PRENTICE-HALL SERIES IN
ELECTRICAL AND COMPUTER ENGINEERING

LEON O. CHUA, *editor*

Chua and Lin *Computer-Aided Analysis of Electronic Circuits: Algorithms and Computational Techniques*

COMPUTER-AIDED ANALYSIS OF ELECTRONIC CIRCUITS

COMPUTER-AIDED ANALYSIS OF ELECTRONIC CIRCUITS

ALGORITHMS AND COMPUTATIONAL TECHNIQUES

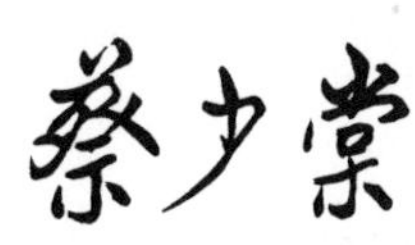

Leon O. Chua

Professor of Electrical Engineering
and Computer Sciences
University of California
Berkeley, California

Pen-Min Lin

Professor of Electrical Engineering
Purdue University
Lafayette, Indiana

PRENTICE-HALL, INC., Englewood Cliffs, New Jersey

Library of Congress Cataloging in Publication Data

CHUA, LEON O.
Computer-aided analysis of electronic circuits.

(Prentice-Hall series in electrical and computer engineering)
Bibliography: p.
Includes index.
1. Electronic data processing—Electronic circuit design. 2. Electronic circuits—Mathematical models. I. Lin, Pen-Min joint author
II. Title.
TK7867.C49 621.3815'3'02854 75-1388
ISBN 0-13-165415-2

10 9 8 7 6 5 4 3 2 1

Printed in the United States of America

PRENTICE-HALL INTERNATIONAL, INC., *London*
PRENTICE-HALL OF AUSTRALIA, PTY. LTD., *Sydney*
PRENTICE-HALL OF CANADA, LTD., *Toronto*
PRENTICE-HALL OF INDIA PRIVATE LIMITED, *New Delhi*
PRENTICE-HALL OF JAPAN, INC., *Tokyo*
PRENTICE-HALL OF SOUTHEAST ASIA (PTE.) LTD., *Singapore*

To our wives:

Diana and Louise

Contents

Preface

Traditionally, the design of most electronic circuits starts with paper-and-pencil work by an engineer who, besides his basic training, is armed with a wealth of design charts, tables, and monograms. He relies very heavily on his intuition, past experience, and knowledge to make reasonable approximations. Then comes the "breadboarding stage," where the result of the preliminary design is confirmed, and perhaps improved, by adjusting circuit element values in a trial-and-error fashion.

The advent of integrated circuits, however, has greatly changed the picture. Not only are the circuits much larger but the specifications are also much tighter. The paper-and-pencil method is no longer adequate when we consider the required accuracy of the results and the time to complete a design. Breadboarding is also of little help because it is impossible to duplicate an integrated circuit with discrete components. Actual production of a mask for an integrated circuit is very costly. Aside from the cost consideration, neither method permits a tolerance or worst-case study. It is in such an environment that the digital computer emerges as an important design tool. Instead of simulating a circuit via breadboarding, a computer program is developed to simulate and analyze the circuit. Such computer-aided circuit analysis is the first step toward an automated circuit design. The other important ingredient is an efficient optimization technique. Today, circuit simulation programs are generally recognized as indispensable tools in any sophisticated circuit design.

This book is devoted to *computer-aided circuit analysis, with emphasis on computational algorithms and techniques*. It is intended as a *textbook* for senior and first-year graduate students in electrical engineering. We hope that the student, after finishing the book, will have not only a thorough understanding of the basic principles and algorithms used in many existing computer simulation programs, but also the capability to develop a small, special-purpose program for his own use. The *prerequisites* for the study of this book are (1) elementary circuit analysis (sinusoidal steady-state analysis, Laplace transform method) and (2) elementary matrix algebra (multiplication, inverse, partitioning). Such background is common among all senior students of electrical engineering.

The preparation of the present text was motivated by an unfulfilled need. At the time of the writing of this book, which started about four years ago, there were already about a dozen texts on the market dealing with various aspects of computer-aided circuit analysis. We examined these and found that they all fell under one of the following categories:

1. *Very narrow coverage.* Some books discuss in great detail how to use certain specific programs, with little or no emphasis on the basic theory. Although useful for training designers whose company happens to have these programs, such texts obviously are unsuitable for college students.
2. *Very elementary.* Such books touch on many important topics superficially, never deep enough to be useful. A case in point is the discussion of the formulation of state equations which is restricted to linear *RLC* networks without controlled sources. Such texts cannot satisfy the demand of serious students.
3. *Very advanced.* These are the books written for those who are already fairly knowledgeable in the field.
4. *Edited volume.* This category seems most prevalent. Because the efforts of a group of experts are pooled, it is possible to cover topics that, at the time, represent the state of the art. On the negative side, such texts usually suffer from notational difficulties and poor coordination among the chapters. These books are good references for researchers, but rarely good textbooks for students.

It was under such circumstances that the authors started four years ago to prepare the present book. It is the outgrowth of class notes, portions of which have been used at the University of California at Berkeley and at Purdue University. The book leads the students step by step from very elementary to fairly advanced topics. Where the frontier of knowledge is reached, and the level is beyond that assumed by the book, we give references to literature. We emphasize algorithms rather than actual programming details, since the former possess "universality," while the latter vary with the programming languages used (FORTRAN, APL, etc.). We do not advocate any *one* approach, to the extent of ignoring the others. The selection of topics is balanced, as evidenced from the table of contents and the following brief synopsis of the chapters.

In the first three sections of *Chapter 1*, we define the problems to be solved with computer simulation programs, describe the main ingredients of a computer simulation program, and give some actual examples of the use of these programs. In the remaining sections of Chapter 1, we give extremely elementary explanations of some basic techniques, idiosyncrasies, and numerical problems associated with many computer simulation programs, with the hope of motivating the students to an in-depth study of the subsequent chapters.

In *Chapter 2* we discuss the principles of modeling electrical components and devices commonly used in electronic circuits. Models for junction diodes and bipolar transistors are described. Models for other semiconductor devices are merely referenced, since a detailed study of this topic is clearly beyond the scope of the present book. A potentially powerful unified *black-box approach* for synthesizing dc circuit models for three-terminal devices is presented with more detail, since it is not available elsewhere. Among other things, a solid understanding of this approach will permit a circuit designer to transform many circuit elements previously not allowed by a particular computer simulation program into an acceptable equivalent circuit. Also included in this chapter is a black-box "macro" circuit model of an *operational amplifier* that is capable of simulating not only the frequency and phase characteristics but also various important nonlinear effects, such as the op amp's slew-rate limitation. The principle used for deriving this model is presented in some detail, not only because it demonstrates the usefulness of the black-box modeling approach but also because the same approach could conceivably be applied to the modeling of other *IC* modules.

In *Chapter 3* we present the fundamentals of graph theory which are applicable to circuit analysis. The results of this chapter are used extensively in the subsequent chapters: in hybrid matrix formulation (Chapter 6), state equation formulation (Chapters 8 and 10), symbolic network analysis (Chapter 14), and adjoint-network sensitivity analysis (Chapter 15). Particular attention is paid to computer generation of various topological matrices.

In *Chapter 4* we present a detailed study of the nodal analysis of linear networks, with emphasis on the use of digital computational techniques. The content of this chapter is practically self-contained. A skeleton program called NODAL is included in the appendix to this chapter so that it may be used as the starting point of a student project to expand it into a full-scale program. Our experience shows that such a project is extremely beneficial for learning the material in this book.

In *Chapter 5* we extend the nodal-analysis method to nonlinear resistive networks. The *fixed-point algorithm* is introduced here as a unifying concept from which several latter algorithms—Newton–Raphson (Section 5-4), predictor-corrector (Section 11-5-2), and periodic solution (Section 17-5-1)—can be considered as special cases. The most commonly used Newton–Raphson method for solving nonlinear functional equations is discussed in detail. The solution of nonlinear resistive networks is actually the cornerstone of most computer simulation programs, and the solution process is usually referred to as *dc analysis*. The use of "discretized *linear* resistive circuit models" for implementing the Newton–Raphson method for *nonlinear* resistive networks is given a detailed treatment because this important technique is usually presented elsewhere without rigorous justification.

In *Chapter 6* we describe the computer formulation of *hybrid matrices* for linear resistive *n*-ports. Like Chapter 3, the material from this chapter is used in several subsequent chapters (Chapters 7, 8, 10, and 15). The general analysis technique that depends on this material is usually referred to as *hybrid analysis*. Although not as widely known as nodal analysis, there are many circuits for which it is computationally advantageous, if not necessary, to resort to hybrid analysis. For example, this approach is particularly suited for analyzing circuits containing nonmonotonic voltage-controlled and current-controlled nonlinear resistors. It is also quite useful for analyzing nonlinear circuits containing many *linear resistors* because the number of nonlinear equations that need to be solved repeatedly would then be equal to the number of nonlinear resistors in the circuit.

In *Chapter 7* we apply the hybrid-matrix method of Chapter 6 to the analysis of nonlinear resistive networks. Several recent algorithms for solving piecewise-linear networks are described. Although computationally less efficient than the Newton–Raphson algorithm of Chapter 5, the piecewise-linear approach has at least two advantageous features. First, for a large class of nonlinear networks, the piecewise-linear approach presented in Section 7-3 is guaranteed to converge, regardless of the initial guess. Second, it is still the only approach capable of finding all solutions of a nonlinear resistive network.

The formulation of state equations is thoroughly discussed in *Chapter 8* for dynamic linear networks, and in *Chapter 10* for dynamic nonlinear networks. The concept of hybrid matrices from Chapter 6 plays a fundamental role in these two chapters. Although it is true that from a programming and computational point of view, the *state-equation approach* is not as appealing as the *tableau approach* (Section 17-2), in so far as developing *large* simulation programs is concerned, the concept and formulation of state equations are of basic importance in many other respects. For example, any qualitative analysis concerning stability, transient decay, bifurcation behavior, etc., requires the formulation of the network's state equations.

Numerical integration techniques for state equations and the associated stability and time-constant problems are given a thorough treatment in *Chapters 11, 12, and 13.* These chapters form a self-contained package containing an *up-to-date* and *in-depth* study of the numerical integration of *ordinary differential equations* and could be virtually lifted out as supplementary text material for a course on *system analysis, modeling*, or *simulation.* Although these chapters are much more mathematical than the rest, all results are derived by elementary methods. A unifying approach is used to derive the three important families of integration methods: the *explicit Adams–Bashforth algorithm*, the *implicit Adams–Moulton algorithm*, and the *implicit Gear's algorithm.* Using a novel approach, a simple yet *generalized formula* which gives the *local truncation error* for all multi-step algorithms is derived. This analysis leads to the formulation of an efficient *variable order*, *variable step-size* predictor–corrector algorithm. (The explicit Adams–Bashforth algorithm serves admirably here as the predictor.) Instead of using the well-known z-transform approach to perform the *stability analysis*, a *difference-equations approach* is adopted here, since it is a more direct approach and it provides additional insights. In particular, it shows vividly how the *parasitic terms* could destroy the stability of an otherwise accurate numerical integration algorithm.

Several special numerical techniques that are applicable only to linear networks are discussed in *Chapter 9*. Also included in this chapter is the computer determination of the transfer function $H(s)$ of a linear circuit. This subject is also encountered in Chapter 14, albeit from a different point of view. Finally, the highly efficient ***QR* algorithm** for computing *eigenvalues* of the associated *A* matrix is presented with many illustrative examples.

The remaining four chapters of the book cover somewhat specialized, and yet very important, topics in computer-aided circuit analysis. As is well known, analysis is the first step to design. The automated design of electronic circuits, a subject of extensive research at present, requires a good analysis program and a good optimization strategy. Most optimization techniques require the efficient computation of partial derivatives of network functions, and these are often obtained from a computer-aided sensitivity analysis. In *Chapter 15* we describe several methods of sensitivity analysis. The adjoint network method receives full attention here because of its generality and efficiency. However, the potential of the symbolic method is also pointed out as a natural sequel to *Chapter 14*, in which a unified treatment of the computer generation of symbolic network function is presented. Although symbolic analysis has some serious shortcomings, it is nevertheless useful for carrying out a comprehensive sensitivity study of small-size networks.

As the network size becomes larger (one hundred nodes or more), it becomes a necessity to apply sparse-matrix techniques to solve the large systems of equations. Although many papers have appeared in the literature, it is difficult to find one written for the novice in this area. *Chapter 16* is written to fill this need. Since sparse-matrix techniques are still an active field of research, new algorithms much more efficient than those presented in this chapter will no doubt be forthcoming, and the interested reader should consult the recent literature on this subject.

Chapter 17, the final chapter, is devoted to a number of recent results from computer-aided circuit analysis. Among other things, the recent *tableau approach* for analyzing large-scale networks and a new efficient algorithm for solving differential-algebraic systems using implicit backward differentiation formulas are presented. Also included in this chapter are efficient algorithms for computing the steady-state periodic response of networks driven by periodic sources, as well as efficient algorithms for computing the periodic solutions of nonlinear oscillators. This final chapter ends with the presentation of the latest algorithm for carrying out a spectrum and distortion analysis of nonlinear communication circuits.

A flowchart depicting the relationships among the chapters is given here, and immediately following the Preface is a complete listing of all algorithms and computational techniques presented in this book. We believe that most of the important results and methods relevant to computer-aided circuit analysis are covered in this book. We have decided not to include any material on *optimization techniques* and *tolerance analysis* because to do justice to these important subjects would have increased the length of this book considerably. Moreover, these subjects are really more relevant to computer-aided *design* and are therefore outside the scope of this book.

The book contains enough material for a one-year course at senior level. As a rule, the sections in each chapter are organized in the order of increasing level of sophistication. In fact, the last two sections in each chapter often contain advanced materials that can be omitted without loss of continuity. Such sections are coded with a star. With judicious selection of topics (not necessarily in the order of the chapters),

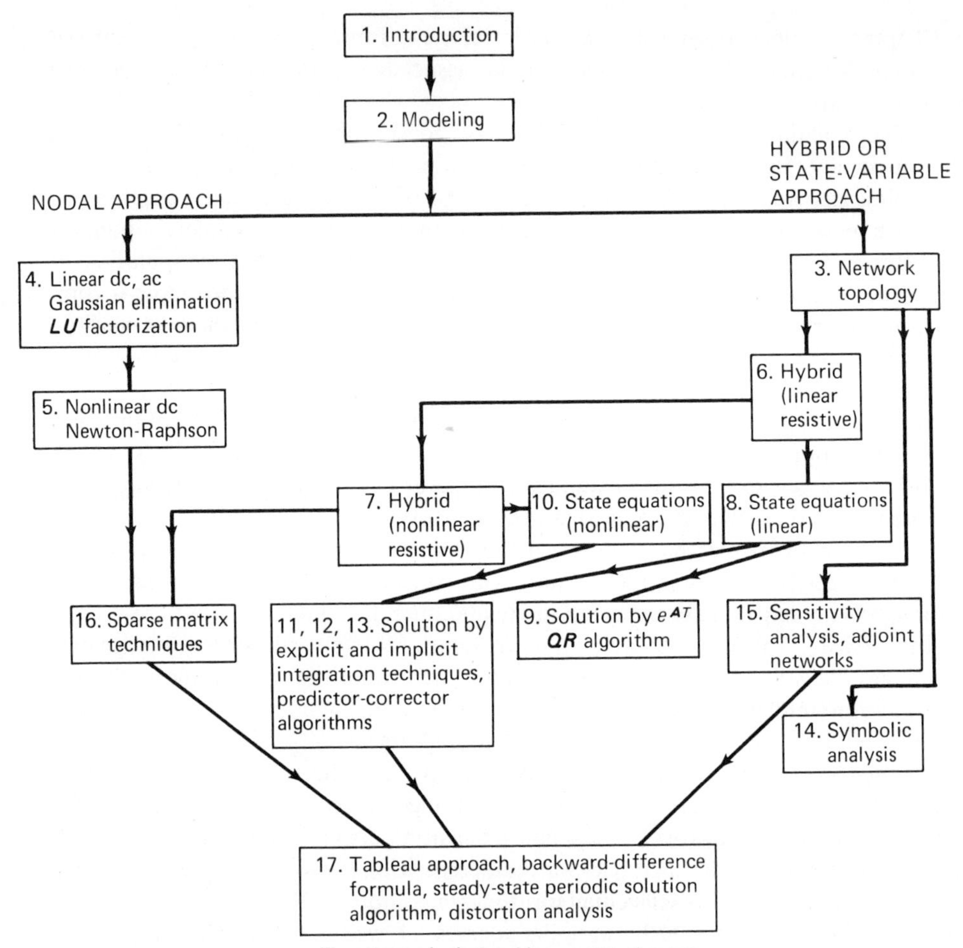

Flowchart of relationships among chapters

it is possible to use the book for a one-semester, a one-quarter, a two-semester, or a two-quarter course. The table on page xxiii illustrates a possible organization of topics for six typical courses in computer-aided circuit analysis. In each course, we recommend that only the first three sections of Chapter 1 be covered initially in class for motivational purposes. The remaining sections may be assigned for self-reading in order to give the students a bird's eye view of the many important ideas which will be covered in depth later. Alternately, each of the remaining sections may be used as a lead-in *synopsis* for a subsequent chapter at the appropriate point in time. Although mathematical proofs are given for most propositions and theorems in this book, they are mainly intended for research-oriented students and may be omitted in any introductory course where the emphasis is on the interpretation and application of these results.

Problems are included at the end of each chapter. Most of these problems have been *class-tested* to ensure that their level of difficulty and degree of complexity are proper for the students. When the book is used for undergraduate classes, those prob-

COURSE	LENGTH OF COURSE AND LEVEL	CHAPTERS AND SECTIONS COVERED	REMARKS
I	One semester, 15 weeks, 3 hours/week, senior	Chaps. 1–6 (omit Sections 4-4 and 6-6); Sections 7-1, 7-2, 8-1, 8-2, 8-4, 9-5; Chaps. 10–13; and selected topics from Chaps. 15 and 16	Covers dc, ac, and transient analysis using both nodal and state-variable methods; emphasizes implicit integration techniques
II	Two semesters, 30 weeks, 3 hours/week, senior	Practically the whole book except Chap. 17	Special-purpose computer simulation programs could be assigned as student group projects
III	One semester, 15 weeks, 3 hours/week, first-year graduate	Quick review: Chaps. 1, 2, 3, 4, 9; Lectures: Chaps. 5, 6, 8, 10, 11, 12, 13, 15, 16; and selected topics from Chap. 17	Graduate students having prior undergraduate backgrounds in computer-aided circuit analysis could skip Chaps. 1, 2, 3, and 4
IV	One quarter, 10 weeks, 3 hours/week, senior	Chaps. 1, 2, 4, 5; Chaps. 11–13; and selected topics from Chaps. 15 and 16	Confined to dc, ac, and transient analysis using the nodal method; emphasizes implicit integration techniques
V	Two quarters, 20 weeks, 3 hours/week, senior	Plan IV, plus Chaps. 3, 6, 7; Sections 8-1, 8-2, 8-4, 9-5; Chap. 10; and selected topics from Chaps. 15 and 16	Second quarter covers hybrid and state variable methods, adjoint-network, and sparse-matrix techniques
VI	One quarter, 10 weeks, 3 hours/week, first-year graduate	Quick review: Chaps. 1, 2, 3, 4, 9; Lectures: Chaps. 5, 6, 8, 10, 11, 12, 13; and selected topics from Chaps. 15, 16, and 17	The star sections dealing with programming details may be assigned as term projects

lems requiring more than elementary matrix algebra should be omitted. This judgment is left to the instructor. To avoid assigning the same problems over successive years, many problems contain several similar parts that differ only in numerical parameters or other trivial details. In these cases, the instructor should assign only the parts that do not duplicate each other.

Although the emphasis of the book is on algorithms rather than programming details, we have included three FORTRAN programs for good reasons. The NODAL program in the appendix of Chapter 4 makes use of the formulation and solution techniques described in that chapter. The program is small enough so that each student may be issued a deck of the source program. He may then use it as a starting point to expand into a full-fledged dc, ac, and transient-analysis program. This avoids much of the frustration he will encounter if he writes the program from scratch.

The HYBRID program in the appendix of Chapter 6 makes use of the concepts and techniques discussed in Chapter 3 and Section 6-6. It is an excellent example of putting theory into practice. Moreover, since no program for obtaining hybrid matrices in the most general case (all four types of controlled sources are allowed) has yet been reported in the literature, the inclusion of this program should be of value to both students and researchers in computer-aided design.

The program SPARSE in the appendix of Chapter 16 is included for the same pedagogical reasons. It is very difficult to find in any textbook a FORTRAN program that is simple enough, and yet illustrates clearly the essence of the sparse-matrix technique.

The authors would like to thank Dr. Y-F Lam and Dr. S. M. Kang of the University of California, Berkeley, and Professor L. P. Huelsman of the University of Arizona for offering many useful suggestions for improving the preliminary version of this book. The authors would also like to acknowledge the contributions of the following colleagues who have read portions of the final manuscript: Professor A. Dervisoglu of the Technical University of Istanbul, Turkey; Dr. T. Roska of the Institute of Telecommunication in Budapest, Hungary; Professor A. Ushida of the Technical College of Tokushima University, Japan; and Professors B. J. Leon and S. C. Bass of Purdue University, Indiana. Two of the computer programs used in Chapter 1 were made available by the following colleagues: SPICE by Professor D. O. Pederson and Dr. L. W. Nagel of the University of California, Berkeley, and CORNAP by Professor C. Pottle of Cornell University. A number of the authors' students have also contributed in various aspects during the evolution of this book. They are G. E. Alderson, L. K. Chen, E. Cohen, W. J. Kim, D. J. Reinagel, T. E. Richardson, C. Stewart, J. Stockman, N. N. Wang, and B. C. K. Wong. They would also like to acknowledge the support provided by the National Science Foundation under Grant 32236. The second author wishes to thank Professor C. L. Coates and J. C. Hancock of Purdue University for their support and encouragement. Finally, the authors would like to thank Mary Ann Ratch, Beth Harris, Linda Stovall, June Harner, and Helen Hancock for their expert typing of various portions of this book.

LEON O. CHUA
Berkeley, California

PEN-MIN LIN
Lafayette, Indiana

List of Algorithms and Chapters Where Found

COMPUTER-AIDED ANALYSIS OF ELECTRONIC CIRCUITS

CHAPTER 1

Once Over Lightly

1-1 BREADBOARDING VERSUS COMPUTER SIMULATION

Most circuit-analysis problems are solved in two steps. The first step consists of formulating the equilibrium equations in an appropriate form, making use of the two Kirchhoff laws and the element characteristics. The second step consists of solving these equations by suitable analytical or numerical techniques. Before the advent of computers, these equations were usually solved by analytical techniques; this approach imposes severe limitations on the size and type of circuits that can be analyzed economically. Large *linear* circuits (say those containing more than 50 elements), or even small *nonlinear* circuits, are seldom analyzed exactly. Instead, engineers often rely heavily on intuition and "seat-of-the-pants" methods to obtain *approximate* analyses of such circuits. Invariably, the final analysis is obtained by *breadboarding* the circuit and measuring the desired variables of interest. Even this breadboarding approach is now inadequate for analyzing integrated circuits, because it is impossible to duplicate an integrated circuit with discrete components. Certainly, parasitic effects and element-matching characteristics between integrated devices cannot be accurately reproduced. Moreover, a tolerance or worst-case analysis cannot be implemented via breadboarding since it would be nearly impossible to vary the device parameters.

Instead of simulating a circuit via breadboarding, a computer program can be developed to perform the analysis automatically. Such a general-purpose analysis program is often called a *computer simulator*. Since devices in an integrated circuit can often be modeled much more accurately by a circuit model than by discrete physical components, the results obtained from the computer simulator are often much more accurate than those obtained via breadboarding. Moreover, the cost of computer simulation is often a small fraction of the breadboarding cost.

1-2 EXAMPLES OF CIRCUIT ANALYSIS VIA COMPUTER SIMULATION

The advantages of computer-aided circuit analysis over breadboarding are so overwhelming that many computer simulators have been developed in the last decade, and many more improved versions are continuing to be developed. Each computer simulator is designed for solving a certain class of circuit problems. Typical problems commonly solved with computer simulators are listed in Table 1-1.

We shall now present examples of the application of computer simulators to solve some of the problems listed in Table 1-1. The name of the computer program

TABLE 1-1. Typical Circuit-Analysis Problems

TYPE OF NETWORK	PROBLEM DESCRIPTION
I. *Linear resistive* (no capacitors or inductors) and *Linear dynamic* (contains at least one capacitor or inductor)	1. DC analysis (find the dc solution of a linear resistive network) 2. AC analysis (find the frequency response of a linear dynamic network) 3. Transient analysis (find the transient response of a linear dynamic network) 4. Noise analysis (ac or transient analysis with noise sources as input) 5. Tolerance analysis (sensitivity or worst-case analysis) 6. Determination of pole–zero locations of transfer functions 7. Generation of symbolic network functions
II. *Nonlinear resistive* (no capacitors or inductors)	1. Operating-point analysis (find the dc solutions of a nonlinear resistive network) 2. Driving-point characteristic determination (find the relationship between driving-point current and driving-point voltage) 3. Transfer characteristic determination (find the relationship between an output voltage or current versus an input voltage or current) 4. Find output waveform due to input time functions
III. *Nonlinear dynamic* (contains at least one capacitor or inductor)	1. Initial condition, bias, or equilibrium state analysis (operating-point analysis with all capacitors replaced by open circuits and inductors by short circuits) 2. Transient analysis (find output waveforms under user- or program-specified initial condition, with or without inputs) 3. Steady-state analysis (find steady-state periodic solution, with or without inputs) 4. Nonlinear distortion analysis (find harmonic, modulation, and intermodulation distortions)

used in each example will be identified along with a complete printout of input data and computer outputs. Our choice of a particular computer program is often influenced by its wide availability and our personal familiarity with its use. It is not our intention here to endorse any one program over another. Nor is it our intention to describe the full capability and usage of any particular program. Such information can be obtained from the user's manual that comes with the program. Our main objective in this section is to impress the reader with the importance of computer simulation as an analysis tool, thereby providing the proper setting for learning the algorithms and techniques to be presented in subsequent chapters.

EXAMPLE 1-1. Figure 1-1(a) shows a single-stage transistor amplifier. We wish to find the operating point, frequency response, and transient response due to a rectangular pulse input with a 10-mV amplitude and 1-s duration.

The program SPICE [1] is chosen for solving this problem. The transistor Q_1 is described by the Ebers–Moll model (see Section 2-4-2) with some minor modifications. First, dc analysis is performed to determine the operating point. This is a necessary prelude to both ac (frequency response) and transient analysis. Once the operating point is determined by the Newton–Raphson method (see Chap. 5), the parameters associated with the small-signal transistor model [see Fig. 1-1(c)] are evaluated by the program. AC nodal analysis (see Chap. 4) is then performed to obtain the frequency response. The results obtained from dc analysis also provide the initial conditions for transient analysis, which makes use of implicit numerical integration techniques (see Chap. 13).

The input data cards are shown in Fig. 1-1(b). The input language is free-format.[1] The following is a brief description of the function of each card in Fig. 1-1(b):

The first card is a *title card*, which will be printed verbatim in each section of the output (dc, ac, etc.). The second card, with an asterisk in the first column, is a *comment card* whose content will be printed in the input listing. The third card means: the branch named *VCC* is connected from node 6 to node 0 (the datum node) and is a 10-V dc voltage source (indicated by the first character *V* in the branch name *VCC*). The fourth card means: the branch named *VIN* (hence a voltage source) is connected from node 1 to node 0. During ac sinusoidal steady-state analysis, *VIN* has value $1 \angle 0^\circ$. During transient analysis, *VIN* is a rectangular pulse of 10-mV amplitude and 1-s duration. The fifth card means: a resistor named *RS* (the first character *R* in *RS* indicates a resistance) is connected from node 1 to node 2 and has a value of 1000 Ω. The next six branch cards are interpreted similarly. The twelfth card means: the three-terminal device *Q*1 is a bipolar junction transistor (indicated by the first character *Q* in the name *Q*1) with collector connected to node 5, base to node 3, and emitter to node 4. The model name for *Q*1 is *BNPN*. The thirteenth card specifies the output to be the voltage from node 5 to node 0. The magnitude versus frequency curve for ac analysis is to be tabulated and plotted. The transient response is to be plotted only.

[1]With *free-format* input, fields in a card are separated by *delimiters* such as blanks, commas, etc. In contrast, a *fixed-format* input has specified columns in a card for each field.

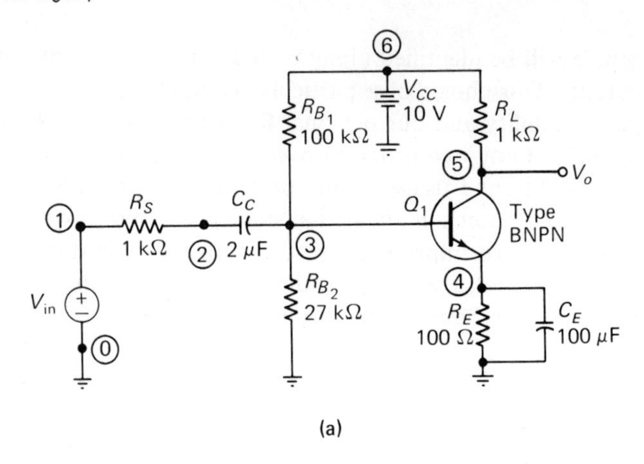

(a)

```
SINGLE STAGE COMMON EMITTER AMPLIFIER
* REFER TO FIG 1-1(A) FOR THE CIRCUIT DIAGRAM
VCC 6 0 DC 10
VIN 1 0 AC 1 PULSE 0. 1UMV 0. 0. 0. 1.
RS 1 2 1K
RL 5 6 1K
RB1 3 6 100K
RB2 3 0 27K
CC  2 3 2UFD
CE 4 0 100UFD
RE 4 0 100
Q1 5 3 4 BNPN
.OUT V5 5 0 PRINT MA PLOT MA PLOT TR
.AC DEC 5 1HZ 100MEGHZ
.TRAN 5NS 300NS
.MODEL BNPN NPN BF=100 BR=0.001 RB=50 VA=100 CCS=2PF CJE=3PF CJC=2PF
.END
```

(b)

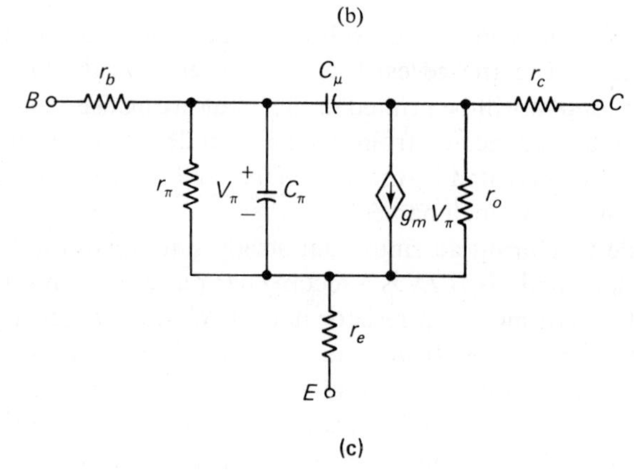

(c)

Fig. 1-1. A single-stage transistor amplifier.

The fourteenth card specifies that in ac analysis the frequency response is to be calculated with logarithmic frequency scale, 5 points per decade, from 1 to 10^8 Hz. The fifteenth card specifies that the transient response is to be calculated from 0 to 300 ns at 5-ns steps. The sixteenth card describes the Ebers–Moll model parameters for the model named *BNPN* (second field). It is of *NPN* type (third field). The forward beta *BF* is 100 (fourth field). The base ohmic resistance *RB* is 50 Ω. The Early voltage *VA* is 100 V. The collector-substrate capacitance *CCS* is 2 pF. The zero-bias B–E junction capacitance *CJE* is 3 pF. The zero-bias B–C junction capacitance *CJC* is 2 pF. For a complete description of the Ebers–Moll model, many additional parameters must be specified (see Chap. 2). The program SPICE requires a total of 15 parameters for the description of a bipolar junction transistor. Only six of them are specified in the sixteenth card. The remaining parameters are given the default values that are set within the program (see SPICE manual for the default values used). The seventeenth card signals the end of the data cards.

To save space, only portions of the output from a computer run of this example are shown. In Fig. 1-2, we first see the 15 Ebers–Moll model parameters used by the program, some of which are specified in the input data deck; the remaining are default values. Next in Fig. 1-2 are the results of dc analysis which consist of all node-to-datum voltages, voltage source currents, and the total power dissipation. The transistor operating point and the corresponding small-signal parameters computed by the program are shown in the last row. The frequency response and transient response are shown in Figs. 1-3 and 1-4, respectively.

For linear networks, the transfer function $H(s)$ provides extremely valuable information about the networks. $H(s)$ may be expressed as the ratio of two polynomials in s, or in terms of pole–zero locations and a gain constant. The following example illustrates the use of CORNAP [2] for obtaining the poles and zeros of a network function.

Example 1-2. Figure 1-5(a) shows the linearized model of the amplifier circuit of Example 1-1. Note that the transistor model is that determined in Example 1-1. Our problem is to find the poles and zeros of the transfer function $H(s) \triangleq V_o(s)/V_i(s)$, the frequency response $H(j\omega)$, the step response, and the impulse response. The complete input card deck for this problem is shown in Fig. 1-5(b). CORNAP uses fixed-format input. We shall not describe here what data should be entered in which columns of the data cards, as this information is available from the user's manual [2]. Instead, we shall briefly explain the function of each card contained in Fig. 1-5(b).

The first card is a title card (see Example 1-1). The second card says that a resistor (indicated by *R* in column 1) branch named *RS* is connected from node 1 to node 2 and has a value of 1000 Ω. The next 12 cards describing the *R* and *C* branches are interpreted similarly. The $+V$ in the ninth card informs the program that the voltage across this branch (from node 5 to node 0) is the desired output. The fifteenth card describes the controlled source. The branch named *ID* is a dependent current source connected from node 5 to node 4, and is controlled by the voltage of the branch named *RPI*, with a controlling coefficient (proportionality constant) equal to 0.181. The sixteenth card specifies that the branch named *VS* is an independent voltage source connected from node 1 to node 0. Independent sources are always considered as inputs of the network. The seventeenth card contains the constants for

```
**** EBERS-MOLL MODEL PARAMETERS

NAME   TYPE    BF      BR      RB     RC    RE     CCS       TF      TR       CJE       CJC       IS       PE    PC    VA      EG
BNPN    N    100.00  00.001   50.0   0.0   0.0 2.00E-12 0.E+00   0.E+00   3.00E-12 2.00E-12 1.00E-14  1.00  1.00  100.00  1.11


*******************************************************************************************************************

SINGLE STAGE COMMON EMITTER AMPLIFIER                                                              ----- SPICE -----

                SMALL SIGNAL BIAS SOLUTION                                  TEMPERATURE      27.000  DEG C

*******************************************************************************************************************


NODE    VOLTAGE     NODE    VOLTAGE     NODE    VOLTAGE     NODE    VOLTAGE     NODE    VOLTAGE

(  1)     0.0000    (  2)     0.0000    (  3)     1.1691    (  4)    00.4732    (  5)     5.3126
(  6)    10.0000


    VOLTAGE SOURCE CURRENTS
       NAME          CURRENT

       VCC         -4.776E-03   AMPS
       VIN          0.E+00      AMPS

    TOTAL POWER DISSIPATION    4.78E-02  WATTS


    TRANSISTOR OPERATING POINTS

NAME    MODEL      IB         IC       VBE      VBC      VCE      GM       RPI      RO       CPI      CMU      BETADC  BETAAC   FT

Q1      BNPN    4.50E-05  4.69E-03  00.696  -4.143   4.839  1.81E-01 5.74E+02 2.22E+04 4.04E-12 8.82E-13   104.1   104.1 5.86E+09
```

Fig. 1-2. DC analysis of the amplifier of Fig. 1-1.

```
*************************************************************************************************************
SINGLE STAGE COMMON EMITTER AMPLIFIER                                                     -----  SPICE  -----

                 AC ANALYSIS                                      TEMPERATURE     27.000 DEG C

*************************************************************************************************************

FREQUENCY    MAGNITUDE OF V5

                       1.000E-01          1.000E+00          1.000E+01          1.000E+02          1.000E+03

1.000E+00    8.244E-01
1.585E+00    1.292E+00
2.512E+00    1.995E+00
3.981E+00    2.985E+00
6.310E+00    4.244E+00
1.000E+01    5.736E+00
1.585E+01    7.672E+00
2.512E+01    1.065E+01
3.981E+01    1.544E+01
6.310E+01    2.268E+01
1.000E+02    3.229E+01
1.585E+02    4.261E+01
2.512E+02    5.093E+01
3.981E+02    5.595E+01
6.310E+02    5.842E+01
1.000E+03    5.950E+01
1.585E+03    5.995E+01
2.512E+03    6.013E+01
3.981E+03    6.020E+01
6.310E+03    6.023E+01
1.000E+04    6.024E+01
1.585E+04    6.025E+01
2.512E+04    6.025E+01
3.981E+04    6.024E+01
6.310E+04    6.023E+01
1.000E+05    6.021E+01
1.585E+05    6.014E+01
2.512E+05    5.998E+01
3.981E+05    5.957E+01
6.310E+05    5.860E+01
1.000E+06    5.633E+01
1.585E+06    5.164E+01
2.512E+06    4.362E+01
3.981E+06    3.326E+01
6.310E+06    2.322E+01
1.000E+07    1.536E+01
```

Fig. 1-3. AC analysis of the amplifier of Fig. 1-1.

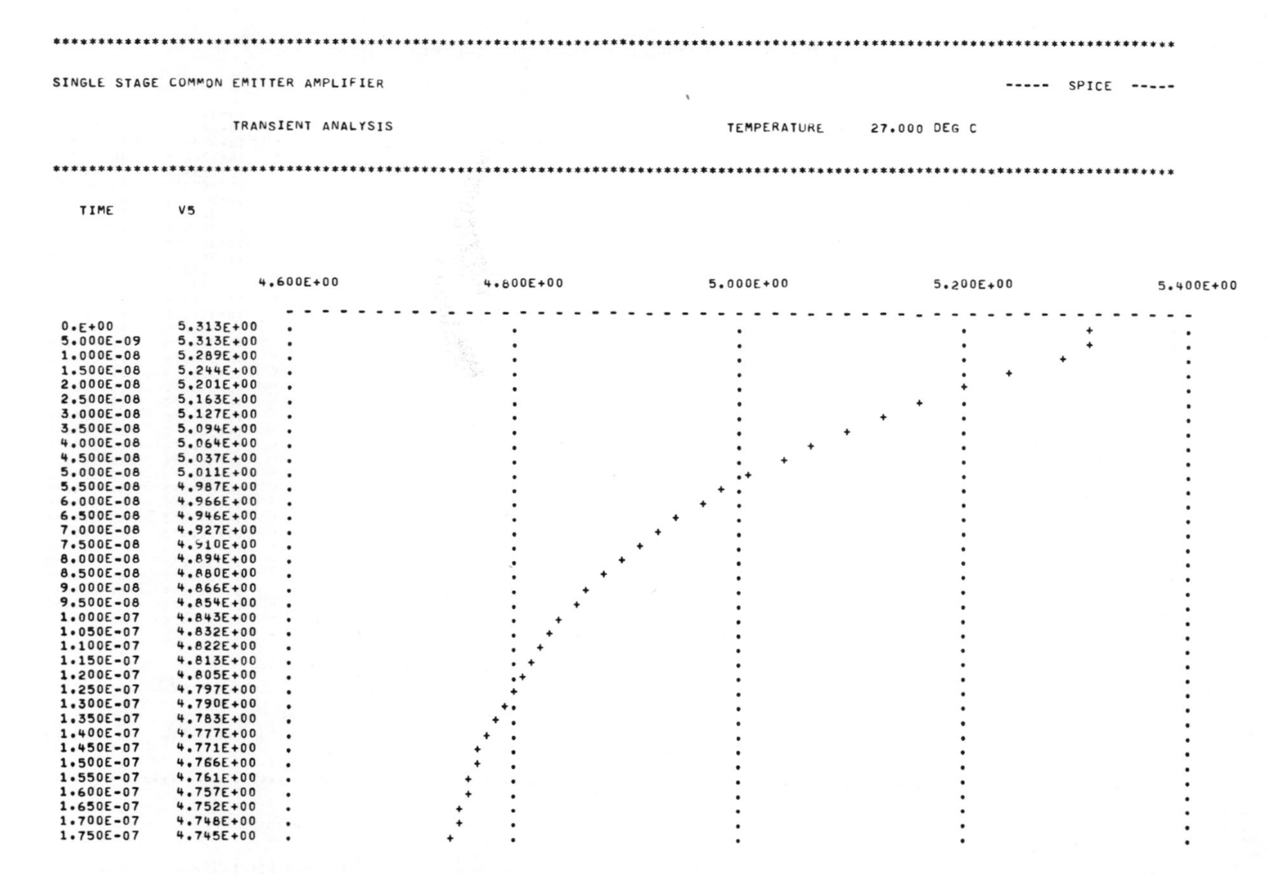

```
***************************************************************************************

SINGLE STAGE COMMON EMITTER AMPLIFIER                                     ----- SPICE -----

            TRANSIENT ANALYSIS                        TEMPERATURE    27.000 DEG C

***************************************************************************************

  TIME        V5

                     4.600E+00        4.800E+00        5.000E+00        5.200E+00        5.400E+00

0.E+00       5.313E+00
5.000E-09    5.313E+00
1.000E-08    5.289E+00
1.500E-08    5.244E+00
2.000E-08    5.201E+00
2.500E-08    5.163E+00
3.000E-08    5.127E+00
3.500E-08    5.094E+00
4.000E-08    5.064E+00
4.500E-08    5.037E+00
5.000E-08    5.011E+00
5.500E-08    4.987E+00
6.000E-08    4.966E+00
6.500E-08    4.946E+00
7.000E-08    4.927E+00
7.500E-08    4.910E+00
8.000E-08    4.894E+00
8.500E-08    4.880E+00
9.000E-08    4.866E+00
9.500E-08    4.854E+00
1.000E-07    4.843E+00
1.050E-07    4.832E+00
1.100E-07    4.822E+00
1.150E-07    4.813E+00
1.200E-07    4.805E+00
1.250E-07    4.797E+00
1.300E-07    4.790E+00
1.350E-07    4.783E+00
1.400E-07    4.777E+00
1.450E-07    4.771E+00
1.500E-07    4.766E+00
1.550E-07    4.761E+00
1.600E-07    4.757E+00
1.650E-07    4.752E+00
1.700E-07    4.748E+00
1.750E-07    4.745E+00
```

Fig. 1-4. Transient analysis of the amplifier of Fig. 1-1.

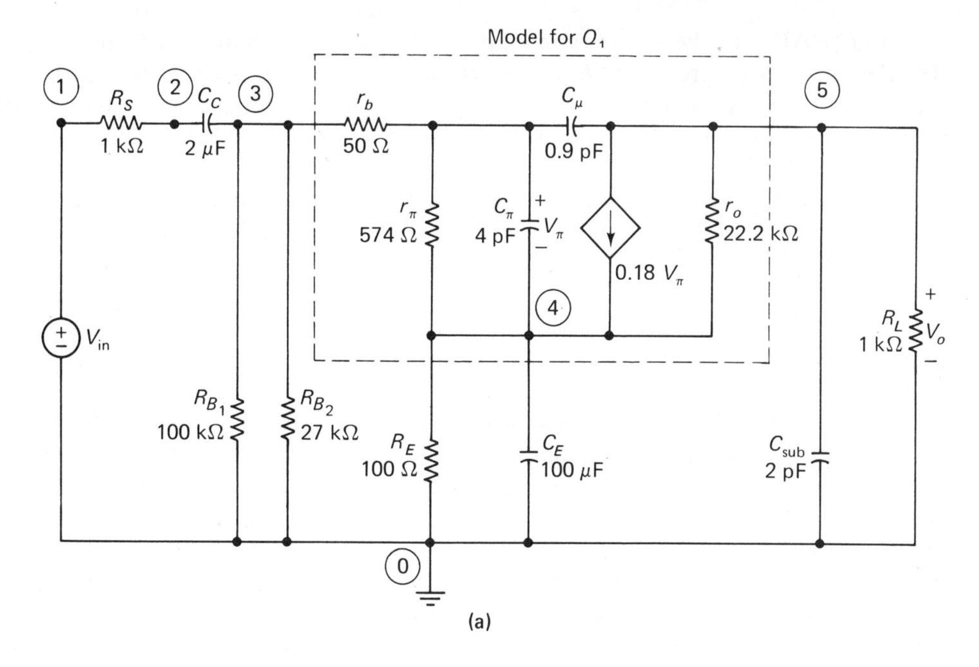

(a)

```
SINGLE STAGE COMMON EMITTER AMPLIFIER
R  RS    1  2 1000.
R  RB1   3  0 100000.
R  RB2   3  0 27000.
R  RE    4  0 100.
R  RB    3  6 50.
R  RO    5  4 22200.
R  RPI   6  4 574.
R  RL    5  0 1000.                             +V
C  CC    2  3             2.E-06
C  CE    4  0           100.E-06
C  CPI   6  4           4.04E-12
C  CMU   6  5          0.882E-12
C  CCS   5  0             2.E-12
I  ID    5  4 0.181                       RPI V
V  VS    1  0 0.
              1.                                    1.
F        7  5                                       1.
T         100   0.000000005
CANCEL
```

(b)

Fig. 1-5. Linear incremental model of the amplifier of Fig. 1-1.

frequency scaling and impedance scaling (see CORNAP manual [2] for explanation). The eighteenth card specifies that the frequency response is to be calculated over 7 decades with 5 points per decade, starting with 1 Hz. The nineteenth card specifies that the step and impulse responses are to be calculated for 100 points at 5-ns steps, starting with $t = 0$. The last card with the word CANCEL signifies the end of data for the present problem and also tells the computer that this is the last problem in the job submitted for execution. (More than one problem can be run in a single job if CANCEL is replaced by MORE.)

CORNAP uses the state-variable approach (see Chaps. 8 and 9). Included in the computer output are the $\boldsymbol{A}$, $\boldsymbol{B}$, $\boldsymbol{C}$, and $\boldsymbol{D}$ matrices. To save space, these matrices are not shown. Figure 1-6 shows portions of the computer output, which contain the pole–zero locations, frequency response, step response, and impulse response.

CORNAP is strictly for *linear* networks. Linearized models for active devices (BJT, JFET, MOSFET, etc.) must be supplied *by the user* before CORNAP is used for networks containing such devices. The program SLIC [3], on the other hand, will perform dc analysis (much the same as SPICE does), determine the operating point, derive linear models for active devices, and then proceed to determine the poles and zeros of $H(s)$. The current version of SLIC calculates frequency response but has no transient response capability.

In Example 1-2 we used CORNAP to obtain $H(s)$. In $H(s)$, the complex frequency variable s is the only symbol. Sometimes it is desirable to obtain *symbolic* network functions in which network parameters besides s are also represented by symbols, e.g., $R = R_1R_2/(R_1 + R_2)$ or $Z = R/(sRC + 1)$. The applications of symbolic network functions and the methods for deriving them with digital computers are discussed in Chapter 14. Here, we shall illustrate how the program SNAP [4] is used to solve such problems.

EXAMPLE 1-3. An amplifier with RC phase-shift network is shown in Fig. 1-7(a). It is desired to obtain the expression for the voltage gain $H(s) \triangleq V_o(s)/E(s)$.

The complete input data for solving the problem with SNAP are shown in Fig. 1-7(b). SNAP uses fixed-format input. We shall describe briefly the function of each card.

The first card is a title card (see Example 1-1). The second card says that the network has six nodes, nine branches, and the basis used in coding symbols is 16 (see SNAP manual [4] for an explanation of the coding scheme). The third card indicates that the input is branch 9, and the output variable is associated with branch 4. The fourth card describes branch 1, which is a resistor (R in column 1) connected from node 1 to node 5 and is represented by a variable R (R in column 19). The next six cards are interpreted similarly. The eleventh card says that branch 8, connected from node 1 to node 5, is a voltage-controlled current source (VC in columns 1 and 2); the controlling variable is associated with branch 9, and the controlling coefficient is GM. The last card says that branch 9, connected from node 6 to node 5, is an independent voltage source (indicated by E in column 1).

SNAP uses the signal-flow-graph approach (see Chap. 14). The complete computer output will contain this information on the signal-flow graph. To save space, only portions of the output that give the final results are shown in Fig. 1-8. Note that each term in the numerator and denominator polynomials is specified by a symbol combination (such as C^2R^2), a power of s, and a real constant. From Fig. 1-8, we obtain the desired gain function, as shown in Fig. 1-7(c).

Most existing nonlinear simulation programs use the Newton–Raphson method for dc analysis (e.g., the program SPICE used in Example 1-1). This is satisfactory for networks having a unique operating point. The method will not yield *all* solutions

SINGLE STAGE COMMON EMITTER AMPLIFIER

TRANSFER FUNCTION CRITICAL FREQUENCIES (SCALED)

OUTPUT VARIABLE - V RL
SOURCE VARIABLE - VS

GAIN CONSTANT IS -2.1286597D+19

POLE POSITIONS			ZERO POSITIONS		
REAL PART	IMAGINARY PART	ORDER	REAL PART	IMAGINARY PART	ORDER
-6.7750186D+08	0.D+00	1	0.D+00	0.D+00	1
-5.2252265D+08	0.D+00	1	-9.9995664D+01	0.D+00	1
-1.0021913D+03	0.D+00	1			
-4.5447759D+01	0.D+00	1			

THIS NETWORK HAS BEEN SCALED FOR COMPUTATION BY THE FOLLOWING FACTORS
FREQUENCY 1.0000000D+00 RADIANS/SEC. IMPEDANCE 1.0000000D+00 OHMS

UNSCALED FREQ RESPONSE

FREQUENCY HZ	MAGNITUDE	MAGNITUDE DB	REAL PART	IMAGINARY PART	PHASE FRAC OF PI	DELAY SEC
1.0000E+00	8.2324E-01	-1.6895E+00	-6.6525E-02	-8.2054E-01	-0.52575	1.2627E-02
1.5849E+00	1.2904E+00	2.2145E+00	-1.6265E-01	-1.2801E+00	-0.54023	1.2091E-02
2.5119E+00	1.9923E+00	5.9871E+00	-3.8300E-01	-1.9551E+00	-0.56158	1.0876E-02
3.9811E+00	2.9811E+00	9.4875E+00	-8.3236E-01	-2.8625E+00	-0.59007	8.4733E-03
6.3096E+00	4.2392E+00	1.2546E+01	-1.5700E+00	-3.9377E+00	-0.62076	4.8496E-03
1.0000E+01	5.7299E+00	1.5163E+01	-2.4727E+00	-5.1689E+00	-0.64203	1.3820E-03
1.5849E+01	7.6651E+00	1.7690E+01	-3.3916E+00	-6.8739E+00	-0.64590	-2.3990E-04
2.5119E+01	1.0641E+01	2.0540E+01	-4.5379E+00	-9.6250E+00	-0.64024	-2.0602E-04
3.9811E+01	1.5429E+01	2.3767E+01	-6.6438E+00	-1.3925E+01	-0.64170	2.6450E-04
6.3096E+01	2.2663E+01	2.7106E+01	-1.1056E+01	-1.9783E+01	-0.66222	5.5004E-04
1.0000E+02	3.2258E+01	3.0173E+01	-1.9409E+01	-2.5765E+01	-0.70551	5.8376E-04
1.5849E+02	4.2551E+01	3.2578E+01	-3.1591E+01	-2.8506E+01	-0.76632	4.4800E-04
2.5119E+02	5.0841E+01	3.4124E+01	-4.3834E+01	-2.5757E+01	-0.83089	2.6498E-04

UNSCALED TIME RESPONSE

INTEGRATION STEP SIZE 1.4706E-10 SEC

TIME	STEP	IMPULSE
0.D+00	0.D+00	0.D+00
5.0000D-09	-4.7702D+01	-5.4322D+09
1.0000D-08	-5.8947D+01	-5.8192D+08
1.5000D-08	-6.0033D+01	-4.8819D+07
2.0000D-08	-6.0122D+01	-3.7261D+06
2.5000D-08	-6.0128D+01	-2.1631D+05
3.0000D-08	-6.0128D+01	4.7950D+04
3.5000D-08	-6.0128D+01	6.7562D+04
4.0000D-08	-6.0128D+01	6.9008D+04
4.5000D-08	-6.0128D+01	6.9114D+04
5.0000D-08	-6.0127D+01	6.9122D+04
5.5000D-08	-6.0127D+01	6.9122D+04
6.0000D-08	-6.0127D+01	6.9122D+04
6.5000D-08	-6.0126D+01	6.9122D+04
7.0000D-08	-6.0126D+01	6.9121D+04
7.5000D-08	-6.0126D+01	6.9121D+04
8.0000D-08	-6.0126D+01	6.9121D+04
8.5000D-08	-6.0125D+01	6.9120D+04
9.0000D-08	-6.0125D+01	6.9120D+04
9.5000D-08	-6.0125D+01	6.9120D+04
1.0000D-07	-6.0124D+01	6.9120D+04
1.0500D-07	-6.0124D+01	6.9119D+04
1.1000D-07	-6.0124D+01	6.9119D+04
1.1500D-07	-6.0124D+01	6.9119D+04
1.2000D-07	-6.0123D+01	6.9118D+04
1.2500D-07	-6.0123D+01	6.9118D+04
1.3000D-07	-6.0123D+01	6.9118D+04
1.3500D-07	-6.0122D+01	6.9118D+04
1.4000D-07	-6.0122D+01	6.9117D+04
1.4500D-07	-6.0122D+01	6.9117D+04
1.5000D-07	-6.0122D+01	6.9117D+04
1.5500D-07	-6.0121D+01	6.9116D+04
1.6000D-07	-6.0121D+01	6.9116D+04

Fig. 1-6. Transfer function poles and zeros, frequency response, and transient response of the amplifier of Fig. 1-1.

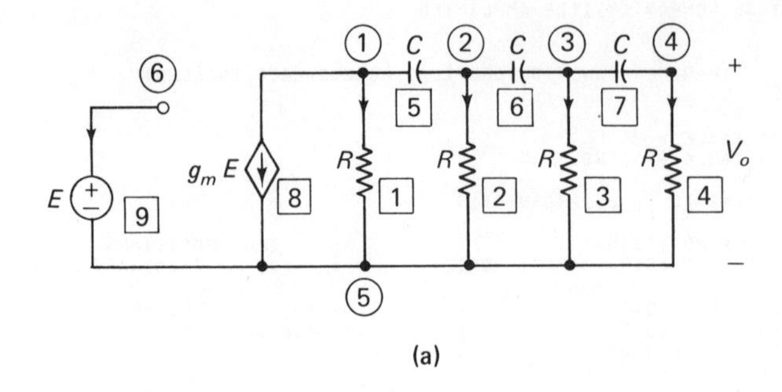

(a)

```
AMPLIFIER WITH RC PHASE SHIFT NETWORK
      6     9    16        11
      9     4
R     1     1     5    R
R     2     2     5    R
R     3     3     5    R
R     4     4     5    R
C     5     1     2    C
C     6     2     3    C
C     7     3     4    C
VC    8     1     5   GM                   9
E     9     6     5
```

(b)

Answer obtained from SNAP

$$\frac{V_o}{E} = \frac{-g_m R(sCR)^3}{4(sCR)^3 + 10(sCR)^2 + 6sCR + 1}$$

(c)

Fig. 1-7. Amplifier with *RC* phase shift network.

when there are more than one. Examples of such circuits are bistable multivibrators (flip-flops) and Schmitt triggers. In such cases, a program based on *combinatorial analysis* of a piecewise-linear resistive network model (see Chaps. 5 and 7) will be very useful. We shall illustrate how the program MECA [5] is used to solve such problems.

EXAMPLE 1-4. A Schmitt trigger circuit is shown in Fig. 1-9(a). After representing the transistors by piecewise-linear models, as shown in Fig. 1-9(b), the complete network is redrawn as shown in Fig. 1-9(c). The diodes are characterized by the piecewise-linear curve $i = 0$, $v \leq 0$ and $v = 0$, $i \geq 0$. It is desired to find the transfer characteristic V_o versus V_i.

The complete input deck for solving this problem using MECA is shown in Fig. 1-9(d). MECA uses fixed-format input. We shall describe briefly the function of each card in the input deck.

The first card contains the title of the problem in columns 1 to 12. The remaining data on the first card say that there are 17 branches in the network; a transfer characteristic is desired (coded 2); the input is branch 1; and one output is specified. The second card specifies the output to be the voltage from node 7 to the datum node (always numbered 0). The third card says that branch 1 is named *VIN*, connected from node 2 to node 0, and is an independent voltage source (indicated by the code

```
AMPLIFIER WITH RC PHASE SHIFT NETWORK


 NUMBER OF NODES=  6
 NUMBER OF BRANCHES=  9
 ELEMENT NUMBER OF SOURCE=  9
 ELEMENT NUMBER ASSOCIATED WITH OUTPUT=  4
 BASE FOR SYMBOL CODES=  16


*************************************************


                          NUMERATOR POLYNOMIAL


COLUMN                 SYMBOL FOR GIVEN COLUMN
    1                          C**3  R**4 GM     /  1


  POWER
  OF S                          CONSTANT COEFS. IN THE POLYNOMIAL
         COLUMN 1               COLUMN
  3    -1.00000E+00
  1     0.
  2     0.
  0     0.
*************************************************


                          DENOMINATOR POLYNOMIAL


COLUMN                 SYMBOL FOR GIVEN COLUMN
    1                          C       R    /  1
    2                          C**2   R**2  /  1
    3                          C**3   R**3  /  1
    4                        1      /  1


  POWER
  OF S                          CONSTANT COEFS. IN THE POLYNOMIAL
         COLUMN 1               COLUMN 2          COLUMN 3           COLUMN 4
  3     0.                     0.              4.00000E+00          0.
  1     6.00000E+00            0.              0.                   0.
  2     0.                     1.00000E+01     0.                   0.
  0     0.                     0.              0.                   1.00000E+00

 EXECUTION TIME IN SECONDS,               .474

AUGUST 1970 VERSION OF SNAP
```

Fig. 1-8. Symbolic analysis of the amplifier of Fig. 1-7.

11 in columns 21 and 22). The fourth card describes the fact that branch 2 is connected from node 1 to node 0 and is a resistor (coded 01) of 1000 Ω. The fifth card describes the fact that branch 3 is connected from node 3 to node 1 and is a nonlinear resistor (coded 22) characterized by three points on its piecewise-linear curve. The next three cards then give the coordinates of these points (voltage first, current second). Branch 5 is

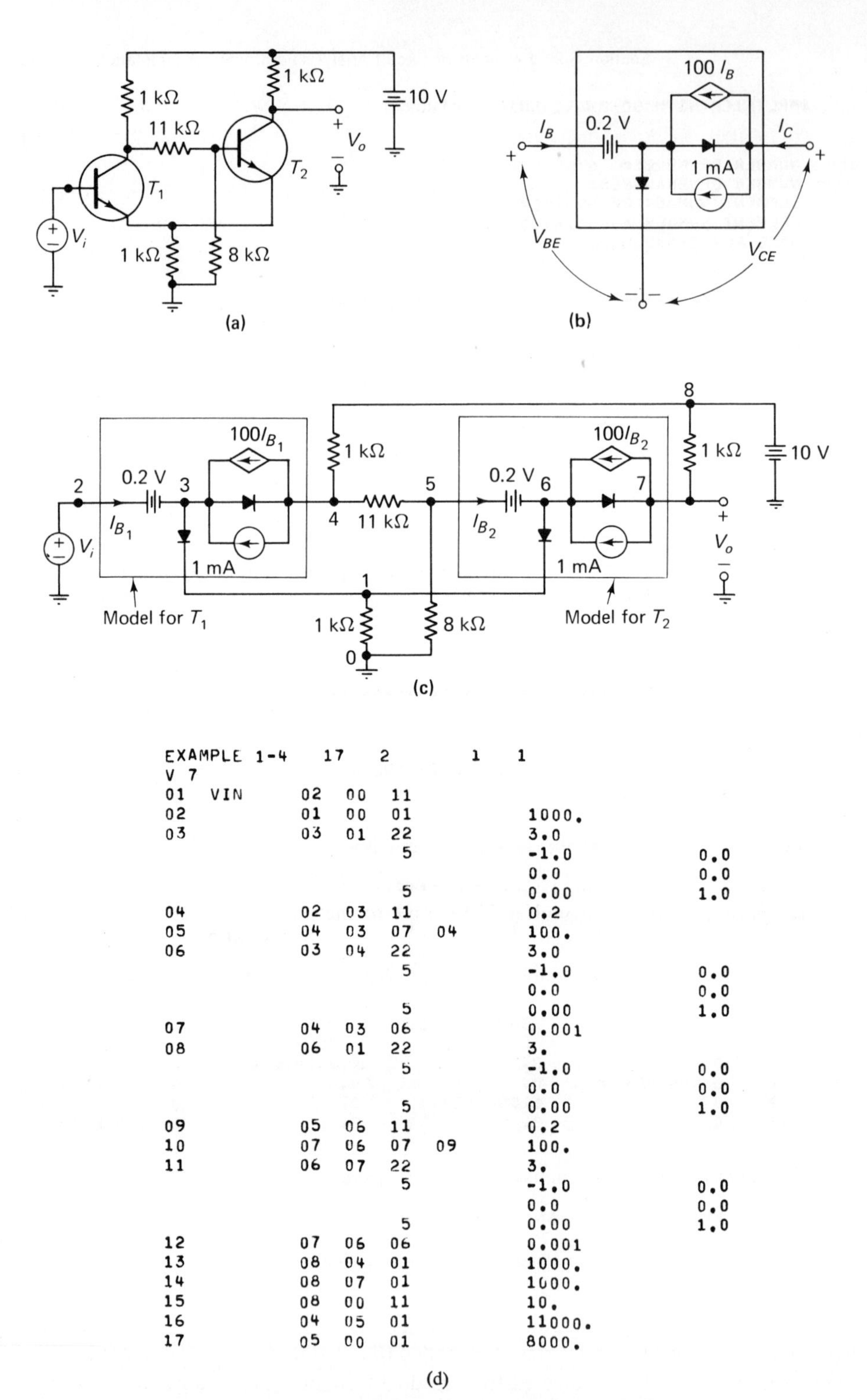


```
EXAMPLE 1-4     17    2          1    1
V 7
01   VIN      02   00   11
02            01   00   01              1000.
03            03   01   22              3.0
                         5              -1.0             0.0
                                        0.0              0.0
                         5              0.00             1.0
04            02   03   11              0.2
05            04   03   07    04        100.
06            03   04   22              3.0
                         5              -1.0             0.0
                                        0.0              0.0
                         5              0.00             1.0
07            04   03   06              0.001
08            06   01   22              3.
                         5              -1.0             0.0
                                        0.0              0.0
                         5              0.00             1.0
09            05   06   11              0.2
10            07   06   07    09        100.
11            06   07   22              3.
                         5              -1.0             0.0
                                        0.0              0.0
                         5              0.00             1.0
12            07   06   06              0.001
13            08   04   01              1000.
14            08   07   01              1000.
15            08   00   11              10.
16            04   05   01              11000.
17            05   00   01              8000.
```

(d)

Fig. 1-9. Piecewise-linear model for a Schmitt trigger circuit.

a current-controlled current source (coded 07). For a controlled source the controlling branch number is entered in columns 21–26, which is 04 for the present case. A similar interpretation is given to branch 10.

MECA performs combinatorial analysis of the piecewise-linear network (see Chap. 5). The complete computer output contains all *valid* segment combinations and the corresponding solutions of voltages and currents (two extreme points are given for each segment of the piecewise-linear-solution curve). To save space, these are not reproduced here. Only the portions of the computer output that are important to the user are shown in Fig. 1-10. Note that the last part of Fig. 1-10 describes the desired transfer characteristic (TC plot). A line printer plot from MECA is shown in Fig. 1-11, with an expanded, more accurate plot shown in the insert. The computed TC plot agrees very well with the measured characteristics. For the range of input voltage $2.95 < V_i < 3.86$ V, there are three possible solutions of V_o for each given V_i. Thus, any program (such as SPICE) based on iterative method will not be able to give the complete solution to this problem.

In Example 1-1, the Ebers–Moll model for the transistor is used for determining the operating point and the small-signal parameters at that point. The frequency response curve of Fig. 1-3 is obtained via linear network analysis. The transient response of Fig. 1-4, however, is obtained via a nonlinear transient analysis subroutine. In view of the small amplitude of the input voltage (10 mV), the effect of nonlinearity is not appreciable. We can obtain essentially the same result by the use of a linear transient analysis program and the hybrid-pi model [Fig. 1-1(c)]. In our final example, we shall use SPICE to obtain the transient waveform of a transistor switch under large-signal operating conditions.

EXAMPLE 1-5. Figure 1-12(a) shows a transistor switch operating normally in the "off" state owing to the -4-V negative voltage applied to node 7. A 12-V positive pulse is applied through a 50-Ω cable to the input of the circuit (node 2) to drive the transistor into saturation. The pulse has a rise time of 2 ns, a pulse width of 100 ns, and a fall time of 2 ns, as shown in Fig. 1-13. We wish to find the waveforms of I_C and V_{CE}.

The input data cards for SPICE are shown in Fig. 1-12(b). The function of each card has been explained in Example 1-1. The only important difference is in the MODEL card. Observe that we have specified the following four additional parameters:

Forward transit time	$T_f = 0.5$ ns
Reverse transit time	$T_r = 10$ ns
Collector ohmic resistance	$R_c = 10\ \Omega$
Emitter ohmic resistance	$R_e = 1\ \Omega$

These are given the default value of zero in Example 1-1. In the present case, because we are interested in the accurate waveform of I_C and V_{CE}, these parameters become very important. Figure 1-14 shows the dc solution from SPICE, which also serves as

```
MECA68-4                                      DP.TC    NO. OF                                                                  NO. OF
                 NO. OF   ANALYSIS ERROR     INPUT   SPECIFIED                                                       DP.TC   CURVES
NETWORK NAME   ELEMENTS    MODE    CODE     SOURCE    OUTPUTS  STARTING TIME    FINAL TIME     TIME INCREMENT      ERROR    /PLOT
EXAMPLE 1-4       17        2       -0         1         1      -0.E+00         -0.E+00         -0.E+00             -0        -0

SPECIFIED OUTPUTS
V  7

.................................................................................................................................

NONLINEAR RESISTORS

FOR EACH SEGMENT, THE VOLTAGE INTERVAL LIES BETWEEN VA AND VB WHILE THE CURRENT INTERVAL LIES BETWEEN IA AND IB.

  ELEMENT      SEG-
NUMBER NAME    MENT        VA              VB                IA                IB             CONDUCTANCE        INTERCEPT
   3            1        - INF.          0.E+00            0.E+00            0.E+00           0.E+00             0.E+00
                2     0.E+00             0.E+00            0.E+00            + INF.                              0.E+00        *
   6            1        - INF.          0.E+00            0.E+00            0.E+00           0.E+00             0.E+00
                2     0.E+00             0.E+00            0.E+00            + INF.                              0.E+00        *
   8            1        - INF.          0.E+00            0.E+00            0.E+00           0.E+00             0.E+00
                2     0.E+00             0.E+00            0.E+00            + INF.                              0.E+00        *
  11            1        - INF.          0.E+00            0.E+00            0.E+00           0.E+00             0.E+00
                2     0.E+00             0.E+00            0.E+00            + INF.                              0.E+00        *
* INDICATES A VOLTAGE INTERCEPT FOR A ZERO RESISTANCE SEGMENT. OTHER INTERCEPTS ARE CURRENT.


.................................................................................................................................
.................................................................................................................................

CHARACTERISTICS

POINTS ARE LISTED FOR EACH CURVE OF THE CHARACTERISTIC
* PRECEEDS A POINT ON A SEMI-INFINITE SEGMENT

.................................................................................................................................

TRANSFER CHARACTERISTIC
OUTPUT IS V  7

CURVE  1
  POINT             INPUT          OUTPUT                INPUT          OUTPUT                 INPUT          OUTPUT
  NUMBER           VOLTAGE         VOLTAGE              VOLTAGE         VOLTAGE               VOLTAGE         VOLTAGE
  1-   3     *( 3.4822684E+00, 6.3572404E+00)    ( 3.8691671E+00, 6.3572404E+00)    ( 2.9531915E+00, 9.9900990E+00)
  4-   6      ( 5.0926020E+00, 9.9900990E+00)    ( 2.4292574E+01, 9.9900990E+00)   *( 2.6721832E+01, 1.0171726E+01)
.................................................................................................................................

   1 CHARACTERISTIC PLOTS GIVEN
```

Fig. 1-10. Transfer characteristic of the Schmitt trigger circuit of Fig. 1-9.

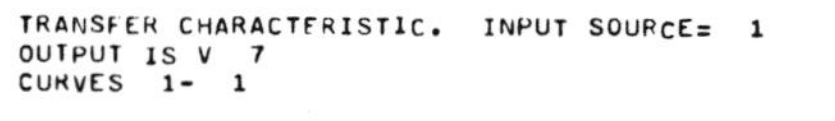

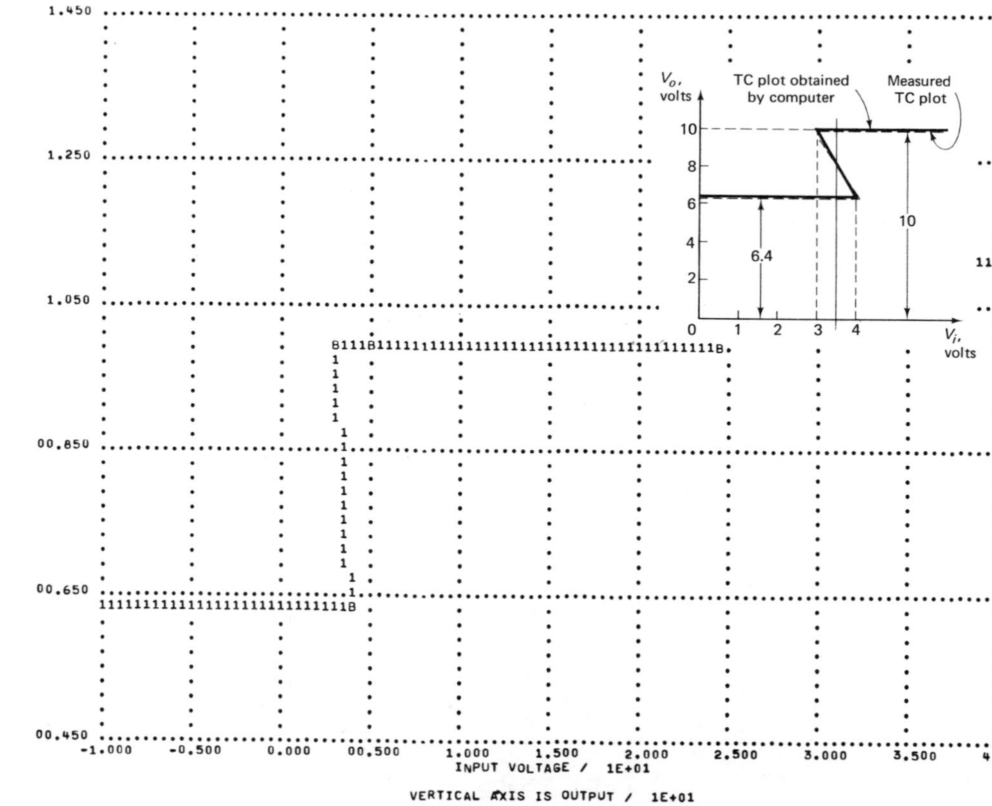

Fig. 1-11. Plot of the transfer characteristic of the Schmitt trigger circuit.

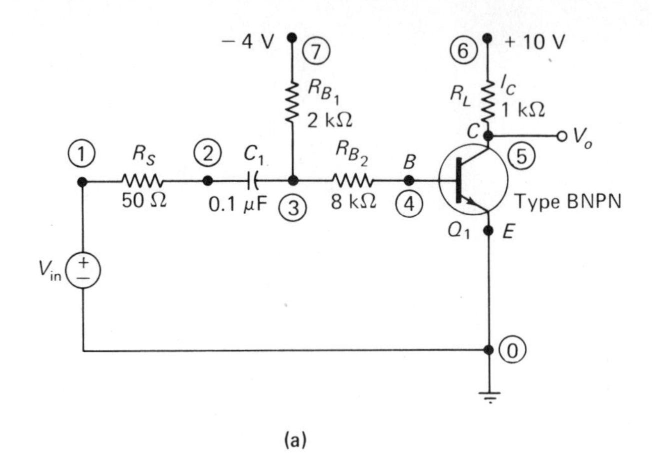

(a)

```
TRANSISTOR SWITCHING TIME
VCC 6 0 DC 10
VBB 7 0 DC -4
VIN 1 0 PULSE 0. 12 2NS 2NS 2NS 100NS
RS 1 2 50
C1  2 3 0.1UF
RB1 7 3 2K
RB2 3 4 8K
RL 6 5 1K
Q1 5 4 0 BNPN
.OUTPUT IVCC VCC PLOT TR
.OUT V5 5 0 PLOT TR
.TRAN 2NS 200NS
.MODEL BNPN NPN BF=100 RB=50 VA=100 CCS=2PF CJE=3PF CJC=2PF TF=0.5NS
+ TR=10NS RC=10 RE=1
.END
```

(b)

Fig. 1-12. Transistor switch.

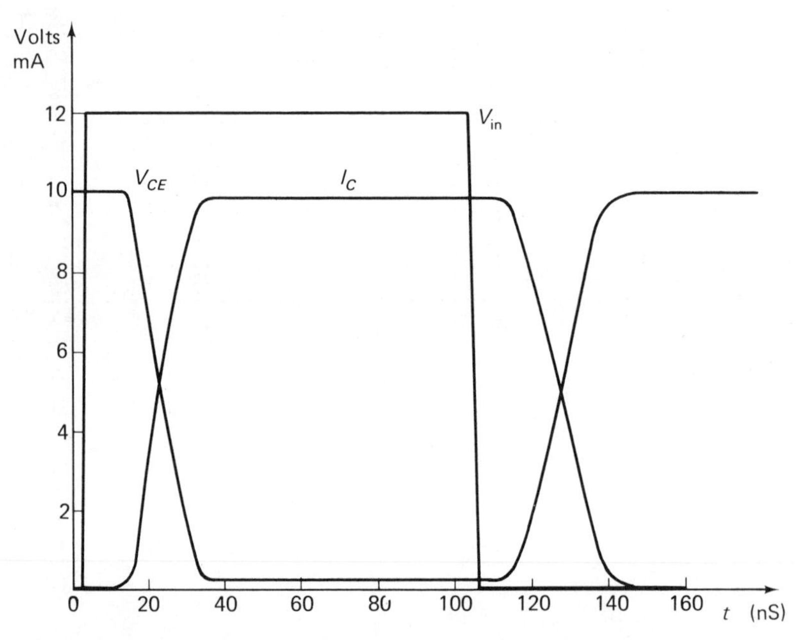

Fig. 1-13. Current and voltage waveforms during switching.

```
*****************************************************************************************************************
TRANSISTOR SWITCHING TIME                                                                       ----- SPICE -----

                  INITIAL TRANSIENT SOLUTION                          TEMPERATURE     27.000  DEG C

*****************************************************************************************************************

 NODE    VOLTAGE     NODE    VOLTAGE     NODE    VOLTAGE     NODE    VOLTAGE     NODE    VOLTAGE

(  1)     0.0000    (  2)     0.0000    (  3)    -4.0000    (  4)    -4.0000    (  5)    10.0000
(  6)    10.0000    (  7)    -4.0000



**** EBERS-MOLL MODEL PARAMETERS
NAME  TYPE   BF      BR      RB     RC     RE     CCS      TF       TR       CJE      CJC      IS      PE   PC    VA     EG
BNPN    N  100.00  1.000   50.0   10.0    1.0 2.00E-12 5.00E-10 1.00E-08 3.00E-12 2.00E-12 1.00E-14 1.00 1.00 100.00 1.11



    TRANSISTOR OPERATING POINTS

NAME   MODEL      IB         IC       VBE     VBC     VCE    BETADC

Q1     BNPN   -2.34E-11  2.38E-11  -4.000 -14.000  10.000    -1.0
```

Fig. 1-14. DC analysis of the transistor switch of Fig. 1-12.

```
****************************************************************************************************************

TRANSISTOR SWITCHING TIME                                                                      ----- SPICE -----

                    TRANSIENT ANALYSIS                                       TEMPERATURE     27.000 DEG C

****************************************************************************************************************

   TIME        V5


                        0.E+00              5.000E+00           1.000E+01           1.500E+01           2.000E+01

0.E+00        1.000E+01
2.000E-09     1.000E+01
4.000E-09     1.011E+01
6.000E-09     1.023E+01
8.000E-09     1.025E+01
1.000E-08     1.023E+01
1.200E-08     1.019E+01
1.400E-08     1.013E+01
1.600E-08     9.460E+00
1.800E-08     8.169E+00
2.000E-08     6.819E+00
2.200E-08     5.555E+00
2.400E-08     4.395E+00
2.600E-08     3.341E+00
2.800E-08     2.399E+00
3.000E-08     1.576E+00
3.200E-08     8.827E-01
3.400E-08     3.315E-01
3.600E-08     2.236E-01
3.800E-08     2.089E-01
4.000E-08     2.000E-01
4.200E-08     1.953E-01
4.400E-08     1.912E-01
4.600E-08     1.891E-01
4.800E-08     1.866E-01
5.000E-08     1.856E-01
5.200E-08     1.840E-01
5.400E-08     1.835E-01
5.600E-08     1.823E-01
5.800E-08     1.822E-01
6.000E-08     1.812E-01
6.200E-08     1.813E-01
6.400E-08     1.805E-01
6.600E-08     1.807E-01
6.800E-08     1.801E-01
7.000E-08     1.803E-01
```

```
7.200E-08   1.797E-01
7.400E-08   1.801E-01
7.600E-08   1.795E-01
7.800E-08   1.799E-01
8.000E-08   1.794E-01
8.200E-08   1.798E-01
8.400E-08   1.793E-01
8.600E-08   1.797E-01
8.800E-08   1.792E-01
9.000E-08   1.796E-01
9.200E-08   1.792E-01
9.400E-08   1.796E-01
9.600E-08   1.792E-01
9.800E-08   1.796E-01
1.000E-07   1.791E-01
1.020E-07   1.795E-01
1.040E-07   1.791E-01
1.060E-07   1.787E-01
1.080E-07   1.856E-01
1.100E-07   2.008E-01
1.120E-07   2.226E-01
1.140E-07   3.300E-01
1.160E-07   7.371E-01
1.180E-07   1.280E+00
1.200E-07   1.898E+00
1.220E-07   2.589E+00
1.240E-07   3.340E+00
1.260E-07   4.143E+00
1.280E-07   4.988E+00
1.300E-07   5.868E+00
1.320E-07   6.773E+00
1.340E-07   7.690E+00
1.360E-07   8.576E+00
1.380E-07   9.258E+00
1.400E-07   9.628E+00
1.420E-07   9.792E+00
1.440E-07   9.862E+00
1.460E-07   9.893E+00
1.480E-07   9.906E+00
1.500E-07   9.914E+00
1.520E-07   9.919E+00
1.540E-07   9.924E+00
1.560E-07   9.929E+00
```

Fig. 1-15. Transient analysis of the transistor switch of Fig. 1-12.

the initial condition for the transient analysis. Included in Fig. 1-14 are the dc node-to-datum voltages, Ebers–Moll model parameters, and the transistor operating point. (The output of the program contains much more information not shown here.) The waveform of V_{CE} is shown in Fig. 1-15. For the purpose of comparison, the waveforms of V_{in}, I_C and V_{CE} are plotted in Fig. 1-13. Note that I_C does not start to rise until about 10 ns after the pulse is applied, and that I_C does not return to zero until about 28 ns after the termination of the input pulse. Also note that the V_{CE} waveform has a rise time of about 20 ns and a fall time of about 25 ns. All these place a limit on the speed at which the transistor can satisfactorily perform as a switch. An explanation of these phenomena can be found in most textbooks on semiconductor physics. In this book we are concerned only with the computer algorithms and numerical techniques that produce accurate solutions, once the device models are properly chosen.

1-3 AN ANATOMY OF COMPUTER SIMULATION PROGRAMS

Most general-purpose computer simulation programs are made up of five main stages: (1) an *input stage*, (2) a *device model retrieval* and *replacement stage*, (3) an *equilibrium equation formulation stage*, (4) a *numerical solution stage*, and (5) an *output stage*. A flow diagram depicting the relationships among these stages is shown in Fig. 1-16, with a summary of the pertinent features in Fig. 1-17 for the first two stages and in

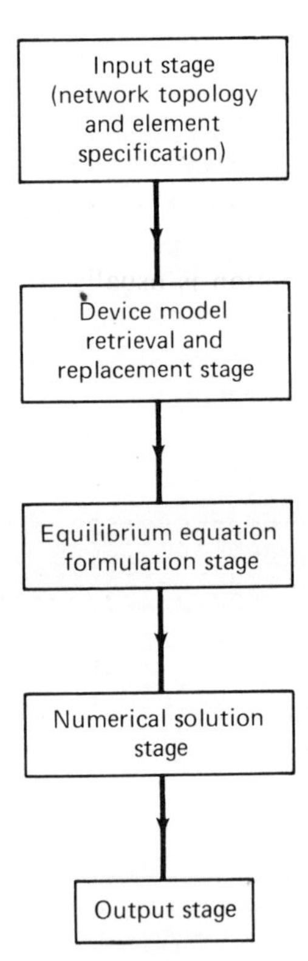

Fig. 1-16. Main stages of computer simulation programs.

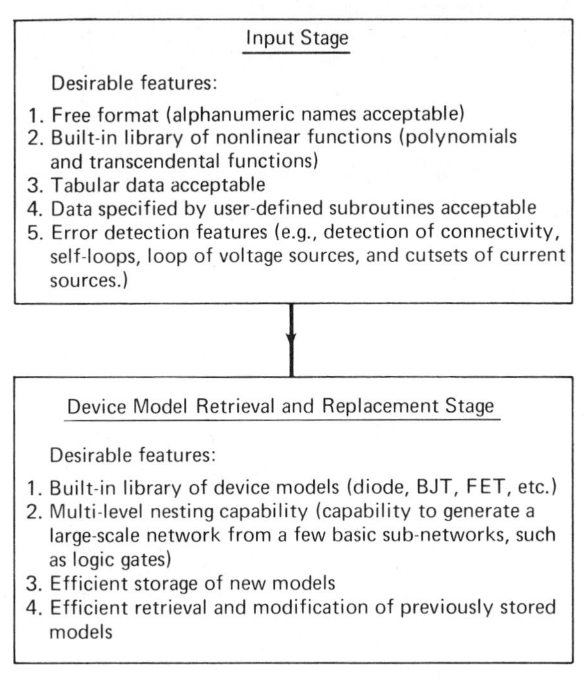

Fig. 1-17. Features of stages 1 and 2.

Fig. 1-18 for the remaining three stages. Here we shall give a brief description of each stage and the chapters in which detailed discussions may be found.

In the input stage the computer receives information from the user with regard to the network configuration, element characteristics, and types of analysis to be performed. Such information is usually conveyed through a deck of punched cards (or the equivalent), as shown in Examples 1-1 to 1-5. The "languages," or the rules, for these input cards vary tremendously in the degree of user convenience. At one extreme, we have the very primitive input language shown in Example 1-4, which uses fixed format and has integer codes for different types of elements (e.g., 22 for a non-linear resistor) and different types of outputs (e.g., 2 for a transfer characteristic). At the other extreme, we have the quite sophisticated input language shown in Example 1-1, which uses free format and has statements that almost read like ordinary English. Some other desirable features of user convenience in the input stage are listed in the first block of Fig. 1-17. To provide such features increases the programming effort considerably. On the average, in any *user-oriented* computer simulation program, about 40 per cent of the codes are used to process the input language and to provide extensive diagnostic messages. Some people tend to regard most of these features as conveniences rather than essentials to the simulation programs. However, if any simulation program is to gain wide acceptance, especially among engineers, user convenience is among the topmost factors to be considered.

The second stage *handles* device models. This stage is usually not needed for small simulation programs and for programs written for instructional uses, but

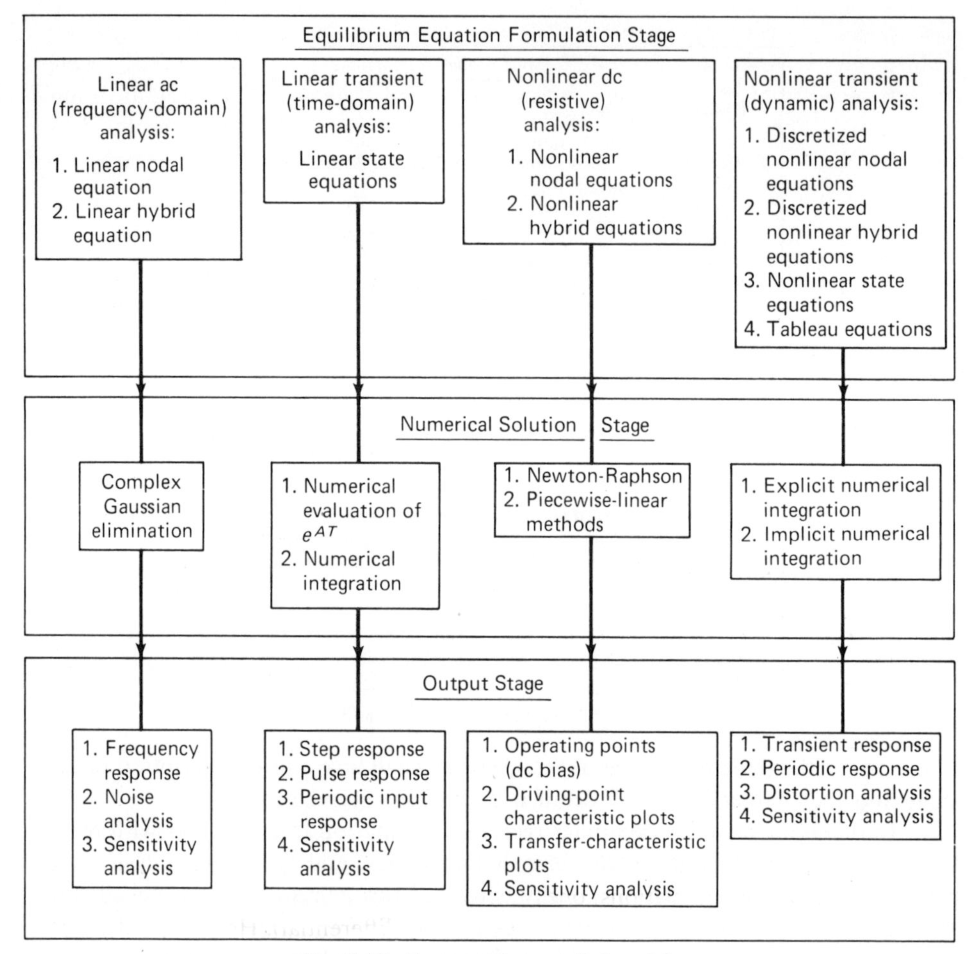

Fig. 1-18. Features of stages 3, 4, and 5.

becomes very important in programs used for designing electronic circuits. Whenever a device, e.g., transistor type 2N3705, appears very often in some circuits, it is a tedious job to describe the device parameters repeatedly to the computer; some devices may require as many as 36 parameters for complete characterization. The model-library feature is very useful in such cases. For example, to describe a complex logic circuit, each basic transistor need only be described once, be assigned a model name, have its external nodes specified, and then be stored in the library. Later, whenever that particular transistor type is needed, all one has to do is to call for that transistor type by its model name, at the same time indicating where the terminals are to be connected. Models can be nested to a number of levels (e.g., 20 levels). In this manner, very complex circuits can be described by the user with fairly simple statements. Note that the model library and nested models do not make a complex circuit simple for the computer to analyze. They only save the user's effort when describing the circuit.

Besides the provisions for defining and retrieving a model, the model library should also permit modification, renaming, and restorage of any model. For those built-in models in the library (i.e., models whose configurations have been predetermined, such as the Ebers–Moll model for a bipolar transistor), the program should also provide a reasonable set of "default" parameter values. In this way, a beginning user need not be bothered with all the details of a model if they are of no importance to the solution of his problem.

A detailed study of the implementation of the first two stages requires considerable knowledge of computer programming and of the operation of peripheral devices, such as disks and magnetic tapes. Such a study is beyond the scope of this book.

In the third stage, the program formulates equilibrium equations for the network whose configuration and element values have been completely specified. A summary of the methods used for this stage is shown in the first block of Fig. 1-18. There are three methods widely used in computer simulation programs: (1) the nodal method, (2) the hybrid method, and (3) the state-variable method. Each method has its advantages and disadvantages, as will become clear in the subsequent chapters.

For dc and ac analysis of linear networks, we discuss the formulation of nodal equations in Chapter 4 and hybrid equations in Chapter 6. For transient analysis (as well as ac analysis) of linear networks, we discuss the formulation of state equations in great detail in Chapter 8. For dc analysis of nonlinear *resistive* networks, we discuss the formulation of nodal equations in Chapter 5 and hybrid equations in Chapter 7. For transient analysis of nonlinear *dynamic* networks, the formulation of state equations is discussed in Chapter 10. The discretized nonlinear nodal equations and hybrid equations are described in Section 1-9.

Any lumped network is governed by three types of constraint equations arising from the Kirchhoff voltage law, Kirchhoff current law, and element characteristics. These constraints constitute a system of equations; some are algebraic equations, and the remaining are, in general, nonlinear differential equations. In formulating the nodal, hybrid, or state equations, one motive has been to reduce the number of equations to be solved *simultaneously* (algebraic and/or differential). However, with recent advances in sparse-matrix techniques, we no longer consider a large number of simultaneous equations as necessarily a computational woe, provided that the system is sparse. As a consequence, the tableau method is emerging as an important method in computer simulation programs. The tableau method, which includes all network information in nonreduced form, is discussed in Chapter 17.

In the fourth stage, the equilibrium equations are solved *numerically* (as opposed to analytic solutions). For the solution of linear algebraic equations, whether resulting from nodal or hybrid analysis, with real or complex coefficients, we discuss the Gaussian elimination method and the ***LU*** decomposition method in Chapter 4. The latter method is again considered in Chapter 16, with necessary restructuring of the procedure to deal with sparse matrices. For the solution of nonlinear algebraic equations, which may be the result of nodal or hybrid analysis, we discuss the Newton–Raphson method in Chapter 5 and piecewise-linear methods in Chapter 7. For linear state equations, the solution in the time domain may be obtained with the aid of the matrix exponential e^{At}. Such a method is discussed in Chapter 9. For nonlinear state equations, analytic solution is in general not possible, and we have to apply numerical integration techniques. The explicit and implicit numerical integration techniques are discussed in Chapters 11 to 13.

Last, but not the least important in a computer simulation program, is the output stage, for this is where the user obtains his answers. A variety of output capabilities is shown in the third block of Fig. 1-18, some of which have been illustrated in Examples 1-1 to 1-5.

Most general-purpose computer simulation programs contain the basic subroutines for dc, ac, and transient analyses. When other types of outputs are desired, special subroutines have to be written. A few such special topics have been included in our discussions. In Chapter 14 we discuss the generation of symbolic network functions (for linear networks). In Chapter 15, we present different methods of sensitivity analysis. Recent advances in the solution of periodic response and distortion analysis are described in Chapter 17.

From the preceding discussions it is clear that our main concern in this book is computer-aided circuit *analysis*, which is the first step toward computer-aided circuit *design*. The automated design of electronic circuits, a subject of extensive research at present, requires a good analysis program and a good optimization strategy. Good references are now available on various aspects of computer-aided circuit design [6–9].

1-4 A GLIMPSE AT EQUATION FORMULATION VIA THE LINEAR *n*-PORT HYBRID-ANALYSIS APPROACH

Among the many distinct methods for formulating network equilibrium equations for resistive networks, the n-port hybrid-analysis approach will be emphasized in this book (see Chapters 6 and 7); not only is it more general, but this method has not been adequately treated in existing textbooks. The hybrid approach is particularly useful for nonlinear networks containing a large percentage of *linear* resistors and controlled sources. The basic philosophy is to form an n-port $\mathfrak{N}$ from the given network by *extracting* an appropriate set of two-terminal elements so that the resultant n-port contains only linear resistors, independent sources, and linear controlled sources. The extracted elements, which are usually nonlinear resistors or energy storage elements, are considered as external loads across the ports. The n-port $\mathfrak{N}$ can be characterized by a hybrid equation of the form

$$\begin{bmatrix} \boldsymbol{i}_a \\ \boldsymbol{v}_b \end{bmatrix} = \boldsymbol{H}\begin{bmatrix} \boldsymbol{v}_a \\ \boldsymbol{i}_b \end{bmatrix} + \boldsymbol{s}(t) = \begin{bmatrix} \boldsymbol{H}_{aa} & \boldsymbol{H}_{ab} \\ \boldsymbol{H}_{ba} & \boldsymbol{H}_{bb} \end{bmatrix}\begin{bmatrix} \boldsymbol{v}_a \\ \boldsymbol{i}_b \end{bmatrix} + \begin{bmatrix} \boldsymbol{s}_a \\ \boldsymbol{s}_b \end{bmatrix} \tag{1-1}$$

where $\boldsymbol{H}$ is an $n \times n$ real matrix, called the *hybrid matrix*, and $\boldsymbol{s}$ an $n \times 1$ *source vector* to account for the presence of independent sources inside $\mathfrak{N}$. The term *hybrid analysis* arises from the mixture of both voltages and currents as independent variables in (1-1). Several efficient computer algorithms for generating $\boldsymbol{H}$ and $\boldsymbol{s}$ via topological techniques are presented in Chapter 6. Once Eq. (1-1) is obtained, the internal elements of the n-port $\mathfrak{N}$ are no longer needed in any subsequent computations. Hence, for linear or nonlinear dynamic networks, this approach enables one to iterate only on the external port variables. Since only one computation needs to be done on the internal variables, the hybrid-analysis approach becomes more advantageous as the percentage of linear resistive elements increases.

To illustrate the essence of the n-port hybrid-analysis approach, consider the transistor amplifier circuit shown in Fig. 1-19(a). We wish to derive the state equations for the associated *linear incremental network* shown in Fig. 1-19(b), which is obtained by replacing the transistor with its hybrid-pi model; for the sake of simplicity, we neglect the 20-μF coupling capacitor. A two-port $\mathfrak{N}$ is then formed by extracting the two capacitors as shown in Fig. 1-19(c). Our next task is a simple one: Call the sub-

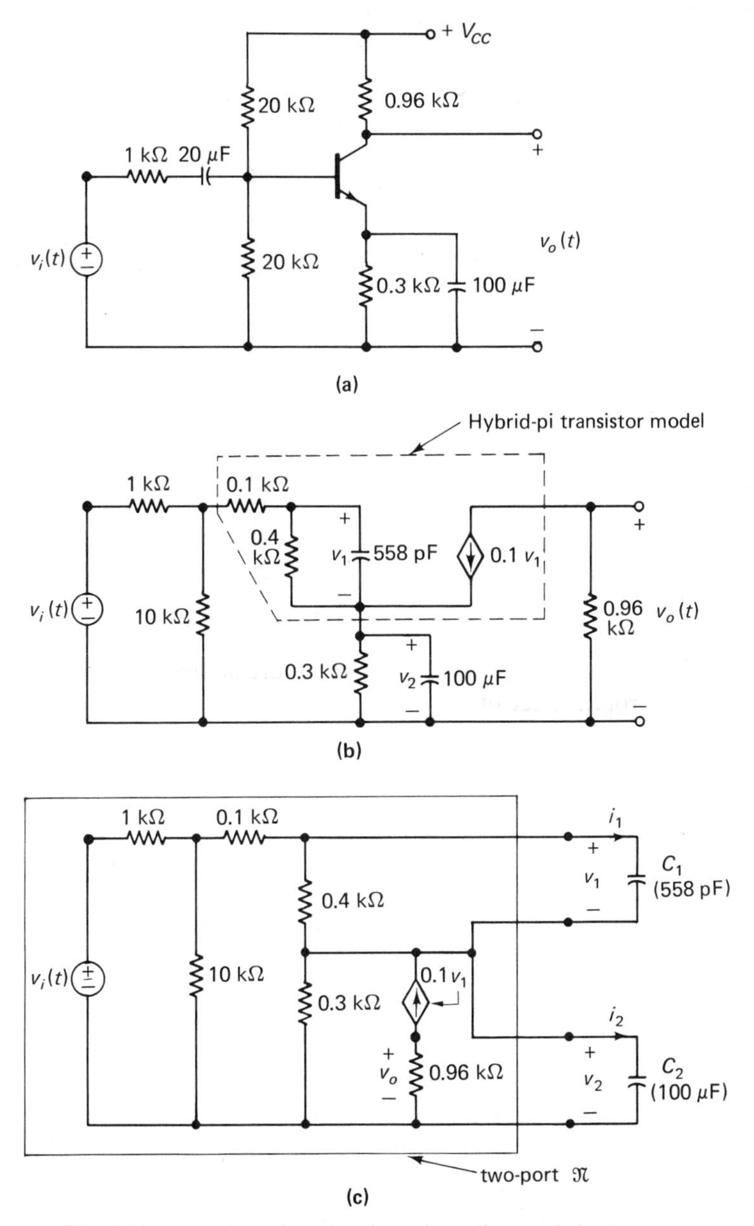

Fig. 1-19. A transistor circuit is redrawn into a linear resistive two-port terminated by two capacitors.

routine HYBRID (see Chapter 6) and obtain

$$\begin{bmatrix} i_1 \\ i_2 \end{bmatrix} = \begin{bmatrix} -0.003491 & -0.000991 \\ 0.09901 & -0.004324 \end{bmatrix} \begin{bmatrix} v_1 \\ v_2 \end{bmatrix} + \begin{bmatrix} 0.0009009 & v_i(t) \\ 0.0009009 & v_i(t) \end{bmatrix} \tag{1-2}$$

Equation (1-2) is a special case of Eq. (1-1) since the independent variables happen to be both voltages. As will be shown later (Chapters 6, 7, and 10), there are situations in which both voltages and currents must be chosen as independent variables.

The two capacitors are characterized by

$$i_1 = C_1 \frac{dv_1}{dt} \tag{1-3}$$

$$i_2 = C_2 \frac{dv_2}{dt} \tag{1-4}$$

To derive the state equations with v_1 and v_2 as state variables, we substitute Eqs. (1-3) and (1-4) for i_1 and i_2 in Eq. (1-2). After some simple operations, we obtain

$$\begin{bmatrix} \dot{v}_1 \\ \dot{v}_2 \end{bmatrix} = \begin{bmatrix} -6.28 \times 10^7 & -1.792 \times 10^7 \\ 990 & -43 \end{bmatrix} \begin{bmatrix} v_1 \\ v_2 \end{bmatrix} + \begin{bmatrix} 1.613 \times 10^7 \\ 9.009 \end{bmatrix} v_i(t) \tag{1-5}$$

Equation (1-5) is of the form

$$\dot{\boldsymbol{x}} = \boldsymbol{A}\boldsymbol{x} + \boldsymbol{B}\boldsymbol{u} \tag{1-6}$$

and is the desired state equation.

An examination of the $\boldsymbol{A}$ matrix in (1-5) shows that the elements differ by several orders of magnitude. In fact, if we calculate the eigenvalues of this matrix, or the natural frequencies of the network, by solving the equation

$$\det(\lambda \mathbf{1} - \boldsymbol{A}) = 0 \tag{1-7}$$

we find that

$$\lambda_1 = -6.28 \times 10^7 \tag{1-8}$$

$$\lambda_2 = -3.26 \times 10^2 \tag{1-9}$$

The two natural frequencies have a ratio of about 200,000. We shall demonstrate in Section 1-6 that this situation is responsible for a very serious numerical problem.

The preceding example shows very clearly that the crux of the problem in deriving state equations is to obtain the hybrid matrix of a resistive n-port. Detailed discussions of this approach will be given in Chapters 8 and 10.

As a final example illustrating the advantage of the n-port hybrid-analysis approach, let us replace capacitors C_1 and C_2 in Fig. 1-19(b) by nonlinear resistors characterized by

$$i_1 = g_1(v_1) \tag{1-10}$$

$$i_2 = g_2(v_2) \tag{1-11}$$

Substituting these two relationships into Eq. (1-2), we obtain the following two nonlinear equations with v_1 and v_2 as the unknown variables.

$$g_1(v_1) + 0.003491 v_1 + 0.000991 v_2 = 0.0009009 v_i(t) \tag{1-12}$$

$$g_2(v_2) - 0.09901 v_1 + 0.004324 v_2 = 0.0009009 v_i(t) \tag{1-13}$$

At each time $t = t_j$, Eqs. (1-12) and (1-13) can be solved numerically on the computer via the *Newton–Raphson method* (Chapter 5), provided that the nonlinear functions $g_1(v_1)$ and $g_2(v_2)$ are such that a *unique* solution exists. On the other hand, if the network has more than one solution, each nonlinear function could be approximated by a piecewise-linear curve, and the resultant network could then be solved by the combinatorial technique to be presented in Chapter 7. In any case, observe that we have only two nonlinear equations using the n-port hybrid-analysis approach. Since the network in Fig. 1-19(b) has six nodes, the more common *nodal-analysis method* (Chapter 5) would have required the solution of five nonlinear equations.

Another advantage of the hybrid-analysis approach is that it allows the nonlinearities to be either *voltage-controlled* (e.g., a tunnel diode) or *current-controlled* (e.g., a glow lamp), a flexibility not shared by the nodal-analysis method. For example, in Fig. 1-19(b) let C_1 and C_2 be replaced by nonlinear resistors characterized by $i_1 = g_1(v_1)$ and $v_2 = f_2(i_2)$, respectively. We assume that i_2 cannot be expressed as a single-valued function of v_2. Then we have no choice but to consider i_2 as one of the two unknown variables. In this case, nodal analysis is no longer applicable since only node voltages are permissible unknowns in nodal equations. Using hybrid analysis, on the other hand, we have no difficulty at all. As before, we can obtain the following hybrid equation for the two-port with the aid of the subroutine HYBRID:

$$\begin{bmatrix} i_1 \\ v_2 \end{bmatrix} = \begin{bmatrix} -26{,}180 & 0.2292 \\ 22.90 & -231.3 \end{bmatrix} \begin{bmatrix} v_1 \\ i_2 \end{bmatrix} + \begin{bmatrix} -0.0002064v_i(t) \\ 0.2084v_i(t) \end{bmatrix} \tag{1-14}$$

Substituting $i_1 = g_1(v_1)$ and $v_2 = f_2(i_2)$ into Eq. (1-14), we obtain the following two nonlinear equations:

$$g_1(v_1) + 21{,}680v_1 - 0.2292i_2 = -0.0002064v_i(t) \tag{1-15}$$

$$f_2(i_2) - 22.90v_1 + 231.3i_2 = 0.2084v_i(t) \tag{1-16}$$

Once again, Eqs. (1-15) and (1-16) can be solved either by the Newton–Raphson method of Chapter 5 or the combinatorial method of Chapter 7.

1-5 A GLIMPSE AT SOME NUMERICAL INTEGRATION ALGORITHMS AND THEIR NUMERICAL STABILITY CHARACTERISTICS

Since the weakest link in any computer simulation program is the numerical integration algorithm used during transient analysis, it is instructive to examine the properties of a few common algorithms. Although these algorithms are intended for obtaining numerical solutions of a system of nonlinear differential equations of the form

$$\dot{\mathbf{x}} = \mathbf{f}(\mathbf{x}, t) \tag{1-17}$$

it is standard practice for most numerical analysts to apply each algorithm to the "test equation"

$$\dot{x} = f(x) = -\lambda x \tag{1-18}$$

associated with a *first-order* (single time constant) *linear circuit*. The exact analytical solution to Eq. (1-18) is given by

$$x(t) = x_0 e^{-\lambda t}, \qquad t \geq 0 \tag{1-19}$$

where $x_0 \triangleq x(0)$ is the initial condition. The reason for choosing Eq. (1-18) as the test equation is not only because it has an *exact* analytical solution to which other numerical solutions can be compared, but also because, *incrementally* speaking, most solutions to a differential equation can be approximated by a portion of an exponential. Consequently, if an algorithm fails to solve Eq. (1-18), chances are that it will also fail when applied to other differential equations.

To solve Eq. (1-17) using numerical integration algorithms, we locate on the time axis a set of points $t_0, t_1, \ldots$, and determine the corresponding values $x_0, x_1, \ldots$. A typical set of solution points is shown in Fig. 1-20(a) along with the exact solution. The difference $\Delta t_j \triangleq t_j - t_{j-1}$ is called the *step size*. For simplicity, we assume a *uniform* step size "h." The concept and principle used to derive numerical integration algorithms will be presented in Chapters 11 to 13. Some special integration algorithms applicable only to linear networks will be discussed in Chapter 9. For the present discussion, let us examine the following three simple algorithms for solving $\dot{x} = f(x)$:

Forward Euler algorithm

$$x_{n+1} = x_n + hf(x_n) \tag{1-20}$$

Backward Euler algorithm

$$x_{n+1} = x_n + hf(x_{n+1}) \tag{1-21}$$

Trapezoidal algorithm

$$x_{n+1} = x_n + \frac{h}{2}[f(x_n) + f(x_{n+1})] \tag{1-22}$$

Applying each algorithm to solve Eq. (1-18), we obtain the following simplified expressions:

Forward Euler solution of $\dot{x} = -\lambda x$

$$x_{n+1} = (1 - h\lambda)x_n \tag{1-23}$$

Backward Euler solution of $\dot{x} = -\lambda x$

$$x_{n+1} = \frac{x_n}{1 + h\lambda} \tag{1-24}$$

Trapezoidal solution of $\dot{x} = -\lambda x$

$$x_{n+1} = \left(\frac{1 - \dfrac{h\lambda}{2}}{1 + \dfrac{h\lambda}{2}} \right) x_n \tag{1-25}$$

Since the solution obtained by numerical integration methods is never exact, it is important that we know how much error is incurred at each time step, as well as

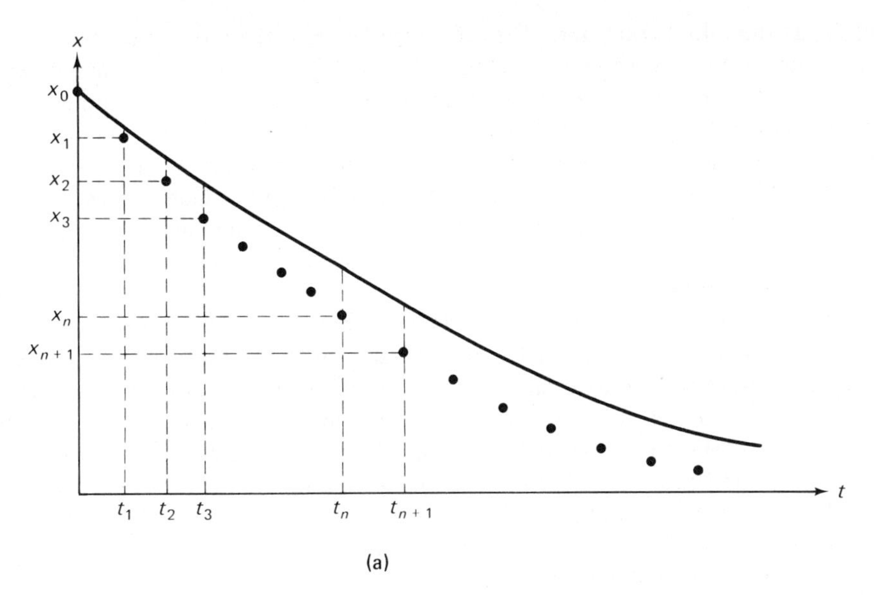

(a)

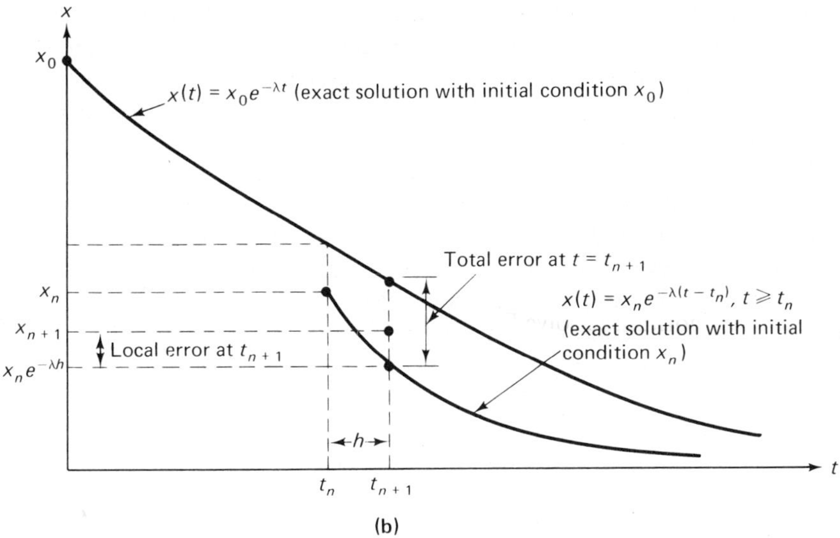

(b)

Fig. 1-20. Difference between local and total errors.

how much error is accumulated after some prescribed time interval. Measures of these errors for the test equation $\dot{x} = -\lambda x$ are given by

$$\text{Local error at } t = t_{n+1} \triangleq (x_n e^{-\lambda h}) - x_{n+1} \tag{1-26}$$

$$\text{Total error at } t = t_{n+1} \triangleq (x_0 e^{-\lambda t_{n+1}}) - x_{n+1} \tag{1-27}$$

The local error defined by Eq. (1-26) can be interpreted as the error occurring at $t = t_{n+1}$, assuming that x_n is the initial condition. In contrast, the total error defined by Eq. (1-27) can be interpreted as the actual error accumulated from $t = 0$ to $t = t_{n+1}$, with x_0 as the initial condition. The difference between these two measures of error is illustrated in Fig. 1-20(b). Observe that since the local error at $t = t_{n+1}$ may be either positive or negative, the total error which accounts in part for the accumulation of local errors may or may not grow with time. It is entirely conceivable that after a few time steps, the local errors may partially cancel out each other owing to variation in sign. *Roughly speaking*, a numerical integration algorithm which has the desirable property that its total error does not get amplified but actually decreases with time is said to be *numerically stable*. Algorithms that do not possess this property are said to be numerically unstable. Clearly, even if the local error is small, the total error of an unstable algorithm will eventually be magnified sufficiently to make the resulting solution useless. One major goal in Chapter 12 is to uncover the origin of numerical instability so that we can determine which algorithms are numerically stable. In this section, we shall only run some numerical experiments. In particular, let us compute Eqs. (1-23)–(1-25) with $\lambda = 40$ and $x_0 = 0.01$. The results corresponding to a uniform step size $h = 0.01$ are shown in Fig. 1-21(a), (b), and (c) for the forward Euler, backward Euler, and trapezoidal algorithms, respectively. In each case, the *exact solution* (neglecting round-off errors) is tabulated in the second column and the *computed solution* in the third column. The local and total errors at each time step are also tabulated in the last two columns. A comparison of these results shows that the *trapezoidal algorithm has the least error* among the three algorithms, and is therefore the best choice for this example *when the step size is chosen to be* $h = 0.01$. Since the error is really negligible in each case, any one of the three algorithms would have been acceptable for all practical purposes. However, this is not true if a larger step size is used. To illustrate this point, let us repeat the above numerical experiment with a step size ten times larger: $h = 0.1$. The corresponding solutions are shown in Fig. 1-22(a), (b), and (c), respectively. Observe that whereas the backward Euler and trapezoidal algorithms continue to give reasonably small local and total errors, the errors for the forward Euler algorithm become exceedingly large, thereby rendering the solution completely meaningless. This example demonstrates without any doubt that the forward Euler algorithm is numerically unstable when *large step sizes* are taken. This observation is not surprising if we examine Eq. (1-23) and see that

$$
\begin{aligned}
x_1 &= (1 - \lambda h)x_0 \\
x_2 &= (1 - \lambda h)x_1 = (1 - \lambda h)^2 x_0 \\
&\;\vdots \\
x_n &= (1 - \lambda h)^n x_0
\end{aligned}
$$

Clearly, if $|1 - \lambda h| > 1$, then $x_n \longrightarrow \infty$ as $n \longrightarrow \infty$. Hence, to ensure numerical stability of the forward Euler algorithm, we must require that $|1 - \lambda h| < 1$ or, equivalently, $0 < \lambda h < 2$. Since λ is a positive real number in the present case, the condition becomes

$$h < \frac{2}{\lambda} \tag{1-28}$$

In the preceding example, we have $\lambda = 40$, and hence the step size h must be smaller

STEP SIZE = .01

TIME	EXACT SOLUTION	FORWARD EULER	LOCAL ERROR	TOTAL ERROR
0.	.10000E−01	.10000E−01	0.	0.
.10000E−01	.67032E−02	.60000E−02	.70320E−03	.70320E−03
.20000E−01	.44933E−02	.36000E−02	.42192E−03	.89329E−03
.30000E−01	.30119E−02	.21600E−02	.25315E−03	.85194E−03
.40000E−01	.20190E−02	.12960E−02	.15189E−03	.72297E−03
.50000E−01	.13534E−02	.77760E−03	.91135E−04	.57575E−03
.60000E−01	.90718E−03	.46656E−03	.54681E−04	.44062E−03
.70000E−01	.60810E−03	.27994E−03	.32809E−04	.32816E−03
.80000E−01	.40762E−03	.16796E−03	.19685E−04	.23966E−03
.90000E−01	.27324E−03	.10078E−03	.11811E−04	.17246E−03
.10000E+00	.18316E−03	.60466E−04	.70866E−05	.12269E−03
.11000E+00	.12277E−03	.36280E−04	.42520E−05	.86494E−04
.12000E+00	.82297E−04	.21768E−04	.25512E−05	.60530E−04
.13000E+00	.55166E−04	.13061E−04	.15307E−05	.42105E−04
.14000E+00	.36979E−04	.78364E−05	.91843E−06	.29142E−04
.15000E+00	.24788E−04	.47018E−05	.55106E−06	.20086E−04
.16000E+00	.16616E−04	.28211E−05	.33063E−06	.13794E−04
.17000E+00	.11138E−04	.16927E−05	.19838E−06	.94451E−05
.18000E+00	.74659E−05	.10156E−05	.11903E−06	.64503E−05
.19000E+00	.50045E−05	.60936E−06	.71417E−07	.43952E−05
.20000E+00	.33546E−05	.36562E−06	.42850E−07	.29890E−05

(a)

TIME	EXACT SOLUTION	BACKWARD EULER	LOCAL ERROR	TOTAL ERROR
0.	.10000E−01	.10000E−01	0.	0.
.10000E−01	.67032E−02	.71429E−02	−.43966E−03	−.43966E−03
.20000E−01	.44933E−02	.51020E−02	−.31404E−03	−.60875E−03
.30000E−01	.30119E−02	.36443E−02	−.22431E−03	−.63237E−03
.40000E−01	.20190E−02	.26031E−02	−.16022E−03	−.58412E−03
.50000E−01	.13534E−02	.18593E−02	−.11445E−03	−.50599E−03
.60000E−01	.90718E−03	.13281E−02	−.81747E−04	−.42092E−03
.70000E−01	.60810E−03	.94865E−03	−.58391E−04	−.34054E−03
.80000E−01	.40762E−03	.67760E−03	−.41708E−04	−.26998E−03
.90000E−01	.27324E−03	.48400E−03	−.29791E−04	−.21077E−03
.10000E+00	.18316E−03	.34572E−03	−.21279E−04	−.16256E−03
.11000E+00	.12277E−03	.24694E−03	−.15200E−04	−.12417E−03
.12000E+00	.82297E−04	.17639E−03	−.10857E−04	−.94088E−04
.13000E+00	.55166E−04	.12599E−03	−.77549E−05	−.70824E−04
.14000E+00	.36979E−04	.89993E−04	−.55392E−05	−.53014E−04
.15000E+00	.24788E−04	.64281E−04	−.39566E−05	−.39493E−04
.16000E+00	.16616E−04	.45915E−04	−.28261E−05	−.29299E−04
.17000E+00	.11138E−04	.32796E−04	−.20187E−05	−.21658E−04
.18000E+00	.74659E−05	.23426E−04	−.14419E−05	−.15960E−04
.19000E+00	.50045E−05	.16733E−04	−.10299E−05	−.11728E−04
.20000E+00	.33546E−05	.11952E−04	−.73567E−06	−.85973E−05

(b)

TIME	EXACT SOLUTION	TRAPEZOIDAL	LOCAL ERROR	TOTAL ERROR
0.	.10000E−01	.10000E−01	0.	0.
.10000E−01	.67032E−02	.66667E−02	.36534E−04	.36534E−04
.20000E−01	.44933E−02	.44444E−02	.24356E−04	.48845E−04
.30000E−01	.30119E−02	.29630E−02	.16237E−04	.48979E−04
.40000E−01	.20190E−02	.19753E−02	.10825E−04	.43657E−04
.50000E−01	.13534E−02	.13169E−02	.72166E−05	.36480E−04
.60000E−01	.90718E−03	.87791E−03	.48110E−05	.29265E−04
.70000E−01	.60810E−03	.58528E−03	.32074E−05	.22824E−04
.80000E−01	.40762E−03	.39018E−03	.21382E−05	.17438E−04
.90000E−01	.27324E−03	.26012E−03	.14255E−05	.13114E−04
.10000E+00	.18316E−03	.17342E−03	.95033E−06	.97411E−05
.11000E+00	.12277E−03	.11561E−03	.63355E−06	.71632E−05
.12000E+00	.82297E−04	.77073E−04	.42237E−06	.52240E−05
.13000E+00	.55166E−04	.51382E−04	.28158E−06	.37833E−05
.14000E+00	.36979E−04	.34255E−04	.18772E−06	.27238E−05
.15000E+00	.24788E−04	.22837E−04	.12515E−06	.19509E−05
.16000E+00	.16616E−04	.15224E−04	.83431E−07	.13912E−05
.17000E+00	.11138E−04	.10150E−04	.55620E−07	.98816E−06
.18000E+00	.74659E−05	.67664E−05	.37080E−07	.69946E−06
.19000E+00	.50045E−05	.45109E−05	.24720E−07	.49358E−06
.20000E+00	.33546E−05	.30073E−05	.16480E−07	.34734E−06

(c)

Fig. 1-21. Comparison of errors in forward Euler, backward Euler, and trapezoidal solutions of $x = -40x$, with $x(0) = 0.1$ and $h = 0.01$.

STEP SIZE = .10

TIME	EXACT SOLUTION	FORWARD EULER	LOCAL ERROR	TOTAL ERROR
0.	.10000E-01	.10000E-01	0.	0.
.10000E+00	.18316E-03	-.30000E-01	.30183E-01	.30183E-01
.20000E+00	.33546E-05	.90000E-01	-.90549E-01	-.89997E-01
.30000E+00	.61442E-07	-.27000E+00	.27165E+00	.27000E+00
.40000E+00	.11254E-08	.81000E+00	-.81495E+00	-.81000E+00
.50000E+00	.20612E-10	-.24300E+01	.24448E+01	.24300E+01
.60000E+00	.37751E-12	.72900E+01	-.73345E+01	-.72900E+01
.70000E+00	.69144E-14	-.21870E+02	.22004E+02	.21870E+02
.80000E+00	.12664E-15	.65610E+02	-.66011E+02	-.65610E+02
.90000E+00	.23195E-17	-.19683E+03	.19803E+03	.19683E+03
.10000E+01	.42484E-19	.59049E+03	-.59410E+03	-.59049E+03
.11000E+01	.77811E-21	-.17715E+04	.17823E+04	.17715E+04
.12000E+01	.14252E-22	.53144E+04	-.53469E+04	-.53144E+04
.13000E+01	.26103E-24	-.15943E+05	.16041E+05	.15943E+05
.14000E+01	.47809E-26	.47830E+05	-.48122E+05	-.47830E+05
.15000E+01	.87565E-28	-.14349E+06	.14437E+06	.14349E+06
.16000E+01	.16038E-29	.43047E+06	-.43310E+06	-.43047E+06
.17000E+01	.29375E-31	-.12914E+07	.12993E+07	.12914E+07
.18000E+01	.53802E-33	.38742E+07	-.38979E+07	-.38742E+07
.19000E+01	.98542E-35	-.11623E+08	.11694E+08	.11623E+08
.20000E+01	.18049E-36	.34868E+08	-.35081E+08	-.34868E+08

(a)

TIME	EXACT SOLUTION	BACKWARD EULER	LOCAL ERROR	TOTAL ERROR
0.	.10000E-01	.10000E-01	0.	0.
.10000E+00	.18316E-03	.20000E-02	-.18168E-02	-.18168E-02
.20000E+00	.33546E-05	.40000E-03	-.36337E-03	-.39665E-03
.30000E+00	.61442E-07	.80000E-04	-.72674E-04	-.79939E-04
.40000E+00	.11254E-08	.16000E-04	-.14535E-04	-.15999E-04
.50000E+00	.20612E-10	.32000E-05	-.29069E-05	-.32000E-05
.60000E+00	.37751E-12	.64000E-06	-.58139E-06	-.64000E-06
.70000E+00	.69144E-14	.12800E-06	-.11628E-06	-.12800E-06
.80000E+00	.12664E-15	.25600E-07	-.23256E-07	-.25600E-07
.90000E+00	.23195E-17	.51200E-08	-.46511E-08	-.51200E-08
.10000E+01	.42484E-19	.10240E-08	-.93022E-09	-.10240E-08
.11000E+01	.77811E-21	.20480E-09	-.18604E-09	-.20480E-09
.12000E+01	.14252E-22	.40960E-10	-.37209E-10	-.40960E-10
.13000E+01	.26103E-24	.81920E-11	-.74418E-11	-.81920E-11
.14000E+01	.47809E-26	.16384E-11	-.14884E-11	-.16384E-11
.15000E+01	.87565E-28	.32768E-12	-.29767E-12	-.32768E-12
.16000E+01	.16038E-29	.65536E-13	-.59534E-13	-.65536E-13
.17000E+01	.29375E-31	.13107E-13	-.11907E-13	-.13107E-13
.18000E+01	.53802E-33	.26214E-14	-.23814E-14	-.26214E-14
.19000E+01	.98542E-35	.52429E-15	-.47627E-15	-.52429E-15
.20000E+01	.18049E-36	.10486E-15	-.95255E-16	-.10486E-15

(b)

TIME	EXACT SOLUTION	TRAPEZOIDAL	LOCAL ERROR	TOTAL ERROR
0.	.10000E-01	.10000E-01	0.	0.
.10000E+00	.18316E-03	-.33333E-02	.35165E-02	.35165E-02
.20000E+00	.33546E-05	.11111E-02	-.11722E-02	-.11078E-02
.30000E+00	.61442E-07	-.37037E-03	.39072E-03	.37043E-03
.40000E+00	.11254E-08	.12346E-03	-.13024E-03	-.12346E-03
.50000E+00	.20612E-10	-.41152E-04	.43413E-04	.41152E-04
.60000E+00	.37751E-12	.13717E-04	-.14471E-04	-.13717E-04
.70000E+00	.69144E-14	-.45725E-05	.48237E-05	.45725E-05
.80000E+00	.12664E-15	.15242E-05	-.16079E-05	-.15242E-05
.90000E+00	.23195E-17	-.50805E-06	.53597E-06	.50805E-06
.10000E+01	.42484E-19	.16935E-06	-.17866E-06	-.16935E-06
.11000E+01	.77811E-21	-.56450E-07	.59552E-07	.56450E-07
.12000E+01	.14252E-22	.18817E-07	-.19851E-07	-.18817E-07
.13000E+01	.26103E-24	-.62723E-08	.66169E-08	.62723E-08
.14000E+01	.47809E-26	.20908E-08	-.22056E-08	-.20908E-08
.15000E+01	.87565E-28	-.69692E-09	.73521E-09	.69692E-09
.16000E+01	.16038E-29	.23231E-09	-.24507E-09	-.23231E-09
.17000E+01	.29375E-31	-.77435E-10	.81690E-10	.77435E-10
.18000E+01	.53802E-33	.25812E-10	-.27230E-10	-.25812E-10
.19000E+01	.98542E-35	-.86039E-11	.90767E-11	.86039E-11
.20000E+01	.18049E-36	.28680E-11	-.30256E-11	-.28680E-11

(c)

Fig. 1-22. Repeating the comparison of Fig. 1-21 with a step size ten times larger ($h = 0.1$).

than $\frac{2}{40} = 0.05$ to avoid numerical instability. This explains why numerical instability occurs in Fig. 1-22(a) where we have used $h = 0.1 > 0.05$.

In a similar manner, we see from Eq. (1-24) that the backward Euler algorithm gives

$$x_n = \frac{1}{(1 + h\lambda)^n} x_0$$

and from Eq. (1-25) that the trapezoidal algorithm gives

$$x_n = \left(\frac{1 - \dfrac{h\lambda}{2}}{1 + \dfrac{h\lambda}{2}} \right)^n x_0$$

In both cases, we have $x_n \longrightarrow 0$ as $n \longrightarrow \infty$, regardless of the step size h. Hence, we conclude that both the *backward Euler and the trapezoidal algorithms are numerically stable.* A much more comprehensive study of this phenomenon will be given in Chapters 12 and 13.

1-6 A GLIMPSE AT A STIFF DIFFERENTIAL EQUATION AND ITS ASSOCIATED TIME-CONSTANT PROBLEM

A fundamental numerical problem that severely limits the usefulness of many computer simulation programs is the *time-constant problem* associated with *stiff* differential equations. An example of this type of equation is given by Eq. (1-5) for the transistor amplifier circuit shown earlier in Fig. 1-19(c). The numerical integration algorithms considered in the preceding section can be generalized for a *system of differential equations* $\dot{\boldsymbol{x}} = \boldsymbol{f}(\boldsymbol{x}, t)$ as follows:

Forward Euler algorithm

$$\boldsymbol{x}_{n+1} = \boldsymbol{x}_n + h\boldsymbol{f}(\boldsymbol{x}_n, t_n) \tag{1-29}$$

Backward Euler algorithm

$$\boldsymbol{x}_{n+1} = \boldsymbol{x}_n + h\boldsymbol{f}(\boldsymbol{x}_{n+1}, t_{n+1}) \tag{1-30}$$

Trapezoidal algorithm

$$\boldsymbol{x}_{n+1} = \boldsymbol{x}_n + \frac{h}{2}[\boldsymbol{f}(\boldsymbol{x}_n, t_n) + \boldsymbol{f}(\boldsymbol{x}_{n+1}, t_{n+1})] \tag{1-31}$$

Applying Eqs. (1-29)–(1-31) to the special case of a system of *linear* differential equations given by

$$\dot{\boldsymbol{x}} = \boldsymbol{A}\boldsymbol{x} + \boldsymbol{u}(t) \tag{1-32}$$

we obtain the following simplified expressions:

Forward Euler solution of $\dot{\boldsymbol{x}} = \boldsymbol{A}\boldsymbol{x} + \boldsymbol{u}(t)$

$$\boldsymbol{x}_{n+1} = (\boldsymbol{1} + h\boldsymbol{A})\boldsymbol{x}_n + h\boldsymbol{u}(t_n) \tag{1-33}$$

Backward Euler solution of $\dot{\boldsymbol{x}} = \boldsymbol{A}\boldsymbol{x} + \boldsymbol{u}(t)$

$$\boldsymbol{x}_{n+1} = (\boldsymbol{1} - h\boldsymbol{A})^{-1}[\boldsymbol{x}_n + h\boldsymbol{u}(t_n)] \tag{1-34}$$

Trapezoidal solution of $\dot{\mathbf{x}} = A\mathbf{x} + \mathbf{B}\mathbf{u}(t)$

$$\mathbf{x}_{n+1} = \left(\mathbf{1} - \frac{1}{2}hA\right)^{-1}\left\{\left(\mathbf{1} + \frac{h}{2}A\right)\mathbf{x}_n + \frac{h}{2}[\mathbf{u}(t_n) + \mathbf{u}(t_{n+1})]\right\} \tag{1-35}$$

Let us now use Eqs. (1-33)–(1-35) to solve the differential equation (1-5) with

$$A = \begin{bmatrix} -6.28 \times 10^7 & -1.792 \times 10^7 \\ 990 & -43 \end{bmatrix} \tag{1-36}$$

and

$$\mathbf{u}(t) = \begin{bmatrix} 1.613 \times 10^7 \\ 9.009 \end{bmatrix} v_i(t) \tag{1-37}$$

After the solutions of $v_1(t)$ and $v_2(t)$ have been obtained, the desired output $v_o(t)$ is simply given by [see Fig. 1-19(b)]

$$v_o(t) = -960 \times 0.1 \times v_1(t) \tag{1-38}$$

Let us assume that the input voltage $v_i(t)$ is a 1-mV stepwise waveform, as shown in Fig. 1-23(a). The computer solution for $v_o(t)$ with an initial condition $v_1(0) = v_2(0) = 0$ and a *step size* $h = 0.02\ \mu s$ is shown in Fig. 1-24 for the time interval from $t = 0$ to $t = 0.01$ s. To save space, only portions of the solution from $t = 0$ to $t = 10\ \mu s$ and $t = 0.01$ ms to $t = 0.01$ s are shown. The exact solution (neglecting round-off error) to this problem is shown in the second column; the numerical solutions obtained from the forward Euler, backward Euler, and trapezoidal algorithms are shown in the third, fourth, and fifth columns, respectively. Observe that the solutions agree very well in all three cases when we use a *sufficiently small* step size, such as $h = 0.02\ \mu s$ in the present case. However, with such a small step size the number of iterations needed to cover the entire 0.01-s time interval is

$$N = \frac{0.01}{0.02 \times 10^{-6}} = 500{,}000$$

which is excessive for the amount of information it produces. Now, if we examine the exact solution plotted in Fig. 1-23(b), we see that the output waveform $v_o(t)$ remains almost constant after $t = 2.0\ \mu s$. Hence, it would seem only prudent for us to enlarge our step size after this time in order to reduce the number of iterations. This is indeed what we have done in Fig. 1-24. The computer solution for $v_o(t)$ shown in Fig. 1-25 corresponds to an initial step size of $0.02\ \mu s$ up to $t = 2\ \mu s$. Thereafter, the step size is doubled to $t = 0.04\ \mu s$. Again, to save space, only the portions covering the time intervals (in microseconds) $0 \leq t \leq 0.16$, $1.76 \leq t \leq 2$, $2.04 \leq t \leq 2.2$, and $16.96 \leq t \leq 18$ are shown in Fig. 1-25. Examination of this figure shows that the solution obtained by the forward Euler algorithm is meaningless because it begins to oscillate violently shortly after $t = 2\ \mu s$, as shown in Fig. 1-23(c). It is interesting that this explosive situation does not occur with the solution obtained from the backward Euler algorithm or the trapezoidal algorithm.

The reason behind the numerical instability shown in Fig. 1-23(c) will be shown in Chapter 13 to be due to the widely different magnitudes of the eigenvalues of the A matrix as given by Eqs. (1-8)–(1-9). Since the reciprocals of the magnitudes of

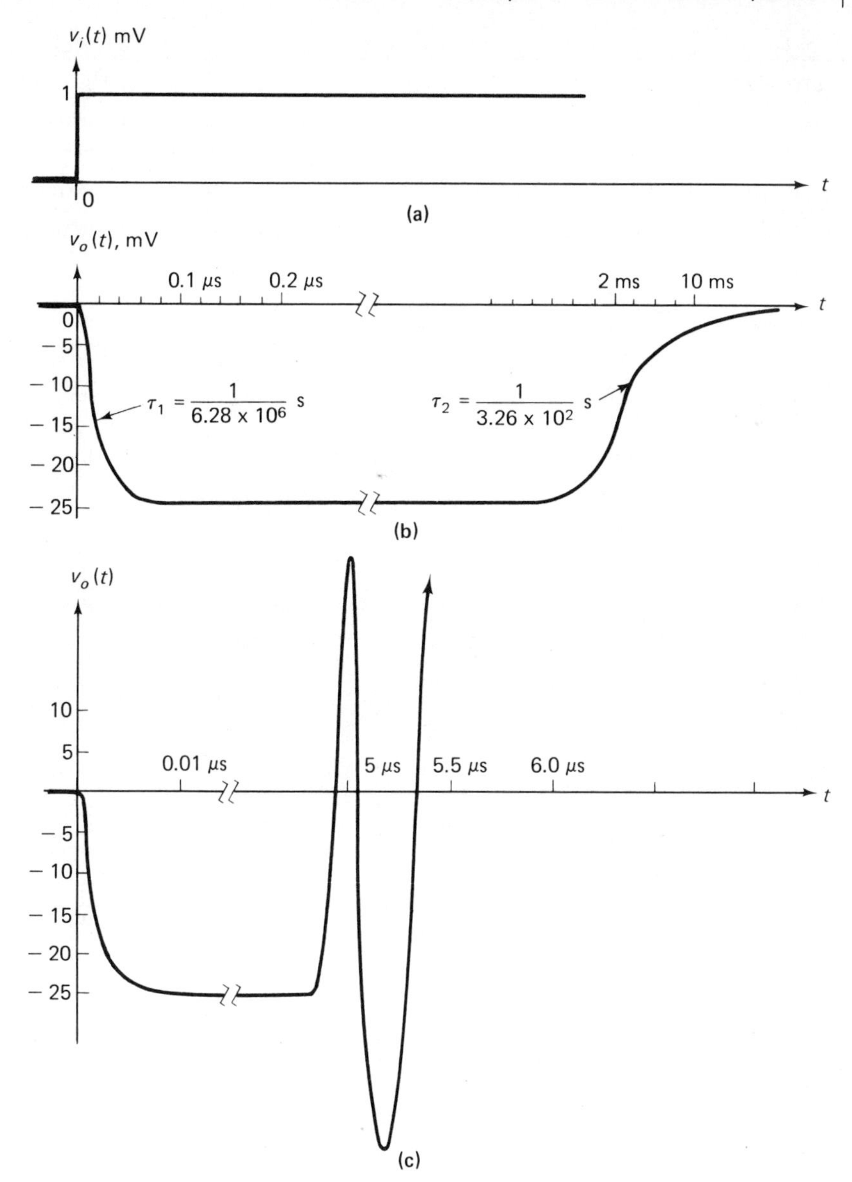

Fig. 1-23. Time-constant problem: (a) input waveform; (b) exact solution; (c) instability occurs when the step size in the forward Euler algorithm is too large.

eigenvalues are the time constants associated with the exponentials in the transient response, we find the two time constants to be

$$\tau_1 = \frac{1}{|\lambda_1|} = \frac{1}{6.28 \times 10^7} \approx 0.016\ \mu\text{s} \tag{1-39}$$

$$\tau_2 = \frac{1}{|\lambda_2|} = \frac{1}{3.26 \times 10^2} \approx 3.1\ \text{ms} \tag{1-40}$$

STEP SIZE = .200E − 07				
TIME	EXACT SOLUTION	FORWARD EULER	BACKWARD EULER	TRAPEZOIDAL
0.	0.	0.	0.	0.
.20000E−07	−.17821E+02	−.31296E+02	−.13872E+02	−.19224E+02
.40000E−07	−.22896E+02	−.23284E+02	−.20021E+02	−.23616E+02
.60000E−07	−.24341E+02	−.25335E+02	−.22747E+02	−.24620E+02
.80000E−07	−.24753E+02	−.24810E+02	−.23955E+02	−.24849E+02
.10000E−06	−.24870E+02	−.24944E+02	−.24490E+02	−.24901E+02
.12000E−06	−.24903E+02	−.24909E+02	−.24727E+02	−.24913E+02
.14000E−06	−.24913E+02	−.24918E+02	−.24833E+02	−.24916E+02
.16000E−06	−.24915E+02	−.24916E+02	−.24879E+02	−.24916E+02
.18000E−06	−.24916E+02	−.24916E+02	−.24900E+02	−.24916E+02
.20000E−06	−.24916E+02	−.24916E+02	−.24909E+02	−.24916E+02
.22000E−06	−.24916E+02	−.24916E+02	−.24913E+02	−.24916E+02
.24000E−06	−.24916E+02	−.24916E+02	−.24914E+02	−.24916E+02
.26000E−06	−.24915E+02	−.24915E+02	−.24915E+02	−.24916E+02
.28000E−06	−.24915E+02	−.24915E+02	−.24915E+02	−.24915E+02
.30000E−06	−.24915E+02	−.24915E+02	−.24915E+02	−.24915E+02
.32000E−06	−.24915E+02	−.24915E+02	−.24915E+02	−.24915E+02
.34000E−06	−.24915E+02	−.24915E+02	−.24915E+02	−.24915E+02
.36000E−06	−.24915E+02	−.24915E+02	−.24915E+02	−.24915E+02
.38000E−06	−.24915E+02	−.24915E+02	−.24915E+02	−.24915E+02
.40000E−06	−.24914E+02	−.24914E+02	−.24914E+02	−.24915E+02
.42000E−06	−.24914E+02	−.24914E+02	−.24914E+02	−.24914E+02
.44000E−06	−.24914E+02	−.24914E+02	−.24914E+02	−.24914E+02
.46000E−06	−.24914E+02	−.24914E+02	−.24914E+02	−.24914E+02
.48000E−06	−.24914E+02	−.24914E+02	−.24914E+02	−.24914E+02
.50000E−06	−.24914E+02	−.24914E+02	−.24914E+02	−.24914E+02
.52000E−06	−.24914E+02	−.24913E+02	−.24914E+02	−.24914E+02
.54000E−06	−.24913E+02	−.24913E+02	−.24913E+02	−.24913E+02
.56000E−06	−.24913E+02	−.24913E+02	−.24913E+02	−.24913E+02
.58000E−06	−.24913E+02	−.24913E+02	−.24913E+02	−.24913E+02
.60000E−06	−.24913E+02	−.24913E+02	−.24913E+02	−.24913E+02
.62000E−06	−.24913E+02	−.24913E+02	−.24913E+02	−.24913E+02
.64000E−06	−.24913E+02	−.24913E+02	−.24913E+02	−.24913E+02
.66000E−06	−.24912E+02	−.24912E+02	−.24913E+02	−.24913E+02
.68000E−06	−.24912E+02	−.24912E+02	−.24912E+02	−.24912E+02
.70000E−06	−.24912E+02	−.24912E+02	−.24912E+02	−.24912E+02
.72000E−06	−.24912E+02	−.24912E+02	−.24912E+02	−.24912E+02
.74000E−06	−.24912E+02	−.24912E+02	−.24912E+02	−.24912E+02
.76000E−06	−.24912E+02	−.24912E+02	−.24912E+02	−.24912E+02
.78000E−06	−.24912E+02	−.24912E+02	−.24912E+02	−.24912E+02
.80000E−06	−.24911E+02	−.24911E+02	−.24912E+02	−.24912E+02
.82000E−06	−.24911E+02	−.24911E+02	−.24911E+02	−.24911E+02
.84000E−06	−.24911E+02	−.24911E+02	−.24911E+02	−.24911E+02
.86000E−06	−.24911E+02	−.24911E+02	−.24911E+02	−.24911E+02
.88000E−06	−.24911E+02	−.24911E+02	−.24911E+02	−.24911E+02
.90000E−06	−.24911E+02	−.24911E+02	−.24911E+02	−.24911E+02
.92000E−06	−.24911E+02	−.24911E+02	−.24911E+02	−.24911E+02
.94000E−06	−.24910E+02	−.24910E+02	−.24911E+02	−.24911E+02
.96000E−06	−.24910E+02	−.24910E+02	−.24910E+02	−.24910E+02
.98000E−06	−.24910E+02	−.24910E+02	−.24910E+02	−.24910E+02
.10000E−05	−.24910E+02	−.24910E+02	−.24910E+02	−.24910E+02

TIME	EXACT SOLUTION	FORWARD EULER	BACKWARD EULER	TRAPEZOIDAL
.10000E−02	−.18689E+02	−.18699E+02	−.18699E+02	−.18699E+02
.20000E−02	−.14194E+02	−.14208E+02	−.14208E+02	−.14208E+02
.30000E−02	−.10949E+02	−.10964E+02	−.10964E+02	−.10964E+02
.40000E−02	−.86072E+01	−.86222E+01	−.86223E+01	−.86222E+01
.50000E−02	−.69166E+01	−.69308E+01	−.69308E+01	−.69308E+01
.60000E−02	−.56964E+01	−.57093E+01	−.57093E+01	−.57093E+01
.70000E−02	−.48156E+01	−.48272E+01	−.48272E+01	−.48272E+01
.80000E−02	−.41799E+01	−.41901E+01	−.41902E+01	−.41901E+01
.90000E−02	−.37210E+01	−.37301E+01	−.37301E+01	−.37301E+01
.10000E−01	−.33898E+01	−.33978E+01	−.33979E+01	−.33978E+01

Fig. 1-24. Numerical solution obtained with step size $h = 0.02\ \mu s$ for $0 \leq t \leq 2\ \mu s$.

STEP SIZE = .200E − 07

TIME	EXACT SOLUTION	FORWARD EULER	BACKWARD EULER	TRAPEZOIDAL
0.	0.	0.	0.	0.
.20000E−07	−.17821E+02	−.31296E+02	−.13872E+02	−.19224E+02
.40000E−07	−.22896E+02	−.23284E+02	−.20021E+02	−.23616E+02
.60000E−07	−.24341E+02	−.25335E+02	−.22747E+02	−.24620E+02
.80000E−07	−.24753E+02	−.24810E+02	−.23955E+02	−.24849E+02
.10000E−06	−.24870E+02	−.24944E+02	−.24490E+02	−.24901E+02
.12000E−06	−.24903E+02	−.24909E+02	−.24727E+02	−.24913E+02
.14000E−06	−.24913E+02	−.24918E+02	−.24833E+02	−.24916E+02
.16000E−06	−.24915E+02	−.24916E+02	−.24879E+02	−.24916E+02

TIME	EXACT SOLUTION	FORWARD EULER	BACKWARD EULER	TRAPEZOIDAL
.17600E−05	−.24904E+02	−.24904E+02	−.24905E+02	−.24905E+02
.17800E−05	−.24904E+02	−.24904E+02	−.24904E+02	−.24904E+02
.18000E−05	−.24904E+02	−.24904E+02	−.24904E+02	−.24904E+02
.18200E−05	−.24904E+02	−.24904E+02	−.24904E+02	−.24904E+02
.18400E−05	−.24904E+02	−.24904E+02	−.24904E+02	−.24904E+02
.18600E−05	−.24904E+02	−.24904E+02	−.24904E+02	−.24904E+02
.18800E−05	−.24904E+02	−.24904E+02	−.24904E+02	−.24904E+02
.19000E−05	−.24903E+02	−.24903E+02	−.24904E+02	−.24904E+02
.19200E−05	−.24903E+02	−.24903E+02	−.24903E+02	−.24903E+02
.19400E−05	−.24903E+02	−.24903E+02	−.24903E+02	−.24903E+02
.19600E−05	−.24903E+02	−.24903E+02	−.24903E+02	−.24903E+02
.19800E−05	−.24903E+02	−.24903E+02	−.24903E+02	−.24903E+02
.20000E−05	−.24903E+02	−.24903E+02	−.24903E+02	−.24903E+02

STEP SIZE IS CHANGED TO h = 0.04E − 06

TIME	EXACT SOLUTION	FORWARD EULER	BACKWARD EULER	TRAPEZOIDAL
.20400E−05	−.24902E+02	−.24902E+02	−.38775E+02	−.34514E+02
.20800E−05	−.24902E+02	−.24902E+02	−.44924E+02	−.33003E+02
.21200E−05	−.24902E+02	−.24902E+02	−.47649E+02	−.33240E+02
.21600E−05	−.24902E+02	−.24902E+02	−.48857E+02	−.33202E+02
.22000E−05	−.24901E+02	−.24901E+02	−.49392E+02	−.33208E+02

TIME	EXACT SOLUTION	FORWARD EULER	BACKWARD EULER	TRAPEZOIDAL
.16960E−04	−.24794E+02	−.93426E+58	−.49711E+02	−.33100E+02
.17000E−04	−.24794E+02	.14126E+59	−.49711E+02	−.33100E+02
.17040E−04	−.24793E+02	−.21359E+59	−.49711E+02	−.33099E+02
.17080E−04	−.24793E+02	.32295E+59	−.49710E+02	−.33099E+02
.17120E−04	−.24793E+02	−.48831E+59	−.49710E+02	−.33099E+02
.17160E−04	−.24792E+02	.73833E+59	−.49710E+02	−.33098E+02
.17200E−04	−.24792E+02	−.11164E+60	−.49710E+02	−.33098E+02
.17240E−04	−.24792E+02	.16880E+60	−.49709E+02	−.33098E+02
.17280E−04	−.24792E+02	−.25522E+60	−.49709E+02	−.33098E+02
.17320E−04	−.24791E+02	.38590E+60	−.49709E+02	−.33097E+02
.17360E−04	−.24791E+02	−.58349E+60	−.49708E+02	−.33097E+02
.17400E−04	−.24791E+02	.88225E+60	−.49708E+02	−.33097E+02
.17440E−04	−.24790E+02	−.13340E+61	−.49708E+02	−.33096E+02
.17480E−04	−.24790E+02	.20170E+61	−.49708E+02	−.33096E+02
.17520E−04	−.24790E+02	−.30497E+61	−.49707E+02	−.33096E+02
.17560E−04	−.24790E+02	.46112E+61	−.49707E+02	−.33096E+02
.17600E−04	−.24789E+02	−.69723E+61	−.49707E+02	−.33095E+02
.17640E−04	−.24789E+02	.10542E+62	−.49706E+02	−.33095E+02
.17680E−04	−.24789E+02	−.15940E+62	−.49706E+02	−.33095E+02
.17720E−04	−.24788E+02	.24102E+62	−.49706E+02	−.33094E+02
.17760E−04	−.24788E+02	−.36442E+62	−.49706E+02	−.33094E+02
.17800E−04	−.24788E+02	.55101E+62	−.49705E+02	−.33094E+02
.17840E−04	−.24787E+02	−.83313E+62	−.49705E+02	−.33094E+02
.17880E−04	−.24787E+02	.12597E+63	−.49705E+02	−.33093E+02
.17920E−04	−.24787E+02	−.19047E+63	−.49704E+02	−.33093E+02
.17960E−04	−.24787E+02	.28800E+63	−.49704E+02	−.33093E+02
.18000E−04	−.24786E+02	−.43545E+63	−.49704E+02	−.33092E+02

Fig. 1-25. Numerical solution obtained with step size h = 0.02 μs for $0 \leq t \leq 2$ μs, and step size h = 0.04 μs for $t > 2$ μs.

corresponding to the two exponentials shown in Fig. 1-23(b). (Notice the change in time scale!) It will be shown in Chapter 13 that the largest step size for the forward Euler algorithm must be smaller than twice the *smallest* time constant in order to avoid numerical instability.[2] Hence, for our present example, we must have $h < 0.032$ μs. It is not surprising, therefore, that our choice of $h = 0.04$ μs results in numerical instability. This is an extremely serious restriction since the second time constant, τ_2, is much larger, thereby resulting in a *slowly* decaying transient. Thus, an excessive number of iterations, each with a step size no larger than τ_1, must be carried out if we want to compute the complete waveform until it reaches steady state. In other words, the maximum allowable step size in the *forward Euler algorithms is determined by the smallest time constant, but the number of iterations required to reach steady state is determined by the largest time constant.* Hence, whenever the time constants differ by many orders of magnitude, the forward Euler algorithm would run into a serious bottleneck problem. We shall show in Chapter 13 that this *widely separated time-constant* problem can be overcome by using *implicit integration algorithms*, of which the backward Euler and the trapezoidal algorithms are special cases.

Although both backward Euler and trapezoidal algorithms are numerically stable, they differ in several important aspects. One aspect will be investigated briefly in the next section, and in depth in Chapter 12. Another aspect is clearly demonstrated by the numerical results indicated under the columns labelled "local and total errors" in Figs. 1-22(b) and (c). Observe that whereas the error decreases monotonically in the backward Euler algorithm, the corresponding error in the trapezoidal algorithm tends to alternate in sign even though its magnitude also decreases monotonically. This "ringing behavior" is typical with the trapezoidal algorithm when implemented with a large step size. This phenomenon is strictly a *numerical property* of the trapezoidal algorithm and is clearly undesirable because it could easily mislead a circuit designer into concluding erroneously the presence of an oscillatory response. The reason behind this phenomenon will be clear after the reader studies Chapter 12 and works out Problems 12-24 and 12-25.

1-7 A GLIMPSE AT ERROR ANALYSIS FOR NUMERICAL INTEGRATION ALGORITHMS

No numerical integration algorithm is exact. We have already presented examples in Section 1-5 illustrating this basic fact of life. Consequently, no full-fledged, computer-aided-circuit-analysis expert can afford not to understand the basic techniques required to carry out a meaningful error analysis. The foundation and methods for error analysis will be presented in Chapters 11 and 12. Here, we merely wish to whet the reader's appetite for a subsequent in-depth study of these somewhat "mathematical" chapters by considering some simple examples and showing that no more than a basic knowledge of "Taylor's formula with a remainder" is needed to perform an error analysis. In particular, we shall show in detail the derivation of *explicit* expressions

[2]We have already derived this fundamental step-size constraint in Eq. (1-28) for a single first-order equation.

for *local* truncation error for two of the numerical algorithms presented in Section 1-5, the *forward Euler* and the *backward Euler* algorithms.

We define the *local truncation error* ϵ_T of a numerical integration algorithm for solving the initial-value problem

$$\dot{x} = f(x, t), \qquad x(0) = x_0 \tag{1-41}$$

as follows:

$$\epsilon_T \triangleq \hat{x}(t_{n+1}) - \hat{x}_{n+1} \tag{1-42}$$

where $\hat{x}(t_{n+1})$ is the *exact solution* $\hat{x}(t)$ to Eq. (1-41) evaluated at $t = t_{n+1}$, and $\hat{x}_{n+1}$ is the corresponding numerical solution obtained at the same time $t = t_{n+1} \triangleq t_n + h$, provided that in using the numerical integration algorithm we assume that $\hat{x}_n = \hat{x}(t_n)$, the exact solution at $t = t_n$.

Since the value of $\hat{x}_n$ at the beginning of the time interval (t_n, t_{n+1}) is assumed to be the exact solution $\hat{x}(t_n)$ when we compute $\hat{x}(t_{n+1})$ in Eq. (1-42), it is clear that ϵ_T, as defined by Eq. (1-42), gives only the error occurring after one *time step;* hence the name *local truncation error.*

Let us now proceed to derive explicit expressions for ϵ_T for the algorithms discussed in Section 1-5. The basis for the error analysis is Taylor's formula with a remainder. Taylor's formula with a remainder after two terms may be expressed as[3]

$$g(t) = g(a) + g'(a)(t - a) + \tfrac{1}{2}g''(\hat{\tau})(t - a)^2 \tag{1-43}$$

where either $a < \hat{\tau} < t$ when $a < t$ or $t < \hat{\tau} < a$ when $t < a$.

The forward Euler algorithm for solving a first-order differential equation $\dot{x} = f(x, t)$ can be expressed as

$$\hat{x}_{n+1} = \hat{x}_n + h\hat{x}'(t_n) \tag{1-44}$$

To derive the local truncation error ϵ_T, we let $\hat{x}_n = x(t_n)$ in Eq. (1-44) to obtain

$$\hat{x}_{n+1} = \hat{x}(t_n) + h\hat{x}'(t_n) \tag{1-45}$$

where $\hat{x}(t)$ is the exact solution. Applying Taylor's formula (1-43) with $g(t) = \hat{x}(t)$, $a = t_n$, and $t = t_{n+1} \triangleq t_n + h$, we obtain

$$\hat{x}(t_{n+1}) = \hat{x}(t_n) + h\hat{x}'(t_n) + \frac{h^2}{2}\hat{x}''(\hat{\tau}), \qquad t_n < \hat{\tau} < t_{n+1} \tag{1-46}$$

From Eq. (1-45), the sum of the first two terms on the right-hand side of Eq. (1-46) is simply $\hat{x}_{n+1}$ as computed from the forward Euler algorithm. Subtracting $\hat{x}_{n+1}$ from both sides of Eq. (1-46), we immediately obtain the *local truncation error* for the forward Euler algorithm:

$$\epsilon_T \doteq \hat{x}(t_{n+1}) - \hat{x}_{n+1} = \frac{h^2}{2}\hat{x}''(\hat{\tau}), \qquad t_n < \hat{\tau} < t_{n+1} \tag{1-47}$$

Notice that ϵ_T in this case is proportional to h^2 as well as the second-order derivative of the exact solution evaluated at some time $\hat{\tau}$ lying between t_n and t_{n+1}.

Next, consider the local truncation error for the backward Euler algorithm when

[3]Throughout this book, we denote $dx(t)/dt$ by either $\dot{x}$ or $x'(t)$, $d^2x(t)/dt^2$ by either $\ddot{x}(t)$ or $x''(t)$, and $d^nx(t)/dt^n$ by $x^{(n)}(t)$ when $n > 2$.

used to solve a first-order differential equation $\dot{x} = f(x, t)$. The backward Euler algorithm can be expressed as

$$\hat{x}_{n+1} = \hat{x}_n + h\hat{x}'(t_{n+1}) \tag{1-48}$$

We proceed as before by letting $\hat{x}_n = \hat{x}(t_n)$ in Eq. (1-48) to obtain

$$\hat{x}_{n+1} = \hat{x}(t_n) + h\hat{x}'(t_{n+1}) \tag{1-49}$$

Applying Taylor's formula with a remainder given by Eq. (1-43), with $g(t) = \hat{x}(t)$, $t = t_n$, and $a = t_{n+1} \triangleq t_n + h$, we obtain

$$\hat{x}(t_n) = \hat{x}(t_{n+1}) - h\hat{x}'(t_{n+1}) + \frac{h^2}{2}\hat{x}''(\hat{\tau}), \qquad t_n < \hat{\tau} < t_{n+1} \tag{1-50}$$

Solving for $\hat{x}(t_{n+1})$ from Eq. (1-50), we obtain

$$\hat{x}(t_{n+1}) = \hat{x}(t_n) + h\hat{x}'(t_{n+1}) - \frac{h^2}{2}\hat{x}''(\hat{\tau}), \qquad t_n < \hat{\tau} < t_{n+1} \tag{1-51}$$

The sum of the first two terms on the right-hand side of Eq. (1-51) is simply $\hat{x}_{n+1}$ as computed from Eq. (1-49). Subtracting $\hat{x}_{n+1}$ from both sides of Eq. (1-51), we immediately obtain the local truncation error for the backward Euler algorithm:

$$\epsilon_T = \hat{x}(t_{n+1}) - \hat{x}_{n+1} = \frac{-h^2}{2}\hat{x}''(\hat{\tau}), \qquad t_n < \hat{\tau} < t_{n+1} \tag{1-52}$$

Except for a change in sign, observe that the *expressions* for the local truncation errors for both the forward and backward Euler algorithms are the same. However, the values for $\hat{\tau}$ in the two expressions are in general different.

Using the preceding approach, we can derive the expression for the local truncation error for the trapezoidal algorithm. The derivation, however, will require the use of Taylor's formula with a remainder after *three* terms. The details of this analysis will be given in Chapter 11. Here we shall only state the result; the local truncation error for the trapezoidal algorithm is

$$\epsilon_T = -\frac{h^3}{12}\hat{x}'''(\hat{\tau}), \qquad t_n < \hat{\tau} < t_{n+1} \tag{1-53}$$

Comparing Eq. (1-53) with Eq. (1-52) or (1-47), we see that, if the step size h is small, then the local truncation error for the trapezoidal algorithm will be much smaller than for both Euler algorithms, assuming that both $\hat{x}''(\hat{\tau})$ and $\hat{x}'''(\hat{\tau})$ are of the same order of magnitude for any $\hat{\tau}$ in the time interval (t_n, t_{n+1}). For example, if we assume that $\hat{x}''(\tau) \approx \hat{x}'''(\tau) \triangleq K$, the *local truncation error* ϵ_T for each of the three algorithms is as listed in the accompanying table for four different values of the step size h.

	STEP SIZE h			
ALGORITHM	0.01	0.10	1.0	10.0
Forward Euler	$\epsilon_T = 5 \times 10^{-5}K$	$\epsilon_T = 5 \times 10^{-3}K$	$\epsilon_T = 0.5K$	$\epsilon_T = 50K$
Backward Euler	$\epsilon_T = -5 \times 10^{-5}K$	$\epsilon_T = -5 \times 10^{-3}K$	$\epsilon_T = -0.5K$	$\epsilon_T = -50K$
Trapezoidal	$\epsilon_T = -8 \times 10^{-8}K$	$\epsilon_T = -8 \times 10^{-5}K$	$\epsilon_T = -0.08K$	$\epsilon_T = -83K$

An examination of this table shows that the trapezoidal algorithm has a smaller local truncation error ϵ_T for $h = 0.01$, 0.1, and 1.0. It has a greater local truncation error for $h = 10$.

Using a unified approach, the preceding analysis will be generalized in Section 11-5-1 where a simple formula will be derived which gives, by direct substitution, the *local truncation error* of any multistep numerical integration algorithm.

1-8 ON THE EFFECT OF CHOICE OF STATE VARIABLES OVER TOTAL ERROR

The error analysis presented in the preceding section is concerned with "local error" incurred at the end of each time step. In practice, we are of course more concerned about the *total error* accumulated at the end of the time interval of interest. Unfortunately, very little is yet known about how to analyze this total error. Until a sound theory analogous to that presented in Section 1-7 is available, one can only be guided by insight and experience in trying to minimize the total error. Our objective in this section is to offer one method by which the total error can be greatly reduced for the case when the circuit to be analyzed contains *nonlinear* capacitors and inductors. In such a situation, it turns out to be advantageous to choose the *capacitor charge q* and the *inductor flux linkage φ* as the state variables, as compared to the more common choice of capacitor voltage and inductor current. The detailed analysis necessary to arrive at this conclusion is given in Chapter 10. Here we shall merely illustrate the conclusion with a simple example.

Consider the one-capacitor circuit shown in Fig. 1-26. We assume that the one-port $\mathfrak{N}$ is purely resistive (no inductors, nor capacitors) and is characterized by a driving-point v-i curve $i = g(v)$. Let the nonlinear capacitor be characterized by a voltage-controlled *incremental capacitance* function $C(v) \triangleq dq/dv$. Then, in Fig. 1-26, the nonlinear capacitor gives the following relationship:

$$-i = \frac{dq}{dt} = \frac{dq}{dv} \cdot \frac{dv}{dt} = C(v) \cdot \dot{v} \tag{1-54}$$

On the other hand, the resistive one-port has the constraint

$$i = g(v) \tag{1-55}$$

From Eqs. (1-54) and (1-55), the state equation with v as the state variable is easily seen to be

$$\dot{v} = -\frac{i}{C(v)} = \frac{-g(v)}{C(v)} \triangleq f(v) \tag{1-56}$$

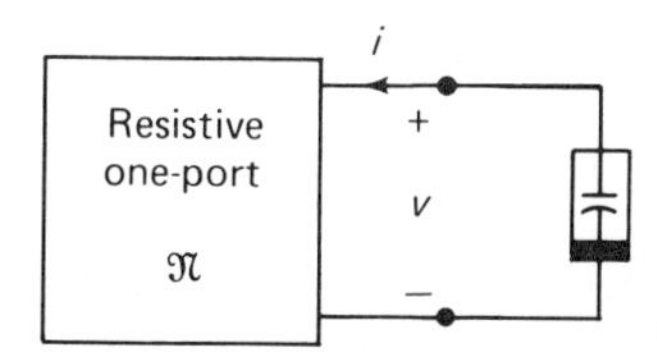

Fig. 1-26. Resistive one-port terminated in a nonlinear capacitor.

Evaluating the derivative of $f(v)$ with respect to v, we obtain

$$\frac{df(v)}{dv} = -\frac{1}{C(v)}g'(v) + \frac{C'(v)}{[C(v)]^2}g(v) \tag{1-57}$$

Examination of (1-57) shows that if $g'(v)C'(v) > 0$, then

$$\boxed{\frac{df(v)}{dv} > 0, \qquad \text{whenever } \frac{g(v)}{g'(v)} > \frac{C(v)}{C'(v)}} \tag{1-58}$$

Now suppose that we choose the *capacitor charge* q, instead of v, as the state variable. The corresponding state equation can be written very simply as follows:

$$\frac{dq}{dt} = -i = -g[\hat{v}(q)] \triangleq f(q) \tag{1-59}$$

where $v = \hat{v}(q)$ is the voltage-versus-charge characteristic of the nonlinear capacitor. Evaluating the derivative of $f(q)$ with respect to q, we obtain

$$\begin{aligned}\frac{df(q)}{dq} &= -\frac{d}{dq}\{g[\hat{v}(q)]\} = -\frac{dg(v)}{dv}\cdot\frac{dv(q)}{dq} \\ &= -\frac{g'(v)}{\frac{dq}{dv}} = -\frac{g'(v)}{C(v)}\end{aligned} \tag{1-60}$$

Examination of Eq. (1-60) shows that

$$\boxed{\frac{df(q)}{dq} < 0, \qquad \text{whenever } \frac{g'(v)}{C(v)} > 0} \tag{1-61}$$

Now in the usual case when both the one-port $\mathfrak{N}$ and the nonlinear capacitor are characterized by strictly *monotonically increasing* v-i and q-v curves, respectively, we have $g'(v) > 0$ and $C(v) > 0$. Under this condition, Eq. (1-61) shows that $df(q)/dq < 0$ at all times. This property is not necessarily true with $df(v)/dv$ because of Eq. (1-58). We shall now show that an error in the initial condition tends to get *magnified* whenever $df(v)/dv > 0$, but *attenuated* whenever $df(q)/dq < 0$. A demonstration of this fact would establish the advantage of choosing q as the state variable.

Consider the first-order initial-value problem

$$\dot{x} = f(x), \qquad x(0) = x_0 \tag{1-62}$$

Let $\hat{x}(t)$ be the *exact solution* when the initial condition is exactly x_0. Now suppose that an error is made so that the initial condition becomes $x_0 + \delta_0$. Let the corresponding exact solution be $\hat{x}(t) + \delta x(t)$. Substituting this solution into Eq. (1-62), we have

$$\dot{\hat{x}} + \delta\dot{x} = f(\hat{x} + \delta x) \tag{1-63}$$

If we assume δx is small and expand the right side of Eq. (1-63) by *Taylor's series* about the point $\hat{x}$, we obtain, when neglecting higher-order terms, the relation

$$\dot{\hat{x}} + \delta\dot{x} \approx f(\hat{x}) + \left.\frac{df(x)}{dx}\right|_{x=\hat{x}} \cdot \delta x \tag{1-64}$$

From Eqs. (1-62) and (1-64), we have

$$\delta\dot{x}(t) = \left.\frac{df(x)}{dx}\right|_{x=\hat{x}(t)} \cdot \delta x(t), \qquad \delta x(0) = \delta_0 \tag{1-65}$$

It is convenient to rewrite Eq. (1-65) as follows:

$$\delta\dot{x} = \lambda(t)\delta x, \qquad \delta x(0) = \delta_0 \tag{1-66}$$

where

$$\lambda(t) \triangleq \left.\frac{df(x)}{dx}\right|_{x=\hat{x}(t)} \tag{1-67}$$

Equation (1-66) is a first-order linear *time-varying* equation and has the explicit solution

$$\delta x(t) = \delta_0 e^{\int_0^t \lambda(\tau)\, d\tau} \tag{1-68}$$

To verify that Eq. (1-68) is indeed a solution of Eq. (1-66), we simply differentiate both sides of Eq. (1-68) with respect to t to see that Eq. (1-66) is satisfied:

$$\begin{aligned} \delta\dot{x}(t) &= [\delta_0 e^{\int_0^t \lambda(\tau)\, d\tau}] \cdot \frac{d}{dt}\left[\int_0^t \lambda(\tau)\, d\tau\right] \\ &= (\delta x)[\lambda(t)] = \lambda(t)\delta x \end{aligned} \tag{1-69}$$

Moreover,

$$\delta x(0) = \delta_0 e^{\int_0^0 \lambda(\tau)\, d\tau} = \delta_0 \tag{1-70}$$

Hence, Eq. (1-68) is a solution of Eq. (1-66).

Examination of Eq. (1-68) shows that $\delta x(t)$ increases if $\lambda(t) > 0$, and decreases if $\lambda(t) < 0$. But $\lambda(t)$ as defined by Eq. (1-67) is simply $df(x)/dx$ evaluated at $x = \hat{x}(t)$. If indeed $df(x)/dx < 0$ for all x, then $\delta x(t)$ will always decrease with time. Now $\delta x(t)$ represents the *error accumulated* at time t in the *exact solution* due to an error δ_0 in the initial condition. It has nothing to do with the numerical algorithm used. However, if we interpret the local truncation error $\boldsymbol{\epsilon}_T$ at $t = t_n$ as the *initial error* δ_0, the preceding analysis shows that the corresponding *exact solution* would contain an error term $\delta x(t)$ which will either get *amplified* if $df(x)/dx > 0$, or attenuated if $df(x)/dx < 0$, as long as the error $\lambda(t)$ remains small enough for Eq. (1-65) to hold. Hence, we conclude that *regardless of the numerical integration algorithm used in solving the non-linear circuit shown in Fig. 1-26, the choice of q as the state variable will result in less error propagation $\delta x(t)$ owing to the presence of local truncation error.*

Examination of Eqs. (1-57) and (1-61) shows that if the capacitor is *linear*, and hence $C'(v) = 0$, we have both $df(v)/dv < 0$ and $df(q)/dq < 0$ as long as $g'(v) > 0$ and $C > 0$. Hence, if the capacitor is linear, in so far as error propagation due to an initial error is concerned, there is no apparent advantage in choosing q instead of v as the state variable. For an in-depth investigation of the choice of state variables, the reader is referred to Section 10-5.

1-9 ASSOCIATED DISCRETE CIRCUIT MODELS FOR CAPACITORS AND INDUCTORS

From the numerical integration point of view, the differential equation characterizing a capacitor or inductor can be approximated by a *resistive circuit* associated with the integration algorithm, henceforth called the *associated discrete circuit model.*[4] The adjective "associated" is used to emphasize that the models associated with different integration algorithms are different. The adjective "discrete" is used to emphasize that the parameters in the model are discrete in nature; i.e., they differ from one time step to another. The use of the associated discrete circuit model actually reduces the transient analysis of a dynamic network to the dc analysis of a resistive network.

An associated discrete circuit model can be easily derived from any *implicit* numerical integration formula (see Chapter 13). In this section, we shall illustrate the basic procedure with three commonly used implicit algorithms: the *backward Euler algorithm*, *trapezoidal algorithm*, and *Gear's second-order algorithm*. This procedure will be generalized in Section 17-1 for any *implicit integration algorithm.*

1-9-1 Deriving Associated Discrete Circuit Models for Linear Capacitors

The *backward Euler algorithm* for solving the first-order differential equation $v' = f(v)$ with a *step size* h is given by

$$\begin{aligned} v_{n+1} &= v_n + hf(v_{n+1}) \\ &= v_n + hv'_{n+1} \end{aligned} \tag{1-71}$$

where the prime denotes time differentiation. Now since

$$i(t) = C\frac{dv}{dt} = Cv'(t)$$

for a *linear* capacitor, it follows that

$$v'(t_{n+1}) = \frac{1}{C}i(t_{n+1}) \tag{1-72}$$

Observe that Eq. (1-72) is an exact relationship, whereas Eq. (1-71) represents an "approximate" solution. If we approximate the *exact solutions* $v'(t_{n+1})$ and $i(t_{n+1})$ by v'_{n+1} and i_{n+1}, respectively, then Eq. (1-72) becomes

$$v'_{n+1} = \frac{1}{C}i_{n+1} \tag{1-73}$$

Substituting Eq. (1-73) into Eq. (1-71) and solving for i_{n+1}, we obtain

$$i_{n+1} = \frac{C}{h}v_{n+1} - \frac{C}{h}v_n \tag{1-74}$$

[4]Also called a *companion model* in the literature.

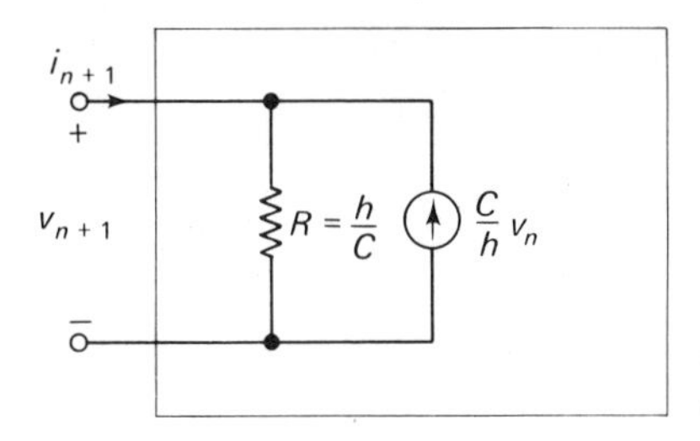

Fig. 1-27. Discrete circuit model associated with backward Euler algorithm for a linear capacitor.

This relationship can be represented by the *equivalent linear one-port* shown in Fig. 1-27, with port current i_{n+1} and port voltage v_{n+1}. Observe that, whereas the value of the resistor R remains fixed (assuming fixed step size), the current source depends on the voltage v_n, which is determined during the *preceding* time step. Clearly, if the only energy storage elements in a circuit are *linear* capacitors, and if each capacitor is replaced by its associated discrete circuit model shown in Fig. 1-27, then the resulting circuit becomes *purely resistive*. By assigning the value $(C_j/h)v_o^{(j)}$ to the current source associated with the capacitor C_j, with $v_o^{(j)}$ being the initial voltage across capacitor C_j, we can solve for the voltage $v_1^{(j)}$ across each model by *any* efficient method, such as *nodal analysis*. The voltage $v_1^{(j)}$ is then the capacitor voltage across C_j at $t = t_1 \triangleq h$ as computed by the backward Euler algorithm. The associated discrete circuit model is then "updated." By this we mean the value of the current source in the model for C_j is changed from $(C_j/h)v_o^{(j)}$ to $(C_j/h)v_1^{(j)}$. This updated resistive network is then solved to yield $v_2^{(j)}$, which is the voltage across C_j at $t = t_2 \triangleq t_1 + h$. Clearly, this procedure can be iterated as many times as necessary.

If we choose another numerical integration algorithm such as the trapezoidal algorithm for solving $v' = f(v)$, we have

$$\begin{aligned} v_{n+1} &= v_n + \frac{h}{2}[f(v_{n+1}) + f(v_n)] \\ &= v_n + \frac{h}{2}v'_{n+1} + \frac{h}{2}v'_n \end{aligned} \tag{1-75}$$

Although the *exact* v-i relationship for a capacitor is Eq. (1-72) at $t = t_{n+1}$, and

$$v'(t_n) = \frac{1}{C}i(t_n) \tag{1-76}$$

at $t = t_n$, we can approximate these two exact relationships by

$$v'_{n+1} = \frac{1}{C}i_{n+1} \tag{1-77}$$

$$v'_n = \frac{1}{C}i_n \tag{1-78}$$

Substituting Eqs. (1-77) and (1-78) into Eq. (1-75) and solving for i_{n+1}, we obtain

$$i_{n+1} = \frac{2C}{h}v_{n+1} - \left(\frac{2C}{h}v_n + i_n\right) \tag{1-79}$$

Equation (1-79) can be represented by the *equivalent linear* one-port shown in Fig.

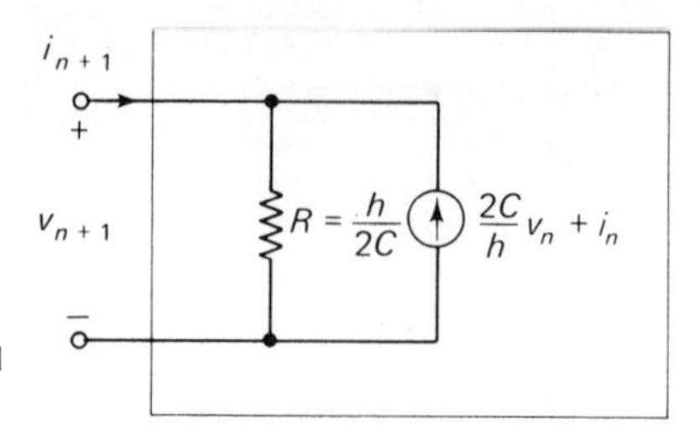

Fig. 1-28. Discrete circuit model associated with trapezoidal algorithm for a linear capacitor.

1-28. This circuit is therefore the discrete circuit model associated with the *trapezoidal algorithm.*

As a final example, consider Gear's second-order algorithm (to be derived in Chapter 13) for solving $\dot{v} = f(v)$:

$$\begin{aligned} v_{n+1} &= \tfrac{4}{3}v_n - \tfrac{1}{3}v_{n-1} + h[\tfrac{2}{3}f(v_{n+1})] \\ &= \tfrac{4}{3}v_n - \tfrac{1}{3}v_{n-1} + \tfrac{2}{3}hv'_{n+1} \end{aligned} \tag{1-80}$$

Substituting Eq. (1-77) into Eq. (1-80) and solving for i_{n+1}, we obtain

$$i_{n+1} = \frac{3C}{2h}v_{n+1} - \left(\frac{2C}{h}v_n - \frac{C}{2h}v_{n-1}\right) \tag{1-81}$$

Equation (1-81) can be represented by the *equivalent linear* one-port shown in Fig. 1-29. This circuit is therefore the discrete circuit model associated with *Gear's second-order algorithm.*

Examination of the three associated discrete circuit models for a linear capacitor shows that the circuit topology remains unchanged. Only the circuit parameters vary. If *nodal* analysis is to be used for solving the resistive network, the models associated with the backward Euler or Gear's second-order algorithms have some computational advantage over the one associated with the *trapezoidal algorithm*, because the current source in Figs. 1-27 and 1-29 depends only on the *voltage* at the preceding time step, as compared to the source in Fig. 1-28, which requires both voltage v_n and current i_n. In the latter case, additional effort must be made in the nodal analysis to obtain i_n after all nodal voltages have been found.

Finally, note that each circuit in Figs. 1-27 to 1-29 can be transformed into its *Thévenin equivalent* circuit consisting of a resistor and a voltage source. However, there is little motivation for choosing the alternative model, because voltage sources are not desirable in nodal analysis.

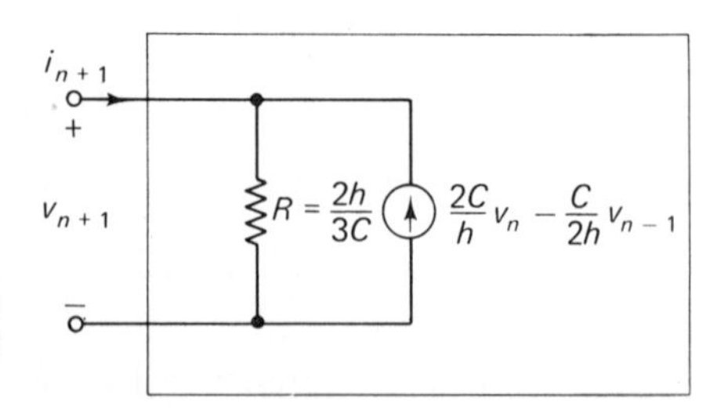

Fig. 1-29. Discrete circuit model associated with Gear's second-order algorithm for a linear capacitor.

1-9-2 Deriving Associated Discrete Circuit Models for Linear Inductors

The derivation given next completely parallels that presented in Section 1-9-1. The *backward Euler algorithm* for solving the first-order differential equation $i' = f(i)$ with a *step size* h is given by

$$\begin{aligned} i_{n+1} &= i_n + hf(i_{n+1}) \\ &= i_n + hi'_{n+1} \end{aligned} \tag{1-82}$$

The exact relationship $v(t) = Li'(t)$ for a *linear* inductor can be approximated by

$$i'_{n+1} = \frac{1}{L} v_{n+1} \tag{1-83}$$

Substituting Eq. (1-83) into Eq. (1-82), we obtain

$$i_{n+1} = \frac{h}{L} v_{n+1} + i_n \tag{1-84}$$

Equation (1-84) can be represented by the *linear equivalent* one-port shown in Fig. 1-30. This circuit is therefore the discrete circuit model associated with the backward Euler algorithm for a linear inductor.

The *trapezoidal algorithm* for solving $i' = f(i)$ is given by

$$\begin{aligned} i_{n+1} &= i_n + \frac{h}{2}[f(i_{n+1}) + f(i_n)] \\ &= i_n + \frac{h}{2} i'_{n+1} + \frac{h}{2} i'_n \end{aligned} \tag{1-85}$$

Following the previous procedure, we make the substitutions $i'_{n+1} \approx (1/L)v_{n+1}$, and $i'_n \approx (1/L)v_n$ to obtain

$$i_{n+1} = \frac{h}{2L} v_{n+1} + \left(\frac{h}{2L} v_n + i_n\right) \tag{1-86}$$

The discrete equivalent circuit for Eq. (1-86) is shown in Fig. 1-31, which is therefore the model for a linear inductor associated with the *trapezoidal algorithm*.

Finally, for Gear's second-order algorithm,

$$\begin{aligned} i_{n+1} &= \tfrac{4}{3} i_n - \tfrac{1}{3} i_{n-1} + h[\tfrac{2}{3} f(i_{n+1})] \\ &= \tfrac{4}{3} i_n - \tfrac{1}{3} i_{n-1} + \tfrac{2}{3} h i'_{n+1} \end{aligned} \tag{1-87}$$

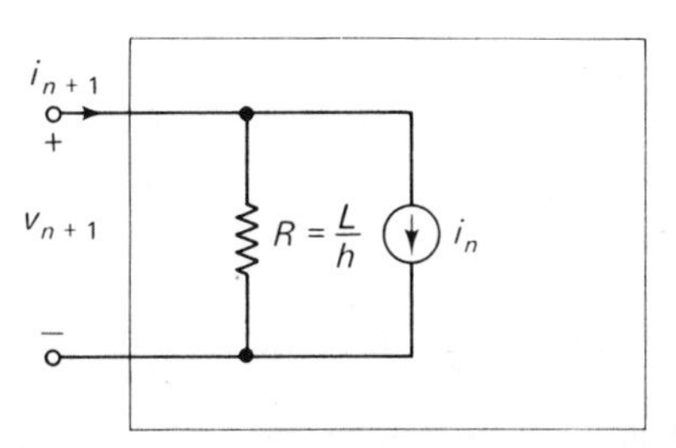

Fig. 1-30. Discrete circuit model associated with backward Euler algorithm for a linear inductor.

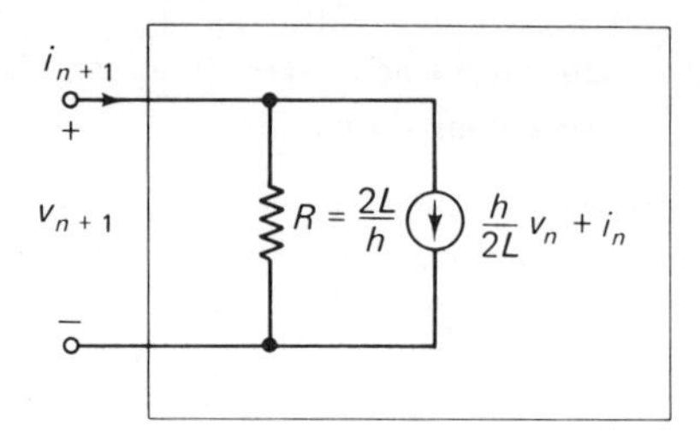

Fig. 1-31. Discrete circuit model associated with trapezoidal algorithm for a linear inductor.

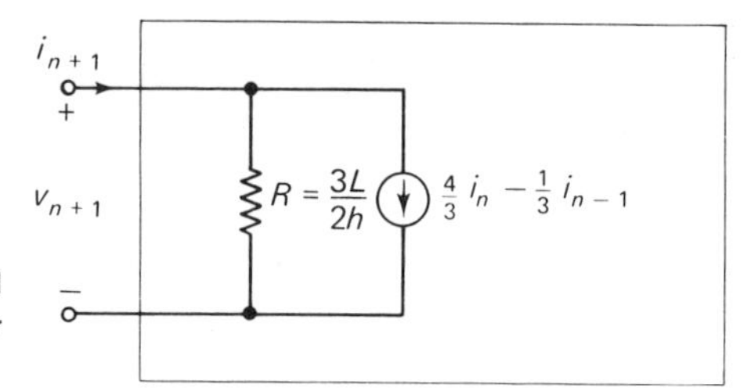

Fig. 1-32. Discrete circuit model associated with Gear's second-order algorithm for a linear inductor.

Following the same procedure, we obtain the equivalent representation

$$i_{n+1} = \frac{2h}{3L} v_{n+1} + \left(\frac{4}{3} i_n - \frac{1}{3} i_{n-1}\right) \tag{1-88}$$

The discrete equivalent circuit for Eq. (1-88) associated with Gear's second-order algorithm is shown in Fig. 1-32.

1-10 A GLIMPSE AT SENSITIVITY ANALYSIS

Consider the sinusoidal steady-state behavior of a linear network. Let $T(j\omega)$ be any network function of interest and $(x_1, \ldots, x_n)$ a set of element parameters subject to changes. The partial derivative $\partial T/\partial x_i$ is called the *unnormalized sensitivity* and $(\partial T/\partial x_i) \cdot (x_i/T)$ the *normalized sensitivity* of T with respect to x_i.

Sensitivity information is important in several applications. It is used to determine the tolerances of network components. In the optimization of linear networks, partial derivatives $\partial T/\partial x_i$ indicate the relative magnitudes of the element changes that will improve the network performance most efficiently.

A brute-force method for determining $\partial T/\partial x_i$ is as follows: perform an analysis of the network with nominal values for $(x_1, \ldots, x_n)$. For each $\partial T/\partial x_i$, perform another analysis with x_i changed *slightly* to $x_i + \Delta x_i$, and obtain the new value of the network function $T + \Delta T$. Then the partial derivative $\partial T/\partial x_i$ is approximated by $\partial T/\partial x_i \approx \Delta T/\Delta x_i$. Such a method has two serious drawbacks. First, the accuracy is poor, because as $\Delta x_i \rightarrow 0$ we are taking the difference between two nearly equal numbers $T + \Delta T$ and T. Second, for each x_i and at each frequency ω, the calculation of $\partial T/\partial x_i$ requires one *analysis of the network*.

Both drawbacks will be avoided if we can obtain the symbolic expression for T containing $(x_1, \ldots, x_n)$ as variables.[5] The function T may then be differentiated to

[5]For certain types of circuits (e.g., crystal filters), the accuracy of the coefficients in the expression poses another serious problem.

obtain the expressions for $\partial T/\partial x_i$. Once this has been done, we only have to *evaluate these expressions* for any given set of nominal parameters $(x_1, \ldots, x_n)$ and frequency ω to obtain all partial derivatives $\partial T/\partial x_i$. This is of course much simpler than analyzing the network repeatedly.

As a simple example of the symbolic procedure, consider the network of Fig. 1-33 in which the nominal element values have been indicated. We wish to find $\partial Y_{\text{in}}/\partial y_1$, $\partial Y_{\text{in}}/\partial y_2$, $\partial Y_{\text{in}}/\partial y_3$, and $\partial Y_{\text{in}}/\partial g_m$. (Since the network is resistive, the result will be independent of the frequency ω.) We first obtain the expression for Y_{in} by any standard method of circuit analysis. The result is

$$Y_{\text{in}} = y_1 + y_3 + \frac{y_3(g_m - y_3)}{y_2 + y_3} \tag{1-89}$$

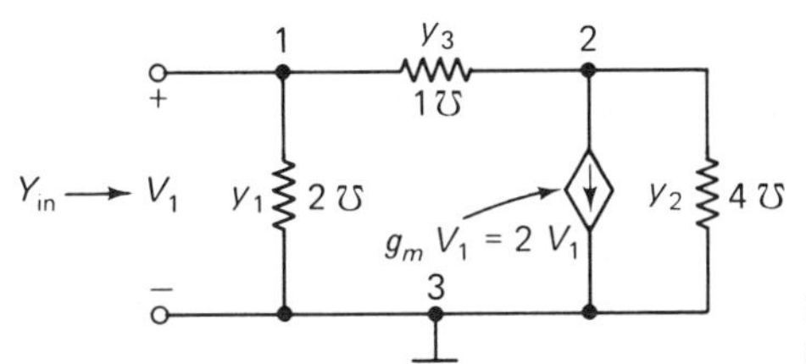

Fig. 1-33. Sensitivity of Y_{in} of this circuit is to be calculated.

Differentiating Eq. (1-89) with respect to y_i, we obtain

$$\frac{\partial Y_{\text{in}}}{\partial y_1} = 1 \tag{1-90}$$

$$\frac{\partial Y_{\text{in}}}{\partial y_2} = -\frac{y_3(g_m - y_3)}{(y_2 + y_3)^2} \tag{1-91}$$

$$\frac{\partial Y_{\text{in}}}{\partial y_3} = \frac{y_2(g_m + y_2)}{(y_2 + y_3)^2} \tag{1-92}$$

$$\frac{\partial Y_{\text{in}}}{\partial g_m} = \frac{y_3}{y_2 + y_3} \tag{1-93}$$

Substituting the nominal values $y_1 = 2$, $y_2 = 4$, $y_3 = 1$, and $g_m = 2$ into the preceding equations, we have

$$\frac{\partial Y_{\text{in}}}{\partial y_1} = 1, \quad \frac{\partial Y_{\text{in}}}{\partial y_2} = \frac{-1}{25}, \quad \frac{\partial Y_{\text{in}}}{\partial y_3} = \frac{24}{25}, \quad \frac{\partial Y_{\text{in}}}{\partial g_m} = \frac{1}{5}$$

If another set of nominal parameter values is given, we only have to reevaluate expressions (1-90)–(1-93). The network need not be analyzed again.

In this simple example, we have derived the expression for Y_{in} by hand. As the network becomes more complex, manual derivation of network functions in symbolic form is extremely error prone (any student can attest to this statement!), if it can be managed at all. Computer programs will be needed to obtain the expressions for the network function T and all partial derivatives $\partial T/\partial x_i$. Chapter 14 presents a number of computer methods of symbolic network analysis. At present, all such computer programs can handle only small- to medium-sized (15-node, 30-branch range) networks with not too many elements represented by variables (on the order of ten).

Thus, for the sensitivity analysis of large networks, we again have to resort to the more conventional, nonsymbolic programs, such as ASTAP [10].

A powerful method for sensitivity analysis and noise analysis of large networks, called the *adjoint-network method,* is now available. With the adjoint-network method, we have to perform only *two* analyses, one for the given network and one for its adjoint network, in order to obtain the partial derivatives $\partial T/\partial x_i$ at some specified frequency with respect to all parameters $(x_1, \ldots, x_n)$. The details of this method will be given in Chapter 15. In this section, we merely wish to illustrate the essence of the method with a very simple example of determining $\partial Y_{\text{in}}/\partial x$. The nodal method (Chapter 4) will be used to analyze the network.

Let Y_{in} be the admittance measured across node 1 and the reference node of the network N. We assume that the network N is modeled with *RLC* elements and voltage-controlled current sources. The adjoint-network method of determining $\partial Y_{\text{in}}/\partial x$ consists of four steps.

Step 1. With a voltage source $V_1 = 1$ V applied to N, perform a nodal analysis and obtain the node voltages $\mathbf{V}_n$.

Step 2. Construct the adjoint network $\hat{N}$ whose nodes are numbered in the same manner as N, and whose element characteristics are such that its nodal admittance matrix $\hat{\mathbf{Y}}_n$ is the transpose of that of N; i.e.,

$$\hat{\mathbf{Y}}_n = \mathbf{Y}_n^t \tag{1-94}$$

Step 3. With a voltage source $\hat{V}_1 = 1$ V applied to $\hat{N}$, perform a nodal analysis and obtain the node voltages $\hat{\mathbf{V}}_n$.

Step 4. Determine $\partial Y_{\text{in}}/\partial x_i$ from the relationship

$$\boxed{dY_{\text{in}} = \hat{\mathbf{V}}_n^t(d\mathbf{Y}_n)\mathbf{V}_n} \tag{1-95}$$

Let us apply the procedure to the previous example. The result of the analysis in step 1 for the circuit shown in Fig. 1-33 is

$$\mathbf{V}_n = [1, \quad -\tfrac{1}{5}]^t \tag{1-96}$$

and is shown in Fig. 1-34(a). The nodal admittance matrix $\mathbf{Y}_n$ can be constructed directly from the given network configuration and element values (see Section 4-6). From Fig. 1-33, we have

$$\mathbf{Y}_n = \begin{bmatrix} y_1 + y_3 & -y_3 \\ g_m - y_3 & y_2 + y_3 \end{bmatrix} = \begin{bmatrix} 3 & -1 \\ 1 & 5 \end{bmatrix} \tag{1-97}$$

Therefore,

$$\hat{\mathbf{Y}}_n = \mathbf{Y}_n^t = \begin{bmatrix} 3 & 1 \\ -1 & 5 \end{bmatrix}$$

The adjoint network $\hat{N}$ constructed in step 2 to have the preceding $\hat{\mathbf{Y}}_n$ is shown in Fig. 1-34(b). Observe that the controlled source is simply "turned around," while all

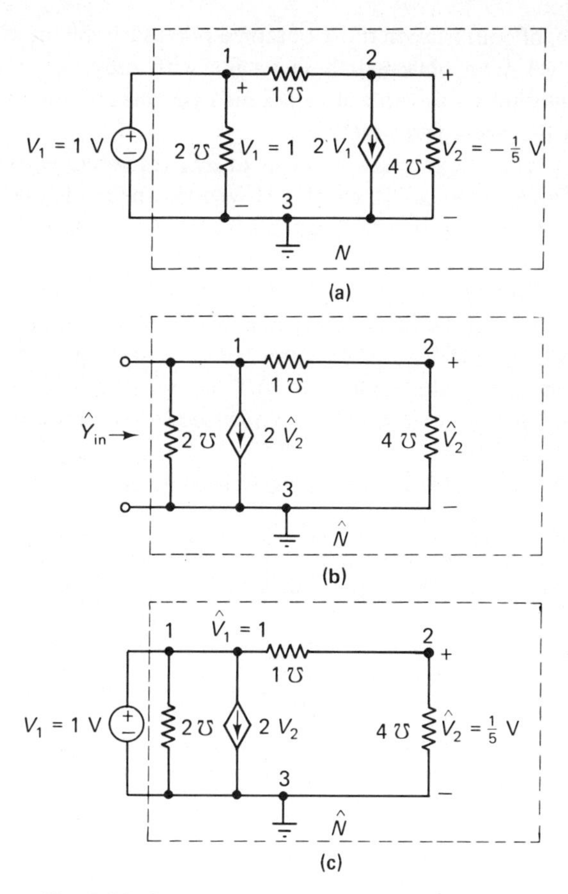

Fig. 1-34. Sensitivity analysis by the adjoint-network method.

conductance elements remain unchanged from N to $\hat{N}$. The result of the analysis in step 3 is shown in Fig. 1-34(c), with

$$\hat{\mathbf{V}}_n = [1, \ \tfrac{1}{5}]^t \tag{1-98}$$

Finally, in step 4, let us first calculate $\partial Y_{\text{in}}/\partial g_m$. For this case, $dy_1 = dy_2 = dy_3 = 0$, since g_m is the only parameter that varies. From Eq. (1-97), we have

$$d\mathbf{Y}_n = \begin{bmatrix} 0 & 0 \\ dg_m & 0 \end{bmatrix} \tag{1-99}$$

Substituting Eqs. (1-96), (1-98), (1-99) into Eq. (1-95), we have

$$dY_{\text{in}} = [1, \ \tfrac{1}{5}] \begin{bmatrix} 0 & 0 \\ dg_m & 0 \end{bmatrix} \begin{bmatrix} 1 \\ -\frac{1}{5} \end{bmatrix} = \tfrac{1}{5} dg_m$$

Hence, $\partial Y_{\text{in}}/\partial g_m = \frac{1}{5}$. In a similar manner, we find that

$$\frac{\partial Y_{\text{in}}}{\partial y_1} = 1, \qquad \frac{\partial Y_{\text{in}}}{\partial y_2} = \frac{-1}{25}, \qquad \frac{\partial Y_{\text{in}}}{\partial y_3} = \frac{24}{25}$$

These results agree, of course, with those obtained previously by the symbolic method. But observe that we have obtained these results with only two nodal analyses of *numerically specified* networks, and have avoided the inaccuracy of the "numerical differentiation" in the brute-force method.

In Eq. (1-95), dY_{in} is expressed in terms of $d\boldsymbol{Y}_n$ (the change in the *nodal admittance matrix*) and nodal voltages $\boldsymbol{V}_n$ and $\hat{\boldsymbol{V}}_n$. How does one relate these to the changes in the *branch admittance* matrix and branch voltages? For the case of a multiport, what are the expressions for dz_{ij} and dy_{ij} by the adjoint-network method? How does one construct the adjoint network $\hat{N}$ if all four types of controlled sources appear in N? All these questions will be investigated and answered in Chapter 15. In doing so, we make extensive use of Tellegen's theorem. Here we shall show that for the special case depicted by Eq. (1-95), the adjoint-network method of sensitivity analysis can be explained with no more than elementary knowledge of nodal analysis and matrix algebra.

Referring to Fig. 1-35 for notations, we write the nodal equations for N as

$$\boldsymbol{Y}_n\boldsymbol{V}_n = \left[\begin{array}{c|ccc} y_{11} & y_{12} & \cdots & y_{1n} \\ \hline y_{21} & y_{22} & \cdots & y_{2n} \\ \cdot & \cdot & & \cdot \\ \cdot & \cdot & & \cdot \\ \cdot & \cdot & & \cdot \\ y_{n1} & y_{n2} & \cdots & y_{nn} \end{array}\right]\left[\begin{array}{c} V_1 \\ \hline V_2 \\ \cdot \\ \cdot \\ \cdot \\ V_n \end{array}\right] = \left[\begin{array}{c|c} y_{11} & \boldsymbol{Y}_{12} \\ \hline \boldsymbol{Y}_{21} & \boldsymbol{Y}_{22} \end{array}\right]\left[\begin{array}{c} V_1 \\ \hline \boldsymbol{V}_2 \end{array}\right] = \left[\begin{array}{c} I_1 \\ \hline \boldsymbol{0} \end{array}\right] \tag{1-100}$$

where the meanings of the submatrices $\boldsymbol{Y}_{12}$, $\boldsymbol{Y}_{21}$, $\boldsymbol{Y}_{22}$, and $\boldsymbol{V}_2$ are indicated by the partitionings of $\boldsymbol{Y}_n$ and $\boldsymbol{V}_n$. We can rewrite Eq. (1-100) as two equations:

$$y_{11}V_1 + \boldsymbol{Y}_{12}\boldsymbol{V}_2 = I_1 \tag{1-101}$$

$$\boldsymbol{Y}_{21}V_1 + \boldsymbol{Y}_{22}\boldsymbol{V}_2 = \boldsymbol{0} \tag{1-102}$$

Assuming that $\boldsymbol{Y}_{22}^{-1}$ exists for the moment, we can solve for $\boldsymbol{V}_2$ from Eq. (1-102), and use the result in Eq. (1-101) to obtain

$$Y_{\text{in}} = \frac{I_1}{V_1} = y_{11} - \boldsymbol{Y}_{12}\boldsymbol{Y}_{22}^{-1}\boldsymbol{Y}_{21} \tag{1-103}$$

If $\boldsymbol{Y}_{22}$ is singular, then Y_{in} does not exist (as a finite number), and neither does any derivative of Y_{in}. Applying the well-known formula $d(xy) = (dx)y + x\,dy$ to Eq.

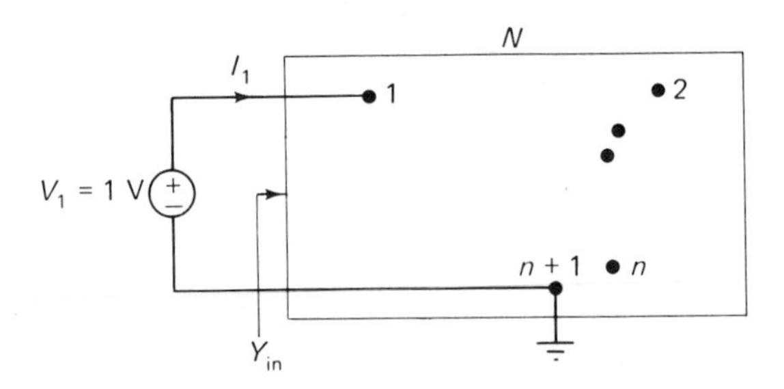

Fig. 1-35. Derivation of the sensitivity expression (1-95).

(1-103), we have

$$dY_{\text{in}} = dy_{11} - (dY_{12})Y_{22}^{-1}Y_{21} - Y_{12}(dY_{22}^{-1})Y_{21} - Y_{12}Y_{22}^{-1}(dY_{21}) \tag{1-104}$$

In step 1 described previously, we have $V_1 = 1$. Then, from Eq. (1-102), we obtain

$$V_2 = -Y_{22}^{-1}Y_{21}V_1 = -Y_{22}^{-1}Y_{21}$$

Therefore,

$$V_n = \begin{bmatrix} V_1 \\ V_2 \end{bmatrix} = \begin{bmatrix} 1 \\ -Y_{22}^{-1}Y_{21} \end{bmatrix} \tag{1-105}$$

By exactly the same reasoning, the voltage vector obtained in step 3 for the adjoint network $\hat{N}$ is

$$\hat{V}_n = \begin{bmatrix} \hat{V}_1 \\ \hat{V}_2 \end{bmatrix} = \begin{bmatrix} 1 \\ -\hat{Y}_{22}^{-1}\hat{Y}_{21} \end{bmatrix} \tag{1-106}$$

But in step 2 we construct $\hat{N}$ such that $\hat{Y}_n^t = Y_n$. This implies that

$$\begin{aligned} \hat{y}_{11} &= y_{11} & \hat{Y}_{12}^t &= Y_{21} \\ \hat{Y}_{21}^t &= Y_{12} & \hat{Y}_{22}^t &= Y_{22} \end{aligned} \tag{1-107}$$

Substituting Eqs. (1-105) and (1-106) into the right side of Eq. (1-95) and making use of Eq. (1-107), we have

$$\begin{aligned} \hat{V}_n^t(dY_n)V_n &= [1, -\hat{Y}_{21}^t(\hat{Y}_{22}^{-1})^t] \begin{bmatrix} dy_{11} & dY_{12} \\ dY_{21} & dY_{22} \end{bmatrix} \begin{bmatrix} 1 \\ -Y_{22}^{-1}Y_{21} \end{bmatrix} \\ &= dy_{11} - (dY_{12})Y_{22}^{-1}Y_{21} - Y_{12}Y_{22}^{-1}(dY_{21}) \\ &\quad + Y_{12}Y_{22}^{-1}(dY_{22})Y_{22}^{-1}Y_{21} \end{aligned} \tag{1-108}$$

Now, for any nonsingular matrix P, we can write $PP^{-1} = \mathbf{1}$. Then $(dP)P^{-1} + P \cdot d(P^{-1}) = \mathbf{0}$; therefore, $P^{-1}(dP)P^{-1} = -d(P^{-1})$. Applying this relationship to the last term of Eq. (1-108), we obtain

$$\hat{V}_n^t(dY_n)V_n = dy_{11} - (dY_{12})Y_{22}^{-1}Y_{21} - Y_{12}Y_{22}^{-1}(dY_{21}) - Y_{12}d(Y_{22}^{-1})Y_{21} \tag{1-109}$$

But according to Eq. (1-104), the right side of Eq. (1-109) is simply dY_{in}. Therefore, we have proved the relationship (1-95) used in the adjoint-network method:

$$dY_{\text{in}} = \hat{V}_n^t(dY_n)V_n$$

1-11 A GLIMPSE AT SPARSE-MATRIX TECHNIQUES FOR CIRCUIT ANALYSIS

The analysis of linear networks mainly consists of the formulation and solution of a set of n linear simultaneous equations of the form

$$Ax = \boldsymbol{\mu} \tag{1-110}$$

where A, x, and $\boldsymbol{\mu}$ are real for dc analysis and complex for ac analysis (sinusoidal steady state). In the case of a *nonlinear* resistive network, the nonlinear simultaneous equations

$$f(x, t) = \mathbf{0} \tag{1-111}$$

may be solved by any iterative method, e.g., the Newton–Raphson algorithm (Chapter 5). Such a method requires *repeated solutions* of linear simultaneous equations. Therefore, it is of extreme importance in computer-aided circuit analysis to consider efficient methods of solving Eq. (1-110).

Any method of solving Eq. (1-110) that pays no attention to whether each element a_{ij} of A is zero or not is called a *full-matrix* method. On the other hand, a method that exploits the sparsity of the coefficient matrix A to reduce the computational effort will be called a *sparse-matrix* method. A matrix A is usually considered sparse if only a small fraction, say less than 30 percent, of its elements are nonzero. The nodal admittance matrices $\boldsymbol{Y}_n$ for most large electronic circuits are sparse. For example, a typical 100-node circuit has about 5 percent nonzero elements in $\boldsymbol{Y}_n$. More specifically, the six-node active circuit of Fig. 1-36 has the following $\boldsymbol{Y}_n$ at $\omega = 1$ (see Section 4-6 for a direct construction of $\boldsymbol{Y}_n$):

$$\boldsymbol{Y}_n = \begin{bmatrix} 1+j & -j & 0 & 0 & 0 & 80 \\ -j & 1+j2 & -j & 0 & 0 & 0 \\ 0 & -j & 1+j2 & -j & 0 & 0 \\ 0 & 0 & -j & 1+j2 & -j & 0 \\ 0 & 0 & 0 & -j & 1+j2 & -j \\ 0 & 0 & 0 & 0 & -j & 1+j \end{bmatrix} \tag{1-112}$$

Observe that only $\frac{17}{36} \times 100 = 47$ percent of the elements of $\boldsymbol{Y}_n$ are nonzero.

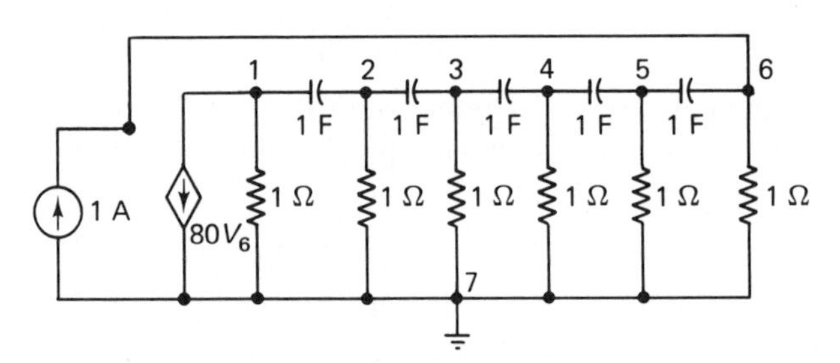

Fig. 1-36. Network with sparse nodal admittance matrix.

One commonly used method for solving Eq. (1-110) is the Gaussian elimination method (Chapter 4). It can be shown that in the case of full matrices the number of operations[6] required by the Gaussian elimination method for solving Eq. (1-110) is

$$\frac{n^3 + 3n^2 - n}{3}$$

which is approximately $n^3/3$ when n is very large (see Problem 4-5). Clearly, as n increases, the required number of operations increases at a much greater rate. For a 200-node circuit, $n^3/3 \approx 2.6 \times 10^6$ operations will be needed to solve Eq. (1-110)

[6]In this section, the count of operations includes only multiplications and divisions.

just once. Even with present-day high-speed digital computers the computing time is prohibitive, since in most cases Eq. (1-110) has to be solved repeatedly with different coefficient matrices. Thus, when large networks (100-node range) are to be analyzed, it becomes an absolute necessity to employ sparse-matrix techniques.

A sparse-matrix method reduces the computational effort for solving Eq. (1-110) mainly by avoiding trivial operations such as $a \times 0 = 0$ and $a + 0 = a$. We shall illustrate with some numerical examples the reduction in the number of operations that can be achieved with a sparse-matrix technique. Consider first the following matrix equation whose coefficient matrix is full:

$$\begin{bmatrix} 2 & 3 & -1 & -1 \\ 4 & 12 & 8 & 4 \\ 2 & -2 & 4 & 1 \\ 3 & -1 & 1 & 2 \end{bmatrix} \begin{bmatrix} x_1 \\ x_2 \\ x_3 \\ x_4 \end{bmatrix} = \begin{bmatrix} 1 \\ -8 \\ 3 \\ 8 \end{bmatrix} \tag{1-113}$$

Let us solve Eq. (1-113) by the Gaussian elimination method. Then the number of multiplications *and* divisions required to obtain the solution is

$$\frac{n^3 + 3n^2 - n}{3} = \frac{64 + 48 - 4}{3} = 36$$

and the solution is

$$\boldsymbol{x} = [2, -1, -1, 1]^t$$

Next, consider the following matrix equation whose coefficient matrix is only 50 percent full:

$$\begin{bmatrix} 2 & 0 & 0 & 6 \\ 0 & 4 & 0 & 8 \\ 0 & 0 & 2 & 4 \\ 4 & 0 & 0 & 6 \end{bmatrix} \begin{bmatrix} x_1 \\ x_2 \\ x_3 \\ x_4 \end{bmatrix} = \begin{bmatrix} 8 \\ 4 \\ 8 \\ 10 \end{bmatrix} \tag{1-114}$$

If a full-matrix technique is used to solve Eq. (1-114), operations such as $4 \times 0 = 0$ will be carried out (by the computer) just the same as $4 \times 3 = 12$. No saving in computational effort is achieved by the presence of the eight zero elements in the coefficient matrix. The number of operations required to solve Eq. (1-114) is 36, the same as that for Eq. (1-113).

Now let us solve Eq. (1-114) again by the Gaussian elimination method, but this time we incorporate a sparse-matrix technique to avoid all trivial operations such as $a \times 0 = 0$. The matrix equation (1-114) actually represents four scalar equations. The solution proceeds as follows.

Normalize a_{11} in the first equation with two divisions:[7] $a_{14} \leftarrow 6 \div 2$ and $\mu_1 \leftarrow 8 \div 2$.

Eliminate x_1 from the fourth equation with two multiplications: $a_{44} \leftarrow (a_{44} - 4 \times a_{14})$ and $\mu_4 \leftarrow (10 - 4 \times \mu_1)$.

[7]The notation $p \leftarrow qr$ means to replace p by the result of qr.

At this time, Eq. (1-114) is reduced to the following equivalent form:

$$\begin{bmatrix} 1 & 0 & 0 & 3 \\ 0 & 4 & 0 & 8 \\ 0 & 0 & 2 & 4 \\ 0 & 0 & 0 & -6 \end{bmatrix} \begin{bmatrix} x_1 \\ x_2 \\ x_3 \\ x_4 \end{bmatrix} = \begin{bmatrix} 4 \\ 4 \\ 8 \\ -6 \end{bmatrix} \tag{1-115}$$

Now refer to Eq. (1-115). Normalize a_{22}, a_{33}, and a_{44} with five divisions: $a_{24} \leftarrow 8 \div 4$, $\mu_2 \leftarrow 4 \div 4$, $a_{34} \leftarrow 4 \div 2$, $\mu_3 \leftarrow 8 \div 2$, and $\mu_4 \leftarrow (-6) \div (-6)$.

At this time, Eq. (1-115) is reduced to

$$\begin{bmatrix} 1 & 0 & 0 & 3 \\ 0 & 1 & 0 & 2 \\ 0 & 0 & 1 & 2 \\ 0 & 0 & 0 & 1 \end{bmatrix} \begin{bmatrix} x_1 \\ x_2 \\ x_3 \\ x_4 \end{bmatrix} = \begin{bmatrix} 4 \\ 1 \\ 4 \\ 1 \end{bmatrix} \tag{1-116}$$

Now refer to Eq. (1-116). Eliminate x_4 from the first three equations with three multiplications: $\mu_3 \leftarrow 4 + (-2) \times 1$, $\mu_2 \leftarrow 1 + (-2) \times 1$, and $\mu_1 \leftarrow 4 + (-3) \times 1$. At this time, Eq. (1-116) is reduced to

$$\begin{bmatrix} 1 & 0 & 0 & 0 \\ 0 & 1 & 0 & 0 \\ 0 & 0 & 1 & 0 \\ 0 & 0 & 0 & 1 \end{bmatrix} \begin{bmatrix} x_1 \\ x_2 \\ x_3 \\ x_4 \end{bmatrix} = \begin{bmatrix} x_1 \\ x_2 \\ x_3 \\ x_4 \end{bmatrix} = \begin{bmatrix} 1 \\ -1 \\ 2 \\ 1 \end{bmatrix} \tag{1-117}$$

which gives the solution of $\boldsymbol{x}$. From this detailed analysis, we see that a total of $2 + 2 + 5 + 3 = 12$ operations are needed to solve Eq. (1-114). The number of operations is only one third that required by the use of a full-matrix technique. The reduction of the number of operations will be much more significant when the coefficient matrix $\boldsymbol{A}$ is larger and sparser. In one example of a 28-node operational-amplifier circuit having about 15 percent nonzero elements in $\boldsymbol{Y}_n$, the operation count is reduced from 7308 by a full-matrix method to 232 by a sparse-matrix method!

When a sparse-matrix method is used, the ordering of the equations and unknowns in Eq. (1-110) has an important effect on the number of operations required in the solution. For example, consider the following equation:

$$\begin{bmatrix} 2 & 4 & -2 & -6 \\ 3 & 9 & 0 & 0 \\ -2 & 0 & 8 & 0 \\ 2 & 0 & 0 & -12 \end{bmatrix} \begin{bmatrix} x \\ y \\ z \\ w \end{bmatrix} = \begin{bmatrix} 6 \\ -6 \\ 14 \\ 26 \end{bmatrix} \tag{1-118}$$

The coefficient matrix has six zero elements. One would expect that the number of operations required for solving Eq. (1-118) should be somewhat less than 36, the number required for solving Eq. (1-113) whose coefficient matrix has no zero elements at all. This turns out to be untrue. Because of the particular distribution of the zero elements in Eq. (1-118), exactly the same number of operations is needed for solving

Eq. (1-118). Even though we intend to omit trivial operations, they never occur in solving Eq. (1-118) if the unknowns are eliminated in the order of x, y, and z.

Now suppose that the unknowns and equations in Eq. (1-118) are reordered as follows:

$$\begin{bmatrix} -12 & 0 & 0 & 2 \\ 0 & 9 & 0 & 3 \\ 0 & 0 & 8 & -2 \\ -6 & 4 & -2 & 2 \end{bmatrix} \begin{bmatrix} w \\ y \\ z \\ x \end{bmatrix} = \begin{bmatrix} 26 \\ -6 \\ 14 \\ 6 \end{bmatrix} \tag{1-119}$$

Then using the Gaussian elimination method, but omitting trivial operations such as $a \times 0 = 0$, it can be shown that only 16 operations, instead of 36, are needed to obtain the solution $w = -2$, $y = -1$, $z = 2$, and $x = 1$.

A question immediately arises: What is the best way to reorder the unknowns and equations so that a sparse-matrix technique is most effective? Furthermore, if $b = 0$, how does a computer algorithm avoid trivial operations such as $ab = 0$ and $a + b = a$ *without* fetching the number b and checking its value? For if a computer program still has to get b and check its value, then the advantage of a sparse-matrix technique will be greatly reduced. These questions will be investigated and answered in Chapter 16.

REFERENCES

1. NAGEL, L. W., and D. O. PEDERSON. *SPICE (Simulation Program with Integrated Circuit Emphasis)*. Berkeley, Calif.: University of California, Electronics Research Laboratory. Memorandum ERL-M382, Apr. 12, 1973.
2. POTTLE, C. *CORNAP User Manual*. Ithaca, N.Y.: Cornell University, School of Electrical Engineering, 1968.
3. IDLEMAN, T. E., F. S. JENKINS, W. J. MCCALLA, and D. O. PEDERSON. "SLIC—A Simulator for Linear Integrated Circuits." *IEEE J. Solid State Circuits* (Special Issue on Computer-Aided Circuit Analysis and Device Modeling), Vol. SC-6, pp. 188–203, Aug. 1971.
4. LIN, P. M., and G. E. ALDERSON. *SNAP—A Computer Program for Generating Symbolic Network Functions*. Lafayette, Ind.: Purdue University, School of Electrical Engineering. Rept. TR-EE70-16, Aug. 1970.
5. CHUA, L. O., and P. A. MEDLOCK. *MECA—A User Oriented Computer Program for Analyzing Resistive Nonlinear Networks*. Vol. 1, User's Manual. Lafayette, Ind.: Purdue University, School of Electrical Engineering. Rept. TR-EE69-7, Apr. 1969.
6. HERSKOWITZ, G. J., ed. *Computer-Aided Integrated Circuit Design*. New York: McGraw-Hill Book Company, 1968.
7. KUO, F. F., and W. G. MAGNUSON, Jr., eds. *Computer-Oriented Circuit Design*. Englewood Cliffs, N. J.: Prentice-Hall, Inc., 1969.
8. CALAHAN, D. A. *Computer-Aided Network Design*. New York: McGraw-Hill Book Company, 1972.
9. DIRECTOR, S. W., ed. *Computer-Aided Circuit Design, Simulation, and Optimization*. Stroudsburg. Pa.: Dowden, Hutchinson & Ross, Inc., 1973.

10. WEEKS, W. T., A. J. JIMENEZ, G. W. MAHONEY, D. METHA, H. QASSEMZADEH, and T. R. SCOTT. "Algorithms for ASTAP—A Network-Analysis Program." *IEEE Trans. Circuit Theory*, Vol. CT-20, pp. 628–634, Nov. 1973.

PROBLEMS

1-1. Find the hybrid matrix $\boldsymbol{H}$ and source vector $\boldsymbol{s}$ in Eq. (1-15) and show that the associated nonlinear equations given by Eqs. (1-16) and (1-17) are equivalent to Eqs. (1-13) and (1-14) in the sense that they have identical solutions.

1-2. Consider the initial-value problem

$$\dot{x} = -80x$$
$$x(0) = 0.005$$

(a) Find the *maximum* step size, $h_{\max}$, to avoid numerical instability when using the forward Euler algorithm.
(b) Write a computer program to find the solution from $t = 0$ to 0.10 s using a step size $h = \frac{1}{4}h_{\max}$.
(c) Repeat part (b) using a step size $h = 2h_{\max}$.
(d) Find the local and total errors in parts (b) and (c).

1-3. Write a computer program to find the solution to

$$\dot{x} = -80x$$
$$x(0) = 0.005$$

using the following numerical integration algorithms:
(a) Backward Euler algorithm with step sizes of $h = 0.005$, 0.1, and 1.0. Find the local and total errors in each case.
(b) Trapezoidal algorithm with step sizes of $h = 0.05$, 0.1, and 1.0. Find the local and total errors in each case.

1-4. The matrix $\boldsymbol{A}$ associated with the state equation of a linear pulse forming circuit is given by

$$\boldsymbol{A} = \begin{bmatrix} -10^6 & -2 \times 10^7 \\ 10^3 & -50 \end{bmatrix}$$

(a) Find the eigenvalues of $\boldsymbol{A}$.
(b) Find the maximum step size, $h_{\max}$, to guarantee numerical stability when the forward Euler algorithm is used to solve the state equation numerically.
(c) Assuming that each time-step iteration in a computer requires 1 μs, find the approximate computing time needed to obtain the complete solution (transient and steady state) of the circuit using a unit step input.

1-5. Consider the initial-value problem

$$\dot{x} = -80x$$
$$x(0) = 0.005$$

(a) Find the solution at $t = 0.01$ using forward Euler algorithm with a step size $h = 0.002$.
(b) Compute the local truncation error ϵ_T.

(c) Find the exact solution $\hat{x}(t)$ and determine the value of $\hat{\tau}$ so that the answer from part (b) is equal to that given by Eq. (1-47).

1-6. Repeat Problem 1-5 using the backward Euler algorithm and find $\hat{\tau}$ so that Eq. (1-52) will give the exact value of the local truncation error at $t = 0.02$ s.

1-7. Repeat Problem 1-5 using the trapezoidal algorithm and find $\hat{\tau}$ so that Eq. (1-66) will give the exact value of the local truncation error at $t = 0.02$ s.

1-8. Construct a simple numerical example to demonstrate the conclusion that under certain conditions the state variable q will lead to less total error than the state variable v for a nonlinear capacitor.

1-9. Use a development dual to that given in Section 1-8 to show that under certain conditions the state variable φ (flux linkage) will lead to less total error than the more usual state variable i for a nonlinear inductor.

1-10. Derive the discrete circuit model for a *nonlinear capacitor* associated with the backward Euler algorithm, the trapezoidal algorithm, and Gear's second-order algorithm.

1-11. Derive the discrete circuit model for a *nonlinear inductor* associated with the backward Euler algorithm, the trapezoidal algorithm, and Gear's second-order algorithm.

1-12. In Fig. 1-34(a), let the controlled current source be replaced by a conductance $g_4 = 1\ \mho$. Find the partial derivatives $\partial Y_{\text{in}}/\partial g_i$, $i = 1, 2, 3, 4$, by
(a) The symbolic network function method.
(b) The adjoint-network method.
Note that in the present case the adjoint network and the original network are identical.

1-13. Verify that the total number of operations (multiplications and divisions) required for solving Eq. (1-119) is 16 if the Gaussian elimination method is used, and trivial operations $a \times 0 = 0$ are avoided.

1-14. Verify that the number of operations (multiplications and divisions) required for solving Eq. (1-118), which is equivalent to Eq. (1-119), is 36, when the same method of solution is used.

1-15. In the circuit shown in Fig. P1-15, the input current is a ramp and the capacitor is initially uncharged; i.e., $v_1(0) = 0$. We wish to find the solution waveform of $v_1(t)$ for $t \geq 0$. This transient analysis problem can be reduced to a repeated dc analysis by replacing the capacitor with its "associated discrete circuit model." The resulting resistive circuit can then be solved for $v_1(t)$ by any convenient method, such as the nodal method of Chapter 4.
(a) Use the model of Fig. 1-27 for the capacitor and find $v_1(t)$ for $t = 0.1, 0.2$, and 0.3 ($h = 0.1$).
(b) Compare the answers from part (a) with the exact solution $v_1(t) = t - 1 + e^{-t}$.

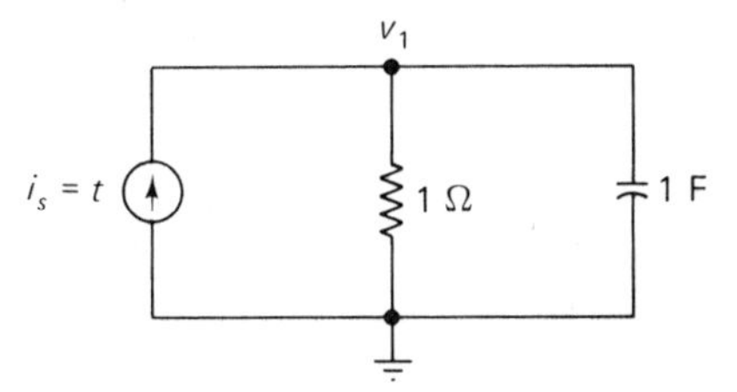

Fig. P1-15

CHAPTER 2

Computer Circuit Models of Electronic Devices and Components

2-1 CIRCUIT MODELS AND THEIR BUILDING BLOCKS—THE BASIC SET

Any computer circuit analysis program will recognize and allow only a *basic set* of circuit elements for which it was designed. For example, if a program was designed to analyze networks containing only resistors and batteries, the basic set for this program would consist of only resistors and batteries. Hence, the larger the basic set allowed in a program, the more versatile it becomes. Such a program would of course be more difficult to develop and would in general require much larger computer storage. Indeed, the maximum number of circuit element types in a basic set is often dictated by the memory size of the class of computer for which the program was intended.

The subject of this chapter is best motivated by asking the following question: How could a computer program be used to analyze a circuit containing elements not allowed in its basic set? The simplest approach—if it could be done—is to replace each disallowed element by an "equivalent circuit" made up of only elements from the basic set. Unfortunately, this is often not possible because, *by definition*, a circuit is equivalent to a given element if, and only if, they are indistinguishable when measured from their external terminals. This strong requirement is never satisfied by such practical devices as diodes, transistors, etc. However, in most practical cases, it has been found possible to replace each disallowed element by an "approximately equivalent circuit,"

TABLE 2-1. The Minimal Basic Set for Lumped Time-Invariant Elements

TYPE OF ELEMENT	SYMBOL AND CHARACTERIZATION	
1. Resistor	(a) Linear resistor 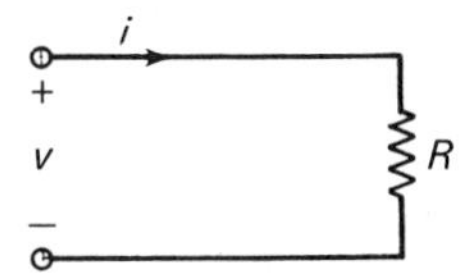	$v = Ri$ $i = Gv$ R = resistance G = conductance
	(b) Nonlinear resistor 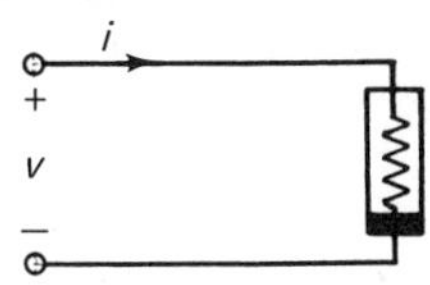	Current-controlled resistor: $v = \hat{v}(i)$ Voltage-controlled resistor: $i = \hat{i}(v)$ A strictly monotonic resistor can be characterized by both.
2. Capacitor	(a) Linear capacitor 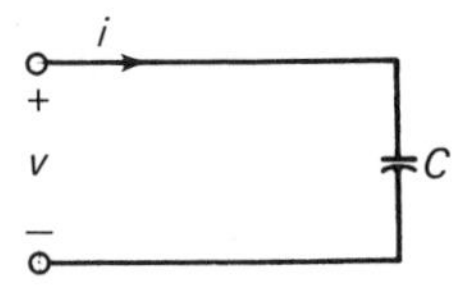	$i = C\dfrac{dv(t)}{dt}$ $v = \dfrac{1}{C}\int_{-\infty}^{t} i(\tau)\,d\tau$
	(b) Nonlinear capacitor 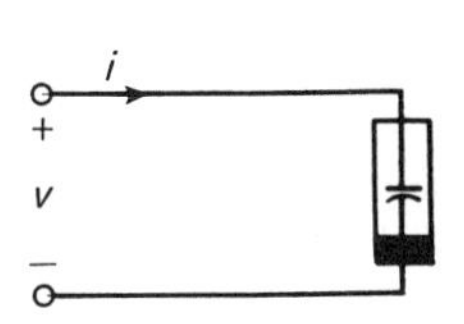	Charge-controlled capacitor: $v = \hat{v}(q)$, $i(t) = dq(t)/dt$ Voltage-controlled capacitor: $q = \hat{q}(v)$, $i(t) = dq(t)/dt$ or $i(t) = C(v)\,dv(t)/dt$ where $C(v) \triangleq dq(v)/dv$ is called the *incremental capacitance*. A strictly monotonic capacitor can be characterized by both.
3. Inductor	(a) Linear inductor 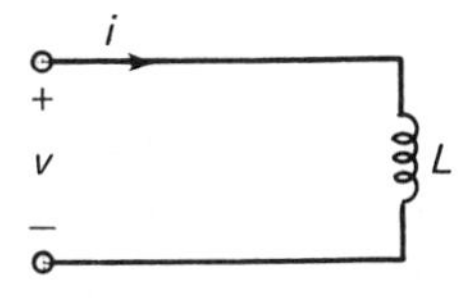	$v = L\dfrac{di(t)}{dt}$ $i = \dfrac{1}{L}\int_{-\infty}^{t} v(\tau)\,d\tau$
	(b) Nonlinear inductor 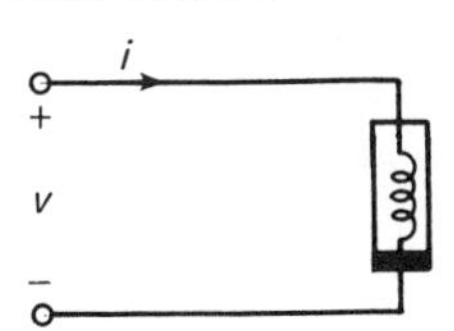	Flux-controlled inductor: $i = \hat{i}(\phi)$, $v(t) = d\phi(t)/dt$ Current-controlled inductor: $\phi = \hat{\phi}(i)$, $v(t) = d\phi(t)/dt$ or $v(t) = L(i)\,di(t)/dt$ where $L(i) \triangleq d\phi(i)/di$ is called the *incremental inductance.* A strictly monotonic inductor can be characterized by both.

TABLE 2-1 (cont.)

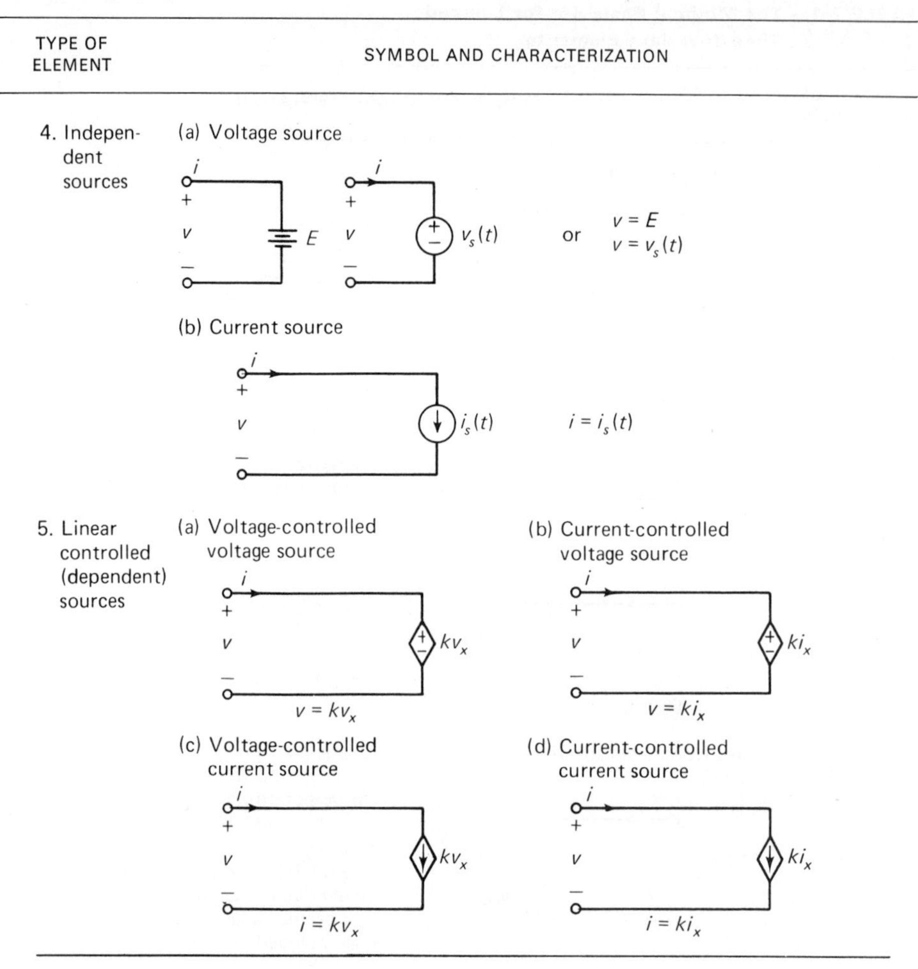

henceforth called a *circuit model*, and obtain computer answers that represent a good approximation to those actually measured. In other words, the usefulness of a computer program is greatly enhanced by the use of *realistic* circuit models.

To synthesize realistic circuit models for most practical devices and components of interest, the basic set must contain at least the five classes of lumped time-invariant circuit elements listed in Table 2-1. A *resistor* is a two-terminal element characterized by a curve in the voltage (v) versus current (i) plane. If the v-i curve consists of simply a straight line through the origin, the resistor is said to be *linear*; otherwise, it is *nonlinear*. If the v-i curve can be expressed as a function of current (voltage), it is said to be current (voltage) controlled. A strictly monotonically increasing v-i curve is both current-controlled and voltage-controlled.

A *capacitor* is a two-terminal element characterized by a curve in the charge (q) versus voltage (v) plane. If the v-q curve consists of simply a straight line through the origin, the capacitor is said to be *linear*; otherwise, it is *nonlinear*. If the v-q curve can be expressed as a function of charge (voltage), it is said to be charge (voltage) controlled. A strictly monotonically increasing v-q curve is both charge-controlled and voltage-controlled.

An *inductor* is a two-terminal element characterized by a curve in the flux linkage (φ) versus current (i) plane. If the i-φ curve consists of simply a straight line through the origin, the inductor is said to be *linear*; otherwise, it is *nonlinear*. If the i-φ curve can be expressed as a function of flux (current), it is said to be flux (current)-controlled. A strictly monotonically increasing i-φ curve is both flux-controlled and current-controlled.

An *independent* voltage source is a two-terminal element whose terminal voltage $v_s(t)$ at any instant of time is prescribed a priori and is therefore independent of its terminal current. An *independent* current source is a two-terminal element whose terminal current $i_s(t)$ at any instant of time is prescribed a priori and is therefore independent of its terminal voltage.

A *linear* controlled (dependent) source is a two-terminal element whose terminal voltage or current at any instant of time is proportional to the voltage v_x or the current i_x in another part of the circuit. If the controlling voltage v_x or the controlling current i_x pertains to the voltage or current of an element x, then this element is called the *controlling element*.

Since most *lumped* and *time-invariant* devices and components can be modeled reasonably well using only the five types of circuit elements listed in Table 2-1, we shall henceforth refer to this set of basic elements as the *minimal basic set*. We do not consider distributed elements in this book because no general and efficient computer-analysis algorithm is currently available for distributed networks. However, many distributed circuit elements, such as transmission lines, can often be modeled by a lumped circuit [1, 7, 8].

2-2 HIERARCHY AND TYPES OF CIRCUIT MODELS

A two-terminal, three-terminal, and $(n + 1)$-terminal device are shown in Fig. 2-1(a), (b), and (c), respectively. Our main objective is to synthesize a network with the same number of external terminals using only the elements from the minimal basic set such that the resulting networks shown in Fig. 2-1(d), (e), and (f) will approximate the characteristics of the corresponding devices to within acceptable accuracy. We refer to the resulting *black box* as a *circuit model* of the physical device. In general, a given device may give rise to many distinct circuit models. Although one model may be better than another in a specific circuit, there is no such thing as the "best model" or a "unique model." It is up to the engineer to choose the best model for his par-

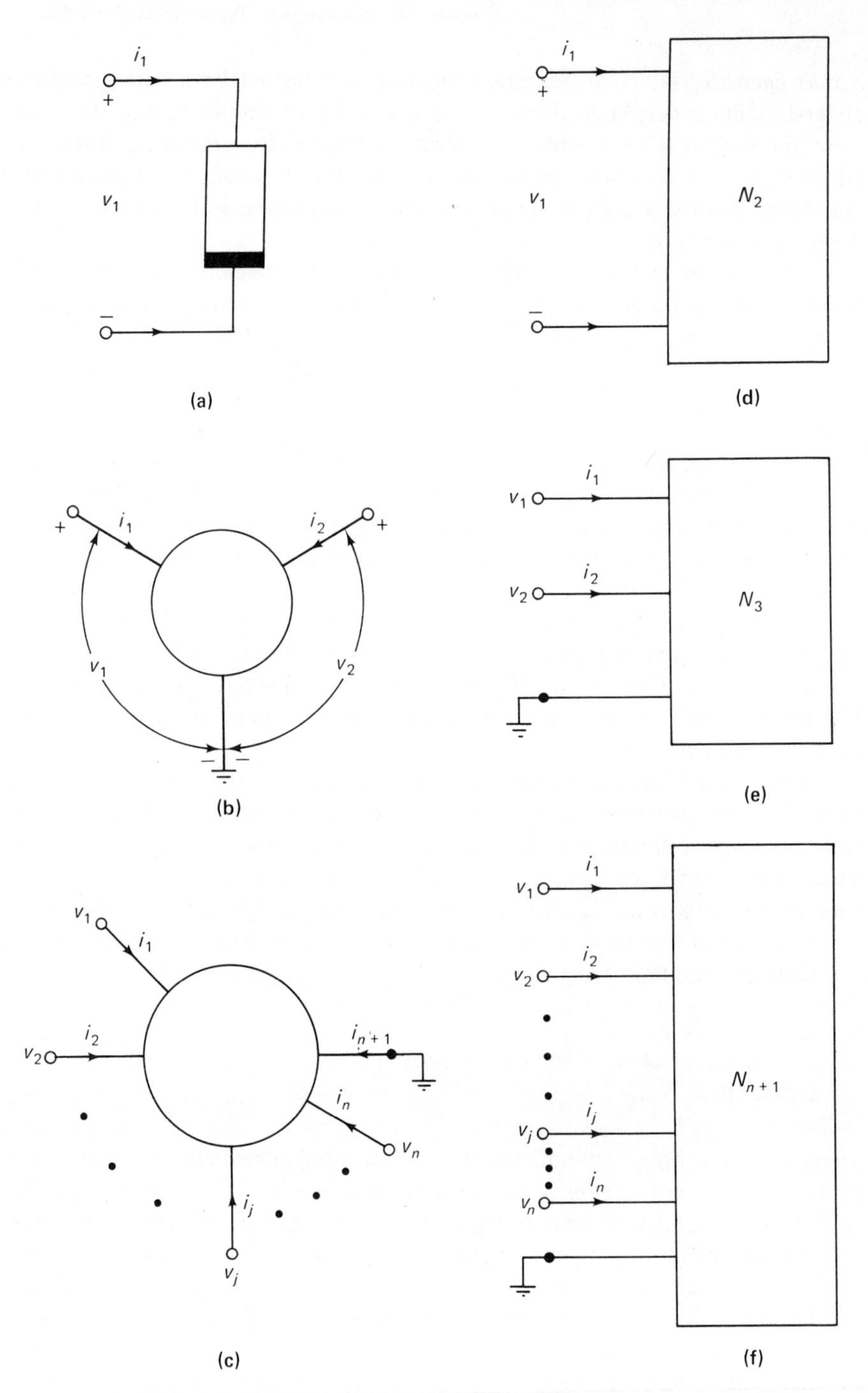

Fig. 2-1. Symbols for a two-terminal, three-terminal, and $(n+1)$-terminal device.

ticular circuits. The ability to select the most appropriate model depends on a firm understanding of the hierarchy and different types of circuit models.

Perhaps the most significant distinction among various types of models is based on the class of signals a model is designed to handle. From the point of view of modeling, the two most important "qualities" of a signal are its *amplitude range* and *bandwidth*. The amplitude range corresponds to the maximum and minimum instantaneous voltage and current swings to which the device will be subjected. The bandwidth corresponds to the lowest- and the highest-frequency components of the signals.

2-2-1 Classification of Models in Terms of Signal Amplitude Range

Depending on the amplitude range of the operating signals, a model may be classified as a *global*, *local*, or *linear incremental model*. A *global model* is designed to simulate a given device over all measurable ranges of terminal voltages and currents. A *local model* is designed to simulate accurately over only some prescribed regions of the device's operating ranges. A *linear incremental* model is a local model made up of only *linear* elements from the minimal basic set (no independent sources and no nonlinear elements). A global model of a physical device is invariably *nonlinear*; i.e., it contains at least one nonlinear resistor, capacitor, or inductor. A local model may or may not be nonlinear, depending on over how large a region it is expected to operate. A linear incremental model simulates the device's characteristics in a small neighborhood of the device's operating point. It can be interpreted geometrically as a translation of the origin to the operating point of the device such that the characteristics about a sufficiently small neighborhood of the "translated" origin are approximately linear.

As an illustration of the differences between various types of models, consider the typical input–output characteristics associated with an *n-p-n* transistor in the common-emitter configuration, as shown in Fig. 2-2(a), (b), and (c). The output characteristic curves predicted by a typical global model[1] are shown in Fig. 2-2(d). Notice that, although these curves are not an exact replica of those shown in Fig. 2-2(c), they represent a fair approximation over all regions of the curves shown in these two figures. On the other hand, the output characteristic curves predicted by a typical local model might be as shown in Fig. 2-2(e). Notice that the approximation in this case is good only over the crosshatched region. Clearly, if the device is known to operate only over this region, the computer solutions obtained using this local model will be as good as those obtained using the global model corresponding to the characteristics shown in Fig. 2-2(d). Since the network corresponding to a local model is in general much simpler than the network corresponding to a global model of comparable accuracy, it makes good sense to use a local model whenever it is applicable.

Finally, if only relatively small signals are involved—such as in class A amplifiers—the operating region will center around a small neighborhood of the operat-

[1]See Chapter 11 of reference 2.

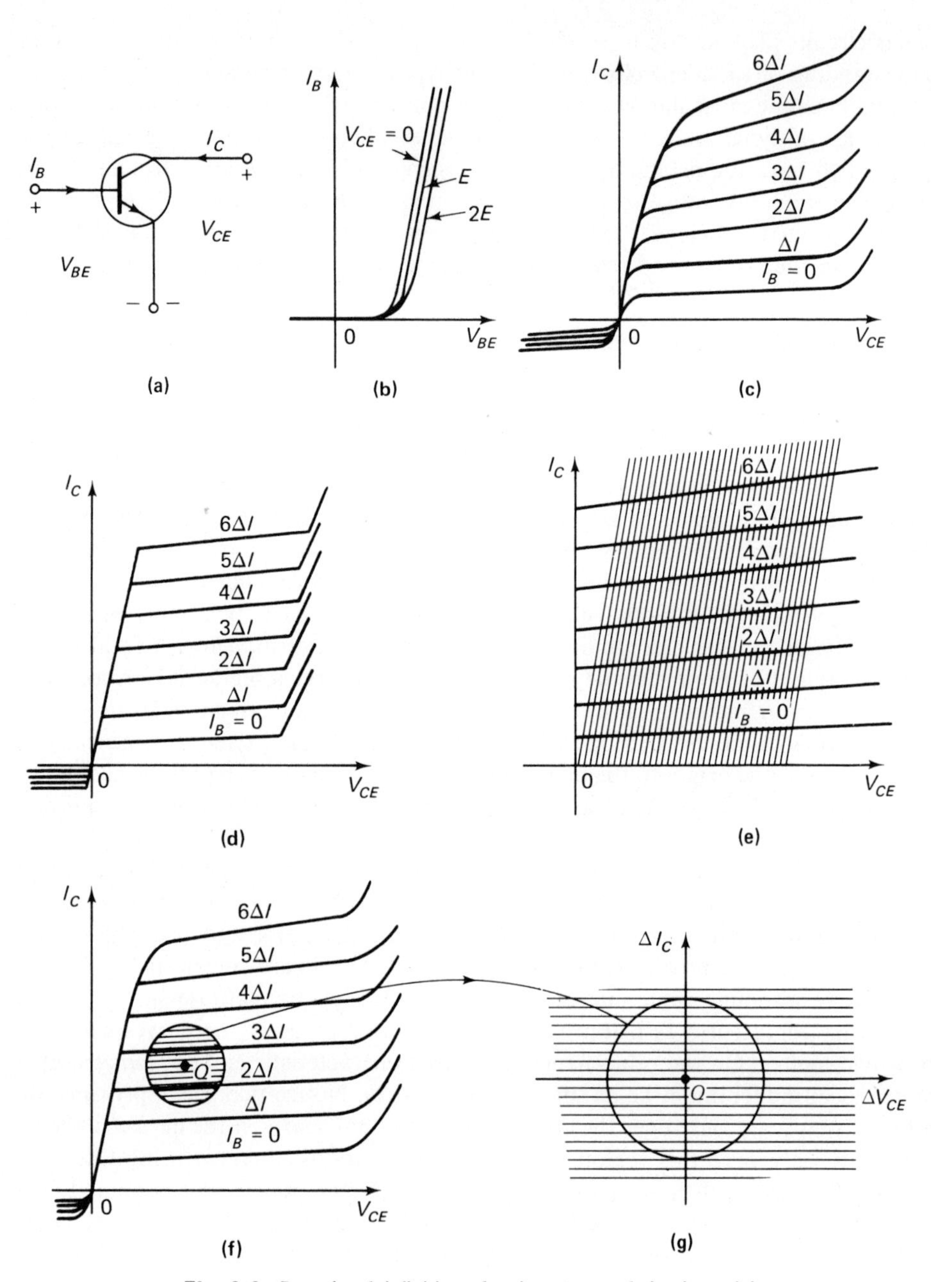

Fig. 2-2. Domain of definition of various types of circuit models.

ing point Q, such as shown in Fig. 2-2(f), where this neighborhood is *magnified* for clarity. Observe that all characteristics within this region are parallel straight lines. The output characteristics predicted by a typical *linear incremental model* are shown in Fig. 2-2(g), where the origin has been translated to coincide with the operating point Q. Observe that, whereas the slopes of the characteristic curves (over a small region) predicted by the global model need only be an *average* value, the slope of the linear characteristics must be quite exact to be accurate. In other words, a realistic linear incremental model must be capable of predicting the "fine structures." In this sense, a global or a local model is analogous to looking at a specimen through a magnifying glass; whereas an incremental model is analogous to looking through a high-power microscope focused over a small neighborhood of the specimen.

2-2-2 Classification of Models in Terms of Signal Bandwidth

Depending on the frequency range over which a model is designed to simulate accurately, it is convenient to classify a model into various types. Roughly speaking, a model is either a *dc model* or an *ac model.* A dc model differs from an ac model in that it is strictly *resistive*; i.e., it contains neither capacitors nor inductors.

A dc model is capable of simulating the device's characteristics from dc to some very low frequency, typically under 10 KHz. As the frequency increases, the reactive elements—capacitors and inductors—begin to play an important role. In fact, for frequencies above 1 GHz, even the small parasitic lead inductance and stray capacitance must be incorporated into the model in order to achieve the desired accuracy.

As an example of the importance of the operating frequency on the validity of a model, consider the following models associated with a *tantalum nitride thin-film resistor* [8]. At low frequencies, say below 10 KHz, a realistic model consists of the single-resistor dc model shown in Fig. 2-3(a). Above 10 KHz, the lead inductance L_P begins to affect the performance of the resistor; and the ac model shown in Fig. 2-3(b) is needed to achieve the desired accuracy. Above 1 GHz, the stray capacitance C_P must be added to maintain accuracy. In fact, above 100 GHz, the leads begin to behave like a transmission line; and several sections of the lumped approximation to a transmission line must be added, as shown in Fig. 2-3(d).

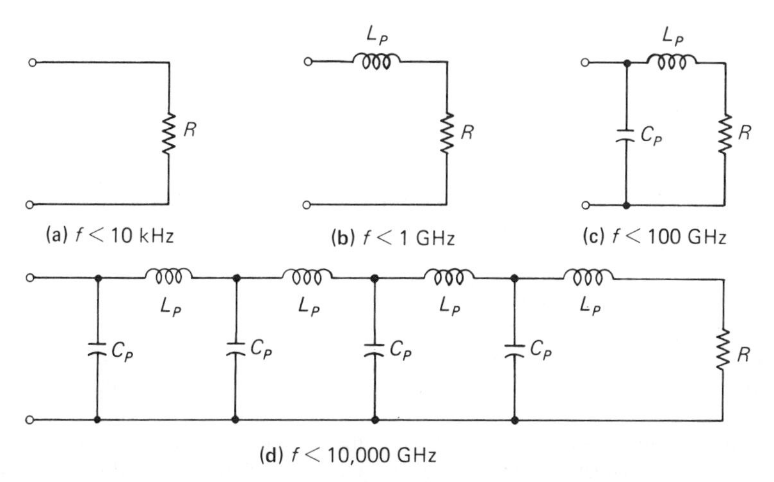

Fig. 2-3. Circuit model of a typical thin-film resistor at different operating frequency ranges.

In view of the preceding observations, it is sometimes convenient to further divide ac models into *low-frequency*, *medium-frequency*, and *high-frequency models.* Although there is yet no standard agreement on the frequency range covered by each model, it is usually assumed that a low-frequency model should be adequate for all frequencies in the audio range, i.e., from 1 to 20 KHz. The medium-frequency model would normally be used for frequencies ranging from 20 KHz to about 50 MHz. Beyond this, the high-frequency model must be used.

2-2-3 Hierarchy of Models

To help visualize the relationships among various models, consider Fig. 2-4. The boxes denote the various types of models and the arrowheads point toward the direction of simplified models. For example, starting from the most complete and accurate ac global model, we can strip away some nonlinear elements to obtain a simplified ac local model; or we can strip away all capacitors and inductors to obtain a dc global model.[2] Similarly, stripping away some nonlinear elements from the dc global model, or stripping away all capacitors and inductors from the ac local model, would give us a dc local model. Likewise, the removal of all dc sources along with the replacement of all nonlinear elements by equivalent linear elements from the ac local model would result in an ac linear incremental model. Applying a similar procedure to a dc local model would result in a dc linear incremental model. Finally, the removal of all capacitors and inductors from the ac linear incremental model would produce a dc linear incremental model.

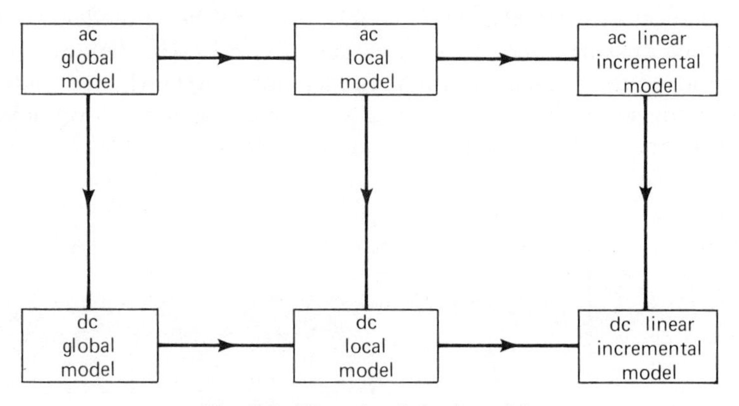

Fig. 2-4. Hierarchy of circuit models.

2-3 FOUNDATION OF MODEL MAKING

No matter what approach one uses to construct a model for a physical device, its validity depends ultimately on how well the model does indeed predict the actual

[2]By stripping or removing capacitors and inductors, we mean open-circuiting all capacitors and short-circuiting all inductors. Our terminology in this section differs somewhat from that usually found in the literature. In particular, our "dc global model" is often called a "dc model," our "ac global model" is often called a "transient model," our "ac linear incremental model" is often called an "ac model," and our "ac local model" is often called a "distortion model."

solutions. Since it is impractical to test a model's validity by subjecting it to all possible excitation waveforms and comparing its computed response with the measured response, it is highly desirable to find some general theorem that guarantees the model's validity on the basis of only a "finite" number of measurements. The following theorem is a case in point.

> ***Theorem 2.1.*** *(Representation Theorem for Linear Systems). Let N_a and N_b be two $(n + 1)$-terminal* ***linear*** *black boxes; i.e., both N_a and N_b obey the principle of superposition. Suppose that both N_a and N_b are driven by n arbitrarily prescribed* ***sinusoidal*** *sources, each having a nonzero amplitude, and suppose that the sources connected to corresponding terminals of N_a and N_b are identical. If the complete responses (both transient and steady state) of the corresponding terminals of N_a and N_b are identical under the described excitation scheme, then N_a and N_b are equivalent to each other in the sense that they will always possess identical responses under any other corresponding excitations.*

The proof of this representation theorem can be found in a number of advanced textbooks in linear system theory.[3] The theorem is a direct consequence of the *linearity assumption* of the two black boxes, and its importance cannot be overemphasized. Because of this theorem, one could establish the validity of both the ac and dc *linear incremental models* in Fig. 2-4 by comparing the model predicted response with the experimentally measured response corresponding to the same set of sinusoidal excitations.

To establish the validity of the remaining models in Fig. 2-4, an analogous representation theorem for *nonlinear* systems must be used. Unfortunately, no such theorem is currently available.[4] Consequently, the only nonlinear models whose validity could be established are the dc global models and the dc local models. Since these models do not contain capacitors and inductors, one only needs to compare the characteristic curves generated by the model against the measured curves. In the case of nonlinear ac global or ac local models, the characteristic curves would depend on the frequency of measurement and are therefore not unique. In view of this observation, it is not yet possible to establish the validity of nonlinear ac global and ac local models. Indeed, until a practical representation theorem for nonlinear systems is discovered, there is little hope that the validity of these models can ever be established rigorously.

In the absence of a firm foundation for constructing nonlinear ac models, two approaches are currently available for synthesizing these models which agree qualitatively with the behavior of real devices: the *physical approach* and the *black-box approach.*

In the physical approach, one tries to translate the physical structure and operating mechanisms of a given device into a circuit model. In this case, the elements of the model usually bear a one-to-one relationship with the device's internal structure.

[3] See Chapter 6 of reference 9.

[4] The only known representation theorem for nonlinear systems is due to Wiener. However, an infinite set of coefficients is required for a complete characterization; it is therefore impractical for modeling purposes.

This approach is quite sound and should yield realistic results provided the physical operating mechanisms of the device are well understood. Unfortunately, the physics of many devices, especially those operating at high power or at high microwave frequencies, is not yet fully understood. Moreover, even when the physical operating principles are sufficiently well known, it is usually necessary to make many simplifying assumptions and idealizations in order to identify the internal structure of the device. Consequently, the physical approach is not always applicable.

In the black-box approach, one first arrives at a valid dc global model, and then augments the model with parasitic capacitors and inductors at one or more *strategic* locations. These locations are usually selected to guarantee that the state equations for the network may be written. For example, it is usually a sound practice to add a parasitic inductor and a parasitic capacitor to a voltage-controlled nonlinear resistor as shown in Fig. 2-5(a), and to a current-controlled nonlinear resistor as shown in Fig. 2-5(b). Observe that the parasitic capacitor is placed to the right of the parasitic inductor in Fig. 2-5(a) and to its left in Fig. 2-5(b). To show that the location of these parasitics is important, consider what happens when the parasitic capacitor C_P in Fig. 2-5(a) is moved to the left of L_P, as shown in Fig. 2-6, where a biasing resistor and battery are added to complete the circuit. Let the voltage-controlled nonlinear resistor be characterized by

$$i_R = v_R^3 - 3v_R \tag{2-1}$$

The equations of motion can be obtained by inspection:

$$\frac{dv_C}{dt} = \frac{1}{C_P}\left(\frac{E - v_C}{R} - i_L\right) \tag{2-2}$$

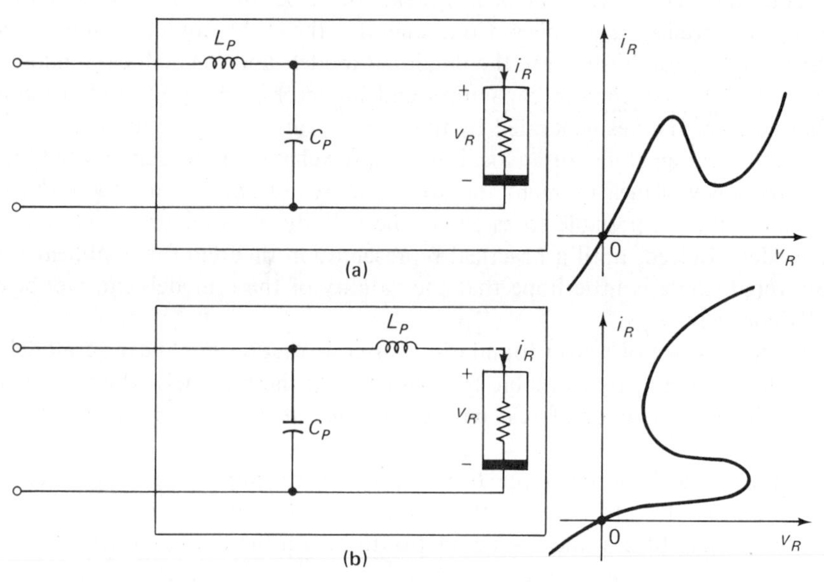

Fig. 2-5. Two sound principles for adding parasitic inductors and capacitors to a dc global model, thereby transforming it into an ac global model.

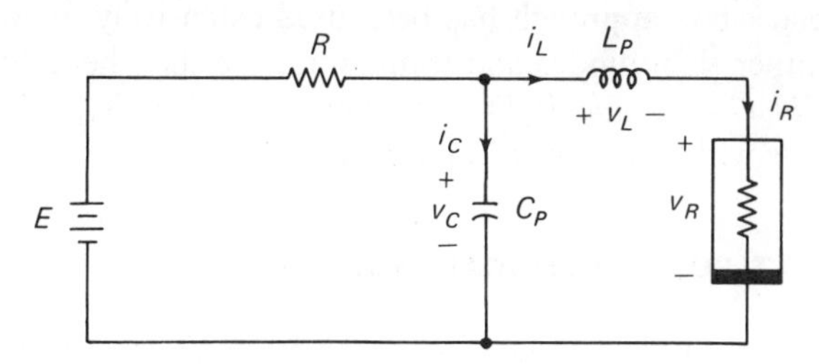

Fig. 2-6. An example illustrating the importance of introducing the correct type of parasitic element at a given location.

$$\frac{di_L}{dt} = \frac{1}{L_P}(v_C - v_R) \tag{2-3}$$

To obtain the *state equation* for this network, it is necessary to express v_R in terms of $i_L = i_R$. But this is impossible, because Eq. (2-1) shows that the v_R-i_R curve is not one-to-one. Since any reasonably modeled physical device when imbedded in a circuit must give rise to a well-defined state equation,[5] these parasitics clearly violated this requirement. This example demonstrates that, although there is no rigorous basis for choosing the appropriate locations for adding a particular type of parasitic element, guidelines do exist as to which location is likely to be associated with a parasitic capacitor or a parasitic inductor.

The advantages of the black-box approach are the following:

1. One does not have to understand the physical operating mechanisms internal to a device.
2. In the case of dc models, a systematic method is available for synthesizing a resistive model to simulate the characteristic curves of a given device.[6] Since this method is device-independent, it is useful for synthesizing models of not only existing devices, but also those yet to be discovered.
3. Since each element in a model synthesized by the black-box approach always performs a specific function, the user knows exactly the role played by each element; consequently, it is comparatively easy for him to identify the parameters and characteristics of the elements in the model from the measured characteristic curves. Moreover, it is a simple matter to obtain local models from the global model, because the user knows exactly which elements are negligible in a given operating region.

Both the physical approach and the black-box approach to device modeling have been widely used in industry. For devices such as diodes and transistors, the physical approach has been quite successful because the physics of these two devices is well understood. We shall henceforth refer to this class of models as *physical models*. Two widely used physical models will be presented in the next section. On the

[5]Otherwise, the solution will contain discontinuous jumps, which is impossible for physical circuits.

[6]See Chapter 11 of reference 2.

other hand, the black-box approach has been used extensively for modeling devices and components other than diodes and transistors. We shall henceforth refer to this class of models as *black-box models*. Due to its greater generality, a detailed treatment of black-box models will be presented in Section 2-5.

2-4 A GLIMPSE AT SOME PHYSICAL MODELS[7]

Since, by definition, a physical model is derived from physical principles governing the device, any rigorous treatment of this subject must, by necessity, delve into the physics of the materials from which the device is made. Such a treatment, although very important in its own right, is outside the mainstream of this book. Consequently, our objective in this section is simply to present some widely used physical models that have stood the test of time. We shall be content with giving the circuit model of a device and a discussion of the parameters and characteristics associated with elements in the model. The physical principles and reasonings that led to the model will not be presented. However, an extensive list of references is provided at the end of this chapter for readers interested in this information.

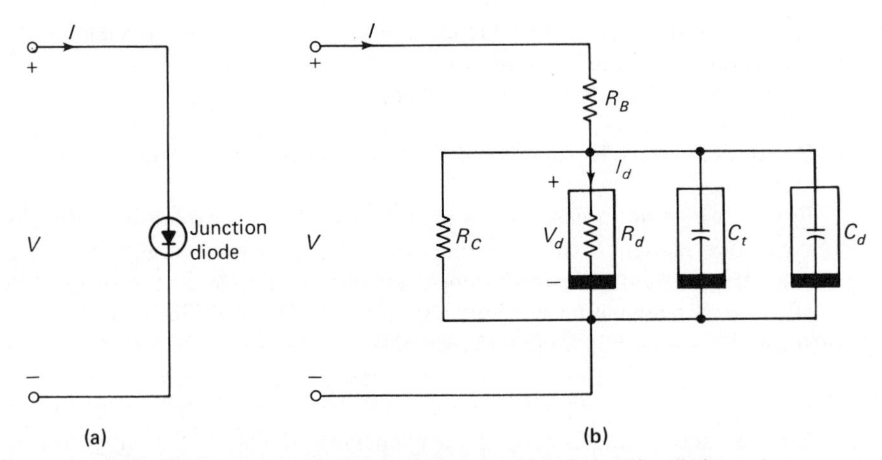

Fig. 2-7. "Physical" ac global circuit model of a junction diode.

2-4-1 Physical Model for Junction Diodes

The symbol of a *pn* junction diode is shown in Fig. 2-7(a) and a widely used ac global model is shown in Fig. 2-7(b). There are five elements in this model:

1. R_B: the semiconductor bulk and contact resistance (typically, $0 < R_B < 100\ \Omega$).
2. R_C: the junction ohmic leakage resistance (typically, $R_C > 1\ \text{M}\Omega$).

[7]See reference 7 for an extensive treatment of physical models of solid-state devices.

3. R_d: the *nonlinear* resistor representing the diode junction. Its V_d-I_d curve is given by

$$I_d = I_s\left[\exp\left(\frac{qV_d}{MkT}\right) - 1\right] \tag{2-4}$$

where

I_s = diode saturation current (typically, $10^{-12} < I_s < 10^{-6}$ mA for a silicon diode and $10^{-8} < I_s < 10^{-2}$ mA for a germanium diode)
q = electron charge = 1.6×10^{-19} C
k = Boltzmann's constant = 1.38×10^{-23} J/°K
T = junction absolute temperature in degrees Kelvin
M = emission constant (a correction factor by which the dynamic resistance and junction voltage are greater than the ideal values given by qV_d/kT (typically, $1.0 < M < 2.5$)

4. C_t: the *nonlinear* junction transition capacitance of the depletion layer, given by

$$C_t = \frac{D}{(V_z - V_d)^n}, \qquad V_d < V_z \tag{2-5}$$

where

D = proportionality constant (typically, $0.5 \times 10^{-12} < D < 5 \times 10^{-12}$)
V_z = junction contact potential (typically, $0.2 < V_z < 0.9$ V)
n = junction grading constant (typically, $0 < n < 1$)

5. C_d: the nonlinear junction diffusion capacitance whose value depends on the value of the junction current I_d in accordance with[8]

$$C_d = \frac{q}{2\pi MkTF}(I_d + I_s) \tag{2-6}$$

where I_s, q, k, T, and M are as defined previously, and where F is the intrinsic diode cutoff frequency (depending on the fabrication process, F may vary from a few megahertz to several hundred gigahertz).

Observe that *a total of ten "physical parameters" must be specified for the diode model* shown in Fig. 2-7(b). These parameters are determined in practice by performing a series of measurements on the given device. A detailed discussion of the measurement procedure is given in [10].

Observe that the model shown in Fig. 2-7(b) is an ac global model. To obtain the dc global model, we simply remove the two capacitors, in which case the three resistors can be combined into a single nonlinear resistor whose v-i curve would coincide with that measured across the diode at low frequencies.

The diode model shown in Fig. 2-7(b) will not simulate accurately the diode "storage" and "fall" times under transient step inputs. A more realistic *pn* junction diode model is given in [15].

[8] Observe that C_d is a nonlinear capacitor because it depends on I_d, which in turn depends on V_d as given by Eq. (2-4).

2-4-2 Physical Model for Transistors

The symbol of a *pnp* transistor is shown in Fig. 2-8(a) and a widely used ac global model—the Ebers–Moll model[9]—is shown in Fig. 2-8(b). The corresponding symbol and model for an *npn* transistor are shown in Fig. 2-8(c) and (d).[10] There are 13 elements in each model:

1. R_{EE}: the emitter bulk and contact resistance (typically, $0 < R_{EE} < 10\ \Omega$).
2. R_{BB}: the base spreading, bulk, and contact resistance (typically, $0 < R_{BB} < 100\ \Omega$).
3. R_{CC}: the collector bulk and contact resistance (typically, $0 < R_{CC} < 10\ \Omega$).
4. R_E: the emitter–base junction ohmic leakage resistance (typically, $R_E > 1\ \text{M}\Omega$).
5. R_C: the collector–base junction ohmic leakage resistance (typically, $R_C > 1\ \text{M}\Omega$).
6. I_e: the collector current-controlled current source:

$$I_e = \alpha_I I_{CF} \qquad (\text{typically, } \alpha_I = 0.5) \tag{2-7}$$

where

α_I = common-base inverted-mode dc current gain
I_{CF} = collector junction diode current

7. I_C: the emitter current-controlled current source:

$$I_C = \alpha_N I_{EF} \qquad (\text{typically, } \alpha_N = 0.99) \tag{2-8}$$

where

α_N = common-base normal-mode dc current gain
I_{EF} = emitter junction diode current

8. R_{de}: the emitter–base junction diode nonlinear resistor:

$$I_{EF} = \frac{I_{ES}}{1 - \alpha_N \alpha_I}\left[\exp\left(\frac{qv_1}{M_E kT}\right) - 1\right] \tag{2-9}$$

where

I_{ES} = emitter–base diode saturation current (typically, $10^{-14} < I_{ES} < 10^{-6}$ mA for silicon and $10^{-8} < I_{ES} < 10^{-2}$ mA for germanium)
q = electron charge = 1.6×10^{-19} C
k = Boltzmann's constant = 1.38×10^{-23} J/°K
T = junction absolute temperature in degrees Kelvin
M_E = emission constant for emitter–base diode (a correction factor by which the emitter dynamic resistance and the junction voltage are

[9]The circuit shown in Fig. 2-8 is called the *Ebers–Moll injection model*. An equivalent form, called the *Ebers–Moll transport model*, has recently been described by Logan [6]; it has the advantage that the parameters are more easily measured. A generalization of the Ebers–Moll model to include various high-injection effects has been given by Gummell and Poon [4]. Another variant of the Ebers–Moll model is given in [16] along with methods for measuring the model parameters.

[10]Notice that these two models differ only in the polarity of the two nonlinear resistors R_{de} and R_{dc} and the polarity of the two nonlinear capacitors C_{te} and C_{tc}. Indeed, they are *complementary models* (see Chapter 7 of reference 2).

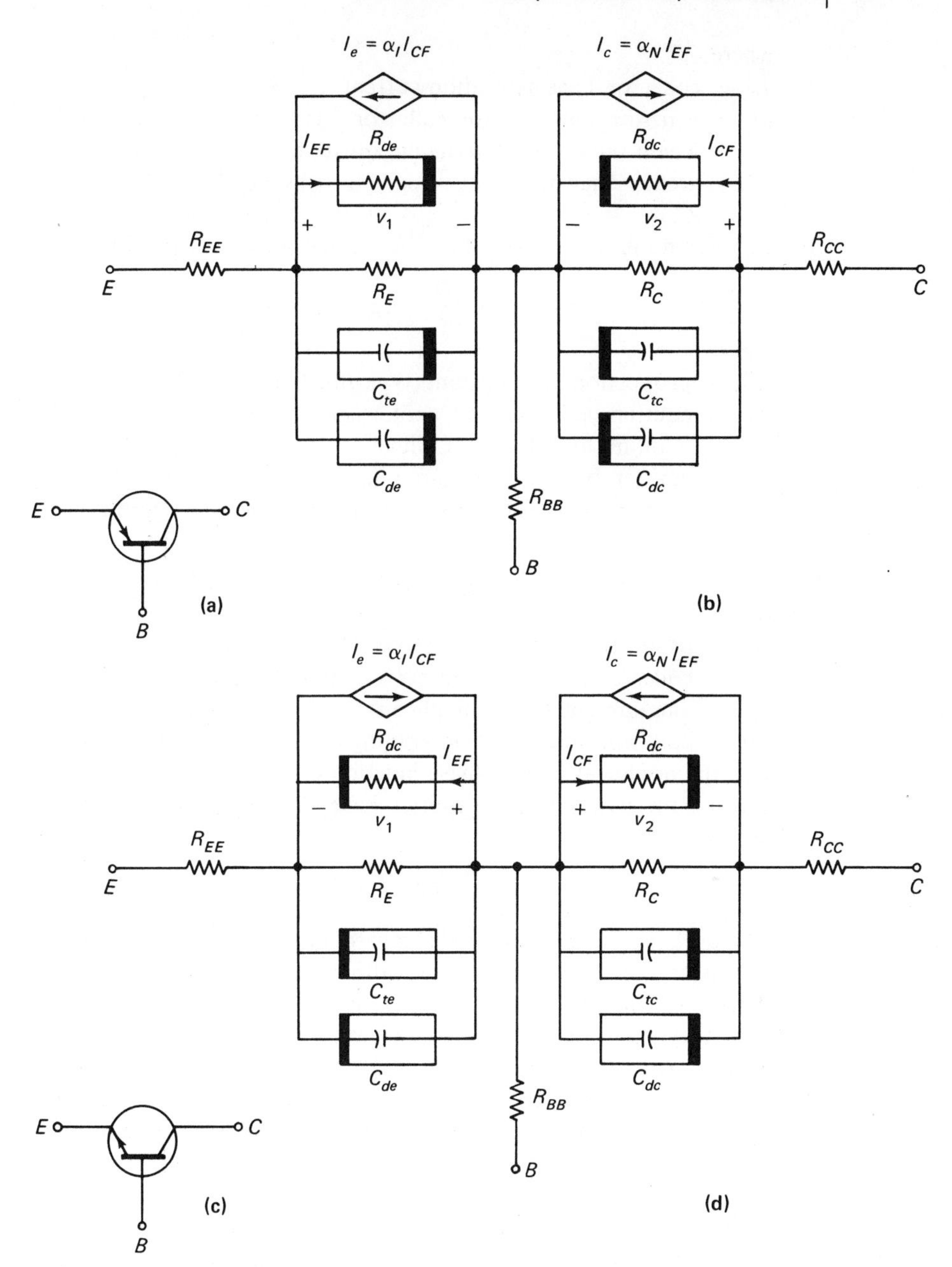

Fig. 2-8. Ebers–Moll model for a *pnp* and an *npn* transistor.

greater than the ideal values given by qv_1/kT; typically, $1 < M_E < 2.5$)

9. R_{dc}: the collector–base junction diode nonlinear resistor:

$$I_{CF} = \frac{I_{CS}}{1 - \alpha_N \alpha_I}\left[\exp\left(\frac{qv_2}{M_C kT}\right) - 1\right] \tag{2-10}$$

where

I_{CS} = collector–base saturation current $[I_{CS} \approx I_{ES}(\alpha_N/\alpha_I)]$

M_C = emission constant for collector–base diode (a correction factor by which the collector dynamic resistance and the junction voltage are greater than the ideal values given by qv_2/kT; typically, $1 < M_C < 2.5$)

10. C_{te}: the nonlinear emitter–base junction transition capacitance, given by

$$C_{te} = \frac{D_1}{(V_{ZE} - v_1)^{n_E}}, \qquad v_1 < V_{ZE} \tag{2-11}$$

where

D_1 = proportionality constant (typically, $0.5 \times 10^{-12} < D_1 < 5 \times 10^{-12}$)

V_{ZE} = emitter–base junction contact potential (typically, 0.4 V for germanium and 0.9 V for silicon)

n_E = emitter–base junction grading constant (typically, $0.3 < n_E < 0.5$)

11. C_{tc}: the nonlinear collector–base transition capacitance, given by

$$C_{tc} = \frac{D_2}{(V_{ZC} - v_2)^{n_C}}, \qquad v_2 < V_{ZC} \tag{2-12}$$

where

D_2 = proportionality constant (typically, $0.5 \times 10^{-12} < D_2 < 5 \times 10^{-12}$)

V_{ZC} = collector–base junction contact potential (typically, 0.4 V for germanium and 0.9 V for silicon)

n_C = collector–base junction grading constant (typically, $0.1 < n_C < 0.5$)

12. C_{de}: the nonlinear emitter–base junction diffusion capacitance whose value depends on the value of the emitter–base junction current I_{EF} in accordance with

$$C_{de} = \frac{q}{2\pi M_E kTF_n}\left[I_{EF} + \frac{I_{ES}}{(1 - \alpha_N\alpha_I)}\right] \tag{2-13}$$

where F_n is the normal-mode gain–bandwidth product of the intrinsic transistor. This is the gain–bandwidth product that would be measured if it were possible to eliminate the transition capacitances and the extrinsic and stray resistances and capacitances (depending on the fabrication process, F_n may vary from a few megahertz to several hundred gigahertz).

13. C_{dc}: the nonlinear collector–base junction diffusion capacitance whose value depends on the value of the collector–base junction current I_{CF} in accordance with

$$C_{dc} = \frac{q}{2\pi M_C kTF_i}\left[I_{CF} + \frac{I_{CS}}{(1 - \alpha_N\alpha_I)}\right] \tag{2-14}$$

where F_i is the inverted-mode gain–bandwidth product of the intrinsic transistor (depending on the fabrication process, F_i may vary from a few megahertz to several hundred gigahertz).

Observe that *a total of 22 physical parameters must be specified for the Ebers–Moll model* shown in Fig. 2-8(b) and (d). Clearly, these parameters are determined in practice by performing a series of measurements on the given device.

As before, the dc global Ebers–Moll model for transistors can be readily obtained by removing the four capacitors from the models in Fig. 2-8(b) and (d).

2-4-3 High-Frequency Linear Incremental Transistor Physical Model

Although in principle a *linear incremental model* could be derived from a *global model* by replacing the nonlinear elements by equivalent linear elements and by stripping away unnecessary components, it may not possess the accuracy demanded of a linear incremental model. A global model tends to smooth things out; hence, the fine details are seldom observed from such models. This inaccuracy is obviously inherited by any linear incremental model derived from it. Thus, one usually constructs linear incremental models from scratch. Fortunately, because of the *linearity* property, one can find an infinite variety of equivalent models from an existing model. Indeed, it follows from *Theorem 2.1* that, incrementally speaking, any three-terminal device can be modeled by a *linear two-port*. Such a network is completely characterized at dc by 2 × 2 constant matrix

$$\begin{bmatrix} \Delta w_1 \\ \Delta w_2 \end{bmatrix} = \begin{bmatrix} \rho_{11} & \rho_{12} \\ \rho_{21} & \rho_{22} \end{bmatrix} \begin{bmatrix} \Delta x_1 \\ \Delta x_2 \end{bmatrix} \tag{2-15}$$

where Δx_i represents either an incremental voltage source or current source, and Δw_i represents the corresponding response. The incremental sources are of course applied with respect to an *operating point*; i.e., the transistor must be biased first at the desired operating point before the measurements are taken. Clearly,

$$\rho_{jk} = \left. \frac{\Delta w_j}{\Delta x_k} \right|_{\Delta x_j = 0} \tag{2-16}$$

As the measurement frequency increases, the parameters ρ_{jk} also change. Since the network is linear, a sinusoidal excitation $\Delta x_i(t)$ will give rise to a corresponding sinusoidal response $\Delta w_i(t)$. In this case, it is more convenient to work in the frequency domain and consider

$$\hat{\rho}_{jk}(\omega) = \left. \frac{\Delta W_j(\omega)}{\Delta X_k(\omega)} \right|_{\Delta X_j(\omega) = 0} \tag{2-17}$$

where the capital letter denotes the associated *complex amplitude*, or *phasor*. Hence, in general, one expects that the two-port parameters $\hat{\rho}_{jk}(\omega)$ will be a function of *not only the operating point but also the frequency*.

Depending on whether ΔX_i is a current or voltage source, one arrives at various distinct *but equivalent* incremental parameters. Three of the most common parameter matrices are the following:

1. *Open-circuit impedance matrix* $\boldsymbol{Z}(\omega)$

$$\begin{bmatrix} \Delta V_1(\omega) \\ \Delta V_2(\omega) \end{bmatrix} = \begin{bmatrix} z_{11}(\omega) & z_{12}(\omega) \\ z_{21}(\omega) & z_{22}(\omega) \end{bmatrix} \begin{bmatrix} \Delta I_1(\omega) \\ \Delta I_2(\omega) \end{bmatrix} \tag{2-18}$$

The open-circuit impedance parameters are given by

$$z_{11}(\omega) = \left.\frac{\Delta V_1(\omega)}{\Delta I_1(\omega)}\right|_{\Delta I_2(\omega)=0} = \text{input impedance (with port 2 open-circuited)}$$

$$z_{21}(\omega) = \left.\frac{\Delta V_2(\omega)}{\Delta I_1(\omega)}\right|_{\Delta I_2(\omega)=0} = \text{forward transfer impedance (with port 2 open-circuited)}$$

$$z_{12}(\omega) = \left.\frac{\Delta V_1(\omega)}{\Delta I_2(\omega)}\right|_{\Delta I_1(\omega)=0} = \text{reverse transfer impedance (with port 1 open-circuited)}$$

$$z_{22}(\omega) = \left.\frac{\Delta V_2(\omega)}{\Delta I_2(\omega)}\right|_{\Delta I_1(\omega)=0} = \text{output impedance (with port 1 open-circuited)}$$

2. *Short-circuit admittance matrix* $\boldsymbol{Y}(\omega)$

$$\begin{bmatrix}\Delta I_1(\omega)\\ \Delta I_2(\omega)\end{bmatrix} = \begin{bmatrix} y_{11}(\omega) & y_{12}(\omega)\\ y_{21}(\omega) & y_{22}(\omega)\end{bmatrix}\begin{bmatrix}\Delta V_1(\omega)\\ \Delta V_2(\omega)\end{bmatrix}$$

The short-circuit admittance parameters are given by

$$y_{11}(\omega) = \left.\frac{\Delta I_1(\omega)}{\Delta V_1(\omega)}\right|_{\Delta V_2(\omega)=0} = \text{input admittance (with port 2 short-circuited)}$$

$$y_{21}(\omega) = \left.\frac{\Delta I_2(\omega)}{\Delta V_1(\omega)}\right|_{\Delta V_2(\omega)=0} = \text{forward transfer admittance (with port 2 short-circuited)}$$

$$y_{12}(\omega) = \left.\frac{\Delta I_1(\omega)}{\Delta V_2(\omega)}\right|_{\Delta V_1(\omega)=0} = \text{reverse transfer admittance (with port 1 short-circuited)}$$

$$y_{22}(\omega) = \left.\frac{\Delta I_2(\omega)}{\Delta V_2(\omega)}\right|_{\Delta V_1(\omega)=0} = \text{output admittance (with port 1 short-circuited)}$$

3. *Hybrid matrix* $\boldsymbol{H}(\omega)$

$$\begin{bmatrix}\Delta V_1(\omega)\\ \Delta I_2(\omega)\end{bmatrix} = \begin{bmatrix} h_{11}(\omega) & h_{12}(\omega)\\ h_{21}(\omega) & h_{22}(\omega)\end{bmatrix}\begin{bmatrix}\Delta I_1(\omega)\\ \Delta V_2(\omega)\end{bmatrix}$$

The hybrid parameters are given by

$$h_{11}(\omega) = \left.\frac{\Delta V_1(\omega)}{\Delta I_1(\omega)}\right|_{\Delta V_2(\omega)=0} = \text{input impedance (with port 2 short-circuited)}$$

$$h_{21}(\omega) = \left.\frac{\Delta I_2(\omega)}{\Delta I_1(\omega)}\right|_{\Delta V_2(\omega)=0} = \text{forward transfer current ratio (with port 2 short-circuited)}$$

$$h_{12}(\omega) = \left.\frac{\Delta V_1(\omega)}{\Delta V_2(\omega)}\right|_{\Delta I_1(\omega)=0} = \text{reverse transfer voltage ratio (with port 1 open-circuited)}$$

$$h_{22}(\omega) = \left.\frac{\Delta I_2(\omega)}{\Delta V_1(\omega)}\right|_{\Delta I_1(\omega)=0} = \text{output admittance (with port 1 open-circuited)}$$

It is sometimes convenient to be able to convert one set of parameters in terms of another set. Standard relationships for this purpose are tabulated in Table 2-2.

Once the incremental parameters are known, it is possible to synthesize an infinite variety of two-ports with the prescribed parameters. Occasionally, a par-

TABLE 2-2. Linear Incremental Two-Port, Black-Box Model of a Three-Terminal Device

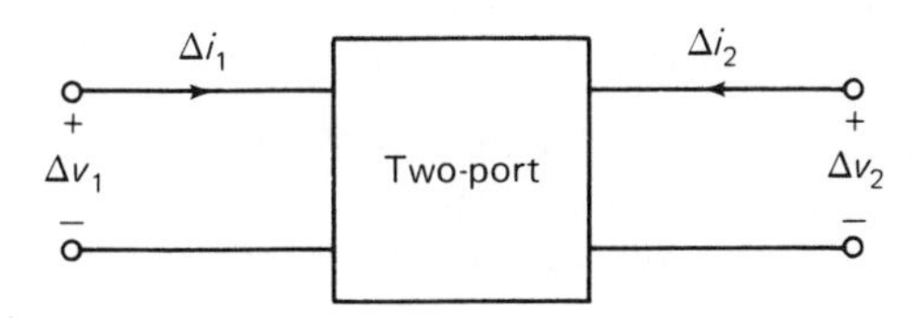

TO	FROM $Z(\omega)$	FROM $Y(\omega)$	FROM $H(\omega)$
$Z(\omega)$	$\begin{bmatrix} z_{11}(\omega) & z_{12}(\omega) \\ z_{21}(\omega) & z_{22}(\omega) \end{bmatrix}$	$\begin{bmatrix} \frac{y_{22}(\omega)}{D_y(\omega)} & \frac{-y_{12}(\omega)}{D_y(\omega)} \\ \frac{-y_{21}(\omega)}{D_y(\omega)} & \frac{y_{11}(\omega)}{D_y(\omega)} \end{bmatrix}$	$\begin{bmatrix} \frac{D_h(\omega)}{h_{22}(\omega)} & \frac{h_{12}(\omega)}{h_{22}(\omega)} \\ \frac{-h_{21}(\omega)}{h_{22}(\omega)} & \frac{1}{h_{22}(\omega)} \end{bmatrix}$
$Y(\omega)$	$\begin{bmatrix} \frac{z_{22}(\omega)}{D_z(\omega)} & \frac{-z_{12}(\omega)}{D_z(\omega)} \\ \frac{-z_{21}(\omega)}{D_z(\omega)} & \frac{z_{11}(\omega)}{D_z(\omega)} \end{bmatrix}$	$\begin{bmatrix} y_{11}(\omega) & y_{12}(\omega) \\ y_{21}(\omega) & y_{22}(\omega) \end{bmatrix}$	$\begin{bmatrix} \frac{1}{h_{11}(\omega)} & \frac{-h_{12}(\omega)}{h_{11}(\omega)} \\ \frac{h_{21}(\omega)}{h_{11}(\omega)} & \frac{D_h(\omega)}{h_{11}(\omega)} \end{bmatrix}$
$H(\omega)$	$\begin{bmatrix} \frac{D_z(\omega)}{z_{22}(\omega)} & \frac{z_{12}(\omega)}{z_{22}(\omega)} \\ \frac{-z_{21}(\omega)}{z_{22}(\omega)} & \frac{1}{z_{22}(\omega)} \end{bmatrix}$	$\begin{bmatrix} \frac{1}{y_{11}(\omega)} & \frac{-y_{12}(\omega)}{y_{11}(\omega)} \\ \frac{y_{21}(\omega)}{y_{11}(\omega)} & \frac{D_y(\omega)}{y_{11}(\omega)} \end{bmatrix}$	$\begin{bmatrix} h_{11}(\omega) & h_{12}(\omega) \\ h_{21}(\omega) & h_{22}(\omega) \end{bmatrix}$

$D_z(\omega) \triangleq z_{11}(\omega)z_{22}(\omega) - z_{12}(\omega)z_{21}(\omega)$, $\quad D_y(\omega) \triangleq y_{11}(\omega)y_{22}(\omega) - y_{12}(\omega)y_{21}(\omega)$
$D_h(\omega) \triangleq h_{11}(\omega)h_{22}(\omega) - h_{12}(\omega)h_{21}(\omega)$

ticular model stands out from among these infinite varieties of equivalent networks because of its *simplicity*, *accuracy over wide frequency range*, and *ease of parameter determination*. One such model is the *common-emitter hybrid-pi* transistor model shown in Fig. 2-9. The parameters corresponding to the seven elements *at a particular* frequency f are given by

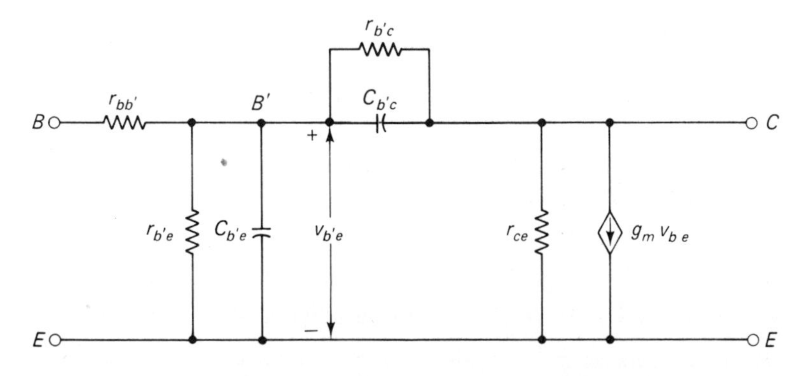

Fig. 2-9. Hybrid-pi ac incremental circuit model of a transistor.

1. $r_{bb'}$: the base spreading resistance. The value of this resistance is practically a constant for a given transistor (typically $5 < r_{bb'} < 100\ \Omega$).
2. $r_{b'e}$: the base–emitter resistance. This resistance is related to the common-emitter forward[11] current transfer ratio h_{fe}

$$r_{b'e} = r_e(1 + h_{fe}) \approx r_e h_{fe} \tag{2-19}$$

   where $r_e = kT/qI_E$ (I_E = emitter current).
3. $C_{b'e}$: the emitter-to-base capacitance. This parameter is responsible for the decrease in gain and the increase in phase shift with frequency in transistors. The time constant $C_{b'e}r_{b'e}$ can be shown to coincide with the 3-dB point on the current gain versus frequency curve. Hence, $C_{b'e}$ can be written as

$$C_{b'e} = \frac{1}{2\pi f_T r_e} \tag{2-20}$$

   where f_T is the gain–bandwidth product and r_e is as defined previously.
4. $r_{b'c}$: the feedback resistance, given by

$$r_{b'c} = \frac{r_e(1 + h_{fe})}{h_{re}} \approx \frac{r_{b'e}}{h_{re}} \tag{2-21}$$

   Typically, $r_{b'c} > 1\ \text{M}\Omega$.
5. $C_{b'c}$: the collector-to-base junction capacitance, given approximately by

$$C_{b'c} = A\left(\frac{k}{V_{CE}}\right)^{1/3} \tag{2-22}$$

   where A is the area of the collector–base junction, k is a constant related to the dielectric constant of the silicon and the impurity profile of the junction, and V_{CE} is the collector-to-emitter dc voltage. This capacitance is usually given as C_{ob} on the manufacturer's data sheet.
6. g_m: current gain of the hybrid-pi, given by

$$g_m = \frac{h_{fe}}{r_e(1 + h_{fe})} \approx \frac{1}{r_e} \tag{2-23}$$

7. r_{ce}: the common-emitter output impedance, given by

$$r_{ce} = \frac{1}{h_{oe} - g_m h_{re}} \tag{2-24}$$

   This parameter is usually very large and may be neglected in most applications.

The hybrid-pi model has been found to yield accurate results over a relatively wide frequency band, say from 1 KHz to 100 MHz.

2-5 SYNTHESIS OF DC GLOBAL BLACK-BOX MODELS OF THREE-TERMINAL DEVICES

Our objective in this section is to present a systematic procedure for synthesizing a dc global model for a large class of three-terminal devices, using the black-box approach. Since the family of input or output characteristic curves of most three-

[11]The hybrid parameters for transistors in the common-emitter configuration are usually denoted by the following symbols: $h_{11} = h_{ie}$, $h_{21} = h_{fe} = \beta$, $h_{12} = h_{re}$, and $h_{22} = h_{oe}$.

terminal devices is located either in the first or third quadrant, and since the characteristic curves in the first quadrant can be transformed into characteristic curves in the third quadrant, or vice versa, by an interchange of terminals, it suffices to consider the synthesis of a global model to simulate the first-quadrant characteristic curves. Throughout this section, we assume that the family of curves is specified in *piecewise-linear form.* Since any curve can be approximated arbitrarily closely by choosing a sufficient number of segments, there is little loss of generality in this assumption. In fact, in most large-signal applications, only two or three segments are usually adequate.

Since our models are *dc* by assumption, the capacitor and inductor in the *minimal basic set* listed in Table 2-1 are not needed. Although the remaining elements in this table are sufficient for our purpose, great simplification in the resulting model can be achieved by introducing three additional *secondary model-building blocks*:[12] *nonlinear controlled sources, controlled linear resistances or conductances,* and *controlled concave or convex resistors.*

1. *Nonlinear controlled sources.* These elements are generalizations of the four *linear* controlled sources listed in Table 2-1. Consequently, the same symbols are used with the terminal voltage or current defined as a *nonlinear* function of the controlling variable, as shown in Fig. 2-10. To show that these elements are secondary building blocks, two *equivalent circuits* using only elements from the minimal set are given in Fig. 2-10 for each nonlinear controlled source. For computer programs that do not recognize nonlinear controlled sources, these equivalent circuits can be substituted in their places.
2. *Controlled linear resistances or conductances.* We define an x-controlled linear resistance by the relationship

$$v = R(x)i \tag{2-25}$$

and an x-controlled linear conductance by the relationship

$$i = G(x)v \tag{2-26}$$

For each *fixed* value of the controlling variable x, Eqs. (2-25) and (2-26) reduce to a linear resistance and conductance, respectively. Although in principle only one of these two relationships is necessary, there are devices in which $R(x)$ can be more accurately measured than $G(x)$, and vice versa. The symbols for these two elements are shown in Fig. 2-11 along with two equivalent circuits containing nonlinear resistors and controlled sources. The nonlinear controlled source can of course be replaced by the equivalent circuits shown in Fig. 2-10. Hence, these two elements are also secondary building blocks.
3. *Controlled concave or convex resistors.* We define a controlled concave resistor by the voltage-controlled relationship

$$i = \tfrac{1}{2}G(x)[|v - E(x)| + v - E(x)] \tag{2-27}$$

[12]These building blocks are *secondary* elements in the sense that they in turn can be modeled using only elements from the minimal set in Table 2-1.

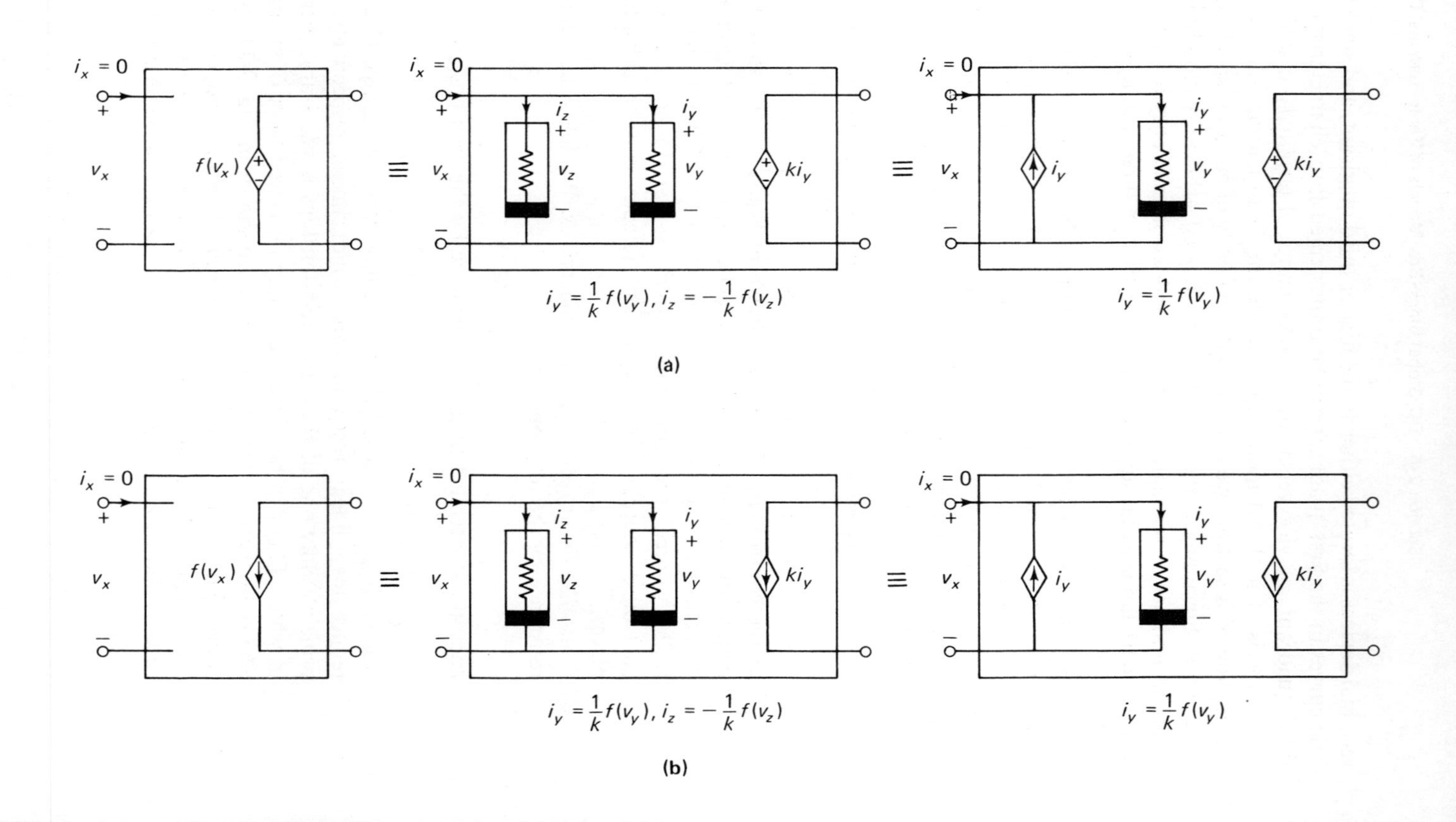

$i_x = 0$
+
v_x
−
$f(v_x)$
≡
i_z
v_z
i_y
v_y
ki_y
$i_y = \frac{1}{k} f(v_y),\ i_z = -\frac{1}{k} f(v_z)$
$i_y = \frac{1}{k} f(v_y)$

(a)

(b)

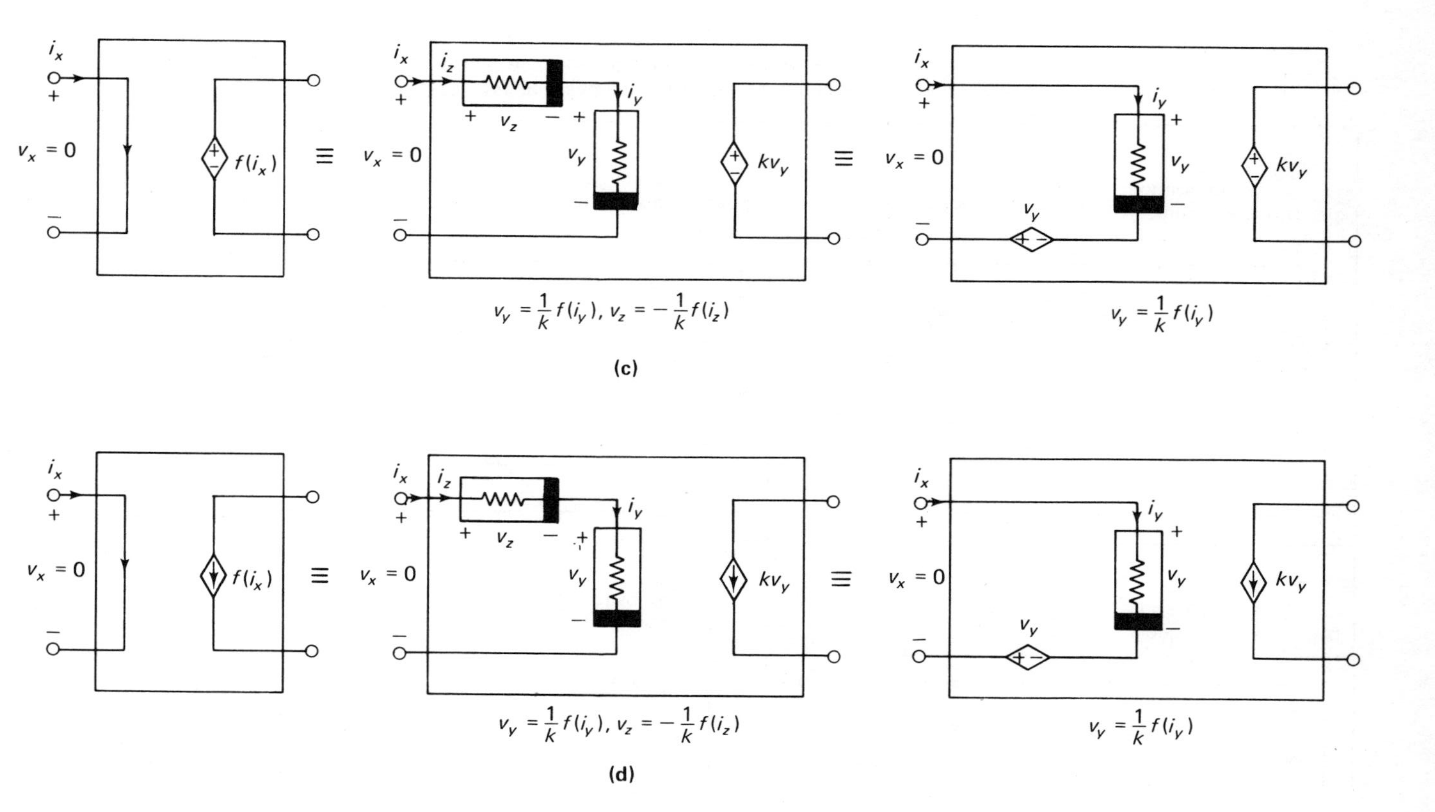

Fig. 2-10. Symbols and equivalent circuits for nonlinear controlled sources.

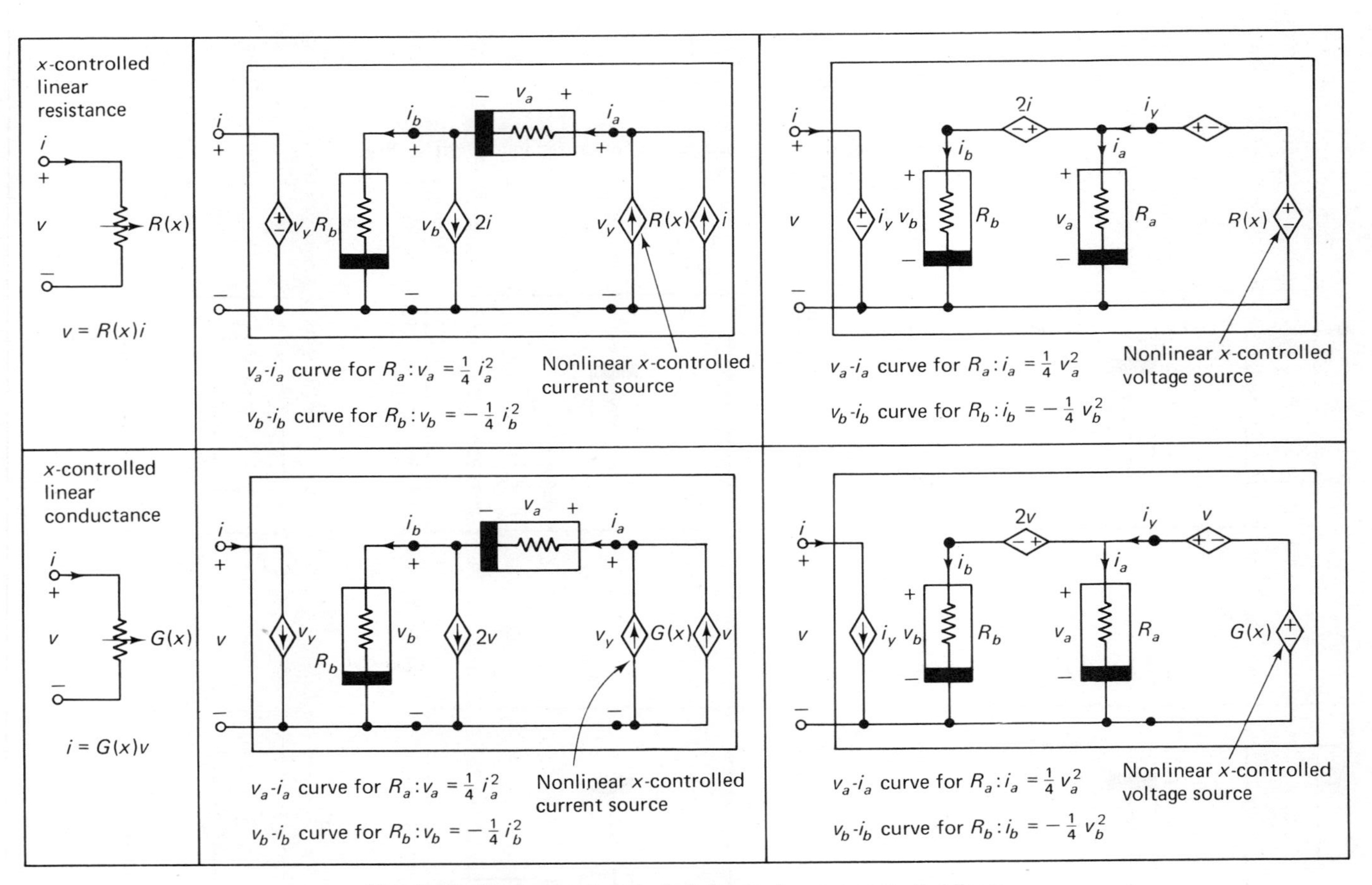

Fig. 2-11. Symbols and equivalent circuits for an x-controlled linear resistance and an x-controlled linear conductance.

where $G(x)$ and $E(x)$ are any function of the controlling variable x. The symbol for this element is shown in Fig. 2-12 along with two typical v-i curves for a fixed value of x. Also shown are an equivalent mathematical representation of Eq. (2-27) in terms of a *unit ramp* function and an equivalent circuit model [valid only if $G(x) \geq 0$] containing an *ideal* diode, an x-controlled linear conductance, and a nonlinear x-controlled voltage source.[13] Observe that a controlled concave resistor is completely specified by the two functions $G(x)$ and $E(x)$, henceforth referred to as the *conductance* and *voltage-intercept functions*, respectively.

We define a controlled convex resistor by the current-controlled relationship

$$v = \tfrac{1}{2}R(x)[|i - I(x)| + i - I(x)] \tag{2-28}$$

where $R(x)$ and $I(x)$ are any function of the controlling variable x. The symbol for this element is shown in Fig. 2-12 along with two typical v-i curves for a fixed value of x. Also shown are an equivalent mathematical representation of Eq. (2-28) in terms of a unit ramp function and an equivalent circuit model (valid only if $R(x) \geq 0$) containing an x-controlled linear resistance and a nonlinear x-controlled current source. Observe that a controlled convex resistor is completely specified by the two functions $R(x)$ and $I(x)$, henceforth referred to as the *resistance* and *current-intercept functions*, respectively.

2-5-1 Canonic Black-Box Models for Families of Two-Segment Paralleled *v-i* Curves

We are now in a position to present a number of canonic black-box models for simulating a prescribed family of two-segment, piecewise-linear v-i curves. We require the corresponding segments to be *parallel* to each other. No restriction, however, is imposed on the relative spacing between segments.

Canonic Black-Box Model I. Typical families of input characteristic curves i_1 versus v_1 (with i_2 as the parameter) and output characteristic curves i_2 versus v_2 (with i_1 as the parameter) for a three-terminal device with a *horizontal breakline*[14] are shown in Fig. 2-13(a). We assume that all segments in each family have identical slope. However, the spacing need not be uniform. The simplest model for simulating this type of characteristic curve is shown in Fig. 2-13(b). The v_{x_1}-i_{x_1} curve for R_{x_1} and the v_{x_2}-i_{x_2} curve for R_{x_2} are chosen to be any one of the v-i curves in the respective family. A convenient choice would be the curve corresponding to $i_2 = 0$ for R_{x_1} and

[13] Equivalent circuits for x-controlled concave and convex resistors that are valid for both positive and negative values of $G(x)$ and $R(x)$ are given in reference 14. However, in view of the complexity of these circuits, it is highly desirable that computer simulation programs be developed which will accept these basic controlled resistors directly.

[14] A curve passing through corresponding breakpoints on the family of v-i curves is called a *breakline*.

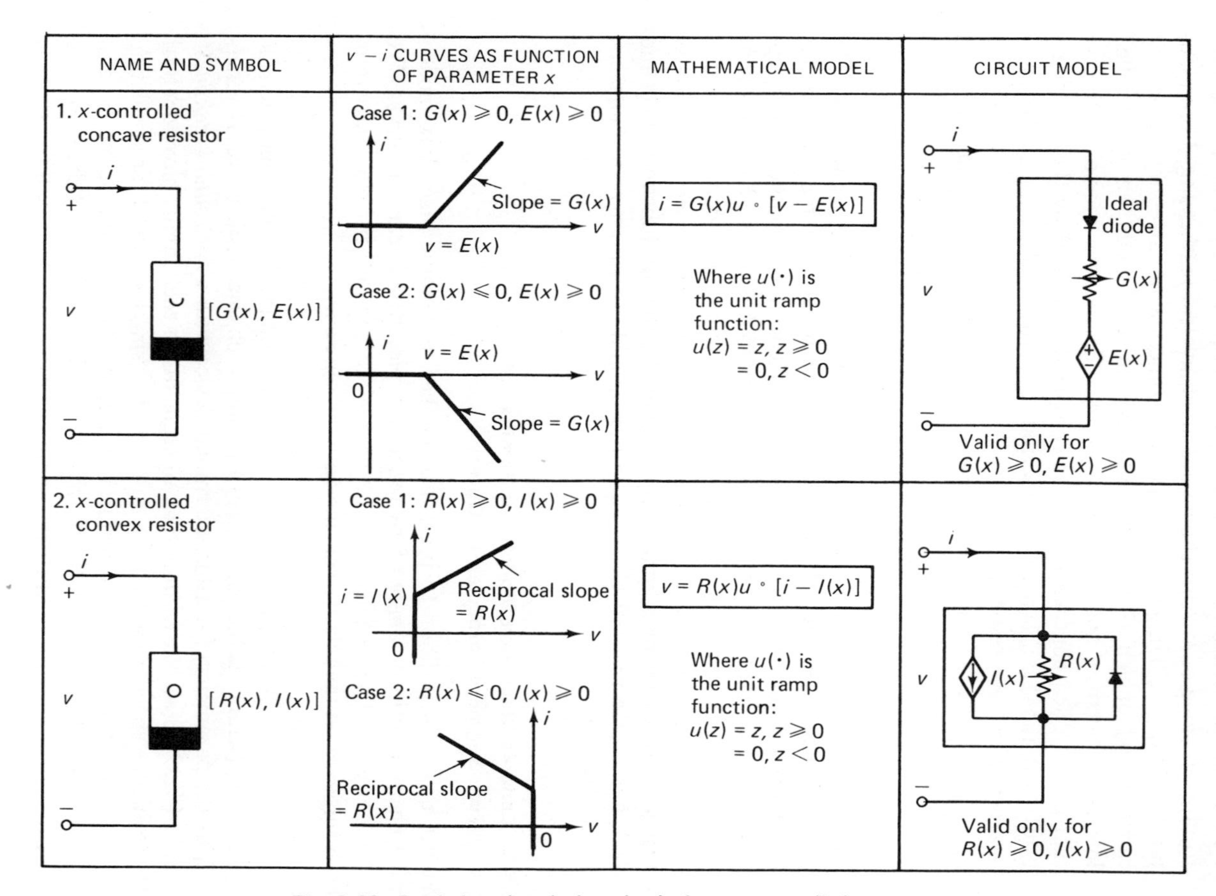

Fig. 2-12. Symbols and equivalent circuits for an x-controlled concave resistor and an x-controlled convex resistor.

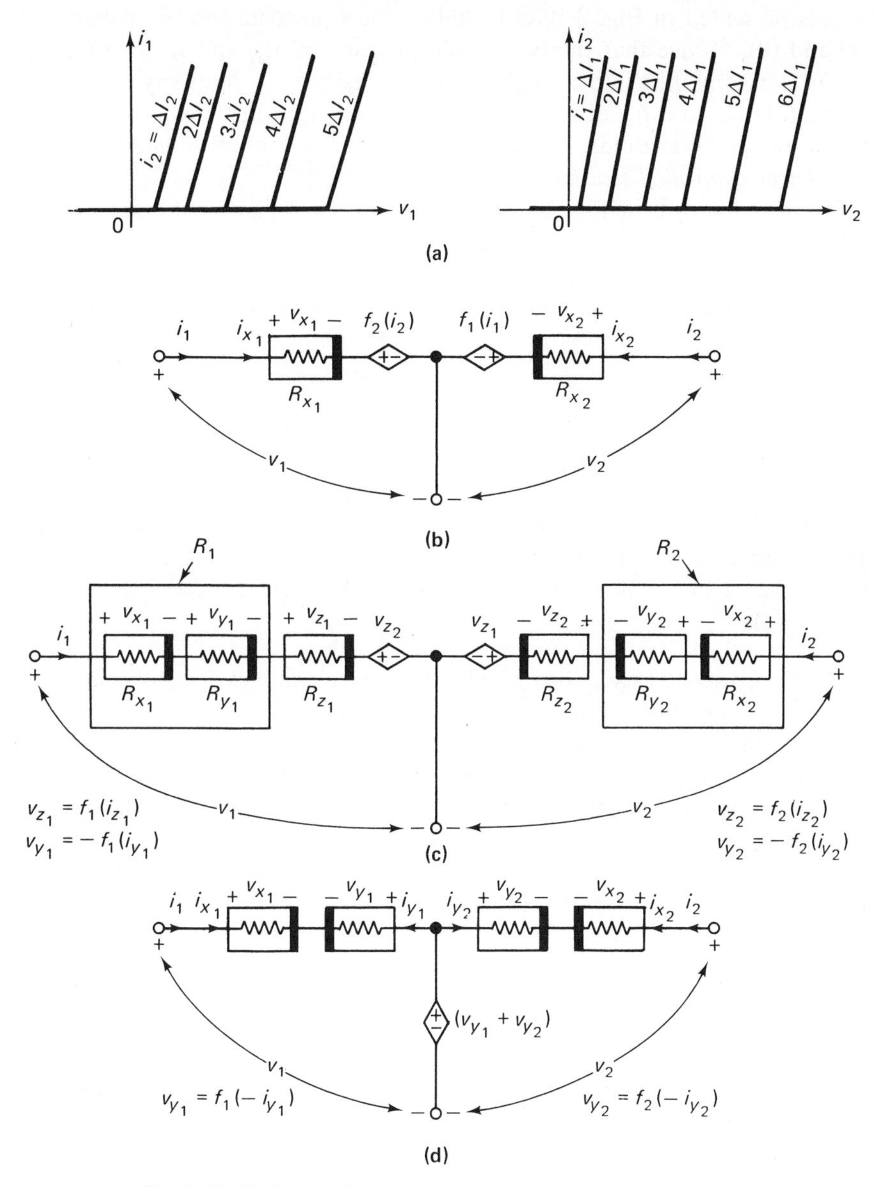

Fig. 2-13. Various equivalent forms of the canonic black-box model I.

$i_1 = 0$ for R_{x_2}. Once the v_{x_1}-i_{x_1} and v_{x_2}-i_{x_2} curves are prescribed, the nonlinear functions $f_2(i_2)$ and $f_1(i_1)$ are determined by the voltage intercepts of the remaining curves in each family. Observe that in the special case where the spacings are uniform, the nonlinear controlled sources reduce to *linear* controlled sources.

To transform the nonlinear controlled sources into an equivalent circuit containing only linear controlled sources and nonlinear resistors, we apply the two techniques presented in Fig. 2-10(c) to obtain the equivalent models shown in Fig. 2-13(c) and (d). Notice that the two nonlinear resistors R_{x_1} and R_{y_1} in Fig. 2-13(c) can be further combined into a single nonlinear resistor R_1. Similarly, the two nonlinear resistors R_{x_2} and R_{y_2} can be combined into an equivalent resistor R_2. In other words, the model shown in Fig. 2-13(c) actually consists of only *four nonlinear resistors* and *two linear controlled sources.*[15]

The model shown in Fig. 2-13(d) is obtained by replacing the nonlinear controlled sources in Fig. 2-13(b) by the second equivalent circuit shown in Fig. 2-10(c) and then applying the *v-shift theorem* [2]. Observe that this time the two nonlinear resistors in series cannot be combined into an equivalent nonlinear resistor because the voltage across one of the two resistors is a controlling variable for the linear controlled source.

CANONIC BLACK-BOX MODEL II. Typical families of input characteristic curves i_1 versus v_1 (with v_2 as the parameter) and output characteristic curves i_2 versus v_2 (with v_1 as the parameter) for a three-terminal device with a horizontal breakline are shown in Fig. 2-14(a). The simplest model consisting of two nonlinear resistors and two nonlinear controlled sources is shown in Fig. 2-14(b). By using the equivalent circuits shown in Fig. 2-10(a), we obtain the two equivalent models shown in Fig. 2-14(c) and (d).

CANONIC BLACK-BOX MODEL III. Typical families of input characteristic curves i_1 versus v_1 (with i_2 as the parameter) and output characteristic curves i_2 versus v_2 (with i_1 as the parameter) for a three-terminal device with a vertical breakline are shown in Fig. 2-15(a). The simplest model consisting of two nonlinear resistors and two nonlinear controlled sources is shown in Fig. 2-15(b). By using the equivalent circuits shown in Fig. 2-10(d), we obtain the two equivalent models shown in Fig. 2-15(c) and (d).

CANONIC BLACK-BOX MODEL IV. Typical families of input characteristic curves i_1 versus v_1 (with v_2 as the parameter) and output characteristic curves i_2 versus v_2 (with v_1 as the parameter) for a three-terminal device with a vertical breakline are shown in Fig. 2-16(a). The simplest model consisting of two nonlinear resistors and two nonlinear controlled sources is shown in Fig. 2-16(b). By using the equivalent circuits shown in Fig. 2-10(b), we obtain the two equivalent models shown in Fig. 2-16(c) and (d). As usual, the two parallel resistors on each arm of the model in Fig. 2-16(c) can be replaced by an equivalent nonlinear resistor.

CANONIC BLACK-BOX MODEL V. Typical families of input characteristic curves i_1 versus v_1 (with v_2 as the parameter) and output characteristic curves i_2 versus v_2 (with i_1 as the parameter) for a three-terminal device with a horizontal breakline for the input characteristics and a vertical breakline for the output characteristics

[15]Observe that it is not possible to further combine the two nonlinear resistors R_1 and R_{z_1} or R_2 and R_{z_2} into one equivalent resistor because the voltages across R_{z_1} and R_{z_2} are controlling variables for the two controlled sources.

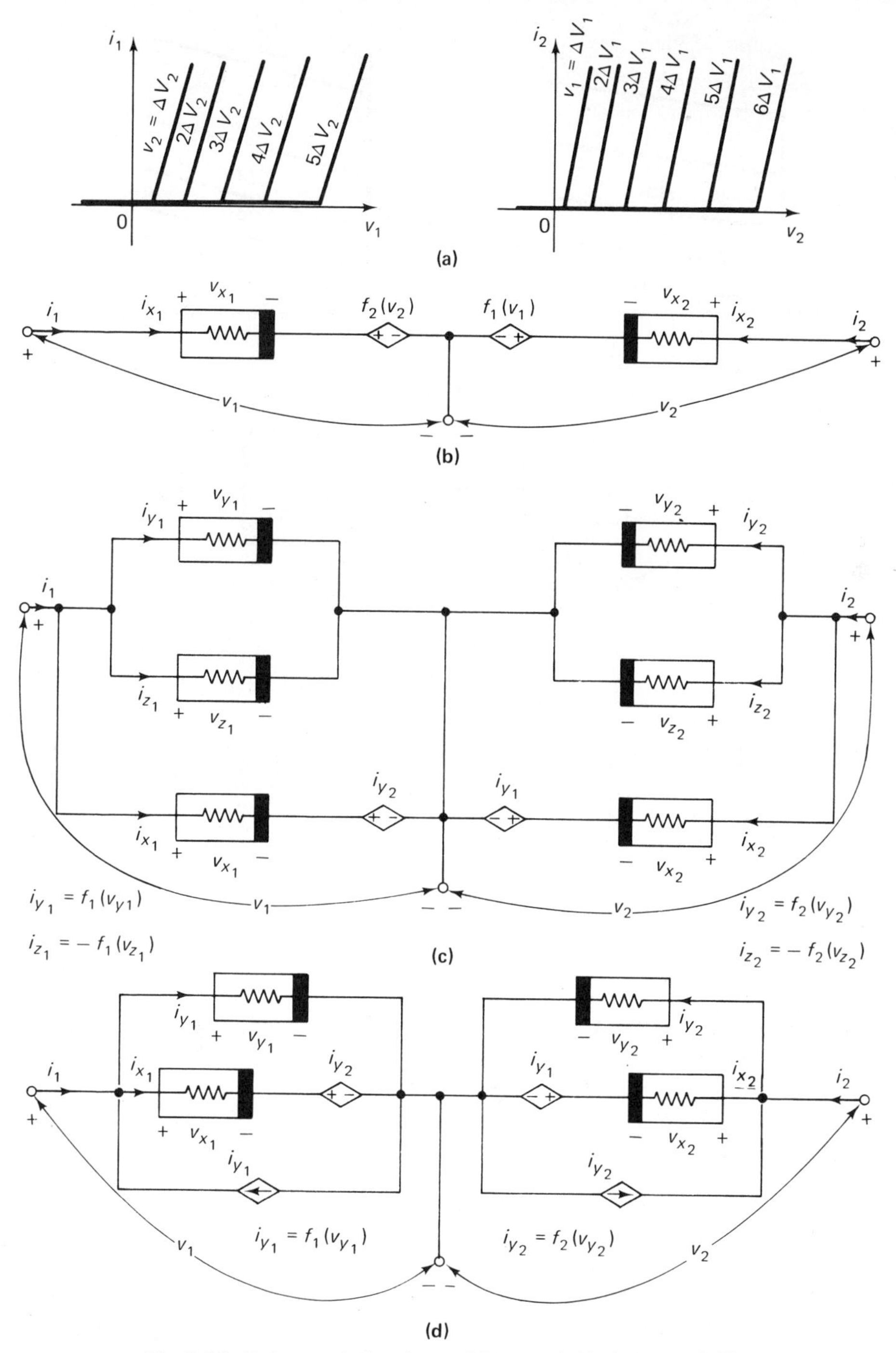

Fig. 2-14. Various equivalent forms of the canonic black-box model II.

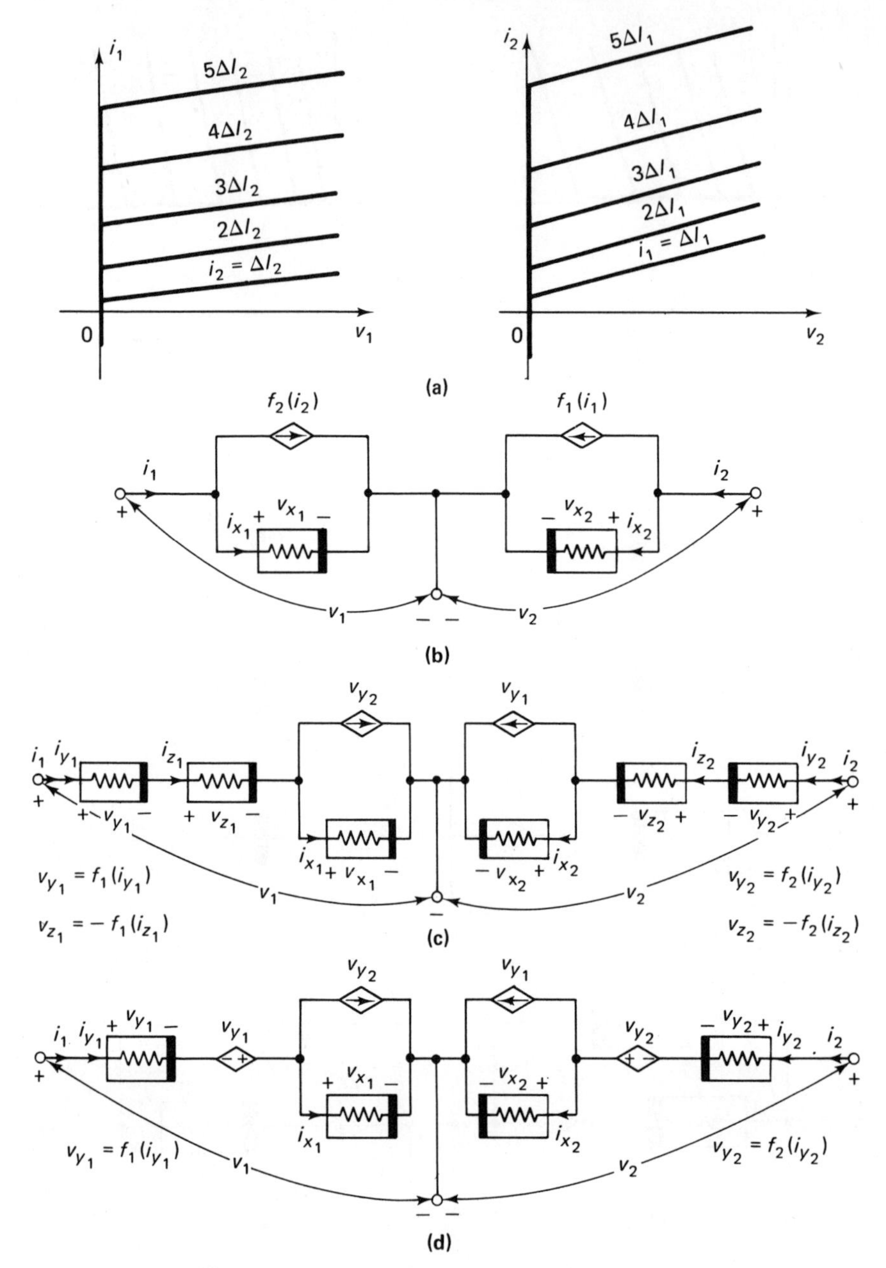

Fig. 2-15. Various equivalent forms of the canonic black-box model III.

are shown in Fig. 2-17(a). The simplest model consisting of two nonlinear resistors and two nonlinear controlled sources is shown in Fig. 2-17(b). By using the equivalent circuits shown in Fig. 2-10(a) and (d), we obtain the two equivalent models shown in Fig. 2-17(c) and (d). In the case of Fig. 2-17(c), the two series resistors on the left arm

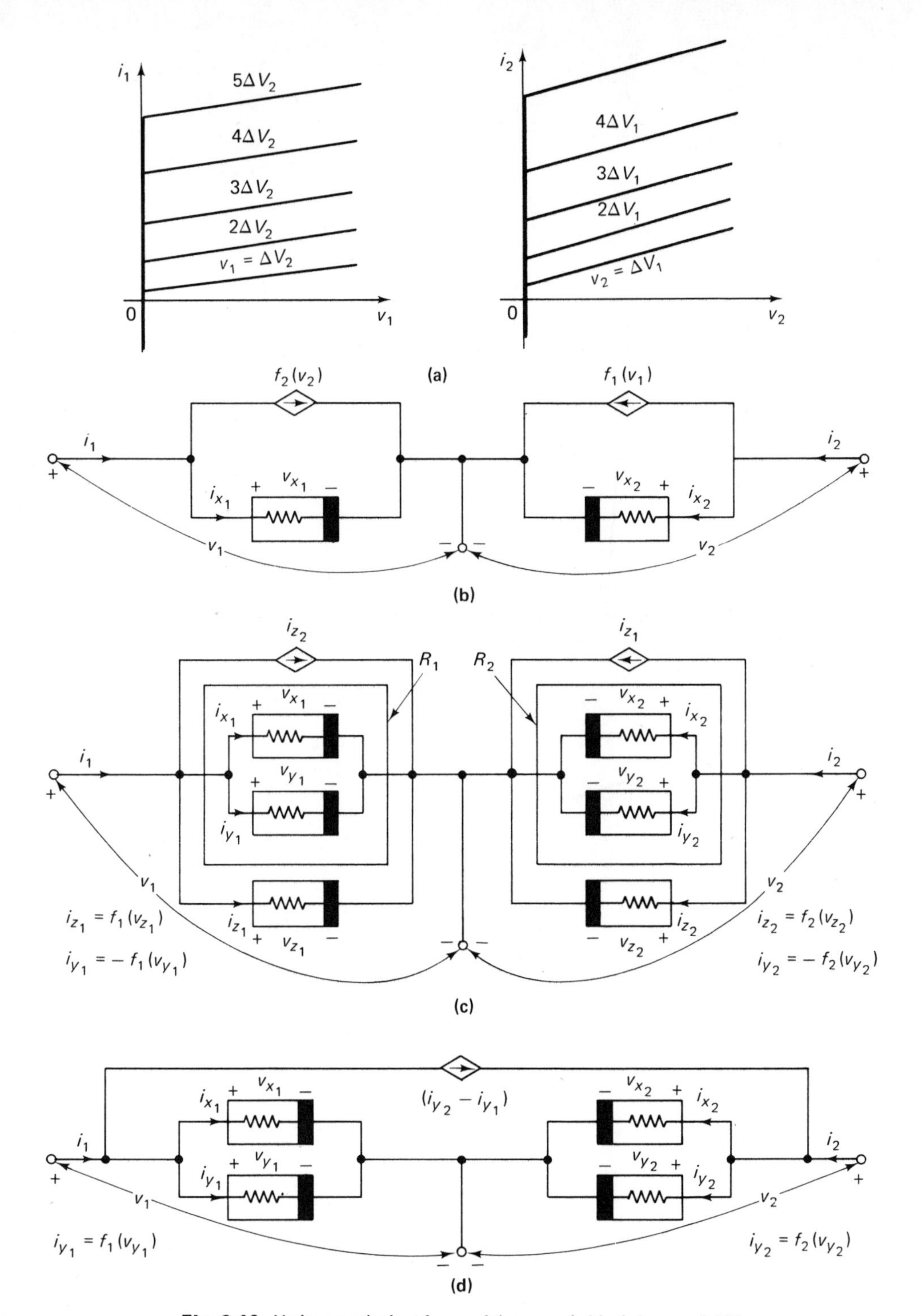

Fig. 2-16. Various equivalent forms of the canonic black-box model IV.

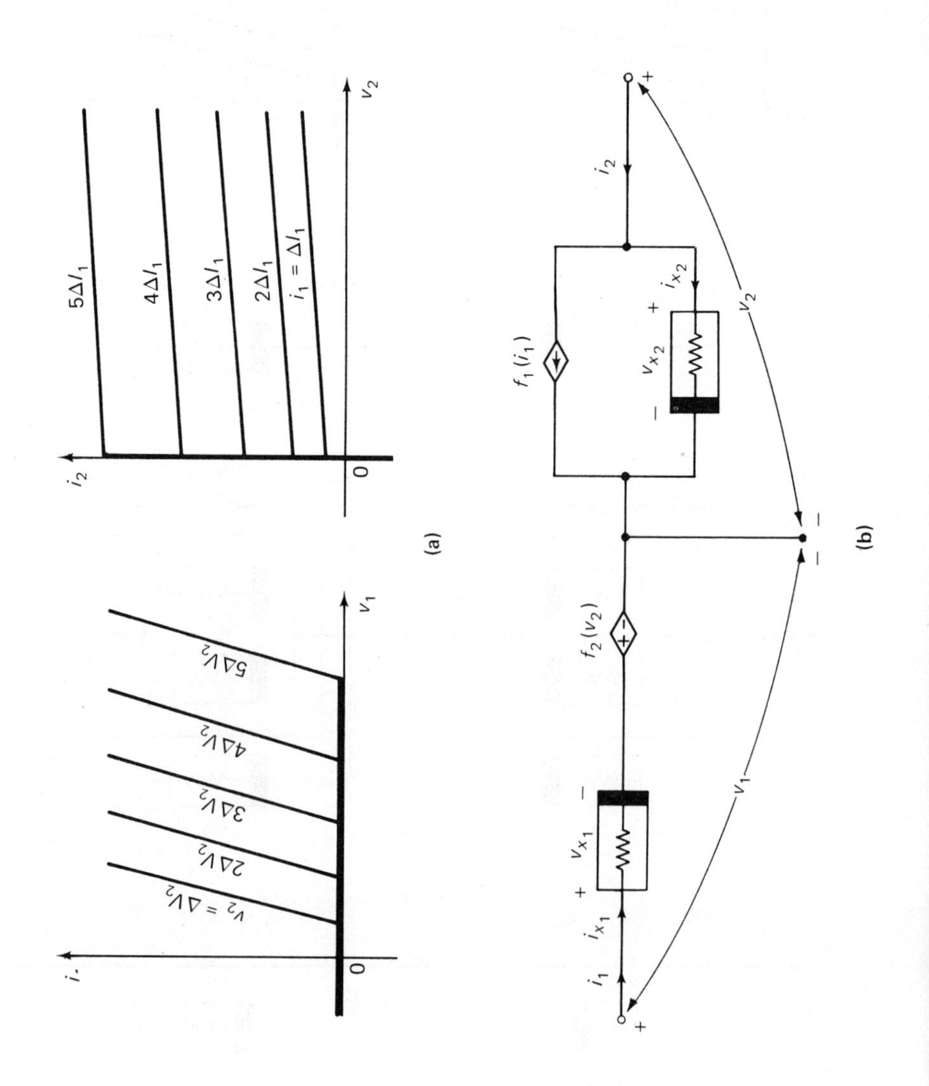

i_1
0
v_1
$v_2 = \Delta V_2$
$2\Delta V_2$
$3\Delta V_2$
$4\Delta V_2$
$5\Delta V_2$
i_2
0
v_2
$5\Delta I_1$
$4\Delta I_1$
$3\Delta I_1$
$2\Delta I_1$
$i_1 = \Delta I_1$
i_1
i_{x_1}
v_{x_1}
$f_2(v_2)$
v_1
$f_1(i_1)$
v_{x_2}
i_{x_2}
i_2
v_2

(a)

(b)

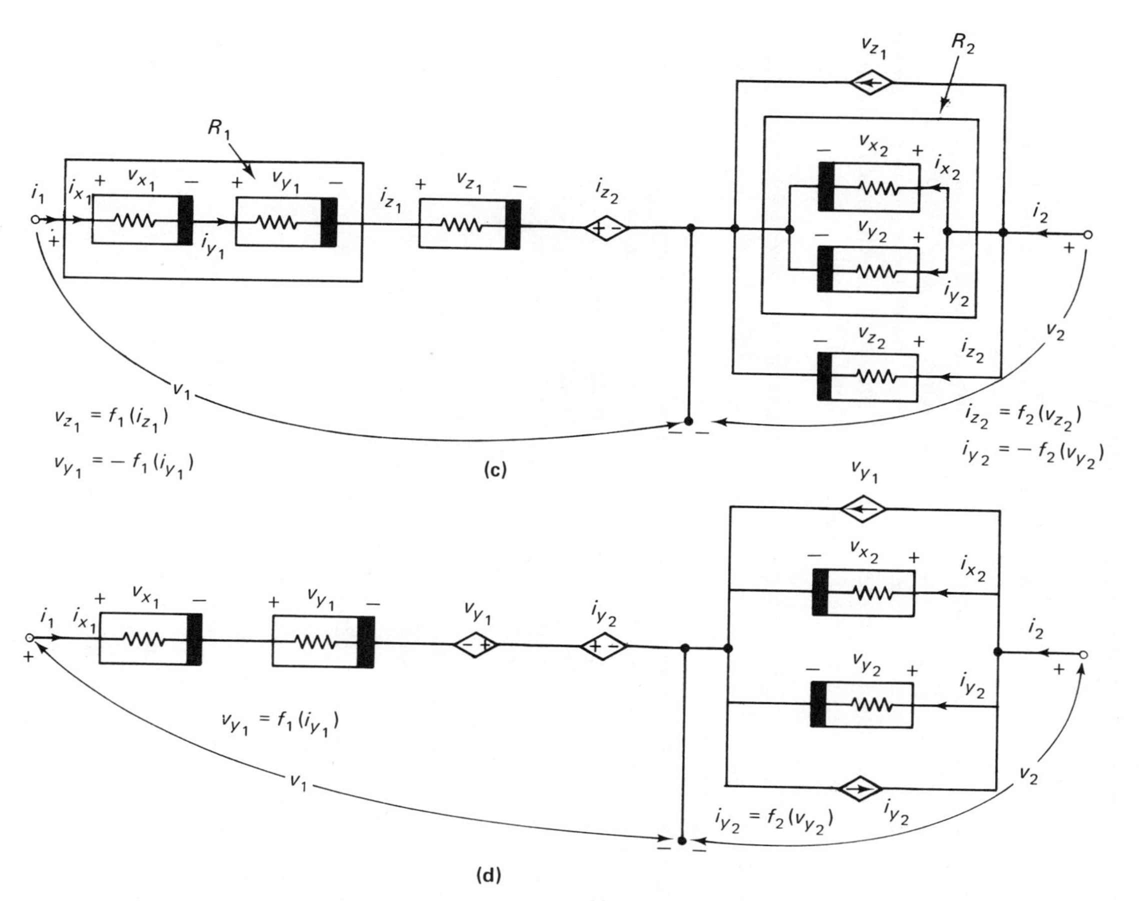

Fig. 2-17. Various equivalent forms of the canonic black-box model V.

and the two parallel resistors on the right arm can be replaced by an equivalent nonlinear resistor.

2-5-2 Canonic Black-Box Models for Arbitrary Families of *v-i* Curves

Consider now the general case in which no restriction is imposed on either the slope or the breakpoint of the family of piecewise-linear v-i curves. In this case, a *unified procedure* can be formulated, provided two additional building blocks are allowed, i.e., *controlled linear resistances or conductances*, and *controlled concave or convex resistors*. The basic approach is a simple extension of the *segment-by-segment method* for synthesizing a prescribed v-i curve as presented in reference 2. The procedure is best described with the help of a hypothetical example.

EXAMPLE 2-1. Synthesize a circuit model for realizing exactly the prescribed input family of curves $i_1 = g_1(v_1, v_2)$ in Fig. 2-18(a) and the output family of curves $i_2 = g_2(v_2, i_1)$ in Fig. 2-18(b).[16]

Solution: (a) *Modeling the input characteristic curves* $i_1 = g_1(v_1, v_2)$. Since each v_1-i_1 curve has four breakpoints, it would be necessary to use a combination of four v_2-controlled concave and convex resistors. The resulting model is shown in the left portion of Fig. 2-18(c). The v_2-controlled linear resistor is used to realize the four leftmost segments (through the origin) in Fig. 2-18(a). Its conductance function $G_1(\cdot)$ is shown in Fig. 2-19(a). An inspection of this function shows that $G_1(0) = 2.0$ m℧, $G_1(2) = 0.5$ m℧, $G_1(4) = 0.3$ m℧, and $G_1(6) = 0.2$ m℧. These values are taken directly from the slopes of the four leftmost segments in Fig. 2-18(a). The function of $G_1(\cdot)$ in Fig. 2-19(a) is obtained by connecting these four data points with straight lines. This is equivalent to assuming that the slope varies uniformly between each pair of segments in Fig. 2-18(a). Clearly, a nonuniform variation could also be simulated by drawing an appropriate smooth curve through the four data points.

The second segment associated with each of the four v_1-i_1 curves in Fig. 2-18(a) is realized by connecting a v_2-controlled concave resistor in parallel with the v_2-controlled linear resistor $G_1(v_2)$. The conductance function $G_2(\cdot)$ and the voltage intercept function $E_2(\cdot)$ for this element are shown in Fig. 2-19(b) and (c). The value of $G_2(v_2)$ for $v_2 = 0, 2, 4$, and 6 is obtained by subtracting the corresponding slope of the second segment from the slope of the first segment. The value of $E_2(v_2)$ for $v_2 = 0, 2, 4$, and 6 is equal to the voltage coordinate v_1 at each of the four breakpoints connecting the first two segments.

The third segment of each of the four prescribed v_1-i_1 curves is characterized by a negative slope and can be realized by connecting a v_2-controlled convex resistor with a *negative* resistance function in series with the two-element network realized so far. The resistance function $R_3(\cdot)$ and current-intercept function $I_3(\cdot)$ for this element

[16]Since it is convenient to assume both i_1 and i_2 as the vertical axis, we abuse our notation slightly by writing $i_1 = g_1(v_1, v_2)$ even though $g_1(\cdot, \cdot)$ is not a single-valued function in Fig. 2-18(a).

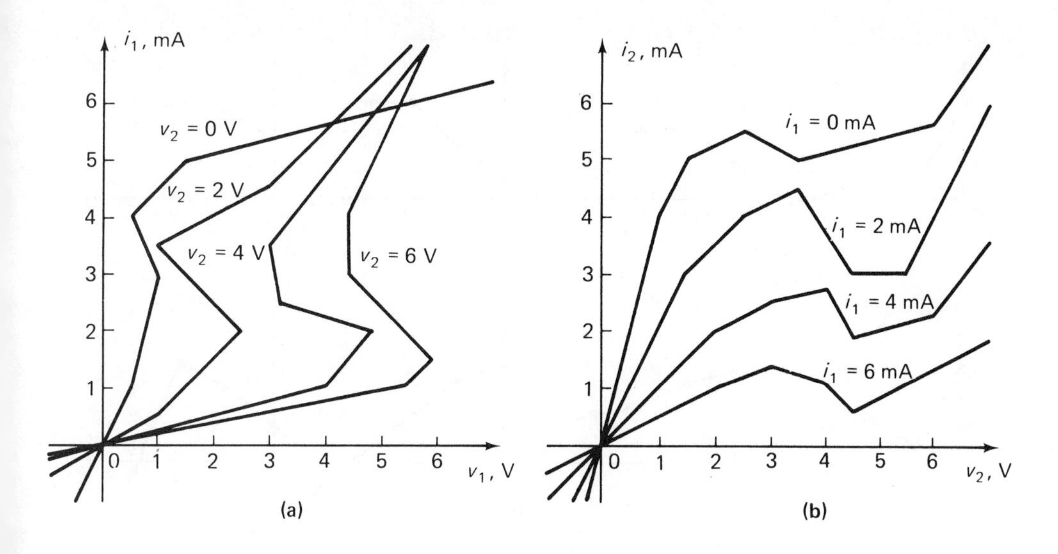

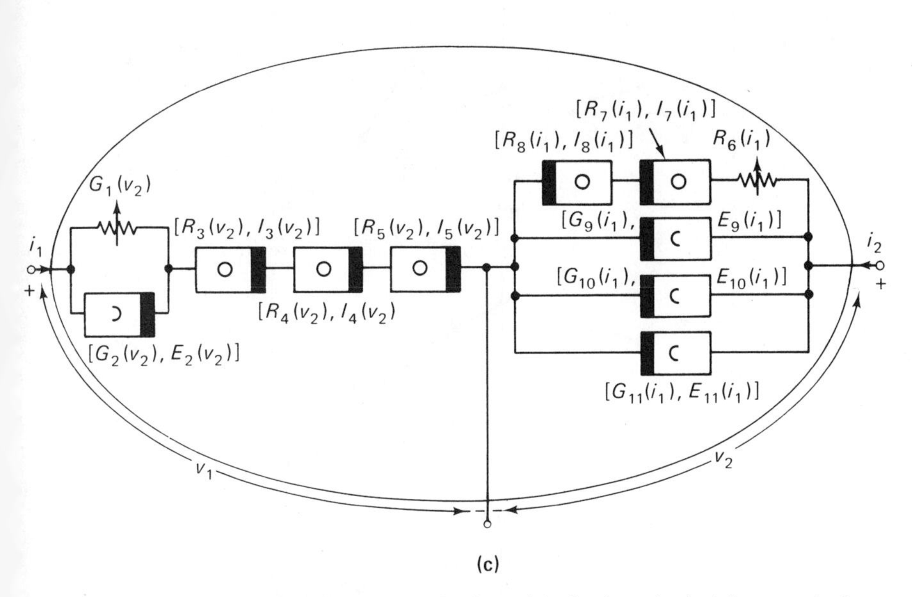

Fig. 2-18. Global dc circuit model of a hypothetical three-terminal device.

are shown in Fig. 2-19(d) and (e). The value of $R_3(v_2)$ for $v_2 = 0, 2, 4$, and 6 is obtained by subtracting the reciprocal slope of the third segment associated with each v_1-i_1 curve from the reciprocal slope associated with the second segment. The value of $I_3(v_2)$ for $v_2 = 0, 2, 4$, and 6 is equal to the current coordinate i_1 at each of the four breakpoints connecting the second and the third segments.

The last two segments associated with each of the four v_1-i_1 curves can be realized

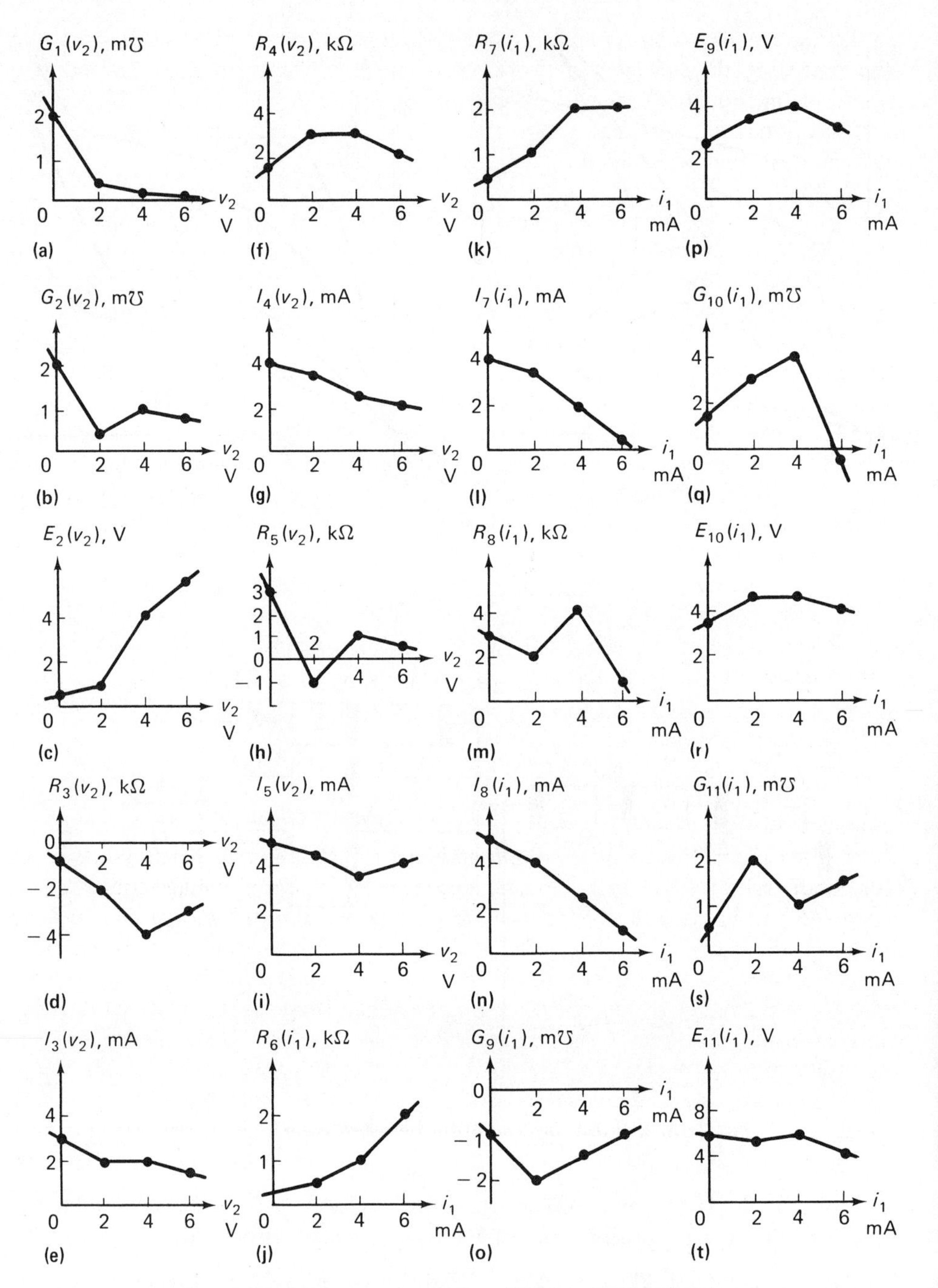

Fig. 2-19. Nonlinear functions for specifying the hypothetical three-terminal device model of Fig. 2-18.

in the same manner upon connecting two v_2-controlled convex resistors in series with the preceding three-element network. The resistance function $R_4(\cdot)$ and current-intercept function $I_4(\cdot)$ are easily computed as before and are given in Fig. 2-19(f) and (g). Similarly, the resistance function $R_5(\cdot)$ and current-intercept function $I_5(\cdot)$ are shown in Fig. 2-19(h) and (i).

(b) *Modeling the output characteristic curves* $i_2 = g_2(v_2, i_1)$. Since each v_2-i_2 curve has five breakpoints, it would be necessary to use a combination of five i_1-controlled concave and convex resistors. The resulting model is shown in the right portion of Fig. 2-18(c). Again, the four leftmost segments through the origin in Fig. 2-18(b) are realized by the i_1-controlled linear resistor with a resistance function $R_6(\cdot)$ as shown in Fig. 2-19(j). The next two segments can be realized by connecting two i_1-controlled convex resistors in series with this resistor, as shown in Fig. 2-18(c). The resistance function $R_7(\cdot)$ and current-intercept function $I_7(\cdot)$ are obtained from the slope and intercept of the second segment relative to the first segment and are shown in Fig. 2-19(k) and (l). Similarly, the resistance function $R_8(\cdot)$ and current-intercept function $I_8(\cdot)$ are obtained from the slope and intercept of the third segment relative to the second segment and are shown in Fig. 2-19(m) and (n).

The remaining three segments in each of the four v_2-i_2 curves in Fig. 2-18(b) can be realized by connecting three appropriate i_1-controlled concave resistors in parallel with the three-element network realized so far, as shown in Fig. 2-18(c). The characterizing conductance and voltage-intercept functions $G_9(\cdot), E_9(\cdot); G_{10}(\cdot), E_{10}(\cdot)$; and $G_{11}(\cdot)$, $E_{11}(\cdot)$ are easily computed from the prescribed curves in Fig. 2-18(b) and are shown in Figs. 2-19(o) to (t), respectively.

The complete model shown in Fig. 2-18(c) clearly simulates the prescribed characteristic curves in Fig. 2-18(a) and (b) exactly. Since no restriction has so far been imposed, it follows that our synthesis procedure is indeed a unified approach for modeling three-terminal devices characterized by either voltage-controlled or current-controlled curves. *In general, a combination of one x-controlled linear resistance or conductance and n x-controlled concave resistors will be required to realize each family of characteristic curves, where n is the total number of breakpoints in each* v_j-i_j *curve in the family.* Observe that the number of elements in the model does not depend on the number of v_j-i_j curves prescribed for each family of characteristic curves. Of course it is up to the circuit designer to specify a sufficient number of curves, since the unspecified continuum of curves will be generated automatically by the model on the basis of a linear *interpolation* between each pair of prescribed curves. *The piecewise-linear functions characterizing each of the x-controlled concave and convex resistors will contain the same number of segments as the number of curves prescribed in each family of characteristic curves.*

We shall now apply the black-box approach presented in Example 2-1 to synthesize several nonlinear dc circuit models of *bipolar transistors.* Only *npn* transistors will be considered here since the same models would apply to *pnp* transistors with minor modifications [2]. The most versatile model to be proposed for the *npn* transistor in Fig. 2-20(a) is the model A shown in Fig. 2-20(b), where the "ideal diode" R_4 is introduced to force $I_C = 0$ for $V_{CE} < 0$. The relationship between the slope and breakpoint of each segment and the characteristic function associated with the building

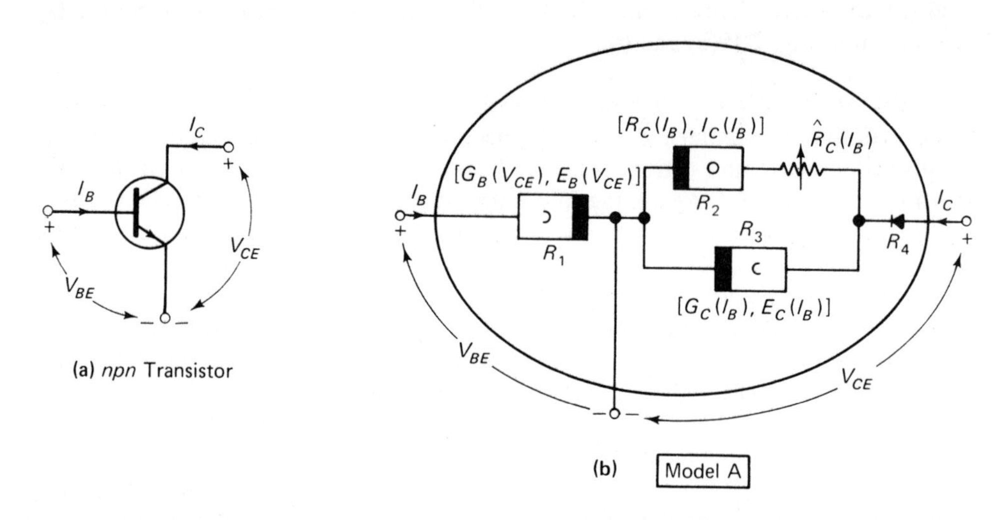

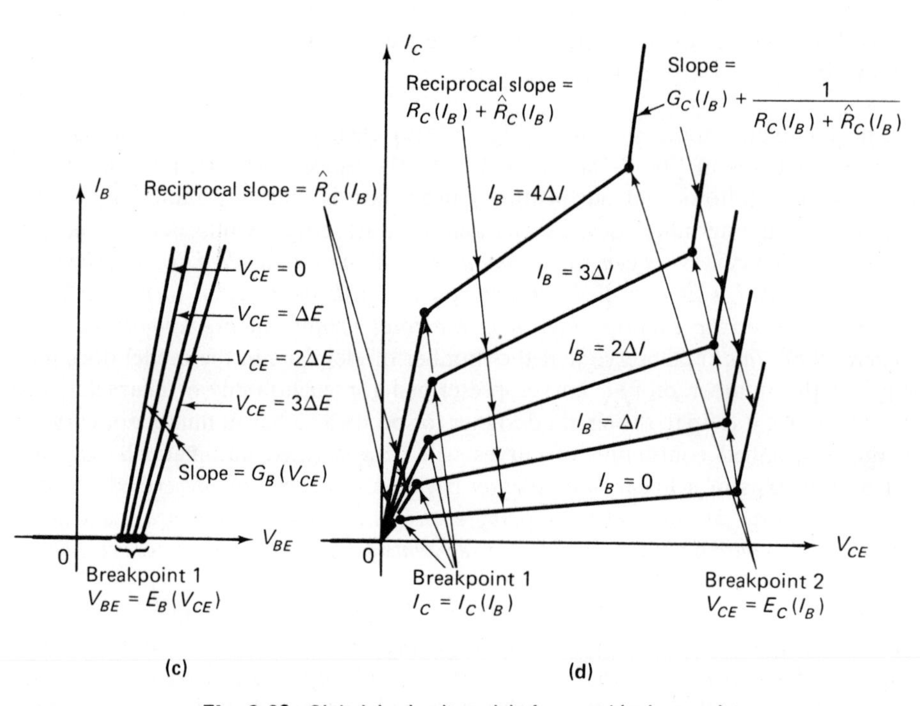

Fig. 2-20. Global dc circuit model of an *npn* bipolar transistor.

blocks are indicated directly on top of the prescribed characteristic curves in Fig. 2-20(c) and (d). Observe that model A is capable of simulating not only the *nonlinear current gain* in the active region, but also various high-injection nonlinear effects, such as the variation of collector output conductance in different regions of the I_C versus V_{CE} plane. Among other things, model A is capable of simulating all the features possessed by the *dc version* of the Gummel and Poon model [4]. However, the parameters associated with model A can be determined much more easily than those required by Gummel and Poon.

Since the dynamic range of operation of most bipolar transistors usually covers only a small portion of the first quadrant, model A may be further simplified under certain assumptions. For example, model A reduces to model B[17] in Fig. 2-21(a) if we approximate the saturation region of the output characteristic curves in Fig. 2-20(d) by a single straight line with slope G_C, as shown in Fig. 2-21(c), and if we delete the third segments, which normally are needed only in the region of high collector voltage. A further reduction from model B to model C in Fig. 2-21(d) is possible, provided the family of input characteristic curves in Fig. 2-21(b) is approximated by a single curve, as shown in Fig. 2-21(e), and provided the collector output conductance in the active region is assumed to be a constant, as shown in Fig. 2-21(f). Model C still contains a *nonlinear* current-controlled current source. This element may be replaced by a *linear* voltage-controlled current source upon decomposing the concave resistor R_1 in Fig. 2-21(d) into two nonlinear resistors R'_B and R''_B in the equivalent model D shown in Fig. 2-21(g). The v'_B-i'_B curve for R'_B is characterized by the relation $v'_B = (1/k)I_C(i'_B)$, where k is any convenient scaling constant [Fig. 2-21(h)]. The v''_B-i''_B curve for R''_B is chosen to take up the "slack" [Fig. 2-21(i)] in order that the composite v_1-i_1 curve [Fig. 2-21(j)] be identical to that shown in Fig. 2-21(e).

Another variant of model C is model E, shown in Fig. 2-21(k). The validity of this model is based on the observation that for a large class of bipolar transistors, the collector curves in the active region when projected backward all meet at a focal point, as shown in Fig. 2-21(l) [4]. Model E takes advantage of this property and consequently requires only the evaluation of $R_C(I_B)$ and E_C. The remaining parameters G_B, E_B, and G_C are known for various types of transistors. With the help of a curve tracer and a straight edge, $R_C(I_B)$ and E_C can be determined almost by inspection! Hence, model E seems to be an excellent compromise between accuracy and simplicity in the determination of the model parameters.

It should now be clear that the same procedure is applicable to any three-terminal device whose first-quadrant v-i curves have been approximated by piecewise-linear functions. One decisive advantage of this black-box approach over the physical approach is the ease with which the model parameters and functions are determined. Another advantage is the relative simplicity of the model itself. Of course, this simplicity is realized only if the computer program includes the three additional building blocks in its repertoire of allowable elements.

[17]The two diode-like symbols in Fig. 2-21 denote *fixed* concave and convex resistors and are defined in reference 2.

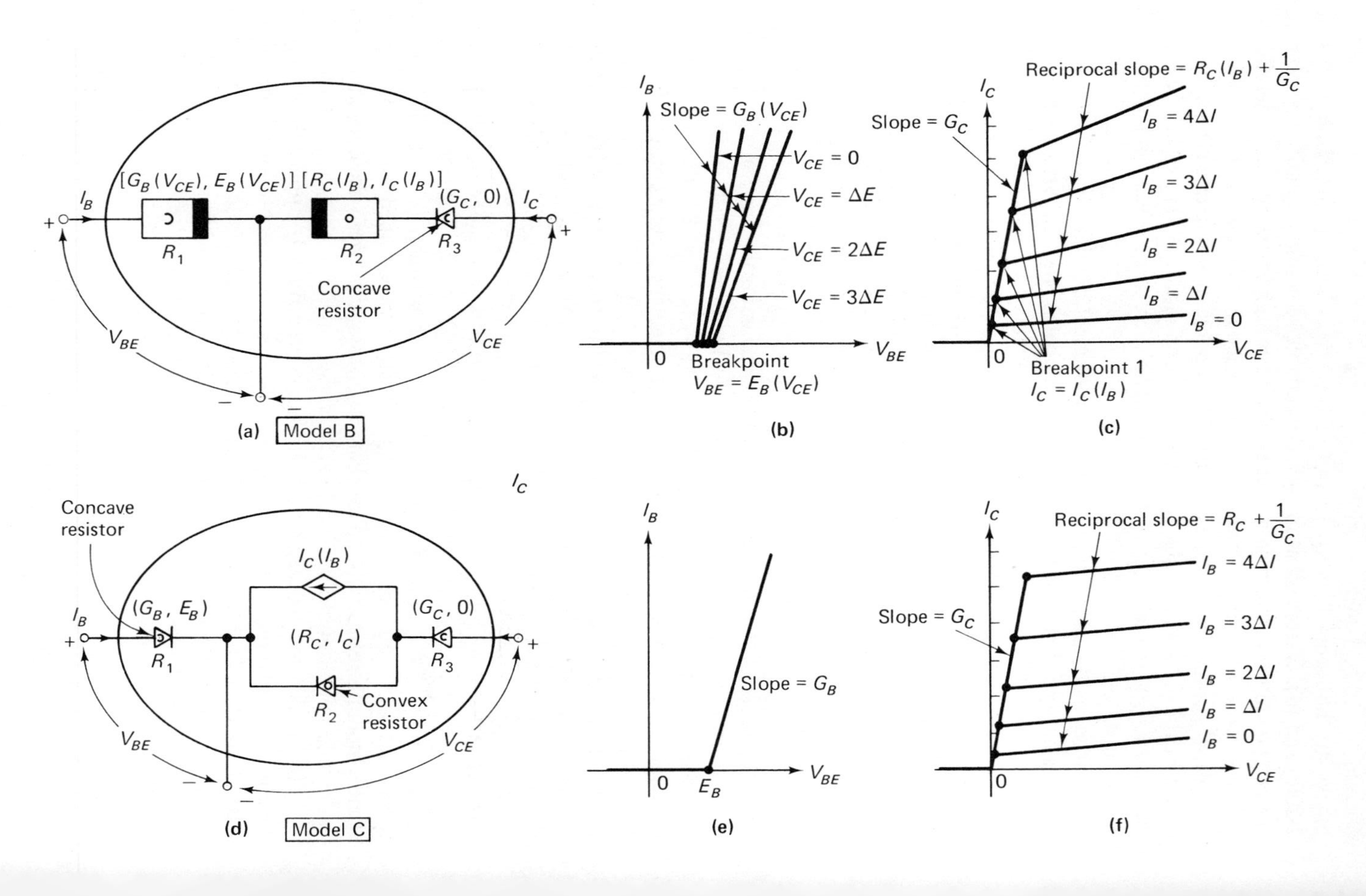
[$G_B(V_{CE})$, $E_B(V_{CE})$] [$R_C(I_B)$, $I_C(I_B)$]
$(G_C, 0)$
I_B
I_C
R_1
R_2
R_3
Concave resistor
V_{BE}
V_{CE}
(a) Model B
I_B
Slope = $G_B(V_{CE})$
$V_{CE} = 0$
$V_{CE} = \Delta E$
$V_{CE} = 2\Delta E$
$V_{CE} = 3\Delta E$
0
Breakpoint
$V_{BE} = E_B(V_{CE})$
V_{BE}
(b)
I_C
Reciprocal slope = $R_C(I_B) + \frac{1}{G_C}$
Slope = G_C
$I_B = 4\Delta I$
$I_B = 3\Delta I$
$I_B = 2\Delta I$
$I_B = \Delta I$
$I_B = 0$
0
Breakpoint 1
$I_C = I_C(I_B)$
V_{CE}
(c)
I_C
Concave resistor
$I_C(I_B)$
(G_B, E_B)
(R_C, I_C)
$(G_C, 0)$
I_B
R_1
R_2
R_3
Convex resistor
V_{BE}
V_{CE}
(d) Model C
I_B
Slope = G_B
0
E_B
V_{BE}
(e)
I_C
Reciprocal slope = $R_C + \frac{1}{G_C}$
Slope = G_C
$I_B = 4\Delta I$
$I_B = 3\Delta I$
$I_B = 2\Delta I$
$I_B = \Delta I$
$I_B = 0$
0
V_{CE}
(f)

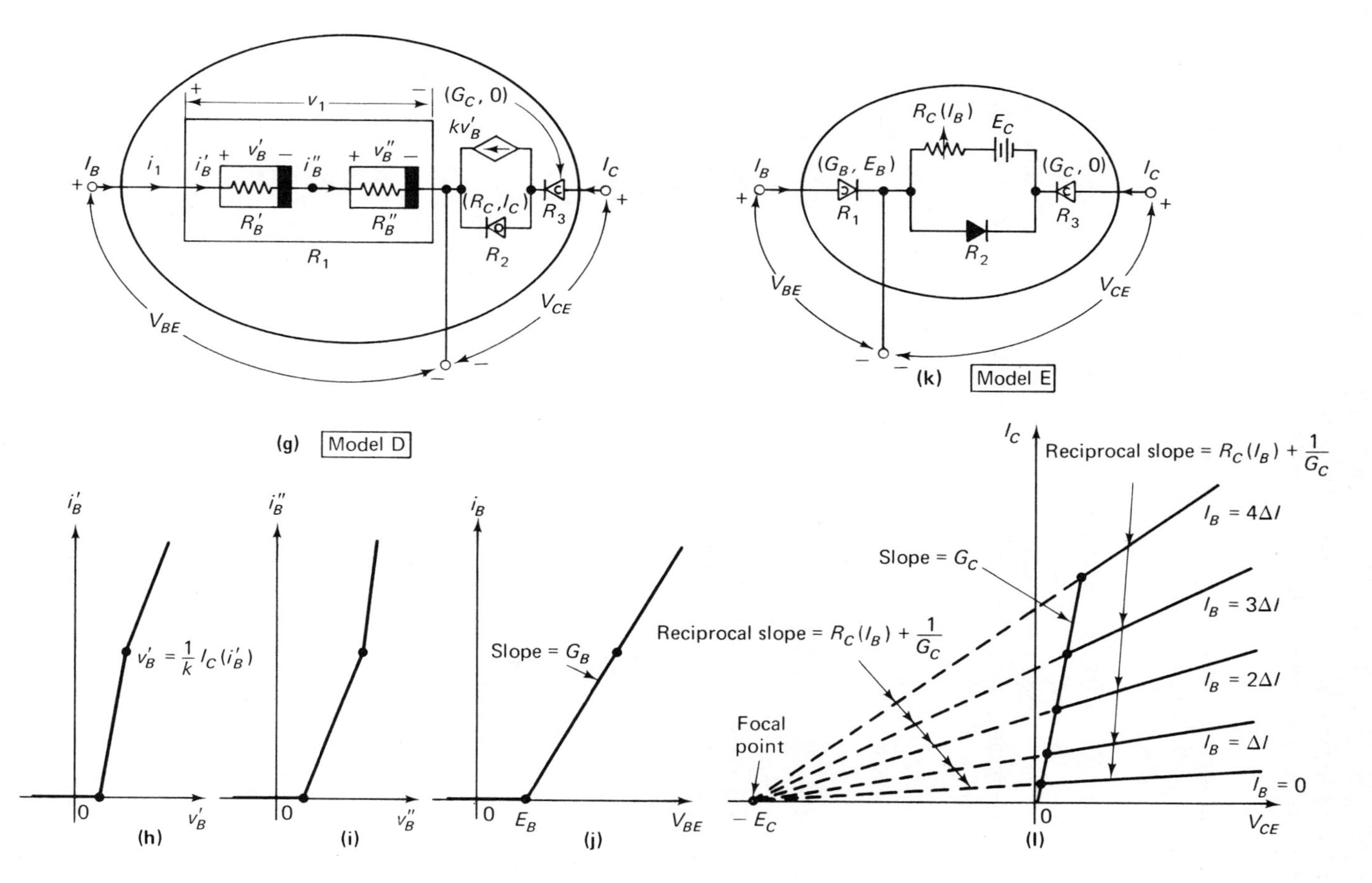

Fig. 2-21. Simplified black-box models of an *npn* bipolar transistor.

2-6 TRANSFORMING A DC GLOBAL BLACK-BOX MODEL INTO AN AC GLOBAL BLACK-BOX MODEL

The canonic models presented in Section 2-5 are dc global black-box models. The task of transforming these into ac models is still a major research problem. Consequently, we shall limit our discussion to a certain class of devices for which a reasonable solution is available.

> ***Definition:*** Quasi-Resistive Devices. *An n-terminal device is said to be quasi-resistive if the energy-storage mechanism of the device can be accounted for by the addition of* ***small linear*** *inductors, henceforth called* ***parasitic inductors****, and* ***small linear capacitors****, henceforth called* ***parasitic capacitors****, to an appropriately chosen dc model of the device.*

Fortunately, a large number of practical devices that exist today are quasi-resistive. For these classes of devices, the following techniques are available for transforming a dc model into an ac model:

1. Add parasitic inductances to simulate the actual *lead inductance* of the terminals.
2. Add parasitic capacitances to simulate the *packaging capacitance* of the device's enclosure.
3. Add a parasitic inductance in series with any *current-controlled* nonlinear resistor that is characterized by a nonmonotonic v-i curve.
4. Add a parasitic capacitance across any *voltage-controlled* nonlinear resistor that is characterized by a nonmonotonic v-i curve.

2-6-1 Lead Inductances and Packaging Capacitances

To understand why the parasitics in items 1 and 2 are generally needed, consider the typical enclosure of a three-terminal device shown in Fig. 2-22(a). As the operating frequency increases, the wires that connect the *intrinsic device* to the outside world behave more and more like inductances. Consequently, it is reasonable to account for these effects by adding three *inner* lead inductances, L_{1a}, L_{2a}, and L_{3a}, and three *outer* lead inductances, L_{1b}, L_{2b}, and L_{3b}, in series with the three terminals $1'$, $2'$, and $3'$ of the dc model, as shown in Fig. 2-22(b).

The next item to consider is the enclosure that is invariably used to protect the intrinsic device from moisture and other undesirable effects of the external environment. This enclosure is usually a combination of metals and insulators. Consequently, it is reasonable to account for these capacitive coupling effects by introducing three *packaging capacitances,* C_{12}, C_{13}, and C_{23}, as shown in Fig. 2-22(b). It is important to observe that the three inner lead inductances and the three outer lead inductances each form an *inductor-only cutset.* Similarly, observe that the three packaging capacitances form a *capacitor-only loop.*[18]

[18]As will be shown in the sequel, the presence of *inductor-only cutsets* and *capacitor-only loops* usually complicates the development of an automatic circuit-analysis program.

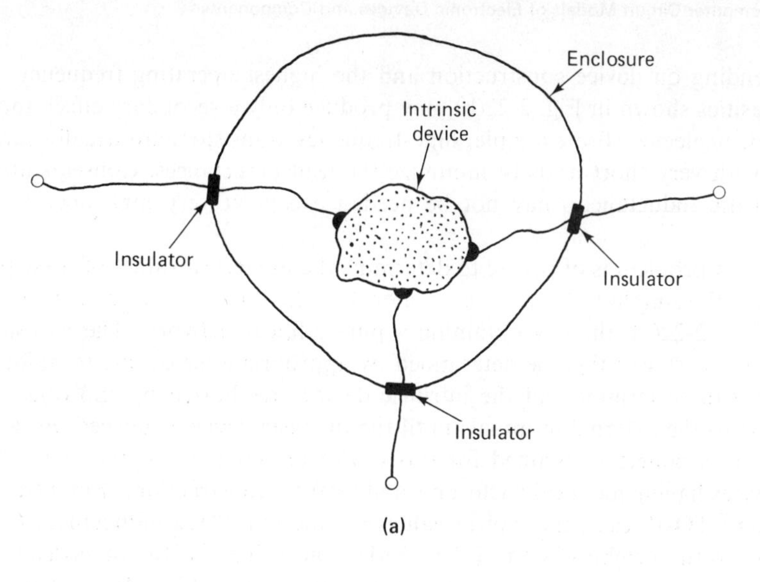

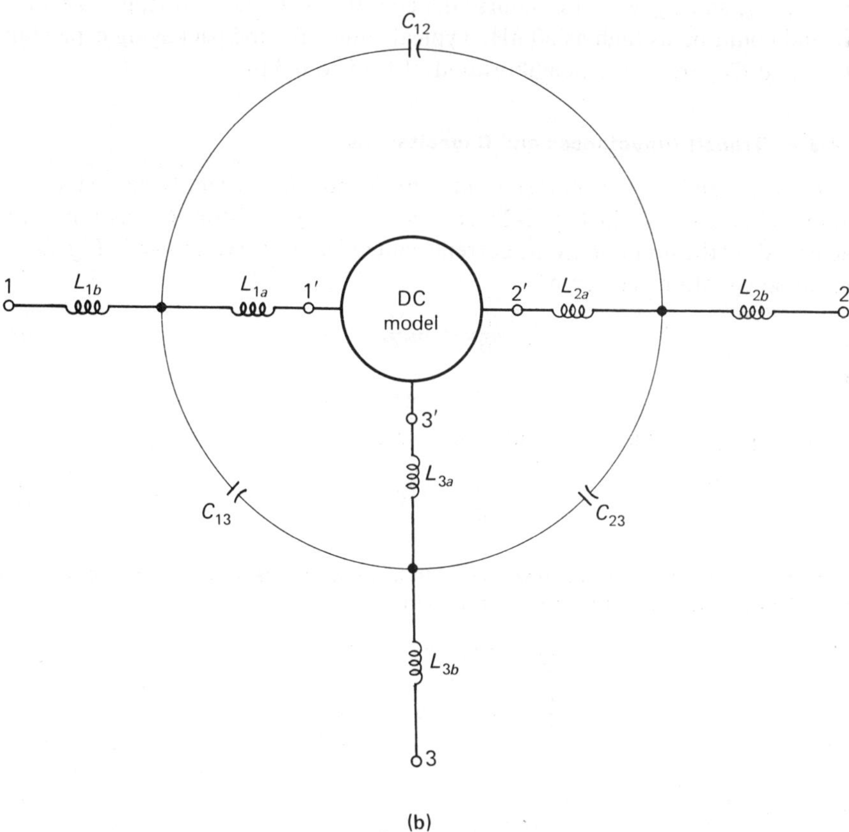

Fig. 2-22. Typical ac circuit model incorporating both parasitic lead inductances and packaging capacitances.

Depending on device construction and the highest operating frequency, some of the parasitics shown in Fig. 2-22(b) may produce only a secondary effect and may therefore be neglected. For example, high-frequency transistors are usually carefully packaged with very short leads to minimize the lead inductances. Consequently, the inner parasitic inductances may not be needed except at very high operating frequencies.

The next problem is of course to determine the numerical values of these parasitics. Perhaps the simplest approach is to short out the three terminals of the intrinsic device in Fig. 2-22(a), thereby obtaining a purely linear network. The parasitics of this linear network can then be determined by appropriate measuring techniques. In practice, the three terminals of the intrinsic device are shorted by applying a high-power pulse to the external terminals until the intrinsic device is burned out and destroyed. The parameters obtained for this device are then considered as typical for similar devices having identical enclosures and internal constructions. For a transistor in a standard TO-18 case, the typical values for the inner lead inductances L_{1a}, L_{2a}, and L_{3a} are in the neighborhood of 2 to 5 nH. The values for the outer lead inductances L_{1b}, L_{2b}, and L_{3b} would of course depend on the actual lead length used in the circuit and could be as high as 50 nH. Typical values for the packaging capacitances C_{12}, C_{13}, and C_{23} are in the neighborhood of 0.15 to 0.3 pF.

2-6-2 Transit Inductances and Capacitances

To understand why the parasitics in items 3 and 4 are generally needed, consider the simple circuit shown in Fig. 2-23(a). The nonlinear resistor R is assumed to be characterized by the nonmonotonic current-controlled v-i curve shown in Fig. 2-23(b). The terminal equation is given by

$$\frac{dv}{dt} = -\frac{i}{C} \tag{2-29}$$

$$v = v(i) \tag{2-30}$$

To reduce Eqs. (2-29) and (2-30) into the *normal form*[19]

$$\frac{dv_c}{dt} = f(v_c) \tag{2-31}$$

[19]A family of differential equations is said to be in the normal form if it can be written as a system of *first-order* differential equations of the form

$$\frac{dx_1}{dt} = f_1(x_1, x_2, \ldots, x_n, t)$$

$$\frac{dx_2}{dt} = f_2(x_1, x_2, \ldots, x_n, t)$$

$$\vdots$$

$$\frac{dx_n}{dt} = f_n(x_1, x_2, \ldots, x_n, t)$$

The independent variables $x_1, x_2, \ldots, x_n$ are then said to be *state variables.*

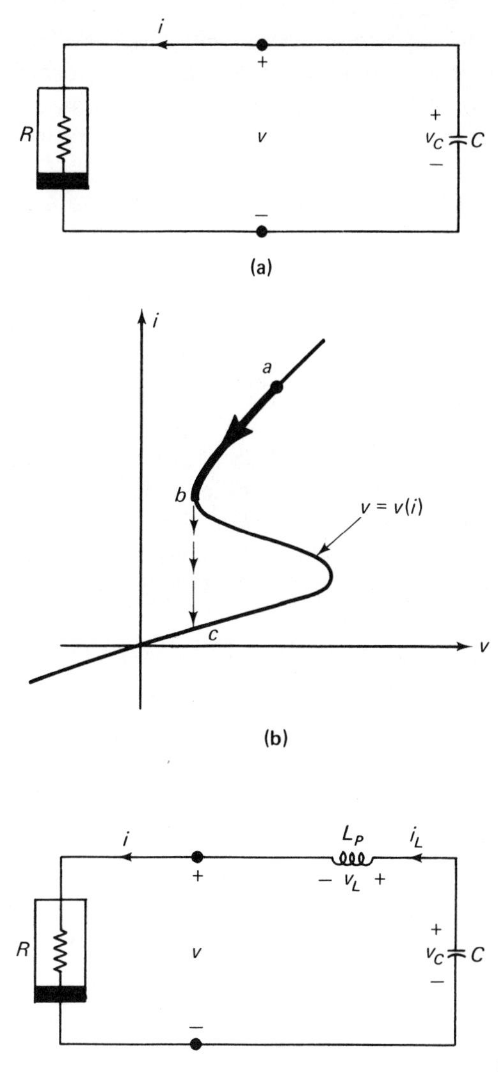

Fig. 2-23. An example illustrating the necessity for adding a transit inductance in series with any current-controlled device.

it would be necessary to solve for i in Eq. (2-30) in terms of v. But this is impossible, because the function $v(i)$ is not a one-to-one function and therefore admits no inverse. Therefore, we conclude that the *normal-form* equation for this network cannot be written. A direct consequence of this is that the solution waveform may become discontinuous. For example, starting with an initial condition corresponding to point a on the v-i curve in Fig. 2-23(b), we find that the motion must follow the curve in a downward direction, because Eq. (2-29) requires that $dv/dt < 0$ for all $i > 0$. When the motion arrives at point b, Eq. (2-29) forbids it to continue into the negative resistance region of the v-i curve; for otherwise the voltage would be increasing $[(dv/dt) >$

0]. To resolve this dilemma, it is necessary to postulate an *instantaneous jump* from point b to c, as shown by the arrows in Fig. 2-23(b). Hence, the current waveform in this case will be discontinuous at this point.[20] Since no physical waveforms can ever be discontinuous, it is clear that the circuit shown in Fig. 2-23(a) is not a realistic model of a physical circuit.

Let us see what happens if we insert a parasitic inductance L_P in series with the nonlinear resistor, as shown in Fig. 2-23(c). The normal-form equations can now be written as

$$\frac{dv_C}{dt} = -\frac{i_L}{C} \tag{2-32}$$

$$\frac{di_L}{dt} = \frac{v_C - v(i_L)}{L_P} \tag{2-33}$$

As will be shown later, the solution of Eqs. (2-32) and (2-33) will always be a continuous function of time, provided that $L_P > 0$. It will also be clear later that the smaller the value of this parasitic inductance, the faster the transition from point b to point c. In other words, the value of this parasitic inductance determines the nonzero transition time T_t in going from point b to point c in Fig. 2-23(b). Consequently, we shall refer to it as a *transit inductance*.

Our example demonstrates the necessity for including a transit inductance in series with any nonmonotonic current-controlled resistor. Using an exactly dual argument, it can be shown that a *transit capacitance* should be added in parallel with any nonmonotonic voltage-controlled resistor.

2-7 BLACK-BOX MODELS OF COMMON MULTIPORT CIRCUIT ELEMENTS AND DEVICES

We shall conclude this chapter with a collection of black-box models for some of the more commonly encountered multiport circuit elements and devices.[21] The name of each of these elements and its *ideal* port characterizations are listed in column 1 of Table 2-3. The corresponding symbols for these circuit elements and devices are shown in the second column, and the models are shown in the third column. It must be emphasized that the port characterizations listed in column 1 of Table 2-3 are idealized mathematical representations of the corresponding physical device. If a device in this table is to be operated at frequencies above the audio range, a more realistic circuit model will be needed to obtain an accurate computer simulation. We shall illustrate this point by deriving a more realistic circuit model for two of the devices listed in Table 2-3, the *two-port transformer* and the *op amp*.

[20]For more details on the resolution of this dilemma, see Chapter 14 of reference 2.

[21]A *multiterminal* element is said to be a *multiport* if the terminals can be arranged into *pairs* such that current entering one terminal of each pair is constrained to leave the second terminal. Every grounded $(n + 1)$-terminal element can be considered an n-port with the ground serving as the *common* second terminal of each port.

TABLE 2-3. Black-Box Circuit Models of Several Multiport Circuit Elements and Devices

ELEMENT NAME AND PORT CHARACTERIZATIONS	ELEMENT SYMBOL	MODELS 1	MODELS 2
1. Ideal two-port transformer $v_2 = nv_1$ $i_2 = -\frac{1}{n} i_1$ where $n = n_2/n_1$	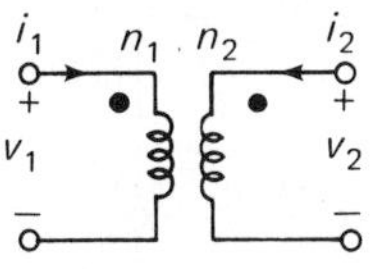 i_1, n_1, n_2, i_2, v_1, v_2	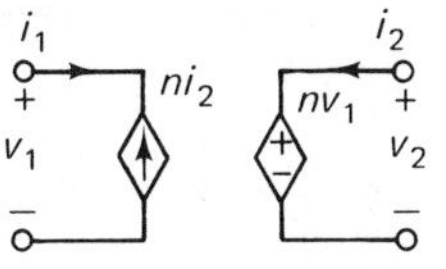 i_1, ni_2, nv_1, i_2, v_1, v_2	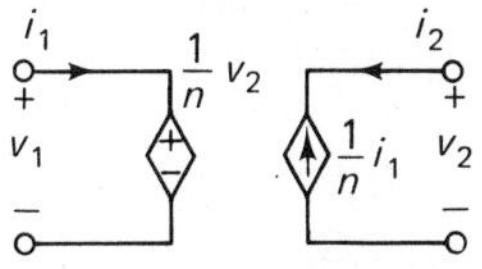 i_1, $\frac{1}{n} v_2$, $\frac{1}{n} i_1$, i_2, v_1, v_2
2. Pair of floating mutual inductors $v_1 = L_1 \frac{di_1}{dt} + M \frac{di_2}{dt}$ $v_2 = M \frac{di_1}{dt} + L_2 \frac{di_2}{dt}$	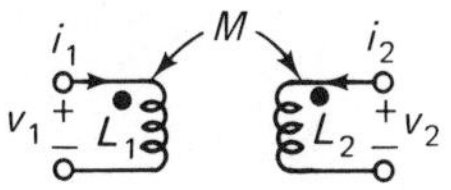 M, i_1, i_2, v_1, L_1, L_2, v_2	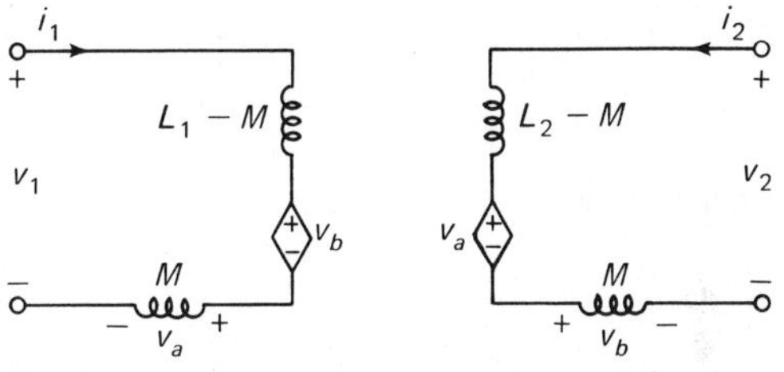 i_1, $L_1 - M$, v_1, v_b, M, v_a	i_2, $L_2 - M$, v_2, v_a, M, v_b
3. Pair of grounded mutual inductors	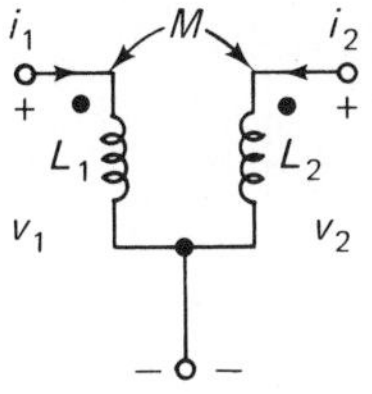 i_1, M, i_2, L_1, L_2, v_1, v_2	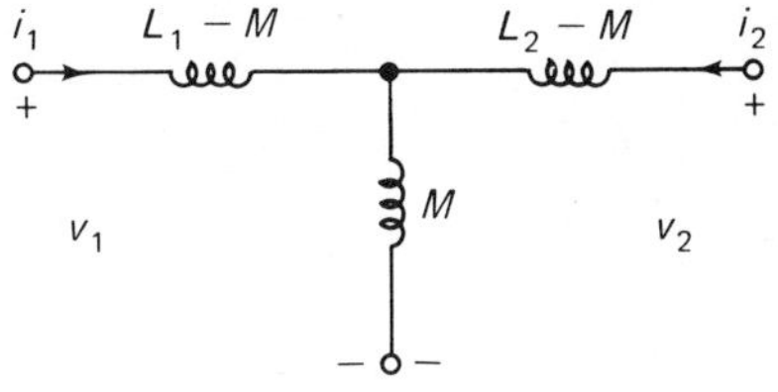 i_1, $L_1 - M$, $L_2 - M$, i_2, M, v_1, v_2	
4. Ideal gyrator $i_1 = Gv_2$ $i_2 = -Gv_1$ where G = gyration conductance	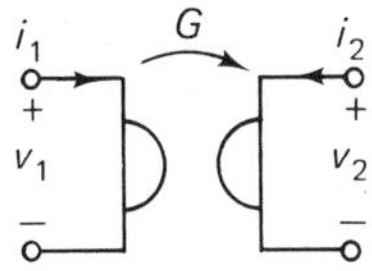 G, i_1, i_2, v_1, v_2	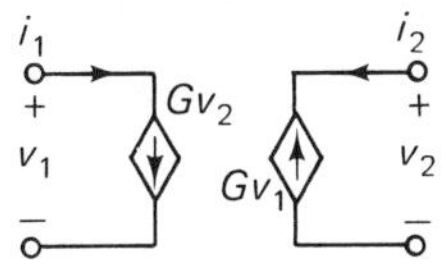 i_1, Gv_2, Gv_1, i_2, v_1, v_2	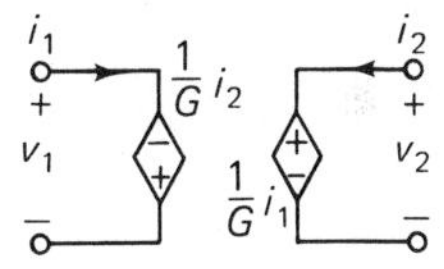 i_1, $\frac{1}{G} i_2$, $\frac{1}{G} i_1$, i_2, v_1, v_2
5. Voltage-inversion negative-impedance converter (VNIC) $v_1 = -v_2$ $i_1 = -i_2$	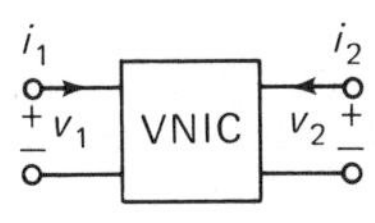 i_1, v_1, VNIC, v_2, i_2	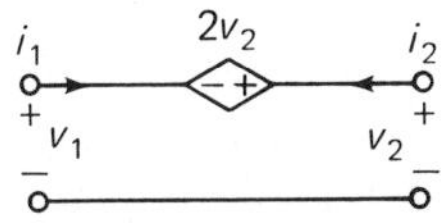 i_1, $2v_2$, i_2, v_1, v_2	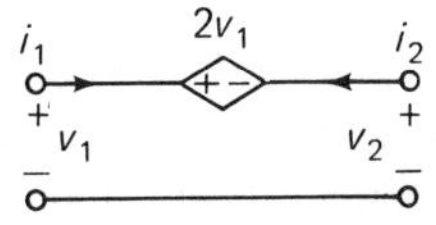 i_1, $2v_1$, i_2, v_1, v_2
6. Current-inversion negative-impedance converter (INIC) $v_1 = v_2$ $i_1 = i_2$	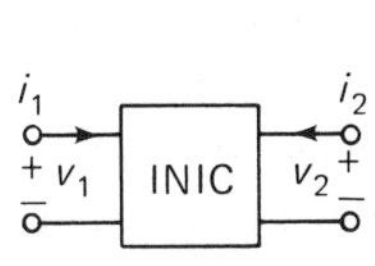 i_1, v_1, INIC, v_2, i_2	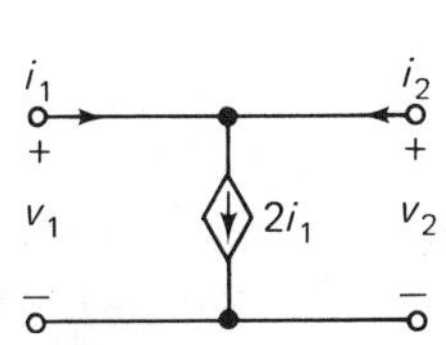 i_1, $2i_1$, i_2, v_1, v_2	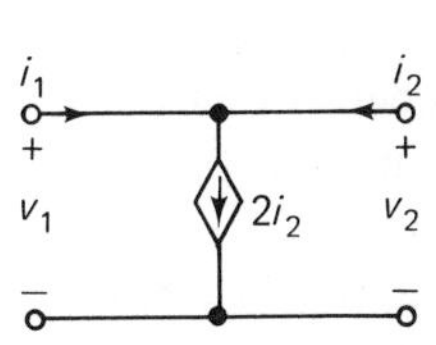 i_1, $2i_2$, i_2, v_1, v_2

TABLE 2-3 (cont.)

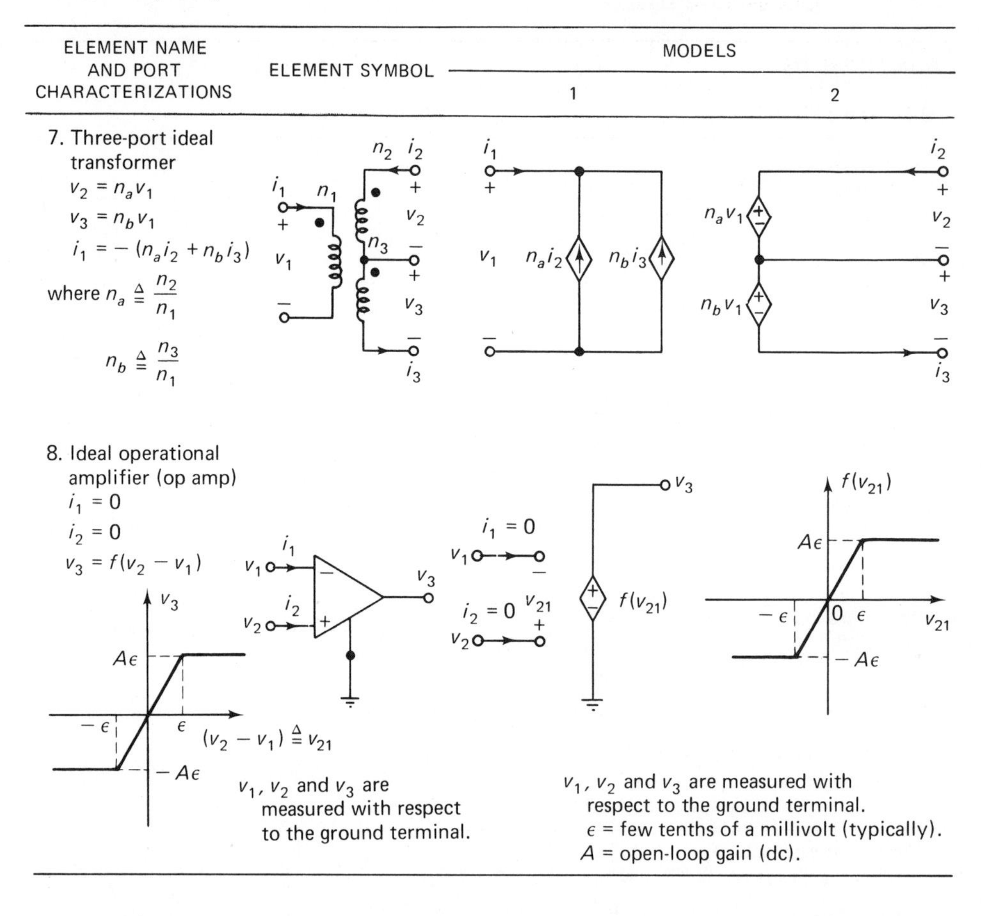

ELEMENT NAME AND PORT CHARACTERIZATIONS
ELEMENT SYMBOL
MODELS
1
2
7. Three-port ideal transformer
$v_2 = n_a v_1$
$v_3 = n_b v_1$
$i_1 = -(n_a i_2 + n_b i_3)$
where $n_a \triangleq \frac{n_2}{n_1}$
$n_b \triangleq \frac{n_3}{n_1}$
$n_a i_2$
$n_b i_3$
$n_a v_1$
$n_b v_1$
8. Ideal operational amplifier (op amp)
$i_1 = 0$
$i_2 = 0$
$v_3 = f(v_2 - v_1)$
$A\epsilon$
$-A\epsilon$
$-\epsilon$
ϵ
$(v_2 - v_1) \triangleq v_{21}$
$f(v_{21})$
v_1, v_2 and v_3 are measured with respect to the ground terminal.
v_1, v_2 and v_3 are measured with respect to the ground terminal.
ϵ = few tenths of a millivolt (typically).
A = open-loop gain (dc).

2-7-1 Circuit Model for a Nonideal Two-Port Transformer

Above the audio frequency range, a nonideal two-port transformer behaves like a pair of "floating" mutual inductors with self-inductances equal to L_1 and L_2, respectively; and a mutual inductance equal to M. Such a transformer can therefore be modeled by the six-element equivalent circuit of a mutual inductor, as shown in the second row of Table 2-3. If the "ideal" two-port transformer is included in the repertoire of allowed circuit elements of a particular computer simulation program, a nonideal two-port transformer can be modeled by the four-element equivalent circuit shown in Fig. 2-24(a), or by the three-element equivalent circuit shown in Fig. 2-24(b). We leave it as a simple exercise for the reader to show that these two-ports are equivalent to a pair of floating mutual inductors.

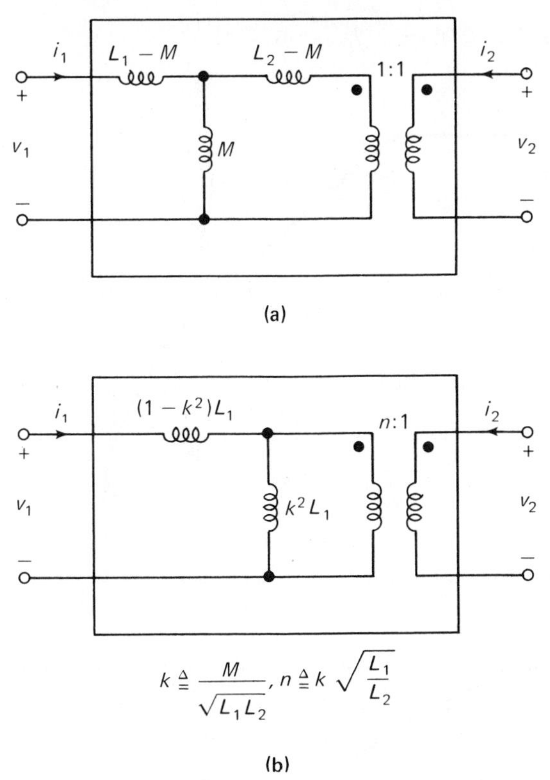

Fig. 2-24. Two equivalent circuit models for a nonideal two-port transformer in terms of inductors and an ideal two-port transformer.

2-7-2 Circuit Model for a Nonideal Op Amp

A nonideal op amp differs from the ideal op amp as defined in Table 2-3 in many respects. A much more accurate op-amp model could of course be derived by simply replacing each transistor and diode in the op-amp circuit by their respective circuit models. However, a typical op amp generally contains at least 20 transistors. If we choose the 13-element Ebers–Moll model shown in Fig. 2-8 for the transistors, the resulting circuit model of the op amp would contain at least 260 elements in addition to the other elements in the op-amp circuit! It is possible, however, to develop a much simpler op-amp circuit model—often called a *macro model* [17]—capable of mimicking realistically the following nonideal op-amp behaviors:

1. *Finite* input resistance $R_i \neq \infty$.
2. *Finite* output resistance $R_o \neq 0$.
3. *Frequency-dependent* open-loop voltage gain $G(\omega)$. A typical Bode plot for $G(\omega)$ is shown in Fig. 2-25(a). The first corner frequency ω_1 is usually called

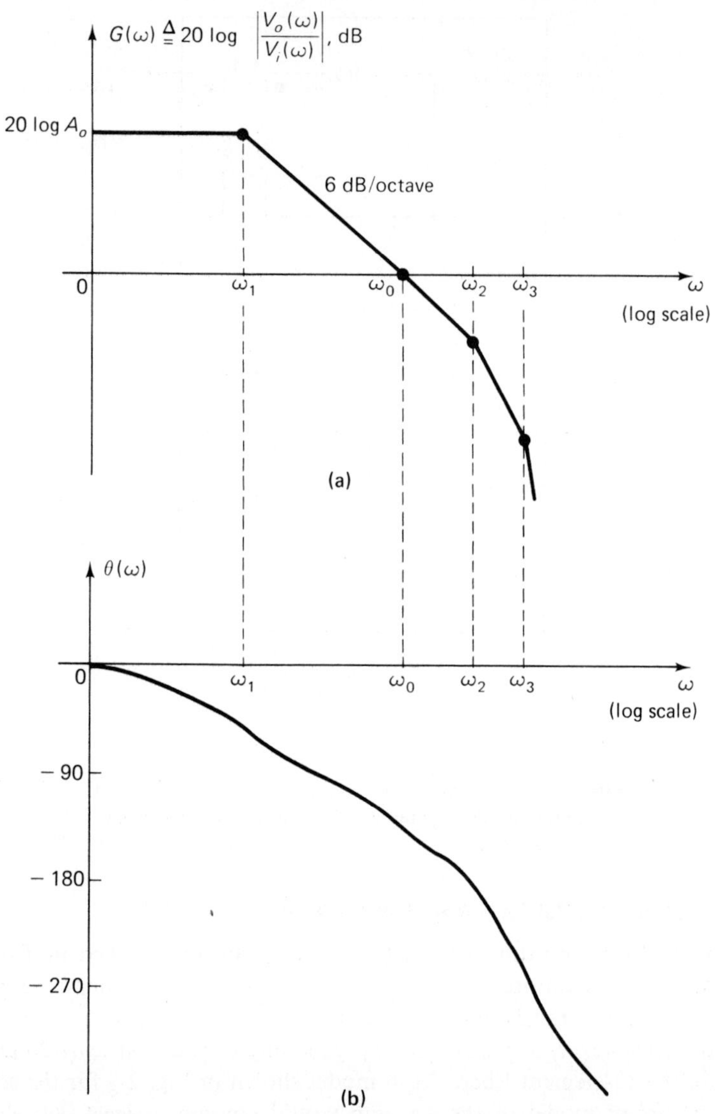

Fig. 2-25. Voltage gain (Bode plot) $G(\omega)$ and phase-shift characteristic $\theta(\omega)$ for a typical nonideal op amp.

the *dominant-pole frequency*; and the frequency ω_0, where the curve $G(\omega)$ crosses the 0-dB axis, is usually called the *unity-gain frequency*.

4. *Frequency-dependent* phase shift $\theta(\omega)$. A typical curve for $\theta(\omega)$ is shown in Fig. 2-25(b).
5. Output voltage-limiting behavior: $|v_o| < V_{o\max}$.
6. Finite *slew-rate limitation M*, where M is defined to be the maximum "time rate of change" of the output voltage that can be attained by the op amp under

the least favorable external circuitry.[22] The slew rate is an important op-amp parameter in the design of high-speed circuits, such as A/D and D/A converters, because it imposes an absolute limit on the speed attainable.

To develop a circuit model for mimicking these nonideal op-amp characteristics, we observe that the first four are typical of those found in linear circuits and should be relatively straightforward to simulate. The fifth characteristic is also typical of that found in voltage limiters and could be simulated with a *nonlinear* resistor. To simulate the last characteristic, we propose the nonlinear circuit building block shown in Fig. 2-26(a), where the *nonlinear* "voltage-controlled current source" is characterized by the "saturation" curve $i_c = f(v_i)$, as shown in Fig. 2-26(b). Observe that since the output current of the controlled source cannot exceed I_m, we have

$$\left|\frac{dv_c(t)}{dt}\right| = \frac{1}{C}|i_c| \leq \frac{I_m}{C} \triangleq S_r \tag{2-34}$$

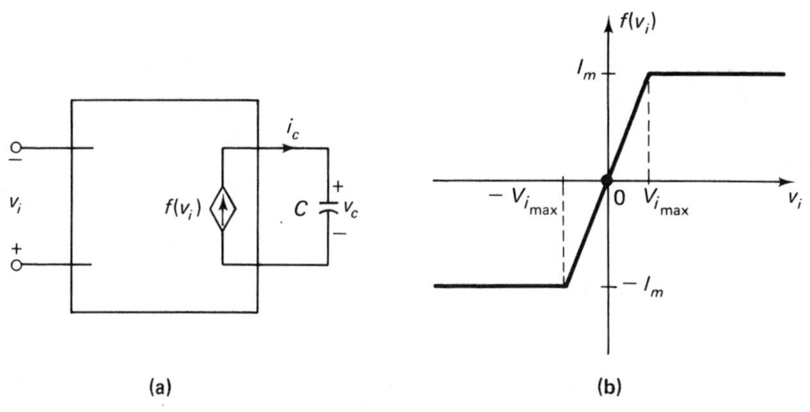

Fig. 2-26. Simple nonlinear circuit building block for simulating slew-rate limitation.

It follows from Eq. (2-34) that the slew rate of the circuit shown in Fig. 2-26(a) is equal to S_r.

Making use of the preceding observations, we now propose the circuit shown in Fig. 2-27(a) as a possible model for a nonideal op amp; the nonlinear function $f(v_i)$ characterizing the nonlinear controlled source is shown in Fig. 2-27(b), and the v-i curve characterizing the nonlinear resistor is shown in Fig. 2-27(c). Observe that this circuit is made up of five stages:

1. Input stage N_1.
2. Slew-rate limiting and gain stage N_2.

[22]Unlike the other parameters, the op-amp slew rate is not a unique number but depends on the particular feedback configuration to which the op amp is connected. Since the worst case usually occurs when the op amp is connected as a *unity-gain voltage follower*, this configuration is usually used in measuring the slew rate.

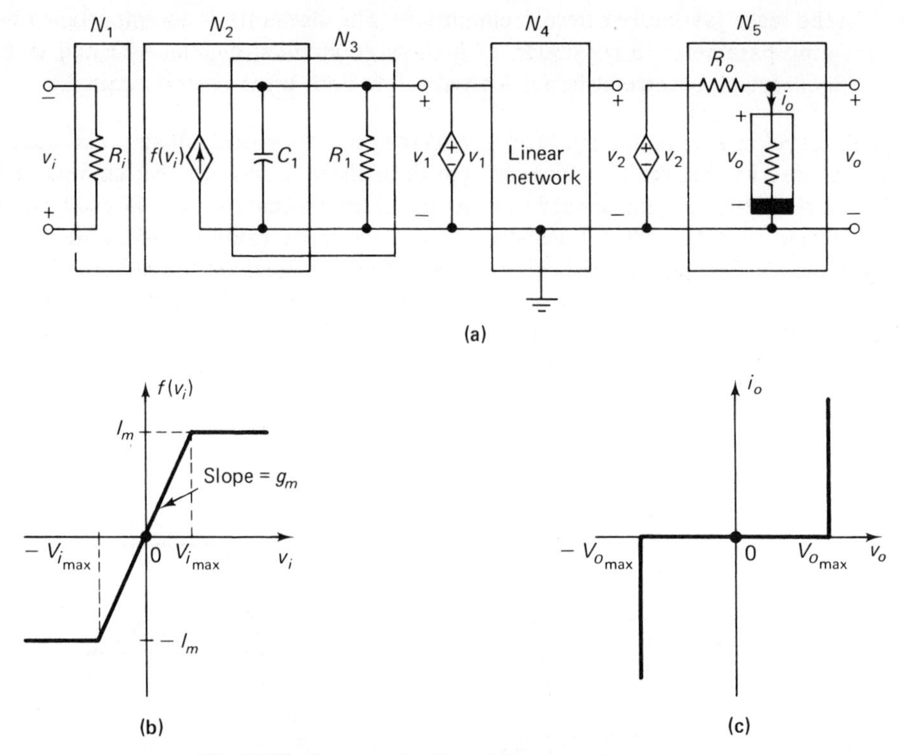

Fig. 2-27. A macro circuit model for a nonideal op amp.

3. Dominant-pole stage N_3.
4. Unity-gain, higher-pole-frequency stage N_4.
5. Output stage N_5.

The *input stage* consists of a single resistor R_i whose resistance is assigned equal to the prescribed op-amp differential mode "input resistance." The *second stage* is simply "transplanted" from Fig. 2-26(a) and consists of a nonlinear voltage-controlled current source in parallel with a linear capacitor C_1. If we choose a resistance for R_1 sufficiently high to offer negligible loading effects on N_2, then the second stage can be used to simulate any prescribed slew rate S_r by choosing I_m and C_1 such that

$$S_r = \frac{I_m}{C_1} \tag{2-35}$$

Since there are two parameters, I_m and C_1, at our disposal, we shall show in the following development that one parameter can be chosen so that the *macro* circuit model in Fig. 2-27(a) will have a prescribed dc open-loop voltage gain A_o.

The *third stage* consists of the R_1C_1 parallel combination and is used to simulate the dominant-pole corner frequency $\omega_1 = 1/R_1C_1$ of the prescribed Bode plot $G(\omega)$, assuming that the plot falls off at 6 dB/octave (20 dB/decade) which is typical of most op amps. Observe that the capacitor C_1 is shared in this case by both stages N_2 and

N_3. Again, since we have two free parameters here, R_1 and C_1, we can assume any convenient value for R_1 (provided it is high enough to avoid loading down stage N_2) and then determine the value of C_1 from

$$C_1 = \frac{1}{\omega_1 R_1} \tag{2-36}$$

If the dominant-pole frequency ω_1 is not available from the manufacturer's data sheet, it can be determined (approximately) from the unity-gain frequency ω_0 (which is usually given in the data sheet) by using the relationship

$$\omega_1 = \frac{\omega_0}{A_0} \tag{2-37}$$

To derive this useful relationship, observe that if $\omega_2 > \omega_0$, then at $\omega = \omega_0$ most of the current from $f(v_i)$ flows into C_1, and we have

$$V_1 \cong g_m V_i \frac{1}{j\omega_0 C_1} \tag{2-38}$$

where V_1 and V_i are the *phasors* associated with $v_1(t)$ and $v_i(t)$, respectively. Hence,

$$\left|\frac{V_1}{V_i}\right| \cong \frac{g_m}{\omega_0 C_1} = 1 \tag{2-39}$$

Now substituting the value $g_m = A_o/R_1$ [to be derived in Eq. (2-40)] and the value $C_1 = 1/\omega_1 R_1$ [from Eq. (2-36)] into Eq. (2-39) and solving for ω_1, we obtain Eq. (2-37).

The *fourth stage* is a grounded *unity-gain* (dc) linear two-port designed to simulate the higher-pole-frequency breakpoints of the prescribed Bode plot for $G(\omega)$ and the phase-shift characteristic for $\theta(\omega)$. Many *RC* circuits can be chosen for this purpose, and the associated circuit parameters can be determined by standard approximation and optimization techniques. In fact, if N_4 is chosen to be a *minimum-phase* linear two-port,[23] then it follows from the well-known Hilbert transform relationship [12] that $G(\omega)$ and $\theta(\omega)$ are not independent of each other. Indeed, given $G(\omega)$, we can calculate the "uniquely associated" $\theta(\omega)$, and vice versa. This observation is highly significant, because the transfer function of most op amps is minimum phase; hence, we need only simulate either the gain or the phase characteristic, whichever is more convenient, knowing that the associated phase or gain characteristic would automatically follow from the Hilbert transform relationship.

The *last stage* consists of a linear and a nonlinear resistor. The value of R_o is chosen equal to the prescribed op-amp output resistance. The *v-i* curve for the nonlinear resistor is chosen as shown in Fig. 2-27(c), where the breakpoint voltage is assigned a value equal to the peak op-amp output voltage $V_{o_{max}}$, which is usually given in the data sheet. Observe that the nonlinear resistor behaves like an open circuit for all output voltages $|v_o| < V_{o_{max}}$. Hence, the last stage is used for simulating both the output resistance and the output voltage-limiting behavior of the nonideal op amp.

It remains to determine the parameters I_m and g_m for the nonlinear voltage-

[23]A linear two-port is said to be *minimum phase* if its transfer function $T(s)$ has no zeros in the open right half-plane.

controlled current source in Fig. 2-27(b).[24] Observe that since the dc gain of N_4 is assumed to be unity, we have $v_o(\text{dc}) = v_1(\text{dc}) = (g_m v_i) R_1$ in the region of interest; i.e., $-V_{o_{\max}} \leq v_o \leq V_{o_{\max}}$. Hence,

$$g_m = \frac{v_o(\text{dc})/v_i(\text{dc})}{R_1} = \frac{A_o}{R_1} \tag{2-40}$$

where A_o is the prescribed dc open-loop gain. The value of I_m is obtained by substituting Eq. (2-36) for C_1 into Eq. (2-35):

$$I_m = \frac{S_r}{\omega_1 R_1} \tag{2-41}$$

Finally, upon dividing Eq. (2-41) by Eq. (2-40) we obtain

$$V_{i_{\max}} = \frac{I_m}{g_m} \tag{2-42}$$

Since the two voltage-controlled voltage sources connected to N_4 and N_5 have unity controlling coefficients (they are used only for isolation purposes), all parameters for the op-amp circuit model shown in Fig. 2-27(a), (b), and (c) can be determined either from the op-amp data sheet or from experimental measurements. To demonstrate the usefulness of this model, let us consider a specific example.

EXAMPLE 2-2. Find a *macro circuit model* for the popular μA741 op amp with the following prescribed parameters (these are typical parameters obtained from the manufacturer's data sheet):

DC open-loop gain: $A_o = 8.35 \times 10^5 = 118$ dB
Input resistance (differential mode): $R_i = 2$ MΩ
Output resistance: $R_o = 75\ \Omega$
Peak output voltage: $V_{o_{\max}} = 15$ V
Slew rate: $S_r = 0.5$ V/μs $= 0.5 \times 10^6$ V/s
Dominant-pole frequency: $\omega_1 = 2\pi(1.4) = 8.8$ rad/s

Solution: To determine the parameters for the op-amp circuit model, let us choose arbitrarily (as long as it is sufficiently large)[25]

$$R_1 = 835\ \text{K}\Omega$$

$$g_m = \frac{A_o}{R_1} = \frac{8.35 \times 10^5}{835\ \text{k}\Omega} = 1\ \mho$$

$$I_m = \frac{S_r}{\omega_1 R_1} = \frac{0.5 \times 10^6}{(8.8)(835\ \text{k}\Omega)} = 68\ \text{mA}$$

$$V_{i_{\max}} = \frac{S_r}{\omega_1 A_o} = \frac{0.5 \times 10^6}{(8.8)(8.35 \times 10^5)} = 68\ \text{mV}$$

[24]Since the circuit shown in Fig. 2-27(a) is strictly a "black-box" model, the parameters I_m and g_m bear no relationship with the transistors inside the op amp. In fact, the choice of these parameters is not unique but depends on the choice of the free parameter R_1.

[25]Observe that we have chosen R_1 so that $g_m = 1$ to simplify computation.

It remains to design the unity-gain (dc) linear two-port N_4 for simulating the frequency-dependent gain and phase characteristics for $\omega > \omega_2$. For this example, we have chosen the four-section RC ladder shown in Fig. 2-28(a). Since this is a minimum-phase network, it suffices to determine its parameters so that its gain or phase characteristics match those of the prescribed or measured op-amp characteristics (apart from a constant dc gain, which is already taken care of by stage N_2). For this example, we have chosen to match the phase-shift characteristic $\theta(\omega)$, and the resulting circuit is shown in Fig. 2-28(a). It must be emphasized that this circuit is by no means unique. The reader with a good background in filter circuit design and computer optimization techniques can no doubt come up with other circuits that are as good if not better. The circuit shown in Fig. 2-28 was designed to have a fourth-order pole at $\omega = (2\pi)(9 \times 10^6)$, which effectively adds a total of 180° phase shift at this frequency. To avoid using computer optimization, each capacitor value is chosen to be one tenth of the preceding capacitor, thereby causing negligible loading effects on each subsequent section. The resulting transfer function is therefore approximately equal to the product of the transfer function of each single RC section; i.e.,

$$\frac{V_2(j\omega)}{V_1(j\omega)} \cong \frac{1}{(1 + j\omega R_2C_2)(1 + j\omega R_3C_3)(1 + j\omega R_4C_4)(1 + j\omega R_5C_5)} \tag{2-43}$$

Observe that at $\omega = 0$, $V_2(0)/V_1(0) = 1$; hence, we have unity dc gain. By assuming $C_2 = 100$, $C_3 = 10$, $C_4 = 1$, and $C_5 = 0.1$ pF, we can add the fourth-order pole at 9 MHz by choosing

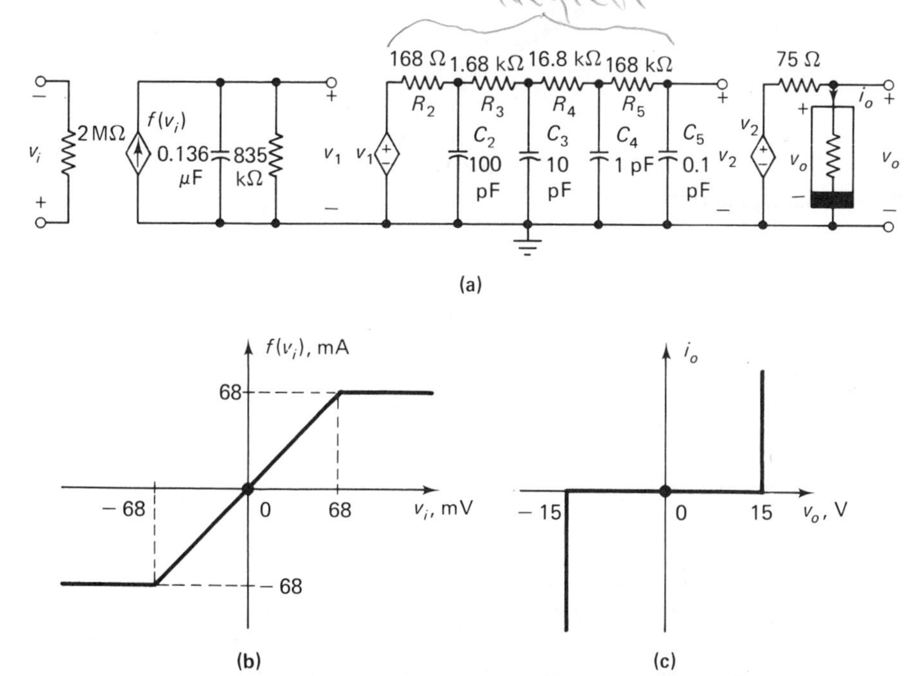

Fig. 2-28. A macro circuit model for the μA741 op amp.

$$R_2 = \frac{1}{\omega C_2} = \frac{1}{(2\pi)(9 \times 10^6)(10^{-10})} = 168\ \Omega$$

$$R_3 = \frac{1}{\omega C_3} = \frac{1}{(2\pi)(9 \times 10^6)(10^{-11})} = 1.68\ \text{k}\Omega$$

$$R_4 = \frac{1}{\omega C_4} = \frac{1}{(2\pi)(9 \times 10^6)(10^{-12})} = 16.8\ \text{k}\Omega$$

$$R_5 = \frac{1}{\omega C_5} = \frac{1}{(2\pi)(9 \times 10^6)(10^{-13})} = 168\ \text{k}\Omega$$

This "seat-of-the-pants" design did not make use of any sophisticated computer optimization techniques. To evaluate the accuracy of the resulting circuit model shown in Fig. 2-28, the gain and phase characteristics have been simulated on a computer

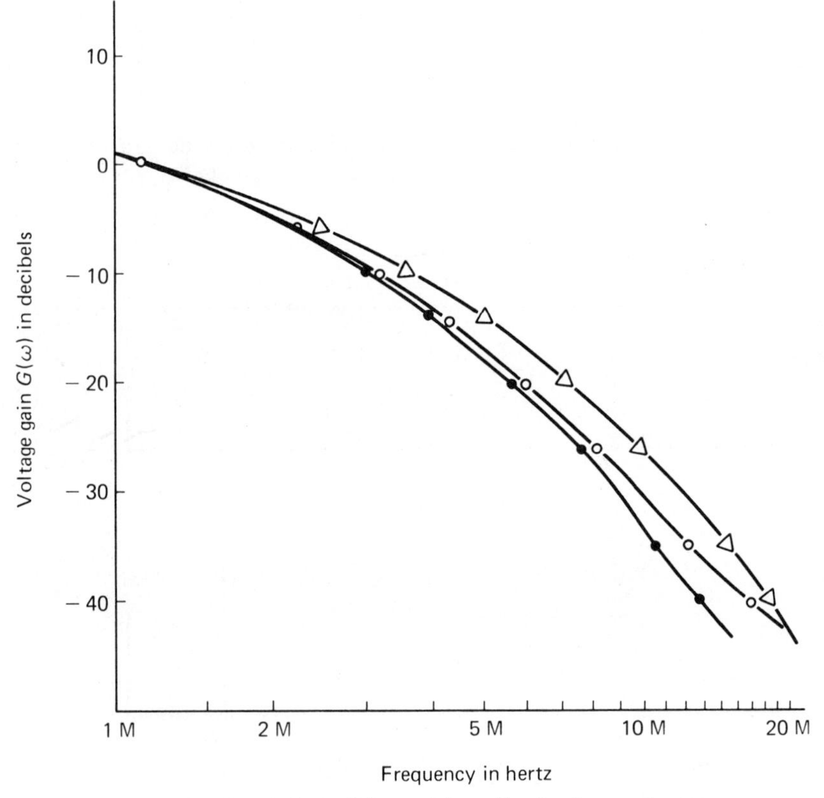

Fig. 2-29. Comparison of the model-*predicted* voltage-gain characteristic with the computer-simulated result using the entire op-amp circuit and with experimentally measured results.

● Computer-simulated result using the op-amp circuit model in Fig. 2-28.

○ Computer-simulated result using the complete op-amp circuit.

△ Experimentally measured result.

and the results are shown in Figs. 2-29 and 2-30, respectively. Also shown are two additional curves obtained respectively by experimental measurement and by simulating the complete op-amp circuit (with the transistors replaced by their Ebers–Moll models). The resulting agreement is seen to be quite good. As a final check, the op-amp circuit model is connected in a unity-gain voltage follower configuration, as shown in Fig. 2-31(a), and driven with a 10-V peak-to-peak square wave. The corresponding computer-simulated output voltage transient response is shown in Fig. 2-31(b). Observe that the output voltage $v_o(t)$ lags the input voltage, and the *maximum* magnitude of the slope of the rising and falling portions of $v_o(t)$ is approximately equal to 0.49 V/μs.[26] Again, the comparison with the prescribed slew rate of 0.5 V/μs is remarkably close!

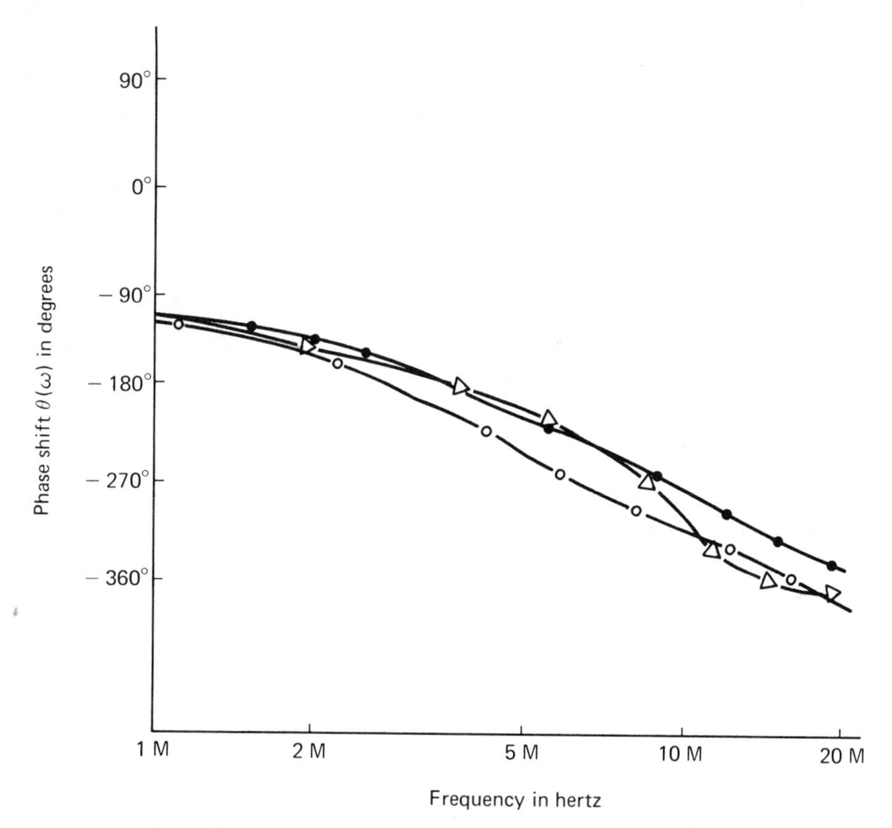

Fig. 2-30. Comparison of the model-*predicted* phase-shift characteristic with the computer-simulated result using the entire op-amp circuit and with experimentally measured results.

● Computer-simulated result using the op-amp circuit model in Fig. 2-28.
○ Computer-simulated result using the complete op-amp circuit.
△ Experimentally measured result.

[26]Additional details on the development of the op-amp circuit model shown in Fig. 2-28 and the computer-simulation results are given in reference 13.

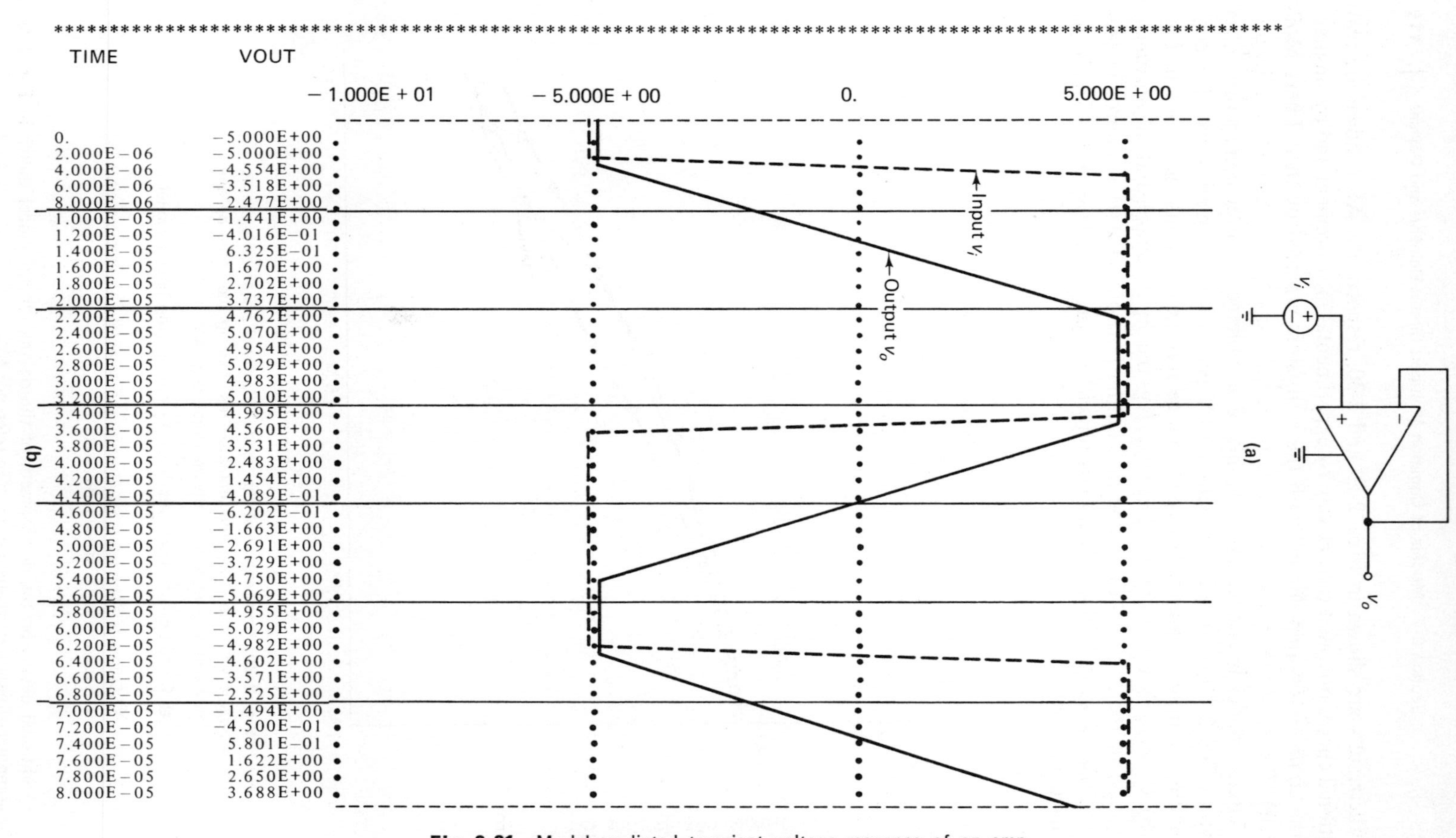

TIME	VOUT
0.	−5.000E+00
2.000E−06	−5.000E+00
4.000E−06	−4.554E+00
6.000E−06	−3.518E+00
8.000E−06	−2.477E+00
1.000E−05	−1.441E+00
1.200E−05	−4.016E−01
1.400E−05	6.325E−01
1.600E−05	1.670E+00
1.800E−05	2.702E+00
2.000E−05	3.737E+00
2.200E−05	4.762E+00
2.400E−05	5.070E+00
2.600E−05	4.954E+00
2.800E−05	5.029E+00
3.000E−05	4.983E+00
3.200E−05	5.010E+00
3.400E−05	4.995E+00
3.600E−05	4.560E+00
3.800E−05	3.531E+00
4.000E−05	2.483E+00
4.200E−05	1.454E+00
4.400E−05	4.089E−01
4.600E−05	−6.202E−01
4.800E−05	−1.663E+00
5.000E−05	−2.691E+00
5.200E−05	−3.729E+00
5.400E−05	−4.750E+00
5.600E−05	−5.069E+00
5.800E−05	−4.955E+00
6.000E−05	−5.029E+00
6.200E−05	−4.982E+00
6.400E−05	−4.602E+00
6.600E−05	−3.571E+00
6.800E−05	−2.525E+00
7.000E−05	−1.494E+00
7.200E−05	−4.500E−01
7.400E−05	5.801E−01
7.600E−05	1.622E+00
7.800E−05	2.650E+00
8.000E−05	3.688E+00

Fig. 2-31. Model-predicted transient voltage response of op-amp voltage follower circuit for calculating slew rate.

REFERENCES

1. CALAHAN, D. A. *Computer-Aided Network Design.* New York: McGraw-Hill Book Company, 1972.
2. CHUA, L. O. *Introduction to Nonlinear Network Theory.* New York: McGraw-Hill Book Company, 1969.
3. GRAY, P. E., D. DEWITT, A. R. BOOTHROYD, and J. F. GIBBONS. *Physical Electronics and Circuit Models of Transistors.* New York: John Wiley & Sons, Inc., 1964.
4. GUMMEL, H. K., and H. C. POON. "An Integral Charge-Control Model of Bipolar Transistors." *Bell System Tech. J.*, Vol. 49, No. 5, pp. 827–850, May–June 1970.
5. LINVILL, J. G. *Models of Transistors and Diodes.* New York: McGraw-Hill Book Company, 1963.
6. LOGAN, J. "Characterization and Modeling for Statistical Design." *Bell System Tech. J.*, Vol. 50, No. 4, pp. 1105–1147, Apr. 1971.
7. HAMILTON, D. J., F. A. LINDHOLM, and A. H. MARSHAK. *Principles and Applications of Semiconductor Device Modeling.* New York: Holt, Rinehart and Winston, Inc., 1971.
8. HERSKOWITZ, G. J., and R. B. SCHILLING. *Semiconductor Device Modeling for Computer-Aided Design.* New York: McGraw-Hill Book Company, 1972.
9. SESHU, S., and N. BALABANIAN. *Linear Network Analysis.* New York: John Wiley & Sons, Inc., 1959.
10. SOKAL, N. O., J. J. SIERAKOWSKI, and J. J. SIROTA. "Modeling Transistors and Diodes for Computer-Aided Nonlinear Circuit Analysis." *Electron. Design*, Part 1, pp. 54–59, June 7, 1967; Part 2, pp. 60–66, June 21, 1967; and Part 3, pp. 80–83, July 5, 1967.
11. SPARKES, J. J. "Device Modeling." *IEEE Trans. Electron. Devices*, Vol. ED-14, No. 5, pp. 229–232, May 1967.
12. BALABANIAN, N., T. A. BICKART, and S. SESHU. *Electrical Network Theory.* New York: John Wiley & Sons, Inc., 1969.
13. WONG, B. C. K. *Operational Amplifier Model Parameters Determination and Measurement.* Berkeley, Calif.: University of California, Master of Science, Plan II Research Report, Sept. 1973.
14. LAM, Y. F. "Comment on Modeling of Three-Terminal Devices: A Black Box Approach." *IEEE Trans. Circuits and Systems*, Vol. CAS-21, No. 6, pp. 807–809, Nov. 1974.
15. CHUA, L. O., and C-W TSENG. "A Memristive Circuit Model for P-N Junction Diodes." *International Journal of Circuit Theory and Applications,* Vol. 2, No. 4, pp. 367–389, Dec. 1974.
16. GETREU, I. "Modeling the Bipolar Transistor." *Electronics*, Part 1, pp. 114–120, September 19, 1974; Part 2, pp. 71–75, October 31, 1974; and Part 3, pp. 137–143, November 14, 1974.
17. BOYLE, G. R., B. M. COHN, D. O. PEDERSON, and J. E. SOLOMON. "Macromodeling of Integrated Circuit Operational Amplifiers," *IEEE Jour. of Solid State Circuits,* Vol. SC-9, pp. 353–364, Dec. 1974.

PROBLEMS

2-1. From the definition

$$C(v) = \frac{dq(v)}{dv}$$

it is clear that the incremental capacitance $C(v)$ associated with a nonlinear capacitor must necessarily be a function of the capacitor voltage v. Referring to Eq. (2-6), however, we observe the *diffusion capacitance*

$$C_d = \frac{q}{2\pi MkTF}(I_d + I_s) = C_d(I_d)$$

associated with the junction diode is a function of the diode current I_d. Resolve this paradox by deriving an equivalent expression for C_d in terms of v_d.

2-2. From the ac global junction diode model in Fig. 2-7, derive the following special cases:
(a) ac local diode model when $V \geq 0$.
(b) ac local diode model when $V \leq 0$.
(c) ac linear incremental diode model about the operating point $V = V_Q > 0$.
(d) ac linear incremental diode model about the operating point $V = V_Q < 0$.
(e) dc global diode model in terms of a single V-I curve.

2-3. An equivalent circuit for the Ebers–Moll "injection" *npn* transistor model shown in Fig. 2-8(d) is the Ebers–Moll "transport" model shown in Fig. P2-3.

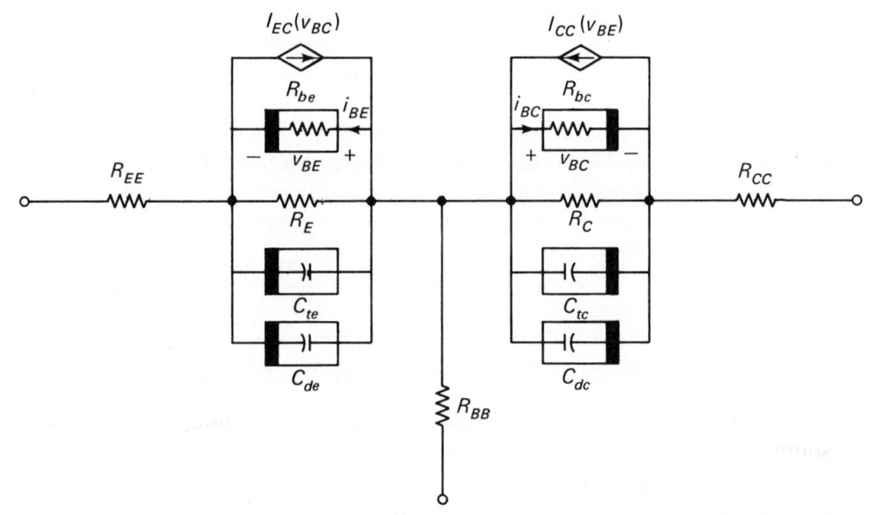

Fig. P2-3. Ebers–Moll *npn* transistor transport model.

(a) Show that the two nonlinear controlled sources must be characterized respectively by

$$I_{CC}(v_{BE}) = \alpha_F I_{ES}\left[\exp\left(\frac{qv_{BE}}{M_E kT}\right) - 1\right]$$

$$I_{EC}(v_{BC}) = \alpha_R I_{CS}\left[\exp\left(\frac{qv_{BC}}{M_C kT}\right) - 1\right]$$

(b) Show that the two nonlinear resistors must be characterized respectively by

$$i_{BE} = \frac{I_{CC}(v_{BE})}{\alpha_F}$$

$$i_{BC} = \frac{I_{EC}(v_{BC})}{\alpha_R}$$

(c) Investigate the advantages of the "transport" model over the "injection" model from the model parameter identification point of view.

2-4. A three-terminal device characterized by

$$i_1 = i_1(v_1, v_2)$$
$$i_2 = i_2(v_1, v_2)$$

is said to be *reciprocal* if

$$\frac{\partial i_1(v_1, v_2)}{\partial v_2} = \frac{\partial i_2(v_1, v_2)}{\partial v_1} \tag{1}$$

for all values of v_1 and v_2. It is said to be *passive* if $i_1 v_1 + i_2 v_2 \geq 0$ for all v_1 and v_2, and *locally active* if $(\delta i_1)(\delta v_1) + (\delta i_2)(\delta v_2) < 0$ about *some* operating point $(V_{1Q}, V_{2Q}, I_{1Q}, I_{2Q})$, where $(\delta i_1, \delta i_2)$ are incremental currents about (I_{1Q}, I_{2Q}) and $(\delta v_1, \delta v_2)$ are incremental voltages about (V_{1Q}, V_{2Q}).

(a) With the help of the Ebers–Moll model, show that although transistors are not reciprocal there are many operating points in which Eq. (1) is satisfied. Specify *all* such operating points. You may assume the well-known relationship $\alpha_I I_{CE} = \alpha_N I_{EF}$.

(b) Show that the Ebers–Moll model predicts that transistors are *locally active* over a wide range of operating points. What is the significance of this property?

(c) Show that some parameters α_I and α_N exist for which the Ebers–Moll model is not passive.

2-5. The dc linear incremental model of a three-terminal device can be characterized by a *constant* (frequency independent) 2×2 *open-circuit impedance matrix* ***Z***, short-circuit admittance matrix ***Y***, or *hybrid matrix* ***H***. Use the Ebers–Moll model to derive these three matrices for a transistor about the operating point $(V_{1Q}, V_{2Q}, I_{1Q}, I_{2Q})$.

2-6. Verify that the two circuits shown on the right side of Fig. 2-10(a), (b), (c), and (d) are equivalent to the *nonlinear* controlled sources shown on the left of the corresponding figures.

2-7. Verify that the two circuits shown in the upper and lower part of Fig. 2-11 are equivalent to the *x-controlled linear resistance* and the *x-controlled linear conductance*, respectively.

2-8. Find a circuit equivalent to an *x-controlled concave resistor* using only *linear* controlled sources and nonlinear resistors.

2-9. Find a circuit equivalent to an *x-controlled convex resistor* using only *linear* controlled sources and nonlinear resistors.

2-10. Derive the *canonic black-box model II* shown in Fig. 2-14(d) and illustrate with a hypothetical numerical example.

2-11. Derive the *canonic black-box model III* shown in Fig. 2-15(d) and illustrate with a hypothetical numerical example.

2-12. Derive the *canonic black-box model IV* shown in Fig. 2-16(d) and illustrate with a hypothetical numerical example.

2-13. Derive the *canonic black-box model V* shown in Fig. 2-17(d) and illustrate with a hypothetical numerical example.

2-14. Referring to the hypothetical model shown in Fig. 2-18(c) along with the model parameters in Fig. 2-19, sketch the two families of input and output characteristic curves corresponding to the controlling voltages $v_2 = 1, 3, 5,$ and 7 V and controlling currents $i_1 = 1, 3, 5,$ and 7 mA, respectively.

2-15. Construct a numerical example illustrating transistor models A, B, C, D, and E shown respectively in Figs. 2-20(b) and 2-21(a), (d), (g), and (k).

2-16. Consider the three-terminal device shown in Fig. P2-16(a) with the two families of input and output characteristic curves restricted to the first and third quadrants, as shown in Fig. P2-16(b) and (c), respectively.

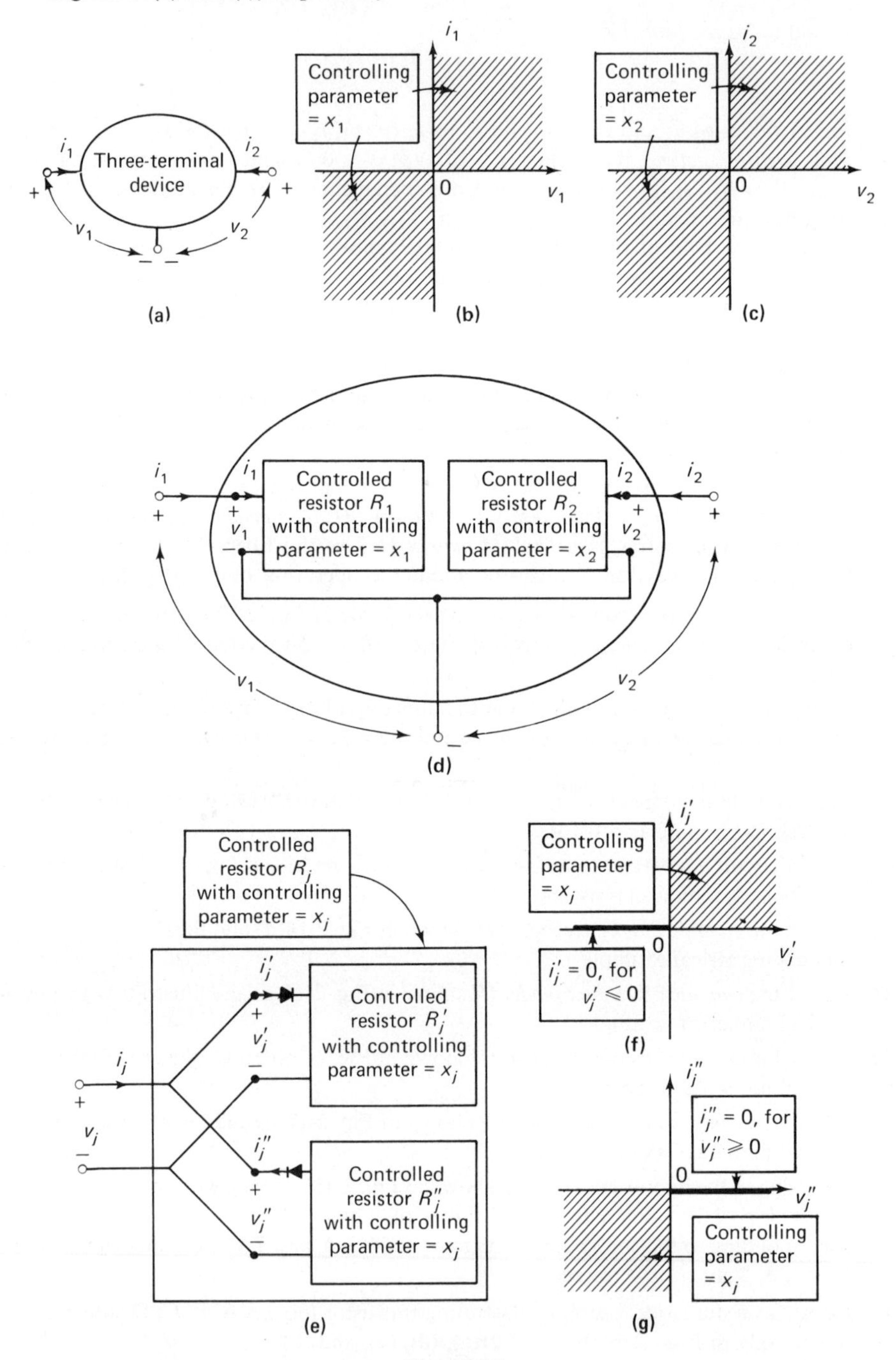

Fig. P2-16

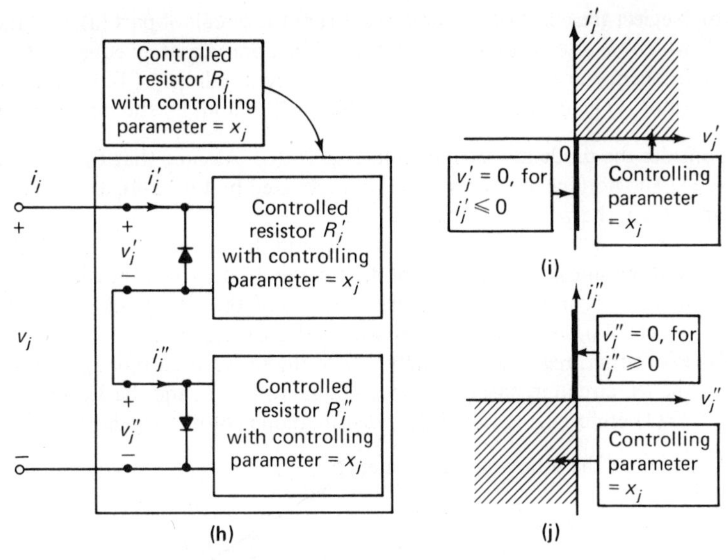

Fig. P2-16 (cont.)

(a) Show how the input and output characteristic curves can be modeled by two controlled resistors R_1 and R_2, as shown in Fig. P2-16(d).
(b) Show how the controlled resistor R_j ($j = 1$ or 2) can be modeled by the canonic circuit shown in Fig. P2-16(e) with the characteristics of R_j' and R_j'', as shown in Fig. P2-16(f) and (g), respectively.
(c) Show how the controlled resistor R_j ($j = 1$ or 2) can be modeled by the canonic circuit shown in Fig. P2-16(h) with the characteristics of R_j' and R_j'' shown in Fig. P2-16(i) and (j), respectively.

2-17. Prove that the two circuits shown in Fig. 2-24 are equivalent to a pair of mutual inductors with self-inductances L_1 and L_2 and mutual inductance M.

2-18. The op-amp circuit model shown in Fig. 2-27 cannot be used with many computer simulation programs based on nodal analysis (such as SPICE) because these programs will not allow voltage-controlled voltage sources, *nonlinear* voltage-controlled current sources, and nonlinear resistors characterized by arbitrary *v-i* curves. Show that these disallowed elements can be modeled in turn by voltage-controlled current sources and junction diodes characterized by the ideal exponential junction law, thereby permitting their use with the above-mentioned computer simulation programs.

2-19. Simulate the results shown in Figs. 2-29, 2-30, and 2-31 for the μA741 op amp using the computer simulation program SPICE.

2-20. A certain junction transistor has the following measured values at the operating point $V_{CE} = 10$ V and $I_E = 2.5$ mA.

$$r_{b'b} = 25\ \Omega \qquad f_T = 1\text{ MHz}$$
$$h_{fe} = 100 \qquad r_{b'c} = 2\text{ M}\Omega$$
$$C_{b'c} = 5\text{ pF} \qquad r_{ce} = 1\text{ M}\Omega$$

(a) Calculate all element values for the hybrid-pi model shown in Fig. 2-9. Refer to Section 2-4 for the expressions needed.

(b) Neglect the effect of $C_{b'c}$ and $r_{b'c}$. From the circuit of part (a), find the frequency at which $|h_{21}| = 1$. (The answer should be approximately equal to f_T.) Also find the frequency at which the short-circuit current gain $|h_{21}|$ is 3 dB down from the dc value. (The product of this 3-dB bandwidth and h_{fe} should be equal to f_T.)

2-21. Consider the diode model shown in Fig. 2-7(b). Assume that $C_t = 0$, $R_B = 0$, and $R_C = \infty$. The nonlinear resistor R_d is characterized by Eq. (2-4), and the diffusion capacitance C_d is characterized by Eq. (2-6).

(a) If $I_s = 10^{-10}$ A, $T = 300$ °K, $M = 1$, and the value of F in Eq. (2-6) is 5×10^7 Hz, find the linear incremental model for the diode, when the diode is forward biased to $I_D = 10$ mA. Find the time constant of the RC circuit. Find the frequency at which $|Z_{RC}|$ is 3 dB down from the value of $|Z_{RC}(0)|$.

(b) Prove that, regardless of the value of the forward current I_D, the time constant of the RC circuit in part (a) is always equal to $1/2\pi F$, and the 3-dB frequency is always F. (Therefore, F is called the *intrinsic* diode cutoff frequency.)

2-22. A diode has the following parameters:

$$R_B = 10\ \Omega \qquad I_s = 10^{-10}\ \text{A}$$
$$M = 1.0 \qquad V_z = 0.8\ \text{V}$$
$$n = 0.5 \qquad C_t(0) = 2\ \text{pF}$$
$$F = 5 \times 10^7\ \text{Hz} \qquad T = 300\ ^\circ\text{K}$$

[Refer to Fig. 2-7 and Eqs. (2-4)–(2-6) for notations.] Construct a linear incremental model for the diode for each of the following operating conditions. Indicate all resistance and capacitance values.

(a) Forward biased: $I_D = 20$ mA.

(b) Reversed biased: $V_D = -10$ V.

Note: Assume that Eq. (2-5) is valid up to $V_D \leq \frac{1}{2}V_z$. For $V_D > \frac{1}{2}V_z$, a linear extrapolation of C_t is made using the slope of $C_t(V_D)$ versus V_D at $V_D = \frac{1}{2}V_z$.

2-23. The *auto transformer* shown in Fig. P2-23(a) is constructed by connecting together one terminal of the primary and secondary windings.

(a) Show that the two circuit models shown in Fig. P2-23(b) and (c) are both equivalent to the auto transformer. *Hint:* Figure P2-23(b) consists of a pair of mutual inductors with one terminal of each winding connected together.

(b) An auto transformer does not possess the usually desirable *isolation* property of the two-port transformer. What then are the reasons for using an auto transformer?

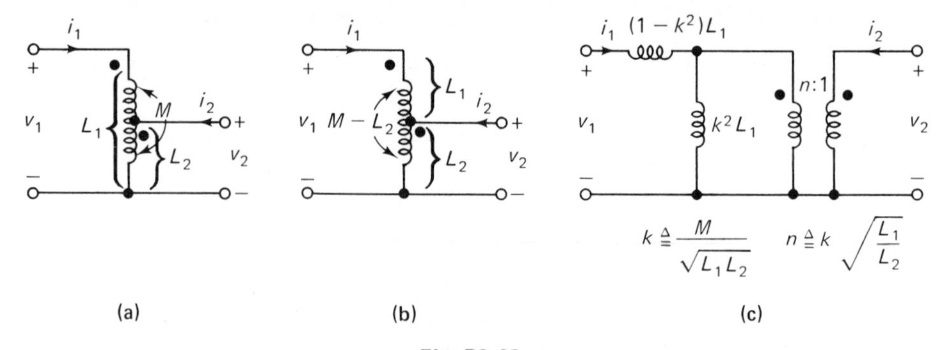

Fig. P2-23

2-24. A nonideal circuit model of a three-port transformer consisting of an interconnection of three mutual inductances is shown in Fig. P2-24.

(a) Find the conditions under which this circuit model reduces to the ideal three-port transformer defined in Table 2-3.

(b) Derive a circuit model for the three-port transformer shown in Fig. P2-24 without using mutual inductances.

Hint: See Problem 2-23.

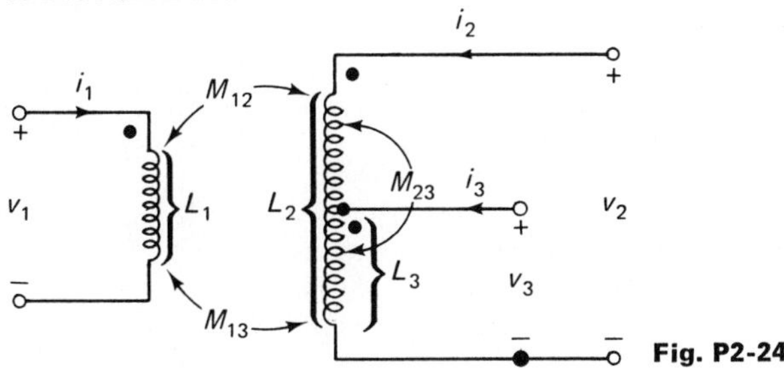

Fig. P2-24

2-25. The *ideal n-port transformer* shown in Fig. P2-25(a) is defined by the hybrid matrix

$$\begin{bmatrix} v_1 \\ v_2 \\ \vdots \\ v_{n-1} \\ i_n \end{bmatrix} = \begin{bmatrix} 0 & 0 & \cdots & \dfrac{n_1}{n_n} \\ 0 & 0 & \cdots & \dfrac{n_2}{n_n} \\ \vdots & \vdots & & \vdots \\ 0 & 0 & \cdots & \dfrac{n_{n-1}}{n_n} \\ \dfrac{-n_1}{n_n} & \dfrac{-n_2}{n_n} & \cdots & 0 \end{bmatrix} \begin{bmatrix} i_1 \\ i_2 \\ \vdots \\ i_{n-1} \\ v_n \end{bmatrix}$$

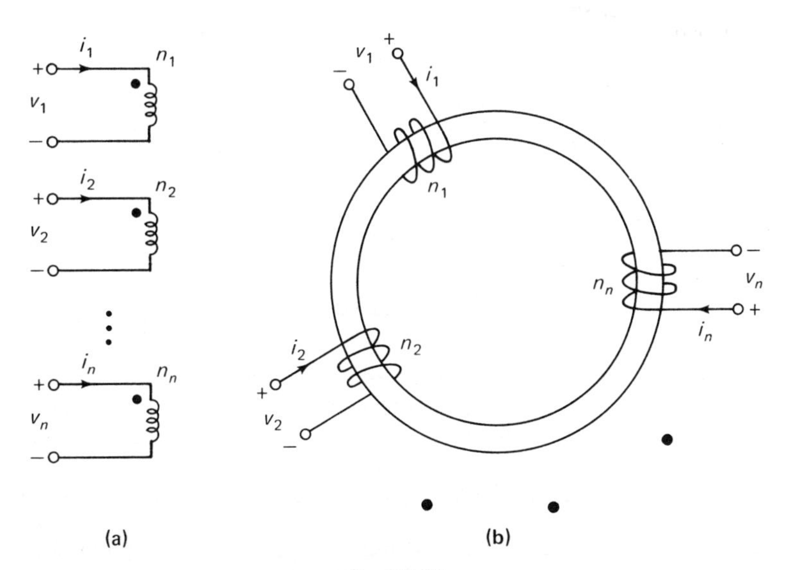

Fig. P2-25

(a) Derive a circuit model for the ideal *n-port transformer*.
(b) A real n-port transformer is usually constructed by winding n coils around an iron ring, as shown in Fig. P2-25(b). Derive a nonideal circuit model without using mutual inductances.
(c) Find the conditions under which your nonideal model reduces to the ideal model.

2-26. The *ideal n-winding transformer* shown in Fig. P2-26(a) is defined by the hybrid matrix

$$\begin{bmatrix} i_1 \\ i_2 \\ \vdots \\ i_{n-1} \\ v_n \end{bmatrix} = \begin{bmatrix} 0 & 0 & \cdots & \frac{n_n}{n_1} \\ 0 & 0 & \cdots & \frac{n_n}{n_2} \\ \vdots & \vdots & & \vdots \\ 0 & 0 & \cdots & \frac{n_n}{n_{n-1}} \\ \frac{-n_n}{n_1} & \frac{-n_n}{n_2} & \cdots & 0 \end{bmatrix} \begin{bmatrix} v_1 \\ v_2 \\ \vdots \\ v_{n-1} \\ i_n \end{bmatrix}$$

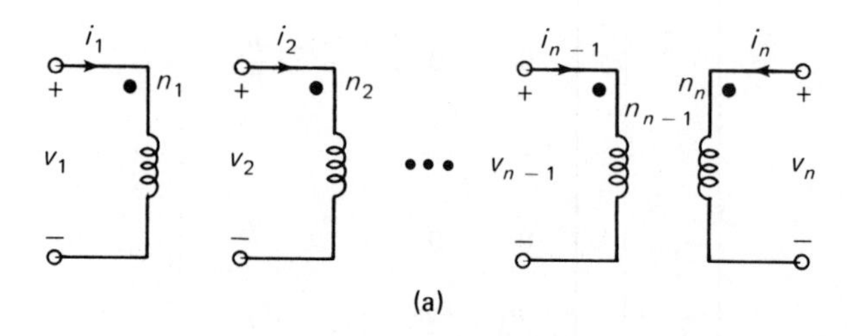

(a)

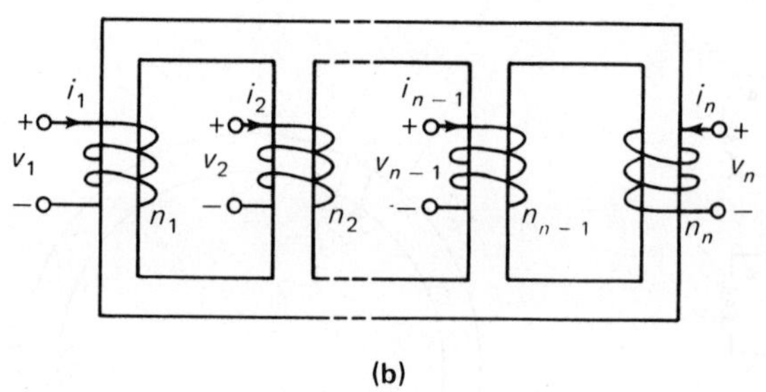

(b)

Fig. P2-26

(a) Derive a circuit model for the ideal *n-winding transformer*.
(b) A real n-winding transformer is usually constructed by winding n coils around n parallel iron legs, as in the structure shown in Fig. P2-26(b). Derive a nonideal circuit model without using mutual inductances.
(c) Find the conditions under which your nonideal model reduces to the ideal model.

2-27. A *circulator* is a microwave device with interesting power transfer properties. The ideal three-port circulator shown in Fig. P2-27 is defined by the open-circuit resistance matrix

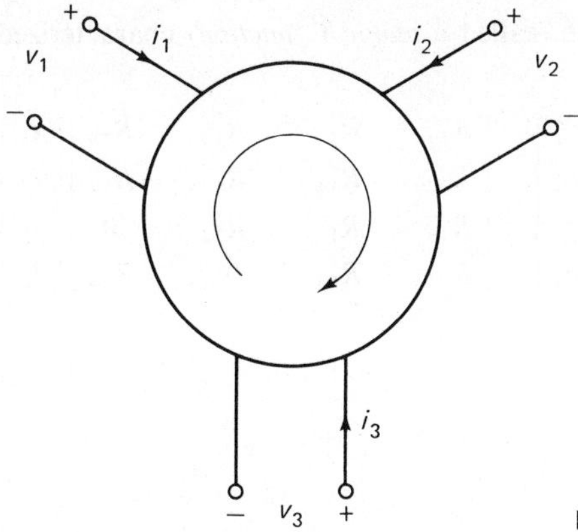

Fig. P2-27

$$\begin{bmatrix} v_1 \\ v_2 \\ v_3 \end{bmatrix} = \begin{bmatrix} 0 & R & -R \\ -R & 0 & R \\ R & -R & 0 \end{bmatrix} \begin{bmatrix} i_1 \\ i_2 \\ i_3 \end{bmatrix}$$

(a) Derive a circuit model of the ideal three-port circulator using controlled sources.

(b) Show that all power entering port 1 is delivered to port 2 if ports 2 and 3 are terminated in an R-Ω linear resistor. Similarly, show that all power entering port 2 is delivered to port 3 if ports 3 and 1 are terminated in an R-Ω linear resistor, and all power entering port 3 is delivered to port 1 if ports 1 and 2 are terminated in an R-Ω linear resistor. Assume the input consists of a voltage source in series with an R-Ω linear resistor.

2-28. Prove that the resistive linear four-port shown in Fig. P2-28 can be used to model an

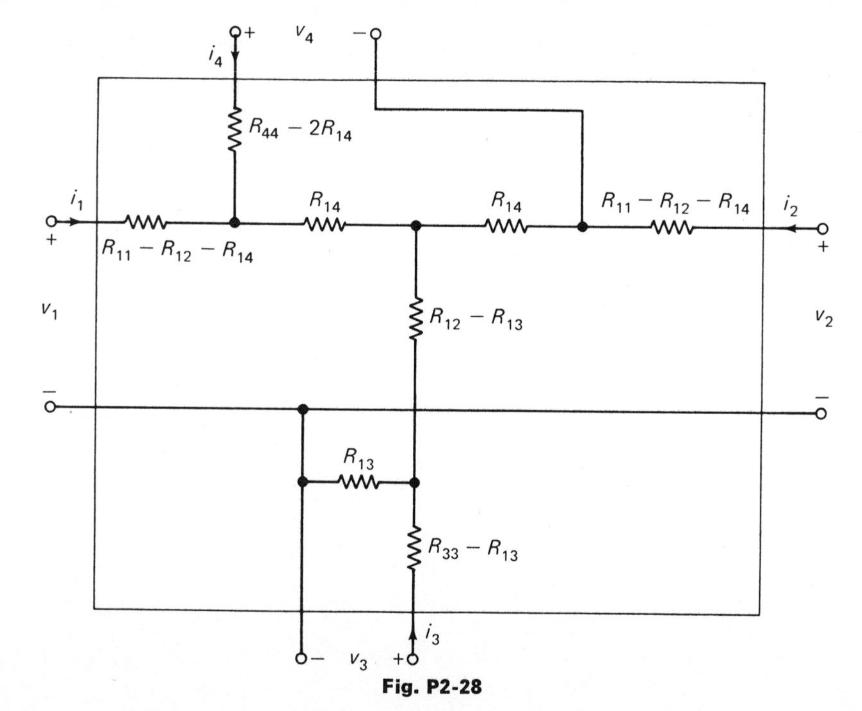

Fig. P2-28

ideal microwave device (called a *magic T junction*) characterized by the open-circuit resistance matrix

$$\begin{bmatrix} v_1 \\ v_2 \\ v_3 \\ v_4 \end{bmatrix} = \begin{bmatrix} R_{11} & R_{12} & R_{13} & R_{14} \\ R_{12} & R_{11} & R_{13} & -R_{14} \\ R_{13} & R_{13} & R_{33} & 0 \\ R_{14} & -R_{14} & 0 & R_{44} \end{bmatrix} \begin{bmatrix} i_1 \\ i_2 \\ i_3 \\ i_4 \end{bmatrix}.$$

CHAPTER 3

Network Topology: The Key to Computer Formulation of the Kirchhoff Laws

3-1 WHAT IS NETWORK TOPOLOGY?

Any lumped network obeys three basic laws: the Kirchhoff voltage law (KVL), the Kirchhoff current law (KCL), and the elements' law (branch characteristics). The first two laws, KVL and KCL, are linear algebraic constraints on branch voltages and currents, arising from the interconnection of branches, and are independent of the branch characteristics.

Network topology deals with those properties of lumped networks which are related to the interconnection of branches only. It is a topic in a powerful branch of mathematics known as graph theory. We do not intend in this chapter to delve in network topology. Good references with emphasis on circuit applications are available ([1]–[5]). Instead, we shall merely present those essential concepts and properties which are required in the development of general-purpose computer simulation programs. Actual applications of the results of this chapter for computer formulation of KVL and KCL equations will appear in later chapters.

As shown in Chapter 2, a large class of lumped networks, both linear and nonlinear, can be modeled as an interconnection of two-terminal elements with specified element characteristics. A complete description of the network model must then contain the following information:

1. How the branches are connected.
2. The reference directions for branch currents and voltages.
3. The branch characteristics.

One natural and simple way to depict items 1 and 2 is to draw a *directed graph* G_d associated with the given network N, according to the following rule: replace each two-terminal element by a line segment, called a branch, with an arrow in the same direction as the assumed positive current through that branch. This arrow also serves as the branch voltage reference: the + voltage terminal is assumed to be at the tail of the arrow. By adopting such an "associated reference," there is no need to carry two sets of references, one for the currents and the other for the voltages. Thus, the directed graph G_d gives complete information for items 1 and 2. For example, Fig. 3-1(a) shows a network N, and Fig. 3-1(b) shows the directed graph G_d associated with N.

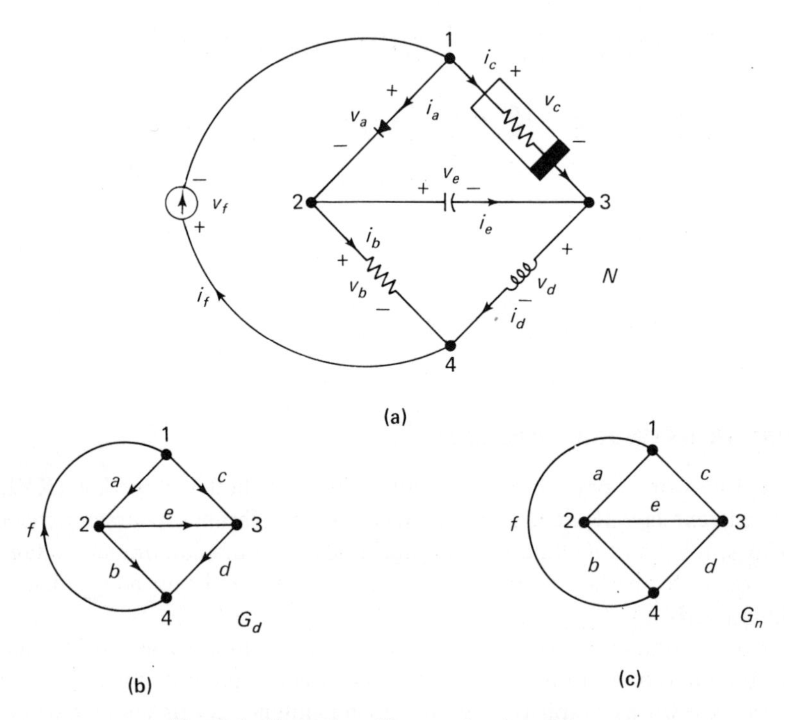

Fig. 3-1. Network and its associated graphs.

When we are not concerned with the reference directions of branch voltages and currents, all the arrows in G_d may be removed. The resultant simpler graph is called the *undirected graph* (nonoriented graph) associated with the network N and is denoted by G_n. For example, the nonoriented graph G_n associated with Fig. 3-1(a) is shown in Fig. 3-1(c). One should regard the network model N, the directed graph G_d, and the undirected graph G_n as three closely related entities. For the sake of brevity, we shall often refer to either G_d or G_n as the *network graph* of N. The context should make it clear which is meant.

We shall next introduce some basic concepts in network topology. In doing so, we find it more convenient to use the undirected graph G_n instead of the directed graph G_d, even though the latter is used in formulating KCL and KVL equations.

Definitions:

Path. *A set of branches*

$$b_1, b_2, \ldots, b_n$$

in G_n is called a path between two nodes V_j and V_k if the branches can be labeled such that

1. *Consecutive branches b_i and b_{i+1} always have a common endpoint.*
2. *No node of G_n is the endpoint of more than two branches in the set.*
3. *V_j is the the endpoint of exactly one branch in the set, and so is V_k.*

In plain language, a path is just a route between two nodes without excursions enroute. For example, in Fig. 3-2, branches (d h i b) form a path between nodes 1 and 2; branches (e g j f) do not form a path because condition 2 is violated; branches (e f h i c) do not form a path because condition 1 is violated.

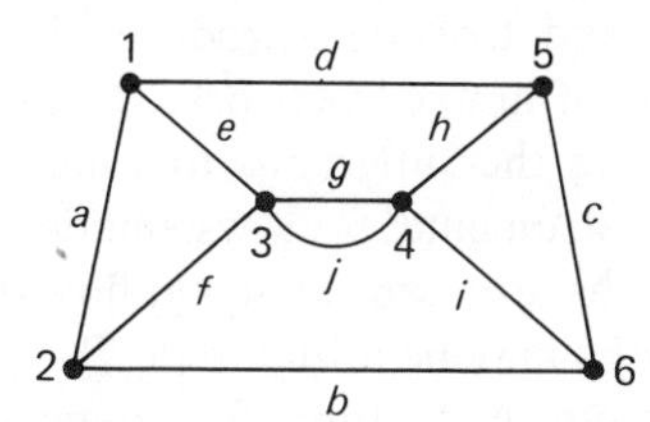

Fig. 3-2. Concepts of path, loop, tree, and cutset.

Connected Graph. *An undirected graph G_n is said to be connected if there exists a path between any two nodes of the graph. A network N or a directed graph G_d is said to be connected if the associated graph G_n is connected.*

Loop (Circuit). *A subgraph G_s of a graph G_n is called a loop if*

1. *G_s is connected.*
2. *Every node of G_s has exactly two branches of G_s incident at it.*

For example, in Fig. 3-2 the branches (a b c d) form a loop; branches (a e f g j) do not form a loop because condition 2 is violated; branches (a e f h i c) do not form a loop because condition 1 is violated.

Tree. *A subgraph G_s of a connected graph G_n is called a tree if*

1. *G_s is connected.*
2. *G_s contains all nodes of G_n.*
3. *G_s has no loops.*

For example, in Fig. 3-2, the branches (a e d g i) form a tree; branches (a e h c) do not form a tree because condition 1 is violated; branches (a b c h) do not form a tree because condition 2 is violated; branches (a b c d e g) do not form a tree because condition 3 is violated.

The branches that belong to a tree T are called *tree branches*, and those which do not belong to a tree T are called *links* (chords). All the links of a given tree T form what is called a *cotree* T_c with respect to the tree T.

It can be shown that for a connected graph G_n with n nodes, any tree T has exactly $n - 1$ tree branches. Furthermore, if $n - 1$ branches of G_n can be chosen such that no loop is formed, then the $n - 1$ branches constitute a tree of G_n [1].

Cutset. *A set of branches of a connected graph G_n is said to be a cutset if*

1. *The removal of the set of branches (but not their endpoints) results in a graph that is not connected.*
2. *After the removal of the set of branches, the restoration of any one branch from the set will result in a connected graph again.*

For example, in Fig. 3-2, branches ($a \quad e \quad d$) form a cutset; so do branches ($d \quad g \quad j \quad b$). However, branches ($e \quad h \quad g \quad f \quad b$) do not form a cutset because condition 1 is violated; and branches ($a \quad e \quad d \quad c$) do not form a cutset because condition 2 is violated. Note that the number of branches in a cutset is not fixed, as it is for a tree.

The concepts of path, loop, cutset, and tree are introduced here because of their important applications in the analysis of lumped networks. Loops are the subgraphs to which we apply KVL. Cutsets are the subgraphs to which we apply the generalized KCL. The concept of a tree is instrumental for a systematic formulation of independent KCL and KVL equations. Paths are used for signal-flow-graph analysis of linear systems. Other applications of paths may be found in [6, 7].

In the next few sections, we shall introduce the three fundamental matrices $\boldsymbol{A}$, $\boldsymbol{B}$, and $\boldsymbol{D}$ associated with a directed graph.

3-2 INCIDENCE MATRIX

Although the directed graph G_d defined in Section 3-1 completely describes the interconnection and the reference directions of the branches of a lumped network, it is not in a form suitable for storing in a digital computer. We must resort to other means. Matrices have been found most useful for such purposes.

The information contained in a directed graph G_d can be completely stored in a matrix called an *incidence matrix*. Although several kinds of incidence matrices exist, the one most widely used in lumped-network analysis is the *node-branch* incidence matrix, or simply the incidence matrix. A precise definition follows.

Definition: Incidence Matrix $\boldsymbol{A}_a$. *For a directed graph G_d with n nodes and b branches, we define the incidence matrix to be an $n \times b$ matrix*

$$\boldsymbol{A}_a = [a_{ij}]$$

where

$a_{ij} = 1$ *if branch j is incident at node i, and the arrow is pointing away from node i*

$a_{ij} = -1$ *if branch j is incident at node i, and the arrow is pointing toward node i*

$a_{ij} = 0$ *if branch j is not incident at node i*

For example, for the directed graph of Fig. 3-1(b)

$$A_a = \begin{array}{c} \\ \text{node } 1 \\ 2 \\ 3 \\ 4 \end{array} \begin{array}{c} \begin{array}{cccccc} a & b & c & d & e & f \end{array} \\ \begin{bmatrix} 1 & 0 & 1 & 0 & 0 & -1 \\ -1 & 1 & 0 & 0 & 1 & 0 \\ 0 & 0 & -1 & 1 & -1 & 0 \\ 0 & -1 & 0 & -1 & 0 & 1 \end{bmatrix} \end{array} \begin{array}{c} \text{incident branches} \\ (acf) \\ (abe) \\ (cde) \\ (bdf) \end{array}$$

Since in this book we consider only directed graphs without self-loops,[1] every branch is connected to two distinct nodes. It follows that every column of A_a has exactly two nonzero elements, a 1 and a -1, with the rest being zeros. We can actually delete any one row of A_a without losing information, because this deleted row may be restored correctly whenever necessary by observing the rule that every column of A_a must add up to zero.

A matrix obtained from A_a by deleting any one row is called a *reduced incidence matrix* and is denoted by A. For distinction, A_a will be called the *complete incidence matrix*. When no confusion can arise, either A_a or A will be referred to simply as the incidence matrix.

Let the branch currents of the network N be represented by a column vector $i(t)$ of order $b \times 1$. Let the columns of A_a and the rows of i be arranged in the same branch order, that is, the kth column of A_a and kth row of i correspond to the same branch of G_d. Then KCL, when applied to all nodes, can be expressed very compactly as one matrix equation

$$A_a i = \mathbf{0} \tag{3-1}$$

For example, writing Eq. (3-1) for Fig. 3-1(b), we have

$$\begin{array}{c} \\ 1 \\ 2 \\ 3 \\ 4 \end{array} \begin{array}{c} \begin{array}{cccccc} a & b & c & d & e & f \end{array} \\ \begin{bmatrix} 1 & 0 & 1 & 0 & 0 & -1 \\ -1 & 1 & 0 & 0 & 1 & 0 \\ 0 & 0 & -1 & 1 & -1 & 0 \\ 0 & -1 & 0 & -1 & 0 & 1 \end{bmatrix} \end{array} \begin{bmatrix} i_a \\ i_b \\ i_c \\ i_d \\ i_e \\ i_f \end{bmatrix} = \begin{bmatrix} 0 \\ 0 \\ 0 \\ 0 \end{bmatrix}$$

By writing out the equations in scalar form, we obtain four equations, which are indeed the KCL applied to nodes 1, 2, 3, and 4.

The set of equations represented by (3-1) is not linearly independent. In fact, any one equation in (3-1) is implied by the remaining $n - 1$ equations. Since in network analysis we want to use only independent equations, we ask whether the remaining $n - 1$ equations are linearly independent. The answer is given by Theorem 3-1.

Theorem 3-1. *For a* ***connected*** *graph G_d, the set of all rows of any reduced incidence matrix A is linearly independent.*

[1]A branch is called a *self-loop* if the two endpoints of the branch are the same.

The proof of this important theorem is given in Appendix 3A. As a direct consequence of Theorem 3-1, we have Corollary 3-1.

Corollary 3-1. *The maximum set of independent KCL equations, obtained from the nodes of a connected network N, can be expressed as*

$$\boxed{A i = 0} \tag{3-2}$$

Since A contains the complete information about G_d and G_n, we should be able to determine whether a set of branches forms a tree or not by suitable matrix operations on A. Theorem 3-2 provides a basis for such matrix operations.

Theorem 3-2. *Let A be the reduced incidence matrix of a connected graph G_d with n nodes. Then $n-1$ columns of A are linearly independent if and only if the branches corresponding to these columns form a tree in G_n.*

The proof of this theorem is given in Appendix 3B. Corollary 3-2 is a direct consequence of Theorem 3-2.

Corollary 3-2. *Let A be partitioned as*

$$A = [A_T \mid A_L]$$

where the columns of A_T correspond to the tree branches of a chosen tree T, and the columns of A_L correspond to the links. Then $\det A_T \neq 0$.

For example, in Fig. 3-1(b), if a tree T is chosen to consist of branches $(d \quad e \quad f)$, and A is obtained from A_a by deleting the row corresponding to node 4, we have

$$A = [A_T \mid A_L]$$

$$= \begin{array}{c} \\ 1 \\ 2 \\ 3 \end{array} \begin{array}{c} \begin{array}{cccccc} d & e & f & a & b & c \end{array} \\ \left[\begin{array}{ccc|ccc} 0 & 0 & -1 & 1 & 0 & 1 \\ 0 & 1 & 0 & -1 & 1 & 0 \\ 1 & -1 & 0 & 0 & 0 & -1 \end{array}\right] \end{array}$$

and

$$|A| \triangleq \det A_T = \begin{vmatrix} 0 & 0 & -1 \\ 0 & 1 & 0 \\ 1 & -1 & 0 \end{vmatrix} = 1 \neq 0$$

where the symbols $|M|$ and $\det M$ will be used interchangeably in this book to denote the *determinant* of the matrix M.

In this book, whenever the columns (rows) of a matrix are partitioned into two blocks corresponding to a tree and its cotree, we shall always write the tree first.

3-3 LOOP MATRIX

To express KVL equations compactly as a single matrix equation, we need to introduce another matrix, the *loop matrix* (*circuit matrix*) associated with a directed graph.

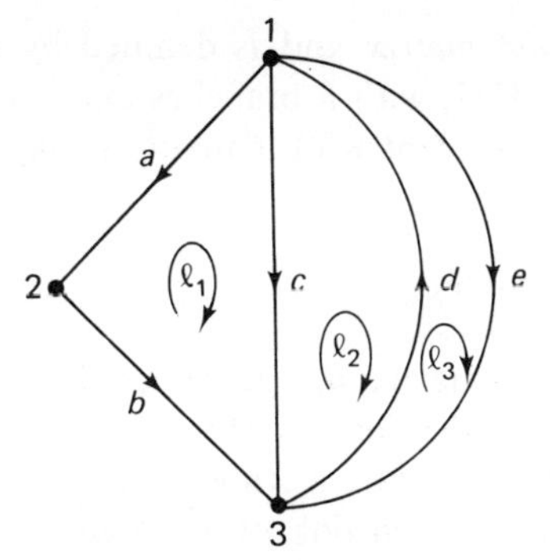

Fig. 3-3. Construction of the loop matrix for a directed graph.

Consider first the undirected graph G_n. Let there be n_l loops in G_n. Each loop is now assigned one of two possible orientations, also indicated by an arrow, as shown in Fig. 3-3. These are now called *oriented loops.*

Definition: Loop Matrix $\boldsymbol{B}_a$. *For a directed graph G_d with b branches and n_l oriented loops, we define the loop matrix to be an $n_l \times b$ matrix*

$$\boldsymbol{B}_a = [b_{ij}]$$

where

$$b_{ij} = 1 \textit{ if branch } j \textit{ is in loop } i, \textit{ and their directions agree}$$

$$b_{ij} = -1 \textit{ if branch } j \textit{ is in loop } i, \textit{ and their directions oppose}$$

$$b_{ij} = 0 \textit{ if branch } j \textit{ is not in loop } i$$

For example, there are six oriented loops in Fig. 3-3 (only loops 1, 2, and 3 are shown), and the loop matrix is

$$\boldsymbol{B}_a = \begin{array}{r} \\ \text{loop } 1 \\ 2 \\ 3 \\ 4 \\ 5 \\ 6 \end{array} \begin{array}{c} \begin{array}{ccccc} a & b & c & d & e \end{array} \\ \begin{bmatrix} -1 & -1 & 1 & 0 & 0 \\ 0 & 0 & -1 & -1 & 0 \\ 0 & 0 & 0 & 1 & 1 \\ -1 & -1 & 0 & -1 & 0 \\ -1 & -1 & 0 & 0 & 1 \\ 0 & 0 & -1 & 0 & 1 \end{bmatrix} \end{array} \begin{array}{c} \text{branches in the loop} \\ (abc) \\ (cd) \\ (de) \\ (abd) \\ (abe) \\ (ce) \end{array}$$

Recall that KVL states that *the algebraic sum of voltages around any loop of a lumped network is zero at all times.* If we represent the branch voltages by a $b \times 1$ column vector $\boldsymbol{v}(t)$ such that rows of $\boldsymbol{v}$ are in the same branch order as columns of $\boldsymbol{B}_a$, then KVL when applied to all loops may be expressed very compactly as

$$\boldsymbol{B}_a\boldsymbol{v} = \boldsymbol{0} \tag{3-3}$$

The reader may easily verify the validity of this assertion. When written out in scalar form, Eq. (3-3) leads to a set of n_l equations; n_l may be very large even for a network of moderate size. However, in network analysis, we do not need all these n_l equations. Any maximum set of independent equations will suffice. This leads to a more useful submatrix of $\boldsymbol{B}_a$. Any submatrix of $\boldsymbol{B}_a$ that consists of the maximum number of independent rows of $\boldsymbol{B}_a$ is called a *basic loop matrix* and is denoted by $\boldsymbol{B}_b$. We shall presently show that for a connected graph G_d with b branches and n nodes, $\boldsymbol{B}_b$ has $b - n + 1$ rows. Thus, the $b - n + 1$ independent KVL equations may be expressed very compactly as

$$\boxed{\boldsymbol{B}_b \boldsymbol{v} = \boldsymbol{0}} \tag{3-4}$$

For connected planar networks, the matrix $\boldsymbol{B}_b$ can be easily constructed by choosing the $b - n + 1$ loops to be the meshes or "windows." For nonplanar networks, this method is not applicable. A systematic method of constructing a basic loop matrix is through the aid of a tree T. Each link of a cotree T_c together with the unique path through the tree T forms a loop, the *fundamental loop* for that link, with respect to the chosen tree T. The orientation of the loop is arbitrarily chosen to coincide with that of the link. For a connected graph G_d with n nodes and b branches, there are $b - n + 1$ links, and hence $b - n + 1$ fundamental loops. A submatrix of $\boldsymbol{B}_a$ constructed by the use of these $b - n + 1$ fundamental loops is called a *fundamental loop matrix*, and is denoted by $\boldsymbol{B}_f$. Since in this book we shall be using $\boldsymbol{B}_f$ practically all the time, the subscript f will be omitted to make the notation simpler.

For example, in Fig. 3-3, if T is chosen to consist of branches (a, d), the fundamental loop matrix is

$$\boldsymbol{B}_f = \boldsymbol{B} = \begin{array}{c} \begin{matrix} a & d & b & c & e \end{matrix} \\ \left[\begin{array}{cc:ccc} 1 & 1 & 1 & 0 & 0 \\ 0 & 1 & 0 & 1 & 0 \\ 0 & 1 & 0 & 0 & 1 \end{array}\right] \end{array} \tag{3-5}$$

From the way the $\boldsymbol{B}$ matrix is constructed, it should be obvious that any $\boldsymbol{B}$ matrix can be partitioned as

$$\boldsymbol{B} = [\boldsymbol{B}_T \mid \boldsymbol{1}] \tag{3-6}$$

where the columns of $\boldsymbol{B}_T$ correspond to tree branches, and the columns of the identity matrix $\boldsymbol{1}$, of order $b - n + 1$, correspond to links. That the rows of $\boldsymbol{B}$ are linearly independent is obvious from the presence of $\boldsymbol{1}$ in Eq. (3-6). We now ask: Can any other rows of $\boldsymbol{B}_a$ be appended to $\boldsymbol{B}$ while still maintaining linear independence? Or, equivalently, can we write more than $b - n + 1$ independent equations with KVL? The next two theorems provide a definite answer to the question.

Theorem 3-3. *If the columns of $\boldsymbol{A}_a$ and $\boldsymbol{B}_a$ are arranged in the same branch order, then for all i and j*

$$[\text{row } i \text{ of } \boldsymbol{B}_a] \cdot [\text{row } j \text{ of } \boldsymbol{A}_a]^t = 0 \tag{3-7}$$

Proof: The left side of Eq. (3-7) may be written as

$$S = \sum_{k=1}^{b} b_{ik} a_{jk}$$

We increment k ($k = 1, 2, \ldots, b$) to evaluate this sum of products. The first nonzero product is encountered when some branch is incident at node j and is in loop i, and the product $b_{ik} a_{jk}$ is ± 1. However, when node j is in loop i, there will be exactly one more branch that is incident at node j and is in loop i. By exhausting all possible orientations of these two branches and loop i, we find that the other nonzero product is always ∓ 1. Then $S = 0 + \ldots \pm 1 + 0 + \ldots \mp 1 + 0 + \ldots = 0$. Q.E.D.

Corollary 3-3. *All the following relationships are valid:*

$$\boldsymbol{B}_a\boldsymbol{A}_a^t = \boldsymbol{0}, \qquad \boldsymbol{A}_a\boldsymbol{B}_a^t = \boldsymbol{0}$$
$$\boldsymbol{B}_a\boldsymbol{A}^t = \boldsymbol{0}, \qquad \boldsymbol{A}\boldsymbol{B}_a^t = \boldsymbol{0}$$
$$\boldsymbol{B}\boldsymbol{A}^t = \boldsymbol{0}, \qquad \boldsymbol{A}\boldsymbol{B}^t = \boldsymbol{0}$$

Theorem 3-4. *For a connected graph G_a with n nodes and b branches, the maximum number of rows of $\boldsymbol{B}_a$ that are linearly independent is $b - n + 1$.*

Proof: A connected graph has at least one tree T. For a tree T, we can construct the fundamental loop matrix $\boldsymbol{B}$. $\boldsymbol{B}$ has $b - n + 1$ rows that are linearly independent. Partition $\boldsymbol{B}_a$ matrix as

$$\boldsymbol{B}_a = \begin{bmatrix} \boldsymbol{B} \\ \hline \boldsymbol{B}_2 \end{bmatrix} = \left[\begin{array}{c|c} \boldsymbol{B}_T & \boldsymbol{1} \\ \hline \boldsymbol{B}_{2T} & \boldsymbol{B}_{2L} \end{array}\right]$$

We shall prove the theorem by showing that every row of $\boldsymbol{B}_2$ is a linear combination of the rows of $\boldsymbol{B}$.

According to Corollary 3-3,

$$\boldsymbol{B}_a\boldsymbol{A}^t = \left[\begin{array}{c|c} \boldsymbol{B}_T & \boldsymbol{1} \\ \hline \boldsymbol{B}_{2T} & \boldsymbol{B}_{2L} \end{array}\right] [\boldsymbol{A}_T \mid \boldsymbol{A}_L]^t$$
$$= \left[\begin{array}{c|c} \boldsymbol{B}_T & \boldsymbol{1} \\ \hline \boldsymbol{B}_{2T} & \boldsymbol{B}_{2L} \end{array}\right] \begin{bmatrix} \boldsymbol{A}_T^t \\ \hline \boldsymbol{A}_L^t \end{bmatrix} = \boldsymbol{0}$$

from which we obtain two matrix equations,

$$\boldsymbol{B}_T\boldsymbol{A}_T^t + \boldsymbol{A}_L^t = \boldsymbol{0} \tag{3-8}$$
$$\boldsymbol{B}_{2T}\boldsymbol{A}_T^t + \boldsymbol{B}_{2L}\boldsymbol{A}_L^t = \boldsymbol{0} \tag{3-9}$$

Solve for $\boldsymbol{A}_L^t$ from Eq. (3-8), and substitute the result into Eq. (3-9). We obtain

$$(\boldsymbol{B}_{2T} - \boldsymbol{B}_{2L}\boldsymbol{B}_T)\boldsymbol{A}_T^t = \boldsymbol{0} \tag{3-10}$$

From Corollary 3-2, $|\boldsymbol{A}_T^t| \neq 0$; therefore, the inverse of $\boldsymbol{A}_T^t$ exists. Postmultiplying Eq. (3-10) by $(\boldsymbol{A}_T^t)^{-1}$, we obtain

$$\boldsymbol{B}_{2T} = \boldsymbol{B}_{2L}\boldsymbol{B}_T$$

Hence,

$$\begin{aligned} \boldsymbol{B}_2 &= [\boldsymbol{B}_{2T} \mid \boldsymbol{B}_{2L}] = [\boldsymbol{B}_{2L}\boldsymbol{B}_T \mid \boldsymbol{B}_{2L}] \\ &= \boldsymbol{B}_{2L}[\boldsymbol{B}_T \mid \boldsymbol{1}] = \boldsymbol{B}_{2L}\boldsymbol{B} \end{aligned} \tag{3-11}$$

In matrix algebra, it is shown that every row of the product $\boldsymbol{PQ}$ is a linear combination of the rows of $\boldsymbol{Q}$. Equation (3-11) shows that every row of $\boldsymbol{B}_2$ in $\boldsymbol{B}_a$ is a linear combination of the rows of $\boldsymbol{B}$. Hence, the theorem. Q.E.D.

Every fundamental loop matrix $\boldsymbol{B}$ is always a basic loop matrix $\boldsymbol{B}_b$. But the converse is not true. For example, in Fig. 3-3, a basic loop matrix can be constructed using loops l_1, l_2, and l_3. However, no tree T exists in the graph whose fundamental loops are l_1, l_2, and l_3.

Although $\boldsymbol{B}_b$ includes $\boldsymbol{B}$ as a special case, and therefore is more general, the $\boldsymbol{B}$ matrix has the advantage of simple digital computer implementation (see Section 3-6). For this reason, we shall use $\boldsymbol{B}$ exclusively in this book. The maximum number of independent KVL equations obtained with the fundamental loops may be expressed compactly as

$$\boxed{\boldsymbol{Bv} = \boldsymbol{0}} \tag{3-12}$$

3-4 CUTSET MATRIX

The Kirchhoff current law is traditionally stated as follows: *The algebraic sum of currents entering (leaving) any node of any lumped network is zero at all times.*

A more general form of the KCL states that *the algebraic sum of all currents through a cutset, from one part to the other, is zero at all times.*

The second form of the KCL is more general because a cutset may or may not consist of all the branches incident at a node. For example, applying the generalized KCL to Fig. 3-1(b), we can write

$$i_a + i_c - i_f = 0$$
$$i_c + i_e + i_b - i_f = 0$$
$$\text{etc.}$$

The second equation is not the result of applying the KCL to any one node.

To express the generalized KCL equations compactly as a single matrix equation, we introduce another matrix, the *cutset matrix* associated with a directed graph.

Consider first the undirected graph G_n associated with G_d. Let there be n_c cutsets in G_n. Figuratively speaking, the branches of any cutset are the "bridges" joining two "islands" separated by a borderline as depicted in Fig. 3-4(a). For the directed graph G_d, we shall be interested in the "traffic" from island 1 to island 2, or vice versa. Once the "traffic" direction is chosen, it is indicated by an arrow, the cutset reference arrow, on the borderline, as shown in Fig. 3-4(b). The cutset is now called an *oriented cutset.*

Definition: Cutset Matrix $\boldsymbol{D}_a$. *For a directed graph G_d with b branches and n_c oriented cutsets, we define the cutset matrix to be an $n_c \times b$ matrix*

$$\boldsymbol{D}_a = [d_{ij}]$$

where

$$d_{ij} = 1 \textit{ if branch } j \textit{ is in cutset } i\textit{, and their directions agree}$$
$$d_{ij} = -1 \textit{ if branch } j \textit{ is in cutset } i\textit{, and their directions oppose}$$
$$d_{ij} = 0 \textit{ if branch } j \textit{ is not in cutset } i$$

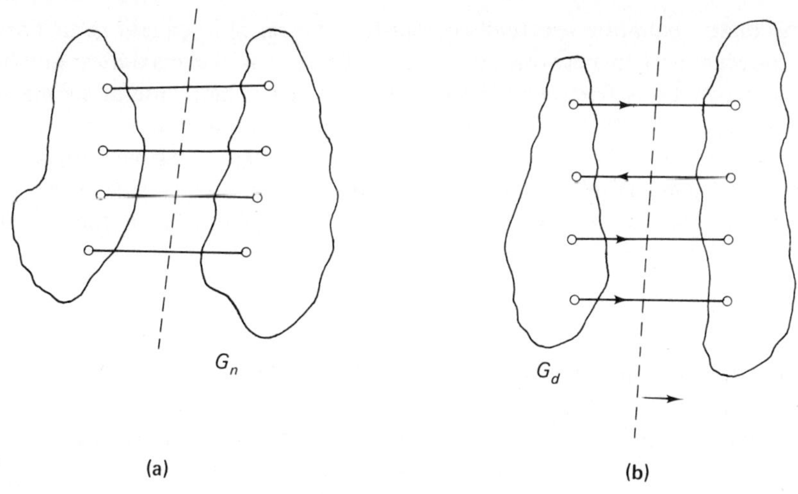

Fig. 3-4. Orientation of a cutset.

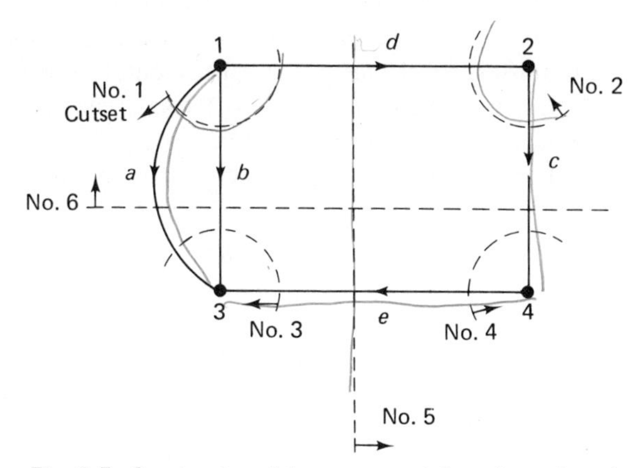

Fig. 3-5. Construction of the cutset matrix for a directed graph.

For example, there are six cutsets in Fig. 3-5, and the cutset matrix is

$$
\boldsymbol{D}_a = \begin{array}{r} \\ \text{cutset } 1 \\ 2 \\ 3 \\ 4 \\ 5 \\ 6 \end{array}
\begin{array}{c} \begin{array}{ccccc} a & b & c & d & e \end{array} \\
\left[\begin{array}{ccccc}
1 & 1 & 0 & 1 & 0 \\
0 & 0 & -1 & 1 & 0 \\
1 & 1 & 0 & 0 & 1 \\
0 & 0 & 1 & 0 & -1 \\
0 & 0 & 0 & 1 & -1 \\
-1 & -1 & -1 & 0 & 0
\end{array}\right] \end{array}
\begin{array}{c} \text{branches in the cutset} \\ (abd) \\ (cd) \\ (abe) \\ (ce) \\ (de) \\ (abc) \end{array}
$$

If the rows of $\boldsymbol{i}$ and columns of $\boldsymbol{D}_a$ are arranged in the same branch order, the generalized KCL, when applied to all cutsets, may be expressed as

$$\boldsymbol{D}_a \boldsymbol{i} = \boldsymbol{0} \tag{3-13}$$

This matrix equation is actually comprised of a set of n_c equations that are not linearly independent. In network analysis, all we need is the maximum number of independent equations from Eq. (3-13). This leads to a more useful submatrix of $\boldsymbol{D}_a$. Any submatrix of $\boldsymbol{D}_a$ that consists of the maximum number of independent rows of $\boldsymbol{D}_a$ is called a *basic cutset matrix* and is denoted by $\boldsymbol{D}_b$. It will be shown shortly that for a connected graph G_d with n nodes any basic cutset matrix $\boldsymbol{D}_b$ has $n - 1$ rows. The maximum number of independent equations obtained from the generalized KCL may be expressed as

$$\boxed{\boldsymbol{D}_b\boldsymbol{i} = \boldsymbol{0}} \qquad (3\text{-}14)$$

The reader should compare Eqs. (3-14) and (3-2), the latter being obtained from the conventional form of the KCL.

A systematic method of constructing a basic cutset matrix is through the aid of a tree T. Each tree branch of T together with some (possibly none) links in the associated cotree T_c forms a cutset, the *fundamental cutset* for that tree branch, with respect to the chosen tree T. The reference arrow for the cutset is arbitrarily chosen to agree with the reference arrow of the tree branch. For a connected graph with n nodes, there are $n - 1$ tree branches, and hence $n - 1$ fundamental cutsets for each chosen tree. A submatrix of $\boldsymbol{D}_a$ constructed with the $n - 1$ fundamental cutsets is called a *fundamental cutset matrix*, and is denoted by $\boldsymbol{D}_f$, or simply $\boldsymbol{D}$.

For example, in Fig. 3-5, if T is chosen to consist of branches a e c, then the fundamental cutset matrix is

$$\begin{array}{c} \\ \boldsymbol{D} = \end{array} \begin{array}{c} \begin{array}{ccccc} a & c & e & b & d \end{array} \\ \left[\begin{array}{ccc|cc} 1 & 0 & 0 & 1 & 1 \\ 0 & 1 & 0 & 0 & -1 \\ 0 & 0 & 1 & 0 & -1 \end{array}\right] \end{array}$$

From the way the $\boldsymbol{D}$ matrix is defined, it is obvious that any $\boldsymbol{D}$ matrix can be partitioned as

$$\boldsymbol{D} = [\boldsymbol{1} \mid \boldsymbol{D}_L] \qquad (3\text{-}15)$$

where the columns of $\boldsymbol{1}$ correspond to tree branches and the columns of $\boldsymbol{D}_L$ correspond to links.

That the rows of $\boldsymbol{D}$ are linearly independent is obvious from the presence of $\boldsymbol{1}$ in Eq. (3-15). We now ask whether it is possible to write more than $n - 1$ independent equations from the generalized KCL. The next two theorems provide the answer.

Theorem 3-5. *If the columns of $\boldsymbol{B}_a$ and $\boldsymbol{D}_a$ are arranged in the same branch order, then for all i and j*

$$[\text{row } i \text{ of } \boldsymbol{B}_a]\cdot[\text{row } j \text{ of } \boldsymbol{D}_a]^t = 0 \qquad [3\text{-}16]$$

Proof: The left side of Eq. (3-16) may be written as

$$S = \sum_{k=1}^{b} b_{ik}d_{jk}$$

We increment k ($k = 1, 2, \ldots, b$) to evaluate this sum of products. A nonzero product is encountered only if some branch is in loop i and is also in cutset j. The product is then equal to 1 or -1. Now if a loop and a cutset have any branch in common, the number of common branches must be an even number $2m$ [1]. Furthermore, exactly m of these common branches have reference arrows agreeing with the cutset reference arrow, and the remaining m branches have reference arrows opposing the cutset reference arrow. Then

$$S = 0 + 0 + \cdots + \underbrace{(1 + 1 + \cdots + 1)}_{m \text{ terms}} + \underbrace{(-1 - 1 - \cdots - 1)}_{m \text{ terms}} = 0$$

Hence, the theorem is proved.

Corollary 3-4. *All the following relationships are valid:*

$$\boldsymbol{D}_a\boldsymbol{B}_a^t = \boldsymbol{0}, \qquad \boldsymbol{B}_a\boldsymbol{D}_a^t = \boldsymbol{0}$$

$$\boldsymbol{D}\boldsymbol{B}^t = \boldsymbol{0}, \qquad \boldsymbol{B}\boldsymbol{D}^t = \boldsymbol{0}$$

$$\boldsymbol{D}_a\boldsymbol{B}^t = \boldsymbol{0}$$

Theorem 3-6. *For a connected graph G_a with n nodes, the maximum number of rows of $\boldsymbol{D}_a$ that are linearly independent is $n - 1$.*

Proof: A connected graph has at least one tree. For a tree T, we can construct the fundamental cutset matrix $\boldsymbol{D}$, which has $n - 1$ rows. Partition $\boldsymbol{D}_a$ as follows:

$$\boldsymbol{D}_a = \begin{bmatrix} \boldsymbol{D} \\ \hline \boldsymbol{D}_2 \end{bmatrix} = \left[\begin{array}{c|c} \boldsymbol{1} & \boldsymbol{D}_L \\ \hline \boldsymbol{D}_{2T} & \boldsymbol{D}_{2L} \end{array}\right]$$

We shall prove the theorem by showing that every row of $\boldsymbol{D}_2$ is a linear combination of rows of $\boldsymbol{D}$.

According to Corollary 3-4,

$$\boldsymbol{D}_a\boldsymbol{B}^t = \left[\begin{array}{c|c} \boldsymbol{1} & \boldsymbol{D}_L \\ \hline \boldsymbol{D}_{2T} & \boldsymbol{D}_{2L} \end{array}\right] [\boldsymbol{B}_T \mid \boldsymbol{1}]^t$$

$$= \left[\begin{array}{c|c} \boldsymbol{1} & \boldsymbol{D}_L \\ \hline \boldsymbol{D}_{2T} & \boldsymbol{D}_{2L} \end{array}\right] \begin{bmatrix} \boldsymbol{B}_T^t \\ \hline \boldsymbol{1} \end{bmatrix} = \boldsymbol{0}$$

from which we obtain two matrix equations:

$$\boldsymbol{B}_T^t + \boldsymbol{D}_L = \boldsymbol{0} \tag{3-17}$$

$$\boldsymbol{D}_{2T}\boldsymbol{B}_T^t + \boldsymbol{D}_{2L} = \boldsymbol{0} \tag{3-18}$$

It follows from Eqs. (3-17) and (3-18) that

$$\boldsymbol{D}_{2L} = \boldsymbol{D}_{2T}\boldsymbol{D}_L$$

Then

$$\begin{aligned} \boldsymbol{D}_2 &= [\boldsymbol{D}_{2T} \mid \boldsymbol{D}_{2L}] \\ &= [\boldsymbol{D}_{2T} \mid \boldsymbol{D}_{2T}\boldsymbol{D}_L] \\ &= \boldsymbol{D}_{2T}[\boldsymbol{1} \mid \boldsymbol{D}_L] = \boldsymbol{D}_{2T}\boldsymbol{D} \end{aligned} \tag{3-19}$$

Since $\boldsymbol{D}_2$ is equal to the product $\boldsymbol{D}_{2T}\boldsymbol{D}$, every row of $\boldsymbol{D}_2$ is a linear combination of the rows of $\boldsymbol{D}$. This completes the proof of Theorem 3-6. Q.E.D.

Every fundamental cutset matrix $\boldsymbol{D}$ is always a basic cutset matrix $\boldsymbol{D}_b$. The converse is not true. For example, in Fig. 3-5 we can construct a basic cutset matrix $\boldsymbol{D}_b$ using cutsets 1, 2, and 3. However, no tree T exists in the graph whose fundamental cutsets are 1, 2, and 3.

Although $\boldsymbol{D}_b$ includes $\boldsymbol{D}$ as a special case, the $\boldsymbol{D}$ matrix lends itself better to digital computer formulation (see Section 3-6). For this reason, $\boldsymbol{D}$ is used in this book whenever the maximum number of independent equations from the generalized KCL is desired. These equations may be expressed compactly as

$$\boxed{\boldsymbol{Di} = \boldsymbol{0}} \tag{3-20}$$

3-5 FUNDAMENTAL RELATIONSHIPS AMONG BRANCH VARIABLES

The Kirchhoff current and voltage laws impose constraints on branch currents and voltages, respectively, of a lumped network. Because of these constraints, only smaller sets of branch currents and voltages are independent, and the remaining branch variables may be expressed in terms of these. We shall now discuss several basic relationships among branch variables.

For a connected network N with n nodes and b branches, we first choose a tree T and partition various matrices as follows:

$$\boldsymbol{A} = [\boldsymbol{A}_T \mid \boldsymbol{A}_L]$$

$$\boldsymbol{B} = [\boldsymbol{B}_T \mid \boldsymbol{1}_\mu]$$

$$\boldsymbol{D} = [\boldsymbol{1}_\rho \mid \boldsymbol{D}_L]$$

$$\boldsymbol{v} = \begin{bmatrix} \boldsymbol{v}_T \\ \hline \boldsymbol{v}_L \end{bmatrix}, \qquad \boldsymbol{i} = \begin{bmatrix} \boldsymbol{i}_T \\ \hline \boldsymbol{i}_L \end{bmatrix}$$

where subscript T stands for tree and L for link, $\rho = n - 1$, and $\mu = b - n + 1$. It is assumed that the rows of $\boldsymbol{v}$ and $\boldsymbol{i}$ and the columns of $\boldsymbol{A}$, $\boldsymbol{B}$, and $\boldsymbol{D}$ are arranged in the same branch order.

First, we observe that link voltages are expressible as linear combinations of tree branch voltages. To find the explicit relationship, we use Eq. (3-12).

$$\boldsymbol{Bv} = [\boldsymbol{B}_T \mid \boldsymbol{1}_\mu]\begin{bmatrix} \boldsymbol{v}_T \\ \hline \boldsymbol{v}_L \end{bmatrix} = \boldsymbol{B}_T\boldsymbol{v}_T + \boldsymbol{v}_L = \boldsymbol{0}$$

Hence,

$$\boxed{\boldsymbol{v}_L = -\boldsymbol{B}_T\boldsymbol{v}_T} \tag{3-21}$$

Similarly, tree branch currents are expressible as linear combinations of link currents. To find the explicit relationship, we use Eq. (3-20).

$$Di = [1_\rho \mid D_L]\begin{bmatrix} i_T \\ \hline i_L \end{bmatrix} = i_T + D_L i_L = 0$$

Hence,

$$\boxed{i_T = -D_L i_L} \tag{3-22}$$

The two matrices B_T and D_L are very simply related. From Corollary 3-4,

$$\begin{aligned} DB^t &= [1_\rho \mid D_L][B_T \mid 1_\mu]^t = [1_\rho \mid D_L]\begin{bmatrix} B_T^t \\ \hline 1_\mu \end{bmatrix} \\ &= B_T^t + D_L = 0 \end{aligned}$$

Therefore,

$$\boxed{D_L = -B_T^t \quad \text{or} \quad B_T = -D_L^t} \tag{3-23}$$

Because of Eq. (3-23), once either D or B has been constructed, the other may be obtained in a very simple manner involving just the transpose of a matrix.

With the aid of Eq. (3-23), we can now express all branch currents in terms of link currents as follows:

$$\begin{aligned} i &= \begin{bmatrix} i_T \\ \hline i_L \end{bmatrix} = \begin{bmatrix} -D_L i_L \\ \hline i_L \end{bmatrix} = \begin{bmatrix} -D_L \\ \hline 1_\mu \end{bmatrix} i_L = \begin{bmatrix} B_T^t \\ \hline 1_\mu \end{bmatrix} i_L \\ &= B^t i_L \end{aligned} \tag{3-24}$$

Similarly, we can express all branch voltages in terms of tree branch voltages as follows:

$$\begin{aligned} v &= \begin{bmatrix} v_T \\ \hline v_L \end{bmatrix} = \begin{bmatrix} v_T \\ \hline -B_T v_T \end{bmatrix} = \begin{bmatrix} 1_\rho \\ \hline -B_T \end{bmatrix} v_T = \begin{bmatrix} 1_\rho \\ \hline D_L^t \end{bmatrix} v_T \\ &= D^t v_T \end{aligned} \tag{3-25}$$

Equation (3-24) is actually a special case of a more general transformation, *loop transformation*, given by

$$i = B_b^t i_m \tag{3-26}$$

where B_b is any basic loop matrix, and i_m is a corresponding set of μ independent currents. Some elements of i_m may not be identifiable as branch currents. However, all elements of i_m may be interpreted as fictitious circulating currents called *loop currents*. Even though Eq. (3-26) is more general than Eq. (3-24), it is not as useful as Eq. (3-24) in digital computer formulation. Therefore, we shall not explore Eq. (3-26) further.

In like manner, Eq. (3-25) is a special case of a more general transformation, *cutset transformation*, given by

$$v = D_b^t v_p \tag{3-27}$$

where D_b is any basic circuit matrix, and elements of v_p may be interpreted as *cutset voltages*. Even though Eq. (3-27) is more general than Eq. (3-24), it is not convenient for digital computer formulation. Therefore, we shall not explore Eq. (3-27) further.

Another extremely useful transformation, *node transformation*, makes use of the reduced incidence matrix A. Consider a connected network N with n nodes. Without any loss of generality, we may choose node n to be the reference or datum node. There are $n - 1$ node-to-datum voltages, which we represent by a column vector

$$\boldsymbol{v}_n \triangleq \begin{bmatrix} v_{1n} \\ v_{2n} \\ \cdot \\ \cdot \\ \cdot \\ v_{(n-1)n} \end{bmatrix}$$

Let A be the reduced incidence matrix whose $n - 1$ rows correspond successively to nodes $1, 2, \ldots, n - 1$. (Note that the datum node is not represented in A.) With such conventions, the branch voltages are related to the node-to-datum voltages explicitly by

$$\boldsymbol{v} = A^t \boldsymbol{v}_n \tag{3-28}$$

Proof: Consider the kth branch. There are only three possible ways of connecting this branch:

1. From node i to node n.
2. From node n to node i.
3. From node i to node j, $i \neq n$, $j \neq n$.

For case 1, the only nonzero element in column k of A is $a_{ik} = 1$. The kth scalar equation from Eq. (3-28) is

$$v_k = v_{in}$$

which is the correct relationship. Case 2 follows similarly, except for $a_{ik} = -1$.

For case 3, the only two nonzero elements in column k of A are $a_{ik} = 1$ and $a_{jk} = -1$. The kth scalar equation from (3-28) is

$$v_k = v_{in} - v_{jn}$$

which again is the correct relationship. Hence, Eq. (3-28) is valid for all cases.

For example, applying the node-transformation Eq. (3-28) to the network graph of Fig. 3-5, we have

$$\begin{bmatrix} v_a \\ v_b \\ v_c \\ v_d \\ v_e \end{bmatrix} = \begin{bmatrix} 1 & 0 & -1 \\ 1 & 0 & -1 \\ 0 & 1 & 0 \\ 1 & -1 & 0 \\ 0 & 0 & -1 \end{bmatrix} \begin{bmatrix} v_{14} \\ v_{24} \\ v_{34} \end{bmatrix}$$

or, in scalar form,

$$v_a = v_{14} - v_{34}$$

$$v_b = v_{14} - v_{34}$$

$$v_c = v_{24}$$
$$v_d = v_{14} - v_{24}$$
$$v_e = -v_{34}$$

which are indeed the correct relationships.

Unlike the fundamental-cutset transformation Eq. (3-25), in which all voltages involved are branch voltages, some of the voltages in $\boldsymbol{v}_n$ may not be identifiable as branch voltages; v_{14} in our example illustrates this point.

3-6 COMPUTER GENERATION OF TOPOLOGICAL MATRICES *A*, *B*, AND *D*

In the preceding sections we have shown that independent KVL equations may be expressed as $\boldsymbol{Bv} = \boldsymbol{0}$, and KCL equations as $\boldsymbol{Ai} = \boldsymbol{0}$ or $\boldsymbol{Di} = \boldsymbol{0}$. In constructing $\boldsymbol{B}$ and $\boldsymbol{D}$ matrices, we rely on the inspection of the network graph to find fundamental loops and cutsets. We shall now show how the $\boldsymbol{A}$, $\boldsymbol{B}$, and $\boldsymbol{D}$ matrices may be generated by a digital computer. This is an essential step in the development of some user-oriented, network-analysis programs.

The generation of $\boldsymbol{A}_a$ (and hence $\boldsymbol{A}$) is extremely simple. We assign consecutive integers, called branch numbers (usually starting from 1), to the branches of a network graph. Similarly, the nodes of a network graph are assigned node numbers, which are usually consecutive integers starting from 1 or 0. If branch k is connected between nodes i and j, with the reference arrow pointing toward node j, we describe the information by a triplet (k, i, j). Each triplet of integers may be punched on part of a data card and read into the computer. Two nonzero elements of $\boldsymbol{A}_a$ are then generated: $a_{ik} = 1$ and $a_{jk} = -1$. (Initially, $\boldsymbol{A}_a = \boldsymbol{0}$.)

Storing the information of a directed graph G_d by means of $\boldsymbol{A}_a$ is very inefficient in regard to the use of computer memory. For example, if G_d has 50 nodes and 200 branches, then $\boldsymbol{A}_a$ will have $50 \times 200 = 10{,}000$ elements and therefore will require 10,000 memory locations (assuming that no additional "packing" is done). Considerable saving in computer storage is possible if $\boldsymbol{A}_a$ is not needed in the form of an $n \times b$ matrix. For example, in the formulation of the nodal admittance matrix (see Chapter 4) the triple product $\boldsymbol{AY}_b\boldsymbol{A}^t$ is needed. In such a case, it is much better to store all the triplets (k, i, j) in the form of a *connection table*, and perform operations on this table. For the 50-node, 200-branch example, the storage needed for the connection table is $200 \times 3 = 600$ locations, as compared with 10,000 locations for the $\boldsymbol{A}_a$ matrix. Taking advantage of the sparsity of a matrix is an important topic, which we shall discuss in greater detail in Chapter 16.

To generate $\boldsymbol{B}$ and $\boldsymbol{D}$, the computer must choose a tree T first. Very often T must be chosen with some preference as to the order of the different types of network elements included in the tree. For example, the particular tree used in formulating the state equations (Chapter 8) is required to have the following preferred order of element types: independent voltage sources, controlled voltage sources, capacitances, resistances, and inductances. In other words, the tree T is to be chosen by including all independent voltage sources[2] as the tree branches first, followed by the controlled voltage sources, capacitances, resistances, and inductances, in this order until a tree is obtained. In this case, the tree may not contain all types of elements. For example, the tree may contain only controlled voltage sources, capacitances, and resistances if the network does not contain independent voltage sources and if a tree can be obtained without the use of inductances.

Therefore, we have two problems to consider:

1. Find a tree T with a given preference of network element types for inclusion in the tree.
2. Find the fundamental loop matrix $\boldsymbol{B}$ and the fundamental cutset matrix $\boldsymbol{D}$ relative to the chosen tree T.

At least two approaches are available for the computer solution of these two problems:

1. Elementary row operations on $\boldsymbol{A}$ or $\boldsymbol{A}_a$.
2. Path search through a connection table.

The details of the first approach will be explained in the next two sections. (See [6] and [9] for a discussion of the second approach.)

3-6-1 Finding a Tree

Consider first the method that makes use of the reduced incidence matrix $\boldsymbol{A}$. Let the columns of $\boldsymbol{A}$ be arranged from left to right in the order corresponding to the desired preference of element types. For example,

$$\boldsymbol{A} = \begin{array}{c} \\ 1 \\ 2 \\ \vdots \\ n-1 \end{array} \overset{\begin{array}{cccccccccc} E1 & C1 & C2 & R1 & R2 & R3 & L1 & L2 & L3 & I1 \end{array}}{\left[\begin{array}{c|c} \\ \quad \boldsymbol{A}_T \quad & \quad \boldsymbol{A}_L \quad \\ \\ \end{array} \right]}$$

According to Theorem 3-2, any $n - 1$ linearly independent columns of $\boldsymbol{A}$ or $\boldsymbol{A}_a$ correspond to a tree of the network graph (assuming a connected graph with n nodes). Our problem is then simply to pick an independent set of $n - 1$ columns, starting from the leftmost column and moving successively to the right.

The recognition of a set of linearly independent columns is made easier by

[2]We assume that no loops are made up of only voltage sources. Otherwise, the network would be either inconsistent or redundant in the sense that some voltage sources could be removed without affecting the solution of the network. Similarly, we assume that no cutsets are made up of only independent current sources. (See Section 6-3 for further discussions of these assumptions.) As a result, independent current sources are never needed as tree branches.

reducing A to the *echelon form* (see Appendix 3C) through a series of elementary row operations. By an elementary row operation, we mean any one of the following three operations:

Type 1. Interchange of two rows.
Type 2. Multiplication of any row by a nonzero scalar constant.
Type 3. Replacement of the jth row by the sum of the jth row and α times the kth row, where $k \neq j$ and α is any scalar constant.

The precise definition of an echelon matrix and the sequence of elementary row operations required to reduce an arbitrary matrix to echelon form are given in Appendix 3C. Here we shall merely illustrate its distinctive features with an example. The following is a typical echelon matrix.

$$\begin{bmatrix} \textcircled{1} & 1 & -1 & 0 & 0 & 1 & -1 \\ 0 & \textcircled{1} & 2 & 1 & 3 & 1 & 0 \\ 0 & 0 & \textcircled{1} & 0 & 2 & 0 & 2 \\ 0 & 0 & 0 & 0 & \textcircled{1} & -1 & 1 \\ 0 & 0 & 0 & 0 & 0 & \textcircled{1} & 1 \\ 0 & 0 & 0 & 0 & 0 & 0 & 0 \end{bmatrix}$$

Note that the dashed line has the form of a stair. Below the stair, all elements are zero. Each element above the stair, and immediately to the right of the vertical dashed lines is always a $+1$ (encircled). Other elements above the stair may be any scalar.

For the present case, the matrix A is $(n - 1) \times b$ and has $n - 1$ independent columns. Following the algorithm described in Appendix 3C, A may be reduced to the following typical echelon form:

$$A_{ech} = \begin{bmatrix} \textcircled{1} & \times & \cdots & \cdots & \cdots & \cdots & \cdots & \cdots & \times \\ 0 & \textcircled{1} & \times & \times & \cdots & \cdots & \cdots & \cdots & \times \\ 0 & 0 & \textcircled{1} & \times & \cdots & \cdots & \cdots & \cdots & \times \\ 0 & 0 & 0 & 0 & \textcircled{1} & \times & \cdots & \cdots & \times \\ 0 & 0 & 0 & 0 & 0 & \textcircled{1} & \times & \cdots & \times \\ 0 & 0 & 0 & 0 & 0 & 0 & \textcircled{1} & \cdots & \times \end{bmatrix}$$

where each $\times$ indicates a $+1$, -1, or 0. Note that in the A_{ech} there are $(n - 1)$ ①'s each of which is the first nonzero element of some row. The $n - 1$ columns of A_{ech} corresponding to these ①'s form an upper triangular matrix and therefore are linearly independent. Since elementary row operations do not affect the linear independence or dependence of a set of columns [8], the corresponding $n - 1$ columns of A are also linearly independent. According to Theorem 3-2, then, the corresponding $n - 1$ branches form a tree. In choosing the $n - 1$ columns from A_{ech}, we have favored columns to the left. Thus, the requirement of the specified element-type preference has been taken care of automatically.

EXAMPLE 3-1. A network graph has the following reduced incidence matrix A. Find a tree using branches corresponding to columns of A as close as possible to the left.

$$A = \begin{array}{c} \\ 1 \\ 2 \\ 3 \\ 4 \end{array} \begin{array}{cccccccc} E1 & C1 & C2 & C3 & L1 & L2 & L3 & L4 \\ \end{array}$$
$$\begin{bmatrix} 1 & 1 & 0 & 1 & 0 & 0 & 0 & 0 \\ 0 & -1 & 1 & 0 & 1 & 0 & 1 & 0 \\ 0 & 0 & 0 & 0 & 0 & 0 & -1 & 1 \\ -1 & 0 & -1 & 0 & 0 & 1 & 0 & 0 \end{bmatrix}$$

Solution: From Appendix 3C, the following sequence of elementary row operations is needed to reduce A to echelon form (row numbers refer to the *updated* matrix after operations):

Add row 1 to row 4.

Multiply row 2 by -1.

Add $-1 \times$ (row 2) to row 4.

Interchange rows 3 and 4.

Multiply row 4 by -1.

The result is

$$A_{\text{ech}} = \begin{bmatrix} \textcircled{1} & 1 & 0 & 1 & 0 & 0 & 0 & 0 \\ 0 & \textcircled{1} & -1 & 0 & -1 & 0 & -1 & 0 \\ 0 & 0 & 0 & \textcircled{1} & 1 & 1 & 1 & 0 \\ 0 & 0 & 0 & 0 & 0 & 0 & \textcircled{1} & -1 \end{bmatrix}$$

(columns: $E1$, $C1$, $C2$, $C3$, $L1$, $L2$, $L3$, $L4$)

from which we see that branches $E1$, $C1$, $C3$, and $L3$ form the required tree. The reader may construct the network graph from the given A matrix and verify that this solution is correct.

3-6-2 Generation of *B* and *D*

Assume that a tree T has been generated and that the matrices $\boldsymbol{A}$, $\boldsymbol{B}$, and $\boldsymbol{D}$ are partitioned as

$$\boldsymbol{A} = [\boldsymbol{A}_T \mid \boldsymbol{A}_L]$$
$$\boldsymbol{B} = [\boldsymbol{B}_T \mid \boldsymbol{1}_\mu]$$
$$\boldsymbol{D} = [\boldsymbol{1}_\rho \mid \boldsymbol{D}_L]$$

The problem at hand is to generate $\boldsymbol{B}_T$ and $\boldsymbol{D}_L$ from the incidence matrix $\boldsymbol{A}$. By Corollary 3-3

$$\boldsymbol{A}\boldsymbol{B}^t = [\boldsymbol{A}_T \mid \boldsymbol{A}_L]\begin{bmatrix} \boldsymbol{B}_T^t \\ \boldsymbol{1}_\mu \end{bmatrix} = \boldsymbol{A}_T\boldsymbol{B}_T^t + \boldsymbol{A}_L = \boldsymbol{0}$$

from which we have

$$B_T^t = -A_T^{-1}A_L \tag{3-29}$$

Equation (3-29), together with Eq. (3-23), enables us to write

$$D = [\mathbf{1}_\rho \mid D_L] = A_T^{-1}[A_T \mid A_L] = A_T^{-1}A \tag{3-30}$$

$$B = [B_T \mid \mathbf{1}_\mu] = [-D_L^t \mid \mathbf{1}_\mu] \tag{3-31}$$

Equation (3-30) is the key to the matrix method of generating D and B. It says that if we first find A_T^{-1}, and then premultiply A by A_T^{-1}, the result is D. Once D is found, we obtain B_T by $B_T = -D_L^t$. Although this method is correct, it is rather inefficient, because every element of A_T^{-1} has to be found explicitly.[3] We shall now digress to discuss two basic theorems in matrix algebra that will lead to an alternative and more efficient method of generating D and B.

Any matrix obtained by performing a single elementary row operation on the identity matrix is called an *elementary matrix*. There are three types of elementary matrices corresponding to the three types of elementary row operations defined earlier. We shall use the symbol $\mathcal{E}$ for an elementary matrix and $\mathcal{E}(j)$ for a type j elementary matrix ($j = 1, 2,$ or 3). The following two theorems on elementary matrices can be proved easily [8].

Theorem 3-7. *Every elementary matrix has an inverse, which is also elementary.*

Theorem 3-8. *To perform an elementary row operation on a matrix Q, calculate the product $\mathcal{E}\, Q$, where $\mathcal{E}$ is the elementary matrix obtained by performing the intended row operation on the identity matrix.*

For example, given the following matrix Q, we wish to add -2 times the second row to the first row; we do the intended operations on $\mathbf{1}$, finding $\mathcal{E}$, and then calculate $\mathcal{E}\, Q$ to obtain the desired result

$$Q = \begin{bmatrix} -2 & 1 & 0 & 1 \\ 1 & 0 & 0 & 3 \\ 0 & 0 & 2 & 5 \end{bmatrix}$$

$$\mathcal{E} = \begin{bmatrix} 1 & -2 & 0 \\ 0 & 1 & 0 \\ 0 & 0 & 1 \end{bmatrix}$$

$$\mathcal{E}Q = \begin{bmatrix} -4 & 1 & 0 & -5 \\ 1 & 0 & 0 & 3 \\ 0 & 0 & 2 & 5 \end{bmatrix}$$

[3] A method which reduces the inversion of A_T to a path-searching problem is given in reference 6.

Following the algorithm of Appendix 3C, if Q is a nonsingular square matrix, we can always find a sequence of m elementary row operations that reduce Q to an identity matrix. By Theorem 3-8, we can always write

$$\mathcal{E}_m \quad \dots \quad \mathcal{E}_3\mathcal{E}_2\mathcal{E}_1 Q = \mathbf{1} \tag{3-32}$$

where $\mathcal{E}_k$ may be either type 1, 2, or 3. From Eq. (3-32) we have

$$Q^{-1} = \mathcal{E}_m \quad \dots \quad \mathcal{E}_3\mathcal{E}_2\mathcal{E}_1$$

and

$$Q = \mathcal{E}_1^{-1}\mathcal{E}_2^{-1}\mathcal{E}_3^{-1} \quad \dots \quad \mathcal{E}_m^{-1} \tag{3-33}$$

Equation (3-33) shows that any nonsingular matrix is expressible as the product of some elementary matrices. (These elementary matrices may be different, but their products are all equal to Q.)

Now, let us return to Eq. (3-30). We may write, as in Eq. (3-33),

$$A_T^{-1} = \mathcal{E}_m \quad \dots \quad \mathcal{E}_3\mathcal{E}_2\mathcal{E}_1$$

The equation

$$D = A_T^{-1}A = (\mathcal{E}_m \quad \dots \quad \mathcal{E}_3\mathcal{E}_2\mathcal{E}_1)A$$

means that D may be obtained from A by performing a sequence of elementary row operations on A. The equation

$$A_T^{-1}A_T = (\mathcal{E}_m \quad \dots \quad \mathcal{E}_3\mathcal{E}_2\mathcal{E}_1)A_T = \mathbf{1}$$

shows that the required sequence of row operations is just those operations which when performed on A_T will result in an identity matrix. Putting these two facts together, we have the following alternative method of generating D and B from A.

Perform elementary row operations on A to reduce A_T to an identity matrix. (See Appendix 3C for an algorithm.) When this state is attained, the resulting matrix is D. Finally, obtain B by the use of $B_T = -D_L^t$. Example 3-2 will illustrate in detail this procedure.

Example 3-2. Given

$$A = [A_T \mid A_L] = \begin{array}{c} \\ 1 \\ 2 \\ 3 \\ 4 \end{array} \begin{array}{c} \begin{array}{ccccccc} a & b & c & d & e & f & g \end{array} \\ \left[\begin{array}{cccc:ccc} 1 & 1 & 0 & 0 & 0 & 0 & -1 \\ -1 & 0 & 0 & 1 & -1 & 0 & 0 \\ 0 & 0 & 0 & -1 & 0 & 1 & 0 \\ 0 & 0 & -1 & 0 & 1 & -1 & 1 \end{array}\right] \end{array}$$

find D and B matrices relative to the tree T $(abcd)$.

Solution: Let ⓚ denote the kth row of the *updated* matrix. Following the algorithm of Appendix 3C, we have

$$A \xrightarrow{\text{add (1) to (2)}} \left[\begin{array}{cccc|ccc} 1 & 1 & 0 & 0 & 0 & 0 & -1 \\ 0 & 1 & 0 & 1 & -1 & 0 & -1 \\ 0 & 0 & 0 & -1 & 0 & 1 & 0 \\ 0 & 0 & -1 & 0 & 1 & -1 & -1 \end{array}\right]$$

$$\xrightarrow{\text{add }-(2)\text{ to (1)}} \left[\begin{array}{cccc|ccc} 1 & 0 & 0 & -1 & 1 & 0 & 0 \\ 0 & 1 & 0 & 1 & -1 & 0 & -1 \\ 0 & 0 & 0 & -1 & 0 & 1 & 0 \\ 0 & 0 & -1 & 0 & 1 & -1 & 1 \end{array}\right]$$

$$\xrightarrow{\text{interchange (4) and (3), multiply (3) by } -1} \left[\begin{array}{cccc|ccc} 1 & 0 & 0 & -1 & 1 & 0 & 0 \\ 0 & 1 & 0 & 1 & -1 & 0 & -1 \\ 0 & 0 & 1 & 0 & -1 & 1 & -1 \\ 0 & 0 & 0 & -1 & 0 & 1 & 0 \end{array}\right]$$

$$\xrightarrow{\text{multiply (4) by } -1,\ \text{add (4) to (1), add }-(4)\text{ to (2)}} \begin{array}{c} \begin{array}{ccccccc} a & b & c & d & e & f & g \end{array} \\ \left[\begin{array}{cccc|ccc} 1 & 0 & 0 & 0 & 1 & -1 & 0 \\ 0 & 1 & 0 & 0 & -1 & 1 & -1 \\ 0 & 0 & 1 & 0 & -1 & 1 & -1 \\ 0 & 0 & 0 & 1 & 0 & -1 & 0 \end{array}\right] \end{array} = D$$

Finally,

$$B = \begin{array}{c} \begin{array}{ccccccc} a & b & c & d & e & f & g \end{array} \\ \left[\begin{array}{cccc|ccc} -1 & 1 & 1 & 0 & 1 & 0 & 0 \\ 1 & -1 & -1 & 1 & 0 & 1 & 0 \\ 0 & 1 & 1 & 0 & 0 & 0 & 1 \end{array}\right] \end{array}$$

The reader should construct the directed graph from the given A matrix, and then verify that the D matrix obtained is correct.

Note that in the preceding reduction process the identity matrix makes its appearance *column by column*. That is, we have

$$\text{first} \begin{bmatrix} 1 \\ 0 \\ 0 \\ 0 \end{bmatrix}, \text{ then} \begin{bmatrix} 1 & 0 \\ 0 & 1 \\ 0 & 0 \\ 0 & 0 \end{bmatrix}, \text{ then} \begin{bmatrix} 1 & 0 & 0 \\ 0 & 1 & 0 \\ 0 & 0 & 1 \\ 0 & 0 & 0 \end{bmatrix}, \text{ finally} \begin{bmatrix} 1 & 0 & 0 & 0 \\ 0 & 1 & 0 & 0 \\ 0 & 0 & 1 & 0 \\ 0 & 0 & 0 & 1 \end{bmatrix}$$

There is an alternative reduction procedure that creates the identity matrix *row by row*. This procedure is most conveniently carried out by row operations on the A_a instead of A matrix. The row appended to A to obtain A_a is just $-$(sum of all rows of A). Any linear combination of rows of A_a is also a linear combination of A. Therefore, the basis of this alternative procedure is still Eq. (3-30).

We shall illustrate this alternative procedure with the same example as before.

Alternative Solution:

$$A_a = [A_{aT} \mid A_{aL}] = \begin{matrix} \\ 1 \\ 2 \\ 3 \\ 4 \\ 5 \end{matrix} \begin{matrix} a & b & c & d & e & f & g \\ \end{matrix} \left[\begin{array}{cccc|ccc} 1 & 1 & 0 & 0 & 0 & 0 & -1 \\ -1 & 0 & 0 & 1 & -1 & 0 & 0 \\ 0 & 0 & 0 & -1 & 0 & 1 & 0 \\ 0 & 0 & -1 & 0 & 1 & -1 & 1 \\ 0 & -1 & 1 & 0 & 0 & 0 & 0 \end{array}\right]$$

In A_a,

$$\xrightarrow[\text{add 4 to 1}]{\text{add 5 to 1,}} [1 \quad 0 \quad 0 \quad 0 \mid 1 \quad -1 \quad 0]$$

$$\xrightarrow[\text{add 3 to 1}]{\text{add 2 to 1,}} [0 \quad 1 \quad 0 \quad 0 \mid -1 \quad 1 \quad -1]$$

$$\xrightarrow{\text{multiply 4 by } -1} [0 \quad 0 \quad 1 \quad 0 \mid -1 \quad 1 \quad -1]$$

$$\xrightarrow{\text{multiply 3 by } -1} [0 \quad 0 \quad 0 \quad 1 \mid 0 \quad -1 \quad 0]$$

Putting the four generated rows together, we have

$$D = \left[\begin{array}{cccc|ccc} 1 & 0 & 0 & 0 & 1 & -1 & 0 \\ 0 & 1 & 0 & 0 & -1 & 1 & -1 \\ 0 & 0 & 1 & 0 & -1 & 1 & -1 \\ 0 & 0 & 0 & 1 & 0 & -1 & 0 \end{array}\right]$$

By going through this example in detail, the reader may discover the strategy used in the reduction process:

To create a row with all zero elements in the tree section, except for a $+1$ in the kth column, we first select the row that has $+1$ in its kth column. Let this row be the mth row. If the mth row of A_a has some other nonzero elements in the tree section, each of them can be reduced to zero by adding some row of A_a to the mth row. (This is always possible because each column of A_a has exactly one $+1$ and one -1.) Repeated applications of such row additions, as illustrated in the example, will lead to a row in the A_{aT} block with only one $+1$ in the kth column and zeros elsewhere.[4]

[4]We have only presented the matrix method of finding a tree T and the topological matrices B and D in this chapter. Other methods are available [6]. No comparative study has been made on the efficiency of each method in digital computer implementation, where the main concerns are execution time and storage requirements. One reason for the lack of this information is probably that the time required for finding T, D, and B is insignificant (usually on the order of a few tenths of a second) as compared to the time required in the other segments of computer simulation programs, such as numerical integration, piecewise-linear iteration, etc.

APPENDIX 3A PROOF OF THEOREM 3-1

Theorem 3-1. *For a **connected** graph G_d, the set of all rows of any reduced incidence matrix A is linearly independent.*

Proof: For a graph G_d with n nodes and b branches, a reduced incidence matrix A is of order $(n-1) \times b$. Let the rows of A be denoted by

$$\boldsymbol{\alpha}_1 = (a_{11}, a_{12}, \dots, a_{1b})$$
$$\vdots$$
$$\boldsymbol{\alpha}_{n-1} = (a_{n-1,1}, a_{n-1,2}, \dots, a_{n-1,b})$$

and the row that has to be appended to A to obtain A_a be denoted by

$$\boldsymbol{\alpha}_n = (a_{n1}, a_{n2}, \dots, a_{nb})$$

We shall now prove that $\boldsymbol{\alpha}_1, \boldsymbol{\alpha}_2, \dots, \boldsymbol{\alpha}_{n-1}$ are linearly independent by the method of contradiction.

Suppose that $\boldsymbol{\alpha}_1, \boldsymbol{\alpha}_2, \dots, \boldsymbol{\alpha}_{n-1}$ are linearly dependent. Then there exist scalars $k_1, k_2, \dots, k_{n-1}$, not all zero, such that

$$k_1\boldsymbol{\alpha}_1 + k_2\boldsymbol{\alpha}_2 + \cdots + k_{n-1}\boldsymbol{\alpha}_{n-1} = \mathbf{0} \tag{3A-1}$$

Without loss of generality, we may assume that

$$k_1, k_2, \dots, k_m, \qquad n-1 \geq m \geq 1$$

are nonzero and $k_{m+1}, k_{m+2}, \dots, k_{n-1}$ are zero. Then Eq. (3A-1) becomes

$$k_1\boldsymbol{\alpha}_1 + k_2\boldsymbol{\alpha}_2 + \cdots + k_m\boldsymbol{\alpha}_m = \mathbf{0} \tag{3A-2}$$

Let us partition the complete incidence matrix A_a as follows:

$$A_a = \begin{bmatrix} A_m \\ \hdashline A_r \end{bmatrix} = \begin{bmatrix} \boldsymbol{\alpha}_1 \\ \vdots \\ \boldsymbol{\alpha}_m \\ \hdashline \boldsymbol{\alpha}_{m+1} \\ \vdots \\ \boldsymbol{\alpha}_{n-1} \\ \boldsymbol{\alpha}_n \end{bmatrix} \tag{3A-3}$$

where A_m and A_r each have at least one row because $m \geq 1$, and A_r contains the row $\boldsymbol{\alpha}_n$. Because of Eq. (3A-2), no column of A_m can have exactly one nonzero element. Thus, there are three possible cases to consider:

1. Every column of A_m is a zero column. Then each of the first m nodes of G_d is not connected to any other node. Therefore, G_d is unconnected.

2. Every column of A_m has exactly one $+1$ and one -1. Then A_r will be a zero matrix, and hence G_d is not connected.
3. Some columns of A_m have one $+1$ and one -1, while some columns of A_m are zero columns. Then A_a may be further partitioned (after some permutation of columns) as follows:[5]

$$A_a = \begin{bmatrix} A_m \\ \hdashline A_r \end{bmatrix} = \left[\begin{array}{c:c} \mathbf{0} & (\pm 1, 0) \\ \hdashline (\pm 1, 0) & \mathbf{0} \end{array}\right] \tag{3A-4}$$

which shows that there is no path between, say, node 1 and node n. Hence, G_d is unconnected.

We have shown that the assumption of linear dependence among $\boldsymbol{\alpha}_1, \boldsymbol{\alpha}_2, \ldots, \boldsymbol{\alpha}_{n-1}$ leads to a conclusion that G_d is unconnected. This contradicts the hypothesis. Therefore, $\boldsymbol{\alpha}_1, \boldsymbol{\alpha}_2, \ldots, \boldsymbol{\alpha}_{n-1}$ are linearly independent, and the theorem is valid.

APPENDIX 3B PROOF OF THEOREM 3-2

Theorem 3-2. *Let A be the reduced incidence matrix of a connected graph G_d with n nodes. Then $n - 1$ columns of A are linearly independent if and only if the branches corresponding to these columns form a tree in G_n.*

Proof:

PART I. SUFFICIENCY. Without loss of generality, let us assume that the first $n - 1$ columns of A correspond to a tree T of the connected graph. Denote by A_T the submatrix consisting of these $n - 1$ columns. A_T is a square matrix of order $n - 1$. Now A_T is simply a reduced incidence matrix for the subgraph T. Since T is connected, the $n - 1$ rows of A_T are linearly independent by Theorem 3-1. Consequently, the $n - 1$ columns of A_T are also linearly independent. (Recall that for any $n \times m$ matrix the maximum numbers of linearly independent rows and linearly independent columns are equal.)

PART II. NECESSITY. Let A_s be a square submatrix consisting of $n - 1$ linearly independent columns of A. We shall show that the $n - 1$ branches corresponding to the columns of A_s form a tree.

Construct a directed graph G_s, using A_s as its reduced incidence matrix. Since A_s has $n - 1$ rows and $n - 1$ columns, then G_s has n nodes and $n - 1$ branches. Now the $n - 1$ columns of A_s are linearly independent by hypothesis. It follows that the $n - 1$ rows of A_s are also linearly independent. Then G_s is a connected graph (see Problem 3-13). The three properties of G_s (connected, n nodes, $n - 1$ branches) imply that G_s contains no loops [1]. Therefore, G_s is a tree.

This completes the proof of Theorem 3-2.

[5]The notation $(\pm 1, 0)$ here denotes a matrix whose elements consist of 1, —1, and 0 only.

APPENDIX 3C AN ALGORITHM FOR REDUCING A RECTANGULAR MATRIX TO AN ECHELON MATRIX

In this book, we emphasize the development of algorithms for solving network problems. We present algorithms rather than actual computer programs. The reason is that algorithms possess universality, whereas computer programs are extremely language dependent. It is generally agreed that once an algorithm is established, the conversion into a computer program is a task that any person with reasonable background in computer programming can perform.

An algorithm is just a step-by-step procedure with the following essential features:

1. A clearly defined starting point (input).
2. A clearly defined end result (output).
3. Finiteness. The algorithm should terminate in a finite number of steps.
4. Definiteness. At each step, there should be no ambiguity as to what action to take.

An algorithm may be presented in several ways: (1) a step-by-step prose account, (2) a flow chart, or (3) an informal explanation of the "how" *and* "why" of the steps. From the pedagogical point of view, method 3 is the most effective since it not only shows how but also explains why. Most algorithms in this book will be presented using method 3. For the purpose of illustrating the styles only, in this appendix we shall present one algorithm first by method 1 and then by method 2.

Definition: Echelon Matrix. *An* $m \times n$ *rectangular matrix* A *is said to be a row echelon matrix, or simply an echelon matrix, if* A *has the following structure:*

1. *The first* k *rows,* $k \geq 0$, *are nonzero and all remaining rows, if any, are zero.*
2. *In the ith row* ($i = 1, 2, \ldots, k$), *the first nonzero element is equal to unity, the column in which it occurs being numbered* c_i.
3. $c_1 < c_2 < c_3 < \ldots < c_k$.

For example, the following matrix A is a row echelon matrix.

$$A = \begin{bmatrix} 0 & 1 & a_{13} & a_{14} & a_{15} & a_{16} & a_{17} & a_{18} \\ 0 & 0 & 0 & 0 & 1 & a_{26} & a_{27} & a_{28} \\ 0 & 0 & 0 & 0 & 0 & 1 & a_{37} & a_{38} \\ 0 & 0 & 0 & 0 & 0 & 0 & 1 & a_{48} \\ 0 & 0 & 0 & 0 & 0 & 0 & 0 & 0 \end{bmatrix}$$

Any $m \times n$ rectangular matrix A can be reduced to echelon form by a series of elementary row operations.[6] We now describe an algorithm for achieving the reduction.

ALGORITHM RE (REDUCTION TO ECHELON MATRIX). Starting with an $m \times n$ matrix A, this algorithm performs elementary row operations on submatrices of A. When the algorithm terminates, A is reduced to echelon form.

1. Let A_s be a submatrix of A. Initially, $A_s = A$.
2. Scan A_s, column by column from the left, until a nonzero column, column c, is encountered. If no such column exists, then $A_s = \mathbf{0}$, and the algorithm terminates.
3. Scan column c, element by element from the top, until a nonzero element in row r is encountered.
4. Interchange row 1 and row r of A_s. (Note that A_s has been changed from the original one. All operations are performed on the latest A_s.)
5. Multiply row 1 of A_s by the reciprocal of $(1, c)$ element of A_s.
6. Reduce all elements below row 1 and in column c of A_s to zero by type 3 elementary row operation.
7. Let the new A_s be a submatrix of the old A_s. The new A_s consists of the elements below row 1 and to the right of column c of the old A_s. If this new A_s does not exist, the algorithm terminates. Otherwise, go to step 2.

As an example, let us apply the algorithm to reduce the following matrix A to echelon form.

$$A_{5\times 6} = \begin{bmatrix} 0 & 0 & -1 & 2 & 4 & 1 \\ 0 & 0.5 & 1 & 2 & 1 & 2 \\ 0 & -1 & -2 & 1 & 1 & 0 \\ 0 & 0 & 3 & 2 & 1 & 2 \\ 0 & -1 & 1 & 3 & 2 & 2 \end{bmatrix}$$

First pass through the steps: Initially $A_s = A$. Steps 1–6 result in

$$A_s = \left[\begin{array}{cc|cccc} 0 & 1 & 2 & 4 & 2 & 4 \\ \hline 0 & 0 & -1 & 2 & 4 & 1 \\ 0 & 0 & 0 & 5 & 3 & 4 \\ 0 & 0 & 3 & 2 & 1 & 2 \\ 0 & 0 & 3 & 7 & 4 & 6 \end{array}\right]$$

Step 7 results in

$$A_s = \begin{bmatrix} -1 & 2 & 4 & 1 \\ 0 & 5 & 3 & 4 \\ 3 & 2 & 1 & 2 \\ 3 & 7 & 4 & 6 \end{bmatrix}$$

[6]The result is not unique, depending on the algorithms used. However, if we use the slightly different definition of row echelon matrix given in reference 8, then the result is unique.

Second pass: Steps 2–6 result in

$$A_s = \begin{bmatrix} 1 & -2 & -4 & -1 \\ 0 & 5 & 3 & 4 \\ 0 & 8 & 13 & 5 \\ 0 & 13 & 16 & 9 \end{bmatrix}$$

Step 7 results in

$$A_s = \begin{bmatrix} 5 & 3 & 4 \\ 8 & 13 & 5 \\ 13 & 16 & 9 \end{bmatrix}$$

Third pass: Steps 2–6 result in

$$A_s = \begin{bmatrix} 1 & \frac{3}{5} & \frac{4}{5} \\ 0 & \frac{41}{5} & -\frac{7}{5} \\ 0 & \frac{41}{5} & -\frac{7}{5} \end{bmatrix}$$

Step 7 results in

$$A_s = \begin{bmatrix} \frac{41}{5} & -\frac{7}{5} \\ \frac{41}{5} & -\frac{7}{5} \end{bmatrix}$$

Fourth pass: Steps 2–6 result in

$$A_s = \begin{bmatrix} 1 & -\frac{7}{41} \\ 0 & 0 \end{bmatrix}$$

Step 7 results in

$$A_s = 0$$

Fifth pass: The algorithm terminates at step 2.

Finally, recording all the changes that have been made to the elements of A in our procedure, we obtain the desired echelon matrix:

$$A_{\text{ech}} = \begin{bmatrix} 0 & 1 & 2 & 4 & 2 & 4 \\ 0 & 0 & 1 & -2 & -4 & -1 \\ 0 & 0 & 0 & 1 & \frac{3}{5} & \frac{4}{5} \\ 0 & 0 & 0 & 0 & 1 & -\frac{7}{41} \\ 0 & 0 & 0 & 0 & 0 & 0 \end{bmatrix}$$

It should be emphasized that there is no short-cut to the full understanding of any algorithm, owing to its precise and detailed nature, except by working out examples with pencil and paper.

In our presentation of the algorithm, we have concentrated on strategy rather than details. The algorithm is described with a flowchart in Fig. 3C-1, which embodies all the detailed operations required by a computer program. The conversion into a *FORTRAN* program, based on this flowchart, is shown in the subroutine *RAECH* in Appendix 6B.

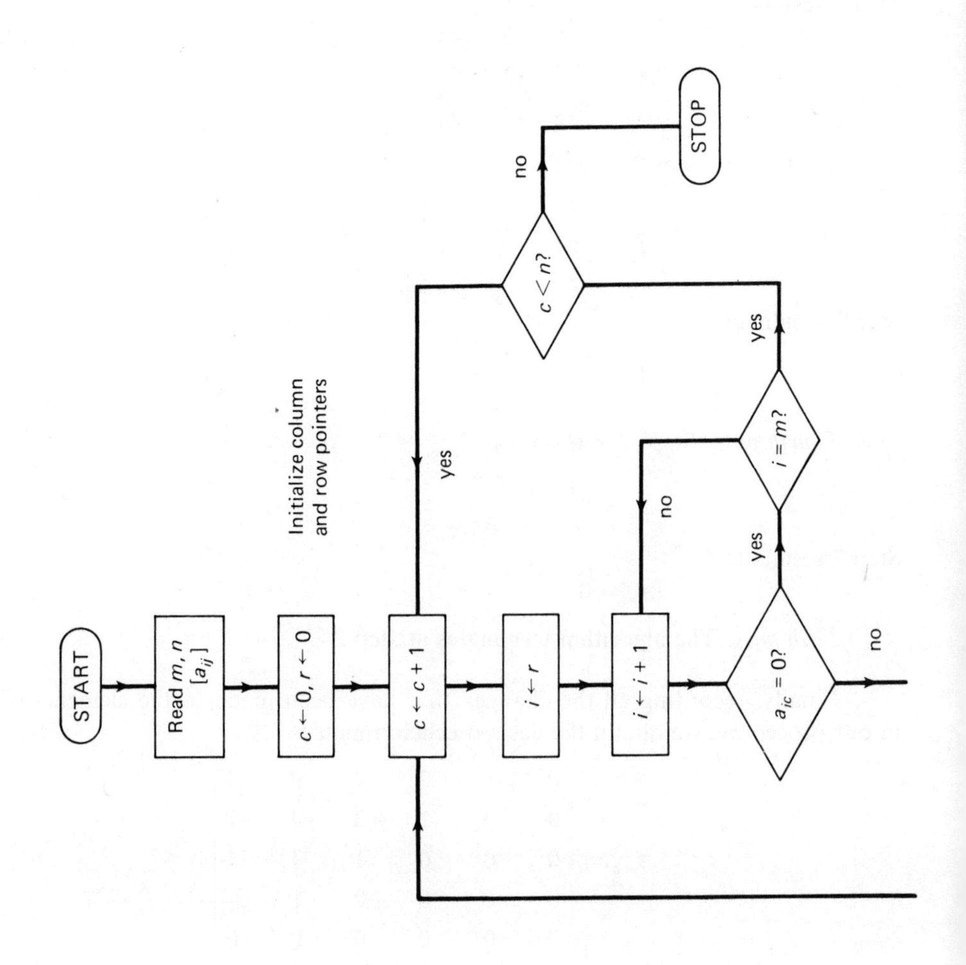
START
Read m, n $[a_{ij}]$
$c \leftarrow 0, r \leftarrow 0$
Initialize column and row pointers
$c \leftarrow c + 1$
$i \leftarrow r$
$i \leftarrow i + 1$
$a_{ic} = 0?$
yes
no
$i = m?$
yes
no
$c < n?$
yes
no
STOP

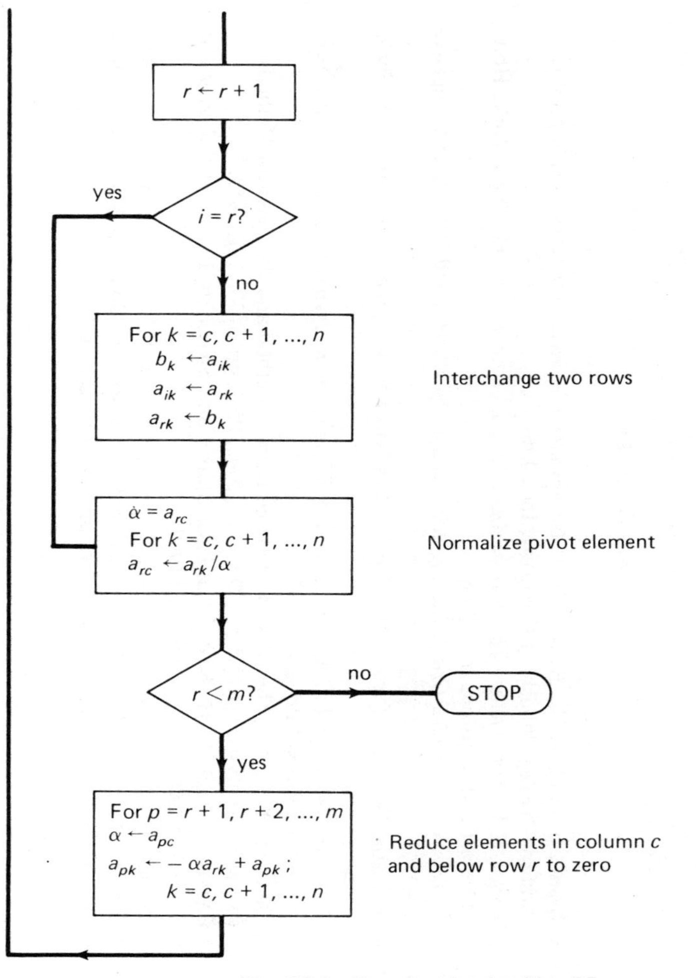

Fig. 3C-1. Flowchart for algorithm RE.

REDUCTION TO AN IDENTITY MATRIX. Any matrix $\boldsymbol{Q}$ may be reduced to echelon form by elementary row operations. If $\boldsymbol{Q}$ is a nonsingular square matrix, its echelon form will be an *upper triangular matrix* with unity diagonal elements. (An upper triangular matrix is a square matrix with all elements below the main diagonal equal to zero.) For each k ($k = n, n-1, \ldots, 3, 2, 1$), the unity diagonal element a_{kk} may be used to reduce $a_{k-1,k}, a_{k-2,k}, \ldots, a_{2k}, a_{1k}$ to zeros by type 3 elementary row operations. Thus, any nonsingular matrix $\boldsymbol{Q}$ is reducible to an identity matrix by elementary row operations.

If $\boldsymbol{Q}$ is a submatrix of an incidence matrix $\boldsymbol{A}_a$, then in reducing to echelon form, the constants required in type 2 and type 3 elementary row operations are $+1$'s and -1's. This, together with the fact that $\boldsymbol{A}_a$ has only $+1$, -1, and 0 as its elements, assures that round-off error never occurs in the reduction of an incidence matrix to its echelon form.

REFERENCES

1. SESHU, S., and M. B. REED. *Linear Graphs and Electrical Networks*. Reading, Mass.: Addison-Wesley Publishing Company, Inc., 1961.
2. CHAN, S. P. *Introductory Topological Analysis of Electrical Networks*. New York: Holt, Rinehart and Winston, Inc., 1968.
3. CHEN, W. K. *Applied Graph Theory*. Amsterdam: North-Holland Publishing Company, 1971.
4. MAYEDA, W. *Graph Theory*. New York: John Wiley & Sons, Inc. (Interscience Division), 1972.
5. JOHNSON, D. E., and J. R. JOHNSON. *Graph Theory with Engineering Applications*. New York: The Ronald Press Co., 1972.
6. BRANIN, F. H., JR. "The Relation between Kron's Method and the Classical Methods of Network Analysis." *IRE WESCON Conv. Rec.*, Pt. 2, pp. 3–28, Aug. 1959.
7. WING, O., and W. H. KIM. "The Path Matrix and Switching Functions." *J. Franklin Inst.*, Vol. 268, pp. 251–259, Oct. 1959.
8. NOBLE, B. *Applied Linear Algebra*. Englewood Cliffs, N.J.: Prentice-Hall, Inc., 1969.
9. LIN, P-M, and G. E. ALDERSON. "Symbolic Network Functions by a Single Pathfinding Algorithm." *Proc. 7th Allerton Conf.*, pp. 196–205, Oct. 1969.

PROBLEMS

3-1. Consider the nonlinear network of Fig. P3-1 in which the branches have been labeled and the reference directions assigned. (v_k means voltage drop across branch k in the direction of the arrow.)

(a) Draw the corresponding directed graph G_d.

(b) Write $\boldsymbol{A}_a$, the incidence matrix, and $\boldsymbol{A}$, the reduced incidence matrix, with node 6 as the datum.

(c) Choose a tree T. By inspection of G_d, write the fundamental loop matrix $\boldsymbol{B}$ in the form of Eq. (3-6), and the fundamental cutset matrix $\boldsymbol{D}$ in the form of Eq. (3-15).

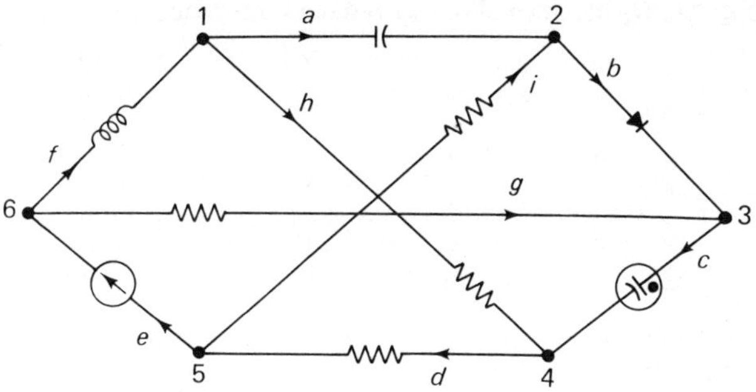

Fig. P3-1

(d) For each of the following matrix equations

$$\boldsymbol{Ai} = \boldsymbol{0}$$

$$\boldsymbol{Di} = \boldsymbol{0}$$

$$\boldsymbol{Bv} = \boldsymbol{0}$$

write out at least two scalar equations and verify that they are indeed Kirchhoff current or voltage equations.

3-2. In the network graph of Fig. P3-2, a tree T is chosen to consist of branches $a\ b\ c\ d$.

(a) Show all entries of the matrices $\boldsymbol{i}_T$, $\boldsymbol{i}_L$, $\boldsymbol{v}_T$, $\boldsymbol{v}_L$, $\boldsymbol{i}$, and $\boldsymbol{v}$.

(b) Write the $\boldsymbol{B}$ matrix.

(c) Show the relationship between $\boldsymbol{v}_L$ and $\boldsymbol{v}_T$ (the entries of all matrices should be shown).

(d) Repeat part (c) for $\boldsymbol{i}_T$ and $\boldsymbol{i}_L$.

(e) Repeat part (c) for $\boldsymbol{i}$ and $\boldsymbol{i}_L$.

(f) Repeat part (c) for $\boldsymbol{v}$ and $\boldsymbol{v}_T$.

(g) If $\boldsymbol{A}$ is obtained from $\boldsymbol{A}_a$ by deleting the row corresponding to node 5 and the following transformation is used,

$$\boldsymbol{v} = \boldsymbol{A}^t\boldsymbol{v}_n$$

write this matrix equation in scalar form. From these equations, interpret the physical meaning of the entries of $\boldsymbol{v}_n$.

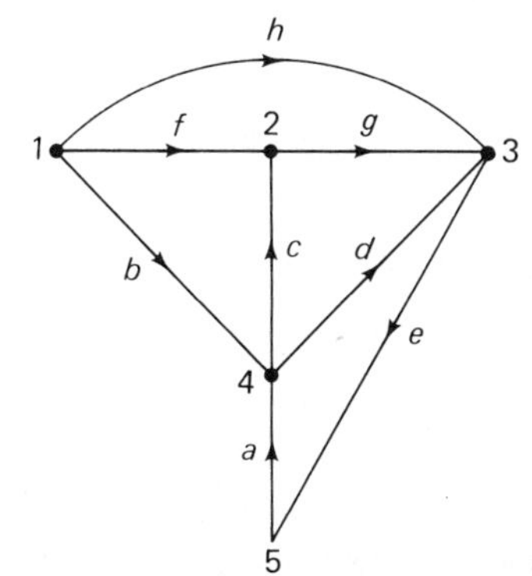

Fig. P3-2

3-3. A directed graph G_d has the following reduced incidence matrix:

$$A = \begin{array}{c} \\ 1 \\ 2 \\ 3 \\ 4 \\ 5 \end{array}\begin{array}{c} \begin{matrix} a & b & c & d & e & f & g & h & i \end{matrix} \\ \begin{bmatrix} 1 & 1 & 0 & 0 & 0 & 0 & 0 & 0 & 0 \\ -1 & 0 & -1 & 0 & 1 & 1 & 0 & 0 & 0 \\ 0 & -1 & 1 & 1 & 0 & 0 & 0 & 1 & 0 \\ 0 & 0 & 0 & 0 & 0 & -1 & 1 & 0 & 1 \\ 0 & 0 & 0 & 0 & 0 & 0 & -1 & -1 & -1 \end{bmatrix} \end{array}$$

By reducing A to echelon form, find a tree corresponding to columns of A as close as possible to the left.

3-4. A directed graph has the following incidence matrix:

$$A = [A_T \mid A_L] = \begin{array}{c} \\ 1 \\ 2 \\ 3 \\ 4 \end{array}\begin{array}{c} \begin{matrix} a & b & c & d & e & f & g & h & i & j \end{matrix} \\ \left[\begin{array}{cccc|cccccc} -1 & 1 & 0 & 0 & 0 & 0 & 0 & 0 & 1 & -1 \\ 0 & -1 & 1 & 1 & 0 & 0 & 1 & 0 & 0 & 0 \\ 0 & 0 & -1 & 0 & 1 & 0 & 0 & -1 & -1 & 0 \\ 0 & 0 & 0 & -1 & -1 & 1 & 0 & 0 & 0 & 1 \end{array}\right] \end{array}$$

(a) Following the matrix method discussed in the text, find the fundamental cutset matrix $\boldsymbol{D}$.

(b) Using the result of part (a), write the fundamental loop matrix $\boldsymbol{B}$.

3-5. Prove that the determinant of any square submatrix of A_a is 1, -1, or 0. A matrix with such a property is said to be *unimodular*. (It can be shown that both $\boldsymbol{B}$ and $\boldsymbol{D}$ matrices are unimodular.)

3-6. List all trees, loops, and cutsets for the graph shown in Fig. 3-1(c).

3-7. For an arbitrary matrix $\boldsymbol{P}$ of order $m \times n$, a *major determinant* of $\boldsymbol{P}$ is a determinant of order min (m, n). The Binet–Cauchy theorem states that if $\boldsymbol{P}$ is an $m \times n$ matrix and $\boldsymbol{Q}$ is an $n \times m$ matrix, where $m \leq n$, then

$$|\boldsymbol{PQ}| = \sum_i \text{(products of corresponding major determinants of } \boldsymbol{P} \text{ and } \boldsymbol{Q})$$

where the summation is over all such major determinants.

Check the theorem with the following numerical example (see p. 226 of [8] for a proof of the theorem).

$$\boldsymbol{P} = \begin{bmatrix} 1 & 2 & 1 \\ -3 & 0 & 1 \end{bmatrix}, \quad \boldsymbol{Q} = \begin{bmatrix} 2 & 1 \\ 0 & -1 \\ 1 & 3 \end{bmatrix}$$

3-8. Let A be the reduced incidence matrix of a directed connected graph G_d. Prove that

$$\text{number of trees of } G_n = |AA^t|$$

Hint: Make use of the Binet–Cauchy theorem, the result of Problem 3-5, and Theorem 3-2.

3-9. An undirected graph is called a *complete graph* if every pair of nodes is joined by exactly one branch. Prove that for a complete graph of n nodes the number of trees is $n^{(n-2)}$. *Hint:* Use the result of Problem 3-8.

3-10. (a) Prove that any set of branches $(b_1, \ldots, b_m)$ of a connected graph G can be made part of a tree of G if and only if the set of branches contains no loop.

(b) Prove that any set of branches $(b_1, \ldots, b_m)$ of a connected graph G can be made part of a cotree with respect to some tree of G if and only if the set of branches contains no cutset.

3-11. Let S_1 and S_2 be two *disjoint* sets of branches of a connected graph G (i.e., S_1 and S_2 have no branches in common). Prove that, if S_1 contains no loop and S_2 contains no cutset, then a tree T can always be constructed which includes S_1 in the tree and S_2 in the cotree.

3-12. Prove that any m columns of the incidence matrix A_a (or A) of a connected directed graph G_d are linearly independent if and only if the corresponding set of m branches of the associated *undirected* graph G_n contains no loop.

3-13. Prove that if the rows of a reduced incidence matrix for a directed graph G_d are linearly independent, then the graph G_d is connected.

3-14. Consider the problem of constructing the fundamental loop matrix $\boldsymbol{B}$ and the fundamental cutset matrix $\boldsymbol{D}$ for a given directed graph G_d and a given tree T of G_d. We can either (1) find the $\boldsymbol{B}$ matrix first, and then obtain $\boldsymbol{D}$ with the aid of Eq. (3-23); or (2) find the $\boldsymbol{D}$ matrix first, and then obtain $\boldsymbol{B}$ with the aid of Eq. (3-23). If the solution is to be obtained by writing a digital computer program, which way do you prefer? If the solution is to be obtained by inspection of the graph G_d, which way do you prefer? Give the reasons for your choices.

CHAPTER 4

Nodal Linear Network Analysis: Algorithms and Computational Methods

4-1 INTRODUCTORY REMARKS

In Chapter 3 we showed that an independent system of KVL and KCL equations for any network graph can be generated systematically by the computer in terms of topological matrices. The resulting equations relate only the topological interconnection among the branches in the network and contain twice as many branch variables as the number of independent equations. Obviously, the remaining equations must come from the element characteristics. In this chapter, we shall consider only *linear networks* containing *RLC* elements, independent voltage or current sources, and voltage-controlled current sources. There are two reasons for restricting ourselves to this seemingly rather small subset of real-life networks: (1) it is particularly easy to write an automatic circuit-analysis program for analyzing this class of networks; (2) the program can be easily generalized to allow also nonlinear resistors. The procedure for doing this will be presented in Chapter 5.

4-2 COMPUTER FORMULATION OF NODAL EQUATIONS FOR LINEAR RESISTIVE NETWORKS

We shall now show how the computer can be used to generate a system of *nodal equations* for linear resistive networks. To this end, it is convenient to consider each branch "k", Fig. 4-1(a), of a network graph as representing a *composite branch* made up of a two-terminal element b_k, an independent voltage source with terminal voltage E_k, and an independent current source with terminal current J_k, as shown in Fig.

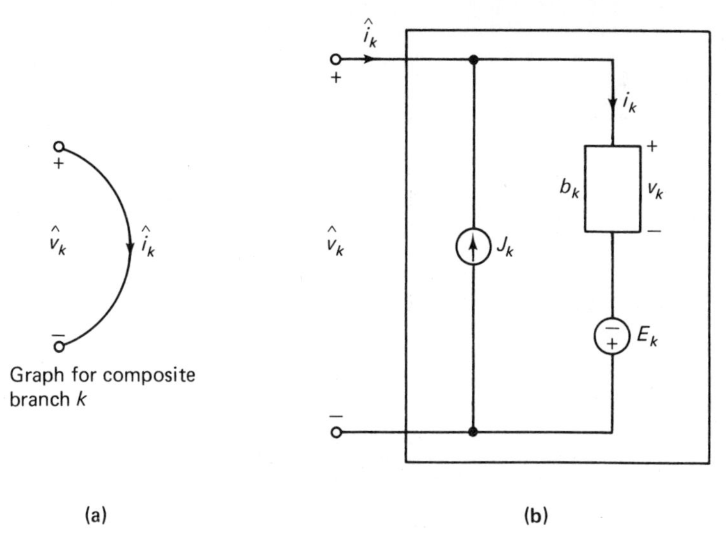

Fig. 4-1. Each branch k is assumed to represent a composite branch.

4-1(b). The two-terminal element b_k can be either a *linear resistor* or a *voltage-controlled current source* that depends linearly on the voltage of another resistor.[1]

Using the preceding notations and definitions, let us consider a *connected* network having b composite branches and $n + 1$ nodes. Let the composite branches be labeled consecutively from 1 to b, and let the nodes be labeled from 0 to n. Without loss of generality, let us choose node 0 as the reference or datum node, and call the voltages from the remaining n nodes with respect to datum the *nodal voltages*, or simply *node voltages*. If we define the voltage vectors

$$\hat{\boldsymbol{v}} \triangleq \begin{bmatrix} \hat{v}_1 \\ \hat{v}_2 \\ \cdot \\ \cdot \\ \cdot \\ \hat{v}_b \end{bmatrix}, \qquad \boldsymbol{v} \triangleq \begin{bmatrix} v_1 \\ v_2 \\ \cdot \\ \cdot \\ \cdot \\ v_b \end{bmatrix}, \qquad \boldsymbol{E} \triangleq \begin{bmatrix} E_1 \\ E_2 \\ \cdot \\ \cdot \\ \cdot \\ E_b \end{bmatrix} \tag{4-1}$$

and the current vectors

$$\hat{\boldsymbol{i}} \triangleq \begin{bmatrix} \hat{i}_1 \\ \hat{i}_2 \\ \cdot \\ \cdot \\ \cdot \\ \hat{i}_b \end{bmatrix}, \qquad \boldsymbol{i} \triangleq \begin{bmatrix} i_1 \\ i_2 \\ \cdot \\ \cdot \\ \cdot \\ i_b \end{bmatrix}, \qquad \boldsymbol{J} \triangleq \begin{bmatrix} J_1 \\ J_2 \\ \cdot \\ \cdot \\ \cdot \\ J_b \end{bmatrix} \tag{4-2}$$

[1]Observe that a *current-controlled current source* which depends linearly on the current of another resistor can be replaced by an *equivalent* voltage-controlled current source. In particular, if $i_k = \beta_{kj} i_j$ is the terminal current of the controlled current source b_k, where i_j is the current of resistor b_j with resistance R_j, then we can replace this controlled source with a voltage-controlled current source with terminal current $i_k = g_{kj} v_j$, where $g_{kj} = \beta_{kj}/R_j$. Consequently, except for this slight inconvenience on the user's part, the *nodal analysis* formulation could allow both voltage- and current-controlled current sources.

then clearly

$$\hat{\boldsymbol{v}} = \boldsymbol{v} - \boldsymbol{E} \tag{4-3}$$

$$\hat{\boldsymbol{i}} = \boldsymbol{i} - \boldsymbol{J} \tag{4-4}$$

The KCL equation from Chapter 3 is given by

$$\boldsymbol{A}\hat{\boldsymbol{i}} = \boldsymbol{0} \tag{4-5}$$

where $\boldsymbol{A}$ is an $n \times b$ reduced incidence matrix obtained from the complete incidence matrix $\boldsymbol{A}_a$ by deleting the row corresponding to the datum node. Substituting Eq. (4-4) into Eq. (4-5) we obtain

$$\boldsymbol{A}\boldsymbol{i} = \boldsymbol{A}\boldsymbol{J} \tag{4-6}$$

Each two-terminal element b_k is characterized by[2]

$$i_k = \frac{1}{R_k} v_k \tag{4-7}$$

if it is a linear resistor with resistance R_k, or by

$$i_k = g_{kj} v_j \tag{4-8}$$

if it is a voltage-controlled current source with controlling voltage v_j and a controlling coefficient equal to g_{kj}. Consequently, the branch element characteristics can be expressed in matrix form by

$$\begin{bmatrix} i_1 \\ i_2 \\ \cdot \\ \cdot \\ \cdot \\ i_b \end{bmatrix} = \begin{bmatrix} y_{11} & y_{12} & \cdots & y_{1b} \\ y_{21} & y_{22} & \cdots & y_{2b} \\ \cdot & \cdot & \cdots & \cdot \\ \cdot & \cdot & \cdots & \cdot \\ \cdot & \cdot & \cdots & \cdot \\ y_{b1} & y_{b2} & \cdots & y_{bb} \end{bmatrix} \begin{bmatrix} v_1 \\ v_2 \\ \cdot \\ \cdot \\ \cdot \\ v_b \end{bmatrix} \tag{4-9}$$

where

$$y_{k\alpha} = \begin{cases} 0, & \alpha \neq k \\ \dfrac{1}{R_k}, & \alpha = k \end{cases} \tag{4-10}$$

if branch k is a linear resistor, and

$$y_{k\alpha} = \begin{cases} 0, & \alpha \neq j \\ g_{kj}, & \alpha = j \end{cases} \tag{4-11}$$

if branch k is a voltage-controlled current source depending on the voltage v_j. Equation (4-9) can be written compactly as

$$\boldsymbol{i} = \boldsymbol{Y}_b \boldsymbol{v} \tag{4-12}$$

where $\boldsymbol{Y}_b$ is called the *branch-admittance matrix.*

Substituting Eq. (4-12) for $\boldsymbol{i}$ in Eq. (4-6), we obtain

$$\boldsymbol{A}\boldsymbol{Y}_b\boldsymbol{v} = \boldsymbol{A}\boldsymbol{J} \tag{4-13}$$

Solving for $\boldsymbol{v}$ from Eq. (4-3) and substituting the resulting expression into Eq. (4-13), we obtain

$$\boldsymbol{A}\boldsymbol{Y}_b\hat{\boldsymbol{v}} = \boldsymbol{A}(\boldsymbol{J} - \boldsymbol{Y}_b\boldsymbol{E}) \tag{4-14}$$

[2]We assume that $R_k \neq 0$ if element b_k is a linear resistor. This assumption is usually met in most practical circuits, and therefore does not appear to be unduly restrictive. See reference 10 for a modified nodal approach that permits $R_k = 0$.

Let us now recall the node transformation relationship derived in Chapter 3:

$$\hat{v} = A^t v_n \tag{4-15}$$

where v_n is the node-to-datum voltage vector. Substituting Eq. (4-15) for $\hat{v}$ in Eq. (4-14), we obtain

$$\boxed{(AY_bA^t)v_n = A(J - Y_bE)} \tag{4-16}$$

which can be written as

$$\boxed{Y_n v_n = J_n} \tag{4-17}$$

where

$$\boxed{Y_n \triangleq AY_bA^t} \tag{4-18}$$

is called the *node-admittance matrix*, and

$$\boxed{J_n \triangleq A(J - Y_bE)} \tag{4-19}$$

is called the *equivalent nodal current source vector*. Solving Eq. (4-17) for v_n, we obtain

$$v_n = Y_n^{-1}J_n \tag{4-20}$$

Once v_n is found, $\hat{v}$ can be computed using Eq. (4-15), and v and i can be computed from Eqs. (4-3) and (4-12). Consequently, Eq. (4-17) completely and uniquely determines the network's solution. It is usually referred to as the *nodal equation*, and the process of solving for v_n from this equation is usually referred to as *nodal analysis*.

EXAMPLE 4-1. For the network shown in Fig. 4-2(a), find Y_n and J_n in the nodal equation (4-17) by the use of Eqs. (4-18) and (4-19).

Solution: The directed graph associated with the network is shown in Fig. 4-2(b), where the directions of the composite branches have been assigned arbitrarily. With node 0 chosen as the datum node, the reduced incidence matrix is

$$A = \begin{matrix} \\ 1 \\ 2 \\ 3 \end{matrix} \begin{matrix} \begin{matrix} 1 & 2 & 3 & 4 & 5 & 6 & 7 \end{matrix} \\ \begin{bmatrix} 1 & 0 & 0 & 1 & 1 & 0 & 1 \\ 0 & 1 & 0 & -1 & -1 & 1 & 0 \\ 0 & 0 & 1 & 0 & 0 & -1 & -1 \end{bmatrix} \end{matrix}$$

The branch-admittance matrix is

$$Y_b = \begin{bmatrix} 2 & 0 & 0 & 0 & 0 & 0 & 0 \\ 0 & 5 & 0 & 0 & 0 & 0 & 0 \\ 0 & 0 & 0 & 0 & 0 & 4 & 0 \\ 0 & 0 & 0 & 0 & 0 & 8 & 0 \\ 0 & 0 & 0 & 0 & 10 & 0 & 0 \\ 0 & 0 & 0 & 0 & 0 & 3 & 0 \\ 0 & 0 & 0 & 0 & 0 & 0 & \frac{1}{2} \end{bmatrix}$$

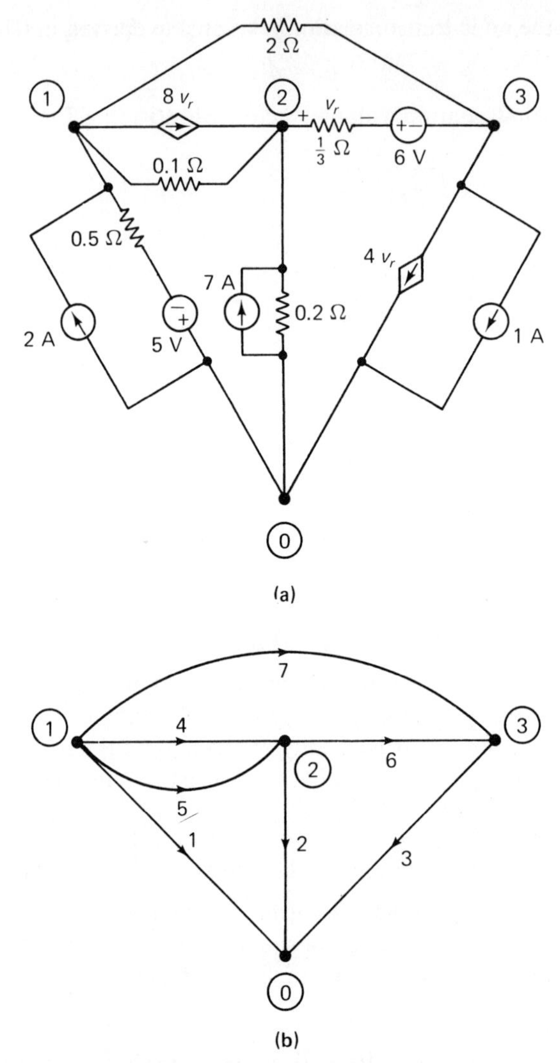

Fig. 4-2. Calculation of $\boldsymbol{Y}_n$ and $\boldsymbol{J}_n$.

The current source and voltage source vectors are, respectively,

$$\boldsymbol{J} = \begin{bmatrix} 2 \\ 7 \\ -1 \\ 0 \\ 0 \\ 0 \\ 0 \end{bmatrix}, \qquad \boldsymbol{E} = \begin{bmatrix} 5 \\ 0 \\ 0 \\ 0 \\ 0 \\ -6 \\ 0 \end{bmatrix}$$

Substituting these expressions into Eqs. (4-18) and (4-19) and carrying out the numerical calculations, we obtain the answers

$$Y_n = AY_bA^t = \begin{bmatrix} 12.5 & -2 & -8.5 \\ -10 & 10 & 5 \\ -0.5 & 1 & -0.5 \end{bmatrix}$$

$$J_n = A(J - Y_bE) = \begin{bmatrix} 40 \\ -23 \\ 5 \end{bmatrix}$$

Since Eq. (4-17) can be readily constructed once the reduced incidence matrix A is known, it can be easily generated by a computer program. The explicit solution given by Eq. (4-20) requires the inversion of an $n \times n$ matrix. It will be shown in Section 4-3 that this operation would require more computer time than that of solving Eq. (4-17) directly. Consequently, from the computational point of view, it is always more efficient to solve a system of linear equations than to compute the inverse of the associated matrix.

4-3 GAUSSIAN ELIMINATION ALGORITHM

Consider a system of n linear equations with real constant coefficients:

$$\begin{aligned} a_{11}x_1 + a_{12}x_2 + \ldots + a_{1n}x_n &= \mu_1 \\ a_{21}x_1 + a_{22}x_2 + \ldots + a_{2n}x_n &= \mu_2 \\ \vdots \qquad \vdots \qquad\quad \vdots \quad &\quad \vdots \\ a_{n1}x_1 + a_{n2}x_2 + \ldots + a_{nn}x_n &= \mu_n \end{aligned} \tag{4-21}$$

Equation (4-21) can be written in the matrix form

$$A\mathbf{x} = \boldsymbol{\mu} \tag{4-22}$$

where

$$\mathbf{x} = (x_1, x_2, \ldots, x_n)^t$$

$$\boldsymbol{\mu} = (\mu_1, \mu_2, \ldots, \mu_n)^t$$

and

$$A = \begin{bmatrix} a_{11} & a_{12} & \cdots & a_{1n} \\ a_{21} & a_{22} & \cdots & a_{2n} \\ \vdots & \vdots & & \vdots \\ a_{n1} & a_{n2} & \cdots & a_{nn} \end{bmatrix}$$

There are many methods for solving Eq. (4-21). The best known, but also the most inefficient, method is by using Cramer's rule. It can be shown that it requires approximately $N = 2(n + 1)!$ multiplications to solve Eq. (4-21) by Cramer's rule.[3] To appreciate how fast this estimate increases with n, observe that $N = 1440$ when $n = 5$, and $N = 79{,}833{,}600$ when $n = 10$. Almost any other method for solving Eq. (4-21) on a computer is more efficient than Cramer's rule. We shall consider only the *most efficient* all-around method in this chapter, the *Gaussian elimination* algorithm.

The Gaussian elimination is based on the extremely elementary idea of eliminating the variables one at a time until there is only one equation in one variable left. This equation is then solved to give the solution for this one variable, say x_n. One then substitutes the value of x_n back into the preceding equations to obtain the remaining solutions. To illustrate this procedure, let us consider the case $n = 4$ and carry the algorithm through to completion; consider

$$a_{11}x_1 + a_{12}x_2 + a_{13}x_3 + a_{14}x_4 = \mu_1 \tag{4-23a}$$

$$a_{21}x_1 + a_{22}x_2 + a_{23}x_3 + a_{24}x_4 = \mu_2 \tag{4-23b}$$

$$a_{31}x_1 + a_{32}x_2 + a_{33}x_3 + a_{34}x_4 = \mu_3 \tag{4-23c}$$

$$a_{41}x_1 + a_{42}x_2 + a_{43}x_3 + a_{44}x_4 = \mu_4 \tag{4-23d}$$

The Gaussian elimination algorithm for solving a system of n equations consists of two major steps: (1) a *forward-elimination step*, and (2) a *back-substitution step.*

FORWARD ELIMINATION. Step 1 is carried out in $n - 1$ *stages*, where n is the number of equations in Eq. (4-21).

Stage 1. Eliminate the variable x_1 from Eqs. (4-23b, c, and d) to obtain

$$a_{11}x_1 + a_{12}x_2 + a_{13}x_3 + a_{14}x_4 = \mu_1 \tag{4-24a}$$

$$a_{22}^{(2)}x_2 + a_{23}^{(2)}x_3 + a_{24}^{(2)}x_4 = \mu_2^{(2)} \tag{4-24b}$$

$$a_{32}^{(2)}x_2 + a_{33}^{(2)}x_3 + a_{34}^{(2)}x_4 = \mu_3^{(2)} \tag{4-24c}$$

$$a_{42}^{(2)}x_2 + a_{43}^{(2)}x_3 + a_{44}^{(2)}x_4 = \mu_4^{(2)} \tag{4-24d}$$

Equation (4-24b) is obtained by multiplying the first equation by $-a_{21}/a_{11}$ and adding the result to Eq. (4-23b). Similarly, Eqs. (4-24c and d) are obtained by multiplying Eq. (4-23a) by $-a_{31}/a_{11}$ and $-a_{41}/a_{11}$, respectively, and adding the result to Eqs. (4-23c and d), respectively. This elimination process is equivalent to premultiplying the matrix equation (4-23) by the following three matrices in the indicated order:

[3]The efficiency of a computational algorithm is based on the total number of arithmetic operations required by the algorithm. However, since the time it takes a computer to perform an addition or subtraction is considerably less than the time it takes to perform a multiplication or division, the computational efficiency of an algorithm is usually figured in practice as the total *number of multiplications* needed to carry the algorithm to completion (a division is counted as a multiplication).

$$\mathcal{E}_1 = \begin{bmatrix} 1 & 0 & 0 & 0 \\ \dfrac{-a_{21}}{a_{11}} & 1 & 0 & 0 \\ 0 & 0 & 1 & 0 \\ 0 & 0 & 0 & 1 \end{bmatrix}, \quad \mathcal{E}_2 = \begin{bmatrix} 1 & 0 & 0 & 0 \\ 0 & 1 & 0 & 0 \\ \dfrac{-a_{31}}{a_{11}} & 0 & 1 & 0 \\ 0 & 0 & 0 & 1 \end{bmatrix},$$

$$\mathcal{E}_3 = \begin{bmatrix} 1 & 0 & 0 & 0 \\ 0 & 1 & 0 & 0 \\ 0 & 0 & 1 & 0 \\ \dfrac{-a_{41}}{a_{11}} & 0 & 0 & 1 \end{bmatrix} \tag{4-25}$$

Observe that each of these three matrices is constructed by adding a number $-a_{j1}/a_{11}$ to an *identity matrix*. Such a matrix is usually called an *elementary matrix*, because it has the effect of carrying out the *elementary operations* required to obtain Eq. (4-24).[4] Observe also that the preceding operation is valid only if the element $a_{11} \neq 0$. This element is called a *pivoting element* or, more simply, a *pivot*. The additional steps required to handle a zero pivot will be discussed later. Meanwhile, let us continue our algorithm.

Stage 2. Eliminate the variable x_2 from Eqs. (4-24c and d) to obtain

$$a_{11}x_1 + a_{12}x_2 + a_{13}x_3 + a_{14}x_4 = \mu_1 \tag{4-26a}$$
$$a_{22}^{(2)}x_2 + a_{23}^{(2)}x_3 + a_{24}^{(2)}x_4 = \mu_2^{(2)} \tag{4-26b}$$
$$a_{33}^{(3)}x_3 + a_{34}^{(3)}x_4 = \mu_3^{(3)} \tag{4-26c}$$
$$a_{43}^{(3)}x_3 + a_{44}^{(3)}x_4 = \mu_4^{(3)} \tag{4-26d}$$

It should now be clear that Eqs. (4-26c and d) can be obtained by premultiplying the matrix equation representing Eq. (4-24) by the following two elementary matrices with $a_{22}^{(2)}$ as a pivot.

$$\mathcal{E}_4 = \begin{bmatrix} 1 & 0 & 0 & 0 \\ 0 & 1 & 0 & 0 \\ 0 & \dfrac{-a_{32}^{(2)}}{a_{22}^{(2)}} & 1 & 0 \\ 0 & 0 & 0 & 1 \end{bmatrix}, \quad \mathcal{E}_5 = \begin{bmatrix} 1 & 0 & 0 & 0 \\ 0 & 1 & 0 & 0 \\ 0 & 0 & 1 & 0 \\ 0 & \dfrac{-a_{42}^{(2)}}{a_{22}^{(2)}} & 0 & 1 \end{bmatrix} \tag{4-27}$$

Stage 3. Eliminate the variable x_3 from Eq. (4-26d) to obtain

$$a_{11}x_1 + a_{12}x_2 + a_{13}x_3 + a_{14}x_4 = \mu_1 \tag{4-28a}$$
$$a_{22}^{(2)}x_2 + a_{23}^{(2)}x_3 + a_{24}^{(2)}x_4 = \mu_2^{(2)} \tag{4-28b}$$
$$a_{33}^{(3)}x_3 + a_{34}^{(3)}x_4 = \mu_3^{(3)} \tag{4-28c}$$
$$a_{44}^{(4)}x_4 = \mu_4^{(4)} \tag{4-28d}$$

[4]See Chapter 3 for definitions of the various types of elementary matrices.

Equation (4-28d) is obtained by premultiplying the matrix equation representing Eq. (4-28) by the following elementary matrix:

$$\boldsymbol{\mathcal{E}}_6 = \begin{bmatrix} 1 & 0 & 0 & 0 \\ 0 & 1 & 0 & 0 \\ 0 & 0 & 1 & 0 \\ 0 & 0 & \dfrac{-a_{43}^{(3)}}{a_{33}^{(3)}} & 1 \end{bmatrix} \tag{4-29}$$

Equation (4-28) can therefore be written as

$$(\boldsymbol{\mathcal{E}}_6\boldsymbol{\mathcal{E}}_5\boldsymbol{\mathcal{E}}_4\boldsymbol{\mathcal{E}}_3\boldsymbol{\mathcal{E}}_2\boldsymbol{\mathcal{E}}_1)\boldsymbol{Ax} \triangleq \boldsymbol{Ux} = \begin{bmatrix} a_{11} & a_{12} & a_{13} & a_{14} \\ 0 & a_{22}^{(2)} & a_{23}^{(2)} & a_{24}^{(2)} \\ 0 & 0 & a_{33}^{(3)} & a_{34}^{(3)} \\ 0 & 0 & 0 & a_{44}^{(4)} \end{bmatrix} \begin{bmatrix} x_1 \\ x_2 \\ x_3 \\ x_4 \end{bmatrix} = \begin{bmatrix} \mu_1 \\ \mu_2^{(2)} \\ \mu_3^{(3)} \\ \mu_4^{(4)} \end{bmatrix} \tag{4-30}$$

Observe that all elements below the main diagonal of the matrix $\boldsymbol{U}$ are zero. Such a matrix is called an *upper triangular matrix*. Consequently, *step 1 of the Gaussian elimination is equivalent to transforming the matrix* $\boldsymbol{A}$ *into an upper triangular matrix.*

Back Substitution. Step 2 of the Gaussian elimination consists of $n - 1$ back-substitution steps. First, we solve for x_4 from Eq. (4-28d) to obtain $x_4 = \mu_4^{(4)}/a_{44}^{(4)}$. This step is equivalent to premultiplying the matrix equation representing Eq. (4-30) by the following elementary matrix:

$$\boldsymbol{S}_1 = \begin{bmatrix} 1 & 0 & 0 & 0 \\ 0 & 1 & 0 & 0 \\ 0 & 0 & 1 & 0 \\ 0 & 0 & 0 & \dfrac{1}{a_{44}^{(4)}} \end{bmatrix} \tag{4-31}$$

This elementary matrix is obtained by multiplying the elements in one row by a scalar constant and differs, therefore, from the elementary matrices $\boldsymbol{\mathcal{E}}_1, \boldsymbol{\mathcal{E}}_2, \ldots, \boldsymbol{\mathcal{E}}_6$ defined earlier. It is convenient to distinguish between these two types of elementary matrices by referring to the matrix constructed by the preceding process as an *elementary matrix of type 2*, and the matrix constructed earlier as an *elementary matrix of type 3.*[5] Premultiplying Eq. (4-30) by $\mathbf{S}_1$, we obtain

$$\boldsymbol{S}_1\boldsymbol{Ux} = \begin{bmatrix} a_{11} & a_{12} & a_{13} & a_{14} \\ 0 & a_{22}^{(2)} & a_{23}^{(2)} & a_{24}^{(2)} \\ 0 & 0 & a_{33}^{(3)} & a_{34}^{(3)} \\ 0 & 0 & 0 & 1 \end{bmatrix} \begin{bmatrix} x_1 \\ x_2 \\ x_3 \\ x_4 \end{bmatrix} = \begin{bmatrix} \mu_1 \\ \mu_2^{(2)} \\ \mu_3^{(3)} \\ \dfrac{\mu_4^{(4)}}{a_{44}^{(4)}} \end{bmatrix} = \begin{bmatrix} \mu_1 \\ \mu_2^{(2)} \\ \mu_3^{(3)} \\ \mu_4^{(5)} \end{bmatrix} \tag{4-32}$$

[5] We shall reserve type 1 for the elementary matrix obtained by an interchange of two rows or two columns. See Section 3-6-1.

The next back-substitution step is to premultiply Eq. (4-32) by the following matrices of type 3 and type 2:

$$\boldsymbol{\mathcal{E}}_7 = \begin{bmatrix} 1 & 0 & 0 & 0 \\ 0 & 1 & 0 & 0 \\ 0 & 0 & 1 & -a_{34}^{(3)} \\ 0 & 0 & 0 & 1 \end{bmatrix}, \qquad \boldsymbol{S}_2 = \begin{bmatrix} 1 & 0 & 0 & 0 \\ 0 & 1 & 0 & 0 \\ 0 & 0 & \dfrac{1}{a_{33}^{(3)}} & 0 \\ 0 & 0 & 0 & 1 \end{bmatrix} \tag{4-33}$$

and obtain

$$\boldsymbol{S}_2\boldsymbol{\mathcal{E}}_7(\boldsymbol{S}_1\boldsymbol{U})\boldsymbol{x} = \begin{bmatrix} a_{11} & a_{12} & a_{13} & a_{14} \\ 0 & a_{22}^{(2)} & a_{23}^{(2)} & a_{24}^{(2)} \\ 0 & 0 & 1 & 0 \\ 0 & 0 & 0 & 1 \end{bmatrix} \begin{bmatrix} x_1 \\ x_2 \\ x_3 \\ x_4 \end{bmatrix} = \begin{bmatrix} \mu_1 \\ \mu_2^{(2)} \\ \mu_3^{(6)} \\ \mu_4^{(5)} \end{bmatrix} \tag{4-34}$$

Next, we premultiply Eq. (4-34) by two type 3 elementary matrices, $\boldsymbol{\mathcal{E}}_8$ and $\boldsymbol{\mathcal{E}}_9$, and a type 2 elementary matrix, $\boldsymbol{S}_3$, where

$$\boldsymbol{\mathcal{E}}_8 = \begin{bmatrix} 1 & 0 & 0 & 0 \\ 0 & 1 & 0 & -a_{24}^{(2)} \\ 0 & 0 & 1 & 0 \\ 0 & 0 & 0 & 1 \end{bmatrix}, \qquad \boldsymbol{\mathcal{E}}_9 = \begin{bmatrix} 1 & 0 & 0 & 0 \\ 0 & 1 & -a_{23}^{(2)} & 0 \\ 0 & 0 & 1 & 0 \\ 0 & 0 & 0 & 1 \end{bmatrix},$$

$$\boldsymbol{S}_3 = \begin{bmatrix} 1 & 0 & 0 & 0 \\ 0 & \dfrac{1}{a_{22}^{(2)}} & 0 & 0 \\ 0 & 0 & 1 & 0 \\ 0 & 0 & 0 & 1 \end{bmatrix} \tag{4-35}$$

The result is

$$\boldsymbol{S}_3\boldsymbol{\mathcal{E}}_9\boldsymbol{\mathcal{E}}_8(\boldsymbol{S}_2\boldsymbol{\mathcal{E}}_7\boldsymbol{S}_1\boldsymbol{U})\boldsymbol{x} = \begin{bmatrix} a_{11} & a_{12} & a_{13} & a_{14} \\ 0 & 1 & 0 & 0 \\ 0 & 0 & 1 & 0 \\ 0 & 0 & 0 & 1 \end{bmatrix} \begin{bmatrix} x_1 \\ x_2 \\ x_3 \\ x_4 \end{bmatrix} = \begin{bmatrix} \mu_1 \\ \mu_2^{(7)} \\ \mu_3^{(6)} \\ \mu_4^{(5)} \end{bmatrix} \tag{4-36}$$

The final step is to premultiply Eq. (4-36) by three type-3 matrices, $\boldsymbol{\mathcal{E}}_{10}$, $\boldsymbol{\mathcal{E}}_{11}$, $\boldsymbol{\mathcal{E}}_{12}$, and a type 2 matrix, $\boldsymbol{S}_4$, where

$$\boldsymbol{\mathcal{E}}_{10} = \begin{bmatrix} 1 & 0 & 0 & -a_{14} \\ 0 & 1 & 0 & 0 \\ 0 & 0 & 1 & 0 \\ 0 & 0 & 0 & 1 \end{bmatrix}, \qquad \boldsymbol{\mathcal{E}}_{11} = \begin{bmatrix} 1 & 0 & -a_{13} & 0 \\ 0 & 1 & 0 & 0 \\ 0 & 0 & 1 & 0 \\ 0 & 0 & 0 & 1 \end{bmatrix}$$

$$\boldsymbol{\mathcal{E}}_{12} = \begin{bmatrix} 1 & -a_{12} & 0 & 0 \\ 0 & 1 & 0 & 0 \\ 0 & 0 & 1 & 0 \\ 0 & 0 & 0 & 1 \end{bmatrix} \tag{4-37}$$

$$\boldsymbol{S}_4 = \begin{bmatrix} \frac{1}{a_{11}} & 0 & 0 & 0 \\ 0 & 1 & 0 & 0 \\ 0 & 0 & 1 & 0 \\ 0 & 0 & 0 & 1 \end{bmatrix} \tag{4-38}$$

The result is

$$\boldsymbol{S}_4\boldsymbol{\mathcal{E}}_{12}\boldsymbol{\mathcal{E}}_{11}\boldsymbol{\mathcal{E}}_{10}(\boldsymbol{S}_3\boldsymbol{\mathcal{E}}_9\boldsymbol{\mathcal{E}}_8\boldsymbol{S}_2\boldsymbol{\mathcal{E}}_7\boldsymbol{S}_1\boldsymbol{U})\boldsymbol{x} = \begin{bmatrix} 1 & 0 & 0 & 0 \\ 0 & 1 & 0 & 0 \\ 0 & 0 & 1 & 0 \\ 0 & 0 & 0 & 1 \end{bmatrix} \begin{bmatrix} x_1 \\ x_2 \\ x_3 \\ x_4 \end{bmatrix} = \begin{bmatrix} \mu_1^{(8)} \\ \mu_2^{(7)} \\ \mu_3^{(6)} \\ \mu_4^{(5)} \end{bmatrix} \tag{4-39}$$

Hence, the final solution is given by $\boldsymbol{x} = (\mu_1^{(8)}, \mu_2^{(7)}, \mu_3^{(6)}, \mu_4^{(5)})^t$.

The preceding developments show that the *Gaussian elimination* can be interpreted as the process of *premultiplying both sides* of the matrix equation $\boldsymbol{A}\boldsymbol{x} = \boldsymbol{\mu}$ by an appropriate sequence of type 2 and type 3 elementary matrices until the matrix $\boldsymbol{A}$ reduces to an identity matrix and the vector $\boldsymbol{\mu}$ to a new vector $\hat{\boldsymbol{\mu}}$, which turns out to be the solution $\boldsymbol{x} = \boldsymbol{A}^{-1}\boldsymbol{\mu}$. If we substitute the product of the elementary matrices $(\boldsymbol{\mathcal{E}}_6\boldsymbol{\mathcal{E}}_5\boldsymbol{\mathcal{E}}_4\boldsymbol{\mathcal{E}}_3\boldsymbol{\mathcal{E}}_2\boldsymbol{\mathcal{E}}_1)\boldsymbol{A}$ from Eq. (4-30) for the matrix $\boldsymbol{U}$ in Eq. (4-39), we would obtain the *complete* sequence of elementary matrices that reduces $\boldsymbol{A}$ to the identity matrix

$$(\boldsymbol{S}_4\boldsymbol{\mathcal{E}}_{12}\boldsymbol{\mathcal{E}}_{11}\boldsymbol{\mathcal{E}}_{10}\boldsymbol{S}_3\boldsymbol{\mathcal{E}}_9\boldsymbol{\mathcal{E}}_8\boldsymbol{S}_2\boldsymbol{\mathcal{E}}_7\boldsymbol{S}_1\boldsymbol{\mathcal{E}}_6\boldsymbol{\mathcal{E}}_5\boldsymbol{\mathcal{E}}_4\boldsymbol{\mathcal{E}}_3\boldsymbol{\mathcal{E}}_2\boldsymbol{\mathcal{E}}_1)\boldsymbol{A} \triangleq \boldsymbol{P}\boldsymbol{A} = \boldsymbol{1} \tag{4-40}$$

It follows from Eq. (4-40) that

$$\boldsymbol{P} \triangleq \boldsymbol{S}_4\boldsymbol{\mathcal{E}}_{12}\boldsymbol{\mathcal{E}}_{11}\boldsymbol{\mathcal{E}}_{10}\boldsymbol{S}_3\boldsymbol{\mathcal{E}}_9\boldsymbol{\mathcal{E}}_8\boldsymbol{S}_2\boldsymbol{\mathcal{E}}_7\boldsymbol{S}_1\boldsymbol{\mathcal{E}}_6\boldsymbol{\mathcal{E}}_5\boldsymbol{\mathcal{E}}_4\boldsymbol{\mathcal{E}}_3\boldsymbol{\mathcal{E}}_2\boldsymbol{\mathcal{E}}_1 = \boldsymbol{A}^{-1} \tag{4-41}$$

In other words, we have shown in detail (for the $n = 4$ case) that *the product of the elementary matrices* which reduces the given matrix $\boldsymbol{A}$ to an identity matrix is actually the *inverse matrix* $\boldsymbol{A}^{-1}$. It is not difficult to show that this is true for a matrix of any order n [see Eq. (3-32)].

Our lengthy derivation was mainly intended to establish the preceding important observation. In actually programming the Gaussian elimination algorithm, we of course do not have to generate the elementary matrices and multiply them as in Eq. (4-41). This would be a lengthy procedure! Instead, we only need to carry out elementary row operations on the matrix $\boldsymbol{A}$ and the vector $\boldsymbol{\mu}$ simultaneously until $\boldsymbol{A}$ reduces to an identity matrix. This is best achieved by augmenting the $n \times n$ matrix $\boldsymbol{A}$ with the column vector $\boldsymbol{\mu}$ to obtain an $n \times (n + 1)$ augmented matrix

$$\boldsymbol{A}_{\text{aug}} \triangleq [\boldsymbol{A} \mid \boldsymbol{\mu}] \tag{4-42}$$

Observe that if we premultiply $\boldsymbol{A}_{\text{aug}}$ by the product $\boldsymbol{P}$ of elementary matrices as defined in (4-41), we obtain

$$\boldsymbol{P}\boldsymbol{A}_{\text{aug}} = [\boldsymbol{P}\boldsymbol{A} \mid \boldsymbol{P}\boldsymbol{\mu}] = [\boldsymbol{1} \mid \hat{\boldsymbol{\mu}}] \tag{4-43}$$

where

$$\hat{\boldsymbol{\mu}} \triangleq \boldsymbol{P}\boldsymbol{\mu} = \boldsymbol{A}^{-1}\boldsymbol{\mu} \tag{4-44}$$

Hence, by performing elementary row operations directly on the augmented matrix $\boldsymbol{A}_{\text{aug}}$ until $\boldsymbol{A}$ reduces to an identity matrix, we automatically recover the solution vector $\boldsymbol{x} = A^{-1}\boldsymbol{\mu} = \hat{\boldsymbol{\mu}}$ in the $(n+1)$th column. It can be shown that the total number of *multiplications* required to implement this reduction process in the Gaussian elimination algorithm is equal *approximately* to $n^3/3$ (see Problem 4-5).

Although this computational algorithm will eventually give us the solution vector $\hat{\boldsymbol{\mu}}$, it does *not* yield the inverse matrix A^{-1}. We shall now show that by an almost trivial extension the matrix A^{-1} can also be obtained by the Gaussian elimination algorithm. To do this, instead of augmenting the matrix A with the vector $\boldsymbol{\mu}$ as in Eq. (4-42), let us augment A with an $n \times n$ identity matrix

$$\hat{A}_{\text{aug}} \triangleq [A \mid \mathbf{1}] \tag{4-45}$$

If we now premultiply $\hat{A}_{\text{aug}}$ by the matrix P in Eq. (4-41), we obtain

$$P\hat{A}_{\text{aug}} = [PA \mid P] = [\mathbf{1} \mid A^{-1}] \tag{4-46}$$

In other words, if we carry out the same elementary row operations on the matrix $\hat{A}_{\text{aug}}$ (formed by attaching an $n \times n$ identity matrix to A) until the matrix A reduces to an identity matrix, the new matrix generated to the right of the identity matrix is precisely the *inverse* matrix A^{-1}. Observe that the augmented matrix $\hat{A}_{\text{aug}}$ in Eq. (4-45) has $n - 1$ more columns than the augmented matrix A_{aug} in Eq. (4-42). Hence, it should now be clear that it requires more multiplications to find the *inverse* of a matrix than to solve for $\boldsymbol{x}$ from the equation $A\boldsymbol{x} = \boldsymbol{\mu}$ by Gaussian elimination. In fact, it can be shown that *inverting a matrix by Gaussian elimination requires approximately n^3 multiplications*, in contrast to $n^3/3$ required when solving for $\boldsymbol{x}$ directly. This is why we do not solve the equation $A\boldsymbol{x} = \boldsymbol{\mu}$ by finding the inverse of A on a computer.

So far we have made the tacit assumption that none of the pivoting elements is zero. What happens if this assumption is violated? To be specific, suppose that $a_{11} = 0$. This means that the variable x_1 is absent in the first equation in Eq. (4-23). In this case, we can *interchange* the first equation and one of the remaining equations, say the jth equation for which $a_{j1} \neq 0$.[6] This same procedure can be carried out in any subsequent stage when a pivot is zero. Fortunately, interchanging two rows on a computer involves only a simple interchange of index.

It occurs frequently in practice that, even if a pivot is not zero, severe error will result if the absolute value of the pivot is very small. This arises because of the round-off error inherent in any computer. To appreciate the seriousness of the problem, consider the following example:

$$\begin{bmatrix} 0.000125 & 1.25 \\ 12.5 & 12.5 \end{bmatrix} \begin{bmatrix} x_1 \\ x_2 \end{bmatrix} = \begin{bmatrix} 6.25 \\ 75 \end{bmatrix} \tag{4-47}$$

The solution correct to five digits is easily found to be $x_1 = 1.0001$ and $x_2 = 4.9999$. Suppose that 3-digit floating arithmetic is used. The solution obtained by Gaussian

[6]There must exist at least one such equation; otherwise, the variable x_1 would be absent in every equation. This is equivalent to saying that the matrix is *singular* and no unique solution is possible. Since we are concerned only with real-life circuits in this book, there is no loss of generality in assuming that the matrix is nonsingular.

elimination without row interchange is given by

$$\begin{aligned} 0.000125x_1 + 1.25x_2 &= 6.25 \\ -1.25 \times 10^5 x_2 &= -6.25 \times 10^5 \end{aligned}$$

Consequently, $x_2 = 5.00$ and $x_1 = 0$, which is clearly unacceptable. Let us now see what happens if we interchange the rows in Eq. (4-47):

$$\begin{bmatrix} 12.5 & 12.5 \\ 0.000125 & 1.25 \end{bmatrix} \begin{bmatrix} x_1 \\ x_2 \end{bmatrix} = \begin{bmatrix} 75 \\ 6.25 \end{bmatrix} \tag{4-48}$$

Applying the Gaussian elimination again, we obtain

$$\begin{aligned} 12.5x_1 + 12.5x_2 &= 75 \\ 1.25x_2 &= 6.25 \end{aligned}$$

Hence, $x_2 = 5.00$ and $x_1 = 1.00$, which is obviously the correct solution.

Our example suggests that too small a pivot is likely to cause severe numerical error. It has been suggested [1] that for most matrices a good strategy is to choose the pivot at the kth stage to be the element with the largest absolute value in column k of all rows from k through n. Testing for size to find the optimum pivot and the subsequent row-interchange operation required to move the desired pivot to the proper position would no doubt require additional computer time. However, the improvement in accuracy is usually worth the extra time. Even further refinement is possible by interchanging not only the rows but also the columns. This strategy is sometimes referred to as *complete pivoting*. For further details concerning this topic, see [3].

*4-4 THE *LU* FACTORIZATION[7]

4-4-1 A Theorem on Factorization

In Section 4-3, we discussed the Gaussian elimination algorithm. It was shown that any *nonsingular* matrix A can be reduced to an upper triangular matrix U, with unit diagonal elements, by a proper sequence of elementary row operations. Or equivalently,

$$\mathcal{E}_m \mathcal{E}_{m-1} \dots \mathcal{E}_2 \mathcal{E}_1 A = U \tag{4-49}$$

where $u_{11} = u_{22} = \dots = u_{nn} = 1$, and $\mathcal{E}_k$ may be any of the three types of elementary matrices.

The type 3 elementary matrices in Eq. (4-49) are lower triangular matrices,[8] because we always add c times one row to a *later* row. Any type 2 elementary matrix is a lower triangular matrix, but this is not so for a type 1 elementary matrix.

If the pivot elements $a_{11}, a_{22}^{(2)}, \dots$ at all stages are nonzero, then type 1 elementary matrices are not needed in Eq. (4-49). Since the product of any two lower triangular matrices is also a lower triangular matrix, it follows that the product

[7]The results of this section are used in Chapter 16 in conjunction with sparse-matrix techniques.

[8]A *lower triangular matrix* L is a square matrix with zero elements above the principal diagonal.

$$(\mathcal{E}_m \mathcal{E}_{m-1} \ldots \mathcal{E}_2 \mathcal{E}_1)$$

in Eq. (4-49) is lower triangular. It can be easily shown that the inverse of any nonsingular lower triangular matrix is also lower triangular. Then we can solve for A in Eq. (4-49) to give

$$\boxed{\begin{aligned} A &= (\mathcal{E}_m \mathcal{E}_{m-1} \ldots \mathcal{E}_2 \mathcal{E}_1)^{-1} \boldsymbol{U} \\ &= \boldsymbol{LU} \end{aligned}} \tag{4-50}$$

where $\boldsymbol{L} = \mathcal{E}_1^{-1} \mathcal{E}_2^{-1} \ldots \mathcal{E}_m^{-1}$ is a lower triangular matrix.

On the other hand, if at any stage the pivot element $a_{kk}^{(k)}$ is zero, then a row interchange is needed for the elimination procedure to continue. In such a case, some of the elementary matrices in Eq. (4-49) will be of type 1. To explain the effect of row interchange, let us introduce the following subscript notations for the three types of elementary matrices:

- $\mathcal{E}_{ij}$: Type 1 elementary matrix obtained from **1** by interchanging the ith and jth rows.
- $\mathcal{E}_i(c)$: Type 2 elementary matrix obtained from **1** by multiplying the ith row by c $(c \neq 0)$.
- $\mathcal{E}_{ij}(c)$: Type 3 elementary matrix obtained from **1** by adding c times the ith row to the jth row.
- $\boldsymbol{P}$: Type 1 elementary matrix, or the product of some type 1 elementary matrices.

Then, for the worst case when a row interchange is needed in every stage of the forward-elimination step, Eq. (4-49) for the case $n = 4$ is of the form[9]

$$\underbrace{\mathcal{E}_4(c_{10})\mathcal{E}_3(c_9)\mathcal{E}_2(c_8)\mathcal{E}_1(c_7)}_{\text{normalizing diagonal elements}}\ \underbrace{\mathcal{E}_{34}(c_6)\boldsymbol{P}_3}_{\text{third stage}}\ \underbrace{\mathcal{E}_{24}(c_5)\mathcal{E}_{23}(c_4)\boldsymbol{P}_2}_{\text{second stage}}\ \underbrace{\mathcal{E}_{14}(c_3)\mathcal{E}_{13}(c_2)\mathcal{E}_{12}(c_1)\boldsymbol{P}_1}_{\text{first stage}}\boldsymbol{A} = \boldsymbol{U} \tag{4-51}$$

where each $\boldsymbol{P}_k$ is an $\mathcal{E}_{ij}$. Each $\boldsymbol{P}_k$ in Eq. (4-51) can be moved to the right of a neighboring $\mathcal{E}_j$ matrix provided that "proper changes" are made on the elements of $\mathcal{E}_j$ to obtain a new elementary $\hat{\mathcal{E}}_j$. For example, if $\boldsymbol{P}_k = \mathcal{E}_{12}$ and $\mathcal{E}_j = \mathcal{E}_{13}(\alpha)$, then

$$\boldsymbol{P}_k \mathcal{E}_j = \begin{bmatrix} 0 & 1 & 0 \\ 1 & 0 & 0 \\ 0 & 0 & 1 \end{bmatrix} \begin{bmatrix} 1 & 0 & 0 \\ 0 & 1 & 0 \\ \alpha & 0 & 1 \end{bmatrix} = \begin{bmatrix} 0 & 1 & 0 \\ 1 & 0 & 0 \\ \alpha & 0 & 1 \end{bmatrix}$$

$$= \begin{bmatrix} 1 & 0 & 0 \\ 0 & 1 & 0 \\ 0 & \alpha & 1 \end{bmatrix} \begin{bmatrix} 0 & 1 & 0 \\ 1 & 0 & 0 \\ 0 & 0 & 1 \end{bmatrix} = \hat{\mathcal{E}}_j \boldsymbol{P}_k$$

[9] Similar arguments hold for the case $n > 4$. But the expression is too lengthy to be given here.

Hence

$$\hat{\mathcal{E}}_j = \mathcal{E}_{23}(\alpha)$$

The general rules for moving P_k one step to the right are as follows (the proof is straightforward). Assuming that $i \neq j \neq k \neq l$, then

$$\begin{aligned}
&\mathcal{E}_{ij}\mathcal{E}_{kl}(c) = \mathcal{E}_{kl}(c)\mathcal{E}_{ij} \\
&\mathcal{E}_{ij}\mathcal{E}_{jk}(c) = \mathcal{E}_{ik}(c)\mathcal{E}_{ij}, \qquad \mathcal{E}_{ij}\mathcal{E}_{ik}(c) = \mathcal{E}_{jk}(c)\mathcal{E}_{ij} \\
&\mathcal{E}_{ij}\mathcal{E}_{kj}(c) = \mathcal{E}_{ki}(c)\mathcal{E}_{ij}, \qquad \mathcal{E}_{ij}\mathcal{E}_{ki}(c) = \mathcal{E}_{kj}(c)\mathcal{E}_{ij} \\
&\mathcal{E}_{ij}\mathcal{E}_{ij}(c) = \mathcal{E}_{ji}(c)\mathcal{E}_{ij}
\end{aligned} \tag{4-52}$$

In other words, the subscripts of $\hat{\mathcal{E}}$ are obtained from the subscripts of $\mathcal{E}$ by making interchanges (possibly none) of the subscripts $j \leftrightarrow i$.

Now consider a typical P_k, say P_2, in Eq. (4-51). Its corresponding row operation is to interchange the second row with some *later* row, say the fourth row, so that a nonzero element is brought to the (2, 2) position. Then $P_2 = \mathcal{E}_{24}$. Applying the rules of Eq. (4-52) to Eq. (4-51), we have

$$\begin{aligned}
P_2\mathcal{E}_{14}(c_3)\mathcal{E}_{13}(c_2)\mathcal{E}_{12}(c_1) &= \mathcal{E}_{24}\mathcal{E}_{14}(c_3)\mathcal{E}_{13}(c_2)\mathcal{E}_{12}(c_1) = \mathcal{E}_{12}(c_3)\mathcal{E}_{24}\mathcal{E}_{13}(c_2)\mathcal{E}_{12}(c_1) \\
&= \mathcal{E}_{12}(c_3)\mathcal{E}_{13}(c_2)\mathcal{E}_{24}\mathcal{E}_{12}(c_1) = \mathcal{E}_{12}(c_3)\mathcal{E}_{13}(c_2)\mathcal{E}_{14}(c_1)\mathcal{E}_{24} \\
&= \mathcal{E}_{12}(c_3)\mathcal{E}_{13}(c_2)\mathcal{E}_{14}(c_1)P_2
\end{aligned}$$

Thus, we have moved P_2 successively to the rightmost position. Although the contents of the other $\mathcal{E}$ matrices change, note that they all remain lower triangular matrices!

From our discussion we see that, after moving all P_k's to the right, Eq. (4-49) may be written as

$$(\hat{\mathcal{E}}_m \ldots \hat{\mathcal{E}}_2\hat{\mathcal{E}}_1)(P_{n-1} \ldots P_2P_1)A = U \tag{4-53}$$

where all $\hat{\mathcal{E}}_k$ are lower triangular matrices. Equation (4-53) directly leads to

$$LU = PA$$

where

$$P = P_{n-1}P_{n-2} \ldots P_2P_1$$

is a *permutation matrix*, and

$$L = (\hat{\mathcal{E}}_m \ldots \hat{\mathcal{E}}_2\hat{\mathcal{E}}_1)^{-1}$$

is a lower triangular matrix. We have, in fact, proved Theorem 4-1.

Theorem 4-1. *Given any* nonsingular *matrix A, there exists some permutation matrix P (possibly $P = 1$) such that*

$$\boxed{LU = PA} \tag{4-54}$$

where U is an upper triangular matrix with unit diagonal elements and L is a *lower triangular matrix with nonzero diagonal elements. Once a proper **P** is chosen, this factorization is unique.*

The factorization of $\boldsymbol{A}$ or $\boldsymbol{PA}$ into the product $\boldsymbol{LU}$ is called ***LU** factorization.* By inserting the product of two diagonal matrices $\boldsymbol{D}_1$ and $\boldsymbol{D}_2$, with $\boldsymbol{D}_1\boldsymbol{D}_2 = \boldsymbol{1}$, between $\boldsymbol{L}$ and $\boldsymbol{U}$, the theorem can take on many different forms, as follows:

$$\boldsymbol{LU} = \boldsymbol{LD}_1\boldsymbol{D}_2\boldsymbol{U} = (\boldsymbol{LD}_1)(\boldsymbol{D}_2\boldsymbol{U}) = \hat{\boldsymbol{L}}\hat{\boldsymbol{U}} = \hat{\boldsymbol{L}}\boldsymbol{D}_2\boldsymbol{U} = \boldsymbol{LD}_1\hat{\boldsymbol{U}} \tag{4-55}$$

where $\hat{\boldsymbol{L}}$ and $\hat{\boldsymbol{U}}$ are lower and upper triangular matrices, respectively, one of which may have arbitrary nonzero diagonal elements.

Once the $\boldsymbol{LU}$ factorization is obtained, by whatever method, the equation

$$\boldsymbol{Ax} = \boldsymbol{LUx} = \boldsymbol{\mu} \tag{4-56}$$

is solved by transforming Eq. (4-56) into

$$\boldsymbol{Ly} = \boldsymbol{\mu} \tag{4-57}$$

and

$$\boldsymbol{Ux} = \boldsymbol{y} \tag{4-58}$$

We first solve for $\boldsymbol{y}$ from Eq. (4-57) and next solve for $\boldsymbol{x}$ from Eq. (4-58). Since both equations are triangular systems of equations, the solutions are easily obtained by back substitutions.

4-4-2 Crout's Algorithm without Row Interchange

We shall now describe a practical method for calculating the $\boldsymbol{L}$ and $\boldsymbol{U}$ factors, which is sometimes referred to as Crout's method [6].

In Gaussian elimination, the forward-elimination step yields $\boldsymbol{U}$ explicitly.[10] The elements of $\boldsymbol{L}$ can be shown to be given by

$$\boldsymbol{L} = \begin{bmatrix} a_{11} & 0 & \cdots & 0 \\ a_{21} & a_{22}^{(2)} & & \\ \cdot & & \cdot & \\ \cdot & & \cdot & \\ \cdot & & & \cdot \\ a_{n1} & a_{n2}^{(2)} & & a_{nn}^{(n)} \end{bmatrix} \tag{4-59}$$

Each element of $\boldsymbol{L}$ appears at a certain stage of the forward-elimination step and may be recorded (usually written over the original matrix to save storage).

Note that in Gaussian elimination many elements of the given matrix are written over quite a number of times before an element in the $\boldsymbol{U}$ matrix is obtained. For example, in Eq. (4-23), the element a_{44} is written over three times:

$$a_{44},\ a_{44}^{(2)},\ a_{44}^{(3)},\ a_{44}^{(4)}$$

The last result is the (4, 4) element of $\boldsymbol{L}$ matrix.

[10] To obtain $\boldsymbol{U}$ with unit diagonal elements, further divide the kth equation in Eq. (4-30) by $a_{kk}^{(k)}$.

The Crout's method, as will be shown presently, is actually a method of calculating the elements of $\boldsymbol{L}$ and $\boldsymbol{U}$ recursively, without writing over previous results. This is a great advantage with desk calculators. With automatic digital computers, this advantage is not an important one. As far as the number of operations is concerned, both Gaussian elimination and Crout's method require approximately $n^3/3$ operations for n large. The real advantage of Crout's method lies in smaller round-off error and in its suitability to incorporate sparse-matrix techniques [7]–[8] (see Chap. 16).

To see how Crout's method generates the elements of $\boldsymbol{L}$ and $\boldsymbol{U}$ recursively, again consider the case $n = 4$, as in Section 4-3. *We assume that no row interchange is necessary in Eq. (4-54).* Then $\boldsymbol{P} = \mathbf{1}$ and

$$\begin{bmatrix} l_{11} & 0 & 0 & 0 \\ l_{21} & l_{22} & 0 & 0 \\ l_{31} & l_{32} & l_{33} & 0 \\ l_{41} & l_{42} & l_{43} & l_{44} \end{bmatrix} \begin{bmatrix} 1 & u_{12} & u_{13} & u_{14} \\ 0 & 1 & u_{23} & u_{24} \\ 0 & 0 & 1 & u_{34} \\ 0 & 0 & 0 & 1 \end{bmatrix} = \begin{bmatrix} a_{11} & a_{12} & a_{13} & a_{14} \\ a_{21} & a_{22} & a_{23} & a_{24} \\ a_{31} & a_{32} & a_{33} & a_{34} \\ a_{41} & a_{42} & a_{43} & a_{44} \end{bmatrix} \tag{4-60}$$

We shall now show how each element l_{jk} of $\boldsymbol{L}$ and each element u_{jk} of $\boldsymbol{U}$ can be determined using the Crout's algorithm.

Define an auxiliary matrix, consisting of elements of $\boldsymbol{L}$ and $\boldsymbol{U}$ in the following manner:

$$\boldsymbol{Q} \triangleq \begin{bmatrix} l_{11} & u_{12} & u_{13} & u_{14} \\ l_{21} & l_{22} & u_{23} & u_{24} \\ l_{31} & l_{32} & l_{33} & u_{34} \\ l_{41} & l_{42} & l_{43} & l_{44} \end{bmatrix} \tag{4-61}$$

Elements of $\boldsymbol{Q}$ are to be calculated one by one in the illustrated order

$$\begin{bmatrix} \textcircled{1} & \textcircled{5} & \textcircled{6} & \textcircled{7} \\ \textcircled{2} & \textcircled{8} & \textcircled{11} & \textcircled{12} \\ \textcircled{3} & \textcircled{9} & \textcircled{13} & \textcircled{15} \\ \textcircled{4} & \textcircled{10} & \textcircled{14} & \textcircled{16} \end{bmatrix} \tag{4-62}$$

where $\textcircled{k}$ indicates the kth element to be calculated. In other words, we calculate

First column of $\boldsymbol{L}$

First row of $\boldsymbol{U}$ (except $u_{11} = 1$)

Second column of $\boldsymbol{L}$

Second row of $\boldsymbol{U}$ (except $u_{22} = 1$)

etc.

The elements of $\boldsymbol{L}$ and $\boldsymbol{U}$ are calculated simply by equating a_{jk} successively, according to the order shown in Eq. (4-62), with the product of the jth row of $\boldsymbol{L}$ and the kth column of $\boldsymbol{U}$. We have

$$
\begin{aligned}
a_{11} &= l_{11}\cdot 1 & \therefore\ l_{11} &= a_{11} \\
a_{21} &= l_{21}\cdot 1 & \therefore\ l_{21} &= a_{21} \\
a_{31} &= l_{31}\cdot 1 & \therefore\ l_{31} &= a_{31} \\
a_{41} &= l_{41}\cdot 1 & \therefore\ l_{41} &= a_{41} \\
a_{12} &= l_{11}u_{12} & \therefore\ u_{12} &= \frac{a_{12}}{l_{11}} \\
a_{13} &= l_{11}u_{13} & \therefore\ u_{13} &= \frac{a_{13}}{l_{11}} \\
a_{14} &= l_{11}u_{14} & \therefore\ u_{14} &= \frac{a_{14}}{l_{11}} \\
a_{22} &= l_{21}u_{12} + l_{22} & \therefore\ l_{22} &= a_{22} - l_{21}u_{12} \\
a_{32} &= l_{31}u_{12} + l_{32} & \therefore\ l_{32} &= a_{32} - l_{31}u_{12} \\
a_{42} &= l_{41}u_{12} + l_{42} & \therefore\ l_{42} &= a_{42} - l_{41}u_{12} \\
a_{23} &= l_{21}u_{13} + l_{22}u_{23} & \therefore\ u_{23} &= \frac{a_{23} - l_{21}u_{13}}{l_{22}} \\
a_{24} &= l_{21}u_{14} + l_{22}u_{24} & \therefore\ u_{24} &= \frac{a_{24} - l_{21}u_{14}}{l_{22}} \\
a_{33} &= l_{31}u_{13} + l_{32}u_{23} + l_{33} & \therefore\ l_{33} &= a_{33} - l_{31}u_{13} - l_{32}u_{23} \\
a_{43} &= l_{41}u_{13} + l_{42}u_{23} + l_{43} & \therefore\ l_{43} &= a_{43} - l_{41}u_{13} - l_{42}u_{23} \\
a_{34} &= l_{31}u_{14} + l_{32}u_{24} + l_{33}u_{34} & \therefore\ u_{34} &= \frac{a_{34} - l_{31}u_{14} - l_{32}u_{24}}{l_{33}} \\
a_{44} &= l_{41}u_{14} + l_{42}u_{24} + l_{43}u_{34} + l_{44} & \therefore\ l_{44} &= a_{44} - l_{41}u_{14} - l_{42}u_{24} - l_{43}u_{34}
\end{aligned}
\tag{4-63}
$$

Note a very important feature of this calculation: the calculation of the element ⓚ of the auxiliary matrix $\boldsymbol{Q}$, which is an element of either $\boldsymbol{L}$ or $\boldsymbol{U}$, involves only the element of $\boldsymbol{A}$ in the same position and some of those elements of $\boldsymbol{Q}$ that have already been calculated. As the element ⓚ is obtained, it is recorded in the $\boldsymbol{Q}$ matrix. (In fact, it may be recorded in the corresponding position in the $\boldsymbol{A}$ matrix, if there is no need to keep the $\boldsymbol{A}$ matrix.) This calculated result need never be written over; it is already one of the elements of $\boldsymbol{L}$ or $\boldsymbol{U}$.

Following the pattern displayed by Eq. (4-63), we can formulate Crout's algorithm for obtaining the auxiliary matrix $\boldsymbol{Q}$ (which contains the components $\boldsymbol{L}$ and $\boldsymbol{U}$) as given below. A rigorous proof of the algorithm may be found in [6].

Crout's Algorithm without Row Interchange

Given $\boldsymbol{A}$, an $n \times n$ matrix. Initially, $\boldsymbol{Q} = \boldsymbol{0}$ of dimension $n \times n$.

Step 1. Obtain the first column of $\boldsymbol{Q}$ by $q_{k1} = a_{k1}$, for $k = 1, 2, \ldots, n$.

Step 2. Complete the first row of $\boldsymbol{Q}$ by $q_{1k} = a_{1k}/q_{11}$, for $k = 2, 3, \ldots, n$.

Step 3. Set $j \leftarrow 2$.

Step 4. Complete the entries of the jth column of $\boldsymbol{Q}$ by $q_{kj} = a_{kj} - q_{k1}q_{1j} - q_{k2}q_{2j} - \cdots - q_{k(j-1)}q_{(j-1)j}$ for $k = j, j+1, \ldots, n$.

Step 5. If $j = n$, terminate the algorithm. Otherwise go to step 6.

Step 6. Complete the entries of the jth row of $\boldsymbol{Q}$ by $q_{jk} = (a_{jk} - q_{j1}q_{1k} - q_{j2}q_{2k} - \cdots - q_{j(j-1)}q_{(j-1)k})/q_{jj}$ for $k = j+1, j+2, \ldots, n$.

Step 7. Set $j \leftarrow j + 1$; go to step 4.

A numerical example will illustrate the application of the algorithm.

EXAMPLE 4-2.

$$\boldsymbol{A} = \begin{bmatrix} 2 & 4 & 4 & 2 \\ 3 & 3 & 12 & 6 \\ 2 & 4 & -1 & 2 \\ 4 & 2 & 1 & 1 \end{bmatrix}$$

Following the algorithm, we have

$$\boldsymbol{Q} = \begin{bmatrix} 2 & 2 & 2 & 1 \\ 3 & -3 & -2 & -1 \\ 2 & 0 & -5 & 0 \\ 4 & -6 & -19 & -9 \end{bmatrix}, \quad \boldsymbol{L} = \begin{bmatrix} 2 & 0 & 0 & 0 \\ 3 & -3 & 0 & 0 \\ 2 & 0 & -5 & 0 \\ 4 & -6 & -19 & -9 \end{bmatrix},$$

$$\boldsymbol{U} = \begin{bmatrix} 1 & 2 & 2 & 1 \\ 0 & 1 & -2 & -1 \\ 0 & 0 & 1 & 0 \\ 0 & 0 & 0 & 1 \end{bmatrix}$$

Elements of $\boldsymbol{Q}$ are calculated as follows:

First column: $2 = 2$, $3 = 3$, $2 = 2$, $4 = 4$.

First row: $2 = \frac{4}{2}$, $2 = \frac{4}{2}$, $1 = \frac{2}{2}$.

Second column: $-3 = 3 - 3 \times 2$, $0 = 4 - 2 \times 2$, $-6 = 2 - 4 \times 2$.

Second row: $-2 = (12 - 3 \times 2)/(-3)$, $-1 = (6 - 3 \times 1)/(-3)$.

Third column: $-5 = -1 - 2 \times 2 - 0 \times (-2)$, $-19 = 1 - 4 \times 2 - (-6) \times (-2)$.

Third row: $0 = 2 - 2 \times 1 - 0 \times (-1)$.

Fourth column: $-9 = 1 - 4 \times 1 - (-6) \times (-1) - (-19) \times 0$.

4-5 SINUSOIDAL STEADY-STATE ANALYSIS OF LINEAR NETWORKS BY NODAL EQUATIONS

The analysis of the behavior of a linear network in sinusoidal steady state is usually referred to as "ac analysis." A great deal of similarity exists between ac analysis and the analysis of linear resistive networks in Section 4-2. In fact, all the relationships discussed there carry over to ac analysis if we make the following transitions in notations and concepts.

ANALYSIS OF LINEAR RESISTIVE NETWORK	AC ANALYSIS
Resistance R	Impedance Z
Conductance G	Admittance Y
$v(t)$, time function	$V(\omega)$, phasor
$i(t)$, time function	$I(\omega)$, phasor
Values of R, G, v, i are real numbers	Values of Z, Y, V, I are complex numbers

Because of such correspondence, we shall not repeat the detailed derivation. Instead, we shall only give the definitions and the main results, and the reader should refer to Section 4-2 for details.

Define the voltage vectors as

$$\hat{V} \triangleq \begin{bmatrix} \hat{V}_1 \\ \hat{V}_2 \\ \cdot \\ \cdot \\ \cdot \\ \hat{V}_b \end{bmatrix}, \quad V \triangleq \begin{bmatrix} V_1 \\ V_2 \\ \cdot \\ \cdot \\ \cdot \\ V_b \end{bmatrix}, \quad E \triangleq \begin{bmatrix} E_1 \\ E_2 \\ \cdot \\ \cdot \\ \cdot \\ E_b \end{bmatrix} \tag{4-64}$$

and the current vectors as

$$\hat{I} \triangleq \begin{bmatrix} \hat{I}_1 \\ \hat{I}_2 \\ \cdot \\ \cdot \\ \cdot \\ \hat{I}_b \end{bmatrix}, \quad I \triangleq \begin{bmatrix} I_1 \\ I_2 \\ \cdot \\ \cdot \\ \cdot \\ I_b \end{bmatrix}, \quad J \triangleq \begin{bmatrix} J_1 \\ J_2 \\ \cdot \\ \cdot \\ \cdot \\ J_b \end{bmatrix} \tag{4-65}$$

Then

$$\hat{\boldsymbol{V}} = \boldsymbol{V} - \boldsymbol{E} \tag{4-66}$$

$$\hat{\boldsymbol{I}} = \boldsymbol{I} - \boldsymbol{J} \tag{4-67}$$

KCL gives

$$\boldsymbol{A}\hat{\boldsymbol{I}} = \boldsymbol{0} \tag{4-68}$$

From Eqs. (4-67) and Eq. (4-68), we have

$$\boldsymbol{A}\boldsymbol{I} = \boldsymbol{A}\boldsymbol{J} \tag{4-69}$$

The element currents $\boldsymbol{I}$ and voltages $\boldsymbol{V}$ are related by

$$\boldsymbol{I} = \boldsymbol{Y}_b\boldsymbol{V} \tag{4-70}$$

or

$$\begin{bmatrix} I_1 \\ I_2 \\ \cdot \\ \cdot \\ \cdot \\ I_b \end{bmatrix} = \begin{bmatrix} Y_{11} & Y_{12} & \cdots & Y_{1b} \\ \cdot & & & \cdot \\ \cdot & & & \cdot \\ \cdot & & & \cdot \\ Y_{b1} & Y_{b2} & \cdots & Y_{bb} \end{bmatrix} \begin{bmatrix} V_1 \\ V_2 \\ \cdot \\ \cdot \\ \cdot \\ V_b \end{bmatrix} \tag{4-71}$$

where, as in Section 4-2, the elements of $\boldsymbol{Y}_b$ are determined by Eqs. (4-10) and (4-11), but with R_k changed to the impedance Z_k. From elementary circuit theory, Z_k is equal to R_k for a linear resistor, $j\omega L_k$ for a linear inductor, or $1/j\omega C_k$ for a linear capacitor.

For controlled sources (strictly speaking, voltage-controlled current sources), we allow the controlling coefficients to be functions of frequency $j\omega$. Of course, as will be clear shortly, the computer analysis will be carried out one frequency at a time; during this analysis, the elements of all matrices and vectors are complex numbers.

The node transformation is expressed by Eq. (3-28)

$$\hat{\boldsymbol{V}} = \boldsymbol{A}^t\boldsymbol{V}_n \tag{4-72}$$

where $\boldsymbol{A}$ is the reduced incidence matrix and $\boldsymbol{V}_n$ is the node-to-datum voltage vector.

Following exactly the same procedure as in Section 4-2, we arrive at the nodal equations in matrix form:

$$\boxed{(\boldsymbol{A}\boldsymbol{Y}_b\boldsymbol{A}^t)\boldsymbol{V}_n = \boldsymbol{A}(\boldsymbol{J} - \boldsymbol{Y}_b\boldsymbol{E})} \tag{4-73}$$

or

$$\boxed{\boldsymbol{Y}_n\boldsymbol{V}_n = \boldsymbol{J}_n} \tag{4-74}$$

where $\boldsymbol{Y}_n = \boldsymbol{A}\boldsymbol{Y}_b\boldsymbol{A}^t$ is the nodal-admittance matrix, and $\boldsymbol{J}_n = \boldsymbol{A}(\boldsymbol{J} - \boldsymbol{Y}_b\boldsymbol{E})$ is the equivalent nodal current source vector.

Equation (4-74) is exactly the same in form as Eq. (4-17), the only difference being that the entries of various matrices are now complex numbers instead of real numbers. To solve for $\boldsymbol{V}_n$ from Eq. (4-74) amounts to solving n simultaneous linear equations with *complex* coefficients. Once $\boldsymbol{V}_n$ is found, $\hat{\boldsymbol{V}}$, $\boldsymbol{V}$, and $\boldsymbol{I}$ can be found successively from the phasor equations (4-72), (4-66), and (4-70). This is actually the procedure used in several computer simulation programs (e.g., ECAP [9]).

The algorithms described in Sections 4-3 and 4-4 for solving simultaneous equations are also applicable in the present case of ac analysis. The only difference is that all quantities are now complex numbers. The total number of multiplications is four times greater than with real coefficients. This is easily seen from the fact that

$$z_1 z_2 = (x_1 + jy_1)(x_2 + jy_2) = (x_1 x_2 - y_1 y_2) + j(x_1 y_2 + y_1 x_2)$$

Thus, to evaluate the product of two complex numbers z_1 and z_2, we need to perform four multiplications of real numbers.

Another difference between dc and ac nodal analysis is the special handling required in the latter for mutual inductances. If a group of m inductances is coupled, then

$$\boldsymbol{V}_L = j\omega \boldsymbol{L} \boldsymbol{I}_L \tag{4-75}$$

where the inductance matrix $\boldsymbol{L}$ is symmetrical and of order $m \times m$. It is necessary to assume that the coupling is less than perfect.[11] Then $|\boldsymbol{L}| \neq 0$, and Eq. (4-75) may be solved to give

$$\boldsymbol{I}_L = \frac{1}{j\omega} \boldsymbol{\Gamma} \boldsymbol{V}_L \tag{4-76}$$

where $\boldsymbol{\Gamma} = \boldsymbol{L}^{-1}$ is the reciprocal inductance matrix. The relationships given by Eq. (4-76) may now be entered into the branch admittance matrix $\boldsymbol{Y}_b$ in Eq. (4-71).

EXAMPLE 4-3. For the network shown in Fig. 4-3(a), find $\boldsymbol{Y}_n$ and $\boldsymbol{J}_n$ in Eq. (4-74) at $\omega = 0.5$ rad/s.

Solution: The directed graph associated with the network is shown in Fig. 4-3(b), where the directions of the composite branches have been assigned arbitrarily. With node 0 chosen as the datum node, the reduced incidence matrix is

$$A = \begin{matrix} \\ 1 \\ 2 \\ 3 \end{matrix} \begin{matrix} \begin{matrix} 1 & 2 & 3 & 4 & 5 & 6 \end{matrix} \\ \begin{bmatrix} 1 & 0 & 0 & 0 & 1 & 0 \\ 0 & 1 & 0 & 0 & -1 & 1 \\ 0 & 0 & 1 & 1 & 0 & -1 \end{bmatrix} \end{matrix}$$

The inverse of the inductance matrix is

$$\boldsymbol{\Gamma} = \boldsymbol{L}^{-1} = \begin{bmatrix} 4 & 6 \\ 6 & 10 \end{bmatrix}^{-1} = \begin{bmatrix} 2.5 & -1.5 \\ -1.5 & 1 \end{bmatrix}$$

The branch-admittance matrix, which contains $(1/j\omega)\boldsymbol{\Gamma} = -j2\boldsymbol{\Gamma}$ as a submatrix, is

$$\boldsymbol{Y}_b = \begin{bmatrix} 1 & 0 & 0 & 0 & 0 & 0 \\ 0 & j3 & 0 & 0 & 0 & 0 \\ 0 & 8 & 0 & 0 & 0 & 0 \\ 0 & 0 & 0 & 2 & 0 & 0 \\ 0 & 0 & 0 & 0 & -j5 & j3 \\ 0 & 0 & 0 & 0 & j3 & -j2 \end{bmatrix}$$

[11]For example, ECAP program requires that the coupling coefficient for a pair of inductances should be less than 0.999995.

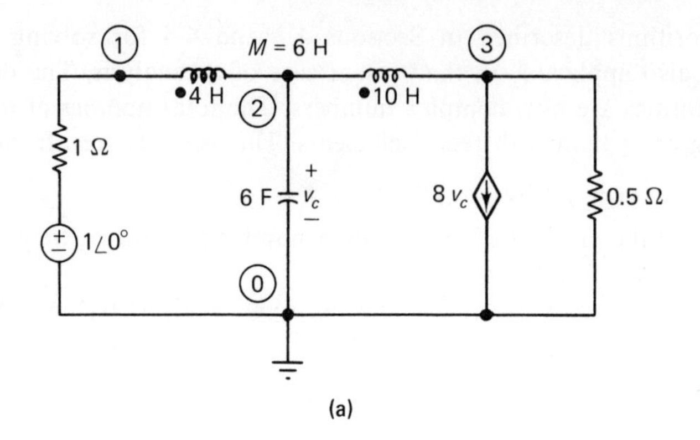

(a)

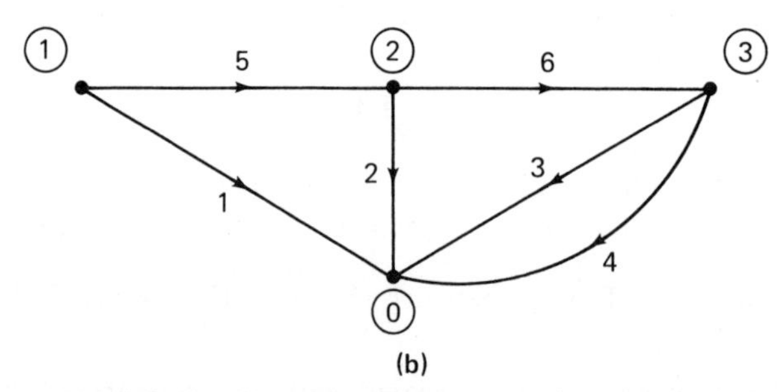

(b)

Fig. 4-3. Handling of mutual inductances in nodal equations.

The current source and voltage source vectors are, respectively,

$$J = \mathbf{0}, \qquad E = [-1 \quad 0 \quad 0 \quad 0 \quad 0 \quad 0]^t$$

Substituting these expressions into the equations for Y_n and J_n and carrying out the numerical calculations, we obtain

$$Y_n = \begin{bmatrix} 1 - j5 & j8 & -j3 \\ j8 & -j10 & j5 \\ -j3 & 8 + j5 & 2 - j2 \end{bmatrix}$$

$$J_n = \begin{bmatrix} 1 \\ 0 \\ 0 \end{bmatrix}$$

4-6 DIRECT CONSTRUCTION OF NODAL-ADMITTANCE MATRIX AND CURRENT SOURCE VECTOR

We have shown in Section 4-2, Eqs. (4-18) and (4-19), that

$$Y_n = AY_bA^t$$
$$J_n = A(J - Y_bE)$$

In the actual implementation of nodal analysis on digital computers, we do not construct the matrices A and Y_b and perform the matrix multiplications. To do so would be extremely wasteful of both computer memory and running time, because A and Y_b are generally very sparse. Instead, we shall devise rules that enable us to "build up" Y_n and J_n bit by bit as the information on each composite branch is received. Thus, in a sense, the method to be described is a sparse-matrix technique for the evaluation of AY_bA^t and $A(J - Y_bE)$.

Suppose that the network under consideration consists of b composite branches and $n + 1$ nodes. As usual, we shall label the nodes from 0 to n, with node 0 as the datum node, and label the composite branches from 1 to b. Y_n is a square matrix of order n, and J_n is an n vector.

We shall now consider how each composite branch contributes to the entries of Y_n and J_n. We distinguish between two types of composite branches: admittance branches and controlled-source branches.

TYPE 1: ADMITTANCE BRANCHES. A typical lth composite branch is directed from node i to node j, with an admittance y_l as the two-terminal element (see Fig. 4-4). Let us write the nodal equations, assuming that neither node is the datum, or calculate AY_bA^t and $A(J - Y_bE)$ just once for this simple configuration to establish the rules. The following results are obtained, which show the places where y_l, J_l, and E_l make their appearances (see Problems 4-15 and 4-17).

$$Y_n = \begin{array}{c} \\ \\ \\ i \\ \\ \\ \\ j \\ \\ \\ \\ \end{array} \begin{bmatrix} \ddots & & \vdots & & \vdots & & \\ & & \vdots & & \vdots & & \\ \cdots & & y_l & \cdots & -y_l & \cdots & \\ & & \vdots & \ddots & \vdots & & \\ & & \vdots & & \vdots & & \\ \cdots & & -y_l & \cdots & y_l & \cdots & \\ & & \vdots & & \vdots & \ddots & \\ & & \vdots & & \vdots & & \ddots \end{bmatrix} \quad \begin{matrix} \\ \\ \\ \end{matrix} \tag{4-77}$$

(columns labeled i and j)

$$J_n = \begin{array}{c} \\ \\ i \\ \\ \\ j \\ \\ \\ \end{array} \begin{bmatrix} \vdots \\ \vdots \\ (J_l - y_lE_l) \\ \vdots \\ \vdots \\ -(J_l - y_lE_l) \\ \vdots \\ \vdots \end{bmatrix} \tag{4-78}$$

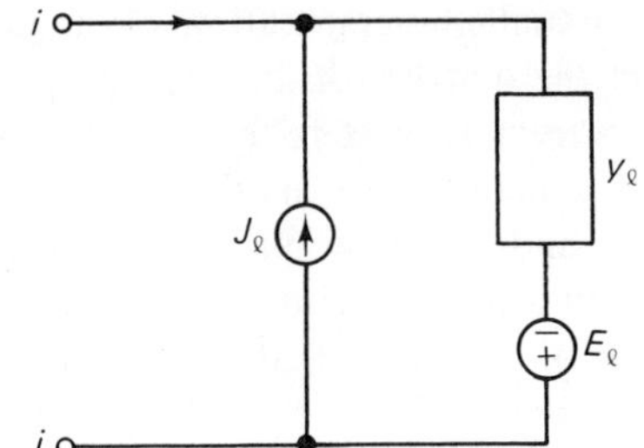

Fig. 4-4. Composite branch containing an admittance.

In general, each y_l will appear in four places in $\boldsymbol{Y}_n$. If $i = 0$, or $j = 0$, demanding entries to be made in "row 0" or "column 0", then the entries are simply not made.[12] This arises when the lth branch is connected to the datum node, with y_l appearing only once on the diagonal of $\boldsymbol{Y}_n$.

TYPE 2: CONTROLLED-SOURCE BRANCHES. A typical lth composite branch is connected from node i to node j, with a voltage-controlled current source as the two-terminal element; the controlling voltage is across the two-terminal element of a composite branch directed from node k to node m. See Fig. 4-5 for the

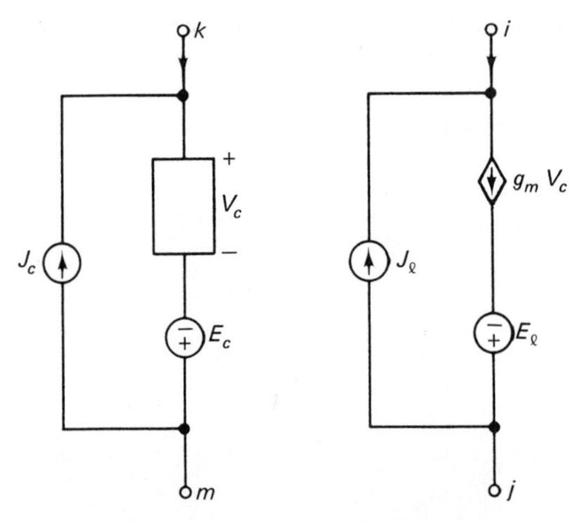

Fig. 4-5. Composite branch containing a voltage-controlled current source together with the controlling branch.

notations and reference directions. As before, we write the nodal equations, assuming that none of the nodes (i, j, k, m) is the datum node, or calculate $\boldsymbol{A}\boldsymbol{Y}_b\boldsymbol{A}^t$ and $\boldsymbol{A}(\boldsymbol{J} - \boldsymbol{Y}_b\boldsymbol{E})$ for Fig. 4-5 just once in order to establish the rules. The following results

[12]These entries could have been made had we added a "row 0" and a "column 0" to $\boldsymbol{Y}_n$. Then the resultant matrix will be an indefinite admittance matrix instead of $\boldsymbol{Y}_n$. See Section 14-3.

are obtained, which show the entries in $\boldsymbol{Y}_n$ and $\boldsymbol{J}_n$ contributed by this lth branch (see Problems 4-16 and 4-18).

$$\boldsymbol{Y}_n = \begin{array}{r} \\ \text{from } i \\ \\ \text{to } j \\ \\ \uparrow \\ \text{controlled source} \end{array} \begin{array}{c} \begin{array}{cc} \text{from } k & \text{to } m \end{array} \leftarrow \text{controlling branch} \\ \begin{bmatrix} & \vdots & & \vdots & \\ \cdots & g_m & \cdots & -g_m & \cdots \\ & \vdots & & \vdots & \\ \cdots & -g_m & \cdots & g_m & \cdots \\ & \vdots & & \vdots & \end{bmatrix} \end{array} \tag{4-79}$$

$$\boldsymbol{J}_n = \begin{array}{r} \\ i \\ \\ j \\ \\ \end{array} \begin{bmatrix} \vdots \\ (J_l - g_m E_c) \\ \vdots \\ -(J_l - g_m E_c) \\ \vdots \end{bmatrix} \tag{4-80}$$

As before, if any one of (i, j, k, m) is 0, demanding entries to be made in "row 0" or "column 0," then the entries are simply not made. Note that it is possible for all four entries of g_m in (4-79) to lie *off the diagonal.*

The procedure for constructing $\boldsymbol{Y}_n$ and $\boldsymbol{J}_n$ without the explicit use of A and $\boldsymbol{Y}_b$ is as follows. Initially, with all branches *conceptually* removed from the network, $\boldsymbol{Y}_n = \boldsymbol{0}$ and $\boldsymbol{J}_n = \boldsymbol{0}$. Composite branches will be added to the network one by one, first the admittance branches and then the controlled-source branches. As the information on each composite branch is received (e.g., by reading in one data card), the entries in $\boldsymbol{Y}_n$ and $\boldsymbol{J}_n$ are updated by *adding* those entries shown in Eqs. (4-77)–(4-80), whichever is appropriate, to the original ones. When all composite branches have been processed this way, the final results are the desired $\boldsymbol{Y}_n$ and $\boldsymbol{J}_n$.

EXAMPLE 4-4. Find $\boldsymbol{Y}_n$ and $\boldsymbol{J}_n$ for the network shown in Fig. 4-6(a) at $\omega = 1$ rad/s. The associated directed graph is shown in Fig. 4-6(b).

Solution: The following table shows the step-by-step updating of $\boldsymbol{Y}_n$ and $\boldsymbol{J}_n$.

BRANCH	CONTRIBUTION IN Y_n	CONTRIBUTION IN J_n
1	2 in (1, 1)	$(2 + j) - 10$ in (1, 1)
2	5 in (2, 2)	7 in (2, 1)
5	10 in (1, 1) and (2, 2) -10 in (1, 2) and (2, 1)	None
6	$j3$ in (2, 2) and (3, 3) $-j3$ in (2, 3) and (3, 2)	$j18$ in (2, 1) $-j18$ in (3, 1)
7	$-j/2$ in (1, 1) and (3, 3) $j/2$ in (1, 3) and (3, 1)	None
3	-4 in (3, 3) 4 in (3, 2)	$-(1 + j) - 4 \times (-6)$ in (3, 1)
4	8 in (1, 2) and (2, 3) -8 in (1, 3) and (2, 2)	48 in (1, 1) -48 in (2, 1)

The final results are

$$Y_n = \begin{bmatrix} 12 - j0.5 & -2 & -8 + j0.5 \\ -10 & 7 + j3 & 8 - j3 \\ j0.5 & 4 - j3 & -4 + j2.5 \end{bmatrix}$$

$$J_n = \begin{bmatrix} 40 + j \\ -41 + j18 \\ 23 - j19 \end{bmatrix}$$

For a network without controlled sources, an alternative way to construct Y_n directly is to make use of the following special properties of Y_n (for an *RLC* network): (1) each diagonal element (i, i) of Y_n is the sum of all admittances connected to node i, either directly or through an independent voltage source in series; (2) each nondiagonal element (i, j) of Y_n is the negative of the sum of all admittances connected between nodes i and j, possibly with a voltage source in series. When Y_n is to be constructed with paper and pencil, these two rules may be more convenient. However, for digital computer implementation of nodal analysis, Eqs. (4-77)–(4-80) are much more general and easy to program. If any controlled source other than the g_m type is encountered, it may be transformed, in most practical cases, into the g_m type, either by the user or by the program (see the footnote in Section 4-2).

A brief remark about mutual inductances is necessary. When a network contains m coupled inductances whose inductance matrix L is given, it is necessary to find $\Gamma = L^{-1}$ first, as discussed in Section 4-5. Once Γ is found, the kth scalar equation in Eq. (4-76) is of the form

$$I_k = \frac{\Gamma_{k1}}{j\omega} V_1 + \cdots + \frac{\Gamma_{kk}}{j\omega} V_k + \cdots + \frac{\Gamma_{km}}{j\omega} V_m \tag{4-81}$$

Thus, we may regard the kth coupled inductance as a new self-inductance of value Γ_{kk} (reciprocal henries) in parallel with $m - 1$ voltage-controlled current sources. The controlling coefficients in the present case are imaginary numbers. The case for a pair of coupled inductances is illustrated in Fig. 4-7. After modeling mutual induc-

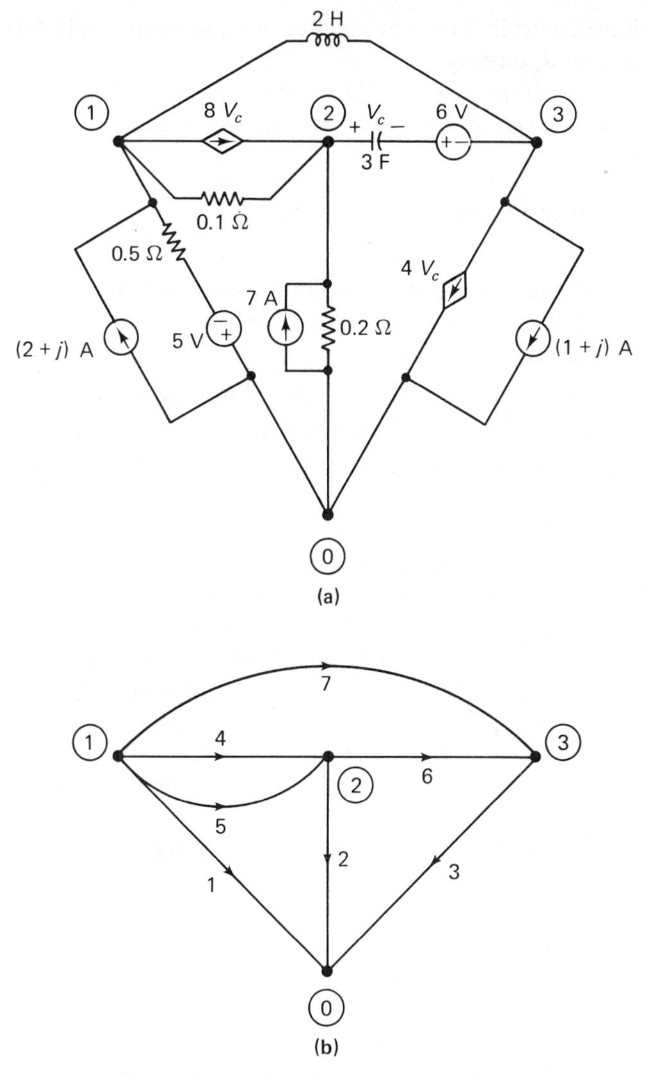

Fig. 4-6. Direct construction of $\boldsymbol{Y}_n$ and $\boldsymbol{J}_n$.

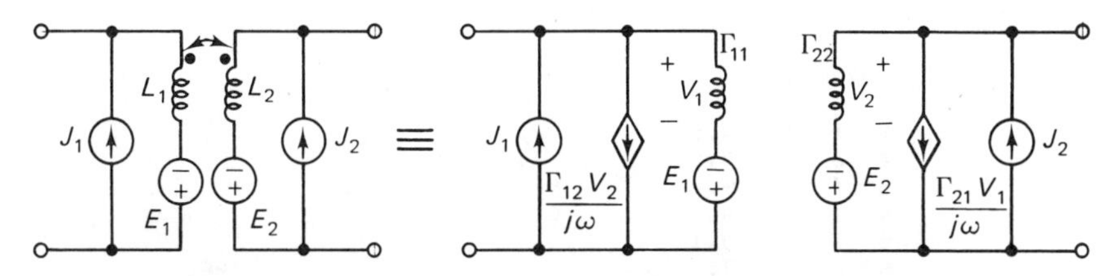

Fig. 4-7. Mutual inductance may be modeled with the aid of voltage-controlled current sources.

tances by voltage-controlled current sources, we can again apply Eqs. (4-77)–(4-80) to construct $\boldsymbol{Y}_n$ and $\boldsymbol{J}_n$ directly.

A complete program called NODAL that embodies all the techniques of Sections 4-3 and 4-6 is given in Appendix 4B. The user's guide to NODAL is given in Appendix 4A.

APPENDIX 4A USER'S GUIDE TO NODAL

The program NODAL performs dc and ac analysis of linear networks by the nodal method. Its output consists of the node-admittance matrix $\boldsymbol{Y}_n$, the equivalent current source vector $\boldsymbol{J}_n$, and the node-to-datum voltages $\boldsymbol{V}_n$.

The data-card setup is as follows:

Card 1. Title card. Cols. 1–80, reproduced exactly in the output.

Card 2. Cols. 1–3, number of branches (I3).

Card 3, 4, . . . , LAST − 1, one card for each branch.

Column	
1–3	Branch number (I3)
4–6	"From node" (I3)
7–9	"To node" (I3) *Datum node is always numbered 0.*
10–11	Type of element (A2); one of the following: ␣R, ␣G, ␣L, ␣C, VC (␣ means blank) where VC stands for voltage-controlled current source.
12–21	Value of R, G, L, C, or g_m (E10.3).
22–24	Branch number of the controlling branch. Leave blank for R, G, L, C elements (I3).
25–34	Re J (J is the independent current source in the composite branch) (E10.3).
35–44	Im J (E10.3).
45–54	Re E (E is the independent voltage source in the composite branch) (E10.3).
55–64	Im E (E10.3).

LAST card. Cols. 1–12, frequency in radians per second (leave blank for dc analysis) (E12.3).

The input and output of a sample problem (Example 4-4) are shown in Figs. 4A-1 and 4A-2.

```
EXAMPLE 4-4
  7
  1  1  0 R 0.5          2.        1.        5.
  2  2  0 R 0.2          7.
  3  3  0VC 4.          6 -1.      -1.
  4  1  2VC 8.          6
  5  1  2 R 0.1
  6  2  3 C 3.                               -6.
  7  1  3 L 2.
1.
```

Fig. 4A-1. Input of Example 4-4 into NODAL program.

```
EXAMPLE 4-4


NO. OF NODES - 1= 3
NUMBER OF BRANCHES=  7
FREQUENCY OF OPERATION= 1.000E+00


         THE NETWORK IS DESCRIBED BY THE FOLLOWING BRANCHES


  I FROM TO TYPE  VALUE    ICONT         J                       E


  1   1   0  R  5.000E-01  -0  2.000E+00   1.000E+00J  5.000E+00  -0.        J

  2   2   0  R  2.000E-01  -0  7.000E+00  -0.        J -0.        -0.        J

  3   3   0 VC  4.000E+00   6 -1.000E+00  -1.000E+00J -0.         -0.        J

  4   1   2 VC  8.000E+00   6 -0.         -0.        J -0.        -0.        J

  5   1   2  R  1.000E-01  -0 -0.         -0.        J -0.        -0.        J

  6   2   3  C  3.000E+00  -0 -0.         -0.        J -6.000E+00 -0.        J

  7   1   3  L  2.000E+00  -0 -0.         -0.        J -0.        -0.        J


  ***NODE ADMITTANCE MATRIX***          ***EQUIVALENT CURRENT SOURCE VECTOR***

COLUMN  1

    ( 1.200E+01  -5.000E-01J)              ( 4.000E+01    1.000E+00J)

    (-1.000E+01   0.        J)             (-4.100E+01    1.800E+01J)

    ( 0.          5.000E-01J)              ( 2.300E+01   -1.900E+01J)

COLUMN  2
                                         **NODE VOLTAGES***
    (-2.000E+00   0.        J)

    ( 7.000E+00   3.000E+00J)            V 1     -1.0424E+00   +   9.4800E-01J
                                         V 2     -2.1595E-01   +   1.9258E-01J
    ( 4.000E+00  -3.000E+00J)            V 3     -6.5071E+00   +   9.0732E-01J

COLUMN  3

    (-8.000E+00   5.000E-01J)

    ( 8.000E+00  -3.000E+00J)

    (-4.000E+00   2.500E+00J)
```

Fig. 4A-2. Output from NODAL program for Example 4-4.

APPENDIX 4B LISTING OF NODAL

```
C THIS PROGRAM PERFORMS AC NODAL ANALYSIS. COMPOSITE BRANCHES ARE USED.
      REAL OMEGA
      COMPLEX YN,Y,J,E
      COMMON YN(40,41),NFROM(200),NTO(200),TYPE(200),
     1VALUE(200),ICONT(200),Y(200),NNODE,NBR,NN,J(200),E(200)
C READ AND WRITE TITLE
1     READ(5,15)(VALUE(K),K=1,80)
      WRITE(6,16)(VALUE(K),K=1,80)
C READ AND WRITE NETWORK INFORMATION
      NNODE=0
      READ(5,6)NBR
      DO 2 K=1,NBR
      READ(5,7)I,NFROM(I),NTO(I),TYPE(I),VALUE(I),ICONT(I),J(I),E(I)
      NNODE=MAX0(NNODE,NFROM(I),NTO(I))
2     CONTINUE
      NN=NNODE+1
      READ(5,8)OMEGA
      WRITE(6,9)NNODE,NBR,OMEGA
      WRITE(6,14)
      DO 3 I=1,NBR
3     WRITE(6,10)I,NFROM(I),NTO(I),TYPE(I),VALUE(I),ICONT(I),J(I),E(I)
C FORMULATE NODAL EQUATIONS AND SOLVE
      CALL FORM(OMEGA)
      WRITE(6,13)
      DO 4  I=1,NNODE
      WRITE(6,18)I
4     WRITE(6,11)(YN(K,I),K=1,NNODE)
      WRITE(6,12)
      WRITE(6,11)(YN(K,NN),K=1,NNODE)
      CALL GAUSS(NNODE)
C PRINT RESULTS
      WRITE (6,17)
      DO 5  I=1,NNODE
5     WRITE(6,19) I,YN(I,NN)
      GO TO 1
C COLLECTION OF FORMAT STATEMENTS
6     FORMAT(I3)
7     FORMAT(3I3,A2,E10.3,I3,4E10.3)
8     FORMAT(E10.3)
9     FORMAT(////1X,'NO. OF NODES - 1=',I2/1X,'NUMBER OF BRANCHES=',I3
     1/1X,'FREQUENCY OF OPERATION=',E10.3//)
10    FORMAT(1H0,3(I3,1X),A2,1X,E10.3,1X,I3,1X,2(1E10.3,2X,1E10.3,1HJ,
     11X))
11    FORMAT(1H0,4X,'(',1E10.3,2X,1E10.3,1HJ ')')
12    FORMAT(1H0//' ***EQUIVALENT CURRENT SOURCE VECTOR***'//)
13    FORMAT(1H1, 2X,'***NODE ADMITTANCE MATRIX***')
     17X,'REAL PART',3X,'IMAG PART'/)
14    FORMAT(10X,'THE NETWORK IS DESCRIBED BY THE FOLLOWING BRANCHES'///
     13X,'I',1X,'FROM',1X,'TO',1X,'TYPE',2X,'VALUE',3X,'ICONT',
     17X,'J',20X,'E'/)
15    FORMAT(80A1)
16    FORMAT(1H1,80A1)
17    FORMAT (1X,//'***NODE VOLTAGES***'//)
18    FORMAT(1H0,'COLUMN ',I2)
19    FORMAT(1X,1HV,I2,3X,1E12.4,2X,1H+,1E12.4,1HJ)
      END
```

```
      SUBROUTINE FORM(OMEGA)
C THIS SUBROUTINE FORMULATES NODE ADMITTANCE MATRIX AND EQUIVALENT
C CURRENT SOURCE VECTOR BY DIRECT CONSTRUCTION.
      REAL OMEGA
```

```
      COMPLEX YN,Y,J,E
      COMMON YN(40,41),NFROM(200),NTO(200),TYPE(200),
     1VALUE(200),ICONT(200),Y(200),NNODE,NBR,NN,J(200),E(200)
      DATA R,L,C,G,VC/2H R,2H L,2H C,2H G,2HVC/
C ZERO OUT YN     MATRIX
      DO 20 I=1,NNODE
      DO 20 K=1,NN
20    YN(I,K)=CMPLX(0.,0.)
      DO 70 I=1,NBR
C DETERMINE ADMITTANCE TYPE AND VALUE
      IF(ICONT(I).NE.0)GO TO 40
      ICONT(I)=I
      IF(TYPE(I).EQ. R)GO TO 30
      IF(TYPE(I).EQ.G)GO TO 22
      IF(TYPE(I).EQ.C )GO TO 25
      Y(I)=CMPLX(0.,-1./(OMEGA*VALUE(I)))
      GO TO 60
22    Y(I)=CMPLX(VALUE(I),0.)
      GO TO 60
25    Y(I)=CMPLX(0.,(OMEGA*VALUE(I)))
      GO TO 60
30    Y(I)=CMPLX(1/VALUE(I),0.)
      GO TO 60
40    IF(TYPE(I).EQ.VC)GO TO 55
C ERROR MESSAGE
      WRITE(6,53)
53    FORMAT(1H0,'ERROR IN ELEMENT TYPE')
      STOP
55    Y(I)=CMPLX(VALUE(I),0.)
60    ICON=ICONT(I)
C ADD CONTRIBUTIONS TO YN FROM ITH BRANCH
      IA=NFROM(I)
      IB=NTO(I)
      IC=NFROM(ICON)
      ID=NTO(ICON)
      IF(IA.NE.0.AND.IC.NE.0)YN(IA,IC)=YN(IA,IC)+Y(I)
      IF(IA.NE.0.AND.ID.NE.0)YN(IA,ID)=YN(IA,ID)-Y(I)
      IF(IB.NE.0.AND.IC.NE.0)YN(IB,IC)=YN(IB,IC)-Y(I)
      IF(IB.NE.0.AND.ID.NE.0)YN(IB,ID)=YN(IB,ID)+Y(I)
C ADD CONTRIBUTIONS TO JN FROM ITH BRANCH
      IF(IA.NE.0)YN(IA,NN)=YN(IA,NN)+J(I)-Y(I)*E(ICON)
      IF(IB.NE.0)YN(IB,NN)=YN(IB,NN)-J(I)+Y(I)*E(ICON)
70    CONTINUE
      RETURN
      END


      SUBROUTINE GAUSS(N)
C THIS SUBROUTINE SOLVES N SIMULTANEOUS LINEAR EQUATIONS BY GAUSSIAN
C ELIMINATION METHOD.  ARRAY A IS THE AUGMENTED COEFFICIENT MATRIX.
C AT EXIT , THE (N+1)TH COLUMN OF A CONTAINS THE SOLUTIONS.
      COMPLEX A,PIVOT,B
      COMMON A(40,41)
      NP1=N+1
      EPS=1.E-30
C FORWARD ELIMINATION
C SEARCH FOR PIVOT ROW
      IC=1
      IR=1
    1 PIVOT=A(IR,IC)
      IPIVOT=IR
      DO 2 I=IR,N
      IF(CABS(A(I,IC)).LE.CABS(PIVOT)) GO TO 2
      PIVOT=A(I,IC)
      IPIVOT=I
    2 CONTINUE
```

```
C INTERCHANGE ROWS
      IF(CABS(PIVOT).LE.EPS) GO TO 8
      IF(IPIVOT.EQ.IR) GO TO 4
      DO 3 K=IC,NP1
      B=A(IPIVOT,K)
      A(IPIVOT,K)=A(IR,K)
      A(IR,K)=B
    3 CONTINUE
    4 CONTINUE
C NORMALIZE PIVOT
      DO 5 K=IC,NP1
    5 A(IR,K)=A(IR,K)/PIVOT
      IF(IR.EQ.N) GO TO 10
      IRP1=IR+1
C COLUMN REDUCTION
      DO 7 IP=IRP1,N
      B=A(IP,IC)
      IF(CABS(B).LE.EPS) GO TO 7
      DO 6 K=IC,NP1
      A(IP,K)=A(IP,K)-A(IR,K)*B
    6 CONTINUE
    7 CONTINUE
      IR=IR+1
      IC=IC+1
      GO TO 1
    8 WRITE(6,9)
    9 FORMAT(47H DETERMINANT EQUAL TO ZERO, NO UNIQUE SOLUTION.)
      STOP
C BACK SUBSTITUTION
   10 NM1=N-1
      DO 12 K=1,NM1
      NMK=N-K
      DO 11 J=1,K
      NP1MJ=N+1-J
   11 A(NMK,NP1)=A(NMK,NP1) -A(NMK,NP1MJ)*A(NP1MJ,NP1)
   12 CONTINUE
      RETURN
      END
```

REFERENCES

1. FORSYTHE, G., and C. B. MOLER. *Computer Solution of Linear Algebraic Systems.* Englewood Cliffs, N.J.: Prentice-Hall, Inc., 1967.
2. CARNAHAN, B., H. A. LUTHER, and J. O. WILKES. *Applied Numerical Methods.* New York: John Wiley & Sons, Inc., 1969.
3. WESTLAKE, J. R. *A Handbook of Numerical Matrix Inversion and Solution of Linear Equations.* New York: John Wiley & Sons, Inc., 1968.
4. HENRICI, P. *Elements of Numerical Analysis.* New York: John Wiley & Sons, Inc., 1964.
5. NOBLE, B. *Applied Linear Algebra.* Englewood Cliffs, N.J.: Prentice-Hall, Inc., 1969.
6. CROUT, P. D. "A Short Method for Evaluating Determinants and Solving Systems of Linear Equations with Real or Complex Coefficients." *AIEE Trans.*, Vol. 69, pp. 1235–1241, 1941.
7. GUSTAVSON, F. G., W. LINIGER, and R. WILLOUGHBY. "Symbolic Generation of an Optimal Crout Algorithm for Sparse System of Linear Equations." *J. ACM*, Vol. 17, No. 1, pp. 87–109, Jan. 1970.
8. CALAHAN, D. A. *Computer-Aided Network Design.* New York: McGraw-Hill Book Company, 1972.
9. JENSEN, R. W., and M. D. LIEBERMAN. *IBM Electronic Circuit Analysis Program.* Englewood Cliffs, N.J.: Prentice-Hall, Inc., 1968.

10. HO, C. W., A. E. RUEHLI, and P. A. BRENNAN. "The Modified Nodal Approach to Network Analysis." *IEEE Trans. Circuits and Systems,* Vol. CAS-22, No. 6, June 1975, pp. 504-509.

PROBLEMS

4-1. Consider the network shown in Fig. P4-1 with nodes and branches already numbered and reference directions assigned.

(a) Write the matrices $\boldsymbol{A}$, $\boldsymbol{Y}_b$, $\boldsymbol{J}$, and $\boldsymbol{E}$.

(b) Calculate $\boldsymbol{Y}_n$ and $\boldsymbol{J}_n$ by the use of Eqs. (4-18) and (4-19), respectively.

(c) Solve the nodal equations by the Gaussian elimination method.

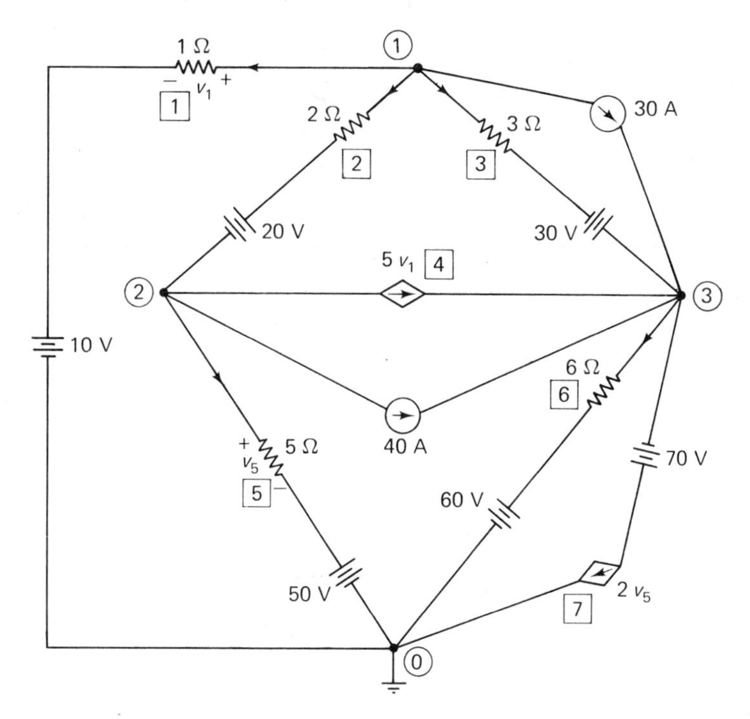

Fig. P4-1

4-2. The network of Fig. P4-2 contains all four types of controlled sources. Assume that only node voltages V_1 and V_6 (with respect to the reference node) are of interest. Show

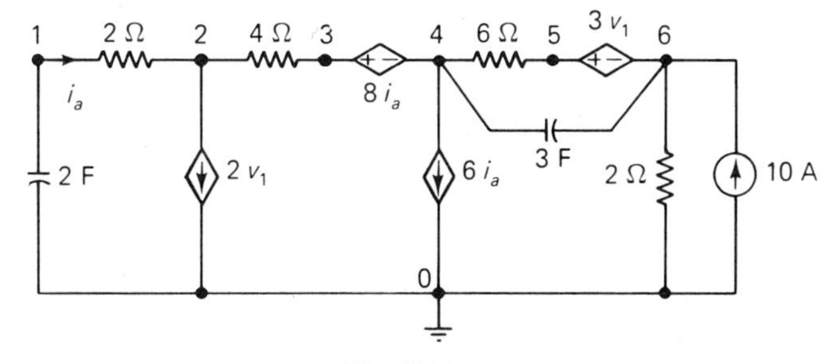

Fig. P4-2

how all controlled sources may be transformed into the g_m type (voltage-controlled current source) without increasing the number of nodes.

4-3. The two-winding ideal transformer shown in Fig. P4-3(a) can be represented by two branches characterized by

$$\begin{bmatrix} v_1 \\ i_2 \end{bmatrix} \begin{bmatrix} 0 & n \\ -\dfrac{1}{n} & 0 \end{bmatrix} \begin{bmatrix} i_1 \\ v_2 \end{bmatrix}$$

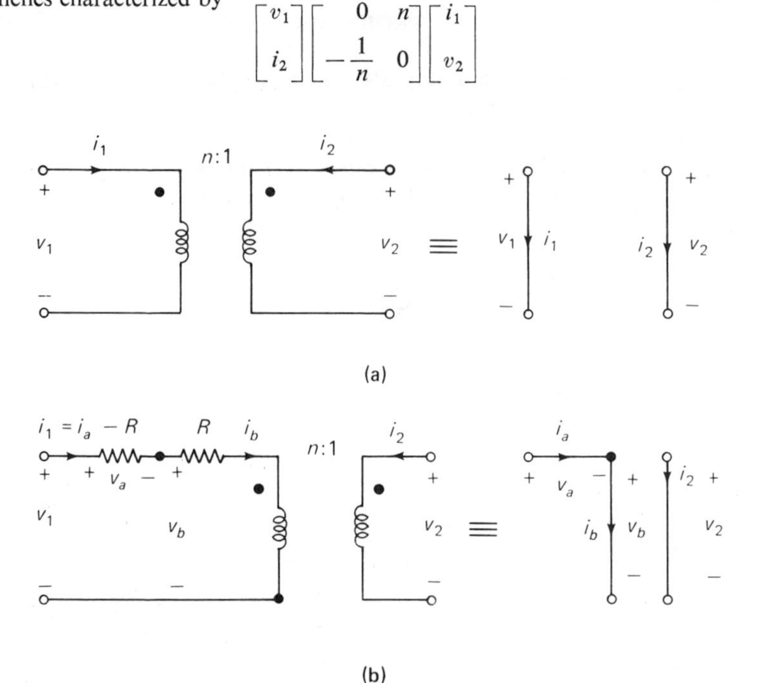

Fig. P4-3

Such coupled branches have no admittance matrix description and consequently are excluded from Eq. (4-9). However, by adding two resistors R and $-R$ in series with one of the windings, as shown in Fig. P4-3(b), we can overcome the difficulty. The ideal transformer is now represented by three branches, two of which are coupled while the third has a negative resistance. Find the branch admittance matrix for the three branches shown in Fig. P4-3(b).

4-4. Given

$$\begin{bmatrix} 2 & 6 & 10 & -2 \\ 1 & 4 & 11 & 0 \\ 3 & 8 & 11 & 6 \\ -2 & -6 & -9 & 4 \end{bmatrix} \begin{bmatrix} x_1 \\ x_2 \\ x_3 \\ x_4 \end{bmatrix} = \begin{bmatrix} 8 \\ -2 \\ 20 \\ -13 \end{bmatrix}$$

show that the number of operations (multiplications and divisions) required for solving the equations by the Gaussian elimination method is

$$\frac{n^3}{3} + n^2 - \frac{n}{3} = \frac{4^3 + 3 \times 4^2 - 4}{3} = 36$$

Use two versions of the Gaussian elimination method to find the values of the x's.

(a) With normalization of the pivot elements.

(b) Without normalization of the pivot elements (as done in Section 4-3).

4-5. (a) Verify that the Gaussian elimination method for solving a general system of n equations in n unknowns requires a total of $(n^3/3) + n^2 - (n/3)$ multiplications (and divisions) and $(n^3/3) + (n^2/2) - \frac{5}{6}n$ additions (and subtractions).

(b) If each multiplication requires 25 μs of computer time and each addition requires 1 μs, find the total computer time needed to solve $\boldsymbol{Ax} = \boldsymbol{b}$, where $\boldsymbol{A}$ is 100×100. Repeat when $\boldsymbol{A}$ is (1000×1000).

4-6. Find $\boldsymbol{Y}_n^{-1}$ by the Gaussian elimination method, where $\boldsymbol{Y}_n = \boldsymbol{A}\boldsymbol{Y}_b\boldsymbol{A}^t$ is obtained from Problem 4-1.

4-7. (a) Verify that to find $\boldsymbol{Y}_n^{-1}$ in Problem 4-6 by Gaussian elimination requires n^3 multiplications and $n^3 - 2n^2 + n$ additions. This is approximately three times the work required for solving $\boldsymbol{Y}_n\boldsymbol{x} = \boldsymbol{b}$, and *not n* times, as might at first be expected. Explain why.

(b) Prove the estimate in part (a) for a general $n \times n$ matrix.

4-8. Given

$$\boldsymbol{A} = \begin{bmatrix} 2 & 6 & 10 & -2 \\ 1 & 4 & 11 & 0 \\ 3 & 8 & 11 & -6 \\ -2 & -6 & -9 & 4 \end{bmatrix}$$

find the $\boldsymbol{LU}$ factors by Crout's method.

4-9. Solve the following system of linear equations by the $\boldsymbol{LU}$ factorization method:

$$\boldsymbol{A}\begin{bmatrix} x_1 \\ x_2 \\ x_3 \\ x_4 \end{bmatrix} = \begin{bmatrix} 8 \\ -2 \\ 20 \\ -13 \end{bmatrix}$$

The matrix $\boldsymbol{A}$ is given in Problem 4-8.

4-10. Solve the following equations by the Gaussian elimination method. Note that row interchange is necessary in this case.

$$\begin{bmatrix} 3 & 9 & 6 \\ 4 & 12 & 12 \\ 1 & -1 & 1 \end{bmatrix}\begin{bmatrix} x_1 \\ x_2 \\ x_3 \end{bmatrix} = \begin{bmatrix} 12 \\ 12 \\ 1 \end{bmatrix}$$

4-11. Repeat Problem 4-10, but with the $\boldsymbol{LU}$ factorization method. From the experience gained in this problem, modify Crout's algorithm in Section 4-4 to include row-interchange operations.

4-12. For each following case, determine whether the factorization $\boldsymbol{A} = \boldsymbol{LU}$ is possible. If so, find the $\boldsymbol{L}$ and $\boldsymbol{U}$ factors. If not, further determine whether a permutation matrix $\boldsymbol{P}$ exists such that the factorization $\boldsymbol{PA} = \boldsymbol{LU}$ is possible.

(a) $\boldsymbol{A} = \begin{bmatrix} 0 & 6 \\ 3 & 8 \end{bmatrix}$ (b) $\boldsymbol{A} = \begin{bmatrix} 4 & 16 & 16 \\ 3 & 12 & 9 \\ 1 & 1 & -1 \end{bmatrix}$

(c) $\boldsymbol{A} = \begin{bmatrix} 10 & 40 & 34 \\ 3 & 12 & 9 \\ 4 & 16 & 16 \end{bmatrix}$

4-13. The linear network shown in Fig. P4-13 is to be analyzed by the nodal method at the angular frequency $\omega = 1$ rad/s.

(a) By the topological rules of Section 4-6, write the nodal admittance matrix Y_n and the equivalent current source vector J_n.

(b) Use the program in Appendix 4A to check your answers to part (a) and to find the node-to-datum voltages.

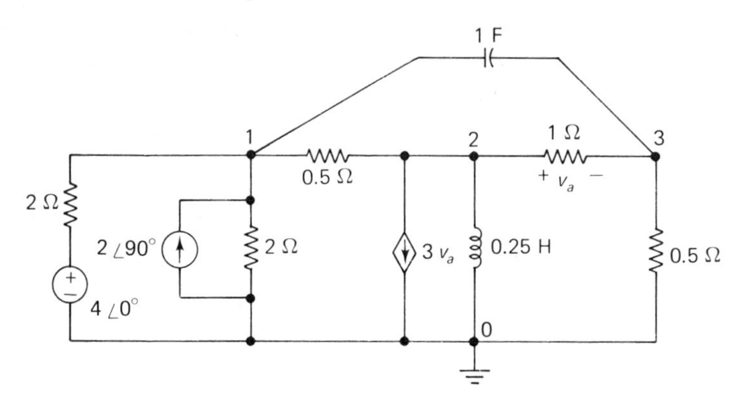

Fig. P4-13

4-14. Consider the network shown in Fig. 4-3(a). Replace the coupled inductors by the model shown in Fig. 4-7; then formulate Y_n and J_n using the topological rules of Section 4-6.

4-15. Consider an $(n + 1)$-node network consisting of only one composite branch, as shown in Fig. 4-4. Write the nodal equations by inspection of this simple network. Assume that neither node i nor j is the reference node. From the Y_n and J_n matrices thus obtained, justify the topological rules given by Eqs. (4-77) and (4-78).

4-16. Consider an $(n + 1)$-node network consisting of only two composite branches as shown in Fig. 4-5. None of the nodes (i, j, k, m) is the reference node. Write nodal equations for this simple network by inspection. From the Y_n and J_n matrices thus obtained, justify the topological rules given by Eqs. (4-79) and (4-80). Can you explain why E_l does not appear in J_n? When will the value of E_l in Fig. 4-5 be needed?

4-17. Consider an $(n + 1)$-node network consisting of only one composite branch, as shown in Fig. 4-4. Neither node i nor j is the reference node.

(a) Write the matrices A, Y_b, E, and J.

(b) Compute $Y_n = AY_bA^t$ and $J_n = A(J - Y_bE)$ by actual matrix multiplications.

(c) Use the results of part (b) to justify the rules given by Eqs. (4-77) and (4-78).

4-18. Consider an $(n + 1)$-node network consisting of only two composite branches, as shown in Fig. 4-5. None of the nodes (i, j, k, m) is the reference node.

(a) Write the matrices A, Y_b, E, and J.

(b) Compute Y_n and J_n from $Y_n = AY_bA^t$ and $J_n = A(J - Y_bE)$.

(c) Use the results of part (b) to justify the rules given by Eqs. (4-79) and (4-80).

4-19. Refer to Table 2-3 for the definition of an ideal gyrator and its model. Show that all four types of linear controlled sources and two-winding ideal transformers can be modeled with g_m type controlled sources and gyrators (and therefore with g_m type controlled sources only).

4-20. In Table 2-3, if the two lower terminals of an ideal gyrator are joined together, the resultant circuit element is called a *grounded gyrator*.

(a) Show that a linear inductor can be modeled as a grounded gyrator terminated in a linear capacitor.

(b) Show that an independent current (voltage) source can be modeled as a grounded gyrator terminated in an independent voltage (current) source.

4-21. In the modified nodal approach of [10], each voltage source (either independent or dependent) is considered as a branch. The currents through voltage-source branches are considered as additional unknowns (besides the node-to-datum voltages), and the corresponding branch constitutive relations as additional equations (besides the KCL equations obtained from all non-datum nodes).

(a) Formulate the modified nodal equations for the network shown in Fig. 4-2, but with the 0.5-Ω resistor replaced by a short circuit.

(b) Repeat part (a) for the network shown in Fig. P4-2, but with the 6-Ω resistor replaced by a short circuit.

CHAPTER 5

Nodal Nonlinear Network Analysis: Algorithms and Computational Methods

5-1 INTRODUCTION

The most basic prerequisite to computer-aided circuit analysis of general *dynamic nonlinear networks* is the ability to analyze *resistive nonlinear networks* efficiently. Just as in the linear case, this task essentially consists of two *independent* problems: (1) formulating the nonlinear equilibrium equations with the help of topological formulas, and (2) solving these nonlinear equations by appropriate numerical techniques. In this chapter, we restrict ourselves to the simpler case in which only *linear resistors, voltage-controlled nonlinear resistors, independent sources, and linear or nonlinear voltage-controlled current sources are allowed.* This restriction will allow a straightforward generalization of the nodal equation formulation algorithm presented in Chapter 4 to the nonlinear case. The more general situation involving both voltage-controlled and current-controlled elements is more complicated and will be presented in Chapter 7, after additional topological machinery has been set up. It must be emphasized, however, that the numerical methods to be presented in this chapter are independent of the topological methods and are therefore applicable to the general case as well.

5-2 TOPOLOGICAL FORMULATION OF NODAL EQUATIONS

Just as in the linear case, we consider the *composite branch* and its *graph* as shown in Fig. 5-1. We assume that the element b_k is either a *voltage-controlled nonlinear resistor* characterized by

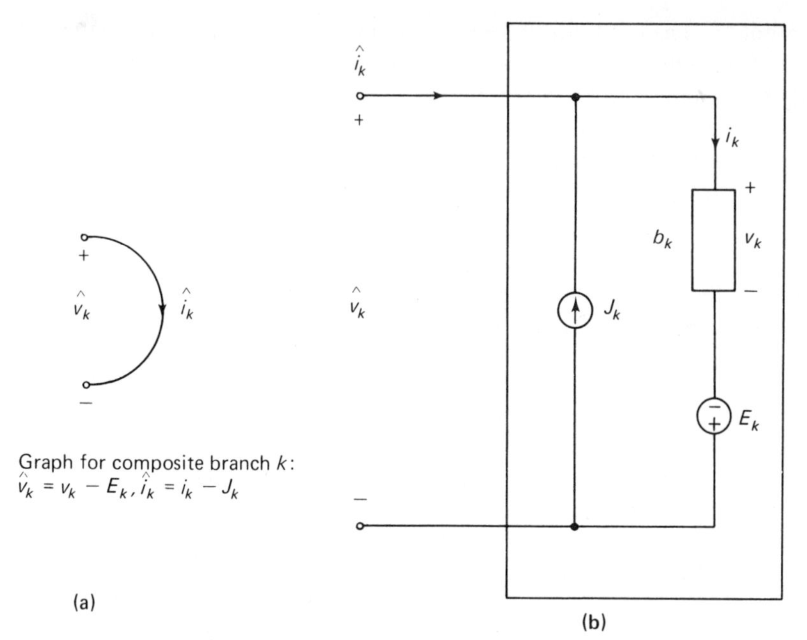

Fig. 5-1. Each branch k in the network graph represents a composite element consisting of an independent current source, an independent voltage source, and a two-terminal resistor, or a voltage-controlled current source.

$$i_k = g_k(v_k) \tag{5-1}$$

or a *voltage-controlled current source* characterized by

$$i_k = g_k(v_j) \tag{5-2}$$

where v_j is the voltage across the controlling element b_j. Notice that the element b_k cannot be a short circuit since it is voltage-controlled by assumption. Equations (5-1) and (5-2) can be combined into the following compact vector form:

$$\boldsymbol{i} = \begin{bmatrix} i_1 \\ i_2 \\ \cdot \\ \cdot \\ \cdot \\ i_b \end{bmatrix} = \begin{bmatrix} g_1(v_\alpha) \\ g_2(v_\beta) \\ \cdot \\ \cdot \\ \cdot \\ g_b(v_\rho) \end{bmatrix} \triangleq \boldsymbol{g}(\boldsymbol{v}) \tag{5-3}$$

where $v_\alpha, v_\beta, \ldots, v_\rho$ may be any branch voltage $v_1, v_2, \ldots, v_b$. Of course, if there were no controlled sources, then $i_k = g_k(v_k)$, and $v_\alpha = v_1, v_\beta = v_2, \ldots, v_\rho = v_b$.

If we substitute Eq. (5-3) for $\boldsymbol{i}$ in Eq. (4-6), we would obtain

$$\boldsymbol{A}\boldsymbol{g}(\boldsymbol{v}) = \boldsymbol{A}\boldsymbol{J} \tag{5-4}$$

Next, substituting $\boldsymbol{v}$ from Eq. (4-3) into Eq. (5-4), we obtain

$$\boldsymbol{A}\boldsymbol{g} \circ (\hat{\boldsymbol{v}} + \boldsymbol{E}) = \boldsymbol{A}\boldsymbol{J} \tag{5-5}$$

The symbol "$\circ$" has been introduced to avoid confusion with multiplication of $\boldsymbol{g}$ with $(\hat{\boldsymbol{v}} + \boldsymbol{E})$. This symbol is called the *composition operation;* whenever it appears, the quantity to its right is to be interpreted as the argument of the function appearing on its left. Let us now substitute $\hat{\boldsymbol{v}}$ from Eq. (4-15) into Eq. (5-5) to obtain the following *nodal equations* for nonlinear resistive networks:

$$\boldsymbol{A}\boldsymbol{g} \circ (\boldsymbol{A}^t\boldsymbol{v}_n + \boldsymbol{E}) = \boldsymbol{A}\boldsymbol{J} \tag{5-6}$$

For an $(n + 1)$-node network, Eq. (5-6) represents a system of n nonlinear nodal equations in terms of the n node-to-datum nodal voltages. If we denote the vector $\boldsymbol{v}_n$ by $\boldsymbol{x}$, Eq. (5-6) can be written in the form

$$\boldsymbol{f}(\boldsymbol{x}) = \boldsymbol{0} \tag{5-7}$$

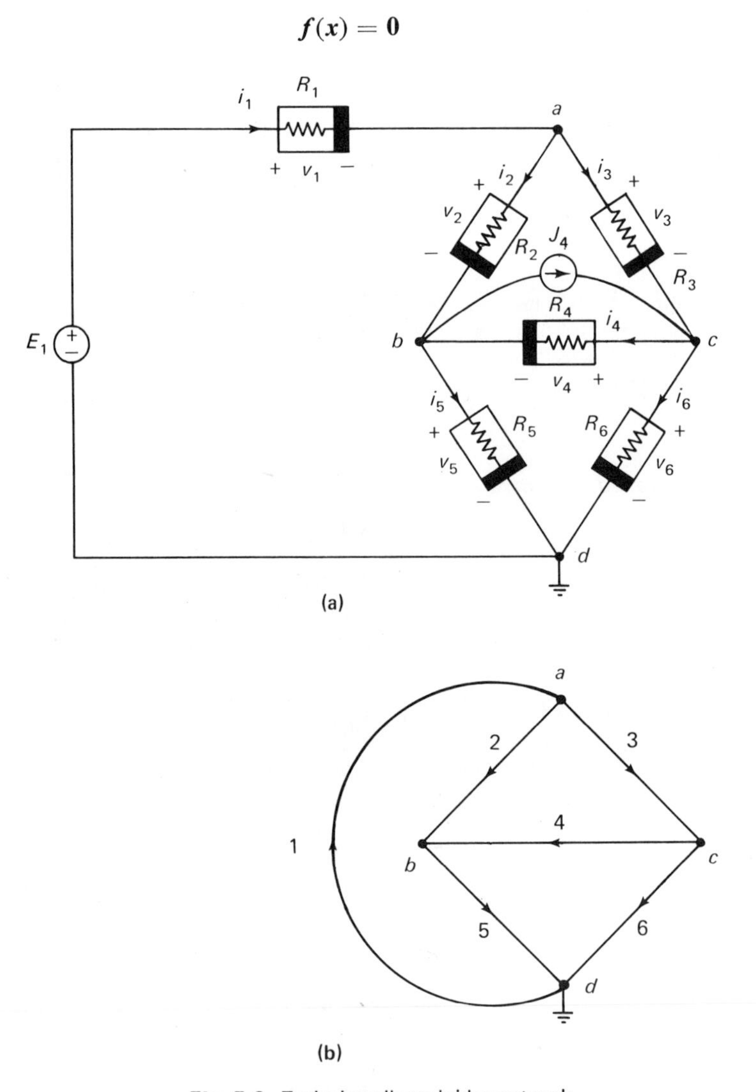

Fig. 5-2. Typical nonlinear bridge network.

where

$$f(x) \triangleq Ag \circ (A^t x + E) - AJ \tag{5-8}$$

Numerical methods for solving Eq. (5-7) on a computer will be presented in the following sections. Meanwhile, let us consider a numerical example.

EXAMPLE 5-1. Consider the nonlinear bridge network shown in Fig. 5-2(a). In terms of the composite branch defined in Fig. 5-1(b), the voltage source E_1 and the resistor R_1 are considered as one composite branch. Similarly, the current source J_4 and the resistor R_4 are also considered as one composite branch. The graph for this network then contains four nodes and six branches, as shown in Fig. 5-2(b). The reduced incidence matrix relative to *datum node d* is readily found to be

$$A = \begin{bmatrix} -1 & 1 & 1 & 0 & 0 & 0 \\ 0 & -1 & 0 & -1 & 1 & 0 \\ 0 & 0 & -1 & 1 & 0 & 1 \end{bmatrix} \tag{5-9}$$

If we let each nonlinear resistor be characterized by a voltage-controlled v_k-i_k curve, as represented in Eq. (5-1), and define the following vectors

$$i = g(v) = \begin{bmatrix} g_1(v_1) \\ g_2(v_2) \\ g_3(v_3) \\ g_4(v_4) \\ g_5(v_5) \\ g_6(v_6) \end{bmatrix}, \quad v = \begin{bmatrix} v_1 \\ v_2 \\ v_3 \\ v_4 \\ v_5 \\ v_6 \end{bmatrix}, \quad i = \begin{bmatrix} i_1 \\ i_2 \\ i_3 \\ i_4 \\ i_5 \\ i_6 \end{bmatrix}, \quad E = \begin{bmatrix} E_1 \\ 0 \\ 0 \\ 0 \\ 0 \\ 0 \end{bmatrix}, \quad J = \begin{bmatrix} 0 \\ 0 \\ 0 \\ J_4 \\ 0 \\ 0 \end{bmatrix} \tag{5-10}$$

$$v_n = \begin{bmatrix} v_a \\ v_b \\ v_c \end{bmatrix} \tag{5-11}$$

we have

$$A^t v_n + E = \begin{bmatrix} -1 & 0 & 0 \\ 1 & -1 & 0 \\ 1 & 0 & -1 \\ 0 & -1 & 1 \\ 0 & 1 & 0 \\ 0 & 0 & 1 \end{bmatrix} \begin{bmatrix} v_a \\ v_b \\ v_c \end{bmatrix} + \begin{bmatrix} E_1 \\ 0 \\ 0 \\ 0 \\ 0 \\ 0 \end{bmatrix} = \begin{bmatrix} E_1 - v_a \\ v_a - v_b \\ v_a - v_c \\ -v_b + v_c \\ v_b \\ v_c \end{bmatrix} \tag{5-12}$$

$$AJ = \begin{bmatrix} -1 & 1 & 1 & 0 & 0 & 0 \\ 0 & -1 & 0 & -1 & 1 & 0 \\ 0 & 0 & -1 & 1 & 0 & 1 \end{bmatrix} \begin{bmatrix} 0 \\ 0 \\ 0 \\ J_4 \\ 0 \\ 0 \end{bmatrix} = \begin{bmatrix} 0 \\ -J_4 \\ J_4 \end{bmatrix} \tag{5-13}$$

$$\boldsymbol{g} \circ (\boldsymbol{A}^t \boldsymbol{v}_n + \boldsymbol{E}) = \begin{bmatrix} g_1 \circ (E_1 - v_a) \\ g_2 \circ (v_a - v_b) \\ g_3 \circ (v_a - v_c) \\ g_4 \circ (-v_b + v_c) \\ g_5 \circ (v_b) \\ g_6 \circ (v_c) \end{bmatrix} \tag{5-14}$$

Substituting Eqs. (5-13) and (5-12) into Eq. (5-6), we obtain

$$\begin{bmatrix} -1 & 1 & 1 & 0 & 0 & 0 \\ 0 & -1 & 0 & -1 & 1 & 0 \\ 0 & 0 & -1 & 1 & 0 & 1 \end{bmatrix} \begin{bmatrix} g_1 \circ (E_1 - v_a) \\ g_2 \circ (v_a - v_b) \\ g_3 \circ (v_a - v_c) \\ g_4 \circ (-v_b + v_c) \\ g_5 \circ (v_b) \\ g_6 \circ (v_c) \end{bmatrix} = \begin{bmatrix} 0 \\ -J_4 \\ J_4 \end{bmatrix} \tag{5-15}$$

Simplifying and rearranging the terms in Eq. (5-15), we obtain

$$\begin{aligned} -g_1 \circ (E_1 - v_a) + g_2 \circ (v_a - v_b) + g_3 \circ (v_a - v_c) &= 0 \\ -g_2 \circ (v_a - v_b) - g_4 \circ (-v_b + v_c) + g_5 \circ (v_b) &= -J_4 \\ -g_3 \circ (v_a - v_c) + g_4 \circ (-v_b + v_c) + g_6 \circ (v_c) &= J_4 \end{aligned} \tag{5-16}$$

Equation (5-16) constitutes the three nodal equations in the three nodal variables v_a, v_b, and v_c. To be more specific, let us assign some definite functions to the v_j-i_j curves. In particular, let $i_1 = g_1(v_1) = v_1^3$, $i_2 = g_2(v_2) = (1/R_2)v_2$, $i_3 = g_3(v_3) = e^{-v_3}$, $i_4 = g_4(v_4) = v_4{}^{1/5}$, $i_5 = g_5(v_5) = v_5 - v_5^3$, and $i_6 = g_6(v_6) = (1/R_6)v_6$. Substituting these functions into Eq. (5-16), we obtain

$$\begin{aligned} -(E_1 - v_a)^3 + \frac{1}{R_2}(v_a - v_b) + e^{-(v_a - v_c)} &= 0 \\ -\frac{1}{R_2}(v_a - v_b) - (-v_b + v_c)^{1/5} + (v_b - v_b^3) &= -J_4 \\ -e^{-(v_a - v_c)} + (-v_b + v_c)^{1/5} + \frac{1}{R_6} v_c &= J_4 \end{aligned} \tag{5-17}$$

Since most practical devices do not have simple mathematical representations, the v_j-i_j curves are usually given graphically. In this case, the nodal equations would be as given by Eq. (5-16), where each function $g_j(\cdot)$ is specified by a curve.[1] Observe that Eq. (5-17) can be recast into the following general form:

$$f_1(v_a, v_b, v_c) \triangleq -(E_1 - v_a)^3 + \frac{1}{R_2}(v_a - v_b) + e^{-(v_a - v_c)} = 0 \tag{5-18a}$$

$$f_2(v_a, v_b, v_c) \triangleq -\frac{1}{R_2}(v_a - v_b) - (-v_b + v_c)^{1/5} + (v_b - v_b{}^3) + J_4 = 0 \tag{5-18b}$$

$$f_3(v_a, v_b, v_c) \triangleq -e^{-(v_a - v_c)} + (-v_b + v_c)^{1/5} + \frac{1}{R_6} v_c - J_4 = 0 \tag{5-18c}$$

[1]It is common practice to let the "dot" inside the parentheses of a function, $g_j(\cdot)$, denote the appropriate *argument* x_j of $g_j(x_j)$.

Equation (5-18) clearly assumes the same general form as Eq. (5-7), i.e., *a system of nonlinear algebraic equations.* Except in very special cases, the solutions of such equations cannot be obtained in *closed analytic form.* Thus, the nodal equations must in general be solved by appropriate *numerical iteration techniques.* The following sections are therefore addressed to the basic idea behind various standard iteration techniques.

5-3 FIXED-POINT ITERATION CONCEPT

Most iteration techniques for solving a system of nonlinear algebraic equations can be considered as special cases of the *fixed-point* iteration algorithm. Perhaps the most intuitive way to introduce the powerful concept of a *fixed point* is through an example. Consider first the simplest case of one equation in one unknown. In particular, consider the equation

$$x = 4 - 2x^{1/3} \triangleq F(x) \tag{5-19}$$

Our objective is to find a number $x = x^*$ that reduces Eq. (5-19) to an identity. Suppose that we start by choosing an arbitrary initial guess $x = x_0$. In all likelihood, our guess is apt to be wrong in the sense that, if we let $F(x_0) \triangleq x_1$, then $x_0 \neq x_1$. If we were to make other guesses arbitrarily, chances are that we would never hit the right value, unless we can find an algorithm which guarantees that each successive guess will be closer to the correct value x^*.

It turns out that, under *appropriate* conditions, the value $x_1 \triangleq F(x_0)$ is a better guess than x_0 in the sense that the error is smaller; i.e., $|x^* - x_1| < |x^* - x_0|$. This observation immediately suggests that, if we make x_1 our next guess, then $x_2 \triangleq F(x_1)$ would be an even better guess than x_1. If we keep repeating this procedure, we eventually arrive at the correct value x^*. Our algorithm consists, therefore, of the following steps:

> Choose an arbitrary initial guess $x = x_0$.
> Calculate $F(x_0)$ and call it x_1.
> Calculate $F(x_1)$ and call it x_2.
> Calculate $F(x_2)$ and call it x_3.
> . . .
> Calculate $F(x_n)$ and call it x_{n+1}.
> Stop whenever the difference between x_{n+1} and x_n becomes less than some prescribed acceptable error.

If we interpret the value of x_n to be a point on the x-axis, as shown in Fig. 5-3, then the algorithm can be interpreted as a *recipe* for moving a point x_n into the next point x_{n+1}. If the distance between each successive pair of points decreases rapidly enough, it is

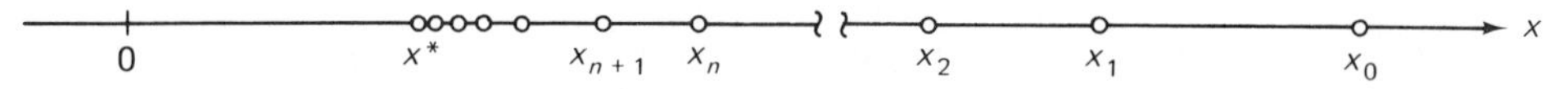

Fig. 5-3. One-dimensional geometrical interpretation of a fixed point.

clear that the algorithm would converge at a point x^* such that each subsequent movement would be onto itself. In view of this interpretation, the point x^* is called a *fixed point* of the function $F(x)$, and the algorithm, the *fixed-point algorithm*. Applying this algorithm to Eq. (5-19), we obtain the results shown in Table 5-1 corresponding to three different initial guesses. A total of 20 iterations is listed, corresponding to the initial guess $x_0 = 27$ (columns 1 and 2), $x_0 = 10$ (columns 3 and 4), and $x_0 = -8$ (columns 5 and 6), respectively. Notice that in each case the iteration converges to the fixed point $x^* = 1.6410$. It is instructive to compare the values of x_n in Table 5-1 corresponding to each initial guess. Observe that a closer initial guess would require a fewer number of iterations.

The fixed-point algorithm has a simple geometrical interpretation. If we plot the curve $y = F(x)$ as shown in Fig. 5-4 (corresponding to our present example), the fixed point of $F(x)$ is simply the intersection of this curve and the 45° straight line through the origin. Since the ordinate (y) and abscissa (x) of each point on this line are equal, the fixed-point algorithm is equivalent to the geometrical constructions shown in Fig. 5-4 (shown for the case $x_0 = 10$). Hence, through $x = 10$, a vertical line is drawn until it intersects the curve $F(x)$ at point a. If we project a horizontal line through a until it intersects the 45° line at point b, the abscissa $x \approx -0.30$ at b gives the next

TABLE 5-1. Computer Output of the Results of the Fixed-Point Iteration of the Equation $F(x) = 4 - 2x^{1/3}$

x_n x	x_{n+1} $F(x)$	x_n x	x_{n+1} $F(x)$	x_n x	x_{n+1} $F(x)$
27.0000	−2.0000	10.0000	−0.3089	−8.0000	8.0000
−2.0000	6.5198	−0.3089	5.3519	8.0000	0.0000
6.5198	0.2637	5.3519	0.5016	0.0000	3.9938
0.2637	2.7175	0.5016	2.4109	3.9938	0.8268
2.7175	1.2090	2.4109	1.3182	0.8268	2.1228
1.2090	1.8694	1.3182	1.8071	2.1228	1.4296
1.8694	1.5363	1.8071	1.5639	1.4296	1.7470
1.5363	1.6923	1.5639	1.6785	1.7470	1.5913
1.6923	1.6167	1.6785	1.6231	1.5913	1.6651
1.6167	1.6527	1.6231	1.6496	1.6651	1.6295
1.6527	1.6354	1.6496	1.6369	1.6295	1.6465
1.6354	1.6437	1.6369	1.6430	1.6465	1.6383
1.6437	1.6397	1.6430	1.6400	1.6383	1.6422
1.6397	1.6416	1.6400	1.6414	1.6422	1.6404
1.6416	1.6407	1.6414	1.6408	1.6404	1.6413
1.6407	1.6411	1.6408	1.6411	1.6413	1.6408
1.6411	1.6409	1.6411	1.6409	1.6408	1.6410
1.6409	1.6410	1.6409	1.6410	1.6410	1.6409
1.6410	1.6410	1.6410	1.6410	1.6409	1.6410
1.6410	1.6410	1.6410	1.6410	1.6410	1.6410

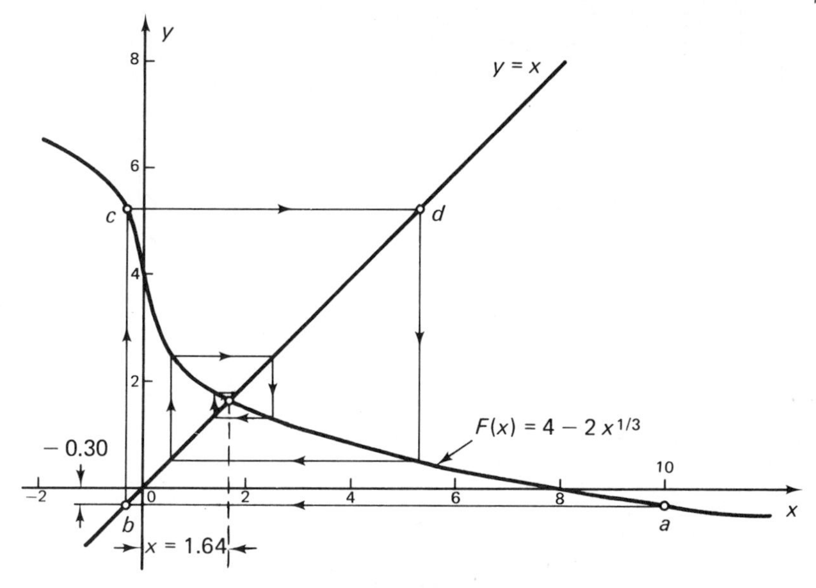

Fig. 5-4. Geometrical interpretation of the fixed-point algorithm.

value x_1. We can repeat the procedure by drawing a vertical line through b until it intersects $F(x)$ at point c. Projecting a horizontal line through c until it intersects the 45° line at point d, we obtain the value of x_2. This construction technique can be repeated a sufficient number of times until the points of intersection converge toward the fixed point.

The reader should verify for this example that the geometrical construction always converges to the desired fixed point, regardless of where we start. Unfortunately, we cannot guarantee that this behavior is true for all cases. There are two problems that one may encounter. The first is best seen from Fig. 5-5(a). Observe that no matter where we start, the iteration diverges. The second problem occurs whenever the curve $F(x)$ has more than one point of intersection with the 45° line, as shown in Fig. 5-5(b). In this case, it is clear that, if the value of our initial guess x_0 is less than x_B, the iteration will converge to point A. Otherwise, it will converge to point C. Since it is not possible to converge to point B, this fixed point is said to be unstable. This example demonstrates a fundamental limitation of all iteration techniques: they cannot account for all solutions.

In the general case of a system of n equations in n unknowns, Eq. (5-19) assumes the standard form

$$\mathbf{x} = \mathbf{F}(\mathbf{x}) \tag{5-20}$$

The fixed-point algorithm for this case can be described compactly by the following recursive recipe:

$$\boxed{\mathbf{x}_{n+1} = \mathbf{F}(\mathbf{x}_n), \qquad n = 0, 1, 2, \ldots} \tag{5-21}$$

Under appropriate conditions, this sequence converges to a fixed point:

$$\lim_{n\to\infty} \mathbf{x}_n = \mathbf{x}^* \tag{5-22}$$

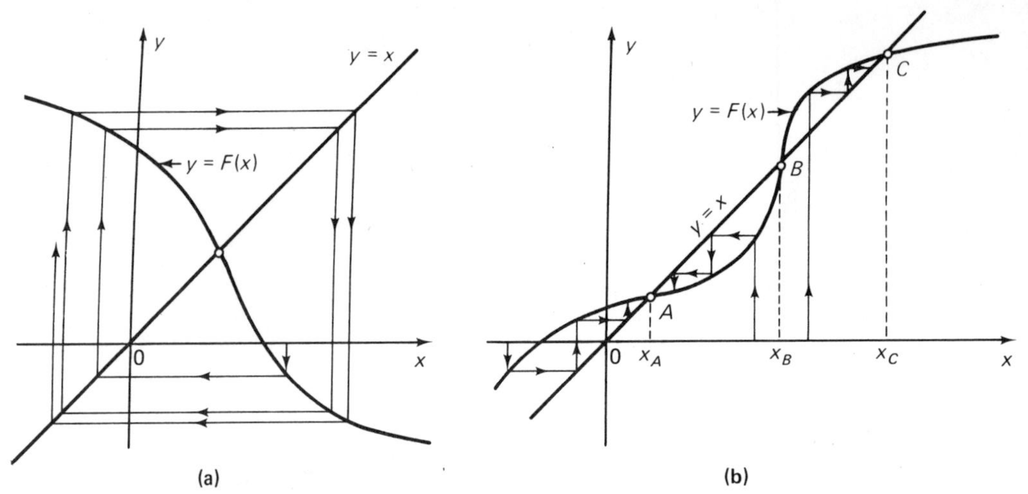

Fig. 5-5. Depending on the shape of $F(x)$, the fixed-point iteration may either diverge, as in part (a), or it may converge to one of several possible stable fixed points, as in part (b).

A simple criterion that guarantees convergence is given by the following theorem:

Theorem 5.1 (Principle of Contraction Mapping). *If* $\mathbf{F}(\mathbf{x})$ *is a* ***contraction*** *from the n-dimensional space* R^n *into* R^n, *i.e., if there exists a constant* $L < 1$ *such that*[2]

$$||\mathbf{F}(\mathbf{x}) - \mathbf{F}(\mathbf{y})|| \leq L\,||\mathbf{x} - \mathbf{y}|| \tag{5-23}$$

for all $\mathbf{x} \in R^n$ *and* $\mathbf{y} \in R^n$, *then* $\mathbf{F}(\mathbf{x})$ *has a* ***unique*** *fixed point* $\mathbf{x} = \mathbf{x}^*$. *Moreover, the sequence of iterates defined by Eq. (5-21)* ***converges*** *to this fixed point. An upper bound on the error for stopping at the jth iteration is given by*

$$||\mathbf{x}^* - \mathbf{x}_j|| \leq \frac{L^j}{(1 - L)}||\mathbf{x}_1 - \mathbf{x}_0|| \tag{5-24}$$

Proof: See Appendix 5A.

Observe that, when L is close to unity, convergence may be very slow. Observe also that the fixed-point algorithm requires that the equations be expressed in the standard form of Eq. (5-20). The nodal equations as defined in Eq. (5-7) can be transformed into the standard "fixed-point form" $\mathbf{x} = \mathbf{F}(\mathbf{x})$ upon defining

$$\boxed{\mathbf{F}(\mathbf{x}) \triangleq \mathbf{x} - \mathbf{K}(\mathbf{x})\mathbf{f}(\mathbf{x})} \tag{5-25}$$

where $\mathbf{K}(\mathbf{x})$ is an $n \times n$ matrix function of $\mathbf{x}$ with the property that $\mathbf{K}(\mathbf{x})$ is *nonsingular* at any solution of $\mathbf{f}(\mathbf{x}) = \mathbf{0}$. Clearly, any solution $\mathbf{x}^*$ of Eq. (5-7) is a fixed point of

[2]The symbol $||\mathbf{x}||$ denotes the *norm* or *length* of the vector $\mathbf{x}$; i.e.,

$$||\mathbf{x}|| = \sqrt{x_1^2 + x_2^2 + \ldots + x_n^2}$$

$\boldsymbol{F}(\boldsymbol{x})$, as can be seen by direct substitution of $\boldsymbol{x} = \boldsymbol{x}^*$ into (5-25). Conversely, if $\boldsymbol{x} = \boldsymbol{x}^*$ is a fixed point of (5-25), then

$$\boldsymbol{x}^* = \boldsymbol{x}^* - \boldsymbol{K}(\boldsymbol{x}^*)\boldsymbol{f}(\boldsymbol{x}^*) \tag{5-26}$$

and hence $\boldsymbol{f}(\boldsymbol{x}^*) = \boldsymbol{0}$ [since $\boldsymbol{K}^{-1}(\boldsymbol{x}^*)$ exists by assumption]. We have therefore shown that $\boldsymbol{x} = \boldsymbol{x}^*$ *is a solution of Eq. (5-7) if, and only if, it is a fixed point of* $\boldsymbol{F}(\boldsymbol{x})$. Observe that the $n \times n$ matrix $\boldsymbol{K}(\boldsymbol{x})$ is arbitrary as long as it is nonsingular at $\boldsymbol{x}^*$. It is clear that different choices of $\boldsymbol{K}(\boldsymbol{x})$ will lead to different convergence behavior. A favorable choice often leads to a more rapid rate of convergence. Indeed, most iteration techniques for solving a system of nonlinear algebraic equations are special cases of Eq. (5-25), each technique being identified with a particular $\boldsymbol{K}(\boldsymbol{x})$. For example, a common choice for $\boldsymbol{K}(\boldsymbol{x})$ is

$$\boldsymbol{K}(\boldsymbol{x}) = \boldsymbol{J}^{-1}(\boldsymbol{x}) \tag{5-27}$$

where

$$\boldsymbol{J}(\boldsymbol{x}) \triangleq \begin{bmatrix} \dfrac{\partial f_1(\boldsymbol{x})}{\partial x_1} & \dfrac{\partial f_1(\boldsymbol{x})}{\partial x_2} & \cdots & \dfrac{\partial f_1(\boldsymbol{x})}{\partial x_n} \\ \dfrac{\partial f_2(\boldsymbol{x})}{\partial x_1} & \dfrac{\partial f_2(\boldsymbol{x})}{\partial x_2} & \cdots & \dfrac{\partial f_2(\boldsymbol{x})}{\partial x_n} \\ \cdot & & & \\ \cdot & & & \\ \cdot & & & \\ \dfrac{\partial f_n(\boldsymbol{x})}{\partial x_1} & \dfrac{\partial f_n(\boldsymbol{x})}{\partial x_2} & \cdots & \dfrac{\partial f_n(\boldsymbol{x})}{\partial x_n} \end{bmatrix} \tag{5-28}$$

is the *Jacobian matrix* of $\boldsymbol{f}(\boldsymbol{x})$. Substituting Eq. (5-27) into Eq. (5-25), we obtain the following recursive formula:

$$\boxed{\boldsymbol{x}_{n+1} = \boldsymbol{x}_n - [\boldsymbol{J}(\boldsymbol{x}_n)]^{-1}\boldsymbol{f}(\boldsymbol{x}_n)} \tag{5-29}$$

Equation (5-29) is therefore a special fixed-point algorithm for solving the system of equations

$$\boxed{\boldsymbol{f}(\boldsymbol{x}) = \boldsymbol{0}} \tag{5-30}$$

It is called the *Newton–Raphson method.* To compare the rate of convergence of this method with that obtained by the original fixed-point method, consider the same example in Eq. (5-19). In view of Eq. (5-30), $\boldsymbol{f}(\boldsymbol{x})$ is now defined by

$$f(x) \triangleq x + 2x^{1/3} - 4 \tag{5-31}$$

Applying Eq. (5-29), we obtain

$$x_{n+1} = x_n - \left[\frac{df(x)}{dx}\bigg|_{x=x_n}\right]^{-1} f(x_n) = x_n - \frac{x_n + 2x_n^{1/3} - 4}{1 + \frac{2}{3}x_n^{-2/3}} \tag{5-32}$$

For comparison purposes, we chose the same initial guesses as in Table 5-1: $x_0 = 27$, 10, and -8. The results corresponding to nine iterations are given in Table 5-2. A comparison between Tables 5-1 and 5-2 clearly shows the superiority of the Newton–Raphson algorithm, at least for this example, since it converges much more rapidly.

TABLE 5-2. Computer Output of the Results of the Newton–Raphson Iteration for the Equation $f(x) = x + 2x^{1/3} - 4$, $f(x) = 0$ at $x = 1.6410$

x_n x	x_{n+1} $F(x)$	x_n x	x_{n+1} $F(x)$	x_n x	x_{n+1} $F(x)$
27.0000	0.0000	10.0000	0.9858	−8.0000	11.2000
0.0000	0.0004	0.9458	1.6977	11.2000	0.8973
0.0004	0.0297	1.5977	1.6409	0.8973	1.5810
0.0297	0.4511	1.6409	1.6410	1.5810	1.6407
0.4511	1.3956	1.6410	1.6410	1.6407	1.6410
1.3956	1.6364	1.6410	1.6410	1.6410	1.6410
1.6364	1.6410	1.6410	1.6410	1.6410	1.6410
1.6410	1.6410	1.6410	1.6410	1.6410	1.6410
1.6410	1.6410	1.6410	1.6410	1.6410	1.6410

5-4 NEWTON–RAPHSON ALGORITHM

The preceding derivation of the Newton–Raphson algorithm as a special case of the general fixed-point algorithm is rather mysterious since no explanation is given as to why $\mathbf{K}(\mathbf{x})$ is chosen in Eq. (5-27). In view of its importance, we shall now rederive this algorithm from a more intuitive point of view.

5-4-1 Newton–Raphson Algorithm for One Equation in One Unknown

Consider the equation

$$f(x) = 0 \tag{5-33}$$

and let $x = x^{(j)}$ be the jth guess.[3] We can safely assume that $x^{(j)}$ is *not* a solution, for if it were a solution, we are done.[4] Hence, $f(x^{(j)}) \neq 0$. Suppose that we obtain the Taylor expansion for $f(x)$ about the point $x = x^{(j)}$:

$$f(x) = f(x^{(j)}) + \left.\frac{df(x)}{dx}\right|_{x=x^{(j)}}(x - x^{(j)}) + \frac{1}{2!}\left.\frac{d^2f(x)}{dx^2}\right|_{x=x^{(j)}}(x - x^{(j)})^2 + \ldots \tag{5-34}$$

[3]Throughout this section, the jth guess is denoted by a *superscript* (j) rather than by a *subscript* as in the preceding section. The change of notation is necessary here to avoid possible confusion with the jth component of a vector, which we shall denote with a subscript.

[4]It follows from Appendix 5A that if the derivative $F'(x) = [f(x)f''(x)]/[f'(x)]^2$ of $F(x) \triangleq x - [f'(x)]^{-1}f(x)$ never vanishes for all values of x, then the Newton–Raphson algorithm $x^{(j+1)} = F(x^{(j)})$ *cannot terminate in a finite number of steps* unless the initial guess $x^{(0)} = x^*$. (This assertion does not apply to linear equations since $F'(x) = 0$ in this case. See also Problem 5-23.) Of course, it will generally take only a *finite* number of iterations to come close to the exact solution x^*. Otherwise, the algorithm would be useless!

If we let $x = x^{(j+1)}$ be the next guess, Eq. (5-34) becomes

$$f(x^{(j+1)}) = f(x^{(j)}) + \left.\frac{df(x)}{dx}\right|_{x=x^{(j)}} (x^{(j+1)} - x^{(j)}) + \frac{1}{2!}\left.\frac{d^2f(x)}{dx^2}\right|_{x=x^{(j)}} \times (x^{(j+1)} - x^{(j)})^2 + \ldots \tag{5-35}$$

Now suppose that we make the *not necessarily valid* assumption that the initial guess is quite *good* in the first place so that $x^{(j+1)} - x^{(j)}$ is a small number. If this assumption is indeed true, we can neglect the higher-order terms in Eq. (5-35), since the nth power of a small number is a much smaller number, for $n \geq 2$. In this case, Eq. (5-35) can be approximated by

$$f(x^{(j+1)}) \approx f(x^{(j)}) + \left.\frac{df(x)}{dx}\right|_{x=x^{(j)}} (x^{(j+1)} - x^{(j)}) \tag{5-36}$$

Our objective is to choose $x^{(j+1)}$ so that it is a solution of Eq. (5-33). Hence, if our preceding assumption is valid, $x^{(j+1)}$ should obviously be chosen so that $f(x^{(j+1)}) = 0$. Equating Eq. (5-36) to zero and solving for $x^{(j+1)}$, we obtain the desired "recipe":

$$\boxed{x^{(j+1)} = x^{(j)} - [J(x^{(j)})]^{-1} f(x^{(j)})} \tag{5-37}$$

where

$$[J(x^{(j)})]^{-1} \triangleq \left[\left.\frac{df(x)}{dx}\right|_{x=x^{(j)}}\right]^{-1} \tag{5-38}$$

Equation (5-37) can now be identified as the *Newton–Raphson equation* derived earlier in (5-29). If our earlier assumption, which led to the derivation of Eq. (5-37) is valid, this algorithm will converge. It is proved in Appendix 5A that, if the initial guess $x^{(0)}$ is sufficiently close to *a correct* solution x^* of Eq. (5-33), then the Newton–Raphson algorithm will *always* converge to x^*. Before proceeding, it will be instructive to present a *geometrical interpretation* of this algorithm. A typical curve representing $y = f(x)$ is shown in Fig. 5-6(a). If we let $P^{(j)}$ represent the point on the curve corresponding to $f(x^{(j)})$, the slope of the tangent drawn through this point is equal to

$$J(x^{(j)}) = \left.\frac{df(x)}{dx}\right|_{x=x^{(j)}} \tag{5-39}$$

Hence, the next guess $x^{(j+1)}$ is simply the distance from the origin to the point of intersection between the x-axis and the tangent from $P^{(j)}$. The next point $P^{(j+1)}$ is therefore the intersection of this curve and the vertical line drawn through $x = x^{(j+1)}$, as shown in Fig. 5-6(a). By repeating this procedure, we find the iteration rapidly approaches point Q, which is the correct solution. To show that the Newton–Raphson algorithm may not converge, consider the curve shown in Fig. 5-6(b). There are two solutions in this case corresponding to points Q_a and Q_b. An initial guess corresponding to point P_a will eventually converge to point Q_a, and an initial guess corresponding to point P_b will eventually converge to point Q_b. However, an initial guess at point P_c will cause the iteration to simply oscillate around the loop without ever

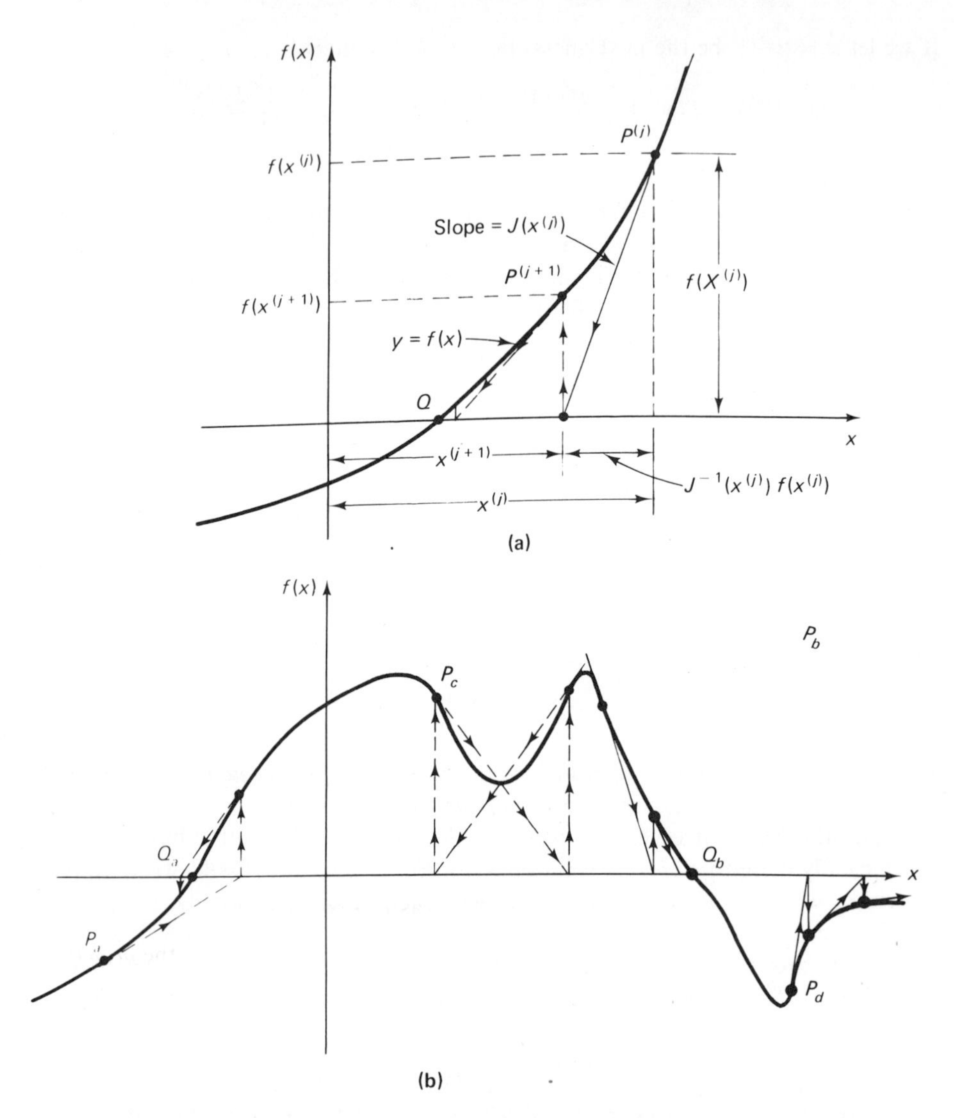

Fig. 5-6. Geometrical interpretation of the Newton–Raphson algorithm.

converging to either point Q_a or Q_b. Similarly, an initial guess at P_d will cause the iteration to diverge toward infinity. This geometrical interpretation clearly demonstrates that, *if the initial guess is far away from a correct solution*, the Newton–Raphson algorithm may not converge at all. However, from this geometrical interpretation it is also reasonable to expect that, if the initial guess is close to a correct solution, the algorithm will always converge. Thus, a program can be written that will detect any divergent behavior and select new guesses until convergence is obtained. Since, for *nonlinear* equations, it takes an infinite number of iterations before the Newton–Raphson algorithm converges to the *exact solution*, in actual programming an acceptable *error* is prescribed so that the program will terminate the iteration as soon as the solution is within this error bound.

One obvious limitation of the Newton–Raphson algorithm is that, even if it converges to a solution, there is no way of knowing whether it is the only solution or whether it converges to a *particular* solution, assuming that the equation has multiple solutions. This limitation is inherent in all known iteration methods. Consequently, another algorithm will be presented in Chapter 7 which guarantees that all solutions will be found. The price we pay for such a more general algorithm is an increase in computation time.

5-4-2 Rate of Convergence

To investigate how rapidly an algorithm converges to a correct solution, let us define the error $\epsilon^{(j)}$ at the jth iteration to be given by

$$\epsilon^{(j)} = |x^* - x^{(j)}| \tag{5-40}$$

where x^* is a correct solution. In terms of $\epsilon^{(j)}$, the one-dimensional case for Eq. (5-24) assumes the form

$$\epsilon^{(j)} \leq \frac{L^j}{(1 - L)} |x_1 - x_0| \tag{5-41}$$

Clearly, for convergence to take place, $\epsilon^{(j)}$ must keep decreasing with j. The rate at which $\epsilon^{(j)}$ decreases is therefore a measure of how fast the algorithm converges and is called the *rate of convergence* of the algorithm. In many iteration algorithms, the following estimate for the rate of convergence can be derived analytically:

$$\epsilon^{(j+1)} \leq k[\epsilon^{(j)}]^m \qquad \text{as } j \to \infty \tag{5-42}$$

where m is an integer and k is a constant. For example, it follows from the *principle of contraction mapping* that the rate of convergence for the *fixed-point algorithm* can be derived as follows:

$$|x^{(j+1)} - x^*| = |f(x^{(j)}) - f(x^*)| \leq L|x^{(j)} - x^*|, \qquad L < 1$$

Hence,

$$\epsilon^{(j+1)} \leq L\epsilon^{(j)}$$

Any iteration algorithm with this property is said to have a *linear rate of convergence*, since $\epsilon^{(j+1)}$ is bounded by a *linear* function of $\epsilon^{(j)}$. Observe that if k is close to unity the convergence would be extremely slow. In the limit $k = 1$, the iteration will fail to converge; when $k > 1$, it will begin to diverge as the error increases with every iteration.

It will be shown in Appendix 5A that the Newton–Raphson algorithm possesses the more desirable property that *its rate of convergence is quadratic;* i.e.,

$$\epsilon^{(j+1)} \leq k[\epsilon^{(j)}]^2 \tag{5-43}$$

If the error $\boldsymbol{\epsilon}^{(j)}$ is small, say $\boldsymbol{\epsilon}^{(j)} \ll 1$, the succeeding error will decrease to *a constant times the square of the previous error*. Thus, the number of correct decimal places is approximately *doubled* after each iteration. It is instructive to examine Tables 5-1 and 5-2 again and compare the rate of convergence in each case. Observe that the Newton–Raphson algorithm indeed converges much more quickly. This property is one reason why the Newton–Raphson algorithm is still the best *general-purpose method* for solving nonlinear equations.

5-4-3 Newton–Raphson Algorithm for Solving Systems of *n* Equations

We are now ready to derive the Newton–Raphson algorithm for solving a system of n nonlinear equations in n unknowns; i.e.,

$$f_i(x_1, x_2, \ldots, x_n) = 0, \qquad i = 1, 2, \ldots, n \tag{5-44}$$

As in the single-variable case, let us suppose that $\boldsymbol{x}^{(j)} = (x_1^{(j)}, x_2^{(j)}, \ldots, x_n^{(j)})^t$ is the value of $\boldsymbol{x}$ obtained at the jth iteration. Again, let us obtain the *Taylor expansion for functions of several variables* [3] about the point $\boldsymbol{x}^{(j)}$:

$$\begin{aligned} f_i(x_1, x_2, \ldots, x_n) = f_i(x_1{}^{(j)}, x_2{}^{(j)}, \ldots x_n{}^{(j)}) \\ + \left.\frac{\partial f_i(\boldsymbol{x})}{\partial x_1}\right|_{\boldsymbol{x}=\boldsymbol{x}^{(j)}} (x_1 - x_1{}^{(j)}) + \left.\frac{\partial f_i(\boldsymbol{x})}{\partial x_2}\right|_{\boldsymbol{x}=\boldsymbol{x}^{(j)}} (x_2 - x_2{}^{(j)}) \\ + \ldots + \left.\frac{\partial f_i(\boldsymbol{x})}{\partial x_n}\right|_{\boldsymbol{x}=\boldsymbol{x}^{(j)}} (x_n - x_n{}^{(j)}) \\ + \text{higher-order terms involving } (x_i - x_i{}^{(j)})^m, \\ m > 1,\ i = 1, 2, \ldots, n \end{aligned} \tag{5-45}$$

Let $\boldsymbol{x} = \boldsymbol{x}^{(j+1)}$ be the value of $\boldsymbol{x}$ at the $(j + 1)$th iteration; substituting this into Eq. (5-45), we obtain

$$\begin{aligned} f_i(x_1{}^{(j+1)}, x_2{}^{(j+1)}, \ldots, x_n{}^{(j+1)}) = f_i(x_1{}^{(j)}, x_2{}^{(j)}, \ldots, x_n{}^{(j)}) \\ + \left.\frac{\partial f_i(\boldsymbol{x})}{\partial x_1}\right|_{\boldsymbol{x}=\boldsymbol{x}^{(j)}} (x_1{}^{(j+1)} - x_1{}^{(j)}) \\ + \left.\frac{\partial f_i(\boldsymbol{x})}{\partial x_2}\right|_{\boldsymbol{x}=\boldsymbol{x}^{(j)}} (x_2{}^{(j+1)} - x_2{}^{(j)}) \\ + \ldots + \left.\frac{\partial f_i(\boldsymbol{x})}{\partial x_n}\right|_{\boldsymbol{x}=\boldsymbol{x}^{(j)}} (x_n{}^{(j+1)} - x_n{}^{(j)}) \\ + \text{higher-order terms involving } (x_i{}^{(j+1)} - x_i{}^{(j)})^m, \\ m > 1,\ i = 1, 2, \ldots, n \end{aligned} \tag{5-46}$$

Under the assumption made earlier, that $\boldsymbol{x}^{(j)}$ is a *good* guess to the correct solution so that each component of the vector $\boldsymbol{x}^{(j+1)} - \boldsymbol{x}^{(j)}$ is a small number, we can neglect all higher-order terms in Eq. (5-46) and obtain the following approximate relationship:

$$
\begin{aligned}
f_i(\mathbf{x}^{(j+1)}) = f_i(\mathbf{x}^{(j)}) &+ \left.\frac{\partial f_i(\mathbf{x})}{\partial x_1}\right|_{\mathbf{x}=\mathbf{x}^{(j)}} (x_1^{(j+1)} - x_1^{(j)}) \\
&+ \left.\frac{\partial f_i(\mathbf{x})}{\partial x_2}\right|_{\mathbf{x}=\mathbf{x}^{(j)}} (x_2^{(j+1)} - x_2^{(j)}) \\
&+ \cdots + \left.\frac{\partial f_i(\mathbf{x})}{\partial x_n}\right|_{\mathbf{x}=\mathbf{x}^{(j)}} (x_n^{(j+1)} - x_n^{(j)}) \qquad i = 1, 2, \ldots, n
\end{aligned}
\tag{5-47}
$$

Since we would like $\mathbf{x}^{(j+1)}$ to be a solution to Eq. (5-44), let us set $f_i(\mathbf{x}^{(j+1)})$ to zero in Eq. (5-47), for $i = 1, 2, \ldots, n$, and obtain the following system of equations:

$$
\begin{aligned}
f_i(\mathbf{x}^{(j)}) &+ \left.\frac{\partial f_i(\mathbf{x})}{\partial x_1}\right|_{\mathbf{x}=\mathbf{x}^{(j)}} (x_1^{(j+1)} - x_1^{(j)}) + \left.\frac{\partial f_i(\mathbf{x})}{\partial x_2}\right|_{\mathbf{x}=\mathbf{x}^{(j)}} (x_2^{(j+1)} - x_2^{(j)}) \\
&+ \ldots + \left.\frac{\partial f_i(\mathbf{x})}{\partial x_n}\right|_{\mathbf{x}=\mathbf{x}^{(j)}} (x_n^{(j+1)} - x_n^{(j)}) = 0 \qquad i = 1, 2, \ldots, n
\end{aligned}
\tag{5-48}
$$

Equation (5-48) represents a system of n equations and can be written in the following matrix form:

$$\mathbf{J}(\mathbf{x}^{(j)})(\mathbf{x}^{(j+1)} - \mathbf{x}^{(j)}) = -\mathbf{f}(\mathbf{x}^{(j)}) \tag{5-49}$$

where $\mathbf{J}(\mathbf{x}^{(j)})$ is the Jacobian matrix of $\mathbf{f}(\mathbf{x})$ evaluated at $\mathbf{x} = \mathbf{x}^{(j)}$, and where

$$
\mathbf{x}^{(j+1)} = \begin{bmatrix} x_1^{(j+1)} \\ x_2^{(j+1)} \\ \cdot \\ \cdot \\ \cdot \\ x_n^{(j+1)} \end{bmatrix}, \qquad
\mathbf{x}^{(j)} = \begin{bmatrix} x_1^{(j)} \\ x_2^{(j)} \\ \cdot \\ \cdot \\ \cdot \\ x_n^{(j)} \end{bmatrix}, \qquad
\mathbf{f}(\mathbf{x}^{(j)}) = \begin{bmatrix} f_1(\mathbf{x}^{(j)}) \\ f_2(\mathbf{x}^{(j)}) \\ \cdot \\ \cdot \\ \cdot \\ f_n(\mathbf{x}^{(j)}) \end{bmatrix}
\tag{5-50}
$$

We can now solve for $\mathbf{x}^{(j+1)}$ from Eq. (5-49) to obtain[5]

$$\boxed{\mathbf{x}^{(j+1)} = \mathbf{x}^{(j)} - [\mathbf{J}(\mathbf{x}^{(j)})]^{-1}\mathbf{f}(\mathbf{x}^{(j)})} \tag{5-51}$$

Equation (5-51) can be identified with Eq. (5-29), the n-dimensional version of the Newton–Raphson equation. A comparison of this equation with Eq. (5-37) for $n = 1$ shows that they are exactly identical in form, except that now all quantities are either vectors or matrices. However, the computational work now greatly increases, because it would be necessary to evaluate *n^2 partial derivatives $\partial f_i(\mathbf{x})/\partial x_j$ at each iteration* to obtain the Jacobian matrix $\mathbf{J}(\mathbf{x}^{(j)})$. This is a major task even with the help of a computer. To overcome this limitation, many Newton-like methods have been proposed that obviate the need to evaluate the Jacobian matrix at each iteration. Some of these

[5]Since inverting a matrix $\mathbf{A}$ by Gaussian elimination requires approximately three times as many multiplications as solving a corresponding system of linear equations $\mathbf{Ax} = \mathbf{b}$, it is computationally more efficient to recast Eq. (5-51) into the form $[\mathbf{J}(\mathbf{x}^{(j)})]\mathbf{x}^{(j+1)} = [\mathbf{J}(\mathbf{x}^{(j)})]\mathbf{x}^{(j)} - \mathbf{f}(\mathbf{x}^{(j)})$ and then solve for $\mathbf{x}^{(j+1)}$ by Gaussian elimination.

methods are known as *quasi-Newton methods.* Although each offers some significant improvement relative to *some class* of nonlinear equations, none can claim to work in the general case. Since an exposition of these methods would divert our attention from the mainstream of this book, and since many of these methods are still in the research state, we refer the interested reader to the published literature for details [2, 4–9].

The n-dimensional Newton–Raphson algorithm suffers from the same basic *divergence problem* alluded to earlier in the one-dimensional case. However, unlike the original fixed-point algorithm, Eq. (5-20), where it is possible for the iteration to diverge regardless of the initial guess [see Fig. 5-5(a)], it can be proved rigorously that, if the initial guess $\boldsymbol{x}^{(0)}$ is close to a correct solution $\boldsymbol{x}^*$ of $\boldsymbol{f}(\boldsymbol{x}) = \boldsymbol{0}$, then the n-dimensional Newton–Raphson algorithm will always converge. Moreover, it can be shown that the rate of convergence is quadratic, as in the single-variable case [1, 2]. In fact, it is even theoretically possible to find how close the initial guess must be to assure convergence. The most general result of this type is given by the following theorem:

Theorem 5.2 (Newton–Raphson–Kantorovich Theorem) [2]. *Let* $\boldsymbol{x}^{(0)} = (x_1^{(0)}, x_2^{(0)}, \ldots, x_n^{(0)})^t$ *be the initial guess and* $\boldsymbol{x}^{(1)} = (x_1^{(1)}, x_2^{(1)}, \ldots, x_n^{(1)})^t$ *be the first iteration obtained from Eq. (5-51). Let the inverse of the Jacobian matrix at* $\boldsymbol{x}^{(0)}$ *be*

$$\boldsymbol{J}^{-1}(\boldsymbol{x}^{(0)}) = \left[\frac{A_{ij}(\boldsymbol{x}^{(0)})}{D(\boldsymbol{x}^{(0)})}\right]^t \tag{5-52}$$

where $A_{ij}(\boldsymbol{x}^{(0)})$ *is the* (i, j)*th cofactor of* $\boldsymbol{J}(\boldsymbol{x}^{(0)})$ *and* $D(\boldsymbol{x}^{(0)})$ *is the determinant of* $\boldsymbol{J}(\boldsymbol{x}^{(0)})$. *Let the following conditions be satisfied:*

$$\sum_{i=1}^{n} (x_i^{(1)} - x_i^{(0)})^2 \leq A^2 \tag{5-53}$$

$$\frac{1}{|D(\boldsymbol{x}^{(0)})|}\left[\sum_{i,j=1}^{n} A_{ij}^2(\boldsymbol{x}^{(0)})\right]^{1/2} \leq B \tag{5-54}$$

$$\sum_{i,j,k=1}^{n} \left[\frac{\partial^2 f_i(\boldsymbol{x})}{\partial x_j \partial x_k}\right]^2_{\boldsymbol{x}=\boldsymbol{x}^{(0)}} \leq K^2 \tag{5-55}$$

where A, B, *and* K *are nonnegative constants such that*

$$ABK \triangleq C_0 \leq \frac{1}{2}$$

Then the sequence defined by Eq. (5-51) converges to a solution $\boldsymbol{x}^* = [x_1^*, x_2^*, \ldots, x_n^*]^t$. *Moreover, we have*

$$||\boldsymbol{x}^* - \boldsymbol{x}^{(0)}|| \triangleq \left[\sum_{i=1}^{n} (x_i^* - x_i^{(0)})^2\right]^{1/2} \leq \frac{1 - [1 - 2C_0]^{1/2}}{C_0} A \tag{5-56}$$

The proof of the Newton–Raphson–Kantorovich theorem can be found in [2]. This theorem is of fundamental *theoretical* importance. Unfortunately, it is not readily amenable to computer programming. In practice, the initial guess is usually chosen arbitrarily. If the iteration diverges, a new guess is chosen, and so on, until the iteration converges. Fortunately, for many electronic circuits the range of the solution can often be determined from the device characteristics, and hence a reasonably good initial guess can usually be specified.

5-5 SOLVING THE NODAL EQUATIONS BY THE NEWTON–RAPHSON ALGORITHM AND ITS ASSOCIATED DISCRETE EQUIVALENT CIRCUIT

Let us now apply the n-dimensional Newton–Raphson algorithm to solve the nonlinear nodal equation (5-6):

$$\boldsymbol{f}(\boldsymbol{v}_n) \triangleq A\boldsymbol{g} \circ (A^t\boldsymbol{v}_n + \boldsymbol{E}) - A\boldsymbol{J} = \boldsymbol{0} \tag{5-57}$$

The Jacobian of $\boldsymbol{f}(\boldsymbol{v}_n)$ is given by

$$J(\boldsymbol{v}_n) = A\frac{\partial \boldsymbol{g}(\boldsymbol{v})}{\partial \boldsymbol{v}}\bigg|_{\boldsymbol{v}=(A^t\boldsymbol{v}_n+\boldsymbol{E})} A^t \triangleq A\frac{\partial \boldsymbol{g} \circ (A^t\boldsymbol{v}_n + \boldsymbol{E})}{\partial \boldsymbol{v}}A^t \tag{5-58}$$

Hence, the Newton–Raphson equation is given by

$$\boxed{\boldsymbol{v}_n^{(j+1)} = \boldsymbol{v}_n^{(j)} - \left[A\frac{\partial \boldsymbol{g} \circ (A^t\boldsymbol{v}_n^{(j)} + \boldsymbol{E})}{\partial \boldsymbol{v}}A^t\right]^{-1}[A\boldsymbol{g} \circ (A^t\boldsymbol{v}_n^{(j)} + \boldsymbol{E}) - A\boldsymbol{J}]} \tag{5-59}$$

To derive a circuit interpretation of Eq. (5-59), let us define

$$\boldsymbol{E}_Q^{(j)} \triangleq A^t\boldsymbol{v}_n^{(j)} + \boldsymbol{E} \tag{5-60}$$

$$\boldsymbol{J}_Q^{(j)} \triangleq \boldsymbol{g} \circ (A^t\boldsymbol{v}_n^{(j)} + \boldsymbol{E}) = \boldsymbol{g}(\boldsymbol{E}_Q^{(j)}) \tag{5-61}$$

$$Y_b^{(j)} \triangleq \frac{\partial \boldsymbol{g} \circ (A^t\boldsymbol{v}_n^{(j)} + \boldsymbol{E})}{\partial \boldsymbol{v}} = \frac{\partial \boldsymbol{g}(\boldsymbol{v})}{\partial \boldsymbol{v}}\bigg|_{\boldsymbol{v}=\boldsymbol{E}_Q^{(j)}} \tag{5-62}$$

Corresponding to the nonlinear bridge circuit shown in Fig. 5-2(a) and the associated nonlinear nodal equations derived earlier in Section 5-2, Eqs. (5-60)–(5-62) assume the form shown in Eqs. (5-63)–(5-65) shown on pages 221–222.

$$\boldsymbol{E}_Q^{(j)} = \begin{bmatrix} E_1 - v_a^{(j)} \\ v_a^{(j)} - v_b^{(j)} \\ v_a^{(j)} - v_c^{(j)} \\ -v_b^{(j)} + v_c^{(j)} \\ v_b^{(j)} \\ v_c^{(j)} \end{bmatrix} \tag{5-63}$$

$$\boldsymbol{J}_Q^{(j)} = \begin{bmatrix} g_1 \circ [E_1 - v_a^{(j)}] \\ g_2 \circ [v_a^{(j)} - v_b^{(j)}] \\ g_3 \circ [v_a^{(j)} - v_c^{(j)}] \\ g_4 \circ [-v_b^{(j)} + v_c^{(j)}] \\ g_5 \circ [v_b^{(j)}] \\ g_6 \circ [v_c^{(j)}] \end{bmatrix} = \begin{bmatrix} [E_1 - v_a^{(j)}]^3 \\ \frac{1}{R_2}[v_a^{(j)} - v_b^{(j)}] \\ e^{-[v_a^{(j)} - v_c^{(j)}]} \\ [-v_b^{(j)} + v_c^{(j)}]^{1/5} \\ v_b^{(j)} - [v_b^{(j)}]^3 \\ \frac{1}{R_6}v_c^{(j)} \end{bmatrix} \tag{5-64}$$

$$\mathbf{Y}_b^{(j)} = \begin{bmatrix} \left.\dfrac{dg_1(v_1)}{dv_1}\right|_{v_1=E_1-v_a^{(j)}} & 0 & 0 & 0 & 0 & 0 \\ 0 & \left.\dfrac{dg_2(v_2)}{dv_2}\right|_{v_2=v_a^{(j)}-v_b^{(j)}} & 0 & 0 & 0 & 0 \\ 0 & 0 & \left.\dfrac{dg_3(v_3)}{dv_3}\right|_{v_3=v_a^{(j)}-v_c^{(j)}} & 0 & 0 & 0 \\ 0 & 0 & 0 & \left.\dfrac{dg_4(v_4)}{dv_4}\right|_{v_4=-v_b^{(j)}+v_c^{(j)}} & 0 & 0 \\ 0 & 0 & 0 & 0 & \left.\dfrac{dg_5(v_5)}{dv_5}\right|_{v_5=v_b^{(j)}} & 0 \\ 0 & 0 & 0 & 0 & 0 & \left.\dfrac{dg_6(v_6)}{dv_6}\right|_{v_6=v_c^{(j)}} \end{bmatrix} \tag{5-65}$$

$$= \begin{bmatrix} 3[E_1 - v_a^{(j)}]^2 & 0 & 0 & 0 & 0 & 0 \\ 0 & \dfrac{1}{R_2} & 0 & 0 & 0 & 0 \\ 0 & 0 & -e^{-[v_a^{(j)} - v_c^{(j)}]} & 0 & 0 & 0 \\ 0 & 0 & 0 & \frac{1}{5}[-v_b^{(j)} + v_c^{(j)}]^{-4/5} & 0 & 0 \\ 0 & 0 & 0 & 0 & 1 - 3[v_b^{(j)}]^2 & 0 \\ 0 & 0 & 0 & 0 & 0 & \dfrac{1}{R_6} \end{bmatrix}$$

Observe that $\boldsymbol{E}_Q^{(j)}$, $\boldsymbol{J}_Q^{(j)}$, and $\boldsymbol{Y}_Q^{(j)}$ have the following simple circuit interpretations: the kth component $E_{Q_k}^{(j)}$ of the $b \times 1$ vector $\boldsymbol{E}_Q^{(j)}$ can be interpreted as the *voltage* across resistor R_k *at the* jth *iteration*. The kth component $J_{Q_k}^{(j)}$ of the $b \times 1$ vector $\boldsymbol{J}_Q^{(j)}$ can be interpreted as the *current* through resistor R_k *at the* jth *iteration*. The kkth element of the $b \times b$ Jacobian matrix $\boldsymbol{Y}_b^{(j)}$ can be interpreted as the *incremental conductance* $G_{Q_k}^{(j)}$ [the slope of the v-i curve $i_k = g_k(v_k)$ of R_k] evaluated at $v_k = E_{Q_k}^{(j)}$; i.e.,

$$Y_{kk}^{(j)} = \frac{dg_k(v_k)}{dv_k}\bigg|_{v_k = E_{Q_k}^{(j)}} \triangleq G_{Q_k}^{(j)} \tag{5-66}$$

Consequently, $\boldsymbol{Y}_b^{(j)}$ can be interpreted as the *incremental branch conductance matrix* associated with the nonlinear resistors. Observe that, if the circuit does not contain controlled sources, $\boldsymbol{Y}_b^{(j)}$ will always be a *diagonal matrix*. Observe also that, in the special case where all resistors are *linear*, $\boldsymbol{Y}_b^{(j)}$ reduces to the familiar *branch conductance matrix* $\boldsymbol{Y}_b$ defined in Eq. (4-12) and is therefore independent of the iteration number j. Viewed from another angle, the kth component $(E_{Q_k}^{(j)}, J_{Q_k}^{(j)})$ of $(\boldsymbol{E}_Q^{(j)}, \boldsymbol{J}_Q^{(j)})$ can be identified as the *operating point* $Q_{Q_k}^{(j)}$ for resistor R_k at the jth iteration, and $G_{Q_k}^{(j)}$ is the incremental conductance of R_k at $Q_k^{(j)}$ [see Fig. 5-7(a)]. Observe that, if we replace the v_k-i_k curve in Fig. 5-7(a) by a straight line drawn *tangent* to the operating point $Q_k^{(j)}$, the resistor R_k can be represented by the equivalent *linearized* circuit shown in Fig. 5-7(b). We shall shortly show that this circuit has far-reaching significance. To

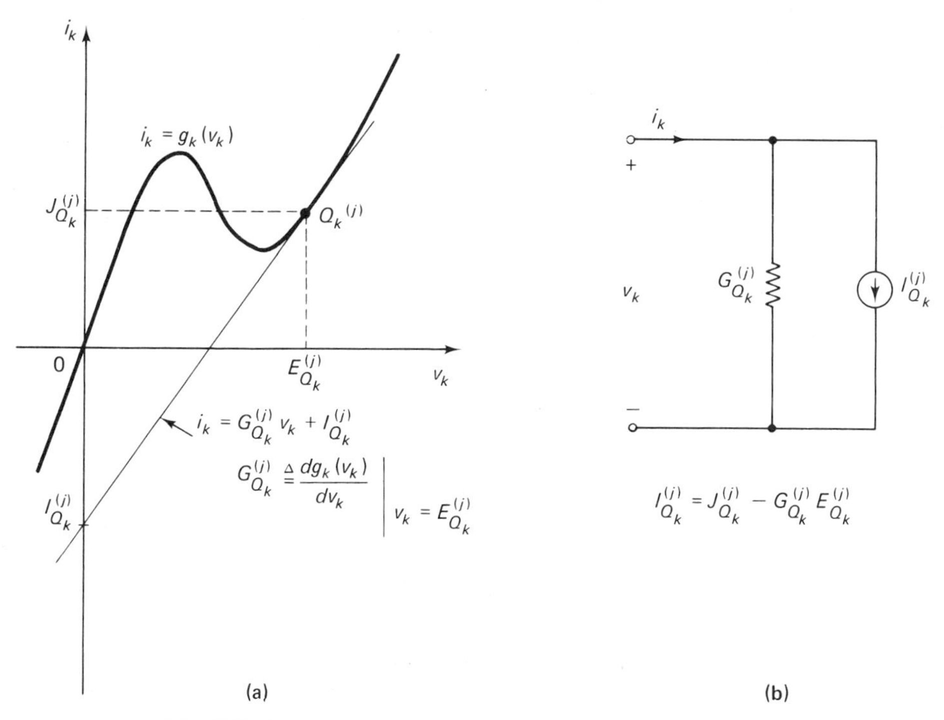

Fig. 5-7. Circuit interpretation of the Newton–Raphson algorithm.

see this, let us substitute Eqs. (5-60)–(5-62) into the Newton–Raphson formula defined by Eq. (5-59):

$$\boldsymbol{v}_n^{(j+1)} = \boldsymbol{v}_n^{(j)} - [\boldsymbol{A}\boldsymbol{Y}_b^{(j)}\boldsymbol{A}^t]^{-1}[\boldsymbol{A}\boldsymbol{J}_Q^{(j)} - \boldsymbol{A}\boldsymbol{J}] \tag{5-67}$$

Equation (5-67) can be rewritten in the form

$$\begin{aligned}[\boldsymbol{A}\boldsymbol{Y}_b^{(j)}\boldsymbol{A}^t]\boldsymbol{v}_n^{(j+1)} &= \boldsymbol{A}[\boldsymbol{J} - \boldsymbol{J}_Q^{(j)} + \boldsymbol{Y}_b^{(j)}\boldsymbol{A}^t\boldsymbol{v}_n^{(j)}] \\ &= \boldsymbol{A}[\boldsymbol{J} - \boldsymbol{J}_Q^{(j)} + \boldsymbol{Y}_b^{(j)}(\boldsymbol{E}_Q^{(j)} - \boldsymbol{E})]\end{aligned} \tag{5-68}$$

If we define an *iterative current source vector* $\boldsymbol{J}^{(j)}$ *at the* jth *iteration* by

$$\boldsymbol{J}^{(j)} \triangleq \boldsymbol{J} - \boldsymbol{J}_Q^{(j)} + \boldsymbol{Y}_b^{(j)}\boldsymbol{E}_Q^{(j)} \tag{5-69}$$

then Eq. (5-68) assumes the following form:

$$\boxed{[\boldsymbol{A}\boldsymbol{Y}_b^{(j)}\boldsymbol{A}^t]\boldsymbol{v}_n^{(j+1)} = \boldsymbol{A}[\boldsymbol{J}^{(j)} - \boldsymbol{Y}_b^{(j)}\boldsymbol{E}]} \tag{5-70}$$

Observe that Eq. (5-70) is still the Newton–Raphson nonlinear nodal analysis equation, but recast in a slightly different form from that of Eq. (5-59). Comparing this equation with Eq. (4-16), we observe that they are *identical* provided we replace the branch conductance matrix $\boldsymbol{Y}_b$ in Eq. (4-16) by the *incremental branch conductance matrix* $\boldsymbol{Y}_b^{(j)}$, and the current source vector $\boldsymbol{J}$ by the *iterative current source vector* $\boldsymbol{J}^{(j)}$ at the jth iteration. If the circuit contains only two-terminal resistors and independent sources, this replacement is equivalent to that of replacing the nonlinear resistor R_k in Fig. 5-8(a) by its *discretized resistor model* shown in Fig. 5-8(b). Observe that this discretized resistor model is identical to the linearized circuit model shown in Fig. 5-7(b). Hence, we have proved the following important result:

Theorem 5-3 (Discrete Circuit Equivalent of Newton–Raphson Nodal Analysis). *The* ***node voltage*** *vector* $\boldsymbol{v}_n^{(j+1)}$ *of a* ***nonlinear*** *resistive network N (containing only two-terminal resistors, linear voltage-controlled current sources, and independent sources) corresponding to the* $(j + 1)$*th Newton–Raphson iteration, as defined by Eq. (5-59), is equal to the node voltage vector of an associated linearized*

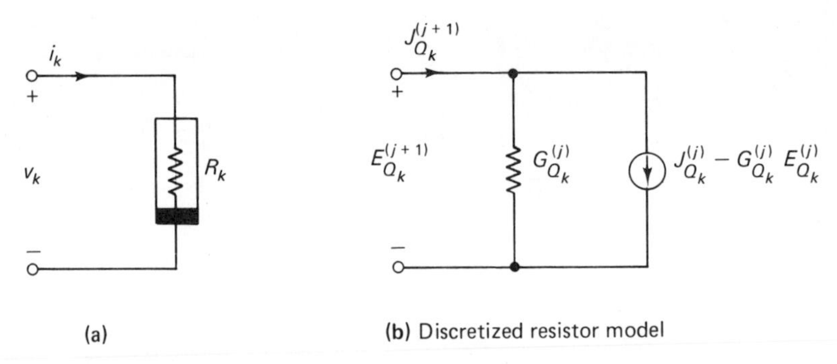

Fig. 5-8. Discretized resistor model associated with a nonlinear resistor in the implementation of the Newton–Raphson algorithm.

network $N^{(j)}$ obtained by replacing each nonlinear resistor R_k in N by its discretized resistor model at the jth iteration step,[6] as shown in Fig. 5-8(b).

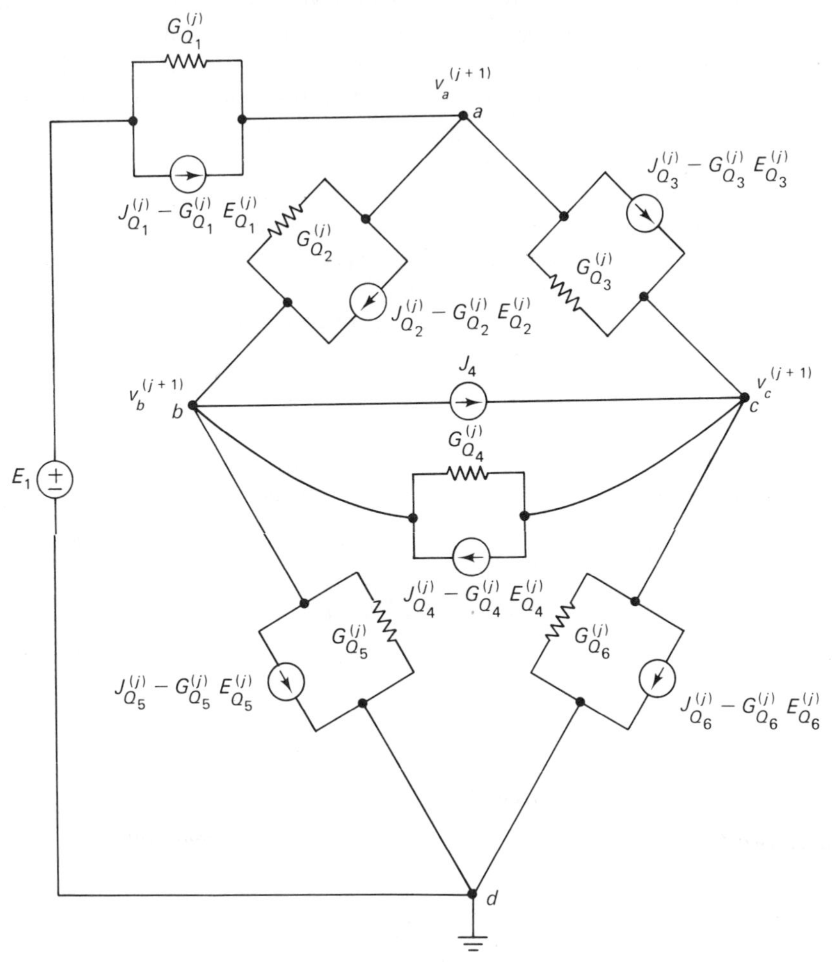

Fig. 5-9. Linearized circuit associated with the nonlinear bridge circuit of Fig. 5-2(a).

To illustrate the application of this theorem, consider again the nonlinear bridge network shown in Fig. 5-2(b), and replace each resistor R_k by its discretized resistor model to obtain the *linearized network* shown in Fig. 5-9. Observe that except for the *original* voltage source E_1 and the original current source J_4—which remain unchanged —the remaining parameters $G_{Q_k}^{(j)}$, $J_{Q_k}^{(j)}$, and $E_{Q_k}^{(j)}$ required to specify the discretized resistor model for R_k *must be updated after each iteration.* One advantage of this *discrete-circuit approach* over that of formulating Eq. (5-59) directly is that any

[6]Also called *companion model* in the literature.

existing *linear circuit-analysis program* can be trivially converted to a *nonlinear circuit-analysis program* capable of analyzing any nonlinear resistive network solvable by nodal analysis. Moreover, the theory of *equivalent circuits* can sometimes be used to simplify the analysis, thereby reducing computation time. One disadvantage of this *discrete-circuit approach* is the increased storage requirements and the slightly more complicated programming efforts needed for coding this algorithm. Observe that the linearized circuit in Fig. 5-9 contains more branches because each composite branch in the original network [recall Fig. 5-1(b)] now requires an additional iterative current source, as shown in Fig. 5-10(a). This current source, however, could be eliminated upon replacing the *original* current source J_k in Fig. 5-10(a) by the equivalent *iterative current source* $J_k^{(j)}$ as shown in Fig. 5-10(b), where $J_k^{(j)}$ is the jth component of the *iterative current source vector* $\boldsymbol{J}^{(j)}$ defined in Eq. (5-69). The two circuits shown in Fig. 5-10(a) and (b) are equivalent because their respective *driving-point characteristic* is given by

$$J_{Q_k}^{(j+1)} = G_{Q_k}^{(j)}[E_{Q_k}^{(j+1)} + E_k] - J_k + J_{Q_k}^{(j)} - G_{Q_k}^{(j)}E_{Q_k}^{(j)} \tag{5-71}$$

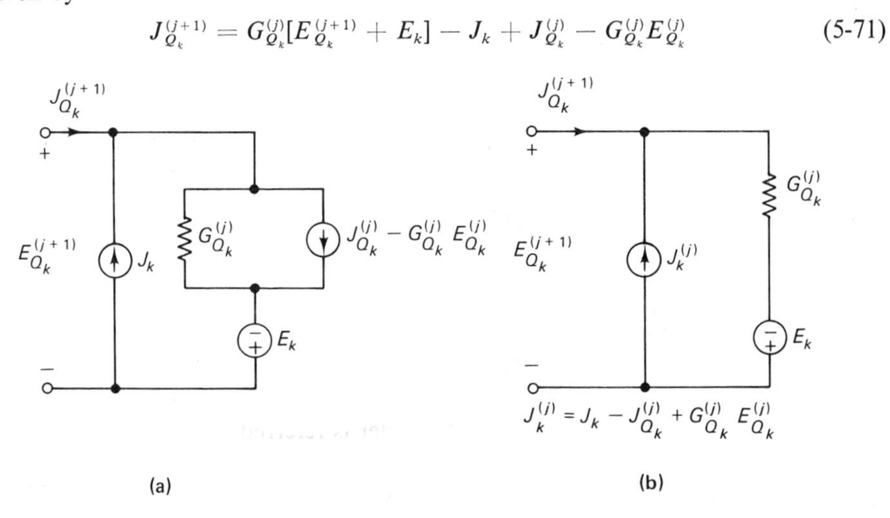

Fig. 5-10. Two equivalent discretized composite branches for implementing the Newton–Raphson algorithm.

The *discrete-circuit approach* for nonlinear nodal analysis by Newton–Raphson iteration is applicable even if the circuit contains voltage-controlled current sources, both linear and nonlinear. Indeed, it is applicable to any circuit containing coupled or multiterminal resistors as long as the branch current vector $\boldsymbol{i}$ is a function of the branch voltage vector $\boldsymbol{v}$; i.e.,

$$\begin{aligned} i_1 &= g_1(v_1, v_2, \ldots, v_b) \triangleq g_1(\boldsymbol{v}) \\ i_2 &= g_2(v_1, v_2, \ldots, v_b) \triangleq g_2(\boldsymbol{v}) \\ &\vdots \\ i_b &= g_b(v_1, v_2, \ldots, v_b) \triangleq g_b(\boldsymbol{v}) \end{aligned} \tag{5-72}$$

In this case, the Jacobian matrix $Y_b^{(j)}$ defined in Eq. (5-62) is no longer a diagonal matrix. Instead, we have

$$Y_b^{(j)} = \begin{bmatrix} \frac{\partial g_1(v)}{\partial v_1} & \frac{\partial g_1(v)}{\partial v_2} & \cdots & \frac{\partial g_1(v)}{\partial v_b} \\ \frac{\partial g_2(v)}{\partial v_1} & \frac{\partial g_2(v)}{\partial v_2} & \cdots & \frac{\partial g_2(v)}{\partial v_b} \\ \cdot & & & \\ \cdot & & & \\ \cdot & & & \\ \frac{\partial g_b(v)}{\partial v_1} & \frac{\partial g_b(v)}{\partial v_2} & \cdots & \frac{\partial g_b(v)}{\partial v_b} \end{bmatrix}_{v = E_Q^{(j)}} \tag{5-73}$$

To apply the discrete-circuit approach, we must develop analogous discretized circuit models for each coupled or multiterminal element in the circuit so that the *node conductance matrix* of the associated linearized circuit is identical to Eq. (5-73). Clearly, this approach is no longer too appealing unless the circuit contains only a few coupled resistors, e.g., transistors. In any event, we can always revert to Eq. (5-59) or Eq. (5-70), the original Newton–Raphson formula.

APPENDIX 5A PROOF OF PRINCIPLES AND PROPERTIES ASSOCIATED WITH THE FIXED-POINT AND NEWTON–RAPHSON ALGORITHMS

In this appendix we present the proofs to a number of assertions and theorems that were only stated in the text. For simplicity, we shall restrict our proofs to one-dimensional cases. The generalization to higher-dimensional cases follows along the same basic reasoning, and the interested reader is referred to [1, 2].

Principle of Contraction Mapping. *If the function $F(x)$ of a single variable x is a* ***contraction****, i.e., if there exists a constant $0 \leq L < 1$ such that*

$$|F(x) - F(y)| \leq L|x - y| \tag{5A-1}$$

for all possible values of x and y, then $F(x)$ has a ***unique*** *fixed point $x = x^*$. Moreover, the sequence of iterates defined by Eq. (5-21)* ***converges*** *to this fixed point. An upper bound on the error for stopping at the jth iteration is given by*

$$|x^* - x_j| \leq \frac{L^j}{1-L}|x_1 - x_0| \tag{5A-2}$$

Proof: First observe that since the function $F(x)$ is a contraction, if there exists a fixed point, it is unique. Now suppose x^* is a fixed point of $F(x)$—whose existence will be established shortly—let us prove that the sequence defined by Eq. (5-21) converges to x^*. It follows from Eq. (5A-1) and the relations $x^* = F(x^*)$ and $x_j = F(x_{j-1})$ that

$$|x^* - x_j| = |F(x^*) - F(x_{j-1})| \leq L|x^* - x_{j-1}| \tag{5A-3}$$

Similarly,

$$|x^* - x_{j-1}| \leq L|x^* - x_{j-2}| \tag{5A-4}$$

$$|x^* - x_{j-2}| \leq L|x^* - x_{j-3}| \tag{5A-5}$$

$$\vdots \qquad \vdots \qquad \vdots$$

$$|x^* - x_1| \leq L|x^* - x_0| \tag{5A-6}$$

Substituting each of these inequalities into the preceding inequality, we obtain

$$|x^* - x_j| \leq L^j|x^* - x_0| \tag{5A-7}$$

Since $0 \leq L < 1$, $L^j \longrightarrow 0$ as $j \longrightarrow \infty$. Hence,

$$\lim_{j \to \infty} x_j = x^* \tag{5A-8}$$

and the algorithm converges to $x = x^*$.

Let us now prove the existence of x^* and derive Eq. (5A-2). Let us first show by *mathematical induction* that

$$|x_{j+1} - x_j| \leq L^j|x_1 - x_0| \tag{5A-9}$$

Clearly, Eq. (5A-9) is true when $j = 0$. Now suppose that Eq. (5A-9) is true for $j = n - 1$, where n is an integer > 0; i.e.,

$$|x_n - x_{n-1}| \leq L^{n-1}|x_1 - x_0| \tag{5A-10}$$

It follows from the induction hypothesis that

$$\begin{aligned} |x_{n+1} - x_n| &= |F(x_n) - F(x_{n-1})| \\ &\leq L|x_n - x_{n-1}| \\ &\leq L \cdot L^{n-1}|x_1 - x_0| \\ &= L^n|x_1 - x_0| \end{aligned} \tag{5A-11}$$

Equation (5A-11) shows that Eq. (5A-9) is true for $j = n$ whenever it is true for $j = 0$ and $j = n - 1$. It is therefore true for all n in view of the *principle of mathematical induction*.

Now let j be a fixed integer and $k > j$. It follows from the *triangle inequality* that

$$\begin{aligned} |x_k - x_j| &= |(x_k - x_{k-1}) + (x_{k-1} - x_{k-2}) + \ldots + (x_{j+1} - x_j)| \\ &\leq |x_k - x_{k-1}| + |x_{k-1} - x_{k-2}| + \ldots + |x_{j+1} - x_j| \end{aligned} \tag{5A-12}$$

It follows from Eqs. (5A-9) and (5A-12) and the fact that $0 \leq L < 1$ that

$$\begin{aligned} |x_k - x_j| &\leq (L^{k-1} + L^{k-2} + \ldots + L^j)|x_1 - x_0| \\ &= L^j(1 + L + L^2 + \ldots + L^{k-j-1})|x_1 - x_0| \\ &\leq L^j(1 + L + L^2 + \ldots + L^{k-j-1} + L^{k-j} + \cdots)|x_1 - x_0| \end{aligned} \tag{5A-13}$$

The expression enclosed in parentheses can be identified as the well-known geometric series that converges to $1/(1 - L)$ whenever $|L| < 1$. Hence, we have

$$|x_k - x_j| \leq \frac{L^j}{1 - L}|x_1 - x_0| \tag{5A-14}$$

Since $L^j \longrightarrow 0$ as $j \longrightarrow \infty$, the sequence $\{x_j\}$ converges to some limit. It follows from $x_{j+1} = F(x_j)$ and the uniform continuity of $F(x)$ that as $j \longrightarrow \infty$, we obtain $x^* = F(x^*)$. Hence, x^* is indeed a fixed point of $F(x)$. Since $k > j$ is any integer, Eq. (5A-14) must hold when $k \longrightarrow \infty$. But

$$\lim_{k \to \infty} x_k = x^* \tag{5A-15}$$

Hence, as $k \longrightarrow \infty$, Eq. (5A-14) reduces to Eq. (5A-2). Q.E.D.

On the Number of Iterations Needed to Achieve Convergence. *Let x^* be the fixed point of a function of a single variable $F(x)$ and let $F'(x) \neq 0$ for all values of x. Then the* fixed-point algorithm *$x_{j+1} = F(x_j)$ cannot terminate in a finite number of steps unless the initial guess $x_0 = x^*$.*

Proof: We shall prove this by contradiction. Suppose that the iteration terminates after n steps; i.e.,

$$x_{n+1} = F(x_n) = x_n, \qquad x_n \neq x_{n-1} \tag{5A-16}$$

It follows from Eq. (5A-16) and $x_n = F(x_{n-1})$ that

$$F(x_{n-1}) - F(x_n) = 0 \tag{5A-17}$$

But the *mean-value theorem* asserts that, if $x_{n-1} < x_n$, then

$$F(x_{n-1}) - F(x_n) = F'(\bar{x})(x_{n-1} - x_n) \tag{5A-18}$$

where $x_{n-1} < \bar{x} < x_n$. It follows from Eqs. (5A-17) and (5A-18) that since $x_n \neq x_{n-1}$ we must have $F'(\bar{x}) = 0$. But this contradicts the hypothesis that $F'(x)$ never vanishes. Hence, it is impossible for the *fixed-point algorithm* to converge to its fixed point x^* in a finite number of steps. Q.E.D.

Convergence Property of the Newton–Raphson Algorithm. *Let x^* be a solution to the equation*

$$f(x) = 0 \tag{5A-19}$$

where $f(x)$ is a continuous function of a single variable with continuous second derivative $f''(x)$. Then the Newton–Raphson algorithm

$$x^{(j+1)} = x^{(j)} - [J(x^{(j)})]^{-1} f(x^{(j)}) \tag{5A-20}$$

is guaranteed to converge to x^* whenever the initial guess $x^{(0)}$ is chosen sufficiently close to x^*. Moreover, if $f'(x^*) \neq 0$, then the error $\epsilon^{(j)} \triangleq |x^* - x^{(j)}|$ decreases quadratically; i.e.,

$$\epsilon^{(j+1)} \leq k[\epsilon^{(j)}]^2 \tag{5A-21}$$

where k is a constant.

Proof: If we define the associated function

$$F(x) \triangleq x - \frac{f(x)}{f'(x)} \tag{5A-22}$$

then applying the fixed-point algorithm on $F(x)$, we obtain

$$x^{(j+1)} = x^{(j)} - \frac{f(x^{(j)})}{f'(x^{(j)})} \tag{5A-23}$$

Observe that Eq. (5A-23) is identical to Eq. (5A-20). Observe also that the solution x^* to Eq. (5A-20) is a *fixed point* of $F(x)$. Taking next the derivative of $F(x)$, we obtain

$$F'(x) = 1 - \frac{f'(x)f'(x) - f(x)f''(x)}{[f'(x)]^2} = \frac{f(x)f''(x)}{[f'(x)]^2} \tag{5A-24}$$

It follows from $f(x^*) = 0$ and $f'(x^*) \neq 0$ that

$$F'(x^*) = 0 \tag{5A-25}$$

Since $F'(x)$ is a *continuous* function of x, there exists an $\varepsilon > 0$ such that

$$|F'(x) - F'(x^*)| = |F'(x)| \leq L < 1 \tag{5A-26}$$

for all x satisfying $|x - x^*| < \varepsilon$. It follows from the *mean-value theorem* that

$$|F(x) - x^*| = |F(x) - F(x^*)| \leq L|x - x^*|, \qquad |x - x^*| < \varepsilon \tag{5A-27}$$

Hence, if we choose the *initial guess* $x^{(0)}$ such that $|x^{(0)} - x^*| < \varepsilon$, it follows from Eq. (5A-27) and $0 \leq L < 1$ that

$$|x^{(1)} - x^*| = |F(x^{(0)}) - x^*| \leq L|x^{(0)} - x^*| < \varepsilon L \tag{5A-28}$$

$$|x^{(2)} - x^*| = |F(x^{(1)}) - x^*| \leq L|x^{(1)} - x^*| \leq L^2|x^{(0)} - x^*| < \varepsilon L^2 \tag{5A-29}$$

$$\vdots$$

$$|x^{(n)} - x^*| = |F(x^{(n-1)}) - x^*| \leq L|x^{(n-1)} - x^*| \leq L^n|x^{(0)} - x^*| < \varepsilon L^n \tag{5A-30}$$

Observe that we can keep applying Eq. (5A-27) to obtain these inequalities since $|x^{(j)} - x^*| < \varepsilon L^j$ in all cases. It follows from Eq. (5A-30) and $0 \leq L < 1$ that

$$\lim_{n \to \infty} x^{(n)} = x^* \tag{5A-31}$$

Hence, we have shown that the Newton–Raphson algorithm always converges as long as the initial guess $x^{(0)}$ satisfies $|x^{(0)} - x^*| < \varepsilon$, where ε is determined by (5A-26).

It remains to prove Eq. (5A-21). Applying *Taylor's formula with a remainder* to $F(x)$ about the fixed point x^*, we obtain (assuming, without loss of generality, $x^* < x$)

$$F(x) = F(x^*) + F'(x^*)(x - x^*) + \tfrac{1}{2}F''(\bar{x})(x - x^*)^2 \tag{5A-32}$$

where $x^* < \bar{x} < x$. Substituting $x = x^{(j)}$ into Eq. (5A-32) and using Eq. (5A-25), we obtain

$$F(x^{(j)}) - F(x^*) = \tfrac{1}{2}F''(\bar{x})(x^{(j)} - x^*)^2 \tag{5A-33}$$

where $x^* < \bar{x} < x^{(j)}$. It follows from Eq. (5A-33) that

$$|x^{(j+1)} - x^*| = \tfrac{1}{2}F''(\bar{x})|x^{(j)} - x^*|^2 \tag{5A-34}$$

Hence, we have

$$\epsilon^{(j+1)} \leq k[\epsilon^{(j)}]^2 \tag{5A-35}$$

where k is some constant.

Q.E.D.

REFERENCES

1. HENRICI, P. *Elements of Numerical Analysis.* New York: John Wiley & Sons, Inc., 1964.
2. ORTEGA, J. M., and W. C. RHEINBOLDT. *Iterative Solution of Nonlinear Equations in Several Variables.* New York: Academic Press, Inc., 1970.
3. WILLIAMSON, R. E., R. H. CROWELL, and H. F. TROTTER. *Calculus of Vector Functions,* 2nd Ed. Englewood Cliffs, N.J.: Prentice-Hall, Inc., 1968.
4. BROYDEN, C. G. "A Class of Methods for Solving Nonlinear Simultaneous Equations." *Math. Computation*, Vol. 19, pp. 577–593, 1965.
5. BROWN, K. M. "A Quadratically Convergent Newton-Like Method Based upon Gaussian Elimination." *SIAM J. Numerical Analysis*, Vol. 6, No. 4, pp. 560–569, Dec. 1969.
6. GILL, P. E., and W. MURRAY. "Quasi-Newton Methods for Unconstrained Optimization." *J. Inst. Math. Appl.* Vol. 9, pp. 91–108, 1972.
7. FLETCHER, R. *FORTRAN Subroutines for Minimization by Quasi-Newton Methods,* Atomic Energy Research Establishment, Harwell, Berkshire, England: Her Majesty's Stationery Office, June, 1972.
8. POLAK, E. *Computational Methods in Optimization: A Unified Approach.* New York: Academic Press, Inc., 1971.
9. CHUA, L. O., and N. N. WANG. "A New Approach to Overcome the Overflow Problem in Computer-Aided Analysis of Nonlinear Resistive Circuits." *International Journal of Circuit Theory and Applications,* Vol. 4, 1976.

PROBLEMS

5-1. Although, strictly speaking, the nodal-analysis method allows only one type of controlled source, show by an example how this restriction can be relaxed to allow all four types of controlled sources through appropriate equivalent circuit transformations.

5-2. Formulate the nonlinear nodal equations for the bridge network shown in Fig. 5-2(a) with the following element characteristics:

(i) R_1 and R_4 are linear 1-kΩ resistors.

(ii) R_3 and R_5 are *junction diodes* characterized by the ideal junction law $i_j = I_s[\exp(v_j/V_T) - 1]$, as shown in Fig. P5-2(a).

(iii) R_2 and R_6 are identical to R_3 and R_5 except that their polarities are oppositely directed; i.e., $i_j = -I_s[\exp(-v_j/V_T) - 1]$, as shown in Fig. P5-2(b).

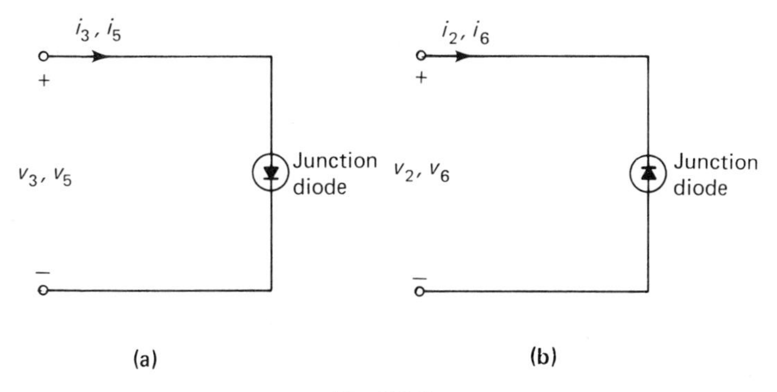

Fig. P5-2

5-3. Use the *fixed-point iteration algorithm* to find the solution of the nodal equations obtained from Problem 5-2. Assume that $I_s = 10^{-12}$ mA, $V_T = 26$ mV, $E_1 = 10$ V, and $J_4 = 0$. Choose an appropriate initial guess to avoid computer overflow problems, which often occur in view of the exponential function in the model [9].

5-4. Repeat Problem 5-3 using the *Newton–Raphson algorithm.* Compare the rate of convergence with that of Problem 5-3 (assuming an identical initial guess).

5-5. Repeat Problem 5-3 using the *discrete-circuit equivalent* of Newton–Raphson nodal analysis.

5-6. Construct a *discretized circuit model* for a bipolar transistor characterized by the Ebers–Moll equation.

5-7. When a nonlinear network contains only nonmonotonic *current-controlled* resistors, nodal analysis is no longer applicable and we must turn to *nonlinear loop analysis.* Formulate this dual method of analysis in terms of topological matrices.

5-8. Construct a simple numerical example illustrating the application of the *Newton–Raphson–Kantorovich theorem* for a system of two nonlinear equations.

5-9. Modify the proof given in Appendix 5A to show that the following weaker version of the *principle of contraction mapping* holds:

Let $f(x)$ be a *continuous* function of a single variable x for all $a \leq x \leq b$. Suppose that (1) $a \leq f(x) \leq b$, for all $a \leq x \leq b$, and (2) there exists $0 \leq L < 1$ such that $[f(x) - f(y)] \leq L|x - y|$, for all $a \leq x \leq b$ and $a \leq y \leq b$. Then the *fixed-point algorithm* converges whenever the initial guess x_0 satisfies $a \leq x_0 \leq b$.

5-10. (a) Show that the function $f(x) = e^{-x}$ satisfies the two conditions given in Problem 5-9 for $0.5 \leq x \leq \log_e 2$.

(b) Apply the fixed-point algorithm to find the solution to the equation $e^{-x} - x = 0$. Choose any initial guess x_0 satisfying $0.5 \leq x_0 \leq \log_e 2$.

5-11. Show that the value of

$$x^* = \sqrt{n + \sqrt{n + \sqrt{n + \sqrt{n + \cdots}}}}$$

is given by the fixed point of the function $f(x) = \sqrt{n + x}$, where n is any positive number. Verify this result numerically as well as graphically for the case $n = 4$.

5-12. Let $f(x)$ be a twice continuously differentiable function for $a \leq x \leq b$. Give a geometrical proof showing that the Newton–Raphson algorithm is guaranteed to converge to the unique solution of $f(x) = 0$ for any initial guess x_0 satisfying $a \leq x_0 \leq b$, provided the following conditions are satisfied:

(i) $f(a)$ and $f(b)$ have opposite signs.

(ii) $f'(x) \neq 0$ for all x satisfying $a \leq x \leq b$.

(iii) Either $f''(x) \geq 0$ or $f''(x) \leq 0$ for all x satisfying $a \leq x \leq b$.

(iv) $\left|\dfrac{f(y)}{f'(y)}\right| \leq b - a$, where $y = a$ if $|f'(a)| < |f'(b)|$, or $y = b$ if $|f'(a)| > |f'(b)|$.

5-13. (a) Show that the positive square root of any real number $n > 0$ can be found from the recursive formula:

$$x_{j+1} = 0.5\left(x_j + \frac{n}{x_j}\right), \qquad x_0 > 0$$

Hint: Use the Newton–Raphson algorithm to solve the equation $x^2 - n = 0$, and apply the result from Problem 5-12 to show the algorithm always converges.

(b) Show that the positive root of order k of any real number $n > 0$ can be found from the recursive formula

$$x_{j+1} = \frac{1}{k}\left[(k-1)x_j + \frac{n}{x_j^{k-1}}\right]$$

5-14. Use the recursive formulas from Problem 5-13 to calculate to 12 decimal places the following quantities:

(a) $\sqrt{e}$.

(b) $(\pi)^{1/3}$.

5-15. (a) Show that the *reciprocal* of any real number $n > 0$ can be found from the recursive relation

$$x_{j+1} = x_j(2 - nx_j)$$

Hint: Apply the Newton–Raphson algorithm to solve the equation $(1/x) - n = 0$.

(b) Apply the result from Problem 5-12 to show that the recursive relation in part (a) converges to $x = 1/n$ whenever the initial guess satisfies $0 < x_0 < 2/n$.

5-16. Use the recursive formulas from Problem 5-15 to calculate to 12 decimal places the following quantities:

(a) $1/e$.

(b) $1/\pi$.

5-17. Show that if the solution x^* to $f(x) = 0$ is a *double root*, i.e., if $f'(x^*) = 0$, then the Newton–Raphson algorithm no longer converges quadratically. In this case, show that, if $f(x)$, $f'(x)$, and $f''(x)$ are continuous and bounded on an interval containing the root $x = x^*$, then the following *modified Newton–Raphson algorithm* converges quadratically:

$$x_{j+1} = x_j - \frac{2f(x_j)}{f'(x_j)} \triangleq F(x_j)$$

Hint: Show $F'(x^*) = 0$ by invoking L'Hôspital's rule.

5-18. Consider the equation

$$f(x) = (\sin x - \tfrac{1}{2}x)^2 = 0$$

(a) Find the root x^* by the Newton–Raphson algorithm with initial guess $x_0 = \frac{1}{2}\pi$.

(b) Show that $f(x)$ has a double root at x^*; i.e., $f'(x^*) = 0$.

(c) Find the root x^* by the *modified* Newton–Raphson algorithm given in Problem 5-17 with $x_0 = \frac{1}{2}\pi$.

(d) Compare the *rate of convergence* between your results from parts (a) and (c) and explain why the latter converges much faster.

5-19. Find the positive root of the equation $x^{20} - 1 = 0$ by the Newton–Raphson algorithm with initial guess $x_0 = 0.5$. Comment on the rate of convergence and explain why.

5-20. (a) Use the Newton–Raphson algorithm to solve for the root of the equation $xe^{-x} = 0$ with initial guess $x_0 = 0.5$.

(b) Sketch the curve $f(x) = xe^{-x}$ and give a geometrical interpretation of your iterations from part (a).

(c) Repeat parts (a) and (b) with initial guess $x_0 = 2.0$.

5-21. Consider the circuit shown in Fig. P5-21. The nonlinear resistors are characterized by

$$i_a = 2v_a^3$$
$$i_b = v_b^3 + 10v_b$$

(a) By inspection of the circuit, write nodal equations in the form

$$f_1(v_1, v_2) = 0$$
$$f_2(v_1, v_2) = 0$$

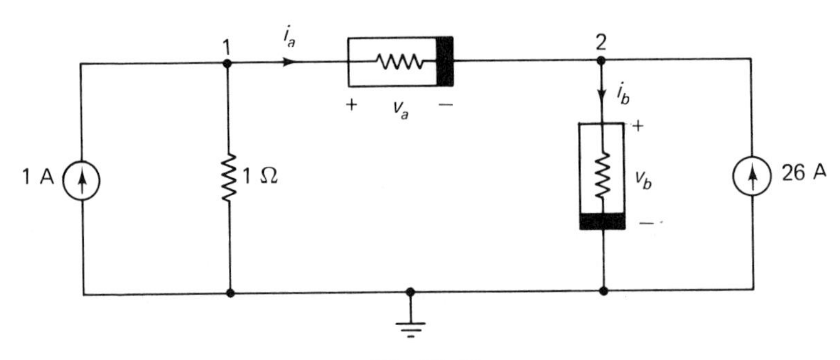

Fig. P5-21

(b) Find the Jacobian matrix $J(v)$.

(c) If the Newton–Raphson algorithm is used to solve the equations and the initial guess is $v^{(0)} = [2, 1]^t$, find $v^{(1)}$ and $v^{(2)}$. Use Eq. (5-49) instead of Eq. (5-51). Why?

5-22. Consider the same circuit given in Problem 5-21. Find the solution of v_1 and v_2 by the Newton–Raphson algorithm with the aid of "discretized" or "linearized" circuit models for the nonlinear resistors as discussed in Section 5-5. Let the initial guess be $v_a^{(0)} = 1$ and $v_b^{(0)} = 1$. More specifically, do the following:

(a) Construct the discretized circuit model suitable for the solution of $v^{(1)}$.

(b) Find $v_1^{(1)}$ and $v_2^{(1)}$ by the nodal method.

(c) Update the discretized circuit model so that it can be used to find $v_1^{(2)}$ and $v_2^{(2)}$.

(d) Find $v_1^{(2)}$ and $v_2^{(2)}$ by the nodal method.

Note: The answers should agree with Problem 5-21.

5-23. Prove that the Newton–Raphson algorithm when applied to solve a system of independent *linear* equations

$$\boldsymbol{Ax} - \boldsymbol{b} = \boldsymbol{0}$$

converges in exactly one step, regardless of the initial guess.

CHAPTER 6

Hybrid Linear Resistive *n*-Port Formulation Algorithms

6-1 WHY HYBRID MATRICES?

The nodal method presented in Chapter 4 is used in several user-oriented circuit-analysis programs, including the widely used SPICE program [1]. For linear networks, the nodal-analysis method is perhaps the most popular because of its simplicity and ease in programming. However, this method does suffer from the following shortcomings:

1. Strictly speaking, only one type of controlled source is allowed, the class of voltage-controlled current sources. All other types of controlled sources have to be converted into voltage-controlled current sources (see Problems 4-2 and 4-19) either by the user or by the program internally.
2. The nodal method cannot handle a 0-Ω resistance as a circuit element, because the branch-conductance matrix, Eq. (4-9), would otherwise contain an infinite entry. Although a short circuit can usually be dispensed with, there are occasions when its inclusion can be very convenient, as we shall shortly see (Section 6-5-1).
3. Multiterminal elements that have no admittance matrix representation cause difficulty in nodal analysis. A well-known example is the *ideal* transformer. Some circuit conversions, performed either by the user or the program, are necessary to force an admittance matrix to exist. Such a procedure usually increases the complexity of the network (see Problem 4-3).

4. Only voltage-controlled nonlinear resistors are allowed. Since there exist practical resistors, such as neon lamps and SCR's, which are current controlled but not voltage controlled, the nodal method is not suitable for analyzing general resistive nonlinear networks.
5. Even though there may be only a few nonlinear resistors in an otherwise linear resistive network, the conventional nodal-analysis method (without the use of sparse-matrix techniques) is unable to take advantage of this fact for the reduction of computing time. On the other hand, the linear n-port hybrid-analysis approach will be very effective in such case, as we have illustrated in Section 1-4.

To overcome these shortcomings, we shall in this chapter introduce the concept of *hybrid n-ports* and describe methods for their characterization. The adjective "hybrid" comes from the fact that *both* currents and voltages appear as unknowns in the equations characterizing the n-port. The n-port under consideration consists of the following types of elements only: positive and negative linear resistors, independent voltage sources, independent current sources, and all four types of controlled sources with real constant controlling coefficients. It should be pointed out that the analysis of such a linear resistive n-port plays a fundamental role in the methods for analyzing nonlinear resistive networks (Chapter 7), linear dynamic networks (Chapter 8), and nonlinear dynamic networks (Chapter 10).

6-2 FORMULATION OF A LINEAR RESISTIVE m-PORT

Given a nonlinear dynamic network, let us extract the following types of two-terminal elements to form an m-port $\hat{N}$ as shown in Fig. 6-1:

Nonlinear resistors
Inductors (linear or nonlinear)
Capacitors (linear or nonlinear)

The m-port $\hat{N}$ is then purely resistive and consists of the following types of elements only:

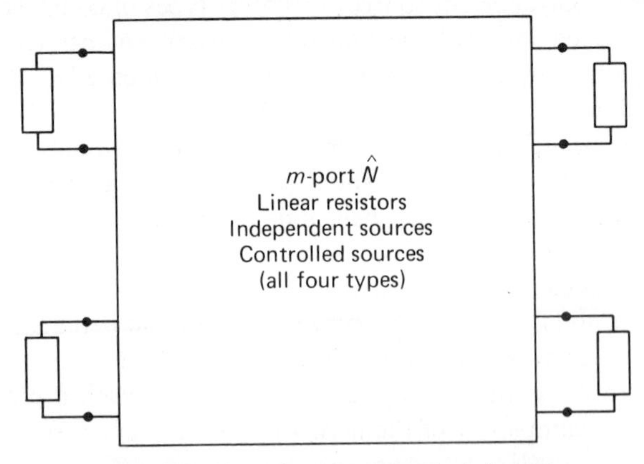

Fig. 6-1. Extraction of two-terminal elements.

Linear resistors
Independent voltage and current sources
All four types of linear controlled sources

Next each extracted two-terminal element is replaced by an independent source (external to $\hat{N}$), as shown in Fig. 6-2. If the replacement source is an independent voltage source, the port is said to be a voltage port. If the replacement source is an independent current source, the port is said to be a current port. The general rules for source replacement are as follows:

1. Replace voltage-controlled nonlinear resistors by independent voltage sources.
2. Replace current-controlled nonlinear resistors by independent current sources.
3. Replace capacitors by independent voltage sources.
4. Replace inductors by independent current sources.

However, there are exceptions to these rules, which will be discussed in detail in Chapters 7, 8, and 10.

Refer to the m-port of Fig. 6-2, where the m_1 voltage ports have been labeled as port 1 to port m_1, and the m_2 current ports have been labeled as port $m_1 + 1$ to port $m_1 + m_2$. (Note that $m = m_1 + m_2$, $m_1 \geq 0$, $m_2 \geq 0$.) Let us define the following port voltage and current vectors:

$$\hat{\mathbf{v}}_a \triangleq \begin{bmatrix} \hat{v}_1 \\ \hat{v}_2 \\ \cdot \\ \cdot \\ \cdot \\ \hat{v}_{m_1} \end{bmatrix}, \quad \hat{\mathbf{v}}_b \triangleq \begin{bmatrix} \hat{v}_{m_1+1} \\ \hat{v}_{m_1+2} \\ \cdot \\ \cdot \\ \cdot \\ \hat{v}_{m_1+m_2} \end{bmatrix}, \quad \hat{\mathbf{i}}_a \triangleq \begin{bmatrix} \hat{i}_1 \\ \hat{i}_2 \\ \cdot \\ \cdot \\ \cdot \\ \hat{i}_{m_1} \end{bmatrix}, \quad \hat{\mathbf{i}}_b \triangleq \begin{bmatrix} \hat{i}_{m_1+1} \\ \hat{i}_{m_2+2} \\ \cdot \\ \cdot \\ \cdot \\ \hat{i}_{m_1+m_2} \end{bmatrix} \tag{6-1}$$

Let $\hat{\mathbf{u}}$ be the vector representing the independent sources *inside* $\hat{N}$. (Note that the assumed reference directions for the port currents are opposite to those used in

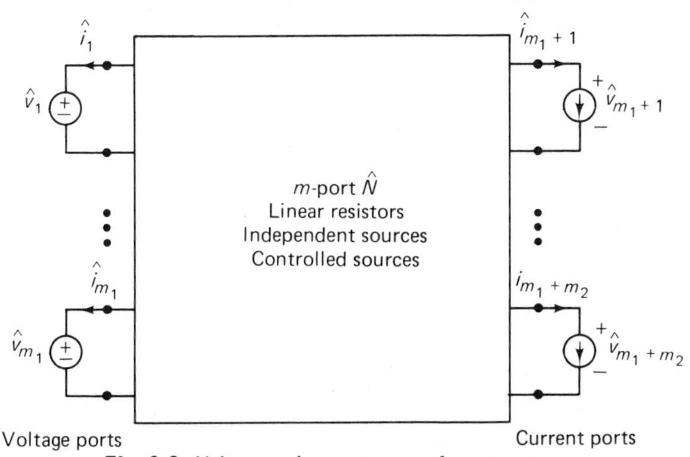

Fig. 6-2. Voltage and current ports of an m-port.

conventional *n*-port theory. Here $\sum_{j=1}^{m} \hat{v}_j \hat{i}_j$ represents the total instantaneous *power delivered by the m-port to the extracted elements.*) Since $\hat{N}$ contains only *linear* resistors, independent sources, and controlled sources, the port voltages and currents may be related by a *hybrid matrix* $\hat{\boldsymbol{H}}$ and a source vector $\hat{\boldsymbol{s}}$ as follows:[1]

$$\boxed{\begin{bmatrix} \hat{\boldsymbol{i}}_a \\ \hat{\boldsymbol{v}}_b \end{bmatrix} = \hat{\boldsymbol{H}} \begin{bmatrix} \hat{\boldsymbol{v}}_a \\ \hat{\boldsymbol{i}}_b \end{bmatrix} + \hat{\boldsymbol{M}}\hat{\boldsymbol{u}} = \hat{\boldsymbol{H}} \begin{bmatrix} \hat{\boldsymbol{v}}_a \\ \hat{\boldsymbol{i}}_b \end{bmatrix} + \hat{\boldsymbol{s}}} \tag{6-2}$$

where $\hat{\boldsymbol{s}} = \hat{\boldsymbol{M}}\hat{\boldsymbol{u}}$.

The matrices $\hat{\boldsymbol{H}}, \hat{\boldsymbol{M}}$, and $\hat{\boldsymbol{s}}$ in Eq. (6-2) can be partitioned according to the dimensions of $\hat{\boldsymbol{i}}_a$ and $\hat{\boldsymbol{v}}_b$ as follows:

$$\begin{bmatrix} \hat{\boldsymbol{i}}_a \\ \hat{\boldsymbol{v}}_b \end{bmatrix} = \begin{bmatrix} \hat{\boldsymbol{H}}_{aa} & \hat{\boldsymbol{H}}_{ab} \\ \hat{\boldsymbol{H}}_{ba} & \hat{\boldsymbol{H}}_{bb} \end{bmatrix} \begin{bmatrix} \hat{\boldsymbol{v}}_a \\ \hat{\boldsymbol{i}}_b \end{bmatrix} + \begin{bmatrix} \hat{\boldsymbol{M}}_a \\ \hat{\boldsymbol{M}}_b \end{bmatrix} \hat{\boldsymbol{u}} = \begin{bmatrix} \hat{\boldsymbol{H}}_{aa} & \hat{\boldsymbol{H}}_{ab} \\ \hat{\boldsymbol{H}}_{ba} & \hat{\boldsymbol{H}}_{bb} \end{bmatrix} \begin{bmatrix} \hat{\boldsymbol{v}}_a \\ \hat{\boldsymbol{i}}_b \end{bmatrix} + \begin{bmatrix} \hat{\boldsymbol{s}}_a \\ \hat{\boldsymbol{s}}_b \end{bmatrix} \tag{6-3}$$

where $\hat{\boldsymbol{s}}_a = \hat{\boldsymbol{M}}_a\hat{\boldsymbol{u}}$ and $\hat{\boldsymbol{s}}_b = \hat{\boldsymbol{M}}_b\hat{\boldsymbol{u}}$. The elements of $\hat{\boldsymbol{H}}$ and $\hat{\boldsymbol{M}}$ are real constants.[2] Conceptually, the simplest method for determining $\hat{\boldsymbol{H}}$ and $\hat{\boldsymbol{s}}$ (or $\hat{\boldsymbol{M}}$) as defined in Eq. (6-2) is to compute each element $\hat{h}_{jk}$ of $\hat{\boldsymbol{H}}$ and each element $\hat{s}_j$ of $\hat{\boldsymbol{s}}$. For example, $\hat{\boldsymbol{s}}_a$ can be found by solving for the current $\hat{\boldsymbol{i}}_a$ when all voltage ports are short-circuited and all current ports are open-circuited. This follows from Eq. (6-3) since

$$\hat{\boldsymbol{s}}_a = \hat{\boldsymbol{i}}_a \big|_{\hat{\boldsymbol{v}}_a = \boldsymbol{0},\, \hat{\boldsymbol{i}}_b = \boldsymbol{0}} \tag{6-4a}$$

Similarly, we can find

$$\hat{\boldsymbol{s}}_b = \hat{\boldsymbol{v}}_b \big|_{\hat{\boldsymbol{v}}_a = \boldsymbol{0},\, \hat{\boldsymbol{i}}_b = \boldsymbol{0}} \tag{6-4b}$$

Observe that $\hat{\boldsymbol{s}} = \boldsymbol{0}$ if $\hat{\boldsymbol{u}} = \boldsymbol{0}$. In other words, the vector $\hat{\boldsymbol{s}}$ is due to the *independent* sources inside $\hat{N}$; hence the name *source vector*.

Each element of $\hat{\boldsymbol{H}}$ can be found by first *setting all independent sources* inside $\hat{N}$ to zero, so that $\hat{\boldsymbol{s}} = \hat{\boldsymbol{M}}\hat{\boldsymbol{u}} = \boldsymbol{0}$, and then obtaining $\hat{h}_{jk}$ by the ratio

$$\hat{h}_{jk} = \frac{\text{response at port } j}{\text{excitation at port } k} \tag{6-5}$$

under the following conditions:

1. Except for port k, all voltage ports are short-circuited and all current ports are open-circuited.
2. The excitation at port k is an independent
   (a) voltage source if port k is a voltage port.
   (b) current source if port k is a current port.
3. The response at port j is considered to be
   (a) the current of port j if port j is a voltage port.
   (b) the voltage of port j if port j is a current port.

[1]Under the assumption that the network of Fig. 6-2 is consistent and has a unique solution. Note that Eq. (6-2) includes the impedance and admittance matrix characterizations of an *n*-port as special cases (see Problems 6-8 and 6-9).

[2]If the independent sources inside $\hat{N}$ are dc sources, then both $\hat{u}$ and $\hat{s} = \hat{M}\hat{u}$ will be *constant* vectors.

Such a procedure may be reasonable if $m_1 + m_2$ is a small number. However, for $m_1 + m_2 > 5$, the computational efforts become excessive, and a more efficient approach for evaluating $\hat{\boldsymbol{H}}$ and $\hat{\boldsymbol{s}}$ is desirable.

In the remaining sections of this chapter, we shall concentrate on the problem of the evaluation of $\hat{\boldsymbol{H}}$ and $\hat{\boldsymbol{s}}$ in Eq. (6-2) for any given linear resistive m-port of Fig. 6-2. Explicit topological formulas for $\hat{\boldsymbol{H}}$ and $\hat{\boldsymbol{s}}$ will be derived. The applications of Eq. (6-2), however, will be the subject of several later chapters (Chaps. 7, 8, and 10).

6-3 LINEAR RESISTIVE n-PORT WITHOUT CONTROLLED SOURCES

Consider first the case of an n-port N consisting of positive linear resistors only.[3] Since there are no independent sources inside N, the source vector $\boldsymbol{s}$ in Eq. (6-2) will be absent. The questions to be answered are: Does the hybrid matrix $\boldsymbol{H}$ exist, and, when it exists, can one compute $\boldsymbol{H}$ by matrix operations? Theorem 6-1 and its proof provide the answers.

Theorem 6-1. *The necessary and sufficient conditions for an n-port N consisting of positive linear resistors only to possess a hybrid matrix* $\boldsymbol{H}$, *as defined in Eq. (6-2), are the following:*

1. *The branches corresponding to voltage ports contain no loops.*
2. *The branches corresponding to current ports contain no cutsets.*

Proof: Necessity. When several independent voltage sources in a linear network form a loop, then either the network is inconsistent, or the currents through these voltage sources cannot be uniquely determined. Therefore, the $\boldsymbol{H}$ matrix, as defined in Eq. (6-2) does not exist. A dual argument can be given for the case where several independent current sources form a cutset.

For example, Fig. 6-3(a) shows a two-port in which the two voltage ports form a loop. If $v_1 \neq v_2$, the network is inconsistent. If $v_1 = v_2 = E$, then $i_1 + i_2 = -E/R$; but it is impossible to uniquely determine i_1 and i_2. In a dual manner, Fig. 6-3(b) shows a two-port in which the two current ports form a cutset (of the entire network). If $-i_2 \neq i_1$, the network is inconsistent. If $-i_2 = i_1 = I$, then $v_2 - v_1 = IR$; but it is impossible to uniquely determine v_1 and v_2. In either case, the two-port cannot be characterized by Eq. (6-2). For if Eq. (6-2) exists, the responses at all ports should be *uniquely* determined when the excitations at all ports are given.

Sufficiency. We shall show that, when the conditions of Theorem 6-1 are met, an explicit matrix expression for $\boldsymbol{H}$ can be derived. Such an explicit formula not only shows the existence of $\boldsymbol{H}$, but also serves as a basis for the evaluation of $\boldsymbol{H}$ by a digital computer.

[3]For this special case, we shall use the notation of Section 6-2, but with all ^ removed and with the subscript m changed to n. See Fig. 6-3(c) for notations.

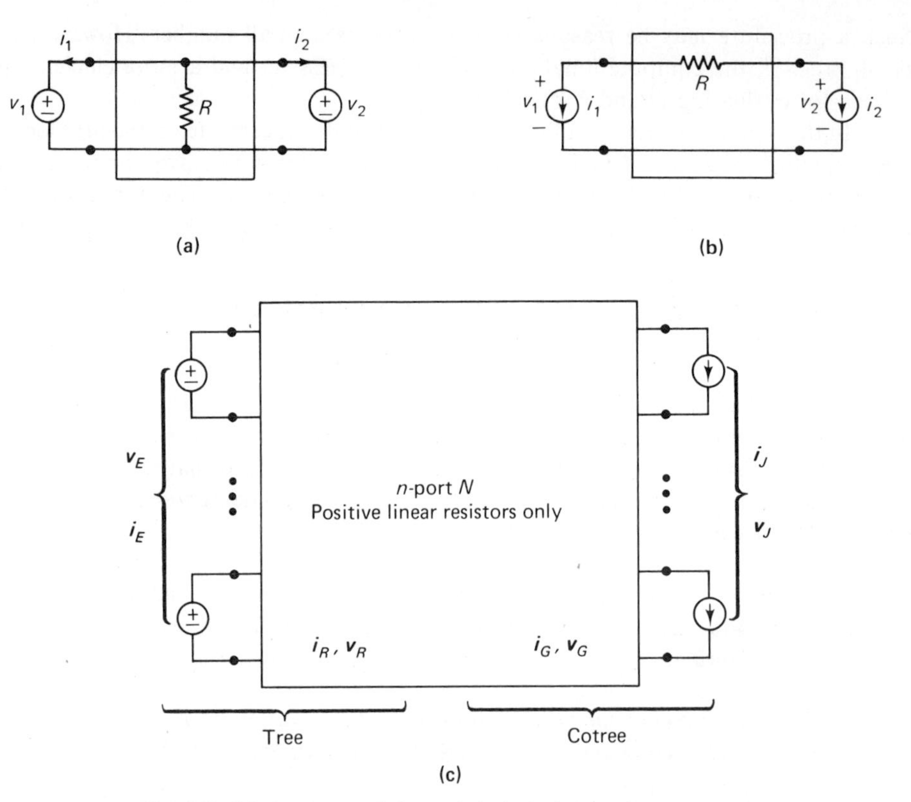

Fig. 6-3. (a) Loop consisting of independent voltage sources. (b) Cutset consisting of independent current sources. (c) Classification of network elements for the proof of Theorem 6-1.

We first select a tree T for which all voltage port branches (indicated by the subscript E) are tree branches, and all current port branches (indicated by the subscript J) are cotree branches. Such a tree T always exists when the conditions of Theorem 6-1 are met (see Problem 3-11). Let those linear resistors inside N which belong to T be characterized by a diagonal resistance matrix $\boldsymbol{Z}_R$, and those which belong to the cotree be characterized by a diagonal conductance matrix $\boldsymbol{Y}_G$. We also use the subscript R for those linear resistors which belong to the tree, and G for those linear resistors in the cotree.

As shown in Chapter 3, the fundamental cutset equations (KCL) for the network of Fig. 6-3(c) can be written as follows:

$$\boldsymbol{D}\boldsymbol{i} = [\boldsymbol{1}_\rho \mid \boldsymbol{D}_L]\boldsymbol{i} = \overset{\begin{matrix} E & \quad R & \quad G & \quad J \end{matrix}}{\underset{\begin{matrix} \text{Tree} & \qquad \text{Cotree} \end{matrix}}{\left[\begin{array}{cc|cc} \boldsymbol{1}_{EE} & \boldsymbol{0}_{ER} & \boldsymbol{D}_{EG} & \boldsymbol{D}_{EJ} \\ \boldsymbol{0}_{RE} & \boldsymbol{1}_{RR} & \boldsymbol{D}_{RG} & \boldsymbol{D}_{RJ} \end{array}\right]}} \begin{bmatrix} \boldsymbol{i}_E \\ \boldsymbol{i}_R \\ \boldsymbol{i}_G \\ \boldsymbol{i}_J \end{bmatrix} = \boldsymbol{0} \tag{6-6}$$

where $\boldsymbol{D}$ is the fundamental cutset matrix. Similarly, the fundamental loop equations

(KVL) can be written as

$$\boldsymbol{Bv} = [\boldsymbol{B}_T \mid \boldsymbol{1}_\mu]\boldsymbol{v} = \begin{array}{c} \begin{array}{cccc} E & R & G & J \end{array} \\ \left[\begin{array}{cc:cc} -\boldsymbol{D}^t_{EG} & -\boldsymbol{D}^t_{RG} & \boldsymbol{1}_{GG} & \boldsymbol{0}_{GJ} \\ -\boldsymbol{D}^t_{EJ} & -\boldsymbol{D}^t_{RJ} & \boldsymbol{0}_{JG} & \boldsymbol{1}_{JJ} \end{array}\right] \end{array} \begin{bmatrix} \boldsymbol{v}_E \\ \boldsymbol{v}_R \\ \boldsymbol{v}_G \\ \boldsymbol{v}_J \end{bmatrix} = \boldsymbol{0} \tag{6-7}$$

where $\boldsymbol{B}$ is the fundamental loop matrix, and where Eq. (3-23) has been used to express $\boldsymbol{B}_T$ in terms of $\boldsymbol{D}_L$.

The linear resistors inside N are characterized by

$$\boldsymbol{v}_R = \boldsymbol{Z}_R\boldsymbol{i}_R \quad \text{or} \quad \boldsymbol{i}_R = \boldsymbol{Y}_R\boldsymbol{v}_R, \quad \text{where} \quad \boldsymbol{Y}_R = \boldsymbol{Z}_R^{-1} \tag{6-8a}$$

and

$$\boldsymbol{i}_G = \boldsymbol{Y}_G\boldsymbol{v}_G \quad \text{or} \quad \boldsymbol{v}_G = \boldsymbol{Z}_G\boldsymbol{i}_G, \quad \text{where} \quad \boldsymbol{Z}_G = \boldsymbol{Y}_G^{-1} \tag{6-8b}$$

Equations (6-6)–(6-8) constitute the complete set of constraints for the network under consideration. To find the $\boldsymbol{H}$ matrix in Eq. (6-2), we must solve for $\boldsymbol{i}_E$ and $\boldsymbol{v}_J$ in terms of $\boldsymbol{v}_E$ and $\boldsymbol{i}_J$. To achieve this, we shall first find $\boldsymbol{v}_R$ and $\boldsymbol{i}_G$ in terms of $\boldsymbol{v}_E$ and $\boldsymbol{i}_J$.

Let us expand Eqs. (6-6) and (6-7) to obtain

$$\boldsymbol{i}_E + \boldsymbol{D}_{EG}\boldsymbol{i}_G + \boldsymbol{D}_{EJ}\boldsymbol{i}_J = \boldsymbol{0} \tag{6-9}$$

$$\boldsymbol{i}_R + \boldsymbol{D}_{RG}\boldsymbol{i}_G + \boldsymbol{D}_{RJ}\boldsymbol{i}_J = \boldsymbol{0} \tag{6-10}$$

$$-\boldsymbol{D}^t_{EG}\boldsymbol{v}_E - \boldsymbol{D}^t_{RG}\boldsymbol{v}_R + \boldsymbol{v}_G = \boldsymbol{0} \tag{6-11}$$

$$-\boldsymbol{D}^t_{EJ}\boldsymbol{v}_E - \boldsymbol{D}^t_{RJ}\boldsymbol{v}_R + \boldsymbol{v}_J = \boldsymbol{0} \tag{6-12}$$

Substituting Eq. (6-8) into Eq. (6-10) to eliminate $\boldsymbol{i}_R$ and $\boldsymbol{i}_G$, we have

$$\boldsymbol{Y}_R\boldsymbol{v}_R + \boldsymbol{D}_{RG}\boldsymbol{Y}_G\boldsymbol{v}_G + \boldsymbol{D}_{RJ}\boldsymbol{i}_J = \boldsymbol{0} \tag{6-13}$$

Solving for $\boldsymbol{v}_G$ in Eq. (6-11) and substituting the resulting expression for $\boldsymbol{v}_G$ in Eq. (6-13), we have, after grouping terms,

$$\boldsymbol{v}_R = [\boldsymbol{Y}_R + \boldsymbol{D}_{RG}\boldsymbol{Y}_G\boldsymbol{D}^t_{RG}]^{-1}[-\boldsymbol{D}_{RG}\boldsymbol{Y}_G\boldsymbol{D}^t_{EG}\boldsymbol{v}_E - \boldsymbol{D}_{RJ}\boldsymbol{i}_J] \tag{6-14}$$

Similarly, we may substitute Eq. (6-8) into Eq. (6-11) to eliminate $\boldsymbol{v}_R$ and $\boldsymbol{v}_G$, and obtain

$$-\boldsymbol{D}^t_{EG}\boldsymbol{v}_E - \boldsymbol{D}^t_{RG}\boldsymbol{Z}_R\boldsymbol{i}_R + \boldsymbol{Z}_G\boldsymbol{i}_G = \boldsymbol{0} \tag{6-15}$$

Solving for $\boldsymbol{i}_R$ from Eq. (6-10), and substituting the result into Eq. (6-15), we obtain

$$\boldsymbol{i}_G = [\boldsymbol{Z}_G + \boldsymbol{D}^t_{RG}\boldsymbol{Z}_R\boldsymbol{D}_{RG}]^{-1}[\boldsymbol{D}^t_{EG}\boldsymbol{v}_E - \boldsymbol{D}^t_{RG}\boldsymbol{Z}_R\boldsymbol{D}_{RJ}\boldsymbol{i}_J] \tag{6-16}$$

Finally, let us substitute Eq. (6-16) for $\boldsymbol{i}_G$ in Eq. (6-9), and Eq. (6-14) for $\boldsymbol{v}_R$ in Eq. (6-12), and then solve for $\boldsymbol{i}_E$ and $\boldsymbol{v}_J$, respectively, from the expressions to obtain

$$\boldsymbol{i}_E = -\boldsymbol{D}_{EG}\boldsymbol{Z}^{-1}\boldsymbol{D}^t_{EG}\boldsymbol{v}_E + [\boldsymbol{D}_{EG}\boldsymbol{Z}^{-1}\boldsymbol{D}^t_{RG}\boldsymbol{Z}_R\boldsymbol{D}_{RJ} - \boldsymbol{D}_{EJ}]\boldsymbol{i}_J \tag{6-17}$$

$$\boldsymbol{v}_J = [\boldsymbol{D}^t_{EJ} - \boldsymbol{D}^t_{RJ}\boldsymbol{Y}^{-1}\boldsymbol{D}_{RG}\boldsymbol{Y}_G\boldsymbol{D}^t_{EG}]\boldsymbol{v}_E - \boldsymbol{D}^t_{RJ}\boldsymbol{Y}^{-1}\boldsymbol{D}_{RJ}\boldsymbol{i}_J \tag{6-18}$$

where

$$\boldsymbol{Z} \triangleq \boldsymbol{Z}_G + \boldsymbol{D}^t_{RG}\boldsymbol{Z}_R\boldsymbol{D}_{RG} \tag{6-19a}$$

$$\boldsymbol{Y} \triangleq \boldsymbol{Y}_R + \boldsymbol{D}_{RG}\boldsymbol{Y}_G\boldsymbol{D}^t_{RG} \tag{6-19b}$$

Equations (6-17) and (6-18) can be rearranged in the following matrix form:

$$\begin{bmatrix} i_E \\ v_J \end{bmatrix} = \begin{bmatrix} -(D_{EG}Z^{-1}D^t_{EG}) & D_{EG}Z^{-1}D^t_{RG}Z_R D_{RJ} - D_{EJ} \\ D^t_{EJ} - D^t_{RJ}Y^{-1}D_{RG}Y_G D^t_{EG} & -(D^t_{RJ}Y^{-1}D_{RJ}) \end{bmatrix} \begin{bmatrix} v_E \\ i_J \end{bmatrix} = \begin{bmatrix} H_{EE} & H_{EJ} \\ H_{JE} & H_{JJ} \end{bmatrix} \begin{bmatrix} v_E \\ i_J \end{bmatrix} \tag{6-20}$$

Equation (6-20) gives the desired H matrix, provided that Z^{-1} and Y^{-1} exist. To show that Y^{-1} exists, we rewrite Eq. (6-19b) as follows:

$$Y \triangleq Y_R + D_{RG}Y_G D^t_{RG} = [\mathbf{1}_{RR} \quad D_{RG}] \begin{bmatrix} Y_R & \mathbf{0} \\ \mathbf{0} & Y_G \end{bmatrix} \begin{bmatrix} \mathbf{1}_{RR} \\ D^t_{RG} \end{bmatrix} \tag{6-21}$$

Since N contains only *positive* linear resistors, Y_R and Y_G are *positive definite* diagonal matrices. It can easily be shown (see Problem 6-2) that Y, as given by Eq. (6-21), is positive definite. Then $|Y| \neq 0$, and Y^{-1} exists. The existence of Z^{-1} can be shown in a completely analogous manner. This completes the proof of Theorem 6-1.

In applying Eq. (6-20), the computation may be simplified somewhat if we take advantage of the following:

1. $H_{EJ} = -H^t_{JE}$. (6-22)
   (see Problem 6-3.)
2. The inverses of Y and Z, as defined in Eq. (6-19), are related as follows (see Problem 6-4):

$$Y^{-1} = Z_R - Z_R D_{RG} Z^{-1} D^t_{RG} Z_R \tag{6-23}$$

   also

$$Z^{-1} = Y_G - Y_G D^t_{RG} Y^{-1} D_{RG} Y_G \tag{6-24}$$

   Thus, we have only to invert either Z or Y, preferably the one of lower order, and then obtain the inverse of the other by Eq. (6-23) or (6-24).

EXAMPLE 6-1. Consider the network shown in Fig. 6-4(a). A four-port N is formed by extracting all independent sources from the given network, as shown in Fig. 6-4(b). Find the hybrid matrix for the four-port N.

Solution: The directed graph associated with the given network is shown in Fig. 6-4(c), where the directions of the resistance branches have been assigned arbitrarily. A tree T is chosen to consist of branches 1, 2, 3, 4, 5, and 6. Note that the voltage port branch 1 is in the tree, while the current port branches 8, 9, and 10 are in the cotree as required by Eqs. (6-6) and (6-7). For this tree, we have

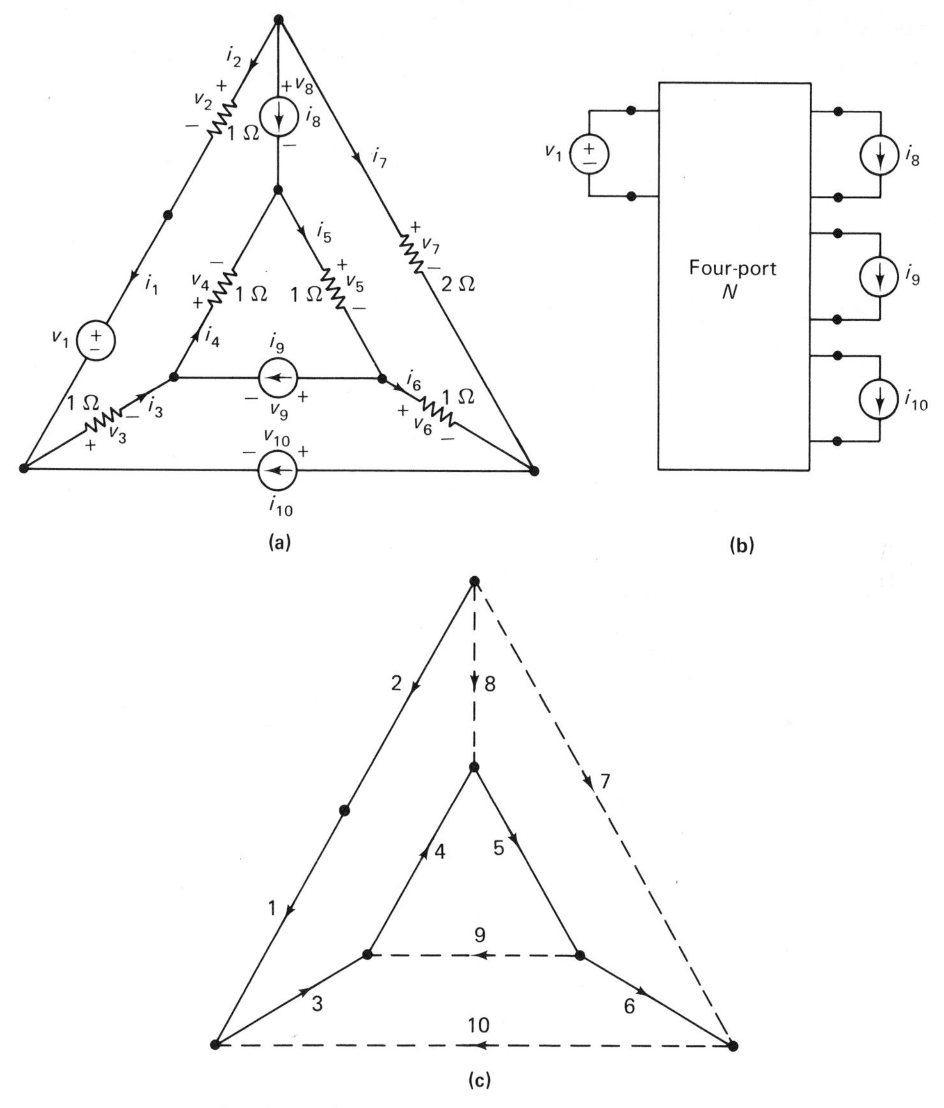

Fig. 6-4. A four-port illustrating the use of Eq. (6-20).

$$
\begin{array}{c} \\ \\ \boldsymbol{B} = \\ \\ \end{array}
\begin{array}{c}
\overset{E}{1} \quad \overbrace{2 \quad 3 \quad 4 \quad 5 \quad 6}^{R} \quad \overset{G}{7} \quad \overbrace{8 \quad 9 \quad 10}^{J} \\
\left[\begin{array}{cccccccccc}
-1 & -1 & -1 & -1 & -1 & -1 & 1 & 0 & 0 & 0 \\
-1 & -1 & -1 & -1 & 0 & 0 & 0 & 1 & 0 & 0 \\
0 & 0 & 0 & 1 & 1 & 0 & 0 & 0 & 1 & 0 \\
0 & 0 & 1 & 1 & 1 & 1 & 0 & 0 & 0 & 1
\end{array}\right]
\end{array}
$$

and

$$\boldsymbol{D} = \left[\begin{array}{c:c|c|c|c|c|c:c:ccc} 1 & 0 & 0 & 0 & 0 & 0 & 1 & 1 & 0 & 0 \\ \hdashline 0 & 1 & 0 & 0 & 0 & 0 & 1 & 1 & 0 & 0 \\ 0 & 0 & 1 & 0 & 0 & 0 & 1 & 1 & 0 & -1 \\ 0 & 0 & 0 & 1 & 0 & 0 & 1 & 1 & -1 & -1 \\ 0 & 0 & 0 & 0 & 1 & 0 & 1 & 0 & -1 & -1 \\ 0 & 0 & 0 & 0 & 0 & 1 & 1 & 0 & 0 & -1 \end{array}\right] \tag{6-25}$$

By comparing Eq. (6-25) with Eq. (6-6), we have

$$\begin{aligned} \boldsymbol{D}_{EG} &= 1 \\ \boldsymbol{D}_{EJ} &= [1 \quad 0 \quad 0] \\ \boldsymbol{D}_{RG} &= [1 \quad 1 \quad 1 \quad 1 \quad 1]^t \\ \boldsymbol{D}_{RJ} &= \begin{bmatrix} 1 & 1 & 1 & 0 & 0 \\ 0 & 0 & -1 & -1 & 0 \\ 0 & -1 & -1 & -1 & -1 \end{bmatrix}^t \end{aligned} \tag{6-26}$$

The linear resistors are characterized by

$$\begin{aligned} \boldsymbol{Z}_G &= 2, \qquad \boldsymbol{Y}_G = 0.5 \\ \boldsymbol{Z}_R &= \mathbf{1}_5, \qquad \boldsymbol{Y}_R = \mathbf{1}_5 \end{aligned} \tag{6-27}$$

In this case, $\boldsymbol{Z}$ is of order 1, and $\boldsymbol{Y}$ of order 5. It will be much easier to compute $\boldsymbol{Z}^{-1}$ from Eq. (6-19a), and then $\boldsymbol{Y}^{-1}$ from Eq. (6-23). The results are

$$\boldsymbol{Z}^{-1} = [2 + 5]^{-1} = \tfrac{1}{7}$$

$$\begin{aligned} \boldsymbol{Y}^{-1} &= \mathbf{1} - \begin{bmatrix} 1 \\ 1 \\ 1 \\ 1 \\ 1 \end{bmatrix} \cdot \tfrac{1}{7} \cdot [1 \quad 1 \quad 1 \quad 1 \quad 1] \\ &= \begin{bmatrix} \frac{6}{7} & -\frac{1}{7} & -\frac{1}{7} & -\frac{1}{7} & -\frac{1}{7} \\ -\frac{1}{7} & \frac{6}{7} & -\frac{1}{7} & -\frac{1}{7} & -\frac{1}{7} \\ -\frac{1}{7} & -\frac{1}{7} & \frac{6}{7} & -\frac{1}{7} & -\frac{1}{7} \\ -\frac{1}{7} & -\frac{1}{7} & -\frac{1}{7} & \frac{6}{7} & -\frac{1}{7} \\ -\frac{1}{7} & -\frac{1}{7} & -\frac{1}{7} & -\frac{1}{7} & \frac{6}{7} \end{bmatrix} \end{aligned} \tag{6-28}$$

Substituting Eqs. (6-26), (6-27), and (6-28) into Eq. (6-20), we obtain

$$\begin{bmatrix} i_1 \\ \hdashline v_8 \\ v_9 \\ v_{10} \end{bmatrix} = \left[\begin{array}{c:ccc} -\frac{1}{7} & -\frac{4}{7} & -\frac{2}{7} & -\frac{4}{7} \\ \hdashline \frac{4}{7} & -\frac{12}{7} & \frac{1}{7} & \frac{2}{7} \\ \frac{2}{7} & \frac{1}{7} & -\frac{10}{7} & -\frac{6}{7} \\ \frac{4}{7} & \frac{2}{7} & -\frac{6}{7} & -\frac{12}{7} \end{array}\right] \begin{bmatrix} v_1 \\ \hdashline i_8 \\ i_9 \\ i_{10} \end{bmatrix} \tag{6-29}$$

which is the desired equation.

6-4 INCLUSION OF INDEPENDENT SOURCES WITHIN AN n-PORT

Having obtained Eq. (6-20) for an n-port consisting of linear resistors only, we can now very easily extend the result to the case in which independent sources are also included within the n-port. This is done by first extracting these independent sources and then reimbedding them. The process of extracting and reimbedding, used in this section for handling independent sources, is also used in Section 6-5 for handling controlled sources.

Let the m-port $\hat{N}$ of Fig. 6-2 be composed of the following:

1. Linear resistors (positive).
2. k_1 independent voltage sources whose voltages and currents are denoted by vectors $\boldsymbol{v}_p$ and $\boldsymbol{i}_p$.
3. k_2 independent current sources whose voltages and currents are denoted by vectors $\boldsymbol{v}_q$ and $\boldsymbol{i}_q$.

To apply Eq. (6-20), we shall extract independent voltage sources $\boldsymbol{v}_p$ and independent current sources $\boldsymbol{i}_q$ and consider them to be additional voltage ports and current ports, respectively, as shown in Fig. 6-5.

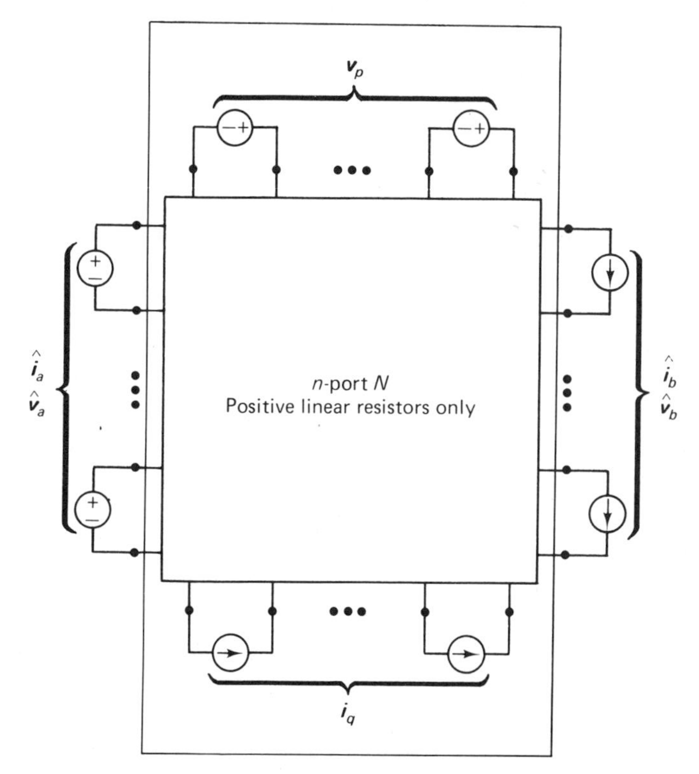

Fig. 6-5. Extraction of independent sources to form ports.

For this new n-port N, with $n = m + k_1 + k_2$, we have

$$\boldsymbol{i}_E = \begin{bmatrix} \hat{\boldsymbol{i}}_a \\ \boldsymbol{i}_p \end{bmatrix}, \quad \boldsymbol{v}_E = \begin{bmatrix} \hat{\boldsymbol{v}}_a \\ \boldsymbol{v}_p \end{bmatrix}, \quad \boldsymbol{i}_J = \begin{bmatrix} \hat{\boldsymbol{i}}_b \\ \boldsymbol{i}_q \end{bmatrix}, \quad \boldsymbol{v}_J = \begin{bmatrix} \hat{\boldsymbol{v}}_b \\ \boldsymbol{v}_q \end{bmatrix}$$

and Eq. (6-20) is applicable.[4] Therefore, we may write Eq. (6-20) as

$$\begin{bmatrix} \hat{\boldsymbol{i}}_a \\ \boldsymbol{i}_p \\ \hat{\boldsymbol{v}}_b \\ \boldsymbol{v}_q \end{bmatrix} = \begin{bmatrix} \boldsymbol{H}_{aa} & \boldsymbol{H}_{ap} & \boldsymbol{H}_{ab} & \boldsymbol{H}_{aq} \\ \boldsymbol{H}_{pa} & \boldsymbol{H}_{pp} & \boldsymbol{H}_{pb} & \boldsymbol{H}_{pq} \\ \boldsymbol{H}_{ba} & \boldsymbol{H}_{bp} & \boldsymbol{H}_{bb} & \boldsymbol{H}_{bq} \\ \boldsymbol{H}_{qa} & \boldsymbol{H}_{qp} & \boldsymbol{H}_{qb} & \boldsymbol{H}_{qq} \end{bmatrix} \begin{bmatrix} \hat{\boldsymbol{v}}_a \\ \boldsymbol{v}_p \\ \hat{\boldsymbol{i}}_b \\ \boldsymbol{i}_q \end{bmatrix} \tag{6-30}$$

Equation (6-30) is for the n-port N shown in Fig. 6-5. To obtain $\hat{N}$ from N, we have to reimbed those ports denoted by $\boldsymbol{v}_p$ and $\boldsymbol{i}_q$, and restore their original roles as independent sources inside $\hat{N}$. The mathematical counterpart of such a reimbedding process is to consider $(\boldsymbol{v}_p, \boldsymbol{i}_q)$ as known quantities, and solve for $(\hat{\boldsymbol{i}}_a, \hat{\boldsymbol{v}}_b)$ in terms of $(\hat{\boldsymbol{v}}_a, \hat{\boldsymbol{i}}_b, \boldsymbol{v}_p, \boldsymbol{i}_q)$. The result is easily shown to be

$$\begin{bmatrix} \hat{\boldsymbol{i}}_a \\ \hat{\boldsymbol{v}}_b \end{bmatrix} = \begin{bmatrix} \boldsymbol{H}_{aa} & \boldsymbol{H}_{ab} \\ \boldsymbol{H}_{ba} & \boldsymbol{H}_{bb} \end{bmatrix} \begin{bmatrix} \hat{\boldsymbol{v}}_a \\ \hat{\boldsymbol{i}}_b \end{bmatrix} + \begin{bmatrix} \boldsymbol{H}_{ap} & \boldsymbol{H}_{aq} \\ \boldsymbol{H}_{bp} & \boldsymbol{H}_{bq} \end{bmatrix} \begin{bmatrix} \boldsymbol{v}_p \\ \boldsymbol{i}_q \end{bmatrix} \tag{6-31}$$

Comparing Eq. (6-31) with Eq. (6-2), we immediately obtain the hybrid matrix $\hat{\boldsymbol{H}}$ and source vector $\hat{\boldsymbol{s}}$ for the m-port $\hat{N}$ as follows:

$$\hat{\boldsymbol{H}} = \begin{bmatrix} \boldsymbol{H}_{aa} & \boldsymbol{H}_{ab} \\ \boldsymbol{H}_{ba} & \boldsymbol{H}_{bb} \end{bmatrix} \tag{6-32}$$

$$\hat{\boldsymbol{s}} = \begin{bmatrix} \boldsymbol{H}_{ap} & \boldsymbol{H}_{aq} \\ \boldsymbol{H}_{bp} & \boldsymbol{H}_{bq} \end{bmatrix} \begin{bmatrix} \boldsymbol{v}_b \\ \boldsymbol{i}_q \end{bmatrix} \tag{6-33}$$

Note that $\hat{\boldsymbol{s}}$ is a constant vector if $\boldsymbol{v}_p$ and $\boldsymbol{i}_q$ are dc independent sources.

6-5 LINEAR RESISTIVE m-PORT WITH CONTROLLED SOURCES

When linear controlled sources are added to the constituents of the m-port of Fig. 6-2, the problem of formulating the hybrid equation (6-2) is no longer as straightforward as the case discussed in the preceding sections. First, from the network topology alone, one cannot determine whether a port should be properly considered to be a voltage port or a current port. The answer depends on network topology *and* element values. To illustrate this point, consider the nonlinear network of Fig. 6-6(a) in which each nonlinear resistor has a monotonically increasing v-i curve. A general analysis method (to be discussed in Chapter 7) requires the extraction of these nonlinear resistors to form a two-port, as shown in Fig. 6-6(b). If $k \neq 2$, each port may be considered either a voltage port or a current port, and Eq. (6-2) exists. Suppose that we consider both ports to be voltage ports; then Eq. (6-2) becomes

$$\begin{bmatrix} i_1 \\ i_2 \end{bmatrix} = \begin{bmatrix} \dfrac{6-k}{3k-6} & \dfrac{k}{3k-6} \\ \dfrac{2}{3k-6} & \dfrac{1}{3k-6} \end{bmatrix} \begin{bmatrix} v_1 \\ v_2 \end{bmatrix} \tag{6-34}$$

[4]Under the conditions stated in Theorem 6-1.

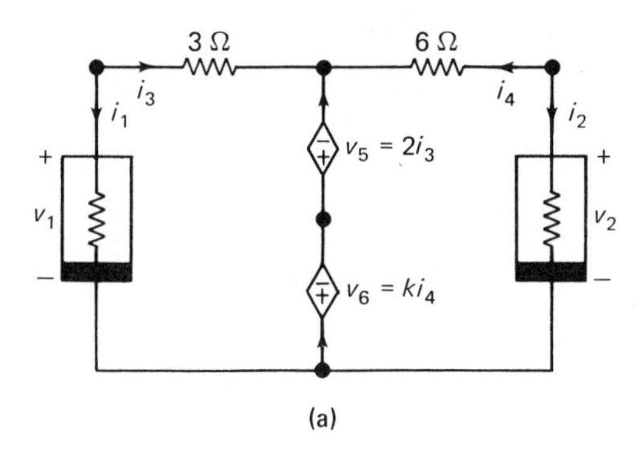

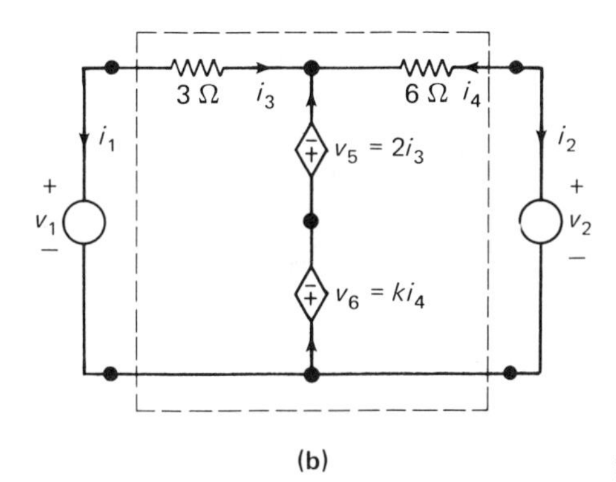

Fig. 6-6. Two-port with no admittance matrix if $k = 2$.

If $k = 2$, the $\boldsymbol{H}$ matrix does not exist when both ports are considered as voltage ports, as can be seen from Eq. (6-34). In this case, however, the hybrid matrices exist for all other port combinations.

Two methods will now be described for formulating the hybrid equation (6-2) for an m-port that may include all four types of controlled sources. We assume that the assignments of voltage ports and current ports are correct. Should the assignment happen to be incorrect, it will result in the singularity of a certain matrix whose inversion is required by the method.

6-5-1 Method of Controlled Source Extraction

Before plunging into the barrage of algebraic manipulations, we shall illustrate the basic idea of this method. Consider the simple one-port network of Fig. 6-7(a). The problem is to formulate Eq. (6-2), which in the present case is of the form $i_1 = -Gv_1$.

If we first extract the controlled current source and replace it by an independent source, as shown in Fig. 6-7(b), we obtain a two-port whose hybrid equation may be computed from Eq. (6-20), or by inspection:

$$\begin{bmatrix} i_1 \\ v_2 \end{bmatrix} = \begin{bmatrix} -\dfrac{1}{R_a} & -1 \\ 1 & -R_b \end{bmatrix} \begin{bmatrix} v_1 \\ i_2 \end{bmatrix} \tag{6-35}$$

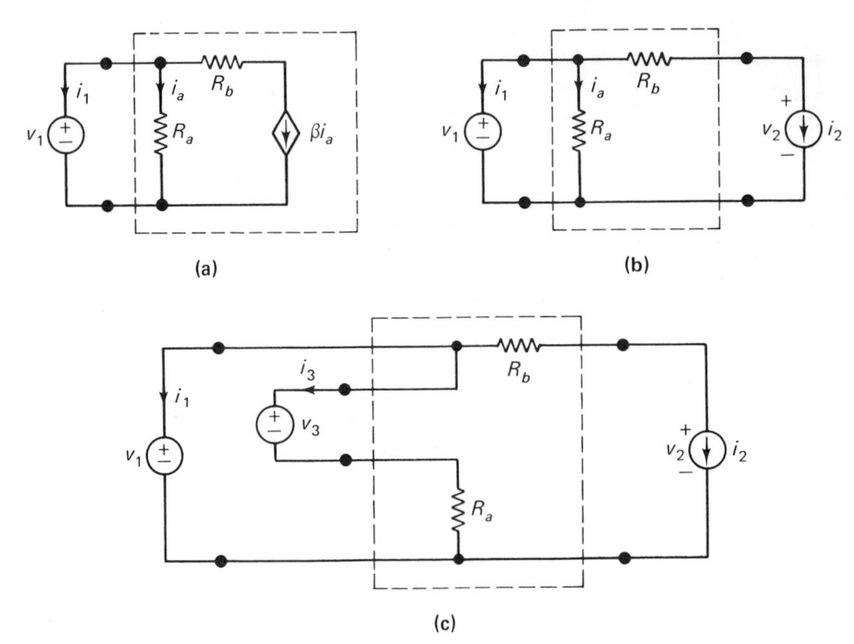

Fig. 6-7. A short-circuit element is extracted as a port.

If to Eq. (6-35) we had added the constraint $i_2 = \beta i_a$, we would have obtained the constraints for the original network of Fig. 6-7(a). Unfortunately, the controlling current i_a does not appear in Eq. (6-35). To remedy this difficulty, we further extract a short-circuit element whose current is the controlling variable of the controlled source to obtain a three-port network, as shown in Fig. 6-7(c). Since a short circuit can be viewed as an independent voltage source whose voltage is identically zero, the extracted element is replaced by a voltage source. For the three-port of Fig. 6-7(c), we can apply Eq. (6-20) to obtain

$$\begin{bmatrix} i_1 \\ v_2 \\ i_3 \end{bmatrix} = \begin{bmatrix} -\frac{1}{R_a} & -1 & \frac{1}{R_a} \\ 1 & -R_b & 0 \\ \frac{1}{R_a} & 0 & -\frac{1}{R_a} \end{bmatrix} \begin{bmatrix} v_1 \\ i_2 \\ v_3 \end{bmatrix} \tag{3-36}$$

For the original network of Fig. 6-7(a), we have to impose two additional constraints:

$$v_3 = 0 \qquad \text{for a short circuit} \tag{6-37}$$

and

$$i_2 = \beta i_3 \qquad \text{for the controlled source} \tag{6-38}$$

From these five equations we can eliminate the four unwanted variables i_2, v_2, i_3, and v_3. The result is

$$i_1 = -\frac{1}{R_a}(\beta + 1)v_1 \tag{6-39}$$

Thus, we see that the essence of this method is to extract a proper set of elements from the given m-port such that Eq. (6-20), which is for a *resistance n-port* only, becomes applicable.

We shall now follow the procedure used in the preceding example and develop an explicit formula for the hybrid matrix for the general case [2]. The m-port under consideration consists of linear resistors and independent and controlled sources (all four types are permissible). Each controlled source is assumed to be either controlled by the current in a *short-circuit* (*sc*) element or by the voltage across an *open-circuit* (*oc*) element. There is no loss of generality in this assumption since an open circuit can always be added in parallel, and a short circuit can always be added in series with any element upon which a controlled source depends. Furthermore, we shall view short-circuit elements as zero-valued independent voltage sources, and open-circuit elements as zero-valued independent current sources. The only restrictions for the present procedure are the following:

Condition 1. Controlled voltage sources and/or independent voltage sources (including voltage ports and short-circuit elements in the present case) do not contain any loops.
Condition 2. Controlled current sources and/or independent current sources (including current ports and open-circuit elements in the present case) do not contain any cutsets.

It should be pointed out that an m-port violating these conditions may still possess a hybrid matrix $\boldsymbol{H}$, which cannot be obtained by the procedure of this section. In such cases, we may apply the procedure of Sections 6-5-2 or 6-6.

Let us form an n-port from the given m-port by the following extraction process:

1. Extract all independent and controlled voltage (current) sources and consider them as voltage (current) ports.[5]
2. Extract all short-circuit (open-circuit) elements whose currents (voltages) are the *controlling* variables for the controlled sources and consider them as voltage (current) ports.

The result will be an n-port as shown in Fig. 6-8, where the subscripts indicate the following:

vc = voltage-controlled current sources
cc = current-controlled current sources
cv = current-controlled voltage sources
vv = voltage-controlled voltage sources
sc = short-circuit elements
oc = open-circuit elements
e and j = independent voltage and current sources, respectively, inside the m-port before extraction

[5]Some controlled source may be equivalent to a linear resistor. When this is apparent, it is desirable to replace the controlled source with a linear resistor before applying the present method.

If the given m-port satisfies conditions 1 and 2, the derived n-port will satisfy the conditions of Theorem 6-1. Consequently, all the equations of Section 6-3 are applicable in the present case, except that we now have

$$\mathbf{v}_E = \begin{bmatrix} \hat{\mathbf{v}}_a \\ \mathbf{v}_e \\ \mathbf{v}_{vv} \\ \mathbf{v}_{cv} \\ \mathbf{v}_{sc} \end{bmatrix}, \quad \mathbf{i}_E = \begin{bmatrix} \hat{\mathbf{i}}_a \\ \mathbf{i}_e \\ \mathbf{i}_{vv} \\ \mathbf{i}_{cv} \\ \mathbf{i}_{sc} \end{bmatrix}, \quad \mathbf{v}_J = \begin{bmatrix} \hat{\mathbf{v}}_b \\ \mathbf{v}_j \\ \mathbf{v}_{cc} \\ \mathbf{v}_{vc} \\ \mathbf{v}_{oc} \end{bmatrix}, \quad \mathbf{i}_J = \begin{bmatrix} \hat{\mathbf{i}}_b \\ \mathbf{i}_j \\ \mathbf{i}_{cc} \\ \mathbf{i}_{vc} \\ \mathbf{i}_{oc} \end{bmatrix} \tag{6-40}$$

Substituting Eq. (6-40) into Eq. (6-20), we may rewrite the equation in the following expanded form:

$$\begin{bmatrix} \hat{\mathbf{i}}_a \\ \mathbf{i}_e \\ \mathbf{i}_{vv} \\ \mathbf{i}_{cv} \\ \mathbf{i}_{sc} \\ \hline \hat{\mathbf{v}}_b \\ \mathbf{v}_j \\ \mathbf{v}_{cc} \\ \mathbf{v}_{vc} \\ \mathbf{v}_{oc} \end{bmatrix} = \left[\begin{array}{ccccc|ccccc} \mathbf{h}_{11} & \mathbf{h}_{12} & \mathbf{h}_{13} & \mathbf{h}_{14} & \mathbf{h}_{15} & \mathbf{h}_{16} & \mathbf{h}_{17} & \mathbf{h}_{18} & \mathbf{h}_{19} & \mathbf{h}_{110} \\ \mathbf{h}_{21} & \mathbf{h}_{22} & \mathbf{h}_{23} & \mathbf{h}_{24} & \mathbf{h}_{25} & \mathbf{h}_{26} & \mathbf{h}_{27} & \mathbf{h}_{28} & \mathbf{h}_{29} & \mathbf{h}_{210} \\ \mathbf{h}_{31} & \mathbf{h}_{32} & \mathbf{h}_{33} & \mathbf{h}_{34} & \mathbf{h}_{35} & \mathbf{h}_{36} & \mathbf{h}_{37} & \mathbf{h}_{38} & \mathbf{h}_{39} & \mathbf{h}_{310} \\ \mathbf{h}_{41} & \mathbf{h}_{42} & \mathbf{h}_{43} & \mathbf{h}_{44} & \mathbf{h}_{45} & \mathbf{h}_{46} & \mathbf{h}_{47} & \mathbf{h}_{48} & \mathbf{h}_{49} & \mathbf{h}_{410} \\ \mathbf{h}_{51} & \mathbf{h}_{52} & \mathbf{h}_{53} & \mathbf{h}_{54} & \mathbf{h}_{55} & \mathbf{h}_{56} & \mathbf{h}_{57} & \mathbf{h}_{58} & \mathbf{h}_{59} & \mathbf{h}_{510} \\ \hline \mathbf{h}_{61} & \mathbf{h}_{62} & \mathbf{h}_{63} & \mathbf{h}_{64} & \mathbf{h}_{65} & \mathbf{h}_{66} & \mathbf{h}_{67} & \mathbf{h}_{68} & \mathbf{h}_{69} & \mathbf{h}_{610} \\ \mathbf{h}_{71} & \mathbf{h}_{72} & \mathbf{h}_{73} & \mathbf{h}_{74} & \mathbf{h}_{75} & \mathbf{h}_{76} & \mathbf{h}_{77} & \mathbf{h}_{78} & \mathbf{h}_{79} & \mathbf{h}_{710} \\ \mathbf{h}_{81} & \mathbf{h}_{82} & \mathbf{h}_{83} & \mathbf{h}_{84} & \mathbf{h}_{85} & \mathbf{h}_{86} & \mathbf{h}_{87} & \mathbf{h}_{88} & \mathbf{h}_{89} & \mathbf{h}_{810} \\ \mathbf{h}_{91} & \mathbf{h}_{92} & \mathbf{h}_{93} & \mathbf{h}_{94} & \mathbf{h}_{95} & \mathbf{h}_{96} & \mathbf{h}_{97} & \mathbf{h}_{98} & \mathbf{h}_{99} & \mathbf{h}_{910} \\ \mathbf{h}_{101} & \mathbf{h}_{102} & \mathbf{h}_{103} & \mathbf{h}_{104} & \mathbf{h}_{105} & \mathbf{h}_{106} & \mathbf{h}_{107} & \mathbf{h}_{108} & \mathbf{h}_{109} & \mathbf{h}_{1010} \end{array}\right] \begin{bmatrix} \hat{\mathbf{v}}_a \\ \mathbf{v}_e \\ \mathbf{v}_{vv} \\ \mathbf{v}_{cv} \\ \mathbf{v}_{sc} \\ \hline \hat{\mathbf{i}}_b \\ \mathbf{i}_j \\ \mathbf{i}_{cc} \\ \mathbf{i}_{vc} \\ \mathbf{i}_{oc} \end{bmatrix} \tag{6-41}$$

We emphasize that the 10×10 block matrix in Eq (6-41) is associated with the n-port of Fig. 6-8, not with the m-port of Fig. 6-2. To obtain the hybrid equation (6-2) for the m-port, we have to eliminate from Eq. (6-41) all voltage and current vectors except $\hat{\mathbf{i}}_a$, $\hat{\mathbf{v}}_a$, $\hat{\mathbf{i}}_b$, $\hat{\mathbf{v}}_b$, $\mathbf{v}_e$, and $\mathbf{i}_j$. To facilitate the elimination process, it is convenient to regroup the voltage and current vectors in Eq. (6-41), and repartition the matrix as follows:

$$\begin{bmatrix} \hat{\mathbf{i}}_a \\ \hat{\mathbf{v}}_b \\ \hline \mathbf{i}_{vv} \\ \mathbf{v}_{vc} \\ \mathbf{i}_{cv} \\ \mathbf{v}_{cc} \\ \hline \mathbf{i}_{sc} \\ \mathbf{v}_{oc} \\ \hline \mathbf{i}_e \\ \mathbf{v}_j \end{bmatrix} = \left[\begin{array}{cc|cccc|cc|cc} \mathbf{h}_{11} & \mathbf{h}_{16} & \mathbf{h}_{13} & \mathbf{h}_{19} & \mathbf{h}_{14} & \mathbf{h}_{18} & \mathbf{h}_{15} & \mathbf{h}_{110} & \mathbf{h}_{12} & \mathbf{h}_{17} \\ \mathbf{h}_{61} & \mathbf{h}_{66} & \mathbf{h}_{63} & \mathbf{h}_{69} & \mathbf{h}_{64} & \mathbf{h}_{68} & \mathbf{h}_{65} & \mathbf{h}_{610} & \mathbf{h}_{62} & \mathbf{h}_{67} \\ \hline \mathbf{h}_{31} & \mathbf{h}_{36} & \mathbf{h}_{33} & \mathbf{h}_{39} & \mathbf{h}_{34} & \mathbf{h}_{38} & \mathbf{h}_{35} & \mathbf{h}_{310} & \mathbf{h}_{32} & \mathbf{h}_{37} \\ \mathbf{h}_{91} & \mathbf{h}_{96} & \mathbf{h}_{93} & \mathbf{h}_{99} & \mathbf{h}_{94} & \mathbf{h}_{98} & \mathbf{h}_{95} & \mathbf{h}_{910} & \mathbf{h}_{92} & \mathbf{h}_{97} \\ \mathbf{h}_{41} & \mathbf{h}_{46} & \mathbf{h}_{43} & \mathbf{h}_{49} & \mathbf{h}_{44} & \mathbf{h}_{48} & \mathbf{h}_{45} & \mathbf{h}_{410} & \mathbf{h}_{42} & \mathbf{h}_{47} \\ \mathbf{h}_{81} & \mathbf{h}_{86} & \mathbf{h}_{83} & \mathbf{h}_{89} & \mathbf{h}_{84} & \mathbf{h}_{88} & \mathbf{h}_{85} & \mathbf{h}_{810} & \mathbf{h}_{82} & \mathbf{h}_{87} \\ \hline \mathbf{h}_{51} & \mathbf{h}_{56} & \mathbf{h}_{53} & \mathbf{h}_{59} & \mathbf{h}_{54} & \mathbf{h}_{58} & \mathbf{h}_{55} & \mathbf{h}_{510} & \mathbf{h}_{52} & \mathbf{h}_{57} \\ \mathbf{h}_{101} & \mathbf{h}_{106} & \mathbf{h}_{103} & \mathbf{h}_{109} & \mathbf{h}_{104} & \mathbf{h}_{108} & \mathbf{h}_{105} & \mathbf{h}_{1010} & \mathbf{h}_{102} & \mathbf{h}_{107} \\ \hline \mathbf{h}_{21} & \mathbf{h}_{26} & \mathbf{h}_{23} & \mathbf{h}_{29} & \mathbf{h}_{24} & \mathbf{h}_{28} & \mathbf{h}_{25} & \mathbf{h}_{210} & \mathbf{h}_{22} & \mathbf{h}_{27} \\ \mathbf{h}_{71} & \mathbf{h}_{76} & \mathbf{h}_{73} & \mathbf{h}_{79} & \mathbf{h}_{74} & \mathbf{h}_{78} & \mathbf{h}_{75} & \mathbf{h}_{710} & \mathbf{h}_{72} & \mathbf{h}_{77} \end{array}\right] \begin{bmatrix} \hat{\mathbf{v}}_a \\ \hat{\mathbf{i}}_b \\ \hline \mathbf{v}_{vv} \\ \mathbf{i}_{vc} \\ \mathbf{v}_{cv} \\ \mathbf{i}_{cc} \\ \hline \mathbf{v}_{sc} \\ \mathbf{i}_{oc} \\ \hline \mathbf{v}_e \\ \mathbf{i}_j \end{bmatrix} \tag{6-42}$$

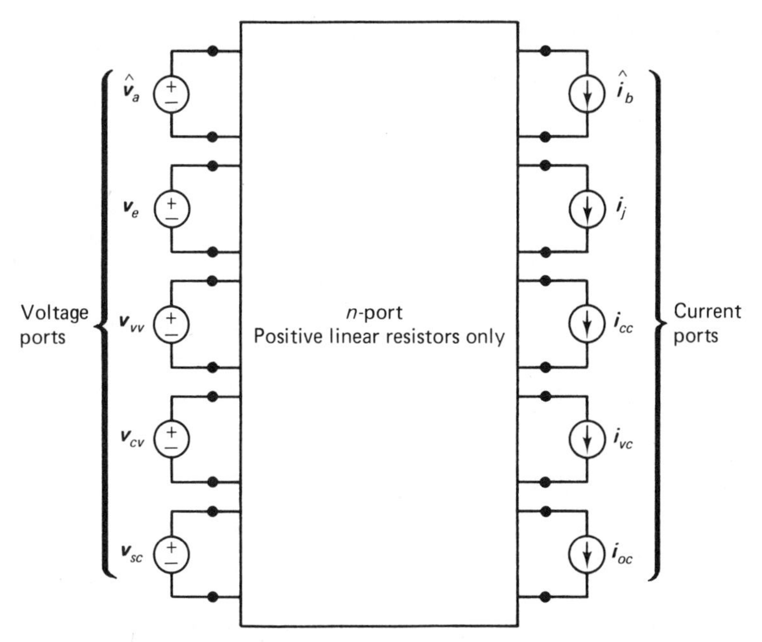

Fig. 6-8. Extraction of independent and controlled sources, short-circuit, and open-circuit elements as ports.

Equation (6-42) can be expressed more compactly as follows:

$$\begin{bmatrix} \boldsymbol{W}_1 \\ \boldsymbol{W}_2 \\ \boldsymbol{W}_3 \\ \boldsymbol{W}_4 \end{bmatrix} = \begin{bmatrix} \boldsymbol{H}_{11} & \boldsymbol{H}_{12} & \boldsymbol{H}_{13} & \boldsymbol{H}_{14} \\ \boldsymbol{H}_{21} & \boldsymbol{H}_{22} & \boldsymbol{H}_{23} & \boldsymbol{H}_{24} \\ \boldsymbol{H}_{31} & \boldsymbol{H}_{32} & \boldsymbol{H}_{33} & \boldsymbol{H}_{34} \\ \boldsymbol{H}_{41} & \boldsymbol{H}_{42} & \boldsymbol{H}_{43} & \boldsymbol{H}_{44} \end{bmatrix} \begin{bmatrix} \boldsymbol{X}_1 \\ \boldsymbol{X}_2 \\ \boldsymbol{X}_3 \\ \boldsymbol{X}_4 \end{bmatrix} \tag{6-43}$$

where the definitions of submatrices[6] $\boldsymbol{W}_i$, $\boldsymbol{X}_i$, and $\boldsymbol{H}_{ij}$ are obtained by pairing off the corresponding blocks in Eqs. (6-42) and (6-43). For example

$$\boldsymbol{W}_1 = \begin{bmatrix} \hat{\boldsymbol{i}}_a \\ \hat{\boldsymbol{v}}_b \end{bmatrix}, \qquad \boldsymbol{X}_1 = \begin{bmatrix} \hat{\boldsymbol{v}}_a \\ \hat{\boldsymbol{i}}_b \end{bmatrix}, \qquad \boldsymbol{X}_4 = \begin{bmatrix} \boldsymbol{v}_e \\ \boldsymbol{i}_j \end{bmatrix}$$

Our aim then is to express $\boldsymbol{W}_1$ in terms of $\boldsymbol{X}_1$ and $\boldsymbol{X}_4$, resulting in Eq. (6-2). We recall that Eq. (6-43) is for the n-port of Fig. 6-8. For the given m-port of Fig. 6-2, two more conditions are available:

1. For the short-circuit and open-circuit elements,

$$\boldsymbol{X}_3 = \boldsymbol{0} \tag{6-44}$$

[6]The notations adopted here are $\boldsymbol{X}_i$ for independent variables, $\boldsymbol{W}_i$ for dependent variables, and $\boldsymbol{H}_{ij}$ for coefficient matrices. Subscript 1 is for the m-port, 2 for controlled sources, 3 for short-circuit and open-circuit elements, and 4 for independent sources originally inside the m-port.

2. For the controlled sources,

$$X_2 = \begin{bmatrix} v_{vv} \\ i_{vc} \\ v_{cv} \\ i_{cc} \end{bmatrix} = \begin{bmatrix} 0 & K_{vv} \\ 0 & K_{vc} \\ K_{cv} & 0 \\ K_{cc} & 0 \end{bmatrix} \begin{bmatrix} i_{sc} \\ v_{oc} \end{bmatrix} = KW_3 \tag{6-45}$$

where the submatrices of K represent the controlling coefficients of the four types of controlled sources.

Substituting Eq. (6-44) into Eq. (6-43) and writing the equations for W_1 and W_3, we have

$$W_1 = H_{11}X_1 + H_{12}X_2 + H_{14}X_4 \tag{6-46a}$$

$$W_3 = H_{31}X_1 + H_{32}X_2 + H_{34}X_4 \tag{6-46b}$$

We next put Eq. (6-46b) in Eq. (6-45), and solve for X_2. The result is

$$X_2 = [1 - KH_{32}]^{-1}[KH_{31}X_1 + KH_{34}X_4] \tag{6-47}$$

Finally, substituting Eq. (6-47) into Eq. (6-46a), we have

$$W_1 = \begin{bmatrix} \hat{i}_a \\ \hat{v}_b \end{bmatrix} = [H_{11} + H_{12}(1 - KH_{32})^{-1}KH_{31}]X_1 + [H_{14} + H_{12}(1 - KH_{32})^{-1}KH_{34}]X_4 \tag{6-48}$$

By comparing Eq. (6-48) with Eq. (6-2), we obtain the expressions for $\hat{H}$ and $\hat{s}$ as follows

$$\hat{H} = H_{11} + H_{12}(1 - KH_{32})^{-1}KH_{31} \tag{6-49a}$$

$$\hat{s} = [H_{14} + H_{12}(1 - KH_{32})^{-1}KH_{34}]\begin{bmatrix} v_e \\ i_j \end{bmatrix} \tag{6-49b}$$

where H_{ij} and K are defined in Eqs. (6-43) and (6-45), respectively. By the use of the identity $(1 + PQ)^{-1}P = P(1 + QP)^{-1}$ (see Problem 6-4), we immediately obtain the following alternative expressions for $\hat{H}$ and $\hat{s}$:

$$\hat{H} = H_{11} + H_{12}K(1 - H_{32}K)^{-1}H_{31} \tag{6-50a}$$

$$\hat{s} = [H_{14} + H_{12}K(1 - H_{32}K)^{-1}H_{34}]\begin{bmatrix} v_e \\ i_j \end{bmatrix} \tag{6-50b}$$

An examination of Eq. (6-49) shows that the hybrid matrix $\hat{H}$ and source vector $\hat{s}$ exist if and only if the matrix[7]

$$F_d \triangleq [1 - KH_{32}]$$

is nonsingular. When F is singular, the given m-port, with the particular combination of voltage ports and current ports, is either inconsistent or has no unique solution,

[7] F_d is sometimes referred to as the *return difference matrix* (see reference 3).

and hence Eq. (6-2) does not exist. In such a case, three different courses of action may be taken. The first is to perturb some element parameters by arbitrarily small quantities within the tolerance of the element. The second is to use a different combination of voltage ports and current ports, for which the hybrid equation may exist. The third is to apply the more general method of Section 6-6.

Example 6-2 will serve to illustrate the method just discussed.

EXAMPLE 6-2. Find the hybrid matrix for the two-port of Fig. 6-6(b) with $k = 2$, port 1 a voltage port, and port 2 a current port.

Solution: Extracting controlled voltage sources and short-circuit elements (branches 7 and 8), we obtain the six-port shown in Fig. 6-9. For the tree T consisting of branches 1, 4, 5, 6, 7 and 8 (see Section 6-3), Eq. (6-6) becomes

$$\mathbf{Di} = \begin{array}{c} \begin{matrix} \quad E \quad & & & & & R & G & J \end{matrix} \\ \left[\begin{array}{ccccc|c|c|c} 1 & 0 & 0 & 0 & 0 & 0 & 1 & 0 \\ 0 & 1 & 0 & 0 & 0 & 0 & 1 & -1 \\ 0 & 0 & 1 & 0 & 0 & 0 & 1 & -1 \\ 0 & 0 & 0 & 1 & 0 & 0 & -1 & 0 \\ 0 & 0 & 0 & 0 & 1 & 0 & 0 & 1 \\ \hline 0 & 0 & 0 & 0 & 0 & 1 & 0 & 1 \end{array}\right] \end{array} \left[\begin{array}{c} i_1 \\ i_5 \\ i_6 \\ i_7 \\ i_8 \\ \hline i_4 \\ \hline i_3 \\ \hline i_2 \end{array}\right] = \mathbf{0}$$

From the partitioning shown, we can identify the submatrices $\boldsymbol{D}_{EG}$, $\boldsymbol{D}_{EJ}$, $\boldsymbol{D}_{RG}$, and $\boldsymbol{D}_{RJ}$. Substituting these matrices into Eq. (6-20), we obtain the following equation, corresponding to Eq. (6-41).

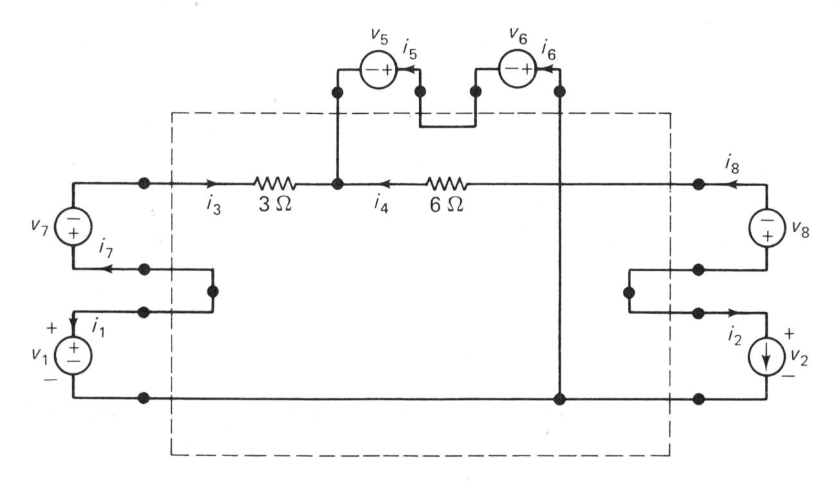

Fig. 6-9. Method of controlled source extraction.

$$\begin{bmatrix} i_1 \\ i_5 \\ i_6 \\ i_7 \\ i_8 \\ \hdashline v_2 \end{bmatrix} = \left[\begin{array}{ccccc:c} -\frac{1}{3} & -\frac{1}{3} & -\frac{1}{3} & \frac{1}{3} & 0 & 0 \\ -\frac{1}{3} & -\frac{1}{3} & -\frac{1}{3} & \frac{1}{3} & 0 & 1 \\ -\frac{1}{3} & -\frac{1}{3} & -\frac{1}{3} & \frac{1}{3} & 0 & 1 \\ \frac{1}{3} & \frac{1}{3} & \frac{1}{3} & -\frac{1}{3} & 0 & 0 \\ 0 & 0 & 0 & 0 & 0 & -1 \\ \hdashline 0 & -1 & -1 & 0 & 1 & -6 \end{array}\right] \begin{bmatrix} v_1 \\ v_5 \\ v_6 \\ v_7 \\ v_8 \\ \hdashline i_2 \end{bmatrix}$$

In the present example, only the following categories of variables appear in this equation:

$$\hat{\boldsymbol{v}}_a = v_1, \qquad \hat{\boldsymbol{i}}_a = i_1$$
$$\boldsymbol{v}_{cv} = [v_5 \ v_6]^t, \qquad \boldsymbol{i}_{cv} = [i_5 \ i_6]^t$$
$$\boldsymbol{v}_{sc} = [v_7 \ v_8]^t, \qquad \boldsymbol{i}_{sc} = [i_7 \ i_8]^t$$
$$\hat{\boldsymbol{i}}_b = i_2, \qquad \hat{\boldsymbol{v}}_b = v_2$$

After regrouping the variables and repartitioning the matrix according to Eq. (6-42), we obtain

$$\begin{bmatrix} i_1 \\ v_2 \\ \hdashline i_5 \\ i_6 \\ \hdashline i_7 \\ i_8 \end{bmatrix} = \left[\begin{array}{cc:cc:cc} -\frac{1}{3} & 0 & -\frac{1}{3} & -\frac{1}{3} & \frac{1}{3} & 0 \\ 0 & -6 & -1 & -1 & 0 & +1 \\ \hdashline -\frac{1}{3} & 1 & -\frac{1}{3} & -\frac{1}{3} & \frac{1}{3} & 0 \\ -\frac{1}{3} & 1 & -\frac{1}{3} & -\frac{1}{3} & \frac{1}{3} & 0 \\ \hdashline \frac{1}{3} & 0 & \frac{1}{3} & \frac{1}{3} & -\frac{1}{3} & 0 \\ 0 & -1 & 0 & 0 & 0 & 0 \end{array}\right] \begin{bmatrix} v_1 \\ i_2 \\ \hdashline v_5 \\ v_6 \\ \hdashline v_7 \\ v_8 \end{bmatrix} \tag{6-51}$$

By comparing Eq. (6-51) with Eq. (6-43), we can readily identify the submatrices $\boldsymbol{H}_{ij}$. Note that in the present case the fourth row and the fourth column in Eq. (6-43) are absent since there is no independent source inside the n-port. The matrix $\boldsymbol{K}$ in Eq. (6-45) is obtained from the following equation describing the controlled sources:

$$\begin{bmatrix} v_5 \\ v_6 \end{bmatrix} = \begin{bmatrix} 2 & 0 \\ 0 & 2 \end{bmatrix} \begin{bmatrix} i_7 \\ i_8 \end{bmatrix} = \boldsymbol{K} \begin{bmatrix} i_7 \\ i_8 \end{bmatrix} \tag{6-52}$$

Substituting $\boldsymbol{K}$ and $\boldsymbol{H}_{ij}$ from Eqs. (6-52) and (6-51), respectively, into Eq. (6-49a), we have

$$\hat{\boldsymbol{H}} = \begin{bmatrix} -\frac{1}{3} & 0 \\ 0 & -6 \end{bmatrix} + \begin{bmatrix} -\frac{1}{3} & -\frac{1}{3} \\ -1 & -1 \end{bmatrix} \left\{ \begin{bmatrix} 1 & 0 \\ 0 & 1 \end{bmatrix} - \begin{bmatrix} 2 & 0 \\ 0 & 2 \end{bmatrix} \begin{bmatrix} \frac{1}{3} & \frac{1}{3} \\ 0 & 0 \end{bmatrix} \right\}^{-1}$$
$$\times \begin{bmatrix} 2 & 0 \\ 0 & 2 \end{bmatrix} \begin{bmatrix} \frac{1}{3} & 0 \\ 0 & -1 \end{bmatrix} = \begin{bmatrix} -1 & 2 \\ -2 & 0 \end{bmatrix}$$

Since there are no internal independent sources, $\hat{\boldsymbol{s}} = \boldsymbol{0}$ by Eq. (6-49b), and the desired hybrid equation is

$$\begin{bmatrix} i_1 \\ v_2 \end{bmatrix} = \begin{bmatrix} -1 & 2 \\ -2 & 0 \end{bmatrix} \begin{bmatrix} v_1 \\ i_2 \end{bmatrix} \tag{6-53}$$

Had we considered both ports 1 and 2 to be voltage ports, we would have the following equation corresponding to Eq. (6-43):

$$\begin{bmatrix} i_1 \\ i_2 \\ \hline i_5 \\ i_6 \\ \hline i_7 \\ i_8 \end{bmatrix} = \left[\begin{array}{cc|cc|cc} -\frac{1}{3} & 0 & -\frac{1}{3} & -\frac{1}{3} & \frac{1}{3} & 0 \\ 0 & -\frac{1}{6} & -\frac{1}{6} & -\frac{1}{6} & 0 & -\frac{1}{6} \\ \hline -\frac{1}{3} & -\frac{1}{6} & -\frac{1}{2} & -\frac{1}{2} & \frac{1}{3} & -\frac{1}{6} \\ -\frac{1}{3} & -\frac{1}{6} & -\frac{1}{2} & -\frac{1}{2} & \frac{1}{3} & -\frac{1}{6} \\ \hline \frac{1}{3} & 0 & \frac{1}{3} & \frac{1}{3} & -\frac{1}{3} & 0 \\ 0 & \frac{1}{6} & \frac{1}{6} & \frac{1}{6} & 0 & \frac{1}{6} \end{array}\right] \begin{bmatrix} v_1 \\ v_2 \\ \hline v_5 \\ v_6 \\ \hline v_7 \\ v_8 \end{bmatrix}$$

and we have

$$\mathbf{1} - \boldsymbol{K}\boldsymbol{H}_{32} = \begin{bmatrix} 1 & 0 \\ 0 & 1 \end{bmatrix} - \begin{bmatrix} 2 & 0 \\ 0 & 2 \end{bmatrix} \begin{bmatrix} \frac{1}{3} & \frac{1}{3} \\ \frac{1}{6} & \frac{1}{6} \end{bmatrix} = \begin{bmatrix} \frac{1}{3} & -\frac{2}{3} \\ -\frac{1}{3} & \frac{2}{3} \end{bmatrix} \tag{6-54}$$

which is singular. Thus, the two-port cannot be characterized by Eq. (6-2) when both ports are voltage ports, which is not surprising when we examine Eq. (6-53) and see that $v_2 = -2v_1$. Since v_1 and v_2 are related to each other by a ratio they cannot both be considered independent voltage sources.

6-5-2 Method of Systematic Elimination[8]

We shall now present a second method by which the hybrid equation (6-2) can be computed even if the two conditions of Section 6-5-1 are violated. As will be seen shortly, the price to be paid for this additional generality is the inversion of a larger matrix than in the previous method.

Again, consider the m-port of Fig. 6-2. Unlike the previous method, we shall not extract controlled sources, and shall not create any short-circuit and open-circuit elements associated with the controlling variables. As usual, we assume that *independent* voltage sources (including voltage ports) contain no loops, and *independent* current sources (including current ports) contain no cutsets. (Otherwise the network is either inconsistent or indeterminate. See the proof of Theorem 6-1.) Then it is always possible to choose a tree T to include all voltage ports as tree branches and all current ports as links. Let the subscripts z and y indicate those nonport branches (including independent and/or controlled sources) which belong to the tree and cotree, respectively.[9] The fundamental cutset equations (KCL) can be written as

$$\begin{array}{cccc} \multicolumn{2}{c}{\overbrace{\qquad\qquad}^{\text{Tree}}} & \multicolumn{2}{c}{\overbrace{\qquad\qquad}^{\text{Cotree}}} \\ a & z & y & b \end{array}$$

$$\begin{bmatrix} \mathbf{1}_{aa} & \mathbf{0}_{az} & \boldsymbol{D}_{ay} & \boldsymbol{D}_{ab} \\ \mathbf{0}_{za} & \mathbf{1}_{zz} & \boldsymbol{D}_{zy} & \boldsymbol{D}_{zb} \end{bmatrix} \begin{bmatrix} \hat{\boldsymbol{i}}_a \\ \boldsymbol{i}_z \\ \boldsymbol{i}_y \\ \hat{\boldsymbol{i}}_b \end{bmatrix} = \begin{bmatrix} \mathbf{0} \\ \mathbf{0} \end{bmatrix} \tag{6-55}$$

[8]This method is actually used in the program MECA (see reference 4).

[9]With the present method, a controlled voltage source may even be included in the cotree and a controlled current source in the tree.

Then, using Eq. (3-23), we can write the fundamental-loop equations as

$$\begin{bmatrix} -D^t_{ay} & -D^t_{zy} & 1_{yy} & 0 \\ -D^t_{ab} & -D_{zb} & 0 & 1_{bb} \end{bmatrix} \begin{bmatrix} \hat{v}_a \\ v_z \\ v_y \\ \hat{v}_b \end{bmatrix} = \begin{bmatrix} 0 \\ 0 \end{bmatrix} \tag{6-56}$$

For the present method, it is more convenient to regroup the variables in Eqs. (6-55) and (6-56) and to rewrite the equations as follows:

$$\begin{bmatrix} i_z \\ v_y \end{bmatrix} = \begin{bmatrix} 0 & -D_{zy} \\ D^t_{zy} & 0 \end{bmatrix} \begin{bmatrix} v_z \\ i_y \end{bmatrix} + \begin{bmatrix} 0 & -D_{zb} \\ D^t_{ay} & 0 \end{bmatrix} \begin{bmatrix} \hat{v}_a \\ \hat{i}_b \end{bmatrix} \tag{6-57}$$

$$\begin{bmatrix} \hat{i}_a \\ \hat{v}_b \end{bmatrix} = \begin{bmatrix} 0 & -D_{ay} \\ D^t_{zb} & 0 \end{bmatrix} \begin{bmatrix} v_z \\ i_y \end{bmatrix} + \begin{bmatrix} 0 & -D_{ab} \\ D^t_{ab} & 0 \end{bmatrix} \begin{bmatrix} \hat{v}_a \\ \hat{i}_b \end{bmatrix} \tag{6-58}$$

Let there be m port branches, and l nonport branches. Then the element laws for the latter may be expressed by l equations in $2 \times (l + m)$ currents and voltages as follows:

$$F \begin{bmatrix} i_z \\ v_y \\ \hat{i}_a \\ \hat{v}_b \\ v_z \\ i_y \\ \hat{v}_a \\ \hat{i}_b \end{bmatrix} = [F_1 \mid F_2 \mid F_3 \mid F_4] \begin{bmatrix} i_z \\ v_y \\ \hdashline \hat{i}_a \\ \hat{v}_b \\ \hdashline v_z \\ i_y \\ \hdashline \hat{v}_a \\ \hat{i}_b \end{bmatrix} = F_1 \begin{bmatrix} i_z \\ v_y \end{bmatrix} + F_2 \begin{bmatrix} \hat{i}_a \\ \hat{v}_b \end{bmatrix} \tag{6-59}$$

$$+ F_3 \begin{bmatrix} v_z \\ i_y \end{bmatrix} + F_4 \begin{bmatrix} \hat{v}_a \\ \hat{i}_b \end{bmatrix} = u$$

where F is an $l \times 2(l + m)$ matrix, and u is an $l \times 1$ vector whose nonzero entries are due to the independent sources within the m-port. Equation (6-59) is an extremely general description of the element laws for a linear resistive m-port. A controlled source may be controlled by the current or voltage of a resistor, a controlled source, or even a port branch. Figure 6-10 and the accompanying equation illustrate the use of Eq. (6-59), where the tree T has been chosen to consist of branches 1, 4, and 5.

The problem at hand is to eliminate from Eq. (6-57)–(6-59) the variables v_z, v_y, i_z, and i_y, and then express $\hat{i}_a$, $\hat{v}_b$ in terms of $\hat{v}_a$, $\hat{i}_b$, and u. To do this, we first substitute Eqs. (6-57) and (6-58) into Eq. (6-59) and solve for v_z and i_y. The result is

$$\begin{bmatrix} v_z \\ i_y \end{bmatrix} = -P^{-1}Q \begin{bmatrix} \hat{v}_a \\ \hat{i}_b \end{bmatrix} + P^{-1}u \tag{6-60a}$$

where

$$P = F_1 \begin{bmatrix} 0 & -D_{zy} \\ D^t_{zy} & 0 \end{bmatrix} + F_2 \begin{bmatrix} 0 & -D_{ay} \\ D^t_{zb} & 0 \end{bmatrix} + F_3 \tag{6-60b}$$

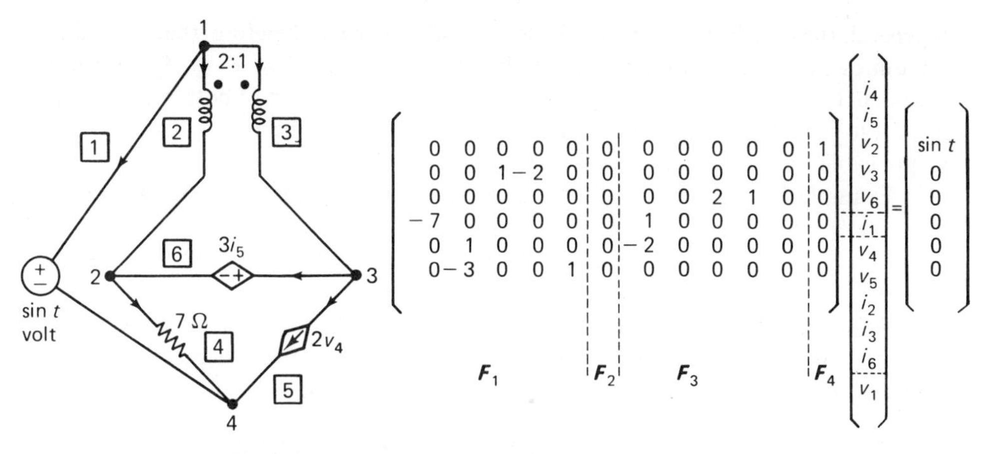

Fig. 6-10. Description of element characteristics by a matrix equation.

and

$$Q = F_1\begin{bmatrix} 0 & -D_{zb} \\ D^t_{ay} & 0 \end{bmatrix} + F_2\begin{bmatrix} 0 & -D_{ab} \\ D^t_{ab} & 0 \end{bmatrix} + F_4 \tag{6-60c}$$

Finally, substituting Eq. (6-60a) into Eq. (6-58), we have

$$\begin{bmatrix} \hat{i}_a \\ \hat{v}_b \end{bmatrix} = \left\{\begin{bmatrix} 0 & D_{ay} \\ -D^t_{zb} & 0 \end{bmatrix} P^{-1}Q - \begin{bmatrix} 0 & D_{ab} \\ -D^t_{ab} & 0 \end{bmatrix}\right\}\begin{bmatrix} \hat{v}_a \\ \hat{i}_b \end{bmatrix} - \begin{bmatrix} 0 & D_{ay} \\ -D^t_{zb} & 0 \end{bmatrix} P^{-1}u = \hat{H}\begin{bmatrix} \hat{v}_a \\ \hat{i}_b \end{bmatrix} + \hat{s} \tag{6-61}$$

From Eq. (6-61), the expressions for $\hat{H}$ and $\hat{s}$ in Eq. (6-2) are immediately obtained.

The existence of the hybrid equation (6-61) depends on the matrix P being nonsingular. If P is singular, the m-port does not possess a hybrid equation for the given voltage ports and current ports. The remarks in Section 6-5-1 about possible remedies are also valid here.

Let us compare the present method with that of Section 6-5-1 paying special attention to the orders of the matrices to be inverted. In Section 6-5-1, the inversions of Z, Y, or F_d are required,[10] where

Order of F_d = number of controlled sources

Order of Z = number of linear resistors in the cotree

Order of Y = number of linear resistors in the tree

For the method of this section, the inversion of P is required. We have

Order of P = number of nonport branches (including controlled sources)

[10]See, however, the remarks in relation to Eqs. (6-23) and (6-24).

In general, the matrix $\boldsymbol{P}$ is of higher order than $\boldsymbol{Z}$, $\boldsymbol{Y}$, or $\boldsymbol{F}_d$; therefore, the inversion of $\boldsymbol{P}$ requires more computational effort than the inversion of $\boldsymbol{Z}$, $\boldsymbol{Y}$, or $\boldsymbol{F}_d$. However, since matrix inversion is only a *part* of the computational effort of both methods, it is difficult to say which method is superior, except to note that the method of this section is applicable to a wider class of m-ports.

Example 6-3 will serve to illustrate the systematic elimination method.

EXAMPLE 6-3. For the one-port of Fig. 6-11, find k in the equation $i_1 = kv_1$.

Solution: The method of Section 6-5-1 is apparently not applicable, because of the loop consisting of the controlled voltage source v_2, the independent voltage source v_1, and the short-circuit element whose current is i_1.

Applying the elimination method, we have for Eq. (6-55), with branch 1 selected as the tree,

$$[1 \;\vdots\; -1 \quad 1]\begin{bmatrix} i_1 \\ \hdashline i_2 \\ i_3 \end{bmatrix} = 0 \tag{6-62}$$

from which $\boldsymbol{D}_{ay} = [-1 \quad 1]$, and $\boldsymbol{D}_{ab}$, $\boldsymbol{D}_{zy}$, and $\boldsymbol{D}_{zb}$ are absent in this case.

The element laws, corresponding to Eq. (6-59), are given by

$$\left[\begin{array}{cc:c:cc:c} -1 & 0 & r & 0 & 0 & 0 \\ 0 & -1 & 0 & 0 & R & 0 \end{array}\right]\begin{bmatrix} v_2 \\ v_3 \\ \hdashline i_1 \\ \hdashline i_2 \\ i_3 \\ \hdashline v_1 \end{bmatrix} = \begin{bmatrix} 0 \\ 0 \end{bmatrix} \tag{6-63}$$

from which

$$\boldsymbol{F}_1 = \begin{bmatrix} -1 & 0 \\ 0 & -1 \end{bmatrix}, \qquad \boldsymbol{F}_2 = \begin{bmatrix} r \\ 0 \end{bmatrix}$$

$$\boldsymbol{F}_3 = \begin{bmatrix} 0 & 0 \\ 0 & R \end{bmatrix}, \qquad \boldsymbol{F}_4 = \begin{bmatrix} 0 \\ 0 \end{bmatrix}$$

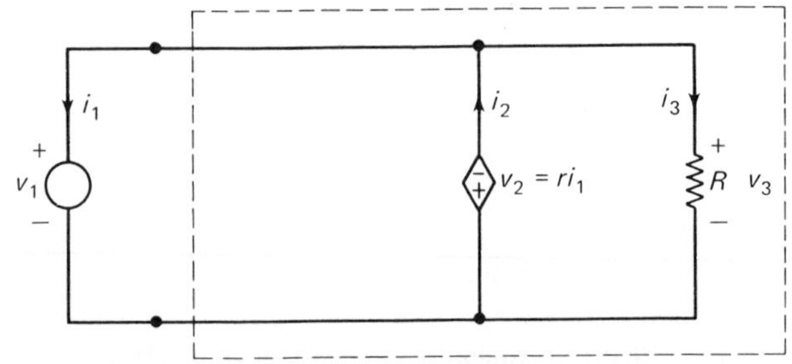

Fig. 6-11. One-port requiring the use of the elimination method.

Then, from Eqs. (6-60b) and (6-60c),

$$P = \begin{bmatrix} r \\ 0 \end{bmatrix} [1 \quad -1] + \begin{bmatrix} 0 & 0 \\ 0 & R \end{bmatrix} = \begin{bmatrix} r & -r \\ 0 & R \end{bmatrix}$$

$$Q = \begin{bmatrix} -1 & 0 \\ 0 & -1 \end{bmatrix} \begin{bmatrix} -1 \\ 1 \end{bmatrix} = \begin{bmatrix} 1 \\ -1 \end{bmatrix}$$

Substituting these submatrices into Eq. (6-61), we have

$$i_1 = [-1 \quad 1] \begin{bmatrix} r & -r \\ 0 & R \end{bmatrix}^{-1} \begin{bmatrix} 1 \\ -1 \end{bmatrix} v_1 = -\frac{1}{r} v_1$$

which is the desired result.

*6-6 FORMULATION OF n-PORT CONSTRAINT MATRICES— THE MOST GENERAL CASE

In Section 6-5 we described two methods for formulating the hybrid matrix of an n-port containing controlled sources. In both methods, the nature of each port (voltage or current port) is predetermined. If the hybrid matrix for a given port combination does not exist, both procedures will abort at some step where an attempt is made to invert a singular matrix [e.g., see Eqs. (6-49) and (6-61)].

In the case of n-ports consisting of positive linear resistors only, the determination of a valid port combination is easy, because the conditions on ports are purely topological, as stated in Theorem 6-1. When an n-port contains controlled sources, it may be difficult (or tedious, at least) to decide whether a port should be considered a voltage or current port. We have illustrated this point with a two-port at the beginning of Section 6-5.

If a particular port combination turns out to be invalid, one natural course is to try other port combinations. For the case of a two-port, this approach may be acceptable since there are only $2^2 = 4$ possible port combinations. However, when the number of ports is large, such an approach is clearly very inefficient, because for each port combination, a complete analysis of the network has to be done. Consider, for example, the four-port of Fig. 6-12. There are $2^4 = 16$ possible port combinations, and yet only one of them is a valid port combination. Unless by sheer good luck we happen to pick the only valid port combination, we shall be spending a great deal of time analyzing possibly up to 15 additional cases. What is more, there may not be any valid port combination and the n-port can only be characterized by a transmission matrix.[11] A well-known example is the nullor [5].

As mentioned in Section 6-5-1, when the matrix $(\mathbf{1} - \boldsymbol{KH}_{32})$ in Eq. (6-49), or the matrix $\boldsymbol{P}$ in Eq. (6-60) is singular, we may seek a remedy by perturbing some element parameters such that the matrix of concern is no longer singular. Such an approach is acceptable for studying practical circuits, because we never know any element value *exactly* in the first place. But then the solution is only *approximate* and

[11]For the case of two-ports, a transmission matrix is also called a *chain matrix* or *ABCD matrix*.

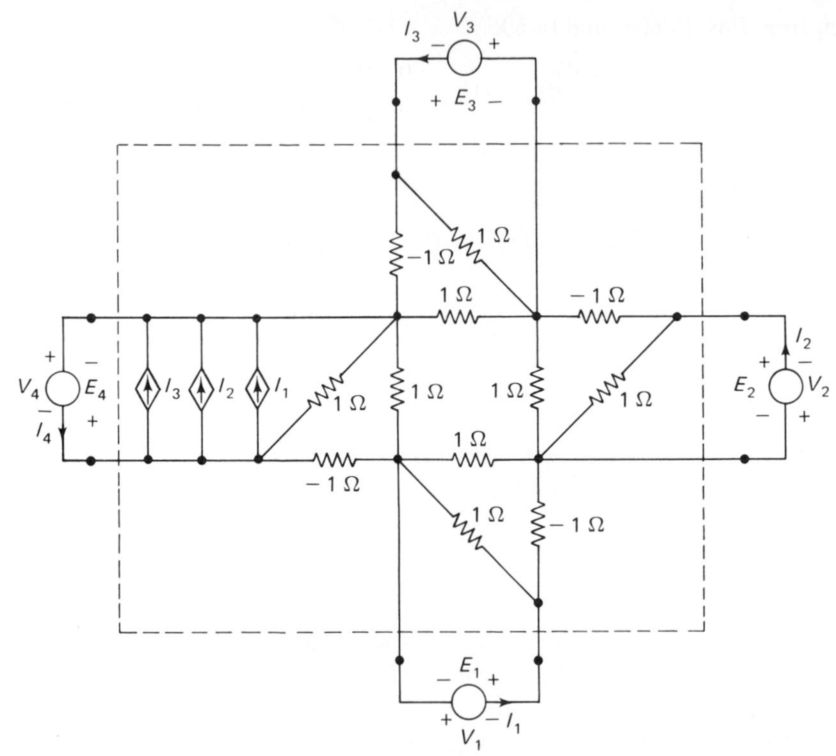

Fig. 6-12. Only one of the 16 possible port combinations of this four-port is valid.

may have a serious accuracy problem if $|\mathbf{1} - \boldsymbol{K}\boldsymbol{H}_{32}|$ or $|\boldsymbol{P}|$ is very close to zero. With a better theory, we should be able to overcome the difficulty without sacrificing the *exactness* of the solution. The method to be described next will achieve this goal [7].

Since the main difficulty lies in not knowing whether a port should be considered to be a voltage or current port, one expedient is to let the nature of each port be "uncommitted" during the analysis. Thus, we introduce the following concept.

Definitions. *An* ***uncommitted independent source*** *is an independent source whose nature (voltage or current source) is not specified. Similarly, an* ***uncommitted port*** *is a port whose nature (voltage or current port) is not specified. In a circuit diagram, an uncommitted source is shown as a circle, but without an arrow or a $\pm$ sign inside the circle.*

For the sake of simpler notations, we shall consider in detail linear resistive n-ports without *internal* independent sources. The extension to include independent sources inside the n-port is straightforward, as described in Section 6-2.

Let all the p ports be uncommitted ports. Instead of seeking the hybrid matrix $\hat{\boldsymbol{H}}$ in Eq. (6-2), which may not exist for a particular port combination, we shall try to find a *maximum* number of m *independent* equations relating the port variables in the form

$$\boldsymbol{C}\begin{bmatrix} \boldsymbol{v}_p \\ \boldsymbol{i}_p \end{bmatrix} = \boldsymbol{0} \tag{6-64}$$

where $\boldsymbol{C}$ is of dimension $m \times 2p$ and is called a *constraint matrix* for the p-port. Normally, the number of constraint equations for a p-port, as given by Eq. (6-64), is the same as the number of ports. However, both $m > p$ and $m < p$ are also possible. Well-known examples are nullators for the former and norators for the latter [5].

To facilitate the formulation of $\boldsymbol{C}$, let us choose an arbitrary tree T. Assume that the network graph is connected and has n nodes. We can construct a tree by picking $n - 1$ branches, paying attention to the rule that they form no loop[12] (see Chap. 3). With respect to this chosen tree T, the network branches can be divided into four categories distinguished by the following subscript notations:

a = port branches in the tree
b = port branches in the cotree
z = nonport branches in the tree
y = nonport branches in the cotree

Then KCL and KVL equations are given by Eqs. (6-55) and (6-56), respectively. The nonport branches may be characterized by

$$\boldsymbol{F}_{iz}\boldsymbol{i}_z + \boldsymbol{F}_{vy}\boldsymbol{v}_y + \boldsymbol{F}_{iy}\boldsymbol{i}_y + \boldsymbol{F}_{vz}\boldsymbol{v}_z + \boldsymbol{F}_{ia}\boldsymbol{i}_a + \boldsymbol{F}_{vb}\boldsymbol{v}_b + \boldsymbol{F}_{ib}\boldsymbol{i}_b + \boldsymbol{F}_{va}\boldsymbol{v}_a = \boldsymbol{0} \tag{6-65}$$

To obtain the constraint matrix $\boldsymbol{C}$, we need to eliminate the four vectors $\boldsymbol{v}_z$, $\boldsymbol{i}_y$, $\boldsymbol{i}_z$, and $\boldsymbol{v}_y$ from Eqs. (6-55), (6-56), and (6-65). In Section 6-5-2, we do this by solving for these vectors *explicitly*. The result is Eq. (6-60), which requires the inversion of $\boldsymbol{P}$. To avoid the kind of difficulty mentioned previously, we shall adopt a different approach that does not require the nonsingularity of any matrix.

Let us rewrite Eqs. (6-55), (6-56), and (6-65) as a single matrix equation. The equation is said to be in the *tableau* form (see Section 17-2).

$$\begin{bmatrix} \boldsymbol{1} & \boldsymbol{0} & \boldsymbol{D}_{zy} & \boldsymbol{0} & \boldsymbol{0} & \boldsymbol{0} & \boldsymbol{D}_{zb} & \boldsymbol{0} \\ \boldsymbol{0} & \boldsymbol{1} & \boldsymbol{0} & -\boldsymbol{D}^t_{zy} & \boldsymbol{0} & \boldsymbol{0} & \boldsymbol{0} & -\boldsymbol{D}^t_{ay} \\ \boldsymbol{0} & \boldsymbol{0} & \boldsymbol{D}_{ay} & \boldsymbol{0} & \boldsymbol{1} & \boldsymbol{0} & \boldsymbol{D}_{ab} & \boldsymbol{0} \\ \boldsymbol{0} & \boldsymbol{0} & \boldsymbol{0} & -\boldsymbol{D}^t_{zb} & \boldsymbol{0} & \boldsymbol{1} & \boldsymbol{0} & -\boldsymbol{D}^t_{ab} \\ \boldsymbol{F}_{iz} & \boldsymbol{F}_{vy} & \boldsymbol{F}_{iy} & \boldsymbol{F}_{vz} & \boldsymbol{F}_{ia} & \boldsymbol{F}_{vb} & \boldsymbol{F}_{ib} & \boldsymbol{F}_{va} \end{bmatrix} \begin{bmatrix} \boldsymbol{i}_z \\ \boldsymbol{v}_y \\ \boldsymbol{i}_y \\ \boldsymbol{v}_z \\ \boldsymbol{i}_a \\ \boldsymbol{v}_b \\ \boldsymbol{i}_b \\ \boldsymbol{v}_a \end{bmatrix} = \boldsymbol{0} \tag{6-66}$$

Note that the first and third rows in Eq. (6-66) are KCL equations, the second and fourth rows are KVL equations, and the last row gives the nonport branch characteristics. The vectors $\boldsymbol{i}_z$ and $\boldsymbol{v}_y$ can be readily eliminated from Eq. (6-66)—by

[12]There is no requirement as to which *type* of network branch is to be picked first as a tree branch.

solving from the first two equations and substituting into the last equation—to yield the following:

$$[F_b \mid F_p]\begin{bmatrix} i_y \\ v_z \\ \hdashline i_a \\ v_b \\ i_b \\ v_a \end{bmatrix} = \left[\begin{array}{cc:cccc} D_{ay} & 0 & 1 & 0 & D_{ab} & 0 \\ 0 & -D^t_{zb} & 0 & 1 & 0 & -D^t_{ab} \\ \hat{F}_{iy} & \hat{F}_{vz} & \hat{F}_{ia} & \hat{F}_{vb} & \hat{F}_{ib} & \hat{F}_{va} \end{array}\right]\begin{bmatrix} i_y \\ v_z \\ \hdashline i_a \\ v_b \\ i_b \\ v_a \end{bmatrix} = 0 \tag{6-67}$$

where

$$\begin{aligned} \hat{F}_{iy} &= F_{iy} - F_{iz}D_{zy} \\ \hat{F}_{vz} &= F_{vz} + F_{vy}D^t_{zy} \\ \hat{F}_{ib} &= F_{ib} - F_{iz}D_{zb} \\ \hat{F}_{va} &= F_{va} + F_{vy}D^t_{ay} \\ \hat{F}_{ia} &= F_{ia} \\ \hat{F}_{vb} &= F_{vb} \end{aligned} \tag{6-68}$$

and submatrices F_b and F_p are partitioned as shown.

Further *explicit* elimination of i_y and v_z from Eq. (6-67) is in general not possible. However, i_y and v_z may be eliminated from Eq. (6-67) by row reduction as follows. Apply elementary row operations to reduce $[F_b \mid F_p]$ to row echelon form (see Chap. 3). In the resultant matrix, those rows in the right block (originally F_p) whose corresponding rows in the left block (originally F_b) are zero rows form the constraint matrix C. Schematically, it looks as follows:

$$[\;F_b \;\mid\; F_p\;] \xrightarrow{\text{elementary row operations}} \left[\begin{array}{c:c} \begin{matrix}\text{Row echelon} \\ \text{form} \\ \\ 0 \;\; 0 \;\; \cdots \;\; 1 \;\; \times \;\; \cdots \end{matrix} & \begin{matrix} \times & \cdots & \times \\ & \cdots & \\ \times & \cdots & \times \end{matrix} \\ \hdashline \mathbf{0} & \begin{matrix}\text{Constraint} \\ \text{matrix}\end{matrix} \end{array}\right] \tag{6-69}$$

Example 6-4 will illustrate this new procedure.

EXAMPLE 6-4. Formulate the constraint equations for the two-port of Fig. 6-6, with $k = 2$.

Solution: There are many possible choices of the tree T. Since the two port branches 1 and 2 contain no loop or cutset, both can be tree branches, both can be cotree branches, or one can be a tree branch and the other a cotree branch. Let us arbitrarily choose a tree to consist of branches 1, 2, 5, and 6. Then

$$i_a = \begin{bmatrix} i_1 \\ i_2 \end{bmatrix}, \quad i_z = \begin{bmatrix} i_5 \\ i_6 \end{bmatrix}, \quad i_y = \begin{bmatrix} i_3 \\ i_4 \end{bmatrix}, \quad i_b\text{—absent}$$

The fundamental cutset matrix is

$$\boldsymbol{D} = \begin{array}{c} \begin{array}{cccccc} 1 & 2 & 5 & 6 & 3 & 4 \end{array} \\ \left[\begin{array}{cc:cc:cc} 1 & 0 & 0 & 0 & 1 & 0 \\ 0 & 1 & 0 & 0 & 0 & 1 \\ \hdashline 0 & 0 & 1 & 0 & 1 & 1 \\ 0 & 0 & 0 & 1 & 1 & 1 \end{array}\right] \end{array} = \begin{bmatrix} \mathbf{1}_{aa} & \mathbf{0} & \boldsymbol{D}_{ay} \\ \mathbf{0} & \mathbf{1}_{zz} & \boldsymbol{D}_{zy} \end{bmatrix} \tag{6-70}$$

The nonport branches are characterized by

$$\left[\begin{array}{cc:cc:cc:cc} 0 & 0 & 1 & 0 & -3 & 0 & 0 & 0 \\ 0 & 0 & 0 & 1 & 0 & -6 & 0 & 0 \\ 0 & 0 & 0 & 0 & -2 & 0 & 1 & 0 \\ 0 & 0 & 0 & 0 & 0 & -2 & 0 & 1 \end{array}\right] \begin{bmatrix} i_5 \\ i_6 \\ \hdashline v_3 \\ v_4 \\ \hdashline i_3 \\ i_4 \\ \hdashline v_5 \\ v_6 \end{bmatrix} = [\boldsymbol{F}_{iz} \mid \boldsymbol{F}_{vy} \mid \boldsymbol{F}_{iy} \mid \boldsymbol{F}_{vz}] \begin{bmatrix} \boldsymbol{i}_z \\ \boldsymbol{v}_y \\ \boldsymbol{i}_y \\ \boldsymbol{v}_z \end{bmatrix} = \mathbf{0} \tag{6-71}$$

From these two equations, we can identify the submatrices $\boldsymbol{D}_{ay}$, $\boldsymbol{D}_{zy}$, $\boldsymbol{F}_{iz}$, $\boldsymbol{F}_{vy}$, $\boldsymbol{F}_{iy}$, and $\boldsymbol{F}_{vz}$. Substituting these submatrices into Eq. (6-67), we obtain

$$[\boldsymbol{F}_b \mid \boldsymbol{F}_p] \begin{bmatrix} \boldsymbol{i}_y \\ \boldsymbol{v}_z \\ \hdashline \boldsymbol{i}_a \\ \boldsymbol{v}_a \end{bmatrix} = \begin{array}{c} \begin{array}{cccc} \boldsymbol{D}_{ay} & \mathbf{0} & \mathbf{1} & \mathbf{0} \end{array} \\ \left[\begin{array}{cc:cc:cc:cc} 1 & 0 & 0 & 0 & 1 & 0 & 0 & 0 \\ 0 & 1 & 0 & 0 & 0 & 1 & 0 & 0 \\ \hdashline -3 & 0 & 1 & 1 & 0 & 0 & 1 & 0 \\ 0 & -6 & 1 & 1 & 0 & 0 & 0 & 1 \\ -2 & 0 & 1 & 0 & 0 & 0 & 0 & 0 \\ 0 & -2 & 0 & 1 & 0 & 0 & 0 & 0 \end{array}\right] \\ \begin{array}{cccc} \hat{\boldsymbol{F}}_{iy} & \hat{\boldsymbol{F}}_{vz} & \hat{\boldsymbol{F}}_{ia} & \hat{\boldsymbol{F}}_{va} \end{array} \end{array} \begin{bmatrix} i_3 \\ i_4 \\ \hdashline v_5 \\ v_6 \\ \hdashline i_1 \\ i_2 \\ \hdashline v_1 \\ v_2 \end{bmatrix} = \mathbf{0} \tag{6-72}$$

We next reduce the coefficient matrix $[\boldsymbol{F}_b \mid \boldsymbol{F}_p]$ to row echelon form. The result is

$$\begin{array}{c} \text{Row echelon form} \\ \left[\begin{array}{cccc:cccc} 1 & 0 & 0 & 0 & 1 & 0 & 0 & 0 \\ 0 & 1 & 0 & 0 & 0 & 1 & 0 & 0 \\ 0 & 0 & 1 & 1 & 3 & 0 & 1 & 0 \\ 0 & 0 & 0 & 1 & 1 & 0 & 1 & 0 \\ \hdashline 0 & 0 & 0 & 0 & 1 & -2 & 1 & 0 \\ 0 & 0 & 0 & 0 & 0 & 0 & 1 & 0.5 \end{array}\right] \\ \begin{array}{cc} \mathbf{0} & \text{Constraint matrix} \end{array} \end{array} \begin{bmatrix} i_3 \\ i_4 \\ v_5 \\ v_6 \\ \hdashline i_1 \\ i_2 \\ v_1 \\ v_2 \end{bmatrix} = \mathbf{0} \tag{6-73}$$

The last two rows in the left block are zero rows. Therefore, by Eq. (6-69), the constraint equation for the two-port is

$$\begin{bmatrix} 1 & -2 & 1 & 0 \\ 0 & 0 & 1 & 0.5 \end{bmatrix} \begin{bmatrix} i_1 \\ i_2 \\ v_1 \\ v_2 \end{bmatrix} = \mathbf{0} \tag{6-74}$$

From Eq. (6-74) it is immediately clear that ports 1 and 2 cannot both be voltage ports, because the columns corresponding to (i_1, i_2) are linearly dependent, and it is not possible to solve for (i_1, i_2) in terms of (v_1, v_2).

On the other hand, if we consider port 1 to be a voltage port and port 2 a current port, we may find (i_1, v_2) from Eq. (6-74) to yield the hybrid matrix

$$\begin{bmatrix} i_1 \\ v_2 \end{bmatrix} = \begin{bmatrix} -1 & 2 \\ -2 & 0 \end{bmatrix} \begin{bmatrix} v_1 \\ i_2 \end{bmatrix} \tag{6-75}$$

which, of course, agrees with the result of Example 6-2.

EXAMPLE 6-5. Find the constraint equations for the four-port shown in Fig. 6-12.

Solution: Using the method of this section as illustrated in Example 6-4, we obtain the following constraint equation:

$$\begin{bmatrix} 0 & 0 & 0 & 1 & 0 & 0 & 0 & 0 \\ 0 & 0 & 0 & 0 & 1 & 0 & 0 & 1 \\ 0 & 0 & 0 & 0 & 0 & 1 & 0 & 1 \\ 0 & 0 & 0 & 0 & 0 & 0 & 1 & 1 \end{bmatrix} \begin{bmatrix} i_1 \\ i_2 \\ i_3 \\ i_4 \\ v_1 \\ v_2 \\ v_3 \\ v_4 \end{bmatrix} = \mathbf{0} \tag{6-76}$$

There are 16 possible port combinations. From Eq. (6-76) it is easily seen that the only valid port-combination is

Ports 1, 2, 3: current ports
Port 4: voltage port

It will be frustrating if we try to find a characterization of this four-port by any method which presumes port nature (e.g., reference 6 or Section 6-5). With the present method using uncommitted ports, we can always obtain the constraint equations first. The task of determining a proper port combination and the corresponding hybrid matrix from the constraint equation is much simpler than the repeated analyses of the entire network. In the present example, with v_1, v_2, v_3, and i_4 as independent sources, we immediately solve from the constraint equation to obtain the hybrid matrix

$$\begin{bmatrix} v_1 \\ v_2 \\ v_3 \\ i_4 \end{bmatrix} = \begin{bmatrix} 0 & 0 & 0 & -1 \\ 0 & 0 & 0 & -1 \\ 0 & 0 & 0 & -1 \\ 0 & 0 & 0 & 0 \end{bmatrix} \begin{bmatrix} i_1 \\ i_2 \\ i_3 \\ v_4 \end{bmatrix} \tag{6-77}$$

6-7 PROGRAM HYBRID AND APPLICATIONS

A FORTRAN program called HYBRID has been written based on the method of Section 6-6. Only resistive n-ports are considered by the program. The user's guide and one sample problem are given in Appendix 6A. The complete listing is given in Appendix 6B.

The application of a hybrid matrix in the analysis of resistive nonlinear networks will be discussed in detail in Chapter 7. The application in the formulation of normal-form equations will be considered in Chapters 8 and 10. Here we shall give some examples of its use in amplifier analysis at midfrequencies where the effect of LC elements is either represented by a short circuit or by an open circuit.

When investigating the performance of an amplifier, we are interested in, among many others, the following pieces of information: V_o/V_i, Z_{in}, Z_{out}, and I_o/I_i. Most of these can be expressed very simply in terms of the parameters of a properly defined n-port. Therefore, the program HYBRID may be used as an aid for the analysis.

EXAMPLE 6-6. For the transistor feedback pair shown in Fig. 6-13(a), with the transistor model shown in Fig. 6-13(b), find the voltage gain E_o/E_i and the input impedance Z_{in}.

Solution: We add a source across the output terminals. The current through this added source will be eventually set to zero. See Fig. 6-13(a) for the notations used and note in particular that $E_i = -V_1$ and $E_o = -V_2$. By the use of the program HYBRID the following constraint equation is obtained:

$$\begin{bmatrix} 1.0 & 0 & 5.605 \times 10^4 & 9.009 \\ 0 & 1.0 & 4.550 \times 10^6 & 909.9 \end{bmatrix} \begin{bmatrix} V_1 \\ V_2 \\ I_1 \\ I_2 \end{bmatrix} = \mathbf{0}$$

Writing in the form of an open-circuit impedance matrix, we have

$$\begin{bmatrix} E_1 \\ E_2 \end{bmatrix} = \begin{bmatrix} z_{11} & z_{12} \\ z_{21} & z_{22} \end{bmatrix} \begin{bmatrix} I_1 \\ I_2 \end{bmatrix} = \begin{bmatrix} 5.605 \times 10^4 & 9.009 \\ 4.550 \times 10^6 & 909.9 \end{bmatrix} \begin{bmatrix} I_1 \\ I_2 \end{bmatrix}$$

Having obtained the z-parameters, we can find E_o/E_i and Z_{in} as follows:

$$\frac{E_o}{E_i} = \left.\frac{E_2}{E_1}\right|_{I_2=0} = \frac{z_{21}}{z_{11}} = \frac{4.550 \times 10^6}{5.605 \times 10^4} = 80.28$$

$$Z_{\text{in}} = \left.\frac{E_1}{I_1}\right|_{I_2=0} = z_{11} = 5.605 \times 10^4\ \Omega$$

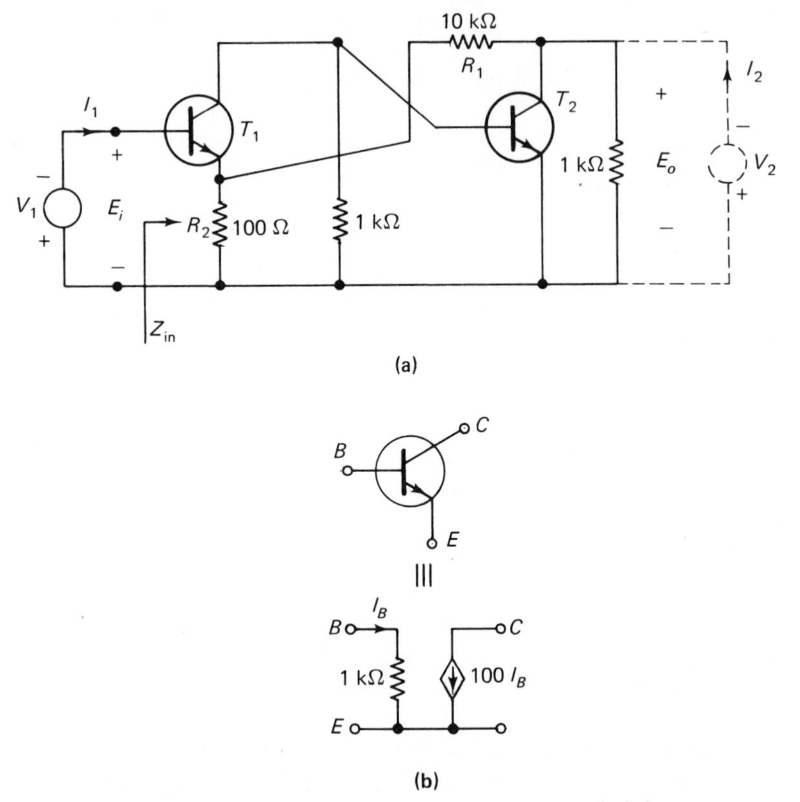

Fig. 6-13. Transforming the determination of Z_{in} and E_o/E_i into an n-port characterization problem.

EXAMPLE 6-7. For the differential amplifier shown in Fig. 6-14(a), with the FET model given by Fig. 6-14(b), find the output impedance Z_o and the differential mode gain A_d.

Solution: As before, we add a source across the output terminals and treat the network as a three-port. By the use of the program HYBRID, the following constraint equation is obtained:

$$\begin{bmatrix} 1 & 0 & 0 & 0 & 0 & 0 \\ 0 & 1 & 0 & 0 & 0 & 0 \\ 0 & 0 & 1 & 0.005 & -0.005 & 6\times 10^{-5} \end{bmatrix} \begin{bmatrix} I_1 \\ I_2 \\ I_3 \\ V_1 \\ V_2 \\ V_3 \end{bmatrix} = \mathbf{0}$$

from which we may write the short-circuit admittance description of the three-port

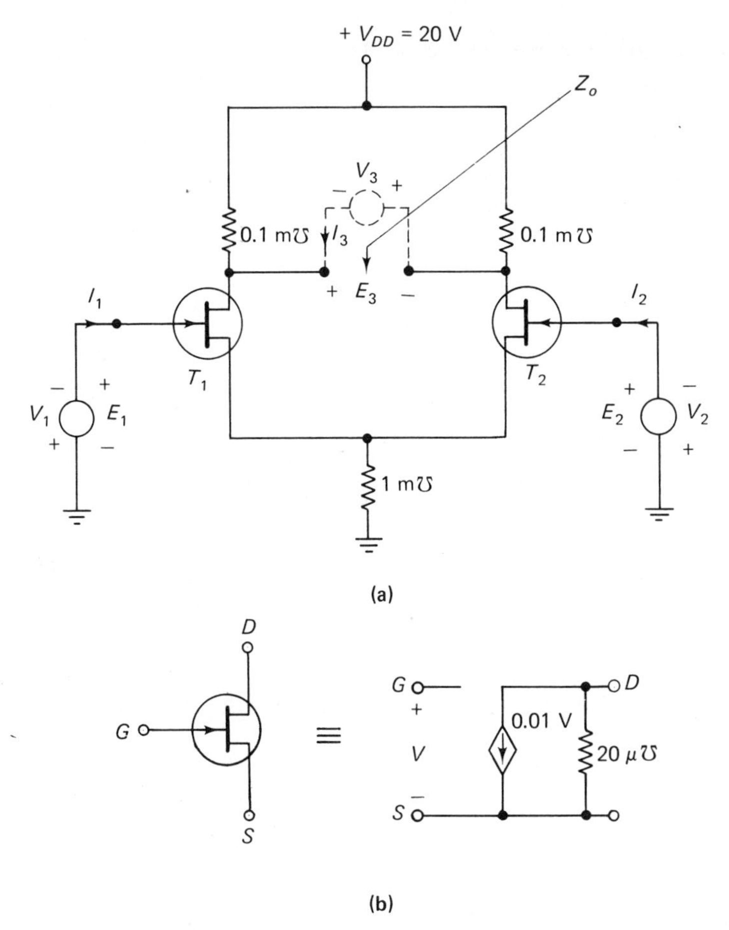

Fig. 6-14. Analysis of a differential amplifier with the aid of the program HYBRID.

as follows:

$$\begin{bmatrix} I_1 \\ I_2 \\ I_3 \end{bmatrix} = \begin{bmatrix} 0 & 0 & 0 \\ 0 & 0 & 0 \\ 0.005 & -0.005 & 6 \times 10^{-5} \end{bmatrix} \begin{bmatrix} E_1 \\ E_2 \\ E_3 \end{bmatrix}$$

Then we have

$$Z_o = \left.\frac{E_3}{I_3}\right|_{E_1=0,\, E_2=0} = \frac{1}{6 \times 10^{-5}} = 16{,}667\ \Omega$$

and

$$A_d = \left.\frac{E_3}{E_1 - E_2}\right|_{I_3=0,\, E_2=-E_1} = -83.33$$

APPENDIX 6A USER'S GUIDE TO HYBRID

PURPOSE. Given a linear n-port consisting of linear resistors (including negative as well as positive resistance values) and all four types of controlled sources, the program HYBRID will find an $n \times 2n$ matrix in echelon form that relates port voltages and currents in the following manner:

$$\begin{bmatrix} n \times 2n \text{ matrix} \\ \text{from HYBRID} \end{bmatrix} \begin{bmatrix} i_a \\ v_b \\ v_a \\ i_b \end{bmatrix} = \mathbf{0}$$

where the subscript a indicates port branches in the tree T (chosen by the program), and the subscript b indicates port branches in the cotree.

DATA CARD SETUP. $(b_t + 2)$ cards are needed for a network of b_t branches. Each independent source is considered as a branch.

Card 1. Title card. Reproduced in the output for problem identification.

Card 2, 3, . . . , $(b_t + 1)$, one card for each branch, entered in the following order:

Group 1. Cards for those port branches that are known or assumed to be *voltage ports*.
Group 2. Cards for nonport branches.
Group 3. Cards for those port branches with undetermined port nature (voltage or current port).
Group 4. Cards for those port branches that are known or assumed to be *current ports*.

For each branch description card, the format is as follows:

Column	
1–3	Branch number (I3)
4–6	"From node" (I3) } (number nodes starting with 0).
7–9	"To node" (I3) }
10–11	Type of branch (A2)

The following entries are permitted:

␣R resistance (␣ for blank)
␣G conductance
␣E voltage port
␣I current port
VC voltage-controlled current source
CC current-controlled current source
CV current-controlled voltage source
VV voltage-controlled voltage source

12–21 value of R, G, or the controlling coefficients (E10.3)
22–24 branch number of the controlling element (I3). Leave blank if the type is R, G, E, or I.

Card $(b_t + 2)$ — last card. Blank.

The input and output of a sample problem (Example 6-6) are shown in Figs. 6A-1 and 6A-2.

```
EXAMPLE 6.5
  3  1   4 R 1000.
  4  3   4 G 0.0
  5  3   4CC 100.          3
  6  3   0 R 1000.
  7  2   0 G 0.0
  8  2   0CC 100.          6
  9  2   4 R 10000.
 10  4   0 R 100.
 11  3   0 R 1000.
 12  2   0 R 1000.
  1  1   0 I
  2  2   0 I

STOP
```

Fig. 6A-1. Input of Example 6-6 into HYBRID program.

```
EXAMPLE 6.6

NETWORK DESCRIPTION


BRANCH      FROM    TO      ELEMENT    ELEMENT     CONTROL
NUMBER      NODE   NODE     TYPE        VALUE      BRANCH
   3          1      4        R       1.000E+03      -0
   4          3      4        G       0.             -0
   5          3      4        CC      1.000E+02       3
   6          3      0        R       1.000E+03      -0
   7          2      0        G       0.             -0
   8          2      0        CC      1.000E+02       6
   9          2      4        R       1.000E+04      -0
  10          4      0        R       1.000E+02      -0
  11          3      0        R       1.000E+03      -0
  12          2      0        R       1.000E+03      -0
   1          1      0        I      -0.             -0
   2          2      0        I      -0.             -0


TREE NON-PORT BRANCHES
 3  4  6  7

LINK NON-PORT BRANCHES
 5  8  9 10 11 12

LINK PORT BRANCHES
 1  2

    V 1            V 2            I 1          I 2

 1.0000E+00   0.              5.6054E+04   9.0090E+00

 0.              1.0000E+00   4.5505E+06   9.0991E+02
```

Fig. 6A-2. Output from HYBRID program for Example 6-6.

METHODS USED. The methods used to find a tree T and to formulate the fundamental cutset and fundamental loop matrices are described in Chapter 3. The method used to derive the constraint matrix is described in Section 6-6.

ADDITIONAL NOTES

1. More problems can be run in the same job simply by stacking the data cards. A title card with the word *STOP* in columns 1-4 will terminate the job.
2. The last blank card of each data deck provides some options. A negative integer in columns 1–3 causes a full output for debugging purposes (of no interest to average users). A real constant may be punched in columns 12–21. Any number smaller than this will then be treated as 0. If left blank, the program considers any number smaller than 10^{-8} as 0.
3. For a network with known voltage ports and current ports, if the data cards are set up as described, the right *half* of the output matrix from HYBRID is the negative of the hybrid matrix defined by Eq. (6-2).

APPENDIX 6B LISTING OF HYBRID

```
      PROGRAM HYBRID(INPUT,OUTPUT,TAPE5=INPUT,TAPE6=OUTPUT)
      INTEGER A,NFROM(80),NTO(80),TYPE(80),ICONT(80),V,E,R,G,VV,CV,CC,
     *VC,FROM,TO,DCOL(80),ICOUNT(2),C,COUN,BEGIN,TEMP,ST,TN,TP,PORT,
     *HEADER(320),BR(80),RBR(80),ANSROW,ANSCOL,DEBUG,ISTP
      REAL VALUE(80),F3(80,40),F6(80,40),ANS(80,160)
      COMMON A(40,80),ANS,HEADER
      DATA E ,IS,R,G,VV,CV,CC,VC,ST,C,V/2H E,2H I,2H R,2H G,
     12HVV,2HCV,2HCC,2HVC,2HST,1HI,1HV/
      DATA ISTP/1HS/
C
C   MAX CIRCUIT CONFIGURATION IS 40 NODES AND 80 BRANCHES
C   MAX OF 40 ELEMENTS IN ANY CATEGORY (TP,TN,LN,LP)
C
1     NBR=0
      NNODE=0
C
C   READ DATA CARDS AND LOAD THE A MATRIX
C
C   READ TITLE CARD
      READ (5,500) (A(40,J),J=1,80)
      IF(A(40,1).EQ.ISTP) STOP
C   WRITE TITLE LINE AND PRINT HEADING OF NETWORK DESCRIPTION
      WRITE (6,501) (A(40,J), J=1,80)
C ZERO OUT A MATRIX
      DO 2 I=1,40
      DO 2 J=1,80
2     A(I,J)=0
C READ IN DATA AND FILL A MATRIX
      DEBUG=0
      DO 3 K=1,80
      READ(5,502)BR(K),NFROM(K),NTO(K),TYPE(K),VALUE(K),ICONT(K)
C   IF BRANCH NUMBER LE ZERO THEN DONE READING IN ELEMENTS
      IF(BR(K).LE.0) GO TO 4
      WRITE(6,503)BR(K),NFROM(K),NTO(K),TYPE(K),VALUE(K),ICONT(K)
C   STORE ENTRIES INTO A MATRIX
      FROM=NFROM(K)
      TO=NTO(K)
      IF(FROM.NE.0)A(FROM,K)=1
```

```
      IF(TO.NE.0)A(TO,K)=-1
      NBR=MAX0(NBR,K)
3     NNODE=MAX0(NNODE,NFROM(K),NTO(K))
C  IF VALUE FIELD LEFT BLANK ON LAST CARD, SET ZERO TO 10**-8, ELSE SET
C  ZERO TO THE VALUE SPECIFIED IN THE VALUE FIELD
4     ZERO=VALUE(K)
      IF(ZERO.EQ.0) ZERO=10.**(-8)
      WRITE(6,504) ZERO
C  IF BRANCH LABEL LT 0, WANT ALL OPTIONAL OUTPUTS FOR THE PROGRAM
C  (SET DEBUG = 1), ELSE SET DEBUG = 0
      IF(BR(K).LT.0) DEBUG=1
      IF(DEBUG.NE.1) GO TO 5
C  PRINT THE A MATRIX FOR DEBUG RUN
      WRITE(6,505)
      DO 6 I=1,NNODE
6     WRITE(6,506)(A(I,J),J=1,NBR)
5     DO 7 I=1,NBR
      DCOL(I)=0
7     RBR(I)=0
C
C  DETERMINE ELEMENTS MAKING UP THE TREE
C
      CALL FTREE(NNODE,NBR,DCOL)
C
C REORDER A MATRIX INTO FOUR CLASSES
C
C    1. TREE PORT BRANCHES(TP)
C    2. TREE NON-PORT BRANCHES(TN)
C    3. LINK NON-PORT BRANCHES(LN)
C    4. LINK PORT BRANCHES(LP)
C
C DCOL CONTAINS ORDERING OF A WITH TREE BRANCHES IN LEFTMOST COLUMNS
      JJ=NNODE+1
      N=1
      DO 8 J=1,NNODE
      M=DCOL(J)
      DO 9 K=N,M
      IF(M.EQ.K)GO TO 8
      DCOL(JJ)=K
      JJ=JJ+1
9     CONTINUE
8     N=M+1
      IF(JJ.EQ.NCOL)GO TO 10
      DO 11 I=N,NBR
11    DCOL(I)=I
C REORDER DCOL INTO FOUR CLASSES
C ICOUNT(1) MARKS LAST PORT COLUMN OF TREE BRANCHES
C ICOUNT(2) MARKS LAST NON-PORT COLUMN OF LINK BRANCHES
10    ICOUNT(1)=1
      IT2=NNODE
      I=1
13    DO 12 M=1,IT2
      MM=(NBR+1)*(I-1)+((3-(2*I))*M)
      ITEM=DCOL(MM)
      IF(TYPE(ITEM).NE.E.AND.TYPE(ITEM).NE.IS)GO TO 12
      ITEM1=ICOUNT(I)
      DCOL(MM)=DCOL(ITEM1)
      DCOL(ITEM1)=ITEM
      ICOUNT(I)=ICOUNT(I)+1-((I-1)*2)
12    CONTINUE
      IF(I.EQ.2) GO TO 14
      ICOUNT(1)=ICOUNT(1)-1
      ICOUNT(2)=NBR
      IT2=NBR-NNODE
      I=2
      GO TO 13
C REORDER THE A MATRIX AND THE ORIGINAL LABEL VECTOR TO CORRESPOND TO
```

```
C  THE REORDERED DCOL
14    NN=2
      N=1
      BEGIN=1
      COUN=0
15    ITEM=DCOL(N)
      IF(ITEM.EQ.BEGIN)GO TO 16
      ITEMP=BR(N)
      BR(N)=BR(ITEM)
      BR(ITEM)=ITEMP
      DO 17 J=1,NNODE
      TEMP=A(J,N)
      A(J,N)=A(J,ITEM)
17    A(J,ITEM)=TEMP
      COUN=COUN+1
      DCOL(N)=-DCOL(N)
      N=ITEM
      GO TO 15
16    DCOL(N)=-DCOL(N)
      IF(COUN.EQ.(NBR-1))GO TO 18
      DO 19 I=NN,NBR
      ITEM=DCOL(I)
      IF(ITEM.EQ.I)GO TO 20
      IF(ITEM.LT.0)GO TO 19
      BEGIN=I
      N=I
      GO TO 15
20    COUN=COUN+1
      DCOL(I)=-DCOL(I)
      NN=I
19    CONTINUE
18    DO 22 N=1,NBR
22    DCOL(N)=IABS(DCOL(N))
C
C  REDUCE REORDERED A MATRIX TO ROW ECHELON FORM
C
      CALL IAECH(NNODE,NBR)
C
C  BACK SUBSTITUTE A MATRIX
C
      DO 23 I=2,NNODE
      LROW=I-1
      DO 23 J=1,LROW
      IFCOL=I
      ITEMP=A(J,IFCOL)
      DO 23 K=I,NBR
23    A(J,K)=A(J,K)-A(I,K)*ITEMP
C
C FORMULATE THE ELEMENT CHARACTERISTICS
C
C TP IS NUMBER OF COLUMNS IN F1 AND F5
C TN IS NUMBER OF COLUMNS IN F2 AND F6
C LN IS NUMBER OF COLUMNS IN F3 AND F7
C LP IS NUMBER OF COLUMNS IN F4 AND F8
      TP=ICOUNT(1)
      TN=NNODE-ICOUNT(1)
      LN=ICOUNT(2)-NNODE
      LP=NBR-ICOUNT(2)
      PORT=TP+LP
      NPORT=TN+LN
      ANSROW=NBR
      ANSCOL=NBR+PORT
      WRITE(6,507)
      IF(TP.EQ.0) GO TO 24
      WRITE(6,508) (BR(I),I=1,TP)
24    J=TP+1
      IF(TN.EQ.0) GO TO 25
```

```
      WRITE(6,509) (BR(I),I=J,NNODE)
25    J=NNODE+1
      JJ=NNODE+LN
      IF(LN.EQ.0) GO TO 26
      WRITE(6,510) (BR(I),I=J,JJ)
26    J=JJ+1
      IF(LP.EQ.0) GO TO 28
      WRITE(6,511) (BR(I),I=J,NBR)
C  ZERO ANS MATRIX
28    DO 27 I=1,ANSROW
      DO 27 J=1,ANSCOL
27    ANS(I,J)=0.0
      DO 29 I=1,NPORT
      DO 30 J=1,TN
30    F6(I,J)=0
      DO 29 J=1,LN
29    F3(I,J)=0
      KOUNT=ICOUNT(1)
      K=0
      J=1
      DO 31 I=1,NBR
      ITEM=BR(I)
31    RBR(ITEM)=I
      IF(DEBUG.NE.1) GO TO 32
      WRITE(6,512) TP,TN,LN,LP
      WRITE(6,513)(BR(I),I=1,NBR)
32    KOUNT=KOUNT+1
      MM=DCOL(KOUNT)
      ITEMP=ICONT(MM)
      ITEMP=RBR(ITEMP)
      IT1=PORT+J
      IF(TYPE(MM).EQ.G.OR.TYPE(MM).EQ.VC.OR.TYPE(MM).EQ.CC)
     1GO TO 33
C VOLTAGE SOURCE TYPE
      IF (KOUNT.GT.NNODE)GO TO 34
C  F2
      IT2=LN+J
      ANS(IT1,IT2)=1.
      IF(TYPE(MM).EQ.CV)GO TO 35
      IF(TYPE(MM).EQ.VV)GO TO 36
      F6(J,J)=-VALUE(MM)
      GO TO 37
34    K=K+1
      F3(J,K)=1.
      IF(TYPE(MM).EQ.CV)GO TO 35
      IF(TYPE(MM).EQ.VV)GO TO 36
C  F7
      ANS(IT1,K)=-VALUE(MM)
      GO TO 37
C CURRENT SOURCE TYPE
33    IF(KOUNT.GT.NNODE)GO TO 38
      F6(J,J)=1.
      IF(TYPE(MM).EQ.VC)GO TO 36
      IF(TYPE(MM).EQ.CC)GO TO 35
C  F2
      IT2=LN+J
      ANS(IT1,IT2)=-VALUE(MM)
      GO TO 37
38    K=K+1
C  F7
      ANS(IT1,K)=1.
      IF(TYPE(MM).EQ.VC)GO TO 36
      IF(TYPE(MM).EQ.CC)GO TO 35
      F3(J,K)=-VALUE(MM)
37    J=J+1
      IF(KOUNT.NE.ICOUNT(2))GO TO 32
      GO TO 39
```

```
C CURRENT CONTROLLED
35     IF(ITEMP.GT.TP)GO TO 40
C  F5
       IT2=NPORT+ITEMP
       ANS(IT1,IT2)=-VALUE(MM)
       GO TO 37
40     IF(ITEMP.GT.NNODE)GO TO 41
       IT=ITEMP-TP
       F6(J,IT)=-VALUE(MM)
       GO TO 37
41     IF(ITEMP.GT.ICOUNT(2))GO TO 42
       IT=ITEMP-NNODE
C  F7
       ANS(IT1,IT)=-VALUE(MM)
       GO TO 37
42     IT=ITEMP-ICOUNT(2)
C  F8
       IT2=NBR+TP+IT
       ANS(IT1,IT2)=-VALUE(MM)
       GO TO 37
C VOLTAGE CONTROLLED
36     IF(ITEMP.GT.TP)GO TO 43
C  F1
       IT2=NBR+ITEMP
       ANS(IT1,IT2)=-VALUE(MM)
       GO TO 37
43     IF(ITEMP.GT.NNODE)GO TO 44
       IT=ITEMP-TP
C  F2
       IT2=LN+IT
       ANS(IT1,IT2)=-VALUE(MM)
       GO TO 37
44     IF(ITEMP.GT.ICOUNT(2))GO TO 45
       IT=ITEMP-NNODE
       F3(J,IT)=-VALUE(MM)
       GO TO 37
45     IT=ITEMP-ICOUNT(2)
C  F4
       IT2=NPORT+TP+IT
       ANS(IT1,IT2)=-VALUE(MM)
       GO TO 37
39     IF(DEBUG.EQ.0) GO TO 46
       IF(LN.EQ.0) GO TO 47
C  WRITE F3 FOR DEBUG RUN
       WRITE(6,514)
       IT1=1
48     IT2=LN
       IF((IT2-IT1).GT.10) GO TO 49
       IF(IT2.EQ.IT1) GO TO 47
       WRITE(6,515)
       DO 50 I=1,NPORT
50     WRITE(6,516) (F3(I,J),J=IT1,IT2)
       GO TO 47
49     IT2=IT1+9
       WRITE(6,515)
       DO 51 I=1,NPORT
51     WRITE(6,516) (F3(I,J),J=IT1,IT2)
       IT1=IT2+1
       GO TO 48
47     IF(TP.EQ.0) GO TO 52
C  WRITE F6 FOR DEBUG RUN
       WRITE(6,517)
       IT1=1
53     IT2=TN
       IF((IT2-IT1).GT.10) GO TO 54
       IF(IT2.EQ.IT1) GO TO 52
```

```
      WRITE(6,515)
      DO 55 I=1,NPORT
55    WRITE(6,516) (F6(I,J),J=IT1,IT2)
      GO TO 52
54    IT2=IT1+9
      WRITE(6,515)
      DO 56 I=1,NPORT
56    WRITE(6,516) (F6(I,J),J=IT1,IT2)
      IT1=IT2+1
      GO TO 53
52    WRITE(6,518)
      CALL PRINT(ANSCOL,ANSROW)
C
C ZERO OUT F6
C
46    IF(TN.EQ.0)GO TO 57
      DO 58 J=1,TN
      KK=TP+J
      DO 58 I=1,NPORT
      IT1=PORT+I
      IF(LN.EQ.0) GO TO 59
C  CHANGE F7
      DO 60 K=1,LN
      LK=NNODE+K
60    ANS(IT1,K)=ANS(IT1,K)-(F6(I,J)*A(KK,LK))
59    IF(LP.EQ.0) GO TO 58
C  CHANGE F8
      DO 61 K=1,LP
      LK=ICOUNT(2)+K
      IT2=NBR+TP+K
61    ANS(IT1,IT2)=ANS(IT1,IT2)-(F6(I,J)*A(KK,LK))
58    CONTINUE
C
C ZERO OUT F3
C
57    IF(LN.EQ.0)GO TO 62
      DO 63 J=1,LN
      LK=NNODE+J
      DO 63 I=1,NPORT
      IT1=PORT+I
      IF(TN.EQ.0) GO TO 64
C  CHANGE F2
      DO 65 K=1,TN
      KK=TP+K
      IT2=LN+K
65    ANS(IT1,IT2)=ANS(IT1,IT2)-(F3(I,J)*(-A(KK,LK)))
64    IF(TP.EQ.0) GO TO 63
C  CHANGE F1
      DO 66 K=1,TP
      IT2=NBR+K
66    ANS(IT1,IT2)=ANS(IT1,IT2)-(F3(I,J)*(-A(K,LK)))
63    CONTINUE
C
C  FILL ANS MATRIX
C
62    IF(DEBUG.EQ.0) GO TO 67
      WRITE(6,519)
      CALL PRINT(ANSCOL,ANSROW)
67    IF(LN.EQ.0.OR.TP.EQ.0)GO TO 68
C  STORE D1
      DO 69 I=1,TP
      DO 69 J=1,LN
      K=NNODE+J
69    ANS(I,J)=A(I,K)
68    LC=LN+1
      ITEMP=TP+1
```

```
      IF(ITEMP.GT.PORT.OR.LC.GT.NPORT)GO TO 70
C   STORE -D4 TRANSPOSE
      DO 71 I=ITEMP,PORT
      JJ=LC+I-ITEMP+NNODE
      DO 71 J=LC,NPORT
      II=J+1-LC+TP
71    ANS(I,J)=-A(II,JJ)
70    IF(TP.EQ.0) GO TO 72
C   STORE UNIT MATRIX ABOVE F5
      DO 73 I=1,TP
      LD=NPORT+I
73    ANS(I,LD)=1.
72    IF(LP.EQ.0) GO TO 74
C   STORE UNIT MATRIX ABOVE F4
      II=TP+1
      DO 75 I=II,PORT
      LD=NPORT+I
75    ANS(I,LD)=1.
74    ITEMP=TP+1
      LF=LD+TP
      LE=LD+1
      IF(ITEMP.GT.PORT.OR.LE.GT.LF)GO TO 76
C   STORE -D2 TRANSPOSE
      DO 77 I=ITEMP,PORT
      JJ=I-ITEMP+ICOUNT(2)+1
      DO 77 J=LE,LF
      II=J+1-LE
77    ANS(I,J)=-A(II,JJ)
76    LE=LF+LP
      LD=LF+1
      IF(TP.EQ.0.OR.LD.GT.LE)GO TO 78
C   STORE D2
      DO 79 I=1,TP
      DO 79 J=LD,LE
      K=ICOUNT(2)+1+J-LD
79    ANS(I,J)=A(I,K)
78    IF(DEBUG.EQ.0) GO TO 80
      WRITE(6,520)
      CALL PRINT(ANSCOL,ANSROW)
C
C   REDUCE ANS MATRIX TO ECHELON FORM
C
80    CALL RAECH(NBR,ANSCOL,ANSCOL,1,1,ZERO)
      IF(DEBUG.EQ.0) GO TO 81
      WRITE(6,521)
      CALL PRINT(ANSCOL,ANSROW)
81    DO 82 I=1,NBR
      DO 82 J=1,NPORT
      II=NBR+1-I
      IF(ABS(ANS(II,J)).LE.ZERO) ANS(II,J)=0.0
      IF (ANS(II,J).NE.0.)GO TO 83
82    CONTINUE
83    II=II+1
C
C   FILL COLUMN HEADING VECTOR FOR FINAL PRINT OUT
C
      J=0
      IF(TP.EQ.0) GO TO 84
      DO 85 I=1,TP
      IT=2*I
      HEADER(IT)=BR(I)
      HEADER(IT-1)=C
      I2=2*(PORT+I)
      HEADER(I2)=BR(I)
85    HEADER(I2-1)=V
84    IF(LP.EQ.0) GO TO 86
      J=TP
```

```
      DO 87 I=1,LP
      J=J+1
      K=I+ICOUNT(2)
      IT=2*J
      HEADER(IT)=BR(K)
      HEADER(IT-1)=V
      I2=2*(PORT+TP+I)
      HEADER(I2)=BR(K)
87    HEADER(I2-1)=C
86    IT=4*PORT
      NPORT1=NPORT+1
      DO 88 I=II,NBR
      DO 88 J=NPORT1,ANSCOL
88    IF(ABS(ANS(I,J)).LE.ZERO) ANS(I,J)=0.0
      IF(DEBUG.EQ.0) GO TO 89
C   PRINT FINAL ANS MATRIX FOR DEBUG RUN
      CALL PRNT1(IT,NPORT1,ANSCOL,II,NBR)
89    IF (II.EQ.NBR) GO TO 90
C
C   BACK SUBSTITUTE FINAL ANSWER MATRIX
C
      IT1=ANSROW-II+1
      IT2=II+1
      DO 91 I=IT2,ANSROW
C   ANS(IRW,ICL) IS PIVOT ELEMENT USING TO ZERO ELEMENTS ABOVE
      IRW=ANSROW+IT2-I
      ICL=NPORT+IT1+IT2-I
      IT3=IRW-1
C   J=ROW ZEROING OUT ABOVE PIVOT
      DO 91 J=II,IT3
      B=ANS(J,ICL)
C   K=COLUMN CHANGING OF JTH ROW
      DO 91 K=ICL,ANSCOL
91    ANS(J,K)=ANS(J,K)-B*ANS(IRW,K)
90    DO 92 I=II,NBR
      DO 92 J=NPORT1,ANSCOL
92    IF(ABS(ANS(I,J)).LE.ZERO) ANS(I,J)=0.0
C
C   PRINT FINAL ANS MATRIX
C
      CALL PRNT1(IT,NPORT1,ANSCOL,II,NBR)
      GO TO 1
500   FORMAT (80A1)
501   FORMAT(1H1/1X,80A1//////1X,'NETWORK DESCRIPTION',///1X,'BRANCH
     * FROM    TO      ELEMENT   ELEMENT   CONTROL'/' NUMBER     NODE  NODE
     *     TYPE      VALUE    BRANCH')
502   FORMAT(3I3,A2,E10.3,I3)
503   FORMAT(2X,I3,6X,I3,3X,I3,7X,A2,4X,E10.3,3X,I3)
504   FORMAT(1H0//,' ZERO = ',E10.3)
505   FORMAT(///' A MATRIX')
506   FORMAT(1H0,40I3)
507   FORMAT(1H0///)
508   FORMAT(1H0'TREE PORT BRANCHES'/50(1X,I2))
509   FORMAT(1H0,'TREE NON-PORT BRANCHES'/50(1X,I2))
510   FORMAT(1H0,'LINK NON-PORT BRANCHES'/50(1X,I2))
511   FORMAT(1H0,'LINK PORT BRANCHES'/50(1X,I2))
512   FORMAT(1H0,'TP = ',I3/' TN = ',I3/' LN = ',I3/' LP = ',I3)
513   FORMAT(1H0,'BR',40(1X,I2))
514   FORMAT(///' F3 BEFORE ZEROING')
515   FORMAT(1X)
516   FORMAT(1X,10(E11.4,1X))
517   FORMAT(///' F6 BEFORE ZEROING')
518   FORMAT(///' ANS MATRIX BEFORE ZEROING')
519   FORMAT(///' ANS MATRIX AFTER ZEROING')
520   FORMAT(///' ANS MATRIX WITH D VALUES FILLED IN')
521   FORMAT(///' ANS MATRIX REDUCED TO ECHELON FORM')
      END
```

```
      SUBROUTINE FTREE(NROW,NCOL,INDCOL)
C
C SUBROUTINE FTREE TAKES THE MATRIX A, APPLIES SUBROUTINE IAECH AND FINDS
C THE INDEPENDENT COLUMNS IN A CLOSEST TO THE LEFT. THESE INDEPENDENT
C COLUMNS MAKE UP THE TREE BRANCHES.
C
      INTEGER A,INDCOL(NROW),COL,TEMP
      COMMON A(40,80)
      L=1
      TEMP=1
      CALL IAECH(NROW,NCOL)
C STEP THROUGH ROWS
      DO 1 K=1,NROW
C STEP THROUGH COLUMNS
      DO 2 J=TEMP,NCOL
C FIND INDEPENDENT COLUMNS
C TEST IF ELEMENT EQUAL TO ONE
      IF (A(K,J).NE.1) GO TO 2
C RECORD INDEPENDENT COLUMN NUMBER
      INDCOL(L)=J
      L=L+1
      TEMP=J+1
      GO TO 1
2     CONTINUE
1     CONTINUE
      RETURN
      END
```

```
      SUBROUTINE IAECH(NROW,NCOL)
C
C SUBROUTINE IAECH MANIPULATES MATRIX A INTO ECHELON FORM
C
      INTEGER A,C,G,GPLUS1,P,B
      COMMON A(40,80)
      C=1
      G=1
2     DO 1 I=G,NROW
      IF(A(I,C).EQ.0)GO TO 1
C INTERCHANGE I AND G ROW TO GET NONZERO PIVOT
      IF(I.EQ.G) GO TO 3
      DO 4 K=C,NCOL
      B=A(I,K)
      A(I,K)=A(G,K)
      A(G,K)=B
4     CONTINUE
C NORMALIZE ROW TO GET POSITIVE NUMBER FOR PIVOT
3     IF(A(G,C).GT.0) GO TO 5
      DO 6 K=C,NCOL
6     A(G,K)=-A(G,K)
5     IF(G.GE.NROW) RETURN
C ZERO COLUMN BELOW PIVOT
      GPLUS1=G+1
      DO 7 P=GPLUS1,NROW
      B=A(P,C)
      IF(B.EQ.0)GO TO 7
      DO 8 K=C,NCOL
8     A(P,K)=-B*A(G,K)+A(P,K)
7     CONTINUE
```

```
      G=G+1
      C=C+1
      GO TO 2
1     CONTINUE
      IF(G.GT.NROW) RETURN
      C=C+1
      GO TO 2
      END
```

```
      SUBROUTINE RAECH(M,N,MARK,ROW1,COL1,ZERO)
      COMMON IA(40,80),A(80,160)
      INTEGER C,G,GPLUS1,P,ROW1,COL1
C
C  RAECH PERFORMS ROW OPERATIONS ON A TO REDUCE A TO ECHELON FORM
C
C  COLUMNS COL1 TO MARK ARE REDUCED TO ROW ECHELON FORM WHILE THE ROW
C  OPERATIONS ARE CARRIED OUT ON THE ROWS FROM MARK + 1 TO N
C  ROWS ROW1 TO M ARE REDUCED TO ROW ECHELON FORM
C  G=ROW DETERMINING PIVOT POINT IN
C  C=COLUMN DETERMINING PIVOT POINT IN
      C=COL1-1
      G=ROW1
1     IF(C.EQ.MARK) RETURN
      C=C+1
C  FIND THE MAX NONZERO ELEMENT IN THE C COLUMN BELOW AND INCLUDING PIVOT
      I=0
      TMZ=0.0
      DO 2 J=G,M
      IF(ABS(A(J,C)).LE.ZERO) A(J,C)=0.0
      IF(ABS(A(J,C)).LE.TMZ) GO TO 2
      TMZ=ABS(A(J,C))
      I=J
2     CONTINUE
      IF(TMZ.EQ.0.0) GO TO 1
C  IF THE NONZERO ELEMENT IS IN THE PIVOT ROW, DO NOT EXCHANGE ROWS
      IF(I.EQ.G)GO TO 3
C  EXCHANGE PIVOT ROW WITH ROW HAVING NONZERO ELEMENT IN PIVOT COLUMN
      DO 4 K=C,N
      B=A(I,K)
      A(I,K)=A(G,K)
4     A(G,K)=B
C  CHECK IF PIVOT POINT ALREADY NORMALIZED TO 1
3     IF(A(G,C).EQ.1.)GO TO 5
C  NORMALIZE PIVOT ROW
      ALPHA=A(G,C)
      DO 6 K=C,N
      A(G,K)=A(G,K)/ALPHA
6     IF(ABS(A(G,K)).LE.ZERO) A(G,K)=0.0
C  CHECK IF JUST NORMALIZED PIVOT IN LAST ROW
5     IF(G.GE.M)RETURN
C  ZERO THE ELEMENTS BELOW THE PIVOT
      GPLUS1=G+1
      DO 7 P=GPLUS1,M
      B=A(P,C)
      IF(ABS(A(P,C)).LE.ZERO) A(P,C)=0.0
      IF(ABS(A(P,C)).EQ.0.0) GO TO 7
      DO 8 K=C,N
8     A(P,K)=-B*A(G,K)+A(P,K)
```

```
7     CONTINUE
      IF(G.GE.M) RETURN
      G=G+1
      GO TO 1
      END
```

```
      SUBROUTINE PRINT(ANSCOL,ANSROW)
C
C  SUBROUTINE PRINT PRINTS THE ENTIRE ANS MATRIX
C
C  PRINTS ANSROW ROWS BY ANSCOL COLUMNS
      INTEGER A(40,80),ANSCOL,ANSROW
      COMMON A,ANS(80,160)
      IT1=1
1     IT2=ANSCOL
      IF((IT2-IT1).GT.9 ) GO TO 2
      IF(IT2.EQ.IT1) RETURN
C  LESS THAN 10 COLUMNS LEFT TO PRINT
      WRITE(6.500)
      DO 3 I=1,ANSROW
3     WRITE(6,501) (ANS(I,J),J=IT1,IT2)
      RETURN
2     IT2=IT1+9
C  MORE THAN 10 COLUMNS LEFT TO PRINT
      WRITE(6,500)
      DO 4 I=1,ANSROW
4     WRITE(6,501) (ANS(I,J),J=IT1,IT2)
      IT1=IT2+1
      GO TO 1
500   FORMAT(1X)
501   FORMAT(1X,10(E11.4,1X))
      END
```

```
      SUBROUTINE PRNT1(HDR,ACL1,ACL2,ARW1,ARW2)
C
C  SUBROUTINE PRNT1 PRINTS ONLY THE DESIRED PART OF THE ANS MATRIX
C  DESCRIBING THE PORT EQUATIONS ALONG WITH THE COLUMN HEADINGS
C
      INTEGER A(40,80),HEADER(320),ACL1,ACL2,ARW1,ARW2,HDR
      COMMON A,ANS(80,160),HEADER
      ITM2=ACL1-1
      IT1=1
1     IT2=HDR
      IF((IT2-IT1).GT.19) GO TO 2
      IF(IT2.EQ.IT1) RETURN
C  LESS OR EQUAL 10 COLUMNS TO PRINT
      WRITE(6,500) (HEADER(I),I=IT1,IT2)
      ITM1=ITM2+1
      DO 3 I=ARW1,ARW2
3     WRITE(6,501) (ANS(I,J),J=ITM1,ACL2)
      RETURN
2     IT2=IT1+19
C  MORE THAN 10 COLUMNS TO PRINT
      WRITE(6,500)(HEADER(I),I=IT1,IT2)
      ITM1=ITM2+1
      ITM2=ITM1+9
```

```
      DO 4 I=ARW1,ARW2
4     WRITE(6,501) (ANS(I,J),J=ITM1,ITM2)
      IT1=IT2+1
      GO TO 1
500   FORMAT(1H0,10(4X,A1,I2,5X))
501   FORMAT(1H0,10(E11.4,1X))
      END
```

REFERENCES

1. NAGEL, L. W., and D. O. PEDERSON. *SPICE (Simulation Program with Integrated Circuit Emphasis)*. Berkeley, Calif.: University of California, Electronics Research Laboratory. Memorandum ERL-M382, Apr. 12, 1973.
2. CHUA, L. O., and D. A. PERREAULT. *Analysis of Nonlinear RLC Networks with Controlled Sources*. Lafayette, Ind.: Purdue University, Tech. Rept. TR-EE-70-29, June 1970.
3. KUH, E. S., and R. A. ROHRER. *Theory of Linear Active Networks*. San Francisco: Holden-Day, Inc., 1967.
4. MEDLOCK, P. A. *Computer-Aided Analysis of Resistive Piecewise Linear Networks*. Lafayette, Ind.: Purdue University, Ph.D. dissertation, June 1971.
5. CARLIN, H. J., and D. C. YOULA. "Network Synthesis with Negative Resistors." *Proc. IRE*, Vol. 49, pp. 907–920, May 1961.
6. SO, H. C. "On the Hybrid Description of a Linear *n*-Port Resulting from Extraction of Arbitrarily Specified Elements." *IEEE Trans. Circuit Theory*, Vol. CT-12, pp. 381–387, Sept. 1965.
7. LIN, P. M. "Formulation of Hybrid Matrices for Linear Multiports Containing Controlled Sources." *IEEE Trans. Circuit Theory*, Vol. CT-21, pp. 169–175, Mar. 1974.

PROBLEMS

6-1. Consider the simple two-port shown in Fig. P6-1.

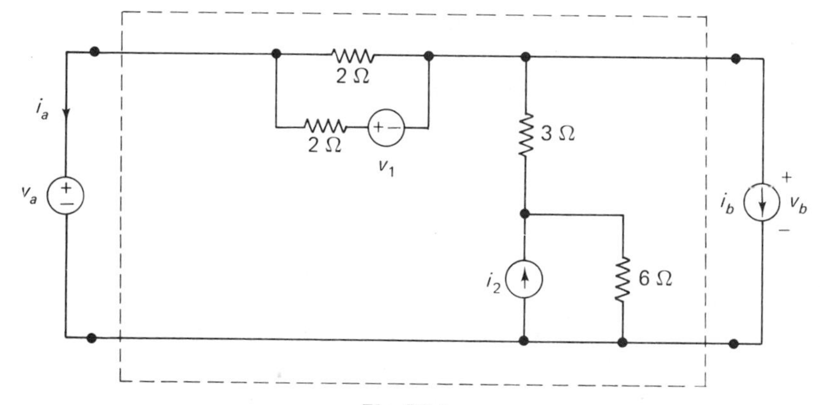

Fig. P6-1

(a) Find the matrices $\hat{H}$ and $\hat{s}$ in Eq. (6-2), using the method of element-by-element calculation, as expressed by Eqs. (6-3)–(6-5).
(b) If the independent source vector is defined as $\hat{u} \triangleq [v_1 \ i_2]^t$, find the matrix $\hat{M}$ in Eq. (6-2).
(c) Repeat part (b) if $\hat{u} \triangleq [i_2 \ v_1]^t$.

6-2. A real $n \times n$ matrix A is said to be positive definite if

$$x^t A x > 0$$

for every real nonzero $n \times 1$ vector x. Prove that the matrices Z and Y, given in Eqs. (6-19a) and (6-19b), respectively, are positive definite provided that Y_R, Y_G, Z_R, and Z_G are positive definite. *Hint:* First write Y in the form of Eq. (6-21).

6-3. The explicit form of the hybrid matrix H has been derived for an n-port consisting of linear resistors only, and the result is shown in Eq. (6-20). From Eq. (6-20), prove that for a resistance n-port

$$H_{EE} = H_{EE}^t$$

$$H_{JJ} = H_{JJ}^t$$

$$H_{EJ} = -H_{JE}^t$$

Hint: For the last part, the crux is to show that

$$Y_G D_{RG}^t Y^{-1} = Z^{-1} D_{RG}^t Z_R$$

or equivalently, $D_{RG} Y_G Z = Y Z_R D_{RG}$. Make use of Eq. (6-19) to establish this last relationship.

6-4. The identities established in this problem will be useful in deriving alternative expressions for the hybrid matrices. Let P be a matrix of dimension $m \times n$, and Q a matrix of dimension $n \times m$.
(a) Prove that $\det(\mathbf{1}_m + PQ) = \det(\mathbf{1}_n + QP)$. *Hint:* Show that both sides are equal to

$$\det\begin{bmatrix} \mathbf{1}_m & P \\ -Q & \mathbf{1}_n \end{bmatrix}$$

(b) Prove that $(\mathbf{1}_m + PQ)^{-1}P = P(\mathbf{1}_n + QP)^{-1}$ if $\det(\mathbf{1}_m + PQ) \neq 0$. *Hint:* First, premultiply both sides by $(\mathbf{1}_m + PQ)$, and then postmultiply both sides by $(\mathbf{1}_n + QP)$.
(c) Prove that if $\det(\mathbf{1}_m + PQ) \neq 0$, then

$$(\mathbf{1}_m + PQ)^{-1} = \mathbf{1}_m - P(\mathbf{1}_n + QP)^{-1}Q$$

Hint: Premultiply both sides by $(\mathbf{1}_m + PQ)$ and then make use of the identity of part (b).
(d) Prove the relationships given by Eqs. (6-23) and (6-24). *Hint:* Make use of the identity of part (c).

6-5. Consider the three-port shown in Fig. P6-5.
(a) Choose a tree to consist of branches 1, 2, and 4. Write the fundamental loop matrix B first and then the fundamental cutset matrix D. Partition D as shown in Eq. (6-6) and identify all submatrices.
(b) Find the matrices Z_R, Y_R, Z_G, and Y_G in Eq. (6-8). Then calculate the matrices Z and Y given by Eq. (6-19).
(c) Find the hybrid matrix H by the use of Eq. (6-20).

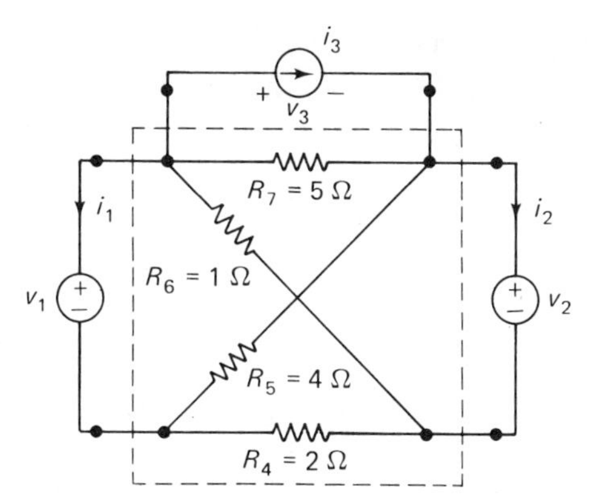

Fig. P6-5

6-6. Equation (6-20) is also applicable to n-ports consisting of linear resistors with negative as well as positive resistance values. However, when negative resistance values are present, the existence of $\boldsymbol{Z}^{-1}$ and $\boldsymbol{Y}^{-1}$ in Eq. (6-20) is not guaranteed. Use Eq. (6-20) to find the hybrid matrix for the two-port of Fig. P6-6. From the result, can you find a well-known component having the same hybrid matrix?

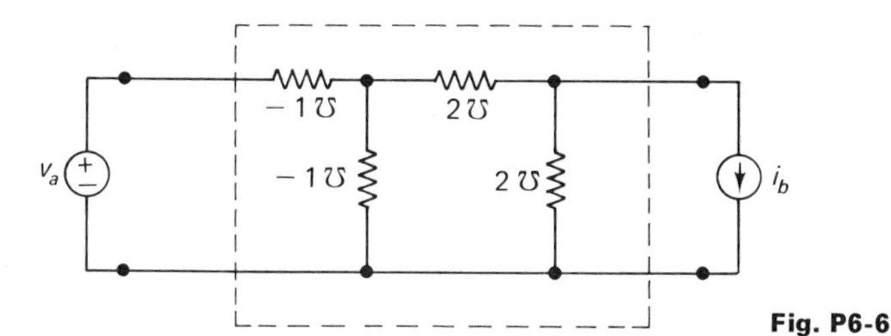

Fig. P6-6

6-7. Use Eq. (6-20) to find the hybrid matrix for the three-port of Fig. P6-7. From the result, can you find a well-known component having the same hybrid matrix?

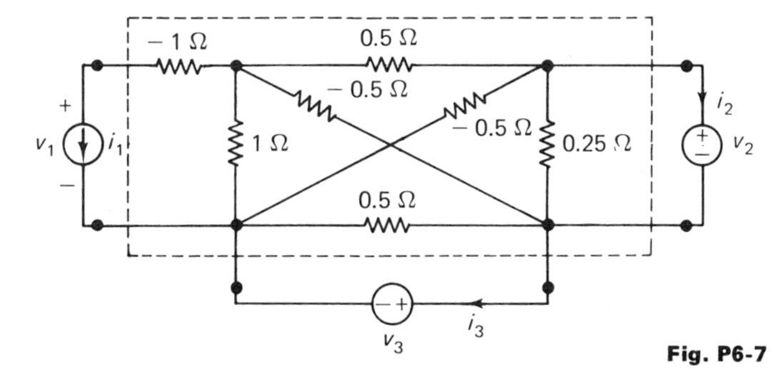

Fig. P6-7

6-8. If all ports of an n-port are voltage ports, Eq. (6-2) becomes

$$\hat{\boldsymbol{i}}_a = \hat{\boldsymbol{H}}_{aa}\hat{\boldsymbol{v}}_a + \hat{\boldsymbol{s}}_a = -\hat{\boldsymbol{Y}}_{sc}\hat{\boldsymbol{v}}_a + \hat{\boldsymbol{s}}_a$$

where $\hat{\boldsymbol{Y}}_{sc}$ is called the short-circuit admittance matrix of the n-port. Thus, $\hat{\boldsymbol{Y}}_{sc}$ is a special case of $\hat{\boldsymbol{H}}$ defined in Eq. (6-2).

(a) Discuss the physical meaning of the elements of $\hat{\boldsymbol{Y}}_{sc}$.

(b) From Eq. (6-20), derive an explicit expression for $\boldsymbol{Y}_{sc}$ for n-ports consisting of linear resistors only.

6-9. If all ports of an n-port are current ports, Eq. (6-2) becomes

$$\hat{\boldsymbol{v}}_b = \hat{\boldsymbol{H}}_{bb}\hat{\boldsymbol{i}}_b + \hat{\boldsymbol{s}}_b = -\hat{\boldsymbol{Z}}_{oc}\hat{\boldsymbol{i}}_b + \hat{\boldsymbol{s}}_b$$

where $\hat{\boldsymbol{Z}}_{oc}$ is called the open-circuit impedance matrix of the n-port. Thus, $\boldsymbol{Z}_{oc}$ is a special case of the $\hat{\boldsymbol{H}}$ matrix of Eq. (6-2).

(a) Discuss the physical meaning of the elements of $\hat{\boldsymbol{Z}}_{oc}$.

(b) From Eq. (6-20), derive an explicit expression for $\boldsymbol{Z}_{oc}$ for n-ports consisting of linear resistors only.

6-10. For each two-port shown in Fig. P6-10, find the short-circuit admittance matrix $\boldsymbol{Y}_{sc}$.

(a) By any method that seems simplest to you.

(b) By the use of the explicit expression of Problem 6-8(b).

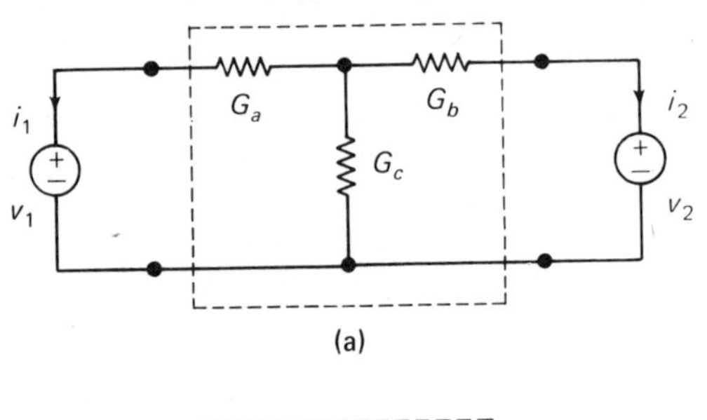

(a)

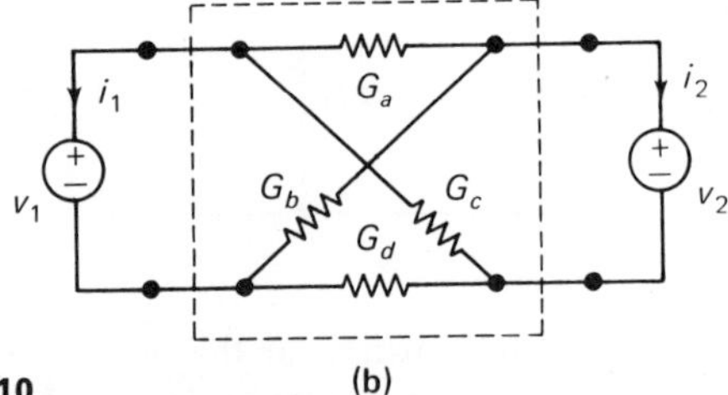

Fig. P6-10 (b)

6-11. For each two-port shown in Fig. P6-11, find the open-circuit impedance matrix $\boldsymbol{Z}_{oc}$.

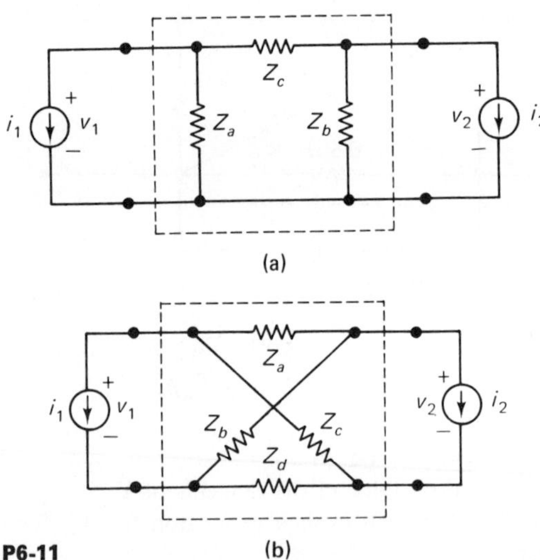

Fig. P6-11 (b)

(a) By any method that seems simplest to you.
(b) By the use of the explicit expression of Problem 6-9(b).

6-12. The two-port shown in Fig. P6-12 is obtained from the three-port of Fig. P6-5 by considering i_3 as an independent current source inside the two-port. Following the method described in Section 6-4, and making use of the result of Problem 6-5, write Eq. (6-2) for the two-port of Fig. P6-12.

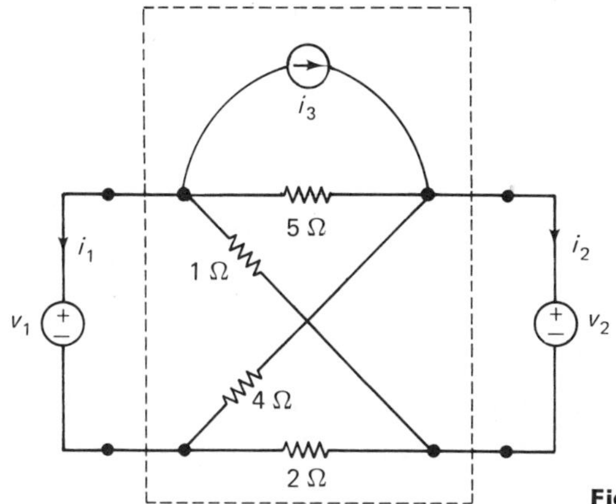

Fig. P6-12

289
6-13. Find a hybrid matrix for the four-port shown in Fig. P6-13. Note that the four-port has neither an open-circuit impedance matrix (why?) nor a short-circuit admittance matrix (why?).

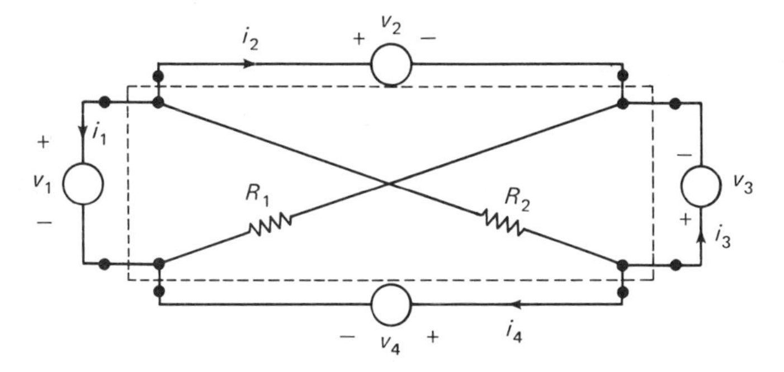

Fig. P6-13

6-14. Consider the two-port network of Fig. 6-6(b), with $k = 2$, and both ports considered current ports.
(a) Extract the ports as described in Section 6-5-1 and find the hybrid equation for the resultant n-port ($n = 6$ in the present case).
(b) Express the result of part (a) in the form of Eq. (6-43) and identify all submatrices $\boldsymbol{H}_{ij}$. Find the matrix $\boldsymbol{K}$ for the controlled sources.
(c) Find the hybrid matrix for the original two-port by the use of Eq. (6-49) and the results of part (b).

6-15. Repeat Problem 6-14 for the case when both ports are considered voltage ports. Does

Eq. (6-2) exist in this case? In exactly which step does the numerical calculation run into difficulty?

6-16. Consider the two-port of Fig. P6-16. Find the hybrid matrix $\hat{\boldsymbol{H}}$ by the method of controlled source extraction described in Section 6-5-1.

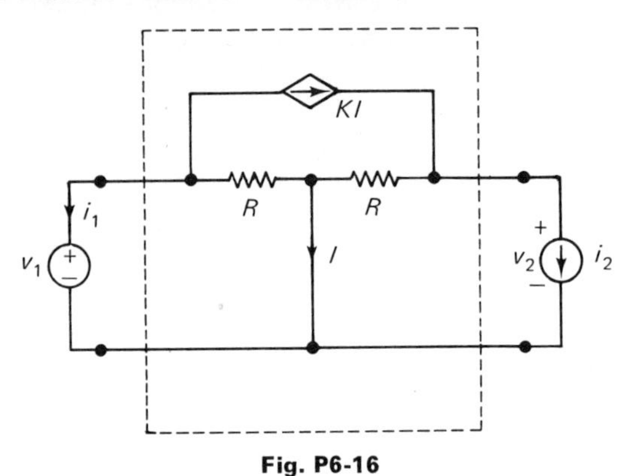

Fig. P6-16

6-17. Repeat Problem 6-16 for the two-port shown in Fig. P6-17.

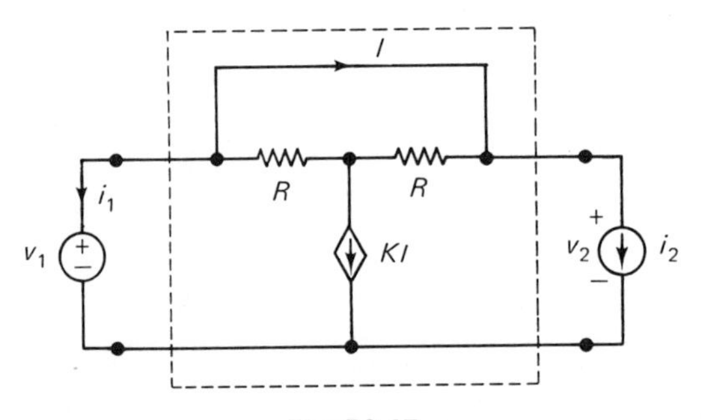

Fig. P6-17

6-18. Find the hybrid matrix $\hat{\boldsymbol{H}}$ for the two-port shown in Fig. P6-16 by the method of systematic elimination described in Section 6-5-2.

6-19. Repeat Problem 6-18 for the two-port shown in Fig. P6-17.

6-20. Find the constraint matrix for the one-port shown in Fig. P6-20 by the method of Section 6-6. Two cases are to be considered: (a) $a = 1, b = 2$; (b) $a = 1, b = 1$.

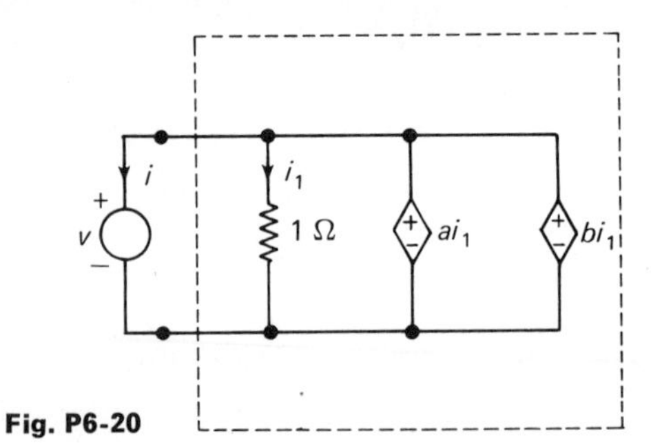

Fig. P6-20

6-21. Find the constraint matrix for the one-port shown in Fig. P6-21 by the method of Section 6-6. Two cases are to be considered: (a) $R_1 = 1$, $R_2 = -1$; (b) $R_1 = -1$, $R_2 = 1$.

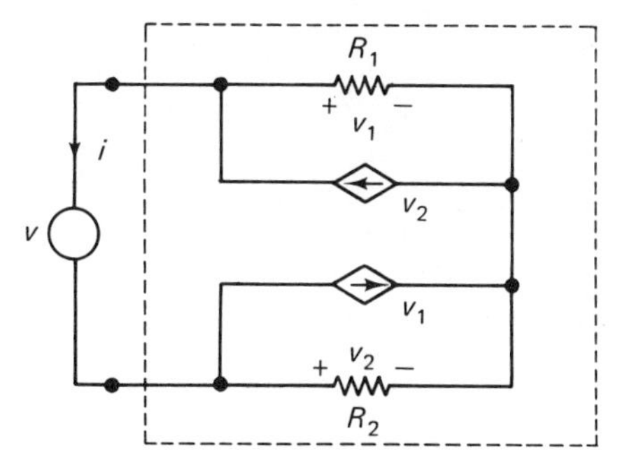

Fig. P6-21

6-22. Find the constraint matrix for the two-port shown in Fig. P6-22 by the method of Section 6-6. From the result obtained, explain why both methods of Section 6-5 will fail if applied to solve the present problem.

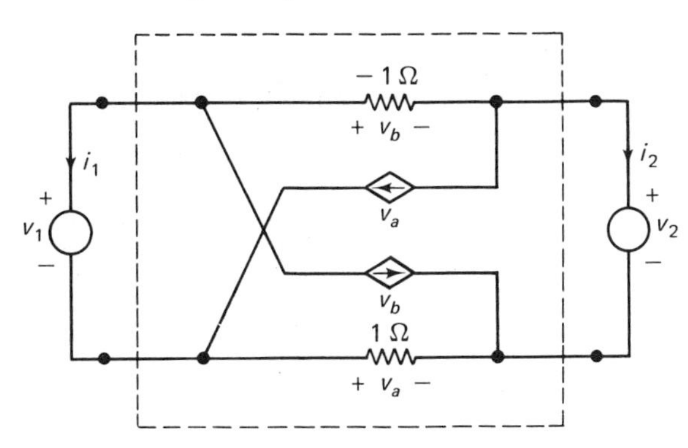

Fig. P6-22

6-23. Consider the transistor feedback amplifier shown in Fig. 6-13. If the transistors have very high forward current gain, the voltage gain of the amplifier is essentially determined by R_1 and R_2, and is given by $E_o/E_i \cong (R_1 + R_2)/R_2$. Use Program HYBRID to find E_o/E_i and Z_i for the amplifier of Fig. 6-13 with the transistor beta changed from 100 to 500. Does the result agree with the predicted value?

359

6-24. Consider the differential amplifier shown in Fig. 6-14(a). Suppose that the two field-effect transistors are slightly mismatched. The parameter values for the model shown in Fig. 6-14(b) are as follows:

$$\text{For } T_1\text{:}\quad g_{m1} = 0.01\ \mho, \qquad g_{o1} = 20\ \mu\mho$$
$$\text{For } T_2\text{:}\quad g_{m2} = 0.02\ \mho, \qquad g_{o2} = 40\ \mu\mho$$

With the aid of the program HYBRID, determine
(a) The differential mode gain A_d.
(b) The common mode gain A_c.
(c) The output impedance Z_o.
Hint: First obtain the short-circuit admittance matrix of the three-port.

6-25. An "ad hoc method" for finding the hybrid matrix of a linear resistive n-port N is to replace the current source at each current port by a grounded gyrator connected to a voltage source (see Problem 4-20). In place of the original current ports, we now have voltage ports.

(a) Show that, with the reference directions for the port voltages and currents properly chosen, the short-circuit admittance matrix of the "gyrator augmented" n-port N_A is equal to the hybrid matrix of the original n-port N.

(b) Making use of the results of part (a) above, Problems 4-19 and 4-20, show how the hybrid matrix of any linear resistive n-port (all four types of controlled sources allowed) can be determined by the *nodal method* of Chapter 4 at the expense of having an increased number of nodes.

CHAPTER 7

Hybrid Nonlinear Network Analysis: Algorithms and Computational Methods

7-1 FORMULATION OF HYBRID EQUATIONS FOR RESISTIVE NONLINEAR NETWORKS

The nodal equation formulation algorithm presented in Chapter 5 does not apply to networks containing *current-controlled* resistors or *controlled* voltage sources. Our objective in this chapter is to formulate a generalized method of analysis—called *hybrid analysis*—that will allow a mixture of both current and voltage-controlled resistors, independent sources, and all four types of *linear* controlled sources. It is clear that such a generalized method of analysis must include both currents and voltages as unknown variables—hence the name *hybrid* analysis. Once the hybrid equations are formulated, they can be solved by a number of iteration techniques, the most general of which is still the *Newton–Raphson algorithm* presented in Chapter 5. A piecewise-linear version of this algorithm will be developed in Section 7-2. Another piecewise-linear algorithm, the *Katzenelson algorithm*, whose convergence is guaranteed under certain conditions will be presented in Section 7-3. All these algorithms are generally capable of finding only *one* solution. For multistate resistive networks, such as flip flops, having *multiple* solutions [7], several *combinatorial algorithms* are presented in Sections 7-4 and 7-5 which guarantee that *all* solutions are found. Such *combinatorial* algorithms are necessarily *less efficient* than *iterative* algorithms; otherwise, there would be no point in studying the less general iterative techniques.

Let N be a nonlinear resistive network containing linear and nonlinear (both voltage- and current-controlled) resistors, constant independent voltage and current sources,[1] and all four types of *linear* controlled sources. We assume that the nonmonotonic *voltage-controlled resistors* do not form *loops*, and the nonmonotonic *current-controlled resistors* do not form *cutsets*. There is little loss of generality in this assumption, since any loop containing only voltage-controlled resistors can be eliminated by replacing one of the resistors in the loop by an equivalent combination of an appropriate *linear* resistor in *series* with another voltage-controlled resistor. Similarly, any cutset containing only current-controlled resistors can be eliminated by replacing one of the resistors in the cutset by an equivalent combination of an appropriate linear resistor in *parallel* with another current-controlled resistor. In view of this assumption, we can create a hybrid m-port $\hat{N}$ as shown in Fig. 7-1 by extracting all voltage-controlled resistors across the m_1 *voltage ports*, and by extracting all current-controlled resistors across the m_2 *current ports*, where $m = m_1 + m_2$. The remaining elements in the m-port $\hat{N}$ consist therefore only of linear resistors, constant independent sources, and linear controlled sources. It follows from Chapter 6 that except for some pathological situations involving exact element

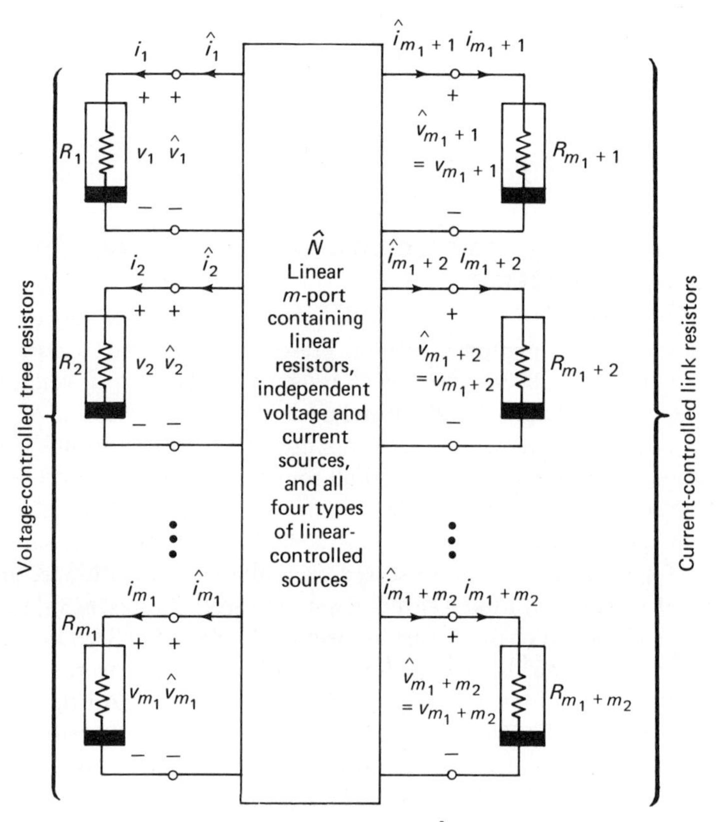

Fig. 7-1. Nonlinear resistors of a network $\hat{N}$ are extracted and shown connected across the $m = m_1 + m_2$ ports of a linear m-port.

[1]Networks containing time-dependent sources are analyzed by repeating the hybrid analysis at different instants of time.

values, or some special topological constraints, the m-port $\hat{N}$ admits a well-defined *hybrid representation* [see Eq. (6-2)]:[2]

$$\begin{bmatrix} \hat{\boldsymbol{i}}_a \\ \hat{\boldsymbol{v}}_b \end{bmatrix} = \begin{bmatrix} \hat{\boldsymbol{H}}_{aa} & \hat{\boldsymbol{H}}_{ab} \\ \hat{\boldsymbol{H}}_{ba} & \hat{\boldsymbol{H}}_{bb} \end{bmatrix} \begin{bmatrix} \hat{\boldsymbol{v}}_a \\ \hat{\boldsymbol{i}}_b \end{bmatrix} + \begin{bmatrix} \hat{\boldsymbol{s}}_a \\ \hat{\boldsymbol{s}}_b \end{bmatrix} \tag{7-1}$$

The *hybrid matrix* $\hat{\boldsymbol{H}}$ and the source vector $\hat{\boldsymbol{s}}$ are defined via *explicit topological formulas* by either Eq. (6-49) or Eq. (6-61). Hence, it is a simple computational task to determine $\hat{\boldsymbol{H}}$ and $\hat{\boldsymbol{s}}$. Moreover, *they need only be determined once.*

To obtain the hybrid equations, we denote the v-i curves of the voltage- and current-controlled resistors, respectively, by

$$\boldsymbol{i}_a \triangleq \begin{bmatrix} i_1 \\ i_2 \\ \cdot \\ \cdot \\ \cdot \\ i_{m_1} \end{bmatrix} = \begin{bmatrix} g_1(v_1) \\ g_2(v_2) \\ \cdot \\ \cdot \\ \cdot \\ g_{m_1}(v_{m_1}) \end{bmatrix} \triangleq \boldsymbol{g}_a(\boldsymbol{v}_a) \tag{7-2}$$

$$\boldsymbol{v}_b \triangleq \begin{bmatrix} v_{m_1+1} \\ v_{m_1+2} \\ \cdot \\ \cdot \\ \cdot \\ v_{m_1+m_2} \end{bmatrix} = \begin{bmatrix} f_{m_1+1}(i_{m_1+1}) \\ f_{m_1+2}(i_{m_1+2}) \\ \cdot \\ \cdot \\ \cdot \\ f_{m_1+m_2}(i_{m_1+m_2}) \end{bmatrix} \triangleq \boldsymbol{f}_b(\boldsymbol{i}_b) \tag{7-3}$$

Substituting $\hat{\boldsymbol{i}}_a = \boldsymbol{i}_a \triangleq \boldsymbol{g}_a(\boldsymbol{v}_a)$, $\hat{\boldsymbol{v}}_b = \boldsymbol{v}_b \triangleq \boldsymbol{f}_b(\boldsymbol{i}_b)$, $\hat{\boldsymbol{v}}_a = \boldsymbol{v}_a$, and $\hat{\boldsymbol{i}}_b = \boldsymbol{i}_b$, as shown in Fig. 7-1, into Eq. (7-1), we obtain

$$\boldsymbol{g}_a(\boldsymbol{v}_a) - \hat{\boldsymbol{H}}_{aa}\boldsymbol{v}_a - \hat{\boldsymbol{H}}_{ab}\boldsymbol{i}_b - \hat{\boldsymbol{s}}_a = \boldsymbol{0} \tag{7-4}$$

$$\boldsymbol{f}_b(\boldsymbol{i}_b) - \hat{\boldsymbol{H}}_{ba}\boldsymbol{v}_a - \hat{\boldsymbol{H}}_{bb}\boldsymbol{i}_b - \hat{\boldsymbol{s}}_b = \boldsymbol{0} \tag{7-5}$$

Equations (7-4) and (7-5) constitute $m = m_1 + m_2$ independent equations in m_1 unknown voltages, $v_1, v_2, \ldots, v_{m_1}$, and m_2 unknown currents, $i_{m_1+1}, i_{m_1+2}, \ldots, i_{m_1+m_2}$, and are called the *hybrid equations* of the nonlinear resistive network N. Once these unknowns are solved, the corresponding solution for the elements inside $\hat{N}$ can be easily determined by the topological formulas developed in Chapter 6. Observe that Eq. (7-5) applies only to the case where the nonlinear resistors form neither loops nor cut sets. A more general form of hybrid equations can be formulated by partitioning an appropriate tree and cotree into two parts in accordance with the algorithm described in *Problem 7-9*. The resulting hybrid equation given in *Problem 7-11* can be easily shown to include Eq. (7-5) as a special case. This generalized *hybrid analysis* approach is of fundamental importance in the analysis of *large-scale* linear and nonlinear networks [11].

[2]Resistors characterized by *strictly monotonically increasing* v-i curves are both voltage- and current-controlled and can be extracted either across the voltage ports or across the current ports.

To solve Eqs. (7-4) and (7-5) by the *Newton–Raphson method* presented in Chapter 5, we apply Eq. (5-51) to obtain

$$\begin{bmatrix} \boldsymbol{v}_a^{(j+1)} \\ \boldsymbol{i}_b^{(j+1)} \end{bmatrix} = \begin{bmatrix} \boldsymbol{v}_a^{(j)} \\ \boldsymbol{i}_b^{(j)} \end{bmatrix} - \begin{bmatrix} \left[\dfrac{\partial \boldsymbol{g}_a(\boldsymbol{v}_a^{(j)})}{\partial \boldsymbol{v}_a} - \hat{\boldsymbol{H}}_{aa}\right] & -\hat{\boldsymbol{H}}_{ab} \\ -\hat{\boldsymbol{H}}_{ba} & \left[\dfrac{\partial \boldsymbol{f}_b(\boldsymbol{i}_b^{(j)})}{\partial \boldsymbol{i}_b} - \hat{\boldsymbol{H}}_{bb}\right] \end{bmatrix}^{-1} \times \begin{bmatrix} \boldsymbol{g}_a(\boldsymbol{v}_a^{(j)}) - \hat{\boldsymbol{H}}_{aa}\boldsymbol{v}_a^{(j)} - \hat{\boldsymbol{H}}_{ab}\boldsymbol{i}_b^{(j)} - \hat{\boldsymbol{s}}_a \\ \boldsymbol{f}_b(\boldsymbol{i}_b^{(j)}) - \hat{\boldsymbol{H}}_{ba}\boldsymbol{v}_a^{(j)} - \hat{\boldsymbol{H}}_{bb}\boldsymbol{i}_b^{(j)} - \hat{\boldsymbol{s}}_b \end{bmatrix} \tag{7-6}$$

where $\boldsymbol{v}_a^{(j+1)}$ and $\boldsymbol{i}_b^{(j+1)}$ denote the values of $\boldsymbol{v}_a$ and $\boldsymbol{i}_b$ at the $(j+1)$th iteration. Recall that the iteration given in Eq. (7-6) *converges quadratically* provided the *initial guess* $\boldsymbol{v}_a^{(0)}, \boldsymbol{i}_b^{(0)}$ is "close" to the correct solution. One serious disadvantage of the Newton–Raphson algorithm is the lack of a systematic procedure for choosing a suitable initial guess. Fortunately, in many electronic circuits, one usually has a priori knowledge on the typical range of solutions; consequently, a successful initial guess can usually be made. Another serious disadvantage of the Newton–Raphson algorithm is the need for evaluating a *Jacobian matrix* at each iteration. Both disadvantages could be overcome to some extent by approximating the *v-i* curve of each nonlinear resistor by a *piecewise-linear curve*. The remaining sections of this chapter are therefore addressed exclusively to piecewise-linear networks [1].

7-2 PIECEWISE-LINEAR VERSION OF THE NEWTON–RAPHSON ALGORITHM [2]

The application of the Newton–Raphson algorithm directly to solve Eqs. (7-4) and (7-5) for $\boldsymbol{v}_a$ and $\boldsymbol{i}_b$ would require the evaluation of a Jacobian matrix *at each iteration*. This is usually a time-consuming and often inaccurate procedure, because finding the derivatives of a function *numerically* is a "noisy" process; i.e., numerical differentiation is a relatively inaccurate operation. It is possible to avoid evaluating the Jacobian if we approximate the v_j-i_j curve of each nonlinear resistor by piecewise-linear segments. A typical piecewise-linear, voltage-controlled v_{a_j}-i_{a_j} curve is shown in Fig. 7-2(a), and a typical piecewise-linear, current-controlled v_{b_j}-i_{b_j} curve is shown in Fig. 7-2(b). Let us label the segments of each curve consecutively from 1 (the leftmost segment) to n_j (the rightmost segment), where n_j is the total number of segments of the jth curve. If we let σ_j be *any* one of the n_j segments of the jth curve, then segment σ_j is completely specified by

$$i_{a_j} = G_{a_j}(\sigma_j)v_{a_j} + I_{a_j}(\sigma_j) \tag{7-7}$$

$$E_{a_j}^-(\sigma_j) \leq v_{a_j} \leq E_{a_j}^+(\sigma_j) \tag{7-8}$$

for a *voltage-controlled* curve, and

$$v_{b_j} = R_{b_j}(\sigma_j)i_{b_j} + E_{b_j}(\sigma_j) \tag{7-9}$$

$$I_{b_j}^-(\sigma_j) \leq i_{b_j} \leq I_{b_j}^+(\sigma_j) \tag{7-10}$$

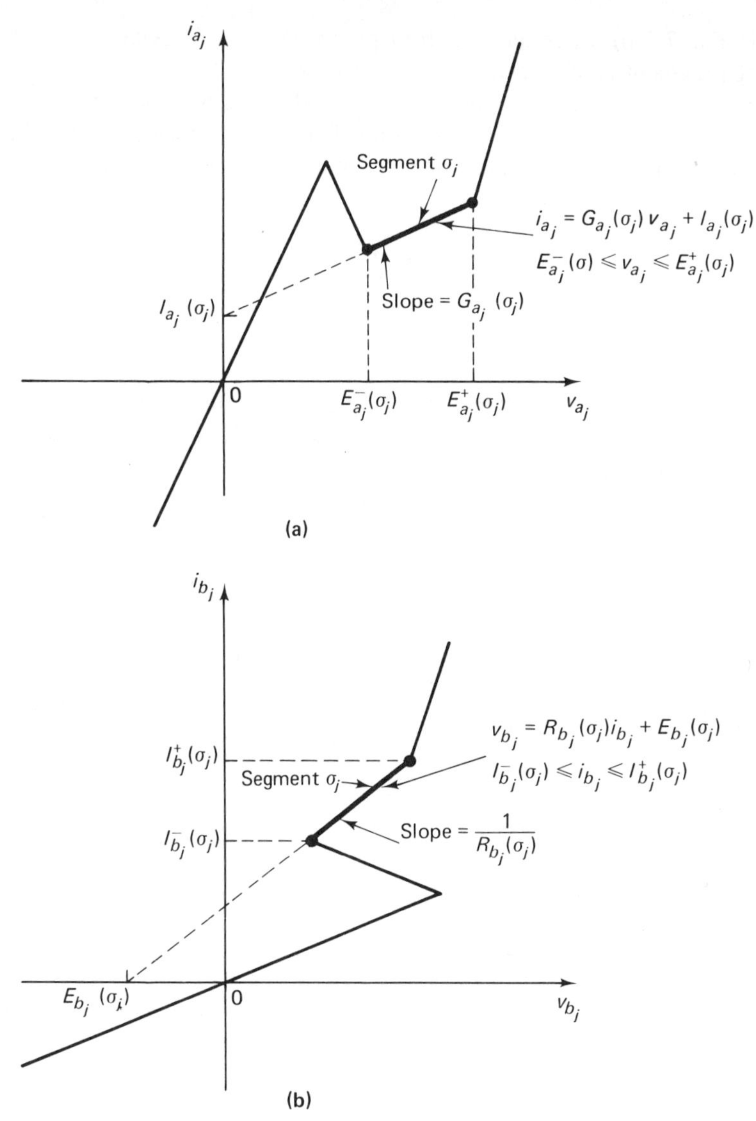

Fig. 7-2. Pertinent parameters defining a piecewise-linear, voltage-controlled v_{a_j}-i_{a_j} curve are shown in (a) and those defining a current-controlled v_{b_j}-i_{b_j} curve are shown in (b).

for a *current-controlled* curve. Geometrically, the parameters $G_{a_j}(\sigma_j)$ and $I_{a_j}(\sigma_j)$ represent the slope (conductance) and the current intercept of segment σ_j, as shown in Fig. 7-2(a). Similarly, the parameters $R_{b_j}(\sigma_j)$ and $E_{b_j}(\sigma_j)$ represent the *reciprocal* slope (resistance) and the voltage intercept of segment σ_j, as shown in Fig. 7-2(b). The interval $[E^-_{a_j}(\sigma_j), E^+_{a_j}(\sigma_j)]$ denotes the *domain of definition* of segment σ_j in Fig. 7-2(a), whereas the interval $[I^-_{b_j}(\sigma_j), I^+_{b_j}(\sigma_j)]$ denotes the domain of definition of segment σ_j in Fig. 7-2(b). In either case, four parameters are necessary and sufficient to specify each segment completely.

Consider now the network shown in Fig. 7-1 containing m_1 voltage-controlled resistors and m_2 current-controlled resistors. If each v_j-i_j curve is approximated by n_j piecewise-linear segments, there will be a total of n_T *distinct segment combinations*, where

$$n_T = \prod_{j=1}^{m_1+m_2} n_j \tag{7-11}$$

Let

$$\sigma = \{\sigma_1, \sigma_2, \ldots, \sigma_{m_1}, \sigma_{m_1+1}, \sigma_{m_1+2}, \ldots, \sigma_{m_1+m_2}\} \tag{7-12}$$

denote *any one* of these n_T segment combinations; i.e., σ is that combination consisting of segments σ_1 for R_1, σ_2 for $R_2, \ldots, \sigma_{m_1}$ for R_{m_1}, σ_{m_1+1} for R_{m_1+1}, σ_{m_1+2} for $R_{m_1+2}, \ldots$, and $\sigma_{m_1+m_2}$ for $R_{m_1+m_2}$.

Corresponding to this segment combination σ, Eqs. (7-7) and (7-8) can be written in the following form for the m_1 voltage-controlled resistors

$$\begin{bmatrix} i_{a_1} \\ i_{a_2} \\ \cdot \\ \cdot \\ \cdot \\ i_{a_{m_1}} \end{bmatrix} = \begin{bmatrix} G_{a_1}(\sigma_1) & 0 & 0 & \cdots & 0 \\ 0 & G_{a_2}(\sigma_2) & 0 & \cdots & 0 \\ \cdot & & \cdot & \cdots & \cdot \\ \cdot & & \cdot & \cdots & \cdot \\ \cdot & & \cdot & \cdots & \cdot \\ 0 & 0 & 0 & \cdots & G_{a_{m_1}}(\sigma_{m_1}) \end{bmatrix} \begin{bmatrix} v_{a_1} \\ v_{a_2} \\ \cdot \\ \cdot \\ \cdot \\ v_{a_{m_1}} \end{bmatrix} + \begin{bmatrix} I_{a_1}(\sigma_1) \\ I_{a_2}(\sigma_2) \\ \cdot \\ \cdot \\ \cdot \\ I_{a_{m_1}}(\sigma_{m_1}) \end{bmatrix} \tag{7-13}$$

$$\begin{bmatrix} E^-_{a_1}(\sigma_1) \\ E^-_{a_2}(\sigma_2) \\ \cdot \\ \cdot \\ \cdot \\ E^-_{a_{m_1}}(\sigma_{m_1}) \end{bmatrix} \leq \begin{bmatrix} v_{a_1} \\ v_{a_2} \\ \cdot \\ \cdot \\ \cdot \\ v_{a_{m_1}} \end{bmatrix} \leq \begin{bmatrix} E^+_{a_1}(\sigma_1) \\ E^+_{a_2}(\sigma_2) \\ \cdot \\ \cdot \\ \cdot \\ E^+_{a_{m_1}}(\sigma_{m_1}) \end{bmatrix} \tag{7-14}$$

where Eq. (7-14) means that $E^-_{a_j}(\sigma_j) \leq v_{a_j} \leq E^+_{a_j}(\sigma_j)$ for $j = 1, 2, \ldots, m_1$. Similarly, Eqs. (7-9) and (7-10) can be written as follows:

$$\begin{bmatrix} v_{b_{m_1+1}} \\ v_{b_{m_1+2}} \\ \cdot \\ \cdot \\ \cdot \\ v_{b_{m_1+m_2}} \end{bmatrix} = \begin{bmatrix} R_{b_{m_1+1}}(\sigma_{m_1+1}) & 0 & \cdots & 0 \\ 0 & R_{b_{m_1+2}}(\sigma_{m_1+2}) & \cdots & 0 \\ \cdot & \cdot & \cdots & \cdot \\ \cdot & \cdot & \cdots & \cdot \\ \cdot & \cdot & \cdots & \cdot \\ 0 & 0 & \cdots & R_{b_{m_1+m_2}}(\sigma_{m_1+m_2}) \end{bmatrix} \times \begin{bmatrix} i_{b_{m_1+1}} \\ i_{b_{m_1+2}} \\ \cdot \\ \cdot \\ \cdot \\ i_{b_{m_1+m_2}} \end{bmatrix} + \begin{bmatrix} E_{b_{m_1+1}}(\sigma_{m_1+1}) \\ E_{b_{m_1+2}}(\sigma_{m_1+2}) \\ \cdot \\ \cdot \\ \cdot \\ E_{b_{m_1+m_2}}(\sigma_{m_1+m_2}) \end{bmatrix} \tag{7-15}$$

$$\begin{bmatrix} I^-_{b_{m_1+1}}(\sigma_{m_1+1}) \\ I^-_{b_{m_1+2}}(\sigma_{m_1+2}) \\ \cdot \\ \cdot \\ \cdot \\ I^-_{b_{m_1+m_2}}(\sigma_{m_1+m_2}) \end{bmatrix} \leq \begin{bmatrix} i_{b_{m_1+1}} \\ i_{b_{m_1+2}} \\ \cdot \\ \cdot \\ \cdot \\ i_{b_{m_1+m_2}} \end{bmatrix} \leq \begin{bmatrix} I^+_{b_{m_1+1}}(\sigma_{m_1+1}) \\ I^+_{b_{m_1+2}}(\sigma_{m_1+2}) \\ \cdot \\ \cdot \\ \cdot \\ I^+_{b_{m_1+m_2}}(\sigma_{m_1+m_2}) \end{bmatrix} \tag{7-16}$$

where Eq. (7-16) means that $I^-_{b_{m+j}}(\sigma_{m_1+j}) \leq i_{b_{m_1+j}} \leq I^+_{b_{m_1+j}}(\sigma_{m_1+j})$ for $j = 1, 2, \ldots, m_2$. It will be convenient for us to rewrite Eqs. (7-13)–(7-16) into the following compact forms:

$$\mathbf{i}_a = \mathbf{G}_a(\sigma)\mathbf{v}_a + \mathbf{I}_a(\sigma) \triangleq \mathbf{g}_a(\mathbf{v}_a) \tag{7-17}$$

$$\mathbf{E}^-_a(\sigma) \leq \mathbf{v}_a \leq \mathbf{E}^+_a(\sigma) \tag{7-18}$$

$$\mathbf{v}_b = \mathbf{R}_b(\sigma)\mathbf{i}_b + \mathbf{E}_b(\sigma) \triangleq \mathbf{f}_b(\mathbf{i}_b) \tag{7-19}$$

$$\mathbf{I}^-_b(\sigma) \leq \mathbf{i}_b \leq \mathbf{I}^+_b(\sigma) \tag{7-20}$$

where it is understood that $\mathbf{G}_a(\sigma)$, $\mathbf{I}_a(\sigma)$, $\mathbf{E}^-_a(\sigma)$, and $\mathbf{E}^+_a(\sigma)$ as well as $\mathbf{R}_b(\sigma)$, $\mathbf{E}_b(\sigma)$, $\mathbf{I}^-_b(\sigma)$, and $\mathbf{I}^+_b(\sigma)$ will assume different parameter values corresponding to different segment combinations σ.

Let us now substitute Eqs. (7-17) and (7-19) into Eqs. (7-4) and (7-5) to obtain

$$\mathbf{G}_a(\sigma)\mathbf{v}_a + \mathbf{I}_a(\sigma) - \hat{\mathbf{H}}_{aa}\mathbf{v}_a - \hat{\mathbf{H}}_{ab}\mathbf{i}_b - \hat{\mathbf{s}}_a = \mathbf{0} \tag{7-21}$$

$$\mathbf{R}_b(\sigma)\mathbf{i}_b + \mathbf{E}_b(\sigma) - \hat{\mathbf{H}}_{ba}\mathbf{v}_a - \hat{\mathbf{H}}_{bb}\mathbf{i}_b - \hat{\mathbf{s}}_b = \mathbf{0} \tag{7-22}$$

Equations (7-21) and (7-22) are valid only for those $\mathbf{v}_a$ and $\mathbf{i}_b$ satisfying

$$\mathbf{E}^-_a(\sigma) \leq \mathbf{v}_a \leq \mathbf{E}^+_a(\sigma) \tag{7-23}$$

$$\mathbf{I}^-_b(\sigma) \leq \mathbf{i}_b \leq \mathbf{I}^+_b(\sigma) \tag{7-24}$$

Let us rewrite Eqs. (7-21) and (7-22) into the following forms:

$$\mathbf{F}(\mathbf{x}) \triangleq \begin{bmatrix} [\hat{\mathbf{H}}_{aa} - \mathbf{G}_a(\sigma)] & \hat{\mathbf{H}}_{ab} \\ \hat{\mathbf{H}}_{ba} & [\hat{\mathbf{H}}_{bb} - \mathbf{R}_b(\sigma)] \end{bmatrix}\mathbf{x} - \begin{bmatrix} \mathbf{I}_a(\sigma) - \hat{\mathbf{s}}_a \\ \mathbf{E}_b(\sigma) - \hat{\mathbf{s}}_b \end{bmatrix} = \mathbf{0} \tag{7-25}$$

where

$$\mathbf{x} = \begin{bmatrix} \mathbf{v}_a \\ \mathbf{i}_b \end{bmatrix} \tag{7-26}$$

To apply the Newton–Raphson algorithm, we start with an initial guess $\mathbf{x}^{(0)}$ corresponding to $\mathbf{v}^{(0)}_a$ and $\mathbf{i}^{(0)}_b$. This initial guess can be used to identify the *initial segment combination* $\sigma^{(0)}$ to be that which satisfies Eqs. (7-23) and (7-24); i.e.,

$$\mathbf{E}^-_a(\sigma^{(0)}) \leq \mathbf{v}^{(0)}_a \leq \mathbf{E}^+_a(\sigma^{(0)}) \tag{7-27}$$

$$\mathbf{I}^-_b(\sigma^{(0)}) \leq \mathbf{i}^{(0)}_b \leq \mathbf{I}^+_b(\sigma^{(0)}) \tag{7-28}$$

The first iteration is then given by

$$\begin{aligned} \begin{bmatrix} \mathbf{v}^{(1)}_a \\ \mathbf{i}^{(1)}_b \end{bmatrix} = \begin{bmatrix} \mathbf{v}^{(0)}_a \\ \mathbf{i}^{(0)}_b \end{bmatrix} &- \begin{bmatrix} [\hat{\mathbf{H}}_{aa} - \mathbf{G}_a(\sigma^{(0)})] & \hat{\mathbf{H}}_{ab} \\ \hat{\mathbf{H}}_{ba} & [\hat{\mathbf{H}}_{bb} - \mathbf{R}_b(\sigma^{(0)})] \end{bmatrix}^{-1} \\ &\times \begin{bmatrix} [\hat{\mathbf{H}}_{aa} - \mathbf{G}_a(\sigma^{(0)})] & \hat{\mathbf{H}}_{ab} \\ \hat{\mathbf{H}}_{ba} & [\hat{\mathbf{H}}_{bb} - \mathbf{R}_b(\sigma^{(0)})] \end{bmatrix} \begin{bmatrix} \mathbf{v}^{(0)}_a \\ \mathbf{i}^{(0)}_b \end{bmatrix} \\ &+ \begin{bmatrix} [\hat{\mathbf{H}}_{aa} - \mathbf{G}_a(\sigma^{(0)})] & \hat{\mathbf{H}}_{ab} \\ \hat{\mathbf{H}}_{ba} & [\hat{\mathbf{H}}_{bb} - \mathbf{R}_b(\sigma^{(0)})] \end{bmatrix}^{-1} \begin{bmatrix} \mathbf{I}_a(\sigma^{(0)}) - \hat{\mathbf{s}}_a \\ \mathbf{E}_b(\sigma^{(0)}) - \hat{\mathbf{s}}_b \end{bmatrix} \end{aligned} \tag{7-29}$$

Observe that the first two terms on the right of Eq. (7-29) cancel each other out, and we have the simple formula

$$\begin{bmatrix} \boldsymbol{v}_a^{(1)} \\ \boldsymbol{i}_b^{(1)} \end{bmatrix} = \begin{bmatrix} [\hat{\boldsymbol{H}}_{aa} - \boldsymbol{G}_a(\boldsymbol{\sigma}^{(0)})] & \hat{\boldsymbol{H}}_{ab} \\ \hat{\boldsymbol{H}}_{ba} & [\hat{\boldsymbol{H}}_{bb} - \boldsymbol{R}_b(\boldsymbol{\sigma}^{(0)})] \end{bmatrix}^{-1} \begin{bmatrix} \boldsymbol{I}_a(\boldsymbol{\sigma}^{(0)}) - \hat{\boldsymbol{s}}_a \\ \boldsymbol{E}_b(\boldsymbol{\sigma}^{(0)}) - \hat{\boldsymbol{s}}_b \end{bmatrix} \tag{7-30}$$

This relationship gives the *first iterated* value of $\boldsymbol{v}_a = \boldsymbol{v}_a^{(1)}$ and $\boldsymbol{i}_b = \boldsymbol{i}_b^{(1)}$ in terms of the parameters of the segment combination $\boldsymbol{\sigma}^{(0)}$ corresponding to the initial guess $\boldsymbol{v}_a^{(0)}$ and $\boldsymbol{i}_b^{(0)}$. It is important to observe that, unlike the usual Newton–Raphson formula, the value of $\boldsymbol{v}_a^{(0)}$ and $\boldsymbol{i}_b^{(0)}$ is not needed in Eq. (7-30); only the corresponding segment combination $\boldsymbol{\sigma}^{(0)}$ is needed. If we take $\boldsymbol{v}_a^{(1)}$ and $\boldsymbol{i}_b^{(1)}$ as the next guess, we can identify a new segment combination $\boldsymbol{\sigma}^{(1)}$ such that

$$\boldsymbol{E}_a^-(\boldsymbol{\sigma}^{(1)}) \leq \boldsymbol{v}_a^{(1)} \leq \boldsymbol{E}_a^+(\boldsymbol{\sigma}^{(1)}) \tag{7-31}$$

$$\boldsymbol{I}_b^-(\boldsymbol{\sigma}^{(1)}) \leq \boldsymbol{i}_b^{(1)} \leq \boldsymbol{I}_b^+(\boldsymbol{\sigma}^{(1)}) \tag{7-32}$$

Applying again the Newton–Raphson formula, we obtain

$$\begin{bmatrix} \boldsymbol{v}_a^{(2)} \\ \boldsymbol{i}_b^{(2)} \end{bmatrix} = \begin{bmatrix} [\hat{\boldsymbol{H}}_{aa} - \boldsymbol{G}_a(\boldsymbol{\sigma}^{(1)})] & \hat{\boldsymbol{H}}_{ab} \\ \hat{\boldsymbol{H}}_{ba} & [\hat{\boldsymbol{H}}_{bb} - \boldsymbol{R}_b(\boldsymbol{\sigma}^{(1)})] \end{bmatrix}^{-1} \begin{bmatrix} \boldsymbol{I}_a(\boldsymbol{\sigma}^{(1)}) - \hat{\boldsymbol{s}}_a \\ \boldsymbol{E}_b(\boldsymbol{\sigma}^{(1)}) - \hat{\boldsymbol{s}}_b \end{bmatrix} \tag{7-33}$$

Again we observe that the value of $\boldsymbol{v}_a^{(1)}$ and $\boldsymbol{i}_b^{(1)}$ is not needed in Eq. (7-33); only its corresponding segment combination $\boldsymbol{\sigma}^{(1)}$ is needed. This iteration procedure can obviously be repeated as many times as necessary until the process either *diverges* or *converges* to a *solution* $\boldsymbol{v}_a^*$, $\boldsymbol{i}_b^*$. In particular, the value of $\boldsymbol{v}_a$ and $\boldsymbol{i}_b$ at the $(n+1)$th iteration is clearly given by

$$\boxed{\begin{bmatrix} \boldsymbol{v}_a^{(n+1)} \\ \boldsymbol{i}_b^{(n+1)} \end{bmatrix} = \begin{bmatrix} [\hat{\boldsymbol{H}}_{aa} - \boldsymbol{G}_a(\boldsymbol{\sigma}^{(n)})] & \hat{\boldsymbol{H}}_{ab} \\ \hat{\boldsymbol{H}}_{ba} & [\hat{\boldsymbol{H}}_{bb} - \boldsymbol{R}_b(\boldsymbol{\sigma}^{(n)})] \end{bmatrix}^{-1} \begin{bmatrix} \boldsymbol{I}_a(\boldsymbol{\sigma}^{(n)}) - \hat{\boldsymbol{s}}_a \\ \boldsymbol{E}_b(\boldsymbol{\sigma}^{(n)}) - \hat{\boldsymbol{s}}_b \end{bmatrix}} \tag{7-34}$$

where $\boldsymbol{\sigma}^{(n)}$ is that segment combination satisfying

$$\boldsymbol{E}_a^-(\boldsymbol{\sigma}^{(n)}) \leq \boldsymbol{v}_a^{(n)} \leq \boldsymbol{E}_a^+(\boldsymbol{\sigma}^{(n)}) \tag{7-35}$$

$$\boldsymbol{I}_b^-(\boldsymbol{\sigma}^{(n)}) \leq \boldsymbol{i}_b^{(n)} \leq \boldsymbol{I}_b^+(\boldsymbol{\sigma}^{(n)}) \tag{7-36}$$

Equation (7-34) is the *piecewise-linear version* of the *Newton–Raphson iteration formula* for solving Eqs. (7-21) and (7-22). Observe that since the hybrid submatrices $\hat{\boldsymbol{H}}_{aa}$, $\hat{\boldsymbol{H}}_{ab}$, $\hat{\boldsymbol{H}}_{ba}$, and $\hat{\boldsymbol{H}}_{bb}$ and the two source vectors $\hat{\boldsymbol{s}}_a$ and $\hat{\boldsymbol{s}}_b$ are fixed, only $\boldsymbol{G}_a(\boldsymbol{\sigma}^{(n)})$, $\boldsymbol{R}_b(\boldsymbol{\sigma}^{(n)})$, $\boldsymbol{I}_a(\boldsymbol{\sigma}^{(n)})$, and $\boldsymbol{E}_b(\boldsymbol{\sigma}^{(n)})$ need be changed in each iteration. Since $\boldsymbol{G}_a(\boldsymbol{\sigma}^{(n)})$ and $\boldsymbol{R}_b(\boldsymbol{\sigma}^{(n)})$ can be obtained trivially from the "slope" of the appropriate segment of the v-i curves, the *Jacobian matrix* in Eq. (7-34) is known for each iteration and therefore does not have to be evaluated numerically. This is one major advantage of the present algorithm. Comparing Eqs. (7-34) and (7-25), we observe that they are the same equation written in different forms. Hence, we have the following theorem

Theorem 7-1. *The process of solving for* $\boldsymbol{v}_a$ *and* $\boldsymbol{i}_b$ *of a piecewise-linear network corresponding to segment combination* $\boldsymbol{\sigma}$ *is equivalent to carrying out a Newton–Raphson iteration with the initial guess falling within segment combination* $\boldsymbol{\sigma}$.

Since the Newton–Raphson algorithm may not converge if the initial "segment combination" guess is not close to the correct one, it is highly desirable that we devise an algorithm which guarantees the iteration to converge. Before we develop such an algorithm, it is instructive to study the typical divergence behavior geometrically via the simple circuit shown in Fig. 7-3(a) and its Thevenin equivalent circuit shown in Fig. 7-3(b). In this one-dimensional case, the piecewise-linear Newton–Raphson

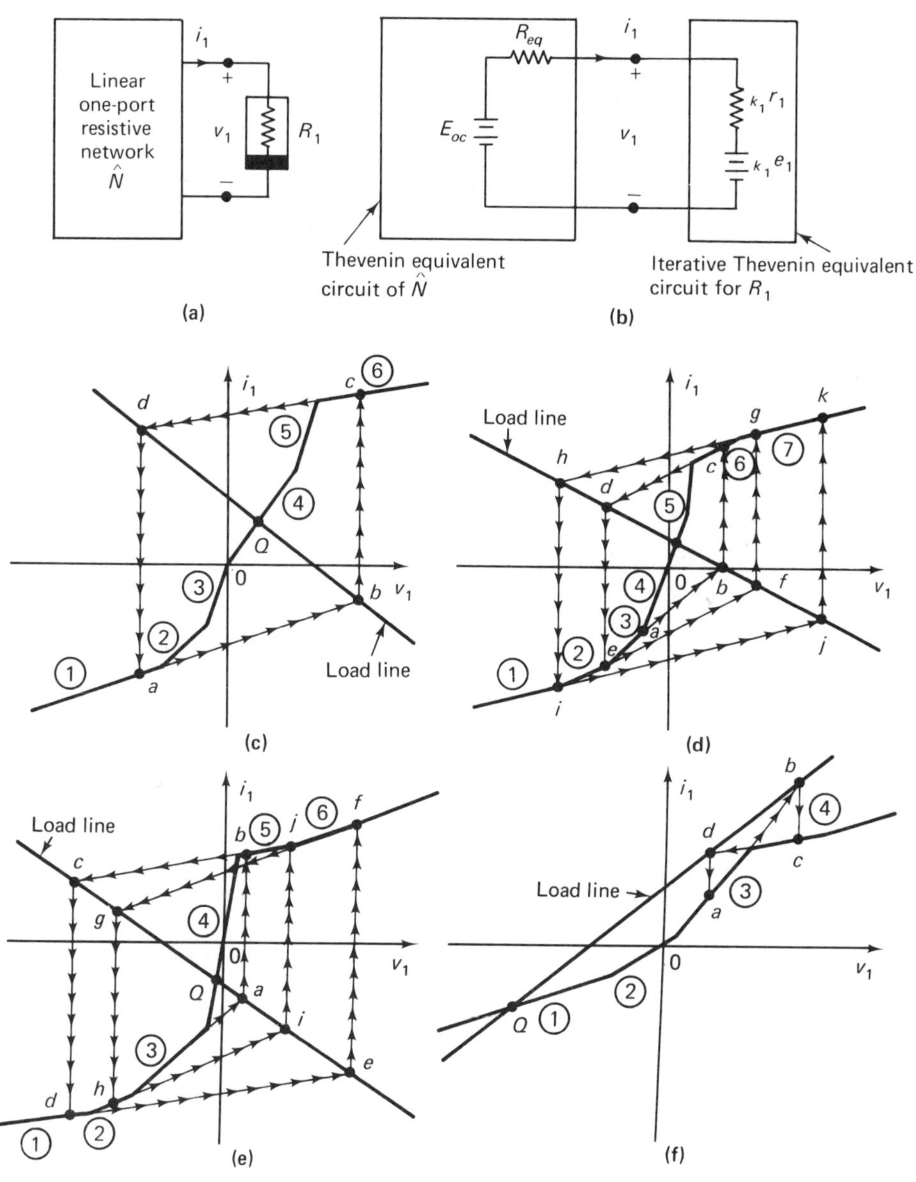

Fig. 7-3. Geometrical interpretation of the divergence mechanism in the piecewise-linear Newton–Raphson algorithm.

algorithm has a simple interpretation in terms of the load line representing the linear one-port $\hat{N}$. For example, consider the four typical monotonically increasing v_1-i_1 curves shown in Fig. 7-3(c), (d), (e), and (f). A load line is drawn on top of each v_1-i_1 curve to demonstrate the four typical divergence behaviors. If we apply the preceding algorithm to the v_1-i_1 curve in Fig. 7-3(c) with an arbitrary initial point *a* on segment 1, we find that the iteration oscillates between the two segments 1 and 6, following the path *a b c d a*. If we repeat the procedure on the v_1-i_1 curve in Fig. 7-3(d), we find that the iteration follows the cyclic path *a b c d e f g h i j k h* corresponding to segments 3 6 2 7 1 7. Similarly, the iteration in Fig. 7-3(e) follows the path *a b c d e f g h i j c* corresponding to segments 3 5 1 6 2 5. Observe that even though this path crosses itself once it eventually becomes cyclic. Finally, the iteration in Fig. 7-3(f) follows along a bow-tie-shaped path *a b c d a* corresponding to segments 3 4 3.

These examples clearly demonstrate that, if the piecewise-linear Newton–Raphson algorithm diverges, it must assume the form of one or more *cyclic paths*. Hence, unlike the usual Newton–Raphson algorithm, it is impossible for our present algorithm to diverge to infinity. This observation leads to the following algorithm for avoiding divergence.

Cyclic Path Detection Algorithm

1. Store all previously iterated segment combinations in the form of an *index table*.
2. Before implementing an iteration corresponding to a new segment combination, check the index table and verify whether this segment combination is in the table or not. If it is, a cyclic path is detected and a new segment combination that does not belong to the index table is chosen as the new initial guess.

Since there is a maximum of n_T distinct segment combinations, where n_T is defined by Eq. (7-11), and since the *cyclic path detection algorithm* rules out the repetition of any previously iterated segment combination, it is clear that, if the network has at least one solution, the combination of the piecewise-linear Newton–Raphson and the cyclic path detection algorithms must necessarily converge to a solution. Moreover, the *final phase* of the algorithm must eventually converge *quadratically* to a solution.[3]

One disadvantage associated with the present algorithm is that the computer storage needed for recording the *index table* could become excessive for *very large networks*. Another is that the "computer time" needed to verify whether a new segment combination belongs to the index table could also become quite significant. Both disadvantages can be partially overcome by judicious programming techniques.

[3]The *final phase* of the algorithm corresponds to the last sequence of iterations that does not contain a cyclic path, and which converges to a solution.

★7-3 PIECEWISE-LINEAR KATZENELSON ALGORITHM [3-6]

Let us now consider another piecewise-linear algorithm whose convergence can be assured for certain classes of networks and that does not require excessive computer storage. This algorithm is due to Katzenelson and applies to systems of piecewise-linear equations of the form

$$f(x) = y \tag{7-37}$$

where $f(x)$ represents a system of m *continuous piecewise-linear equations* in the m variables $x_1, x_2, \ldots, x_m$. The vector $y = [y_1\ y_2 \ldots y_m]^t$ is a *constant* vector called the *source vector*. The basic assumption of the Katzenelson algorithm is that the function $f(x)$ is *one-to-one* and *onto*; i.e., given *any* constant source vector $\hat{y}$, there exists one, and only one, solution $\hat{x}$. To develop this algorithm, let us recast Eqs. (7-21)–(7-24) into the form of Eq. (7-37):

$$G_a(\sigma)v_a + I_a(\sigma) - \hat{H}_{aa}v_a - \hat{H}_{ab}i_b = \hat{s}_a \tag{7-38}$$

$$R_b(\sigma)i_b + E_b(\sigma) - \hat{H}_{ba}v_a - \hat{H}_{bb}i_b = \hat{s}_b \tag{7-39}$$

$$E_a^-(\sigma) \leq v_a \leq E_a^+(\sigma) \tag{7-40}$$

$$I_b^-(\sigma) \leq i_b \leq I_b^+(\sigma) \tag{7-41}$$

If we define

$$x \triangleq \begin{bmatrix} v_a \\ i_b \end{bmatrix}, \qquad \hat{y} \triangleq \begin{bmatrix} \hat{s}_a \\ \hat{s}_b \end{bmatrix} \tag{7-42}$$

then Eqs. (7-38)–(7-41) assume the form

$$A^{(k)}x + w^{(k)} = \hat{y} \tag{7-43}$$

$$x^-(\sigma^{(k)}) \leq x \leq x^+(\sigma^{(k)}) \tag{7-44}$$

where

$$A^{(k)} \triangleq \begin{bmatrix} [G_a(\sigma^{(k)}) - \hat{H}_{aa}] & -\hat{H}_{ab} \\ -\hat{H}_{ba} & [R_b(\sigma^{(k)}) - \hat{H}_{bb}] \end{bmatrix}, \qquad w^{(k)} \triangleq \begin{bmatrix} I_a(\sigma^{(k)}) \\ E_b(\sigma^{(k)}) \end{bmatrix} \tag{7-45}$$

$$x^-(\sigma^{(k)}) \triangleq \begin{bmatrix} E_a^-(\sigma^{(k)}) \\ I_b^-(\sigma^{(k)}) \end{bmatrix}, \qquad x^+(\sigma^{(k)}) \triangleq \begin{bmatrix} E_a^+(\sigma^{(k)}) \\ I_b^+(\sigma^{(k)}) \end{bmatrix} \tag{7-46}$$

are the pertinent parameters defining the piecewise-linear equation corresponding to segment combination $\sigma = \sigma^{(k)}$.

Let $\sigma = \sigma^{(0)}$ be an arbitrary initial segment combination. For example, for a network with three nonlinear resistors, we could choose our initial segment combination to be $\sigma^{(0)} = (1, 3, 5)$, corresponding to segment 1 of resistor 1, segment 3 of resistor 2, and segment 5 of resistor 3. Let $x^{(0)}$ be *any* point belonging to the *hypercube* corresponding to $\sigma^{(0)}$; i.e.,[4]

$$x^-(\sigma^{(0)}) \leq x^{(0)} \leq x^+(\sigma^{(0)}) \tag{7-47}$$

[4]In the two-dimensional case, Eq. (7-47) defines a *rectangular* region bounded by a pair of parallel horizontal lines and a pair of parallel vertical lines.

Substituting $k = 0$ and $\mathbf{x} = \mathbf{x}^{(0)}$ in Eq. (7-43), we obtain

$$A^{(0)}\mathbf{x}^{(0)} + \mathbf{w}^{(0)} \triangleq \mathbf{y}^{(0)} \tag{7-48}$$

Observe that if $\mathbf{y}^{(0)} = \hat{\mathbf{y}}$, then $\mathbf{x} = \mathbf{x}^{(0)}$ is a solution and we are done. Of course, in general, we expect $\mathbf{y}^{(0)} \neq \hat{\mathbf{y}}$, and $\mathbf{x}^{(0)} \neq \hat{\mathbf{x}}^{(0)}$, where $\hat{\mathbf{x}}^{(0)}$ is the solution satisfying Eq. (7-43); i.e.,

$$A^{(0)}\hat{\mathbf{x}}^{(0)} + \mathbf{w}^{(0)} = \hat{\mathbf{y}} \tag{7-49}$$

If $\hat{\mathbf{x}}^{(0)}$ lies within the hypercube defined by Eq. (7-47), it is a solution and we are done. In general, however, $\hat{\mathbf{x}}^{(0)}$ will be outside of the hypercube, and we call $\hat{\mathbf{x}}^{(0)}$ a *virtual solution*. If we subtract Eq. (7-48) from Eq. (7-49), we obtain

$$\hat{\mathbf{x}}^{(0)} = \mathbf{x}^{(0)} + A^{(0)^{-1}}(\hat{\mathbf{y}} - \mathbf{y}^{(0)}) \tag{7-50}$$

Consider the straight line connecting $\mathbf{x}^{(0)}$ to $\hat{\mathbf{x}}^{(0)}$ in the $\mathbf{x}$-space as shown in Fig. 7-4(a). This straight line can be defined mathematically via the parametric equation

$$\tilde{\mathbf{x}}^{(0)}(\lambda) = \lambda\hat{\mathbf{x}}^{(0)} + (1 - \lambda)\mathbf{x}^{(0)}, \qquad 0 \leq \lambda \leq 1 \tag{7-51}$$

Observe that $\tilde{\mathbf{x}}^{(0)}(0) = \mathbf{x}^{(0)}$ and $\tilde{\mathbf{x}}^{(0)}(1) = \hat{\mathbf{x}}^{(0)}$. Substituting Eq. (7-50) into Eq. (7-51), we obtain the following alternative representation of the straight line from $\mathbf{x}^{(0)}$ to $\hat{\mathbf{x}}^{(0)}$:

$$\tilde{\mathbf{x}}^{(0)}(\lambda) = \mathbf{x}^{(0)} + \lambda A^{(0)^{-1}}(\hat{\mathbf{y}} - \mathbf{y}^{(0)}), \qquad 0 \leq \lambda \leq 1 \tag{7-52}$$

It is instructive to consider the *image* of points of this straight line in the $\mathbf{y}$-space under the mapping defined by Eq. (7-37). Since $\mathbf{x}^{(0)}$ is mapped into $\mathbf{y}^{(0)}$ and $\hat{\mathbf{x}}^{(0)}$ is mapped into $\hat{\mathbf{y}}$, it is clear that Eq. (7-52) maps the straight line from $\mathbf{x}^{(0)}$ to $\hat{\mathbf{x}}^{(0)}$ into the straight line in the $\mathbf{y}$-space connecting $\mathbf{y}^{(0)}$ to $\hat{\mathbf{y}}$, as shown in Fig. 7-4(b). Indeed, the point $\tilde{\mathbf{x}}^{(0)}(\lambda)$ is mapped into the point

$$\tilde{\mathbf{y}}^{(0)}(\lambda) \triangleq A^{(0)}\mathbf{x}^{(0)}(\lambda) + \mathbf{w}^{(0)} \tag{7-53}$$

If we substitute Eq. (7-52) into Eq. (7-53), we obtain with the help of Eq. (7-48) the equation

$$\tilde{\mathbf{y}}^{(0)}(\lambda) = \lambda\hat{\mathbf{y}} + (1 - \lambda)\mathbf{y}^{(0)} \tag{7-54}$$

which is a parametric equation of the straight line from $\mathbf{y}^{(0)}$ to $\hat{\mathbf{y}}$.

Now returning to Eq. (7-52), let us locate the point $\tilde{\mathbf{x}} = \mathbf{x}^{(0)}(\lambda_0)$ where this straight line intersects the *boundary*[5] between the hypercube corresponding to the initial combination $\sigma^{(0)}$ and an adjacent hypercube corresponding to a *new* segment combination $\sigma^{(1)}$. Let this point correspond to a value $\lambda = \lambda_0 > 0$. In view of Eq. (7-54), this corresponds to the point $\tilde{\mathbf{y}}^{(0)}(\lambda_0) \triangleq \mathbf{y}^{(1)}$ in the $\mathbf{y}$-space.

Consider now the new segment combination $\sigma^{(1)}$ and choose $\mathbf{x}^{(1)} \triangleq \tilde{\mathbf{x}}^{(0)}(\lambda_0)$ to be our next guess. Let us repeat the algorithm using the *updated* parameters $A^{(1)}$, $\mathbf{w}^{(1)}$, $\mathbf{x}^{-}(\sigma^{(1)})$, and $\mathbf{x}^{+}(\sigma^{(1)})$ *corresponding to the new segment combination* $\sigma = \sigma^{(1)}$. Corresponding to Eqs. (7-52) and (7-54), we now obtain

$$\tilde{\mathbf{x}}^{(1)}(\lambda) = \mathbf{x}^{(1)} + \lambda A^{(1)^{-1}}(\hat{\mathbf{y}} - \mathbf{y}^{(1)}) \tag{7-55a}$$

$$\tilde{\mathbf{y}}^{(1)}(\lambda) = \lambda\hat{\mathbf{y}} + (1 - \lambda)\mathbf{y}^{(1)} \tag{7-55b}$$

[5]We assume for simplicity here that the point of intersection does not occur at the "corner" between two boundaries. This special case leads to some complication since it is no longer clear which segment combination to consider next. In practice, this "corner problem" can be avoided by perturbing the initial guess by a small increment.

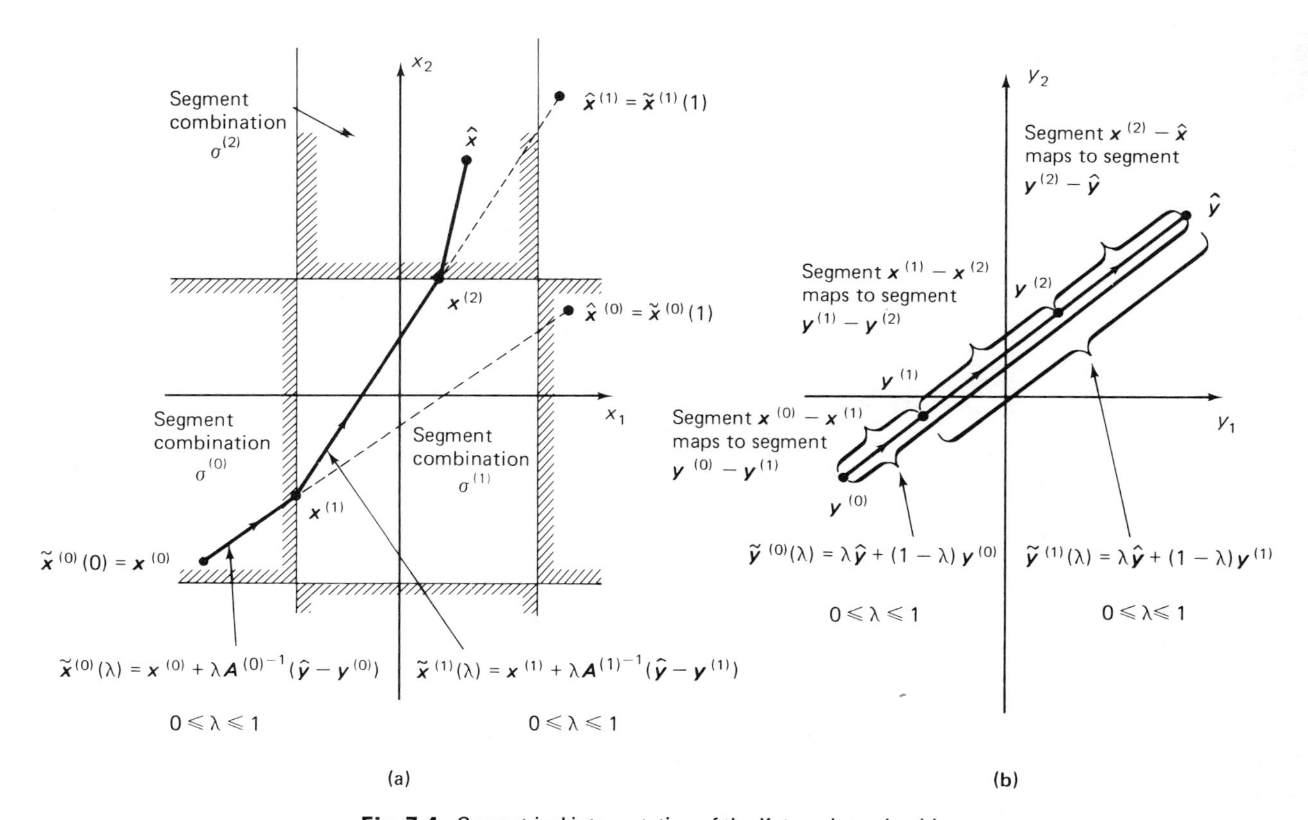

Fig. 7-4. Geometrical interpretation of the Katzenelson algorithm.

where $0 \le \lambda \le 1$. Observe that the line segment represented by $\tilde{\boldsymbol{y}}^{(1)}(\lambda)$ with $0 \le \lambda \le 1$ is a portion of the previously defined line segment represented by $\tilde{\boldsymbol{y}}^{(0)}(\lambda)$ with $0 \le \lambda \le 1$. Observe also that the variable λ defined in Eqs. (7-52) and (7-54) is not the same λ as used in Eq. (7-55).

Let $\lambda = \lambda_1 > 0$ be the point where the straight line defined by Eq. (7-55a) intersects the boundary between the hypercube corresponding to segment combination $\boldsymbol{\sigma}^{(1)}$ and an adjacent hypercube corresponding to another segment combination $\boldsymbol{\sigma}^{(2)}$. Since $\lambda_1 > \boldsymbol{0}$ (by assumption), this line segment maps into a corresponding segment from $\boldsymbol{y}^{(1)}$ to $\tilde{\boldsymbol{y}}^{(1)}(\lambda_1) \triangleq \boldsymbol{y}^{(2)}$, as shown in Fig. 7-4(b).

Let $\boldsymbol{\sigma}^{(2)}$ be the new segment combination and let $\boldsymbol{x}^{(2)} \triangleq \tilde{\boldsymbol{x}}^{(1)}(\lambda_1)$ be the new initial guess. We can clearly iterate the preceding algorithm each time a new value of λ_j is determined along with a new segment combination $\boldsymbol{\sigma}^{(j)}$. For the nth iteration, we obtain, corresponding to Eqs. (7-52) and (7-55a), the following equations for the corresponding straight lines in the $\boldsymbol{x}$-space and $\boldsymbol{y}$-space, respectively:

$$\tilde{\boldsymbol{x}}^{(n)}(\lambda) = \boldsymbol{x}^{(n)} + \lambda \boldsymbol{A}^{(n)^{-1}}(\hat{\boldsymbol{y}} - \boldsymbol{y}^{(n)}) \tag{7-56a}$$

$$\tilde{\boldsymbol{y}}^{(n)}(\lambda) = \lambda \hat{\boldsymbol{y}} + (1 - \lambda)\boldsymbol{y}^{(n)} \tag{7-56b}$$

The solution $\hat{\boldsymbol{x}}^{(n)}$ corresponding to segment combination $\boldsymbol{\sigma}^{(n)}$ is obtained by setting $\lambda = 1$ in Eq. (7-56a):

$$\boxed{\hat{\boldsymbol{x}}^{(n)} \triangleq \tilde{\boldsymbol{x}}^{(n)}(1) = \boldsymbol{x}^{(n)} + \boldsymbol{A}^{(n)^{-1}}(\hat{\boldsymbol{y}} - \boldsymbol{y}^{(n)})} \tag{7-56c}$$

To find the point where the *straight line*

$$\tilde{\boldsymbol{x}}^{(n)}(\lambda) = \lambda \hat{\boldsymbol{x}}^{(n)} + (1 - \lambda)\boldsymbol{x}^{(n)} \tag{7-57a}$$

connecting $\boldsymbol{x}^{(n)}$ to $\hat{\boldsymbol{x}}^{(n)}$ crosses the boundary of the hypercube corresponding to segment combination $\boldsymbol{\sigma}^{(n)}$, we impose the *boundary condition*

$$\begin{aligned} \tilde{x}_j^{(n)}(\lambda) &= x_j^{+}(\boldsymbol{\sigma}^{(n)}), \quad \text{if } \hat{x}_j^{(n)} > x_j^{(n)} \\ &= x_j^{-}(\boldsymbol{\sigma}^{(n)}), \quad \text{if } \hat{x}_j^{(n)} < x_j^{(n)} \end{aligned}, \qquad \text{for } j = 1, 2, \ldots, m \tag{7-57b}$$

on Eq. (7-57a), where $x_j^{\pm}(\boldsymbol{\sigma}^{(n)}) = E_j^{\pm}(\boldsymbol{\sigma}^{(n)})$ if x_j represents the voltage v_j of resistor R_j, or $x_j^{\pm}(\boldsymbol{\sigma}^{(n)}) = I_j^{\pm}(\boldsymbol{\sigma}^{(n)})$ if x_j represents the current i_j of resistor R_j [See Eqs. (7-40) and (7-41)]. Hence,

$$x_j^{\pm}(\boldsymbol{\sigma}^{(n)}) = \lambda \hat{x}_j^{(n)} + (1 - \lambda)x_j^{(n)} \tag{7-57c}$$

Solving for λ, we obtain

$$\boxed{\lambda = \frac{x_j^{\pm}(\boldsymbol{\sigma}^{(n)}) - x_j^{(n)}}{\hat{x}_j^{(n)} - x_j^{(n)}} \triangleq \lambda_j^{(n)}} \tag{7-58a}$$

where the superscript $^+$ is chosen if $\hat{x}_j^{(n)} > x_j^{(n)}$ and the superscript $^-$ is chosen if $\hat{x}_j^{(n)} < x_j^{(n)}$. The value of $\lambda_j^{(n)}$ computed from Eq. (7-58a) represents the value of λ where the straight line crosses the boundary of that segment $\sigma_j^{(n)}$ of resistor R_j corresponding to segment combination $\boldsymbol{\sigma}^{(n)}$. Clearly, the value of λ where this straight line *first* crosses a boundary is given by

$$\lambda(\sigma^{(n)}) = \min\{\lambda_1^{(n)}, \lambda_2^{(n)}, \ldots, \lambda_{m_1+m_2}^{(n)}\} \tag{7-58b}$$

The point $P(\sigma^{(n)})$ where the straight line defined by Eq. (7-57a) intersects the boundary of the hypercube defined by segment combination $\sigma^{(n)}$ is therefore given by

$$\tilde{\mathbf{x}}^{(n+1)} \triangleq \tilde{\mathbf{x}}^{(n)}(\lambda)|_{\lambda=\lambda(\sigma^{(n)})} \tag{7-58c}$$

Assuming that $P(\sigma^{(n)})$ is not a *corner point* (common breakpoint for two or more resistor v-i curves), then Eq. (7-58c) can be used to identify the next segment combination $\sigma^{(n+1)}$. Moreover, $\mathbf{x}^{(n+1)}$ becomes the *initial point* for our next iteration:

$$\mathbf{y}^{(n+1)} = \mathbf{A}^{(n+1)}\mathbf{x}^{(n+1)} + \mathbf{w}^{(n+1)} \tag{7-58d}$$

$$\tilde{\mathbf{x}}^{(n+1)}(\lambda) = \mathbf{x}^{(n+1)} + \lambda \mathbf{A}^{(n+1)^{-1}}(\hat{\mathbf{y}} - \mathbf{y}^{(n+1)}) \tag{7-58e}$$

The algorithm then repeats itself upon substituting $\mathbf{x}^{(n+1)}$ and $\mathbf{y}^{(n+1)}$ for $\mathbf{x}^{(n)}$ and $\mathbf{y}^{(n)}$ in Eq. (7-56). Since $\lambda_j > 0$ (by assumption) in each iteration, each straight line segment in the $\mathbf{x}$-space when mapped into the $\mathbf{y}$-space will dovetail into one another (in the same direction) along the line connecting $\mathbf{y}^{(0)}$ and $\hat{\mathbf{y}}$. Hence, we shall eventually arrive at the point $\hat{\mathbf{y}}$ in a *finite* number of iterations. The value $\mathbf{x} = \hat{\mathbf{x}}$ that maps into the endpoint $\hat{\mathbf{y}}$ is therefore the desired solution. We illustrate this convergence phenomenon in Fig. 7-4 using three iterations. Observe that the polygonal curve from $\mathbf{x}^{(0)}$ to $\hat{\mathbf{x}}$ in the $\mathbf{x}$-space—the *solution curve*—is mapped into the entire straight line from $\mathbf{y}^{(0)}$ to $\hat{\mathbf{y}}$ in the $\mathbf{y}$-space.

The convergence of the algorithm depends critically on the *assumption* that $\lambda_j > 0$ in each iteration so that a segment of finite length is added *along* the direction of the $\mathbf{y}^{(0)} - \hat{\mathbf{y}}$ straight line from $\mathbf{y}^{(0)}$ to $\mathbf{y}^{(1)}$, then from $\mathbf{y}^{(1)}$ to $\mathbf{y}^{(2)}$, etc., until at the nth iteration, where we have $\mathbf{y}^{(n)} = \hat{\mathbf{y}}$. Observe that this assumption will be satisfied if the function $\mathbf{f}(\mathbf{x}) = \mathbf{y}$ is one-to-one and onto. For suppose that $\lambda_j < 0$ at the jth iteration. Then the corresponding segment will be mapped into the line from $\mathbf{y}^{(0)}$ to $\hat{\mathbf{y}}$ in the opposite direction. Thus, there are at least two points in the $\mathbf{x}$-space that are mapped into the same point in the $\mathbf{y}$-space, contradicting the hypothesis that $\mathbf{f}(\mathbf{x})$ is one-to-one.

An equivalent and sometimes more enlightening way to visualize this algorithm is to consider the straight line connecting $\mathbf{y}^{(0)}$ to $\hat{\mathbf{y}}$ and its corresponding points in the $\mathbf{x}$-space under the function $\mathbf{f}(\mathbf{x})$. Since $\mathbf{f}(\mathbf{x})$ is assumed to be continuous, one-to-one, and onto, it is clear that each point on the $\mathbf{y}^{(0)} - \hat{\mathbf{y}}$ straight line corresponds to *exactly* one point in the $\mathbf{x}$-space, and vice versa. Hence, we must have a *connected curve* in the $\mathbf{x}$-space. Moreover, since $\mathbf{f}(\mathbf{x})$ is a piecewise-linear function, corresponding points in the $\mathbf{x}$-space must also be straight line segments, and we conclude that *the straight line* connecting $\mathbf{y}^{(0)}$ to $\hat{\mathbf{y}}$ in the $\mathbf{y}$-space must be mapped onto a unique *polygonal curve*. Hence, our algorithm for finding intersections of this polygonal curve with the boundaries is indeed a sound strategy.

If the function $\mathbf{f}(\mathbf{x})$ is *not* one-to-one and onto, the Katzenelson algorithm generally does not converge, since some λ_j could be negative. However, it can be proved [4–5] that, if the determinant of $\mathbf{A}^{(k)}$ has the same sign for all segment combinations, the Katzenelson algorithm is also guaranteed to converge. In fact, this algorithm can be modified such that the determinant of only those segment combinations located in the unbounded region need have the same sign [6].

★7-4 PIECEWISE-LINEAR COMBINATORIAL ALGORITHM FOR FINDING MULTIPLE SOLUTIONS

We have already mentioned that, if a network possesses *multiple* solutions, then the Newton–Raphson algorithm or the Katzenelson algorithm will generally find only one of several possible solutions. Indeed, all existing iterative methods suffer from this serious disadvantage [7]. Our objective in this section is to present a noniterative but *combinatorial* algorithm that is capable of finding all solutions.

We proceed as before by extracting all m nonlinear resistors and writing the equations of the network as in Eq. (7-25), as follows:[6]

$$\begin{bmatrix} [\hat{\boldsymbol{H}}_{aa} - \boldsymbol{G}_a(\sigma)] & \hat{\boldsymbol{H}}_{ab} \\ \hat{\boldsymbol{H}}_{ba} & [\hat{\boldsymbol{H}}_{bb} - \boldsymbol{R}_b(\sigma)] \end{bmatrix} \begin{bmatrix} \boldsymbol{v}_a \\ \boldsymbol{i}_b \end{bmatrix} = \begin{bmatrix} \boldsymbol{I}_a(\sigma) - \hat{\boldsymbol{s}}_a \\ \boldsymbol{E}_b(\sigma) - \hat{\boldsymbol{s}}_b \end{bmatrix} \tag{7-59}$$

Corresponding to each segment combination σ, Eq. (7-59) represents a system of *linear algebraic equations* and, therefore, has a solution that can be obtained by the Gaussian elimination method presented in Chapter 4. Let this solution be written as follows:

$$\begin{bmatrix} \boldsymbol{v}_a \\ \boldsymbol{i}_b \end{bmatrix} = \begin{bmatrix} [\hat{\boldsymbol{H}}_{aa} - \boldsymbol{G}_a(\sigma)] & \hat{\boldsymbol{H}}_{ab} \\ \hat{\boldsymbol{H}}_{ba} & [\hat{\boldsymbol{H}}_{bb} - \boldsymbol{R}_b(\sigma)] \end{bmatrix}^{-1} \begin{bmatrix} \boldsymbol{I}_a(\sigma) - \hat{\boldsymbol{s}}_a \\ \boldsymbol{E}_b(\sigma) - \hat{\boldsymbol{s}}_b \end{bmatrix} \tag{7-60}$$

Clearly, $\boldsymbol{v}_a$ and $\boldsymbol{i}_b$ would represent a *valid* solution if, and only if, each component of $\boldsymbol{v}_a$ and $\boldsymbol{i}_b$ falls within the domain of definition of each segment corresponding to the *assumed* segment combination $\boldsymbol{\sigma}$. By repeating this process for all possible segment combinations, we shall clearly obtain all solutions of the network. Since there are a total of n_T distinct segment combinations, where n_T is given in Eq. (7-11), the *combinatorial piecewise-linear* method would call for solving Eq. (7-59) a total of n_T times, each time corresponding to a different segment combination. Hence, the *combinatorial piecewise-linear* method in its present form could be extremely inefficient. We can measure its efficiency by defining the following *combinatorial efficiency index*:

$$\eta = \frac{n_A}{n_\sigma} \tag{7-61}$$

where n_A is the actual number of valid solutions and n_σ is the number of times that Eq. (7-59) has to be solved to sort out these n_A solutions. In the "unmodified" combinatorial method to be presented in this section, we have $n_\sigma = n_T$; i.e., all possible segment combinations must be tested. In the next section, we shall show that it is possible

[6]We remark that even though Eq. (7-59) was derived earlier for m-ports terminated only by voltage- and current-controlled resistors, we can now allow any nonlinear resistor characterized by arbitrary piecewise-linear curves, including multivalued or disconnected v-i curves. In fact, instead of using the hybrid matrix, we may just as well use the open-circuit resistance matrix, provided the matrix exists, i.e., if it contains no infinite entries.

to modify the combinatorial method so that $n_\sigma \ll n_T$, thereby greatly increasing the combinatorial efficiency index. We observe that it is possible for this index to assume the maximum value of "unity" even in the unmodified method. For example, consider a simple resistor-tunnel diode series circuit biased in such a manner that the load line intersects all n_T segments of the tunnel diode v-i curve. In most practical cases, however, the more nonlinear resistors there are, the smaller usually is the combinatorial index, because the solutions corresponding to a large percentage of the n_T segment combinations are usually *virtual* solutions, i.e., solutions which fall outside the domain of definition of the segment combination.

We shall now show that even in the present unmodified combinatorial piecewise-linear method, it is possible to save considerable computation time by taking advantage of the fact that only the *diagonal* matrices $\boldsymbol{G}_a(\sigma)$ and $\boldsymbol{R}_b(\sigma)$ are altered in going from one segment combination to another.

Suppose that we have two matrices $\boldsymbol{A}$ and $\boldsymbol{B}$ which differ from each other by only one diagonal element, say the kkth element. In particular, let

$$(\boldsymbol{B})_{jk} = (\boldsymbol{A})_{jk}, \qquad j \neq k \tag{7-62}$$

and

$$(\boldsymbol{B})_{kk} = (\boldsymbol{A})_{kk} + (\Delta \boldsymbol{A})_{kk} \tag{7-63}$$

where $(\boldsymbol{B})_{jk}$ and $(\boldsymbol{A})_{jk}$ represent the jkth element of the matrices $\boldsymbol{B}$ and $\boldsymbol{A}$, respectively. Now suppose that $\boldsymbol{A}^{-1}$ has been computed, and let α_{jk} be its jkth element. A well-known method from numerical analysis [8] provides us with the following useful relationship:

$$\boldsymbol{B}^{-1} = \boldsymbol{A}^{-1} - \left[\frac{(\Delta \boldsymbol{A})_{kk}}{1 + (\Delta \boldsymbol{A})_{kk}\alpha_{kk}}\right]\boldsymbol{M}_k \tag{7-64}$$

where $\boldsymbol{M}_k$ is an $m \times m$ matrix defined by

$$\boldsymbol{M}_k = \begin{bmatrix} \alpha_{1k} \\ \alpha_{2k} \\ \cdot \\ \cdot \\ \cdot \\ \alpha_{mk} \end{bmatrix} [\alpha_{k1} \quad \alpha_{k2} \quad \cdots \quad \alpha_{km}] = \begin{bmatrix} \alpha_{1k}\alpha_{k1} & \alpha_{1k}\alpha_{k2} & \cdots & \alpha_{1k}\alpha_{km} \\ \alpha_{2k}\alpha_{k1} & \alpha_{2k}\alpha_{k2} & \cdots & \alpha_{2k}\alpha_{km} \\ \cdot & \cdot & \cdots & \cdot \\ \cdot & \cdot & \cdots & \cdot \\ \cdot & \cdot & \cdots & \cdot \\ \alpha_{mk}\alpha_{k1} & \alpha_{mk}\alpha_{k2} & \cdots & \alpha_{mk}\alpha_{km} \end{bmatrix} \tag{7-65}$$

This relationship shows that the inverse of $\boldsymbol{B}$ can be found by *modifying* the *inverse* of $\boldsymbol{A}$, which is already known, and is generally referred to as *matrix inversion by modification.*[7] Observe that this method of finding $\boldsymbol{B}^{-1}$ requires only $m^2 + m + 2$ *multiplication* operations, as compared to m^3 operations if $\boldsymbol{B}^{-1}$ were to be computed from scratch using the already efficient Gaussian elimination method.

It is now clear that only one of the n_T matrices in the combinatorial method needs to be inverted from scratch. After that, each new segment combination can be chosen sequentially so that *only one* nonlinear resistor changes segment at a time. This would give us a sequence of matrices for Eq. (7-60) such that successive pairs of matrices differ from each other by one and only one *diagonal element*. The method of

[7]This powerful method appears to have been discovered independently by several researchers. Consequently, it is also known as Kron's method or as Woodbury's formula.

matrix inversion by modification can therefore be used to find the inverse of the remaining $n_T - 1$ matrices. The computational time saved as a result of this approach clearly increases with n_T, and is quite significant for most networks of moderate complexity.

The combinatorial method remains applicable if the hybrid matrix characterizing the linear m-port is an *open-circuit resistance matrix* when all ports are driven with current sources, or a *short-circuit conductance matrix* when all ports are driven with voltage sources. In fact, *any* hybrid matrix corresponding to an arbitrary combination of voltage- and current-driven ports is acceptable, provided that the matrix exists; i.e., all elements of the matrix are finite numbers. To emphasize this observation, let us redraw Fig. 7-1 into the nonlinear resistor terminated m-port shown in Fig. 7-5(a). To characterize the linear m-port $\hat{N}$, let us connect m independent sources $x_1, x_2, \ldots, x_m$ across its ports, as shown in Fig. 7-5(b), where x_i could be either a

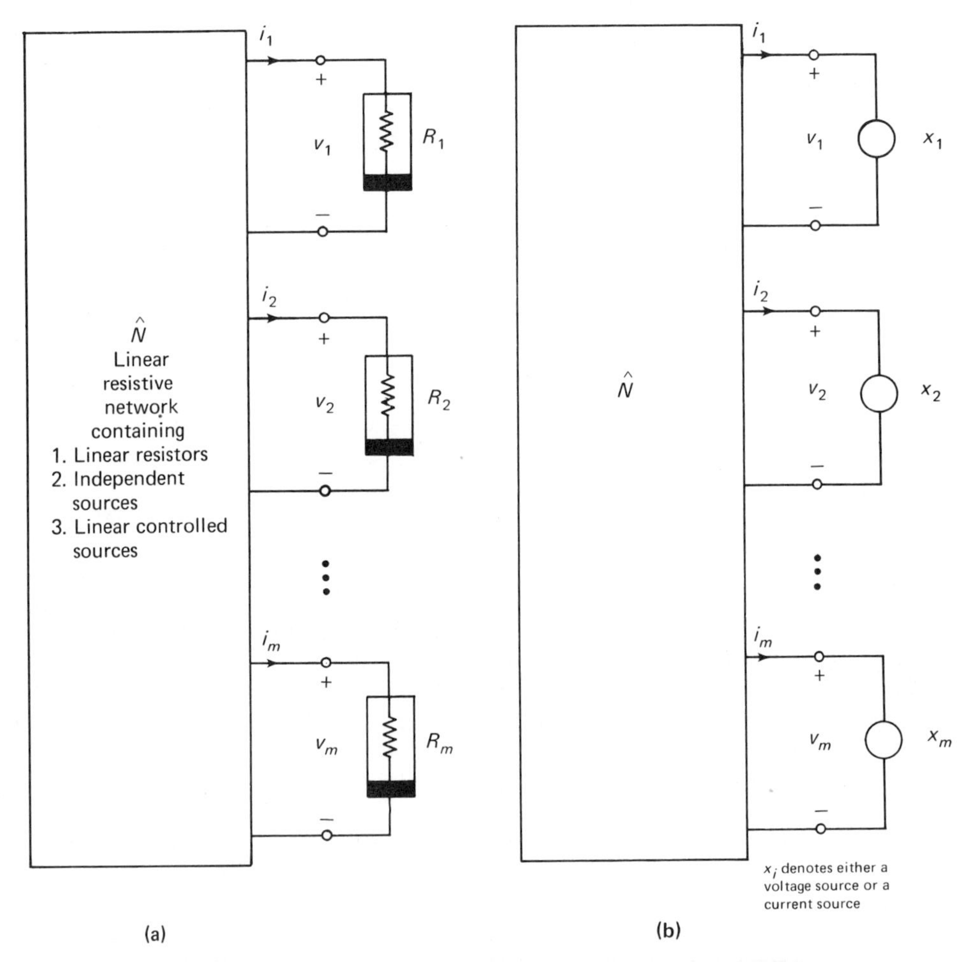

Fig. 7-5. A resistive nonlinear network can always be redrawn into a linear m-port with the m nonlinear resistors connected across the ports.

voltage or a current source. We let y_i denote the associated port variable. For example, if x_5 is a current source, then port 5 is driven by a current source with terminal current $i_5 = x_5$, and its associated port voltage v_5 is then equal to y_5. Unlike Fig. 7-1, where the first m_1 ports must be voltage ports and the remaining m_2 ports must be current ports, we now allow an arbitrary mixture of voltage and current ports. If we let $\boldsymbol{x} = (x_1, x_2, \ldots, x_m)^t$, and $\boldsymbol{y} = (y_1, y_2, \ldots, y_m)^t$, then the linear m-port $\hat{N}$ can be characterized by

$$\boldsymbol{y} = \hat{\boldsymbol{H}}\boldsymbol{x} + \hat{\boldsymbol{s}} \tag{7-66}$$

where $\hat{\boldsymbol{H}}$ is the $m \times m$ *hybrid matrix* of $\hat{N}$, and $\hat{\boldsymbol{s}}$ is an $m \times 1$ *source vector* representing the independent sources *inside* $\hat{N}$. Observe that if we choose $\boldsymbol{x} = \boldsymbol{i}$, then $\boldsymbol{Z}_{oc} \triangleq -\hat{\boldsymbol{H}}$ can be identified as the open-circuit resistance matrix. Similarly, if $\boldsymbol{x} = \boldsymbol{v}$, then $\boldsymbol{Y}_{sc} \triangleq -\hat{\boldsymbol{H}}$ can be identified as the short-circuit conductance matrix. The hybrid matrix corresponding to the $(m_1 + m_2)$-port $\hat{N}$ in Fig. 7-1 would correspond to $\boldsymbol{x} = (v_1, v_2, \ldots, v_{m_1}, i_{m_1+1}, i_{m_1+2}, \ldots, i_{m_1+m_2})^t$ and $\boldsymbol{y} = (i_1, i_2, \ldots, i_{m_1}, v_{m_1+1}, v_{m_1+2}, \ldots, v_{m_1+m_2})^t$. Since there are 2^m distinct combinations of these port variables, there are as many distinct combinations of hybrid matrices and source vectors.

As usual, we shall let $\boldsymbol{\sigma} = \{\sigma_1, \sigma_2, \ldots, \sigma_m\}$ denote the segment combination consisting of segment σ_1 of R_1, σ_2 of R_2, . . . , and σ_m of R_m. Hence, corresponding to each segment combination $\boldsymbol{\sigma}$, segment σ_j of the v_j-i_j curve of R_j across port j is characterized by

$$y_j = h_j(\sigma_j)x_j + s_j(\sigma_j) \tag{7-67}$$

$$x_j^-(\sigma_j) \leq x_j \leq x_j^+(\sigma_j) \tag{7-68}$$

where $h_j(\sigma_j)$ and $s_j(\sigma_j)$ denote, respectively, the slope and the intercept of segment σ_j of R_j, and where $[x_j^-(\sigma_j), x_j^+(\sigma_j)]$ represents the domain of definition of segment σ_j of R_j, $j = 1, 2, \ldots, m$. The terminal constraint imposed by the m nonlinear resistors can therefore be expressed by

$$\boldsymbol{y} = \boldsymbol{H}(\boldsymbol{\sigma})\boldsymbol{x} + \boldsymbol{s}(\boldsymbol{\sigma}) \tag{7-69}$$

$$\boldsymbol{x}^-(\boldsymbol{\sigma}) \leq \boldsymbol{x} \leq \boldsymbol{x}^+(\boldsymbol{\sigma}) \tag{7-70}$$

where $\boldsymbol{H}(\boldsymbol{\sigma})$ is an $m \times m$ *diagonal* matrix whose jjth diagonal element is equal to $h_j(\sigma_j)$, and where $\boldsymbol{s}(\boldsymbol{\sigma})$, $\boldsymbol{x}^-(\boldsymbol{\sigma})$, and $\boldsymbol{x}^+(\boldsymbol{\sigma})$ are $m \times 1$ vectors whose jth component is equal to $s_j(\sigma_j)$, $x_j^-(\sigma_j)$, and $x_j^+(\sigma_j)$. If we substitute Eq. (7-69) for $\boldsymbol{y}$ in Eq. (7-66) and solve for $\boldsymbol{x}$, we obtain the following solution corresponding to each segment combination $\boldsymbol{\sigma}$:

$$\boxed{\boldsymbol{x} = [\hat{\boldsymbol{H}} - \boldsymbol{H}(\boldsymbol{\sigma})]^{-1}[\boldsymbol{s}(\boldsymbol{\sigma}) - \hat{\boldsymbol{s}}]} \tag{7-71}$$

This solution is then checked to see if it indeed satisfies Eq. (7-70). If it does, a valid solution is found. If it does not, the solution must correspond to a *virtual segment combination*[8] and is therefore discarded. The "unmodified" combinatorial piecewise-linear method presented in this section when recast in terms of an arbitrary hybrid matrix $\hat{\boldsymbol{H}}$ and source vector $\hat{\boldsymbol{s}}$ would call for solving Eq. (7-71) for all n_T distinct

[8]A segment combination that does not give rise to a valid solution is called a *virtual* segment combination.

segment combinations. It is desirable therefore to develop algorithms that are capable of bypassing some of the virtual segment combinations.

*7-5 ALGORITHMS FOR IMPROVING THE COMBINATORIAL EFFICIENCY INDEX

The linear m-port network $\hat{N}$ shown in Fig. 7-5(a) imposes certain constraints on the possible values of the port current vector $\boldsymbol{i}$ and the port voltage vector $\boldsymbol{v}$. The constraint imposed upon each nonlinear resistor R_j is given by Eq. (7-66); i.e.,

$$y_j = \hat{h}_{j1}x_1 + \hat{h}_{j2}x_2 + \ldots + \hat{h}_{jj}x_j + \ldots + \hat{h}_{jm}x_m + \hat{s}_j, \qquad j = 1, 2, \ldots, m \tag{7-72}$$

where the variables y_k and x_k may represent either the voltage or the current associated with the resistor R_k. From Eq. (7-72) we obtain two inequalities:

$$y_j \geq \hat{s}_j, \quad \text{if } \hat{h}_{jk}x_k \geq 0, \qquad k = 1, 2, \ldots, m \tag{7-73}$$

and

$$y_j \leq \hat{s}_j, \quad \text{if } \hat{h}_{jk}x_k \leq 0, \qquad k = 1, 2, \ldots, m \tag{7-74}$$

Each inequality $\hat{h}_{jk}x_k \geq 0$ represents a region in the y_k-x_k plane corresponding to the half-plane $x_k \geq 0$, if $\hat{h}_{jk} \geq 0$, or the half-plane $x_k \leq 0$, if $\hat{h}_{jk} \leq 0$. Similarly, each inequality $\hat{h}_{jk}x_k \leq 0$ represents a region in the y_k-x_k plane corresponding to the half-plane $x_k \leq 0$, if $\hat{h}_{jk} \geq 0$, or the half-plane $x_k \geq 0$, if $\hat{h}_{jk} \leq 0$. Therefore, if we divide the v_j-i_j plane of each resistor R_j into a right half-plane (corresponding to all points with $v_j \geq 0$), a left half-plane (corresponding to all points with $v_j \leq 0$), an upper half-plane (corresponding to all points with $i_j \geq 0$), and a lower half-plane (corresponding to all points with $i_j \leq 0$), then only the points in certain half-plane combinations can satisfy the inequalities given by Eqs. (7-73) and (7-74) simultaneously. Those segment combinations that do not satisfy Eqs. (7-73) and (7-74) are, therefore, virtual segment combinations and can be systematically eliminated in accordance with two elimination criteria:

Elimination Criteria 1. All combinations consisting of those segments of the v_j-i_j curve of R_j in the rectangular region bounded by $y_j \leq \hat{s}_j$ and $\hat{h}_{jj}x_j \geq 0$ *and* those segments of the v_k-i_k curve of the remaining resistors R_k, $k = 1, 2, \ldots, m$, $k \neq j$, in the half-planes bounded by $\hat{h}_{jk}x_k \geq 0$ are virtual segment combinations.

Elimination Criteria 2. All combinations consisting of those segments of the v_j-i_j curve of R_j in the rectangular region bounded by $y_j \geq \hat{s}_j$ and $\hat{h}_{jj}x_j \leq 0$ *and* those segments of the v_k-i_k curve of the remaining resistors R_k, $k = 1, 2, \ldots, m$, $k \neq j$, in the half-plane bounded by $\hat{h}_{jk}x_k \leq 0$ are virtual segment combinations.

Observe that these elimination criteria are valid for *each* of the m equations in Eq. (7-72). Therefore, depending on the "sign" of the hybrid parameters $\hat{h}_{jk}$, each equation in Eq. (7-72) can be used to eliminate those virtual segment combinations satisfying criteria 1 and 2. Observe also that the elimination procedure pertains to only one hybrid representation of the linear m-port network $\hat{N}$. Since there is a total of 2^m distinct hybrid representations, the elimination procedure may be repeated with *all* 2^m hybrid representations to eliminate as many virtual segment combinations as possible.

EXAMPLE 7-1. To illustrate the steps for implementing the preceding algorithm, consider the network shown in Fig. 7-6(a), which is redrawn in Fig. 7-6(b) in the form of a two-port $\hat{N}$ terminated by the Thevenin equivalent circuits associated with the two nonlinear resistors. To demonstrate that the algorithm is applicable to arbitrary nonlinear resistors—including those characterized by multivalued v_j-i_j curves[9]—we have chosen the v_1-i_1 and v_2-i_2 curves shown in Fig. 7-7(a) and (b) to represent two nonlinear resistors R_1 and R_2. The parameters associated with each segment of these v_j-i_j curves are tabulated in Fig. 7-6(c).

The first step is to derive the four distinct hybrid representations associated with the linear two-port $\hat{N}$. This is easily found and the result is tabulated in column 1 of Table 7-1. The next step is to look at each of the two terminal equations associated with each hybrid representation and identify the virtual segment combinations in accordance with the two elimination criteria presented earlier. For example, applying the elimination criteria 1 to the first terminal equation,

$$v_1 = -2i_1 - i_2 + 2 \tag{7-75}$$

we identify $y_1 = v_1, x_1 = i_1, x_2 = i_2$, and $\hat{s}_1 = 2$. Therefore, all combinations consisting of those segments of the v_1-i_1 curve of R_1 in the rectangular region bounded by $v_1 \leq 2$ and $i_1 \leq 0$ (or $-2i_1 \geq 0$), and those segments of the v_2-i_2 curve of R_2 in the half-plane $i_2 \leq 0$ (or $-i_2 \geq 0$) are virtual segment combinations. Examination of the v_1-i_1 curve in Fig. 7-7(a) shows that segments 1, 5, and 6 are located in the rectangular region bounded by $v_1 \leq 2$ and $i_1 \leq 0$. Similarly, examination of the v_2-i_2 curve in Fig. 7-7(b) shows that segments 1, 2, and 3 are located in the lower half-plane $i_2 \leq 0$. Therefore, elimination criteria 1 can be used to eliminate *nine* virtual segment combinations associated with Eq. (7-75): (1, 1), (1, 2), (1, 3), (5, 1), (5, 2), (5, 3), (6, 1), (6, 2), and (6, 3).[10] These results are summarized in the first row of Table 7-1. If we next apply elimination criteria 2 to Eq. (7-75), we find that all combinations consisting of segments of the v_1-i_1 curve of R_1 in the rectangular region bounded by $v_1 \geq 2$ and $i_1 \geq 0$ (or $-2i_1 \leq 0$), and those segments of the v_2-i_2 curve of R_2 in the half-plane $i_2 \geq 0$ (or $-i_2 \leq 0$) are virtual segment combinations. Examination of the v_1-i_1 curve in Fig. 7-7(a) shows that segments 9 and 10 are located in the rectangular region bounded by $v_1 \geq 2$ and $i_1 \geq 0$. Similarly, examination of the v_2-i_2 curve in Fig. 7-7(b) shows that segments 4, 5, and 6 are located in the upper half-plane $i_2 \geq 0$. Consequently, elimination criteria 2 can be used to eliminate *six* virtual segment combinations associated with Eq. (7-75): (9, 4), (9, 5), (9, 6), (10, 4), (10, 5), and (10, 6). These results are tabulated in row 2 of Table 7-1.

[9] A multivalued v_j-i_j curve is usually encountered in practice when a one-port resistive network containing nonmonotonic resistors is replaced by an equivalent nonlinear resistor.

[10] The notation (σ_1, σ_2) denotes segment σ_1 of R_1 and segment σ_2 of R_2.

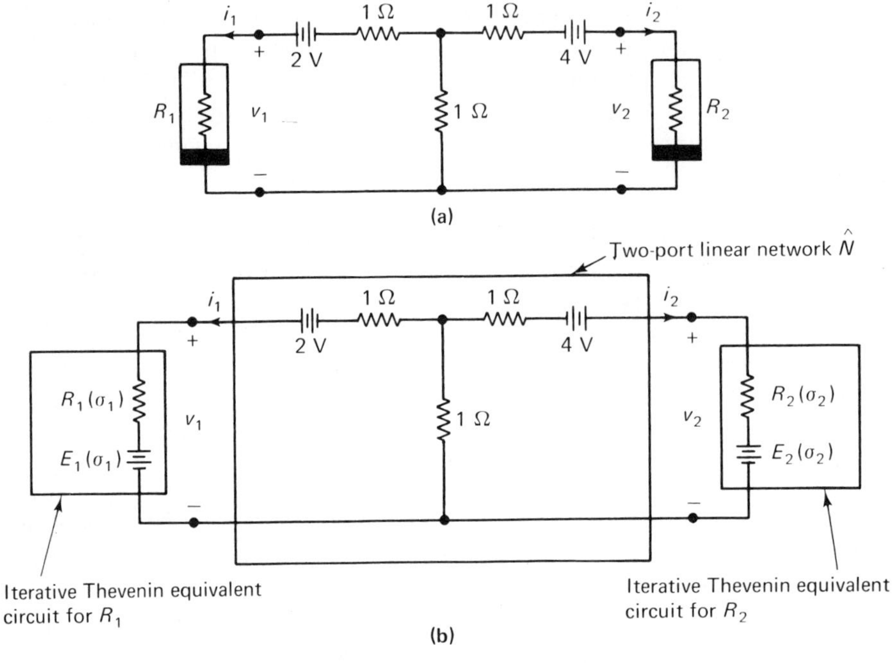

RESISTOR R_j	SEGMENT σ_j	$R_j(\sigma_j)(\Omega)$	$E_j(\sigma_j)(\mathrm{V})$	INTERVAL OF DEFINITION $I_j^-(\sigma_j) \leqslant i_j \leqslant I_j^+(\sigma_j)$,	$E_j^-(\sigma_j) \leqslant v_j \leqslant E_j^+(\sigma_j)$
R_1	1	0.50	− 4.5	$-\infty \leqslant i_1 \leqslant -1$,	$-\infty \leqslant v_1 \leqslant -5$
	2	1.334	− 3.75	$-1 \leqslant i_1 \leqslant 1$,	$-5 \leqslant v_1 \leqslant -3$
	3	2.00	− 4.00	$1 \leqslant i_1 \leqslant 4$,	$-3 \leqslant v_1 \leqslant 4$
	4	1.00	0.00	$0 \leqslant i_1 \leqslant 4$,	$0 \leqslant v_1 \leqslant 4$
	5	3.00	0.00	$-1 \leqslant i_1 \leqslant 0$,	$-3 \leqslant v_1 \leqslant 0$
	6	− 0.50	− 3.50	$-3 \leqslant i_1 \leqslant -1$,	$-3 \leqslant v_1 \leqslant -2$
	7	5.00	13.00	$-3 \leqslant i_1 \leqslant -2$,	$-2 \leqslant v_1 \leqslant 3$
	8	1.00	5.00	$-2 \leqslant i_1 \leqslant 1$,	$3 \leqslant v_1 \leqslant 6$
	9	− 0.50	6.50	$1 \leqslant i_1 \leqslant 3$,	$5 \leqslant v_1 \leqslant 6$
	10	1.00	2.00	$3 \leqslant i_1 \leqslant \infty$,	$5 \leqslant v_1 \leqslant \infty$
R_2	1	4.00	10.00	$-\infty \leqslant i_2 \leqslant -3$,	$-\infty \leqslant v_2 \leqslant -2$
	2	− 1.00	− 5.00	$-3 \leqslant i_2 \leqslant -1$,	$-4 \leqslant v_2 \leqslant -2$
	3	4.00	0.00	$-1 \leqslant i_2 \leqslant 0$,	$-4 \leqslant v_2 \leqslant 0$
	4	1.00	0.00	$0 \leqslant i_2 \leqslant 5$,	$0 \leqslant v_2 \leqslant 5$
	5	− 0.50	7.50	$3 \leqslant i_2 \leqslant 5$,	$5 \leqslant v_2 \leqslant 6$
	6	1.00	3.00	$3 \leqslant i_2 \leqslant \infty$,	$6 \leqslant v_2 \leqslant \infty$

(c)

Fig. 7-6. Segment elimination algorithm in terms of the "sign" of the hybrid parameters.

The next step is to apply the two elimination criteria to the second terminal equation

$$v_2 = -i_1 - 2i_2 + 4 \tag{7-76}$$

associated with the first hybrid representation. For example, elimination criteria 1

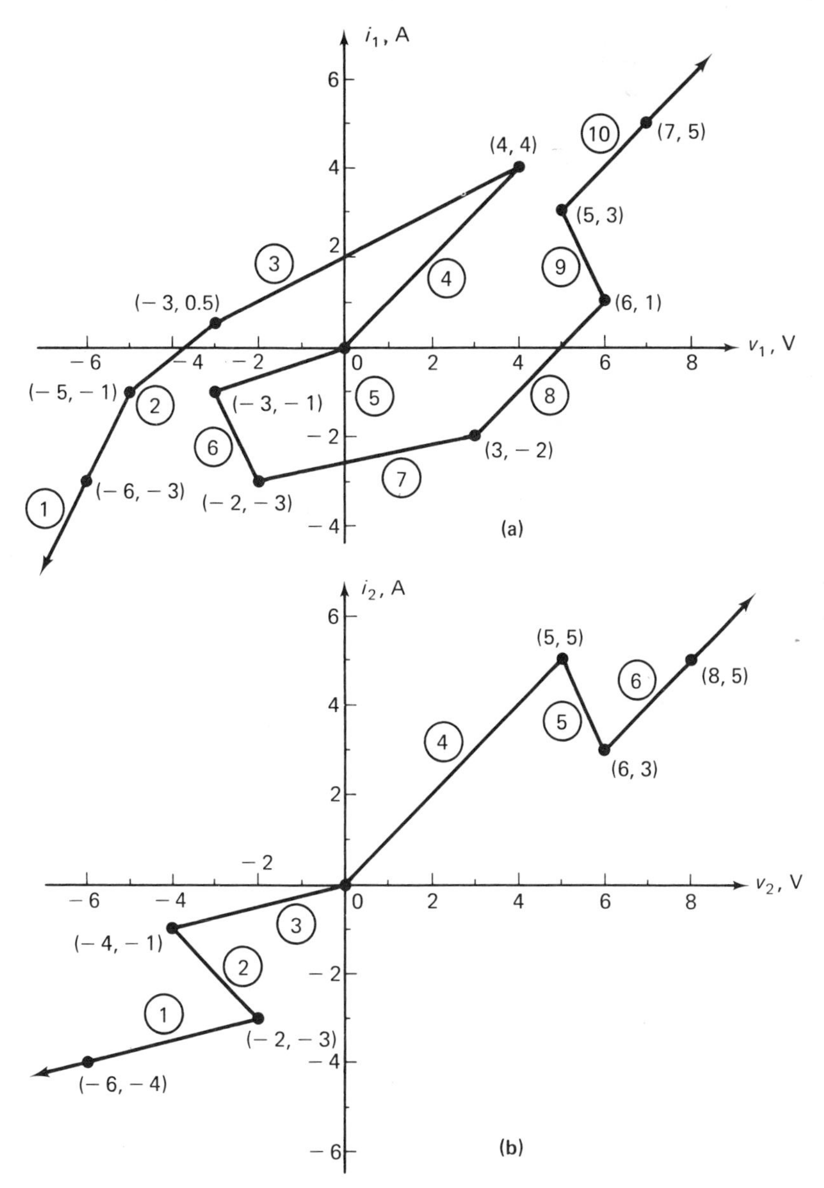

Fig. 7-7. The v_j-i_j curves associated with the two nonlinear resistors in Fig. 7-6.

can be used to eliminate *twelve* segment combinations corresponding to segments 1, 5, 6, and 7 for R_1 and segments 1, 2, and 3 for R_2. Observe, however, that nine of these twelve combinations coincided with those already eliminated in row 1. Therefore, only *three* new virtual segment combinations are eliminated in this case: (7, 1), (7, 2), and (7, 3). These results are tabulated in row 3 of Table 7-1. By a similar procedure, elimination criteria 2 can be used to eliminate *eight* segment combinations corresponding to segments 3, 4, 9 and 10 for R_1 and segments 5 and 6 for R_2. This

TABLE 7-1. Summary of Steps for Implementing the Efficient Algorithm for Eliminating the Virtual Segment Combinations of the Network in Fig. 7-6

HYBRID REPRESENTATION	TERMINAL EQUATION	ELIMINATION CRITERIA	REGION CONTAINING VIRTUAL SEGMENT COMBINATIONS	VIRTUAL SEGMENTS FOR R_1	VIRTUAL SEGMENTS FOR R_2	VIRTUAL SEGMENT COMBINATIONS (SEGMENT COMBINATIONS THAT HAVE ALREADY BEEN ELIMINATED ARE NOT SHOWN)	NUMBER OF VIRTUAL SEGMENT COMBINATIONS
$\begin{bmatrix} v_1 \\ v_2 \end{bmatrix} = \begin{bmatrix} -2 & -1 \\ -1 & -2 \end{bmatrix} \begin{bmatrix} i_1 \\ i_2 \end{bmatrix} + \begin{bmatrix} 2 \\ 4 \end{bmatrix}$	$v_1 = -2i_1 - i_2 + 2$	(1)	$v_1 \leq 2, i_1 \leq 0, i_2 \leq 0$	1, 5, 6	1, 2, 3	(1, 1), (1, 2), (1, 3), (5, 1), (5, 2), (5, 3), (6, 1), (6, 2), (6, 3)	9
		(2)	$v_1 \geq 2, i_1 \geq 0, i_2 \geq 0$	9, 10	4, 5, 6	(9, 4), (9, 5), (9, 6), (10, 4), (10, 5), (10, 6)	6
	$v_2 = -i_1 - 2i_2 + 4$	(1)	$v_2 \leq 4, i_2 \leq 0, i_1 \leq 0$	1, 5, 6, 7	1, 2, 3	(7, 1), (7, 2), (7, 3)	3
		(2)	$v_2 \geq 4, i_2 \geq 0, i_1 \geq 0$	3, 4, 9, 10	5, 6	(3, 5), (3, 6), (4, 5), (4, 6)	4
$\begin{bmatrix} i_1 \\ i_2 \end{bmatrix} = \begin{bmatrix} -\frac{2}{3} & \frac{1}{3} \\ \frac{1}{3} & -\frac{2}{3} \end{bmatrix} \begin{bmatrix} v_1 \\ v_2 \end{bmatrix} + \begin{bmatrix} 0 \\ 2 \end{bmatrix}$	$i_1 = -\frac{2}{3}v_1 + \frac{1}{3}v_2$	(1)	$i_1 \leq 0, v_1 \leq 0, v_2 \geq 0$	1, 5, 6	4, 5, 6	(1, 4), (1, 5), (1, 6), (5, 4), (5, 5), (5, 6), (6, 4), (6, 5), (6, 6)	9
		(2)	$i_1 \geq 0, v_1 \geq 0, v_2 \leq 0$	4, 9, 10	1, 2, 3	(4, 1), (4, 2), (4, 3), (9, 1), (9, 2), (9, 3), (10, 1), (10, 2), (10, 3)	9
	$i_2 = \frac{1}{3}v_1 - \frac{2}{3}v_2 + 2$	(1)	$i_2 \leq 2, v_2 \leq 0, v_1 \geq 0$	4, 8, 9, 10	1, 2, 3	(8, 1), (8, 2), (8, 3)	3
		(2)	$i_2 \geq 2, v_2 \geq 0, v_1 \leq 0$	1, 2, 5, 6	5, 6	(2, 5), (2, 6)	2

$\begin{bmatrix} v_1 \\ i_2 \end{bmatrix} = \begin{bmatrix} -\frac{3}{2} & \frac{1}{2} \\ -\frac{1}{2} & -\frac{1}{2} \end{bmatrix} \begin{bmatrix} i_1 \\ v_2 \end{bmatrix} + \begin{bmatrix} 0 \\ 2 \end{bmatrix}$	$v_1 = -\frac{3}{2}i_1 + \frac{1}{2}v_2$	(1)	$v_1 \leq 0, i_1 \leq 0, v_2 \geq 0$	1, 5, 6	4, 5, 6		0
		(2)	$v_1 \geq 0, i_1 \geq 0, v_2 \leq 0$	4, 9, 10	1, 2, 3		0
	$i_2 = -\frac{1}{2}i_1 - \frac{1}{2}v_2 + 2$	(1)	$i_2 \leq 2, v_2 \leq 0, i_1 \leq 0$	1, 5, 6, 7	1, 2, 3		0
		(2)	$i_2 \geq 2, v_2 \geq 0, i_1 \geq 0$	3, 4, 9, 10	5, 6		0
$\begin{bmatrix} i_1 \\ v_2 \end{bmatrix} = \begin{bmatrix} -\frac{1}{2} & -\frac{1}{2} \\ \frac{1}{2} & -\frac{3}{2} \end{bmatrix} \begin{bmatrix} v_1 \\ i_2 \end{bmatrix} + \begin{bmatrix} 1 \\ 3 \end{bmatrix}$	$i_1 = -\frac{1}{2}v_1 - \frac{1}{2}i_2 + 1$	(1)	$i_1 \leq 1, v_1 \leq 0, i_2 \leq 0$	1, 2	1, 2, 3	(2, 1), (2, 2), (2, 3)	3
		(2)	$i_1 \geq 1, v_1 \geq 0, i_2 \geq 0$	9, 10	4, 5, 6		0
	$v_2 = \frac{1}{2}v_1 - \frac{3}{2}i_2 + 3$	(1)	$v_2 \leq 3, i_2 \leq 0, v_1 \geq 0$	4, 8, 9, 10	1, 2, 3		0
		(2)	$v_2 \geq 3, i_2 \geq 0, v_1 \leq 0$	1, 2, 5, 6	5, 6		0

Total number of virtual segment combinations eliminated = 48

Total number of segment combinations = 10 × 6 = 60
Total number of virtual segment combinations eliminated = −48
Total number of iterations = 12

time only *four* new segment combinations are eliminated: (3, 5), (3, 6), (4, 5), and (4, 6). These results are tabulated in row 4 of Table 7-1.

The algorithm is then repeatedly applied to each of the remaining three hybrid representations. The results are tabulated in the remaining 12 rows of Table 7-1. Observe that a total of 48 virtual segment combinations have been identified, and hence eliminated. Since there is a total of 60 distinct segment combinations for this network, only 12 segment combinations need actually be analyzed. They are (2, 4), (3, 1), (3, 2), (3, 3), (3, 4), (4, 4), (7, 4), (7, 5), (7, 6), (8, 4), (8, 5), and (8, 6). To obtain the solutions, we must substitute the appropriate parameters associated with each of these 12 segment combinations into Eq. (7-71), where $\hat{\boldsymbol{H}}$ and $\hat{\boldsymbol{s}}$ may correspond to any one of the four hybrid representations.

From the preceding example, it should be clear that the combinatorial efficiency will increase as the number of segments per v_j-i_j curve increases, because the algorithm eliminates not just one segment at a time, but an entire rectangular region or half-plane. In fact, this algorithm would not be efficient if each nonlinear resistor contained only two segments—as in the case of an ideal diode—because then in each row of Table 7-1 we could eliminate at most one segment combination. Therefore, we can now afford to approximate each v_j-i_j curve with more piecewise-linear segments and still not run the risk of excessive computer time.

An examination of Table 7-1 shows that out of the four hybrid representations, there is a total of $4^2 = 16$ elimination steps (corresponding to the 16 rows in Table 7-1). Some of these elimination steps do not furnish new information because the associated virtual segment combinations have been found earlier.[11] It is possible to find additional algorithms to detect this situation beforehand, provided additional restrictions are imposed on the location of the v_j-i_j curves, such as restricting all v_j-i_j curves to the first and third quadrants.

Finally, it is important to observe that a computer program can be easily written to implement the preceding algorithm automatically. The computer must, of course, be programmed to evaluate the 2^m distinct hybrid matrices $\hat{\boldsymbol{H}}$ and the 2^m distinct source vectors $\hat{\boldsymbol{s}}$ associated with the linear m-port $\hat{N}$. This would have been a very time-consuming task had it not been for the development of a very efficient technique to be described in Section 7-5-1. The remaining tasks to be performed by the computer are essentially logic operations, which are comparatively much faster than arithmetic operations. Observe that the algorithm for finding the regions containing the virtual segment combinations (column 4 in Table 7-1) depend only on the *sign* of the hybrid parameters $\hat{h}_{jk}$ and the *value* of the source constant $\hat{s}_j$. Since the values of the hybrid parameters are not used at all in the elimination steps, the hybrid matrices $\hat{\boldsymbol{H}}$ need not be calculated with precision. Moreover, to save storage space, only the *sign* of the $\hat{h}_{jk}$ need be stored.

Since the segment combinations to be analyzed usually can be ordered consecutively so that two adjacent segment combinations differ by only one segment, only one diagonal element of $\boldsymbol{R}(\sigma)$ needs to be changed. For example, the remaining 12 segment combinations from the preceding example can be arranged in the following

[11]Since there exist examples where *all* elimination steps will yield at least one new virtual segment combination, it is generally necessary to go through all $(m)(2^m)$ distinct elimination steps to eliminate as many virtual segment combinations as possible.

order: (3, 1), (3, 2), (3, 3), (3, 4), (4, 4), (7, 4), (7, 5), (7, 6), (8, 6), (8, 5), (8, 4), and (2, 4). In view of this observation, much computation time can be saved if the matrix $[\hat{\boldsymbol{H}} - \boldsymbol{H}(\sigma)]$ is inverted by the *method of matrix inversion by modification* described in Section 7-4.

Let us illustrate the application of matrix inversion by modification with the example tabulated in Table 7-1. The first segment combination shown in this table corresponds to the *open-circuit resistance matrix* representation:[12]

$$\boldsymbol{i} = [\hat{\boldsymbol{R}} - \boldsymbol{R}(\sigma)]^{-1}[\boldsymbol{E}(\sigma) - \hat{\boldsymbol{s}}] \tag{7-77}$$

Out of the 12 segment combinations isolated earlier, we found that only three segment combinations result in a valid solution. They are (3, 4), (4, 4), and (8, 4). To save space, only the computations corresponding to these three segment combinations are shown here:

Segment combination 1: $\sigma^{(1)} = (3, 4)$

$$[\hat{\boldsymbol{R}} - \boldsymbol{R}(\sigma^{(1)})]^{-1} = \left\{\begin{bmatrix} -2.0 & -1.0 \\ -1.0 & -2.0 \end{bmatrix} - \begin{bmatrix} 2.0 & 0 \\ 0 & 1.0 \end{bmatrix}\right\}^{-1}$$

$$= \begin{bmatrix} -4.0 & -1.0 \\ -1.0 & -3.0 \end{bmatrix}^{-1} = \begin{bmatrix} -\frac{3}{11} & \frac{1}{11} \\ \frac{1}{11} & -\frac{4}{11} \end{bmatrix}$$

$$[\boldsymbol{E}(\sigma^{(1)}) - \hat{\boldsymbol{s}}] = \begin{bmatrix} -4.0 \\ 0.0 \end{bmatrix} - \begin{bmatrix} 2.0 \\ 4.0 \end{bmatrix} = \begin{bmatrix} -6.0 \\ -4.0 \end{bmatrix}$$

$$\therefore \begin{bmatrix} i_1 \\ i_2 \end{bmatrix} = \begin{bmatrix} -\frac{3}{11} & \frac{1}{11} \\ \frac{1}{11} & -\frac{4}{11} \end{bmatrix}\begin{bmatrix} -6.0 \\ -4.0 \end{bmatrix} = \begin{bmatrix} \frac{14}{11} \\ \frac{10}{11} \end{bmatrix} = \begin{bmatrix} 1.272 \\ 0.909 \end{bmatrix}$$

Segment combination 2: $\sigma^{(2)} = (4, 4)$

$$[\hat{\boldsymbol{R}} - \boldsymbol{R}(\sigma^{(2)})] = \begin{bmatrix} -2.0 & -1.0 \\ -1.0 & -2.0 \end{bmatrix} - \begin{bmatrix} 1.0 & 0.0 \\ 0.0 & 1.0 \end{bmatrix} = \begin{bmatrix} -3.0 & -1.0 \\ -1.0 & -3.0 \end{bmatrix}$$

$$[\boldsymbol{E}(\sigma^{(2)}) - \hat{\boldsymbol{s}}] = \begin{bmatrix} 0.0 \\ 0.0 \end{bmatrix} - \begin{bmatrix} 2.0 \\ 4.0 \end{bmatrix} = \begin{bmatrix} -2.0 \\ -4.0 \end{bmatrix}$$

If we let $\boldsymbol{A} \triangleq [\hat{\boldsymbol{R}} - \boldsymbol{R}(\sigma^{(1)})]$ and $\boldsymbol{B} \triangleq [\hat{\boldsymbol{R}} - \boldsymbol{R}(\sigma^{(2)})]$, then only the first diagonal elements of $\boldsymbol{A}$ and $\boldsymbol{B}$ are different; i.e., $(\boldsymbol{B})_{11} = -4.0 + (\Delta \boldsymbol{A})_{11} = -3.0$, or $(\Delta \boldsymbol{A})_{11} = 1.0$. Hence, using Eq. (7-64), we obtain

$$[\hat{\boldsymbol{R}} - \boldsymbol{R}(\sigma^{(2)})]^{-1} = \boldsymbol{B}^{-1} = \boldsymbol{A}^{-1} - \left[\frac{(\Delta \boldsymbol{A})_{11}}{1 + (\Delta \boldsymbol{A})_{11}\alpha_{11}}\right]\boldsymbol{M}_1$$

$$= \begin{bmatrix} -\frac{3}{11} & \frac{1}{11} \\ \frac{1}{11} & -\frac{4}{11} \end{bmatrix} - \left[\frac{1}{1 + (-\frac{3}{11})}\right]\begin{bmatrix} -\frac{3}{11} \\ \frac{1}{11} \end{bmatrix}[-\tfrac{3}{11} \quad \tfrac{1}{11}]$$

$$= \begin{bmatrix} -\frac{3}{11} & \frac{1}{11} \\ \frac{1}{11} & -\frac{4}{11} \end{bmatrix} - \begin{bmatrix} \frac{9}{88} & -\frac{3}{88} \\ -\frac{3}{88} & \frac{1}{88} \end{bmatrix} = \begin{bmatrix} -\frac{3}{8} & \frac{1}{8} \\ \frac{1}{8} & -\frac{3}{8} \end{bmatrix}$$

$$\therefore \begin{bmatrix} i_1 \\ i_2 \end{bmatrix} = \begin{bmatrix} -\frac{3}{8} & \frac{1}{8} \\ \frac{1}{8} & -\frac{3}{8} \end{bmatrix}\begin{bmatrix} -2.0 \\ -4.0 \end{bmatrix} = \begin{bmatrix} \frac{1}{4} \\ \frac{5}{4} \end{bmatrix} = \begin{bmatrix} 0.25 \\ 1.25 \end{bmatrix}$$

[12]Since there is no apparent advantage for starting with any particular hybrid matrix representation, we shall choose the open-circuit resistance matrix for our initial segment combination since this matrix can be most easily computed by the topological formulas presented in Chapter 6.

Segment combination 3: $\sigma^{(3)} = (8, 4)$

$$[\hat{\boldsymbol{R}} - \boldsymbol{R}(\sigma^{(3)})] = \begin{bmatrix} -2.0 & -1.0 \\ -1.0 & -2.0 \end{bmatrix} - \begin{bmatrix} 1.0 & 0.0 \\ 0.0 & 1.0 \end{bmatrix} = \begin{bmatrix} -3.0 & -1.0 \\ -1.0 & -3.0 \end{bmatrix}$$

$$[\boldsymbol{E}(\sigma^{(3)}) - \hat{\boldsymbol{s}}] = \begin{bmatrix} 5.0 \\ 0.0 \end{bmatrix} - \begin{bmatrix} 2.0 \\ 4.0 \end{bmatrix} = \begin{bmatrix} 3.0 \\ -4.0 \end{bmatrix}$$

Since the matrix $[\hat{\boldsymbol{R}} - \boldsymbol{R}(\sigma^{(3)})]$ is identical to the matrix $[\hat{\boldsymbol{R}} - \boldsymbol{R}(\sigma^{(2)})]$, the inverse is already known

$$\therefore \begin{bmatrix} i_1 \\ i_2 \end{bmatrix} = \begin{bmatrix} -\frac{3}{8} & \frac{1}{8} \\ \frac{1}{8} & -\frac{3}{8} \end{bmatrix} \begin{bmatrix} 3.0 \\ -4.0 \end{bmatrix} = \begin{bmatrix} -\frac{13}{8} \\ \frac{15}{8} \end{bmatrix} = \begin{bmatrix} -1.625 \\ 1.875 \end{bmatrix}$$

These three solutions represent valid operating points of the network, as can be easily verified against the interval of definition of the respective segments in Fig. 7-6(c). The remaining nine solutions represent *virtual operating points* of the network, because in all cases either the value of i_1 or i_2 (or both) falls outside the interval of definition of the corresponding segments.

To summarize, we have shown an example where 48 of the 60 iterations required by the combinatorial algorithm can be eliminated by the preceding method. Hence, the combinatorial efficiency increases from $\eta = \frac{3}{60}$ to $\eta = \frac{3}{12}$.

7-5-1 A Simple Method for Generating All Hybrid Representations

To implement the preceding algorithm efficiently, it is necessary to derive a simple relationship between the hybrid representations so that, as soon as one representation is computed, the remaining $2^m - 1$ representations can be obtained quickly. Our main objective in this section is to derive the desired relationships.

Consider first any two hybrid representations out of the 2^m distinct representations: e.g.,

$$\boldsymbol{y} = \hat{\boldsymbol{H}}\boldsymbol{x} + \hat{\boldsymbol{s}} \tag{7-78}$$

$$\boldsymbol{y}^* = \hat{\boldsymbol{H}}^*\boldsymbol{x}^* + \hat{\boldsymbol{s}}^* \tag{7-79}$$

Since $y_j = v_j$ if $x_j = i_j$, and $y_j = i_j$ if $x_j = v_j$, these two sets of port vectors are always related by

$$\boldsymbol{y}^* = \boldsymbol{A}\boldsymbol{y} + \boldsymbol{B}\boldsymbol{x} \tag{7-80}$$

$$\boldsymbol{x}^* = \boldsymbol{B}\boldsymbol{y} + \boldsymbol{A}\boldsymbol{x} \tag{7-81}$$

where $\boldsymbol{A}$ and $\boldsymbol{B}$ are $m \times m$ *diagonal* matrices with diagonal elements

$$a_{jj} = 1, \qquad b_{jj} = 0, \qquad \text{if } y_j^* = y_j \tag{7-82}$$

$$a_{jj} = 0, \qquad b_{jj} = 1, \qquad \text{if } y_j^* = x_j \tag{7-83}$$

We can now prove the following theorem.[13]

[13]This theorem is valid for arbitrary $m \times m$ matrices $\boldsymbol{A}$ and $\boldsymbol{B}$, and is not restricted by Eqs. (7-82) and (7-83).

Theorem 7-2. *The two hybrid representations are related to each other by the following relationships:*

$$\hat{\boldsymbol{H}}^* = (\boldsymbol{A}\hat{\boldsymbol{H}} + \boldsymbol{B})(\boldsymbol{B}\hat{\boldsymbol{H}} + \boldsymbol{A})^{-1} \tag{7-84}$$

$$\hat{\boldsymbol{s}}^* = (\boldsymbol{A} - \hat{\boldsymbol{H}}^*\boldsymbol{B})\hat{\boldsymbol{s}} \tag{7-85}$$

Proof: Substituting Eq. (7-78) into (7-80) and (7-81), we obtain

$$\boldsymbol{y}^* = \boldsymbol{A}(\hat{\boldsymbol{H}}\boldsymbol{x} + \hat{\boldsymbol{s}}) + \boldsymbol{B}\boldsymbol{x} = (\boldsymbol{A}\hat{\boldsymbol{H}} + \boldsymbol{B})\boldsymbol{x} + \boldsymbol{A}\hat{\boldsymbol{s}} \tag{7-86}$$

$$\boldsymbol{x}^* = \boldsymbol{B}(\hat{\boldsymbol{H}}\boldsymbol{x} + \hat{\boldsymbol{s}}) + \boldsymbol{A}\boldsymbol{x} = (\boldsymbol{B}\hat{\boldsymbol{H}} + \boldsymbol{A})\boldsymbol{x} + \boldsymbol{B}\hat{\boldsymbol{s}} \tag{7-87}$$

Solving for $\boldsymbol{x}$ in Eq. (7-87), we obtain

$$\boldsymbol{x} = (\boldsymbol{B}\hat{\boldsymbol{H}} + \boldsymbol{A})^{-1}\boldsymbol{x}^* - (\boldsymbol{B}\hat{\boldsymbol{H}} + \boldsymbol{A})^{-1}\boldsymbol{B}\hat{\boldsymbol{s}} \tag{7-88}$$

Substituting Eq. (7-88) into Eq. (7-86), we obtain after rearrangement

$$\boldsymbol{y}^* = [(\boldsymbol{A}\hat{\boldsymbol{H}} + \boldsymbol{B})(\boldsymbol{B}\hat{\boldsymbol{H}} + \boldsymbol{A})^{-1}]\boldsymbol{x}^* + [\boldsymbol{A} - (\boldsymbol{A}\hat{\boldsymbol{H}} + \boldsymbol{B})(\boldsymbol{B}\hat{\boldsymbol{H}} + \boldsymbol{A})^{-1}\boldsymbol{B}]\hat{\boldsymbol{s}} \tag{7-89}$$

Q.E.D.

Assuming that the first hybrid matrix $\hat{\boldsymbol{H}}$ and its associated source vector $\hat{\boldsymbol{s}}$ have been computed, Eqs. (7-84) and (7-85) can be used to generate the remaining $2^m - 1$ hybrid representations. These representations exist if, and only if, the matrix $(\boldsymbol{B}\hat{\boldsymbol{H}} + \boldsymbol{A})$ is nonsingular. Since the preceding algorithm requires that all 2^m hybrid representations be found, we can generate these systematically by interchanging only one port current and one port voltage at a time. Under this condition Theorem 7-2 can be restated as follows:

Theorem 7-3. *If the vectors* $\mathbf{y}$ *and* $\mathbf{y}^*$ *(and hence also* $\mathbf{x}$ *and* $\mathbf{x}^*$*) differ in only one element* α*, i.e., if* $y_j^* = y_j$ *and* $x_j^* = x_j$*, for all* $j \neq \alpha$*, then the elements* $\hat{h}_{jk}^*$ *of* $\hat{\boldsymbol{H}}^*$ *can be found directly from the elements* $\hat{h}_{jk}$ *of* $\hat{\boldsymbol{H}}$ *in accordance with the following relationships:*

$$\begin{aligned}
\hat{h}_{jk}^* &= \hat{h}_{jk} - \hat{h}_{j\alpha}\frac{\hat{h}_{\alpha k}}{\hat{h}_{\alpha\alpha}}, &&\text{for all } j \neq \alpha,\ k \neq \alpha \\
&= \frac{\hat{h}_{j\alpha}}{\hat{h}_{\alpha\alpha}}, &&\text{for all } j \neq \alpha,\ k = \alpha \\
&= -\frac{\hat{h}_{\alpha k}}{\hat{h}_{\alpha\alpha}}, && j = \alpha,\ k \neq \alpha \\
&= \frac{1}{\hat{h}_{\alpha\alpha}}, && j = k = \alpha
\end{aligned} \tag{7-90}$$

Similarly, the elements $\hat{s}_j^*$ *of* $\hat{\boldsymbol{s}}^*$ *can be found directly from the elements of* $\hat{\boldsymbol{s}}$ *and* $\hat{\boldsymbol{H}}$ *in accordance with the following relationships:*

$$\begin{aligned}
\hat{s}_j^* &= \hat{s}_j - \hat{h}_{j\alpha}^*\hat{s}_\alpha, &&\text{for all } j \neq \alpha \\
&= -\hat{h}_{j\alpha}^*\hat{s}_\alpha, && j = \alpha
\end{aligned} \tag{7-91}$$

Proof: It is easily shown that if $\boldsymbol{y}^*$ differs from $\boldsymbol{y}$ only in one element α, then

$$(\boldsymbol{B}\hat{\boldsymbol{H}}+\boldsymbol{A})^{-1}=\begin{bmatrix} 1 & 0 & \cdots & 0 \\ 0 & 1 & \cdots & 0 \\ 0 & 0 & \cdots & 0 \\ \cdot & \cdot & \cdots & \cdot \\ \cdot & \cdot & \cdots & \cdot \\ \cdot & \cdot & \cdots & \cdot \\ \hat{h}_{\alpha 1} & \hat{h}_{\alpha 2}\cdots & \hat{h}_{\alpha\alpha}\cdots & \hat{h}_{\alpha m} \\ \cdot & \cdot & \cdots & \cdot \\ \cdot & \cdot & \cdots & \cdot \\ \cdot & \cdot & \cdots & \cdot \\ 0 & 0 & \cdots & 1 \end{bmatrix}^{-1}=\begin{bmatrix} 1 & 0 & \cdots & 0 \\ 0 & 1 & \cdots & 0 \\ 0 & 0 & \cdots & 0 \\ \cdot & \cdot & \cdots & \cdot \\ \cdot & \cdot & \cdots & \cdot \\ \cdot & \cdot & \cdots & \cdot \\ -\dfrac{\hat{h}_{\alpha 1}}{\hat{h}_{\alpha\alpha}} & -\dfrac{\hat{h}_{\alpha 2}}{\hat{h}_{\alpha\alpha}}\cdots & \dfrac{1}{\hat{h}_{\alpha\alpha}}\cdots & -\dfrac{\hat{h}_{\alpha m}}{\hat{h}_{\alpha\alpha}} \\ \cdot & \cdot & \cdots & \cdot \\ \cdot & \cdot & \cdots & \cdot \\ \cdot & \cdot & \cdots & \cdot \\ 0 & 0 & \cdots & 0 \end{bmatrix} \tag{7-92}$$

Substituting Eq. (7-92) into Eq. (7-84), we obtain

$$\hat{\boldsymbol{H}}^*=\begin{bmatrix} \hat{h}_{11} & \hat{h}_{12} & \cdots & \hat{h}_{1m} \\ \hat{h}_{21} & \hat{h}_{22} & \cdots & \hat{h}_{2m} \\ \cdot & \cdot & \cdots & \cdot \\ \cdot & \cdot & \cdots & \cdot \\ \cdot & \cdot & \cdots & \cdot \\ 0 & 0 & 1 & 0 \\ \hat{h}_{\alpha+1\,1} & \hat{h}_{\alpha+1\,2} & \cdots & \hat{h}_{\alpha+1\,m} \\ \cdot & \cdot & \cdots & \cdot \\ \cdot & \cdot & \cdots & \cdot \\ \cdot & \cdot & \cdots & \cdot \\ \hat{h}_{m1} & \hat{h}_{m2} & \cdots & \hat{h}_{mm} \end{bmatrix}\begin{bmatrix} 1 & 0 & \cdots & 0 \\ 0 & 1 & \cdots & 0 \\ 0 & 0 & \cdots & 0 \\ \cdot & \cdot & \cdots & \cdot \\ \cdot & \cdot & \cdots & \cdot \\ \cdot & \cdot & \cdots & \cdot \\ -\dfrac{\hat{h}_{\alpha 1}}{\hat{h}_{\alpha\alpha}} & -\dfrac{\hat{h}_{\alpha 2}}{\hat{h}_{\alpha\alpha}}\cdots & \dfrac{1}{\hat{h}_{\alpha\alpha}}\cdots & -\dfrac{\hat{h}_{\alpha m}}{\hat{h}_{\alpha\alpha}} \\ & & \cdots & \\ \cdot & \cdot & \cdots & \cdot \\ \cdot & \cdot & \cdots & \cdot \\ \cdot & \cdot & \cdots & \cdot \\ 0 & 0 & \cdots & 0 \end{bmatrix} \tag{7-93}$$

If we carry out this multiplication, we obtain the relationships given in Eq. (7-90). By a similar procedure, we obtain the relationships given in Eq. (7-91). Q.E.D.

Observe that Theorem 7-3 implies that the sequence of hybrid representations will exist if, and only if, the element $\hat{h}_{\alpha\alpha}$ associated with the preceding representation is not zero.

As an illustration, let us assume the first hybrid representation in Table 7-1 has been found, and we would like to obtain the third representation from it. In this case, we must assign

$$\boldsymbol{A}=\begin{bmatrix} 1 & 0 \\ 0 & 0 \end{bmatrix} \quad \text{and} \quad \boldsymbol{B}=\begin{bmatrix} 0 & 0 \\ 0 & 1 \end{bmatrix} \tag{7-94}$$

Substituting Eq. (7-94) into Eqs. (7-84) and (7-85), we obtain

$$\begin{aligned} \hat{\boldsymbol{H}}^* &= \left\{\begin{bmatrix} 1 & 0 \\ 0 & 0 \end{bmatrix}\begin{bmatrix} -2 & -1 \\ -1 & -2 \end{bmatrix}+\begin{bmatrix} 0 & 0 \\ 0 & 1 \end{bmatrix}\right\}\left\{\begin{bmatrix} 0 & 0 \\ 0 & 1 \end{bmatrix}\begin{bmatrix} -2 & -1 \\ -1 & -2 \end{bmatrix}+\begin{bmatrix} 1 & 0 \\ 0 & 0 \end{bmatrix}\right\}^{-1} \\ &= \begin{bmatrix} -\frac{3}{2} & \frac{1}{2} \\ -\frac{1}{2} & -\frac{1}{2} \end{bmatrix} \end{aligned} \tag{7-95}$$

$$\hat{\boldsymbol{s}}^* = \left\{\begin{bmatrix}1 & 0\\0 & 0\end{bmatrix} - \begin{bmatrix}-\frac{3}{2} & \frac{1}{2}\\-\frac{1}{2} & -\frac{1}{2}\end{bmatrix}\begin{bmatrix}0 & 0\\0 & 1\end{bmatrix}\right\}\begin{bmatrix}2\\4\end{bmatrix} = \begin{bmatrix}0\\2\end{bmatrix} \tag{7-96}$$

Alternatively, we can obtain the elements of $\hat{\boldsymbol{H}}^*$ and $\hat{\boldsymbol{s}}^*$ from Eqs. (7-90) and (7-91) directly:

$$\hat{h}_{11}^* = \hat{h}_{11} - \hat{h}_{12}\frac{\hat{h}_{21}}{\hat{h}_{22}} = -\frac{3}{2}, \qquad \hat{h}_{12}^* = \frac{\hat{h}_{12}}{\hat{h}_{22}} = \frac{1}{2}$$

$$\hat{h}_{21}^* = -\frac{\hat{h}_{21}}{\hat{h}_{22}} = -\frac{1}{2}, \qquad \hat{h}_{22}^* = \frac{1}{\hat{h}_{22}} = -\frac{1}{2}$$

$$\hat{s}_1^* = \hat{s}_1 - \hat{h}_{12}^*\hat{s}_2 = 0, \qquad \hat{s}_2^* = -\hat{h}_{22}^*\hat{s}_2 = 2$$

7-5-2 Modified Combinatorial Piecewise-Linear Algorithm

The algorithm presented in Section 7-5 makes use of only the "sign" of the elements of the hybrid matrices. If the "value" of each element of the hybrid matrices is used instead of just its sign, it is possible to eliminate more virtual segment combinations, which may not have been eliminated by the preceding combinatorial algorithm. We shall now present this modified combinatorial algorithm [10].

Let us proceed as before and derive the port equations of the m-port $\hat{N}$ in terms of its hybrid parameters, as shown earlier in Eq. (7-72). Let us rewrite this equation into the following more convenient form:

$$(y_j - \hat{h}_{jj}x_j) - \sum_{\substack{k=1\\k\neq j}}^{m} \hat{h}_{jk}x_k - \hat{s}_j = 0, \qquad j = 1, 2, \ldots, m \tag{7-97}$$

The v_j-i_j relationship for the nonlinear resistor R_j connected across port j corresponding to any segment combination σ is given by

$$y_j = f_j(x_j) = h_j(\sigma_j)x_j + s_j(\sigma_j) \tag{7-98}$$

where

$$x_j^-(\sigma_j) \leq x_j \leq x_j^+(\sigma_j) \tag{7-99}$$

If we let $y_j^-(\sigma_j)$ and $y_j^+(\sigma_j)$ be the lower and upper limit for the value of y_j corresponding to segment combination σ_j, then

$$\begin{aligned} y_j^-(\sigma_j) \leq y_j \leq y_j^+(\sigma_j), &\quad \text{when } h_j(\sigma_j) \geq 0\\ y_j^+(\sigma_j) \leq y_j \leq y_j^-(\sigma_j), &\quad \text{when } h_j(\sigma_j) < 0 \end{aligned} \tag{7-100}$$

Now observe that, when Eq. (7-98) is substituted for y_j in Eq. (7-97), we obtain a system of m linear equations in the unknowns $x_1, x_2, \ldots, x_m$, *where each unknown x_j is restricted to lie between the interval given in Eq. (7-99).* A solution obtained by solving for $x_1, x_2, \ldots, x_m$ from this system of equations is therefore valid if, and only if, it satisfies Eq. (7-99). Since each equation in Eq. (7-97) is *linear*, it is easy to compute the *maximum value* (*upper bound*) $U_j(\sigma)$ and the *minimum value* (*lower bound*) $L_j(\sigma)$ of

$$z_j(x_1, x_2, \ldots, x_m) \triangleq y_j - \hat{h}_{jj}x_j - \sum_{\substack{k=1\\k\neq j}}^{m} \hat{h}_{jk}x_k - \hat{s}_j, \qquad j = 1, 2, \ldots, m \tag{7-101}$$

when x_j is allowed to vary *over all values* within the interval prescribed by Eq. (7-99); i.e.,[14]

$$U_j(\sigma) = \max[y_j^+(\sigma_j) - \hat{h}_{jj}x_j^+(\sigma_j), y_j^-(\sigma_j) - \hat{h}_{jj}x_j^-(\sigma_j)] + \sum_{\substack{k=1 \\ k \neq j}}^{m} \max[-\hat{h}_{jk}x_k^+(\sigma_k), -\hat{h}_{jk}x_k^-(\sigma_k)] - \hat{s}_j \tag{7-102}$$

and

$$L_j(\sigma) = \min[y_j^+(\sigma_j) - \hat{h}_{jj}x_j^+(\sigma_j), y_j^-(\sigma_j) - \hat{h}_{jj}x_j^-(\sigma_j)] + \sum_{\substack{k=1 \\ k \neq j}}^{m} \min[-\hat{h}_{jk}x_k^+(\sigma_k), -\hat{h}_{jk}x_k^-(\sigma_k)] - \hat{s}_j \tag{7-103}$$

where $\sigma = (\sigma_1, \sigma_2, \ldots, \sigma_m)$. Now observe that, if $(\hat{x}_1, \hat{x}_2, \ldots, \hat{x}_m)$ is a solution of Eq. (7-97) which also satisfies Eq. (7-99), then Eq. (7-101) must reduce to zero; i.e.,

$$z_j(\hat{x}_1, \hat{x}_2, \ldots, \hat{x}_m) = 0, \qquad j = 1, 2, \ldots, m \tag{7-104}$$

For this to be possible, however, we must have $U_j(\sigma) > 0$ and $L_j(\sigma) < 0$. Hence, we have the following elimination criterion:

Elimination Criterion Based on Upper and Lower Bounds

A segment combination σ gives rise to a valid solution if, and only if, $U_j(\sigma) > 0$ and $L_j(\sigma) < 0$ for all $j = 1, 2, \ldots, m$.

In view of this criterion, our present algorithm simply calls for computing the value of $U_j(\sigma)$ and $L_j(\sigma)$ corresponding to each segment combination σ and then checking the sign of $U_j(\sigma)$ and $L_j(\sigma)$. This procedure is then repeated for all possible segment combinations corresponding to a given hybrid matrix.

The present algorithm clearly entails a lot more steps than that of the preceding algorithm, because *only one segment combination is eliminated at a time.* However, it is still more efficient than the unmodified combinatorial algorithm presented in Section 7-4, because it takes a lot less time to compute $U_j(\sigma)$ and $L_j(\sigma)$ than to solve Eq. (7-59) repeatedly. In the special but rather important case when the hybrid matrix contains many *zero entries*, this algorithm can be rendered much more efficient by reordering the equations so that those containing the fewest number of nonzero variables are considered first. For example, suppose that we have a system of 10 equations in 10 unknowns. Suppose that the third equation contains only two nonzero variables, say

$$(y_5 - \hat{h}_{55}x_5) - \hat{h}_{57}x_7 - \hat{s}_5 = 0 \tag{7-105}$$

[14]The notations max $[A, B]$ and min $[A, B]$ denote, respectively, the *greater* and the *lesser* between the two quantities. Hence,

$$\max[A, B] = A, \quad \text{if } A > B$$
$$= B, \quad \text{if } A \leq B$$

and

$$\min[A, B] = B, \quad \text{if } A > B$$
$$= A, \quad \text{if } A \leq B$$

We consider this equation first, and suppose that for some initial segment combination $\sigma = (1, 1, 1, 1, 1, 1, 1, 1, 1, 1)$ we find that σ is a virtual segment combination because $U_5 < 0$ or $L_5 > 0$. Then clearly the segment combinations $(\sigma_1, \sigma_2, \sigma_3, \sigma_4, 1, \sigma_6, 1, \sigma_8, \sigma_9, \sigma_{10})$, $\sigma_j = 1, 2, \ldots, n_j$ for $j = 1, 2, 3, 4, 6, 8, 9$, and 10, where n_j is the number of piecewise-linear segments in the R_j curve, must also be virtual because Eq. (7-105) does not contain the remaining variables and is therefore unaffected in value by changes in segment combinations, as long as $\sigma_5 = 1$ and $\sigma_7 = 1$. Hence, the present algorithm can be rendered much more efficient in cases when the hybrid matrix has many zero entries. Let us illustrate the application of this algorithm with an example.

EXAMPLE 7-2. Consider the network shown in Fig. 7-8. By the topological formulas presented in Chapter 6, we derive the following hybrid matrix and source vector representation:

$$\begin{bmatrix} i_1 \\ i_3 \\ v_2 \end{bmatrix} = \begin{bmatrix} 1 & 1 & 1 \\ 0 & -1 & 0 \\ 0 & 1 & 1 \end{bmatrix} \begin{bmatrix} v_1 \\ v_3 \\ i_2 \end{bmatrix} + \begin{bmatrix} -5 \\ 5 \\ -5 \end{bmatrix} \tag{7-106}$$

Since row 2 contains two zero entries and row 3 contains one zero entry, we reorder the equations in the order 2, 3, and 1. The reordered equation now takes the form

$$\begin{bmatrix} i_3 \\ v_2 \\ i_1 \end{bmatrix} = \begin{bmatrix} -1 & 0 & 0 \\ 1 & 1 & 0 \\ 1 & 1 & 1 \end{bmatrix} \begin{bmatrix} v_3 \\ i_2 \\ v_1 \end{bmatrix} + \begin{bmatrix} 5 \\ -5 \\ -5 \end{bmatrix} \tag{7-107}$$

Observe that the reordered matrix will always contain zeros in the upper-right corner. We can now apply the elimination criterion based on upper and lower bounds to Eq. (7-107). The sequence of segment elimination takes place as shown in Table 7-2.[15]

TABLE 7-2. Sequence of Segment Elimination

EQUATION NUMBER	SEGMENT COMBINATION $(\sigma_3, \sigma_2, \sigma_1)$	UPPER BOUND	LOWER BOUND	BOUNDS SATISFIED
1	1, 1, 1	−3	−∞	No
1	2, 1, 1	−3	−3	No
1	3, 1, 1	1	−3	Yes
2	3, 1, 1	∞	2	No
2	3, 2, 1	3	−2	Yes
3	3, 2, 1	∞	2	No
3	3, 2, 2	9	−8	Yes, a possible solution
3	3, 2, 3	−1	−∞	No
2	3, 3, 1	−1	−∞	No
1	4, 1, 1	∞	1	No
	End of iteration			

[15]The notation (a, b, c) in Table 7-2 denotes segment a for R_3, b for R_2, and c for R_1.

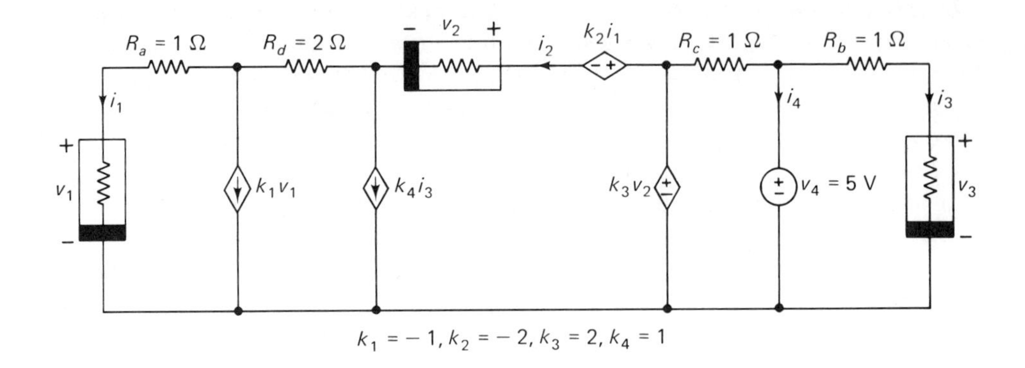

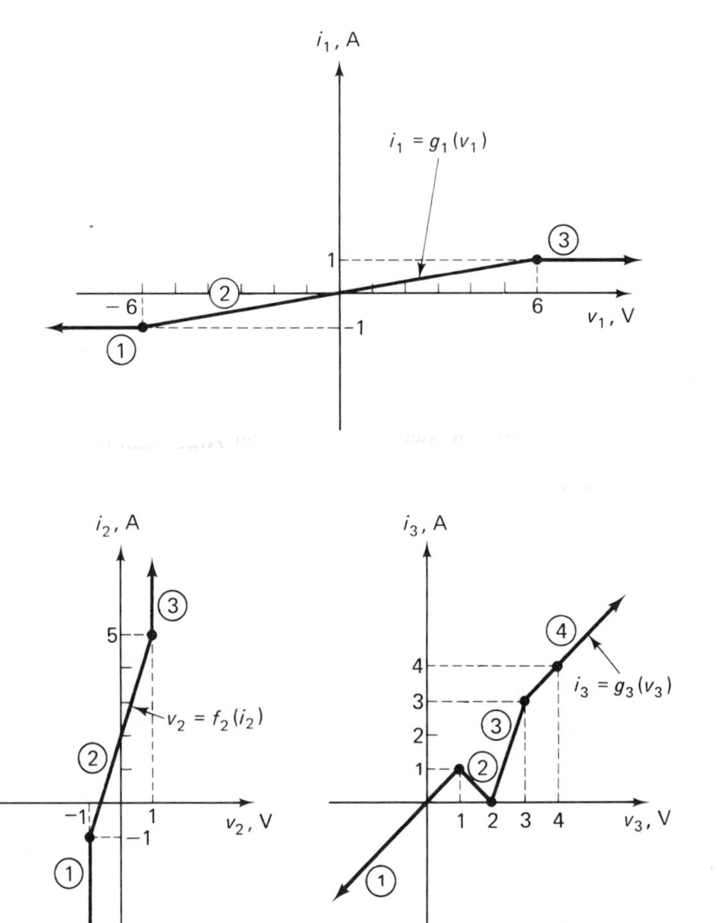

Fig. 7-8. Segment elimination algorithm in terms of upper and lower bounds.

Observe that out of 36 possible segment combinations, only eight were evaluated and only one can possibly yield a valid solution: segment combination (3, 2, 2). The bounds were computed on 10 equations to perform the elimination. Finally, it remains to determine the solution corresponding to segment combinations (3, 2, 2):

$$\begin{bmatrix} v_1 \\ v_3 \\ i_2 \end{bmatrix} = \left\{ \begin{bmatrix} \frac{1}{6} & 0 & 0 \\ 0 & 3 & 0 \\ 0 & 0 & \frac{1}{3} \end{bmatrix} + \begin{bmatrix} -1 & -1 & -1 \\ 0 & 1 & 0 \\ 0 & -1 & -1 \end{bmatrix} \right\}^{-1} \left\{ \begin{bmatrix} -5 \\ 5 \\ -5 \end{bmatrix} - \begin{bmatrix} 0 \\ -6 \\ -\frac{2}{3} \end{bmatrix} \right\} = \begin{bmatrix} -\frac{3}{20} \\ \frac{11}{4} \\ \frac{19}{8} \end{bmatrix} \tag{7-108}$$

Since $[v_1, v_3, i_2] = [-\frac{3}{20}, \frac{11}{4}, \frac{19}{8}]$ falls within the domain of definition for segment 3 for R_3, 2 for R_2, and 2 for R_1, it is indeed a solution.

REFERENCES

1. CHUA, L. O. *Introduction to Nonlinear Network Theory*. New York: McGraw-Hill Book Company, 1969.
2. CHUA, L. O. "Efficient Computer Algorithm for Piecewise-Linear Analysis of Resistive Nonlinear Networks." *IEEE Trans. Circuit Theory*, Vol. CT-18, pp. 73–85, Jan. 1971.
3. KATZENELSON, J. "An Algorithm for Solving Nonlinear Resistive Networks." *Bell System Tech. J.*, Vol. 44, pp. 1605–1620, Oct. 1965.
4. FUJISAWA, T., and E. S. KUH. "Piecewise-Linear Theory of Nonlinear Networks." *SIAM J. Appl. Math.*, Vol. 22, pp. 307–328, Mar. 1972.
5. FUJISAWA, T., E. S. KUH, and T. OHTSUKI. "A Sparse Matrix Method for Analysis of Piecewise-Linear Resistive Networks." *IEEE Trans. Circuit Theory*, Vol. CT-19, pp. 571–584, Nov. 1972.
6. CHIEN, M. J., and E. S. KUH. "Solving Piecewise-Linear Equations for Resistive Networks." Berkeley, Calif.: University of California, Electronics Research Laboratory, Memorandum ERL-M471, Sept. 10, 1974.
7. CHUA, L. O., and A. USHIDA. "A Switching-Parameter Algorithm for Finding Multiple Solutions of Nonlinear Resistive Circuits." Berkeley, Calif.: University of California, Electronics Research Laboratory, Memorandum ERL-M497, Feb. 12, 1975.
8. HOUSEHOLDER, A. S. *Principles of Numerical Analysis*. New York: McGraw-Hill Book Company, 1953.
9. OHTSUKI, T., Y. ISHIZAKI, and H. WATANABE. "Topological Degrees of Freedom and Mixed Analysis of Electrical Networks." *IEEE Trans. Circuit Theory*, Vol. CT-17, No. 4, pp. 491–499, Nov. 1970.
10. PERREAULT, D. A. *Analysis of Nonlinear RLC Networks with Controlled Sources*. Lafayette, Ind.: Purdue University, Ph.D. dissertation, Aug. 1970.
11. CHUA, L. O., and L. K. CHEN. "Nonlinear Diakoptics." *Proceedings of the International Symposium on Circuits and Systems*. Boston, Apr. 21-23, 1975.

PROBLEMS

7-1. Redraw the circuit shown in Fig. 7-8 into the form shown in Fig. 7-1 and derive the *hybrid representation*

$$\begin{bmatrix} i_1 \\ i_3 \\ v_2 \end{bmatrix} = \begin{bmatrix} \hat{h}_{11} & \hat{h}_{12} & \hat{h}_{13} \\ \hat{h}_{21} & \hat{h}_{22} & \hat{h}_{23} \\ \hat{h}_{31} & \hat{h}_{32} & \hat{h}_{33} \end{bmatrix} \begin{bmatrix} v_1 \\ v_3 \\ i_2 \end{bmatrix} + \begin{bmatrix} \hat{s}_1 \\ \hat{s}_2 \\ \hat{s}_3 \end{bmatrix}$$

using the following methods:
(a) Conventional method; i.e., determine $\hat{h}_{ij}$ and $\hat{s}_i$ one at a time through open-circuiting and short-circuiting appropriate ports.
(b) Controlled source extraction method (review Section 6-5-1).
(c) Systematic elimination method (review Section 6-5-2).
(d) N-port constraint matrix method (review Section 6-6).
Warning: Be careful with the reference direction and polarity of controlled sources that depend on port variables.

7-2. Consider the circuit shown in Fig. 7-8 and denote the v-i curves of R_1, R_2, and R_3 by $i_1 = g_1(v_1)$, $v_2 = f_2(i_2)$, and $i_3 = g_3(v_3)$, respectively.
(a) Derive the *hybrid equations* in terms of the unknowns v_1, v_3, and i_2. Use Eq. (7-106) to represent the n-port.
(b) Solve the hybrid equations from part (a) using the Newton–Raphson method.
(c) Solve the hybrid equations from part (a) using the *piecewise-linear* version of the Newton–Raphson method (Section 7-2), and verify that this method coincides with that of solving a sequence of linear networks.

7-3. Solve the hybrid equations from Problem 7-2(a) using Katzenelson's method. Compare the rate of convergence with that of the Newton–Raphson method using several initial guesses.

7-4. Starting with the hybrid representation given in Eq. (7-106), derive the remaining seven hybrid representations using the method described in Section 7-5-1.

7-5. Using the hybrid representations obtained from Problem 7-4, solve the hybrid equations from Problem 7-2(a) by the *piecewise-linear combinatorial method* described in Section 7-4.

7-6. Use the piecewise-linear combinatorial method to obtain the solutions of the circuit shown in Fig. P7-6, where the v_j-i_j curves for R_1 and R_2 are given respectively by the *duals* of the v_j-i_j curves shown in Fig. 7-7. (The dual v-i curve is obtained simply by interchanging the voltage and current axes.)

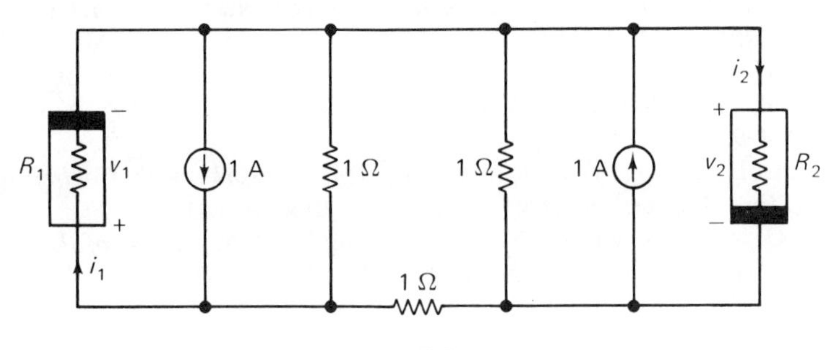

Fig. P7-6

7-7. Use the piecewise-linear combinatorial method to obtain the solutions of the twin tunnel-diode circuit shown in Fig. P7-7(a) where the v-i curve of the two identical tunnel diodes is shown in Fig. P7-7(b).

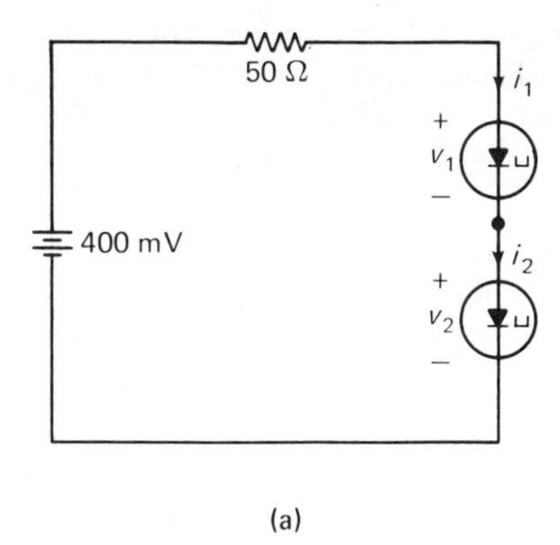

(a)

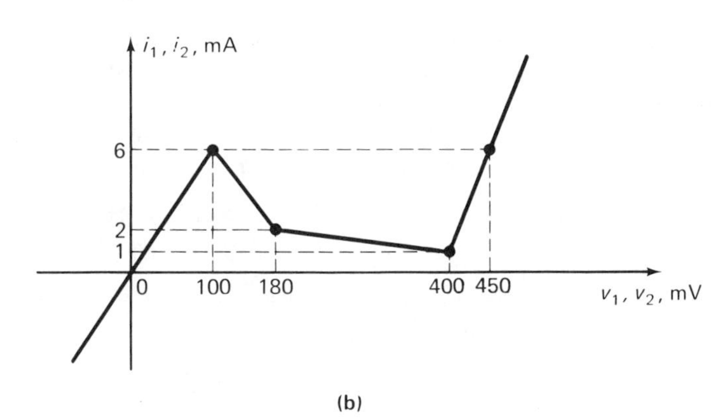

(b)

Fig. P7-7

7-8. Verify the sequence of segment elimination listed in Table 7-2 for the circuit shown in Fig. 7-8.

7-9. A mixed set of voltage and current variables

$$S = \{v_1, v_2, \ldots, v_p, i_{p+1}, i_{p+2}, \ldots, i_q\}$$

associated with a network graph N is said to be *complete* if these variables do not violate KCL or KVL, *and* if the voltage or the current (or both) of *every* branch in N can be expressed as a linear combination of the variables in S via KVL and KCL.

(a) Let $\mathfrak{T}$ be a tree of N and let $\mathfrak{T}$ be partitioned into two parts, $\mathfrak{T}_1$ and $\mathfrak{T}_2$. Let $\mathfrak{L}_1$ be those links which form closed loops with $\mathfrak{T}_1$, and let $\mathfrak{L}_2$ be the remaining links. Show that the voltages associated with branches in $\mathfrak{T}_1$ and the currents associated with the branches in $\mathfrak{L}_2$ form a complete set.

(b) Show that if we label the branches consecutively in the order in which they are found in $\mathfrak{T}_1$, $\mathfrak{T}_2$, $\mathfrak{L}_1$, and $\mathfrak{L}_2$, then the *fundamental loop matrix* $\boldsymbol{B}$ always assumes the form

$$\boldsymbol{B} = \begin{bmatrix} \boldsymbol{B}_{\mathfrak{L}_1\mathfrak{T}_1} & \mathbf{0}_{\mathfrak{L}_1\mathfrak{T}_2} & \mathbf{1}_{\mathfrak{L}_1\mathfrak{L}_1} & \mathbf{0}_{\mathfrak{L}_1\mathfrak{L}_2} \\ \boldsymbol{B}_{\mathfrak{L}_2\mathfrak{T}_1} & \boldsymbol{B}_{\mathfrak{L}_2\mathfrak{T}_2} & \mathbf{0}_{\mathfrak{L}_2\mathfrak{L}_1} & \mathbf{1}_{\mathfrak{L}_2\mathfrak{L}_2} \end{bmatrix}$$

where **1** denotes the unity matrix and **0** the zero matrix.

7-10. Referring to Problem 7-9 and denoting the voltage and current vectors associated respectively with branches in $\mathfrak{L}_1$, $\mathfrak{L}_2$, $\mathfrak{T}_1$, and $\mathfrak{T}_2$ by $\boldsymbol{v}_{\mathfrak{L}_1}$, $\boldsymbol{i}_{\mathfrak{L}_1}$, $\boldsymbol{v}_{\mathfrak{L}_2}$, $\boldsymbol{i}_{\mathfrak{L}_2}$, $\boldsymbol{v}_{\mathfrak{T}_1}$, $\boldsymbol{i}_{\mathfrak{T}_1}$, $\boldsymbol{v}_{\mathfrak{T}_2}$, and $\boldsymbol{i}_{\mathfrak{T}_2}$, show that KVL and KCL lead to the following important relationships:

$$\hat{\boldsymbol{v}}_{\mathfrak{L}_1} = -\boldsymbol{B}_{\mathfrak{L}_1\mathfrak{T}_1}\hat{\boldsymbol{v}}_{\mathfrak{T}_1}$$

$$\hat{\boldsymbol{i}}_{\mathfrak{T}_2} = \boldsymbol{B}^t_{\mathfrak{L}_2\mathfrak{T}_2}\hat{\boldsymbol{i}}_{\mathfrak{L}_2}$$

$$\hat{\boldsymbol{v}}_{\mathfrak{L}_2} + \boldsymbol{B}_{\mathfrak{L}_2\mathfrak{T}_1}\hat{\boldsymbol{v}}_{\mathfrak{T}_1} + \boldsymbol{B}_{\mathfrak{L}_2\mathfrak{T}_2}\hat{\boldsymbol{v}}_{\mathfrak{T}_2} = \boldsymbol{0}_{\mathfrak{L}_2}$$

$$-\boldsymbol{B}^t_{\mathfrak{L}_1\mathfrak{T}_1}\hat{\boldsymbol{i}}_{\mathfrak{L}_1} - \boldsymbol{B}^t_{\mathfrak{L}_2\mathfrak{T}_1}\hat{\boldsymbol{i}}_{\mathfrak{L}_2} + \hat{\boldsymbol{i}}_{\mathfrak{T}_1} = \boldsymbol{0}_{\mathfrak{T}_1}$$

7-11. Let N be a network containing nonmonotonic voltage-controlled resistors, nonmonotonic current-controlled resistors, independent voltage sources, and independent current sources. Form a tree $\mathfrak{T}$ by including as many voltage-controlled resistors as possible. Let these voltage-controlled resistors be grouped into $\mathfrak{T}_1$ and the remaining current-controlled tree resistors be grouped into $\mathfrak{T}_2$. Let all remaining voltage-controlled resistors (which must necessarily form closed loops with the branches in $\mathfrak{T}_1$) be grouped into $\mathfrak{L}_1$ and all remaining current-controlled resistors be grouped into $\mathfrak{L}_2$.

(a) Using the composite branch notation shown in Fig. 4-1, derive the following relationships:

$$\boldsymbol{v}_{\mathfrak{L}_1} = -\boldsymbol{B}_{\mathfrak{L}_1\mathfrak{T}_1}\boldsymbol{v}_{\mathfrak{T}_1} + [\boldsymbol{E}_{\mathfrak{L}_1}(t) + \boldsymbol{B}_{\mathfrak{L}_1\mathfrak{T}_1}\boldsymbol{E}_{\mathfrak{T}_1}(t)]$$

$$\boldsymbol{i}_{\mathfrak{T}_2} = \boldsymbol{B}^t_{\mathfrak{L}_2\mathfrak{T}_2}\boldsymbol{i}_{\mathfrak{L}_2} + [\boldsymbol{J}_{\mathfrak{T}_2}(t) - \boldsymbol{B}^t_{\mathfrak{L}_2\mathfrak{T}_2}\boldsymbol{J}_{\mathfrak{L}_2}(t)]$$

$$\boldsymbol{v}_{\mathfrak{L}_2} + \boldsymbol{B}_{\mathfrak{L}_2\mathfrak{T}_1}\boldsymbol{v}_{\mathfrak{T}_1} + \boldsymbol{B}_{\mathfrak{L}_2\mathfrak{T}_2}\boldsymbol{v}_{\mathfrak{T}_2} = \boldsymbol{E}_{\mathfrak{L}_2}(t) + \boldsymbol{B}_{\mathfrak{L}_2\mathfrak{T}_1}\boldsymbol{E}_{\mathfrak{T}_1}(t) + \boldsymbol{B}_{\mathfrak{L}_2\mathfrak{T}_2}\boldsymbol{E}_{\mathfrak{T}_2}(t)$$

$$-\boldsymbol{B}^t_{\mathfrak{L}_1\mathfrak{T}_1}\boldsymbol{i}_{\mathfrak{L}_1} - \boldsymbol{B}^t_{\mathfrak{L}_2\mathfrak{T}_1}\boldsymbol{i}_{\mathfrak{L}_2} + \boldsymbol{i}_{\mathfrak{T}_1} = \boldsymbol{J}_{\mathfrak{T}_1}(t) - \boldsymbol{B}^t_{\mathfrak{L}_1\mathfrak{T}_1}\boldsymbol{J}_{\mathfrak{L}_1}(t) - \boldsymbol{B}^t_{\mathfrak{L}_2\mathfrak{T}_1}\boldsymbol{J}_{\mathfrak{L}_2}(t)$$

(b) If the constitutive relations of the nonlinear resistors are characterized by

$$\boldsymbol{i}_{\mathfrak{L}_1} = \boldsymbol{i}_{\mathfrak{L}_1}(\boldsymbol{v}_{\mathfrak{L}_1}), \qquad \boldsymbol{v}_{\mathfrak{L}_2} = \boldsymbol{v}_{\mathfrak{L}_2}(\boldsymbol{i}_{\mathfrak{L}_2}), \qquad \boldsymbol{i}_{\mathfrak{T}_1} = \boldsymbol{i}_{\mathfrak{T}_1}(\boldsymbol{v}_{\mathfrak{T}_1}), \qquad \boldsymbol{v}_{\mathfrak{T}_2} = \boldsymbol{v}_{\mathfrak{T}_2}(\boldsymbol{i}_{\mathfrak{T}_2})$$

show that the *hybrid equations* of the network assume the following standard form:

$$\boldsymbol{v}_{\mathfrak{L}_2}(\boldsymbol{i}_{\mathfrak{L}_2}) + \boldsymbol{B}_{\mathfrak{L}_2\mathfrak{T}_1}\boldsymbol{v}_{\mathfrak{T}_1} + \boldsymbol{B}_{\mathfrak{L}_2\mathfrak{T}_2}\boldsymbol{v}_{\mathfrak{T}_2} \circ \{\boldsymbol{B}^t_{\mathfrak{L}_2\mathfrak{T}_2}\boldsymbol{i}_{\mathfrak{L}_2} + [\boldsymbol{J}_{\mathfrak{T}_2}(t) - \boldsymbol{B}^t_{\mathfrak{L}_2\mathfrak{T}_2}\boldsymbol{J}_{\mathfrak{L}_2}(t)]\}$$
$$= \boldsymbol{E}_{\mathfrak{L}_2}(t) + \boldsymbol{B}_{\mathfrak{L}_2\mathfrak{T}_1}\boldsymbol{E}_{\mathfrak{T}_1}(t) + \boldsymbol{B}_{\mathfrak{L}_2\mathfrak{T}_2}\boldsymbol{E}_{\mathfrak{T}_2}(t)$$

$$-\boldsymbol{B}^t_{\mathfrak{L}_1\mathfrak{T}_1}\boldsymbol{i}_{\mathfrak{L}_1} \circ \{-\boldsymbol{B}_{\mathfrak{L}_1\mathfrak{T}_1}\boldsymbol{v}_{\mathfrak{T}_1} + [\boldsymbol{E}_{\mathfrak{L}_1}(t) + \boldsymbol{B}_{\mathfrak{L}_1\mathfrak{T}_1}\boldsymbol{E}_{\mathfrak{T}_1}(t)]\} - \boldsymbol{B}^t_{\mathfrak{L}_2\mathfrak{T}_1}\boldsymbol{i}_{\mathfrak{L}_2} + \boldsymbol{i}_{\mathfrak{T}_1}(\boldsymbol{v}_{\mathfrak{T}_1})$$
$$= \boldsymbol{J}_{\mathfrak{T}_1}(t) - \boldsymbol{B}^t_{\mathfrak{L}_1\mathfrak{T}_1}\boldsymbol{J}_{\mathfrak{L}_1}(t) - \boldsymbol{B}^t_{\mathfrak{L}_2\mathfrak{T}_1}\boldsymbol{J}_{\mathfrak{L}_2}(t)$$

where $\circ$ denotes the composition operation. *Hint:* Make use of the relationships from Problems 7-9 and 7-10.

7-12. Let N be a linear resistive network, and let the linear resistors be partitioned arbitrarily into four groups, $\mathfrak{L}_1$, $\mathfrak{L}_2$, $\mathfrak{T}_1$, and $\mathfrak{T}_2$ as in Problem 7-11. Derive the *hybrid equations* in matrix form and illustrate your result with a simple circuit example. Use the composite branch notation of Fig. 4-1. *Note:* Your result should be a special case of the hybrid equations given in Problem 7-11.

7-13. Referring to Problem 7-12, construct a linear network in which the associated hybrid equations contain fewer variables than those associated with either nodal equations, i.e., $n - 1$, or loop equations, i.e., $b - n + 1$, where n is the number of nodes and b is the number of branches.

7-14. Referring to Problem 7-12, show that in general there exist many different sets of hybrid equations for a given linear network. Develop an algorithm for partitioning the linear resistors so that the resulting hybrid equations contain the *minimum* number m of branch variables. (This number m is called the *topological degree* of the network graph.) *Hint:* Study reference 9.

7-15. Referring to the system of piecewise-linear equations $f(x) = y$ defined by Eq. (7-43), show that the necessary and sufficient condition for the piecewise linear function $f(x)$ to be *continuous* for all values of x is that the Jacobian matrices $A^{(p)}$ and $A^{(q)}$ between any two adjacent regions (regions that share a common hyperplane boundary) be related by

$$A^{(p)} = A^{(q)} + cr^t$$

where c is an arbitrary constant vector and r is the unit vector *normal* to the common hyperplane boundary. Illustrate this condition with a two-dimensional example.

7-16. Prove that the Katzenelson algorithm described in Section 7-3 converges in a finite number of steps if the determinants of the Jacobian matrices $A^{(k)}$ are either all positive, or all negative. *Hint:* Study reference 4.

7-17. The hybrid equations from Problem (7-12) can be further generalized by defining a *generalized* voltage vector $\hat{v}_{\mathfrak{I}_1} \triangleq T_Q v_{\mathfrak{I}_1}$ and a *generalized* current vector $\hat{i}_{\mathfrak{L}_2} \triangleq T_B i_{\mathfrak{L}_2}$, where T_Q and T_B are any *nonsingular* square matrices of compatible dimensions. Formulate the *generalized hybrid equations* in terms of $\hat{v}_{\mathfrak{I}_1}$ and $\hat{i}_{\mathfrak{L}_2}$.

7-18. If the tree $\mathfrak{I}$ in Problem (7-12) is partitioned in such a way that $\mathfrak{I}_1$ consists of several *disconnected* subgraphs, show that the associated hybrid equations will assume a computationally desirable *bordered block-diagonal form*, where the equations involving the variable $v_{\mathfrak{I}_1}$ actually consist of several separate subsystems of equations. Show that this method of analysis—also called *diakoptic analysis* [11]—can be interpreted as that of *tearing* the original network into several disconnected pieces and that each subnetwork may be analyzed separately, and concurrently.

CHAPTER 8

Computer Formulation of State Equations for Dynamic Linear Networks

8-1 WHY A STATE-VARIABLE APPROACH?

The nodal method has been used in Chapter 4 for the analysis of linear networks subject to both dc and ac excitations, and again in Chapter 5 for the analysis of nonlinear resistive networks. Among the important information needed concerning the behavior of a linear time-invariant network are the following:

1. Time-domain responses when the excitations are of arbitrary waveforms.
2. The locations of poles and zeros of some desired network functions.

In the case of large linear networks, the *conventional* nodal method is not the best approach for solving these two types of problems. To see why, let us consider the network shown in Fig. 8-1. We wish to find the step response, i.e., $v_o(t)$ when $v_i(t)$ is a unit step function, and the poles and zeros of the transfer function $V_o(s)/V_i(s)$. For these problems, the initial capacitor voltages and inductor currents are assumed to be

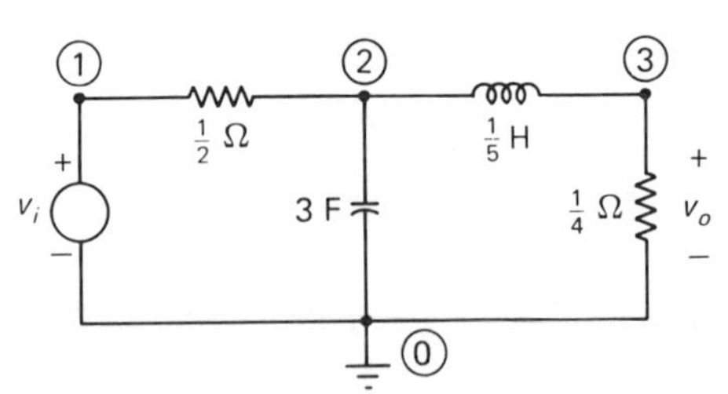

Fig. 8-1. Nodal equations for this network are integrodifferential equations.

zero. Therefore, the nodal equations are

$$\text{At node 2:} \quad 2\Big[v_2(t) - v_i(t)\Big] + 3\frac{dv_2}{dt} + 5\int_0^t [v_2(\tau) - v_0(\tau)]d\tau = 0$$

$$\text{At node 3:} \quad 5\int_0^t [v_0(\tau) - v_2(\tau)]d\tau + 4v_0(t) = 0 \tag{8-1}$$

Note that Eqs. (8-1) contain both derivatives and integrals. In general, nodal analysis, or loop analysis, or cutset analysis of lumped linear networks will all lead to a set of *integrodifferential* equations as exemplified by Eqs. (8-1). A *conventional* method for solving Eqs. (8-1) has been the Laplace transform method. Applying Laplace transform to Eqs. (8-1), we have, after rearranging terms,

$$\Big(3s + 2 + \frac{5}{s}\Big)V_2(s) - \frac{5}{s}V_o(s) = 2V_i(s)$$

$$-\frac{5}{s}V_2(s) + \Big(4 + \frac{5}{s}\Big)V_o(s) = 0 \tag{8-2}$$

where $V_o(s)$, $V_i(s)$, and $V_2(s)$ are the Laplace transforms of $v_o(t)$, $v_i(t)$, and $v_2(t)$, respectively.

By eliminating $V_2(s)$ from Eqs. (8-2), we can solve for the desired network function

$$H(s) \triangleq \frac{V_o(s)}{V_i(s)} = \frac{N(s)}{D(s)} \tag{8-3}$$

where $N(s)$ and $D(s)$ are polynomials in s and have no nontrivial common factors between them. Finding the poles and zeros of $H(s)$ amounts to solving the equations $D(s) = 0$ and $N(s) = 0$. To find the step response, we let $V_i(s) = 1/s$ and perform a partial fraction expansion of $V_o(s)$. Finally, we obtain the time-domain response by taking the inverse Laplace transform of each term in the partial fraction expansion.

This transform method as outlined has the following difficulties when we try to apply the method to *large-scale* networks:

1. Finding the roots of $D(s) = 0$ and $N(s) = 0$ is a serious source of numerical inaccuracy.
2. Partial fraction expansion of $N(s)/D(s)$ is another source of large errors when several poles are quite close or coincide.
3. To obtain $N(s)$ and $D(s)$, we have to calculate some determinants whose elements contain the symbol s. This is not a trivial problem for a digital computer when the equations are in the form of Eq. (8-2).

As the complexity of electronic circuits and systems grows, we have to use more powerful techniques for analyzing circuits, especially those methods which can be adapted for a digital computer. The nodal method becomes very effective for solving the transient problem when we make use of associated discrete circuit models (Sections 1-9 and 17-1), numerical integration techniques (Chaps. 11 and 12), and sparse-matrix techniques (Chap. 16). The nodal method also becomes effective for solving the pole-zero problem when we use the gyrator-capacitor model for an inductor and reduce the problem to a generalized eigenvalue problem (Section 14-4-3).

Since the late 1950s, the state-variable approach has also become popular for analyzing large-scale electronic circuits with the aid of a digital computer. In the state-variable approach, a linear time-invariant network is characterized by two equations of the following forms:

$$\dot{\boldsymbol{x}} = \boldsymbol{A}\boldsymbol{x} + \boldsymbol{B}\boldsymbol{u} \tag{8-4}$$

and

$$\boldsymbol{y} = \boldsymbol{C}\boldsymbol{x} + \boldsymbol{D}\boldsymbol{u} + (\boldsymbol{D}_1\dot{\boldsymbol{u}} + \ldots) \tag{8-5}$$

where

$\boldsymbol{u} = m \times 1$ vector representing the m inputs (independent sources)

$\boldsymbol{y} = p \times 1$ vector representing the p outputs (voltages and/or currents of interest)

$\boldsymbol{x} = n \times 1$ vector consisting of a set of n *independent* auxiliary variables

$\boldsymbol{A}, \boldsymbol{B}, \boldsymbol{C}, \boldsymbol{D}, \boldsymbol{D}_1 =$ constant, real matrices of appropriate dimensions; note that $\boldsymbol{A}$ is always a square matrix of order n

The first equation, Eq. (8-4), is a set of n first-order differential equations (coupled in general), usually referred to as the *state equation* in normal form (other names are *normal form equation* and *differential state equation*). The set of auxiliary variables $x_1, x_2, \ldots, x_n$ are called the *state variables*, and $\boldsymbol{x} = (x_1, x_2, \ldots, x_n)^t$ is called the *state vector*. The second equation, Eq. (8-5), is called the *output equation* (other names are *input-state-output equation* and *read-out map*).

The main attraction of the state-variable approach has been its amenability to many numerical methods of analysis. We shall cite two cases:

1. To find the time-domain response we may apply any of the solution techniques of Chapter 9 to Eq. (8-4). After $\boldsymbol{x}(t)$ is found from Eq. (8-4), it is a trivial matter to obtain $\boldsymbol{y}(t)$ from Eq. (8-5). In this way, we have completely avoided the difficulties inherent in the transform method discussed earlier.
2. The problem of finding the poles and zeros of a transfer function is reduced to that of finding eigenvalues of two real matrices (see Section 9-5) for which many excellent numerical methods are available.

Besides, the number of scalar equations in Eq. (8-4) is often smaller than that in the nodal method (see Eq. (1-5) for example).

Although the state-variable approach offers some advantages over the conventional transform method in linear network analysis, these advantages are not decisive ones. The real value of the state-variable approach becomes apparent when we have to deal with nonlinear dynamic networks (Chapter 10). The Laplace transform method does not apply to nonlinear networks, whereas the state-variable method can be extended without much difficulty to deal with nonlinear networks. Furthermore, many important *qualitative* results of nonlinear networks are formulated only through the state variable approach.

In the next section, we shall give a more detailed discussion of basic concepts, and the remaining sections of this chapter are devoted to the algorithms for formulating Eqs. (8-4) and (8-5) by a digital computer. The time-domain solution of Eq. (8-4) and the calculation of transfer functions will be discussed in Chapter 9.

8-2 STATE VARIABLES, ORDER OF COMPLEXITY, AND INITIAL CONDITIONS

Any lumped network obeys three basic laws: Kirchhoff's voltage law (KVL), Kirchhoff's current law (KCL), and the element laws (branch characteristics). If, from these three types of constraints, we can succeed in reducing the equations to a system of *linearly independent* first-order differential equations

$$\dot{\boldsymbol{x}} = \boldsymbol{f}(\boldsymbol{x}, t) \tag{8-6}$$

where $\boldsymbol{x}$ is a set of n independent, auxiliary variables, we say that the *normal-form equation* (state equation in normal form) for the network exists. In this case, the entries $x_1, x_2, \ldots, x_n$ are called the state variables, and n is defined to be the *order of complexity* of the network.

For a lumped, linear, time-invariant network, if the normal-form equation (8-6) exists, it can be written in the special form given by Eq. (8-4).[1] Furthermore, any set of branch voltages and currents, representing the output vector $\boldsymbol{y}$, is related to $\boldsymbol{x}$ and the input $\boldsymbol{u}$ by Eq. (8-5). For convenience, these two equations are repeated below:

$$\dot{\boldsymbol{x}} = \boldsymbol{Ax} + \boldsymbol{Bu}$$
$$\boldsymbol{y} = \boldsymbol{Cx} + \boldsymbol{Du} + (\boldsymbol{D}_1\dot{\boldsymbol{u}} + \ldots)$$

Even though $\boldsymbol{y}$ may consist of only a few voltages and/or currents of interest, the possibility of $\boldsymbol{y}$ consisting of *all* branch voltages and currents should be kept in mind.

Some obvious questions immediately arise: Does the normal-form equation exist for every linear network? What physical quantities constitute the state variables $x_1, x_2, \ldots, x_n$? How does one determine the order of complexity n? We will answer these questions in this section with the minimum amount of mathematics. Many specific examples will be used to illustrate the concepts.

8-2-1 Significance of the Initial Condition

It is known in elementary differential equations that the most general solution of Eq. (8-4) must contain n arbitrary constants, to be determined by *n initial conditions*. Usually, but not always, these n initial conditions are the values of $x_1, \ldots, x_n$ at $t = 0$. [A brief derivation of the solution of Eq. (8-4) is given in Chapter 9.] Then the order of complexity n of a network is equal to the number of *independent* initial

[1] When written in the form of Eq. (8-4), each x_i is, in general, a linear combination of capacitor voltages, inductor currents, inputs, and derivatives of the inputs. On the other hand, if Eq. (8-6) is written in the form of Eq. (8-48), then the elements of $\boldsymbol{x}$ can be a subset of capacitor voltages and inductor currents.

conditions that can be, and must be, specified in terms of the electrical variables in order to have the complete solution of $\boldsymbol{x}(t)$, and hence $\boldsymbol{y}(t)$.

What electrical variables qualify as the state variables $x_1, \ldots, x_n$? Or, to address the question in a different form, what are the electrical variables whose initial values must be specified in order to have a complete solution of *all* voltages and currents in the network? For linear networks, we shall see (Section 8-4) that capacitor voltages (or charges) and inductor currents (or flux linkages) are *always* among the qualifying variables, whereas capacitor currents and inductor voltages *may or may not* qualify. To see the latter case, consider the networks of Fig. 8-2. In Fig. 8-2(a), complete solutions of all voltages and currents can be obtained for $t \geq 0$ if we know either $v_C(0)$ or $i_C(0)$. On the other hand, in Fig. 8-2(b), complete solutions of all voltages and currents *cannot* be obtained if we are given the initial values $i_{C1}(0) = -i_{C2}(0) = I_0$.

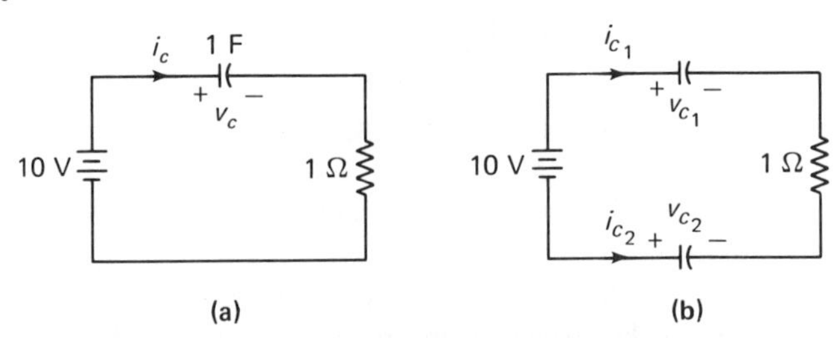

Fig. 8-2. Capacitor current is not suitable as a state variable.

Mathematically, initial conditions represent only arbitrary constants introduced for the purpose of obtaining a unique solution. Physically, initial conditions are introduced because of our ignorance or incomplete knowledge of the past history of the excitations that have been applied to the network. For example, consider the linear capacitor in Fig. 8-2(a). We have

$$i_C = C\frac{dv_C}{dt}$$

$$v_C(t) = \frac{1}{C}\int_{-\infty}^{t} i_C(\tau)\,d\tau = \frac{1}{C}\int_{-\infty}^{0} i_C(\tau)\,d\tau + \frac{1}{C}\int_{0}^{t} i_C(\tau)\,d\tau$$

Therefore,

$$v_C(t) = v_C(0) + \frac{1}{C}\int_{0}^{t} i_C(\tau)\,d\tau, \qquad t \geq 0 \tag{8-7a}$$

where

$$v_C(0) \triangleq \frac{1}{C}\int_{-\infty}^{0} i_C(\tau)\,d\tau \tag{8-7b}$$

Equation (8-7a) signifies that, given the current through the capacitor for $t \geq 0$, the complete determination of the capacitor voltage $v_C(t)$ requires the knowledge of the initial capacitor voltage $v_C(0)$ (not the initial capacitor current). Equation (8-7b) signifies that $v_C(0)$ sums up the effect of the excitation $i_C(t)$ during $-\infty < t < 0$. As far as determining $v_C(t)$ for $t \geq 0$ is concerned, it is sufficient to know $v_C(0)$ without further inquiring into the past history of $i_C(t)$. A dual example can be given for the case of an inductor.

8-2-2 Order of Complexity of *RLC* Networks

From the intuitive discussion of Section 8-2-1, we see that for a lumped, linear, time-invariant network the order of complexity is equal to the number of *independent capacitor voltages and inductor currents* at any t (in particular at $t = 0$). How can capacitor voltages and/or inductor currents become related? For *RLC* networks, such dependence can only arise under the following conditions:

Condition 1. C–E_i loops (the subscript i signifies *independent* sources). The network has some loops consisting of only capacitors and, possibly, independent voltage sources. For example, in Fig. 8-3, three independent C–E_i loops are indicated

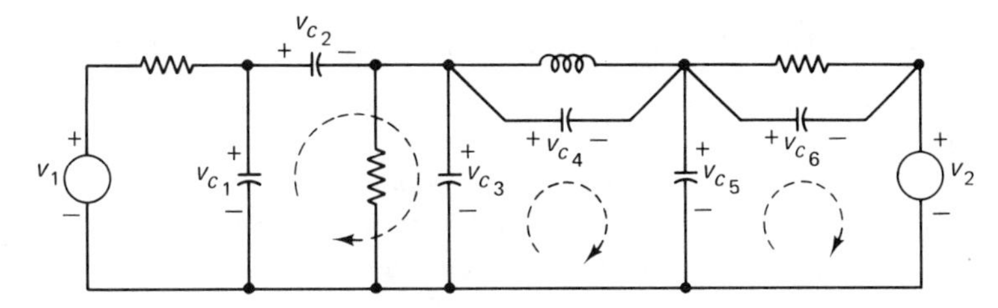

Fig. 8-3. Reduction of order of complexity due to C–E_i loops.

by dashed lines. Therefore, we have three independent relationships:

$$
\begin{aligned}
-v_{C1} + v_{C2} + v_{C3} &= 0 \\
-v_{C3} + v_{C4} + v_{C5} &= 0 \\
-v_{C5} + v_{C6} + v_2 &= 0
\end{aligned}
\tag{8-8}
$$

Although there are six capacitor voltages, only three are independent. For example, if we consider v_{C1}, v_{C3}, and v_{C5} to be independent voltages, then the remaining capacitor voltages v_{C2}, v_{C4}, and v_{C6} can be expressed in terms of v_{C1}, v_{C3}, and v_{C5} by solving Eq. (8-8).

Condition 2. L–J_i cutsets. The network has some cutsets consisting of only inductors and, possibly, independent current sources. For example, in Fig. 8-4, three

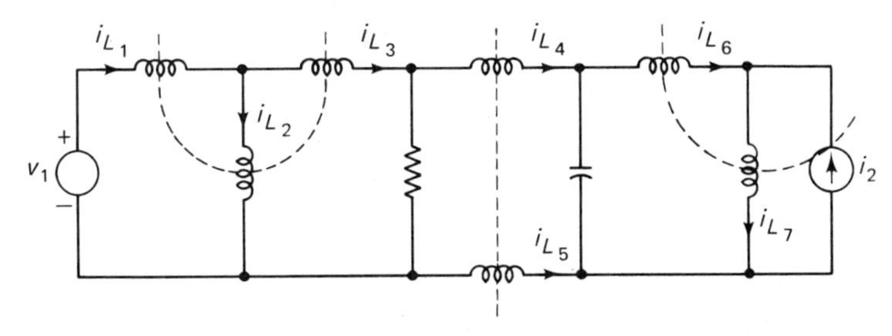

Fig. 8-4. Reduction of order of complexity due to L–J_i cutsets.

independent L–J_i cutsets are indicated by dashed lines. Therefore, we have three independent relationships among the inductor currents:

$$\begin{aligned} i_{L1} - i_{L2} - i_{L3} &= 0 \\ i_{L4} + i_{L5} &= 0 \\ i_{L6} - i_{L7} + i_2 &= 0 \end{aligned} \tag{8-9}$$

Although there are seven inductor currents, only four of them, say i_{L1}, i_{L2}, i_{L4}, and i_{L6}, are independent ones. The remaining inductor currents can be expressed in terms of the chosen independent inductor currents by solving Eq. (8-9).

It should be pointed out that, in the case of C–E_i loops or L–J_i cutsets, dependence among capacitor voltages or inductor currents exists regardless of the exact values of the network elements.

In view of the preceding observations, we can formulate Theorem 8-1.

Theorem 8-1. *The order of complexity of any RLC network is given by*

$$n = b_{LC} - n_C - n_L$$

where

$$\begin{aligned} b_{LC} &= \text{total number of capacitors and inductors} \\ n_C &= \text{total number of \textbf{independent} } C\text{–}E_i \text{ loops} \\ n_L &= \text{total number of \textbf{independent} } L\text{–}J_i \text{ cutsets} \end{aligned}$$

Corollary. *If there are no C–E_i loops and no L–J_i cutsets, the order of complexity of an RLC network is equal to the total number of LC elements.*

8-2-3 Order of Complexity of Linear Active Networks

We shall now consider linear networks containing controlled sources and negative elements. Although the C–E_i loop and L–J_i cutset conditions still give rise to dependency among capacitor voltages and inductor currents, they are now not the only conditions that lead to the dependency. Some "hidden" relationships may exist owing to the action of controlled sources and negative elements. Such dependency occurs when network elements possess some *exact* values and therefore exists only in the idealized model of a network. (No parameter in any *practical* network can have an exact value.) Because the dependency of capacitor voltages and inductor currents due to controlled sources cannot be ascertained from network topology alone, it is not possible to give an explicit formula for the order of complexity n of an active linear network. In such cases, the bounds of n are given by

$$b_{LC} - n_C - n_L \geq n \geq 0$$

We shall give several specific examples to illustrate the intricate situations associated with networks containing controlled sources.

EXAMPLE 8-1. Consider the network of Fig. 8-5(a). Writing the nodal equation, we have

$$i_C - \alpha i_C + \frac{v_C}{R} + \frac{v_C - E}{R} = 0 \tag{8-10}$$

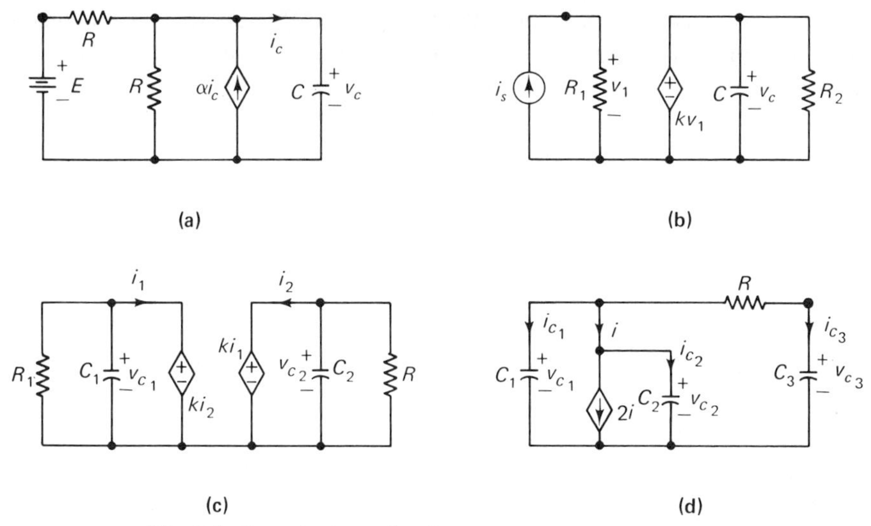

Fig. 8-5. Some intricate situations encountered in determining the order of complexity of linear active networks.

As long as $\alpha \neq 1$, we have

$$i_C = \frac{E - 2v_C}{(1 - \alpha)R}$$

from which the normal-form equation is

$$\dot{v}_C = \frac{i_C}{C} = \frac{-2}{(1 - \alpha)RC} v_C + \frac{E}{(1 - \alpha)RC} \tag{8-11}$$

Hence the order of complexity is 1.

However, if $\alpha = 1$, we have from Eq. (8-10)

$$\frac{v_C}{R} + \frac{v_C - E}{R} = 0$$

or

$$v_C = \frac{E}{2} \tag{8-12}$$

Even though there is no C–E_i loop, the capacitance voltage is related to the voltage source by Eq. (8-12), and the order of complexity is 0, *regardless of the value of C.*

EXAMPLE 8-2. Consider the network of Fig. 8-5(b). There is no C–E_i loop. However, there is a loop consisting of C and a controlled voltage source (C–E_d loop. The subscript d signifies *dependent* sources). Straightforward analysis of the circuit immediately shows that

$$v_C = kR_1 i_s \tag{8-13}$$

Thus, $v_C(0)$ is not an independent capacitor voltage, and the order of complexity is 0. But one should not jump to the conclusion that capacitor and controlled voltage source loops will always reduce the order of complexity. The next example will illustrate this point.

EXAMPLE 8-3. Consider the network of Fig. 8-5(c). There are two independent loops consisting of capacitors and controlled voltage sources only. However, in this case, $v_{C1}(0)$ and $v_{C2}(0)$ are independent voltages. An analysis of the network leads to the normal-form equation

$$\begin{bmatrix} \dot{v}_{C1} \\ \dot{v}_{C2} \end{bmatrix} = \begin{bmatrix} \dfrac{-1}{R_1C_1} & \dfrac{-1}{kC_1} \\ \dfrac{-1}{kC_2} & \dfrac{-1}{R_2C_2} \end{bmatrix} \begin{bmatrix} v_{C1} \\ v_{C2} \end{bmatrix} \tag{8-14}$$

The order of complexity is 2, in spite of the C–E_d loops.

EXAMPLE 8-4. Consider the network of Fig. 8-5(d). Because of the loop consisting of C_1 and C_2, we have

$$v_{C1} = v_{C2}$$

Thus, the order of complexity is at most 2. Suppose that we choose v_{C2} and v_{C3} as the state variables; then we can write

$$\begin{aligned} C_2\dot{v}_{C2} &= i_{C2} = i - 2i = -i \\ &= i_{C1} + \frac{v_{C2} - v_{C3}}{R} \\ &= C_1\dot{v}_{C1} + \frac{v_{C2} - v_{C3}}{R} \\ &= C_1\dot{v}_{C2} + \frac{v_{C2} - v_{C3}}{R} \end{aligned} \tag{8-15}$$

and

$$C_3\dot{v}_{C3} = \frac{v_{C2} - v_{C3}}{R} \tag{8-16}$$

from which the normal-form equation is

$$\begin{bmatrix} \dot{v}_{C2} \\ \dot{v}_{C3} \end{bmatrix} = \begin{bmatrix} \dfrac{1}{R(C_2 - C_1)} & \dfrac{-1}{R(C_2 - C_1)} \\ \dfrac{1}{RC_3} & -\dfrac{1}{RC_3} \end{bmatrix} \begin{bmatrix} v_{C2} \\ v_{C3} \end{bmatrix} \tag{8-17}$$

Provided that $C_1 \neq C_2$, the order of complexity is 2. For the case $C_1 = C_2$, Eq. (8-15) reduces to $v_{C2} = v_{C3}$. Therefore, there is only one independent capacitor voltage! The difference between the present case and that of Example 8-1 is that in the present case the dependency of capacitor voltages requires *exact* values of LC elements.

From these examples it should be clear that the determination of the order of complexity for a linear network containing controlled sources is by no means a simple problem. The most widely known solution to the problem is a recursive method, which will be described in Section 8-5.

Does every linear network possess a normal-form equation? The answer is obviously "no." For if a network can be described by Eqs. (8-4) and (8-5), there is a unique solution of $x(t)$ and $y(t)$ for $t \geq t_0$, given $x(t_0)$ and $u(t)$ for $t \geq t_0$. But it is well known that idealized network models may have inconsistencies that preclude a unique solution of $x(t)$. Obviously, for such a network the normal-form equation does not exist. As a specific example, let us replace the resistor R in Fig. 8-5(d) by an independent current source, and let $C_1 = C_2$. Then the equations of motion for this modified network cannot be reduced to the normal form. The details of this assertion are left as a problem (Problem 8-9).

8-3 COMPUTER FORMULATION OF STATE EQUATIONS FOR *RLCM* NETWORKS

The usefulness of a computer program for carrying out the state-variable analysis of *RLCM* (*M* stands for mutual inductance) networks, without any controlled sources, is undoubtedly very limited. The reason for our discussing the topic is mainly pedagogic. For it is usually a better learning process to go from some simple cases to the more complicated general case. Several excellent treatments on this simpler case of *RLCM* networks are now available [1, 2]. Therefore, our treatment will be brief. We shall concentrate on the *computational* aspect of the problem. The general case of linear networks containing controlled sources will be discussed in great detail in Section 8-4. Readers not interested in the special case of *RLCM* networks may proceed directly to Section 8-4.

The network under consideration consists of *RLC* elements, all with *positive* values. Mutual inductances are allowed, provided that the coefficient of coupling never reaches unity. (Another way of stating the last condition is that the inductance matrix is positive definite.) We assume a network with the following conditions:

1. No loops consisting of independent voltage sources only.
2. No cutsets consisting of independent current sources only.

If either condition is violated, the network either is inconsistent or has no unique solution (see Section 6-3). Such situations are usually due to over idealization of the network model and can be remedied by using more realistic models for some network elements.

The procedure begins with the selection of a tree $\mathfrak{T}$ that contains the following:

1. All independent voltage sources.
2. No independent current sources.
3. As many capacitors as possible.
4. As few inductors as possible.

Such a tree, called a *normal tree* or a *proper tree*, can always be constructed under the two conditions just stated (see Problem 3-11). A matrix procedure for selecting $\mathfrak{T}$ is described in Section 3-6-1. If we use the second-level subscript $\mathfrak{T}$ for the normal tree and $\mathfrak{L}$ for the links (cotree), then the branch voltage and current vectors may be partitioned as follows:[2]

$$v = \begin{bmatrix} v_{E_\mathfrak{J}} \\ v_{C_\mathfrak{J}} \\ v_{R_\mathfrak{J}} \\ v_{L_\mathfrak{J}} \\ v_{J_\mathfrak{L}} \\ v_{L_\mathfrak{L}} \\ v_{R_\mathfrak{L}} \\ v_{C_\mathfrak{L}} \end{bmatrix}, \quad i = \begin{bmatrix} i_{E_\mathfrak{J}} \\ i_{C_\mathfrak{J}} \\ i_{R_\mathfrak{J}} \\ i_{L_\mathfrak{J}} \\ i_{J_\mathfrak{L}} \\ i_{L_\mathfrak{L}} \\ i_{R_\mathfrak{L}} \\ i_{C_\mathfrak{L}} \end{bmatrix} \tag{8-18}$$

The fundamental cutset (KCL) equations for the network can be written as follows:[3]

$$\begin{array}{c} \\ \mathbf{D}i = \end{array} \begin{array}{c} \begin{array}{cccccccc} E_\mathfrak{J} & C_\mathfrak{J} & R_\mathfrak{J} & L_\mathfrak{J} & J_\mathfrak{L} & L_\mathfrak{L} & R_\mathfrak{L} & C_\mathfrak{L} \end{array} \\ \left[\begin{array}{cccc|cccc} \mathbf{1}_{E_\mathfrak{J}} & \mathbf{0} & \mathbf{0} & \mathbf{0} & \mathbf{F}_{11} & \mathbf{F}_{12} & \mathbf{F}_{13} & \mathbf{F}_{14} \\ \mathbf{0} & \mathbf{1}_{C_\mathfrak{J}} & \mathbf{0} & \mathbf{0} & \mathbf{F}_{21} & \mathbf{F}_{22} & \mathbf{F}_{23} & \mathbf{F}_{24} \\ \mathbf{0} & \mathbf{0} & \mathbf{1}_{R_\mathfrak{J}} & \mathbf{0} & \mathbf{F}_{31} & \mathbf{F}_{32} & \mathbf{F}_{33} & \mathbf{0} \\ \mathbf{0} & \mathbf{0} & \mathbf{0} & \mathbf{1}_{L_\mathfrak{J}} & \mathbf{F}_{41} & \mathbf{F}_{42} & \mathbf{0} & \mathbf{0} \end{array}\right] \end{array} i = \mathbf{0} \tag{8-19}$$

Making use of the relationship $\mathbf{B}_T = -\mathbf{D}_L^t$, as proved in Section 3-5, we can write the fundamental loop (KVL) equations as follows:

$$\begin{array}{c} \\ \mathbf{B}v = \end{array} \begin{array}{c} \begin{array}{cccccccc} E_\mathfrak{J} & C_\mathfrak{J} & R_\mathfrak{J} & L_\mathfrak{J} & J_\mathfrak{L} & L_\mathfrak{L} & R_\mathfrak{L} & C_\mathfrak{L} \end{array} \\ \left[\begin{array}{cccc|cccc} -\mathbf{F}_{11}^t & -\mathbf{F}_{21}^t & -\mathbf{F}_{31}^t & -\mathbf{F}_{41}^t & \mathbf{1}_{J_\mathfrak{L}} & \mathbf{0} & \mathbf{0} & \mathbf{0} \\ -\mathbf{F}_{12}^t & -\mathbf{F}_{22}^t & -\mathbf{F}_{32}^t & -\mathbf{F}_{42}^t & \mathbf{0} & \mathbf{1}_{L_\mathfrak{L}} & \mathbf{0} & \mathbf{0} \\ -\mathbf{F}_{13}^t & -\mathbf{F}_{23}^t & -\mathbf{F}_{33}^t & \mathbf{0} & \mathbf{0} & \mathbf{0} & \mathbf{1}_{R_\mathfrak{L}} & \mathbf{0} \\ -\mathbf{F}_{14}^t & -\mathbf{F}_{24}^t & \mathbf{0} & \mathbf{0} & \mathbf{0} & \mathbf{0} & \mathbf{0} & \mathbf{1}_{C_\mathfrak{L}} \end{array}\right] \end{array} v = \mathbf{0} \tag{8-20}$$

The fact that $\mathbf{F}_{34} = \mathbf{0}$, $\mathbf{F}_{43} = \mathbf{0}$, and $\mathbf{F}_{44} = \mathbf{0}$ in Eq. (8-19) is a consequence of the definition of the normal tree. It is perhaps easier to first demonstrate that the three submatrices $-\mathbf{F}_{43}^t$, $-\mathbf{F}_{34}^t$, and $-\mathbf{F}_{44}^t$ in Eq. (8-20) are null matrices. Then we can infer that the three submatrices $\mathbf{F}_{34}$, $\mathbf{F}_{43}$, and $\mathbf{F}_{44}$ in Eq. (8-19) are null matrices from the relationship $\mathbf{B}_T = -\mathbf{D}_L^t$. Recall that we construct a normal tree by using all the independent voltage sources first and then supplementing with capacitors, resistors, and inductors, in the indicated order. We have to put a capacitor in the cotree only when it forms a loop with some E's and C's already chosen as tree branches. Thus, the fundamental loop corresponding to a capacitor in the cotree will contain no resistors or inductors. This accounts for $-\mathbf{F}_{34}^t$ and $-\mathbf{F}_{44}^t$ being null matrices in Eq. (8-20). Similarly, we have to put a resistor in the cotree only when it forms a loop with some

[2]Each independent source (voltage or current source) is considered as an individual branch. This allows more flexibility than the "composite branches" of Chapter 4.

[3]$\mathbf{1}_{C_\mathfrak{J}}$ denotes an identity matrix of order equal to the number of capacitors in the normal tree. Similar interpretations are given to other identity matrices. The **0** matrices in Eqs. (8-19) and (8-20) are in general rectangular matrices whose dimensions are implied by the structure of the **D** and **B** matrices.

E's, C's, and R's already chosen as tree branches. Therefore, the fundamental loop corresponding to a resistor in the cotree will not contain any inductors. This accounts for $-\boldsymbol{F}_{34}^t$ being a null matrix in Eq. (8-20).

The *RLC* branches are characterized by

$$R: \quad \boldsymbol{v}_{R_{\mathfrak{J}}} = \boldsymbol{R}_{\mathfrak{J}}\boldsymbol{i}_{R_{\mathfrak{J}}} \quad \text{or} \quad \boldsymbol{i}_{R_{\mathfrak{J}}} = \boldsymbol{G}_{\mathfrak{J}}\boldsymbol{v}_{R_{\mathfrak{J}}}$$
$$\boldsymbol{i}_{R_{\mathfrak{L}}} = \boldsymbol{G}_{\mathfrak{L}}\boldsymbol{v}_{R_{\mathfrak{L}}} \quad \text{or} \quad \boldsymbol{v}_{R_{\mathfrak{L}}} = \boldsymbol{R}_{\mathfrak{L}}\boldsymbol{i}_{R_{\mathfrak{L}}} \tag{8-21a}$$

$$C: \quad \begin{bmatrix} \boldsymbol{i}_{C_{\mathfrak{J}}} \\ \boldsymbol{i}_{C_{\mathfrak{L}}} \end{bmatrix} = \begin{bmatrix} \boldsymbol{C}_{\mathfrak{J}} & \boldsymbol{0} \\ \boldsymbol{0} & \boldsymbol{C}_{\mathfrak{L}} \end{bmatrix} \frac{d}{dt} \begin{bmatrix} \boldsymbol{v}_{C_{\mathfrak{J}}} \\ \boldsymbol{v}_{C_{\mathfrak{L}}} \end{bmatrix} \tag{8-21b}$$

and

$$L: \quad \begin{bmatrix} \boldsymbol{v}_{L_{\mathfrak{J}}} \\ \boldsymbol{v}_{L_{\mathfrak{L}}} \end{bmatrix} = \begin{bmatrix} \boldsymbol{L}_{\mathfrak{J}\mathfrak{J}} & \boldsymbol{L}_{\mathfrak{J}\mathfrak{L}} \\ \boldsymbol{L}_{\mathfrak{L}\mathfrak{J}} & \boldsymbol{L}_{\mathfrak{L}\mathfrak{L}} \end{bmatrix} \frac{d}{dt} \begin{bmatrix} \boldsymbol{i}_{L_{\mathfrak{J}}} \\ \boldsymbol{i}_{L_{\mathfrak{L}}} \end{bmatrix} \quad \text{or} \quad \frac{d}{dt} \begin{bmatrix} \boldsymbol{i}_{L_{\mathfrak{J}}} \\ \boldsymbol{i}_{L_{\mathfrak{L}}} \end{bmatrix} = \begin{bmatrix} \boldsymbol{\Gamma}_{\mathfrak{J}\mathfrak{J}} & \boldsymbol{\Gamma}_{\mathfrak{J}\mathfrak{L}} \\ \boldsymbol{\Gamma}_{\mathfrak{L}\mathfrak{J}} & \boldsymbol{\Gamma}_{\mathfrak{L}\mathfrak{L}} \end{bmatrix} \begin{bmatrix} \boldsymbol{v}_{L_{\mathfrak{J}}} \\ \boldsymbol{v}_{L_{\mathfrak{L}}} \end{bmatrix} \tag{8-21c}$$

The two square matrices in Eq. (8-21c) are inverses of each other. The inverse exists because the inductance matrix is assumed to be positive definite. (Note that $\boldsymbol{L}_{\mathfrak{J}\mathfrak{J}}$ and $\boldsymbol{L}_{\mathfrak{L}\mathfrak{L}}$ are symmetric, and $\boldsymbol{L}_{\mathfrak{J}\mathfrak{L}}^t = \boldsymbol{L}_{\mathfrak{L}\mathfrak{J}}$.)

Our task is to reduce Eqs. (8-19), (8-20), and (8-21) to the normal-form equation (8-6). As will be demonstrated presently, a suitable choice for the state variables in the present case is $\boldsymbol{v}_{C_{\mathfrak{J}}}$ and $\boldsymbol{i}_{L_{\mathfrak{L}}}$. Then, in Eqs. (8-19)–(8-21) all voltages and currents except $\boldsymbol{v}_{C_{\mathfrak{L}}}$, $\boldsymbol{i}_{L_{\mathfrak{J}}}$, $\boldsymbol{v}_{E_{\mathfrak{J}}}$, and $\boldsymbol{i}_{J_{\mathfrak{L}}}$ must be eliminated. In the present case of *RLCM* networks, this elimination is always possible. The resultant equation, which is of the form of Eq. (8-6) and has *coefficient matrices made up only of the submatrices of Eqs. (8-19)–(8-21)*, is called the *explicit-form* state equation.

We shall now present the elimination procedure in detail. Since the elimination procedure starts with Eqs. (8-19)–(8-21), it is completely general and is applicable to any *RLCM* network. However, the reader may find it much easier to grasp the significance of each step by referring to the simple network shown in Fig. 8-6. This

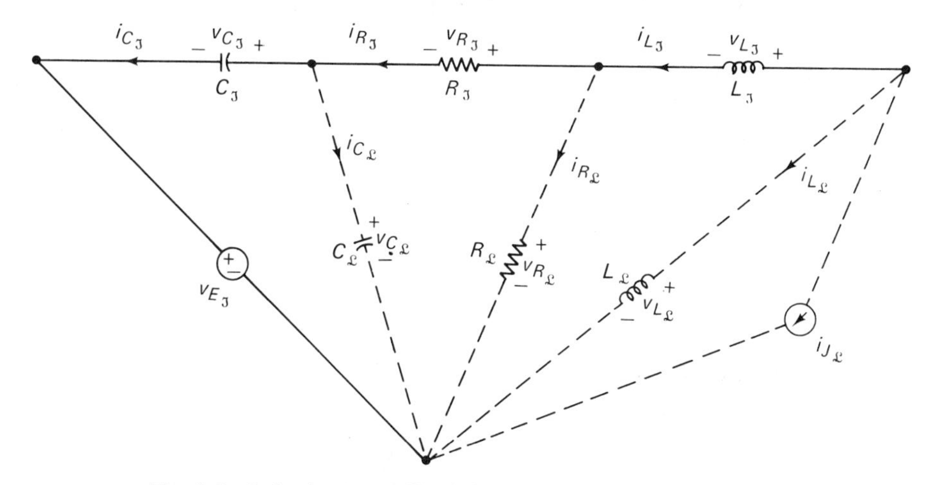

Fig. 8-6. A simple network illustrating the derivation of explicit-form state equations of *RLCM* networks.

network has eight elements, one in each of the eight categories indicated by the subscripts of Eq. (8-18). For the network of Fig. 8-6, with the chosen tree shown in heavy lines, all the submatrices $\boldsymbol{F}_{ij}$ in Eq. (8-19) are scalars having a value equal to 1.

Since we intend to formulate the state equation with $\boldsymbol{v}_{C_\mathfrak{T}}$ and $\boldsymbol{i}_{L_\mathfrak{L}}$ as the state variables, it is natural to write down the element characteristics for these two groups of elements ($C_\mathfrak{T}$ and $L_\mathfrak{L}$) as the starting point. From Eqs. (8-21b) and (8-21c), we have

$$\begin{bmatrix} \boldsymbol{C}_\mathfrak{T} & 0 \\ 0 & \boldsymbol{L}_{\mathfrak{L}\mathfrak{L}} \end{bmatrix} \frac{d}{dt} \begin{bmatrix} \boldsymbol{v}_{C_\mathfrak{T}} \\ \boldsymbol{i}_{L_\mathfrak{L}} \end{bmatrix} = \begin{bmatrix} \boldsymbol{i}_{C_\mathfrak{T}} \\ \boldsymbol{v}_{L_\mathfrak{L}} - \boldsymbol{L}_{\mathfrak{L}\mathfrak{T}} \frac{d}{dt} \boldsymbol{i}_{L_\mathfrak{T}} \end{bmatrix} \tag{8-22}$$

Clearly, our next task is to express $\boldsymbol{i}_{C_\mathfrak{T}}$, $\boldsymbol{v}_{L_\mathfrak{L}}$, and $\boldsymbol{i}_{L_\mathfrak{T}}$ on the right side of Eq. (8-22) in terms of the state variables ($\boldsymbol{v}_{C_\mathfrak{T}}$, $\boldsymbol{i}_{L_\mathfrak{L}}$) and the independent sources ($\boldsymbol{v}_{E_\mathfrak{T}}$, $\boldsymbol{i}_{J_\mathfrak{L}}$). This is done step by step as follows.

Step 1. Express $\boldsymbol{i}_{C_\mathfrak{T}}$ in terms of link currents and $\boldsymbol{v}_{L_\mathfrak{L}}$ in terms of tree branch voltages.

From the second rows of Eqs. (8-19) and (8-20), respectively, we have

$$\boldsymbol{i}_{C_\mathfrak{T}} = -\boldsymbol{F}_{21}\boldsymbol{i}_{J_\mathfrak{L}} - \boldsymbol{F}_{22}\boldsymbol{i}_{L_\mathfrak{L}} - \boldsymbol{F}_{23}\boldsymbol{i}_{R_\mathfrak{L}} - \boldsymbol{F}_{24}\boldsymbol{i}_{C_\mathfrak{L}} \tag{8-23}$$

and

$$\boldsymbol{v}_{L_\mathfrak{L}} = \boldsymbol{F}_{12}^t\boldsymbol{v}_{E_\mathfrak{T}} + \boldsymbol{F}_{22}^t\boldsymbol{v}_{C_\mathfrak{T}} + \boldsymbol{F}_{32}^t\boldsymbol{v}_{R_\mathfrak{T}} + \boldsymbol{F}_{42}^t\boldsymbol{v}_{L_\mathfrak{T}} \tag{8-24}$$

Step 2. Express $\boldsymbol{i}_{L_\mathfrak{T}}$ in terms of ($\boldsymbol{i}_{L_\mathfrak{L}}$, $\boldsymbol{i}_{J_\mathfrak{L}}$), and $\boldsymbol{v}_{C_\mathfrak{L}}$ in terms of ($\boldsymbol{v}_{C_\mathfrak{T}}$, $\boldsymbol{v}_{E_\mathfrak{T}}$).

From the fourth rows of Eqs. (8-19) and (8-20), respectively, we have

$$\boldsymbol{i}_{L_\mathfrak{T}} = -\boldsymbol{F}_{41}\boldsymbol{i}_{J_\mathfrak{L}} - \boldsymbol{F}_{42}\boldsymbol{i}_{L_\mathfrak{L}} \tag{8-25}$$

and

$$\boldsymbol{v}_{C_\mathfrak{L}} = \boldsymbol{F}_{14}^t\boldsymbol{v}_{E_\mathfrak{T}} + \boldsymbol{F}_{24}^t\boldsymbol{v}_{C_\mathfrak{T}} \tag{8-26}$$

Step 3. Express $\boldsymbol{v}_{L_\mathfrak{T}}$ in terms of the *derivatives* of ($\boldsymbol{i}_{J_\mathfrak{L}}$, $\boldsymbol{i}_{L_\mathfrak{L}}$), and $\boldsymbol{i}_{C_\mathfrak{L}}$ in terms of the derivatives of ($\boldsymbol{v}_{E_\mathfrak{T}}$, $\boldsymbol{v}_{C_\mathfrak{T}}$).

From Eq. (8-21c), we have

$$\boldsymbol{v}_{L_\mathfrak{T}} = \boldsymbol{L}_{\mathfrak{T}\mathfrak{T}}\frac{d}{dt}\boldsymbol{i}_{L_\mathfrak{T}} + \boldsymbol{L}_{\mathfrak{T}\mathfrak{L}}\frac{d}{dt}\boldsymbol{i}_{L_\mathfrak{L}} \tag{8-27}$$

Substituting Eq. (8-25) into Eq. (8-27), we have

$$\begin{aligned} \boldsymbol{v}_{L_\mathfrak{T}} &= \boldsymbol{L}_{\mathfrak{T}\mathfrak{T}}\frac{d}{dt}(-\boldsymbol{F}_{41}\boldsymbol{i}_{J_\mathfrak{L}} - \boldsymbol{F}_{42}\boldsymbol{i}_{L_\mathfrak{L}}) + \boldsymbol{L}_{\mathfrak{T}\mathfrak{L}}\frac{d}{dt}\boldsymbol{i}_{L_\mathfrak{L}} \\ &= -\boldsymbol{L}_{\mathfrak{T}\mathfrak{T}}\boldsymbol{F}_{41}\frac{d}{dt}\boldsymbol{i}_{J_\mathfrak{L}} + (\boldsymbol{L}_{\mathfrak{T}\mathfrak{L}} - \boldsymbol{L}_{\mathfrak{T}\mathfrak{T}}\boldsymbol{F}_{42})\frac{d}{dt}\boldsymbol{i}_{L_\mathfrak{L}} \end{aligned} \tag{8-28}$$

Similarly, from Eqs. (8-21b) and (8-26), we obtain

$$\begin{aligned} \boldsymbol{i}_{C_\mathfrak{L}} &= \boldsymbol{C}_\mathfrak{L}\frac{d}{dt}\boldsymbol{v}_{C_\mathfrak{L}} \\ &= \boldsymbol{C}_\mathfrak{L}\boldsymbol{F}_{14}^t\frac{d}{dt}\boldsymbol{v}_{E_\mathfrak{T}} + \boldsymbol{C}_\mathfrak{L}\boldsymbol{F}_{24}^t\frac{d}{dt}\boldsymbol{v}_{C_\mathfrak{T}} \end{aligned} \tag{8-29}$$

Step 4. Express $v_{R_\mathfrak{J}}$ in terms of $(v_{E_\mathfrak{J}}, v_{C_\mathfrak{J}}, i_{J_\mathfrak{L}}, i_{L_\mathfrak{L}})$, and do the same for $i_{R_\mathfrak{L}}$. From the third rows of Eqs. (8-19) and (8-20), we have

$$i_{R_\mathfrak{J}} = -F_{31}i_{J_\mathfrak{L}} - F_{32}i_{L_\mathfrak{L}} - F_{33}i_{R_\mathfrak{L}} \tag{8-30}$$

$$v_{R_\mathfrak{L}} = F_{13}^t v_{E_\mathfrak{J}} + F_{23}^t v_{C_\mathfrak{J}} + F_{33}^t v_{R_\mathfrak{J}} \tag{8-31}$$

Using these two relationships and Eq. (8-21a), we can write

$$G_\mathfrak{J} v_{R_\mathfrak{J}} = i_{R_\mathfrak{J}} = -F_{31}i_{J_\mathfrak{L}} - F_{32}i_{L_\mathfrak{L}} - F_{33}G_\mathfrak{L}(F_{13}^t v_{E_\mathfrak{J}} + F_{23}^t v_{C_\mathfrak{J}} + F_{33}^t v_{R_\mathfrak{J}})$$

which, after regrouping of terms, becomes

$$\begin{aligned}(G_\mathfrak{J} + F_{33}G_\mathfrak{L}F_{33}^t)v_{R_\mathfrak{J}} &= -F_{31}i_{J_\mathfrak{L}} - F_{32}i_{L_\mathfrak{L}} \\ &\quad - F_{33}G_\mathfrak{L}F_{13}^t v_{E_\mathfrak{J}} - F_{33}G_\mathfrak{L}F_{23}^t v_{C_\mathfrak{J}}\end{aligned} \tag{8-32}$$

Define

$$G \triangleq G_\mathfrak{J} + F_{33}G_\mathfrak{L}F_{33}^t \tag{8-33}$$

Since $G_\mathfrak{J}$ and $G_\mathfrak{L}$ are assumed to be positive definite in the present case, G is also positive definite (for a proof, see Section 6-3 and Problem 6-2). Therefore, G is nonsingular, and we obtain the desired expression for $v_{R_\mathfrak{J}}$ as

$$v_{R_\mathfrak{J}} = G^{-1}(-F_{31}i_{J_\mathfrak{L}} - F_{32}i_{L_\mathfrak{L}} - F_{33}G_\mathfrak{L}F_{13}^t v_{E_\mathfrak{J}} - F_{33}G_\mathfrak{L}F_{23}^t v_{C_\mathfrak{J}}) \tag{8-34}$$

The desired expression for $i_{R_\mathfrak{L}}$ can be derived in a completely analogous manner. From Eqs. (8-30), (8-31), and (8-21a), we can write

$$R_\mathfrak{L} i_{R_\mathfrak{L}} = v_{R_\mathfrak{L}} = F_{13}^t v_{E_\mathfrak{J}} + F_{23}^t v_{C_\mathfrak{J}} + F_{33}^t R_\mathfrak{J}(-F_{31}i_{J_\mathfrak{L}} - F_{32}i_{L_\mathfrak{L}} - F_{33}i_{R_\mathfrak{L}}) \tag{8-35}$$

which, after regrouping of terms, becomes

$$(R_\mathfrak{L} + F_{33}^t R_\mathfrak{J} F_{33})i_{R_\mathfrak{L}} = F_{13}^t v_{E_\mathfrak{J}} + F_{23}^t v_{C_\mathfrak{J}} - F_{33}^t R_\mathfrak{J} F_{31} i_{J_\mathfrak{L}} - F_{33}^t R_\mathfrak{J} F_{32} i_{L_\mathfrak{L}} \tag{8-36}$$

Define

$$R \triangleq R_\mathfrak{L} + F_{33}^t R_\mathfrak{J} F_{33} \tag{8-37}$$

Since $R_\mathfrak{L}$ and $R_\mathfrak{J}$ are assumed to be positive definite in the present case, R is also positive definite, and hence nonsingular. Therefore, the desired expression for $i_{R_\mathfrak{L}}$ is

$$i_{R_\mathfrak{L}} = R^{-1}(F_{13}^t v_{E_\mathfrak{J}} + F_{23}^t v_{C_\mathfrak{J}} - F_{33}^t R_\mathfrak{J} F_{31} i_{J_\mathfrak{L}} - F_{33}^t R_\mathfrak{J} F_{32} i_{L_\mathfrak{L}}) \tag{8-38}$$

Step 5. Final substitutions. We are now ready to make use of the expressions obtained in Steps 1–4 to produce the desired state equation. Start with Eq. (8-22). Use Eqs. (8-23), (8-24), and (8-25), respectively, to eliminate $i_{C_\mathfrak{J}}$, $v_{L_\mathfrak{L}}$, and $i_{L_\mathfrak{J}}$ in Eq. (8-22). This substep results in two equations containing the eight variables $(v_{C_\mathfrak{J}}, i_{L_\mathfrak{L}}, v_{E_\mathfrak{J}}, i_{J_\mathfrak{L}}, i_{C_\mathfrak{L}}, v_{L_\mathfrak{J}}, v_{R_\mathfrak{J}}, \text{and } i_{R_\mathfrak{L}})$:

$$C_\mathfrak{J}\frac{d}{dt}v_{C_\mathfrak{J}} = -F_{21}i_{J_\mathfrak{L}} - F_{22}i_{L_\mathfrak{L}} - F_{23}i_{R_\mathfrak{L}} - F_{24}i_{C_\mathfrak{L}}$$

$$\begin{aligned}L_{\mathfrak{L}\mathfrak{L}}\frac{d}{dt}i_{L_\mathfrak{L}} &= F_{12}^t v_{E_\mathfrak{J}} + F_{22}^t v_{C_\mathfrak{J}} + F_{32}^t v_{R_\mathfrak{J}} + F_{42}^t v_{L_\mathfrak{J}} \\ &\quad + L_{\mathfrak{L}\mathfrak{J}}F_{41}\frac{d}{dt}i_{J_\mathfrak{L}} + L_{\mathfrak{L}\mathfrak{J}}F_{42}\frac{d}{dt}i_{L_\mathfrak{L}}\end{aligned}$$

Next, use Eqs. (8-28) and (8-29) to eliminate $v_{L_\mathfrak{J}}$ and $i_{C_\mathfrak{L}}$ from the preceding equations. The resulting equations contain only six variables ($v_{C_\mathfrak{J}}$, $i_{L_\mathfrak{L}}$, $v_{E_\mathfrak{J}}$, $i_{J_\mathfrak{L}}$, $v_{R_\mathfrak{J}}$, and $i_{R_\mathfrak{L}}$):

$$C_\mathfrak{J}\frac{d}{dt}v_{C_\mathfrak{J}} = -F_{21}i_{J_\mathfrak{L}} - F_{22}i_{L_\mathfrak{L}} - F_{23}i_{R_\mathfrak{L}} - F_{24}C_\mathfrak{L}F^t_{14}\frac{d}{dt}v_{E_\mathfrak{J}} - F_{24}C_\mathfrak{L}F^t_{24}\frac{d}{dt}v_{C_\mathfrak{J}}$$

$$L_{\mathfrak{L}\mathfrak{L}}\frac{d}{dt}i_{L_\mathfrak{L}} = F^t_{12}v_{E_\mathfrak{J}} + F^t_{22}v_{C_\mathfrak{J}} + F^t_{32}v_{R_\mathfrak{J}} + (-F^t_{42}L_{\mathfrak{J}\mathfrak{J}} + L_{\mathfrak{L}\mathfrak{J}})F_{41}\frac{d}{dt}i_{J_\mathfrak{L}} + [F^t_{42}(L_{\mathfrak{J}\mathfrak{L}} - L_{\mathfrak{J}\mathfrak{J}}F_{42}) + L_{\mathfrak{L}\mathfrak{J}}F_{42}]\frac{d}{dt}i_{L_\mathfrak{L}}$$

Next, use Eqs. (8-34) and (8-38) to eliminate $v_{R_\mathfrak{J}}$ and $i_{R_\mathfrak{L}}$ in the preceding two equations. The resulting equations contain only four variables ($v_{C_\mathfrak{J}}$, $i_{L_\mathfrak{L}}$, $v_{E_\mathfrak{J}}$, and $i_{J_\mathfrak{L}}$):

$$C_\mathfrak{J}\frac{d}{dt}v_{C_\mathfrak{J}} = (-F_{21} + F_{23}R^{-1}F^t_{33}R_\mathfrak{J}F_{31})i_{J_\mathfrak{L}} + (-F_{22} + F_{23}R^{-1}F^t_{33}R_\mathfrak{J}F_{32})i_{L_\mathfrak{L}} - F_{23}R^{-1}F^t_{13}v_{E_\mathfrak{J}} - F_{23}R^{-1}F^t_{23}v_{C_\mathfrak{J}} - F_{24}C_\mathfrak{L}F^t_{14}\frac{d}{dt}v_{E_\mathfrak{J}} - F_{24}C_\mathfrak{L}F^t_{24}\frac{d}{dt}v_{C_\mathfrak{J}}$$

$$L_{\mathfrak{L}\mathfrak{L}}\frac{d}{dt}i_{L_\mathfrak{L}} = (F^t_{12} - F^t_{32}G^{-1}F_{33}G_\mathfrak{L}F^t_{13})v_{E_\mathfrak{J}} + (F^t_{22} - F^t_{32}G^{-1}F_{33}G_\mathfrak{L}F^t_{23})v_{C_\mathfrak{J}} - F^t_{32}G^{-1}F_{31}i_{J_\mathfrak{L}} - F^t_{32}G^{-1}F_{32}i_{L_\mathfrak{L}} + (-F^t_{42}L_{\mathfrak{J}\mathfrak{J}} + L_{\mathfrak{L}\mathfrak{J}})F_{41}\frac{d}{dt}i_{J_\mathfrak{L}} + [F^t_{42}(L_{\mathfrak{J}\mathfrak{L}} - L_{\mathfrak{J}\mathfrak{J}}F_{42}) + L_{\mathfrak{L}\mathfrak{J}}F_{42}]\frac{d}{dt}i_{L_\mathfrak{L}}$$

Finally, we rewrite the preceding two equations as a single matrix equation, as follows:

$$M^{(0)}\frac{d}{dt}\begin{bmatrix} v_{C_\mathfrak{J}} \\ i_{L_\mathfrak{L}} \end{bmatrix} = A^{(0)}\begin{bmatrix} v_{C_\mathfrak{J}} \\ i_{L_\mathfrak{L}} \end{bmatrix} + B^{(0)}\begin{bmatrix} v_{E_\mathfrak{J}} \\ i_{J_\mathfrak{L}} \end{bmatrix} + B_1^{(0)}\frac{d}{dt}\begin{bmatrix} v_{E_\mathfrak{J}} \\ i_{L_\mathfrak{L}} \end{bmatrix} \tag{8-39}$$

where

$$M^{(0)} = \begin{bmatrix} C_\mathfrak{J} + F_{24}C_\mathfrak{L}F^t_{24} & 0 \\ 0 & (L_{\mathfrak{L}\mathfrak{L}} - F^t_{42}L_{\mathfrak{J}\mathfrak{L}} - L_{\mathfrak{L}\mathfrak{J}}F_{42} + F^t_{42}L_{\mathfrak{J}\mathfrak{J}}F_{42}) \end{bmatrix} \tag{8-40}$$

$$A^{(0)} = \begin{bmatrix} -F_{23}R^{-1}F^t_{23} & (-F_{22} + F_{23}R^{-1}F^t_{33}R_\mathfrak{J}F_{32}) \\ (F^t_{22} - F^t_{32}G^{-1}F_{33}G_\mathfrak{L}F^t_{23}) & -F^t_{32}G^{-1}F_{32} \end{bmatrix} \tag{8-41}$$

$$B^{(0)} = \begin{bmatrix} -F_{23}R^{-1}F_{13} & (-F_{21} + F_{23}R^{-1}F^t_{33}R_\mathfrak{J}F_{31}) \\ (F^t_{12} - F^t_{32}G^{-1}F_{33}G_\mathfrak{L}F^t_{13}) & -F^t_{32}G^{-1}F_{31} \end{bmatrix} \tag{8-42}$$

$$B_1^{(0)} = \begin{bmatrix} -F_{24}C_\mathfrak{L}F^t_{14} & 0 \\ 0 & -F^t_{42}L_{\mathfrak{J}\mathfrak{J}}F_{41} + L_{\mathfrak{L}\mathfrak{J}}F_{41} \end{bmatrix} \tag{8-43}$$

The definitions of all matrices on the right sides of Eqs. (8-40)–(8-43) are to be found in Eqs. (8-19), (8-21), (8-33), and (8-37).

It can be shown that the matrix $M^{(0)}$ in Eq. (8-39) is nonsingular (see Problem 8-10). Then, premultiplying both sides of Eq. (8-39) by $[M^{(0)}]^{-1}$, we obtain an equation of the form

$$\boxed{\dot{\hat{x}} = A\hat{x} + Bu + B_1\dot{u}} \tag{8-44}$$

where

$$A = [M^{(0)}]^{-1}A^{(0)}$$
$$B = [M^{(0)}]^{-1}B^{(0)}$$
$$B_1 = [M^{(0)}]^{-1}B_1^{(0)}$$

We have thus obtained the state equation in *explicit* form, Eq. (8-44), for *RLCM* networks. Equation (8-44) is a special case of Eq. (8-6). Observe that the derivative of the input vector u appears in the equation. But this can be removed by a simple change of variables:

$$x \triangleq \hat{x} - B_1u \tag{8-45}$$

Putting Eq. (8-45) into Eq. (8-44), we have

$$\dot{x} = Ax + (B + AB_1)u \tag{8-46}$$

which is of exactly the same form as Eq. (8-4).

Example 8-5 illustrates the use of the explicit form of the state equation given by Eq. (8-44).

EXAMPLE 8-5. Write the state equation for the RLCM network shown in Fig. 8-7.

Solution: Let us choose a normal tree $\mathfrak{T}$ to consist of branches 1, 2, 3, 4, 5, and 6. Then

$$v_{C_\mathfrak{T}} = \begin{bmatrix} v_2 \\ v_3 \end{bmatrix}, \quad i_{L_\mathfrak{L}} = \begin{bmatrix} i_8 \\ i_9 \end{bmatrix}, \quad v_{E_\mathfrak{T}} = v_1, \quad i_{J_\mathfrak{L}} = i_7$$

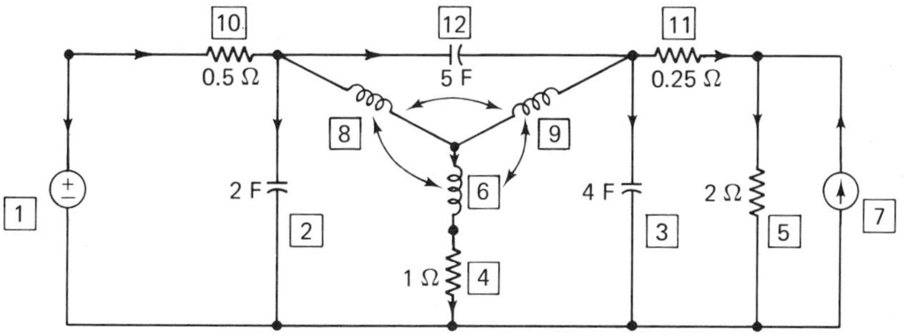

$$\begin{bmatrix} v_6 \\ v_8 \\ v_9 \end{bmatrix} = \begin{bmatrix} 4 & -1 & -1 \\ -1 & 2 & -1 \\ -1 & -1 & 2 \end{bmatrix} \frac{d}{dt} \begin{bmatrix} i_6 \\ i_8 \\ i_9 \end{bmatrix}$$

Fig. 8-7. A typical *RLCM* network.

The fundamental cutset matrix with respect to the tree $\mathfrak{J}$ is found to be

$$
\begin{array}{c|c|cc|cc|c}
\begin{matrix} E_{\mathfrak{J}} & \overbrace{C_{\mathfrak{J}}} & \overbrace{R_{\mathfrak{J}}} & L_{\mathfrak{J}} \\ 1 & 2\ \ 3 & 4\ \ 5 & 6 \end{matrix} & \begin{matrix} J_{\mathfrak{L}} \\ 7 \end{matrix} & \multicolumn{2}{c|}{\begin{matrix} \overbrace{\qquad L_{\mathfrak{L}} \qquad} \\ 8 \qquad 9 \end{matrix}} & \multicolumn{2}{c|}{\begin{matrix} \overbrace{\qquad R_{\mathfrak{L}} \qquad} \\ 10 \qquad 11 \end{matrix}} & \begin{matrix} C_{\mathfrak{L}} \\ 12 \end{matrix}
\end{array}
$$

$$
D = \left[\begin{array}{cccccc|c|cc|cc|c}
 & & & & & & 0 & 0 & 0 & 1 & 0 & 0 \\ \hline
 & & & & & & 0 & -1 & 0 & -1 & 0 & 1 \\
 & & \mathbf{1} & & & & 0 & 0 & -1 & 0 & 1 & -1 \\ \hline
 & & & & & & 0 & 1 & 1 & 0 & 0 & 0 \\
 & & & & & & -1 & 0 & 0 & 0 & -1 & 0 \\ \hline
 & & & & & & 0 & 1 & 1 & 0 & 0 & 0
\end{array}\right]
$$

from which we can identify the following matrices in Eq. (8-19):

$$
F_{11} = 0, \qquad F_{12} = [0 \quad 0], \qquad F_{13} = [1 \quad 0] \qquad F_{14} = 0
$$

$$
F_{21} = \begin{bmatrix} 0 \\ 0 \end{bmatrix}, \qquad F_{22} = \begin{bmatrix} -1 & 0 \\ 0 & -1 \end{bmatrix}, \qquad F_{23} = \begin{bmatrix} -1 & 0 \\ 0 & 1 \end{bmatrix}, \qquad F_{24} = \begin{bmatrix} 1 \\ -1 \end{bmatrix}
$$

$$
F_{31} = \begin{bmatrix} 0 \\ -1 \end{bmatrix}, \qquad F_{32} = \begin{bmatrix} 1 & 1 \\ 0 & 0 \end{bmatrix}, \qquad F_{33} = \begin{bmatrix} 0 & 0 \\ 0 & -1 \end{bmatrix}
$$

$$
F_{41} = 0, \qquad F_{42} = [1 \quad 1]
$$

The matrices describing branch characteristics are [see Eq. (8-21) for definitions] as follows:

$$
R_{\mathfrak{J}} = \begin{bmatrix} R_4 & 0 \\ 0 & R_5 \end{bmatrix} = \begin{bmatrix} 1 & 0 \\ 0 & 2 \end{bmatrix}, \qquad G_{\mathfrak{J}} = \begin{bmatrix} 1 & 0 \\ 0 & 0.5 \end{bmatrix}
$$

$$
R_{\mathfrak{L}} = \begin{bmatrix} R_{10} & 0 \\ 0 & R_{11} \end{bmatrix} = \begin{bmatrix} 0.5 & 0 \\ 0 & 0.25 \end{bmatrix}, \qquad G_{\mathfrak{L}} = \begin{bmatrix} 2 & 0 \\ 0 & 4 \end{bmatrix}
$$

$$
C_{\mathfrak{J}} = \begin{bmatrix} C_2 & 0 \\ 0 & C_3 \end{bmatrix} = \begin{bmatrix} 2 & 0 \\ 0 & 4 \end{bmatrix}, \qquad C_{\mathfrak{L}} = 5
$$

$$
L_{\mathfrak{J}\mathfrak{J}} = L_6 = 4, \qquad L_{\mathfrak{J}\mathfrak{L}} = [M_{68} \quad M_{69}] = [-1 \quad -1]
$$

$$
L_{\mathfrak{L}\mathfrak{J}} = \begin{bmatrix} M_{86} \\ M_{96} \end{bmatrix} = \begin{bmatrix} -1 \\ -1 \end{bmatrix}, \qquad L_{\mathfrak{L}\mathfrak{L}} = \begin{bmatrix} L_8 & M_{89} \\ M_{98} & L_9 \end{bmatrix} = \begin{bmatrix} 2 & -1 \\ -1 & 2 \end{bmatrix}
$$

From Eqs. (8-33) and (8-37), we have

$$
\begin{aligned}
G &= G_{\mathfrak{J}} + F_{33} G_{\mathfrak{L}} F_{33}^t \\
&= \begin{bmatrix} 1 & 0 \\ 0 & 0.5 \end{bmatrix} + \begin{bmatrix} 0 & 0 \\ 0 & -1 \end{bmatrix} \begin{bmatrix} 2 & 0 \\ 0 & 4 \end{bmatrix} \begin{bmatrix} 0 & 0 \\ 0 & -1 \end{bmatrix} = \begin{bmatrix} 1 & 0 \\ 0 & 4.5 \end{bmatrix}
\end{aligned}
$$

$$
\begin{aligned}
R &= R_{\mathfrak{L}} + F_{33}^t R_{\mathfrak{J}} F_{33} \\
&= \begin{bmatrix} 0.5 & 0 \\ 0 & 0.25 \end{bmatrix} + \begin{bmatrix} 0 & 0 \\ 0 & -1 \end{bmatrix} \begin{bmatrix} 1 & 0 \\ 0 & 2 \end{bmatrix} \begin{bmatrix} 0 & 0 \\ 0 & -1 \end{bmatrix} = \begin{bmatrix} 0.5 & 0 \\ 0 & 2.25 \end{bmatrix}
\end{aligned}
$$

Substituting these results into Eqs. (8-40)–(8-43) and carrying out the numerical calculations, we obtain

$$\boldsymbol{M}^{(0)} = \left[\begin{array}{cc|cc} 7 & -5 & 0 & 0 \\ -5 & 9 & 0 & 0 \\ \hline 0 & 0 & 8 & 5 \\ 0 & 0 & 5 & 8 \end{array}\right]$$

$$\boldsymbol{A}^{(0)} = \left[\begin{array}{cc|cc} -2 & 0 & 1 & 0 \\ 0 & -\frac{4}{9} & 0 & 1 \\ \hline -1 & 0 & -1 & -1 \\ 0 & -1 & -1 & -1 \end{array}\right]$$

$$\boldsymbol{B}^{(0)} = \left[\begin{array}{c|c} 2 & 0 \\ 0 & \frac{8}{9} \\ \hline 0 & 0 \\ 0 & 0 \end{array}\right]$$

$$\boldsymbol{B}_1^{(0)} = \mathbf{0}$$

The inverse of $\boldsymbol{M}^{(0)}$ is calculated next.

$$[\boldsymbol{M}^{(0)}]^{-1} = \left[\begin{array}{cc|cc} \frac{9}{38} & \frac{5}{38} & 0 & 0 \\ \frac{5}{38} & \frac{7}{38} & 0 & 0 \\ \hline 0 & 0 & \frac{8}{39} & -\frac{5}{39} \\ 0 & 0 & -\frac{5}{39} & \frac{8}{39} \end{array}\right]$$

Finally, substituting the calculated results of $[\boldsymbol{M}^{(0)}]^{-1}$, $\boldsymbol{A}^{(0)}$, $\boldsymbol{B}^{(0)}$, and $\boldsymbol{B}_1^{(0)}$ into Eq. (8-44), we obtain the state equation for the *RLCM* network of Fig. 8-7, as follows:

$$\frac{d}{dt}\left[\begin{array}{c} v_2 \\ v_3 \\ \hline i_8 \\ i_9 \end{array}\right] = \left[\begin{array}{cc|cc} -\frac{9}{19} & -\frac{10}{171} & \frac{9}{38} & \frac{5}{38} \\ -\frac{5}{19} & -\frac{14}{171} & \frac{5}{38} & \frac{7}{38} \\ \hline -\frac{8}{39} & \frac{5}{39} & -\frac{1}{13} & -\frac{1}{13} \\ \frac{5}{39} & -\frac{8}{39} & -\frac{1}{13} & -\frac{1}{13} \end{array}\right]\left[\begin{array}{c} v_2 \\ v_3 \\ \hline i_8 \\ i_9 \end{array}\right] + \left[\begin{array}{cc} \frac{9}{19} & \frac{20}{171} \\ \frac{5}{19} & \frac{28}{171} \\ \hline 0 & 0 \\ 0 & 0 \end{array}\right]\left[\begin{array}{c} v_1 \\ i_7 \end{array}\right]$$

8-4 COMPUTER FORMULATION OF STATE EQUATIONS FOR LINEAR ACTIVE NETWORKS

The linear networks under consideration consist of *RLCM* elements, independent sources, and all four types of controlled sources. Naturally, the method of this section is applicable to *RLCM* networks, too. However, the method to be described is an iterative procedure. We shall not be able to obtain the normal-form equations in explicit form, such as we have done for *RLCM* networks [see Eq. (8-46)].

The formulation of normal-form equations for linear active networks may be divided into two stages:

Stage 1. Formulation of the "initial state equations" of the form

$$\boxed{\boldsymbol{M}^{(0)}\dot{\boldsymbol{x}}^{(0)} = \boldsymbol{A}^{(0)}\boldsymbol{x}^{(0)} + \boldsymbol{B}^{(0)}\boldsymbol{u}} \tag{8-47}$$

where the vector $\boldsymbol{x}^{(0)}$ consists of *all* capacitor voltages and inductor currents, and $\boldsymbol{u}$ consists of all independent sources. The variables in $\boldsymbol{x}^{(0)}$ may or may not be linearly independent. The dependent variables in $\boldsymbol{x}^{(0)}$ will be uncovered and eliminated in the next stage.

Stage 2. Reduction of Eq. (8-47) first to the form

$$\boxed{\dot{\boldsymbol{x}} = \boldsymbol{A}\boldsymbol{x} + \boldsymbol{B}\boldsymbol{u} + (\boldsymbol{B}_1\dot{\boldsymbol{u}} + \dots)} \tag{8-48}$$

where the elements of $\boldsymbol{x}$ are, in general, a *subset* of $\boldsymbol{x}^{(0)}$, and then to the normal form given by Eq. (8-4).

Stage 1 may be reduced basically to a resistive n-port problem. We shall make extensive use of the hybrid matrices studied in Chapter 6. Stage 2 consists mainly of recursive matrix manipulations which can be investigated without any network concepts. The given network may be inconsistent or indeterminate, and therefore not describable by normal form equations. Such conditions will be uncovered in Stage 1 or Stage 2. The details of these two stages will now be described.[4]

8-4-1 Formulation of the Initial State Equations

The basic strategy here is to extract *all* capacitors and inductors to form an n-port N, while leaving all resistances, independent sources, and controlled sources (of all four types) *inside* N. Each extracted LC element may be replaced by a voltage port or a current port; the only condition is that the port combination should result in the hybrid equation given by Eq. (6-2), which we repeat here for convenience.[5]

$$\begin{bmatrix} \hat{\boldsymbol{i}}_a \\ \hat{\boldsymbol{v}}_b \end{bmatrix} = \begin{bmatrix} \hat{\boldsymbol{H}}_{aa} & \hat{\boldsymbol{H}}_{ab} \\ \hat{\boldsymbol{H}}_{ba} & \hat{\boldsymbol{H}}_{bb} \end{bmatrix} \begin{bmatrix} \hat{\boldsymbol{v}}_a \\ \hat{\boldsymbol{i}}_b \end{bmatrix} + \begin{bmatrix} \hat{\boldsymbol{s}}_a \\ \hat{\boldsymbol{s}}_b \end{bmatrix} \tag{8-49}$$

Once Eq. (8-49) has been obtained, we eliminate all capacitor currents and induc-or voltages by the use of

$$i_C = C\frac{dv_C}{dt}$$

$$v_L = L\frac{di_L}{dt}$$

[4]See reference 7 for a different method of formulating the state equations and determining the solvability of linear active networks. With the method of reference 7, the state variables may include the variables associated with the controlled sources.

[5]If no port combination results in Eq. (6-2), then the network is either inconsistent or indeterminate. Normal form equations do not exist for such a network.

The derivative terms may appear on both sides of Eq. (8-49). Upon moving all derivative terms to the left side and nonderivative terms to the right, we immediately have the initial state equations (8-47).

EXAMPLE 8-6. Formulate the initial state equation for the network shown in Fig. 8-8(a). (See Example 8-9 for a continuation to the problem of obtaining normal-form equations.)

Solution: Extracting LC elements, we obtain a four-port, as shown in Fig. 8-8(b). There are $2^4 = 16$ possible port combinations. But due to the obvious loop consisting of C_1, C_2, and E_a, we see that ports 1 and 2 cannot both be voltage ports. Similarly, because of the cutset consisting of L_3, L_4, and J_b, ports 3 and 4 cannot both be current ports. A total of seven port combinations are eliminated by these considerations.

Suppose that we consider ports 1 and 4 as voltage ports, and ports 2 and 3 as current ports. Then, by any suitable method (e.g., the methods of Section 6-5), the following hybrid equation is obtained [compare with Eq. (8-49)]:

$$\begin{bmatrix} i_1 \\ i_4 \\ v_2 \\ v_3 \end{bmatrix} = \begin{bmatrix} 0 & 0 & -1 & -1 \\ 0 & 0 & 0 & 1 \\ 1 & 0 & 0 & 0 \\ -1 & -1 & 0 & 1 \end{bmatrix} \begin{bmatrix} v_1 \\ v_4 \\ i_2 \\ i_3 \end{bmatrix} + \begin{bmatrix} 0 & 0 \\ 0 & -1 \\ 1 & 0 \\ -1 & 0 \end{bmatrix} \begin{bmatrix} E_a \\ J_b \end{bmatrix} \tag{8-50}$$

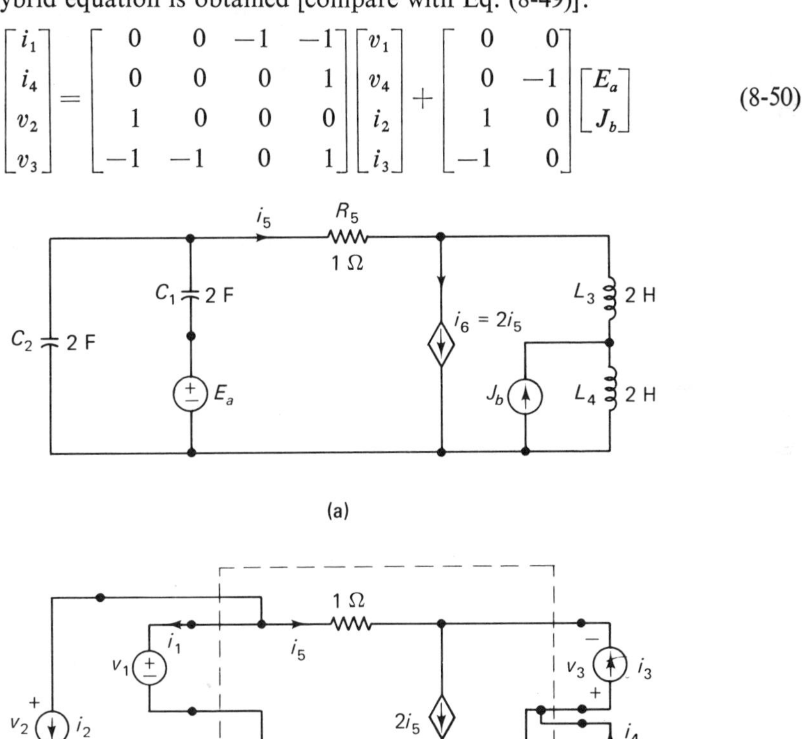

(a)

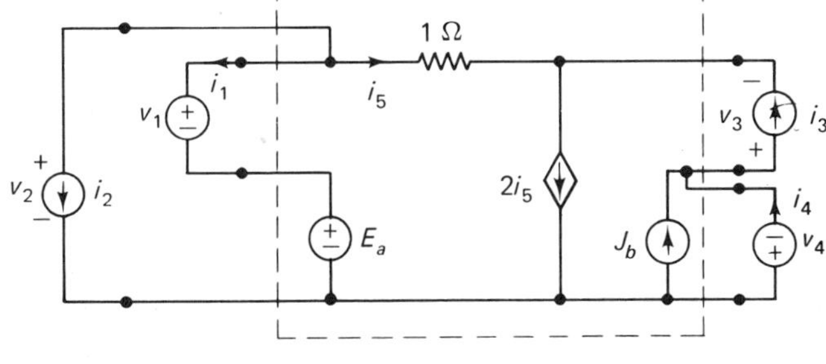

(b)

Fig. 8-8. Formulation of the initial state equation for a linear active network.

Substituting $i_1 = 2\dot{v}_1$, $i_2 = 2\dot{v}_2$, $v_3 = 2\dot{i}_3$, and $v_4 = 2\dot{i}_4$ into Eq. (8-50), we obtain

$$\begin{bmatrix} 2\dot{v}_1 \\ i_4 \\ v_2 \\ 2\dot{i}_3 \end{bmatrix} = \begin{bmatrix} 0 & 0 & -1 & -1 \\ 0 & 0 & 0 & 1 \\ 1 & 0 & 0 & 0 \\ -1 & -1 & 0 & 1 \end{bmatrix} \begin{bmatrix} v_1 \\ 2\dot{i}_4 \\ 2\dot{v}_2 \\ i_3 \end{bmatrix} + \begin{bmatrix} 0 & 0 \\ 0 & -1 \\ 1 & 0 \\ -1 & 0 \end{bmatrix} \begin{bmatrix} E_a \\ J_b \end{bmatrix} \tag{8-51}$$

After rearranging terms, we finally obtain the initial state equation

$$\begin{bmatrix} 2 & 2 & 0 & 0 \\ 0 & 0 & 2 & 2 \\ 0 & 0 & 0 & 0 \\ 0 & 0 & 0 & 0 \end{bmatrix} \begin{bmatrix} \dot{v}_1 \\ \dot{v}_2 \\ \dot{i}_3 \\ \dot{i}_4 \end{bmatrix} = \begin{bmatrix} 0 & 0 & -1 & 0 \\ -1 & 0 & 1 & 0 \\ 1 & -1 & 0 & 0 \\ 0 & 0 & 1 & -1 \end{bmatrix} \begin{bmatrix} v_1 \\ v_2 \\ i_3 \\ i_4 \end{bmatrix} + \begin{bmatrix} 0 & 0 \\ -1 & 0 \\ 1 & 0 \\ 0 & -1 \end{bmatrix} \begin{bmatrix} E_a \\ J_b \end{bmatrix} \tag{8-52}$$

We could also have used other port combinations to obtain different initial state equations. For example, let us consider both ports 1 and 2 as current ports and ports 3 and 4 as voltage ports [this is quite contrary to the common practice of considering as many capacitors (inductors) as possible to be voltage (current) ports]. Then the hybrid equation becomes

$$\begin{bmatrix} v_1 \\ v_2 \\ i_3 \\ i_4 \end{bmatrix} = \begin{bmatrix} -1 & -1 & -1 & -1 \\ -1 & -1 & -1 & -1 \\ -1 & -1 & 0 & 0 \\ -1 & -1 & 0 & 0 \end{bmatrix} \begin{bmatrix} i_1 \\ i_2 \\ v_3 \\ v_4 \end{bmatrix} + \begin{bmatrix} -1 & 0 \\ 0 & 0 \\ 0 & 0 \\ 0 & -1 \end{bmatrix} \begin{bmatrix} E_a \\ J_b \end{bmatrix} \tag{8-53}$$

Upon substituting $i_C = C\dot{v}_C$ and $v_L = L\dot{i}_L$ in Eq. (8-53) and regrouping terms, we obtain the following initial state equation:

$$\begin{bmatrix} 2 & 2 & 2 & 2 \\ 2 & 2 & 2 & 2 \\ 2 & 2 & 0 & 0 \\ 2 & 2 & 0 & 0 \end{bmatrix} \begin{bmatrix} \dot{v}_1 \\ \dot{v}_2 \\ \dot{i}_3 \\ \dot{i}_4 \end{bmatrix} = \begin{bmatrix} -1 & 0 & 0 & 0 \\ 0 & -1 & 0 & 0 \\ 0 & 0 & -1 & 0 \\ 0 & 0 & 0 & -1 \end{bmatrix} \begin{bmatrix} v_1 \\ v_2 \\ i_3 \\ i_4 \end{bmatrix} + \begin{bmatrix} -1 & 0 \\ 0 & 0 \\ 0 & 0 \\ 0 & -1 \end{bmatrix} \begin{bmatrix} E_a \\ J_b \end{bmatrix} \tag{8-54}$$

Note that in both Eqs. (8-52) and (8-54) the matrix $\boldsymbol{M}^{(0)}$ is singular.

As we have just demonstrated, there usually is a great deal of freedom in assigning the nature of an extracted port. From the computer programming point of view, what we are looking for is a systematic and easily programmable method of determining the port combination. One commonly used method [3, 4] is to first choose a tree with the following order of preference for tree branches: independent voltage sources, controlled voltage sources, capacitors, resistors, inductors, controlled current sources, and independent current sources. Such a tree is called a *normal tree* or a *proper tree* for a linear active network. Under the assumptions that independent voltage sources contain no loops and independent current sources contain no cutsets, any normal tree will contain all independent voltage sources and no independent current sources (see Problem 3-11). Those capacitors not in the normal tree are called *excess capacitors*, and those inductors in the normal tree are called *excess inductors*. The rules for extracting ports are as follows:

1. Tree capacitors and tree inductors are extracted as voltage ports.
2. Cotree capacitors and cotree inductors are extracted as current ports.

The hybrid equation may then be formulated by any of the methods discussed in Sections 6-5 and 6-6. For example, a proper tree for the network of Fig. 8-8(a) can be chosen to consist of V_a, C_1, R_5, and L_4. Then C_1 and L_4 will be extracted as voltage ports, and C_2 and L_3 as current ports. The corresponding hybrid equation is given by Eq. (8-50).

Some remarks about the use of a proper tree are necessary.

1. For *RLC* networks, excess-capacitor voltages and excess-inductor currents are not considered as state variables. This is not necessarily so for active networks (see Example 8-8). The adjective "proper" has a much weaker connotation in the case of active networks.
2. If there is any loop consisting of independent and controlled voltage sources, or any cutset consisting of independent and controlled current sources, the method of Section 6-5-1 is not applicable. We can, however, use the method of Section 6-5-2 or Section 6-6.
3. If, after determining a port combination with the aid of a proper tree, the hybrid matrix fails to exist, a revision of the port combination is required, and a new analysis has to be carried out.
4. If the most general method of Section 6-6 is used, one does not need the concept of a proper tree. Simply extract all *LC* elements as uncommitted ports and obtain the constraint equation (6-64) from which we can easily obtain a hybrid equation, if it exists. Substituting $i_C = C\dot{v}_C$ and $v_L = L\dot{i}_L$ into the hybrid equation and regrouping terms, we have the initial state equations.

8-4-2 Reduction to Normal-Form Equations

In Eq. (8-47) the vector $\mathbf{x}^{(0)}$ consists of *all* capacitor voltages and inductor currents, which are not necessarily linearly independent.[6] Our next task is to uncover dependent variables in $\mathbf{x}^{(0)}$ and eliminate them. In the case of *RLCM* networks, we have demonstrated very clearly in Section 8-3 that excess-capacitor voltages and excess-inductor currents may be eliminated [see Eq. (8-39)], and the remaining variables are linearly independent. However, for active networks it is not necessarily true that all excess variables can be eliminated; yet in some cases, even if all excess variables can be eliminated, there may still be linear dependence among the remaining variables due to the action of controlled sources and/or negative-valued resistors [6-7]. The task of discovering such "hidden" relationships among elements of $\mathbf{x}^{(0)}$ is not so straightforward. One procedure for achieving this goal, to be discussed next, is a

[6]One case in which elements of $\mathbf{x}^{(0)}$ are *always* linearly independent is an *RLC* network without C–E_i loops and L–J_i cutsets.

recursive process [3, 4]. This is the reason for using the superscript (k) to indicate the result of successive stages of the recursive process. In the final results, we shall drop the superscripts.

We apply elementary row operations to Eq. (8-47), or more conveniently to the coefficient matrix

$$[\boldsymbol{M}^{(0)}, \boldsymbol{A}^{(0)}, \boldsymbol{B}^{(0)}] \tag{8-55}$$

to reduce it to echelon form (see Appendix 3C). If $\boldsymbol{M}^{(0)}$ is nonsingular, then after reduction to echelon form the result will be an upper triangular matrix with nonzero diagonal elements. Further elementary row operations on (8-55) will reduce $\boldsymbol{M}^{(0)}$ to an identity matrix and (8-55) to the form

$$[\mathbf{1}, \boldsymbol{A}, \boldsymbol{B}] \tag{8-56}$$

which corresponds directly to the desired normal-form equations. Note that the final result of this process is equivalent to premultiplying (8-55) by $[\boldsymbol{M}^{(0)}]^{-1}$.

Next, consider the case where $\boldsymbol{M}^{(0)}$ is singular. Then, in the process of reducing (8-55) to row echelon form, at some step some jth row in the first block [originally $\boldsymbol{M}^{(0)}$] will become a zero row. There are three possible cases to consider.

Case 1. The jth row in the second block [originally $\boldsymbol{A}^{(0)}$] and the third block [originally $\boldsymbol{B}^{(0)}$] are also zero rows. Thus, we have an equation of the form

$$[0, 0, \ldots, 0]\dot{\boldsymbol{x}} = [0, 0, \ldots, 0]\boldsymbol{x} + [0, 0, \ldots, 0]\boldsymbol{u}$$

which is trivial. Obviously, the given network has no unique solution, because there are fewer constraint equations than there are unknowns.

Case 2. The jth row in the second block is a zero row, but the jth row in the third block is not. Thus, we have an equation of the form

$$[0, 0, \ldots, 0]\dot{\boldsymbol{x}} = [0, 0, \ldots, 0]\boldsymbol{x} + [b_{j1}, b_{j2}, \ldots, b_{jm}]\boldsymbol{u}$$

or

$$b_{j1}u_1 + b_{j2}u_2 + \ldots + b_{jm}u_m = 0 \tag{8-57}$$

In this case the given network is inconsistent [unless Eq. (8-57) is identically zero for all t], because we have an equation relating only the elements of $\boldsymbol{u}$, which are supposed to be independent sources. If Eq. (8-57) is identically zero, then case 2 has the same effect as case 1.

Case 3. The jth row in the second block is not a zero row, whereas the jth row in the third block may or may not be a zero row. Thus, we have an equation of the form

$$[0, 0, \ldots, 0]\dot{\boldsymbol{x}} = [a_{j1}, a_{j2} \ldots, a_{jn}]\boldsymbol{x} + [b_{j1}, b_{j2}, \ldots, b_{jm}]\boldsymbol{u}$$

Since there is at least one nonzero a_{jk}, we can always solve for x_k to obtain

$$x_k = -\frac{1}{a_{jk}}\{a_{j1}x_1 + a_{j2}x_2 + \ldots + b_{j1}u_1 + b_{j2}u_2 + \ldots\} \tag{8-58}$$

We may now substitute Eq. (8-58) into Eq. (8-47) to eliminate the variables x_k and $\dot{x}_k$.

The result of this first round of the recursive process is of the form

$$\boldsymbol{M}^{(1)}\dot{\boldsymbol{x}}^{(1)} = \boldsymbol{A}^{(1)}\boldsymbol{x}^{(1)} + \boldsymbol{B}^{(1)}\boldsymbol{u} + \boldsymbol{B}_1^{(1)}\dot{\boldsymbol{u}} \tag{8-59}$$

Note in particular that the $\dot{\boldsymbol{u}}$ term may appear in Eq. (8-59) after the first round, and that $\boldsymbol{x}^{(1)}$ has one fewer elements than $\boldsymbol{x}^{(0)}$.

We continue with the second round of the recursive process. The operations we apply to (8-55) are now applied to Eq. (8-59). This process is applied repeatedly. If there is any dependence among the variables of $\boldsymbol{x}^{(k)}$, the next round of the reduction process will eliminate one dependent variable, accompanied by the possible appearance of the next higher derivative of $\boldsymbol{u}$. When all variables in $\boldsymbol{x}^{(k)}$ are linearly independent, the last round of elementary row operations leads to the form of Eq. (8-48).

A few examples will illustrate the above procedure. The final step of reduction to Eq. (8-4) will be described in Section 8-5.

EXAMPLE 8-7. A certain linear active network contains two capacitors and one inductor. The initial state equation has been found to be

$$\begin{bmatrix} 1 & 3 & 1 \\ 2 & 7 & 3 \\ 1 & 4 & 2 \end{bmatrix} \begin{bmatrix} \dot{v}_{C1} \\ \dot{v}_{C2} \\ \dot{i}_{L3} \end{bmatrix} = \begin{bmatrix} 2 & 3 & 2 \\ 3 & 4 & 5 \\ 5 & 7 & 1 \end{bmatrix} \begin{bmatrix} v_{C1} \\ v_{C2} \\ i_{L3} \end{bmatrix} + \begin{bmatrix} 2 \\ 1 \\ 1 \end{bmatrix} u \tag{8-60}$$

Find the equation in the form of Eq. (8-48).

Solution: Reduce $[\boldsymbol{M}^{(0)}, \boldsymbol{A}^{(0)}, \boldsymbol{B}^{(0)}]$ to row echelon form by the algorithm discussed in Chapter 3. A zero row in the $\boldsymbol{M}^{(0)}$ block first appears when we obtain the following equations:

$$\begin{bmatrix} 1 & 3 & 1 \\ 0 & 1 & 1 \\ 0 & 0 & 0 \end{bmatrix} \begin{bmatrix} \dot{v}_{C1} \\ \dot{v}_{C2} \\ \dot{i}_{L3} \end{bmatrix} = \begin{bmatrix} 2 & 3 & 2 \\ -1 & -2 & 1 \\ 4 & 6 & -2 \end{bmatrix} \begin{bmatrix} v_{C1} \\ v_{C2} \\ i_{L3} \end{bmatrix} + \begin{bmatrix} 2 \\ -3 \\ 2 \end{bmatrix} u \tag{8-61}$$

From the last row, we have

$$4v_{C1} + 6v_{C2} - 2i_{L3} + 2u = 0$$

We may solve for v_{C1} to obtain

$$v_{C1} = -\tfrac{3}{2}v_{C2} + \tfrac{1}{2}i_{L3} - \tfrac{1}{2}u \tag{8-62}$$

Substituting Eq. (8-62) and its derivative equation into Eq. (8-61), and discarding the last scalar equation, which is equivalent to Eq. (8-62), we have

$$\begin{bmatrix} \frac{3}{2} & \frac{3}{2} \\ 1 & 1 \end{bmatrix} \begin{bmatrix} \dot{v}_{C2} \\ \dot{i}_{L3} \end{bmatrix} = \begin{bmatrix} 0 & 3 \\ -\frac{1}{2} & \frac{1}{2} \end{bmatrix} \begin{bmatrix} v_{C2} \\ i_{L3} \end{bmatrix} + \begin{bmatrix} 1 \\ -\frac{5}{2} \end{bmatrix} u + \begin{bmatrix} \frac{1}{2} \\ 0 \end{bmatrix} \dot{u} \tag{8-63}$$

This is the result of the first round of reduction.

Applying the reduction process again to Eq. (8-63), we obtain

$$\begin{bmatrix} \frac{3}{2} & \frac{3}{2} \\ 0 & 0 \end{bmatrix} \begin{bmatrix} \dot{v}_{C2} \\ \dot{i}_{L3} \end{bmatrix} = \begin{bmatrix} 0 & 3 \\ -\frac{1}{2} & -\frac{3}{2} \end{bmatrix} \begin{bmatrix} v_{C2} \\ i_{L3} \end{bmatrix} + \begin{bmatrix} 1 \\ -\frac{19}{6} \end{bmatrix} u + \begin{bmatrix} \frac{1}{2} \\ -\frac{1}{3} \end{bmatrix} \dot{u} \tag{8-64}$$

From the last scalar equation of Eq. (8-64),

$$v_{C2} = -3i_{L3} - \tfrac{19}{3}u - \tfrac{2}{3}\dot{u} \tag{8-65}$$

Substituting Eq. (8-65) and its derivative into Eq. (8-64) to eliminate v_{C2} and $\dot{v}_{C2}$, we obtain

$$\dot{i}_{L3} = -i_{L3} - \tfrac{1}{3}u - \tfrac{10}{3}\dot{u} - \tfrac{1}{3}\ddot{u} \tag{8-66}$$

which is the desired equation.

EXAMPLE 8-8. Formulate the state equations for the linear active network of Fig. 8-9.

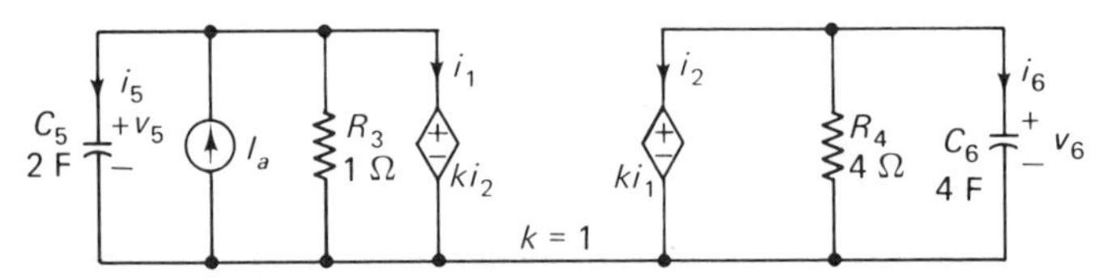

Fig. 8-9. For a linear active network, voltages of "excess capacitors" may be considered as state variables.

Solution: A normal tree in this case will consist of the two controlled voltage sources. C_1 and C_2 will be in the cotree and considered as excess capacitors. Therefore, when C_1 and C_2 are extracted to form a two-port, they are replaced by current sources. This two-port can easily be shown to be characterized by

$$\begin{bmatrix} v_5 \\ v_6 \end{bmatrix} = \begin{bmatrix} \tfrac{1}{3} & -\tfrac{4}{3} \\ -\tfrac{4}{3} & \tfrac{4}{3} \end{bmatrix} \begin{bmatrix} i_5 \\ i_6 \end{bmatrix} + \begin{bmatrix} -\tfrac{1}{3} \\ \tfrac{4}{3} \end{bmatrix} I_a \tag{8-67}$$

Upon substitution of $i_5 = 2\dot{v}_5$ and $i_6 = 4\dot{v}_6$ into Eq. (8-67), we obtain the initial state equation

$$\begin{bmatrix} \tfrac{2}{3} & -\tfrac{16}{3} \\ -\tfrac{8}{3} & \tfrac{16}{3} \end{bmatrix} \begin{bmatrix} \dot{v}_5 \\ \dot{v}_6 \end{bmatrix} = \begin{bmatrix} 1 & 0 \\ 0 & 1 \end{bmatrix} \begin{bmatrix} v_5 \\ v_6 \end{bmatrix} + \begin{bmatrix} \tfrac{1}{3} \\ -\tfrac{4}{3} \end{bmatrix} I_a \tag{8-68}$$

Applying the reduction process to Eq. (8-68), we obtain, after the first round, the desired normal-form equation:

$$\begin{bmatrix} \dot{v}_5 \\ \dot{v}_6 \end{bmatrix} = \begin{bmatrix} -\tfrac{1}{2} & -\tfrac{1}{2} \\ -\tfrac{1}{4} & -\tfrac{1}{16} \end{bmatrix} \begin{bmatrix} v_5 \\ v_6 \end{bmatrix} + \begin{bmatrix} \tfrac{1}{2} \\ 0 \end{bmatrix} I_a$$

Example 8-8 points out one important fact: even though C_1 and C_2 are not in the proper tree, their voltages may constitute state variables. Thus, the concept of "normal tree" or "proper tree" does not have the strong implication as for the case of *RLC* networks. A normal tree in this case is just an aid in deriving the hybrid matrix from which we obtain the initial state equations.

★8-5 COMPUTER FORMULATION OF THE OUTPUT EQUATIONS

Up to now we have discussed only the formulation of the state equation (8-4). In this section we shall show how with some slight modifications the output equation (8-5) can also be formulated by a digital computer program.

The concept of an n-port hybrid matrix is again very useful here. If a desired output voltage is not associated with an LC element, we shall consider it as the voltage across an open-circuit element, which in turn may be considered as a zero-valued independent current source. Similarly, if a desired output current is not associated with an LC element, we consider it to be the current through a short-circuit element, which in turn may be considered as a zero-valued independent voltage source.

In addition to extracting *all* capacitors and inductors, as we have done in Section 8-4 to derive the initial state equations, we now also extract the above-mentioned open-circuit elements and short-circuit elements as current ports and voltage ports, respectively. Then, for this group of ports, $\boldsymbol{i}_{oc} = \mathbf{0}$ and $\boldsymbol{v}_{sc} = \mathbf{0}$. The desired output $\boldsymbol{y}$ consists of $\boldsymbol{v}_{oc}$, $\boldsymbol{i}_{sc}$, and possibly voltages or currents of some LC elements. The resultant resistive multiport consists of resistances, controlled sources, and independent sources (inside the multiport). By the method of Chapter 6, one can obtain the hybrid equation in the following form:

$$\begin{bmatrix} \hat{\boldsymbol{i}}_a \\ \hat{\boldsymbol{v}}_b \\ \boldsymbol{i}_{sc} \\ \boldsymbol{v}_{oc} \end{bmatrix} = \begin{bmatrix} \boldsymbol{H}_{11} & \boldsymbol{H}_{12} & \boldsymbol{H}_{13} & \boldsymbol{H}_{14} \\ \boldsymbol{H}_{21} & \boldsymbol{H}_{22} & \boldsymbol{H}_{23} & \boldsymbol{H}_{24} \\ \boldsymbol{H}_{31} & \boldsymbol{H}_{32} & \boldsymbol{H}_{33} & \boldsymbol{H}_{34} \\ \boldsymbol{H}_{41} & \boldsymbol{H}_{42} & \boldsymbol{H}_{43} & \boldsymbol{H}_{44} \end{bmatrix} \begin{bmatrix} \hat{\boldsymbol{v}}_a \\ \hat{\boldsymbol{i}}_b \\ \boldsymbol{v}_{sc} \\ \boldsymbol{i}_{oc} \end{bmatrix} + \begin{bmatrix} \boldsymbol{s}_1 \\ \boldsymbol{s}_2 \\ \boldsymbol{s}_3 \\ \boldsymbol{s}_4 \end{bmatrix} \tag{8-69}$$

where $\hat{\boldsymbol{v}}_a$ and $\hat{\boldsymbol{i}}_a$ are associated with voltage ports replacing extracted LC elements, $\hat{\boldsymbol{v}}_b$ and $\hat{\boldsymbol{i}}_b$ are associated with current ports replacing LC elements, and the $\boldsymbol{s}_j$ vectors are due to independent sources inside the n-port. From Eq. (8-69) and recalling that $\boldsymbol{v}_{sc} = \mathbf{0}$ and $\boldsymbol{i}_{oc} = \mathbf{0}$, we can write

$$\begin{bmatrix} \hat{\boldsymbol{i}}_a \\ \hat{\boldsymbol{v}}_b \end{bmatrix} = \begin{bmatrix} \boldsymbol{H}_{11} & \boldsymbol{H}_{12} \\ \boldsymbol{H}_{21} & \boldsymbol{H}_{22} \end{bmatrix} \begin{bmatrix} \hat{\boldsymbol{v}}_a \\ \hat{\boldsymbol{i}}_b \end{bmatrix} + \begin{bmatrix} \boldsymbol{s}_1 \\ \boldsymbol{s}_2 \end{bmatrix} \tag{8-70a}$$

$$\begin{bmatrix} \boldsymbol{i}_{sc} \\ \boldsymbol{v}_{oc} \end{bmatrix} = \begin{bmatrix} \boldsymbol{H}_{31} & \boldsymbol{H}_{32} \\ \boldsymbol{H}_{41} & \boldsymbol{H}_{42} \end{bmatrix} \begin{bmatrix} \hat{\boldsymbol{v}}_a \\ \hat{\boldsymbol{i}}_b \end{bmatrix} + \begin{bmatrix} \boldsymbol{s}_3 \\ \boldsymbol{s}_4 \end{bmatrix} \tag{8-70b}$$

Equation (8-70a) is precisely (except for the difference in subscript notations) Eq. (8-49), which we used to derive the initial state equations (8-47).

The output vector $\boldsymbol{y}$ consists of $\boldsymbol{i}_{sc}$, $\boldsymbol{v}_{oc}$, and possibly some elements of $\hat{\boldsymbol{v}}_a$, $\hat{\boldsymbol{i}}_a$, $\hat{\boldsymbol{v}}_b$, and $\hat{\boldsymbol{i}}_b$. Since the list of variables in $(\hat{\boldsymbol{v}}_a, \hat{\boldsymbol{i}}_a, \hat{\boldsymbol{v}}_b, \hat{\boldsymbol{i}}_b)$ consists of *all* capacitor voltages, capacitor currents, inductor voltages, and inductor currents, and since $\hat{\boldsymbol{i}}_C = \boldsymbol{C}\dot{\hat{\boldsymbol{v}}}_C$ and $\hat{\boldsymbol{v}}_L = \boldsymbol{L}\dot{\hat{\boldsymbol{i}}}_L$, it is clear that $\boldsymbol{y}$ can always be expressed in terms of $\hat{\boldsymbol{v}}_C$, $\dot{\hat{\boldsymbol{v}}}_C$, $\hat{\boldsymbol{i}}_L$, $\dot{\hat{\boldsymbol{i}}}_L$, and the input vector $\boldsymbol{u}$. [Note that $\boldsymbol{u}$ is responsible for the presence of the $\boldsymbol{s}_j$ vectors in Eq. (8-69).] Recalling that the vector $\boldsymbol{x}^{(0)}$ in Eq. (8-47) consists of *all* capacitor voltages and inductor currents, we conclude that the output vector $\boldsymbol{y}$ can always be expressed as

$$\boldsymbol{y} = \boldsymbol{C}_1^{(0)}\dot{\boldsymbol{x}}^{(0)} + \boldsymbol{C}^{(0)}\boldsymbol{x}^{(0)} + \boldsymbol{D}^{(0)}\boldsymbol{u} \tag{8-71}$$

which may be called the *initial output equation*.

The reduction of Eq. (8-71) to the final output equation (8-5) is easily accomplished along with the reduction process discussed in the preceding section. Each time a relationship such as Eq. (8-58) (or a more general relationship also involving higher derivatives of $\boldsymbol{u}$) is obtained, we eliminate x_k and $\dot{x}_k$ from Eq. (8-47), and at the same time we may also eliminate x_k and $\dot{x}_k$ from Eq. (8-71). This procedure continues until all dependent variables are eliminated from both Eqs. (8-47) and (8-71). The final results will be two equations of the following form:

$$\dot{\boldsymbol{x}} = \boldsymbol{Ax} + \boldsymbol{Bu} + \boldsymbol{B}_1\frac{d}{dt}\boldsymbol{u} + \ldots + \boldsymbol{B}_p\left(\frac{d}{dt}\right)^p\boldsymbol{u} \tag{8-72}$$

$$\boldsymbol{y} = \boldsymbol{Cx} + \boldsymbol{Du} + \boldsymbol{D}_1\frac{d}{dt}\boldsymbol{u} + \ldots + \boldsymbol{D}_p\left(\frac{d}{dt}\right)^p\boldsymbol{u} \tag{8-73}$$

Note that, if a total of p dependent variables is eliminated from $\boldsymbol{x}^{(0)}$, then the highest-order derivative of $\boldsymbol{u}$ which can possibly appear in Eqs. (8-72) and (8-73) is the pth.

Equation (8-72) is not of the desired form, as given by Eq. (8-4), because of the presence of the derivatives of $\boldsymbol{u}$. We shall now show that by a suitable change of variable all derivatives of the independent sources may be eliminated from Eq. (8-72). Let

$$\boldsymbol{z} = \boldsymbol{x} - \boldsymbol{B}_p\left(\frac{d}{dt}\right)^{p-1}\boldsymbol{u} \tag{8-74}$$

Then

$$\boldsymbol{x} = \boldsymbol{z} + \boldsymbol{B}_p\left(\frac{d}{dt}\right)^{p-1}\boldsymbol{u} \tag{8-75}$$

Substituting Eq. (8-75) into Eqs. (8-72) and (8-73), we have

$$\dot{\boldsymbol{z}} = \boldsymbol{Az} + \boldsymbol{Bu} + \boldsymbol{B}_1\frac{d}{dt}\boldsymbol{u} + \ldots + (\boldsymbol{AB}_p + \boldsymbol{B}_{p-1})\left(\frac{d}{dt}\right)^{p-1}\boldsymbol{u} \tag{8-76}$$

and

$$\boldsymbol{y} = \boldsymbol{Cz} + \boldsymbol{Du} + \boldsymbol{D}_1\frac{d}{dt}\boldsymbol{u} + \ldots + (\boldsymbol{D}_{p-1} + \boldsymbol{CB}_p)\left(\frac{d}{dt}\right)^{p-1}\boldsymbol{u} + \boldsymbol{D}_p\left(\frac{d}{dt}\right)^p\boldsymbol{u} \tag{8-77}$$

Note that, with the new state variables $\boldsymbol{z}$, the highest order of the derivative of $\boldsymbol{u}$ in Eq. (8-76) is the $(p-1)$th, which is one lower than that in Eq. (8-72). Apparently, by repeated application of the transformation similar to Eq. (8-74), we can eventually eliminate all derivatives of $\boldsymbol{u}$ from Eq. (8-72). However, the highest-order derivative of $\boldsymbol{u}$ that can possibly occur in the output equation is still the pth [4].

EXAMPLE 8-9. In Fig. 8-8(a), let the desired outputs be i_3, i_4, and v_6. Find the state equation and the output equation.

Solution: Extract C_1 and L_4 as voltage ports, C_2 and L_3 as current ports, and an open-circuit element in parallel with the controlled current source as a current port. Note that $v_{oc} = v_6$ and $i_{oc} = 0$. For this five-port, we may formulate the hybrid equation using the methods of Chapter 6. The result, corresponding to Eq. (8-69), is

$$\begin{bmatrix} i_1 \\ i_4 \\ v_2 \\ v_3 \\ v_6 \end{bmatrix} = \begin{bmatrix} 0 & 0 & -1 & -1 & 1 \\ 0 & 0 & 0 & 1 & 0 \\ 1 & 0 & 0 & 0 & 0 \\ -1 & -1 & 0 & 1 & -1 \\ 1 & 0 & 0 & -1 & 1 \end{bmatrix} \begin{bmatrix} v_1 \\ v_4 \\ i_2 \\ i_3 \\ i_{oc} \end{bmatrix} + \begin{bmatrix} 0 & 0 \\ 0 & -1 \\ 1 & 0 \\ -1 & 0 \\ 1 & 0 \end{bmatrix} \begin{bmatrix} E_a \\ J_b \end{bmatrix} \tag{8-78}$$

From Eq. (8-78) we can write two equations corresponding to Eqs. (8-70) and (8-71). The former is given by Eq. (8-50), and the latter is

$$\begin{bmatrix} i_3 \\ i_4 \\ v_6 \end{bmatrix} = \begin{bmatrix} 0 & 0 & 0 & 1 \\ 0 & 0 & 0 & 1 \\ 1 & 0 & 0 & -1 \end{bmatrix} \begin{bmatrix} v_1 \\ v_4 \\ i_2 \\ i_3 \end{bmatrix} + \begin{bmatrix} 0 & 0 \\ 0 & -1 \\ 1 & 0 \end{bmatrix} \begin{bmatrix} E_a \\ J_b \end{bmatrix} \tag{8-79}$$

As shown in Example 8-6, Eq. (8-50) leads to the initial state equation (8-52). The last two rows of Eq. (8-52) give

$$v_2 = v_1 + E_a \tag{8-80}$$

$$i_4 = i_3 - J_b \tag{8-81}$$

As we substitute Eqs. (8-80) and (8-81) into Eq. (8-52) to eliminate (v_2, i_4), we do the same to Eq. (8-79). The results are

$$\begin{bmatrix} \dot{v}_1 \\ \dot{i}_3 \end{bmatrix} = \begin{bmatrix} 0 & -\frac{1}{4} \\ -\frac{1}{4} & \frac{1}{4} \end{bmatrix} \begin{bmatrix} v_1 \\ i_3 \end{bmatrix} + \begin{bmatrix} 0 & 0 \\ -\frac{1}{4} & 0 \end{bmatrix} \begin{bmatrix} E_a \\ J_b \end{bmatrix} + \begin{bmatrix} -\frac{1}{2} & 0 \\ 0 & \frac{1}{2} \end{bmatrix} \begin{bmatrix} \dot{E}_a \\ \dot{J}_b \end{bmatrix} \tag{8-82}$$

and

$$\begin{bmatrix} i_3 \\ i_4 \\ v_6 \end{bmatrix} = \begin{bmatrix} 0 & 1 \\ 0 & 1 \\ 1 & -1 \end{bmatrix} \begin{bmatrix} v_1 \\ i_3 \end{bmatrix} + \begin{bmatrix} 0 & 0 \\ 0 & -1 \\ 1 & 0 \end{bmatrix} \begin{bmatrix} E_a \\ J_b \end{bmatrix} \tag{8-83}$$

To remove the derivative terms $\dot{E}_a$ and $\dot{J}_b$ from Eq. (8-82), we define new state variables

$$\begin{bmatrix} z_1 \\ z_2 \end{bmatrix} = \begin{bmatrix} v_1 \\ i_3 \end{bmatrix} - \begin{bmatrix} -\frac{1}{2} & 0 \\ 0 & \frac{1}{2} \end{bmatrix} \begin{bmatrix} E_a \\ J_b \end{bmatrix} \tag{8-84}$$

Substituting Eq. (8-84) into Eqs. (8-82) and (8-83), we obtain

$$\begin{bmatrix} \dot{z}_1 \\ \dot{z}_2 \end{bmatrix} = \begin{bmatrix} 0 & -\frac{1}{4} \\ -\frac{1}{4} & \frac{1}{4} \end{bmatrix} \begin{bmatrix} z_1 \\ z_2 \end{bmatrix} + \begin{bmatrix} 0 & -\frac{1}{8} \\ -\frac{1}{8} & \frac{1}{8} \end{bmatrix} \begin{bmatrix} E_a \\ J_b \end{bmatrix} \tag{8-85}$$

and

$$\begin{bmatrix} i_3 \\ i_4 \\ v_6 \end{bmatrix} = \begin{bmatrix} 0 & 1 \\ 0 & 1 \\ 1 & -1 \end{bmatrix} \begin{bmatrix} z_1 \\ z_2 \end{bmatrix} + \begin{bmatrix} 0 & \frac{1}{2} \\ 0 & -\frac{1}{2} \\ \frac{1}{2} & \frac{1}{2} \end{bmatrix} \begin{bmatrix} E_a \\ J_b \end{bmatrix} \tag{8-86}$$

These two equations, which are exactly of the forms given by Eqs. (8-4) and (8-5), are the desired results.

EXAMPLE 8-10. Consider the network shown in Fig. 8-10(a): (1) Find the initial state equation and initial output equation as given by Eqs. (8-47) and (8-71), respectively. (2) Reduce the equations to the forms of Eqs. (8-72) and (8-73), respectively. (3) Obtain the normal-form equations as given by Eqs. (8-4) and (8-5).

Solution: Extract C_1 as a voltage port and C_2 as a current port, as shown in Fig. 8-10(b). Using any of the methods described in Chapter 6, we obtain the following

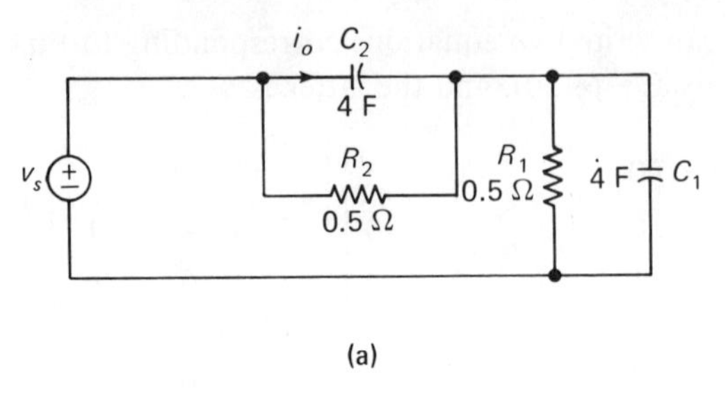

(a)

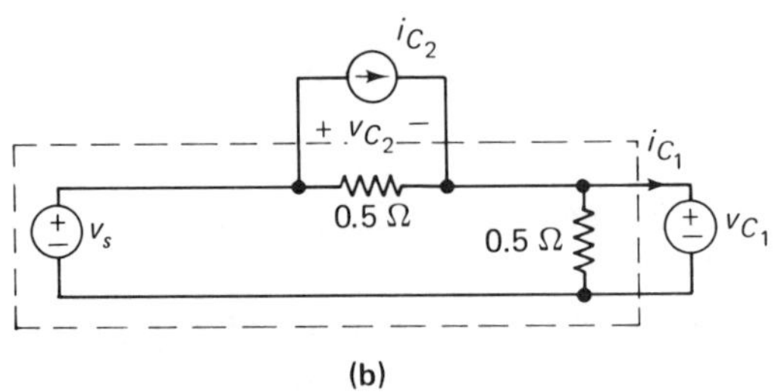

(b)

Fig. 8-10. Formulation of output equations.

hybrid equation for the two-port:

$$\begin{bmatrix} i_{C1} \\ v_{C2} \end{bmatrix} = \begin{bmatrix} -4 & 1 \\ -1 & 0 \end{bmatrix} \begin{bmatrix} v_{C1} \\ i_{C2} \end{bmatrix} + \begin{bmatrix} 2 \\ 1 \end{bmatrix} v_s \tag{8-87}$$

Upon substituting $i_{C1} = 4\dot{v}_{C1}$ and $i_{C2} = 4\dot{v}_{C2}$ into Eq. (8-87), moving all derivative terms to the left side and all nonderivative terms to the right side, we obtain the initial state equation

$$\begin{bmatrix} 4 & -4 \\ 0 & 0 \end{bmatrix} \begin{bmatrix} \dot{v}_{C1} \\ \dot{v}_{C2} \end{bmatrix} = \begin{bmatrix} -4 & 0 \\ -1 & -1 \end{bmatrix} \begin{bmatrix} v_{C1} \\ v_{C2} \end{bmatrix} + \begin{bmatrix} 2 \\ 1 \end{bmatrix} v_s \tag{8-88}$$

For the output equation, we have

$$y = i_o = i_{C2} = 4\dot{v}_{C2} \tag{8-89}$$

which can be written in the form of Eq. (8-71) as follows:

$$y = [0 \quad 4] \begin{bmatrix} \dot{v}_{C1} \\ \dot{v}_{C2} \end{bmatrix} + [0 \quad 0] \begin{bmatrix} v_{C1} \\ v_{C2} \end{bmatrix} + 0v_s \tag{8-90}$$

To reduce the initial state equation (8-88) to the standard form, we normally have to use the algorithm described in Section 8-4. In the present case, the second row of Eq. (8-88) gives

$$-v_{C1} - v_{C2} + v_s = 0 \tag{8-91}$$

Suppose that we solve for v_{C2}:

$$v_{C2} = v_s - v_{C1} \tag{8-92}$$

Substituting Eq. (8-92) into Eqs. (8-88) and (8-90) to eliminate v_{C2}, we obtain

$$\dot{v}_{C1} = -\tfrac{1}{2}v_{C1} + \tfrac{1}{4}v_s + \tfrac{1}{2}\dot{v}_s \tag{8-93}$$

and

$$i_o = 2v_{C1} - v_s + 2\dot{v}_s \tag{8-94}$$

These two equations are in the form of Eqs. (8-47) and (8-71), respectively, as desired.

We can eliminate the $\dot{v}_s$ term in Eq. (8-93) by the transformation of variables given by Eq. (8-74):

$$x = v_{C1} - \tfrac{1}{2}v_s \tag{8-95}$$

Substituting Eq. (8-95) into Eqs. (8-93) and (8-94), we obtain the following equations, which are in the form of Eqs. (8-4) and (8-5), as desired.

$$\dot{x} = -\tfrac{1}{2}x \tag{8-96}$$

$$y = i_o = 2x + 2\dot{v}_s \tag{8-97}$$

We conclude this chapter with additional remarks about the normal-form equation for linear networks.

1. Although $\boldsymbol{x}^{(0)}$ in the initial state equation consists of *all* capacitor voltages and inductor currents, the state variables are, in the general case, a subset of $\boldsymbol{x}^{(0)}$. Furthermore, whenever a transformation such as Eq. (8-74) is introduced, the new variables in $\boldsymbol{z}$ are, in general, linear combinations of capacitor voltages, inductor currents, independent sources $\boldsymbol{u}$ and their derivatives $\boldsymbol{u}^{(k)}$.
2. For linear networks, no derivatives of $\boldsymbol{u}$ appear in the initial output equation (8-71), and at most the first derivative of $\boldsymbol{u}$ appears in the initial state equation (8-47). After the reduction process, we obtain Eq. (8-48). For *RLC* networks without any voltage sources in C–E_i loops nor any current sources in L–J_i cutsets, no derivative of $\boldsymbol{u}$ will appear in Eq. (8-48); and $\dot{\boldsymbol{u}}$ (but no higher derivatives) will appear in Eq. (8-48) if there is any C–E_i loop or L–J_i cutset. These facts are rigorously established by the explicit formulation described in Section 8-3. For linear active networks, however, it is possible to have higher derivatives of $\boldsymbol{u}$ appearing in Eq. (8-48).
3. For *RLC* networks, by the use of the transformation (8-74), it is always possible to have the state equation and output equation without any derivatives of $\boldsymbol{u}$. For linear active networks, however, the derivatives of $\boldsymbol{u}$ can always be eliminated from Eq. (8-72); this is not necessarily so with the output equation (8-73).
4. After eliminating the derivatives $\boldsymbol{u}^{(k)}$ from the state equation, if some derivative $\boldsymbol{u}^{(k)}$ appears in the output equation, the system is said to be improper. It is unadvisable to solve such equations by a digital computer, because the differentiation of the input $\boldsymbol{u}$, when performed on a discrete basis, can lead to very serious accuracy problems. In fact, all existing programs [4, 5] using the state-variable approach preclude such a system and will give some error message when an improper system is encountered.

REFERENCES

1. Bryant, P. R. "The Explicit Form of Bashkow's A Matrix." *IRE Trans. Circuit Theory*, Vol. CT-9, pp. 303–306, Sept. 1962.
2. Rohrer, R. *Circuit Theory: An Introduction to the State Variable Approach to Network Theory*. New York: McGraw-Hill Book Company, 1969.
3. Dervisoglu, A. *State Models of Active RLC Networks*. Urbana, Ill.: Coordinated Science Laboratory, University of Illinois, Report R-237, Dec. 1964.
4. Pottle, C. "State Space Techniques for General Active Network Analysis." Ch. 7 in F. F. Kuo and J. F. Kaiser, eds., *System Analysis by Digital Computer*. New York: John Wiley & Sons, Inc., 1966.
5. Bowers, J. C., and S. R. Sedore. *SCEPTRE: A Computer Program for Circuit and System Analysis*. Englewood Cliffs, N.J.: Prentice-Hall, Inc., 1971.
6. Parker, S. R., and V. T. Barmes. "Existence of Numerical Solution and the Order of Linear Circuits with Dependent Sources." *IEEE Trans. Circuit Theory*, Vol. CT-18, pp. 368–374, May 1971.
7. Purslow, E. J. "Solvability and Analysis of Linear Active Networks by Use of the State Equations." *IEEE Trans. Circuit Theory*, Vol. CT-17, pp. 469–475, Nov. 1970.

PROBLEMS

8-1. Consider the network shown in Fig. P8-1. All element values are assumed to be positive. (Exact values are not needed.)
(a) Determine the order of complexity.
(b) Show a suitable set of state variables.

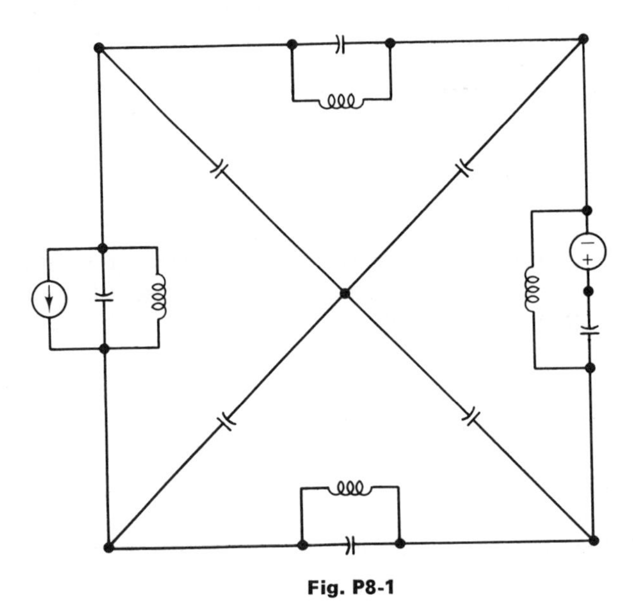

Fig. P8-1

8-2. Repeat Problem 8-1 for the network shown in Fig. P8-2.

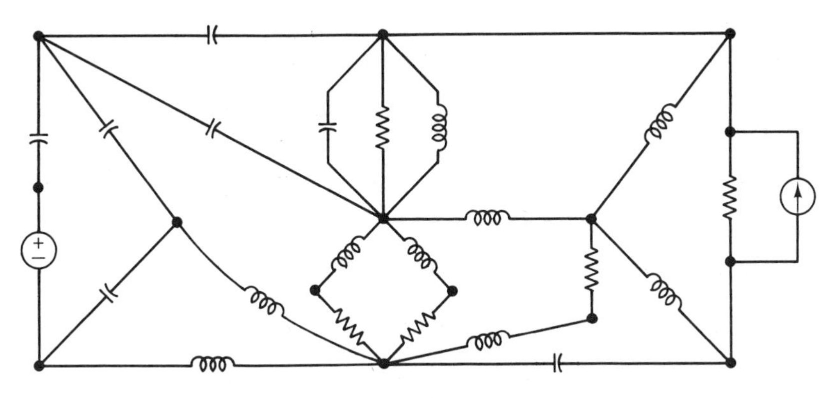

Fig. P8-2

360

8-3. Formulate the state equations for the *RLC* network shown in Fig. P8-3 by the following procedure:

(i) Select a proper tree.

(ii) Extract independent voltage sources, capacitors, and inductors in the tree as voltage ports. Extract independent current sources, inductors, and capacitors in the cotree as current ports.

(iii) Find the hybrid matrix for the resultant n-port. This step may be done (a) by inspection, for simple networks, or (b) by the use of Eq. (6-20), or (c) by the use of the program HYBRID.

(iv) Substitute

$$i_C = C\dot{v}_C$$
$$v_L = L\dot{i}_L$$

in the hybrid equation. Manipulate to get the state equations.

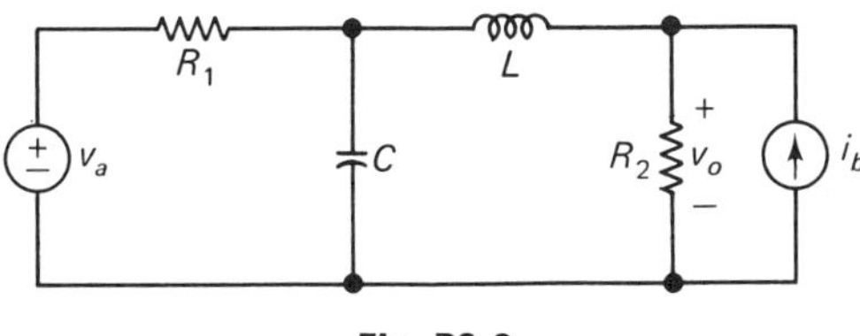

Fig. P8-3

8-4. Find the state equations for the network of Fig. P8-3 by the use of the explicit formula (8-44). This involves the following steps:

(i) Choose a proper tree and identify all submatrices $\boldsymbol{F}_{ij}$ in Eq. (8-19).

(ii) Obtain appropriate $\boldsymbol{R}$, $\boldsymbol{G}$, $\boldsymbol{C}$, and $\boldsymbol{L}$ matrices in Eq. (8-21).

(iii) Substitute in Eq. (8-44) to obtain the state equations.

8-5. Repeat Problem 8-3 for the network shown in Fig. P8-5 (diagram on next page).

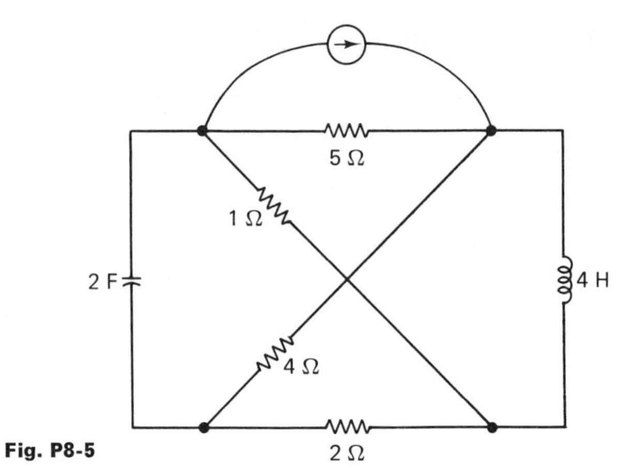

Fig. P8-5

8-6. Repeat Problem 8-3 for the network shown in Fig. P8-6.

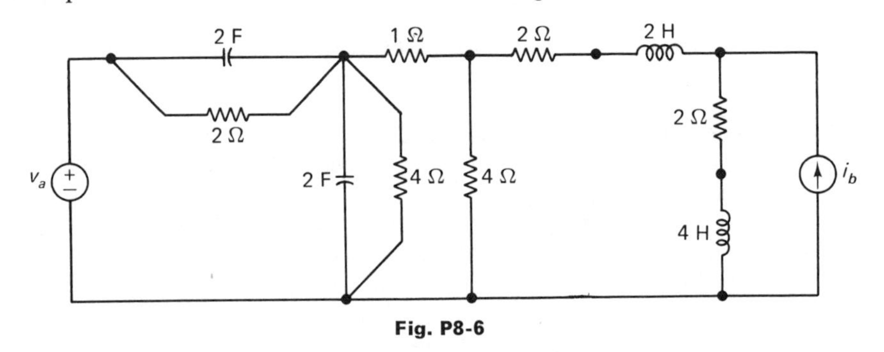

Fig. P8-6

8-7. The three-port shown in Fig. P8-7 is characterized by

$$\begin{bmatrix} i_1 \\ i_2 \\ v_3 \end{bmatrix} = \begin{bmatrix} 0 & 1 & 1 \\ -1 & 0 & 1 \\ 1 & -1 & 0 \end{bmatrix} \begin{bmatrix} v_1 \\ v_2 \\ i_3 \end{bmatrix}$$

(a) Write the initial state equations.
(b) Following the iterative reduction of Section 8-4, find the normal form equation $\dot{\boldsymbol{x}} = \boldsymbol{A}\boldsymbol{x} + \boldsymbol{B}\boldsymbol{u}$.

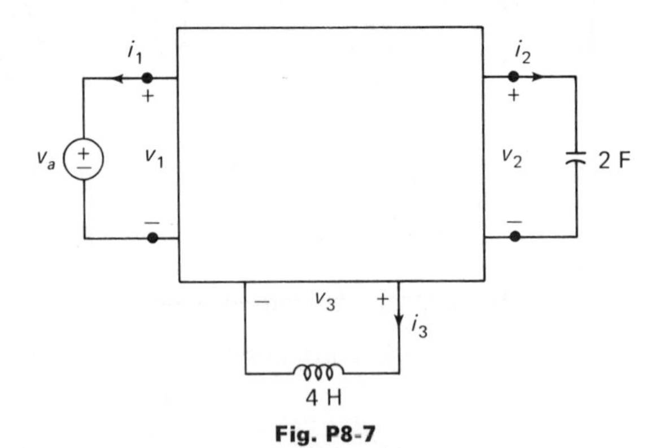

Fig. P8-7

8-8. The two-port shown in Fig. P8-8 is characterized by

$$\begin{bmatrix} i_1 \\ v_2 \end{bmatrix} = \begin{bmatrix} -1 & 2 \\ -2 & 0 \end{bmatrix} \begin{bmatrix} v_1 \\ i_2 \end{bmatrix}$$

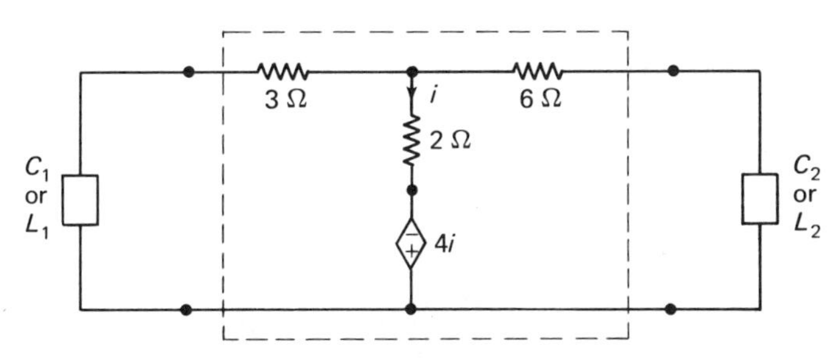

Fig. P8-8

Formulate the state equations $\dot{\boldsymbol{x}} = \boldsymbol{A}\boldsymbol{x}$ for each of the following cases by first writing down the initial state equations and then reducing to normal-form equations by the method described in Section 8-4.

(a) $C_1 = 2$ F, $L_2 = 4$ H.
(b) $L_1 = 4$ H, $C_2 = 2$ F.
(c) $C_1 = 2$ F, $C_2 = 4$ F.
(d) $L_1 = 4$ H, $L_2 = 2$ H.

8-9. Prove that the network of Fig. P8-9 is inconsistent, and therefore cannot be characterized by normal-form equations.

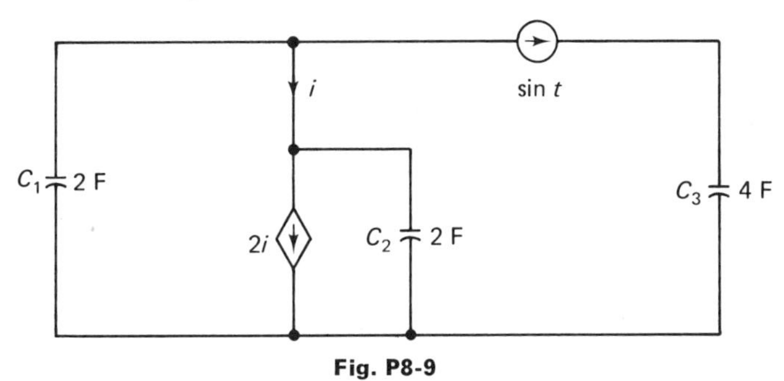

Fig. P8-9

8-10. Prove that the leading matrix on the left side of Eq. (8-39) is nonsingular under the following conditions:

(a) $\boldsymbol{C}_{\mathfrak{J}}$ and $\boldsymbol{C}_{\mathfrak{L}}$ are positive definite.

(b) $\begin{bmatrix} \boldsymbol{L}_{\mathfrak{J}\mathfrak{J}} & \boldsymbol{L}_{\mathfrak{J}\mathfrak{L}} \\ \boldsymbol{L}_{\mathfrak{L}\mathfrak{J}} & \boldsymbol{L}_{\mathfrak{L}\mathfrak{L}} \end{bmatrix}$ is positive definite.

8-11. The outputs have been indicated for the circuit shown in Fig. P8-11. Formulate the state equation and the output equation in the form of Eqs. (8-4) and (8-5), respectively. Clearly identify the $\boldsymbol{A}$, $\boldsymbol{B}$, $\boldsymbol{C}$, and $\boldsymbol{D}$ matrices and the $\boldsymbol{x}$ vector. Although the circuit is simple and consequently can be solved by inspection, you should use the result obtained this way only as a check. Follow the algorithms discussed in Sections 8-4 and 8-5 to learn the way the problem is solved with a digital computer.

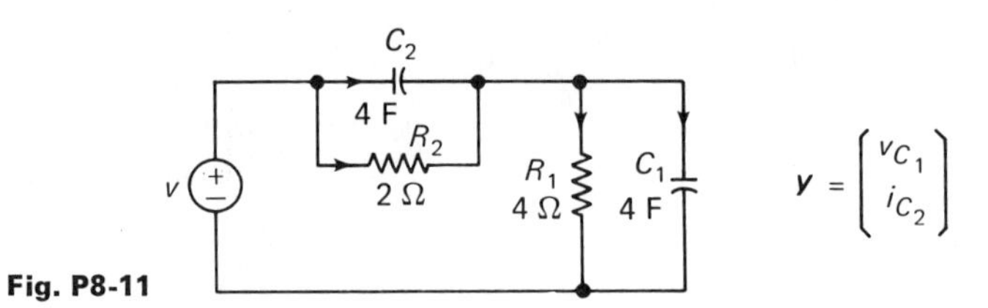

Fig. P8-11

8-12. Repeat Problem 8-11 for the circuit of Fig. P8-12.

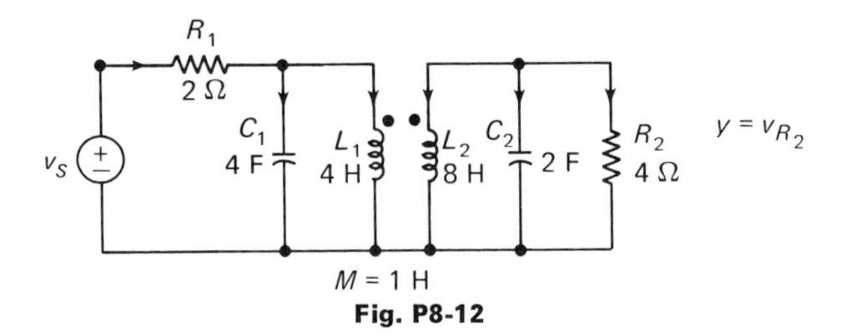

Fig. P8-12

8-13. Repeat Problem 8-11 for the circuit of Fig. P8-13.

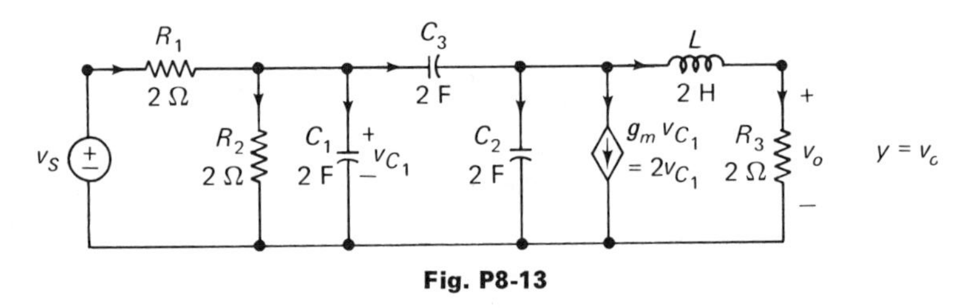

Fig. P8-13

8-14. Repeat Problem 8-11 for the circuit of Fig. P8-14.

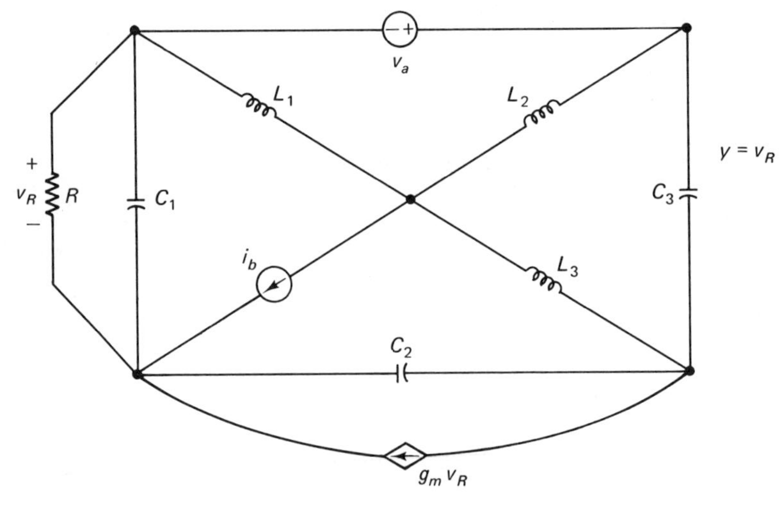

$C_1 = C_2 = C_3 = L_1 = L_2 = L_3 = R = g_m = 2$
(in farads, henries, ohms, and mhos)

Fig. P8-14

8-15. Repeat Problem 8-11 for the circuit of Fig. P8-15.

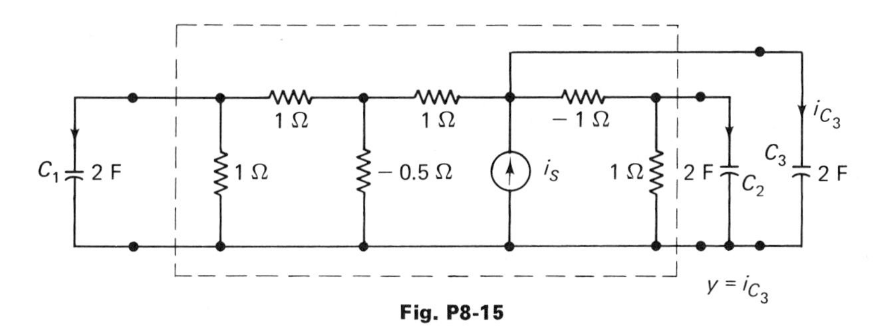

Fig. P8-15

8-16. Repeat Problem 8-11 for the circuit of Fig. P8-16.

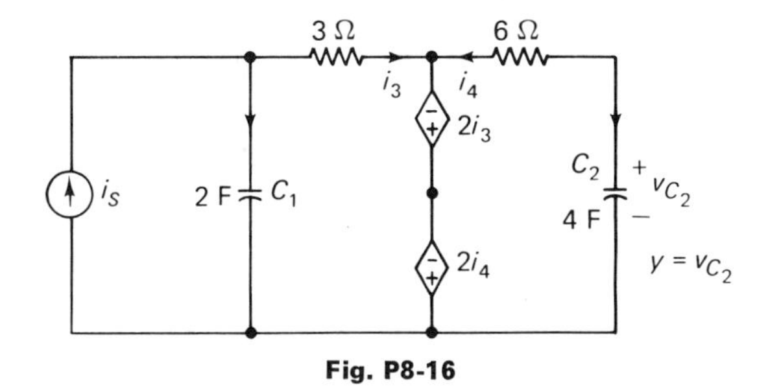

Fig. P8-16

8-17. Repeat Problem 8-11 for the circuit of Fig. P8-17.

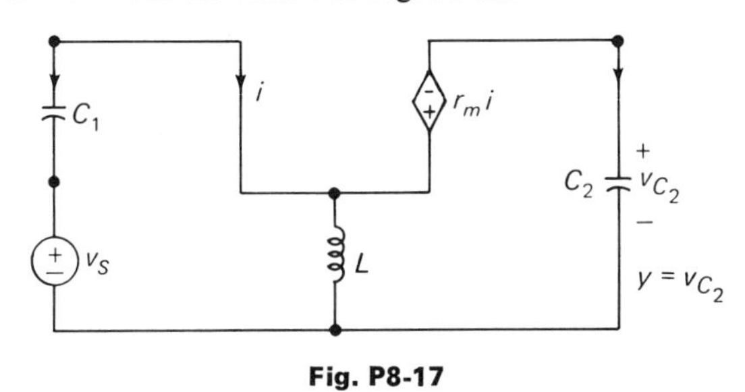

Fig. P8-17

CHAPTER 9

Numerical Solution of State Equations for Dynamic Linear Networks

In Chapter 8 we described in detail the formulation of state equations for linear dynamic networks in the form of Eqs. (8-4) and (8-5). For the sake of convenience, these two equations are repeated here:

$$\dot{\boldsymbol{x}} = \boldsymbol{A}\boldsymbol{x} + \boldsymbol{B}\boldsymbol{u} \tag{9-1}$$

$$\boldsymbol{y} = \boldsymbol{C}\boldsymbol{x} + \boldsymbol{D}\boldsymbol{u} + (\boldsymbol{D}_1\dot{\boldsymbol{u}} + \ldots) \tag{9-2}$$

As pointed out in Section 8-1, one important reason for choosing the state-variable approach is its amenability to a digital computer solution. In this chapter, we shall show how some elegant numerical methods are applied to obtain the solution of Eqs. (9-1) and (9-2). Both time-domain and frequency-domain problems will be considered.

9-1 TIME-DOMAIN SOLUTION OF THE STATE EQUATION

Given a dynamic linear network characterized by Eqs. (9-1) and (9-2), the initial state $\boldsymbol{x}(t_0)$, and the input $\boldsymbol{u}(t)$ for $t \geq t_0$, we wish to find the output $\boldsymbol{y}(t)$ for $t \geq t_0$. Since the solution is to be obtained with digital techniques, both $\boldsymbol{x}(t)$ and $\boldsymbol{y}(t)$ will be determined only for some discrete values of t, for example,

$$t = t_0, t_0 + T, t_0 + 2T, \ldots$$

where T is some chosen "time step." The input $\boldsymbol{u}(t)$ may be expressed explicitly as functions of t or may be available only in the form of sampled data.

The crux of the problem is obviously to solve Eq. (9-1); this is the problem we shall concentrate on. We shall first review the solution of a first-order linear differential equation by the method of *variation of parameters.* Then after a digression to discuss some important properties of matrix exponentials, we obtain the solution of Eq. (9-1) in a completely analogous manner. [This is, in fact, one advantage of writing the equation of motion in the form of Eq. (9-1).]

9-1-1 Method of Variation of Parameters

Given a first-order linear differential equation

$$\dot{x} = ax + bu \tag{9-3}$$

and $x(t_0)$, there are many ways of finding $x(t)$ for $t \geq t_0$. The method of *variation of parameters* will be used here.

First consider the special case when $u(t) \equiv 0$, and Eq. (9-3) is thus homogeneous. The most general solution is easily shown to be

$$x(t) = e^{at}K \tag{9-4}$$

where K is an arbitrary constant. (We write $e^{at}K$, instead of the more conventional Ke^{at}, to show the striking similarity between the equations of this section and those of Section 9-1-3.)

For the case $u(t) \not\equiv 0$, we try to obtain a solution of Eq. (9-3) by letting K be a function of t, so that

$$x(t) = e^{at}K(t) \tag{9-5}$$

Since Eq. (9-5) is *assumed* to be a solution of Eq. (9-3), we have, upon substituting Eq. (9-5) and its derivative into Eq. (9-3),

$$ae^{at}K(t) + e^{at}\dot{K}(t) = ae^{at}K(t) + bu(t) \tag{9-6}$$

Therefore,

$$\begin{aligned} \dot{K}(t) &= (e^{at})^{-1}bu(t) \\ &= e^{-at}bu(t) \end{aligned} \tag{9-7}$$

Integrating both sides of Eq. (9-7) from $t = t_0$ to t, we have

$$\int_{t_0}^{t} \dot{K}(\tau)\,d\tau = \int_{t_0}^{t} e^{-a\tau}bu(\tau)\,d\tau \tag{9-8}$$

or

$$K(t) = \int_{t_0}^{t} e^{-a\tau}bu(\tau)\,d\tau + K(t_0) \tag{9-9}$$

From Eqs. (9-9) and (9-5), we have

$$x(t) = e^{at}\left[\int_{t_0}^{t} e^{-a\tau}bu(\tau)\,d\tau + K(t_0)\right] \tag{9-10}$$

To determine $K(t_0)$, we let $t = t_0$ in Eq. (9-5)

$$x(t_0) = e^{at_0}K(t_0)$$

or

$$K(t_0) = e^{-at_0}x(t_0) \tag{9-11}$$

Therefore, the solution of Eq. (9-3) is

$$\boxed{x(t) = e^{at}\int_{t_0}^{t} e^{-a\tau}bu(\tau)\,d\tau + e^{a(t-t_0)}x(t_0)} \tag{9-12}$$

9-1-2 Some Properties of e^{At}

The solution of a linear equation $ax = k$ may be expressed as $x = a^{-1}k$. Similarly, the solution of a set of simultaneous equations $\boldsymbol{Ax} = \boldsymbol{k}$ may be expressed as $\boldsymbol{x} = \boldsymbol{A}^{-1}\boldsymbol{k}$. Except for a change from scalars to matrices the solutions in the two cases are of exactly the same form. However, since matrix multiplication is in general not commutative ($\boldsymbol{AB} \neq \boldsymbol{BA}$ in general), care should be exercised in keeping various matrices in their proper order. For example, in the scalar equation, we can write the solution as either $x = a^{-1}k$ or $x = ka^{-1}$, whereas in the matrix equation we can only write $\boldsymbol{x} = \boldsymbol{A}^{-1}\boldsymbol{k}$, not $\boldsymbol{x} = \boldsymbol{k}\boldsymbol{A}^{-1}$.

To obtain the solution to Eq. (9-1) in a manner analogous to that described in Section 9-1-1, we have to define e^{At} and explore some of its basic properties. It turns out that e^{At} and e^{at} have many (but not all) properties in common. Therefore, very little effort is needed to become familiar with the properties of e^{At}.

The matrix exponential e^{At}, where A is a constant square matrix of order n, is defined by the following infinite series:

$$e^{At} = \mathbf{1} + At + \frac{1}{2!}(At)^2 + \dots + \frac{1}{n!}(At)^n + \dots \tag{9-13}$$

It can be shown that this infinite series converges for all values of t [1]. From Eq. (9-13), e^{At} is a square matrix of order n, whose elements are functions of t.

EXAMPLE 9-1. If $A = \begin{bmatrix} a & 0 \\ 0 & b \end{bmatrix}$, then

$$e^{At} = \begin{bmatrix} 1 & 0 \\ 0 & 1 \end{bmatrix} + \begin{bmatrix} a & 0 \\ 0 & b \end{bmatrix} t + \frac{1}{2!}\begin{bmatrix} a & 0 \\ 0 & b \end{bmatrix}^2 t^2 + \dots$$

$$= \begin{bmatrix} 1 + at + \frac{1}{2}(at)^2 + \dots & 0 \\ 0 & 1 + bt + \frac{1}{2!}(bt)^2 + \dots \end{bmatrix} = \begin{bmatrix} e^{at} & 0 \\ 0 & e^{bt} \end{bmatrix}$$

EXAMPLE 9-2. If $A = \begin{bmatrix} 0 & a \\ -b & 0 \end{bmatrix}$, $a > 0$ and $b > 0$, then

$$A^{n+2} = \begin{bmatrix} 0 & a \\ -b & 0 \end{bmatrix}\begin{bmatrix} 0 & a \\ -b & 0 \end{bmatrix} A^n = -abA^n$$

Therefore, we have

$$
\begin{aligned}
e^{\begin{bmatrix} 0 & a \\ -b & 0 \end{bmatrix}t} &= \mathbf{1} + At + \frac{1}{2!}(At)^2 + \frac{1}{3!}(At)^3 + \dots \\
&= \begin{bmatrix} 1 & 0 \\ 0 & 1 \end{bmatrix}\left\{1 - \frac{abt^2}{2!} + \frac{1}{4!}a^2b^2t^4 - \frac{1}{6!}a^3b^3t^6 + \dots\right\} \\
&\quad + \begin{bmatrix} 0 & a \\ -b & 0 \end{bmatrix}\left\{t - \frac{1}{3!}abt^3 + \frac{1}{5!}a^2b^2t^5 - \dots\right\} \\
&= \begin{bmatrix} 1 & 0 \\ 0 & 1 \end{bmatrix} \cdot \cos\sqrt{ab}\,t + \begin{bmatrix} 0 & a \\ -b & 0 \end{bmatrix}\frac{1}{\sqrt{ab}} \cdot \sin\sqrt{ab}\,t \\
&= \begin{bmatrix} \cos\sqrt{ab}\,t & \sqrt{\dfrac{a}{b}}\sin\sqrt{ab}\,t \\ -\sqrt{\dfrac{b}{a}}\sin\sqrt{ab}\,t & \cos\sqrt{ab}\,t \end{bmatrix}
\end{aligned}
$$

In general, it is very difficult to obtain the elements of e^{At} in closed form by summing up the terms in Eq. (9-13). However, if t is set to a constant $t = T$, then Eq. (9-13) is a very practical way of evaluating e^{At} (see Section 9-3).

Some important properties of e^{At} that we shall need for solving Eq. (9-1) are the following:

Property 1. $e^{A0} = \mathbf{1}$. (9-14)

Property 2. $e^{At} \cdot e^{Bt} = e^{(A+B)t}$ *if and only if* $AB = BA$. (9-15)

Property 3. $[e^{At}]^{-1} = e^{-At}$. (9-16)

Property 4. $\dfrac{d}{dt}e^{At} = Ae^{At} = e^{At}A$. (9-17)

Property 5.
$$
\begin{aligned}
\int_0^t e^{A\tau}\,d\tau &= A^{-1}(e^{At} - \mathbf{1}) \\
&= (e^{At} - \mathbf{1})A^{-1}, \qquad \text{if } A \text{ is nonsingular.}
\end{aligned}
\tag{9-18}
$$

Equation (9-14) follows directly from Eq. (9-13) by letting $t = 0$. Equation (9-15) may be verified by expanding all matrix exponentials in infinite series (see Problem 9-1). Equation (9-16) follows from Eq. (9-15). Equation (9-17) is obtained by differentiating both sides of Eq. (9-13) and then factoring A from the infinite series. Equation (9-18) is a direct consequence of Eq. (9-17).

9-1-3 Solution of the State Equation

We now have all the tools needed to solve Eq. (9-1). It is interesting to note that, except for a change from scalar to matrix notations, the following derivation is an exact replica of that given in Section 9-1-1.

The most general solution of $\dot{x} = Ax$ is

$$x(t) = e^{At}K \tag{9-19}$$

where K is an arbitrary $n \times 1$ vector with constant elements. This assertion may be verified by the use of property 4:

$$\dot{x} = \frac{d}{dt}(e^{At}K) = Ae^{At}K = Ax$$

For Eq. (9-1), we try to obtain a solution by letting K be a function of t, so that

$$x(t) = e^{At}K(t) \tag{9-20}$$

Substituting Eq. (9-20) into Eq. (9-1) and making use of Eq. (9-17), we have

$$Ae^{At}K(t) + e^{At}\dot{K}(t) = Ae^{At}K(t) + Bu(t) \tag{9-21}$$

From Eqs. (9-21) and (9-16), we have

$$\dot{K}(t) = e^{-At}Bu(t) \tag{9-22}$$

Integrating Eq. (9-22) from $t = t_0$ to t, we have

$$\int_{t_0}^{t} \dot{K}(\tau)\,d\tau = \int_{t_0}^{t} e^{-A\tau}Bu(\tau)\,d\tau \tag{9-23}$$

or

$$K(t) = \int_{t_0}^{t} e^{-A\tau}Bu(\tau)\,d\tau + K(t_0) \tag{9-24}$$

Then, from Eqs. (9-24) and (9-20),

$$x(t) = e^{At}\left[\int_{t_0}^{t} e^{-A\tau}Bu(\tau)\,d\tau + K(t_0)\right] \tag{9-25}$$

$$x(t_0) = e^{At_0}K(t_0)$$

or

$$K(t_0) = e^{-At_0}x(t_0) \tag{9-26}$$

Therefore, the solution of Eq. (9-1) is

$$\boxed{x(t) = e^{At}\int_{t_0}^{t} e^{-A\tau}Bu(\tau)\,d\tau + e^{A(t-t_0)}x(t_0)} \tag{9-27}$$

Putting Eq. (9-27) in Eq. (9-2), we obtain

$$\boxed{y(t) = Ce^{A(t-t_0)}x(t_0) + \left\{Ce^{At}\int_{t_0}^{t} e^{-A\tau}Bu(\tau)\,d\tau + Du(t) + D_1\dot{u}(t) + \ldots\right\}} \tag{9-28}$$

The first term in Eq. (9-28) is the *natural response* or the *zero-input response*; the second term is the *forced response* or the *zero-state response*.

9-2 CONVERSION TO DIFFERENCE EQUATIONS

Although Eq. (9-27) is the *exact* solution of Eq. (9-1), it is not in a form suitable for digital processing. With a digital computer, we can only calculate $x(t)$ for some discrete values of t. Usually, we calculate $x(t)$ for $t = kT$, where k is an integer and T is a suitably chosen time interval. Since the input $u(kT)$ is assumed to be known for all k, what we need is an equation relating $\boldsymbol{x}[(k+1)T]$ to $\boldsymbol{u}(kT)$ and $\boldsymbol{x}(kT)$. Such an equation is a special case of a difference equation. Once the difference equation is obtained, $\boldsymbol{x}(kT)$ may be calculated successively for all k. For example, if

$$\begin{aligned} x[(k+1)T] &= \tfrac{1}{2}x(kT) + u(kT) \\ u(kT) &= (\tfrac{1}{4})^k, \qquad \text{for all } k \\ x(0) &= 1 \end{aligned} \tag{9-29}$$

then from $x(0) = 1$ and $u(0) = 1$, we have $x(T) = \frac{3}{2}$ by Eq. (9-29). Next, from $x(T) = \frac{3}{2}$ and $u(T) = \frac{1}{4}$, we have $x(2T) = \frac{3}{4} + \frac{1}{4} = 1$. The process goes on for $k = 1, 2, \ldots$.

We shall now show two basic methods for converting Eq. (9-27) into a difference equation. Since our starting point is Eq. (9-27), *the results will be applicable to linear time-invariant networks only*. Methods for converting the state equation for a nonlinear network into a difference equation will be discussed in Chapters 11 and 12. Calculation of $\boldsymbol{x}(kT)$ through a difference equation is one form of numerical integration, the details of which may be found in many textbooks on numerical analysis [2].

In Eq. (9-27), let $t_0 = kT$ and $t = (k+1)T$; we have

$$\boldsymbol{x}[(k+1)T] = e^{\boldsymbol{A}T}\boldsymbol{x}(kT) + e^{\boldsymbol{A}(k+1)T}\int_{kT}^{(k+1)T} e^{-\boldsymbol{A}\tau}\boldsymbol{B}\boldsymbol{u}(\tau)\,d\tau \tag{9-30}$$

For the following special cases, the integral in Eq. (9-30) can be evaluated exactly.

Case 1. $\boldsymbol{u}(t)$ is piecewise constant, such that

$$\boldsymbol{u}(t) = \boldsymbol{u}(kT), \qquad \text{for } kT \le t < (k+1)T,\ k = 0, 1, 2, \ldots \tag{9-31}$$

An example of such a forcing function is shown in Fig. 9-1(a). Then, making use of Eq. (9-18), we have

$$\begin{aligned} \int_{kT}^{(k+1)T} e^{-\boldsymbol{A}\tau}\boldsymbol{B}\boldsymbol{u}(\tau)\,d\tau &= \int_{kT}^{(k+1)T} e^{-\boldsymbol{A}\tau}\,d\tau \cdot \boldsymbol{B}\boldsymbol{u}(kT) \\ &= [-e^{-\boldsymbol{A}\tau}]_{kT}^{(k+1)T} \cdot \boldsymbol{A}^{-1}\boldsymbol{B}\boldsymbol{u}(kT) \\ &= [-e^{-\boldsymbol{A}(k+1)T} + e^{-\boldsymbol{A}kT}] \cdot \boldsymbol{A}^{-1}\boldsymbol{B}\boldsymbol{u}(kT) \end{aligned}$$

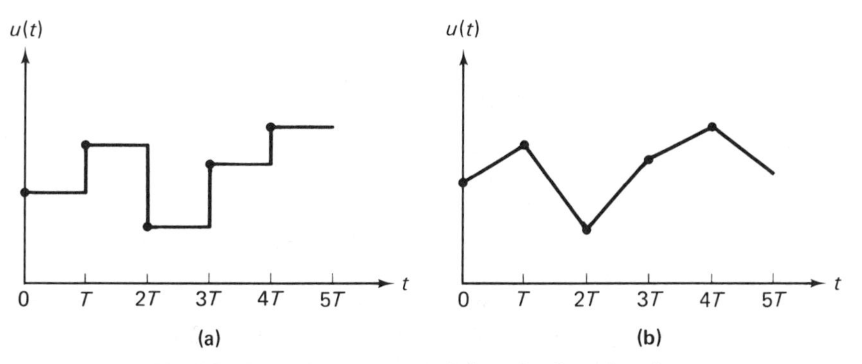

Fig. 9-1. Piecewise-constant and piecewise-linear functions.

Therefore, Eq. (9-30) becomes

$$\boxed{\boldsymbol{x}[(k+1)T] = e^{\boldsymbol{A}T}\boldsymbol{x}(kT) + [e^{\boldsymbol{A}T} - \mathbf{1}]\boldsymbol{A}^{-1}\boldsymbol{B}\cdot\boldsymbol{u}(kT)} \tag{9-32}$$

which is the desired difference equation. Equation (9-32), which is in effect a recursive formula, can be programmed very easily on a digital computer. At first, the presence of $\boldsymbol{A}^{-1}$ in Eq. (9-32) may seem to require that $\boldsymbol{A}$ be nonsingular. But such is not the case, as we actually calculate $e^{\boldsymbol{A}T}$ by summing up a number of terms of the infinite series (9-13). Thus,

$$\begin{aligned}[e^{\boldsymbol{A}T} - \mathbf{1}]\boldsymbol{A}^{-1} &= [(\mathbf{1} + \boldsymbol{A}T + \tfrac{1}{2}\boldsymbol{A}^2T^2 + \dots) - \mathbf{1}]\boldsymbol{A}^{-1} \\ &= \mathbf{1}\cdot T + \frac{1}{2!}\boldsymbol{A}T^2 + \frac{1}{3!}\boldsymbol{A}^2T^3 + \dots\end{aligned} \tag{9-33}$$

Note that there is no need to calculate $\boldsymbol{A}^{-1}$ in Eq. (9-33).

Case 2. $\boldsymbol{u}(t)$ is continuous and piecewise linear, such that

$$\boldsymbol{u}(t) = \boldsymbol{u}(kT) + \frac{\boldsymbol{u}[(k+1)T] - \boldsymbol{u}(kT)}{T}(t - kT) \tag{9-34}$$

for $kT \leq t \leq (k+1)T$, $k = 0, 1, 2, \dots$ An example of such a forcing function is shown in Fig. 9-1(b).

Putting Eq. (9-34) in Eq. (9-30), we obtain, after some manipulations, the following results (the details are left as an exercise; see Problem 9-2):

$$\boxed{\boldsymbol{x}[(k+1)T] = \boldsymbol{F}\boldsymbol{x}(kT) + \boldsymbol{G}\boldsymbol{u}(kT) + \boldsymbol{H}\boldsymbol{u}[(k+1)T]} \tag{9-35}$$

where

$$\begin{aligned}\boldsymbol{F} &= e^{\boldsymbol{A}T} = \sum_{n=0}^{\infty}\frac{1}{n!}(\boldsymbol{A}T)^n \\ \boldsymbol{G} &= [e^{\boldsymbol{A}T}(-\mathbf{1} + \boldsymbol{A}T) + \mathbf{1}](\boldsymbol{A}^2T^2)^{-1}\boldsymbol{B}T \\ &= \sum_{n=0}^{\infty}\frac{1}{n!(n+2)}(\boldsymbol{A}T)^n\cdot\boldsymbol{B}T \\ \boldsymbol{H} &= [e^{\boldsymbol{A}T} - \mathbf{1} - \boldsymbol{A}T](\boldsymbol{A}^2T^2)^{-1}\boldsymbol{B}T \\ &= \sum_{n=0}^{\infty}\frac{1}{(n+2)!}(\boldsymbol{A}T)^n\cdot\boldsymbol{B}T\end{aligned}$$

Equations (9-32) and (9-35) provide *exact* solutions of $\boldsymbol{x}(kT)$ only for some special cases of the forcing function $\boldsymbol{u}(t)$. In general, if $\boldsymbol{u}(t)$ is continuous, Eqs. (9-32) and (9-35) give *approximate* solutions of $\boldsymbol{x}(kT)$. Since the source of error lies in approximating the original forcing function by Eqs. (9-31) or (9-34), as illustrated in Fig. 9-2(a) and (b), respectively, it should be obvious that the error may be reduced by using a smaller time step T. A smaller T means a longer computer run time. Such a trade-off exists in every type of numerical method of solving differential equations. Also, from Fig. 9-2 it is clear that Eq. (9-34) is better than Eq. (9-31) as an approximation to the original continuous function.

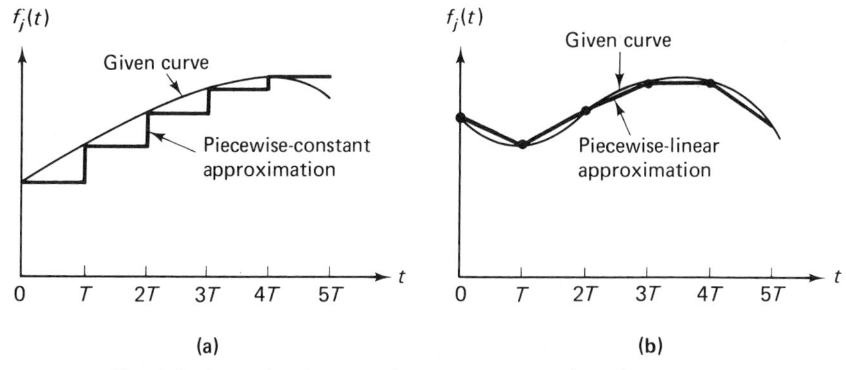

Fig. 9-2. Approximating a continuous curve by a piecewise-constant or a piecewise-linear curve.

So far we have only considered the *exact* integration of Eq. (9-30) when the forcing function $\boldsymbol{u}(t)$ is a piecewise-constant or piecewise-linear function. We can also obtain an *approximate* difference equation by assuming $\dot{\boldsymbol{x}}(t)$ or the integrand in Eq. (9-30) to be piecewise constant, piecewise linear, or piecewise parabolic. A summary of these approximate difference equations is given below. Their derivations are left as exercises [see Problems 9-3 to 9-8. Also see Chapters 11 and 12].

1. $\dot{\boldsymbol{x}}(t)$ is assumed to be piecewise constant; i.e., $f_j(t) = \dot{x}_j(t)$ in Fig. 9-2(a). The result is what is commonly called the forward Euler formula:

$$\boldsymbol{x}[(k+1)T] = (\mathbf{1} + \boldsymbol{A}T)\boldsymbol{x}(kT) + T\boldsymbol{B}\boldsymbol{u}(kT) \tag{9-36}$$

Note that if we use $e^{\boldsymbol{A}T} \cong \mathbf{1} + \boldsymbol{A}T$ as a very crude approximation to $e^{\boldsymbol{A}T}$, then Eq. (9-32) reduces to Eq. (9-36).

2. $\dot{\boldsymbol{x}}(t)$ is assumed to be piecewise linear; i.e., $f_j(t) = \dot{x}_j(t)$ in Fig. 9-2(b). The result is commonly called the trapezoidal formula:

$$\begin{aligned}\boldsymbol{x}[(k+1)T] &= \left(\mathbf{1} - \frac{T}{2}\boldsymbol{A}\right)^{-1}\left(\mathbf{1} + \frac{T}{2}\boldsymbol{A}\right)\boldsymbol{x}(kT) \\ &\quad + \left(\mathbf{1} - \frac{T}{2}\boldsymbol{A}\right)^{-1}\frac{T}{2}\cdot \boldsymbol{B}[\boldsymbol{u}(kT) + \boldsymbol{u}(kT+T)]\end{aligned} \tag{9-37}$$

3. The integrand in Eq. (9-30)

$$\boldsymbol{f}(t) = e^{-\boldsymbol{A}T}\boldsymbol{B}\boldsymbol{u}(t)$$

is assumed to be piecewise constant, as shown in Fig. 9-2(a). We obtain

$$\boldsymbol{x}[(k+1)T] = e^{\boldsymbol{A}T}\boldsymbol{x}(kT) + e^{\boldsymbol{A}T}\cdot T\boldsymbol{B}\boldsymbol{u}(kT) \tag{9-38}$$

4. The integrand in Eq. (9-30) is assumed to be piecewise linear, as shown in Fig. 9-2(b). We obtain

$$\boldsymbol{x}[(k+1)T] = e^{\boldsymbol{A}T}\boldsymbol{x}(kT) + e^{\boldsymbol{A}T}\boldsymbol{B}\frac{T}{2}\boldsymbol{u}(kT) + \boldsymbol{B}\frac{T}{2}\boldsymbol{u}[(k+1)T] \tag{9-39}$$

5. The integrand of Eq. (9-30) is assumed to be a second-degree polynomial in t. Evaluation of the integral under such an assumption leads to Simpson's

rules (see Problem 9-8 for details). We obtain

$$\begin{aligned}\boldsymbol{x}[(2n+2)T] = e^{2AT}\boldsymbol{x}(2nT) + e^{2AT} \cdot \frac{T}{3} \boldsymbol{Bu}(2nT) \\ + e^{AT} \cdot \frac{4}{3} T\boldsymbol{Bu}[(2n+1)T] + \frac{T}{3} \boldsymbol{Bu}[(2n+2)T]\end{aligned} \tag{9-40}$$

Note that in Eq. (9-40) $\boldsymbol{u}(t)$ is required every T seconds apart, while $\boldsymbol{x}(t)$ is computed every $2T$ apart.

The accuracy and stability problems associated with these formulas will be discussed in Section 9-4 after a digression to the problem of calculating e^{AT}.

9-3 EVALUATION OF e^{AT}

In Section 9-2, we gave several methods of converting the state equation (9-1) into approximate difference equations. The matrix e^{AT} appears in several of these difference equations. Literature abounds with methods for calculating e^{At}, the *state transition matrix* for a time-invariant linear system, as functions of t in closed form [3]. One can of course obtain e^{AT} by first calculating e^{At} and then letting $t = T$. Such an approach is unnecessarily complicated when a digital computer is used. Since we are really interested in e^{At} for one particular value of t, it is much easier to use the series expansion of e^{AT}. We *approximate* e^{AT} by the first $K + 1$ terms of Eq. (9-13):

$$e^{AT} \cong \mathbf{1} + AT + \frac{1}{2!}(AT)^2 + \dots + \frac{1}{K!}(AT)^K \triangleq \boldsymbol{M} = [m_{ij}] \tag{9-41}$$

Then the error matrix $\boldsymbol{R}$ is given by the difference between Eq. (9-13) and Eq. (9-41)

$$\boldsymbol{R} = \frac{1}{(K+1)!}(AT)^{(K+1)} + \frac{1}{(K+2)!}(AT)^{(K+2)} + \dots = \sum_{k=K+1}^{\infty} \frac{(AT)^k}{k!} \tag{9-42}$$

An upper bound U can be found for the magnitudes of elements of $\boldsymbol{R}$. Let $||A||$ be one of the following *norms* of the square matrix A of order n

$$||A|| \triangleq \max_i \sum_{j=1}^{n} |a_{ij}| \tag{9-43}$$

or

$$||A|| \triangleq \max_j \sum_{i=1}^{n} |a_{ij}| \tag{9-44}$$

The norm $||A||$ given by Eq. (9-43) is the maximum among the n values, each of which is the sum of $|a_{ij}|$ in the same row. Equation (9-44) has a similar meaning, but with "row" replaced by "column." Obviously, $|a_{ij}| \leq ||A||$ for all i and j.

It can be proved [1] that

$$||\boldsymbol{AB}|| \leq ||\boldsymbol{A}|| \cdot ||\boldsymbol{B}|| \tag{9-45}$$

Applying Eq. (9-45) to Eq. (9-42), we obtain a bound for the elements of $\boldsymbol{R} = [r_{ij}]$:

$$
\begin{aligned}
|r_{ij}| &\leq \sum_{k=K+1}^{\infty} \frac{1}{k!} |(i, j) \text{ element of } (AT)^k| \\
&\leq \sum_{k=K+1}^{\infty} \frac{1}{k!} \| AT \|^k \\
&\leq \frac{\| AT \|^{(K+1)}}{(K+1)!} \{1 + \| AT \| + \| AT \|^2 + \dots\}
\end{aligned}
\tag{9-46}
$$

Provided that $\| AT \| < 1$, the infinite series in Eq. (9-46) converges, leading to the following result [4]:

$$
|r_{ij}| \leq \frac{\| AT \|^{(k+1)}}{(K+1)!} \frac{1}{1 - \| AT \|} \triangleq U \tag{9-47}
$$

Equation (9-47) establishes an upper bound of the error when $\boldsymbol{M}$, given by Eq. (9-41), is used to approximate e^{AT}. Obviously, we shall use the smaller of Eq. (9-43) and Eq. (9-44) in Eq. (9-47). Once an upper bound U is established, the accuracy of $\boldsymbol{M}$ becomes evident by comparing U with all m_{ij}. More accurate results may be obtained by using more terms in Eq. (9-41). Some examples will illustrate this process.

EXAMPLE 9-3. Given

$$
A = \begin{bmatrix} -2 & 1 & 0 \\ 0 & -2 & 1 \\ 0 & 0 & -2 \end{bmatrix}, \qquad T = 0.1
$$

find e^{AT}, accurate to eight significant figures.

Solution:

$$
\| A \| = 3, \qquad \| AT \| = 0.3 < 1
$$

If 11 ($K = 10$) terms are used in Eq. (9-41), we obtain

$$
e^{AT} \cong \boldsymbol{M} = \begin{bmatrix} 8.1873075 \times 10^{-1} & 8.1873075 \times 10^{-2} & 4.0936538 \times 10^{-3} \\ 0 & 8.1873075 \times 10^{-1} & 8.1873075 \times 10^{-2} \\ 0 & 0 & 8.1873075 \times 10^{-1} \end{bmatrix} \tag{9-48}
$$

The error upper bound, as given by Eq. (9-47) is

$$
U = 6.33986 \times 10^{-14}
$$

By comparing U with all elements of $\boldsymbol{M}$, we see that the largest relative error is approximately $6.33986 \times 10^{-14}/4.09365 \times 10^{-3} = 1.549 \times 10^{-11}$. In general, if the magnitude of the relative error of a given number x is less than 0.5×10^{-n}, then the number x is accurate to n digits. The result given by Eq. (9-48) is therefore accurate to 10 significant figures. The reader can verify the correctness of this assertion by comparing Eq. (9-48) with the following exact solution:

$$
e^{AT} = [e^{At}]_{t=T} = \begin{bmatrix} e^{-2t} & te^{-2t} & \frac{t^2}{2} e^{-2t} \\ 0 & e^{-2t} & te^{-2t} \\ 0 & 0 & e^{-2t} \end{bmatrix}_{t=0.1} = \begin{bmatrix} e^{-0.2} & 0.1e^{-0.2} & 0.005e^{-0.2} \\ 0 & e^{-0.2} & 0.1e^{-0.2} \\ 0 & 0 & e^{-0.2} \end{bmatrix}
$$

9-4 A COMPLETE EXAMPLE OF TRANSIENT RESPONSE CALCULATION

EXAMPLE 9-4. Figure 9-3 shows a second-order system characterized by

$$\dot{\boldsymbol{x}} = \begin{bmatrix} 0 & 1 \\ -1 & -1 \end{bmatrix} \boldsymbol{x} + \begin{bmatrix} 0 \\ 1 \end{bmatrix} u \qquad (9\text{-}49)$$
$$y = [1 \quad 0]\boldsymbol{x}$$

The system is initially at rest, i.e., $\boldsymbol{x}(0) = \boldsymbol{0}$, and the input is

$$u(t) = 50(1 - e^{-t})u_0(t) - 50[1 - e^{-(t-1)}]u_0(t-1)$$

where $u_0(t)$ represents the unit step function. Find the response $y(t)$ for $0 \leq t \leq 8$ s. Use $T = 0.1$ s as the integration time step.

Solution: Calculate e^{AT} by Eq. (9-41), with $K = 10$; we have

$$e^{AT} = \begin{bmatrix} 0.995166 & 0.0950041 \\ -0.000950041 & 0.900016 \end{bmatrix}$$

The upper bound of error, given by Eq. (9-47), is

$$U = \frac{(0.2)^{11}}{11!} \cdot \frac{1}{1 - 0.2} = 6.41334 \times 10^{-16}$$

Therefore, the calculated e^{AT} is correct to the last digit shown. For comparison, $y(t)$ is calculated by five different methods and labeled as follows:

$y_{EX}(t)$ — exact solution obtained by the Laplace transform method

$y_{EU}(t)$ — approximate solution by the forward Euler method, Eq. (9-36)

$y_{PC}(t)$ — with piecewise-constant approximation to the input, Eq. (9-32)

$y_{PL}(t)$ — with piecewise-linear approximation to the input, Eq. (9-35)

$y_{SI}(t)$ — evaluation of the integral in Eq. (9-30) by Simpson's rule, Eq. (9-40)

Portions of the calculated results are shown in Table 9-1 (with six significant figures), from which it is seen that the forward Euler method produces too large an error. The output $y_{PC}(t)$, which should be exact when the input is piecewise constant, has appreciable error in the present case of exponential pulse input. The output $y_{SI}(t)$, calculated by Simpson's rule, is accurate to five significant figures in the present case.

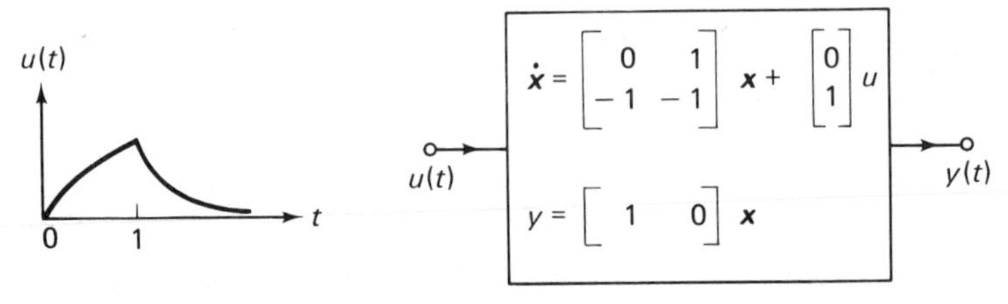

Fig. 9-3. Calculating the transient response of a second-order system.

TABLE 9-1. Results of Example 9-4

t	$y_{EX}(t)$	$y_{EU}(t)$	$y_{PC}(t)$	$y_{PL}(t)$	$y_{SI}(t)$
0.	0.	0.	0.	0.	0.
1.0	4.93067	3.97455	4.32917	4.92060	4.93070
2.0	17.3386	17.4830	16.9138	17.3318	17.3386
3.0	18.5793	19.8984	18.8166	18.5833	18.5792
4.0	10.7122	11.4089	11.1837	10.7199	10.7121
5.0	2.49951	2.08848	2.83012	2.50487	2.49951
6.0	−1.63955	−2.44095	−1.53577	−1.63790	−1.63954
7.0	−2.08409	−2.53799	−2.11774	−2.08466	−2.08408
8.0	−0.989119	−0.941643	−1.05139	−0.990137	−0.989118

The output $y_{PL}(t)$, which approximates the input with a piecewise linear function, is accurate to four significant figures in the present example.

We have given only the geometrical interpretations of the approximations used to obtain various difference equations. The whole process can also be explained with the aid of the Taylor series; such an approach is necessary when the errors are to be evaluated (see Chap. 11). In the previous cases of piecewise constant, piecewise linear, and Simpson's approximations, we are actually approximating some function over some time interval by one term, two terms, and three terms, respectively, of the Taylor series expansion. Thus, one naturally expects that, among the three methods, Simpson's rule should give the best result. (It can be shown that with Simpson's rule the error of the integration is proportional to the fifth power of the time step, when the time step is very small.) However, this is not necessarily the case when we compare Eq. (9-35) with Eq. (9-40). We derive Eq. (9-35) from the exact solution (9-30) by assuming that the input function is piecewise linear. Thus, for the case when the input *is* piecewise linear, Eq. (9-35) will give the *exact* solution, whereas Eq. (9-40) still remains a very good approximate solution. Example 9-5 will bear out this point.

EXAMPLE 9-5. Repeat Example 9-4 with the following changes: $u(t) = t$, $T = 0.5$ s, and $K = 15$.

Solution:

$$e^{AT} = \begin{bmatrix} 0.895594 & 0.377345 \\ -0.377345 & 0.518249 \end{bmatrix}$$

Again, the exact solution $y_{EX}(t)$ is obtained by the Laplace transform method. Other solutions are obtained from Eqs. (9-36), (9-32), (9-35), and (9-40). From the results given in Table 9-2 (with six significant figures), it is seen that $y_{PL}(t)$ is more accurate than $y_{SI}(t)$ in the present case.

Besides the accuracy problem, there is another important problem that must be considered in using the difference equations: the stability problem. A rigorous treatment of this topic is given in Chapters 12 and 13. Here we shall illustrate the basic facts with a simple example (see also Section 1-5). Consider an unforced first-order

TABLE 9-2. Results of Example 9-5

t	$y_{EX}(t)$	$y_{EU}(t)$	$y_{PC}(t)$	$y_{PL}(t)$	$y_{SI}(t)$
0.	0.	0.	0.	0.	0.
1.0	0.126193	0.	0.0522027	0.126193	0.125782
2.0	0.731295	0.437500	0.527599	0.731295	0.730322
3.0	1.74240	1.57812	1.46399	1.742240	1.74113
4.0	2.89641	3.00000	2.60699	2.89641	2.89511
5.0	4.01335	4.23730	3.74278	4.01335	4.01213
6.0	5.04860	5.17798	4.79688	5.04860	5.04745
7.0	6.03328	6.00000	5.78945	6.03328	6.03214
8.0	7.00828	6.89989	6.76371	7.00828	7.00712

system characterized by

$$\dot{x} = -ax, \qquad a > 0 \tag{9-50}$$

and

$$x(0) = X_0$$

It is readily shown that the exact solution is

$$x(t) = X_0 e^{-at}$$

or

$$x(kT) = X_0 e^{-akT} = X_0(e^{-aT})^k \tag{9-51}$$

which is finite and approaches 0 as $k \longrightarrow \infty$. If the forward Euler method is used, we have, from Eq. (9-36),

$$x[(k+1)T] = (1 - aT)x(kT) \tag{9-52}$$

Clearly, if the integration time step T is too large, such that $|1 - aT| > 1$, then $x[(k+1)T]$, as given by Eq. (9-52), becomes infinite as $k \longrightarrow \infty$. In such a case we say that the numerical integration is unstable. Note that the original system is stable. It is the use of an improper numerical method that leads to $x(\infty) \longrightarrow \infty$. We must seek a more suitable method.

Suppose that Eq. (9-32) is used. We have

$$x[(k+1)T] = e^{-aT}x(kT) \tag{9-53}$$

Since $a > 0$, the magnitude of e^{-aT} is less than 1 for any time step T. There is no instability problem with Eq. (9-53), no matter how large the time step T may be, provided that e^{-aT} is evaluated *exactly*. For first- and second-order systems, e^{AT} may be evaluated very easily by the Laplace transform method or any other method. For higher-order systems, the usual method for calculating e^{AT} with a computer program is by the series expansion (9-41) with a finite number of terms. When such a method is used, the numerical instability problem appears again when the time step is too large. For the present example of Eq. (9-53), if $aT = 4$, then the value of M, calculated from Eq. (9-41) as a function of K, is as follows:

K	1	2	3	4	5	6	7	8
M	−3.000	5.000	−5.667	5.000	−3.533	2.156	−1.095	0.5302

It is seen that, if less than nine terms are used for approximating e^{-aT}, Eq. (9-53) is unstable.

In general, to guarantee numerical stability when one of Eqs. (9-32), (9-35), (9-38) or (9-40) is used, the time step T and the number of terms $(K + 1)$ in Eq. (9-41) must satisfy the following condition (see [3]):

All roots $z_1, \ldots, z_n$ of the equation

$$|z\mathbf{1} - \boldsymbol{M}| = 0$$

should have magnitudes not greater than 1, where $\boldsymbol{M}$ is given by Eq. (9-41).

Other numerical methods that are applicable to nonlinear as well as to linear dynamic networks, the estimate of errors, the numerical stability problem, and the choice of optimum time steps will be studied in detail in Chapters 11–13.

9-5 FREQUENCY-DOMAIN SOLUTION OF STATE EQUATIONS

A linear system characterized by Eqs. (9-1) and (9-2) is said to be a *proper system* if the terms $(\boldsymbol{D}_1 \dot{\boldsymbol{u}} + \ldots)$ are not needed for its characterization. Otherwise, the system is said to be improper. Only proper systems will be considered in this section. Thus, the system under consideration is characterized by

$$\dot{\boldsymbol{x}} = \boldsymbol{A}\boldsymbol{x} + \boldsymbol{B}\boldsymbol{u} \tag{9-54}$$

$$\boldsymbol{y} = \boldsymbol{C}\boldsymbol{x} + \boldsymbol{D}\boldsymbol{u} \tag{9-55}$$

By definition, the transfer function matrix $\boldsymbol{H}(s)$ of a linear time-invariant system is a matrix that relates the Laplace transforms of the output and input, with the system initially at rest [i.e., $\boldsymbol{x}(0) = \boldsymbol{0}$], in the following manner:

$$\boldsymbol{Y}(s) = \boldsymbol{H}(s)\boldsymbol{U}(s) \tag{9-56}$$

There are many reasons for our interest in the transfer function matrix $\boldsymbol{H}(s)$. Two important ones are the following:

1. From the poles of the elements of $\boldsymbol{H}(s)$, we can immediately determine the stability of the system.
2. When the frequency response (for sinusoidal steady state) is desired, we simply let $s = j\omega$, and evaluate

$$\boldsymbol{Y}(j\omega) = \boldsymbol{H}(j\omega)\boldsymbol{U}(j\omega) \tag{9-57}$$

at the desired frequencies. This should be compared with the nodal method discussed in Chapter 4, where for each frequency ω one has to solve a set of simultaneous equations with complex coefficients. The advantage of using the transfer function $\boldsymbol{H}(j\omega)$ should be obvious.

Given the state equations (9-54) and (9-55), we can derive an explicit formula for $\boldsymbol{H}(s)$ very easily, as follows. Applying the Laplace transform to Eqs. (9-54) and (9-55), and recalling that $\boldsymbol{x}(0) = \boldsymbol{0}$, we have

$$s\boldsymbol{X} = \boldsymbol{A}\boldsymbol{X} + \boldsymbol{B}\boldsymbol{U} \tag{9-58}$$

$$\boldsymbol{Y} = \boldsymbol{C}\boldsymbol{X} + \boldsymbol{D}\boldsymbol{U} \tag{9-59}$$

where $\boldsymbol{X}$, $\boldsymbol{Y}$, and $\boldsymbol{U}$ are transforms of $\boldsymbol{x}$, $\boldsymbol{y}$, and $\boldsymbol{u}$, respectively. From Eq. (9-58),

$$(s\mathbf{1} - \boldsymbol{A})\boldsymbol{X} = \boldsymbol{B}\boldsymbol{U}$$

and

$$X = (s\mathbf{1} - A)^{-1}BU \tag{9-60}$$

Substituting Eq. (9-60) into Eq. (9-59), we have

$$Y = [C(s\mathbf{1} - A)^{-1}B + D]U \tag{9-61}$$

Comparing Eq. (9-61) with Eq. (9-56), we obtain

$$\boxed{H(s) = C(s\mathbf{1} - A)^{-1}B + D} \tag{9-62}$$

For low-order systems ($n \leq 3$), we can evaluate Eq. (9-62) by hand with the aid of a well-known formula [1]:

$$[s\mathbf{1} - A]^{-1} = \frac{\text{adj}[s\mathbf{1} - A]}{|s\mathbf{1} - A|} \tag{9-63}$$

When the system is of order 4 or higher, we generally evaluate Eq. (9-62) with a digital computer program. The next two sections describe algorithms for the evaluation of Eq. (9-62).

9-5-1 Souriau-Frame Algorithm [5]

The crux in the evaluation of Eq. (9-62) is to find the inverse of $(s\mathbf{1} - A)$. For the case when A is $n \times n$, we can always write

$$\begin{aligned}(s\mathbf{1} - A)^{-1} &= \frac{\text{adj}(s\mathbf{1} - A)}{|s\mathbf{1} - A|} \\ &= \frac{P_1 s^{n-1} + P_2 s^{n-2} + \dots + P_{n-1}s + P_n}{s^n + q_1 s^{n-1} + \dots + q_{n-1}s + q_n}\end{aligned} \tag{9-64}$$

where q_i are constants and P_i are $n \times n$ matrices.

The Souriau–Frame algorithm is a recursive method for calculating $P_1, q_1, \dots,$ using only additions and multiplications of real constant matrices. A proof of the method may be found in many textbooks in linear algebra [5]. Hence, we shall only describe the method, and then illustrate it with a numerical example.

We need the definition of "trace." The *trace* of a square matrix A, denoted by $\text{tr}(A)$, is the algebraic sum of the main diagonal elements of A; i.e., $\text{tr}(A) \triangleq \sum_{i=1}^{n} a_{ii}$.

The Souriau–Frame algorithm proceeds as follows:

$$\boxed{\begin{aligned} P_1 &= \mathbf{1} \\ q_1 &= -\text{tr}(A) \\ P_2 &= P_1 A + q_1 \mathbf{1} \\ q_2 &= -\tfrac{1}{2}\text{tr}(P_2 A) \\ &\vdots \\ P_k &= P_{k-1} A + q_{k-1} \mathbf{1} \\ q_k &= -\frac{1}{k}\text{tr}(P_k A) \end{aligned}}$$

The process terminates when $\boldsymbol{P}_n$ and q_n have been evaluated and substituted into Eq. (9-64).

EXAMPLE 9-6. Given

$$\dot{\boldsymbol{x}} = \begin{bmatrix} -4 & 1 \\ -6 & 1 \end{bmatrix} \boldsymbol{x} + \begin{bmatrix} 3 \\ 7 \end{bmatrix} u$$

$$y = [1 \quad 1]\boldsymbol{x} + 2u$$

find the transfer function $\boldsymbol{H}(s)$ (a scalar in the present case).

Solution:

$$\boldsymbol{P}_1 = \mathbf{1}$$

$$q_1 = -(-4 + 1) = 3$$

$$\boldsymbol{P}_2 = \begin{bmatrix} -4 & 1 \\ -6 & 1 \end{bmatrix} + \begin{bmatrix} 3 & 0 \\ 0 & 3 \end{bmatrix} = \begin{bmatrix} -1 & 1 \\ -6 & 4 \end{bmatrix}$$

$$q_2 = -\tfrac{1}{2}\text{tr}\left\{\begin{bmatrix} -1 & 1 \\ -6 & 4 \end{bmatrix}\begin{bmatrix} -4 & 1 \\ -6 & 1 \end{bmatrix}\right\} = -\tfrac{1}{2}\text{tr}\begin{bmatrix} -2 & 0 \\ 0 & -2 \end{bmatrix} = 2$$

From Eq. (9-64), we have

$$(s\mathbf{1} - \boldsymbol{A})^{-1} = \frac{\begin{bmatrix} s-1 & 1 \\ -6 & s+4 \end{bmatrix}}{s^2 + 3s + 2}$$

Putting this result in Eq. (9-62), we have

$$\boldsymbol{H}(s) = \frac{2s^2 + 16s + 18}{s^2 + 3s + 2}$$

Even though the method is easy to program on a digital computer, it has a serious accuracy problem because of the subtractions of nearly equal numbers that may occur at various steps of the evaluation. Furthermore, the final answers are always in the form of polynomials in s. Although we can find poles and zeros of the network functions by the use of a complex roots subroutine, it will be much better if the poles and zeros can be obtained directly. The next section describes a method for this purpose.

9-5-2 Transfer Functions as Eigenvalue Problems

As may be easily seen from Eq. (9-62), each element $h_{ij}(s)$ of the transfer function matrix $\boldsymbol{H}(s)$ has $|s\mathbf{1} - \boldsymbol{A}|$ as its denominator (before cancellation of common factors with the numerator). The roots of the characteristic equation

$$|s\mathbf{1} - \boldsymbol{A}| = (s - s_1)(s - s_2)\dots(s - s_n) = 0 \tag{9-65}$$

i.e., $s_1, s_2, \dots, s_n$, are called *natural frequencies* of the system. In linear algebra terminology, they are called the eigenvalues of the matrix $\boldsymbol{A}$. Calculation of eigenvalues is a very important topic in numerical analysis, and many methods are now available. The ***QR*** algorithm [6], considered as the best by most numerical analysts, is described in Section 9-6.

Once the natural frequencies are found, we know the denominator of the transfer function to within a multiplicative constant. These natural frequencies may or may not be the poles of the transfer function, depending on whether cancellation of common factors with the numerator takes place.

It would be very nice if we could also reduce the problem of finding the numerator of a transfer function to the eigenvalue problem associated with certain matrices. This has been made possible for single-input single-output systems through the aid of a feedback system to be described next [7]. A rigorous, algebraic treatment of the problem may be found in [8].

Figure 9-4(a) shows a single-input single-output system characterized by state equations

$$\dot{\mathbf{x}} = \mathbf{A}\mathbf{x} + \mathbf{B}u \tag{9-66}$$

$$y = \mathbf{C}\mathbf{x} + du \tag{9-67}$$

The transfer function $H(s)$ may be expressed as

$$\begin{aligned}\frac{Y}{U} = H(s) &= \mathbf{C}(s\mathbf{1} - \mathbf{A})^{-1}\mathbf{B} + d \\ &= \frac{N(s)}{|s\mathbf{1} - \mathbf{A}|}\end{aligned} \tag{9-68}$$

We have, in terms of transformed variables,

$$|s\mathbf{1} - \mathbf{A}|\,Y = N(s)U \tag{9-69}$$

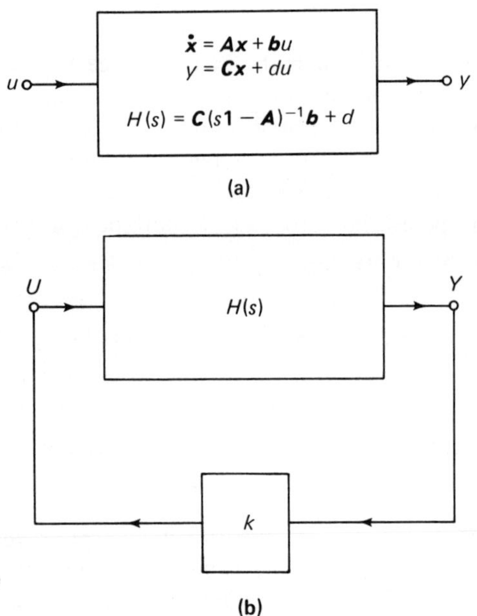

Fig. 9-4. Relating the zeros of $H(s)$ to the natural frequencies of a feedback system.

In considering the natural frequencies, we may let $U = 0$, thus leading to

$$|s\mathbf{1} - A|\, Y = 0 \tag{9-70}$$

Equation (9-70) indicates, from the frequency-domain point of view, that the natural frequencies of the systems are the zeros of the polynomial $\Delta(s) \triangleq |s\mathbf{1} - A|$, where $\Delta(s)\, Y = 0$ characterizes the unforced system.

Next, consider the system shown in Fig. 9-4(b), which is constructed by amplifying the output y by a gain k ($k > 0$) and using the resultant output signal as the sole input to the original system. For this system, we have frequency-domain relationships as follows:

$$Y = H(s)U = \frac{N(s)}{|s\mathbf{1} - A|} U \tag{9-71}$$

and

$$U = kY \tag{9-72}$$

From Eqs. (9-71) and (9-72), we have

$$\left[-\frac{|s\mathbf{1} - A|}{k} + N(s) \right] Y = 0 \tag{9-73}$$

As $k \longrightarrow \infty$, Eq. (9-73) reduces to

$$N(s)Y = 0 \tag{9-74}$$

Thus, by comparing Eq. (9-74) with Eq. (9-70), it is seen that the system of Fig. 9-4(b), with $k \longrightarrow \infty$, has natural frequencies which are just the roots of $N(s) = 0$.

Our remaining task is to derive the state equation for the system of Fig. 9-4(b). As will be shown shortly, the initial state equation for Fig. 9-4(b), under the condition $k \longrightarrow \infty$, can always be written as

$$\hat{\boldsymbol{M}}^{(0)}\dot{\boldsymbol{x}}^{(0)} = \hat{\boldsymbol{A}}^{(0)}\boldsymbol{x}^{(0)} \tag{9-75}$$

The reduction algorithm of Section 8-4 is then applied to obtain the final state equation,

$$\dot{\hat{\boldsymbol{x}}} = \hat{\boldsymbol{A}}\hat{\boldsymbol{x}} \tag{9-76}$$

where $\hat{\boldsymbol{x}}$ in general is a subset of $\boldsymbol{x}^{(0)}$. Once the $\hat{\boldsymbol{A}}$ matrix has been found, its eigenvalues may be evaluated by the $\boldsymbol{QR}$ algorithm or any of a number of other methods.

In the time domain, the feedback system of Fig. 9-4(b) is characterized by the following three equations:

$$\dot{\boldsymbol{x}} = \boldsymbol{Ax} + \boldsymbol{B}u \tag{9-77}$$

$$y = \boldsymbol{Cx} + du \tag{9-78}$$

and

$$u = ky \tag{9-79}$$

If we substitute Eq. (9-78) into Eq. (9-79), we have

$$u = ky = k\boldsymbol{Cx} + k\,du$$

or

$$(1 - kd)u = k\boldsymbol{Cx} \tag{9-80}$$

Depending on the value of d, we have to distinguish between two cases.

Case 1. $d \neq 0$. Then, as $k \longrightarrow \infty$, Eq. (9-80) yields

$$u = -\frac{\boldsymbol{C}}{d}\boldsymbol{x} \tag{9-81}$$

Substituting Eq. (9-81) into Eq. (9-77), we obtain

$$\dot{\boldsymbol{x}} = \left(\boldsymbol{A} - \frac{\boldsymbol{BC}}{d}\right)\boldsymbol{x} \tag{9-82}$$

Since Eq. (9-82) is already in the form of Eq. (9-76), or (9-75) with $\hat{\boldsymbol{M}}^{(0)} = \mathbf{1}$, we have for this case, $\hat{\boldsymbol{x}} = \boldsymbol{x}$ and

$$\hat{\boldsymbol{A}} = \boldsymbol{A} - \frac{\boldsymbol{BC}}{d} \tag{9-83}$$

Case 2. $d = 0$. Then Eq. (9-80) yields

$$u = k\boldsymbol{Cx} \tag{9-84}$$

Substituting Eq. (9-84) into Eq. (9-77), we have

$$\mathbf{1}\dot{\boldsymbol{x}} = (\boldsymbol{A} + k\boldsymbol{BC})\boldsymbol{x} \tag{9-85}$$

Care must be exercised in passing k to infinity. Note that $\boldsymbol{B}$ is a column vector,

$$\boldsymbol{B} = [b_1, b_2, \ldots, b_n]^t$$

in the present case. If $b_i = 0$, the ith equation in Eq. (9-85) is not affected by the constant k at all. If $b_i \neq 0$, we divide the ith equation in Eq. (9-85) by k, and then let $k \longrightarrow \infty$. These steps lead to the following two rules for obtaining the initial state equation (9-75):

> *Rule 1.* If $b_i = 0$, then the ith row of $[\hat{\boldsymbol{M}}^{(0)}, \hat{\boldsymbol{A}}^{(0)}]$ is the same as the ith row of $[\mathbf{1}, \boldsymbol{A}]$.
>
> *Rule 2.* If $b_i \neq 0$, then the ith row of $[\hat{\boldsymbol{M}}^{(0)}, \hat{\boldsymbol{A}}^{(0)}]$ is the same as the ith row of $[\mathbf{0}, \boldsymbol{BC}]$. Or, equivalently, the ith row of $\hat{\boldsymbol{M}}^{(0)}$ is a zero row, and the ith row of $\hat{\boldsymbol{A}}^{(0)}$ is equal to b_i times $\boldsymbol{C}$.

In terms of $\boldsymbol{A}$ and $\hat{\boldsymbol{A}}$, the desired transfer function may be expressed as

$$\frac{Y}{U} = H(s) = K\frac{|s\mathbf{1} - \hat{\boldsymbol{A}}|}{|s\mathbf{1} - \boldsymbol{A}|} \tag{9-86}$$

The gain constant K is usually determined by evaluating both Eqs. (9-86) and (9-62) at some convenient real value of s. A comparison of the results will reveal the value of K.

Some numerical examples will illustrate the preceding methods.

EXAMPLE 9-7. Find the transfer function of the system in Example 9-6 by the eigenvalue method.

Solution:

$$\boldsymbol{A} = \begin{bmatrix} -4 & 1 \\ -6 & 1 \end{bmatrix}$$

$$|s\mathbf{1} - \boldsymbol{A}| = (s+1)(s+2)$$

From Eq. (9-83)

$$\hat{\boldsymbol{A}} = \begin{bmatrix} -4 & 1 \\ -6 & 1 \end{bmatrix} - \frac{1}{2}\begin{bmatrix} 3 \\ 7 \end{bmatrix}[1 \quad 1] = \begin{bmatrix} -5.5 & -0.5 \\ -9.5 & -2.5 \end{bmatrix}$$

$$|s\mathbf{1} - \hat{\boldsymbol{A}}| = s^2 + 8s + 9$$

To determine K, let $s \longrightarrow \infty$. Then from Eq. (9-62), $H(\infty) = 2$. But from Eq. (9-86), $H(\infty) = K$. Therefore, $K = 2$ and

$$H(s) = 2\frac{s^2 + 8s + 9}{(s+1)(s+2)}$$

EXAMPLE 9-8. Find the transfer function for the following system by the eigenvalue method.

$$\dot{\boldsymbol{x}} = \begin{bmatrix} -4 & 1 \\ -6 & 1 \end{bmatrix}\boldsymbol{x} + \begin{bmatrix} 0 \\ 2 \end{bmatrix}u$$

$$y = [1 \quad 2]\boldsymbol{x}$$

Solution:

$$\boldsymbol{A} = \begin{bmatrix} -4 & 1 \\ -6 & 1 \end{bmatrix}$$

$$|s\mathbf{1} - \boldsymbol{A}| = (s+1)(s+2)$$

According to the rules given in this section, we have

$$[\hat{\boldsymbol{M}}^{(0)} \mid \hat{\boldsymbol{A}}^{(0)}] = \left[\begin{array}{cc|cc} 1 & 0 & -4 & 1 \\ 0 & 0 & 2 & 4 \end{array}\right]$$

Applying the reduction algorithm of Section 8-4, we obtain the final state equations of the feedback system of Fig. 9-4(b), as follows:

$$\dot{\hat{x}} = -\tfrac{9}{2}\hat{x}$$

$$\hat{\boldsymbol{A}} = -\tfrac{9}{2}$$

$$|s\mathbf{1} - \hat{\boldsymbol{A}}| = s + \tfrac{9}{2}$$

Therefore,

$$H(s) = K\frac{s + \frac{9}{2}}{(s+1)(s+2)}$$

To determine K, let $s = 0$. From the preceding equation, we have

$$H(0) = K\tfrac{9}{4}$$

But from Eq. (9-62)

$$H(0) = [1 \quad 2]\begin{bmatrix} 4 & -1 \\ 6 & -1 \end{bmatrix}^{-1}\begin{bmatrix} 0 \\ 2 \end{bmatrix} = 9$$

Therefore,

$$K = 4$$

and

$$H(s) = \frac{4(s + 4.5)}{(s + 1)(s + 2)}$$

*9-6 THE *QR* ALGORITHM

In Section 9-5 we showed that the determination of the poles and zeros of a transfer function can be treated as an eigenvalue problem. One natural way to find the eigenvalues of a general matrix A is to obtain the characteristic polynomial $|s\mathbf{1} - A|$ (e.g., by the method of Section 9-5-1) and then find the roots of this polynomial. Such a method has a serious accuracy problem when the matrix is of an order greater than 4 or 5. Of all methods presently available for finding eigenvalues, the QR algorithm [6] is regarded by numerical analysts as the best for digital computer implementation. A complete proof of the QR algorithm is beyond the scope of this book. We shall only describe the essential concepts of the method, and illustrate them with some simple numerical examples. Readers interested in the rigorous theory behind the method should consult [6, 9, 10].

All matrices under consideration in the present section are assumed to be real. The given matrix A may be asymmetric and singular. Before describing the QR algorithm, let us examine a few special cases in which the eigenvalues of A can be determined very easily.

Case 1. Matrix A is triangular. Then a direct expansion of the determinant $|s\mathbf{1} - A|$ yields (details left as an exercise)

$$|s\mathbf{1} - A| = (s - a_{11})(s - a_{22}) \dots (s - a_{nn}) \tag{9-87}$$

Therefore, the eigenvalues of A are just the diagonal elements of A.

Case 2. Matrix A is of order 2. Then we have

$$|s\mathbf{1} - A| = \begin{vmatrix} s - a_{11} & -a_{12} \\ -a_{21} & s - a_{22} \end{vmatrix} = s^2 - (a_{11} + a_{22})s + a_{11}a_{22} - a_{12}a_{21}$$

and the eigenvalues of A are

$$s_{1,2} = \frac{(a_{11} + a_{22}) \pm \sqrt{(a_{11} + a_{22})^2 - 4|A|}}{2} \tag{9-88}$$

Case 3. Matrix A is *block triangular.* An upper block triangular matrix is of the form

$$A = \begin{bmatrix} A_{11} & A_{12} & \cdots & A_{1m} \\ \mathbf{0} & A_{22} & \cdots & A_{2m} \\ \cdot & & & \\ \cdot & & & \\ \cdot & & & \\ \mathbf{0} & \mathbf{0} & \cdots & A_{mm} \end{bmatrix} \tag{9-89}$$

where each A_{ii} is square, and $A_{ij} = \mathbf{0}$ for all $i > j$. Note that, if all A_{ii} are 1×1 matrices, then (9-89) becomes a triangular matrix.

Applying Laplace expansion [1] to the determinant $|s\mathbf{1} - \boldsymbol{A}|$, we obtain

$$|s\mathbf{1} - \boldsymbol{A}| = |s\mathbf{1}_{11} - \boldsymbol{A}_{11}|\cdot|s\mathbf{1}_{22} - \boldsymbol{A}_{22}|\ldots|s\mathbf{1}_{mm} - \boldsymbol{A}_{mm}| \tag{9-90}$$

where $\mathbf{1}_{kk}$ is an identity matrix of the same order as $\boldsymbol{A}_{kk}$. From the above equation, it is seen that the eigenvalues of a block triangular matrix $\boldsymbol{A}$ consist of the eigenvalues of $\boldsymbol{A}_{11}, \boldsymbol{A}_{22}, \ldots, \boldsymbol{A}_{mm}$.

EXAMPLE 9-9. Find the eigenvalues of the following 6×6 matrix:

$$\boldsymbol{A} = \left[\begin{array}{cc|c|c|cc} -4 & 1 & 2 & 1 & 6 & 2 \\ -6 & 1 & 5 & -1 & 8 & 3 \\ \hline 0 & 0 & 3 & 4 & 7 & 0 \\ \hline 0 & 0 & 0 & -2 & 1 & 0 \\ \hline 0 & 0 & 0 & 0 & 1 & -1 \\ 0 & 0 & 0 & 0 & 1 & 1 \end{array}\right] \tag{9-91}$$

Solution: Applying Eqs. (9-87)–(9-90) to the present case, we immediately obtain the six eigenvalues, as follows:

$$\begin{array}{lll} s_1 = -1, & s_2 = -2 & \text{eigenvalues of } \boldsymbol{A}_{11} \\ s_3 = 3, & s_4 = -2, & \text{eigenvalues of } \boldsymbol{A}_{22} \text{ and } \boldsymbol{A}_{33} \\ s_5 = 1 + j, & s_6 = 1 - j, & \text{eigenvalues of } \boldsymbol{A}_{44} \end{array}$$

9-6-1 Essence of the *QR* Algorithm

From Example 9-9, it is clear that the eigenvalues of $\boldsymbol{A}$ are very simply determined if $\boldsymbol{A}$ is block triangular *with* 1×1 *or* 2×2 *diagonal blocks.* A general matrix $\boldsymbol{A}$ of course will not be of such form. The ***QR*** algorithm is a method that attempts to generate another matrix $\boldsymbol{B}$ having the same eigenvalues as $\boldsymbol{A}$ and the above-mentioned desirable structure.

It is easy to derive from $\boldsymbol{A}$ another matrix $\boldsymbol{B}$ having the same eigenvalues as $\boldsymbol{A}$. Let $\boldsymbol{P}$ be any *nonsingular* matrix of the same order as $\boldsymbol{A}$. Let

$$\boldsymbol{B} = \boldsymbol{P}^{-1}\boldsymbol{A}\boldsymbol{P} \tag{9-92}$$

Then we say the two matrices $\boldsymbol{A}$ and $\boldsymbol{B}$ are *similar*, and the transformation given by Eq. (9-92) is called a *similarity transformation.* The most important property of a similarity transformation is that it preserves the eigenvalues; i.e., $\boldsymbol{B}$ and $\boldsymbol{A}$ have the same eigenvalues. This is easy to see, as follows:

$$\begin{aligned} |s\mathbf{1} - \boldsymbol{B}| &= |s\boldsymbol{P}^{-1}\boldsymbol{P} - \boldsymbol{P}^{-1}\boldsymbol{A}\boldsymbol{P}| = |\boldsymbol{P}^{-1}(s\mathbf{1} - \boldsymbol{A})\boldsymbol{P}| \\ &= |\boldsymbol{P}^{-1}||s\mathbf{1} - \boldsymbol{A}||\boldsymbol{P}| = |s\mathbf{1} - \boldsymbol{A}| \end{aligned} \tag{9-93}$$

Thus, the characteristic polynomials of $\boldsymbol{A}$ and $\boldsymbol{B}$ are the same; and so are the eigenvalues.

In the ***QR*** algorithm, the nonsingular matrix $\boldsymbol{P}$ used in Eq. (9-92) is required to have the additional property of being *orthogonal.* An $n \times n$ real matrix $\boldsymbol{Q}$ is said to be an *orthogonal matrix* if

$$Q^tQ = QQ^t = \mathbf{1} \tag{9-94}$$

Note that the nonsingularity of Q is implied in Eq. (9-94). Hereafter, we shall use Q to denote an orthogonal matrix. Examples of orthogonal matrices are

$$Q_1 = \begin{bmatrix} 0 & 0 & 1 \\ 0 & 1 & 0 \\ 1 & 0 & 0 \end{bmatrix}, \qquad Q_2 = \begin{bmatrix} \cos\theta & -\sin\theta \\ \sin\theta & \cos\theta \end{bmatrix}$$

It is easy to prove the following properties of an orthogonal matrix:

1. $|Q| = \pm 1$.
2. $Q^{-1} = Q^t$.
3. *The product of any number of orthogonal matrices is also an orthogonal matrix.*

Given *any* matrix A, we can always factor A as

$$A = QU \tag{9-95}$$

where Q is an orthogonal matrix and U is an upper triangular matrix. (One method of QU factorization is described in Section 9-6-3.) After a QU factorization of A, if we form another matrix B such that

$$B = UQ \tag{9-96}$$

then

$$B = UQ = (Q^{-1}A)Q \tag{9-97}$$

Thus, B is similar to A. However, with just one application of Eqs. (9-95) and (9-96), B in general does not yet have the desired block triangular structure mentioned earlier. The basic idea of the QR algorithm is to repeat this transformation until the desired matrix structure appears. Let A_1 be the given matrix. The standard QR algorithm[1] generates a sequence of matrices $(A_1, A_2, \ldots)$ as follows:

For $k = 1, 2, \ldots$, factor A_k as

$$A_k = Q_k U_k \tag{9-98a}$$

and then form

$$A_{k+1} \triangleq U_k Q_k \tag{9-98b}$$

A study of the convergence conditions of the sequence as $k \longrightarrow \infty$ is beyond our scope (see [6, 9, 10]). It suffices to state here that, as $k \longrightarrow \infty$, in *nearly all cases* A_k tends toward a block triangular matrix whose diagonal blocks are 1×1 or 2×2 matrices. Then the eigenvalues are easily determined, as illustrated by Example 9-9. In some rare cases where the convergence of the sequence $(A_1, A_2, \ldots)$ to the abovementioned structure fails, the difficulty can be overcome by a "shift of origin," to be described in Section 9-6-5.

[1]An upper triangular matrix is sometimes called a right triangular matrix. Equation (9-98a) is then written as $A_k = Q_k R_k$; hence the name QR algorithm.

When the computations are done by hand, the $\boldsymbol{QR}$ algorithm is impractical because of the large amount of operations involved. But as a digital computer method, the $\boldsymbol{QR}$ algorithm is very powerful, because it is very stable in the sense that the round-off error does not seriously affect the accuracy of the calculated eigenvalues.

In actual computer implementation of the $\boldsymbol{QR}$ algorithm, we always perform a preliminary transformation of $\boldsymbol{A}$ into a Hessenberg matrix $\boldsymbol{H}$ having the same eigenvalues and then apply the $\boldsymbol{QR}$ algorithm to $\boldsymbol{H}$. A matrix $\boldsymbol{H}$ is said to be an (upper) *Hessenberg matrix* if $h_{ij} = 0$ for all $i > j + 1$. A 5×5 upper Hessenberg matrix has the following form:

$$\boldsymbol{H} = \begin{bmatrix} \times & \times & \times & \times & \times \\ \times & \times & \times & \times & \times \\ 0 & \times & \times & \times & \times \\ 0 & 0 & \times & \times & \times \\ 0 & 0 & 0 & \times & \times \end{bmatrix}$$

— diagonal
— subdiagonal

Thus, it is essentially an upper triangular matrix, but with one extra set of subdiagonal elements, as shown. The reason for reducing $\boldsymbol{A}$ to a Hessenberg matrix is that this step greatly reduces the total number of operations of the $\boldsymbol{QR}$ algorithm. For a full matrix, the number of multiplications and additions in going from $\boldsymbol{A}_k$ to $\boldsymbol{A}_{k+1}$ is proportional to n^3; with a Hessenberg matrix the number is proportional to only n^2. Furthermore, it can be shown that, if $\boldsymbol{A}_k$ is a Hessenberg matrix, so is $\boldsymbol{A}_{k+1}$ [9]. Thus, the simpler form of the matrix $\boldsymbol{A}_k$ is preserved throughout the $\boldsymbol{QR}$ algorithm. Since a long sequence of matrices $(\boldsymbol{A}_1, \boldsymbol{A}_2, \ldots)$ has to be generated, the savings in computational effort by going to the Hessenberg matrix is quite significant. One method of reducing $\boldsymbol{A}$ to a Hessenberg matrix while preserving the eigenvalues is described in Section 9-6-2.

Assume that the given matrix $\boldsymbol{A}$ is already in Hessenberg form. The application of the $\boldsymbol{QR}$ algorithm can be modified as follows. Generate a sequence of Hessenberg matrices $(\boldsymbol{A}_1, \boldsymbol{A}_2, \ldots)$ by Eq. (9-98). As soon as any one (or more) subdiagonal elements of $\boldsymbol{A}_k$ becomes zero,[2] $\boldsymbol{A}_k$ becomes a block triangular matrix with two (or more) diagonal blocks, each of which is also a Hessenberg matrix. For example, if the (4, 3) element on the subdiagonal of a 6×6 matrix $\boldsymbol{A}_k$ becomes zero, we have

$$\boldsymbol{A}_k = \left[\begin{array}{ccc|ccc} \times & \times & \times & \times & \times & \times \\ \times & \times & \times & \times & \times & \times \\ 0 & \times & \times & \times & \times & \times \\ \hline 0 & 0 & 0 & \times & \times & \times \\ 0 & 0 & 0 & \times & \times & \times \\ 0 & 0 & 0 & 0 & \times & \times \end{array}\right] = \begin{bmatrix} \boldsymbol{A}_{k11} & \boldsymbol{A}_{k12} \\ \boldsymbol{0} & \boldsymbol{A}_{k22} \end{bmatrix}$$

[2] In floating-point calculations, we rarely get an exact zero. A small positive number ε is selected as an effective zero. Then we consider $h_{ij} = 0$ whenever $|h_{ij}| \leq \varepsilon$.

We can invoke Eq. (9-90) to reduce the original problem to several eigenvalue problems of smaller Hessenberg matrices. This process is repeated until we encounter a diagonal 1×1 or 2×2 block whose eigenvalues are easily determined by Eqs. (9-87) and (9-88).

Several examples of finding eigenvalues by the $\boldsymbol{QR}$ algorithm are given in Sections 9-6-4 and 9-6-5, after a digression to describe the reduction to Hessenberg form and the $\boldsymbol{QU}$ factorization in the next two sections.

9-6-2 Reduction to Hessenberg Matrix

Two methods, the Householder method and the Givens method, are commonly used for this purpose. We shall not attempt to present here a rigorous development of these methods and their geometrical interpretations, as they can be found in most textbooks on linear algebra (e.g., see [1]). Instead, we shall only describe the Givens method without proof and then illustrate the method with a numerical example.

For any $n \times n$ real matrix $\boldsymbol{A}$, we can always find an orthogonal matrix $\boldsymbol{Q}$ such that $\boldsymbol{Q}^{-1}\boldsymbol{A}\boldsymbol{Q} = \boldsymbol{H}$ is an upper Hessenberg matrix. $\boldsymbol{Q}$ is usually built up as the product of some simple orthogonal matrices. Let us first consider the simpler problem of finding $\boldsymbol{Q}$ such that the (i, j) element in $\boldsymbol{H} = \boldsymbol{Q}^{-1}\boldsymbol{A}\boldsymbol{Q}$ is reduced to zero ($i > j + 1$ in the present case). The following rule can be proved [1]:

Rule 1. *To reduce the* (i, j) *element of* $\boldsymbol{Q}^{-1}\boldsymbol{A}\boldsymbol{Q}$ *to zero, we may use the following* orthogonal *matrix* $\boldsymbol{Q}$:

$$\boldsymbol{Q} = \begin{array}{c} \\ j+1 \\ \\ i \\ \\ \end{array} \begin{bmatrix} \boxed{\mathbf{1}} & & & & \\ & c & & -s & \\ & & \boxed{\mathbf{1}} & & \\ & s & & c & \\ & & & & \boxed{\mathbf{1}} \end{bmatrix} \quad \text{(columns } j+1 \text{ and } i\text{)} \tag{9-99}$$

where

$$c = a_{j+1,j}(a_{j+1,j}^2 + a_{ij}^2)^{-1/2}$$
$$s = a_{ij}(a_{j+1,j}^2 + a_{ij}^2)^{-1/2}$$

and, except for the four elements $(j + 1, j + 1)$, (i, i), $(j + 1, i)$, *and* $(i, j + 1)$, *as just defined, other elements of* $\boldsymbol{Q}$ *are the same as those of an identity matrix.*

(The notations c and s are used here because they can be considered as $\cos\theta$ and $\sin\theta$, respectively.)

Suppose that rule 1 is used to reduce the (4, 2) element of $\boldsymbol{A}_2 = \boldsymbol{Q}_1^{-1}\boldsymbol{A}_1\boldsymbol{Q}_1$ to zero. We may apply rule 1 again to $\boldsymbol{A}_2$, with a suitable $\boldsymbol{Q}_2$, so as to reduce another element, say the (3, 1) element of $\boldsymbol{A}_3 = \boldsymbol{Q}_2^{-1}\boldsymbol{A}_2\boldsymbol{Q}_2$, to zero. But then the (4, 2) element, which has been reduced to zero in $\boldsymbol{A}_2$, will in general no longer be zero in $\boldsymbol{A}_3$. However, if the following rule is observed, all elements that have been reduced to zero will *remain* zero (see [1] for a proof).

Rule 2. *In generating the sequence of matrices $A_{k+1} = Q_k^{-1}A_kQ_k$, $k = 1, 2, \ldots,$ to preserve the zeros created by the use of rule 1, choose Q_k to reduce successively the elements below the subdiagonal to zero in the following order:*

First column	$(3, 1), (4, 1), \ldots, (n, 1)$
Second column	$(4, 2), (5, 2), \ldots, (n, 2)$
.	
.	
.	
$(n-3)$th *column*	$(n-1, n-3), (n, n-3)$
$(n-2)$th *column*	$(n, n-2)$

Givens method for obtaining $H = Q^{-1}AQ$ is simply repeated applications of rule 1, with the order of zeroing out elements given by rule 2. Let the sequence of orthogonal matrices used in the process be

$$Q_1, Q_2, \ldots, Q_p$$

It is easy to show that

$$p \leq \tfrac{1}{2}(n-2)(n-1)$$

and that

$$H = Q_p^{-1} \ldots Q_2^{-1}Q_1^{-1}AQ_1Q_2 \ldots Q_p \tag{9-100}$$

$$Q = Q_1Q_2 \ldots Q_p \tag{9-101}$$

Since Q_k is an orthogonal matrix, $Q_k^{-1} = Q_k^t$, Eq. (9-100) can also be written as

$$H = Q_p^t \ldots Q_2^tQ_1^tAQ_1Q_2 \ldots Q_p \tag{9-102}$$

From Eq. (9-102) it is easily seen that, if A is symmetric, the corresponding matrix H is also symmetric. A symmetric Hessenberg matrix is called a *tridiagonal* matrix because of its obvious structure. Tridiagonal matrices are of interest in circuit analysis because the nodal admittance matrix of an *RLC ladder* network is of such form.

EXAMPLE 9-10. Reduce the following matrix A_1 to a Hessenberg matrix H by $H = Q^{-1}A_1Q$.

$$A_1 = \begin{bmatrix} 2 & 1 & 4 & 1 \\ 3 & 3 & 2 & -2 \\ 4 & 2 & -1 & 3 \\ 5 & -1 & 4 & 2 \end{bmatrix}$$

Solution: According to rule 2, three elements will be reduced to zero in the order (3, 1), (4, 1), and (4, 2). By rule 1, we have

$$Q_1 = \begin{bmatrix} 1 & 0 & 0 & 0 \\ 0 & c_1 & -s_1 & 0 \\ 0 & s_1 & c_1 & 0 \\ 0 & 0 & 0 & 1 \end{bmatrix}, \qquad c_1 = \frac{3}{(3^2 + 4^2)^{1/2}} = \frac{3}{5} = 0.6 \qquad s_1 = \frac{4}{5} = 0.8$$

and $A_2 = Q_1^{-1}A_1Q_1 = Q_1^tA_1Q_1$. (*Note:* s_1 here is not to be confused with one root of $|s\mathbf{1} - A| = 0$.)

$$A_2 = \begin{bmatrix} 2 & 3.8 & 1.6 & 1 \\ 5 & 2.36 & -2.48 & 1.2 \\ 0 & -2.48 & -0.36 & 3.4 \\ 5 & 2.6 & 3.2 & 2 \end{bmatrix}$$

Continuing the process, and retaining four significant figures in all answers, we have

$$Q_2 = \begin{bmatrix} 1 & 0 & 0 & 0 \\ 0 & c_2 & 0 & -s_2 \\ 0 & 0 & 1 & 0 \\ 0 & s_2 & 0 & c_2 \end{bmatrix}, \qquad c_2 = \frac{5}{(5^2 + 5^2)^{1/2}} = \frac{1}{\sqrt{2}} = 0.7071$$

$$s_2 = \frac{1}{\sqrt{2}} = 0.7071$$

$$A_3 = Q_2^{-1}A_2Q_2 = \begin{bmatrix} 2.000 & 3.394 & 1.600 & -1.980 \\ 7.071 & 4.080 & 0.5091 & -0.8800 \\ 0 & 0.6505 & -0.3600 & 4.157 \\ 0 & 0.5200 & 4.016 & 0.2800 \end{bmatrix}$$

$$Q_3 = \begin{bmatrix} 1 & 0 & 0 & 0 \\ 0 & 1 & 0 & 0 \\ 0 & 0 & c_3 & -s_3 \\ 0 & 0 & s_3 & c_3 \end{bmatrix}, \qquad c_3 = \frac{0.6505}{(0.6505^2 + 0.52^2)^{1/2}} = \frac{0.6505}{0.8328} = 0.7811$$

$$s_3 = \frac{0.52}{0.8328} = 0.6244$$

and finally

$$H = A_4 = Q_3^{-1}A_3Q_3 = \begin{bmatrix} 2.000 & 3.394 & 0.01358 & -2.545 \\ 7.071 & 4.080 & -0.1518 & -1.005 \\ 0 & 0.8328 & 3.876 & 1.283 \\ 0 & 0 & 1.142 & -3.956 \end{bmatrix}$$

9-6-3 The QU Factorization

Given any $n \times n$ real matrix A, we can always express A as the product $A = QU$, where Q is an orthogonal matrix and U is an upper triangular matrix. By writing the relationship as $Q^{-1}A = U$, we see that the problem is the same as finding an orthogonal matrix Q, such that $Q^{-1}A$ is an upper triangular matrix. The matrix Q is usually built up as the product of some simple orthogonal matrices. We shall describe Givens method for QU factorization without proof and then illustrate the method with a numerical example. For a detailed discussion of several other methods, the reader is referred to [1].

Consider first the simpler problem of finding Q such that the (i, j) element of $Q^{-1}A$ is reduced to zero ($i > j$ in the present case). The following rule can be proved [1].

Rule 1. *To reduce the (i, j) element of $\boldsymbol{Q}^{-1}\boldsymbol{A}$ to zero, we may use the following orthogonal matrix $\boldsymbol{Q}$:*

$$\begin{array}{cc} & \begin{array}{ccccc} \phantom{\boxed{1}} & j & \phantom{\boxed{1}} & i & \phantom{\boxed{1}} \end{array} \\ \boldsymbol{Q} = \begin{array}{c} \phantom{\boxed{1}} \\ j \\ \phantom{\boxed{1}} \\ i \\ \phantom{\boxed{1}} \end{array} & \left[\begin{array}{ccccc} \boxed{1} & & & & \\ & c & & -s & \\ & & \boxed{1} & & \\ & s & & c & \\ & & & & \boxed{1} \end{array}\right] \end{array} \tag{9-103}$$

where

$$c = a_{jj}(a_{jj}^2 + a_{ij}^2)^{-1/2}$$
$$s = a_{ij}(a_{jj}^2 + a_{ij}^2)^{-1/2}$$

and, except for the four elements (j, j), (i, i), (i, j) and (j, i), as just defined, other elements of $\boldsymbol{Q}$ are the same as those of an identity matrix.

Suppose that rule 1 is used to reduce the (4, 2) elements of $\boldsymbol{A}_2 = \boldsymbol{Q}_1^{-1}\boldsymbol{A}_1$ to zero. We may apply rule 1 again with a suitable $\boldsymbol{Q}_2$, so as to reduce another element, say the (3, 1) element of $\boldsymbol{A}_3 = \boldsymbol{Q}_2^{-1}\boldsymbol{A}_2$, to zero. But then the (4, 2) element, which has been reduced to zero in $\boldsymbol{A}_2$, will in general no longer be zero in $\boldsymbol{A}_3$. However, if the following rule is observed, all elements that have been reduced to zero will *remain* zero (see [1] for a proof).

Rule 2. *In generating the sequence of matrices $\boldsymbol{A}_{k+1} = \boldsymbol{Q}_k^{-1}\boldsymbol{A}_k$, $k = 1, 2, \ldots,$ to preserve the zeros created by the use of rule 1, choose $\boldsymbol{Q}_k$ to reduce successively the elements below the diagonal to zero in the following order:*

$$\begin{array}{rl} \textit{First column} & (2, 1), (3, 1), \ldots, (n, 1) \\ \textit{Second column} & (3, 2), (4, 2), \ldots, (n, 2) \\ \vdots & \\ (n-1)\text{th } \textit{column} & (n, n-1) \end{array}$$

Givens method of $\boldsymbol{QU}$ factorization is simply repeated applications of rule 1, with the order of zeroing out elements given by rule 2. Let the sequence of orthogonal matrices used in the process be

$$\boldsymbol{Q}_1, \boldsymbol{Q}_2, \ldots, \boldsymbol{Q}_p$$

It is easy to show that for a full matrix $\boldsymbol{A}_1$,

$$p \leq \tfrac{1}{2}(n-1)n$$

and that if $\boldsymbol{A}_1$ is a Hessenberg matrix, then

$$p \leq n - 1$$

The complete reduction process yields

$$\boldsymbol{U} = \boldsymbol{Q}_p^{-1} \ldots \boldsymbol{Q}_2^{-1}\boldsymbol{Q}_1^{-1}\boldsymbol{A}_1 \tag{9-104}$$

from which

$$A_1 = (Q_1 Q_2 \dots Q_p)U \tag{9-105}$$

and therefore

$$Q = Q_1 Q_2 \dots Q_p \tag{9-106}$$

EXAMPLE 9-11. Express the following matrix A_1 as the product QU:

$$A_1 = \begin{bmatrix} 3 & 3 & 2 \\ 4 & 2 & -1 \\ 2 & 1 & 4 \end{bmatrix}$$

Solution: According to rule 2, we shall reduce three elements to zero in the order (2, 1), (3, 1), and (3, 2). By rule 1, we have

$$Q_1 = \begin{bmatrix} c_1 & -s_1 & 0 \\ s_1 & c_1 & 0 \\ 0 & 0 & 1 \end{bmatrix}, \qquad c_1 = \frac{3}{(3^2 + 4^2)^{1/2}} = \frac{3}{5} = 0.6 \qquad s_1 = \frac{4}{5} = 0.8$$

$$A_2 = Q_1^{-1} A_1 = Q_1^t A_1 = \begin{bmatrix} 5 & 3.4 & 0.4 \\ 0 & -1.2 & -2.2 \\ 2 & 1 & 4 \end{bmatrix}$$

Continuing the process, and retaining four significant figures in all answers, we have

$$Q_2 = \begin{bmatrix} c_2 & 0 & -s_2 \\ 0 & 1 & 0 \\ s_2 & 0 & c_2 \end{bmatrix}, \qquad c_2 = \frac{5}{(5^2 + 2^2)^{1/2}} = \frac{5}{5.385} = 0.9285 \qquad s_2 = \frac{2}{5.385} = 0.3714$$

$$A_3 = Q_2^{-1} A_2 = Q_2^t A_2 = \begin{bmatrix} 5.385 & 3.528 & 1.857 \\ 0 & -1.200 & -2.200 \\ 0 & -0.3342 & 3.565 \end{bmatrix}$$

$$Q_3 = \begin{bmatrix} 1 & 0 & 0 \\ 0 & c_3 & -s_3 \\ 0 & s_3 & c_3 \end{bmatrix}, \qquad c_3 = \frac{-1.2}{(1.2^2 + 0.3342^2)^{1/2}} = \frac{-1.2}{1.246} = -0.9633$$

$$s_3 = \frac{-0.3342}{1.246} = -0.2682$$

and finally

$$U = A_4 = Q_3^{-1} A_3 = Q_3^t A_3 = \begin{bmatrix} 5.385 & 3.528 & 1.857 \\ 0 & 1.246 & 1.163 \\ 0 & 0 & -4.025 \end{bmatrix}$$

$$Q = Q_1 Q_2 Q_3 = \begin{bmatrix} 0.5571 & 0.8304 & 0 \\ 0.7428 & -0.4982 & 0.4474 \\ 0.3714 & -0.2491 & -0.8944 \end{bmatrix}$$

9-6-4 Numerical Examples of Calculating Eigenvalues by the *QR* Algorithm

The QR algorithm is described in Section 9-6-1. Since any matrix can be transformed into a Hessenberg matrix while preserving the eigenvalues (Section 9-6-2), we assume that the transformation has been done. Since the procedure for QU factorization has been clearly described in Section 9-6-3 we shall omit all such details in the following examples.

EXAMPLE 9-12. Find the eigenvalues of the following matrix by the QR algorithm.

$$A_1 = \begin{bmatrix} -5 & -5 & -4 \\ 1 & 0 & 0 \\ 0 & 1 & 0 \end{bmatrix}$$

Solution: Following the method of Section 9-6-3, we obtain, to four significant figures,

$$Q_1 = \begin{bmatrix} -0.9806 & -0.1373 & 0.1400 \\ 0.1961 & -0.6865 & 0.7001 \\ 0 & 0.7140 & 0.7001 \end{bmatrix}$$

$$U_1 = \begin{bmatrix} 5.099 & 4.903 & 3.922 \\ 0 & 1.400 & 0.5492 \\ 0 & 0 & -0.5601 \end{bmatrix}$$

$$A_2 = U_1 Q_1 = \begin{bmatrix} -4.038 & -1.265 & 6.893 \\ 0.2746 & -0.5694 & 1.365 \\ 0 & -0.3999 & -0.3921 \end{bmatrix}$$

Note that the (2, 1) element of A_2 is already close to zero, indicating that one of the eigenvalues of A_2 (and hence A_1) is approximately -4.038. Also note that A_2 is again a Hessenberg matrix. The Hessenberg form will be preserved throughout the QR algorithm. The second iteration yields the following matrices:

$$Q_2 = \begin{bmatrix} -0.9977 & -0.05789 & -0.3540 \\ 0.06787 & -0.8511 & -0.5205 \\ 0 & -0.5217 & 0.8531 \end{bmatrix}$$

$$U_2 = \begin{bmatrix} 4.048 & 1.224 & -6.784 \\ 0 & 0.7665 & -1.356 \\ 0 & 0 & -1.289 \end{bmatrix}$$

$$A_3 = U_2 Q_2 = \begin{bmatrix} \boxed{-4.038} & 2.263 & -6.568 \\ 0.05201 & \boxed{\begin{matrix} 0.05520 & -1.556 \\ 0.6726 & -1.100 \end{matrix}} \\ 0 & & \end{bmatrix}$$

If the number 0.05201 can be considered as an *effective zero*, then A_3 becomes a block triangular matrix with one 1×1 and one 2×2 diagonal block. (See boxes in the preceding equation.) By Eqs. (9-87)–(9-89), the three eigenvalues of the given matrix $\boldsymbol{A}_1$ are easily found to be

$$-4.038, \qquad 0.5456 + j0.8445, \qquad 0.5456 - j0.8445$$

These are to be compared with the exact eigenvalues

$$-4, \qquad 0.5 \pm j\frac{\sqrt{3}}{2} = 0.5 \pm j0.8660$$

As more iterations are carried out (by a digital computer, of course), the result will be much more accurate.

EXAMPLE 9-13. Find the eigenvalues of the following matrix by the $\boldsymbol{QR}$ algorithm:

$$\boldsymbol{A}_1 = \begin{bmatrix} 0 & 0 & 1 \\ 1 & 0 & 0 \\ 0 & 1 & 0 \end{bmatrix}$$

Solution: Following the method of Section 9-6-3, we have

$$\boldsymbol{A}_1 = \boldsymbol{Q}_1\boldsymbol{U}_1 = \underset{\boldsymbol{Q}_1}{\begin{bmatrix} 0 & 0 & 1 \\ 1 & 0 & 0 \\ 0 & 1 & 0 \end{bmatrix}} \underset{\boldsymbol{U}_1}{\begin{bmatrix} 1 & 0 & 0 \\ 0 & 1 & 0 \\ 0 & 0 & 1 \end{bmatrix}}$$

Therefore, $\boldsymbol{A}_2 = \boldsymbol{U}_1\boldsymbol{Q}_1 = \boldsymbol{A}_1$.

Continuing the process, we obtain $\boldsymbol{A}_k = \boldsymbol{A}_1$ for $k = 2, 3, 4, \ldots$. Thus, none of the subdiagonal elements of $\boldsymbol{A}_k$ will converge to zero. The standard $\boldsymbol{QR}$ algorithm, as described in Section 9-6-1, will not terminate. One way to overcome this difficulty is to apply the "shifted $\boldsymbol{QR}$ algorithm," a brief description of which is given in the next section.

9-6-5 Shift of Origin

The standard $\boldsymbol{QR}$ algorithm described in Section 9-6-1 may fail in some rare cases, as illustrated by Example 9-13. The difficulty can be overcome by a *shift-of-origin* technique. The $\boldsymbol{QR}$ algorithm with shift of origin may be stated as follows:

For $k = 1, 2, \ldots,$ factor $\boldsymbol{A}_k - p_k\mathbf{1}$ as

$$\boldsymbol{A}_k - p_k\mathbf{1} = \boldsymbol{Q}_k\boldsymbol{U}_k \tag{9-107a}$$

and then form

$$\boldsymbol{A}_{k+1} \triangleq \boldsymbol{U}_k\boldsymbol{Q}_k + p_k\mathbf{1} \tag{9-107b}$$

where p_k is an arbitrary constant. Note that, if $p_k = 0$ for all k, then Eq. (9-107) reduces to the standard $\boldsymbol{QR}$ algorithm given by Eq. (9-98). A sequence of similar matrices is generated by Eq. (9-107), and hence they have the same eigenvalues. This is easy to see, as we have

$$\begin{aligned} A_{k+1} &= U_k Q_k + p_k \mathbf{1} = Q_k^{-1}(A_k - p_k \mathbf{1})Q_k + p_k \mathbf{1} \\ &= Q_k^{-1} A_k Q_k - p_k \mathbf{1} + p_k \mathbf{1} = Q_k^{-1} A_k Q_k \end{aligned} \tag{9-108}$$

A "lock-of-pattern," such as that occurring in Example 9-13, can usually be broken with a random shift of origin.

EXAMPLE 9-14. Find the eigenvalues of the matrix of Example 9-13 by QR algorithm with origin shift.

Solution: Arbitrarily let $p_1 = 1$. Then

$$A_1 - p_1\mathbf{1} = \begin{bmatrix} -1 & 0 & 1 \\ 1 & -1 & 0 \\ 0 & 1 & -1 \end{bmatrix} = Q_1 U_1$$

$$= \underbrace{\begin{bmatrix} -\frac{1}{\sqrt{2}} & -\frac{1}{\sqrt{6}} & \frac{1}{\sqrt{3}} \\ \frac{1}{\sqrt{2}} & -\frac{1}{\sqrt{6}} & \frac{1}{\sqrt{3}} \\ 0 & \frac{2}{\sqrt{6}} & \frac{1}{\sqrt{3}} \end{bmatrix}}_{Q_1} \underbrace{\begin{bmatrix} \sqrt{2} & -\frac{1}{\sqrt{2}} & -\frac{1}{\sqrt{2}} \\ 0 & \sqrt{1.5} & -\sqrt{1.5} \\ 0 & 0 & 0 \end{bmatrix}}_{U_1}$$

$$A_2 = U_1 Q_1 + p_1 \mathbf{1} = \begin{bmatrix} -\frac{1}{2} & -\frac{\sqrt{3}}{2} & 0 \\ \frac{\sqrt{3}}{2} & -\frac{1}{2} & 0 \\ 0 & 0 & 1 \end{bmatrix}$$

One subdiagonal element of A_2 is already zero. By the use of Eqs. (9-87)–(9-89), we obtain the three eigenvalues as follows:

$$1, \qquad -\frac{1}{2} + j\frac{\sqrt{3}}{2}, \qquad -\frac{1}{2} - j\frac{\sqrt{3}}{2}$$

We have only demonstrated the use of shift of origin when the QR algorithm fails to yield any 1×1 or 2×2 diagonal submatrix. In fact, even if such difficulty does not occur, the speed of convergence may be greatly improved by a suitable shift of origin. Furthermore, it is sometimes advantageous to use conjugate complex values for p_k and p_{k+1} in the origin shift. It appears that complex arithmetics will have to be used. But with a technique called the *double QR algorithm*, computations with complex numbers can be avoided even if p_k and p_{k+1} are complex. For a detailed discussion of suitable choices of p_k and the double QR algorithm, the reader is referred to [9]. Discussions on the digital computer implementation of the QR algorithm may be found in [10].

REFERENCES

1. NOBLE, B. *Applied Linear Algebra.* Englewood Cliffs, N.J.: Prentice-Hall, Inc., 1969.
2. HAMMING, R. W. *Numerical Methods for Scientists and Engineers.* New York: McGraw-Hill Book Company, 1962.
3. DERUSSO, P. M., R. J. ROY, and C. M. CLOSE. *State Variables for Engineers.* New York: John Wiley & Sons, Inc., 1966.
4. LIOU, M. L. "Time and Frequency Domain Analysis of Linear Time-Invariant Systems." In F. F. Kuo and J. F. Kaiser, eds., *System Analysis by Digital Computer.* New York: John Wiley & Sons, Inc., 1966.
5. HILDERBRAND, F. B. *Introduction to Numerical Analysis.* New York: McGraw-Hill Book Company, 1956.
6. FRANCIS, J. G. F. "The *QR* Transformation—A Unitary Analogue to the *LR* Transformation." *Computer J.*, Part 1, pp. 265–271, 1961; Part 2, pp. 332–345, 1962.
7. POTTLE, C. "A 'Textbook' Computerized State-Space Network Analysis Algorithm." *IEEE Trans. Circuit Theory*, Vol. CT-16, pp. 566–568, Nov. 1969.
8. SANDBERG, I. W., and H. C. SO. "A Two-Sets-of-Eigenvalues Approach to the Computer Analysis of Linear Systems." *IEEE Trans. Circuit Theory*, Vol. CT-16, pp. 509–517, Nov. 1969.
9. WILKINSON, J. H. *The Algebraic Eigenvalue Problem.* New York: Oxford University Press, Inc., 1965.
10. PARLETT, B. N. "The *LU* and *QR* Algorithms." in A. Ralston and H. S. Wilf, eds., *Numerical Methods for Digital Computers.* New York: John Wiley & Sons, Inc., 1967. Vol. II, pp. 116–130.

PROBLEMS

9-1. Prove that $e^{At} \cdot e^{Bt} = e^{(A+B)t}$ if and only if $AB = BA$.

9-2. Prove the relationship given by Eq. (9-35) for the case of piecewise linear-input $u(t)$ given by Eq. (9-34).

9-3. Derive the difference equation (9-36) from the state equation

$$\dot{x} = Ax + Bu$$

by two methods:

(a) *Method 1.* Assume that the derivative is approximated by

$$\dot{x}(t) = \frac{x(t+T) - x(t)}{T}$$

(b) *Method 2.* Assume that $\dot{x}(t)$ is piecewise constant, such that

$$\dot{x}(t) = \dot{x}(kT), \qquad kT \leq t < (k+1)T,\ k = 0, 1, 2, \ldots$$

9-4. If $\dot{x}(t)$ is assumed to be piecewise constant such that

$$\dot{x}(t) = \dot{x}[(k+1)T], \qquad kT < t \leq (k+1)T,\ k = 0, 1, 2, \ldots$$

show that the state equation converts to the backward Euler formula

$$x[(k+1)T] = (1 - AT)^{-1}x(kT) + (1 - AT)^{-1}BTu[(k+1)T]$$

Note very carefully the difference between the assumption made on $\dot{x}(t)$ in this problem and that in Problem 9-3.

9-5. If $\dot{x}(t)$ is assumed to be piecewise linear, as shown in Fig. 9-2(b), show that the state equation converts to Eq. (9-37). *Hint:* Recall the following:

(i) $\int_{kT}^{(k+1)T} \dot{\phi}(t)\,dt = \phi[(k+1)T] - \phi(kT)$

(ii) The area under the $\dot{x}(t)$ curve, bounded by $t = kT$ and $t = (k+1)T$, is given by the formula

$$\frac{\dot{x}(kT) + \dot{x}[(k+1)T]}{2} \cdot T$$

9-6. Prove the relationship given by Eq. (9-38). *Hint:* Start with Eq. (9-30).

9-7. Prove the relationship given by Eq. (9-39).

9-8. Consider the integral

$$\int_{kT}^{(k+2)T} f(t)\,dt$$

If, in the interval from kT to $(k+2)T$, the arc of the $f(t)$ versus t curve is very nearly coincident with the arc of the parabola through the midpoint and the terminal points of the arc, then the integral is given approximately by

$$\int_{kT}^{(k+2)T} f(t)\,dt = \frac{T}{3}[f(kT+2T) + 4f(kT+T) + f(kT)]$$

Using this formula as the starting point, derive the difference equation given by Eq. (9-40).

9-9. Let $A = \begin{bmatrix} -4 & 1 \\ -6 & 1 \end{bmatrix}$, $\quad T = 0.1$

(a) Calculate e^{AT} by

$$e^{At} = \mathcal{L}^{-1}[s\mathbf{1} - A]^{-1}$$

and

$$e^{AT} = [e^{At}]_{t=0.1}$$

(b) Calculate e^{AT} by using four terms of the series expansion [$K = 3$ in Eq. (9-41)]. What is an upper bound of error, as predicted by Eq. (9-47)?

9-10. A linear system is characterized by

$$\begin{bmatrix} \dot{x}_1 \\ \dot{x}_2 \end{bmatrix} = \begin{bmatrix} -3 & 1 \\ 1 & -3 \end{bmatrix} \begin{bmatrix} x_1 \\ x_2 \end{bmatrix} + \begin{bmatrix} 1 \\ 1 \end{bmatrix} u \qquad (1)$$

$$y = [1 \quad 1] \begin{bmatrix} x_1 \\ x_2 \end{bmatrix} + 2u \qquad (2)$$

Convert the state equation (1) into a difference equation by Euler's methods. Assume that $T = 0.1$.

9-11. A first-order linear system is characterized by

$$\dot{x} = -2x + 3u(t)$$

For each of the following cases, find $x(t)$ for $t \geq 0$ by the use of Eq. (9-12).

(a) $x(0) = 5$, $u(t) = e^{-10t}$.
(b) $x(0) = 5$, $u(t) = 0$.
(c) $x(0) = 0$, $u(t) = e^{-10t}$.

9-12. (a) Convert the differential equation of Problem 9-11 into approximate difference equations, with $T = 0.1$, by the use of Eqs. (9-31), (9-32), (9-36), (9-37), and (9-38).

(b) If these difference equations are used to calculate $x(kT)$ for cases (a), (b), and (c) of Problem 9-11, which difference equation is *exact*, and for which case?

9-13. Consider the linear system given in Problem 9-10.

(a) Find the eigenvalues of the A matrix.
(b) Find $[s\mathbf{1} - A]^{-1}$ by the use of

$$[s\mathbf{1} - A]^{-1} = \frac{\text{adj}[s\mathbf{1} - A]}{|s\mathbf{1} - A|}$$

where

$$\text{adj}\,\boldsymbol{P} \triangleq [\text{cofactors of } p_{ij}]^t$$

(c) Repeat part (b), but use the Souriau–Frame algorithm.
(d) Find the transfer function for the system.

9-14. (a) Find a matrix A whose eigenvalues are the zeros of the transfer function $H(s)$ of the system given in Problem 9-10.
(b) Repeat part (a), but with the output equation of the system changed to

$$y = [1 \quad 1]\begin{bmatrix} x_1 \\ x_2 \end{bmatrix}$$

and the state equation changed to

$$\dot{\boldsymbol{x}} = \begin{bmatrix} -3 & 1 \\ 1 & -3 \end{bmatrix} \boldsymbol{x} + \begin{bmatrix} 0 \\ 3 \end{bmatrix} u$$

9-15. Consider a single-input linear system characterized by

$$\dot{\boldsymbol{x}} = \boldsymbol{Ax} + \boldsymbol{Bu}$$

Find the *exact* difference equation relating $x[(k+1)T]$ and $x(kT)$ for each of the following cases:
(a) $u(t)$ is a unit step function.
(b) $u(t)$ is a unit impulse function.

9-16. The circuit shown in Fig. P9-16 is characterized by

$$\dot{\boldsymbol{x}} = \begin{bmatrix} \dot{v}_C \\ \dot{i}_L \end{bmatrix} = \begin{bmatrix} -\frac{1}{RC} & -\frac{1}{C} \\ \frac{1}{L} & 0 \end{bmatrix} \begin{bmatrix} v_C \\ i_L \end{bmatrix} + \begin{bmatrix} \frac{1}{C} \\ 0 \end{bmatrix} i_s$$

The input $i_s(t)$ is a zero-mean, 0.5-Hz square wave with a peak-to-peak value of 20 mA.
(a) Find the *exact* relationship between $\boldsymbol{x}[(k+1)T]$ and $\boldsymbol{x}(kT)$ when $T = 1$ s.
(b) If $\boldsymbol{x}(0) = \boldsymbol{0}$, find the value of $v_C(t)$ from the result of part (a) for $t = 0, 1, 2, 3, 4, 5$ s.
(c) Write a FORTRAN program to carry out the calculation of part (b) until the steady state is reached. What is the peak-to-peak value of $v_C(t)$ in steady state?

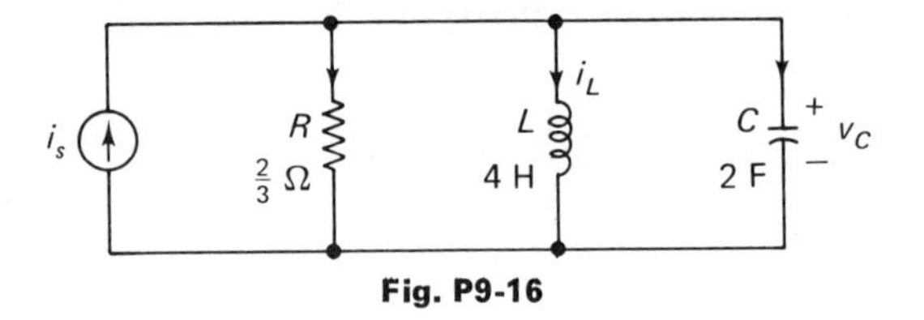

Fig. P9-16

9-17. The circuit shown in Fig. P9-17 is characterized by

$$\dot{\boldsymbol{x}} = \begin{bmatrix} \dot{v}_C \\ \dot{i}_L \end{bmatrix} = \begin{bmatrix} 0 & 1 \\ -1 & 0 \end{bmatrix} \begin{bmatrix} v_C \\ i_L \end{bmatrix} + \begin{bmatrix} 0 \\ 1 \end{bmatrix} v_s$$

The input voltage $v_s(t)$ is a rectangular pulse given by $v_s(t) = 10[u(t) - u(t - 2\pi)]$ V.
(a) Find the exact relationship between $\boldsymbol{x}[(k + 1)T]$ and $\boldsymbol{x}(kT)$ when $T = \pi/2$ s.
(b) If $\boldsymbol{x}(0) = \boldsymbol{0}$ and $T = \pi/2$, find $v_C(kT)$ for $k = 0, 1, 2, \ldots, 50$.

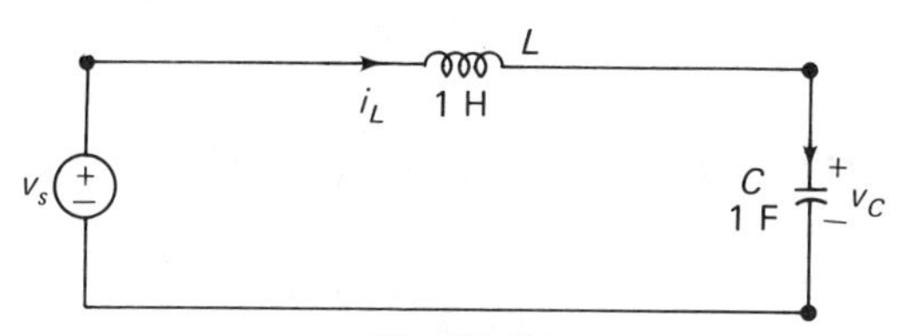

Fig. P9-17

9-18. A linear system is characterized by $\dot{x} = (-\log_e 3)x + 3$. Find the *exact* relationship between $x[(k + 1)T]$ and $x(kT)$ when $T = 0.1$ s.

9-19. Reduce the matrix

$$A = \begin{bmatrix} 100 & 600 & -80 \\ 10 & -100 & 8 \\ 10 & 8 & 5 \end{bmatrix}$$

to an upper Hessenberg matrix $\boldsymbol{H}$ by $\boldsymbol{H} = \boldsymbol{Q}^{-1}\boldsymbol{A}\boldsymbol{Q}$, where $\boldsymbol{Q}$ is an orthogonal matrix. Perform all calculations correct to four significant figures.

9-20. Find the eigenvalues of the following matrix by the $\boldsymbol{QR}$ algorithm:

$$A = \begin{bmatrix} -1 & 0 & 2 \\ 3.5 & 0 & 3 \\ 0 & -1 & 0 \end{bmatrix}$$

Consider the effective zero to be $h = 0.1$ for hand solution, and $h = 0.01$ for computer solution.

CHAPTER 10

Computer Formulation of State Equations for Dynamic Nonlinear Networks

10-1 INTRODUCTION

We have seen in Chapter 9 that the solution of a dynamic linear network can always be expressed *analytically* in terms of the state transition matrix e^{At}, as in Eq. (9-27). In sharp contrast to this, the solution of a dynamic nonlinear network seldom exists in analytic form. Consequently, the solution of most practical nonlinear networks must be obtained either by graphical or numerical methods. For first- and second-order nonlinear networks, e.g., networks containing one or two energy-storage elements, graphical techniques are usually superior to numerical methods, because both quantitative and qualitative behaviors of the network can be obtained. A detailed description of the graphical techniques of solution can be found in [1]. The most serious shortcoming of graphical methods is that they are usually not applicable for nonlinear networks of order greater than 2. By contrast, the numerical methods of analysis are valid for nonlinear networks of any order of complexity. In view of their greater generality and ease of implementation on a digital computer, only numerical methods will be presented in this book.

Since the majority of numerical methods that are readily implemented on a digital computer pertain to a system of first-order differential equations in the normal form

$$\begin{aligned} \dot{x}_1 &= f_1(x_1, x_2, \ldots, x_n, t) \\ \dot{x}_2 &= f_2(x_1, x_2, \ldots, x_n, t) \\ &\vdots \\ \dot{x}_n &= f_n(x_1, x_2, \ldots, x_n, t) \end{aligned} \tag{10-1}$$

or in compact vector notation,

$$\dot{\boldsymbol{x}} = \boldsymbol{f}(\boldsymbol{x}, t) \tag{10-2}$$

our main objective in this chapter is to learn how to formulate the equations of motion of a given network in this form, which incidentally coincide with the *state equations* already introduced in Chapter 8. The numerical methods for solving the state equations of dynamic nonlinear networks will be presented in Chapters 11, 12, and 13.

10-2 EXISTENCE OF NORMAL-FORM EQUATIONS FOR DYNAMIC NONLINEAR NETWORKS

We have seen in Section 8-2 that the normal-form equations for some linear networks containing controlled sources may not exist at all. This class of networks has been dismissed in Chapter 8 as being pathological, because it arises only when a certain parameter assumes a precise value. Unfortunately, this excuse is no longer valid in the case of nonlinear networks, for there exist many examples of practical nonlinear networks whose equations of motion cannot be formulated in the normal form with respect to a certain set of state variables. To illustrate the various possibilities that may arise in formulating the normal-form equations for nonlinear networks, consider the single-loop nonlinear *RC* network shown in Fig. 10-1. The two candidates for the state variable associated with this network are the capacitor voltage v_2 and the capacitor charge q_2. The state equation, if it exists, will be of the form

$$\dot{v}_2 = f(v_2, t) \tag{10-3}$$

if v_2 is chosen to be the state variable, or

$$\dot{q}_2 = g(q_2, t) \tag{10-4}$$

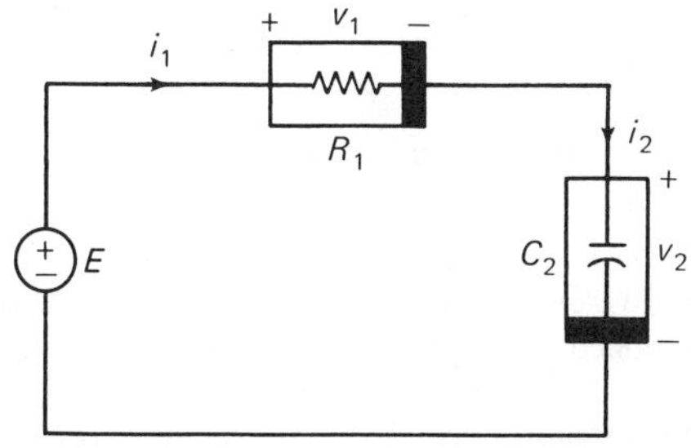

Fig. 10-1. Single-loop *RC* network.

if q_2 is chosen to be the state variable. The KCL and KVL equations are given by

$$i_1 = i_2 = \dot{q}_2 \tag{10-5}$$

$$v_1 = E - v_2 \tag{10-6}$$

To obtain Eq. (10-3) or (10-4), it will be necessary to make use of the v_1-i_1 curve of the nonlinear resistor R_1 and the v_2-q_2 curve of the nonlinear capacitor C_2. We shall now attempt to write the state equation for this network corresponding to different types of characteristic curves for R_1 and C_2. We shall assume that both the v_1-i_1 curve for R_1 and the v_2-q_2 curve for C_2 *represent functions.*[1]

Case 1. Both v_1-i_1 and v_2-q_2 curves are strictly monotonically increasing.[2] In particular, let these characteristic curves be given by

$$R_1\colon \quad i_1 = v_1^3 \tag{10-7}$$

$$C_2\colon \quad v_2 = \ell n(q_2 + 1) \tag{10-8}$$

To derive Eq. (10-3), we make use of Eqs. (10-5) and (10-7) and the chain rule to obtain

$$v_1^3 = \dot{q}_2 = \frac{dq_2}{dv_2}\frac{dv_2}{dt} \tag{10-9}$$

Substituting Eq. (10-6) for v_1 in Eq. (10-9), we obtain

$$\dot{v}_2 = \frac{(E - v_2)^3}{dq_2/dv_2} \tag{10-10}$$

To arrive at the desired normal form, it would be necessary to solve for q_2 in terms of v_2 from Eq. (10-8) before we could obtain dq_2/dv_2. Since the v_2-q_2 curve is strictly monotonically increasing, we know the inverse function exists and is given by

$$q_2 = e^{v_2} - 1 \tag{10-11}$$

Differentiating Eq. (10-11) with respect to v_2 and then substituting the resulting expression into Eq. (10-10), we obtain the state equation:

$$\dot{v}_2 = \frac{(E - v_2)^3}{e^{v_2}} \tag{10-12}$$

To derive Eq. (10-4), we make use of Eqs. (10-5)–(10-7) to obtain

$$\dot{q}_2 = v_1^3 = (E - v_2)^3 \tag{10-13}$$

Substituting Eq. (10-8) for v_2 in Eq. (10-13), we obtain the alternative state equation

$$\dot{q}_2 = [E - \ell n(q_2 + 1)]^3 \tag{10-14}$$

Hence, the state equation for this case exists with either v_2 or q_2 as the state variable.

[1]A curve in the x-y plane is said to represent a function $y = f(x)$ [$x = g(y)$] if each value of x determines a unique value of y [if each value of y determines a unique value of x]. In this case, the function $y = f(x)$ is said to represent an "x-controlled curve" [the function $x = g(y)$ is said to represent a "y-controlled curve"].

[2]A function $y = f(x)$ is said to be strictly monotonically increasing if $f(x_2) > f(x_1)$ whenever $x_2 > x_1$.

Case 2. R_1 is nonmonotonic voltage-controlled and C_2 is nonmonotonic charge-controlled. In particular, let

$$R_1: \quad i_1 = 1 - v_1^2 + v_1^3 \tag{10-15}$$

$$C_2: \quad v_2 = q_2^3 - 3q_2 \tag{10-16}$$

To derive Eq. (10-3), we proceed as before and obtain, corresponding to Eqs. (10-9) and (10-10),

$$1 - v_1^2 + v_1^3 = \dot{q}_2 = \frac{dq_2}{dv_2}\frac{dv_2}{dt} \tag{10-17}$$

$$\dot{v}_2 = \frac{1 - (E - v_2)^2 + (E - v_2)^3}{dq_2/dv_2} \tag{10-18}$$

The next step is to solve for q_2 as a function of v_2 from Eq. (10-16). But this is impossible, because Eq. (10-16) is not a *one-to-one* function.[3] Thus, q_2 is a *multivalued* function of v_2, and hence so is dq_2/dv_2. We conclude, therefore, that the normal-form equation for this network does not exist with v_2 as the state variable. Our only alternative will be to choose q_2 as the state variable and attempt to derive Eq. (10-4). Substituting Eqs. (10-5) and (10-6) into Eq. (10-15) and making use of Eq. (10-16), we obtain the state equation

$$\dot{q}_2 = 1 - [E - (q_2^3 - 3q_2)]^2 + [E - (q_2^3 - 3q_2)]^3 \tag{10-19}$$

We conclude, therefore, that in this case the state equation exists only if q_2 is chosen to be the state variable.

Case 3. Both R_1 and C_2 are nonmonotonic but voltage-controlled. In particular, let

$$R_1: \quad i_1 = 1 - v_1^2 + v_1^3 \tag{10-20}$$

$$C_2: \quad q_2 = v_2^3 - 3v_2 \tag{10-21}$$

Since the v_1-i_1 curve for this case is identical to Eq. (10-15), to derive Eq. (10-3), we proceed as in case 2 and obtain Eq. (10-18). Differentiating Eq. (10-21) with respect to v_2 and substituting the resulting expression into Eq. (10-18), we obtain the state equation

$$\dot{v}_2 = \frac{1 - (E - v_2)^2 + (E - v_2)^3}{3v_2^2 - 3} \tag{10-22}$$

To derive Eq. (10-4), we obtain from Eqs. (10-5), (10-6), and (10-20)

$$\dot{q}_2 = 1 - (E - v_2)^2 + (E - v_2)^3 \tag{10-23}$$

To arrive at the desired normal form, it would be necessary to solve for v_2 as a function of q_2 from Eq. (10-21). But, again, this is impossible, because Eq. (10-21) is not one-

[3] A function $y = f(x)$ is said to be one-to-one if to each value $\hat{y}$ there corresponds one and only one value of $\hat{x}$ such that $\hat{y} = f(\hat{x})$. A continuous function $y = f(x)$ has an *inverse* $x = f^{-1}(y)$ that is also a function if, and only if, $f(x)$ is strictly monotonically increasing or decreasing.

to-one; hence v_2 is a multivalued function of q_2. We conclude that in this case the state equation exists only if the state variable is chosen to be v_2.

Case 4. R_1 is nonmonotonic current-controlled and C_2 is a linear capacitor. In particular, let

$$R_1: \quad v_1 = i_1^3 - 3i_1 \tag{10-24}$$

$$C_2: \quad q_2 = Cv_2 \tag{10-25}$$

To derive Eq. (10-3), we proceed as in case 1 to obtain

$$\dot{v}_2 = \frac{i_1}{dq_2/dv_2} = \frac{i_1}{C} \tag{10-26}$$

To transform Eq. (10-26) into the normal-form Eq. (10-3), it would be necessary to express i_1 first in terms of v_1. Once again, this is impossible since Eq. (10-24) is not a one-to-one function. Turning next to the alternative normal-form Eq. (10-4), we observe that, since

$$\dot{q}_2 = i_1 \tag{10-27}$$

and since it is impossible to express i_1 in terms of v_1, the state equation also does not exist with q_2 as the state variable. Hence, here is a simple example of a circuit whose equation of motion cannot be reduced to the normal form at all. In such cases, we say the network does not possess a state equation, or that the normal-form equation of the network does not exist.

The nonexistence of state equations for a nonlinear network is often the result of a poor choice of network models to represent the physical circuit. In Chapter 2, we learned that good modeling techniques require each current-controlled nonlinear resistor to be associated with a parasitic inductance. Hence, if we insert a linear inductor with a small inductance in series with the nonlinear resistor R_1 in Fig. 10-2 to simulate the lead inductance, we can derive the following equations:

$$\text{KCL:} \quad i_1 = i_2 = i_3 = \dot{q}_2 \tag{10-28}$$

$$\text{KVL:} \quad v_3 = E - v_1 - v_2 = L\dot{i}_3 \tag{10-29}$$

Substituting Eq. (10-25) into Eq. (10-28), and Eq. (10-24) into Eq. (10-29), we obtain

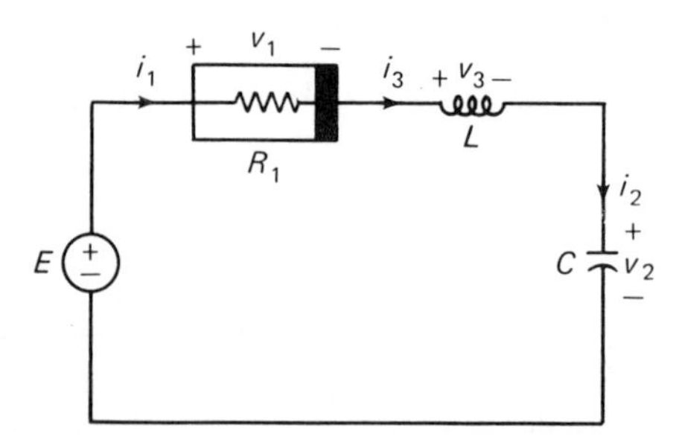

Fig. 10-2. Single-loop nonlinear *RLC* network obtained by inserting a small parasitic inductance in series with a nonmonotonic current-controlled resistor R_1.

the state equations

$$\dot{v}_2 = \frac{1}{C} i_3 \tag{10-30}$$

$$\dot{i}_3 = \frac{E - (i_3^3 - 3i_3) - v_2}{L} \tag{10-31}$$

From these examples, it is clear that the nonexistence of the normal-form equation arises invariably because certain nonmonotonic functions fail to possess an inverse. Hence, *the presence of nonlinear elements characterized by nonmonotonic characteristic curves in a network may lead to the nonexistence of the normal-form equation.* This observation, however, does not preclude the possibility of the nonexistence of normal forms for networks containing nonlinear elements with strictly monotonically increasing characteristic curves and *negative resistors* or *controlled sources.* For example, the nonmonotonic resistor R_1 in case 4 can be replaced by an equivalent network consisting of a strictly monotonically increasing nonlinear resistor R'_1 characterized by $v'_1 = i'^3_1$ in series with a -3Ω negative resistor. Since a negative resistor could be modeled by a voltage-controlled current source with terminal current $i = -Gv$, or a current-controlled voltage source with terminal voltage $v = -Ri$, it follows that the presence of even one controlled source may lead to the nonexistence of normal-form equations for a network whose nonlinear elements are all characterized by strictly monotonically increasing curves. Since the model of almost all three-terminal devices, such as transistors, invariably contains controlled sources, it is entirely possible for nonlinear networks containing three-terminal devices to fail to possess normal-form equations. For example, the transistor network shown in Fig. 10-3(a) does not have a normal-form equation, because, by assigning appropriate values for the resistors, the v-i curve of the two-terminal resistive black box N to the right of terminals a and b can be shown to be given by the nonmonotonic voltage-controlled driving-point characteristic curve shown in Fig. 10-3(b). Similarly, the normal-form equation fails to exist for the transistor network shown in Fig. 10-4(a) because, by an appropriate choice of resistor values, the driving-point characteristic curve of the resistive black box N is given by the nonmonotonic current-controlled v-i curve shown in Fig. 10-4(b).

Our objective in the following sections is to develop a general algorithm for formulating the state equations of a very large class of nonlinear networks. Several approaches for deriving state equations exist. The simplest allows only *strictly monotonically increasing* resistors, inductors, and capacitors, but no controlled sources [2, 3]. More general approaches that allow *nonmonotonic* resistors but still no controlled sources are given in [4–8]. The approach we choose in this chapter is based on the hybrid matrix formulation in Chapter 6 and allows both nonmonotonic elements and linear controlled sources. This approach is motivated by the fact that a large percentage of elements in many practical nonlinear electronic circuits consist of *linear resistors* and *linear controlled sources.* Consequently, these elements can be represented by a *linear hybrid m-port.* Since the hybrid matrix can be determined by explicit topological formulas, it can be calculated very efficiently. Moreover, the calculation needs to be carried out only once for each given network.

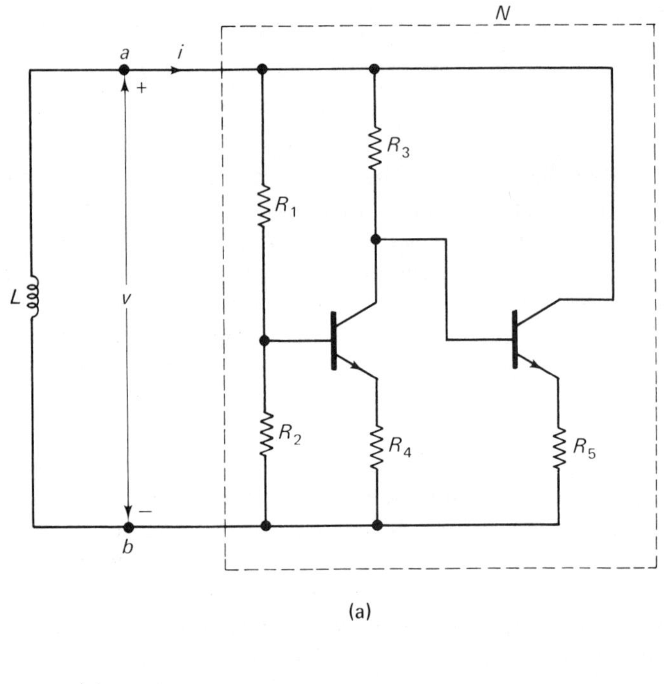

(a)

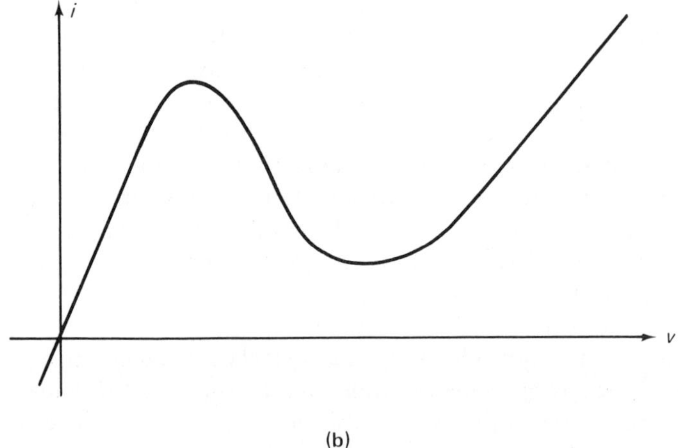

(b)

Fig. 10-3. Transistor circuit consisting of an inductor connected across a black box characterized by a nonmonotonic voltage-controlled v-i curve.

10-3 TOPOLOGICAL FORMULATION OF STATE EQUATIONS FOR DYNAMIC NONLINEAR NETWORKS

The examples in Section 10-2 demonstrated that the possibility of formulating the equations of motion for nonlinear networks in the normal form depends strongly on the nature of the characteristic curves of the nonlinear elements. We also noted

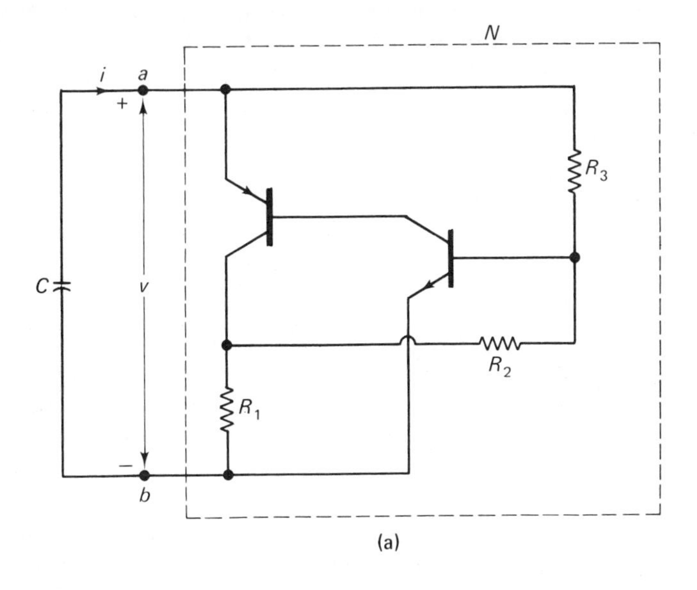

(a)

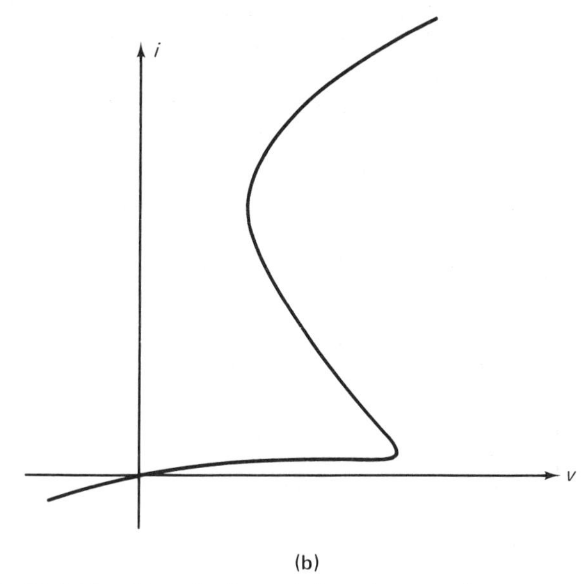

(b)

Fig. 10-4. Transistor circuit consisting of a capacitor connected across a black box characterized by a nonmonotonic current-controlled v-i curve.

that the presence of negative resistors and controlled sources could lead to difficulty in formulating normal-form equations. Our objective in this section is to develop an algorithm so that a computer program can be written to formulate the normal-form equations of a large class of nonlinear networks containing both linear and nonlinear resistors, inductors, and capacitors, independent voltage and current sources, and the four types of linear controlled sources. To avoid dealing with certain types of networks whose normal-form equations either do not exist or are pathological, in the sense that their order of complexity depends on the *precise* value of some element parameters, we shall assume that our networks meet the following requirements:

10-3-1 Standing Assumptions on the Class of Allowable Networks

1. *Consistency Assumptions*
   (a) There does not exist any "loop" made up of only independent and/or controlled voltage sources, henceforth called an *E loop*.
   (b) There does not exist any "cutset" made up of only independent and/or controlled current sources, henceforth called a *J cutset*.
2. *Assumptions on Controlling Variables*
   (a) The controlled sources do not depend on the branch current of any element of a "loop" made up of only capacitors and independent and/or controlled voltage sources, henceforth called a *C-E loop*.
   (b) The controlled sources do not depend on the branch voltage of any element of a "cutset" made up of only inductors and independent and/or controlled current sources, henceforth called an *L-J cutset*.
3. *Normal-Tree Assumptions*
   With each controlled source considered as a two-terminal element, let us choose a special tree, henceforth called the *normal tree*, whose elements are selected according to the following priority:[4] (1) *all* independent and controlled voltage sources, (2) as many capacitors as possible, (3) as many resistors as possible, and (4) inductors. If we let $\mathfrak{T}$ be a normal tree and $\mathfrak{L}$ the corresponding cotree, the following additional requirements must be satisfied with respect to $\mathfrak{T}$:
   (a) Any controlled source belonging to a *C-E* loop or an *L-J* cutset can depend only on the voltage of a tree capacitor or the current of a cotree inductor.
   (b) All *cotree capacitors* are voltage-controlled; i.e., the charge q of each cotree capacitor is a function of its voltage; i.e., $q = \hat{q}(v)$.
   (c) All *tree inductors* are current-controlled; i.e., the flux linkage φ of each tree inductor is a function of its current; i.e., $\varphi = \hat{\varphi}(i)$.

To illustrate the nature of the preceding standing assumptions, we have contrived two networks that violate the various stipulated requirements. The network shown in Fig. 10-5(a) contains an E loop, thereby violating assumption 1(a). It also contains a J cutset, thereby violating assumption 1(b). The network shown in Fig. 10-5(b) contains a C-E loop and an L-J cutset. This network violates assumption 2(a)

[4] In view of the consistency assumptions, such a normal tree will always contain all independent and controlled voltage sources and will not contain any independent or controlled current sources. In general, the choice of a normal tree is *not unique*, since many different trees may qualify.

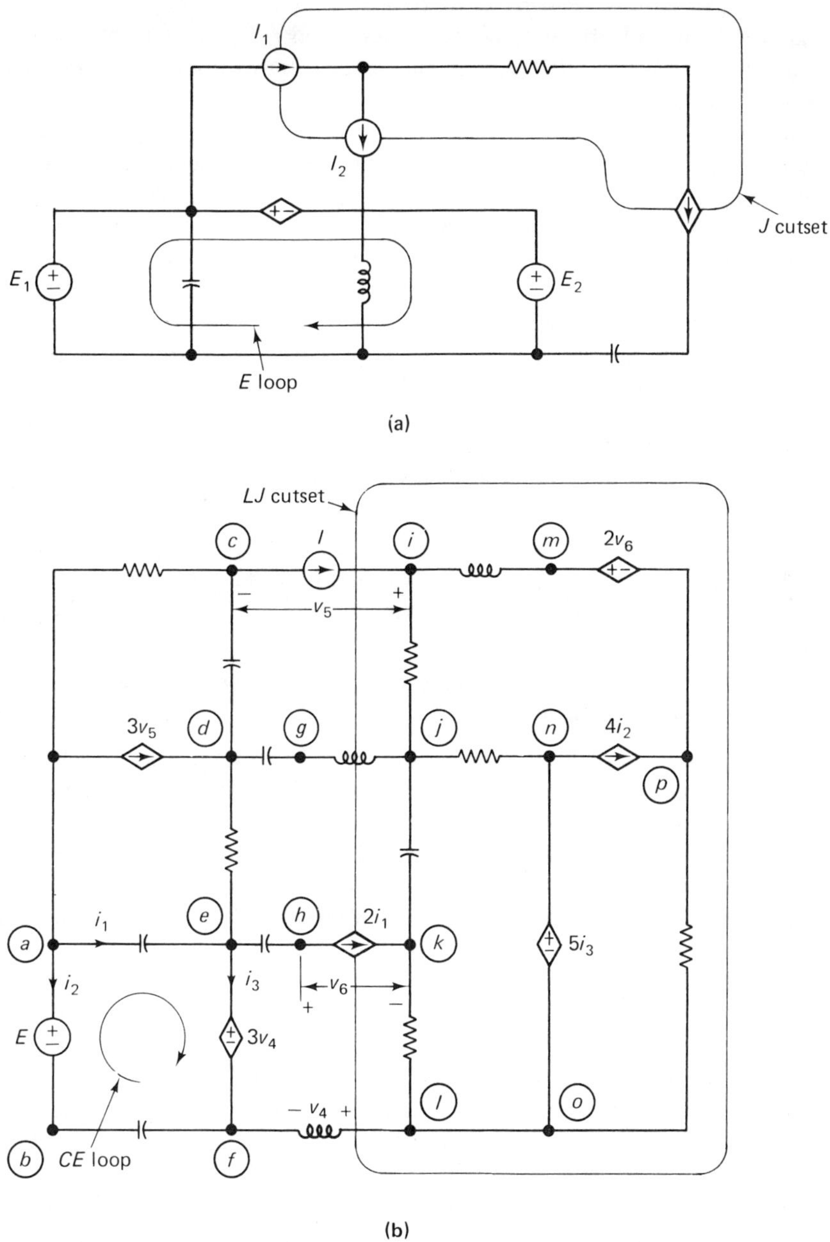

Fig. 10-5. Two contrived networks that violate one or more standing assumptions.

on three counts: the controlled source across nodes h and k depends on the current i_1 of a capacitor in the C-E loop; the controlled source across nodes n and p depends on the current i_2 of an independent voltage source in the C-E loop; and finally, the controlled source across nodes n and o depends on the current i_3 of a controlled voltage source in the C-E loop. This network also violates assumption 2(b) on three counts: the controlled source across nodes e and f depends on the voltage v_4 of an inductor in the L-J cutset; the controlled source across nodes a and d depends on the voltage v_5 of an independent current source in the L-J cutset; and, finally, the controlled source across nodes m and p depends on the voltage v_6 of a controlled current source in the L-J cutset. Observe that this network also violates assumption 3(a) because the controlled voltage source across nodes e and f in the C-E loop depends on an inductor voltage v_4, which is not the current of a cotree inductor. Likewise, the controlled source across nodes h and k in the L-J cutset violates assumption 3(a) because it depends on the current i_1, which is neither a voltage of a tree capacitor, nor a current of a cotree inductor.

Of course, these networks are specially contrived to clarify the standing assumptions. Most practical networks do not violate these assumptions. In fact, even in those rare cases when a network violates one or more of these assumptions, we could easily overcome the difficulty by introducing small parasitics to the network. For example, a small resistance can be inserted in series with a capacitor or a voltage source belonging to a C-E loop, thereby eliminating it. Similarly, a small conductance can be connected in parallel with an inductor or a current source belonging to an L-J cutset to eliminate it. In either case, the augmented parasitic could be used to simulate the lead resistance or the shunt conductance invariably present in any physical element.

10-3-2 Step 1: Formation and Characterization of a Hybrid m-Port $\hat{N}$

Let N be a network that satisfies the standing assumptions, and let $\mathfrak{I}$ be a normal tree and $\mathfrak{L}$ its corresponding cotree. Our first task is to form a *linear* m-port $\hat{N}$ obtained by extracting from N all *independent* sources, capacitors (both linear and nonlinear), inductors (both linear and nonlinear), and all *nonlinear* resistors, as shown in Fig. 10-6.[5] In view of our procedure for selecting the normal tree $\mathfrak{I}$, all extracted elements on the left of $\hat{N}$ are tree branches and therefore constitute a part of the normal tree $\mathfrak{I}$. Similarly, all elements on the right of $\hat{N}$ are cotree branches and therefore constitute a part of the cotree $\mathfrak{L}$. The remaining elements of $\mathfrak{I}$ and $\mathfrak{L}$ are either controlled sources or linear resistors inside $\hat{N}$, as shown in Fig. 10-6. Without loss of generality, each controlled source is assumed to depend on either the current in a short circuit or the voltage across an open circuit. To show this explicitly, we have further extracted the controlled sources, the short circuits, and the open circuits, thereby forming an augmented n-port N_A. By our procedure for constructing the normal tree, all controlled voltage sources are tree branches and all controlled current sources are cotree branches. For the same reason given in Chapter 6, each short circuit is considered as a distinct tree branch of the normal tree $\mathfrak{I}$, and each open circuit is treated as a distinct cotree branch of $\mathfrak{L}$. The augmented n-port N_A and the m-port $\hat{N}$ are redrawn, as shown in Figs. 10-7 and 10-8, with all extracted tree branches replaced by voltage sources and all extracted cotree branches replaced by current sources. Since N_A and $\hat{N}$ contain neither nonlinear elements nor independent sources, they are completely

[5]Each extracted element in Fig. 10-6 represents a "group" of elements of the same kind. Hence, if N contains five independent voltage sources, the single voltage source $\boldsymbol{v}_{E_3}$ would represent five voltage sources.

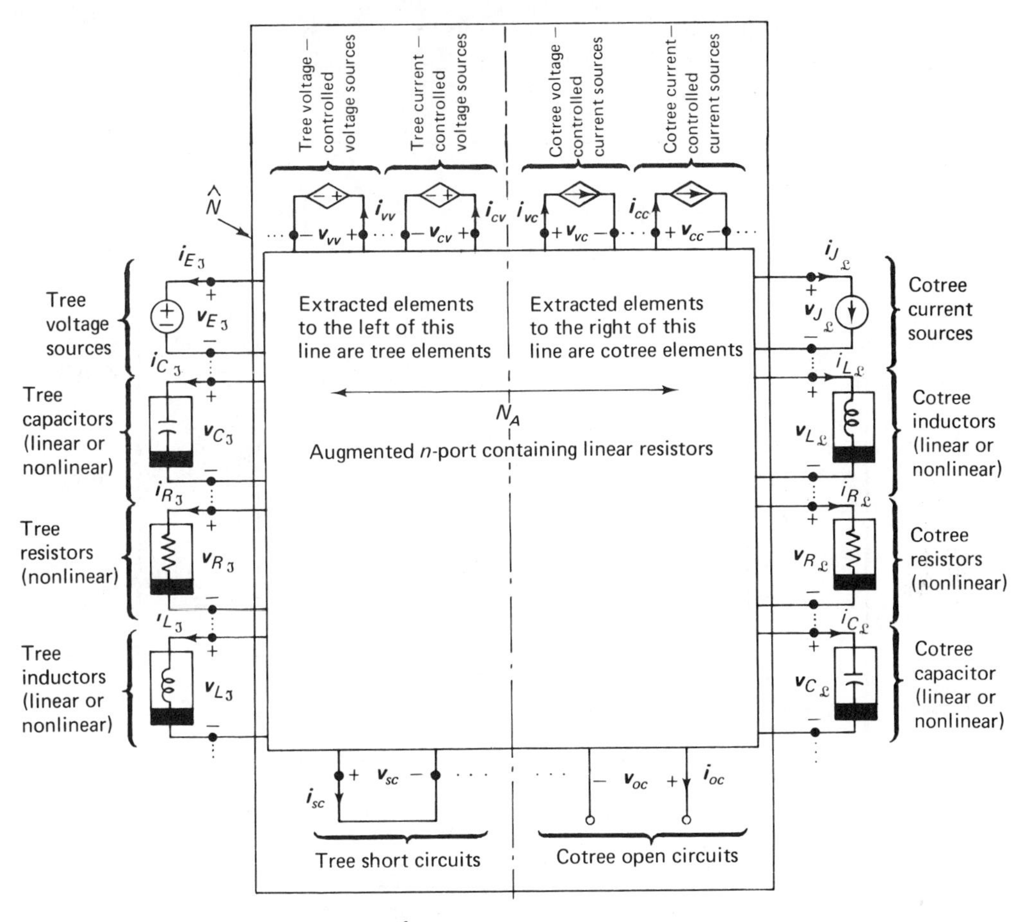

Fig. 10-6. An m-port $\hat{N}$ created by extracting all independent sources and nonlinear elements.

characterized by hybrid matrices $\boldsymbol{H}$ and $\hat{\boldsymbol{H}}$, respectively. Our ultimate objective is to find the hybrid matrix $\hat{\boldsymbol{H}}$ of $\hat{N}$. However, as in Chapter 6, to determine $\hat{\boldsymbol{H}}$ by an explicit topological formula, it will be necessary to determine first the hybrid matrix $\boldsymbol{H}$ of the augmented n-port N_A. To do this, let us proceed as in Chapter 6 and define the following port vectors:

$$\boldsymbol{v}_{\mathfrak{T}} \triangleq \begin{bmatrix} \boldsymbol{v}_{E_{\mathfrak{T}}} \\ \boldsymbol{v}_{C_{\mathfrak{T}}} \\ \boldsymbol{v}_{R_{\mathfrak{T}}} \\ \boldsymbol{v}_{L_{\mathfrak{T}}} \\ \hdashline \boldsymbol{v}_{vv} \\ \boldsymbol{v}_{cv} \\ \boldsymbol{v}_{sc} \end{bmatrix}, \qquad \boldsymbol{i}_{\mathfrak{T}} \triangleq \begin{bmatrix} \boldsymbol{i}_{E_{\mathfrak{T}}} \\ \boldsymbol{i}_{C_{\mathfrak{T}}} \\ \boldsymbol{i}_{R_{\mathfrak{T}}} \\ \boldsymbol{i}_{L_{\mathfrak{T}}} \\ \hdashline \boldsymbol{i}_{vv} \\ \boldsymbol{i}_{cv} \\ \boldsymbol{i}_{sc} \end{bmatrix}, \qquad \boldsymbol{v}_{\mathfrak{L}} \triangleq \begin{bmatrix} \boldsymbol{v}_{J_{\mathfrak{L}}} \\ \boldsymbol{v}_{L_{\mathfrak{L}}} \\ \boldsymbol{v}_{R_{\mathfrak{L}}} \\ \boldsymbol{v}_{C_{\mathfrak{L}}} \\ \hdashline \boldsymbol{v}_{cc} \\ \boldsymbol{v}_{vc} \\ \boldsymbol{v}_{oc} \end{bmatrix}, \qquad \boldsymbol{i}_{\mathfrak{L}} \triangleq \begin{bmatrix} \boldsymbol{i}_{J_{\mathfrak{L}}} \\ \boldsymbol{i}_{L_{\mathfrak{L}}} \\ \boldsymbol{i}_{R_{\mathfrak{L}}} \\ \boldsymbol{i}_{C_{\mathfrak{L}}} \\ \hdashline \boldsymbol{i}_{cc} \\ \boldsymbol{i}_{vc} \\ \boldsymbol{i}_{oc} \end{bmatrix} \tag{10-32}$$

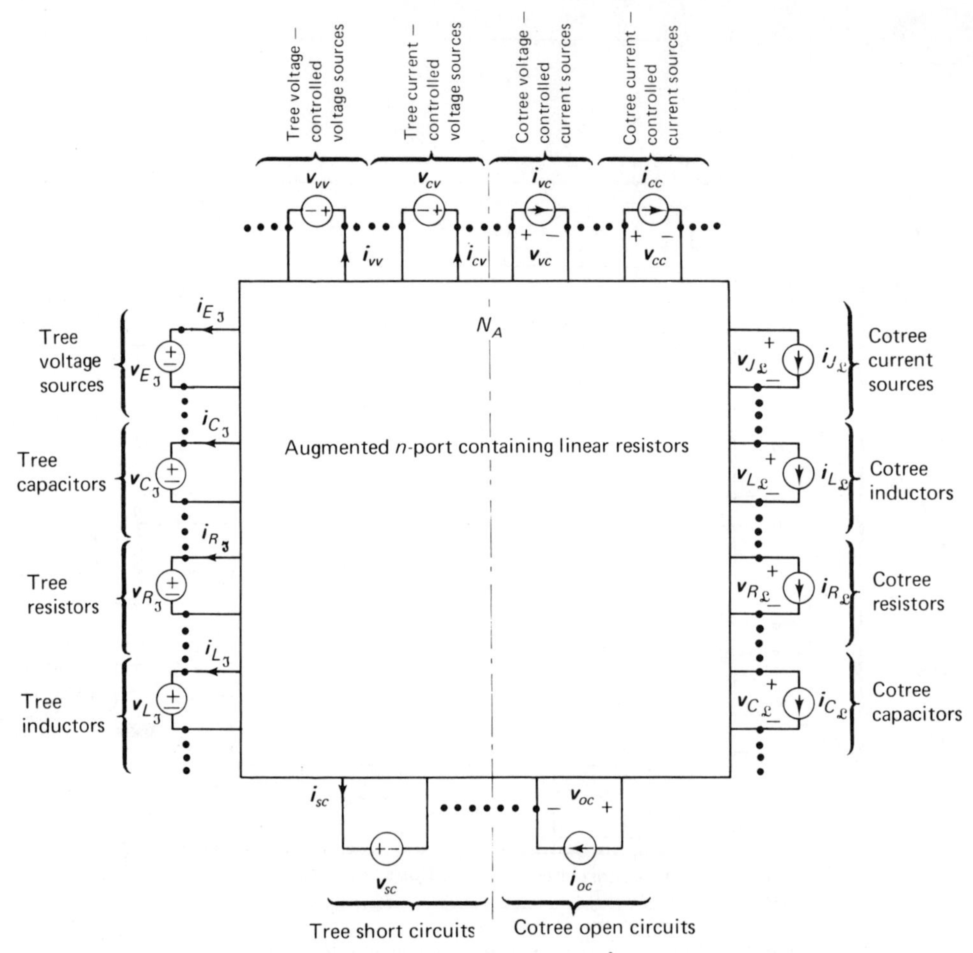

Fig. 10-7. Augmented n-port N_A created from $\hat{N}$ by further extracting all linear controlled sources and the controlling short circuits and open circuits.

Next let us define the vectors $\boldsymbol{v}_\mathcal{R}$ and $\boldsymbol{i}_\mathcal{R}$ to be the voltage and current vectors associated with the internal *tree branch* linear resistors inside N_A. Finally, let us define the vectors $\boldsymbol{v}_\mathcal{G}$ and $\boldsymbol{i}_\mathcal{G}$ to be the voltage and current vectors associated with the internal *cotree branch* linear resistors inside N_A. We assume that the branches are labeled consecutively, starting with the extracted tree branches and followed by the internal tree branches, the internal cotree branches, and finally the extracted cotree branches. The fundamental cutset (KCL) equations for the complete network in Fig. 10-7 can be written as follows:

$$\boldsymbol{D}\boldsymbol{i} = \left[\begin{array}{cc|cc} \mathbf{1}_{\mathcal{T}\mathcal{T}} & \mathbf{0}_{\mathcal{T}\mathcal{R}} & \boldsymbol{D}_{\mathcal{T}\mathcal{G}} & \boldsymbol{D}_{\mathcal{T}\mathcal{L}} \\ \mathbf{0}_{\mathcal{R}\mathcal{T}} & \mathbf{1}_{\mathcal{R}\mathcal{R}} & \boldsymbol{D}_{\mathcal{R}\mathcal{G}} & \boldsymbol{D}_{\mathcal{R}\mathcal{L}} \end{array}\right] \left[\begin{array}{c} \boldsymbol{i}_\mathcal{T} \\ \boldsymbol{i}_\mathcal{R} \\ \hline \boldsymbol{i}_\mathcal{G} \\ \boldsymbol{i}_\mathcal{L} \end{array}\right] = \mathbf{0} \tag{10-33}$$

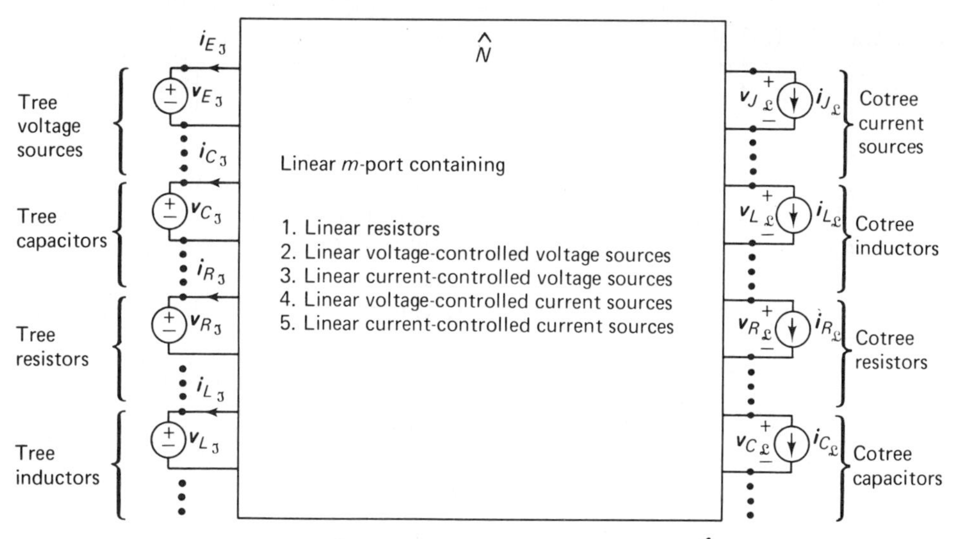

Fig. 10-8. Simplified representation of the hybrid m-port $\hat{N}$ with the tree voltage ports appearing on the left and the cotree current ports appearing on the right.

Using the same procedure for deriving Eq. (6-20) in Chapter 6, we obtain the following hybrid representation for N_A:

$$\begin{bmatrix} \boldsymbol{i}_{\mathfrak{I}} \\ \boldsymbol{v}_{\mathcal{L}} \end{bmatrix} = \begin{bmatrix} -\boldsymbol{D}_{\mathfrak{I}\mathcal{G}}\boldsymbol{Z}^{-1}\boldsymbol{D}^t_{\mathfrak{I}\mathcal{G}} & \boldsymbol{D}_{\mathfrak{I}\mathcal{G}}\boldsymbol{Z}^{-1}(\boldsymbol{D}^t_{\mathcal{R}\mathcal{G}}\boldsymbol{Z}_{\mathcal{R}}\boldsymbol{D}_{\mathcal{R}\mathcal{L}}) - \boldsymbol{D}_{\mathfrak{I}\mathcal{L}} \\ \boldsymbol{D}^t_{\mathfrak{I}\mathcal{L}} - \boldsymbol{D}^t_{\mathcal{R}\mathcal{L}}\boldsymbol{Y}^{-1}(\boldsymbol{D}_{\mathcal{R}\mathcal{G}}\boldsymbol{Y}_{\mathcal{G}}\boldsymbol{D}^t_{\mathfrak{I}\mathcal{G}}) & -\boldsymbol{D}^t_{\mathcal{R}\mathcal{L}}\boldsymbol{Y}^{-1}\boldsymbol{D}_{\mathcal{R}\mathcal{L}} \end{bmatrix} \begin{bmatrix} \boldsymbol{v}_{\mathfrak{I}} \\ \boldsymbol{i}_{\mathcal{L}} \end{bmatrix} \tag{10-34}$$

where

$$\boldsymbol{Z} \triangleq \boldsymbol{Z}_{\mathcal{G}} + \boldsymbol{D}^t_{\mathcal{R}\mathcal{G}}\boldsymbol{Z}_{\mathcal{R}}\boldsymbol{D}_{\mathcal{R}\mathcal{G}} \tag{10-35a}$$

$$\boldsymbol{Y} \triangleq \boldsymbol{Y}_{\mathcal{R}} + \boldsymbol{D}_{\mathcal{R}\mathcal{G}}\boldsymbol{Y}_{\mathcal{G}}\boldsymbol{D}^t_{\mathcal{R}\mathcal{G}} \tag{10-35b}$$

$$\boldsymbol{Z}_{\mathcal{R}} \triangleq \boldsymbol{Y}^{-1}_{\mathcal{R}} = \text{diagonal } \textit{branch resistance matrix} \text{ associated with internal tree resistors} \tag{10-35c}$$

$$\boldsymbol{Z}_{\mathcal{G}} \triangleq \boldsymbol{Y}^{-1}_{\mathcal{G}} = \text{diagonal } \textit{branch resistance matrix} \text{ associated with internal cotree resistors} \tag{10-35d}$$

We can now identify the hybrid matrix $\boldsymbol{H}$ from Eq. (10-34):

$$\boxed{\begin{aligned} \boldsymbol{H} &= \begin{bmatrix} -\boldsymbol{D}_{\mathfrak{I}\mathcal{G}}\boldsymbol{Z}^{-1}\boldsymbol{D}^t_{\mathfrak{I}\mathcal{G}} & \boldsymbol{D}_{\mathfrak{I}\mathcal{G}}\boldsymbol{Z}^{-1}(\boldsymbol{D}^t_{\mathcal{R}\mathcal{G}}\boldsymbol{Z}_{\mathcal{R}}\boldsymbol{D}_{\mathcal{R}\mathcal{L}}) - \boldsymbol{D}_{\mathfrak{I}\mathcal{L}} \\ \boldsymbol{D}^t_{\mathfrak{I}\mathcal{L}} - \boldsymbol{D}^t_{\mathcal{R}\mathcal{L}}\boldsymbol{Y}^{-1}(\boldsymbol{D}_{\mathcal{R}\mathcal{G}}\boldsymbol{Y}_{\mathcal{G}}\boldsymbol{D}^t_{\mathfrak{I}\mathcal{G}}) & -\boldsymbol{D}^t_{\mathcal{R}\mathcal{L}}\boldsymbol{Y}^{-1}\boldsymbol{D}_{\mathcal{R}\mathcal{L}} \end{bmatrix} \\ &\triangleq \begin{bmatrix} \boldsymbol{H}_{\mathfrak{I}\mathfrak{I}} & \boldsymbol{H}_{\mathfrak{I}\mathcal{L}} \\ \boldsymbol{H}_{\mathcal{L}\mathfrak{I}} & \boldsymbol{H}_{\mathcal{L}\mathcal{L}} \end{bmatrix} \end{aligned}} \tag{10-36}$$

As in Chapter 6 [recall Eq. (6-22)], it is important to recognize that

$$\boxed{\boldsymbol{H}_{\mathcal{L}\mathfrak{I}} = -\boldsymbol{H}^t_{\mathfrak{I}\mathcal{L}}} \tag{10-37}$$

and [recall Eq. (6-23)]

$$\boxed{\boldsymbol{H}_{\mathcal{L}\mathcal{L}} = (\boldsymbol{D}^t_{\mathcal{R}\mathcal{L}}\boldsymbol{Z}_{\mathcal{R}}\boldsymbol{D}_{\mathcal{R}\mathcal{G}})\boldsymbol{Z}^{-1}(\boldsymbol{D}^t_{\mathcal{R}\mathcal{G}}\boldsymbol{Z}_{\mathcal{R}}\boldsymbol{D}_{\mathcal{R}\mathcal{L}}) - \boldsymbol{D}^t_{\mathcal{R}\mathcal{L}}\boldsymbol{Z}_{\mathcal{R}}\boldsymbol{D}_{\mathcal{R}\mathcal{L}}} \tag{10-38}$$

In view of Eqs. (10-37) and (10-38), we only need to compute the inverse of one matrix, $\boldsymbol{Z}^{-1}$, and only $\boldsymbol{H}_{\mathcal{J}\mathcal{J}}$, $\boldsymbol{H}_{\mathcal{J}\mathcal{L}}$, and $\boldsymbol{H}_{\mathcal{L}\mathcal{L}}$ must be evaluated. Moreover, we observe that the hybrid matrix $\boldsymbol{H}$ of the augmented n-port N_A exists if, and only if, the matrix $\boldsymbol{Z}$ defined in Eq. (10-35a) is nonsingular. Now recall from Chapter 6 that, if N_A contains only *positive resistances*, then $\boldsymbol{Z}^{-1}$ always exists.

Now that we have found $\boldsymbol{H}$ for N_A, we can proceed to find $\hat{\boldsymbol{H}}$ for $\hat{N}$. Let us define

$$\hat{\boldsymbol{v}}_{\mathcal{J}} \triangleq \begin{bmatrix} \boldsymbol{v}_{E_{\mathcal{J}}} \\ \boldsymbol{v}_{C_{\mathcal{J}}} \\ \boldsymbol{v}_{R_{\mathcal{J}}} \\ \boldsymbol{v}_{L_{\mathcal{J}}} \end{bmatrix}, \qquad \hat{\boldsymbol{i}}_{\mathcal{J}} \triangleq \begin{bmatrix} \boldsymbol{i}_{E_{\mathcal{J}}} \\ \boldsymbol{i}_{C_{\mathcal{J}}} \\ \boldsymbol{i}_{R_{\mathcal{J}}} \\ \boldsymbol{i}_{L_{\mathcal{J}}} \end{bmatrix}, \qquad \hat{\boldsymbol{v}}_{\mathcal{L}} \triangleq \begin{bmatrix} \boldsymbol{v}_{J_{\mathcal{L}}} \\ \boldsymbol{v}_{L_{\mathcal{L}}} \\ \boldsymbol{v}_{R_{\mathcal{L}}} \\ \boldsymbol{v}_{C_{\mathcal{L}}} \end{bmatrix}, \qquad \hat{\boldsymbol{i}}_{\mathcal{L}} \triangleq \begin{bmatrix} \boldsymbol{i}_{J_{\mathcal{L}}} \\ \boldsymbol{i}_{L_{\mathcal{L}}} \\ \boldsymbol{i}_{R_{\mathcal{L}}} \\ \boldsymbol{i}_{C_{\mathcal{L}}} \end{bmatrix} \tag{10-39}$$

Using this notation for the partitioned subvectors in Eq. (10-32), we can expand Eq. (10-34) as follows:

$$\begin{bmatrix} \hat{\boldsymbol{i}}_{\mathcal{J}} \\ \boldsymbol{i}_{vv} \\ \boldsymbol{i}_{cv} \\ \boldsymbol{i}_{sc} \\ \hline \hat{\boldsymbol{v}}_{\mathcal{L}} \\ \boldsymbol{v}_{cc} \\ \boldsymbol{v}_{vc} \\ \boldsymbol{v}_{oc} \end{bmatrix} = \left[\begin{array}{cccc|cccc} \boldsymbol{h}_{11} & \boldsymbol{h}_{12} & \boldsymbol{h}_{13} & \boldsymbol{h}_{14} & \boldsymbol{h}_{15} & \boldsymbol{h}_{16} & \boldsymbol{h}_{17} & \boldsymbol{h}_{18} \\ \boldsymbol{h}_{21} & \boldsymbol{h}_{22} & \boldsymbol{h}_{23} & \boldsymbol{h}_{24} & \boldsymbol{h}_{25} & \boldsymbol{h}_{26} & \boldsymbol{h}_{27} & \boldsymbol{h}_{28} \\ \boldsymbol{h}_{31} & \boldsymbol{h}_{32} & \boldsymbol{h}_{33} & \boldsymbol{h}_{34} & \boldsymbol{h}_{35} & \boldsymbol{h}_{36} & \boldsymbol{h}_{37} & \boldsymbol{h}_{38} \\ \boldsymbol{h}_{41} & \boldsymbol{h}_{42} & \boldsymbol{h}_{43} & \boldsymbol{h}_{44} & \boldsymbol{h}_{45} & \boldsymbol{h}_{46} & \boldsymbol{h}_{47} & \boldsymbol{h}_{48} \\ \hline \boldsymbol{h}_{51} & \boldsymbol{h}_{52} & \boldsymbol{h}_{53} & \boldsymbol{h}_{54} & \boldsymbol{h}_{55} & \boldsymbol{h}_{56} & \boldsymbol{h}_{57} & \boldsymbol{h}_{58} \\ \boldsymbol{h}_{61} & \boldsymbol{h}_{62} & \boldsymbol{h}_{63} & \boldsymbol{h}_{64} & \boldsymbol{h}_{65} & \boldsymbol{h}_{66} & \boldsymbol{h}_{67} & \boldsymbol{h}_{68} \\ \boldsymbol{h}_{71} & \boldsymbol{h}_{72} & \boldsymbol{h}_{73} & \boldsymbol{h}_{74} & \boldsymbol{h}_{75} & \boldsymbol{h}_{76} & \boldsymbol{h}_{77} & \boldsymbol{h}_{78} \\ \boldsymbol{h}_{81} & \boldsymbol{h}_{82} & \boldsymbol{h}_{83} & \boldsymbol{h}_{84} & \boldsymbol{h}_{85} & \boldsymbol{h}_{86} & \boldsymbol{h}_{87} & \boldsymbol{h}_{88} \end{array}\right] \begin{bmatrix} \hat{\boldsymbol{v}}_{\mathcal{J}} \\ \boldsymbol{v}_{vv} \\ \boldsymbol{v}_{cv} \\ \boldsymbol{v}_{sc} \\ \hline \hat{\boldsymbol{i}}_{\mathcal{L}} \\ \boldsymbol{i}_{cc} \\ \boldsymbol{i}_{vc} \\ \boldsymbol{i}_{oc} \end{bmatrix} \tag{10-40}$$

Since our objective is to find $\hat{\boldsymbol{H}}$ such that

$$\begin{bmatrix} \hat{\boldsymbol{i}}_{\mathcal{J}} \\ \hline \hat{\boldsymbol{v}}_{\mathcal{L}} \end{bmatrix} = \hat{\boldsymbol{H}} \begin{bmatrix} \hat{\boldsymbol{v}}_{\mathcal{J}} \\ \hline \hat{\boldsymbol{i}}_{\mathcal{L}} \end{bmatrix} \tag{10-41}$$

let us extract from Eq. (10-40) the following relationships:[6]

$$\begin{bmatrix} \hat{\boldsymbol{i}}_{\mathcal{J}} \\ \hline \hat{\boldsymbol{v}}_{\mathcal{L}} \end{bmatrix} = \left[\begin{array}{c|c} \boldsymbol{h}_{11} & \boldsymbol{h}_{15} \\ \hline \boldsymbol{h}_{51} & \boldsymbol{h}_{55} \end{array}\right] \begin{bmatrix} \hat{\boldsymbol{v}}_{\mathcal{J}} \\ \hline \hat{\boldsymbol{i}}_{\mathcal{L}} \end{bmatrix} + \left[\begin{array}{cc|cc} \boldsymbol{h}_{12} & \boldsymbol{h}_{17} & \boldsymbol{h}_{13} & \boldsymbol{h}_{16} \\ \hline \boldsymbol{h}_{52} & \boldsymbol{h}_{57} & \boldsymbol{h}_{53} & \boldsymbol{h}_{56} \end{array}\right] \begin{bmatrix} \boldsymbol{v}_{vv} \\ \boldsymbol{i}_{vc} \\ \hline \boldsymbol{v}_{cv} \\ \boldsymbol{i}_{cc} \end{bmatrix} \tag{10-42}$$

$$\begin{bmatrix} \boldsymbol{i}_{sc} \\ \hline \boldsymbol{v}_{oc} \end{bmatrix} = \left[\begin{array}{c|c} \boldsymbol{h}_{41} & \boldsymbol{h}_{45} \\ \hline \boldsymbol{h}_{81} & \boldsymbol{h}_{85} \end{array}\right] \begin{bmatrix} \hat{\boldsymbol{v}}_{\mathcal{J}} \\ \hline \hat{\boldsymbol{i}}_{\mathcal{L}} \end{bmatrix} + \left[\begin{array}{cc|cc} \boldsymbol{h}_{42} & \boldsymbol{h}_{47} & \boldsymbol{h}_{43} & \boldsymbol{h}_{46} \\ \hline \boldsymbol{h}_{82} & \boldsymbol{h}_{87} & \boldsymbol{h}_{83} & \boldsymbol{h}_{86} \end{array}\right] \begin{bmatrix} \boldsymbol{v}_{vv} \\ \boldsymbol{i}_{vc} \\ \hline \boldsymbol{v}_{cv} \\ \boldsymbol{i}_{cc} \end{bmatrix} \tag{10-43}$$

[6]We set $\boldsymbol{v}_{sc} = \boldsymbol{0}$ and $\boldsymbol{i}_{oc} = \boldsymbol{0}$ since they correspond to the voltage across a short circuit and the current in an open circuit.

We continue as in Chapter 6 to denote the relationships between the controlled sources and their controlling variables as follows:

$$\begin{bmatrix} v_{vv} \\ i_{vc} \\ \hline v_{cv} \\ i_{cc} \end{bmatrix} = \left[\begin{array}{c|c} 0 & K_{vv} \\ 0 & K_{vc} \\ \hline K_{cv} & 0 \\ K_{cc} & 0 \end{array}\right] \begin{bmatrix} i_{sc} \\ \hline v_{oc} \end{bmatrix} \tag{10-44}$$

where K_{vv}, K_{vc}, K_{cv}, and K_{cc} are matrices representing the controlling coefficients of the voltage-controlled voltage sources (vv), voltage-controlled current sources (vc), current-controlled voltage sources (cv), and current-controlled current sources (cc), respectively. Substituting Eq. (10-43) into Eq. (10-44), we obtain after simplification

$$\begin{bmatrix} v_{vv} \\ i_{vc} \\ \hline v_{cv} \\ i_{cc} \end{bmatrix} = M^{-1}P \begin{bmatrix} \hat{v}_3 \\ \hat{i}_{\mathcal{L}} \end{bmatrix} \tag{10-45}$$

where

$$M \triangleq \begin{bmatrix} (1 - K_{vv}h_{82}) & -K_{vv}h_{87} & -K_{vv}h_{83} & -K_{vv}h_{86} \\ -K_{vc}h_{82} & (1 - K_{vc}h_{87}) & -K_{vc}h_{83} & -K_{vc}h_{86} \\ -K_{cv}h_{42} & -K_{cv}h_{47} & (1 - K_{cv}h_{43}) & -K_{cv}h_{46} \\ -K_{cc}h_{42} & -K_{cc}h_{47} & -K_{cc}h_{43} & (1 - K_{cc}h_{46}) \end{bmatrix} \tag{10-46}$$

and

$$P \triangleq \begin{bmatrix} K_{vv}h_{81} & K_{vv}h_{85} \\ K_{vc}h_{81} & K_{vc}h_{85} \\ K_{cv}h_{41} & K_{cv}h_{45} \\ K_{cc}h_{41} & K_{cc}h_{45} \end{bmatrix} \tag{10-47}$$

Substituting Eq. (10-45) into Eq. (10-42), we obtain

$$\begin{bmatrix} \hat{i}_3 \\ \hat{v}_{\mathcal{L}} \end{bmatrix} = \left\{ \begin{bmatrix} h_{11} & h_{15} \\ h_{51} & h_{55} \end{bmatrix} + \begin{bmatrix} h_{12} & h_{17} & h_{13} & h_{16} \\ h_{52} & h_{57} & h_{53} & h_{56} \end{bmatrix} M^{-1}P \right\} \begin{bmatrix} \hat{v}_3 \\ \hat{i}_{\mathcal{L}} \end{bmatrix} \tag{10-48}$$

Comparing Eq. (10-48) with Eq. (10-41), we identify the hybrid matrix $\hat{H}$ for the linear m-port $\hat{N}$ to be given by

$$\boxed{\hat{H} = \begin{bmatrix} h_{11} & h_{15} \\ h_{51} & h_{55} \end{bmatrix} + \begin{bmatrix} h_{12} & h_{17} & h_{13} & h_{16} \\ h_{52} & h_{57} & h_{53} & h_{56} \end{bmatrix} M^{-1}P} \tag{10-49}$$

In view of our earlier remark concerning the existence of the hybrid matrix H for N_A, we can now state the following important theorem:

Theorem 10-1 (Existence of Hybrid Matrix for $\hat{N}$). *The hybrid matrix $\hat{H}$ for the linear m-port $\hat{N}$ exists if the following two conditions are satisfied:*

1. *The matrix Z defined in Eq. (10-35a) is nonsingular.*
2. *The matrix M defined in Eq. (10-46) is nonsingular.*

For most practical circuits, both conditions are satisfied; hence, the hybrid matrix $\hat{\boldsymbol{H}}$ for $\hat{N}$ will usually exist. In the rare situation where $\hat{\boldsymbol{H}}$ fails to exist because either $\boldsymbol{Z}$ or $\boldsymbol{M}$ is singular, one can overcome the problem by perturbing some element parameters slightly. Consequently, this situation is indeed pathological and need not concern us.

Now that we have found $\hat{\boldsymbol{H}}$, let us partition Eq. (10-48) as shown in Eq. (10-50):

$$\left[\begin{array}{c} \boldsymbol{i}_{E_\mathcal{T}} \\ \boldsymbol{i}_{C_\mathcal{T}} \\ \boldsymbol{i}_{R_\mathcal{T}} \\ \boldsymbol{i}_{L_\mathcal{T}} \\ \hline \boldsymbol{v}_{J_\mathcal{L}} \\ \boldsymbol{v}_{L_\mathcal{L}} \\ \boldsymbol{v}_{R_\mathcal{L}} \\ \boldsymbol{v}_{C_\mathcal{L}} \end{array}\right] = \left[\begin{array}{cccc|cccc} \boldsymbol{H}_{E_\mathcal{T}E_\mathcal{T}} & \boldsymbol{H}_{E_\mathcal{T}C_\mathcal{T}} & \boldsymbol{H}_{E_\mathcal{T}R_\mathcal{T}} & \boldsymbol{H}_{E_\mathcal{T}L_\mathcal{T}} & \boldsymbol{H}_{E_\mathcal{T}J_\mathcal{L}} & \boldsymbol{H}_{E_\mathcal{T}L_\mathcal{L}} & \boldsymbol{H}_{E_\mathcal{T}R_\mathcal{L}} & \boldsymbol{H}_{E_\mathcal{T}C_\mathcal{L}} \\ \boldsymbol{H}_{C_\mathcal{T}E_\mathcal{T}} & \boldsymbol{H}_{C_\mathcal{T}C_\mathcal{T}} & \boldsymbol{H}_{C_\mathcal{T}R_\mathcal{T}} & \boldsymbol{H}_{C_\mathcal{T}L_\mathcal{T}} & \boldsymbol{H}_{C_\mathcal{T}J_\mathcal{L}} & \boldsymbol{H}_{C_\mathcal{T}L_\mathcal{L}} & \boldsymbol{H}_{C_\mathcal{T}R_\mathcal{L}} & \boldsymbol{H}_{C_\mathcal{T}C_\mathcal{L}} \\ \boldsymbol{H}_{R_\mathcal{T}E_\mathcal{T}} & \boldsymbol{H}_{R_\mathcal{T}C_\mathcal{T}} & \boldsymbol{H}_{R_\mathcal{T}R_\mathcal{T}} & \boldsymbol{H}_{R_\mathcal{T}L_\mathcal{T}} & \boldsymbol{H}_{R_\mathcal{T}J_\mathcal{L}} & \boldsymbol{H}_{R_\mathcal{T}L_\mathcal{L}} & \boldsymbol{H}_{R_\mathcal{T}R_\mathcal{L}} & \boldsymbol{H}_{R_\mathcal{T}C_\mathcal{L}} \\ \boldsymbol{H}_{L_\mathcal{T}E_\mathcal{T}} & \boldsymbol{H}_{L_\mathcal{T}C_\mathcal{T}} & \boldsymbol{H}_{L_\mathcal{T}R_\mathcal{T}} & \boldsymbol{H}_{L_\mathcal{T}L_\mathcal{T}} & \boldsymbol{H}_{L_\mathcal{T}J_\mathcal{L}} & \boldsymbol{H}_{L_\mathcal{T}L_\mathcal{L}} & \boldsymbol{H}_{L_\mathcal{T}R_\mathcal{L}} & \boldsymbol{H}_{L_\mathcal{T}C_\mathcal{L}} \\ \hline \boldsymbol{H}_{J_\mathcal{L}E_\mathcal{T}} & \boldsymbol{H}_{J_\mathcal{L}C_\mathcal{T}} & \boldsymbol{H}_{J_\mathcal{L}R_\mathcal{T}} & \boldsymbol{H}_{J_\mathcal{L}L_\mathcal{T}} & \boldsymbol{H}_{J_\mathcal{L}J_\mathcal{L}} & \boldsymbol{H}_{J_\mathcal{L}L_\mathcal{L}} & \boldsymbol{H}_{J_\mathcal{L}R_\mathcal{L}} & \boldsymbol{H}_{J_\mathcal{L}C_\mathcal{L}} \\ \boldsymbol{H}_{L_\mathcal{L}E_\mathcal{T}} & \boldsymbol{H}_{L_\mathcal{L}C_\mathcal{T}} & \boldsymbol{H}_{L_\mathcal{L}R_\mathcal{T}} & \boldsymbol{H}_{L_\mathcal{L}L_\mathcal{T}} & \boldsymbol{H}_{L_\mathcal{L}J_\mathcal{L}} & \boldsymbol{H}_{L_\mathcal{L}L_\mathcal{L}} & \boldsymbol{H}_{L_\mathcal{L}R_\mathcal{L}} & \boldsymbol{H}_{L_\mathcal{L}C_\mathcal{L}} \\ \boldsymbol{H}_{R_\mathcal{L}E_\mathcal{T}} & \boldsymbol{H}_{R_\mathcal{L}C_\mathcal{T}} & \boldsymbol{H}_{R_\mathcal{L}R_\mathcal{T}} & \boldsymbol{H}_{R_\mathcal{L}L_\mathcal{T}} & \boldsymbol{H}_{R_\mathcal{L}J_\mathcal{L}} & \boldsymbol{H}_{R_\mathcal{L}L_\mathcal{L}} & \boldsymbol{H}_{R_\mathcal{L}R_\mathcal{L}} & \boldsymbol{H}_{R_\mathcal{L}C_\mathcal{L}} \\ \boldsymbol{H}_{C_\mathcal{L}E_\mathcal{T}} & \boldsymbol{H}_{C_\mathcal{L}C_\mathcal{T}} & \boldsymbol{H}_{C_\mathcal{L}R_\mathcal{T}} & \boldsymbol{H}_{C_\mathcal{L}L_\mathcal{T}} & \boldsymbol{H}_{C_\mathcal{L}J_\mathcal{L}} & \boldsymbol{H}_{C_\mathcal{L}L_\mathcal{L}} & \boldsymbol{H}_{C_\mathcal{L}R_\mathcal{L}} & \boldsymbol{H}_{C_\mathcal{L}C_\mathcal{L}} \end{array}\right] \left[\begin{array}{c} \boldsymbol{v}_{E_\mathcal{T}} \\ \boldsymbol{v}_{C_\mathcal{T}} \\ \boldsymbol{v}_{R_\mathcal{T}} \\ \boldsymbol{v}_{L_\mathcal{T}} \\ \hline \boldsymbol{i}_{J_\mathcal{L}} \\ \boldsymbol{i}_{L_\mathcal{L}} \\ \boldsymbol{i}_{R_\mathcal{L}} \\ \boldsymbol{i}_{C_\mathcal{L}} \end{array}\right] \tag{10-50}$$

Each of the 64 submatrices in Eq. (10-50) can be given a simple circuit interpretation in Fig. 10-8. For example, the submatrix $\boldsymbol{H}_{E_\mathcal{T}E_\mathcal{T}}$ can be interpreted to be the *driving-point admittance matrix* of the p-port formed by the tree voltage sources when all other tree ports are short-circuited and when all cotree ports are open-circuited; i.e.,

$$\boldsymbol{i}_{E_\mathcal{T}} = \boldsymbol{H}_{E_\mathcal{T}E_\mathcal{T}}\boldsymbol{v}_{E_\mathcal{T}}\Big|_{\boldsymbol{v}_{C_\mathcal{T}}=\boldsymbol{0},\,\boldsymbol{v}_{R_\mathcal{T}}=\boldsymbol{0},\,\boldsymbol{v}_{L_\mathcal{T}}=\boldsymbol{0},\,\boldsymbol{i}_{J_\mathcal{L}}=\boldsymbol{0},\,\boldsymbol{i}_{L_\mathcal{L}}=\boldsymbol{0},\,\boldsymbol{i}_{R_\mathcal{L}}=\boldsymbol{0},\,\boldsymbol{i}_{C_\mathcal{L}}=\boldsymbol{0}} \tag{10-51}$$

Similarly, the submatrix $\boldsymbol{H}_{R_\mathcal{L}C_\mathcal{T}}$ can be interpreted as the *voltage transfer matrix* from the tree capacitor ports to the cotree resistor ports when all tree ports (except the tree capacitor ports) are short-circuited and all cotree ports are open-circuited; i.e.,

$$\boldsymbol{v}_{R_\mathcal{L}} = \boldsymbol{H}_{R_\mathcal{L}C_\mathcal{T}}\boldsymbol{v}_{C_\mathcal{T}}\Big|_{\boldsymbol{v}_{E_\mathcal{T}}=\boldsymbol{0},\,\boldsymbol{v}_{R_\mathcal{T}}=\boldsymbol{0},\,\boldsymbol{v}_{L_\mathcal{T}}=\boldsymbol{0},\,\boldsymbol{i}_{J_\mathcal{L}}=\boldsymbol{0},\,\boldsymbol{i}_{L_\mathcal{L}}=\boldsymbol{0},\,\boldsymbol{i}_{R_\mathcal{L}}=\boldsymbol{0},\,\boldsymbol{i}_{C_\mathcal{L}}=\boldsymbol{0}} \tag{10-52}$$

Observe that some of the submatrices in Eq. (10-50) may be absent, i.e., of dimension 0×0 for certain networks. For example, if the network does not contain any *C-E* loop, there are no cotree capacitor ports, and all submatrices in the last row of Eq. (10-50) are absent. Similarly, if the network does not contain any *L-J* cutset, there are no tree inductor ports, and all submatrices in the fourth row of Eq. (10-50) are absent. It is also possible that for certain networks some of the submatrices may be *null matrices* (with zero entries).[7] In fact, in view of our procedure for constructing the normal tree and our standing assumptions, there are a number of submatrices in Eq. (10-50) that are always null matrices or matrices with zero dimension for all networks which satisfy the standing assumptions. This property turns out to be

[7]It is important to distinguish between a 0-dimensional submatrix and a $p \times q$ null submatrix with zero entries. A 0-dimensional submatrix arises only if a certain group of elements in Fig. 10-8 is completely absent. A null submatrix may occur in a particular network when certain degeneracies, such as short circuits, exist between certain ports.

crucial in the formulation of the normal-form equations, and is important enough to be stated in the form of a theorem.

> ***Theorem 10-2.*** *For any network that satisfies the standing assumptions, the following submatrices of Eq. (10-50) are either null matrices or matrices with zero dimension:*
>
> (a) $\boldsymbol{H}_{L_\mathfrak{T}E_\mathfrak{T}}, \boldsymbol{H}_{L_\mathfrak{T}R_\mathfrak{T}}, \boldsymbol{H}_{L_\mathfrak{T}L_\mathfrak{T}}, \boldsymbol{H}_{L_\mathfrak{T}R_\mathfrak{L}}, \boldsymbol{H}_{L_\mathfrak{T}C_\mathfrak{L}}$
> (b) $\boldsymbol{H}_{C_\mathfrak{L}R_\mathfrak{T}}, \boldsymbol{H}_{C_\mathfrak{L}L_\mathfrak{T}}, \boldsymbol{H}_{C_\mathfrak{L}J_\mathfrak{L}}, \boldsymbol{H}_{C_\mathfrak{L}R_\mathfrak{L}}, \boldsymbol{H}_{C_\mathfrak{L}C_\mathfrak{L}}$
> (c) $\boldsymbol{H}_{R_\mathfrak{T}L_\mathfrak{T}}, \boldsymbol{H}_{R_\mathfrak{T}C_\mathfrak{L}}$
> (d) $\boldsymbol{H}_{R_\mathfrak{L}L_\mathfrak{T}}, \boldsymbol{H}_{R_\mathfrak{L}C_\mathfrak{L}}$
> (e) $\boldsymbol{H}_{C_\mathfrak{T}L_\mathfrak{T}}, \boldsymbol{H}_{L_\mathfrak{L}C_\mathfrak{L}}, \boldsymbol{H}_{E_\mathfrak{T}L_\mathfrak{T}}, \boldsymbol{H}_{J_\mathfrak{L}C_\mathfrak{L}}$
>
> *Proof:* (a) From Eq. (10-50), we obtain

$$\begin{aligned}\boldsymbol{i}_{L_\mathfrak{T}} = \boldsymbol{H}_{L_\mathfrak{T}E_\mathfrak{T}}\boldsymbol{v}_{E_\mathfrak{T}} + \boldsymbol{H}_{L_\mathfrak{T}C_\mathfrak{T}}\boldsymbol{v}_{C_\mathfrak{T}} + \boldsymbol{H}_{L_\mathfrak{T}R_\mathfrak{T}}\boldsymbol{v}_{R_\mathfrak{T}} + \boldsymbol{H}_{L_\mathfrak{T}L_\mathfrak{T}}\boldsymbol{v}_{L_\mathfrak{T}} + \boldsymbol{H}_{L_\mathfrak{T}J_\mathfrak{L}}\boldsymbol{i}_{J_\mathfrak{L}} \\ + \boldsymbol{H}_{L_\mathfrak{T}L_\mathfrak{L}}\boldsymbol{i}_{L_\mathfrak{L}} + \boldsymbol{H}_{L_\mathfrak{T}R_\mathfrak{L}}\boldsymbol{i}_{R_\mathfrak{L}} + \boldsymbol{H}_{L_\mathfrak{T}C_\mathfrak{L}}\boldsymbol{i}_{C_\mathfrak{L}}\end{aligned} \tag{10-53}$$

Since the tree inductors form cutsets with the cotree inductors and independent and/or controlled current sources, we can write

$$\boldsymbol{i}_{L_\mathfrak{T}} = \boldsymbol{\alpha}\boldsymbol{i}_{L_\mathfrak{L}} + \boldsymbol{\beta}\boldsymbol{i}_{J_\mathfrak{L}} + \boldsymbol{\gamma}\boldsymbol{i}_{vc} + \boldsymbol{\delta}\boldsymbol{i}_{cc} \tag{10-54}$$

But by assumption 3(a),

$$\boldsymbol{i}_{vc} = \boldsymbol{\gamma}_1\boldsymbol{v}_{C_\mathfrak{T}} \tag{10-55}$$

$$\boldsymbol{i}_{cc} = \boldsymbol{\delta}_1\boldsymbol{i}_{L_\mathfrak{L}} \tag{10-56}$$

Substituting Eqs. (10-55) and (10-56) into Eq. (10-54) and rearranging the terms, we obtain

$$\boldsymbol{i}_{L_\mathfrak{T}} = (\boldsymbol{\alpha} + \boldsymbol{\delta}\boldsymbol{\delta}_1)\boldsymbol{i}_{L_\mathfrak{L}} + \boldsymbol{\gamma}\boldsymbol{\gamma}_1\boldsymbol{v}_{C_\mathfrak{T}} + \boldsymbol{\beta}\boldsymbol{i}_{J_\mathfrak{L}} \tag{10-57}$$

Comparing Eqs. (10-57) and (10-53), we identify

$$\boldsymbol{H}_{L_\mathfrak{T}L_\mathfrak{L}} = (\boldsymbol{\alpha} + \boldsymbol{\delta}\boldsymbol{\delta}_1), \qquad \boldsymbol{H}_{L_\mathfrak{T}C_\mathfrak{T}} = \boldsymbol{\gamma}\boldsymbol{\gamma}_1, \qquad \boldsymbol{H}_{L_\mathfrak{T}J_\mathfrak{L}} = \boldsymbol{\beta}$$

$$\boldsymbol{H}_{L_\mathfrak{T}E_\mathfrak{T}} = \boldsymbol{0}, \quad \boldsymbol{H}_{L_\mathfrak{T}R_\mathfrak{T}} = \boldsymbol{0}, \quad \boldsymbol{H}_{L_\mathfrak{T}L_\mathfrak{T}} = \boldsymbol{0}, \quad \boldsymbol{H}_{L_\mathfrak{T}R_\mathfrak{L}} = \boldsymbol{0}, \quad \boldsymbol{H}_{L_\mathfrak{T}C_\mathfrak{L}} = \boldsymbol{0}$$

(b) By a dual procedure, we find

$$\boldsymbol{H}_{C_\mathfrak{L}R_\mathfrak{T}} = \boldsymbol{0}, \quad \boldsymbol{H}_{C_\mathfrak{L}L_\mathfrak{T}} = \boldsymbol{0}, \quad \boldsymbol{H}_{C_\mathfrak{L}J_\mathfrak{L}} = \boldsymbol{0}, \quad \boldsymbol{H}_{C_\mathfrak{L}R_\mathfrak{L}} = \boldsymbol{0}, \quad \boldsymbol{H}_{C_\mathfrak{L}C_\mathfrak{L}} = \boldsymbol{0}$$

(c) From Eq. (10-50), we obtain

$$\boldsymbol{i}_{R_\mathfrak{T}} = \boldsymbol{H}_{R_\mathfrak{T}L_\mathfrak{T}}\boldsymbol{v}_{L_\mathfrak{T}} + \boldsymbol{H}_{R_\mathfrak{T}C_\mathfrak{L}}\boldsymbol{i}_{C_\mathfrak{L}}\Big|_{\boldsymbol{v}_{E_\mathfrak{T}}=\boldsymbol{0},\, \boldsymbol{v}_{C_\mathfrak{T}}=\boldsymbol{0},\, \boldsymbol{v}_{R_\mathfrak{T}}=\boldsymbol{0},\, \boldsymbol{i}_{J_\mathfrak{L}}=\boldsymbol{0},\, \boldsymbol{i}_{L_\mathfrak{L}}=\boldsymbol{0},\, \boldsymbol{i}_{R_\mathfrak{L}}=\boldsymbol{0}} \tag{10-58}$$

Since the m-port $\hat{N}$ is linear, we can apply the principle of superposition and set all components of $(\boldsymbol{v}_{L_\mathfrak{T}}, \boldsymbol{i}_{C_\mathfrak{L}})$ except one to zero, and compute for the corresponding $\boldsymbol{i}_{R_\mathfrak{T}}$. The total current is then obtained by adding the individual contributions. To be specific, let us investigate the effect of driving the cotree capacitor port with a current source $i_{C_{\mathfrak{L}_j}}$ while short-circuiting all tree ports and open-circuiting all cotree ports. Since each cotree capacitor forms a loop with the tree capacitors and independent and/or controlled voltage sources, the current source $i_{C_{\mathfrak{L}_j}}$ must form a loop with only short circuits because $\boldsymbol{v}_{E_{\mathfrak{T}}} = \mathbf{0}$, $\boldsymbol{v}_{C_{\mathfrak{T}}} = \mathbf{0}$, and $\boldsymbol{i}_{L_{\mathfrak{L}}} = \mathbf{0}$, and the terminal voltage of each controlled voltage source depends only on $\boldsymbol{v}_{C_{\mathfrak{T}}}$ or $\boldsymbol{i}_{L_{\mathfrak{L}}}$ by assumption 3(a). Hence, the current $i_{C_{\mathfrak{L}_j}}$ will circulate around this short-circuited path. Now, in view of assumption 2(a), no controlled sources can depend on the current of any short-circuited branch for which the current $i_{C_{\mathfrak{L}_j}}$ flows. Therefore, all controlled voltage sources must have zero terminal voltage and all controlled current sources must have zero terminal current, since $\boldsymbol{v}_{E_{\mathfrak{T}}} = \mathbf{0}$ and $\boldsymbol{i}_{J_{\mathfrak{L}}} = \mathbf{0}$. Moreover, the resistor $R_{\mathfrak{T}}$, being a *tree* resistor, cannot appear in parallel with any element in the *C-E* loop; therefore, it is impossible for the current $i_{C_{\mathfrak{L}_j}}$ to shunt through the short-circuited port across $R_{\mathfrak{T}}$. Consequently, all branches not lying in the closed loop through which $i_{C_{\mathfrak{L}_j}}$ circulates must have zero current. A dual analysis shows that, if we connect a voltage source $v_{L_{\mathfrak{T}_j}}$ across one tree inductor port while shorting all tree ports and opening all cotree ports, no voltage can exist across any branch that does not belong to the cutset containing this tree inductor. By superposition, it follows that $\boldsymbol{i}_{R_{\mathfrak{T}}} = \mathbf{0}$, independent of $\boldsymbol{v}_{L_{\mathfrak{T}}}$ and $\boldsymbol{i}_{C_{\mathfrak{L}}}$. But from Eq. (10-58), this is possible only if $\boldsymbol{H}_{R_{\mathfrak{T}}L_{\mathfrak{T}}} = \mathbf{0}$ and $\boldsymbol{H}_{R_{\mathfrak{T}}C_{\mathfrak{L}}} = \mathbf{0}$.

(d) By a dual analysis, we can conclude that $\boldsymbol{H}_{R_{\mathfrak{L}}L_{\mathfrak{T}}} = \mathbf{0}$ and $\boldsymbol{H}_{R_{\mathfrak{L}}C_{\mathfrak{L}}} = \mathbf{0}$.

(e) By an analysis similar to part (c), we find that $\boldsymbol{H}_{C_{\mathfrak{T}}L_{\mathfrak{T}}} = \mathbf{0}$, $\boldsymbol{H}_{L_{\mathfrak{L}}C_{\mathfrak{L}}} = \mathbf{0}$, $\boldsymbol{H}_{E_{\mathfrak{T}}L_{\mathfrak{T}}} = \mathbf{0}$, and $\boldsymbol{H}_{J_{\mathfrak{L}}C_{\mathfrak{L}}} = \mathbf{0}$.

This completes the proof of Theorem 10-2. Q.E.D.

10-3-3 Step 2: Solving the Resistive Nonlinear Subnetwork

In view of Theorem 10-2, we can now extract the following two equations from Eq. (10-50):

$$\begin{bmatrix} \boldsymbol{i}_{R_{\mathfrak{T}}} \\ \boldsymbol{v}_{R_{\mathfrak{L}}} \end{bmatrix} = \begin{bmatrix} \boldsymbol{H}_{R_{\mathfrak{T}}E_{\mathfrak{T}}} & \boldsymbol{H}_{R_{\mathfrak{T}}C_{\mathfrak{T}}} & \boldsymbol{H}_{R_{\mathfrak{T}}R_{\mathfrak{T}}} & \boldsymbol{H}_{R_{\mathfrak{T}}J_{\mathfrak{L}}} & \boldsymbol{H}_{R_{\mathfrak{T}}L_{\mathfrak{L}}} & \boldsymbol{H}_{R_{\mathfrak{T}}R_{\mathfrak{L}}} \\ \boldsymbol{H}_{R_{\mathfrak{L}}E_{\mathfrak{T}}} & \boldsymbol{H}_{R_{\mathfrak{L}}C_{\mathfrak{T}}} & \boldsymbol{H}_{R_{\mathfrak{L}}R_{\mathfrak{T}}} & \boldsymbol{H}_{R_{\mathfrak{L}}J_{\mathfrak{L}}} & \boldsymbol{H}_{R_{\mathfrak{L}}L_{\mathfrak{L}}} & \boldsymbol{H}_{R_{\mathfrak{L}}R_{\mathfrak{L}}} \end{bmatrix} \begin{bmatrix} \boldsymbol{v}_{E_{\mathfrak{T}}} \\ \boldsymbol{v}_{C_{\mathfrak{T}}} \\ \boldsymbol{v}_{R_{\mathfrak{T}}} \\ \boldsymbol{i}_{J_{\mathfrak{L}}} \\ \boldsymbol{i}_{L_{\mathfrak{L}}} \\ \boldsymbol{i}_{R_{\mathfrak{L}}} \end{bmatrix} \tag{10-59}$$

Equation (10-59) can be written in a more convenient form:

$$\begin{bmatrix} \boldsymbol{i}_{R_{\mathfrak{T}}} \\ \hline \boldsymbol{v}_{R_{\mathfrak{L}}} \end{bmatrix} = \left[\begin{array}{c|c} \boldsymbol{H}_{R_{\mathfrak{T}}R_{\mathfrak{T}}} & \boldsymbol{H}_{R_{\mathfrak{T}}R_{\mathfrak{L}}} \\ \hline \boldsymbol{H}_{R_{\mathfrak{L}}R_{\mathfrak{T}}} & \boldsymbol{H}_{R_{\mathfrak{L}}R_{\mathfrak{L}}} \end{array}\right] \begin{bmatrix} \boldsymbol{v}_{R_{\mathfrak{T}}} \\ \hline \boldsymbol{i}_{R_{\mathfrak{L}}} \end{bmatrix} + \begin{bmatrix} \boldsymbol{s}_{R_{\mathfrak{T}}} \\ \hline \boldsymbol{s}_{R_{\mathfrak{L}}} \end{bmatrix} \tag{10-60}$$

where

$$\begin{bmatrix} \boldsymbol{s}_{R_{\mathfrak{T}}} \\ \hline \boldsymbol{s}_{R_{\mathfrak{L}}} \end{bmatrix} \triangleq \left[\begin{array}{cc|cc} \boldsymbol{H}_{R_{\mathfrak{T}}C_{\mathfrak{T}}} & \boldsymbol{H}_{R_{\mathfrak{T}}L_{\mathfrak{L}}} & \boldsymbol{H}_{R_{\mathfrak{T}}E_{\mathfrak{T}}} & \boldsymbol{H}_{R_{\mathfrak{T}}J_{\mathfrak{L}}} \\ \hline \boldsymbol{H}_{R_{\mathfrak{L}}C_{\mathfrak{T}}} & \boldsymbol{H}_{R_{\mathfrak{L}}L_{\mathfrak{L}}} & \boldsymbol{H}_{R_{\mathfrak{L}}E_{\mathfrak{T}}} & \boldsymbol{H}_{R_{\mathfrak{L}}J_{\mathfrak{L}}} \end{array}\right] \begin{bmatrix} \boldsymbol{v}_{C_{\mathfrak{T}}} \\ \boldsymbol{i}_{L_{\mathfrak{L}}} \\ \hline \boldsymbol{v}_{E_{\mathfrak{T}}} \\ \boldsymbol{i}_{J_{\mathfrak{L}}} \end{bmatrix} \tag{10-61}$$

Equation (10-60) brings into focus the port vectors $\boldsymbol{i}_{R_{\mathfrak{T}}}$ and $\boldsymbol{v}_{R_{\mathfrak{T}}}$ associated with the nonlinear tree resistors, and the port vectors $\boldsymbol{i}_{R_\mathfrak{L}}$ and $\boldsymbol{v}_{R_\mathfrak{L}}$ associated with the nonlinear cotree resistors. For reasons that will soon be obvious, the contribution due to the remaining four port vectors $\boldsymbol{v}_{C_\mathfrak{J}}$, $\boldsymbol{i}_{L_\mathfrak{L}}$, $\boldsymbol{v}_{E_\mathfrak{J}}$, and $\boldsymbol{i}_{J_\mathfrak{L}}$ is lumped together by the vector $[\boldsymbol{s}_{R_\mathfrak{J}}, \boldsymbol{s}_{R_\mathfrak{L}}]^t$ as defined in Eq. (10-61).[8] Equation (10-60) can be interpreted as the hybrid matrix representation of a new $(p+q)$ — port $\hat{N}_R$ whose ports consist of only the p *nonlinear tree resistors* and the q *nonlinear cotree resistors*, as shown in Fig. 10-9. Observe that $\hat{N}_R$ now contains both linear resistors and controlled *and* independent sources, as in Chapter 6. A comparison of Eq. (10-60) with the hybrid representation derived in Chapter 6 [Eq. (6-2)] shows that the vector $[\boldsymbol{s}_{R_\mathfrak{J}}, \boldsymbol{s}_{R_\mathfrak{L}}]^t$ on the right side of Eq. (10-60) is simply the "source vector" due to the *independent* sources $\boldsymbol{v}_{C_\mathfrak{J}}$, $\boldsymbol{i}_{L_\mathfrak{L}}$, $\boldsymbol{v}_{E_\mathfrak{J}}$, and $\boldsymbol{i}_{J_\mathfrak{L}}$ inside $\hat{N}_R$. The network shown in Fig. 10-9 is a *nonlinear resistive network*. Since the nonlinear resistors may be either voltage- or current-controlled, there are several possible combinations; we shall investigate only the combination when all nonlinear tree resistors are voltage-controlled and all nonlinear cotree resistors are current-controlled. To be specific, let

$$\boldsymbol{i}_{R_\mathfrak{J}} = \begin{bmatrix} g_1(v_{R_{\mathfrak{J}_1}}) \\ g_2(v_{R_{\mathfrak{J}_2}}) \\ \vdots \\ g_p(v_{R_{\mathfrak{J}_p}}) \end{bmatrix} \triangleq \boldsymbol{g}(\boldsymbol{v}_{R_\mathfrak{J}}), \qquad \boldsymbol{v}_{R_\mathfrak{L}} = \begin{bmatrix} f_1(i_{R_{\mathfrak{L}_1}}) \\ f_2(i_{R_{\mathfrak{L}_2}}) \\ \vdots \\ f_q(i_{R_{\mathfrak{L}_q}}) \end{bmatrix} \triangleq \boldsymbol{f}(\boldsymbol{i}_{R_\mathfrak{L}}) \tag{10-62}$$

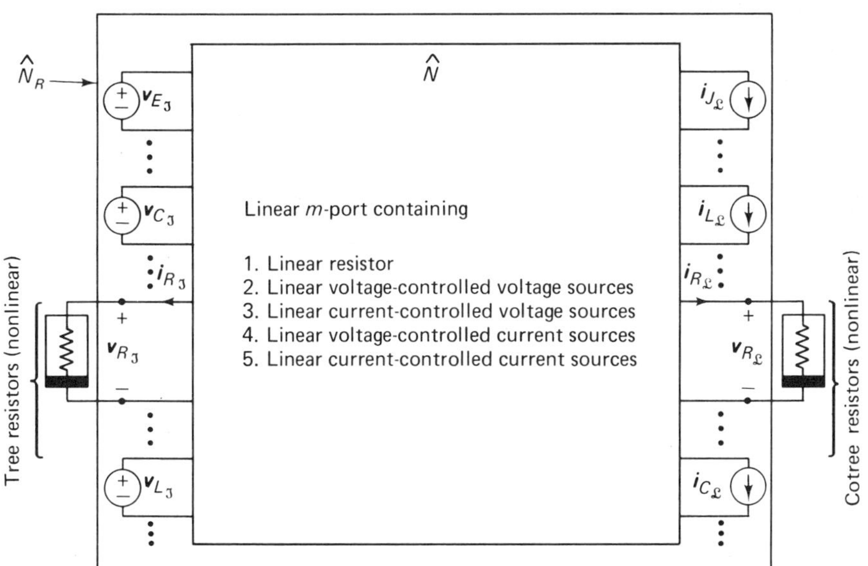

Fig. 10-9. A $(p+q)$-port $\hat{N}_R$ obtained from $\hat{N}$ by imbedding all ports except the p nonlinear tree resistors and the q nonlinear cotree resistors.

[8]Throughout this chapter, as well as in subsequent chapters, we will often abuse our notation by deleting the "transpose" symbol associated with a column vector and write $[\boldsymbol{x}^t\, \boldsymbol{y}^t \ldots \boldsymbol{z}^t]^t$ simply as $[\boldsymbol{x}\, \boldsymbol{y} \ldots \boldsymbol{z}]^t$.

If we substitute Eq. (10-62) into Eq. (10-60), we obtain the following equations of motion of the resistive network in Fig. 10-9:[9]

$$\begin{bmatrix} g(\boldsymbol{v}_{R_\mathfrak{J}}) \\ \hdashline f(\boldsymbol{i}_{R_\mathfrak{L}}) \end{bmatrix} - \begin{bmatrix} \boldsymbol{H}_{R_\mathfrak{J}R_\mathfrak{J}} & \boldsymbol{H}_{R_\mathfrak{J}R_\mathfrak{L}} \\ \hdashline \boldsymbol{H}_{R_\mathfrak{L}R_\mathfrak{J}} & \boldsymbol{H}_{R_\mathfrak{L}R_\mathfrak{L}} \end{bmatrix} \begin{bmatrix} \boldsymbol{v}_{R_\mathfrak{J}} \\ \hdashline \boldsymbol{i}_{R_\mathfrak{L}} \end{bmatrix} - \begin{bmatrix} \boldsymbol{s}_{R_\mathfrak{J}} \\ \hdashline \boldsymbol{s}_{R_\mathfrak{L}} \end{bmatrix} = \begin{bmatrix} \boldsymbol{0} \\ \boldsymbol{0} \end{bmatrix} \tag{10-63}$$

which is in exactly the same form as the system of nonlinear equations (5-30). Consequently, corresponding to any *specified* value of the source vector $[\boldsymbol{s}_{R_\mathfrak{J}}, \boldsymbol{s}_{R_\mathfrak{L}}]^t$, Eq. (10-63) can be solved for $\boldsymbol{v}_{R_\mathfrak{J}}$ and $\boldsymbol{i}_{R_\mathfrak{L}}$ by any of the methods described in Chapter 5. Since the source vector $[\boldsymbol{s}_{R_\mathfrak{J}}, \boldsymbol{s}_{R_\mathfrak{L}}]^t$ defined by Eq. (10-61) depends on the vectors $\boldsymbol{v}_{C_\mathfrak{J}}$, $\boldsymbol{i}_{L_\mathfrak{L}}$, $\boldsymbol{v}_{E_\mathfrak{J}}$, and $\boldsymbol{i}_{J_\mathfrak{L}}$, it follows that, corresponding to any *specified* value of $\boldsymbol{v}_{C_\mathfrak{J}}$, $\boldsymbol{i}_{L_\mathfrak{L}}$, $\boldsymbol{v}_{E_\mathfrak{J}}$, and $\boldsymbol{i}_{J_\mathfrak{L}}$, we can solve Eq. (10-63) for $\boldsymbol{v}_{R_\mathfrak{J}}$ and $\boldsymbol{i}_{R_\mathfrak{L}}$. We have already learned that Eq. (10-63) may possess more than one solution. However, for the normal-form equations to exist, we must at this point stipulate the following assumption, in addition to the standing assumptions already made:

> *Uniqueness Assumption*
> For *any* specified $\boldsymbol{v}_{C_\mathfrak{J}}$, $\boldsymbol{i}_{L_\mathfrak{L}}$, $\boldsymbol{v}_{E_\mathfrak{J}}$, and $\boldsymbol{i}_{J_\mathfrak{L}}$, the corresponding source vector $[\boldsymbol{s}_{R_\mathfrak{J}}, \boldsymbol{s}_{R_\mathfrak{L}}]^t$ in Eq. (10-63) gives rise to a unique solution $\boldsymbol{v}_{R_\mathfrak{J}} = \hat{\boldsymbol{v}}_{R_\mathfrak{J}}$ and $\boldsymbol{i}_{R_\mathfrak{L}} = \hat{\boldsymbol{i}}_{R_\mathfrak{L}}$.

To emphasize the fact that the solution $[\hat{\boldsymbol{v}}_{R_\mathfrak{J}}, \hat{\boldsymbol{i}}_{R_\mathfrak{L}}]^t$ depends on the specified value of $\boldsymbol{v}_{C_\mathfrak{J}}$, $\boldsymbol{i}_{L_\mathfrak{L}}$, $\boldsymbol{v}_{E_\mathfrak{J}}$, and $\boldsymbol{i}_{J_\mathfrak{L}}$, we write

$$\begin{bmatrix} \boldsymbol{v}_{R_\mathfrak{J}} \\ \boldsymbol{i}_{R_\mathfrak{L}} \end{bmatrix} = \begin{bmatrix} \hat{\boldsymbol{v}}_{R_\mathfrak{J}}(\boldsymbol{v}_{C_\mathfrak{J}}, \boldsymbol{i}_{L_\mathfrak{L}}, \boldsymbol{v}_{E_\mathfrak{J}}, \boldsymbol{i}_{J_\mathfrak{L}}) \\ \hat{\boldsymbol{i}}_{R_\mathfrak{L}}(\boldsymbol{v}_{C_\mathfrak{J}}, \boldsymbol{i}_{L_\mathfrak{L}}, \boldsymbol{v}_{E_\mathfrak{J}}, \boldsymbol{i}_{J_\mathfrak{L}}) \end{bmatrix} \tag{10-64}$$

10-3-4 Step 3: Solving the *C-E* Loops and *L-J* Cutsets

Our next step is to express the tree inductor voltages $\boldsymbol{v}_{L_\mathfrak{J}}$ and the cotree capacitor currents $\boldsymbol{i}_{C_\mathfrak{L}}$ in terms of the tree capacitor voltages $\boldsymbol{v}_{C_\mathfrak{J}}$ and the cotree inductor currents $\boldsymbol{i}_{L_\mathfrak{L}}$. In view of assumptions 3(b) and 3(c), the voltage across each tree inductor and the current in each cotree capacitor can be written as

$$i_{C_{\mathfrak{L}_j}} = C_{\mathfrak{L}_j}(v_{C_{\mathfrak{L}_j}})\frac{dv_{C_{\mathfrak{L}_j}}}{dt}, \qquad j = 1, 2, \ldots, \rho \tag{10-65}$$

$$v_{L_{\mathfrak{J}_j}} = L_{\mathfrak{J}_j}(i_{L_{\mathfrak{J}_j}})\frac{di_{L_{\mathfrak{J}_j}}}{dt}, \qquad j = 1, 2, \ldots, \sigma \tag{10-66}$$

where $C_{\mathfrak{L}_j}$ $(v_{C_{\mathfrak{L}_j}})$ is the *incremental capacitance* of the jth cotree capacitor, ρ is the total number of cotree capacitors, $L_{\mathfrak{J}_j}$ $(i_{L_{\mathfrak{J}_j}})$ is the *incremental inductance* of the jth

[9]There is little loss of generality in assuming $g(\cdot)$ and $f(\cdot)$ are one-to-one functions, since a linear *negative* resistor can always be extracted, if necessary; the linear resistor can then be imbedded within $\hat{N}$. Observe also that Eq. (10-63) still holds even if the nonlinear resistors are *coupled* to each other.

tree inductor, and σ is the total number of tree inductors. Equations (10-65) and (10-66) can be rewritten into the following compact matrix form:

$$\begin{bmatrix} i_{C_\mathfrak{L}} \\ v_{L_\mathfrak{J}} \end{bmatrix} = \begin{bmatrix} C_\mathfrak{L}(v_{C_\mathfrak{L}}) & 0 \\ 0 & L_\mathfrak{J}(i_{L_\mathfrak{J}}) \end{bmatrix} \begin{bmatrix} \dot{v}_{C_\mathfrak{L}} \\ \dot{i}_{L_\mathfrak{J}} \end{bmatrix} \tag{10-67}$$

where

$$C_\mathfrak{L}(v_{C_\mathfrak{L}}) \triangleq \begin{bmatrix} C_{\mathfrak{L}_1}(v_{C_{\mathfrak{L}_1}}) & & & \\ & C_{\mathfrak{L}_2}(v_{C_{\mathfrak{L}_2}}) & & \\ & & \ddots & \\ & & & C_{\mathfrak{L}_\rho}(v_{C_{\mathfrak{L}_\rho}}) \end{bmatrix} \tag{10-68}$$

and

$$L_\mathfrak{J}(i_{L_\mathfrak{J}}) \triangleq \begin{bmatrix} L_{\mathfrak{J}_1}(i_{L_{\mathfrak{J}_1}}) & & & \\ & L_{\mathfrak{J}_2}(i_{L_{\mathfrak{J}_2}}) & & \\ & & \ddots & \\ & & & L_{\mathfrak{J}_\sigma}(i_{L_{\mathfrak{J}_\sigma}}) \end{bmatrix} \tag{10-69}$$

are *diagonal* incremental capacitance and inductance matrices of the cotree capacitors and tree inductors, respectively, and the "dot" denotes the time derivatives of the vectors $i_{L_\mathfrak{J}}$ and $v_{C_\mathfrak{L}}$, respectively. The relationship between these two vectors in terms of $v_{C_\mathfrak{J}}$, $i_{L_\mathfrak{L}}$, $v_{E_\mathfrak{J}}$, and $i_{J_\mathfrak{L}}$ can be extracted from Eq. (10-50) upon setting the submatrices specified in Theorem 10-2 to zero:

$$\begin{bmatrix} v_{C_\mathfrak{L}} \\ i_{L_\mathfrak{J}} \end{bmatrix} = \begin{bmatrix} H_{C_\mathfrak{L}C_\mathfrak{J}} & H_{C_\mathfrak{L}L_\mathfrak{L}} & H_{C_\mathfrak{L}E_\mathfrak{J}} & 0_{C_\mathfrak{L}J_\mathfrak{L}} \\ H_{L_\mathfrak{J}C_\mathfrak{J}} & H_{L_\mathfrak{J}L_\mathfrak{L}} & 0_{L_\mathfrak{J}E_\mathfrak{J}} & H_{L_\mathfrak{J}J_\mathfrak{L}} \end{bmatrix} \begin{bmatrix} v_{C_\mathfrak{J}} \\ i_{L_\mathfrak{L}} \\ v_{E_\mathfrak{J}} \\ i_{J_\mathfrak{L}} \end{bmatrix} \tag{10-70}$$

If we differentiate both sides of Eq. (10-70) with respect to time and substitute the resulting expression for $[\dot{v}_{C_\mathfrak{L}}, \dot{i}_{L_\mathfrak{J}}]^t$ in Eq. (10-67), we obtain the following important relationship:

$$\begin{bmatrix} i_{C_\mathfrak{L}} \\ v_{L_\mathfrak{J}} \end{bmatrix} = \left\{ \begin{bmatrix} C_\mathfrak{L}(v_{C_\mathfrak{L}}) & 0 \\ 0 & L_\mathfrak{J}(i_{L_\mathfrak{J}}) \end{bmatrix} \begin{bmatrix} H_{C_\mathfrak{L}C_\mathfrak{J}} & H_{C_\mathfrak{L}L_\mathfrak{L}} & H_{C_\mathfrak{L}E_\mathfrak{J}} & 0_{C_\mathfrak{L}J_\mathfrak{L}} \\ H_{L_\mathfrak{J}C_\mathfrak{J}} & H_{L_\mathfrak{J}L_\mathfrak{L}} & 0_{L_\mathfrak{J}E_\mathfrak{J}} & H_{L_\mathfrak{J}J_\mathfrak{L}} \end{bmatrix} \right\} \begin{bmatrix} \dot{v}_{C_\mathfrak{J}} \\ \dot{i}_{L_\mathfrak{L}} \\ \dot{v}_{E_\mathfrak{J}} \\ \dot{i}_{J_\mathfrak{L}} \end{bmatrix} \tag{10-71}$$

10-3-5 Step 4: The Homestretch

We are now ready to collect the relationships we have derived so far for the purpose of formulating the normal-form equations for the network in Fig. 10-6. In Section 10-2, we have already seen that the normal-form equations may exist for some choice of state variables, but may not exist for another. Since there are at least two possible state variables that might be chosen for each capacitor (its voltage or charge) and each inductor (its current or flux linkage), there are many possible combinations of state variables for a given network. In this book, we shall consider only two of the most common combinations.

A. State Variable Combination 1: $\boldsymbol{v}_{C_\mathfrak{J}}, \boldsymbol{i}_{L_\mathfrak{L}}$. Let the tree capacitor voltage $\boldsymbol{v}_{C_\mathfrak{J}}$ and the cotree inductor current $\boldsymbol{i}_{L_\mathfrak{L}}$ be the state variables. In this case, it would be necessary to stipulate the following additional assumption:

> *Assumption A*
>
> All tree capacitors are voltage-controlled and all cotree inductors are current-controlled.

Under this assumption, the incremental capacitance of each tree capacitor and the incremental inductance of each cotree inductor are well defined, and we can write, as in Eq. (10-67), the following relationship:

$$\begin{bmatrix} \boldsymbol{i}_{C_\mathfrak{J}} \\ \boldsymbol{v}_{L_\mathfrak{L}} \end{bmatrix} = \begin{bmatrix} \boldsymbol{C}_\mathfrak{J}(\boldsymbol{v}_{C_\mathfrak{J}}) & \boldsymbol{0} \\ \boldsymbol{0} & \boldsymbol{L}_\mathfrak{L}(\boldsymbol{i}_{L_\mathfrak{L}}) \end{bmatrix} \begin{bmatrix} \dot{\boldsymbol{v}}_{C_\mathfrak{J}} \\ \dot{\boldsymbol{i}}_{L_\mathfrak{L}} \end{bmatrix} \tag{10-72}$$

where

$$\boldsymbol{C}_\mathfrak{J}(\boldsymbol{v}_{C_\mathfrak{J}}) \triangleq \begin{bmatrix} C_{\mathfrak{J}_1}(v_{C_{\mathfrak{J}_1}}) & & & & \\ & C_{\mathfrak{J}_2}(v_{C_{\mathfrak{J}_2}}) & & & \\ & & \cdot & & \\ & & & \cdot & \\ & & & & \cdot \\ & & & & C_{\mathfrak{J}_\alpha}(v_{C_{\mathfrak{J}_\alpha}}) \end{bmatrix} \tag{10-73}$$

and

$$\boldsymbol{L}_\mathfrak{L}(\boldsymbol{i}_{L_\mathfrak{L}}) \triangleq \begin{bmatrix} L_{\mathfrak{L}_1}(i_{L_{\mathfrak{L}_1}}) & & & & \\ & L_{\mathfrak{L}_2}(i_{L_{\mathfrak{L}_2}}) & & & \\ & & \cdot & & \\ & & & \cdot & \\ & & & & \cdot \\ & & & & L_{\mathfrak{L}_\beta}(i_{L_{\mathfrak{L}_\beta}}) \end{bmatrix} \tag{10-74}$$

are *diagonal* incremental capacitance and inductance matrices of the tree capacitors and cotree inductors, respectively, α is the number of tree capacitors, and β is the number of cotree inductors. To transform Eq. (10-72) into the normal form, we must express the vectors $\boldsymbol{i}_{C_\mathfrak{J}}$ and $\boldsymbol{v}_{L_\mathfrak{L}}$ on the left side in terms of $\boldsymbol{v}_{C_\mathfrak{J}}, \boldsymbol{i}_{L_\mathfrak{L}}, \boldsymbol{v}_{E_\mathfrak{J}}$, and $\boldsymbol{i}_{J_\mathfrak{L}}$. This relationship has already been obtained in Eq. (10-50):

$$\begin{bmatrix} \boldsymbol{i}_{C_\mathfrak{J}} \\ \boldsymbol{v}_{L_\mathfrak{L}} \end{bmatrix} = \begin{bmatrix} \boldsymbol{H}_{C_\mathfrak{J}C_\mathfrak{J}} & \boldsymbol{H}_{C_\mathfrak{J}L_\mathfrak{L}} & \boldsymbol{H}_{C_\mathfrak{J}E_\mathfrak{J}} & \boldsymbol{H}_{C_\mathfrak{J}J_\mathfrak{L}} \\ \boldsymbol{H}_{L_\mathfrak{L}C_\mathfrak{J}} & \boldsymbol{H}_{L_\mathfrak{L}L_\mathfrak{L}} & \boldsymbol{H}_{L_\mathfrak{L}E_\mathfrak{J}} & \boldsymbol{H}_{L_\mathfrak{L}J_\mathfrak{L}} \end{bmatrix} \begin{bmatrix} \boldsymbol{v}_{C_\mathfrak{J}} \\ \boldsymbol{i}_{L_\mathfrak{L}} \\ \boldsymbol{v}_{E_\mathfrak{J}} \\ \boldsymbol{i}_{J_\mathfrak{L}} \end{bmatrix} + \begin{bmatrix} \boldsymbol{H}_{C_\mathfrak{J}R_\mathfrak{J}} & \boldsymbol{H}_{C_\mathfrak{J}R_\mathfrak{L}} \\ \boldsymbol{H}_{L_\mathfrak{L}R_\mathfrak{J}} & \boldsymbol{H}_{L_\mathfrak{L}R_\mathfrak{L}} \end{bmatrix} \begin{bmatrix} \boldsymbol{v}_{R_\mathfrak{J}} \\ \boldsymbol{i}_{R_\mathfrak{L}} \end{bmatrix}$$

$$+ \begin{bmatrix} \boldsymbol{H}_{C_\mathfrak{J}C_\mathfrak{L}} & \boldsymbol{0}_{C_\mathfrak{J}L_\mathfrak{J}} \\ \boldsymbol{0}_{L_\mathfrak{L}C_\mathfrak{L}} & \boldsymbol{H}_{L_\mathfrak{L}L_\mathfrak{J}} \end{bmatrix} \begin{bmatrix} \boldsymbol{i}_{C_\mathfrak{L}} \\ \boldsymbol{v}_{L_\mathfrak{J}} \end{bmatrix} \tag{10-75}$$

Substituting Eqs. (10-64) and (10-71) into Eq. (10-75), we obtain

$$\begin{aligned}
\begin{bmatrix} i_{C_\mathfrak{J}} \\ v_{L_\mathfrak{L}} \end{bmatrix} &= \begin{bmatrix} H_{C_\mathfrak{J}C_\mathfrak{J}} & H_{C_\mathfrak{J}L_\mathfrak{L}} & H_{C_\mathfrak{J}E_\mathfrak{J}} & H_{C_\mathfrak{J}J_\mathfrak{L}} \\ H_{L_\mathfrak{L}C_\mathfrak{J}} & H_{L_\mathfrak{L}L_\mathfrak{L}} & H_{L_\mathfrak{L}E_\mathfrak{J}} & H_{L_\mathfrak{L}J_\mathfrak{L}} \end{bmatrix} \begin{bmatrix} v_{C_\mathfrak{J}} \\ i_{L_\mathfrak{L}} \\ v_{E_\mathfrak{J}} \\ i_{J_\mathfrak{L}} \end{bmatrix} \\
&+ \begin{bmatrix} H_{C_\mathfrak{J}R_\mathfrak{J}} & H_{C_\mathfrak{J}R_\mathfrak{L}} \\ H_{L_\mathfrak{L}R_\mathfrak{J}} & H_{L_\mathfrak{L}R_\mathfrak{L}} \end{bmatrix} \begin{bmatrix} \hat{v}_{R_\mathfrak{J}}(v_{C_\mathfrak{J}}, i_{L_\mathfrak{L}}, v_{E_\mathfrak{J}}, i_{J_\mathfrak{L}}) \\ \hat{i}_{R_\mathfrak{L}}(v_{C_\mathfrak{J}}, i_{L_\mathfrak{L}}, v_{E_\mathfrak{J}}, i_{J_\mathfrak{L}}) \end{bmatrix} \\
&+ \left\{ \begin{bmatrix} H_{C_\mathfrak{J}C_\mathfrak{L}} & 0_{C_\mathfrak{J}L_\mathfrak{J}} \\ 0_{L_\mathfrak{L}C_\mathfrak{L}} & H_{L_\mathfrak{L}L_\mathfrak{J}} \end{bmatrix} \begin{bmatrix} C_\mathfrak{L}(v_{C_\mathfrak{L}}) & 0 \\ 0 & L_\mathfrak{J}(i_{L_\mathfrak{J}}) \end{bmatrix} \begin{bmatrix} H_{C_\mathfrak{L}C_\mathfrak{J}} & H_{C_\mathfrak{L}L_\mathfrak{L}} \\ H_{L_\mathfrak{J}C_\mathfrak{J}} & H_{L_\mathfrak{J}L_\mathfrak{L}} \end{bmatrix} \right\} \begin{bmatrix} \dot{v}_{C_\mathfrak{J}} \\ \dot{i}_{L_\mathfrak{L}} \end{bmatrix} \\
&+ \left\{ \begin{bmatrix} H_{C_\mathfrak{J}C_\mathfrak{L}} & 0_{C_\mathfrak{J}L_\mathfrak{J}} \\ 0_{L_\mathfrak{L}C_\mathfrak{L}} & H_{L_\mathfrak{L}L_\mathfrak{J}} \end{bmatrix} \begin{bmatrix} C_\mathfrak{L}(v_{C_\mathfrak{L}}) & 0 \\ 0 & L_\mathfrak{J}(i_{L_\mathfrak{J}}) \end{bmatrix} \begin{bmatrix} H_{C_\mathfrak{L}E_\mathfrak{J}} & 0_{C_\mathfrak{L}J_\mathfrak{L}} \\ 0_{L_\mathfrak{J}E_\mathfrak{J}} & H_{L_\mathfrak{J}J_\mathfrak{L}} \end{bmatrix} \right\} \begin{bmatrix} \dot{v}_{E_\mathfrak{J}} \\ \dot{i}_{J_\mathfrak{L}} \end{bmatrix}
\end{aligned} \tag{10-76}$$

Substituting Eq. (10-76) into Eq. (10-72) and rearranging terms, we obtain

$$\begin{aligned}
&\left\{ \begin{bmatrix} C_\mathfrak{J}(v_{C_\mathfrak{J}}) & 0 \\ 0 & L_\mathfrak{L}(i_{L_\mathfrak{L}}) \end{bmatrix} - \begin{bmatrix} H_{C_\mathfrak{J}C_\mathfrak{L}} & 0_{C_\mathfrak{J}L_\mathfrak{J}} \\ 0_{L_\mathfrak{L}C_\mathfrak{L}} & H_{L_\mathfrak{L}L_\mathfrak{J}} \end{bmatrix} \begin{bmatrix} C_\mathfrak{L}(v_{C_\mathfrak{L}}) & 0 \\ 0 & L_\mathfrak{J}(i_{L_\mathfrak{J}}) \end{bmatrix} \begin{bmatrix} H_{C_\mathfrak{L}C_\mathfrak{J}} & H_{C_\mathfrak{L}L_\mathfrak{L}} \\ H_{L_\mathfrak{J}C_\mathfrak{J}} & H_{L_\mathfrak{J}L_\mathfrak{L}} \end{bmatrix} \right\} \begin{bmatrix} \dot{v}_{C_\mathfrak{J}} \\ \dot{i}_{L_\mathfrak{L}} \end{bmatrix} \\
&= \begin{bmatrix} H_{C_\mathfrak{J}C_\mathfrak{J}} & H_{C_\mathfrak{J}L_\mathfrak{L}} & H_{C_\mathfrak{J}E_\mathfrak{J}} & H_{C_\mathfrak{J}J_\mathfrak{L}} \\ H_{L_\mathfrak{L}C_\mathfrak{J}} & H_{L_\mathfrak{L}L_\mathfrak{L}} & H_{L_\mathfrak{L}E_\mathfrak{J}} & H_{L_\mathfrak{L}J_\mathfrak{L}} \end{bmatrix} \begin{bmatrix} v_{C_\mathfrak{J}} \\ i_{L_\mathfrak{L}} \\ v_{E_\mathfrak{J}} \\ i_{J_\mathfrak{L}} \end{bmatrix} \\
&+ \begin{bmatrix} H_{C_\mathfrak{J}R_\mathfrak{J}} & H_{C_\mathfrak{J}R_\mathfrak{L}} \\ H_{L_\mathfrak{L}R_\mathfrak{J}} & H_{L_\mathfrak{L}R_\mathfrak{L}} \end{bmatrix} \begin{bmatrix} \hat{v}_{R_\mathfrak{J}}(v_{C_\mathfrak{J}}, i_{L_\mathfrak{L}}, v_{E_\mathfrak{J}}, i_{J_\mathfrak{L}}) \\ \hat{i}_{R_\mathfrak{L}}(v_{C_\mathfrak{J}}, i_{L_\mathfrak{L}}, v_{E_\mathfrak{J}}, i_{J_\mathfrak{L}}) \end{bmatrix} \\
&+ \left\{ \begin{bmatrix} H_{C_\mathfrak{J}C_\mathfrak{L}} & 0_{C_\mathfrak{J}L_\mathfrak{J}} \\ 0_{L_\mathfrak{L}C_\mathfrak{L}} & H_{L_\mathfrak{L}L_\mathfrak{J}} \end{bmatrix} \begin{bmatrix} C_\mathfrak{L}(v_{C_\mathfrak{L}}) & 0 \\ 0 & L_\mathfrak{J}(i_{L_\mathfrak{J}}) \end{bmatrix} \begin{bmatrix} H_{C_\mathfrak{L}E_\mathfrak{J}} & 0_{C_\mathfrak{L}J_\mathfrak{L}} \\ 0_{L_\mathfrak{J}E_\mathfrak{J}} & H_{L_\mathfrak{J}J_\mathfrak{L}} \end{bmatrix} \right\} \begin{bmatrix} \dot{v}_{E_\mathfrak{J}} \\ \dot{i}_{J_\mathfrak{L}} \end{bmatrix}
\end{aligned} \tag{10-77}$$

Finally, the normal-form equations for the network in Fig. 10-6 in terms of the state variables $v_{C_\mathfrak{J}}$ and $i_{L_\mathfrak{L}}$ are obtained from Eq. (10-77) as follows:

$$\begin{aligned}
&\begin{bmatrix} \dot{v}_{C_\mathfrak{J}} \\ \dot{i}_{L_\mathfrak{L}} \end{bmatrix} \\
&= \left\{ \begin{bmatrix} C_\mathfrak{J}(v_{C_\mathfrak{J}}) & 0 \\ 0 & L_\mathfrak{L}(i_{L_\mathfrak{L}}) \end{bmatrix} - \begin{bmatrix} H_{C_\mathfrak{J}C_\mathfrak{L}} & 0_{C_\mathfrak{J}L_\mathfrak{J}} \\ 0_{L_\mathfrak{L}C_\mathfrak{L}} & H_{L_\mathfrak{L}L_\mathfrak{J}} \end{bmatrix} \begin{bmatrix} C_\mathfrak{L}(v_{C_\mathfrak{L}}) & 0 \\ 0 & L_\mathfrak{J}(i_{L_\mathfrak{J}}) \end{bmatrix} \begin{bmatrix} H_{C_\mathfrak{L}C_\mathfrak{J}} & H_{C_\mathfrak{L}L_\mathfrak{L}} \\ H_{L_\mathfrak{J}C_\mathfrak{J}} & H_{L_\mathfrak{J}L_\mathfrak{L}} \end{bmatrix} \right\}^{-1} \\
&\left\{ \begin{bmatrix} H_{C_\mathfrak{J}C_\mathfrak{J}} & H_{C_\mathfrak{J}L_\mathfrak{L}} \\ H_{L_\mathfrak{L}C_\mathfrak{J}} & H_{L_\mathfrak{L}L_\mathfrak{L}} \end{bmatrix} \begin{bmatrix} v_{C_\mathfrak{J}} \\ i_{L_\mathfrak{L}} \end{bmatrix} + \begin{bmatrix} H_{C_\mathfrak{J}R_\mathfrak{J}} & H_{C_\mathfrak{J}R_\mathfrak{L}} \\ H_{L_\mathfrak{L}R_\mathfrak{J}} & H_{L_\mathfrak{L}R_\mathfrak{L}} \end{bmatrix} \begin{bmatrix} \hat{v}_{R_\mathfrak{J}}(v_{C_\mathfrak{J}}, i_{L_\mathfrak{L}}, v_{E_\mathfrak{J}}, i_{J_\mathfrak{L}}) \\ \hat{i}_{R_\mathfrak{L}}(v_{C_\mathfrak{J}}, i_{L_\mathfrak{L}}, v_{E_\mathfrak{J}}, i_{J_\mathfrak{L}}) \end{bmatrix} \right. \\
&+ \begin{bmatrix} H_{C_\mathfrak{J}E_\mathfrak{J}} & H_{C_\mathfrak{J}J_\mathfrak{L}} \\ H_{L_\mathfrak{L}E_\mathfrak{J}} & H_{L_\mathfrak{L}J_\mathfrak{L}} \end{bmatrix} \begin{bmatrix} v_{E_\mathfrak{J}} \\ i_{J_\mathfrak{L}} \end{bmatrix} \\
&\left. + \begin{bmatrix} H_{C_\mathfrak{J}C_\mathfrak{L}} & 0_{C_\mathfrak{J}L_\mathfrak{J}} \\ 0_{L_\mathfrak{L}C_\mathfrak{L}} & H_{L_\mathfrak{L}L_\mathfrak{J}} \end{bmatrix} \begin{bmatrix} C_\mathfrak{L}(v_{C_\mathfrak{L}}) & 0 \\ 0 & L_\mathfrak{J}(i_{L_\mathfrak{J}}) \end{bmatrix} \begin{bmatrix} H_{C_\mathfrak{L}E_\mathfrak{J}} & 0_{C_\mathfrak{L}J_\mathfrak{L}} \\ 0_{L_\mathfrak{J}E_\mathfrak{J}} & H_{L_\mathfrak{J}J_\mathfrak{L}} \end{bmatrix} \begin{bmatrix} \dot{v}_{E_\mathfrak{J}} \\ \dot{i}_{J_\mathfrak{L}} \end{bmatrix} \right\}
\end{aligned} \tag{10-78}$$

where the two terms $\hat{\boldsymbol{v}}_{R_{\mathfrak{Z}}}(\boldsymbol{v}_{C_{\mathfrak{Z}}}, \boldsymbol{i}_{L_{\mathfrak{L}}}, \boldsymbol{v}_{E_{\mathfrak{Z}}}, \boldsymbol{i}_{J_{\mathfrak{L}}})$ and $\hat{\boldsymbol{i}}_{R_{\mathfrak{L}}}(\boldsymbol{v}_{C_{\mathfrak{Z}}}, \boldsymbol{i}_{L_{\mathfrak{L}}}, \boldsymbol{v}_{E_{\mathfrak{Z}}}, \boldsymbol{i}_{J_{\mathfrak{L}}})$ represent the solution of the *resistive nonlinear* subnetwork defined earlier by Eq. (10-64), and where the incremental capacitance and inductance matrices are redefined as follows for handy reference [observe that $\boldsymbol{v}_{C_{\mathfrak{L}}}$ and $\boldsymbol{i}_{L_{\mathfrak{Z}}}$ in Eq. (10-78) are defined in terms of the state variables $\boldsymbol{v}_{C_{\mathfrak{Z}}}$ and $\boldsymbol{i}_{L_{\mathfrak{L}}}$ in Eq. (10-70)]:

$$\boldsymbol{C}_{\mathfrak{Z}}(\boldsymbol{v}_{C_{\mathfrak{Z}}}) \triangleq \begin{bmatrix} \dfrac{\partial \hat{q}_{C_{\mathfrak{Z}_1}}(v_{C_{\mathfrak{Z}_1}})}{\partial v_{C_{\mathfrak{Z}_1}}} & & & \\ & \dfrac{\partial \hat{q}_{C_{\mathfrak{Z}_2}}(v_{C_{\mathfrak{Z}_2}})}{\partial v_{C_{\mathfrak{Z}_2}}} & & \\ & & \ddots & \\ & & & \dfrac{\partial \hat{q}_{C_{\mathfrak{Z}_\alpha}}(v_{C_{\mathfrak{Z}_\alpha}})}{\partial v_{C_{\mathfrak{Z}_\alpha}}} \end{bmatrix} \tag{10-79}$$

$$\boldsymbol{C}_{\mathfrak{L}}(\boldsymbol{v}_{C_{\mathfrak{L}}}) \triangleq \begin{bmatrix} \dfrac{\partial \hat{q}_{C_{\mathfrak{L}_1}}(v_{C_{\mathfrak{L}_1}})}{\partial v_{C_{\mathfrak{L}_1}}} & & & \\ & \dfrac{\partial \hat{q}_{C_{\mathfrak{L}_2}}(v_{C_{\mathfrak{L}_2}})}{\partial v_{C_{\mathfrak{L}_2}}} & & \\ & & \ddots & \\ & & & \dfrac{\partial \hat{q}_{C_{\mathfrak{L}_\rho}}(v_{C_{\mathfrak{L}_\rho}})}{\partial v_{C_{\mathfrak{L}_\rho}}} \end{bmatrix} \tag{10-80}$$

$$\boldsymbol{L}_{\mathfrak{L}}(\boldsymbol{i}_{L_{\mathfrak{L}}}) \triangleq \begin{bmatrix} \dfrac{\partial \hat{\varphi}_{L_{\mathfrak{L}_1}}(i_{L_{\mathfrak{L}_1}})}{\partial i_{L_{\mathfrak{L}_1}}} & & & \\ & \dfrac{\partial \hat{\varphi}_{L_{\mathfrak{L}_2}}(i_{L_{\mathfrak{L}_2}})}{\partial i_{L_{\mathfrak{L}_2}}} & & \\ & & \ddots & \\ & & & \dfrac{\partial \hat{\varphi}_{L_{\mathfrak{L}_\beta}}(i_{L_{\mathfrak{L}_\beta}})}{\partial i_{L_{\mathfrak{L}_\beta}}} \end{bmatrix} \tag{10-81}$$

$$\boldsymbol{L}_{\mathfrak{Z}}(\boldsymbol{i}_{L_{\mathfrak{Z}}}) \triangleq \begin{bmatrix} \dfrac{\partial \hat{\varphi}_{L_{\mathfrak{Z}_1}}(i_{L_{\mathfrak{Z}_1}})}{\partial i_{L_{\mathfrak{Z}_1}}} & & & \\ & \dfrac{\partial \hat{\varphi}_{L_{\mathfrak{Z}_2}}(i_{L_{\mathfrak{Z}_2}})}{\partial i_{L_{\mathfrak{Z}_2}}} & & \\ & & \ddots & \\ & & & \dfrac{\partial \hat{\varphi}_{L_{\mathfrak{Z}_\sigma}}(i_{L_{\mathfrak{L}_\sigma}})}{\partial i_{L_{\mathfrak{Z}_\sigma}}} \end{bmatrix} \tag{10-82}$$

Observe that the incremental capacitance and inductance matrices defined in Eqs. (10-79)–(10-82) are all *diagonal* matrices, because we have so far assumed for convenience that there are no *mutual couplings* among the capacitors or the inductors. If one or more mutual couplings are present, as in a pair of mutual inductances, we could easily take care of this situation by either including the mutual couplings with the help of controlled sources, as shown in Fig. 10-10, or by including the mutual couplings into the incremental matrices. In the latter case, the only modification needed in Eq. (10-78) is that the incremental matrices defined are no longer diagonal, and additional off-diagonal coupling terms must be introduced to take care of the mutual couplings.

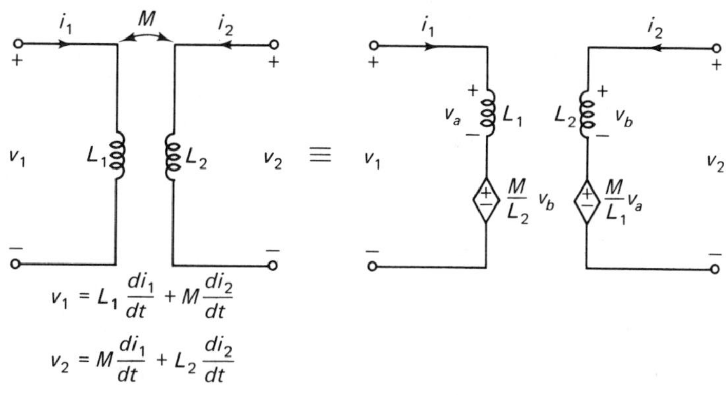

Fig. 10-10. Equivalent circuit for representing a pair of mutual inductances in terms of a pair of self-inductances and a pair of linear controlled sources.

B. STATE VARIABLE COMBINATION 2: $\mathbf{q}_{C_{\mathfrak{J}}}$, $\boldsymbol{\varphi}_{L_{\mathfrak{L}}}$. Consider now the case when the charge $\mathbf{q}_{C_{\mathfrak{J}}}$ of the tree capacitors and the flux linkage $\boldsymbol{\varphi}_{L_{\mathfrak{L}}}$ of the cotree inductors are chosen to be the state variables. In this case, it would be necessary to stipulate the following additional assumption:

> *Assumption B*
>
> All tree capacitors are charge-controlled and all cotree inductors are flux-controlled.

Under this assumption, we can write

$$v_{C_{\mathfrak{J}_j}} = \hat{v}_{C_{\mathfrak{J}_j}}(q_{C_{\mathfrak{J}_j}}), \qquad j = 1, 2, \ldots, \alpha \tag{10-83}$$

$$i_{L_{\mathfrak{L}_j}} = \hat{i}_{L_{\mathfrak{L}_j}}(\varphi_{L_{\mathfrak{L}_j}}), \qquad j = 1, 2, \ldots, \beta \tag{10-84}$$

Let us rewrite Eqs. (10-83) and (10-84) into the following compact vector form:

$$\begin{bmatrix} \mathbf{v}_{C_{\mathfrak{J}}} \\ \mathbf{i}_{L_{\mathfrak{L}}} \end{bmatrix} = \begin{bmatrix} \hat{\mathbf{v}}_{C_{\mathfrak{J}}}(\mathbf{q}_{C_{\mathfrak{J}}}) \\ \hat{\mathbf{i}}_{L_{\mathfrak{L}}}(\boldsymbol{\varphi}_{L_{\mathfrak{L}}}) \end{bmatrix} \tag{10-85}$$

If we differentiate Eq. (10-85) with respect to time, we obtain with the help of the chain rule the following relationship:

$$\begin{bmatrix} \dot{\boldsymbol{v}}_{C_{\mathfrak{T}}} \\ \dot{\boldsymbol{i}}_{L_{\mathfrak{L}}} \end{bmatrix} = \begin{bmatrix} \dfrac{\partial \hat{\boldsymbol{v}}_{C_{\mathfrak{T}}}(\boldsymbol{q}_{C_{\mathfrak{T}}})}{\partial \boldsymbol{q}_{C_{\mathfrak{T}}}} & \boldsymbol{0} \\ \boldsymbol{0} & \dfrac{\partial \hat{\boldsymbol{i}}_{L_{\mathfrak{L}}}(\boldsymbol{\varphi}_{L_{\mathfrak{L}}})}{\partial \boldsymbol{\varphi}_{L_{\mathfrak{L}}}} \end{bmatrix} \begin{bmatrix} \dot{\boldsymbol{q}}_{C_{\mathfrak{T}}} \\ \dot{\boldsymbol{\varphi}}_{L_{\mathfrak{L}}} \end{bmatrix} \tag{10-86}$$

where $\partial \hat{\boldsymbol{v}}_{C_{\mathfrak{T}}}(\boldsymbol{q}_{C_{\mathfrak{T}}})/\partial \boldsymbol{q}_{C_{\mathfrak{T}}}$ is a diagonal matrix whose jjth element is $\partial \hat{v}_{C_{\mathfrak{T}_j}}(q_{C_{\mathfrak{T}_j}})/\partial q_{C_{\mathfrak{T}_j}}$, and $\partial \hat{\boldsymbol{i}}_{L_{\mathfrak{L}}}(\boldsymbol{\varphi}_{L_{\mathfrak{L}}})/\partial \boldsymbol{\varphi}_{L_{\mathfrak{L}}}$ is a diagonal matrix whose jjth element is $\partial \hat{i}_{L_{\mathfrak{L}_j}}(\varphi_{L_{\mathfrak{L}_j}})/\partial \varphi_{L_{\mathfrak{L}_j}}$. Equation (10-85) expresses the tree capacitor voltages $\boldsymbol{v}_{C_{\mathfrak{T}}}$ and the cotree inductor currents $\boldsymbol{i}_{L_{\mathfrak{L}}}$ in terms of the state variables $\boldsymbol{q}_{C_{\mathfrak{T}}}$ and $\boldsymbol{\varphi}_{L_{\mathfrak{L}}}$. It remains now to do the same thing for the cotree capacitor voltages $\boldsymbol{v}_{C_{\mathfrak{L}}}$ and the tree inductor currents $\boldsymbol{i}_{L_{\mathfrak{T}}}$. This is easily achieved by substituting Eq. (10-85) into Eq. (10-70) to obtain

$$\boldsymbol{v}_{C_{\mathfrak{L}}} = \boldsymbol{H}_{C_{\mathfrak{L}}C_{\mathfrak{T}}}\hat{\boldsymbol{v}}_{C_{\mathfrak{T}}}(\boldsymbol{q}_{C_{\mathfrak{T}}}) + \boldsymbol{H}_{C_{\mathfrak{L}}L_{\mathfrak{L}}}\hat{\boldsymbol{i}}_{L_{\mathfrak{L}}}(\boldsymbol{\varphi}_{L_{\mathfrak{L}}}) + \boldsymbol{H}_{C_{\mathfrak{L}}E_{\mathfrak{T}}}\boldsymbol{v}_{E_{\mathfrak{T}}} \tag{10-87}$$

$$\boldsymbol{i}_{L_{\mathfrak{T}}} = \boldsymbol{H}_{L_{\mathfrak{T}}C_{\mathfrak{T}}}\hat{\boldsymbol{v}}_{C_{\mathfrak{T}}}(\boldsymbol{q}_{C_{\mathfrak{T}}}) + \boldsymbol{H}_{L_{\mathfrak{T}}L_{\mathfrak{L}}}\hat{\boldsymbol{i}}_{L_{\mathfrak{L}}}(\boldsymbol{\varphi}_{L_{\mathfrak{L}}}) + \boldsymbol{H}_{L_{\mathfrak{T}}J_{\mathfrak{L}}}\boldsymbol{i}_{J_{\mathfrak{L}}} \tag{10-88}$$

Since

$$\begin{bmatrix} \dot{\boldsymbol{q}}_{C_{\mathfrak{T}}} \\ \dot{\boldsymbol{\varphi}}_{L_{\mathfrak{L}}} \end{bmatrix} = \begin{bmatrix} \boldsymbol{i}_{C_{\mathfrak{T}}} \\ \boldsymbol{v}_{L_{\mathfrak{L}}} \end{bmatrix} \tag{10-89}$$

we can substitute Eqs. (10-85), (10-86), and (10-89) into Eq. (10-76) and rearrange the terms to solve for $\dot{\boldsymbol{q}}_{C_{\mathfrak{T}}}$ and $\dot{\boldsymbol{\varphi}}_{L_{\mathfrak{L}}}$. The resulting expression is the following system of normal-form equations for the network in Fig. 10-6 in terms of the state variables $\boldsymbol{q}_{C_{\mathfrak{T}}}$ and $\boldsymbol{\varphi}_{L_{\mathfrak{L}}}$:

$$\begin{aligned}
\begin{bmatrix} \dot{\boldsymbol{q}}_{C_{\mathfrak{T}}} \\ \dot{\boldsymbol{\varphi}}_{L_{\mathfrak{L}}} \end{bmatrix} = \Bigg\{ & \begin{bmatrix} \boldsymbol{1} & \boldsymbol{0} \\ \boldsymbol{0} & \boldsymbol{1} \end{bmatrix} - \begin{bmatrix} \boldsymbol{H}_{C_{\mathfrak{T}}C_{\mathfrak{L}}} & \boldsymbol{0}_{C_{\mathfrak{T}}L_{\mathfrak{T}}} \\ \boldsymbol{0}_{L_{\mathfrak{L}}C_{\mathfrak{L}}} & \boldsymbol{H}_{L_{\mathfrak{L}}L_{\mathfrak{T}}} \end{bmatrix} \begin{bmatrix} \boldsymbol{C}_{\mathfrak{L}}(\boldsymbol{v}_{C_{\mathfrak{L}}}) & \boldsymbol{0} \\ \boldsymbol{0} & \boldsymbol{L}_{\mathfrak{T}}(\boldsymbol{i}_{L_{\mathfrak{T}}}) \end{bmatrix} \begin{bmatrix} \boldsymbol{H}_{C_{\mathfrak{L}}C_{\mathfrak{T}}} & \boldsymbol{H}_{C_{\mathfrak{L}}L_{\mathfrak{L}}} \\ \boldsymbol{H}_{L_{\mathfrak{T}}C_{\mathfrak{T}}} & \boldsymbol{H}_{L_{\mathfrak{T}}L_{\mathfrak{L}}} \end{bmatrix} \\
& \begin{bmatrix} \boldsymbol{S}_{\mathfrak{T}}(\boldsymbol{q}_{C_{\mathfrak{T}}}) & \boldsymbol{0} \\ \boldsymbol{0} & \boldsymbol{\Gamma}_{\mathfrak{L}}(\boldsymbol{\varphi}_{L_{\mathfrak{L}}}) \end{bmatrix} \Bigg\}^{-1} \Bigg\{ \begin{bmatrix} \boldsymbol{H}_{C_{\mathfrak{T}}C_{\mathfrak{T}}} & \boldsymbol{H}_{C_{\mathfrak{T}}L_{\mathfrak{L}}} \\ \boldsymbol{H}_{L_{\mathfrak{L}}C_{\mathfrak{T}}} & \boldsymbol{H}_{L_{\mathfrak{L}}L_{\mathfrak{L}}} \end{bmatrix} \begin{bmatrix} \hat{\boldsymbol{v}}_{C_{\mathfrak{T}}}(\boldsymbol{q}_{C_{\mathfrak{T}}}) \\ \hat{\boldsymbol{i}}_{L_{\mathfrak{L}}}(\boldsymbol{\varphi}_{L_{\mathfrak{L}}}) \end{bmatrix} \\
& + \begin{bmatrix} \boldsymbol{H}_{C_{\mathfrak{T}}R_{\mathfrak{T}}} & \boldsymbol{H}_{C_{\mathfrak{T}}R_{\mathfrak{L}}} \\ \boldsymbol{H}_{L_{\mathfrak{L}}R_{\mathfrak{T}}} & \boldsymbol{H}_{L_{\mathfrak{L}}R_{\mathfrak{L}}} \end{bmatrix} \begin{bmatrix} \hat{\boldsymbol{v}}_{R_{\mathfrak{T}}}(\boldsymbol{v}_{C_{\mathfrak{T}}}(\boldsymbol{q}_{C_{\mathfrak{T}}}), \boldsymbol{i}_{L_{\mathfrak{L}}}(\boldsymbol{\varphi}_{L_{\mathfrak{L}}}), \boldsymbol{v}_{E_{\mathfrak{T}}}, \boldsymbol{i}_{J_{\mathfrak{L}}}) \\ \hat{\boldsymbol{i}}_{R_{\mathfrak{L}}}(\boldsymbol{v}_{C_{\mathfrak{T}}}(\boldsymbol{q}_{C_{\mathfrak{T}}}), \boldsymbol{i}_{L_{\mathfrak{L}}}(\boldsymbol{\varphi}_{L_{\mathfrak{L}}}), \boldsymbol{v}_{E_{\mathfrak{T}}}, \boldsymbol{i}_{J_{\mathfrak{L}}}) \end{bmatrix} \\
& + \begin{bmatrix} \boldsymbol{H}_{C_{\mathfrak{T}}E_{\mathfrak{T}}} & \boldsymbol{H}_{C_{\mathfrak{T}}J_{\mathfrak{L}}} \\ \boldsymbol{H}_{L_{\mathfrak{L}}E_{\mathfrak{T}}} & \boldsymbol{H}_{L_{\mathfrak{L}}J_{\mathfrak{L}}} \end{bmatrix} \begin{bmatrix} \boldsymbol{v}_{E_{\mathfrak{T}}} \\ \boldsymbol{i}_{J_{\mathfrak{L}}} \end{bmatrix} \\
& + \begin{bmatrix} \boldsymbol{H}_{C_{\mathfrak{T}}C_{\mathfrak{L}}} & \boldsymbol{0}_{C_{\mathfrak{T}}L_{\mathfrak{T}}} \\ \boldsymbol{0}_{L_{\mathfrak{L}}C_{\mathfrak{L}}} & \boldsymbol{H}_{L_{\mathfrak{L}}L_{\mathfrak{T}}} \end{bmatrix} \begin{bmatrix} \boldsymbol{C}_{\mathfrak{L}}(\boldsymbol{v}_{C_{\mathfrak{L}}}) & \boldsymbol{0} \\ \boldsymbol{0} & \boldsymbol{L}_{\mathfrak{T}}(\boldsymbol{i}_{L_{\mathfrak{T}}}) \end{bmatrix} \begin{bmatrix} \boldsymbol{H}_{C_{\mathfrak{L}}E_{\mathfrak{T}}} & \boldsymbol{0}_{C_{\mathfrak{L}}J_{\mathfrak{L}}} \\ \boldsymbol{0}_{L_{\mathfrak{T}}E_{\mathfrak{T}}} & \boldsymbol{H}_{L_{\mathfrak{T}}J_{\mathfrak{L}}} \end{bmatrix} \begin{bmatrix} \dot{\boldsymbol{v}}_{E_{\mathfrak{T}}} \\ \dot{\boldsymbol{i}}_{J_{\mathfrak{L}}} \end{bmatrix} \Bigg\}
\end{aligned} \tag{10-90}$$

where the various incremental inverse capacitance and inductance matrices are redefined here for handy reference:

$$\boldsymbol{S}_{\mathfrak{J}}(\boldsymbol{q}_{C_{\mathfrak{J}}}) \triangleq \begin{bmatrix} \frac{\partial \hat{v}_{C_{\mathfrak{J}_1}}(q_{C_{\mathfrak{J}_1}})}{\partial q_{C_{\mathfrak{J}_1}}} & & & \\ & \frac{\partial \hat{v}_{C_{\mathfrak{J}_2}}(q_{C_{\mathfrak{J}_2}})}{\partial q_{C_{\mathfrak{J}_2}}} & & \\ & & \ddots & \\ & & & \frac{\partial \hat{v}_{C_{\mathfrak{J}_\alpha}}(q_{C_{\mathfrak{J}_\alpha}})}{\partial q_{C_{\mathfrak{J}_\alpha}}} \end{bmatrix} \tag{10-91}$$

$$\boldsymbol{\Gamma}_{\mathfrak{L}}(\boldsymbol{\varphi}_{L_{\mathfrak{L}}}) \triangleq \begin{bmatrix} \frac{\partial \hat{i}_{L_{\mathfrak{L}_1}}(i_{L_{\mathfrak{L}_1}})}{\partial \varphi_{L_{\mathfrak{L}_1}}} & & & \\ & \frac{\partial \hat{i}_{L_{\mathfrak{L}_2}}(i_{L_{\mathfrak{L}_2}})}{\partial \varphi_{L_{\mathfrak{L}_2}}} & & \\ & & \ddots & \\ & & & \frac{\partial \hat{i}_{L_{\mathfrak{L}_\beta}}(i_{L_{\mathfrak{L}_\beta}})}{\partial \varphi_{L_{\mathfrak{L}_\beta}}} \end{bmatrix} \tag{10-92}$$

$$\boldsymbol{C}_{\mathfrak{L}}(\boldsymbol{v}_{C_{\mathfrak{L}}}) \triangleq \begin{bmatrix} \frac{\partial \hat{q}_{C_{\mathfrak{L}_1}}(v_{C_{\mathfrak{L}_1}})}{\partial v_{C_{\mathfrak{L}_1}}} & & & \\ & \frac{\partial \hat{q}_{C_{\mathfrak{L}_2}}(v_{C_{\mathfrak{L}_2}})}{\partial v_{C_{\mathfrak{L}_2}}} & & \\ & & \ddots & \\ & & & \frac{\partial \hat{q}_{C_{\mathfrak{L}_\rho}}(v_{C_{\mathfrak{L}_\rho}})}{\partial v_{C_{\mathfrak{L}_\rho}}} \end{bmatrix} \tag{10-93}$$

where $v_{C_{\mathfrak{L}_1}}, v_{C_{\mathfrak{L}_2}}, \ldots, v_{C_{\mathfrak{L}_\rho}}$ are in turn defined in terms of the state variables $\boldsymbol{q}_{C_{\mathfrak{J}}}$ and $\boldsymbol{\varphi}_{L_{\mathfrak{L}}}$ in Eq. (10-87),

$$\boldsymbol{L}_{\mathfrak{J}}(\boldsymbol{i}_{L_{\mathfrak{J}}}) \triangleq \begin{bmatrix} \frac{\partial \hat{\varphi}_{L_{\mathfrak{J}_1}}(i_{L_{\mathfrak{J}_1}})}{\partial i_{L_{\mathfrak{J}_1}}} & & & \\ & \frac{\partial \hat{\varphi}_{L_{\mathfrak{J}_2}}(i_{L_{\mathfrak{J}_2}})}{\partial i_{L_{\mathfrak{J}_2}}} & & \\ & & \ddots & \\ & & & \frac{\partial \hat{\varphi}_{L_{\mathfrak{J}_\sigma}}(i_{L_{\mathfrak{J}_\sigma}})}{\partial i_{L_{\mathfrak{J}_\sigma}}} \end{bmatrix} \tag{10-94}$$

where $i_{L_{\mathfrak{T}_1}}, i_{L_{\mathfrak{T}_2}}, \ldots, i_{L_{\mathfrak{T}_\sigma}}$ are in turn defined in terms of the state variables $\boldsymbol{q}_{C_\mathfrak{T}}$ and $\boldsymbol{\varphi}_{L_\mathfrak{L}}$ in Eq. (10-88). These are diagonal matrices if there are no mutual couplings among the capacitors or the inductors. Our earlier remarks concerning the modification needed to allow mutual couplings also apply here.

The only terms left that may need some clarification are the tree resistor voltages $\hat{\boldsymbol{v}}_{R_\mathfrak{T}}$ and the cotree resistor currents $\hat{\boldsymbol{i}}_{R_\mathfrak{L}}$ in Eq. (10-90). These two terms represent the solutions of the *resistive nonlinear subnetwork* defined earlier by Eq. (10-64), where the tree capacitor voltages $\boldsymbol{v}_{C_\mathfrak{T}}$ are defined in terms of $\hat{\boldsymbol{v}}_{C_\mathfrak{T}}(\boldsymbol{q}_{C_\mathfrak{T}})$ and the cotree inductor currents $\boldsymbol{i}_{L_\mathfrak{L}}$ are defined in terms of $\hat{\boldsymbol{i}}_{L_\mathfrak{L}}(\boldsymbol{\varphi}_{L_\mathfrak{L}})$. Computationally, this means that given any value of the state variables $(\boldsymbol{q}_{C_\mathfrak{T}}, \boldsymbol{\varphi}_{L_\mathfrak{L}})$, we compute $\boldsymbol{v}_{C_\mathfrak{T}} = \hat{\boldsymbol{v}}_{C_\mathfrak{T}}(\boldsymbol{q}_{C_\mathfrak{T}})$ and $\boldsymbol{i}_{L_\mathfrak{L}} = \hat{\boldsymbol{i}}_{L_\mathfrak{L}}(\boldsymbol{\varphi}_{L_\mathfrak{L}})$. These numbers are then assigned to be the terminal voltages for the tree capacitor voltage sources and the terminal currents for the cotree inductor current sources in Fig. 10-9. The resulting solutions for $\boldsymbol{v}_{R_\mathfrak{T}}$ and $\boldsymbol{i}_{L_\mathfrak{L}}$ then give us the value of $\hat{\boldsymbol{v}}_{R_\mathfrak{T}}$ and $\hat{\boldsymbol{i}}_{L_\mathfrak{L}}$ required by Eq. (10-90). To emphasize the dependence of these two variables upon the state variables $\boldsymbol{q}_{C_\mathfrak{T}}$ and $\boldsymbol{\varphi}_{L_\mathfrak{L}}$ as well as upon the independent voltage sources $\boldsymbol{v}_{E_\mathfrak{T}}$ and the independent current sources $\boldsymbol{i}_{J_\mathfrak{L}}$, we write symbolically

$$\boldsymbol{v}_{R_\mathfrak{T}} = \hat{\boldsymbol{v}}_{R_\mathfrak{T}}(\hat{\boldsymbol{v}}_{C_\mathfrak{T}}(\boldsymbol{q}_{C_\mathfrak{T}}), \hat{\boldsymbol{i}}_{L_\mathfrak{L}}(\boldsymbol{\varphi}_{L_\mathfrak{L}}), \boldsymbol{v}_{E_\mathfrak{T}}, \boldsymbol{i}_{J_\mathfrak{L}}) \tag{10-95}$$

$$\boldsymbol{i}_{R_\mathfrak{L}} = \hat{\boldsymbol{i}}_{R_\mathfrak{L}}(\hat{\boldsymbol{v}}_{C_\mathfrak{T}}(\boldsymbol{q}_{C_\mathfrak{T}}), \hat{\boldsymbol{i}}_{L_\mathfrak{L}}(\boldsymbol{\varphi}_{L_\mathfrak{L}}), \boldsymbol{v}_{E_\mathfrak{T}}, \boldsymbol{i}_{J_\mathfrak{L}}) \tag{10-96}$$

10-3-6 Topological Equations for Determining the Nonstate Variables

The state equations given by Eq. (10-78) or Eq. (10-90) can be solved *numerically* for $\boldsymbol{v}_{C_\mathfrak{T}}(t)$ and $\boldsymbol{i}_{L_\mathfrak{L}}(t)$ for $t \geq t_0$, once the *initial states* $\boldsymbol{v}_{C_\mathfrak{T}}(t_0)$ and $\boldsymbol{i}_{L_\mathfrak{L}}(t_0)$ are specified. The numerical methods for solving these equations are given in Chapters 11, 12, and 13. Once the solutions $\boldsymbol{v}_{C_\mathfrak{T}}(t)$ and $\boldsymbol{i}_{L_\mathfrak{L}}(t)$ are obtained, it is a trivial matter to obtain the voltage or current of *any* element in the network. For example, the *nonstate* variables $\boldsymbol{i}_{R_\mathfrak{T}}(t)$ and $\boldsymbol{v}_{R_\mathfrak{L}}(t)$ associated with the *nonlinear resistors* can be obtained from Eq. (10-59) by *direct substitution*. The corresponding variables $\boldsymbol{v}_{R_\mathfrak{T}}(t)$ and $\boldsymbol{i}_{R_\mathfrak{L}}(t)$ are then obtained from their respective constitutive relations. Similarly, the *nonstate* variables $\boldsymbol{v}_{C_\mathfrak{L}}$ and $\boldsymbol{i}_{L_\mathfrak{T}}$ can be obtained from Eq. (10-70) by *direct substitution*. The corresponding variables $\boldsymbol{i}_{C_\mathfrak{L}}$ and $\boldsymbol{v}_{L_\mathfrak{T}}$ are then obtained from Eq. (10-67). Likewise, the variable $\boldsymbol{i}_{C_\mathfrak{T}}(t)$ and $\boldsymbol{v}_{L_\mathfrak{L}}(t)$ can be obtained from Eq. (10-72). Finally, the voltages and currents associated with the *linear resistors* can be obtained by *direct substitution* into the appropriate topological formulas presented in Chapter 6 for deriving the hybrid matrix $\boldsymbol{H}$.

If we let $\boldsymbol{y}$ denote the *nonstate* variables, the preceding algorithm shows that $\boldsymbol{y}$ can be obtained by *direct substitution* once the *state variables* $\boldsymbol{v}_{C_\mathfrak{T}}(t)$ and $\boldsymbol{i}_{L_\mathfrak{L}}(t)$ and the *independent* sources $\boldsymbol{v}_{E_\mathfrak{T}}(t)$ and $\boldsymbol{i}_{J_\mathfrak{L}}(t)$ are given. Hence, we can write symbolically

$$\boxed{\boldsymbol{y} = \hat{\boldsymbol{y}}(\boldsymbol{v}_{C_\mathfrak{T}}, \boldsymbol{i}_{L_\mathfrak{L}}; \boldsymbol{v}_{E_\mathfrak{T}}, \boldsymbol{i}_{J_\mathfrak{L}}; \dot{\boldsymbol{v}}_{E_\mathfrak{T}}, \dot{\boldsymbol{i}}_{J_\mathfrak{L}})} \tag{10-97}$$

Equation (10-97) represents the *output equations*[10] in a state-variable analysis. A

[10]Also known as the *readout function*.

complete analysis of a dynamic nonlinear network using the *state-variable approach* will therefore call for two sets of equations:

$$\dot{x} = f(x, u, \dot{u}) \tag{10-98}$$

$$y = \hat{y}(x, u, \dot{u}) \tag{10-99}$$

where Eq. (10-98) represents either Eq. (10-78) or (10-90), and Eq. (10-99) represents Eq. (10-97). Observe that, unlike most control systems, *the state* and *output equations* of a dynamic nonlinear network may contain terms representing the *time derivative of the forcing functions.*

10-4 FORMULATION OF STATE EQUATIONS FOR NETWORKS CONTAINING NEITHER *C-E* LOOPS NOR *L-J* CUTSETS— THE AD HOC APPROACH

The *explicit* state equations given by either Eq. (10-78) or (10-90) in Section 10-3 are applicable to any dynamic nonlinear network satisfying the standing assumptions stipulated in Section 10-3-1. Networks that violate one or more of these assumptions are rare, if not pathological. Consequently, our preceding state equation formulation algorithm should cover almost all dynamic nonlinear networks of interest. Such broad generality certainly justifies the long sequence of involved algebraic substitutions and manipulations used to derive Eqs. (10-78) and (10-90). It is also refreshing to recognize that, since these equations are already expressed in *explicit* form,[11] the intermediate steps are no longer needed in any analysis, either qualitative or quantitative. In fact, a general computer simulation program could be easily developed by simply coding either Eq. (10-78) or Eq. (10-90).

To demonstrate the generality of Eq. (10-78), consider the special case when the dynamic network contains neither *C-E* loops (loops made up exclusively of capacitors and independent and/or controlled voltage sources) nor *L-J* cutsets (cutsets made up exclusively of inductors and independent and/or controlled current sources). In this case, standing assumptions 2(a), 2(b), and 3(a) are automatically satisfied and Eq. (10-78) simplifies to

$$\begin{bmatrix} \dot{v}_{C_\mathfrak{I}} \\ \dot{i}_{L_\mathfrak{L}} \end{bmatrix} = \begin{bmatrix} C_\mathfrak{I}^{-1}(v_{C_\mathfrak{I}}) & 0 \\ 0 & L_\mathfrak{L}^{-1}(i_{L_\mathfrak{L}}) \end{bmatrix} \left\{ \begin{bmatrix} H_{C_\mathfrak{I}C_\mathfrak{I}} & H_{C_\mathfrak{I}L_\mathfrak{L}} \\ H_{L_\mathfrak{L}C_\mathfrak{I}} & H_{L_\mathfrak{L}L_\mathfrak{L}} \end{bmatrix} \begin{bmatrix} v_{C_\mathfrak{I}} \\ i_{L_\mathfrak{L}} \end{bmatrix} + \begin{bmatrix} H_{C_\mathfrak{I}R_\mathfrak{I}} & H_{C_\mathfrak{I}R_\mathfrak{L}} \\ H_{L_\mathfrak{L}R_\mathfrak{I}} & H_{L_\mathfrak{L}R_\mathfrak{L}} \end{bmatrix} \begin{bmatrix} \hat{v}_{R_\mathfrak{I}}(v_{C_\mathfrak{I}}, i_{L_\mathfrak{L}}, v_{E_\mathfrak{I}}, i_{J_\mathfrak{L}}) \\ \hat{i}_{R_\mathfrak{L}}(v_{C_\mathfrak{I}}, i_{L_\mathfrak{L}}, v_{E_\mathfrak{I}}, i_{J_\mathfrak{L}}) \end{bmatrix} + \begin{bmatrix} H_{C_\mathfrak{I}E_\mathfrak{I}} & H_{C_\mathfrak{I}J_\mathfrak{L}} \\ H_{L_\mathfrak{L}E_\mathfrak{I}} & H_{L_\mathfrak{L}J_\mathfrak{L}} \end{bmatrix} \begin{bmatrix} v_{E_\mathfrak{I}} \\ i_{J_\mathfrak{L}} \end{bmatrix} \right\} \tag{10-100}$$

[11]Equations (10-78) and (10-90) can be programmed *explicitly* since the two terms $\hat{v}_{R_\mathfrak{I}}(v_{C_\mathfrak{I}}, i_{L_\mathfrak{L}}, v_{E_\mathfrak{I}}, i_{J_\mathfrak{L}})$ and $\hat{i}_{R_\mathfrak{L}}(v_{C_\mathfrak{I}}, i_{L_\mathfrak{L}}, v_{E_\mathfrak{I}}, i_{J_\mathfrak{L}})$ representing the resistive subnetwork can be solved *numerically* from Eq. (10-63) at each time step. We assume of course that the constitutive relations of all *nonlinear* resistors are such that a *unique* solution exists. Some sufficient conditions are given in the problems. Observe that Eq. (10-63) is still valid even when there are *mutual couplings*—such as those defining *nonlinear controlled sources*—among the resistors.

where $\hat{v}_{R_{\mathfrak{Z}}}(v_{C_{\mathfrak{Z}}}, i_{L_{\mathfrak{L}}}, v_{E_{\mathfrak{Z}}}, i_{J_{\mathfrak{L}}})$ and $\hat{i}_{R_{\mathfrak{L}}}(v_{C_{\mathfrak{Z}}}, i_{L_{\mathfrak{L}}}, v_{E_{\mathfrak{Z}}}, i_{J_{\mathfrak{L}}})$ are the solution of Eq. (10-63), which we reproduce here for convenience:

$$\begin{bmatrix} g(v_{R_{\mathfrak{Z}}}) \\ \hline f(i_{R_{\mathfrak{L}}}) \end{bmatrix} - \left[\begin{array}{c|c} H_{R_{\mathfrak{Z}}R_{\mathfrak{Z}}} & H_{R_{\mathfrak{Z}}R_{\mathfrak{L}}} \\ \hline H_{R_{\mathfrak{L}}R_{\mathfrak{Z}}} & H_{R_{\mathfrak{L}}R_{\mathfrak{L}}} \end{array}\right] \begin{bmatrix} v_{R_{\mathfrak{Z}}} \\ \hline i_{R_{\mathfrak{L}}} \end{bmatrix} - \begin{bmatrix} s_{R_{\mathfrak{Z}}} \\ \hline s_{R_{\mathfrak{L}}} \end{bmatrix} = \begin{bmatrix} 0 \\ \hline 0 \end{bmatrix} \tag{10-101}$$

If we assume further that all capacitors and inductors are *linear*, Eq. (10-100) reduces to

$$\dot{x} = Ax + B_r u_r + B_s u_s \tag{10-102}$$

where

$$x \triangleq \begin{bmatrix} v_{C_{\mathfrak{Z}}} \\ i_{L_{\mathfrak{L}}} \end{bmatrix}, \quad u_r \triangleq \begin{bmatrix} \hat{v}_{R_{\mathfrak{Z}}}(v_{C_{\mathfrak{Z}}}, i_{L_{\mathfrak{L}}}, v_{E_{\mathfrak{Z}}}, i_{J_{\mathfrak{L}}}) \\ \hat{i}_{R_{\mathfrak{L}}}(v_{C_{\mathfrak{Z}}}, i_{L_{\mathfrak{L}}}, v_{E_{\mathfrak{Z}}}, i_{J_{\mathfrak{L}}}) \end{bmatrix}, \quad u_s \triangleq \begin{bmatrix} v_{E_{\mathfrak{Z}}} \\ i_{J_{\mathfrak{L}}} \end{bmatrix} \tag{10-103}$$

$$A \triangleq \begin{bmatrix} C^{-1} & 0 \\ 0 & L^{-1} \end{bmatrix} \begin{bmatrix} H_{C_{\mathfrak{Z}}C_{\mathfrak{Z}}} & H_{C_{\mathfrak{Z}}L_{\mathfrak{L}}} \\ H_{L_{\mathfrak{L}}C_{\mathfrak{Z}}} & H_{L_{\mathfrak{L}}L_{\mathfrak{L}}} \end{bmatrix} \tag{10-104}$$

$$B_r \triangleq \begin{bmatrix} C^{-1} & 0 \\ 0 & L^{-1} \end{bmatrix} \begin{bmatrix} H_{C_{\mathfrak{Z}}R_{\mathfrak{Z}}} & H_{C_{\mathfrak{Z}}R_{\mathfrak{L}}} \\ H_{L_{\mathfrak{L}}R_{\mathfrak{Z}}} & H_{L_{\mathfrak{L}}R_{\mathfrak{L}}} \end{bmatrix} \tag{10-105}$$

$$B_s \triangleq \begin{bmatrix} C^{-1} & 0 \\ 0 & L^{-1} \end{bmatrix} \begin{bmatrix} H_{C_{\mathfrak{Z}}E_{\mathfrak{Z}}} & H_{C_{\mathfrak{Z}}J_{\mathfrak{L}}} \\ H_{L_{\mathfrak{L}}E_{\mathfrak{Z}}} & H_{L_{\mathfrak{L}}J_{\mathfrak{L}}} \end{bmatrix} \tag{10-106}$$

Observe that, at each time step $t = t_k$, the vector $u_r(t_k)$ can be computed from Eq. (10-101) once $s_{R_{\mathfrak{Z}}}(t_k)$ and $s_{R_{\mathfrak{L}}}(t_k)$, as defined in Eq. (10-61), are given. Hence, once the state variables $v_{C_{\mathfrak{Z}}}(t_k)$ and $i_{L_{\mathfrak{L}}}(t_k)$ and the independent sources $v_{E_{\mathfrak{Z}}}(t_k)$ and $i_{J_{\mathfrak{L}}}(t_k)$ are given at $t = t_k$, $u_r(t_k)$ can be computed and represented as a *constant vector*. In other words, from the *computational* point of view, Eq. (10-102) is equivalent to a system of *linear* state equations of the form

$$\dot{x} = Ax + Bu \tag{10-107}$$

where

$$B \triangleq [B_r \mid B_s] \tag{10-108}$$

and

$$u \triangleq \begin{bmatrix} u_r \\ \hline u_s \end{bmatrix} \tag{10-109}$$

Hence, for the special class of dynamic *nonlinear* network under consideration here, the solution of the associated state equations can be obtained by the same numerical methods presented in Chapter 9 for solving *linear* state equations. In fact, Eq. (10-109) shows that any computer program for formulating *state equations* for *linear networks* can be easily modified to include this class of nonlinear networks by simply extracting all *nonlinear resistors* and representing them as if they were *independent sources* u_r. The only computational difference between u_r and u_s is that the value of $u_r(t_k)$ at any time step $t = t_k$ is obtained by solving a system of *nonlinear algebraic equations*—Eq.

(10-101), whereas the value of $\boldsymbol{u}_s(t_k)$ is obtained by direct substitution of the value of the *independent* sources $\boldsymbol{v}_{E_\mathfrak{J}}(t_k)$ and $\boldsymbol{i}_{J_\mathfrak{L}}(t_k)$ at $t = t_k$.

10-5 CHOICE OF STATE VARIABLES

The preceding developments show that the choice of state variables depends on the nature of the nonlinear elements. For example, we found that *capacitor voltage* must be chosen as the state variable if the capacitor is characterized by a nonmonotonic *voltage-controlled q-v* curve. On the other hand, *capacitor charge* must be chosen as the state variable if the capacitor is characterized by a nonmonotonic *charge-controlled* q-v curve. A dual situation applies to the case of inductors characterized by nonmonotonic φ-i curves. If the capacitors and inductors are characterized by *strictly monotonic* curves, however, the state equation can often be written with either v or q as the state variable associated with capacitors, and with either i or φ as the state variable associated with inductors. In this case, we shall now show that, from the computational point of view, it is advantageous to choose q and φ as the state variables [9].

To simplify notation, let us consider a resistive nonlinear n-port N characterized by

$$\boldsymbol{i} = \hat{\boldsymbol{i}}(\boldsymbol{v}, t) \tag{10-110}$$

where the variable t accounts for the presence of time-dependent sources. Let N be terminated by n *nonlinear* capacitors characterized by strictly monotonic q-v curves. The state equation of the resulting network is then given trivially by

$$\dot{\boldsymbol{v}} = -\boldsymbol{C}^{-1}(\boldsymbol{v})\hat{\boldsymbol{i}}(\boldsymbol{v}, t) \triangleq \boldsymbol{f}(\boldsymbol{v}, t) \tag{10-111}$$

where $\boldsymbol{C}(\boldsymbol{v})$ is an $n \times n$ diagonal matrix whose jjth element is given by the capacitance $C_j(\boldsymbol{v}_j)$ of the jth capacitor. If we evaluate the Jacobian matrix $\boldsymbol{J}_f(\boldsymbol{v}, t)$ of $\boldsymbol{f}(\boldsymbol{v}, t)$ with respect to $\boldsymbol{v}$, we obtain with the help of the *chain rule* the expression

$$\boxed{\boldsymbol{J}_f(\boldsymbol{v}, t) \triangleq \frac{\partial \boldsymbol{f}(\boldsymbol{v}, t)}{\partial \boldsymbol{v}} = \boldsymbol{A}_1(\boldsymbol{v}, t) + \boldsymbol{A}_2(\boldsymbol{v}, t)} \tag{10-112}$$

where

$$\boldsymbol{A}_1(\boldsymbol{v}, t) \triangleq -\boldsymbol{C}^{-1}(\boldsymbol{v})\frac{\partial \hat{\boldsymbol{i}}(\boldsymbol{v}, t)}{\partial \boldsymbol{v}} \tag{10-113}$$

and

$$\boldsymbol{A}_2(\boldsymbol{v}, t) \triangleq \sum_{i=1}^{n}\left\{\frac{\partial}{\partial v_i}[-\boldsymbol{C}^{-1}(\boldsymbol{v})][\hat{\boldsymbol{i}}(\boldsymbol{v}, t)]\boldsymbol{e}_i\right\} \tag{10-114}$$

are $n \times n$ matrices. The matrix $\partial/\partial v_i\,[-\boldsymbol{C}^{-1}(\boldsymbol{v})]$ in Eq. (10-114) is obtained by differentiating each element of $\boldsymbol{C}^{-1}(\boldsymbol{v})$ with respect to v_i; the vector

$$\boldsymbol{e}_i \triangleq [0\,0\ldots0\,1\,0\ldots0] \tag{10-115}$$

denotes a $1 \times n$ unit row vector whose ith element is equal to unity. An examination of Eq. (10-113) shows that $\boldsymbol{A}_1(\boldsymbol{v},t)$ can be interpreted as the Jacobian matrix associated with an *incremental linear network* obtained by replacing all nonlinear elements by their incremental *linear* equivalent circuit about the operating point at time t, and by setting all independent sources to zero. From classical linear circuit theory, we know, for any fixed t, the *eigenvalues* of $A_1(v, t)$ are the natural frequencies of the associated linear network. Hence, if the network contains only *passive* elements and independent sources, all eigenvalues of $A_1(v, t)$ have a negative real part; i.e.,

$$\lambda_k = \alpha_k + j\beta_k, \qquad \alpha_k < 0 \tag{10-116}$$

where $k = 1, 2, \ldots, n$. This property no longer holds, however, for the Jacobian matrix $J_f(v, t)$, as defined in Eq. (10-112), because the presence of the second matrix $A_2(v, t)$ could shift the eigenvalues of $A_1(v, t)$ from the left half-plane into the right half-plane. An example illustrating this mechanism was given in Section 1-8 for the case $n = 1$, where it is shown that such a situation is undesirable because any initial error tends to get amplified when the eigenvalue has a positive real part. The same argument used there can be invoked here to show that this undesirable initial-error-amplification phenomenon also occurs for $n > 1$ whenever one or more eigenvalues are in the open right half-plane. In Section 1-8 this difficulty was overcome by choosing the capacitor charge q, instead of voltage v, as the state variable. We shall now show that this technique also applies in the general case.

If we choose q as the state variable for our example, the state equation assumes the form

$$\dot{q} = -\hat{i}(\hat{v}(q), t) \triangleq f(q, t) \tag{10-117}$$

where $v = \hat{v}(q)$ is an $n \times 1$ vector whose jth element is given by $v_j = \hat{v}_j(q_j)$, the q-v curve of the jth capacitor. The Jacobian matrix $J_f(q, t) \triangleq \partial f(q, t)/\partial q$ can be expressed in terms of v as follows:

$$J_f(v, t) \triangleq \left.\frac{\partial f(q, t)}{\partial q}\right|_{q=\hat{q}(v)} = -\frac{\partial \hat{i}(v, t)}{\partial v} \cdot [C^{-1}(v)] \tag{10-118}$$

where $q = \hat{q}(v)$ is the *inverse function* of $v = \hat{v}(q)$, and where

$$C(v) \triangleq \frac{\partial \hat{q}(v)}{\partial v} \tag{10-119}$$

Equation (10-118) can be recast into the following equivalent form:

$$\begin{aligned} J_f(v, t) &= C(v)\left[-C^{-1}(v)\frac{\partial \hat{i}(v, t)}{\partial v}\right]C^{-1}(v) \\ &= C(v)A_1(v, t)C^{-1}(v) \end{aligned} \tag{10-120}$$

where $A_1(v, t)$ is as defined in Eq. (10-113). Since the two matrices $J_f(v, t)$ and $A_1(v, t)$ are related by a *similarity transformation*, they have identical eigenvalues. Since we have already shown that, for networks containing only *passive* elements and independent sources, all eigenvalues of $A_1(v, t)$ are in the left half-plane, it follows that all eigenvalues of $J_f(v, t)$ are similarly located. Following a similar analysis, we can conclude that *for nonlinear dynamic networks containing passive elements and independent sources it is desirable to choose capacitor charges and inductor flux linkages as the state variables.* A similar criterion seems to apply also for most networks containing active elements and controlled sources, although a rigorous proof has not been found.

REFERENCES

1. CHUA, L. O. *Introduction to Nonlinear Network Theory*. New York: McGraw-Hill Book Company, 1969.
2. LIU, R. W., and L. AUTH. "Qualitative Synthesis of Nonlinear Networks." *Proc. First Annual Allerton Conf. Circuit System Theory*, pp. 330–343, Nov. 1963.
3. CHUA, L. O. "State Variable Formulation of Nonlinear *RLC* Networks in Explicit Normal Form." *Proc. IEEE*, Vol. 53, No. 2, pp. 206–207, Feb. 1965.
4. BRAYTON, R. K., and J. K. MOSER. "A Theory of Nonlinear Networks." *Quart. Appl. Math.*, Vol. 22, No. 1, pp. 1–33, July 1964; and Vol. 22, No. 2, pp. 81–104, July 1964.
5. DESOER, C. A., and J. KATZENELSON. "Nonlinear *RLC* Networks," *Bell System Tech. J.*, Vol. 44, pp. 161–198, Jan. 1965.
6. CHUA, L. O., and R. A. ROHRER. "On the Dynamic Equations of a Class of Nonlinear *RLC* Networks." *IEEE Trans. Circuit Theory*, Vol. CT-12, No. 4, pp. 475–489, Dec. 1965.
7. STERN, T. E. "On the Equations of Nonlinear Networks." *IEEE Trans. Circuit Theory*, Vol. CT-13, No. 1, pp. 74–81, Mar. 1961.
8. PERREAULT, D. A. *Analysis of Nonlinear RLC Networks with Controlled Sources*. Lafayette, Ind.: Purdue University, Ph.D. dissertation, Aug. 1970.
9. CALAHAN, D. A. *Computer-Aided Network Design*. New York: McGraw-Hill Book Company, 1972.

PROBLEMS

10-1. Consider the nonlinear circuit shown in Fig. P10-1. Formulate the state equation with the *inductor current* i_2 or the *inductor flux linkage* φ_2 as the state variable (whenever possible) for the following choices of element characteristics.

(a) $R_1: v_1 = i_1^3; L_2: i_2 = \ln(\varphi_2 + 1)$.

(b) $R_1: v_1 = 1 - i_1^2 + i_1^3; L_2: i_2 = \varphi_2^3 - 3\varphi_2$.

(c) $R_1: v_1 = 1 - i_1^2 + i_1^3; L_2: \varphi_2 = i_2^3 - 3i_2$.

(d) $R_1: i_1 = v_1^3 - 3v_1; L_2: \varphi_2 = Li_2$.

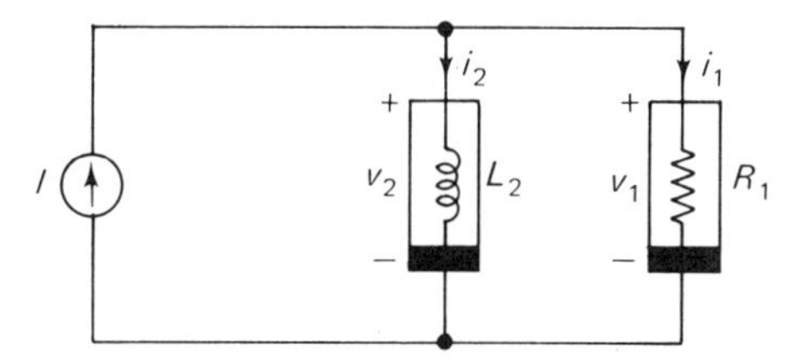

Fig. P10-1

10-2. Show that, by introducing a *parasitic* linear capacitor C across the resistor R_1 in Fig. P10-1, the state equations can be written in terms of two state variables for part (d) of Problem 10-1.

10-3. Show that the state equations of any nonlinear *RLC* network (no controlled sources and coupled elements) which contains neither capacitor voltage source loops nor inductor current source cutsets can always be formulated, without invoking the *unique-*

ness assumption, if all nonlinear elements are characterized by *strictly monotonically increasing* curves which tend to $\pm\infty$. *Hint:* Generalize the algorithm presented in Section 8-3 to derive the *explicit* normal-form equations in terms of topological matrices.

10-4. Show that the state equations of any nonlinear *RLC* network (no controlled sources and coupled elements) which contains neither capacitor voltage source loops nor inductor current source cutsets can always be formulated in explicit form if each *voltage-controlled resistor* is in parallel with a *voltage-controlled capacitor*, and if each *current-controlled resistor* is in *series* with a *current-controlled inductor.*

10-5. Show that the class of networks allowed in Problems 10-3 and 10-4 can be enlarged to include the following additional types of circuit elements:

(a) Mutual inductances.

(b) Controlled sources whose controlling variables represent either capacitor voltages or inductor currents.

10-6. Show that it is possible to eliminate any voltage source $v_s(t)$ which forms a *loop* exclusively with capacitors by applying the equivalent circuit transformation technique shown in Fig. P10-6.

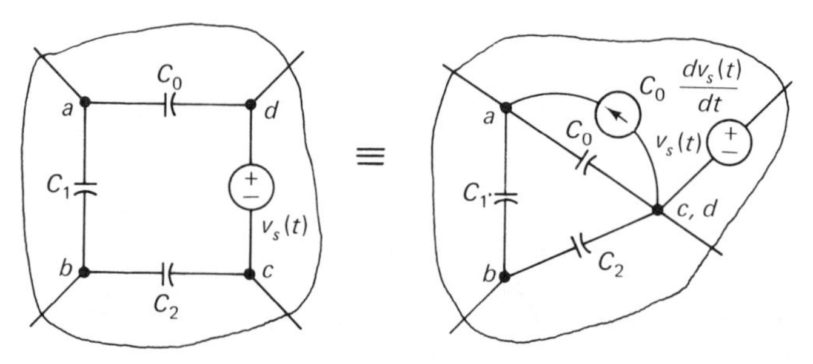

Fig. P10-6

10-7. Show that it is possible to eliminate any current source $i_s(t)$ which forms a *cutset* exclusively with inductors by applying the equivalent circuit transformation technique shown in Fig. P10-7.

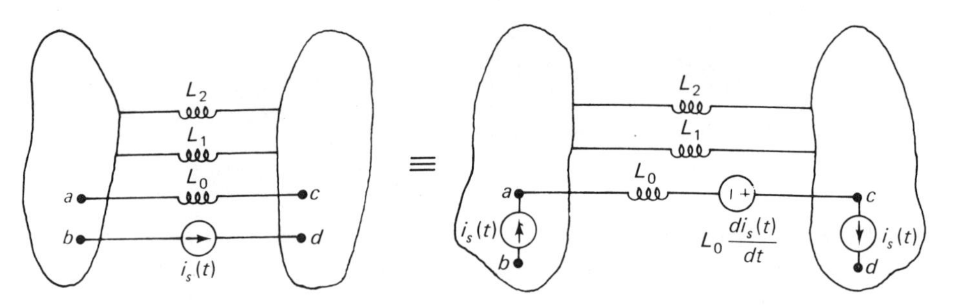

Fig. P10-7

10-8. Show that it is possible to eliminate any *loop* of *linear capacitors* in a nonlinear network by applying the equivalent capacitor-controlled source circuit transformation technique shown in Fig. P10-8.

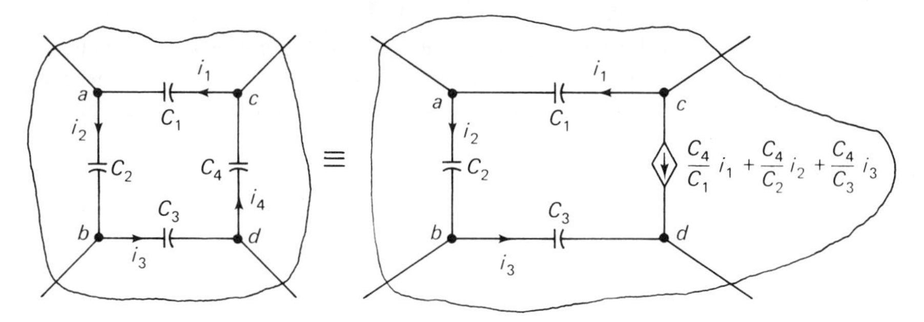

Fig. P10-8

10-9. Show that it is possible to eliminate any *loop* of *linear capacitors* in a nonlinear network by applying the equivalent capacitor-coupled circuit transformation technique shown in Fig. P10-9. The polarities of coupling are such that the system of mutually coupled capacitors is characterized by

$$\begin{bmatrix} i_1 \\ i_2 \\ i_3 \end{bmatrix} = \begin{bmatrix} C_1 + C_4 & C_4 & C_4 \\ C_4 & C_2 + C_4 & C_4 \\ C_4 & C_4 & C_3 + C_4 \end{bmatrix} \begin{bmatrix} \dot{v}_1 \\ \dot{v}_2 \\ \dot{v}_3 \end{bmatrix}$$

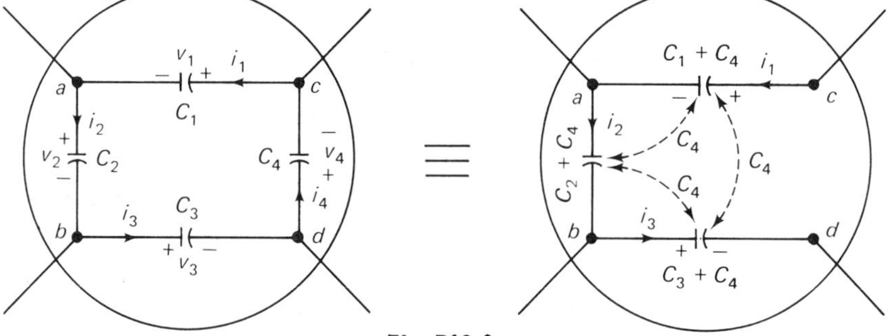

Fig. P10-9

10-10. Show that it is possible to eliminate any *cutset* of *linear inductors* in a nonlinear network by applying the equivalent inductor-controlled source circuit transformation technique shown in Fig. P10-10.

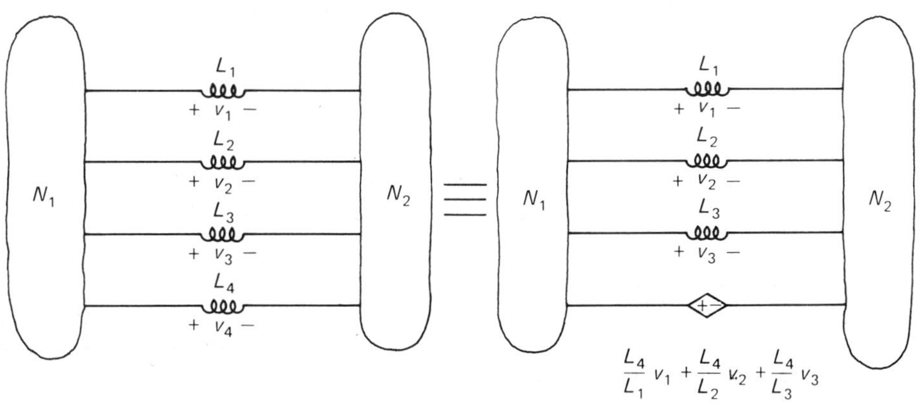

Fig. P10-10

10-11. Show that it is possible to eliminate any *cutset* of *linear inductors* in a nonlinear network by applying the equivalent inductor-coupled circuit transformation technique shown in Fig. P10-11. The polarities of coupling are such that the system of mutually coupled inductors is characterized by

$$\begin{bmatrix} v_1 \\ v_2 \\ v_3 \end{bmatrix} = \begin{bmatrix} L_1 + L_4 & L_4 & L_4 \\ L_4 & L_2 + L_4 & L_4 \\ L_4 & L_4 & L_3 + L_4 \end{bmatrix} \begin{bmatrix} \dot{i}_1 \\ \dot{i}_2 \\ \dot{i}_3 \end{bmatrix}$$

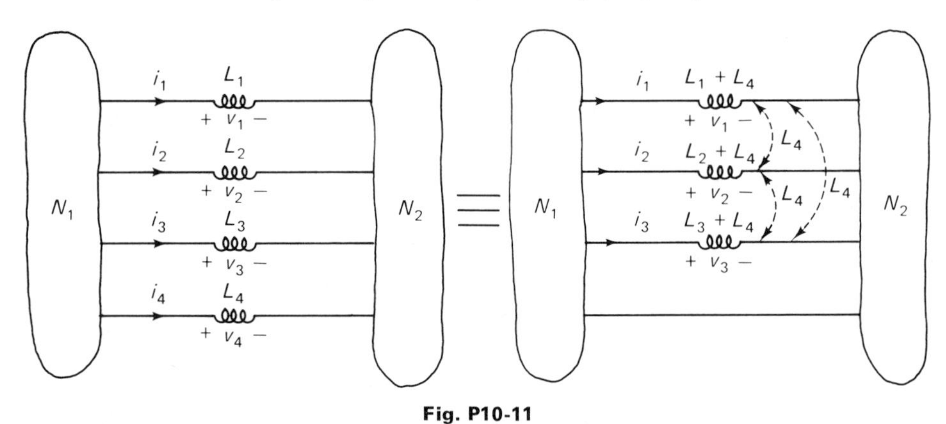

Fig. P10-11

10-12. Construct a nontrivial example illustrating the proof in the text showing $\boldsymbol{H}_{R_{\mathfrak{J}}L_{\mathfrak{J}}} = \mathbf{0}$ and $\boldsymbol{H}_{R_{\mathfrak{J}}C_{\mathfrak{L}}} = \mathbf{0}$.

10-13. Give a detailed proof showing $\boldsymbol{H}_{R_{\mathfrak{L}}L_{\mathfrak{J}}} = \mathbf{0}$ and $\boldsymbol{H}_{R_{\mathfrak{L}}C_{\mathfrak{L}}} = \mathbf{0}$. Illustrate your proof with a nontrivial example.

10-14. Show that the *consistency assumptions* for the formulation of normal-form equations for dynamic nonlinear networks in Section 10-3-1 can be relaxed to that of requiring only "no loops containing capacitors and/or *independent* voltage sources" and "no cutsets containing inductors and/or *independent* current sources." Develop a detailed algorithm in a form analogous to that given in the text. *Hint:* Formulate the hybrid matrix via the method of *systematic elimination* in Section 6-5-2.

10-15. Construct a nontrivial example illustrating how each of the standing assumptions specified in Section 10-3-1 can be violated.

10-16. Develop the specialized form of the normal-form equation formulation in the text for the case when *all* resistors are *nonlinear* so that the n-port N_A contains only connecting wires.

10-17. Formulate the *explicit* state equations similar to those of Eqs. (10-78) and (10-90) for the case when *all nonlinear tree resistors are current-controlled and all nonlinear cotree resistors are voltage-controlled.*

10-18. Formulate the explicit state equations similar to those of Eqs. (10-78) and (10-90) if all nonlinear resistors in subtree $\mathfrak{J}1$ and subcotree $\mathfrak{L}2$ are *voltage-controlled*, and if all nonlinear resistors in subtree $\mathfrak{J}2$ and subcotree $\mathfrak{L}1$ are *current-controlled*, where $\mathfrak{J} = \mathfrak{J}1 \cup \mathfrak{J}2$ and $\mathfrak{L} = \mathfrak{L}1 \cup \mathfrak{L}2$.

10-19. Formulate the *explicit* state equations similar to those of Eqs. (10-78) and (10-90) if all nonlinear resistors in subtree $\mathfrak{J}1$ and subcotree $\mathfrak{L}2$ are *current-controlled*, and if all nonlinear resistors in subtree $\mathfrak{J}2$ and subcotree $\mathfrak{L}1$ are *voltage-controlled*, where $\mathfrak{J} = \mathfrak{J}1 \cup \mathfrak{J}2$ and $\mathfrak{L} = \mathfrak{L}1 \cup \mathfrak{L}2$.

10-20. Find topological conditions which guarantee that Eqs. (10-98) and (10-99) are independent of $\dot{\boldsymbol{u}}$; i.e.,

$$\dot{\boldsymbol{x}} = \boldsymbol{f}(\boldsymbol{x}, \boldsymbol{u})$$

$$\boldsymbol{y} = \hat{\boldsymbol{y}}(\boldsymbol{x}, \boldsymbol{u})$$

10-21. Formulate the *explicit* state equations similar to those of Eqs. (10-78) and (10-90) but with $\boldsymbol{v}_{C_\mathfrak{J}}$ and $\boldsymbol{\varphi}_{L_\mathfrak{L}}$ as the state variables. State all assumptions on the constitutive relations of the capacitors and inductors.

10-22. Formulate the *explicit* state equations similar to those of Eqs. (10-78) and (10-90) but with $\boldsymbol{q}_{C_\mathfrak{J}}$ and $\boldsymbol{i}_{L_\mathfrak{L}}$ as the state variables. State all assumptions on the constitutive relations of the capacitors and inductors.

10-23. Formulate the explicit state equations similar to those of Eqs. (10-78) and (10-90) but with $\boldsymbol{v}_{C_{\mathfrak{J}_1}}$, $\boldsymbol{q}_{C_{\mathfrak{J}_2}}$, $\boldsymbol{i}_{L_{\mathfrak{L}_1}}$, and $\boldsymbol{\varphi}_{L_{\mathfrak{L}_2}}$ as the state variables, and where $C_\mathfrak{J} = C_{\mathfrak{J}_1} \cup C_{\mathfrak{J}_2}$ denotes capacitors in the tree and $L_\mathfrak{L} = L_{\mathfrak{L}_1} \cup L_{\mathfrak{L}_2}$ denotes inductors in the cotree. State all assumptions on the constitutive relations of the capacitors and inductors.

10-24. Derive the *explicit* topological equations for determining the voltages and currents of all *linear* resistors in terms of the state variables $\boldsymbol{v}_{C_\mathfrak{J}}(t)$ and $\boldsymbol{i}_{L_\mathfrak{J}}(t)$ and the independent sources $\boldsymbol{v}_{E_\mathfrak{J}}(t)$ and $\boldsymbol{i}_{J_\mathfrak{L}}(t)$.

10-25. Show that for the class of dynamic networks considered in Section 10-4 the *output equations* assume the *linear* form

$$\boldsymbol{y} = \boldsymbol{P}\boldsymbol{x} + \boldsymbol{Q}_r\boldsymbol{s}_r + \boldsymbol{Q}_s\boldsymbol{s}_s$$

where $\boldsymbol{y}$ is a vector denoting all *nonstate* variables. Specify the matrices $\boldsymbol{P}$, $\boldsymbol{Q}_r$, and $\boldsymbol{Q}_s$, and the vectors $\boldsymbol{s}_r$ and $\boldsymbol{s}_s$.

CHAPTER 11

Numerical Solution of State Equations for Dynamic Nonlinear Networks

11-1 EXISTENCE AND UNIQUENESS OF SOLUTIONS

In Chapter 10, we presented the algorithms for formulating the equations of motion of a large class of dynamic nonlinear networks into the *normal form*

$$\begin{aligned}\dot{x}_1 &= f_1(x_1, x_2, \ldots, x_n, t)\\ \dot{x}_2 &= f_2(x_1, x_2, \ldots, x_n, t)\\ &\cdot\\ &\cdot\\ &\cdot\\ \dot{x}_n &= f_n(x_1, x_2, \ldots, x_n, t)\end{aligned} \tag{11-1}$$

It is convenient to rewrite Eq. (11-1) into the following compact vector form:

$$\dot{\boldsymbol{x}} = \boldsymbol{f}(\boldsymbol{x}, t) \tag{11-2}$$

where $\boldsymbol{f}(\cdot, \cdot)$: $R^{n+1} \longrightarrow R^n$ is a vector-valued function from the $(n + 1)$-dimensional $(\boldsymbol{x}, t)$-Euclidean space R^{n+1} into the n-dimensional $\boldsymbol{x}$-Euclidean space R^n. Although the function $\boldsymbol{f}(\cdot, \cdot)$ is seldom expressible in analytic form, our algorithm can be programmed so that the value of $\boldsymbol{f}(\boldsymbol{x}, t)$ corresponding to *any* vector $\boldsymbol{x}$ and scalar t can be efficiently computed. Our objective in this chapter is to present the basic principles for solving Eq. (11-1) *numerically* using *difference methods*.[1]

[1]Another approach for solving Eq. (11-1) numerically is based on the idea of *orthogonal approximation* of functions. Since this approach is not particularly suited for computer implementation, the interested reader is referred to [1] for details.

Most numerical methods for solving Eq. (11-1) subject to the initial condition $\mathbf{x}(t_0) = \mathbf{x}_0$ *tacitly assume* the solution to this *initial-value problem* exists and is well defined for all times $t \geq t_0$ [2]. In the event that the solution does not exist or is pathological, most numerical methods would still produce a set of numbers, which of course is meaningless. To demonstrate the situations where such difficulties could occur, consider the following examples.

EXAMPLE 11-1. Consider a network consisting of a 1-Ω linear resistor in parallel with a nonlinear capacitor characterized by a square-law incremental capacitance $C(v) = 2v^2$. The associated state equation is readily shown to be

$$\frac{dv}{dt} = -\frac{1}{2v} \triangleq f(v) \tag{11-3}$$

The solution to Eq. (11-3) corresponding to an initial voltage $v(0) = v_0$ is given by

$$v(t) = \sqrt{v_0^2 - t} \tag{11-4}$$

This solution is shown in Fig. 11-1. Observe that if we choose the initial voltage to be $v_0 = 0$, then $v(t) = \sqrt{-t}$ is imaginary for $t > 0$. One might object to this example on the grounds that Eq. (11-3) is undefined at $v = 0$. However, even if we redefine Eq. (11-3) by

$$\begin{aligned} \frac{dv}{dt} &= -\frac{1}{2v}, \qquad v \neq 0 \\ &= 0, \qquad\quad v = 0 \end{aligned} \tag{11-5}$$

Eq. (11-5) still has no solution with $v(0) = 0$. If we choose a *nonzero* initial voltage, the solution would exist only during the time interval $0 \leq t \leq v_0^2$. In either case, we have a highly nonphysical situation. Clearly, the *circuit model* chosen for this particular network is defective, and a more realistic model must be chosen before a numerical solution is attempted.

Example 11-1 motivates the derivation of some criterion that will guarantee the existence of a solution in the general case. The Peano existence theorem shows that the *continuity* of $\mathbf{f}(\mathbf{x}, t)$ is a sufficient condition. For a proof of this classic theorem, see reference 3.

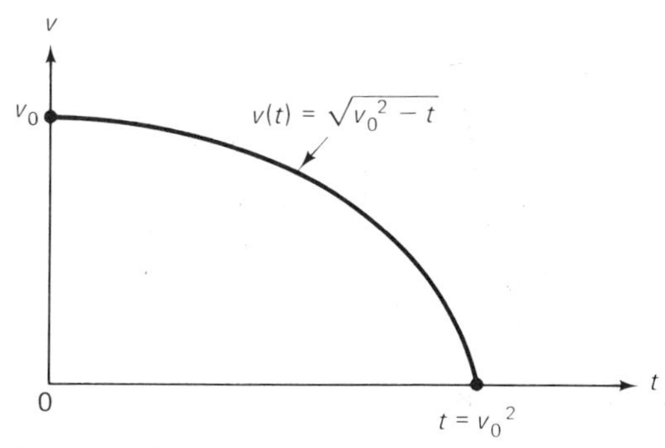

Fig. 11-1. Initial-value problem whose solution does not exist for $t > v_0^2$.

Theorem 11-1 (Peano Existence Theorem). *If $f(\mathbf{x}, t)$ is continuous at $(\mathbf{x}_0, t_0)$, then there exists* at least *one solution of Eq. (11-1) with initial condition $\mathbf{x}(t_0) = \mathbf{x}_0$ over some time interval $t_0 - \varepsilon \leq t \leq t_0 + \varepsilon$, where $\varepsilon > 0$.*

In Example 11-2, we demonstrate that the Peano existence theorem does not guarantee a *unique* solution.

EXAMPLE 11-2. Consider a network consisting of a 1-F linear capacitor in parallel with a nonlinear resistor characterized by $i = -\frac{3}{2}v^{1/3}$. The state equation in this case is given by

$$\frac{dv}{dt} = \frac{3}{2}v^{1/3} \triangleq f(v) \tag{11-6}$$

It is easily verified that a solution of Eq. (11-6) with initial condition $v(0) = 0$ is given by

$$\begin{aligned} v(t) &= 0, \qquad & 0 \leq t \leq k \\ &= (t - k)^{3/2}, \qquad & t > k \end{aligned} \tag{11-7}$$

where k is *any* nonnegative real number. This solution is shown in Fig. 11-2 for three different values of k. Hence, this network has an infinite number of distinct solutions all having the same initial condition. Since $f(v) \triangleq \frac{3}{2}v^{1/3}$ is certainly a continuous function, it is clear that additional conditions must be imposed to guarantee a unique solution. The clue to this condition lies in the observation that the function $f(v)$ has infinite slope at $v = 0$, and is therefore not a continuously differentiable function. This suggests the additional condition that $f(v)$ be continuously differentiable. This condition unfortunately excludes many well-behaved functions of interest to engineers. In particular, it excludes the important class of piecewise-linear functions, since such functions are not differentiable at the breakpoints. Fortunately, a weaker condition, the *Lipschitz condition*, is sufficient to guarantee a unique solution.

Definition. *A function $f(\mathbf{x}, t)$: $R^{n+1} \longrightarrow R^n$ is said to satisfy a* ***Lipschitz condition*** *with respect to $\mathbf{x}$ in a closed and bounded set $D \subset R^{n+1}$ if there exists a* ***finite***

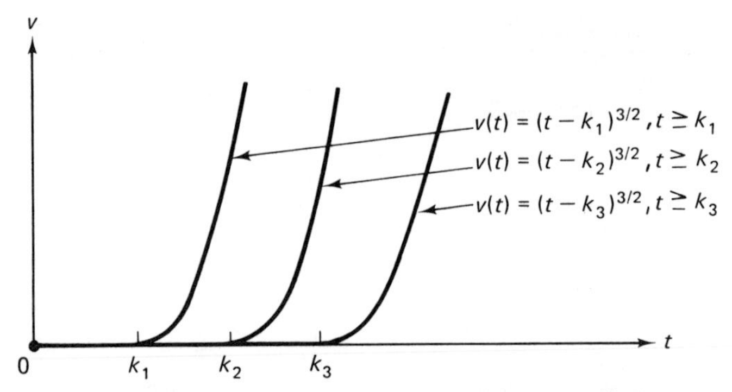

Fig. 11-2. Initial-value problem having infinitely many distinct solutions with identical initial condition $v(0) = 0$.

constant L, the ***Lipschitz constant,*** *such that*

$$\|\boldsymbol{f}(\boldsymbol{x}', t) - \boldsymbol{f}(\boldsymbol{x}'', t)\| \le L \|\boldsymbol{x}' - \boldsymbol{x}''\| \tag{11-8}$$

for all $(\boldsymbol{x}', t) \in D$ *and* $(\boldsymbol{x}'', t) \in D$.

A geometrical interpretation of a Lipschitz condition *at any given instant of time t* is shown in Fig. 11-3 for the case $n = 1$. Since the time t is fixed, we simply write $y = f(x)$, instead of $y = f(x, t)$. Observe that the function $y = f(x)$ shown in Fig. 11-3(a), (b), and (c) is contained entirely within the shaded area bounded by the

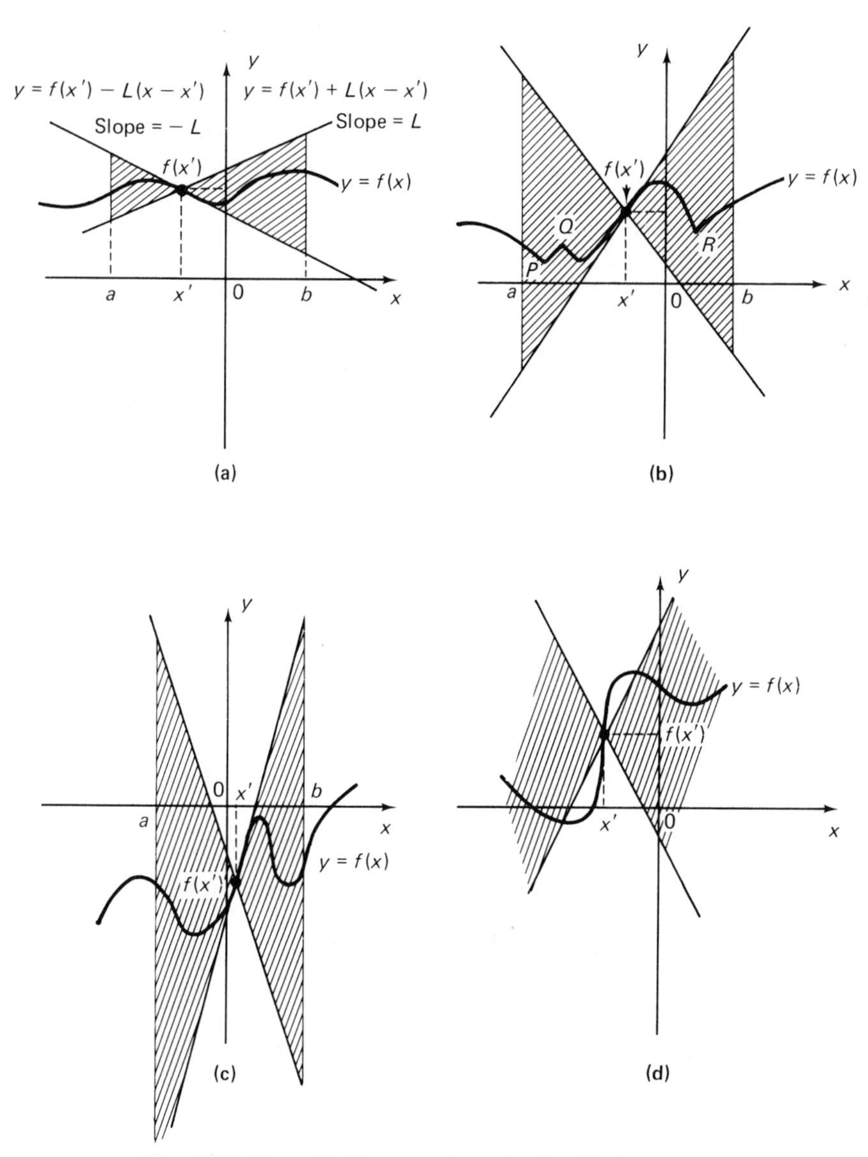

Fig. 11-3. Geometrical interpretations of a Lipschitz function.

pair of straight lines $y = f(x') \pm L(x - x')$ drawn through any point $x' \in (a, b)$, where $L < \infty$. Hence, $f(x)$ is Lipschitz in the domain $D = [a, b]$ in each case. The function $f(x)$ shown in Fig. 11-3(d), however, is *not* Lipschitz in any interval containing the point $x = x'$ since the slope of $f(x)$ is infinite at this point, and no pair of straight lines with a finite slope can possibly bound the curve in the vicinity of $x = x'$. Observe that the function shown in Fig. 11-3(b) is not differentiable at points P, Q, and R, and yet it is Lipschitz. In other words, a Lipschitz function need not be differentiable. It is clear from this geometrical interpretation, however, that a Lipschitz function must be *continuous*. Hence, the Lipschitz condition lies somewhere in between the continuity and the differentiability requirements. *Roughly speaking*, a function is Lipschitz if its value does not change too rapidly to contain points with infinite rate of change. It can be shown that a *continuously differentiable function is automatically a Lipschitz function*. The importance of the Lipschitz condition is given by the following theorem whose proof can be found in reference 3.

Theorem 11-2 (Picard's Existence and Uniqueness Theorem). *If $f(\mathbf{x}, t)$ is continuous in $D \subset R^{n+1}$ and satisfies a Lipschitz condition with respect to $\mathbf{x}$ in D, then for any $(\mathbf{x}_0, t_0) \in D$ there exists a* ***unique*** *solution $\mathbf{x}(t)$ that satisfies the initial condition $\mathbf{x}(t_0) = \mathbf{x}_0$ and which is defined over some time interval $t_0 - \varepsilon \leq t \leq t_0 + \varepsilon$, where $\varepsilon > 0$. Moreover, the solution $\mathbf{x}(t)$ depends* ***continuously*** *on both $\mathbf{x}_0$ and t_0.*

Both the Peano and Picard's theorems are *local theorems*, since they assert the existence of a solution only over some nonzero time interval centered about the initial time t_0. To guarantee the existence of a solution over *all* time, much more severe conditions must be imposed on $\mathbf{f}(\mathbf{x}, t)$, as shown by Example 11-3.

EXAMPLE 11-3. Consider a network consisting of a 1-F linear capacitor in parallel with a nonlinear resistor characterized by $i = -v^2$. The state equation for this network is given by

$$\frac{dv}{dt} = v^2 \tag{11-9}$$

Corresponding to the initial condition $v(0) = 1$, the solution to Eq. (11-9) is given by

$$v(t) = \frac{1}{1 - t} \tag{11-10}$$

Observe that this solution approaches infinity as t approaches 1! A network with this property is said to have a *finite escape time* and is clearly nonphysical. The next theorem provides a sufficient condition that guarantees no finite escape time. The proof can be found in references 4 and 5.

Theorem 11-3 (Wintner's Global Existence Theorem). *Let $\mathbf{f}(\mathbf{x}, t)$ be continuous in the $(n + 1)$-dimensional space R^{n+1}. If it is possible to find a continuous function $L(r)$, where $r \triangleq [x_1^2 + x_2^2 + \ldots + x_n^2]^{1/2}$ such that*

$$|f_i(x_1, x_2, \ldots, x_n, t)| < L(r), \qquad i = 1, 2, \ldots, n \tag{11-11}$$

for all values of $0 \leq r < \infty$, *and if*

$$\int_0^\infty \frac{dr}{L(r)} = \infty \tag{11-12}$$

then all solutions of Eq. (11-1) are defined for all times $-\infty < t < \infty$.

Observe that the conditions required by Wintner's global existence theorem are rather strong. Roughly speaking, these conditions prevent the function $\boldsymbol{f}(\boldsymbol{x}, t)$ from growing too rapidly. For example, these conditions rule out all polynomials with degree greater than 1. However, these conditions are certainly not superfluous; we have already demonstrated the existence of finite escape time with a *second*-degree polynomial $f(x) = x^2$ (Example 11-3). On the other hand, the conditions required by Wintner's theorem do allow a large class of nonelementary functions, such as $f(x) = (1 + x^3)^{-1/2}$, $(1 + x^2)^{1/3}$, $x^3/(1 + x^2)$, and $x \exp(\sin x)$. In fact, we have the following corollary.

Corollary to Wintner's Global Existence Theorem. *If there exists a piecewise-continuous function* $L: R_+ \longrightarrow R_+$ *(*R_+ *denotes the interval* $[0, \infty)$*) such that*

$$||\boldsymbol{f}(\boldsymbol{x}', t) - \boldsymbol{f}(\boldsymbol{x}'', t)|| \leq L(t)\,||\boldsymbol{x}' - \boldsymbol{x}''||$$

for all $t \in R_+$ *and for all* $\boldsymbol{x}', \boldsymbol{x}'' \in R^n$ *(this is called a global Lipschitz condition), then Eq. (11-1) has a unique solution for all time* $0 \leq t < \infty$.

The three theorems presented in this section are introduced to provide a mathematical foundation for the numerical integration of normal-form equations. It must be emphasized that *these theorems provide only sufficient conditions*. Hence, even if $\boldsymbol{f}(\boldsymbol{x}, t)$ does not satisfy Wintner's global existence theorem, the solution may still exist for all time $t \geq t_0$. However, when a computer solution behaves strangely, some kind of ill-conditioning due to the violation of one or more conditions required by the preceding theorems may be suspected. Since these conditions are generally difficult to verify on a computer, the practical approach would be to *assume* that a unique solution does exist and to resort to these theorems only when difficulties are encountered.

11-2 ERROR CONSIDERATIONS IN THE NUMERICAL SOLUTION OF INITIAL-VALUE PROBLEMS

An initial-value problem consists of a prescribed system of first-order ordinary differential equations $\dot{\boldsymbol{x}} = \boldsymbol{f}(\boldsymbol{x}, t)$, a prescribed initial state vector $\boldsymbol{x}_0$, and a prescribed solution time interval T. A vector time function $\boldsymbol{x} = \hat{\boldsymbol{x}}(t)$ is said to be a solution to this initial-value problem if $\hat{\boldsymbol{x}}(t_0) = \boldsymbol{x}_0$ and $\dot{\hat{\boldsymbol{x}}}(t) = \boldsymbol{f}(\hat{\boldsymbol{x}}(t), t)$ for all $t \in [t_0, t_0 + T]$. To find $\hat{\boldsymbol{x}}(t)$ *numerically*, we first divide the solution time interval T into small time increments, as illustrated in Fig. 11-4 for the case $n = 1$. Each time increment $h_i \triangleq (\Delta t)_i$ is called a *step size*. Our objective is to find $\hat{\boldsymbol{x}}(t)$ at $t_k \triangleq t_0 + \sum_{i=1}^{k} h_i$, $k = 1, 2, \ldots, N$, where $t_N = t_0 + T$. Since no numerical method is capable of finding $\hat{\boldsymbol{x}}(t_k)$

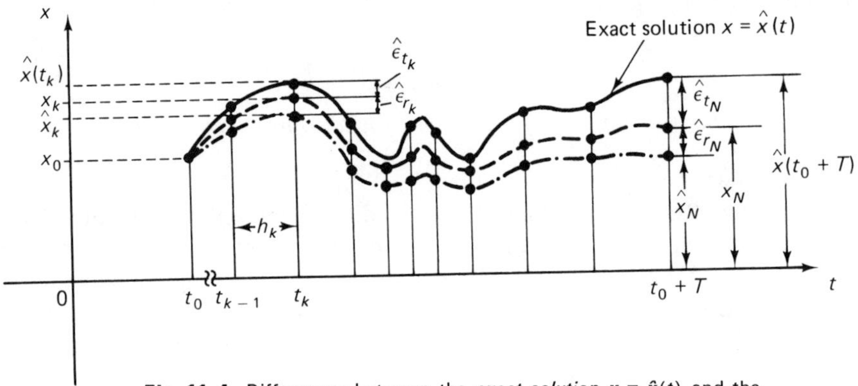

Fig. 11-4. Differences between the *exact solution* $\mathbf{x} = \hat{\mathbf{x}}(t)$ and the *computed* solution x_N without round-off error and $\hat{x}_N$ having both local truncation and round-off errors.

exactly, we let $\hat{\mathbf{x}}_k$ denote the *computed* value at $t = t_k$. The quantity

$$\hat{\boldsymbol{\epsilon}}_k \triangleq \|\hat{\mathbf{x}}(t_k) - \hat{\mathbf{x}}_k\| \tag{11-13}$$

therefore represents the *total error* at $t = t_k$. The total error always consists of two components: a *truncation error* $\hat{\boldsymbol{\epsilon}}_{t_k}$ and a *round-off error* $\hat{\boldsymbol{\epsilon}}_{r_k}$. The truncation error is the error that results when infinite precision arithmetic is used. We refer to this error component as an *algorithmic error*, since it depends on the nature of the numerical algorithm used in computing $\hat{\mathbf{x}}_k$. The *round-off error* is the additional error due to the finite word length in a computer. We refer to this error component as a *machine error*, since it depends only on the arithmetic accuracy of the computer. It is clear from Fig. 11-4 that both truncation and round-off errors tend to accumulate with an increasing number of time steps. Hence, to compare the accuracy between two algorithms, it is necessary to compare them at the same time $t = t_k$ and with the same initial state. For purposes of comparison, however, it suffices to consider the *local error* at $t = t_1$, where there is yet no previous error; i.e.,

$$\boldsymbol{\epsilon}_1 \triangleq \|\hat{\mathbf{x}}(t_1) - \hat{\mathbf{x}}_1\| \tag{11-14}$$

The adjective *local* is used here to emphasize that *it is the error that occurs after one time step, assuming the value of* **x** *at the preceding time step is exact*. The two components of local error are called *local truncation error* and *local round-off error*, and are denoted by $\boldsymbol{\epsilon}_{t_1}$ and $\boldsymbol{\epsilon}_{r_1}$, respectively. If we let $\mathbf{x}_1$ denote the value of $\mathbf{x}(t)$ at $t = t_1 \triangleq t_0 + h_1$, which has been computed *with infinite arithmetic precision*,[2] then the local truncation error at t_1 is given by

$$\boldsymbol{\epsilon}_{t_1} \triangleq \|\hat{\mathbf{x}}(t_1) - \mathbf{x}_1\| \tag{11-15}$$

Since the local round-off error depends crucially on the nature of the computer, it cannot be reduced once a computer is chosen. However, different algorithms react differently with regard to how fast the round-off error is propagated. It is important

[2]It is important to distinguish between the symbols $\hat{\mathbf{x}}_k$ and $\mathbf{x}_k$; the former *includes* round-off errors. See Fig. 11-4.

to recognize that the *total* round-off error at $t = t_k$ is not just the sum of the local round-off error that occurs at each time step, because *local* round-off error may either decay or grow as the computation proceeds. An algorithm with the desirable property that local round-off error decays with an increasing number of time steps is said to be *numerically stable*. Otherwise, it is *numerically unstable* and is of no practical value. The study of *numerical stability* of algorithms is a difficult subject to be investigated in Chapter 12, and we shall be satisfied in this chapter with only a qualitative discussion of this phenomenon.

Although there exist many algorithms for solving initial-value problems, most of them are based on two basic approaches: the *Taylor series expansion* approach and the *polynomial approximation approach*. Algorithms based on the Taylor series expansion approach are usually called *Runge–Kutta* algorithms. Those based on the polynomial approximation approach are usually called *numerical-integration* algorithms. Our objective in the following sections is to illustrate the basic differences between these two approaches. Unless otherwise stated, we shall simplify matters by assuming a *uniform step size* $h_i = h$ in the formulation of the algorithms. Under this assumption, the time step is simplified to

$$t_n = t_0 + nh, \qquad n = 1, 2, \ldots, N \tag{11-16}$$

where $Nh \leq T$ and $(N+1)h > T$.

11-3 NUMERICAL SOLUTION BY TAYLOR SERIES EXPANSION

Let $x = \hat{x}(t)$ be the exact solution to the initial-value problem $\dot{x} = f(x, t), x(t_0) = x_0$.[3] Let us expand $\hat{x}(t)$ in a Taylor series about the point $t = t_n$ and evaluate the expansion at $t = t_{n+1}$:

$$\begin{aligned}\hat{x}(t_{n+1}) = \hat{x}(t_n) &+ \frac{\hat{x}^{(1)}(t_n)}{1!}(t_{n+1} - t_n) + \frac{\hat{x}^{(2)}(t_n)}{2!}(t_{n+1} - t_n)^2 \\ &+ \ldots + \frac{\hat{x}^{(p)}(t_n)}{p!}(t_{n+1} - t_n)^p + \text{higher-order terms}\end{aligned} \tag{11-17}$$

where $\hat{x}^{(j)}(t_n)$ denotes the jth time derivative of $\hat{x}(t)$ at $t = t_n$. Substituting $t_{n+1} - t_n = h$ into Eq. (11-17), we obtain

$$\begin{aligned}\hat{x}(t_{n+1}) = \hat{x}(t_n) &+ \frac{h}{1!}\hat{x}^{(1)}(t_n) + \frac{h^2}{2!}\hat{x}^{(2)}(t_n) + \frac{h^3}{3!}\hat{x}^{(3)}(t_n) \\ &+ \ldots + \frac{h^p}{p!}\hat{x}^{(p)}(t_n) + \text{higher-order terms}\end{aligned} \tag{11-18}$$

If we transpose the higher-order terms to the left of Eq. (11-18) and substitute $f(\hat{x}(t_n), t_n)$ for $\hat{x}^{(1)}(t_n)$, $f^{(1)}(\hat{x}(t_n), t_n)$ for $\hat{x}^{(2)}(t_n), \ldots,$ and $f^{(p-1)}(\hat{x}(t_n), t_n)$ for $\hat{x}^{(p)}(t_n)$, we obtain

$$\begin{aligned}&\hat{x}(t_{n+1}) - \text{higher-order terms} \\ &= \hat{x}(t_n) + \frac{h}{1!}f(\hat{x}(t_n), t_n) + \frac{h^2}{2!}f^{(1)}(\hat{x}(t_n), t_n) + \ldots + \frac{h^p}{p!}f^{(p-1)}(\hat{x}(t_n), t_n)\end{aligned} \tag{11-19}$$

[3]To simplify notation, we consider here only one equation in one state variable x. However, virtually every step applies equally well for a system of n equations.

The right side of Eq. (11-19) is the *exact* truncated Taylor series. The *Taylor algorithm* consists of replacing the expression on the left of Eq. (11-19) by x_{n+1} and replacing the exact value $\hat{x}(t_n)$ on the right of Eq. (11-19) by x_n; thus,

$$x_{n+1} = x_n + \frac{h}{1!}f(x_n, t_n) + \frac{h^2}{2!}f^{(1)}(x_n, t_n) + \ldots + \frac{h^p}{p!}f^{(p-1)}(x_n, t_n) \tag{11-20}$$

It is standard practice to rewrite Eq. (11-20) in the form

$$x_{n+1} = x_n + hT_p(x_n, t_n; h) \tag{11-21}$$

where

$$T_p(x_n, t_n; h) \triangleq f(x_n, t_n) + \frac{h}{2!}f^{(1)}(x_n, t_n) + \cdots + \frac{h^{p-1}}{p!}f^{(p-1)}(x_n, t_n) \tag{11-22}$$

It is important to observe that so far we are assuming no round-off error. Consequently, the actual computed value will be denoted by $\hat{x}_{n+1}$, where

$$\hat{\epsilon}_{r_{n+1}} \triangleq |\hat{x}_{n+1} - x_{n+1}| \tag{11-23}$$

represents the round-off error. Even if we assume that Eq. (11-21) is computed with infinite precision arithmetic, it is still not obvious that the truncation error can be restricted to small values by choosing a sufficiently small step size. The only way to guarantee the accuracy of the computed solution is to derive an *upper bound* for the truncation error. Theorem 11-4 is therefore of crucial importance in our study of Taylor's algorithm.

Theorem 11-4 (Truncation Error Estimate for Taylor's Algorithm). *Let $x = \hat{x}(t)$ be the exact solution to the initial-value problem $\dot{x} = f(x, t)$, $\hat{x}(t_0) = x_0$, $t \in [t_0, t_0 + T]$. Let the following two conditions be satisfied:*

1. $$\left|\frac{d^{(p+1)}\hat{x}(t)}{dt^{p+1}}\right| \leq X_{p+1}, \qquad t \in [t_0, t_0 + T] \tag{11-24}$$

2. $T_p(x_n, t_n; h)$ (11-25)
satisfies a Lipschitz condition with Lipschitz constant L; i.e.,

$$|T_p(x'_n, t_n; h) - T_p(x''_n, t_n; h)| \leq L|x'_n - x''_n| \tag{11-26}$$

Then the ***truncation error***[4] *at $t = t_n$,*

$$\hat{\epsilon}_{t_n} \triangleq |\hat{x}(t_n) - x_n| \tag{11-27}$$

is bounded by

$$\boxed{\hat{\epsilon}_{t_n} \leq \hat{K}_p h^p} \tag{11-28}$$

where

$$\hat{K}_p \triangleq \frac{X_{p+1}}{(p+1)!}\left[\frac{e^{L(t_n - t_0)} - 1}{L}\right] \tag{11-29}$$

is independent of h.

[4]Equation (11-27) is consistent with Eq. (11-15) since we can always let x_{n-1} be the initial value of a new initial-value problem with initial time t_{n-1}. *Local* truncation error at $t = t_n$ is bounded by $\epsilon_{t_n} \leq K_p h^p$, where K_p is $\hat{K}_p$ with $n = 1$ in Eq. (11-29).

Proof: Expand the exact solution $\hat{x}(t)$ using Taylor's formula with a remainder [6] about $t = t_n$ and evaluate the resulting expansion at $t_{n+1} = t_n + h$:

$$\hat{x}(t_{n+1}) = \hat{x}(t_n) + \frac{h}{1!}f(\hat{x}(t_n), t_n) + \frac{h^2}{2!}f^{(1)}(\hat{x}(t_n), t_n) + \cdots + \frac{h^p}{p!}f^{(p-1)}(\hat{x}(t_n), t_n) + \frac{h^{p+1}}{(p+1)!}\hat{x}^{(p+1)}(t_n^*) \tag{11-30}$$

where $t_n^* \in (t_n, t_{n+1})$. We can rewrite Eq. (11-30) using the notation defined in Eq. (11-22):

$$\hat{x}(t_{n+1}) = \hat{x}(t_n) + hT_p(\hat{x}(t_n), t_n; h) + \frac{h^{p+1}}{(p+1)!}\hat{x}^{(p+1)}(t_n^*) \tag{11-31}$$

Subtracting Eq. (11-21) from Eq. (11-31), we obtain

$$\hat{x}(t_{n+1}) - x_{n+1} = \hat{x}(t_n) - x_n + h[T_p(\hat{x}(t_n), t_n; h) - T_p(x_n, t_n; h)] + \frac{h^{p+1}}{(p+1)!}\hat{x}^{(p+1)}(t_n^*) \tag{11-32}$$

Taking the absolute value of both sides of Eq. (11-32) and using Eq. (11-27) and the triangle inequality, we obtain

$$\hat{\epsilon}_{t_{n+1}} \leq \hat{\epsilon}_{t_n} + h|T_p(\hat{x}(t_n), t_n; h) - T_p(x_n, t_n; h)| + \frac{h^{p+1}}{(p+1)!}|\hat{x}^{(p+1)}(t_n^*)| \tag{11-33}$$

This inequality can be simplified with the help of Eqs. (11-24) and (11-26), as follows:

$$\hat{\epsilon}_{t_{n+1}} \leq (1 + hL)\hat{\epsilon}_{t_n} + h^{(p+1)}\left[\frac{X_{p+1}}{(p+1)!}\right] \tag{11-34}$$

It remains to prove that Eq. (11-34) implies

$$\hat{\epsilon}_{t_n} \leq \frac{X_{p+1}}{(p+1)!}\left[\frac{e^{L(t_n - t_0)} - 1}{L}\right]h^p \tag{11-35}$$

We shall prove Eq. (11-35) by *induction.* It is certainly true when $n = 0$ since $\hat{\epsilon}_{t_0} = 0$. Assuming that Eq. (11-35) is true for $n - 1$, i.e.,

$$\hat{\epsilon}_{t_{n-1}} \leq \frac{X_{p+1}}{(p+1)!}\left[\frac{e^{L(t_{n-1} - t_0)} - 1}{L}\right]h^p \tag{11-36}$$

Eq. (11-34) can be written as

$$\hat{\epsilon}_{t_n} \leq (1 + hL)\hat{\epsilon}_{t_{n-1}} + h^{(p+1)}\left[\frac{X_{p+1}}{(p+1)!}\right] \tag{11-37}$$

Substituting Eq. (11-36) for $\hat{\epsilon}_{t_{n-1}}$ in Eq. (11-37) and simplifying, we obtain

$$\hat{\epsilon}_{t_n} \leq h^{p+1}\left[\frac{X_{p+1}}{(p+1)!}\right]\left\{\frac{(1 + hL)e^{L(t_{n-1} - t_0)} - 1}{hL}\right\} \tag{11-38}$$

Now since $(1 + hL) \leq e^{hL}$ and since $(t_{n-1} - t_0) = (n - 1)h$, Eq. (11-38) becomes

$$\hat{\epsilon}_{t_n} \leq h^{p+1}\left[\frac{X_{p+1}}{(p+1)!}\right]\left\{\frac{e^{hL} \cdot e^{L(n-1)h} - 1}{hL}\right\} = \frac{X_{p+1}}{(p+1)!}\left[\frac{e^{nhL} - 1}{L}\right]h^p \tag{11-39}$$

Substituting $t_n - t_0 = nh$ in Eq. (11-39), we obtain Eq. (11-35) and the theorem is proved by induction. Q.E.D.

The error bound given by Eq. (11-28) for the Taylor algorithm is often denoted by the standard mathematical "big O" notation

$$\hat{\boldsymbol{\epsilon}}_{t_n} = O(h^p) \qquad \text{as } h \longrightarrow 0 \tag{11-40}$$

Although Eq. (11-40) is only a notation for Eq. (11-28), it emphasizes the important fact that the truncation error $\hat{\boldsymbol{\epsilon}}_{t_n}$ tends to zero at least as fast as h^p tends to zero. For this reason, Taylor's algorithm is called a *pth-order algorithm*. To appreciate the significance of Eq. (11-28), let us choose a step size $h = \frac{1}{10}$. Then the truncation error corresponding to different values of p is given by

$$\begin{aligned} p &= 1, \qquad \hat{\boldsymbol{\epsilon}}_{t_n} \le 0.1\hat{K}_1 \\ p &= 2, \qquad \hat{\boldsymbol{\epsilon}}_{t_n} \le 0.01\hat{K}_2 \\ p &= 3, \qquad \hat{\boldsymbol{\epsilon}}_{t_n} \le 0.001\hat{K}_3 \\ &\vdots \\ p &= m, \qquad \hat{\boldsymbol{\epsilon}}_{t_n} \le \frac{1}{10^m}\hat{K}_m \end{aligned}$$

Since $\hat{K}_p$ is a constant independent of h, the truncation error decreases dramatically as the *order* p of the algorithm increases. Conversely, given a maximum allowable truncation error $\hat{\boldsymbol{\epsilon}}_{t_n} \le \boldsymbol{\epsilon}_m$, a much larger step size could be chosen when a higher-order algorithm is used. Since the total number of time steps is given by $N = T/h$, it is clear that N decreases as the order of the algorithm increases. Unfortunately, the higher the order, the more derivative terms of $f(x, t)$ with respect to t must be evaluated, as shown in Eq. (11-22). Consequently, Taylor's algorithm is seldom used beyond the fourth order. Let us now study some typical low-order Taylor's algorithms.

11-3-1 First-Order Taylor Algorithm: The Forward Euler Algorithm

In this case, $p = 1$ and Eqs. (11-21) and (11-22) can be combined to give

$$\boxed{x_{n+1} = x_n + hf(x_n, t_n)} \tag{11-41}$$

Equation (11-41) is called the *forward Euler* algorithm and corresponds to the first two terms of a Taylor series expansion. The forward Euler algorithm has the simple geometrical interpretation shown in Fig. 11-5, where $x = \hat{x}(t)$ is the exact solution corresponding to the "new" initial condition x_n at $t = t_n$. The exact solution at $t = t_{n+1}$ is given by $\hat{x}(t_{n+1})$, and the approximate solution computed by the forward Euler algorithm is given by $x_{n+1} = x_n + h\,[d\hat{x}(t)/dt]|_{t=t_n}$. Observe that the local truncation error $\boldsymbol{\epsilon}_{t_{n+1}}$ is quite large. Hence, to obtain reasonable accuracy, the step size in the forward Euler algorithm must be made very small, which is why this algorithm is seldom used in practice.

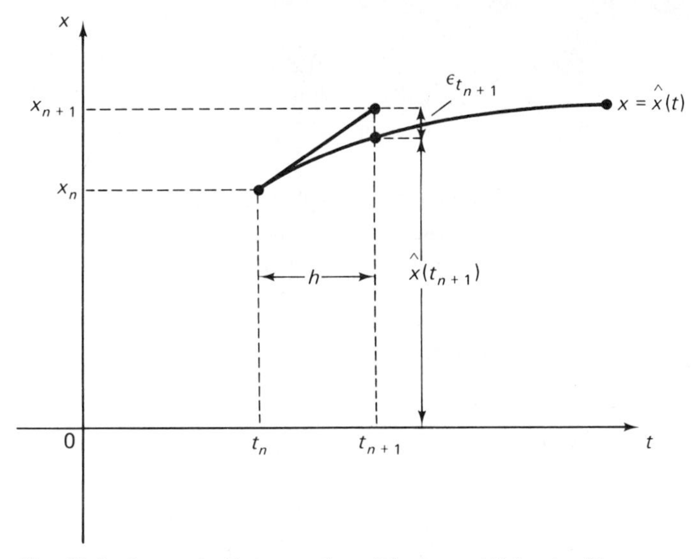

Fig. 11-5. Geometrical interpretation of the forward Euler algorithm.

11-3-2 Second-Order Taylor Algorithm

In this case, $p = 2$ and Eqs. (11-21) and (11-22) become

$$x_{n+1} = x_n + hT_2(x_n, t_n; h) \tag{11-42}$$

where

$$T_2(x_n, t_n; h) = f(x_n, t_n) + \frac{h}{2}[f_x(x_n, t_n)f(x_n, t_n) + f_t(x_n, t_n)] \tag{11-43}$$

where

$$f_x \triangleq \frac{\partial f(x, t)}{\partial x} \quad \text{and} \quad f_t \triangleq \frac{\partial f(x, t)}{\partial t}$$

The expression in the bracket comes from

$$f^{(1)}(x_n, t_n) = \frac{\partial f}{\partial x}\frac{dx}{dt} + \frac{\partial f}{\partial t}\bigg|_{t=t_n} = f_x(x_n, t_n)f(x_n, t_n) + f_t(x_n, t_n) \tag{11-44}$$

11-3-3 Third-Order Taylor Algorithm

In this case $p = 3$ and Eqs. (11-21) and (11-22) become

$$x_{n+1} = x_n + hT_3(x_n, t_n; h) \tag{11-45}$$

where

$$\begin{aligned} T_3(x_n, t_n; h) = {} & f(x_n, t_n) + \frac{h}{2}[f_x(x_n, t_n)f(x_n, t_n) + f_t(x_n, t_n)] \\ & + \frac{h^2}{3!}\{f_{tt}(x_n, t_n) + 2f_{tx}(x_n, t_n)f(x_n, t_n) \\ & + f_{xx}(x_n, t_n)f^2(x_n, t_n) + [f_t(x_n, t_n) \\ & + f_x(x_n, t_n)f(x_n, t_n)]f_x(x_n, t_n)\} \end{aligned} \tag{11-46}$$

where

$$f_{tt} \triangleq \frac{\partial^2 f(x, t)}{\partial t^2}, \qquad f_{tx} \triangleq \frac{\partial^2 f(x, t)}{\partial t\, \partial x}, \qquad f_{xx} \triangleq \frac{\partial^2 f(x, t)}{\partial x^2}$$

The major disadvantage of the Taylor algorithm is now apparent: evaluating the partial derivatives is a formidable task. Moreover, it is extremely error-prone, especially if $f(x, t)$ is not available in explicit analytic form, as in most network problems. Fortunately, the mathematicians Runge and Kutta have found a clever method that obviates the need for evaluating the partial derivatives and still retains the same order of accuracy as the Taylor algorithm. The basic idea consists of replacing the function $T_p(x_n, t_n; h)$ in Eq. (11-22) by another function $K_p(x_n, t_n; h)$ requiring no partial derivatives of $f(x, t)$ such that

$$|K_p(x_n, t_n; h) - T_p(x_n, t_n; h)| \leq Rh^p \tag{11-47}$$

where R is some constant independent of h. In view of Eq. (11-47), the modified algorithm

$$\boxed{x_{n+1} = x_n + hK_p(x_n, t_n; h)} \tag{11-48}$$

has precisely the same order of magnitude of truncation error as the corresponding Taylor's algorithm. Consequently, Eq. (11-48) is called the *pth-order Runge–Kutta algorithm.*

11-4 RUNGE–KUTTA ALGORITHM

To understand how the functions $K_p(x_n, t_n; h)$ are derived, we will consider in detail the case $p = 2$. The higher-order Runge–Kutta algorithms can be similarly derived, but the manipulation becomes extremely involved and tedious. The interested reader is referred to reference 1 for additional details.

11-4-1 Second-Order Runge–Kutta Algorithm

Suppose that instead of

$$T_2(x_n, t_n; h) = f(x_n, t_n) + \frac{h}{2}[f_x(x_n, t_n)f(x_n, t_n) + f_t(x_n, t_n)] \tag{11-49}$$

we define

$$K_2(x_n, t_n; h) \triangleq a_1 f(x_n, t_n) + a_2 f[x_n + \alpha h f(x_n, t_n), t_n + \beta h] \tag{11-50}$$

where the coefficients a_1, a_2, α, and β are to be determined such that

$$|K_2(x_n, t_n; h) - T_2(x_n, t_n; h)| < Rh^2 \tag{11-51}$$

Let us first expand $f(x, t)$ by Taylor series in two variables about (x_n, t_n) and evaluate the resulting expansion at $x = x_n + \alpha h f(x_n, t_n)$ and $t = t_n + \beta h$:

$$f[x_n + \alpha h f(x_n, t_n), t_n + \beta h] = f(x_n, t_n) + \alpha h f(x_n, t_n) f_x(x_n, t_n) + \beta h f_t(x_n, t_n) + O(h^2) \tag{11-52}$$

where the symbol $O(h^2)$ denotes terms involving $h^p, p \geq 2$. Substituting Eq. (11-52) into Eq. (11-50), we obtain

$$K_2(x_n, t_n; h) = (a_1 + a_2)f(x_n, t_n) + h[a_2\alpha f(x_n, t_n)f_x(x_n, t_n) + a_2\beta f_t(x_n, t_n] + O(h^2) \quad (11\text{-}53)$$

Observe next that if we choose

$$a_1 + a_2 = 1 \quad (11\text{-}54)$$

$$a_2\alpha = \tfrac{1}{2} \quad (11\text{-}55)$$

$$a_2\beta = \tfrac{1}{2} \quad (11\text{-}56)$$

then, except for the remainder term $O(h^2)$, Eq. (11-56) is identical to Eq. (11-49)! Hence, by making a_1, a_2, α, and β satisfy Eqs. (11-54)–(11-56), we have indeed the desirable property specified by Eq. (11-51). It follows from Eqs. (11-54), (11-55), and (11-56) that

$$a_1 = 1 - a_2 \quad (11\text{-}57)$$

$$\alpha = \beta = \frac{1}{2a_2} \quad (11\text{-}58)$$

provided $a_2 \neq 0$. Substituting Eqs. (11-57) and (11-58) into (11-50), we obtain the following *second-order Runge–Kutta algorithm*:

$$x_{n+1} = x_n + hK_2(x_n, t_n; h) \quad (11\text{-}59)$$

where

$$K_2(x_n, t_n; h) = (1 - a_2)f(x_n, t_n) + a_2 f\left[x_n + \frac{h}{2a_2}f(x_n, t_n), t_n + \frac{h}{2a_2}\right] \quad (11\text{-}60)$$

Observe that we still have a free parameter $a_2 \neq 0$ in Eq. (11-60). Consequently, an entire family of second-order Runge–Kutta algorithms can be derived by assigning different values to a_2. We consider two common choices:

Case 1. Heun's Algorithm (Modified Trapezoidal Algorithm). In this case, we choose $a_2 = \frac{1}{2}$; Eqs. (11-59) and (11-60) can be combined to give

$$x_{n+1} = x_n + \frac{h}{2}\{f(x_n, t_n) + f[x_n + hf(x_n, t_n), t_n + h]\} \quad (11\text{-}61)$$

Equation (11-61) is usually referred to as either Heun's algorithm or as the modified trapezoidal algorithm.

Case 2. Modified Euler–Cauchy Algorithm. In this case, we choose $a_2 = 1$; Eqs. (11-59) and (11-60) can be combined to give

$$x_{n+1} = x_n + hf\left[x_n + \frac{h}{2}f(x_n, t_n), t_n + \frac{h}{2}\right] \quad (11\text{-}62)$$

Equation (11-62) is sometimes referred to as the *modified Euler–Cauchy algorithm*.

11-4-2 Fourth-Order Runge–Kutta Algorithm

For larger step size and greater accuracy, the *fourth-order Runge–Kutta* algorithm is the most widely used algorithm. It is given by

$$x_{n+1} = x_n + hK_4(x_n, t_n; h) \tag{11-63}$$

where

$$K_4(x_n, t_n; h) = \tfrac{1}{6}[k_1 + 2k_2 + 2k_3 + k_4] \tag{11-64}$$

$$k_1 \triangleq f(x_n, t_n) \tag{11-65}$$

$$k_2 \triangleq f\left[x_n + \frac{h}{2}k_1, t_n + \frac{h}{2}\right] \tag{11-66}$$

$$k_3 \triangleq f\left[x_n + \frac{h}{2}k_2, t_n + \frac{h}{2}\right] \tag{11-67}$$

$$k_4 \triangleq f[x_n + hk_3, t_n + h] \tag{11-68}$$

Observe that the fourth-order Runge–Kutta algorithm computes first for the slope k_1 at x_n. Using this value, we move one half-step $h/2$ forward in time and evaluate the new slope there. Using this modified slope k_2, we move again one half-step forward from the same point (x_n, t_n) and calculate the slope k_3. Finally, we move one full step forward from (x_n, t_n) and calculate the slope k_4. The four slopes are then averaged with respective weights $\frac{1}{6}, \frac{2}{6}, \frac{2}{6}$, and $\frac{1}{6}$. It is interesting that, if $f(x, t)$ does not depend on x, this averaging process reduces to the familiar Simpson's rule.

Since the algorithm is a *fourth-order* algorithm, a relatively large step size could be chosen with a relatively small local truncation error. Unfortunately, because of the involved mathematical manipulations used in deriving Eq. (11-63), it is difficult to estimate the *actual* local error per time step. Consequently, the step size h is usually chosen rather conservatively; i.e., much smaller than is necessary to meet a prescribed maximum error. Another disadvantage of the fourth-order Runge–Kutta algorithm is that the function $f(x_n, t_n)$ must be evaluated four times at each time step. Moreover, these values of the function are not used in any subsequent computation. Hence, computationally speaking, this algorithm is not as efficient as some of the *multistep algorithms* to be presented in Section 11-5, which do make use of previously computed data.

11-5 NUMERICAL SOLUTION BY POLYNOMIAL APPROXIMATION

Suppose that the *exact* solution to an initial-value problem $\dot{x} = f(x, t)$, $x(t_0) = x_0$, is given by a *polynomial of degree* k:

$$\hat{x}(t) = \alpha_0 + \alpha_1 t + \alpha_2 t^2 + \cdots + \alpha_k t^k \tag{11-69}$$

where $\alpha_0, \alpha_1, \alpha_2, \ldots, \alpha_k$ are constants. Suppose that we are given the exact *value* of $\hat{x}(t)$ and its first derivative $\hat{x}'(t)$ at $t = t_n, t_{n-1}, t_{n-2}, \ldots, t_{n-p}$; i.e., $\hat{x}(t_n), \hat{x}(t_{n-1}), \ldots, \hat{x}(t_{n-p})$, and $\hat{x}'(t_n), \hat{x}'(t_{n-1}), \ldots, \hat{x}'(t_{n-p})$. Our objective in this section is to derive numerical algorithms such that the value of $\hat{x}(t_{n+1}) \triangleq \hat{x}(t_n + h)$ can be computed *exactly*. For example, if $k = 1$ in Eq. (11-69), the forward Euler algorithm presented in Section 11-4 would have exactly this property. In general, *any algorithm capable of calculating the exact value $\hat{x}(t_{n+1})$ for an initial-value problem having an exact solution in the form of a* k*th degree polynomial is called a numerical-integration formula of order* k.[5] Of course, if the exact solution is *not* a polynomial, a numerical-integration formula will generally give only an *approximate value* x_{n+1} and not the exact value $\hat{x}(t_{n+1})$. However, in view of the classical *Weierstrass approximation theorem* [6], which asserts that *any continuous function can be approximated arbitrarily closely within any closed interval by a polynomial of sufficiently high degree*, it is clear that even if the solution is not a polynomial a numerical-integration formula of sufficiently high order can, *in principle*, be used to calculate $\hat{x}(t_{n+1})$ to any desired accuracy. In practice, however, the amount of computation and round-off error increases with the order of the integration formula, and only values of $k < 10$ are of practical value.

Unlike Taylor's algorithm, past information from previous time steps is utilized in most numerical-integration formulas to compute x_{n+1}. This is illustrated by the shaded region under the curve $\hat{x}(t)$ in Fig. 11-6 for the case $p = 9$. Consequently, a numerical-integration algorithm with $p > 1$ is called a *multistep algorithm* in contrast to Taylor's algorithm, which is a *single-step algorithm*. A multistep numerical-integration formula is generally of the following form:

Multistep Numerical-Integration Algorithm

$$\begin{aligned} x_{n+1} &= a_0 x_n + a_1 x_{n-1} + \cdots + a_p x_{n-p} + h[b_{-1} f(x_{n+1}, t_{n+1}) \\ &\quad + b_0 f(x_n, t_n) + \cdots + b_p f(x_{n-p}, t_{n-p})] \\ &= \sum_{i=0}^{p} a_i x_{n-i} + h \sum_{i=-1}^{p} b_i f(x_{n-i}, t_{n-i}) \end{aligned} \tag{11-70}$$

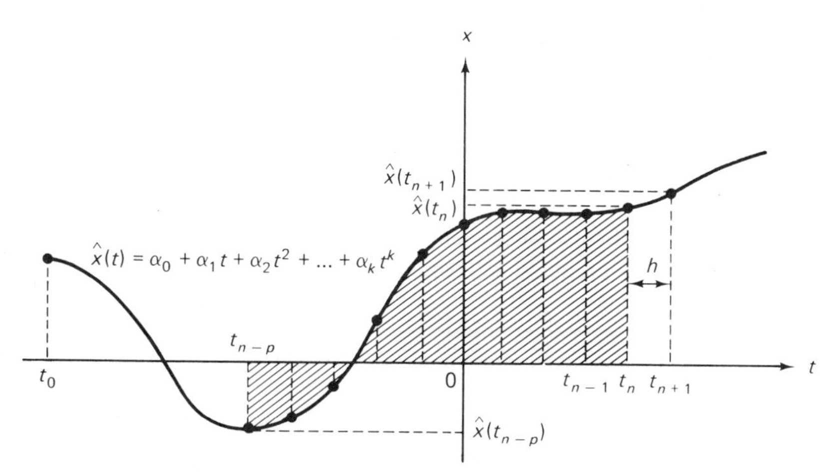

Fig. 11-6. Geometrical interpretation of a multistep algorithm.

[5]The term "numerical integration" comes from the observation that algorithms of this type are similar to those used for numerical integration of functions. The "order" of a numerical-integration algorithm refers to the *maximum*-degree polynomial solution in which the formula would give the *exact* value for $\hat{x}(t_{n+1})$ if there were no round-off error. This term should not be confused with the "order" of Taylor or Runge–Kutta algorithms presented in Sections 11-3 and 11-4.

where $a_0, a_1, \ldots, a_p, b_{-1}$, and $b_0, b_1, \ldots, b_p$ are $2p + 3$ coefficients to be determined such that, if the exact solution is a polynomial and if the previously calculated values $x_n, x_{n-1}, \ldots, x_{n-p}$ and $x'_n, x'_{n-1}, \ldots, x'_{n-p}$ are *assumed* to be exact, then Eq. (11-70) would give the *exact* value of x_{n+1}. To illustrate how these coefficients are determined, let us derive a numerical-integration formula of order $k = 2$. Hence, Eq. (11-69) simplifies to a second-degree polynomial:

$$\hat{x}(t) = \alpha_0 + \alpha_1 t + \alpha_2 t^2 \tag{11-71}$$

Substituting $t = t_{n+1} = t_n + h$ into Eq. (11-71), we obtain

$$\begin{aligned}\hat{x}(t_{n+1}) &= \alpha_0 + \alpha_1(t_n + h) + \alpha_2(t_n + h)^2 \\ &= (\alpha_0 + \alpha_1 t_n + \alpha_2 t_n^2) + h(\alpha_1 + 2\alpha_2 t_n + \alpha_2 h) \\ &= \hat{x}(t_n) + h(\alpha_1 + 2\alpha_2 t_n + \alpha_2 h)\end{aligned} \tag{11-72}$$

Now suppose that we choose the simplest case, where $p = 0$ in Eq. (11-70):

$$x_{n+1} = a_0 x_n + h[b_{-1} f(x_{n+1}, t_{n+1}) + b_0 f(x_n, t_n)] \tag{11-73}$$

Our objective is to find the coefficients a_0, b_{-1}, and b_0 such that if $\hat{x}(t_n) = x_n$, then $\hat{x}(t_{n+1}) = x_{n+1}$.

We can find these coefficients by the following *method of undetermined coefficients*. From Eq. (11-71), we have

$$f(x_{n+1}, t_{n+1}) \triangleq \hat{x}'(t_{n+1}) = \alpha_1 + 2\alpha_2 t_{n+1} \tag{11-74}$$

$$f(x_n, t_n) \triangleq \hat{x}'(t_n) = \alpha_1 + 2\alpha_2 t_n \tag{11-75}$$

Substituting Eqs. (11-74) and (11-75) into Eq. (11-73), we obtain

$$\begin{aligned}x_{n+1} &= a_0 x_n + h\{b_{-1}[\alpha_1 + 2\alpha_2(t_n + h)] + b_0(\alpha_1 + 2\alpha_2 t_n)\} \\ &= a_0 x_n + h[\alpha_1(b_{-1} + b_0) + 2\alpha_2 t_n(b_{-1} + b_0) + \alpha_2 h(2b_{-1})]\end{aligned} \tag{11-76}$$

Comparing Eqs. (11-72) and (11-76) and recalling that $\hat{x}(t_n) = x_n$ and $\hat{x}(t_{n+1}) = x_{n+1}$, we obtain the following constraints:

$$a_0 = 1 \tag{11-77}$$

$$b_{-1} + b_0 = 1 \tag{11-78}$$

$$2b_{-1} = 1 \tag{11-79}$$

Observe that we have three independent equations in three unknowns, and hence the coefficients can be uniquely determined; i.e., $a_0 = 1$, $b_{-1} = \frac{1}{2}$, and $b_0 = \frac{1}{2}$. Hence, Eq. (11-73) becomes

Second-Order Trapezoidal Algorithm

$$x_{n+1} = x_n + \frac{h}{2}[f(x_{n+1}, t_{n+1}) + f(x_n, t_n)] \tag{11-80}$$

This second-order numerical-integration formula is usually called the *trapezoidal algorithm*, since the second term can be interpreted as the *area under a trapezoid*. The trapezoidal algorithm is a two-step formula[6] because the values of $\hat{x}(t)$ at two

[6] Also known as a two-point formula.

instants of time, t_n and t_{n+1}, are needed. Observe that Eq. (11-80) defines x_{n+1} only *implicitly* since the unknown x_{n+1} appears on both sides of the equation. Such algorithms are called *implicit algorithms* in contrast to Taylor's algorithm, which is an *explicit algorithm.*[7] Methods for calculating x_{n+1} from an implicit algorithm will be discussed in Section 11-5-2.

The preceding method of undetermined coefficients can be used to derive numerical-integration formulas of order $k > 2$. In this case, more terms must be chosen from Eq. (11-70) in order to obtain the same number of independent constraint equations as there are unknowns. The resulting algorithm would therefore be a *multistep* (multipoint) formula. If $b_{-1} = 0$, it is an *explicit* (open-type) formula. If $b_{-1} \neq 0$, it is an *implicit* (closed-type) formula. An example of a three-step explicit formula, to be derived in Chapter 12 by the method of undetermined coefficients, is given by

Third-Order Adams–Bashforth Algorithm

$$x_{n+1} = x_n + h[\tfrac{23}{12}f(x_n, t_n) - \tfrac{16}{12}f(x_{n-1}, t_{n-1}) + \tfrac{5}{12}f(x_{n-2}, t_{n-2})] \tag{11-81}$$

An example of a four-step implicit formula, to be derived in Chapter 12, is given by

Fourth-Order Adams–Moulton Algorithm

$$x_{n+1} = x_n + h[\tfrac{9}{24}f(x_{n+1}, t_{n+1}) + \tfrac{19}{24}f(x_n, t_n) - \tfrac{5}{24}f(x_{n-1}, t_{n-1}) + \tfrac{1}{24}f(x_{n-2}, t_{n-2})] \tag{11-82}$$

11-5-1 Local Truncation Error of Numerical-Integration Formulas

To determine the accuracy of numerical-integration formulas, let us derive the *local truncation error.*[8]

$$\epsilon_T \triangleq \hat{x}(t_{n+1}) - x_{n+1} \tag{11-83}$$

Observe that when the exact solution $\hat{x}(t)$ is not a *polynomial* of degree k the local truncation error is generally not zero. To derive ϵ_T, let us first expand the exact solution $\hat{x}(t)$ about the point t_{n-p} using *Taylor's formula with an integral remainder* [6]:

$$\hat{x}(t) = \hat{x}(t_{n-p}) + (t - t_{n-p})\hat{x}^{(1)}(t_{n-p}) + \cdots + \frac{(t - t_{n-p})^k}{k!}\hat{x}^{(k)}(t_{n-p}) + \frac{1}{k!}\int_{t_{n-p}}^{t} \hat{x}^{(k+1)}(\tau)(t - \tau)^k \, d\tau \tag{11-84}$$

[7]Implicit algorithms are also known as algorithms of a *closed type* or *iterative algorithms;* explicit algorithms are also known as algorithms of an *open type* or *forward algorithms.*

[8]Observe that our present definition for *local truncation error* does not require taking the *norm* of $\hat{x}(t_{n+1}) - x_{n+1}$ and therefore differs slightly from the definition given earlier in Eq. (11-15). To emphasize this difference, an uppercase subscript "T" is used in place of the lowercase subscript in Eq. (11-15). Our present definition is more convenient for deriving local truncation error for first-order systems. Moreover, since the *sign* of the error is not suppressed, we can tell whether the *computed* solution is above or below the true solution.

Equation (11-84) can be rewritten into a more convenient form by defining the kth-*degree polynomial*

$$p_k(t) \triangleq \hat{x}(t_{n-p}) + (t - t_{n-p})\hat{x}^{(1)}(t_{n-p}) + \cdots + \frac{(t - t_{n-p})^k}{k!}\hat{x}^{(k)}(t_{n-p}) \tag{11-85}$$

and the *truncated power function*

$$U_k(y) \triangleq \begin{Bmatrix} y^k, & y \geq 0 \\ 0, & y < 0 \end{Bmatrix} \tag{11-86}$$

Equation (11-84) can now be written as follows:

$$\hat{x}(t) = p_k(t) + \frac{1}{k!}\int_{t_{n-p}}^{\infty} \hat{x}^{(k+1)}(\tau)U_k(t - \tau)\,d\tau \tag{11-87}$$

To simplify notation, we let $\mathcal{A}_t$ be an *operator* (algorithm) that converts any differentiable *time function* $F(t)$ into a *number* in accordance with the following formula:

$$\begin{aligned}\mathcal{A}_t\{F(t)\} &= a_0F(t_n) + a_1F(t_{n-1}) + \cdots + a_pF(t_{n-p}) \\ &\quad + h[b_{-1}F'(t_{n+1}) + b_0F'(t_n) + \cdots + b_pF'(t_{n-p})] \\ &= \sum_{i=0}^{p} a_iF(t_{n-i}) + h\sum_{i=-1}^{p} b_iF'(t_{n-i})\end{aligned} \tag{11-88}$$

Observe that the coefficients a_i and b_i defining $\mathcal{A}_t$ in Eq. (11-88) are identical to the coefficients defining the multistep numerical-integration algorithm in Eq. (11-70). If we choose $F(t)$ in Eq. (11-88) to be the time function $\hat{x}(t)$ defined in Eq. (11-87), we obtain

$$\mathcal{A}_t\{\hat{x}(t)\} = \mathcal{A}_t\{p_k(t)\} + \frac{1}{k!}\int_{t_{n-p}}^{\infty} \hat{x}^{(k+1)}(\tau)\mathcal{A}_t\{U_k(t - \tau)\,d\tau\} \tag{11-89}$$

where the operator $\mathcal{A}_t$ is shifted beside the term $U_k(t - \tau)$, since $\mathcal{A}_t$ operates only on *time functions*, and $U_k(t - \tau)$ is the only term inside the integral that is a function of time. It follows from Eqs. (11-87) and (11-89) that

$$\begin{aligned}\hat{x}(t_n + h) - \mathcal{A}_t\{\hat{x}(t)\} &= p_k(t_n + h) + \frac{1}{k!}\int_{t_{n-p}}^{\infty} \hat{x}^{(k+1)}(\tau)U_k(t_n + h - \tau)\,d\tau \\ &\quad - \mathcal{A}_t\{p_k(t)\} - \frac{1}{k!}\int_{t_{n-p}}^{\infty} \hat{x}^{(k+1)}(\tau)\mathcal{A}_t\{U_k(t - \tau)\,d\tau\}\end{aligned} \tag{11-90}$$

Since $p_k(t)$ is a kth-*degree polynomial* and since the multistep numerical-integration algorithm defined by Eq. (11-70) will predict the *exact* value of a kth-*degree* polynomial at $t_{n+1} \triangleq t_n + h$ when the *exact* values of $p_k(t)$ and $p'_k(t)$ are given at $t = t_n, t_{n-1}, \ldots, t_{n-p}$, it follows that

$$\mathcal{A}_t\{p_k(t)\} \triangleq \sum_{i=0}^{p} a_ip_k(t_{n-i}) + h\sum_{i=-1}^{p} b_ip'_k(t_{n-i}) = p_k(t_n + h) \tag{11-91}$$

Since, *by assumption*, $x_n = \hat{x}(t_n)$, $x_{n-1} = \hat{x}(t_{n-1}), \ldots, x_{n-p} = \hat{x}(t_{n-p})$, $x'_n = f(x_n, t_n)$, $x'_{n-1} = f(x_{n-1}, t_{n-1}), \ldots, x'_{n-p} = f(x_{n-p}, t_{n-p})$, we can write

$$\begin{aligned}\mathcal{A}_t\{\hat{x}(t)\} &\triangleq \sum_{i=0}^{p} a_i\hat{x}(t_{n-i}) + h\sum_{i=-1}^{p} b_i\hat{x}'(t_{n-i}) = \sum_{i=0}^{p} a_i\hat{x}_{n-i} \\ &\quad + h\sum_{i=-1}^{p} b_if(x_{n-i}, t_{n-i}) = x_{n+1}\end{aligned} \tag{11-92}$$

Substituting Eqs. (11-91) and (11-92) into Eq. (11-90), we obtain

$$\begin{aligned}\hat{x}(t_n + h) - x_{n+1} &= \frac{1}{k!}\int_{t_{n-p}}^{\infty} \hat{x}^{(k+1)}(\tau)[U_k(t_n + h - \tau) - \mathcal{A}_t\{U_k(t - \tau)\}]\, d\tau \\ &= \frac{1}{k!}\int_{t_{n-p}}^{\infty} \hat{x}^{(k+1)}(\tau)E_k(t_n, \tau)\, d\tau\end{aligned} \tag{11-93}$$

where

$$E_k(t_n, \tau) \triangleq U_k(t_n + h - \tau) - \mathcal{A}_t\{U_k(t - \tau)\} \tag{11-94}$$

It follows from Eq. (11-86) that $U_k(t_n + h - \tau) = 0$ when $\tau > t_n + h$, and $\mathcal{A}_t\{U_k(t - \tau)\} = 0$ when $\tau > t_n$ [recall Eq. (11-88)]. Hence, $E_k(t_n, \tau) = 0$ for $\tau > t_n + h$. On the other hand, when $\tau \leq t_n - ph$, $t_n - \tau > 0$, and $U_k(t_n - \tau) = (t_n - \tau)^k$, a kth-*degree polynomial.* Hence, $\mathcal{A}_t\{U_k(t - \tau)\} = U_k(t_n + h - \tau)$ and $E_k(t_n, \tau) = 0$ when $\tau \leq t_n - ph$. Therefore, Eq. (11-94) is *nonzero* only in the interval $t_n - ph \leq \tau \leq t_n + h$. To simplify notation, let us choose $t_n = 0$ in Eqs. (11-93) and (11-94) and use the fact that $E_k(0, \tau) = 0$, whenever $\tau > h$ and $\tau < -ph$; thus

$$\hat{x}(h) - x_1 = \frac{1}{k!}\int_{-ph}^{h} \hat{x}^{(k+1)}(\tau)E_k(0, \tau)\, d\tau \tag{11-95}$$

where

$$\boxed{E_k(0, \tau) \triangleq U_k(h - \tau) - \mathcal{A}_t\{U_k(t - \tau)\}\big|_{t_n=0}, \qquad -ph \leq \tau \leq h} \tag{11-96}$$

Observe that for each multistep numerical integration algorithm, there corresponds a unique operator $\mathcal{A}_t$ and hence a unique function $E_k(0, \tau)$. For example, consider the *trapezoidal algorithm* in Eq. (11-80). Here, $k = 2$, $p = 0$, $0 \leq \tau \leq h$, and $a_0 = 1$, $b_{-1} = b_0 = \frac{1}{2}$. Equation (11-96) reduces to

$$E_2(0, \tau) = U_2(h - \tau) - \mathcal{A}_t\{U_2(t - \tau)\}\big|_{t_n=0}, \qquad 0 \leq \tau \leq h \tag{11-97}$$

Now

$$U_2(h - \tau) = \begin{cases} (h - \tau)^2, & \tau \leq h \\ 0 \quad, & \text{elsewhere} \end{cases} \tag{11-98}$$

$$\mathcal{A}_t\{U_2(t - \tau)\} = U_2(t_n - \tau) + \frac{h}{2}\left[\frac{\partial U_2(t - \tau)}{\partial t}\bigg|_{t=t_n+h} + \frac{\partial U_2(t - \tau)}{\partial t}\bigg|_{t=t_n}\right] \tag{11-99}$$

Substituting $t_n = 0$ into Eq. (11-99) and noting that $U_2(-\tau) = 0$ whenever $\tau > 0$,

$$\mathcal{A}_t\{U_2(t - \tau)\}\big|_{t_n=0} = U_2(-\tau) + \frac{h}{2}[2(h - \tau) + 0] = h(h - \tau) \tag{11-100}$$

Substituting Eqs. (11-99) and (11-100) into Eq. (11-97), we obtain the following expression for the *trapezoidal algorithm*:

$$E_2(0, \tau) = (h - \tau)^2 - h(h - \tau) = \tau^2 - \tau h, \qquad 0 \leq \tau \leq h \tag{11-101}$$

If we substitute Eq. (11-101) into Eq. (11-95) and set $k = 2$, $p = 0$, we would obtain the following local truncation error for the trapezoidal algorithm:

$$\epsilon_T \triangleq \hat{x}(h) - x_1 = \frac{1}{2}\int_0^h \hat{x}^{(3)}(\tau)E_2(0, \tau)\, d\tau \tag{11-102}$$

Since $E_2(0, \tau)$ does not change sign in $0 \leq \tau \leq h$, it follows from the *mean-value theorem* [6] that there exists $\hat{\tau} \in (0, h)$ such that

$$\epsilon_T = \frac{\hat{x}^{(3)}(\hat{\tau})}{2} \int_0^h E_2(0, \tau)\, d\tau = \frac{\hat{x}^{(3)}(\hat{\tau})}{2} \int_0^h (\tau^2 - \tau h)\, d\tau$$

$$= -\frac{\hat{x}^{(3)}(\hat{\tau})}{12} h^3 = O(h^3)$$

Following the same procedure as above, we will now derive the *main result* in this chapter.

Theorem 11-5 (Local Truncation Error for Multistep Algorithms). *The local truncation error ϵ_T of the* ***kth-order multistep numerical-integration algorithm***

$$x_{n+1} = \sum_{i=0}^{p} a_i x_{n-i} + h \sum_{i=-1}^{p} b_i f(x_{n-i}, t_{n-i})$$

which is exact for polynomial solution of degree less than or equal to k is given by

$$\epsilon_T = C_k \hat{x}^{(k+1)}(\hat{\tau}) h^{k+1} = O(h^{k+1}) \tag{11-103}$$

where

$$-ph < \hat{\tau} < h$$

and

$$C_k \triangleq \frac{1}{(k+1)!} \Big\{ (p+1)^{k+1} - \Big[\sum_{i=0}^{p-1} a_i (p-i)^{k+1} + (k+1) \sum_{i=-1}^{p-1} b_i (p-i)^k \Big] \Big\} \tag{11-104}$$

provided that $E_k(0, \tau)$ defined in Eq. (11-96) does not change sign in the interval $-ph \le \tau \le h$.

Proof. Since $E_k(0, \tau)$ does not change sign in the interval $-ph \le \tau \le h$, it follows from the *mean-value theorem* [6] that there exists $\hat{\tau} \in (-ph, h)$ such that Eq. (11-95) can be rewritten as follows:

$$\epsilon_T \triangleq \hat{x}(h) - x_1 = \frac{\hat{x}^{(k+1)}(\hat{\tau})}{k!} \int_{-ph}^{h} E_k(0, \tau)\, d\tau = C_k \hat{x}^{(k+1)}(\hat{\tau}) h^{k+1}$$

where

$$C_k \triangleq \frac{1}{k!\, h^{k+1}} \int_{-ph}^{h} E_k(0, \tau)\, d\tau \tag{11-105}$$

To evaluate $E_k(0, \tau)$, we recall the definition of the operator $\mathcal{A}_t$ from Eq. (11-88) and write

$$\mathcal{A}_t\{U_k(t - \tau)\}\big|_{t_n = 0} = \sum_{i=0}^{p} a_i U_k(-ih - \tau) + h \sum_{i=-1}^{p} b_i U_k'(-ih - \tau) \tag{11-106}$$

where

$$U_k'(-ih - \tau) = \frac{\partial U_k(t - \tau)}{\partial t}\bigg|_{t=-ih} = \begin{cases} k(-ih - \tau)^{k-1}, & -ph \le \tau \le -ih \\ 0, & \text{elsewhere} \end{cases} \tag{11-107}$$

Substituting Eq. (11-106) into Eq. (11-96) and recalling the definition of $U_k(h-\tau)$ from Eq. (11-86), we obtain

$$\begin{aligned}
\int_{-ph}^{h} E_k(0,\tau)\,d\tau &= \\
&= \int_{-ph}^{h}\left\{U_k(h-\tau)-\left[\sum_{i=0}^{p} a_iU_k(-ih-\tau)+h\sum_{i=-1}^{p} b_iU_k'(-ih-\tau)\right]\right\}d\tau \\
&= \int_{-ph}^{h} U_k(h-\tau)\,d\tau-\sum_{i=0}^{p}\int_{-ph}^{h} a_iU_k(-ih-\tau)d\tau-h\sum_{i=-1}^{p}\int_{-ph}^{h} b_iU_k'(-ih-\tau)\,d\tau \\
&= \int_{-ph}^{h}(h-\tau)^k\,d\tau-\sum_{i=0}^{p}\int_{-ph}^{-ih} a_i(-ih-\tau)^k\,d\tau-h\sum_{i=-1}^{p}\int_{-ph}^{-ih} b_ik(-ih-\tau)^{k-1}\,d\tau \\
&= \int_{-ph}^{h}(h-\tau)^k\,d\tau-\sum_{i=0}^{p-1}\int_{-ph}^{-ih} a_i(-ih-\tau)^k\,d\tau-h\sum_{i=-1}^{p-1}\int_{-ph}^{-ih} b_ik(-ih-\tau)^{k-1}\,d\tau \\
&= \frac{[(p+1)h]^{k+1}}{k+1}-\sum_{i=0}^{p-1}\frac{a_i[(p-i)h]^{k+1}}{k+1}-h\sum_{i=-1}^{p-1} b_i[(p-i)h]^k \\
&= \frac{h^{k+1}}{k+1}\left\{(p+1)^{k+1}-\left[\sum_{i=0}^{p-1} a_i(p-i)^{k+1}+(k+1)\sum_{i=-1}^{p-1} b_i(p-i)^k\right]\right\}
\end{aligned} \tag{11-108}$$

Substituting Eq. (11-108) into Eq. (11-105), we obtain

$$\begin{aligned}
C_k = \frac{1}{k!(k+1)}&\left\{(p+1)^{k+1}\right. \\
&\left.-\left[\sum_{i=0}^{p-1} a_i(p-i)^{k+1}+(k+1)\sum_{i=-1}^{p-1} b_i(p-i)^k\right]\right\}
\end{aligned} \tag{11-109}$$

which can be rewritten as Eq. (11-104). Q.E.D.

11-5-2 Implicit Algorithm via Predictor–Corrector Formulas

Let us now focus our attention on the case of *implicit* numerical-integration algorithms ($b_{-1} \neq 0$) and rewrite Eq. (11-70) in the following more convenient explicit form:

$$\boxed{x_{n+1} = \sum_{i=0}^{p}[a_ix_{n-i}+hb_if(x_{n-i},t_{n-i})]+hb_{-1}f(x_{n+1},t_{n+1})} \tag{11-110}$$

Since the *only unknown quantity* in Eq. (11-110) is x_{n+1}, let us emphasize this fact by writing Eq. (11-110) in the form

$$x_{n+1} = F(x_{n+1}) \tag{11-111}$$

where

$$F(x_{n+1}) \triangleq \sum_{i=0}^{p}[a_ix_{n-i}+hb_if(x_{n-i},t_{n-i})]+hb_{-1}f(x_{n+1},t_{n+1}) \tag{11-112}$$

Observe that Eq. (11-111) is in the standard form considered in Eq. (5-20). The solution x_{n+1} is therefore a *fixed point* of the function $F(x_{n+1})$. Applying the *fixed-point iteration formula* defined by Eq. (5-21), we obtain[9]

$$\boxed{x_{n+1}^{(j+1)} = F(x_{n+1}^{(j)})} \tag{11-113}$$

Hence, starting with an *initial guess* $x_{n+1}^{(0)}$, we obtain $x_{n+1}^{(1)}$; from $x_{n+1}^{(1)}$ we compute $x_{n+1}^{(2)}$, and so on. If this *iterative algorithm* converges, then the solution x_{n+1} is given by

$$x_{n+1} = \lim_{j \to \infty} x_{n+1}^{(j)} \tag{11-114}$$

Our next task is to derive a criterion which assures that the fixed-point iteration always converges. Invoking the *principle of contraction mapping* from Chapter 5, we find that a *sufficient condition* is for $F(x_{n+1})$ to be a *contraction*, i.e., if there exists a constant $L < 1$ such that

$$|F(x'_{n+1}) - F(x''_{n+1})| \leq L|x'_{n+1} - x''_{n+1}| \tag{11-115}$$

where x'_{n+1} and x''_{n+1} are any two points lying within the closed interval $[x_{n+1} - \delta, x_{n+1} + \delta]$ centered about the solution x_{n+1}, where δ is some positive number. Substituting Eq. (11-112) for $F(x'_{n+1})$ and $F(x''_{n+1})$ in Eq. (11-115), we obtain

$$\begin{aligned} |F(x'_{n+1}) - F(x''_{n+1})| &= |hb_{-1}f(x'_{n+1}, t_{n+1}) - hb_{-1}f(x''_{n+1}, t_{n+1})| \\ &= h|b_{-1}|\cdot|f(x'_{n+1}, t_{n+1}) - f(x''_{n+1}, t_{n+1})| \end{aligned} \tag{11-116}$$

Now suppose that the function $f(x, t)$ satisfies a Lipschitz condition with Lipschitz constant L_{n+1} at $t = t_{n+1}$ for $x_{n+1} - \delta \leq x \leq x_{n+1} + \delta$; i.e.,

$$|f(x'_{n+1}, t_{n+1}) - f(x''_{n+1}, t_{n+1})| \leq L_{n+1}|x'_{n+1} - x''_{n+1}| \tag{11-117}$$

Substituting Eq. (11-117) into Eq. (11-116), we obtain

$$|F(x'_{n+1}) - F(x''_{n+1})| \leq h|b_{-1}|\cdot L_{n+1}|x'_{n+1} - x''_{n+1}| \tag{11-118}$$

Comparing Eq. (11-115) with Eq. (11-118), we see that the fixed point iteration will converge if

$$L \triangleq h|b_{-1}|\cdot L_{n+1} < 1 \tag{11-119}$$

or

$$\boxed{h < \frac{1}{|b_{-1}|\cdot L_{n+1}}} \tag{11-120}$$

Equation (11-120) shows that to ensure convergence of Eq. (11-111) the step size h must be chosen small enough. We shall summarize this observation in the form of a theorem, and give a separate proof for it without using the principle of contraction mapping.

Theorem 11-6. *If $f(x, t)$ satisfies a Lipschitz condition with Lipschitz constant L_{n+1} at $t = t_{n+1}$ for all x satisfying $x_{n+1} - \delta \leq x \leq x_{n+1} + \delta$, $\delta > 0$, where*

[9] The iteration index $j + 1$ is now indicated as a superscript to avoid confusion with the subscript in the unknown x_{n+1}.

x_{n+1} is the exact solution of the implicit equation (11-111), then the fixed-point iteration algorithm

$$x_{n+1}^{(j+1)} = \sum_{i=0}^{p} [a_i x_{n-i} + hb_i f(x_{n-i}), t_{n-i})] + hb_{-1} f(x_{n+1}^{(j)}, t_{n+1}) \tag{11-121}$$

*converges to the solution x_{n+1} if the **step size h** satisfies Eq. (11-120).*

Proof: Let us define the following error measures:

$$\Delta^{(0)} \triangleq |x_{n+1}^{(0)} - x_{n+1}|, \qquad \Delta^{(1)} \triangleq |x_{n+1}^{(1)} - x_{n+1}|, \ldots,$$
$$\Delta^{(j)} \triangleq |x_{n+1}^{(j)} - x_{n+1}|, \qquad \Delta^{(j+1)} \triangleq |x_{n+1}^{(j+1)} - x_{n+1}|$$

Now

$$x_{n+1}^{(j+1)} - x_{n+1} = hb_{-1}[f(x_{n+1}^{(j)}, t_{n+1}) - f(x_{n+1}, t_{n+1})] \tag{11-122}$$

It follows from Eq. (11-122) that

$$\begin{aligned}\Delta^{(j+1)} &\leq h|b_{-1}|\cdot L_{n+1}|x_{n+1}^{(j)} - x_{n+1}| \\ &\leq [h|b_{-1}|\cdot L_{n+1}][h|b_{-1}|L_{n+1}]|x_{n+1}^{(j-1)} - x_{n+1}| \\ &\leq \underbrace{[h|b_{-1}|\cdot L_{n+1}]\ldots[h|b_{-1}|\cdot L_{n+1}]}_{(j+1)\text{ times}}|x_{n+1}^{(0)} - x_{n+1}|\end{aligned} \tag{11-123}$$

Therefore,

$$\Delta^{(j+1)} \leq [h|b_{-1}|\cdot L_{n+1}]^{j+1}\Delta^{(0)} \tag{11-124}$$

provided $x_{n+1}^{(1)}, x_{n+1}^{(2)}, \ldots, x_{n+1}^{(j)}$ remain within the interval $x_{n+1} - \delta \leq x_{n+1}^{(j)} \leq x_{n+1} + \delta$, where $f(x, t)$ is Lipschitz with a Lipschitz constant L_{n+1} at $t = t_{n+1}$. This is satisfied because $|x_{n+1}^{(0)} - x_{n+1}| \triangleq \Delta^{(0)} < \delta$ by hypothesis, and $|x_{n+1}^{(1)} - x_{n+1}| = \Delta^{(1)} < \Delta^{(0)} < \delta$, $|x_{n+1}^{(2)} - x_{n+1}| = \Delta^{(2)} < \Delta^{(1)} < \Delta^{(0)} < \delta$, etc. It follows from Eqs. (11-120) and (11-124) that

$$\lim_{j\to\infty} \Delta^{(j+1)} = 0 \tag{11-125}$$

Hence, $x_{n+1}^{(j+1)}$ converges to the fixed point x_{n+1} as $j \to \infty$. Q.E.D.

Examination of Eq. (11-124) shows that, as the constant $L \triangleq h|b_{-1}|\cdot L_{n+1}$ approaches 1, the rate of convergence becomes extremely slow. In fact, when $L = 1$, we have $\Delta^{(j+1)} = \Delta^{(0)}$, and the error may never decrease. Consequently, to hasten convergence of the fixed-point iteration it is desirable to choose a small step size h. Unfortunately, as h decreases, the total number of time steps $N \triangleq T/h$ increases, where T is the length of the time interval in which the solution is desired. A compromise is clearly necessary to satisfy these conflicting requirements.

For a chosen step size h, the number of fixed-point iterations can be reduced if the *initial guess* $x_{n+1}^{(0)}$ can be chosen close enough to the solution x_{n+1}. Rather than choosing $x_{n+1}^{(0)}$ arbitrarily, it is more satisfactory to use an *explicit algorithm* to *predict*

$x_{n+1}^{(0)}$. For example, suppose that we choose the forward Euler algorithm

$$x_{n+1} = x_n + hf(x_n, t_n) \tag{11-126}$$

to predict the initial guess $x_{n+1}^{(0)}$ for the trapezoidal algorithm defined by Eq. (11-80):

$$x_{n+1}^{(1)} = x_n + \frac{h}{2}\{f[x_n + hf(x_n, t_n), t_{n+1}] + f(x_n, t_n)\} \tag{11-127}$$

Equation (11-127) gives the *corrected* value of x_{n+1} after one fixed-point iteration. In general, several iterations may be necessary to assure that the computed value is close to the fixed point. Hence, we iterate as follows:

$$x_{n+1}^{(2)} = x_n + \frac{h}{2}[f(x_{n+1}^{(1)}, t_{n+1}) + f(x_n, t_n)] \tag{11-128}$$

$$x_{n+1}^{(3)} = x_n + \frac{h}{2}[f(x_{n+1}^{(2)}, t_{n+1}) + f(x_n, t_n)] \tag{11-129}$$

It must be emphasized that the values $x_{n+1}^{(0)}, x_{n+1}^{(1)}, x_{n+1}^{(2)}$, etc., are only intermediate values used to compute one from the other and are of no further use once the final iteration $x_{n+1}^{(m)}$ is found. This procedure for computing x_{n+1} from an *implicit algorithm* is called a *predictor–corrector algorithm*. We can summarize this algorithm succinctly as follows:

Predictor

$$x_{n+1}^{(0)} = x_n + hf(x_n, t_n) \tag{11-130}$$

Corrector

$$x_{n+1}^{(j)} = x_n + \frac{h}{2}[f(x_{n+1}^{(j-1)}, t_{n+1}) + f(x_n, t_n)] \tag{11-131}$$

The number m of fixed-point iterations chosen for a predictor–corrector algorithm depends on the desired accuracy and the local truncation error of the predictor. Observe that when $m = 1$ we obtain the Heun algorithm presented earlier in Section 11-4. Hence, the Heun algorithm can be interpreted as the trapezoidal algorithm whose initial guess is predicted by the forward Euler algorithm. Clearly, this is not an accurate algorithm. The following explicit higher-order algorithms to be derived in Chapter 12 will provide more accurate predictions:

Adams–Bashforth Predictors

$$x_{n+1}^{(0)} = x_n + \frac{h}{2}[3f(x_n) - f(x_{n-1})] \tag{11-132}$$

$$x_{n+1}^{(0)} = x_n + \frac{h}{12}[23f(x_n) - 16f(x_{n-1}) + 5f(x_{n-2})] \tag{11-133}$$

With accurate predictors, only one fixed-point iteration ($m = 1$) is often sufficient to obtain reasonably accurate answers. There are many more predictor–corrector algorithms. The reader is referred to [1, 2] for additional details.

11-5-3 Methods for Starting Multistep Numerical-Integration Algorithms

In contrast to single-step algorithms, the multistep numerical-integration algorithm ($p > 0$) defined by Eq. (11-70) is *not self-starting*, since initially only x_0 and t_0 are given. To see this, it suffices to consider the simpler case where $b_{-1} = 0$ in Eq. (11-70), and write out x_1 explicitly as follows:

$$\begin{aligned} x_1 = a_0 x_0 + a_1 x_{-1} + \ldots + a_p x_{-p} + h[b_0 f(x_0, t_0) \\ + b_1 f(x_{-1}, t_{-1}) + \ldots + b_p f(x_{-p}, t_{-p})] \end{aligned} \tag{11-134}$$

Equation (11-134) shows that the values of $x_{-1}, x_{-2}, \ldots, x_{-p}$ must be given, in addition to x_0 and t_0, in order to compute x_1. In general, to compute x_{n+1}, we need to be given $p + 1$ preceding values of x, i.e., $x_n, x_{n-1}, \ldots, x_{n-p}$, assuming a *uniform step size* h. Clearly, to obtain these values, a single-step algorithm must be used at least $p + 1$ times before a multistep algorithm can be initiated. Because of its high degree of accuracy and ease in programming, the *fourth-order Runge–Kutta algorithm* is frequently used to provide these initiating values. Since only a few time steps are needed, i.e., $p + 1$, the difficulty in controlling the error propagation of the Runge–Kutta algorithm—the major flaw of this algorithm—does not present any problem here. Consequently, efficient and accurate numerical algorithms for solving initial-value problems are almost always a combination of single-step and multistep algorithms, with the former used only to obtain the starting values for initiating the latter. Multistep algorithms are used to compute the remaining points because they are often computationally more efficient, and the propagation of both local truncation and round-off errors can be more easily controlled. To see why multistep algorithms are computationally more efficient than single-step algorithms having the same order of local truncation errors, we observe that the fourth-order Runge–Kutta algorithm requires *four* evaluations of the function $f(x, t)$ per time step. A predictor–corrector algorithm having the same order of local truncation error would require only *two* function evaluations. Since the evaluation of $f(x, t)$ requires the most computation time, especially in nonlinear networks where it is not available explicitly in analytic form, it is clear that a predictor–corrector method having the same accuracy as the fourth-order Runge–Kutta algorithm would be almost twice as fast!

*11-6 CANONICAL MATRIX REPRESENTATIONS FOR PREDICTOR–CORRECTOR ALGORITHMS[10]

Several examples of multistep algorithms have been presented without derivation in the preceding discussions to illustrate the fundamental concept of using an explicit algorithm—*the predictor*—to find an initial guess for iterating an implicit algorithm—*the corrector*. These examples are special cases of several important families of

[10]The material in this section represents a simplified version of the *multivalue methods* presented in Chapter 9 of reference 2. To facilitate further study of Gear's book, the same examples are deliberately chosen and worked out in detail here.

multistep algorithms to be derived by a unified approach in the next two chapters. Our objective in this section is to show that any combination of predictor and corrector algorithms can be recast into an equivalent *canonical matrix representation*, which is important for at least three reasons: (1) it is completely general, and hence a unified theory of multistep algorithms can be formulated from this representation; (2) the representation is in a form most suitable for efficient programming; (3) the representation simplifies the difficult task of changing the *step size* and *order* of a multistep algorithm.

Consider the differential equation

$$\dot{x} = f(x, t) \tag{11-135}$$

and the following general predictor–corrector algorithm:

Predictor

$$x_{n+1}^{(0)} = \sum_{i=0}^{p} [\alpha_i^* x_{n-i} + \beta_i^* h x'_{n-i}] \tag{11-136}$$

Corrector

$$\begin{aligned} x_{n+1}^{(j+1)} &= \sum_{i=0}^{p} [\alpha_i x_{n-i} + \beta_i h x'_{n-i}] + h\beta_{-1} f(x_{n+1}^{(j)}, t_{n+1}), \\ j &= 0, 1, 2, \ldots, M-1 \end{aligned} \tag{11-137}$$

where

$$x'_{n-i} \triangleq f(x_{n-i}, t_{n-i}), \qquad 0 \leq i \leq p$$

The number of iterations, M, is chosen such that

$$|x_{n+1} - x_{n+1}^{(M)}| < \varepsilon \tag{11-138}$$

where ε is the maximum allowable error. If we subtract Eq. (11-137) with $j = 0$ from Eq. (11-136), we obtain

$$x_{n+1}^{(1)} = x_{n+1}^{(0)} - \beta_{-1}\left\{\left[\sum_{i=0}^{p} \gamma_i x_{n-i} + \delta_i h x'_{n-i}\right] - hf(x_{n+1}^{(0)}, t_{n+1})\right\} \tag{11-139}$$

where

$$\gamma_i \triangleq \frac{\alpha_i^* - \alpha_i}{\beta_{-1}} \tag{11-140}$$

and

$$\delta_i \triangleq \frac{\beta_i^* - \beta_i}{\beta_{-1}} \tag{11-141}$$

Examination of Eq. (11-137) reveals that the expression enclosed within the bracket remains unchanged during each iteration. Hence, except for the first iteration given by Eq. (11-139), much redundant computation can be obviated by subtracting any two subsequent consecutive iterations from Eq. (11-137) to obtain the following *equivalent corrector formula:*

$$x_{n+1}^{(j+1)} = x_{n+1}^{(j)} - \beta_{-1}[hf(x_{n+1}^{(j-1)}, t_{n+1}) - hf(x_{n+1}^{(j)}, t_{n+1})], \quad j = 1, 2, \dots, M-1 \tag{11-142}$$

To show that Eqs. (11-136), (11-139), and (11-142) can be recast into an equivalent matrix form, let us define the following four $(2p+2) \times 1$ vectors:

$$\boldsymbol{y}_{n+1}^{(0)} \triangleq \begin{bmatrix} x_{n+1}^{(0)} \\ x_n \\ \vdots \\ x_{n-p+1} \\ \hdashline \sum_{i=0}^{p} [\gamma_i x_{n-i} + \delta_i h x'_{n-i}] \\ hx'_n \\ \vdots \\ hx'_{n-p+1} \end{bmatrix}, \qquad \boldsymbol{y}_n \triangleq \begin{bmatrix} x_n \\ x_{n-1} \\ \vdots \\ x_{n-p} \\ \hdashline hx'_n \\ hx'_{n-1} \\ \vdots \\ hx'_{n-p} \end{bmatrix} \tag{11-143}$$

$$\boldsymbol{y}_{n+1}^{(j)} \triangleq \begin{bmatrix} x_{n+1}^{(j)} \\ x_n \\ \vdots \\ x_{n-p+1} \\ \hdashline hf(x_{n+1}^{(j-1)}, t_{n+1}) \\ hx'_n \\ \vdots \\ hx'_{n-p+1} \end{bmatrix}, \qquad \boldsymbol{c}_y \triangleq \begin{bmatrix} \beta_{-1} \\ 0 \\ \vdots \\ 0 \\ \hdashline 1 \\ 0 \\ \vdots \\ 0 \end{bmatrix} \tag{11-144}$$

It is easily verified (by carrying out the following matrix multiplication) that Eqs. (11-136), (11-139), and (11-142) are contained within the following two matrix equations:

> *Predictor Canonical Representation*
>
> $$\boldsymbol{y}_{n+1}^{(0)} = \boldsymbol{Y}\boldsymbol{y}_n \tag{11-145}$$
>
> *Corrector Canonical Representation*
>
> $$\boldsymbol{y}_{n+1}^{(j+1)} = \boldsymbol{y}_{n+1}^{(j)} + F_y(\boldsymbol{y}_{n+1}^{(j)})\boldsymbol{c}_y, \quad j = 0, 1, 2, \dots, M-1 \tag{11-146}$$

where $\boldsymbol{Y}$ is a $(2p+2) \times (2p+2)$ matrix defined by

$$\boldsymbol{Y} \triangleq \left[\begin{array}{cccc|cccc}
\alpha_0^* & \alpha_1^* & \cdots & \alpha_p^* & \beta_0^* & \beta_1^* & \cdots & \beta_p^* \\
1 & 0 & \cdots & 0 & 0 & 0 & \cdots & 0 \\
0 & 1 & \cdots & 0 & 0 & 0 & \cdots & 0 \\
\cdot & \cdot & \cdots & \cdot & \cdot & \cdot & \cdots & \cdot \\
\cdot & \cdot & \cdots & \cdot & \cdot & \cdot & \cdots & \cdot \\
\cdot & \cdot & \cdots & \cdot & \cdot & \cdot & \cdots & \cdot \\
0 & 0 & 1 & 0 & 0 & 0 & \cdots & 0 \\
\hline
\gamma_0 & \gamma_1 & \cdots & \gamma_p & \delta_0 & \delta_1 & \cdots & \delta_p \\
0 & 0 & \cdots & 0 & 1 & 0 & \cdots & 0 \\
0 & 0 & \cdots & 0 & 0 & 1 & 0 & 0 \\
\cdot & \cdot & \cdots & \cdot & \cdot & \cdot & \cdots & \cdot \\
\cdot & \cdot & \cdots & \cdot & \cdot & \cdot & \cdots & \cdot \\
\cdot & \cdot & \cdots & \cdot & \cdot & \cdot & \cdots & \cdot \\
0 & 0 & \cdots & 0 & 0 & 0 & 1 & 0
\end{array}\right] \tag{11-147}$$

where the parameters γ_i and δ_i are defined in Eqs. (11-140) and (11-141). $F_y(\boldsymbol{y}_{n+1}^{(j)})$ is a *scalar* "corrector function" defined by

$$F_y(\boldsymbol{y}_{n+1}^{(0)}) \triangleq -\sum_{i=0}^{p} [\gamma_i x_{n-i} + \delta_i h x'_{n-i}] + h f(x_{n+1}^{(0)}, t_{n+1}) \tag{11-148}$$

and

$$F_y(\boldsymbol{y}_{n+1}^{(j)}) \triangleq -h f(x_{n+1}^{(j-1)}, t_{n+1}) + h f(x_{n+1}^{(j)}, t_{n+1}), \qquad j = 1, 2, \ldots, M-1 \tag{11-149}$$

Observe that the expression $\sum_{i=0}^{p} [\gamma_i x_{n-1} + \delta_i h x'_{n-i}]$ in Eq. (11-148) is equal to the $(p+2)$th component of $\boldsymbol{y}_{n+1}^{(0)}$ and need not be computed separately. An examination of Eqs. (11-143) and (11-144) shows that the vectors $\boldsymbol{y}_n$, $\boldsymbol{y}_{n+1}^{(j)}$, and $\boldsymbol{c}_y$ do not depend on the parameters α_i^*, β_i^*, α_i, and β_i. Hence, the matrix $\boldsymbol{Y}$ in Eq. (11-145) completely defines the predictor algorithm, while the vector $F_y(y_{n+1}^{(j)})\, \boldsymbol{c}_y$ in Eq. (11-146) completely defines the corrector algorithm. Observe also that the terms in Eq. (11-145) do not contain the function $f(x, t)$; hence, the *same* predictor applies to *any equation* $\dot{x} = f(x, t)$. Let us now illustrate the preceding canonical representation by deriving $\boldsymbol{Y}$ and $F_y(\boldsymbol{y}_{n+1}^{(j)})\, \boldsymbol{c}_y$ associated with the following *example*:[11]

Predictor: Third-Order Adams–Bashforth Algorithm

$$x_{n+1}^{(0)} = x_n + h[\tfrac{23}{12} f(x_n, t_n) - \tfrac{16}{12} f(x_{n-1}, t_{n-1}) + \tfrac{5}{12} f(x_{n-2}, t_{n-2})] \tag{11-150}$$

Corrector: Fourth-Order Adams–Moulton Algorithm

$$x_{n+1}^{(j+1)} = x_n + h[\tfrac{9}{24} f(x_{n+1}^{(j)}, t_{n+1}) + \tfrac{19}{24} f(x_n, t_n) - \tfrac{5}{24} f(x_{n-1}, t_{n-1}) + \tfrac{1}{24} f(x_{n-2}, t_{n-2}) \tag{11-151}$$

[11]The Adams–Bashforth algorithms represent a family of *explicit* multistep algorithms (see Table 12-1), whereas the Adams–Moulton algorithms represent a family of *implicit* multistep algorithms (see Table 12-2) to be derived in Chapter 12.

It follows from Eqs. (11-150) and (11-151) that

$$\alpha_0^* = 1, \quad \alpha_1^* = 0, \quad \alpha_2^* = 0, \quad \beta_0^* = \tfrac{23}{12}, \quad \beta_1^* = -\tfrac{16}{12}, \quad \beta_2^* = \tfrac{5}{12}$$

$$\alpha_0 = 1, \quad \alpha_1 = 0, \quad \alpha_2 = 0, \quad \beta_{-1} = \tfrac{9}{24}, \quad \beta_0 = \tfrac{19}{24}, \quad \beta_1 = -\tfrac{5}{24}, \quad \beta_2 = \tfrac{1}{24}$$

Substituting these parameters into Eqs. (11-140) and (11-141), we obtain

$$\gamma_0 = 0, \quad \gamma_1 = 0, \quad \gamma_2 = 0, \quad \delta_0 = 3, \quad \delta_1 = -3, \quad \delta_2 = 1$$

Substituting these parameters into the preceding equations, we obtain

$$\boldsymbol{y}_{n+1}^{(0)} = \begin{bmatrix} x_{n+1}^{(0)} \\ x_n \\ x_{n-1} \\ \hdashline 3hx'_n - 3hx'_{n-1} + hx'_{n-2} \\ hx'_n \\ hx'_{n-1} \end{bmatrix}, \qquad \boldsymbol{y}_n = \begin{bmatrix} x_n \\ x_{n-1} \\ x_{n-2} \\ \hdashline hx'_n \\ hx'_{n-1} \\ hx'_{n-2} \end{bmatrix}$$

$$\boldsymbol{y}_{n+1}^{(j)} = \begin{bmatrix} x_{n+1}^{(j)} \\ x_n \\ x_{n-1} \\ \hdashline hf(x_{n+1}^{(j-1)}, t_{n+1}) \\ hx'_n \\ hx'_{n-1} \end{bmatrix}, \qquad \boldsymbol{c}_y = \begin{bmatrix} \frac{9}{24} \\ 0 \\ 0 \\ \hdashline 1 \\ 0 \\ 0 \end{bmatrix}$$

$$\boldsymbol{Y} = \left[\begin{array}{ccc:ccc} 1 & 0 & 0 & \frac{23}{12} & -\frac{16}{12} & \frac{5}{12} \\ 1 & 0 & 0 & 0 & 0 & 0 \\ 0 & 1 & 0 & 0 & 0 & 0 \\ \hdashline 0 & 0 & 0 & 3 & -3 & 1 \\ 0 & 0 & 0 & 1 & 0 & 0 \\ 0 & 0 & 0 & 0 & 1 & 0 \end{array}\right]$$

$$F_y(\boldsymbol{y}_{n+1}^{(0)}) = -[3hx'_n - 3hx'_{n-1} + hx'_{n-2}] + hf(x_{n+1}^{(0)}, t_{n+1})$$

$$F_y(\boldsymbol{y}_{n+1}^{(j)}) = -hf(x_{n+1}^{(j-1)}, t_{n+1}) + hf(x_{n+1}^{(j)}, t_{n+1})$$

Observe that the expression $3hx'_n - 3hx'_{n-1} + hx'_{n-2}$ in the expression for $F_y(\boldsymbol{y}_{n+1}^{(0)})$ is equal to the fourth component of $\boldsymbol{y}_{n+1}^{(0)}$ and need not be computed separately. Since the values of x_{n-1} and x_{n-2} are not needed to compute $x_{n+1}^{(0)}$ from Eq. (11-150), these two elements represent "excess baggage" for this example and could be deleted to save computer storage. Deleting the corresponding entries from $\boldsymbol{Y}$ and $\boldsymbol{c}_y$, we obtain

$$\boldsymbol{Y} = \begin{bmatrix} 1 & \frac{23}{12} & -\frac{16}{12} & \frac{5}{12} \\ 0 & 3 & -3 & 1 \\ 0 & 1 & 0 & 0 \\ 0 & 0 & 1 & 0 \end{bmatrix}, \qquad \boldsymbol{c}_y = \begin{bmatrix} \frac{3}{8} \\ 1 \\ 0 \\ 0 \end{bmatrix} \tag{11-152}$$

To verify that Eq. (11-152) indeed completely defines the preceding predictor–corrector algorithm, let us substitute $\boldsymbol{Y}$ and $\boldsymbol{c}_y$ in Eqs. (11-145) and (11-146):

$$\begin{bmatrix} x_{n+1}^{(0)} \\ 3hx'_n - 3hx'_{n-1} + hx'_{n-2} \\ hx'_n \\ hx'_{n-1} \end{bmatrix} = \begin{bmatrix} 1 & \frac{23}{12} & -\frac{16}{12} & \frac{5}{12} \\ 0 & 3 & -3 & 1 \\ 0 & 1 & 0 & 0 \\ 0 & 0 & 1 & 0 \end{bmatrix} \begin{bmatrix} x_n \\ hx'_n \\ hx'_{n-1} \\ hx'_{n-2} \end{bmatrix} \tag{11-153}$$

$$\begin{bmatrix} x_{n+1}^{(1)} \\ hf(x_{n+1}^{(0)}, t_{n+1}) \\ hx'_n \\ hx'_{n-1} \end{bmatrix} = \begin{bmatrix} x_{n+1}^{(0)} \\ 3hx'_n - 3hx'_{n-1} + hx'_{n-2} \\ hx'_n \\ hx'_{n-1} \end{bmatrix} + \begin{bmatrix} -\frac{9}{8}hx'_n + \frac{9}{8}hx'_{n-1} - \frac{3}{8}hx'_{n-2} + \frac{3}{8}hf(x_{n+1}^{(0)}, t_{n+1}) \\ -3hx'_n + 3hx'_{n-1} - hx'_{n-2} + hf(x_{n+1}^{(0)}, t_{n+1}) \\ 0 \\ 0 \end{bmatrix} \tag{11-154}$$

$$\begin{bmatrix} x_{n+1}^{(j+1)} \\ hf(x_{n+1}^{(j)}, t_{n+1}) \\ hx'_n \\ hx'_{n-1} \end{bmatrix} = \begin{bmatrix} x_{n+1}^{(j)} \\ hf(x_{n+1}^{(j-1)}, t_{n+1}) \\ hx'_n \\ hx'_{n-1} \end{bmatrix} + \begin{bmatrix} -\frac{3}{8}hf(x_{n+1}^{(j-1)}, t_{n+1}) + \frac{3}{8}hf(x_{n+1}^{(j)}, t_{n+1}) \\ -hf(x_{n+1}^{(j-1)}, t_{n+1}) + hf(x_{n+1}^{(j)}, t_{n+1}) \\ 0 \\ 0 \end{bmatrix} \tag{11-155}$$

$$j = 1, 2, \ldots, M-1$$

If we expand Eq. (11-153), we would obtain Eq. (11-150), in addition to three identities. Similarly, if we expand Eqs. (11-154) and (11-155), we obtain

$$x_{n+1}^{(1)} = x_{n+1}^{(0)} - \tfrac{9}{8}hx'_n + \tfrac{9}{8}hx'_{n-1} - \tfrac{3}{8}hx'_{n-2} + \tfrac{3}{8}hf(x_{n+1}^{(0)}, t_{n+1}) \tag{11-156}$$

$$x_{n+1}^{(j+1)} = x_{n+1}^{(j)} - \tfrac{3}{8}hf(x_{n+1}^{(j-1)}, t_{n+1}) + \tfrac{3}{8}hf(x_{n+1}^{(j)}, t_{n+1}) \qquad j = 1, 2, \ldots, M-1 \tag{11-157}$$

in addition to three identities. Observe that Eqs. (11-156) and (11-157) are simply the equivalent corrector representation obtained by subtracting consecutive iterations between $x_{n+1}^{(j+1)}$ and $x_{n+1}^{(j)}$, $j = 0, 1, 2, \ldots, M-1$. They correspond to Eqs. (11-139) and (11-142). Hence, Eqs. (11-153)–(11-155) indeed contain Eqs. (11-150) and (11-151) as a subset.

Finally, it is important to recognize that once $\boldsymbol{Y}$ and $F_y(\boldsymbol{y}_{n+1}^{(j)})\ \boldsymbol{c}_y$ are given, the predictor–corrector algorithm is specified, and the only *data* that need be stored from one iteration to another iteration is simply $\boldsymbol{y}_n$. Observe also that, at the end of the corrector iteration, $\boldsymbol{y}_{n+1}^{(M)}$ becomes $\boldsymbol{y}_{n+1}$, and the cycle is repeated with $\boldsymbol{y}_{n+1}^{(0)} = \boldsymbol{Y}\boldsymbol{y}_{n+1}$, etc.

*11-7 EQUIVALENT CANONICAL MATRIX REPRESENTATIONS FOR PREDICTOR–CORRECTOR ALGORITHMS

The vector $\boldsymbol{y}_n$ defined in Eq. (11-143) contains $2p + 2$ pieces of information: x_n, $x_{n-1}, \ldots, x_{n-p}, hx'_n, hx'_{n-1}, \ldots$, and hx'_{n-p}. These $(2p + 2)$ quantities correspond to the *ordinate* and *slope* at $p + 1$ points of a $(2p + 1)$th-degree polynomial. In other words, the vector $\boldsymbol{y}_n$ uniquely determines a $(2p+1)$th-degree polynomial. Observe, however, that there are many other methods for determining a polynomial of a given degree. For example, the $(2p+1)$th-degree polynomial could also have been determined by the *ordinates* of $2p+2$ points on the polynomial. In fact, this observation could be generalized by defining a new $(2p+2) \times 1$ vector

$$\boldsymbol{z}_n \triangleq \boldsymbol{T}\boldsymbol{y}_n \tag{11-158}$$

where $\boldsymbol{T}$ is any $(2p+2) \times (2p+2)$ *nonsingular matrix.* To show that $\boldsymbol{z}_n$ uniquely determines the same $(2p+1)$th-degree polynomial, we need only recover first $\boldsymbol{y}_n = \boldsymbol{T}^{-1}\boldsymbol{z}_n$ and then determine the polynomial associated with $\boldsymbol{y}_n$. To derive the predictor–corrector canonical representation in terms of the vector $\boldsymbol{z}_n$, let us premultiply both sides of Eqs. (11-145) and (11-146) by the nonsingular matrix $\boldsymbol{T}$ to obtain

$$\boldsymbol{z}_{n+1}^{(0)} \triangleq \boldsymbol{T}\boldsymbol{y}_{n+1}^{(0)} = \boldsymbol{T}\boldsymbol{Y}\boldsymbol{y}_n \tag{11-159}$$

$$\boldsymbol{z}_{n+1}^{(j+1)} \triangleq \boldsymbol{T}\boldsymbol{y}_{n+1}^{(j+1)} = \boldsymbol{T}\boldsymbol{y}_{n+1}^{(j)} + F_y(\boldsymbol{y}_{n+1}^{(j)})\boldsymbol{T}\boldsymbol{c}_y \tag{11-160}$$

Substituting $\boldsymbol{y}_n = \boldsymbol{T}^{-1}\boldsymbol{z}_n$ and $\boldsymbol{y}_{n+1}^{(j)} = \boldsymbol{T}^{-1}\boldsymbol{z}_{n+1}^{(j)}$ into Eqs. (11-159) and (11-160), we obtain

Predictor Canonical Representation:

$$\boldsymbol{z}_{n+1}^{(0)} = \boldsymbol{Z}\boldsymbol{z}_n \tag{11-161}$$

where

$$\boldsymbol{Z} \triangleq \boldsymbol{T}\boldsymbol{Y}\boldsymbol{T}^{-1} \tag{11-162}$$

Corrector Canonical Representation:

$$\boldsymbol{z}_{n+1}^{(j+1)} = \boldsymbol{z}_{n+1}^{(j)} + F_z(\boldsymbol{z}_{n+1}^{(j)})\boldsymbol{c}_z, \qquad j = 0, 1, 2, \ldots, M-1 \tag{11-163}$$

where

$$F_z(\boldsymbol{z}_{n+1}^{(j)}) \triangleq F_y(\boldsymbol{T}^{-1}\boldsymbol{z}_{n+1}^{(j)}) \tag{11-164}$$

and

$$\boldsymbol{c}_z \triangleq \boldsymbol{T}\boldsymbol{c}_y \tag{11-165}$$

Equations (11-161) and (11-163) are the *equivalent predictor–corrector canonical representation* in terms of the new vector $\boldsymbol{z}_n$. Since each component of $\boldsymbol{z}_n$ could be any nonsingular linear combination of the components of $\boldsymbol{y}_n$, it can no longer be interpreted in terms of some geometrical parameters (such as ordinate or slope) at a point on the polynomial. Consequently, it is more appropriate to call this equivalent predictor–corrector canonical representation a *multivalue algorithm.* If $\boldsymbol{T}$ is chosen to be the identity matrix, a multivalue algorithm (in terms of $\boldsymbol{z}_n$) reduces to a multistep algorithm (in terms of $\boldsymbol{y}_n$). We shall now illustrate the preceding generalized canonical representation by studying two special cases of practical importance: the *backward-difference vector representation* and the *Nordsieck vector representation.*

11-7-1 Predictor–Corrector Algorithm via the Backward-Difference Vector Representation

Consider $\dot{x} = f(x, t)$ and let

$$f_m \triangleq \dot{x}_m = f(x(t_m), t_m) \tag{11-166}$$

where $x_m = x(t_m)$.

We define the *backward-difference operator* $\nabla^j f_m$ as follows:

$$
\begin{aligned}
\nabla^0 f_m &\triangleq f_m \\
\nabla^1 f_m &\triangleq \nabla^0 f_m - \nabla^0 f_{m-1} = f_m - f_{m-1} \\
\nabla^2 f_m &\triangleq \nabla^1 f_m - \nabla^1 f_{m-1} = f_m - 2f_{m-1} + f_{m-2} \\
\nabla^3 f_m &\triangleq \nabla^2 f_m - \nabla^2 f_{m-1} = f_m - 3f_{m-1} + 3f_{m-2} + f_{m-3} \\
&\vdots \\
\nabla^j f_m &\triangleq \nabla^{j-1} f_m - \nabla^{j-1} f_{m-1}
\end{aligned}
\tag{11-167}
$$

It follows from Eq. (11-167) that $\nabla^j(hx'_n)$, the jth-order backward difference of hx'_n, is a *linear combination* of $hx'_n, hx'_{n-1}, \ldots, hx'_{n-j}$. Hence, if we define the *backward-difference vector* as

$$z_n \triangleq [x_n, x_{n-1}, \ldots, x_{n-p}, hx'_n, \nabla^1(hx'_n), \nabla^2(hx'_n), \ldots, \nabla^p(hx'_n)]^t \tag{11-168}$$

then z_n is simply related to

$$y_n \triangleq [x_n, x_{n-1}, \ldots, x_{n-p}, hx'_n, hx'_{n-1}, hx'_{n-2}, \ldots, hx'_{n-p}]^t \tag{11-169}$$

by a *nonsingular linear transformation* $z_n = Ty_n$. Hence, we can derive an equivalent predictor–corrector algorithm in terms of z_n.

EXAMPLE 11-4. Consider the third-order Adams–Bashforth predictor and fourth-order Adams–Moulton corrector algorithm defined earlier by Eq. (11-152). We can transform the vector $y_n \triangleq [x_n, hx'_n, hx'_{n-1}, hx'_{n-2}]^t$ into the backward-difference vector

$$z_n \triangleq [x_n, hx'_n, \nabla^1(hx'_n), \nabla^2(hx'_n)]^t$$

by the following nonsingular linear transformation:

$$
\underbrace{\begin{bmatrix} x_n \\ hx'_n \\ \nabla^1(hx'_n) \\ \nabla^2(hx'_n) \end{bmatrix}}_{z_n}
=
\underbrace{\begin{bmatrix} 1 & 0 & 0 & 0 \\ 0 & 1 & 0 & 0 \\ 0 & 1 & -1 & 0 \\ 0 & 1 & -2 & 1 \end{bmatrix}}_{T}
\underbrace{\begin{bmatrix} x_n \\ hx'_n \\ hx'_{n-1} \\ hx'_{n-2} \end{bmatrix}}_{y_n}
\tag{11-170}
$$

It follows from Eq. (11-170) that

$$
\begin{aligned}
Z \triangleq TYT^{-1} &=
\begin{bmatrix} 1 & 0 & 0 & 0 \\ 0 & 1 & 0 & 0 \\ 0 & 1 & -1 & 0 \\ 0 & 1 & -2 & 1 \end{bmatrix}
\begin{bmatrix} 1 & \frac{23}{12} & -\frac{16}{12} & \frac{5}{12} \\ 0 & 3 & -3 & 1 \\ 0 & 1 & 0 & 0 \\ 0 & 0 & 1 & 0 \end{bmatrix}
\begin{bmatrix} 1 & 0 & 0 & 0 \\ 0 & 1 & 0 & 0 \\ 0 & 1 & -1 & 0 \\ 0 & 1 & -2 & 1 \end{bmatrix} \\
&= \begin{bmatrix} 1 & 1 & \frac{1}{2} & \frac{5}{12} \\ 0 & 1 & 1 & 1 \\ 0 & 0 & 1 & 1 \\ 0 & 0 & 0 & 1 \end{bmatrix}
\end{aligned}
\tag{11-171}
$$

and

$$
\boldsymbol{c}_z \triangleq \boldsymbol{T}\boldsymbol{c}_y = \begin{bmatrix} 1 & 0 & 0 & 0 \\ 0 & 1 & 0 & 0 \\ 0 & 1 & -1 & 0 \\ 0 & 1 & -2 & 1 \end{bmatrix} \begin{bmatrix} \frac{3}{8} \\ 1 \\ 0 \\ 0 \end{bmatrix} = \begin{bmatrix} \frac{3}{8} \\ 1 \\ 1 \\ 1 \end{bmatrix} \tag{11-172}
$$

Hence, in terms of the backward-difference vector z_n, the predictor–corrector algorithm assumes the form

$$
\underbrace{\begin{bmatrix} x_{n+1}^{(0)} \\ hx_{n+1}^{\prime(0)} \\ \nabla^1(hx_{n+1}^{\prime(0)}) \\ \nabla^2(hx_{n+1}^{\prime(0)}) \end{bmatrix}}_{z_{n+1}^{(0)}} = \underbrace{\begin{bmatrix} 1 & 1 & \frac{1}{2} & \frac{5}{12} \\ 0 & 1 & 1 & 1 \\ 0 & 0 & 1 & 1 \\ 0 & 0 & 0 & 1 \end{bmatrix}}_{\boldsymbol{Z}} \underbrace{\begin{bmatrix} x_n \\ hx_n' \\ \nabla^1(hx_n') \\ \nabla^2(hx_n') \end{bmatrix}}_{z_n} \tag{11-173}
$$

$$
\underbrace{\begin{bmatrix} x_{n+1}^{(j+1)} \\ hx_{n+1}^{\prime(j+1)} \\ \nabla^1(hx_{n+1}^{\prime(j+1)}) \\ \nabla^2(hx_{n+1}^{\prime(j+1)}) \end{bmatrix}}_{z_{n+1}^{(j+1)}} = \underbrace{\begin{bmatrix} x_{n+1}^{(j)} \\ hx_{n+1}^{\prime(j)} \\ \nabla^1(hx_{n+1}^{\prime(j)}) \\ \nabla^2(hx_{n+1}^{\prime(j)}) \end{bmatrix}}_{z_{n+1}^{(j)}} + F_z(z_{n+1}^{(j)}) \underbrace{\begin{bmatrix} \frac{3}{8} \\ 1 \\ 1 \\ 1 \end{bmatrix}}_{\boldsymbol{c}_z} \tag{11-174}
$$

where

$$
F_z(z_{n+1}^{(0)}) \triangleq F_y(\boldsymbol{T}^{-1}z_{n+1}^{(0)}) = -[3hx_n' - 3hx_{n-1}' + hx_{n-2}'] + hf(x_{n+1}^{(0)}, t_{n+1}) \tag{11-175}
$$

$$
F_z(z_{n+1}^{(j)}) \triangleq F_y(\boldsymbol{T}^{-1}z_{n+1}^{(j)}) = -hf(x_{n+1}^{(j-1)}, t_{n+1}) + hf(x_{n+1}^{(j)}, t_{n+1}) \tag{11-176}
$$

In terms of the backward-difference vector representation, the preceding predictor–corrector canonical representation is completely characterized by the matrix $\boldsymbol{Z}$ and the vector $\boldsymbol{c}_z$. Observe that, even though $\boldsymbol{Z}$ is a 4×4 matrix, the matrix multiplication indicated in Eq. (11-173) requires only two scalar multiplications, five additions, and two memory locations for storing intermediate values:

Multiplication Operations:

$$
a_1 \triangleq \tfrac{1}{2}[\nabla^1(hx_n')]
$$

$$
a_2 \triangleq \tfrac{5}{12}[\nabla^2(hx_n')]
$$

Addition Operations:

$$
\nabla^1(hx_{n+1}^{\prime(0)}) = \nabla^1(hx_n') + \nabla^2(hx_n')
$$

$$
hx_{n+1}^{\prime(0)} = hx_n' + \nabla^1(hx_{n+1}^{\prime(0)})
$$

$$
x_{n+1}^{(0)} = x_n + hx_n' + a_1 + a_2
$$

Observe also that the expression $[3hx_n' - 3hx_{n-1}' + hx_{n-2}']$ in Eq. (11-175) is equal to the *second* component of $z_{n+1}^{(0)}$, since

$$\begin{aligned} hx'^{(0)}_{n+1} &= hx'_n + \nabla^1(hx'_n) + \nabla^2(hx'_n) \\ &= hx'_n + [hx'_n - hx'_{n-1}] + [hx'_n - 2hx'_{n-1} + hx'_{n-2}] \\ &= 3hx'_n - 3hx'_{n-1} + hx'_{n-2} \end{aligned}$$

The advantages and disadvantages of the backward-difference representation will be investigated in Section 12-8.

11-7-2 Predictor–Corrector Algorithm via the Nordsieck Vector Representation

We have shown in Section 11-7-1 that a $(2p+1)$th-degree polynomial can be uniquely determined by the $(2p+1) \times 1$ *backward-difference vector* as defined by Eq. (11-168). Our objective in this section is to present yet another vector that uniquely determines the same polynomial. This vector, the *Nordsieck vector*, is defined by [2]

$$z_n \triangleq \left[x_n \quad hx'_n \quad \frac{h^2x''_n}{2} \quad \frac{h^3x'''_n}{3!} \quad \cdots \quad \frac{h^{(2p+1)}x_n^{(2p+1)}}{(2p+1)!} \right]^t \tag{11-177}$$

Except for the constants of proportionality, the Nordsieck vector z_n consists simply of the ordinate x_n and the "weighted" first $2p+1$ derivatives $x'_n, x''_n, \ldots, x_n^{(2p+1)}$ of $x(t)$ at $t = t_n$. To show that this vector is related to the vector y_n of Eq. (11-169) by a linear transformation, let us consider the same example (third-order Adams–Bashforth predictor and fourth-order Adams–Moulton corrector algorithm) and determine the matrix T such that $z_n = Ty_n$. Recall that a third-order algorithm is *exact* whenever the solution happens to be a polynomial of degree less than or equal to 3; i.e.,

$$x(t) = a_0 + a_1t + a_2t^2 + a_3t^3 \tag{11-178}$$

It follows from Eq. (11-178) that

$$x'(t) = a_1 + 2a_2t + 3a_3t^2 \tag{11-179}$$

$$x''(t) = 2a_2 + 6a_3t \tag{11-180}$$

$$x'''(t) = 6a_3 \tag{11-181}$$

Assuming a *uniform* step size h for the past three steps, let us choose $t_n = 0, t_{n-1} = -h$, and $t_{n-2} = -2h$, and substitute these values into Eqs. (11-178) and (11-179):

$$x_n = a_0 \tag{11-182}$$

$$x'_n = a_1 \tag{11-183}$$

$$x'_{n-1} = a_1 - 2a_2h + 3a_3h^2 \tag{11-184}$$

$$x'_{n-2} = a_1 - 4a_2h + 12a_3h^2 \tag{11-185}$$

Solving for a_0, a_1, a_2, and a_3 in terms of x_n, x'_n, x'_{n-1}, and x'_{n-2}, we obtain

$$a_0 = x_n \tag{11-186}$$

$$a_1 = x'_n \tag{11-187}$$

$$a_2 = \frac{1}{4h}[3x'_n - 4x'_{n-1} + x'_{n-2}] \tag{11-188}$$

$$a_3 = \frac{1}{6h^2}[x'_n - 2x'_{n-1} + x'_{n-2}] \tag{11-189}$$

If we substitute Eqs. (11-188) and (11-189) into Eqs. (11-180) and (11-181), we obtain

$$\frac{h^2x''}{2} = \frac{3}{4}hx'_n - hx'_{n-1} + \frac{1}{4}hx'_{n-2} \tag{11-190}$$

$$\frac{h^3x'''}{3!} = \frac{1}{6}hx'_n - \frac{1}{3}hx'_{n-1} + \frac{1}{6}hx'_{n-2} \tag{11-191}$$

It follows from Eqs. (11-190) and (11-191) that the Nordsieck vector z_n is related to the vector y_n by a *nonsingular linear* transformation:

$$\underbrace{\begin{bmatrix} x_n \\ hx'_n \\ \frac{h^2x''_n}{2} \\ \frac{h^3x'''_n}{3!} \end{bmatrix}}_{z_n} = \underbrace{\begin{bmatrix} 1 & 0 & 0 & 0 \\ 0 & 1 & 0 & 0 \\ 0 & \frac{3}{4} & -1 & \frac{1}{4} \\ 0 & \frac{1}{6} & -\frac{1}{3} & \frac{1}{6} \end{bmatrix}}_{T} \underbrace{\begin{bmatrix} x_n \\ hx'_n \\ hx'_{n-1} \\ hx'_{n-2} \end{bmatrix}}_{y_n} \tag{11-192}$$

This nonsingular matrix T is valid so long as the step size h has remained unchanged during the preceding three time steps. Substituting the matrix T into Eqs. (11-162) and (11-165), we obtain

$$Z = \underbrace{\begin{bmatrix} 1 & 0 & 0 & 0 \\ 0 & 1 & 0 & 0 \\ 0 & \frac{3}{4} & -1 & \frac{1}{4} \\ 0 & \frac{1}{6} & -\frac{1}{3} & \frac{1}{6} \end{bmatrix}}_{T} \underbrace{\begin{bmatrix} 1 & \frac{23}{12} & -\frac{16}{12} & \frac{5}{12} \\ 0 & 3 & -3 & 1 \\ 0 & 1 & 0 & 0 \\ 0 & 0 & 1 & 0 \end{bmatrix}}_{Y} \underbrace{\begin{bmatrix} 1 & 0 & 0 & 0 \\ 0 & 1 & 0 & 0 \\ 0 & 1 & -2 & 3 \\ 0 & 1 & -4 & 12 \end{bmatrix}}_{T^{-1}}$$

$$= \underbrace{\begin{bmatrix} 1 & 1 & 1 & 1 \\ 0 & 1 & 2 & 3 \\ 0 & 0 & 1 & 3 \\ 0 & 0 & 0 & 1 \end{bmatrix}}_{Z} \tag{11-193}$$

$$c_z = \underbrace{\begin{bmatrix} 1 & 0 & 0 & 0 \\ 0 & 1 & 0 & 0 \\ 0 & \frac{3}{4} & -1 & \frac{1}{4} \\ 0 & \frac{1}{6} & -\frac{1}{3} & \frac{1}{6} \end{bmatrix}}_{T} \underbrace{\begin{bmatrix} \frac{3}{8} \\ 1 \\ 0 \\ 0 \end{bmatrix}}_{c_y} = \underbrace{\begin{bmatrix} \frac{3}{8} \\ 1 \\ \frac{3}{4} \\ \frac{1}{6} \end{bmatrix}}_{c_z} \tag{11-194}$$

It follows from Eqs. (11-193) and (11-194) that in terms of the Nordsieck vector representation, the third-order Adams–Bashforth predictor and fourth-order Adams–

Moulton corrector algorithm assumes the following canonical form:

$$\underbrace{\begin{bmatrix} x_{n+1}^{(0)} \\ hx_{n+1}'^{(0)} \\ \dfrac{h^2 x_{n+1}''^{(0)}}{2} \\ \dfrac{h^3 x_{n+1}'''^{(0)}}{3!} \end{bmatrix}}_{z_{n+1}^{(0)}} = \underbrace{\begin{bmatrix} 1 & 1 & 1 & 1 \\ 0 & 1 & 2 & 3 \\ 0 & 0 & 1 & 3 \\ 0 & 0 & 0 & 1 \end{bmatrix}}_{Z} \underbrace{\begin{bmatrix} x_n \\ hx_n' \\ \dfrac{h^2 x_n''}{2} \\ \dfrac{h^3 x_n'''}{3!} \end{bmatrix}}_{z_n} \tag{11-195}$$

$$\underbrace{\begin{bmatrix} x_{n+1}^{(j+1)} \\ hx_{n+1}'^{(j+1)} \\ \dfrac{h^2 x_{n+1}''^{(j+1)}}{2} \\ \dfrac{h x_{n+1}'''^{(j+1)}}{3!} \end{bmatrix}}_{z_{n+1}^{(j+1)}} = \underbrace{\begin{bmatrix} x_{n+1}^{(j)} \\ hx_{n+1}'^{(j)} \\ \dfrac{h^2 x_{n+1}''^{(j)}}{2} \\ \dfrac{h^3 x_{n+1}'''^{(j)}}{3!} \end{bmatrix}}_{z_{n+1}^{(j)}} + F_z(z_{n+1}^{(j)}) \underbrace{\begin{bmatrix} \frac{3}{8} \\ 1 \\ \frac{3}{4} \\ \frac{1}{6} \end{bmatrix}}_{c_z} \tag{11-196}$$

The matrix Z shown in Eq. (11-195) is called the *Pascal triangle matrix* whose ijth element is given by

$$(Z)_{i+1,\, j+1} = \binom{j}{i} \triangleq \frac{j!}{(j-i)!\,i!}, \qquad 4 > j \geq i \geq 0 \tag{11-197}$$

where i and j vary from 0 to 3. This special property of the matrix Z allows one to compute $z_{n+1}^{(0)}$ from Eq. (11-195) rather efficiently. That Z is simply the Pascal triangle matrix can be derived by expanding $x(t)$ by a Taylor series about $t = t_n$ and truncating all terms of degree greater than 3 in $t - t_n$, since the predictor algorithm is assumed to be exact only for polynomial solutions of degree 3:

$$x(t) = x(t_n) + (t - t_n)x'(t_n) + \frac{(t-t_n)^2}{2!}x''(t_n) + \frac{(t-t_n)^3}{3!}x'''(t_n) + O((t-t_n)^4) \tag{11-198}$$

Differentiating Eq. (11-198) with respect to t, we obtain

$$x'(t) = x'(t_n) + (t - t_n)x''(t_n) + \frac{(t-t_n)^2}{2}x'''(t_n) + O((t-t_n)^3) \tag{11-199}$$

$$x''(t) = x''(t_n) + (t - t_n)x'''(t_n) + O((t-t_n)^2) \tag{11-200}$$

$$x'''(t) = x'''(t_n) + O((t-t_n)) \tag{11-201}$$

If we substitute $h \triangleq t - t_n$ and $x(t) = x(t_n + h) \triangleq x_{n+1}$ into Eqs. (11-198)–(11-201) and premultiply both sides of Eqs. (11-199)–(11-201) by h, $h^2/2$, and $h^3/3!$, respectively, we obtain

$$x_{n+1} = x_n + hx_n' + \frac{h^2 x_n''}{2} + \frac{h^3 x_n'''}{3!} + O(h^4) \tag{11-202}$$

$$hx_{n+1}' = hx_n' + 2\left(\frac{h^2 x_n''}{2}\right) + 3\left(\frac{h^3 x_n'''}{3!}\right) + O(h^4) \tag{11-203}$$

$$\frac{h^2x''_{n+1}}{2} = \frac{h^2x''_n}{2} + 3\left(\frac{h^3x'''_n}{3!}\right) + O(h^4) \tag{11-204}$$

$$\frac{h^3x'''_{n+1}}{3!} = \frac{h^3x'''_n}{3!} + O(h^4) \tag{11-205}$$

Equations (11-202)–(11-205) give the Nordsieck vector

$$z_{n+1} \triangleq \left[x_{n+1} \quad hx'_{n+1} \quad \frac{h^2x''_{n+1}}{2} \quad \frac{h^3x'''_{n+1}}{3!}\right]^t \qquad \text{at } t = t_{n+1}$$

in terms of the Nordsieck vector

$$z_n \triangleq \left[x_n \quad hx'_n \quad \frac{h^2x''_n}{2} \quad \frac{h^3x'''_n}{3!}\right]^t \qquad \text{at } t = t_n$$

If we truncate the terms denoted by $O(h^4)$, the resulting expressions are *exact* only if the solution $x(t)$ is a polynomial of degree less than or equal to 3. But this property is precisely that which characterizes the third-order Adams–Bashforth algorithm. Hence, it follows that the third-order Adams–Bashforth predictor is simply given by

$$x^{(0)}_{n+1} = x_n + hx'_n + \frac{h^2x''_n}{2} + \frac{h^3x'''_n}{3!} \tag{11-206}$$

$$hx'^{(0)}_{n+1} = hx'_n + 2\left(\frac{h^2x''_n}{2}\right) + 3\left(\frac{h^3x'''_n}{3!}\right) \tag{11-207}$$

$$\frac{h^2x''^{(0)}_{n+1}}{2} = \frac{h^2x''_n}{2} + 3\left(\frac{h^3x'''_n}{3!}\right) \tag{11-208}$$

$$\frac{h^3x'''^{(0)}_{n+1}}{3!} = \frac{h^3x'''_n}{3!} \tag{11-209}$$

If we rewrite Eqs. (11-206)–(11-209) in matrix form, we would obtain Eq. (11-195)!

Using the same derivation, it is readily seen that, in terms of the Nordsieck vector representation, the pth-order Adams–Bashforth predictor is given by $z^{(0)}_{n+1} = \boldsymbol{Z}\, z_n$, where $\boldsymbol{Z}$ is again the *Pascal triangle* matrix defined by

$$(\boldsymbol{Z})_{i+1,\, j+1} = \binom{j}{i} \triangleq \frac{j!}{(j-i)!\,i!}\,, \qquad (p+1) > j \geq i \geq 0 \tag{11-210}$$

where i and j vary from 0 to p. We shall see in Section 12-8 that the Nordsieck vector representation is extremely useful for changing step sizes automatically.

Although the matrices $\boldsymbol{T}$ and $\boldsymbol{Z}$ for the preceding Nordsieck vector representation were derived for a third-order algorithm, they can be easily generalized for the nth-order case. To implement the "Nordsieck" canonical representation in Eq. (11-163), however, requires the evaluation of the *scalar* "corrector function" $F_z(z^{(j)}_{n+1})$. This function is defined in Eq. (11-164) via Eqs. (11-148)–(11-149) and requires the past values x_{n-i}, x'_{n-i}, and $x^{(j-1)}_{n+1}$. Unfortunately, these past values are not directly available from the previously stored Nordsieck vector

$$z^{(j)}_{n+1} \triangleq \left[x^{(j)}_{n+1} \quad hx'^{(j)}_{n+1} \quad \frac{h^2x''^{(j)}_{n+1}}{2} \quad \frac{h^3x'''^{(j)}_{n+1}}{3!} \quad \cdots \quad \frac{h^{(2p+1)}x^{(2p+1)(j)}_{n+1}}{(2p+1)!}\right] \tag{11-211}$$

Our final task in this section is to derive the following equivalent formulas for $F_z(z^{(j)}_{n+1})$ which can be computed directly from the previously stored Nordsieck vector:

$$F_z(z_{n+1}^{(0)}) = -hx'^{(0)}_{n+1} + hf(x_{n+1}^{(0)}, t_{n+1}) \tag{11-212}$$

$$F_z(z_{n+1}^{(j)}) = -hx'^{(j)}_{n+1} + hf(x_{n+1}^{(j)}, t_{n+1}) \tag{11-213}$$

To derive Eq. (11-212), observe that the second component of z_n in Eq. (11-177) and the $(p + 2)$th component of $\boldsymbol{y}_n$ in Eq. (11-143) are both equal to hx'_n. Hence, the action of the matrix $\boldsymbol{T}$ in $z_n = \boldsymbol{T}\boldsymbol{y}_n$ is to force the second component of $\boldsymbol{T}\boldsymbol{v}$ to coincide with the $(p + 2)$th component of the vector "$\boldsymbol{v}$", where $\boldsymbol{v}$ is an arbitrary vector. In particular, if we choose $\boldsymbol{v} \triangleq \boldsymbol{y}_{n+1}^{(0)}$ as defined in Eq. (11-143), then the second component of $z_{n+1}^{(0)} \triangleq \boldsymbol{T}\boldsymbol{y}_{n+1}^{(0)}$ must be equal to the $(p + 2)$th component of $\boldsymbol{y}_{n+1}^{(0)}$; namely,

$$hx'^{(0)}_{n+1} = \sum_{i=0}^{p} \{\gamma_i x_{n-i} + \delta_i hx'_{n-i}\} \tag{11-214}$$

Substituting Eq. (11-214) into Eq. (11-148), we obtain

$$F_y(\boldsymbol{y}_{n+1}^{(0)}) = -hx'^{(0)}_{n+1} + hf(x_{n+1}^{(0)}, t_{n+1}) \tag{11-215}$$

Since Eq. (11-215) can be calculated directly from the initial Nordsieck vector $z_{n+1}^{(0)}$, we can replace the symbol "y" in Eq. (11-215) by "z" to obtain Eq. (11-212).

To derive Eq. (11-213), we choose $\boldsymbol{v} = \boldsymbol{c}_y$ as defined in Eq. (11-144). It follows from the preceding property of $\boldsymbol{T}$ that the second component of $\boldsymbol{c}_z = \boldsymbol{T}\boldsymbol{c}_y$ must be equal to the $(p + 2)$th component of $\boldsymbol{c}_y$, namely 1. Hence, equating the second component in both sides of Eq. (11-163) for $j = 0$, we obtain

$$hx'^{(1)}_{n+1} = hx'^{(0)}_{n+1} + F_z(z_{n+1}^{(0)})(1) \tag{11-216}$$

Substituting Eq. (11-212) for $F_z(z_{n+1}^{(0)})$ in Eq. (11-216), we obtain

$$hf(x_{n+1}^{(0)}, t_{n+1}) = hx'^{(1)}_{n+1} \tag{11-217}$$

Now let $j = 1$ in Eq. (11-149) and substitute Eq. (11-217) into the resulting expression, and we obtain

$$F_y(\boldsymbol{y}_{n+1}^{(1)}) = -hx'^{(1)}_{n+1} + hf(x_{n+1}^{(1)}, t_{n+1}) \tag{11-218}$$

Since Eq. (11-218) can be computed from $z_{n+1}^{(1)}$, we can replace the symbol "y" with "z" and conclude that Eq. (11-213) is valid for $j = 1$. To prove that Eq. (11-213) is valid in general, suppose it is valid for $j = k - 1$; namely,

$$F_z(z_{n+1}^{(k-1)}) = -hx'^{(k-1)}_{n+1} + hf(x_{n+1}^{(k-1)}, t_{n+1}) \tag{11-219}$$

Now, repeating the preceding procedure, we equate the second component in both sides of Eq. (11-163) for $j = k - 1$ and obtain

$$hx'^{(k)}_{n+1} = hx'^{(k-1)}_{n+1} + F_z(z_{n+1}^{(k-1)})(1) \tag{11-220}$$

Substituting Eq. (11-219) for $F_z(z_{n+1}^{(k-1)})$ in Eq. (11-220), we obtain

$$hf(x_{n+1}^{(k-1)}, t_{n+1}) = hx'^{(k)}_{n+1} \tag{11-221}$$

Now let $j = k$ in Eq. (11-149) and substitute Eq. (11-221) into the resulting expression, we obtain, upon replacing the symbol "y" with "z," the expression

$$F_z(z_{n+1}^{(k)}) = -hx'^{(k)}_{n+1} + hf(x_{n+1}^{(k)}, t_{n+1}) \tag{11-222}$$

Since by assuming that Eq. (11-213) is valid for $j = k - 1$—the *induction hypothesis*—we have shown that it is valid for $j = k$. It follows from the *principle of mathematical induction* that Eq. (11-213) is valid for all j. This completes the proof.

REFERENCES

1. RALSTON, A. *A First Course in Numerical Analysis.* New York: McGraw-Hill Book Company, 1965.
2. GEAR, C. W. *Numerical Initial Value Problems in Ordinary Differential Equations.* Englewood Cliffs, N.J.: Prentice-Hall, Inc., 1971.
3. STRUBLE, R. A. *Nonlinear Differential Equations.* New York: McGraw-Hill Book Company, 1962.
4. HALE, J. K. *Ordinary Differential Equations.* New York: McGraw-Hill Book Company, 1969.
5. WINTNER, A. "The Non-Local Existence Problem of Ordinary Differential Equations." *Am. J. Math.*, Vols. 67, pp. 277–284, 1945.
6. HILLE, E. *Analysis*, Vols. I and II. Lexington, Mass.: Xerox College Publishing, 1966.

PROBLEMS

11-1. Formulate the *forward Euler algorithm* for a system of n ordinary differential equations $\dot{\mathbf{x}} = \mathbf{f}(\mathbf{x}, t)$.

11-2. Formulate the *second-order Taylor algorithm* for a system of n ordinary differential equations $\dot{\mathbf{x}} = \mathbf{f}(\mathbf{x}, t)$.

11-3. Formulate the *third-order Taylor algorithm* for a system of n ordinary differential equations $\dot{\mathbf{x}} = \mathbf{f}(\mathbf{x}, t)$.

11-4. Formulate the *fourth-order Runge–Kutta algorithm* for a system of n ordinary differential equations $\dot{\mathbf{x}} = \mathbf{f}(\mathbf{x}, t)$.

11-5. Formulate the *multistep numerical-integration algorithm* (Eq. 11-70) for a system of n ordinary differential equations $\dot{\mathbf{x}} = \mathbf{f}(\mathbf{x}, t)$.

11-6. Formulate Theorem 11-5 for a system of n ordinary differential equations $\dot{\mathbf{x}} = \mathbf{f}(\mathbf{x}, t)$.

11-7. Show that the *third-order Adams–Bashforth algorithm* is *exact* for all polynomial solutions of degree 3.

11-8. Show that the *fourth-order Adams–Moulton algorithm* is *exact* for all polynomial solutions of degree 4.

11-9. Calculate the *local truncation error* for the *third-order Adams–Bashforth algorithm.*

11-10. Calculate the *local truncation error* for the *fourth-order Adams-Moulton algorithm.*

11-11. Show that the following constraints must be satisfied by the coefficients of any third-order multistep algorithm:

(a) $a_0 + a_1 + a_2 + \ldots + a_p = 1.$

(b) $-a_1 - 2a_2 - 3a_3 - \ldots - pa_p + b_{-1} + b_0 + b_1 + \ldots + b_p = 1.$

(c) $a_1 + 4a_2 + 9a_3 + \ldots + p^2a_p + 2b_{-1} - 2b_1 - 4b_2 - \ldots - 2pb_p = 1.$

(d) $-a_1 - 8a_2 - 27a_3 - \ldots - p^3a_p + 3b_{-1} + 3b_1 + 12b_2 + \ldots + 3p^2b_p = 1.$

11-12. Specify the general form of a *predictor–corrector algorithm* using a kth-*order explicit multistep algorithm* as predictor for an nth-*order implicit multistep algorithm.* Give an example illustrating your algorithm with $k = 2$ and $n = 4$.

11-13. Consider the initial-value problem

$$\dot{x} = f(x, t) = x + e^t, \qquad x(0) = 1$$

(a) Derive the following total time derivatives of $f(x, t)$ with respect to t:

$$f^{(1)}(x, t), f^{(2)}(x, t), \ldots, f^{(5)}(x, t)$$

(b) Derive the nth-order Taylor algorithm for solving this problem, where $n = 1, 2, \ldots, 5$.

11-14. Use the following algorithms to find the solution to the initial-value problem

$$\dot{x} = f(x, t) = 2 - x, \qquad x(0) = 1$$

and compare your solution with the exact solution at $t = 1$. Use step sizes $h = 0.1$, 0.2, and 0.5, respectively.
(a) Forward Euler algorithm.
(b) Heun's algorithm.
(c) Trapezoidal algorithm.
(d) Fourth-order Runge–Kutta algorithm.

11-15. Find the maximum step size h that can be used with the following algorithms for solving $\dot{x} = x$, $x(0) = 1$ from $t = 0$ to $t = 1$ if the *local truncation error* is not to exceed 0.1, 0.01, and 0.001, respectively.
(a) Forward Euler algorithm.
(b) Second-order Taylor algorithm.
(c) Trapezoidal algorithm.

11-16. Use the following algorithms to find the solution to the initial-value problem

$$\dot{x} = f(x, t) = -x^2 + t, \qquad x(0) = 0$$

from $t = 0$ to $t = 1$. Use the step sizes $h = 0.1$, 0.2, and 0.5, respectively.
(a) Forward Euler algorithm.
(b) Heun's algorithm.
(c) Trapezoidal algorithm.
(d) Fourth-order Runge–Kutta algorithm.

11-17. Assuming the *maximum* local truncation error is not to exceed 0.001, find the *minimum* number of uniform time steps required to solve $\dot{x} = x$, $x(0) = 1$ from $t = 0$ to $t = 1$ using the following algorithms:
(a) Forward Euler algorithm.
(b) Second-order Taylor algorithm.
(c) Trapezoidal algorithm.

11-18. Use Heun's algorithm and the trapezoidal algorithm to solve the initial-value problem

$$\dot{x} = (x - t^3) + 3t^2, \qquad x(0) = 0$$

from $t = 0$ to $t = 1$ with step sizes $h = 2^{-2}, 2^{-4}, 2^{-6}, 2^{-8}, 2^{-10}$, and 2^{-12}. Use log–log paper to plot the relationship between $\log|\epsilon|$ versus $\log h$, where ϵ is the truncation error at $t = 1$. Explain the differences and similarities between the two algorithms, and explain why Heun's algorithm is sometimes referred to as the *modified* trapezoidal algorithm.

11-19. Use the following algorithms to solve the initial-value problem

$$\dot{x} = [1 - 0.2 \cos x]^{-1}, \qquad x(0) = 0$$

from $t = 0$ to $t = 10$.

(a) Modified Euler–Cauchy algorithm with step size $h = 0.5$.
(b) Fourth-order Runge–Kutta algorithm with step size $h = 1$.

11-20. Use the following algorithms to solve the initial-value problem

$$\dot{x}_1 = x_1 - x_2, \qquad x_1(0) = 1$$
$$\dot{x}_2 = x_1 + x_2 + t, \qquad x_2(0) = 2$$

from $t = 0$ to $t = 2$ with step size $h = 0.05$.
(a) Forward Euler algorithm.
(b) Heun's algorithm.
(c) Trapezoidal algorithm.
(d) Fourth-order Runge–Kutta algorithm.

11-21. Repeat Problem 11-20 for the following initial-value problem:

$$\dot{x}_1 = x_1 + 3x_2 + t, \qquad x_1(0) = 1$$
$$\dot{x}_2 = x_1 + 2x_2 + t^2, \qquad x_2(0) = 1$$

11-22. Repeat Problem 11-20 for the following initial-value problem:

$$\dot{x}_1 = -2x_2 + 2t^2, \qquad x_1(0) = -2$$
$$\dot{x}_2 = \tfrac{1}{2}x_1 + 2t, \qquad x_2(0) = 0$$

Compare your answers with the *exact* solution $x_1(t) = -2 \cos t$ and $x_2(t) = -2 \sin t + t^2$.

11-23. Verify that Eqs. (11-136), (11-139), and (11-142) are contained within the two matrix equations given by Eqs. (11-145)–(11-146).

11-24. (a) Derive the expressions defining Eqs. (11-143)–(11-149) corresponding to a *3rd-order Adams-Bashforth predictor* and a *3rd-order Gear's corrector* as defined in Table 13-1 on page 526.
(b) Verify that the expressions obtained from part (a) completely define the predictor-corrector algorithm.
(c) Transform the canonical representation from part (a) into an equivalent *backward-difference* representation.
(d) Transform the canonical representation from part (a) into an equivalent *Nordsieck* representation.

CHAPTER 12

Multistep Numerical-Integration Algorithms

12-1 EXACTNESS CONSTRAINTS FOR MULTISTEP ALGORITHMS

A kth-*order multistep numerical-integration algorithm* for solving initial-value problems assumes the general form

$$x_{n+1} = \sum_{i=0}^{p} a_i x_{n-i} + h \sum_{i=-1}^{p} b_i f(x_{n-i}, t_{n-i}) \tag{12-1}$$

The $2p + 3$ coefficients $\{a_0, a_1, \ldots, a_p, b_{-1}, b_0, b_1, \ldots, b_p\}$ are *usually* chosen such that if the solution $\hat{x}(t)$ of an initial-value problem is a *polynomial* of degree k, then Eq. (12-1) will give the *exact* solution $x_{n+1} = \hat{x}(t_{n+1})$.[1] Since a straight line (polynomial of degree 1) is uniquely determined by two parameters, a parabola (polynomial of degree 2) is uniquely determined by three parameters, and a polynomial of degree k is uniquely determined by $k + 1$ parameters, it is clear that Eq. (12-1) must have at least $k + 1$ coefficients; thus,

$$2p + 3 \geq k + 1 \tag{12-2}$$

In most practical kth-order multistep algorithms, we have $2p + 3 > k + 1$, and the *excess* coefficients can therefore be arbitrarily prescribed. The remaining $k + 1$ coef-

[1]Since polynomials $\hat{x}(t) = \alpha_0 + \alpha_1 t + \alpha_2 t^2 + \cdots + \alpha_k t^k$ of degree k include all polynomials of degree lower than k—by assigning zero value to the higher-order coefficients—Eq. (12-1) must also give the exact solution to all initial-value problems having a polynomial solution of degree less than k.

ficients can be determined by the *method of undetermined coefficients*, as was done in Section 11-5. Our objective in this section is to derive a more systematic method for prescribing the $2p + 3$ coefficients in order that Eq. (12-1) be *exact for all polynomial solutions of degree less than or equal to k*. Clearly, these coefficients are constrained to satisfy certain relations. We shall now derive these constraints by examining the form of initial-value problems

$$\dot{x} = f(x, t) \tag{12-3}$$

which gives rise to a polynomial solution of degree k.

1. $k = 0$, $x = \hat{x}(t) = \alpha_0$

The class of initial-value problems with the solution $\hat{x}(t) = \alpha_0$ is $\dot{x} = f(x, t) = 0$. Hence, the exactness constraint requires that

$$x_{n+1} = \alpha_0, \qquad x_{n-i} = \alpha_0, \qquad f(x_{n-i}, t_{n-i}) = 0 \tag{12-4}$$

Substituting Eq. (12-4) into Eq. (12-1), we obtain

$$\alpha_0 = \sum_{i=0}^{p} a_i \alpha_0 \tag{12-5}$$

Hence, the *exactness constraint for polynomial solutions of degree* 0 is simply

$$\sum_{i=0}^{p} a_i = 1 \tag{12-6}$$

2. $k = 1$, $x = \hat{x}(t) = \alpha_0 + \alpha_1 t$

The class of initial-value problems with the solution $\hat{x}(t) = \alpha_0 + \alpha_1 t$ is $\dot{x} = f(x, t) = \alpha_1$.

For convenience, let us choose $t_n = 0$; hence, $t_{n+1} = h$, $t_{n-1} = -h$, $t_{n-2} = -2h, \ldots, t_{n-p} = -ph$. The exactness constraint in this case requires that

$$x_{n+1} = \alpha_0 + \alpha_1 h, \qquad x_{n-i} = \alpha_0 - \alpha_1(ih), \qquad f(x_{n-i}, t_{n-i}) = \alpha_1 \tag{12-7}$$

Substituting Eq. (12-7) into Eq. (12-1), we obtain

$$\alpha_0 + \alpha_1 h = \sum_{i=0}^{p} a_i[\alpha_0 - \alpha_1(ih)] + h \sum_{i=-1}^{p} b_i \alpha_1 \tag{12-8}$$

Substituting Eq. (12-6) into Eq. (12-8), we obtain

$$\alpha_1 h = \sum_{i=0}^{p} [-a_i \alpha_1(ih)] + h \sum_{i=-1}^{p} b_i \alpha_1 \tag{12-9}$$

Dividing both sides of Eq. (12-9) by $\alpha_1 h$, we obtain the following *exactness constraint for polynomial solutions of degree* 1:

$$\sum_{i=0}^{p} (-i)a_i + \sum_{i=-1}^{p} b_i = 1 \tag{12-10}$$

3. $k = 2$, $x = \hat{x}(t) = \alpha_0 + \alpha_1 t + \alpha_2 t^2$

The class of initial-value problems with the solution $\hat{x}(t) = \alpha_0 + \alpha_1 t + \alpha_2 t^2$ is $\dot{x} = 2\alpha_2 t + \alpha_1$. Again choosing $t_n = 0$ for convenience, the exactness constraint in this case requires that

$$\begin{gathered} x_{n+1} = \alpha_0 + \alpha_1 h + \alpha_2 h^2, \quad x_{n-i} = \alpha_0 - \alpha_1(ih) + \alpha_2(ih)^2, \\ f(x_{n-i}, t_{n-i}) = 2\alpha_2(-ih) + \alpha_1 \end{gathered} \tag{12-11}$$

Substituting Eq. (12-11) into Eq. (12-1), we obtain

$$\alpha_0 + \alpha_1 h + \alpha_2 h^2 = \sum_{i=0}^{p} a_i[\alpha_0 - \alpha_1(ih) + \alpha_2(ih)^2] + h \sum_{i=-1}^{p} b_i[2\alpha_2(-ih) + \alpha_1] \tag{12-12}$$

Substituting Eqs. (12-6) and (12-10) into Eq. (12-12), we obtain

$$\alpha_2 h^2 = \sum_{i=0}^{p} \alpha_2 i^2 h^2 a_i - h^2 \sum_{i=-1}^{p} 2\alpha_2 b_i(i) \tag{12-13}$$

Dividing both sides of Eq. (12-13) by $\alpha_2 h^2$, we obtain the following *exactness constraint for polynomial solutions of degree* 2:

$$\sum_{i=1}^{p} i^2 a_i + 2 \sum_{i=-1}^{p} (-i)b_i = 1 \tag{12-14}$$

4. General case, $x = \hat{x}(t) = \alpha_0 + \alpha_1 t + \alpha_2 t^2 + \ldots + \alpha_k t^k$

The class of initial-value problems with solution $\hat{x}(t) = \alpha_0 + \alpha_1 t + \alpha_2 t^2 + \ldots + \alpha_k t^k$ is given by

$$\dot{x} = k\alpha_k t^{k-1} + (k-1)\alpha_{k-1} t^{k-2} + \ldots + 2\alpha_2 t + \alpha_1 \tag{12-15}$$

Following the same procedure as before, we obtain the following *exactness constraint for polynomial solutions of degree k:*

$$\boxed{\sum_{i=1}^{p} (-i)^k a_i + k \sum_{i=-1}^{p} (-i)^{k-1} b_i = 1} \tag{12-16}$$

Examination of the exactness constraints given by Eqs. (12-10), (12-14), and (12-16) shows that they all have the same form and, in fact, can be considered as special cases of the following:[2]

Exactness Polynomial Constraints for k*th-Order Multistep Algorithm*

$$\sum_{i=0}^{p} a_i = 1 \tag{12-17}$$

$$\sum_{i=1}^{p} (-i)^j a_i + j \sum_{i=-1}^{p} (-i)^{j-1} b_i = 1, \qquad j = 1, 2, \ldots, k$$

Equation (12-17) gives the relationships that must be satisfied by the $2p + 3$ coefficients in Eq. (12-1) in order that this algorithm will give *exact* values for x_{n+1} (assuming no round-off error) whenever the solution of $\dot{x} = f(x, t)$ is *a polynomial* of degree less than or equal to k. One could obviously derive other constraining relationships by demanding the algorithm to be exact whenever the solution belongs to some class of functions—such as exponentials—other than polynomials. However, most multistep algorithms in use today are based on the polynomial-exactness crite-

[2]The first equation in Eq. (12-17) can be included in the second equation by replacing the summation index from $i = 1$ to $i = 0$, and by letting $j = 0, 1, 2, \ldots, k$.

rion, and the *order* of the algorithm is defined to be the *degree* of the *highest-degree polynomial solution* for which it is exact.

Any multistep algorithm in the form of Eq. (12-1) is said to be *consistent* if its $2p + 3$ coefficients satisfy Eqs. (12-6) and (12-10). In other words, a consistent multistep algorithm is one that gives exact numerical solution to an initial-value problem $\dot{x} = f(x, t)$ having a *linear* polynomial solution $\hat{x}(t) = \alpha_0 + \alpha_1 t$. Hence, all multistep algorithms of order $k \geq 1$ are consistent. We shall now derive two important families of consistent multistep algorithms commonly used in practice.

12-2 ADAMS–BASHFORTH ALGORITHM

The kth-order Adams–Bashforth algorithm is an *explicit* multistep algorithm obtained by setting

$$p = k - 1, \qquad a_1 = a_2 = \ldots = a_{k-1} = 0, \qquad b_{-1} = 0 \tag{12-18}$$

in Eq. (12-1); i.e.,

$$\begin{aligned} x_{n+1} = a_0 x_n + h\{&b_0 f(x_n, t_n) + b_1 f(x_{n-1}, t_{n-1}) \\ &+ \ldots + b_{k-1} f(x_{n-k+1}, t_{n-k+1})\} \end{aligned} \tag{12-19}$$

The $k + 1$ coefficients $a_0, b_0, b_1, \ldots, b_{k-1}$ are to be determined so that Eq. (12-19) is exact for all polynomial solutions of degree k. Since a kth-degree polynomial is uniquely determined by $k + 1$ parameters, we have exactly the correct number of unknown coefficients to be determined using Eq. (12-17). It follows immediately from the exactness constraint $\sum_{i=1}^{p} a_i = 0$ that $a_0 = 1$. Since the remaining coefficients $b_0, b_1, \ldots, b_{k-1}$ will depend on the numerical value of k, we let $b_i \triangleq \beta_i(k)$ to emphasize its dependence on k, and rewrite Eq. (12-19) as follows:

$$x_{n+1} = x_n + h \sum_{i=0}^{k-1} \beta_i(k) f(x_{n-i}, t_{n-i}) \tag{12-20}$$

To determine $\beta_0(k), \beta_1(k), \ldots, \beta_{k-1}(k)$, we substitute $a_0 = 1$, $b_i = \beta_i(k)$, and the coefficients defined by Eq. (12-18) into Eq. (12-17):

$$\sum_{i=0}^{k-1} (-i)^{j-1} \beta_i(k) = \frac{1}{j}, \qquad j = 1, 2, \ldots, k \tag{12-21}$$

Equation (12-21), when expanded, consists of a system of k linear equations in the k unknowns $\beta_0(k), \beta_1(k), \ldots, \beta_{k-1}(k)$; thus,

$$\begin{bmatrix} 1 & 1 & 1 & 1 & \cdots & 1 \\ 0 & -1 & -2 & -3 & \cdots & -(k-1) \\ 0 & 1 & 4 & 9 & \cdots & (k-1)^2 \\ 0 & -1 & -8 & -27 & \cdots & -(k-1)^3 \\ \vdots & \vdots & \vdots & \vdots & \cdots & \vdots \\ 0 & (-1)^{k-1} & (-2)^{k-1} & (-3)^{k-1} & \cdots & [-(k-1)]^{k-1} \end{bmatrix} \begin{bmatrix} \beta_0(k) \\ \beta_1(k) \\ \beta_2(k) \\ \beta_3(k) \\ \vdots \\ \beta_{k-1}(k) \end{bmatrix} = \begin{bmatrix} 1 \\ \frac{1}{2} \\ \frac{1}{3} \\ \frac{1}{4} \\ \vdots \\ \frac{1}{k} \end{bmatrix} \tag{12-22}$$

The solution of Eq. (12-22) therefore uniquely specifies the remaining coefficients of the kth-order Adams–Bashforth algorithm. We shall illustrate this by considering three cases.

Case 1: $k = 1$. In this case, Eq. (12-22) reduces trivially to $\beta_0(1) = 1$. Substituting this coefficient into Eq. (12-20), we obtain the following *first-order Adams–Bashforth algorithm:*

$$x_{n+1} = x_n + hf(x_n, t_n) \tag{12-23}$$

which is just the *forward Euler algorithm* defined earlier in Eq. (11-41).

Case 2: $k = 2$. In this case, Eq. (12-22) becomes

$$\begin{bmatrix} 1 & 1 \\ 0 & -1 \end{bmatrix} \begin{bmatrix} \beta_0(2) \\ \beta_1(2) \end{bmatrix} = \begin{bmatrix} 1 \\ \frac{1}{2} \end{bmatrix}$$

The solution is given by $\beta_0(2) = \frac{3}{2}$ and $\beta_1(2) = -\frac{1}{2}$. Substituting these coefficients into Eq. (12-20), we obtain the following *second-order Adams–Bashforth algorithm:*

$$x_{n+1} = x_n + h\{\tfrac{3}{2}f(x_n, t_n) - \tfrac{1}{2}f(x_{n-1}, t_{n-1})\} \tag{12-24}$$

Case 3: $k = 3$. In this case, Eq. (12-22) becomes

$$\begin{bmatrix} 1 & 1 & 1 \\ 0 & -1 & -2 \\ 0 & 1 & 4 \end{bmatrix} \begin{bmatrix} \beta_0(3) \\ \beta_1(3) \\ \beta_2(3) \end{bmatrix} = \begin{bmatrix} 1 \\ \frac{1}{2} \\ \frac{1}{3} \end{bmatrix}$$

The solution is given by $\beta_0(3) = \frac{23}{12}$, $\beta_1(3) = -\frac{16}{12}$, and $\beta_2(3) = \frac{5}{12}$. Substituting these coefficients into Eq. (12-20), we obtain the following *third-order Adams–Bashforth algorithm:*

$$x_{n+1} = x_n + h\{\tfrac{23}{12}f(x_n, t_n) - \tfrac{16}{12}f(x_{n-1}, t_{n-1}) + \tfrac{5}{12}f(x_{n-2}, t_{n-2})\} \tag{12-25}$$

For future reference, the formulas for the first- to sixth-order Adams–Bashforth algorithms are collected in Table 12-1.

Examination of Table 12-1 shows that the kth-order Adams–Bashforth algorithm requires k *starting* values: $x_n, x_{n-1}, x_{n-2}, \ldots, x_{n-k+1}$. Hence, a kth-order Adams–Bashforth algorithm is a k-step algorithm. If we apply Theorem 11-5 to the Adams–Bashforth algorithm, we obtain the following corresponding result:

TABLE 12-1. Adams–Bashforth Algorithms

ORDER	
First	$x_{n+1} = x_n + hf(x_n, t_n)$
Second	$x_{n+1} = x_n + h\{\frac{3}{2}f(x_n, t_n) - \frac{1}{2}f(x_{n-1}, t_{n-1})\}$
Third	$x_{n+1} = x_n + h\{\frac{23}{12}f(x_n, t_n) - \frac{16}{12}f(x_{n-1}, t_{n-1}) + \frac{5}{12}f(x_{n-2}, t_{n-2})\}$
Fourth	$x_{n+1} = x_n + h\{\frac{55}{24}f(x_n, t_n) - \frac{59}{24}f(x_{n-1}, t_{n-1}) + \frac{37}{24}f(x_{n-2}, t_{n-2}) - \frac{9}{24}f(x_{n-3}, t_{n-3})\}$
Fifth	$x_{n+1} = x_n + h\{\frac{1901}{720}f(x_n, t_n) - \frac{2774}{720}f(x_{n-1}, t_{n-1}) + \frac{2616}{720}f(x_{n-2}, t_{n-2}) - \frac{1274}{720}f(x_{n-3}, t_{n-3}) + \frac{251}{720}f(x_{n-4}, t_{n-4})\}$
Sixth	$x_{n+1} = x_n + h\{\frac{4277}{1440}f(x_n, t_n) - \frac{7923}{1440}f(x_{n-1}, t_{n-1}) + \frac{9982}{1440}f(x_{n-2}, t_{n-2}) - \frac{7298}{1440}f(x_{n-3}, t_{n-3}) + \frac{2877}{1440}f(x_{n-4}, t_{n-4}) - \frac{475}{1440}f(x_{n-5}, t_{n-5})\}$

Theorem 12-1. *The local truncation error* ϵ_T *for the kth-order Adams–Bashforth algorithm is given by*

$$\epsilon_T = [C_k \hat{x}^{(k+1)}(\hat{\tau})]h^{k+1} = O(h^{k+1}), \qquad t_n < \hat{\tau} < t_{n+1} \tag{12-26a}$$

$$C_1 = \tfrac{1}{2}, C_2 = \tfrac{5}{12}, C_3 = \tfrac{3}{8}, C_4 = \tfrac{251}{720}, C_5 = \tfrac{95}{288}, C_6 = \tfrac{19{,}087}{60{,}480}, \ldots,$$

$$C_k = \frac{1}{(k+1)!}\left[k^{k+1} - (k-1)^{k+1} - (k+1)\sum_{i=0}^{k-2} b_i(k-1-i)^k\right] \tag{12-26b}$$

12-3 ADAMS–MOULTON ALGORITHM

The kth-order Adams–Moulton algorithm is an *implicit* multistep algorithm obtained by setting[3]

$$p = k - 2, \; a_1 = a_2 = a_3 = \ldots = a_{k-2} = 0 \tag{12-27}$$

in Eq. (12-1); thus,

$$\begin{aligned} x_{n+1} = a_0 x_n + h\{&b_{-1}f(x_{n+1}, t_{n+1}) + b_0 f(x_n, t_n) + b_1 f(x_{n-1}, t_{n-1}) \\ &+ \ldots + b_{k-2} f(x_{n-k+2}, t_{n-k+2})\} \end{aligned} \tag{12-28}$$

The $k + 1$ coefficients $a_0, b_{-1}, b_0, b_1, \ldots, b_{k-2}$ are to be determined so that Eq. (12-28) is exact for all polynomial solutions of degree k. Again observe that we have exactly the correct number of unknown coefficients to be determined using Eq. (12-17). It follows immediately that $a_0 = 1$. To distinguish the remaining coefficients, which depend on the value of k, from those associated with the Adams–Bashforth algorithm, we let $b_i \triangleq \mu_i(k)$ and rewrite Eq. (12-28) as follows:

$$x_{n+1} = x_n + h\sum_{i=-1}^{k-2} \mu_i(k) f(x_{n-i}, t_{n-i}) \tag{12-29}$$

To determine $\mu_{-1}(k), \mu_0(k), \mu_1(k), \ldots, \mu_{k-2}(k)$, we substitute $a_0 = 1$, $b_i = \mu_i(k)$, and the coefficients defined by Eq. (12-27) into Eq. (12-17):

$$\sum_{i=-1}^{k-2} (-i)^{j-1}\mu_i(k) = \frac{1}{j}, \qquad j = 1, 2, \ldots, k \tag{12-30}$$

which, when expanded, consists of a system of k linear equations in the k unknowns $\mu_{-1}(k), \mu_0(k), \mu_1(k), \ldots, \mu_{k-2}(k)$:

$$\begin{bmatrix} 1 & 1 & 1 & 1 & 1 & \cdots & 1 \\ 1 & 0 & -1 & -2 & -3 & \cdots & -(k-2) \\ 1 & 0 & 1 & 4 & 9 & \cdots & (k-2)^2 \\ 1 & 0 & -1 & -8 & -27 & \cdots & -(k-2)^3 \\ \vdots & \vdots & \vdots & \vdots & \vdots & \ddots & \vdots \\ 1 & 0 & (-1)^{k-1} & (-2)^{k-1} & (-3)^{k-1} & \cdots & [-(k-2)]^{k-1} \end{bmatrix} \begin{bmatrix} \mu_{-1}(k) \\ \mu_0(k) \\ \mu_1(k) \\ \mu_2(k) \\ \vdots \\ \mu_{k-2}(k) \end{bmatrix} = \begin{bmatrix} 1 \\ \frac{1}{2} \\ \frac{1}{3} \\ \frac{1}{4} \\ \vdots \\ \frac{1}{k} \end{bmatrix} \tag{12-31}$$

[3] We interpret $\sum_{i=k_1}^{k_2} a_i x_i = a_{k_1} x_{k_1}$ whenever $k_2 < k_1$.

The solution of Eq. (12-31) therefore uniquely specifies the remaining coefficients of the kth-order Adams–Moulton algorithm. We shall illustrate this by considering three cases.

Case 1: $k = 1$. In this case, Eq. (12-31) reduces trivially to $\mu_{-1}(1) = 1$. Substituting this coefficient into Eq. (12-29), we obtain the following *first-order Adams–Moulton algorithm:*

$$x_{n+1} = x_n + hf(x_{n+1}, t_{n+1}) \tag{12-32}$$

In contrast to the *explicit* forward Euler algorithm defined by Eq. (12-23), this *implicit* algorithm is called the *backward Euler algorithm.*

Case 2: $k = 2$. In this case, Eq. (12-31) becomes

$$\begin{bmatrix} 1 & 1 \\ 1 & 0 \end{bmatrix} \begin{bmatrix} \mu_{-1}(2) \\ \mu_0(2) \end{bmatrix} = \begin{bmatrix} 1 \\ \frac{1}{2} \end{bmatrix}$$

The solution is given by $\mu_{-1}(2) = \frac{1}{2}$ and $\mu_0(2) = \frac{1}{2}$. Substituting these coefficients into Eq. (12-29), we obtain the following *second-order Adams–Moulton algorithm:*

$$x_{n+1} = x_n + h\{\tfrac{1}{2}f(x_{n+1}, t_{n+1}) + \tfrac{1}{2}f(x_n, t_n)\} \tag{12-33}$$

which is just the *trapezoidal algorithm* defined earlier in Eq. (11-80).

Case 3: $k = 3$. In this case, Eq. (12-31) becomes

$$\begin{bmatrix} 1 & 1 & 1 \\ 1 & 0 & -1 \\ 1 & 0 & 1 \end{bmatrix} \begin{bmatrix} \mu_{-1}(3) \\ \mu_0(3) \\ \mu_1(3) \end{bmatrix} = \begin{bmatrix} 1 \\ \frac{1}{2} \\ \frac{1}{3} \end{bmatrix}$$

The solution is given by $\mu_{-1}(3) = \frac{5}{12}$, $\mu_0(3) = \frac{8}{12}$, and $\mu_1(3) = -\frac{1}{12}$. Substituting these coefficients into Eq. (12-29), we obtain the following *third-order Adams–Moulton algorithm:*

$$x_{n+1} = x_n + h\{\tfrac{5}{12}f(x_{n+1}, t_{n+1}) + \tfrac{8}{12}f(x_n, t_n) - \tfrac{1}{12}f(x_{n-1}, t_{n-1})\} \tag{12-34}$$

For future reference, the formulas for the first- to sixth-order Adams–Moulton algorithms are collected in Table 12-2.

TABLE 12-2. Adams–Moulton Algorithms

ORDER	
First	$x_{n+1} = x_n + hf(x_{n+1}, t_{n+1})$
Second	$x_{n+1} = x_n + h\{\frac{1}{2}f(x_{n+1}, t_{n+1}) + \frac{1}{2}f(x_n, t_n)\}$
Third	$x_{n+1} = x_n + h\{\frac{5}{12}f(x_{n+1}, t_{n+1}) + \frac{8}{12}f(x_n, t_n) - \frac{1}{12}f(x_{n-1}, t_{n-1})\}$
Fourth	$x_{n+1} = x_n + h\{\frac{9}{24}f(x_{n+1}, t_{n+1}) + \frac{19}{24}f(x_n, t_n) - \frac{5}{24}f(x_{n-1}, t_{n-1}) + \frac{1}{24}f(x_{n-2}, t_{n-2})\}$
Fifth	$x_{n+1} = x_n + h\{\frac{251}{720}f(x_{n+1}, t_{n+1}) + \frac{646}{720}f(x_n, t_n) - \frac{264}{720}f(x_{n-1}, t_{n-1})$ $+ \frac{106}{720}f(x_{n-2}, t_{n-2}) - \frac{19}{720}f(x_{n-3}, t_{n-3})\}$
Sixth	$x_{n+1} = x_n + h\{\frac{475}{1440}f(x_{n+1}, t_{n+1}) + \frac{1427}{1440}f(x_n, t_n) - \frac{798}{1440}f(x_{n-1}, t_{n-1})$ $+ \frac{482}{1440}f(x_{n-2}, t_{n-2}) - \frac{173}{1440}f(x_{n-3}, t_{n-3}) + \frac{27}{1440}f(x_{n-4}, t_{n-4})\}$

Examination of Table 12-2 shows that the kth-order Adams–Moulton algorithm requires only $k - 1$ *starting* values: $x_n, x_{n-1}, x_{n-2}, \ldots, x_{n-k+2}$. Hence, a kth-order Adams–Moulton algorithm is a $(k - 1)$-step algorithm. Observe that this is one less than the kth-order Adams–Bashforth algorithm, which is a k-step algorithm requiring k starting values. If we apply Theorem 11-5 to the Adams–Moulton algorithm, we obtain the following corresponding result:

Theorem 12-2. *The local truncation error ϵ_T for the kth-order Adams–Moulton algorithm is given by*

$$\epsilon_T = [C_k \hat{x}^{(k+1)}(\hat{\tau})]h^{k+1} = O(h^{k+1}), \; t_n < \hat{\tau} < t_{n+1} \tag{12-35a}$$

$$C_1 = -\frac{1}{2}, \quad C_2 = -\frac{1}{12}, \quad C_3 = -\frac{1}{24}, \quad C_4 = -\frac{19}{720},$$

$$C_5 = -\frac{3}{160}, \quad C_6 = -\frac{863}{60{,}480}$$

$$C_k = \frac{1}{(k+1)!}\left[(k-1)^{k+1} - (k-2)^{k+1} - (k+1)\sum_{i=-1}^{k-3} b_i(k-2-i)^k\right] \tag{12-35b}$$

Comparing Eqs. (12-26) and (12-35), we see that the *order of accuracy* of the local truncation error of the kth-order *explicit* Adams–Bashforth algorithm is the same as that of the *implicit* Adams–Moulton algorithm. However, since the latter requires one less starting value than the former, we can say that the *Adams–Moulton algorithm is one order more accurate than the Adams–Bashforth algorithm.* The disadvantage of the Adams–Moulton algorithm is that, being *implicit*, a *predictor* is necessary. The Runge–Kutta algorithm discussed in Section 11-4 or the Adams–Bashforth explicit algorithms of appropriate orders are frequently used as predictors for the Adams–Moulton algorithm. Needless to say, the order of the predictor algorithm *need not* be the same as the order of the Adams–Moulton "corrector" algorithm.

12-4 ANALYSIS OF ERROR PROPAGATION—A CASE STUDY

The Adams–Bashforth and Adams–Moulton algorithms are just two of the infinitely many multistep algorithms having equal if not better numerical accuracies. This is because of the $2p + 3$ coefficients only $k + 1$ are needed to satisfy the exactness constraints prescribed by Eq. (12-17). The remaining coefficients could be assigned arbitrary values. Why then do we single out the Adams–Bashforth and Adams–Moulton algorithms over other algorithms that not only have lower local truncation errors but also possibly require fewer starting values? The answer is that such algorithms, although more accurate to start with, often share the undesirable tendency to amplify the local truncation and round-off errors in each time step such that, over a period of time, the propagated error dominates the solution itself, thereby rendering these algorithms completely useless. To understand this undesirable phenomenon, it is necessary to analyze the error-propagation mechanism of the algorithms—a difficult task to be developed in Section 12-5. To help the reader understand the subsequent stability analysis, we now consider a *specific* example that is simple enough and yet contains most of the salient features of a general multistep algorithm. Consider the following *explicit* multistep algorithm:

$$x_{n+1} = x_{n-3} + h\{\tfrac{8}{3}f(x_n, t_n) - \tfrac{4}{3}f(x_{n-1}, t_{n-1}) + \tfrac{8}{3}f(x_{n-2}, t_{n-2})\} \qquad (12\text{-}36)$$

It is easily verified that the coefficients associated with this algorithm satisfy the exactness constraints prescribed in Eq. (12-17) for $j = 1, 2, 3,$ and 4. Hence, Eq. (12-36) is a *fourth-order explicit algorithm*. Its local truncation error can be derived using Theorem 11-5 and is found to be given by

$$\epsilon_T = \tfrac{28}{90}\hat{x}^{(5)}(\hat{\tau})h^5 = O(h^5) \qquad (12\text{-}37)$$

Hence, this algorithm has the same order of accuracy as the fourth-order Adams–Bashforth formula. However, a simple exercise using this algorithm to solve any initial-value problem will convince the reader that the overall error becomes intolerably unacceptable after a few time steps. To uncover the reason for this behavior, consider the simplest initial-value problem

$$\dot{x} = -\lambda x, \qquad x(0) = 1 \qquad (12\text{-}38)$$

where λ is a positive real number. The exact solution to Eq. (12-38) is

$$x(t) = e^{-\lambda t} \qquad (12\text{-}39)$$

Applying Eq. (12-36) to solve Eq. (12-38), we obtain the *recurrence relation*

$$x_{n+1} = x_{n-3} + h\{-\tfrac{8}{3}\lambda x_n + \tfrac{4}{3}\lambda x_{n-1} - \tfrac{8}{3}\lambda x_{n-2}\} \qquad (12\text{-}40)$$

To analyze the asymptotic behavior of x_{n+1} as $n \longrightarrow \infty$, it is convenient to recast Eq. (12-40) into the following equivalent *linear difference equation:*

$$x_{n+1} + (\tfrac{8}{3}\sigma x_n) - (\tfrac{4}{3}\sigma)x_{n-1} + (\tfrac{8}{3}\sigma)x_{n-2} - x_{n-3} = 0 \qquad (12\text{-}41)$$

where

$$\sigma \triangleq h\lambda \qquad (12\text{-}42)$$

To solve Eq. (12-41), let us try

$$x_n = cz^n \qquad (12\text{-}43)$$

where c and z are constants (possibly complex) that if properly chosen would reduce Eq. (12-41) to an identity. To find c and z, we substitute Eq. (12-43) into Eq. (12-41) and obtain, after simplification,

$$\begin{aligned} &z^{n+1} + (\tfrac{8}{3}\sigma)z^n - (\tfrac{4}{3}\sigma)z^{n-1} + (\tfrac{8}{3}\sigma)z^{n-2} - z^{n-3} \\ &= z^{n-3}[z^4 + (\tfrac{8}{3}\sigma)z^3 - (\tfrac{4}{3}\sigma)z^2 + (\tfrac{8}{3}\sigma)z - 1] = 0 \end{aligned} \qquad (12\text{-}44)$$

Since Eq. (12-44) does not contain the constant c, it follows that Eq. (12-43) is "a" solution of Eq. (12-41) for *any* value of c and for all values of z which satisfies Eq. (12-44); i.e., z is *any root* of the polynomial equation

$$\mathcal{P}(z) = z^4 + (\tfrac{8}{3}\sigma)z^3 - (\tfrac{4}{3}\sigma)z^2 + (\tfrac{8}{3}\sigma)z - 1 = 0 \qquad (12\text{-}45)$$

For each fixed value of σ, it is well known that $\mathcal{P}(z)$ has four roots: z_1, z_2, z_3, and z_4, which are generally complex numbers. Since Eq. (12-41) is *linear*, we can invoke the principle of superposition to assert that

$$x_n = c_1 z_1^n + c_2 z_2^n + c_3 z_3^n + c_4 z_4^n \qquad (12\text{-}46)$$

is the most general solution of Eq. (12-41), provided the roots z_1, z_2, z_3, and z_4 are all distinct. The constants c_1, c_2, c_3, and c_4 are arbitrary, as usual, and can be determined after the starting values x_0, x_1, x_2, and x_3 are given. In particular, they are the solutions of the system of linear equations:

$$
\begin{aligned}
c_1 z_1^0 + c_2 z_2^0 + c_3 z_3^0 + c_4 z_4^0 &= x_0 \\
c_1 z_1^1 + c_2 z_2^1 + c_3 z_3^1 + c_4 z_4^1 &= x_1 \\
c_1 z_1^2 + c_2 z_2^2 + c_3 z_3^2 + c_4 z_4^2 &= x_2 \\
c_1 z_1^3 + c_2 z_2^3 + c_3 z_3^3 + c_4 z_4^3 &= x_3
\end{aligned}
$$

If the four roots z_1, z_2, z_3, and z_4 are not all distinct, say $z_1 = z_2$, then the general solution to Eq. (12-41) assumes the modified form

$$x_n = (c_1 + c_2 n) z_1^n + c_3 z_3^n + c_4 z_4^n \tag{12-47}$$

Examination of Eq. (12-47) shows that if $|z_1| = 1$, then $|x_n| \longrightarrow \infty$ as $n \longrightarrow \infty$. Since we know the exact solution to Eq. (12-38) tends to zero as $t \longrightarrow \infty$, it is clear that the algorithm would be useless if the polynomial equation $\mathcal{P}(z) = 0$ contains a *multiple* root of magnitude 1. The roots of Eq. (12-45) of course depend on the value of the parameter σ. Let us focus our attention on the case where $\sigma = 0$. Equation (12-45) reduces to

$$\mathcal{P}(z) = z^4 - 1 = 0 \tag{12-48}$$

The solutions to Eq. (12-48) are the four roots of unity: $z_1 = 1$, $z_2 = -1$, $z_3 = i$, and $z_4 = -i$, where $i \triangleq \sqrt{-1}$. Observe that these four roots are distinct. We shall now invoke a well-known result from the theory of complex variables that *each simple root of the polynomial equation* $\mathcal{P}(z) = 0$ *is an analytic function of the coefficient* σ *and can therefore be expanded in a Taylor series about* $\sigma = 0$ [1]. Hence, we can write

$$z_1 = 1 + \alpha_1\sigma + \alpha_2\sigma^2 + \alpha_3\sigma^3 + \dots \tag{12-49}$$

$$z_2 = -1 + \beta_1\sigma + \beta_2\sigma^2 + \beta_3\sigma^3 + \dots \tag{12-50}$$

$$z_3 = i + \gamma_1\sigma + \gamma_2\sigma^2 + \gamma_3\sigma^3 + \dots \tag{12-51}$$

$$z_4 = -i + \delta_1\sigma + \delta_2\sigma^2 + \delta_3\sigma^3 + \dots \tag{12-52}$$

where z_1, z_2, z_3, and z_4 are the roots of Eq. (12-45) and are therefore functions of σ. Observe that, when $\sigma = 0$, Eqs. (12-49)–(12-52) revert back to the four roots of unity, as expected. To determine the coefficients $\alpha_1, \alpha_2, \alpha_3, \dots$, we substitute Eq. (12-49) into Eq. (12-45) and equate the coefficients of σ^k to zero, $k = 1, 2, \dots$, etc. Carrying this procedure out, we obtain

$$
\begin{aligned}
&(1 + \alpha_1\sigma + \alpha_2\sigma^2 + \dots)^4 + \tfrac{8}{3}\sigma(1 + \alpha_1\sigma + \alpha_2\sigma^2 + \dots)^3 \\
&\quad - \tfrac{4}{3}\sigma(1 + \alpha_1\sigma + \alpha_2\sigma^2 + \dots)^2 + \tfrac{8}{3}\sigma(1 + \alpha_1\sigma + \alpha_2\sigma^2 + \dots) - 1 = 0
\end{aligned}
$$

Coefficient of σ^1: $4\alpha_1 + \frac{8}{3} - \frac{4}{3} + \frac{8}{3} = 0$. Solving for α_1, we obtain $\alpha_1 = -1$.

Coefficient of σ^2: $6\alpha_1^2 + 4\alpha_2 + 8\alpha_1 - \frac{8}{3}\alpha_1 + \frac{8}{3}\alpha_1 = 0$. Substituting $\alpha_1 = -1$ and solving for α_2, we obtain $\alpha_2 = \frac{1}{2}$.

Coefficient of σ^3: $4\alpha_3 + 4\alpha_1^3 + 12\alpha_1\alpha_2 + 8\alpha_1^2 + 8\alpha_2 - \frac{4}{3}\alpha_1^2 - \frac{8}{3}\alpha_2 + \frac{8}{3}\alpha_2 = 0.$
Substituting $\alpha_1 = -1$, $\alpha_2 = \frac{1}{2}$ and solving for α_3, we obtain

$$\alpha_3 = -\frac{1}{3!}.$$

Repeating the same procedure for the coefficient of σ^4, we obtain

$$\alpha_4 = \frac{1}{4!}$$

Hence, Eq. (12-49) can be written as

$$z_1 = 1 - \sigma + \frac{\sigma^2}{2!} - \frac{\sigma^3}{3!} + \frac{\sigma^4}{4!} + O(\sigma^5) \tag{12-53}$$

where $O(\sigma^5)$ denotes higher-order terms in $\sigma^5, \sigma^6, \ldots$, etc.

By a similar procedure, we can derive the coefficients β_j, γ_j, and δ_j for Eqs. (12-50)–(12-52):

$$z_2 = -1 - \tfrac{5}{3}\sigma + O(\sigma^2) \tag{12-54}$$

$$z_3 = i - \frac{i}{3}\sigma + O(\sigma^2) \tag{12-55}$$

$$z_4 = -i + \frac{i}{3}\sigma + O(\sigma^2) \tag{12-56}$$

Examination of Eq. (12-53) shows that the first five terms are identical to the corresponding terms in the Taylor series expansion of $e^{-\sigma} = e^{-h\lambda}$. In other words, for *small* values of σ, z_1 is a good approximation to $e^{-h\lambda}$; hence,

$$z_1^n \cong e^{-\lambda(nh)} = e^{-\lambda t_n} \tag{12-57}$$

is a good approximation to the exact solution to Eq. (12-38) at $t = t_n \triangleq nh$. According to Eq. (12-53), the error of approximation is given by $O(\sigma^5) = O(\lambda^5 h^5)$, which is consistent with the local truncation error of this algorithm as defined by Eq. (12-37), as it should be. As $\sigma \longrightarrow 0$, we expect z_1^n to tend to the exact solution $x(t_n) = e^{-\lambda t_n}$. But Eq. (12-46) shows that the complete solution x_n contains three other components due to z_2^n, z_3^n, and z_4^n. These are called *parasitic components*, since their presence is undesirable and can only contribute additional errors to what would have been the correct solution z_1^n. We hope, therefore, that as $n \longrightarrow \infty$ these parasitic components will be damped out and become negligible compared to z_1^n. Examination of Eq. (12-54) shows, unfortunately, that as $\sigma \longrightarrow 0$

$$|z_2| \cong |-1 - \tfrac{5}{3}\sigma| > 1 \tag{12-58}$$

Hence, the parasitic component z_2^n actually increases exponentially and, in fact, will eventually dominate z_1^n as n increases, since the latter tends to zero! This unstable behavior is *inherent*, since $|z_2| > 1$ as long as $\sigma \triangleq h\lambda \neq 0$. Thus, no matter how small the step size h is chosen, the use of the fourth-order multistep algorithm defined by Eq. (12-36) to solve the initial-value problem $\dot{x} = -\lambda x$ will always lead to a useless numerical solution, which blows up as $n \longrightarrow \infty$.

Our example points out the necessity for an algorithm to be not only *accurate* but also *stable*, in the sense that the local truncation error should remain *bounded* as $n \longrightarrow \infty$ *for sufficiently small* step size h.[4] It is important therefore that we derive criteria which guarantee that a multistep algorithm is stable. This is the subject of Section 12-5.

Finally, it is important to recognize that a "stable" multistep algorithm can only guarantee that the local truncation and round-off error will not get amplified for a *sufficiently small step size h.* As the step size increases, however, a stable multistep algorithm could become unstable, or even if it remains stable, the local truncation error could oscillate, thereby giving an erroneous "ringing" numerical solution (see Fig. 1-22(c) and Problem 12-25). This undesirable property is one reason why the backward Euler algorithm is sometimes preferred over the trapezoidal algorithm (see Problem 12-24), even though the latter is more accurate.

12-5 STABILITY OF MULTISTEP ALGORITHMS

In comparing and analyzing the stability properties of multistep algorithms, it is standard practice to refer to the initial-value problem

$$\dot{x} = -\lambda x, \qquad x(0) = 1 \tag{12-59}$$

where λ is a *complex* constant. The need for allowing λ to be complex stems from the stability considerations in solving a *system* of linear equations

$$\dot{\boldsymbol{x}} = \boldsymbol{A}\boldsymbol{x} \tag{12-60}$$

where $\boldsymbol{A}$ is an $n \times n$ constant *real* matrix. Assuming that the matrix $\boldsymbol{A}$ is diagonalizable into the form

$$\boldsymbol{\Lambda} = \boldsymbol{M}\boldsymbol{A}\boldsymbol{M}^{-1} \tag{12-61}$$

where $\boldsymbol{\Lambda}$ is an $n \times n$ *diagonal* matrix whose diagonal entries are eigenvalues of $\boldsymbol{A}$, then upon defining $\boldsymbol{y} = \boldsymbol{M}\boldsymbol{x}$, we obtain the following equivalent system of *decoupled* linear equations:

$$\dot{\boldsymbol{y}} = \boldsymbol{\Lambda}\boldsymbol{y} \tag{12-62}$$

Each equation in Eq. (12-62) is in the form

$$\dot{y}_i = \lambda_i y_i \tag{12-63}$$

where λ_i is generally a *complex* eigenvalue of $\boldsymbol{A}$. Hence, to be generally applicable, a multistep algorithm must be stable relative to Eq. (12-63). There are several reasons for choosing Eq. (12-59) as the *standard test equation:* (1) it is the *simplest* equation commonly encountered in practice, and any algorithm that fails to solve this problem is clearly of little value; (2) since it is linear, it is possible to derive some rather meaningful stability criteria in explicit form; (3) even in the case of a general *system* of *nonlinear equations*

$$\dot{\boldsymbol{x}} = \boldsymbol{f}(\boldsymbol{x}) \tag{12-64}$$

its local behavior about some initial point $\boldsymbol{x}_0$ can often be approximated by a *variational linear equation*

$$\delta\dot{\boldsymbol{x}} = \boldsymbol{A}\delta\boldsymbol{x} \tag{12-65}$$

[4]As will be shown in Chapter 13, a *stable* multistep algorithm may become unstable for *large* values of h. Methods for overcoming this problem will also be presented in Chapter 13.

where

$$A \triangleq \left.\frac{\partial f(x)}{\partial x}\right|_{x=x_0} \tag{12-66}$$

is the $n \times n$ *Jacobian matrix* of $f(x)$ evaluated at the initial point x_0. For a small step size h, the solution obtained by any numerical algorithm is necessarily of a local nature, and is roughly exponential in character. Hence, intuitively speaking, an algorithm that works well with the standard test equation ought to work well for most nonlinear equations.

To analyze the stability behavior of a general multistep algorithm, let us apply Eq. (12-1) to the solution of Eq. (12-59); we obtain

$$x_{n+1} = \sum_{i=0}^{p} a_i x_{n-i} + h \sum_{i=-1}^{p} b_i(-\lambda x_{n-i}) \tag{12-67}$$

which can be recast into the following equivalent *linear difference equation:*

$$\boxed{\begin{aligned}(1 + \sigma b_{-1})x_{n+1} - (a_0 - \sigma b_0)x_n - (a_1 - \sigma b_1)x_{n-1} \\ - \ldots - (a_p - \sigma b_p)x_{n-p} = 0\end{aligned}} \tag{12-68}$$

where

$$\sigma \triangleq h\lambda \tag{12-69}$$

By following the same procedure used earlier to solve Eq. (12-41), we find that the *complete solution* to Eq. (12-68) is given by

$$\boxed{x_n = c_1 z_1^n + c_2 z_2^n + \ldots + c_{p+1} z_{p+1}^n} \tag{12-70}$$

where $z_1, z_2, \ldots, z_{p+1}$ are the $p + 1$ roots of the polynomial equation

$$\boxed{\begin{aligned}\mathcal{P}(z) \triangleq (1 + \sigma b_{-1})z^{p+1} - (a_0 - \sigma b_0)z^p - (a_1 - \sigma b_1)z^{p-1} \\ - \ldots - (a_p - \sigma b_p) = 0\end{aligned}} \tag{12-71}$$

provided that these roots are all distinct. The arbitrary constants $c_1, c_2, \ldots, c_{p+1}$ are determined from the starting values $x_{n-p}, x_{n-p+1}, \ldots, x_{n-1}, x_n$. If Eq. (12-71) contains any *multiple* root $z = z_i$ of *multiplicity* m_i, then the corresponding term in Eq. (12-70) will contain the term

$$(c_{i_0} + c_{i_1}n + c_{i_2}n^2 + \ldots + c_{i_{m_i-1}} n^{m_i-1})z_i^n \tag{12-72}$$

Observe that if $|z_i| = 1$ such a term will result in $|x_n| \longrightarrow \infty$ as $n \longrightarrow \infty$. If $|z_i| > 1$, then $|x_n| \longrightarrow \infty$ even if z_i is a distinct root. It will be obvious shortly that in this case we could tolerate a root of multiplicity greater than 1 as long as $|z_i| < |z_1|$. Hence, we have proved the following proposition (see also Problem 12-26).

Proposition 12-1. *A* necessary *condition for the stability of the multistep algorithm defined by Eq. (12-1) is that all nontrivial roots of the polynomial equation* $\mathcal{P}(z) = 0$ *of magnitude 1 be simple.*

Following the procedure of Section 12-4, let us investigate the behavior of the roots of $\mathcal{P}(z) = 0$ in the vicinity of $\sigma = 0$. The crucial result that is needed to expand each root $z = z_i$ in a Taylor series about $\sigma = 0$ is given by the following proposition.

Proposition 12-2. *Let* $\mathcal{P}(z) = 0$ *be the polynomial equation associated with a* ***consistent*** *multistep algorithm.*[5] *Then the value* $z = 1$ *is a simple root of* $\mathcal{P}(z) = 0$ *when* $\sigma = 0$ *if, and only if,*

$$\boxed{b_{-1} + b_0 + b_1 + b_2 + \ldots + b_p \neq 0} \tag{12-73}$$

Proof: When $\sigma = 0$, $\mathcal{P}(z)$ in Eq. (12-71) reduces to

$$\mathcal{P}(z)|_{\sigma=0} = z^{p+1} - a_0 z^p - a_1 z^{p-1} - \ldots - a_{p-1} z - a_p \tag{12-74}$$

Substituting $z = 1$ into Eq. (12-74), we obtain

$$\mathcal{P}(1)|_{\sigma=0} = 1 - a_0 - a_1 - \ldots - a_p = 1 - \sum_{i=0}^{p} a_i = 0 \tag{12-75}$$

where we have made use of the exactness constraint defined by Eq. (12-6). This proves that $z = 1$ is a root of $\mathcal{P}(z)|_{\sigma=0} = 0$. It remains to be proved that it is a *simple* root. To do this, it suffices to prove that

$$\left.\frac{d\mathcal{P}(z)}{dz}\right|_{\sigma=0,\, z=1} \neq 0 \tag{12-76}$$

From Eq. (12-71), we obtain

$$\begin{aligned}
\left.\frac{d\mathcal{P}(z)}{dz}\right|_{\sigma=0,\, z=1} &= [(p+1)z^p - pa_0 z^{p-1} - (p-1)a_1 z^{p-2} \\
&\quad -(p-2)a_2 z^{p-3} - (p-3)a_3 z^{p-4} - \ldots - a_{p-1}]|_{z=1} \\
&= (p+1) - pa_0 - (p-1)a_1 - (p-2)a_2 \\
&\quad - (p-3)a_3 - \ldots - a_{p-1} \\
&= (p+1) - p(1 - a_1 - a_2 - \ldots - a_p) - (p-1)a_1 \\
&\quad - (p-2)a_2 - (p-3)a_3 - \ldots - a_{p-1} \\
&= 1 + a_1 + 2a_2 + 3a_3 + \ldots + pa_p
\end{aligned} \tag{12-77}$$

Invoking next the second exactness constraint defined by Eq. (12-10), we obtain

$$1 + a_1 + 2a_2 + 3a_3 + \ldots + pa_p = b_{-1} + b_0 + b_1 + \cdots + b_p \tag{12-78}$$

It follows from Eqs. (12-77) and (12-78) that

$$\left.\frac{d\mathcal{P}(z)}{dz}\right|_{\sigma=0,\, z=1} \neq 0$$

if, and only if, Eq. (12-73) is satisfied. Q.E.D.

[5]Recall that a multistep algorithm is said to be *consistent* if its coefficients satisfy the two exactness constraints defined by Eqs. (12-6) and (12-10).

Under the mild assumption that the multistep algorithm is consistent and that Eq. (12-73) is satisfied, we are assured that $z = z_1 = 1$ is a simple root of $\mathcal{P}(z) = 0$ when $\sigma = 0$. Hence, z_1 is an *analytic* function of σ and can be expanded in a Taylor series about $\sigma = 0$:

$$z_1 = 1 + \alpha_1\sigma + \alpha_2\sigma^2 + \alpha_3\sigma^3 + \dots \qquad (12\text{-}79)$$

Substituting Eq. (12-79) into Eq. (12-71) and equating the coefficients of σ^k to zero, $k = 1, 2, \dots$, we obtain, as in the previous section, $\alpha_1 = -1$, $\alpha_2 = 1/2!$, $\alpha_3 = -(1/3!)$, $\dots$, $\alpha_k = (-1)^k(1/k!)$. Hence, we can write

$$z_1 = 1 - \sigma + \frac{1}{2!}\sigma^2 - \frac{1}{3!}\sigma^3 + \frac{1}{4!}\sigma^4 - \dots + (-1)^k\frac{1}{k!}\sigma^k + O(\sigma^{k+1}) \qquad (12\text{-}80)$$

Observe that the first $k + 1$ terms in z_1 are identical to the Taylor series expansion for $e^{-\sigma} = e^{-h\lambda}$. Hence, for small values of σ, z_1 is a good approximation to $e^{-h\lambda}$ and

$$z_1^n \cong e^{-\lambda(nh)} = e^{-\lambda t_n} \qquad \text{as } \sigma \longrightarrow 0 \qquad (12\text{-}81)$$

is a good approximation to the exact solution to Eq. (12-59) at $t = t_n \triangleq nh$. It follows from Eq. (12-80) that the error of approximation is given by $O(h^{k+1})$, an estimate consistent with the kth-order Adams–Bashforth or Adams–Moulton algorithm. Since z_1^n alone constitutes the correct solution, the remaining components $z_2^n, z_3^n, \dots, z_{p+1}^n$ in the complete solution defined by Eq. (12-70) can only contribute to the error terms and are therefore called the *parasitic components*. For a multistep algorithm to be stable, it is necessary that these parasitic components be *damped* out as n increases. Hence, we must have

$$|z_j(\sigma)| < |z_1(\sigma)|, \qquad j \geq 2 \qquad (12\text{-}82)$$

Since $z_1(0) = 1$, this means that the *parasitic roots* $z_2(0), z_3(0), \dots, z_{p+1}(0)$ *must all lie strictly inside the unit circle*. Since $z = 1$ is a root of $\mathcal{P}(z)$, we can factor the term $z - 1$ from $\mathcal{P}(z)$, leaving a polynomial equation

$$\hat{\mathcal{P}}_{(z)} \triangleq \left.\frac{\mathcal{P}(z)}{z-1}\right|_{\sigma=0} = \frac{z^{p+1} - a_0 z^p - a_1 z^{p-1} - a_2 z^{p-2} - \dots - a_p}{z - 1} = 0 \qquad (12\text{-}83)$$

whose roots are all parasitic roots and must therefore lie strictly inside the unit circle. We shall summarize the preceding observations as follows:

Theorem 12-3 (Numerical Stability Criterion for Multistep Algorithms). *A consistent multistep algorithm satisfying the constraint*

$$b_{-1} + b_0 + b_1 + b_2 + \dots + b_p \neq 0 \qquad (12\text{-}84)$$

is stable in the sense that the local truncation and round-off errors remain bounded for a sufficiently small step size h if, and only if, all roots of the polynomial equation $\hat{\mathcal{P}}(z)$ defined by Eq. (12-83) lie strictly within the unit circle.

Corollary. *The Adams–Bashforth and Adams–Moulton algorithms are stable in the sense that the local truncation and round-off errors remain bounded for sufficiently small step size h.*

Proof: In both algorithms, we have $a_0 = 1$, $a_1 = a_2 = \ldots = a_p = 0$. Hence,

$$\hat{\mathcal{P}}(z) = \frac{z^{p+1} - z^p}{z - 1} = z^p = 0 \tag{12-85}$$

Since the parasitic roots of Eq. (12-85) are all zero, they certainly lie within the unit circle; hence, these algorithms are stable. Q.E.D.

It is important that except for the rather mild restriction in Eq. (12-84) involving the coefficients b_i, only the coefficients $a_0, a_1, a_2, \ldots, a_p$ of the multistep algorithm are used in determining the *stability* of the algorithm. Intuitively, this makes sense, since the coefficients b_i in Eq. (12-1) are all multiplied by the step size h, which we allow to assume an *arbitrarily small* value in determining the stability of an algorithm. Consequently, the coefficients b_i will have negligible effect on the error as $h \longrightarrow 0$. In practice, however, the step size is usually chosen as large as possible, in which case the b_i coefficients will play a significant role in determining the propagation of local truncation and round-off errors. The effect of a large step size on the stability of an algorithm will be presented in Chapter 13.

To illustrate the utility of Theorem 12-3, consider the consistent multistep algorithm defined by

$$x_{n+1} = -4x_n + 5x_{n-1} + h\{4f(x_n, t_n) + 2f(x_{n-1}, t_{n-1})\} \tag{12-86}$$

This is a *third-order algorithm* and, in fact, is the most accurate two-step explicit algorithm. The coefficients pertinent in forming $\hat{\mathcal{P}}(z)$ are $a_0 = -4$ and $a_1 = 5$. The polynomial equation defined by Eq. (12-83) assumes the form

$$\hat{\mathcal{P}}(z) = \left.\frac{z^{p+1} + 4z^p - 5z^{p-1}}{z - 1}\right|_{p=1} = \frac{(z-1)(z+5)}{z-1} = z + 5 = 0 \tag{12-87}$$

which has a parasitic root $z = -5$ lying outside the unit circle. Hence, it is *unstable*. To demonstrate the unstable nature of this algorithm, consider using it to solve the initial-value problem

$$\dot{x} = 0, \qquad x(0) = 0 \tag{12-88}$$

The exact solution to Eq. (12-88) is $\hat{x}(t) = 0$ for all t. However, suppose that there is an initial round-off error such that $x_1 = \varepsilon$. A continued application of Eq. (12-86) with $x_0 = 0$, $x_1 = \varepsilon$ will give (even assuming no further round-off error) $x_2 = -4\varepsilon$, $x_3 = 21\varepsilon$, $x_4 = -104\varepsilon$, $x_5 = 521\varepsilon$, $x_6 = -2604\varepsilon$, etc. The result will continue to get worse, and eventually $|x_n| \longrightarrow \infty$ as $n \longrightarrow \infty$. Observe that this unstable behavior is independent of the step size h. Hence, in spite of its high accuracy and fewer starting values, this third-order algorithm is useless for all practical purposes.

12-6 CONVERGENCE OF MULTISTEP ALGORITHMS

The *stability* of a multistep algorithm only guarantees that the local truncation and round-off errors are not amplified, but rather remain *bounded* for a sufficiently small step size h, as $n \longrightarrow \infty$. There is no requirement that the local truncation error tend to zero as $h \longrightarrow 0$. For example, an algorithm that is obviously stable is given by

$$x_{n+1} = x_n \tag{12-89}$$

If we use Eq. (12-88) to solve the initial-value problem $\dot{x} = f(x, t)$, $x(0) = x_0$, we would obtain $x_{n+1} = x_0$ for all n. Since the step size h is not even present in Eq. (12-88), it is clear that this algorithm, although stable, will never yield zero error as $h \longrightarrow 0$, except for the trivial problem $\dot{x} = 0$ (assuming no round-off error). This example motivates the following definition.

Definition 12.1 (Convergent Algorithm). *A multistep algorithm is said to be* ***convergent*** *if, for any initial-value problem $\dot{x} = f(x, t)$, $x(0) = x_0$ having a unique solution $x = \hat{x}(t)$, the* ***computed*** *solution x_{n+1}* ***converges*** *to $\hat{x}(t)$ uniformly in $0 \le t \le T$ as $n \longrightarrow \infty$, with $h = T/n$.*

We have already shown that *stability* and *convergence* of an algorithm are two different, unrelated concepts. We close this section with Theorem 12-4, an important result whose proof can be found in reference 2.

Theorem 12-4. *Every* ***stable*** *and* ***consistent*** *multistep algorithm is convergent. Conversely, every convergent multistep algorithm is stable.*

*12-7 STRATEGY FOR CHOOSING OPTIMUM ORDER AND STEP SIZE

So far we have implicitly assumed that, given an initial-value problem, a numerical-integration algorithm of a certain "order" is selected and remains *fixed* during the entire integration process. Under this assumption, the *step size* for each time step may be optimized by choosing the largest possible value of h for which the *local truncation error* ϵ_T remains bounded below the user-specified *maximum allowable error*, and for which the algorithm remains numerically stable.[6] For *large* systems of equations, the amount of computation does not increase substantially when the order of the algorithm is increased. Consequently, it often turns out to be more efficient to vary both the *order* and the *step size* during each time step. Our objective in this section is to present a strategy for choosing the *optimum* order and step size.

Let the solution time interval be from t_0 to $t_0 + T$, and let $E_{\max}$ be the *maximum allowable error* over the solution interval T. The maximum allowable error *per unit time* is given by

$$e_{\max} \triangleq \frac{E_{\max}}{T} \tag{12-90}$$

The maximum allowable error *per time step* (with step size h) is therefore given by $h \cdot e_{\max}$. Now the *local truncation error* for most *multistep* algorithms of *order k* can be expressed in the form

$$\boxed{\epsilon_T = [C_k \hat{x}^{(k+1)}(\hat{\tau})]h^{k+1} = O(h^{k+1})} \tag{12-91}$$

[6]It will be shown in Chapter 13 that, although the step size must be chosen below certain maximum allowable values to ensure stability of an *explicit* algorithm, no such restriction is necessary for *implicit* algorithms.

where $\hat{x}^{(k+1)}(\hat{\tau})$ denotes the $(k+1)$th derivative of the solution $\hat{x}(t)$ at $t = \hat{\tau}$, where $t_n < \hat{\tau} < t_{n+1}$, and where C_k is a *constant* that depends on the *order* of the algorithm. The values of C_k for the *Adams–Bashforth algorithm* have been given in Eq. (12-26) and those for the *Adams–Moulton algorithm* in Eq. (12-35). If we substitute $|\epsilon_T| = h \cdot e_{\max}$ into Eq. (12-91) and solve for $e_{\max}$, we obtain

$$\boxed{e_{\max} = |C_k \hat{x}^{(k+1)}(\tau)| h^k} \qquad (12\text{-}92)$$

A typical family of curves representing the relationship between the *maximum per unit allowable error* $e_{\max}$ versus the *maximum time step h* is shown in Fig. 12-1 for $k = 1, 2, 3$, and 4. Observe that for $e_{\max} > E_a$, a first-order algorithm actually gives the largest permissible *step size* without exceeding $e_{\max}$. On the other hand, a second-, third-, or fourth-order algorithm would give a larger step size if $E_b < e_{\max} < E_a$, $E_c < e_{\max} < E_b$, or $e_{\max} < E_c$, respectively. Since the curves shown in Fig. 12-1 depend on both the value of C_k and $\hat{x}^{(k+1)}(t_n)$, it is clear that given a maximum *per unit* allowable error $e_{\max}$, the optimum *order* and *step size* may vary from one time

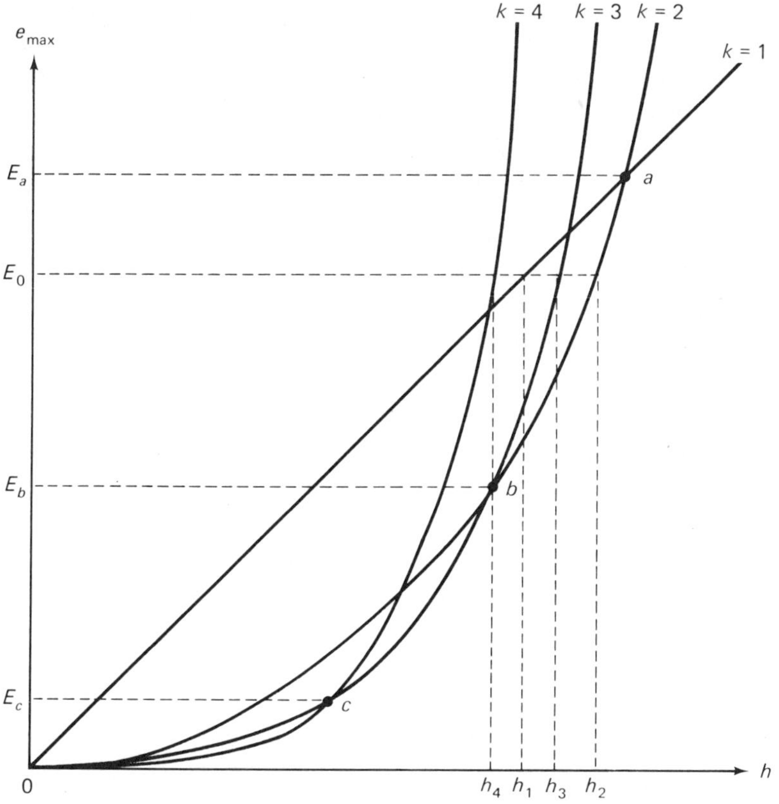

Fig. 12-1. Typical family of curves showing the relationship between the *maximum per unit error*, $e_{\max}$, and the *maximum permissible step size h* for algorithms of order 1, 2, 3, and 4.

step to another. For example, with the curves shown in Fig. 12-1 and with $e_{max} = E_0$, a second-order algorithm would be the optimum choice, and the corresponding maximum time step would be h_2.

A careful inspection shows that optimum order is determined by the *outermost envelope* formed by segments of the respective curves shown in Fig. 12-1. Hence, *increasing the value of* e_{max} *will never increase the order for maximum step size and decreasing the value of* e_{max} *will never decrease the optimum order.* To take full advantage of our strategy for choosing optimum order and step size during each time step, an efficient computation and bookkeeping procedure for accurately calculating the $(k+1)$th *time derivative* $\hat{x}^{(k+1)}(\hat{\tau})$ must be developed.

The preceding development shows that to achieve maximum computational efficiency it is desirable to vary both the *order* and the *step size* of a multistep algorithm during the course of a numerical-integration process. Our next objective therefore is to investigate the mechanics for changing order and step size so that an optimum algorithm may be formulated.

12-7-1 Change of Order

From a programming point of view, changing the order requires only selecting a set of coefficients defining the multistep algorithm of the desired order. Increasing (decreasing) the order would require an increase (decrease) in the number of coefficients with a corresponding increase (decrease) in storage space. In most cases of practical interest, the order may vary from $k = 1$ to $k = 6$. Thus, enough "past" values must be stored so that the *highest-order* algorithm can be implemented whenever called for. Of course, some stored past values may not be needed if a lower-order algorithm is used. In any event, the unused values do not cost anything except possibly a modest amount of storage space.

12-7-2 Change of Step Size

Unlike change of order, which requires little extra programming and computational effort, changing the *step size* could entail considerable additional computation time, because often the previously stored "past" values corresponding to step size h must be interpolated to yield a set of *transformed past values* corresponding to the new step size $\hat{h}$. For example, if

$$\mathbf{y}_n \triangleq [x_n, x_{n-1}, \ldots, x_{n-p}, hx'_n, hx'_{n-1}, \ldots, hx'_{n-p}]^t$$

represents the stored vector at $t = t_n$, then $x_n, x_{n-1}, \ldots, x_{n-p}$ denote the saved values for $x(t)$ at $t = t_n, t_n - h, \ldots, t_n - ph$, respectively; and $hx'_n, hx'_{n-1}, \ldots, hx'_{n-p}$ denote the saved values for $hx'(t)$ at $t = t_n, t_n - h, \ldots, t_n - ph$, respectively [see Fig. 12-2(a)]. If we decide to change the step size at $t = t_{n+1}$ to $\hat{h} \triangleq \alpha h$, where α is a constant, the past values needed are $x(t)$ and $hx'(t)$ at $t = t_n, t_n - \hat{h}, \ldots, t_n - p\hat{h}$. These values represent a new vector

$$\hat{\mathbf{y}}_n \triangleq [\hat{x}_n, \hat{x}_{n-1}, \ldots, \hat{x}_{n-p}, \hat{h}\hat{x}'_n, \hat{h}\hat{x}'_{n-1}, \ldots, \hat{h}\hat{x}'_{n-p}]^t$$

where only $\hat{x}_n = x_n$ [see Fig. 12-2(b)]. The remaining values must be computed by interpolation, which could result in considerable computation cost. It is this difficulty that discourages changing the step size. However, instead of saving the vector $\mathbf{y}_n$, the

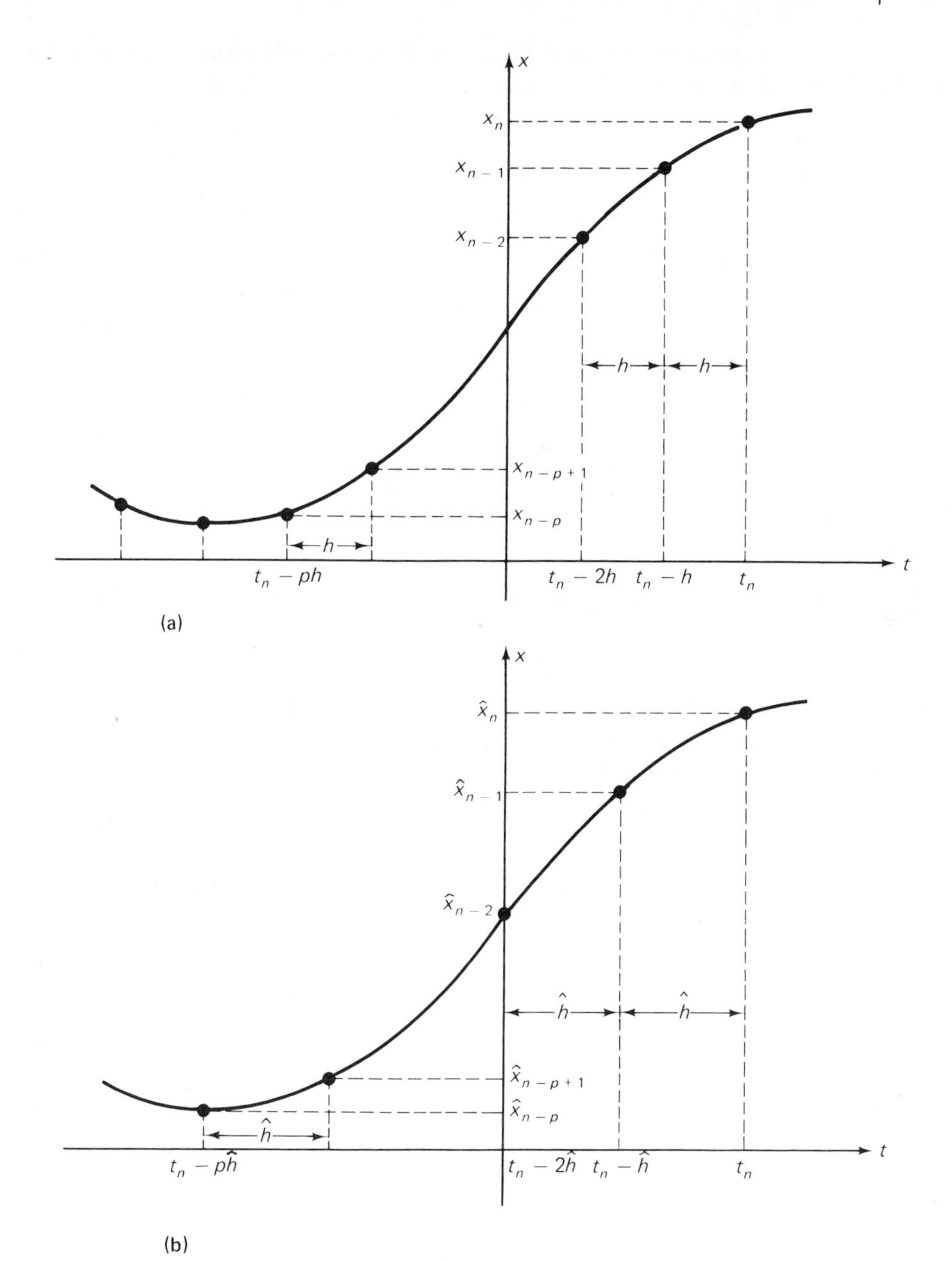

Fig. 12-2. Past values that need to be stored for (a) step size h, and (b) step size $\hat{h}$.

computation time could be greatly reduced if we choose to save some other *equivalent canonical vector* $\mathbf{z}_n$, such as those presented in Section 11-7. In particular, let us investigate the additional computation needed if we choose to store either the *backward-difference vector*, defined by Eq. (11-168), or the *Nordsieck vector*, defined by Eq. (11-177).

CHANGING THE STEP SIZE VIA THE BACKWARD-DIFFERENCE VECTOR. Let the backward-difference vector corresponding to *step size* h be denoted by

$$z_n \triangleq [x_n, x_{n-1}, \ldots, x_{n-p}, hx'_n, \nabla^1(hx'_n), \nabla^2(hx'_n), \ldots, \nabla^p(hx'_n)]^t \tag{12-93}$$

and that corresponding to *step size* $\hat{h} \triangleq \alpha h$ be denoted by

$$\hat{z}_n \triangleq [\hat{x}_n, \hat{x}_{n-1}, \ldots, \hat{x}_{n-p}, \hat{h}\hat{x}'_n, \nabla^1(\hat{h}\hat{x}'_n), \nabla^2(\hat{h}\hat{x}'_n), \ldots, \nabla^p(\hat{h}\hat{x}'_n)]^t \tag{12-94}$$

Given z_n, our problem is to compute for $\hat{z}_n$. To simplify computation, we shall derive the desired relationship between $\hat{z}_n$ and z_n for the *third-order Adams–Bashforth predictor* example presented earlier in Section 11-7-1. Hence, let $\hat{h} \triangleq \alpha h$ and define

$$z_n = [x_n, hx'_n, \nabla^1(hx'_n), \nabla^2(hx'_n)]^t$$

$$\hat{z}_n = [\hat{x}_n, \hat{h}\hat{x}'_n, \nabla^1(\hat{h}\hat{x}'_n), \nabla^2(\hat{h}\hat{x}'_n)]^t$$

Clearly,

$$\boxed{\hat{x}_n = x_n} \tag{12-95}$$

$$\boxed{\hat{h}\hat{x}'_n = \alpha(hx'_n)} \tag{12-96}$$

To determine $\nabla^1(\hat{h}\hat{x}'_n)$, let us expand $x(t)$ via Taylor series about $t = t_n$ and obtain

$$x(t) = x_n + (t - t_n)x'_n + \frac{(t - t_n)^2}{2} x''_n + \frac{(t - t_n)^3}{3!} x'''_n + O((t - t_n)^4) \tag{12-97}$$

where only the first four terms are shown (since the algorithm under consideration is of third order and is therefore exact only for third-degree polynomial solutions). Differentiating Eq. (12-97) with respect to t, we obtain

$$x'(t) = x'_n + (t - t_n)x''_n + \frac{(t - t_n)^2}{2} x'''_n + O((t - t_n)^3) \tag{12-98}$$

Substituting $t = t_n - h$ and $t = t_n - \hat{h}$, respectively, into Eq. (12-98), we obtain, neglecting higher-order terms,

$$x'_{n-1} \triangleq x'(t_n - h) = x'_n - hx''_n + \frac{h^2}{2} x'''_n \tag{12-99}$$

$$\hat{x}'_{n-1} \triangleq x'(t_n - \hat{h}) = x'_n - \alpha hx''_n + \frac{(\alpha h)^2}{2} x'''_n \tag{12-100}$$

It follows from Eqs. (12-99) and (12-100) and the definition of the backward-difference operator $\nabla^j(\cdot)$ in Eq. (11-167) that

$$\nabla^1(hx'_n) \triangleq hx'_n - hx'_{n-1} = h^2x''_n - \frac{h^3x'''_n}{2} \tag{12-101}$$

$$\begin{aligned}\nabla^1(\hat{h}\hat{x}'_n) &\triangleq \hat{h}\hat{x}'_n - \hat{h}\hat{x}'_{n-1} = (\alpha h)^2x''_n - \frac{(\alpha h)^3x'''_n}{2} \\ &= \alpha^2\left[h^2x''_n - \frac{h^3x'''_n}{2}\right] + \frac{\alpha^2h^3x'''_n}{2} - \frac{(\alpha h)^3x'''_n}{2} \\ &= \alpha^2\nabla^1(hx'_n) + \frac{\alpha^2(1 - \alpha)h^3x'''_n}{2}\end{aligned} \tag{12-102}$$

To express x'''_n in Eq. (12-102) in terms of backward differences, substitute $t = t_n - 2h$

into Eq. (12-98) to obtain, neglecting higher-order terms,

$$x'_{n-2} \triangleq x'(t_n - 2h) = x'_n - 2hx''_n + 2h^2x'''_n \tag{12-103}$$

It follows from Eq. (11-167) that

$$\begin{aligned}\nabla^2(hx'_n) &= \nabla^1(hx'_n) - \nabla^1(hx'_{n-1}) \\ &= hx'_n - 2hx'_{n-1} + hx'_{n-2}\end{aligned} \tag{12-104}$$

Substituting Eqs. (12-99) and (12-103) into Eq. (12-104) and simplifying, we obtain

$$\nabla^2(hx'_n) = h^3x'''_n \tag{12-105}$$

Similarly, we obtain

$$\nabla^2(\hat{h}\hat{x}'_n) = \hat{h}^3\hat{x}'''_n = \alpha^3(h^3x'''_n) \tag{12-106}$$

Substituting Eq. (12-105) into Eqs. (12-102) and (12-106), we obtain

$$\boxed{\nabla^1(\hat{h}\hat{x}'_n) = \alpha^2\nabla^1(hx'_n) + \frac{\alpha^2(1-\alpha)\nabla^2(hx'_n)}{2}} \tag{12-107}$$

$$\boxed{\nabla^2(\hat{h}\hat{x}'_n) = \alpha^3\nabla^2(hx'_n)} \tag{12-108}$$

It follows from Eqs. (12-95), (12-96), (12-107), and (12-108) that $\hat{z}_n$ and z_n are related by

$$\underbrace{\begin{bmatrix}\hat{x}_n \\ \hat{h}\hat{x}'_n \\ \nabla^1(\hat{h}\hat{x}'_n) \\ \nabla^2(\hat{h}\hat{x}'_n)\end{bmatrix}}_{\hat{z}_n} = \underbrace{\begin{bmatrix}1 & 0 & 0 & 0 \\ 0 & \alpha & 0 & 0 \\ 0 & 0 & \alpha^2 & \frac{\alpha^2(1-\alpha)}{2} \\ 0 & 0 & 0 & \alpha^3\end{bmatrix}}_{\Lambda(\alpha)} \underbrace{\begin{bmatrix}x_n \\ hx'_n \\ \nabla^1(hx'_n) \\ \nabla^2(hx'_n)\end{bmatrix}}_{z_n} \tag{12-109}$$

In the general case, the two vectors $\hat{z}_n$ and z_n are related by

$$\boxed{\hat{z}_n = \Lambda(\alpha)z_n} \tag{12-110}$$

where $\Lambda(\alpha)$ is a square matrix whose elements are functions of the parameter

$$\alpha \triangleq \frac{\hat{h}}{h} \tag{12-111}$$

Changing the Step Size via the Nordsieck Vector. Let the Nordsieck vector corresponding to *step size h* be denoted by

$$z_n \triangleq \left[x_n, hx'_n, \frac{h^2x''_n}{2}, \frac{h^3x'''_n}{3!}, \ldots, \frac{h^{2p+1}x_n^{(2p+1)}}{(2p+1)!}\right]^t \tag{12-112}$$

and that corresponding to step size $\hat{h} \triangleq \alpha h$ be denoted by

$$\hat{z}_n \triangleq \left[\hat{x}_n, \hat{h}\hat{x}'_n, \frac{\hat{h}^2\hat{x}''_n}{2}, \frac{\hat{h}^3\hat{x}'''_n}{3!}, \ldots, \frac{\hat{h}^{2p+1}\hat{x}_n^{(2p+1)}}{(2p+1)!}\right]^t \tag{12-113}$$

Now observe that

$$\hat{x}_n = x_n, \quad \hat{x}'_n = x'_n, \quad \hat{x}''_n = x''_n, \quad \hat{x}'''_n = x'''_n, \ldots, \quad \hat{x}_n^{(2p+1)} = x_n^{(2p+1)}$$

(since these quantities are all referred to the *same time* $t = t_n$). It follows from this observation and Eq. (12-111) that $\hat{z}_n$ and z_n are related simply by

$$\underbrace{\begin{bmatrix} \hat{x}_n \\ \hat{h}\hat{x}'_n \\ \dfrac{\hat{h}^2\hat{x}''_n}{2} \\ \dfrac{\hat{h}^3\hat{x}'''_n}{3!} \\ \cdot \\ \cdot \\ \cdot \\ \dfrac{\hat{h}^{2p+1}\hat{x}_n^{(2p+1)}}{(2p+1)!} \end{bmatrix}}_{\hat{z}_n} = \underbrace{\begin{bmatrix} 1 & 0 & 0 & 0 & \cdots & 0 \\ 0 & \alpha & 0 & 0 & \cdots & 0 \\ 0 & 0 & \alpha^2 & 0 & \cdots & 0 \\ 0 & 0 & 0 & \alpha^3 & \cdots & 0 \\ \cdot & \cdot & \cdot & \cdot & \cdots & \cdot \\ \cdot & \cdot & \cdot & \cdot & \cdots & \cdot \\ \cdot & \cdot & \cdot & \cdot & \cdots & \cdot \\ 0 & 0 & 0 & 0 & \cdots & \alpha^{2p+1} \end{bmatrix}}_{\Lambda(\alpha)} \underbrace{\begin{bmatrix} x_n \\ hx'_n \\ \dfrac{h^2x''_n}{2} \\ \dfrac{h^3x'''_n}{3!} \\ \cdot \\ \cdot \\ \cdot \\ \dfrac{h^{2p+1}x_n^{(2p+1)}}{(2p+1)!} \end{bmatrix}}_{z_n} \tag{12-114}$$

A comparison between Eqs. (12-109) and (12-114) shows that, in so far as changing *step size* is concerned, the *Nordsieck vector* is preferable since $\hat{z}_n$ is related to z_n by a simple *diagonal* matrix $\Lambda(\alpha)$.

★12-8 AUTOMATIC CONTROL OF ORDER AND STEP SIZE

It is clear from the presentation in Section 12-7 that the optimum strategy for implementing the *predictor–corrector algorithm* should incorporate a scheme for automatically controlling the order and step size. This would require an efficient computing method for estimating the local truncation error ϵ_T at $t = t_n$ and an efficient bookkeeping technique for storing and retrieving the relevant past values.

The *local truncation error* ϵ_T for a multistep numerical-integration algorithm of order k was derived in Theorem 11-5 and is seen to be *proportional* to the $(k+1)$th derivative of the solution $x(t)$. Thus, regardless of which multistep algorithm is used, *it is essential that sufficient information be stored so that the* $(k+1)$th *derivative of* $x(t)$ *can be estimated.* This requirement can be met quite easily if we choose to store either the *backward-difference vector*

$$z_n \triangleq [x_n, x_{n-1}, \ldots, x_{n-k}, hx'_n, \nabla^1(hx'_n), \nabla^2(hx'_n), \ldots, \nabla^k(hx'_n)]^t \tag{12-115}$$

or the *Nordsieck vector*

$$z_n \triangleq \left[x_n, hx'_n, \frac{h^2x''_n}{2}, \frac{h^3x'''_n}{3!}, \ldots, \frac{h^{2k+1}x_n^{(2k+1)}}{(2k+1)!}\right]^t \tag{12-116}$$

If we choose to store the backward-difference vector, it follows from Eq. (11-167) that

$$\nabla^1(hx'_n) \triangleq hx'_n - hx'_{n-1} \cong h^2x''_n \tag{12-117a}$$

$$\nabla^2(hx'_n) \triangleq \nabla^1(hx'_n) - \nabla^1(hx'_{n-1}) = h^2x''_n - h^2x''_{n-1} \cong h^3x'''_n \tag{12-117b}$$

$$\cdot$$
$$\cdot$$
$$\cdot$$

$$\nabla^k(hx'_n) \triangleq \nabla^{k-1}(hx'_n) - \nabla^{k-1}(hx'_{n-1}) = h^kx_n^{(k)} - h^kx_{n-1}^{(k)} \cong h^{k+1}x_n^{(k+1)} \tag{12-118}$$

Hence, an *estimate* for the higher-order derivative terms is already given by the components of the backward-difference vector. Therefore, in so far as estimating the local truncation error is concerned, the *backward-difference vector* is the most convenient choice. Unfortunately, it is not nearly as convenient to change the step size using this vector, as demonstrated by the derivation leading to Eq. (12-109). If we choose instead to store the Nordsieck vector, changing the step size becomes an extremely simple task, as shown by Eq. (12-114). To obtain an estimate for the higher-order derivatives, consider the following Nordsieck vector for a kth-*order* multistep algorithm:

$$z_n \triangleq \left[x_n, hx'_n, \frac{h^2 x''_n}{2}, \frac{h^3 x'''_n}{3!}, \ldots, \frac{h^k x_n^{(k)}}{k!}\right]^t \tag{12-119}$$

If we let

$$(z_n)_{k+1} \triangleq \frac{h^k x_n^{(k)}}{k!} \tag{12-120}$$

denote the *last* component of the Nordsieck vector, then

$$\nabla^1 (z_n)_{k+1} \triangleq \frac{h^k x_n^{(k)}}{k!} - \frac{h^k x_{n-1}^{(k)}}{k!} = \frac{h^k}{k!}[x_n^{(k)} - x_{n-1}^{(k)}] \cong \frac{h^{k+1} x_n^{(k+1)}}{k!} \tag{12-121}$$

where we have made use of the approximation

$$\frac{x_n^{(k)} - x_{n-1}^{(k)}}{h} = x_n^{(k+1)} + O(h^2) \tag{12-122}$$

Similarly, we obtain

$$\nabla^2 (z_n)_{k+1} \triangleq \nabla^1 (z_n)_{k+1} - \nabla^1 (z_{n-1})_{k+1} \cong \frac{h^{k+2} x_n^{(k+2)}}{k!} \tag{12-123}$$

Equations (12-121) and (12-123) show that the higher-order derivatives $x_n^{(k+1)}$ and $x_n^{(k+2)}$ can be estimated quite easily by simply taking the first- and second-order backward difference of the last component of the Nordsieck vector defined in Eq. (12-119). Hence, if our objective is to be able to compute the higher-order derivatives of $x(t)$ *as well as* to change the step size efficiently, the Nordsieck vector would be preferred over the backward-difference vector. We shall close this chapter by formulating an algorithm that will automatically change the *order* and *step size*. To be specific, the algorithm will be formulated for the Adams–Moulton corrector. The same concept of course applies to any other implicit algorithm, including Gear's algorithm to be presented in Chapter 13.

12-8-1 Algorithm for Changing Order and Step Size Automatically

Assuming that a kth-order *Adams–Moulton algorithm* is chosen for the corrector, our objective is to devise an algorithm so that at $t = t_n$ the following two decisions are made automatically:

1. Should the *order* of the algorithm be changed before computing for $x(t)$ at $t = t_{n+1}$? If so, should we increase or decrease the order?

2. Should the *step size* be changed before computing for $x(t)$ at $t = t_{n+1}$? If so, what should be the new step size $\hat{h} \triangleq \alpha h$?

To answer these questions, consider the following *local truncation error* for the kth-order Adams–Moulton algorithm given earlier in Eq. (12-35):

$$\boxed{\begin{aligned} &\epsilon_T = [C_k x^{(k+1)}(\hat{\tau})]h^{k+1}, \qquad t_n < \hat{\tau} < t_{n+1} \\ &C_1 = -\tfrac{1}{2},\; C_2 = -\tfrac{1}{12},\; C_3 = -\tfrac{1}{24},\; C_4 = -\tfrac{19}{720}, \\ &C_5 = -\tfrac{3}{160},\; C_6 = -\tfrac{863}{60{,}480} \end{aligned}} \tag{12-124}$$

If we choose to store the Nordsieck vector defined in Eq. (12-119), we should express the higher-order derivatives of $x(t)$ in terms of these stored values. In particular, let us solve for $h^{k+1}x_n^{(k+1)}$ and $h^{k+2}x_n^{(k+2)}$ from Eqs. (12-121) and (12-123); then substituting the resulting expressions into Eq. (12-124) for ϵ_T, we obtain

$$\epsilon_T = C_k[h^{k+1}x_n^{(k+1)}(\hat{\tau})] \cong C_k(k!)[\nabla^1(z_n)_{k+1}] \tag{12-125}$$

for the kth-order Adams–Moulton algorithm, and

$$\epsilon_T = C_{k+1}[h^{k+2}x_n^{(k+2)}(\hat{\tau})] \cong C_{k+1}(k!)[\nabla^2(z_n)_{k+1}] \tag{12-126}$$

for the $(k + 1)$th-*order Adams–Moulton algorithm*, where $(z_n)_{k+1}$ is the *last component* of the Nordsieck vector. Observe that in deriving Eqs. (12-125) and (12-126), we have tacitly assumed that the solution $x(t)$ is sufficiently smooth to have well-defined higher-order derivatives such that $x_n^{(k+1)}(\hat{\tau}) \cong x_n^{(k+1)}$ and $x_n^{(k+2)}(\hat{\tau}) \cong x_n^{(k+2)}$. If we change the step size from h to

$$\hat{h} \triangleq \alpha h \tag{12-127}$$

where α is a proportionality constant, the local truncation error at $t = t_n$ is given by

$$\epsilon_T \cong C_k\alpha^{k+1}(k!)[\nabla^1(z_n)_{k+1}] \tag{12-128}$$

for the kth-order Adams–Moulton algorithm, and

$$\epsilon_T \cong C_{k+1}\alpha^{k+2}(k!)[\nabla^2(z_n)_{k+1}] \tag{12-129}$$

for the $(k + 1)$th-order Adams–Moulton algorithm. Solving for α from Eqs. (12-128) and (12-129), respectively, we obtain

$$\alpha \cong \left\{\frac{\epsilon_T}{C_k(k!)[\nabla^1(z_n)_{k+1}]}\right\}^{1/(k+1)} \tag{12-130}$$

for the kth-order algorithm, and

$$\alpha \cong \left\{\frac{\epsilon_T}{C_{k+1}(k!)[\nabla^2(z_n)_{k+1}]}\right\}^{1/(k+2)} \tag{12-131}$$

for the $(k + 1)$th-order algorithm. For the $(k - 1)$th-order algorithm, we obtain from Eq. (12-124)

$$\epsilon_T = C_{k-1}(\alpha^k h^k)x_n^{(k)}(\hat{\tau}) \tag{12-132}$$

Solving for α, we obtain

$$\alpha \cong \left\{\frac{\epsilon_T}{C_{k-1}(k!)(\nabla^0 z_n)_{k+1}}\right\}^{1/k} \tag{12-133}$$

where $\nabla^0(z_n)_{k+1} \triangleq (z_n)_{k+1}$

Once the *maximum* allowable local truncation error

$$\epsilon_T \leq \epsilon_{\max} \tag{12-134}$$

is *prescribed*, we can substitute $\epsilon_{\max}$ for ϵ_T in Eqs. (12-130), (12-131), and (12-133) and solve for the α's. The maximum of the three α's will give the maximum allowable step size $\hat{h} = \alpha h$, and the corresponding order would be the *optimum* order to compute for x_{n+1}. The basic step control algorithm is then to execute one step and test whether Eq. (12-134) is satisfied. If it is, the step is accepted. If it is not, the step is rejected and a smaller step size ($\alpha < 1$) should be used. The exact step size to use for executing the next step, or for repeating the rejected step, is given by $\hat{h} = \alpha h$, where α is the *maximum* value computed from Eqs. (12-130), (12-131), and (12-133). To avoid too frequent changes in step size and order, Gear has proposed to impose weights on the three α's to maximize computational efficiency [2]; i.e.,

$$\alpha = \frac{1}{1.2}\left\{\frac{\epsilon_{\max}}{C_k(k!)[\nabla^1(z_n)_{k+1}]}\right\}^{1/(k+1)}, \qquad \text{for order } k \tag{12-135}$$

$$\alpha = \frac{1}{1.3}\left\{\frac{\epsilon_{\max}}{C_{k-1}(k!)(\nabla^0 z_n)_{k+1}}\right\}^{1/k}, \qquad \text{for order } k-1 \tag{12-136}$$

$$\alpha = \frac{1}{1.4}\left\{\frac{\epsilon_{\max}}{C_{k+1}(k!)[\nabla^2(z_n)_{k+1}]}\right\}^{1/(k+2)}, \qquad \text{for order } k+1 \tag{12-137}$$

The weight 1/1.2 in Eq. (12-135) is chosen to increase the likelihood that Eq. (12-134) will be satisfied for the subsequent step in order that the step size need not be changed after each step. The weight for the $(k-1)$th-order algorithm is chosen greater than the $(k+1)$th-order algorithm because the latter would require slightly more computation and should therefore not be chosen unless the improvement in step size is significant. Gear and others have found that increasing the step size before $(k+1)$ steps could result in large accumulated errors, thereby requiring a subsequent reduction in step size [2]. This is because the matrix $\boldsymbol{T}$—such as the one given in Eq. (11-192)—is usually derived under the assumption that the step size h remained unchanged during the preceding k time steps. Consequently, changing h before $(k+1)$ time steps would invalidate the definition $\boldsymbol{c}_z \triangleq \boldsymbol{T}\boldsymbol{c}_y$ as given in Eq. (11-165). Hence, it is neither necessary nor desirable to compute for α from Eqs. (12-135)–(12-137) after every step. Instead, Gear recommends that a new estimate for α be made only under the following circumstances:

1. If the current step size h violates Eq. (12-134).
2. If Eq. (12-134) is satisfied for $k+1$ *consecutive* steps after the last change in either the *order* or the *step size*.
3. If Eq. (12-134) is satisfied for 10 consecutive steps after the α was last estimated and if no increase in step size was made at that time.

In the general case involving a system of N equations

$$\dot{\boldsymbol{x}} = \boldsymbol{f}(\boldsymbol{x}, t) \tag{12-138}$$

N Nordsieck vectors $(z_n)^1, (z_n)^2, \ldots, (z_n)^i, \ldots, (z_n)^N$ would be stored, one for each variable x_i. In this case, the algorithm for controlling order and step size still holds as long as the backward-differences $\nabla^j(z_n)_{k+1}$ in Eqs. (12-135)–(12-137) are replaced by the following backward-difference vector norm [2]:

$$\|\nabla^j(z_n)_{k+1}\| \triangleq \sqrt{\sum_{i=1}^{N} (\nabla^j(z_n)_{k+1})_i^2} \tag{12-139}$$

where $(\nabla^j(z_n)_{k+1})_i$ denotes the jth order backward difference of the last component of the Nordsieck vector corresponding to the ith variable x_i. If it is desirable to control the error for each component x_i differently, Eq. (12-139) can be replaced by the *weighted vector norm*

$$\|\nabla^j(z_n)_{k+1}\| \triangleq \sqrt{\sum_{i=1}^{N} \left[\frac{(\nabla^j(z_n)_{k+1})_i}{w_i}\right]^2} \tag{12-140}$$

where w_i is the weight assigned to the local truncation error associated with the ith variable x_i. It must be emphasized that the algorithm for increasing the step size h tacitly assumes that the multistep algorithm of optimum order k is stable for the chosen step size. It will be shown in Chapter 13 that, for the *Adams–Moulton implicit algorithm*, the step size h cannot be chosen too large even though the maximum allowable local truncation error is not exceeded. In fact, it will be clear that the Adams–Moulton algorithm is not efficient enough for solving a large class of "stiff" nonlinear networks, such as multivibrators, because of its inherent restriction on the maximum allowable step size h in order for the algorithm to be stable. To overcome this problem, Gear has developed a family of "stiffly stable" implicit multistep algorithms in which step size is restricted only by the maximum allowable local truncation error. These algorithms are therefore ideally suited for implementing the preceding automatic order and step size changing algorithm. A highly efficient computer program for integrating a system of N ordinary differential equations, which includes, among many desirable features, the automatic order and step size changing "stiffly stable" algorithm, has been developed by Gear; a complete listing of this program is given in [2].

REFERENCES

1. DIEUDONNE, J. *Foundations of Modern Analysis*. New York: Academic Press, Inc., 1969.
2. GEAR, C. W. *Numerical Initial Value Problems in Ordinary Differential Equations*. Englewood Cliffs, N.J.: Prentice-Hall, Inc., 1971.
3. HENRICI, P. *Discrete Variable Methods in Ordinary Differential Equations*. New York: John Wiley & Sons, Inc., 1962.

PROBLEMS

12-1. Derive the exactness constraint as specified by Eq. (12-16).

12-2. Derive the fourth-order Adams–Bashforth algorithm given in Table 12-1.

12-3. Derive the fifth-order Adams–Bashforth algorithm given in Table 12-1.

12-4. Derive the sixth-order Adams–Bashforth algorithm given in Table 12-1.

12-5 Derive the fourth-order Adams–Moulton algorithm given in Table 12-2.

12-6. Derive the fifth-order Adams–Moulton algorithm given in Table 12-2.

12-7. Derive the sixth-order Adams–Moulton algorithm given in Table 12-2.

12-8. Derive the *local truncation error* for the kth-order Adams–Bashforth algorithm. Specify the explicit formula for $k = 1, 2, \ldots, 6$.

12-9. Derive the *local truncation error* for the kth-order Adams–Moulton algorithm. Specify the explicit formula for $k = 1, 2, \ldots, 6$.

12-10. Derive Eqs. (12-26b) and (12-35b).

12-11. Construct an example demonstrating that a *multiple root* of a *polynomial* equation need not be an analytic function of its coefficients and therefore cannot be expanded in a Taylor series.

12-12. Derive Eqs. (12-54)–(12-56).

12-13. Derive Eq. (12-80).

12-14. Derive the exactness constraints analogous to Eq. (12-17) for a kth-order numerical integration algorithm with *nonuniform* step sizes $h_{k+1} \triangleq t_{k+1} - t_k$, where $k = n, n-1, \ldots, n-p$, and $h_{n+1} \triangleq h$.

12-15. Derive the second-order Adams–Bashforth algorithm with *nonuniform* step sizes $h \triangleq t_{n+1} - t_n$ and $h_n \triangleq t_n - t_{n-1}$.

12-16. Derive the third-order Adams–Bashforth algorithm with *nonuniform* step sizes $h \triangleq t_{n+1} - t_n$, $h_n \triangleq t_n - t_{n-1}$, and $h_{n-1} \triangleq t_{n-1} - t_{n-2}$.

12-17. Derive the second-order Adams–Moulton algorithm with *nonuniform* step sizes $h \triangleq t_{n+1} - t_n$ and $h_n \triangleq t_n - t_{n-1}$.

12-18. Derive the third-order Adams–Moulton algorithm with *nonuniform* step sizes $h \triangleq t_{n+1} - t_n$, $h_n \triangleq t_n - t_{n-1}$, and $h_{n-1} \triangleq t_{n-1} - t_{n-2}$.

12-19. Use the following algorithms to solve the initial-value problem

$$\dot{x} = -x + t^3 + 3t^2, \qquad x(0) = 0$$

from $t = 0$ to $t = 1$ with step sizes $h = 0.05, 0.1, 0.25$, and 0.5. Compare the results with the exact solution $x(t) = t^3$.

(a) Adams–Bashforth algorithm of order $k = 1, 2, \ldots, 6$.
(b) Adams–Moulton algorithm of order $k = 1, 2, \ldots, 6$.

12-20. Repeat Problem 12-19 with $h = 0.005$ for the equation $\dot{x} = -100x + 100t^3 + 3t^2$. Explain the differences observed between the two sets of solutions.

12-21. Show that the *exact* solution to the initial-value problem

$$\dot{x} = -\lambda x + \lambda f(t) + f'(t), \qquad x(0) = k + f(0)$$

is given by

$$x(t) = ke^{-\lambda t} + f(t)$$

Assuming $f(t) = \sin t$, $x(0) = 1$, and a step size $h = 0.1$, find the solution from $t = 0$ to $t = 1$ by the second-order Adams–Bashforth algorithm and the second-order Adams–Moulton algorithm for $\lambda = 0.1, 1.0, 10$, and 100. Plot the local truncation error as a function of λ (assuming zero round-off error) and explain the differences between the two plots.

12-22. Formulate a strategy for choosing the optimal order and step size for the family of Adams–Bashforth algorithms by deriving the associated $e_{\max}$ versus h curves similar to those shown in Fig. 12-1.

12-23. Formulate a strategy for choosing the optimal order and step size for the family of Adams–Moulton algorithms by deriving the associated $e_{\max}$ versus h curves similar to those shown in Fig. 12-1.

12-24. The linear difference equation associated with the backward Euler algorithm is given by $(1 + \sigma)x_{n+1} - x_n = 0$.

(a) Sketch the locus of the root of the associated polynomial equation $\mathcal{P}(z) = (1 + \sigma)z - 1 = 0$ in the $\sigma \triangleq \lambda h$ plane (assume that λ is real and positive) as h increases from 0 to ∞.

(b) Show that the numerical solution $x_{n+1} = c_1 z^{n+1}$ of the test equation $\dot{x} = -\lambda x$ as obtained by the backward Euler algorithm always decreases monotonically with time. Verify this conclusion numerically with several different initial conditions and several different values of λ.

12-25. The linear difference equation associated with the trapezoidal algorithm is given by $(1 + \frac{1}{2}\sigma)x_n - (1 - \frac{1}{2}\sigma) = 0$.

(a) Sketch the locus of the root of the associated polynomial equation $\mathcal{P}(z) = (1 + \frac{1}{2}\sigma)z - (1 - \frac{1}{2}\sigma) = 0$ in the $\sigma \triangleq \lambda h$ plane (assume that λ is real and positive) as h increases from 0 to ∞.

(b) Show that the numerical solution $x_{n+1} = c_1 z^{n+1}$ of the test equation $\dot{x} = -\lambda x$ as obtained by the trapezoidal algorithm decreases monotonically with time only if $h < 2/\lambda$.

(c) If the step size $h > 2/\lambda$, show that the numerical solution always exhibits a decaying oscillatory response even though the exact solution should have been a monotonically decaying exponential.

(d) Verify numerically the conclusions in parts (b) and (c) with several initial conditions and several different values of λ.

12-26. Proposition 12.1 asserts that a *multistep algorithm* is *stable* if the magnitudes of all roots of $\mathcal{P}(z) = 0$ are less than 1. This is true even when $\mathcal{P}(z) = 0$ has multiple roots. Let z_i be a root of $\mathcal{P}(z) = 0$ with multiplicity m_i, $|z_i| < 1$; prove that the expression given by Eq. (12-72) approaches zero as n approaches infinity.

12-27. Sketch a family of $\log e_{\max}$ versus $\log h$ curves for $k = 1, 2, \ldots, 6$ and compare the results with the corresponding curves shown in Fig. 12.1. What conclusions can you infer from these curves?

CHAPTER 13

Implicit Algorithms for Solving Networks Characterized by Stiff State Equations

13-1 REGIONS OF ABSOLUTE STABILITY

The important concept of the stability of a multistep algorithm has been introduced in Sections 12-4 and 12-5. This concept is based on the ability of the algorithm to keep the local and round-off errors within bounds as the step size $h \longrightarrow 0$. The *test equation* used for verifying this property is

$$\dot{x} = -\lambda x, \qquad x(0) = 1 \tag{13-1}$$

where λ is generally a *complex number*.[1] Since the solution to Eq. (13-1) is given by $\hat{x}(t) = e^{-\lambda t}$, the parameter $1/\lambda$ can be identified as the *time constant* of the system. It is important to recognize that the stability of an algorithm guarantees only that the local and round-off errors will be bounded for sufficiently small step size h. If h is too small, a very large number N of time steps will be necessary to cover the specified solution time interval $[t_0, t_0 + T]$, where $Nh = T$. Not only is this operation inefficient and costly in terms of computation time, but the round-off error will increase as N increases, thereby resulting in poor accuracy. In practice, therefore, it is desirable to use as large a step size as possible. Unfortunately, the knowledge that an algorithm is stable does not tell us how small h should be. To answer this question, we must

[1]Recall that the system of linear equations $\dot{\boldsymbol{x}} = -\boldsymbol{A}\boldsymbol{x}$ in Eq. (12-60) when transformed into the equivalent system $\dot{\boldsymbol{y}} = -\boldsymbol{\Lambda}\boldsymbol{y}$ in Eq. (12-62) reduces to a system of n uncoupled linear equations $\dot{y}_i = -\lambda_i y_i$. This equation has the same form as Eq. (13-1), where λ_i is an eigenvalue of $\boldsymbol{A}$, and is generally a complex number.

return to Eq. (12-71), which we reproduce here for convenience:

$$\boxed{\begin{aligned}\mathcal{P}(z) \triangleq (1 + \sigma b_{-1})z^{p+1} &- (a_0 - \sigma b_0)z^p \\ &- (a_1 - \sigma b_1)z^{p-1} - \ldots - (a_p - \sigma b_p)\end{aligned}} \tag{13-2}$$

The solution of Eq. (13-1) using a multistep algorithm

$$x_{n+1} = \sum_{i=0}^{p} a_i x_{n-i} + h \sum_{i=-1}^{p} b_i f(x_{n-i}, t_{n-i}) \tag{13-3}$$

has been shown to be given by

$$x_n = c_1 z_1^n + c_2 z_2^n + \ldots + c_{p+1} z_{p+1}^n \tag{13-4}$$

where $z_1, z_2, \ldots, z_{p+1}$ are the roots of Eq. (13-2). Recall that for *small* $\sigma \triangleq h\lambda$, one root, say z_1, of $\mathcal{P}(z) = 0$ gives a good approximation to the exact solution to Eq. (13-1); i.e.,

$$x_n = z_1^n \approx e^{-\lambda t_n} \tag{13-5}$$

Hence, z_1 is called the *principal root*, and the remaining extraneous roots $z_2, z_3, \ldots, z_{p+1}$ of $\mathcal{P}(z) = 0$ are undesirable and are called *parasitic roots*. Now if Re $\lambda > 0$, then $x_n \longrightarrow 0$ as $t_n \longrightarrow \infty$, and we must require the principal root z and the parasitic roots $z_2, z_3, \ldots, z_{p+1}$ to all lie within the unit circle in order for

$$z_i^n \longrightarrow 0, \qquad i = 1, 2, 3, \ldots, p + 1 \tag{13-6}$$

as $n \longrightarrow \infty$. These conditions are of course satisfied by any stable algorithm as long as $h \longrightarrow 0$. The more interesting case when $h \neq 0$ leads us to Definition 13-1.

Definition 13-1* (*Absolute Stability*).** *A multistep algorithm is said to be* ***absolutely stable *for those values of* $\sigma = h\lambda$ *for which the* $p + 1$ *roots of the polynomial equation* $\mathcal{P}(z) = 0$ *defined in Eq. (13-2) lie within or on the unit circle* $|z| = 1$.

A stable algorithm is clearly absolutely stable for $\sigma = 0$. But it is also generally absolutely stable for other values of $\sigma \neq 0$. *The set of all values* $\sigma \triangleq h\lambda$ *for which a multistep algorithm is absolutely stable is called the region of absolute stability.* Since λ is generally a complex number, so is σ; hence, the region of absolute stability for a multistep algorithm is a region in the complex σ plane. Our next objective therefore is to study methods for determining the region of absolute stability for any multistep algorithm.

13-1-1 Method for Determining Regions of Absolute Stability

Let us rewrite Eq. (13-2) as follows:

$$\begin{aligned}\mathcal{P}(z) = (z^{p+1} - a_0 z^p - a_1 z^{p-1} - \ldots - a_p)& \\ + \sigma(b_{-1} z^{p+1} + b_0 z^p + b_1 z^{p-1} + \ldots + b_p) &= 0\end{aligned} \tag{13-7}$$

If we define

$$\mathcal{P}_a(z) \triangleq z^{p+1} - a_0 z^p - a_1 z^{p-1} - \ldots - a_p \tag{13-8}$$

$$\mathcal{P}_b(z) \triangleq b_{-1} z^{p+1} + b_0 z^p + b_1 z^{p-1} + \ldots + b_p \tag{13-9}$$

then Eq. (13-7) can be recast into the form

$$\sigma = \frac{-\mathcal{P}_a(z)}{\mathcal{P}_b(z)} \tag{13-10}$$

To determine whether an algorithm is absolutely stable for a given value of $\sigma = \sigma_0$, we could solve Eq. (13-10) for z with $\sigma = \sigma_0$ and check whether $|z| \leq 1$. Repeating this procedure over all points in the complex σ plane would therefore identify the region where the algorithm is absolutely stable. Needless to say, this procedure is clearly impractical. A much more practical method is to let z in Eq. (13-10) assume all values on the unit circle $|z| = 1$ and compute the corresponding value $\sigma(z)$. We shall refer to the points $\sigma(z)$ in the complex σ plane corresponding to those z with $|z| = 1$ as the *unity-root locus* $\boldsymbol{\Gamma}_\sigma$. Since any z on the unit circle can be expressed as

$$z = e^{j\theta}, \qquad 0 \leq \theta \leq 2\pi \tag{13-11}$$

we can substitute Eq. (13-11) for z in Eq. (13-10) to obtain

$$\sigma(\theta) = \frac{-\mathcal{P}_a(e^{j\theta})}{\mathcal{P}_b(e^{j\theta})} \tag{13-12}$$

Substituting Eqs. (13-8) and (13-9) into Eq. (13-12) and letting θ vary from $\theta = 0$ to $\theta = 2\pi$, we obtain

$$\boxed{\sigma(\theta) = \frac{-e^{j(p+1)\theta} + a_0 e^{jp\theta} + a_1 e^{j(p-1)\theta} + \ldots + a_{p-1}e^{j\theta} + a_p}{b_{-1}e^{j(p+1)\theta} + b_0 e^{jp\theta} + b_1 e^{j(p-1)\theta} + \ldots + b_{p-1}e^{j\theta} + b_p}, \qquad 0 \leq \theta \leq 2\pi} \tag{13-13}$$

The locus generated by Eq. (13-13) is precisely the *unity-root locus* $\boldsymbol{\Gamma}_\sigma$. This locus can be interpreted geometrically, as shown in Fig. 13-1, as a transformation of the unit circle in the z plane into $\boldsymbol{\Gamma}_\sigma$ in the $\sigma \triangleq h\lambda$ plane. Since the circle is a closed curve, so

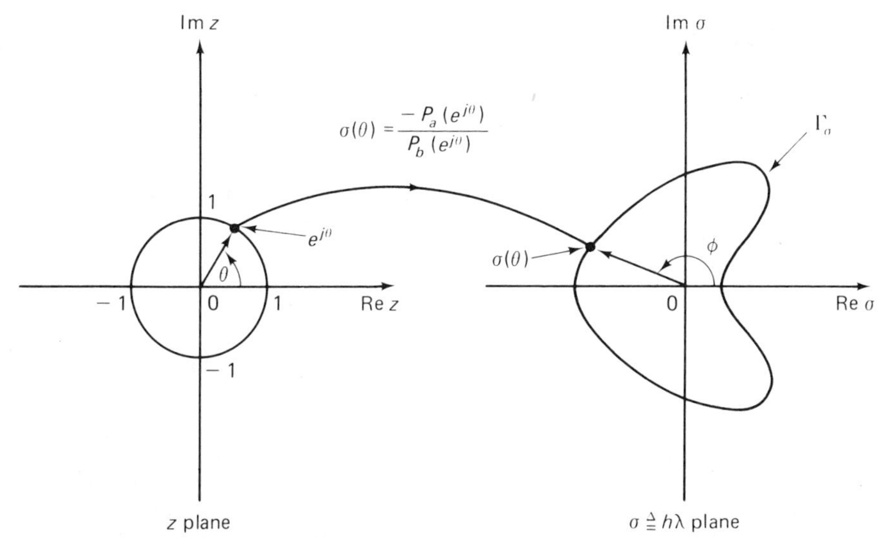

Fig. 13-1. Geometrical construction showing the transformation of the unit circle $z = e^{j\theta}$ in the z plane into the unity-root locus $\boldsymbol{\Gamma}_\sigma$ in the $\sigma \triangleq h\lambda$ plane.

is Γ_σ. The horizontal symmetry of Γ_σ shown in Fig. 13-1 is not by accident, as shown by the following:

Proposition 13-1. *The unity-root locus Γ_σ is symmetric with respect to the horizontal axis Re σ.*

Proof: It follows from the identities $e^{jx} = \cos x + j \sin x$ and $e^{-jx} = \cos x - j \sin x$ that $\text{Re}\, e^{jx} = \text{Re}\, e^{-jx}$, and $\text{Im}\, e^{jx} = -\text{Im}\, e^{-jx}$. Since Eq. (13-13) is made up of terms of the form e^{jkx}, it follows that $\text{Re}\, \sigma(x) = \text{Re}\, \sigma(-x)$, and $\text{Im}\, \sigma(x) = -\text{Im}\, \sigma(-x)$. Thus, the horizontally symmetric points $e^{j\theta}$ and $e^{-j\theta}$ in the z plane are mapped into the horizontally symmetric points $\text{Re}\, \sigma(\theta) + j \,\text{Im}\, \sigma(\theta)$ and $\text{Re}\, \sigma(\theta) - j \,\text{Im}\, \sigma(\theta)$ in the σ plane. Hence, Γ_σ must exhibit horizontal symmetry. Q.E.D.

In view of Proposition 13-1, we need only plot $\sigma(\theta)$ from $\theta = 0$ to $\theta = \pi$. The corresponding locus is then reflected about the horizontal axis to obtain Γ_σ.

Since the simple roots of a polynomial equation are continuous functions of its coefficients, the simple roots of $\mathcal{P}(z) = 0$ must be a continuous function of σ. If Γ_σ is a simple closed curve,[2] it separates the σ plane into an *interior region* and an *exterior region*. Since $z = e^{j\theta}$ is a root of $\mathcal{P}(z) = 0$ for any σ on Γ_σ, it follows from the continuity of the roots z as a function of σ that either (1) $|z| < 1$ for all σ in the interior of Γ_σ, in which case the interior of Γ_σ is the region of absolute stability, or (2) $|z| < 1$ for all σ in the exterior of Γ_σ, in which case the exterior of Γ_σ is the region of absolute stability. Consequently, by choosing any convenient point $\sigma = \sigma_0$ in the interior or the exterior of Γ_σ in the σ plane and testing whether the corresponding roots of $\mathcal{P}(z) = 0$ satisfy $|z| < 1$ or not, we can identify either the interior or the exterior of the *unity-root locus* Γ_σ as the region of absolute stability of the multistep algorithm under consideration.

13-1-2 Regions of Absolute Stability for Explicit Adams–Bashforth Algorithms

Let us now apply the method developed in Section 13-1-1 to determine the regions of absolute stability for the family of explicit Adams–Bashforth algorithms.

FIRST-ORDER ADAMS–BASHFORTH (FORWARD EULER) ALGORITHM

$$k = 1: \quad x_{n+1} = x_n + hf(x_n, t_n)$$

The nonzero coefficients are $a_0 = 1$ and $b_0 = 1$. Substituting these coefficients into Eq. (13-13) and setting $p = k - 1 = 0$, we obtain

$$\sigma(\theta) = \frac{-e^{j\theta} + 1}{1} = 1 - e^{j\theta}, \qquad 0 \leq \theta \leq 2\pi \tag{13-14}$$

The locus of $\sigma(\theta)$ as θ varies from 0 to 2π is therefore a circle Γ_σ with unit radius and with center located at $\sigma = 1 + j0$ in the complex σ-plane, as shown in Fig. 13-2(a).

[2]A closed curve is said to be *simple* if it does not intersect itself.

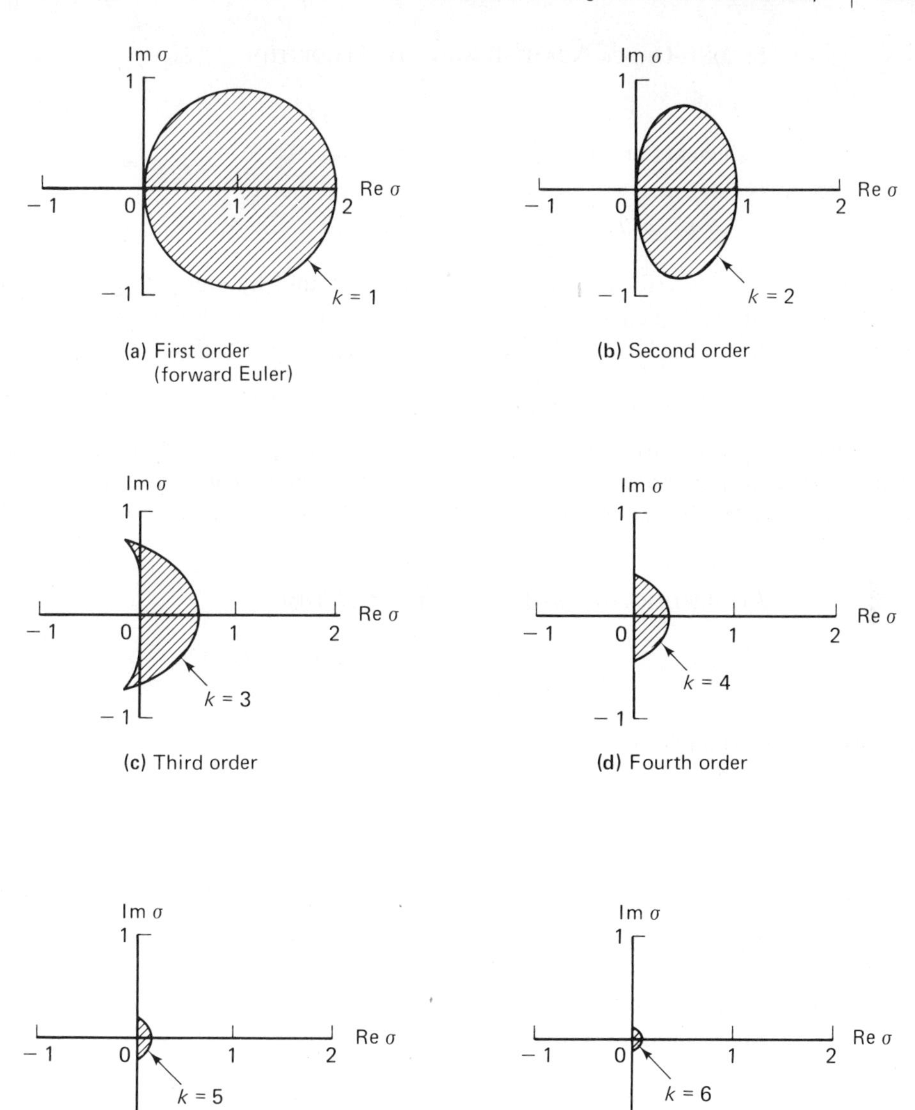

Fig. 13-2. Regions of absolute stability for the family of kth-order Adams–Bashforth algorithms, $k = 1, 2, \ldots, 6$.

To determine whether the interior or exterior is the region of stability, we choose $\sigma = 1 + j0$ in the interior of Γ_σ and substitute this value into Eq. (13-7) to obtain

$$\mathcal{P}(z) = (z - 1) + 1(1) = 0 \tag{13-15}$$

Since the only root $z = 0$ of Eq. (13-15) satisfies $|z| < 1$, it follows that the *forward Euler algorithm is absolutely stable in the interior of the unit circle centered at* $\sigma = 1 + j0$.

SECOND-ORDER ADAMS–BASHFORTH ALGORITHM

$$k = 2: \quad x_{n+1} = x_n + h\{\tfrac{3}{2}f(x_n, t_n) - \tfrac{1}{2}f(x_{n-1}, t_{n-1})\} \tag{13-16}$$

The nonzero coefficients are $a_0 = 1$, $b_0 = \frac{3}{2}$, and $b_1 = -\frac{1}{2}$.

Hence, we have

$$\sigma(\theta) = \frac{-e^{j2\theta} + e^{j\theta}}{\frac{3}{2}e^{j\theta} - \frac{1}{2}}, \qquad 0 \le \theta \le 2\pi \tag{13-17}$$

The locus of $\sigma(\theta)$ described by Eq. (13-17) can be found by computing $\sigma(\theta)$ from $\theta = 0$ to $\theta = \pi$, and then invoking horizontal symmetry to obtain the *unity-root locus* Γ_σ shown in Fig. 13-2(b). To determine the region of absolute stability, let us examine

$$\mathcal{P}(z) = (z^2 - z) + (\tfrac{3}{2}z - \tfrac{1}{2})\sigma = 0 \tag{13-18}$$

Choosing $\sigma = \frac{2}{3}$ for convenience, we find two roots $z = \pm 1/\sqrt{3}$. Since $\sigma = \frac{2}{3}$ lies in the interior of Γ_σ, and since $|z| < 1$, it follows that the *region of absolute stability of the second-order Adams–Bashforth algorithm is the shaded (interior) region shown in Fig. 13-2(b).*

kTH-ORDER ADAMS–BASHFORTH ALGORITHM

$$\begin{aligned} x_{n+1} = a_0 x_n + h\{&b_0 f(x_n, t_n) + b_1 f(x_{n-1}, t_{n-1}) \\ &+ \ldots + b_{k-1} f(x_{n-k+1}, t_{n-k+1})\} \end{aligned} \tag{13-19}$$

Equations (13-13) and (13-7) reduce to ($p = k - 1$)

$$\sigma(\theta) = \frac{-e^{jp\theta} + a_0}{b_0 e^{j(k-1)\theta} + b_1 e^{j(k-2)\theta} + \ldots + b_{k-2}e^{j\theta} + b_{k-1}}, \qquad 0 \le \theta \le 2\pi \tag{13-20}$$

$$\mathcal{P}(z) = (z^k - a_0 z^{k-1}) + \sigma(b_0 z^{k-1} + b_1 z^{k-2} + \ldots + b_{k-1}) = 0 \tag{13-21}$$

The regions of absolute stability for the Adams–Bashforth family of orders 3, 4, 5, and 6 are shown in the shaded regions in Figs. 13-2(c), (d), (e), and (f), respectively.

13-1-3 Regions of Absolute Stability for Implicit Adams–Moulton Algorithms

FIRST-ORDER ADAMS–MOULTON (BACKWARD EULER) ALGORITHM

$$k = 1: \quad x_{n+1} = x_n + hf(x_{n+1}, t_{n+1})$$

The nonzero coefficients are $a_0 = 1$ and $b_{-1} = 1$. Substituting these coefficients into Eq. (13-13) and setting $p = k - 2 = -1$, we obtain

$$\sigma(\theta) = -1 + e^{-j\theta}, \qquad 0 \le \theta \le 2\pi \tag{13-22}$$

The locus of $\sigma(\theta)$ as θ varies from 0 to 2π is therefore a circle Γ_σ with unit radius and with center located at $\sigma = -1 + j0$ in the complex σ-plane, as shown in Fig. 13-3(a). To determine the region of absolute stability, let us examine

$$\mathcal{P}(z) = (1 - z^{-1}) + 1 \cdot \sigma = 0 \tag{13-23}$$

For convenience, let us choose $\sigma = 1$. In this case, the root of $\mathcal{P}(z)$ is given by $z = \frac{1}{2}$.

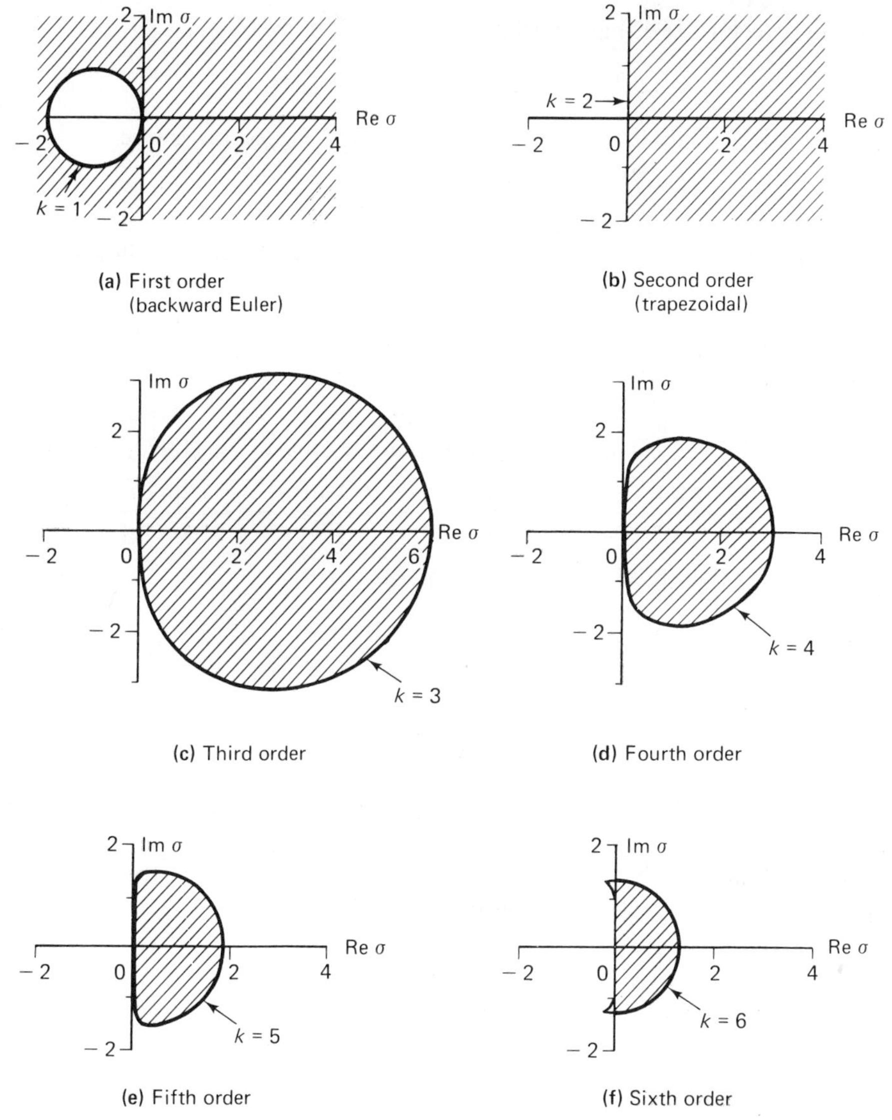

Fig. 13-3. Regions of absolute stability for the family of kth-order Adams–Moulton algorithms, $k = 1, 2, \ldots, 6$.

Since $\sigma = 1$ lies in the *exterior* of Γ_σ, and since the root of $\mathcal{P}(z) = 0$ satisfies $|z| < 1$, it follows that *the backward Euler algorithm is absolutely stable in the exterior of the unit circle centered at* $\sigma = -1 + j0$.

SECOND-ORDER ADAMS–MOULTON (TRAPEZOIDAL) ALGORITHM

$$k = 2: \quad x_{n+1} = x_n + h\{\tfrac{1}{2}f(x_{n+1}, t_{n+1}) + \tfrac{1}{2}f(x_n, t_n)\} \tag{13-24}$$

The nonzero coefficients are $a_0 = 1$, $b_{-1} = \frac{1}{2}$, and $b_0 = \frac{1}{2}$. Substituting these coefficients into Eq. (13-13) and setting $p = k - 2 = 0$, we obtain

$$\sigma(\theta) = \frac{-e^{j\theta} + 1}{\frac{1}{2}e^{j\theta} + \frac{1}{2}}, \qquad 0 \le \theta \le 2\pi \tag{13-25}$$

A straightforward analysis of Eq. (13-25) shows that $\text{Re}\,\sigma(\theta) = 0$; hence,

$$\sigma(\theta) = 0 + j\,\text{Im}\,\sigma(\theta), \qquad 0 \le \theta \le 2\pi \tag{13-26}$$

which shows that the locus of $\sigma(\theta)$ as θ varies from 0 to 2π is simply the imaginary axis—a circle with an infinite radius. To determine the region of absolute stability, let us examine[3]

$$\mathcal{P}(z) = (z - 1) + (\tfrac{1}{2}z + \tfrac{1}{2})\sigma = 0 \tag{13-27}$$

The root of $\mathcal{P}(z) = 0$ is given by

$$z = \frac{1 - \frac{1}{2}\sigma}{1 + \frac{1}{2}\sigma} \tag{13-28}$$

It follows from Eq. (13-28) that the magnitudes of the roots of Eq. (13-27) are given by

$$|z| = \left|\frac{(1 - \frac{1}{2}\,\text{Re}\,\sigma) - j\frac{1}{2}\,\text{Im}\,\sigma}{(1 + \frac{1}{2}\,\text{Re}\,\sigma) + j\frac{1}{2}\,\text{Im}\,\sigma}\right| = \frac{\sqrt{(1 - \frac{1}{2}\,\text{Re}\,\sigma)^2 + (\frac{1}{2}\,\text{Im}\,\sigma)^2}}{\sqrt{(1 + \frac{1}{2}\,\text{Re}\,\sigma)^2 + (\frac{1}{2}\,\text{Im}\,\sigma)^2}} \tag{13-29}$$

Hence,

$$\begin{aligned} |z| &< 1, & \text{Re}\,\sigma &> 0 \\ &= 1, & \text{Re}\,\sigma &= 0 \\ &> 1, & \text{Re}\,\sigma &< 0 \end{aligned} \tag{13-30}$$

It follows from Eq. (13-30) that the *second-order Adams–Moulton* (*trapezoidal*) *algorithm is absolutely stable in the right half-plane*, as shown in Fig. 13-3(b).

kTH-ORDER ADAMS–MOULTON ALGORITHM

$$\begin{aligned} x_{n+1} = a_0 x_n + h\{&b_{-1}f(x_{n+1}, t_{n+1}) + b_0 f(x_n, t_n) + b_1 f(x_{n-1}, t_{n-1}) \\ &+ \dots + b_{k-2}f(x_{n-k+2}, t_{n-k+2})\} \end{aligned} \tag{13-31}$$

Equations (13-13) and (13-7) reduce to ($p = k - 2$)

$$\sigma(\theta) = \frac{-e^{j(k-1)\theta} + a_0 e^{j(k-2)\theta}}{b_{-1}e^{j(k-1)\theta} + b_0 e^{j(k-2)\theta} + b_1 e^{j(k-3)\theta} + \dots + b_{k-3}e^{j\theta} + b_{k-2}}, \quad 0 \le \theta \le 2\pi \tag{13-32}$$

$$\mathcal{P}(z) = (z^{k-1} - a_0 z^{k-2}) + \sigma(b_{-1}z^{k-1} + b_0 z^{k-2} + b_1 z^{k-3} + \dots + b_{k-2}) = 0 \tag{13-33}$$

The regions of absolute stability for the Adams–Moulton family of orders 3, 4, 5, and 6 are shown in the shaded regions in Fig. 13-3(c), (d), (e), and (f), respectively.

13-1-4 Comparison of Regions of Absolute Stability between Adams–Bashforth and Adams–Moulton Algorithms

It is instructive to compare the regions of absolute stability for the *explicit* Adams–Bashforth formulas of order $k = 1, 2, \dots, 6$, as shown in Fig. 13-2, and the corresponding regions for the *implicit* Adams–Moulton formulas, as shown in Fig.

[3]Although we could determine the region of absolute stability by testing a single point $\sigma = \sigma_0$, as in the preceding cases, we shall do it for all values of σ this time because a simple analytic expression can be derived from this case.

13-3. Observe that the latter is much larger for $k = 1, 2$, and approximately 10 times larger for $k = 3, 4, 5$, and 6 (notice that the scales are different). In fact, for any given $\lambda > 0$, the first- and second-order *implicit* (backward Euler and trapezoidal) algorithms will be stable for any step size! Hence, the choice of step size h for the backward Euler and trapezoidal algorithms is restricted only by accuracy—the maximum allowable local truncation error—and not by stability. For orders $k = 3$, 4, 5, and 6, the step size h for the *implicit* Adams–Moulton algorithm can be chosen to be at least *10 times* larger than that for the *explicit* Adams–Bashforth algorithm of the same order. Moreover, it can be shown that even though the local truncation errors for the Adams–Bashforth and the Adams–Moulton algorithms of order k are both proportional to h^{k+1} [see Eqs. (12-26) and (12-35)], the constant of proportionality for the Adams–Moulton algorithm is smaller than that of the Adams–Bashforth algorithm of the same order. Consequently, the possibility of increased accuracy and choice of a greater step size (by at least 10) with the *implicit* Adams–Moulton algorithm usually more than offsets the need to solve the implicit equations. In view of these observations, *the family of Adams–Moulton algorithms is generally considered to be the best family of general-purpose algorithms for solving the initial-value problem* $\dot{\boldsymbol{x}} = \boldsymbol{f}(\boldsymbol{x}, t)$, $\boldsymbol{x}(t_0) = \boldsymbol{x}_0$. In fact, many current network simulation programs make use of only the first-order *backward Euler* algorithm or the second-order *trapezoidal* algorithm.

13-2 STIFF STATE EQUATIONS—AN INTRODUCTION

We shall motivate the subject of this section with two examples.

EXAMPLE 13-1. Consider the first-order linear circuit shown in Fig. 13-4(a). The state equation is readily seen to be given by

$$\dot{v}_c = -\frac{1}{RC}[v_c - v_s(t)] + \frac{dv_s(t)}{dt} \tag{13-34}$$

This equation is of the form

$$\dot{x} = -\lambda_1[x - s(t)] + \frac{ds(t)}{dt} \tag{13-35}$$

where $-\lambda_1 \triangleq -1/RC$ is the natural frequency of the circuit, and $s(t)$ is any continuously differentiable forcing function. The *exact* solution of Eq. (13-35) is readily seen to be given by

$$x(t) = ke^{-\lambda_1 t} + s(t) \tag{13-36}$$

$$k \triangleq x(0) - s(0) \tag{13-37}$$

To be specific, let us choose an exponential signal

$$s(t) = 1 - e^{-\lambda_2 t} \tag{13-38}$$

The exact solution corresponding to this case is shown in Fig. 13-4(b) along with its two components. Suppose the two constants λ_1 and λ_2 differ by several orders of magnitude, say $\lambda_1 = 10^9$ and $\lambda_2 = 1$. Since an exponential waveform decays to a negligible value in about five time constants, the first component in Eq. (13-36) is

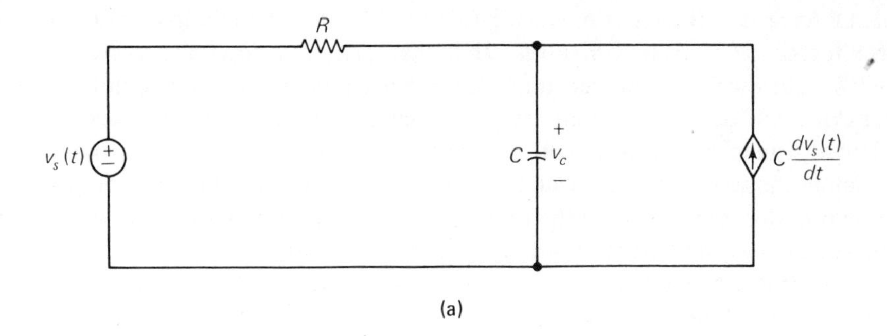

(a)

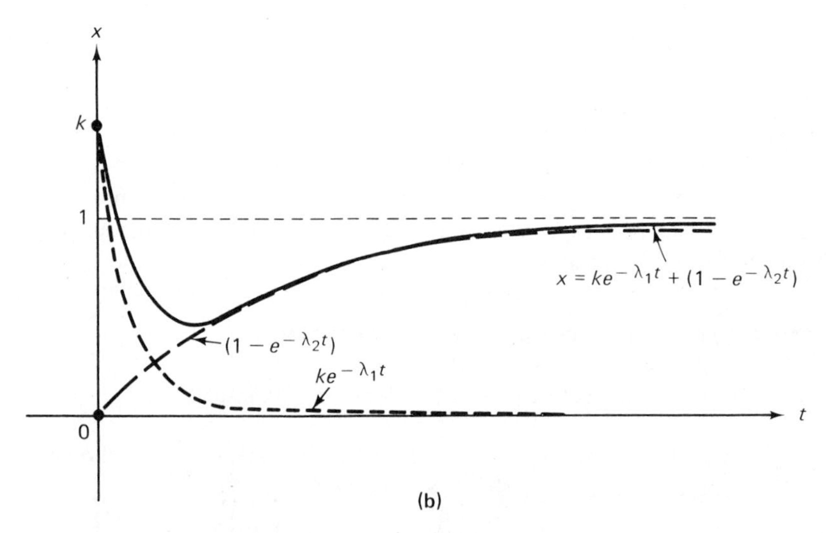

(b)

Fig. 13-4. "Stiff" circuit and its complete solution consisting of a "fast" component $ke^{-\lambda_1 t}$ and a "slow" component $1 - e^{-\lambda_2 t}$.

negligible for $t \geq 5 \times 10^{-9}$ s, whereas the second component will not reach steady state until after $t = 5$ s. Hence, if the numerical solution is to display the entire transient waveform, we must integrate over a time interval of at least 5 s. Consider what happens if we choose the *forward Euler* algorithm to solve this problem. From, Eq. (13-14) and Fig. 13-2(a), we observe that this algorithm is *stable* if, and only if $\sigma \triangleq h\lambda$ lies within the unit circle centered at $\sigma = 1 + j0$. For $\lambda = \lambda_1 = 10^9$, this means that $10^9 h < 2$; hence, we must choose

$$h < 2 \times 10^{-9} \tag{13-39}$$

over the entire time interval $0 \leq t \leq 5$ in order for the local truncation and round-off errors to remain bounded. A simple exercise using a step size $h > 2 \times 10^{-9}$ to solve this problem will show the solution blowing up as time increases, thereby giving an erroneous answer. To integrate with the maximum permissible step size $h = 2 \times 10^{-9}$ over 5 s would require a total of $N = 2.5 \times 10^9$ time steps. Assuming that it takes

10 μs of computer time to integrate over one time step, it will take 2.5×10^4 s, or approximately 7 hours, of computer time to obtain the solution to this problem!

To obtain an intuitive feeling for why this unfortunate restriction to such a small step size is necessary, consider the graphical construction shown in Fig. 13-5 for solving this problem using the forward Euler algorithm with a relatively large step size h. Rather than decaying to the steady waveform defined by Eq. (13-38), observe that the solution actually grows, and will blow up as time increases. To avoid this unstable behavior, the step size must be chosen before the tangent line intersects the envelope of the steady-state waveform at points $a, b, c, d, \ldots$, etc.

EXAMPLE 13-2. Consider the following system of *uncoupled* linear equations:

$$\dot{x}_1 = -\lambda_1 x_1 \tag{13-40}$$

$$\dot{x}_2 = -\lambda_2 x_2 \tag{13-41}$$

The exact solution is given by

$$x_1 = k_1 e^{-\lambda_1 t} \tag{13-42}$$

$$x_2 = k_2 e^{-\lambda_2 t} \tag{13-43}$$

where $s_1 = -\lambda_1$ and $s_2 = -\lambda_2$ are the natural frequencies. If we integrate Eqs. (13-40) and (13-41) by the forward Euler method with $\lambda_1 = 10^9$ and $\lambda_2 = 1$, then to ensure numerical stability we must restrict the step size to $h < 2 \times 10^{-9}$ s for Eq.

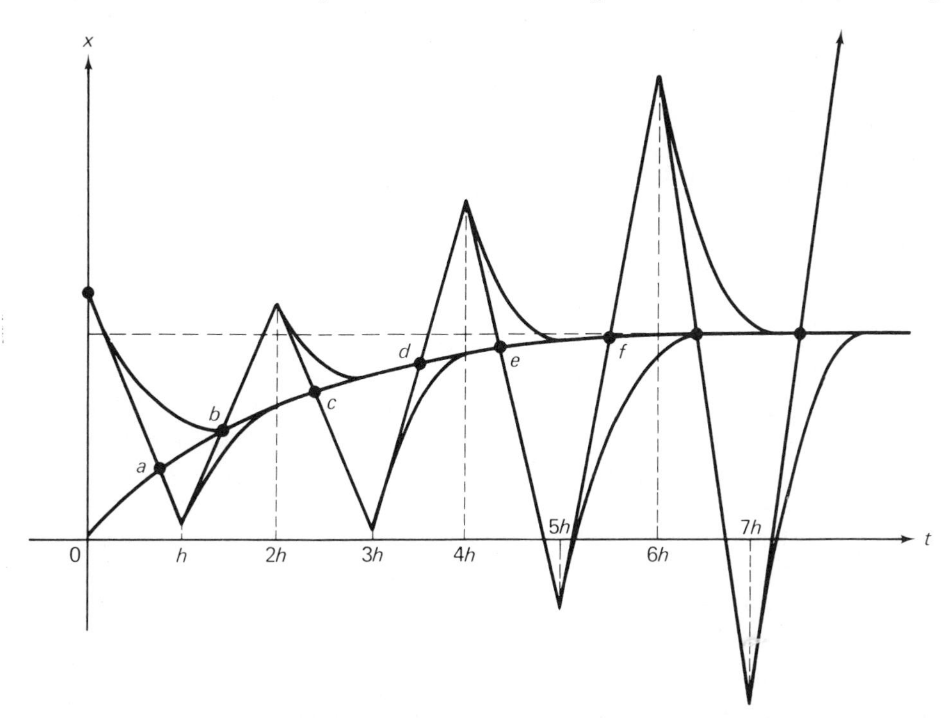

Fig. 13-5. Geometrical illustration of the mechanism leading to numerical instability in the forward Euler algorithm.

(13-42), and $h < 2$ s for Eq. (13-43). For this *uncoupled* system, we could have solved it as two separate problems using two separate step sizes. However, in general, the variables are coupled to one another, and instead of Eqs. (13-40) and (13-41), we have

$$\dot{x}_1 = -a_{11}x_1 - a_{12}x_2 \tag{13-44}$$

$$\dot{x}_2 = -a_{21}x_1 - a_{22}x_2 \tag{13-45}$$

Supposing that $a_{12} = a_{21}$ and assuming that the natural frequencies of this system—i.e., the eigenvalues of the matrix

$$-A = \begin{bmatrix} -a_{11} & -a_{12} \\ -a_{21} & -a_{22} \end{bmatrix} \tag{13-46}$$

are given by $s_1 = -\lambda_1 = -10^9$ and $s_2 = -\lambda_2 = -1$, then Eqs. (13-44) and (13-45) can be transformed into the form of Eqs. (13-40) and (13-41) by the transformation $y = Mx$ [recall Eqs. (12-60)–(12-63)]. Integrating the transformed system by the forward Euler algorithm will therefore lead to the same step size restrictions as before. However, since the complete solution $x_1(t)$ and $x_2(t)$ are *linear sums* of the solution of the transformed variables x_1 and x_2, a *common* step size must be chosen to be the smaller of the two; i.e., $h = 2 \times 10^{-9}$. Once again, we are confronted with the same problem as in Example 13-1 where the transient solution actually decays in about 5 μs, and yet we have to keep inching along with a ridiculously small step size, even though the computed solution varies only by an insignificant increment between any two time steps.

Any system of state equations whose solution contains both "very fast" and "very slow" components—such as the preceding examples—is said to be *stiff*. For linear systems, the stiffness occurs whenever the associated system matrix **A** possesses widely separated eigenvalues. A nonlinear system is stiff if its associated Jacobian matrix exhibits widely separated eigenvalues at the operating points of interest. Most nonlinear electronic circuits that exhibit near stepwise solution waveforms, such as multivibrators, are stiff.

13-3 DESIRABLE REGION OF ABSOLUTE STABILITY FOR SOLVING STIFF STATE EQUATIONS [1, 7]

To solve stiff equations efficiently, one must select an algorithm that will allow the step size to be varied over a wide range of values and yet will remain numerically stable. Such an algorithm would start with a sufficiently small step size commensurate with the desired accuracy for obtaining the initial "fast" transient component, and then would change progressively to larger step sizes after the transient has settled down. Our objective in this section is to identify the desirable regions of absolute stability in the complex $\sigma \triangleq h\lambda$ plane for meeting these requirements.

Consider the system of linear state equations

$$\dot{y} = -Ay \tag{13-47}$$

Assuming that the $n \times n$ matrix A has n distinct eigenvalues

$$\lambda_i \triangleq \alpha_i + j\beta_i, \qquad i = 1, 2, \ldots, n \tag{13-48}$$

and that Eq. (13-47) can be transformed into a system of uncoupled linear equations as in Section 12-5, i.e.,

$$\dot{x}_i = -\lambda_i x_i, \qquad i = 1, 2, \ldots, n \tag{13-49}$$

the *exact* solution to Eq. (13-49) is

$$\hat{x}_i(t) = k_i e^{-\lambda_i t} = (k_i e^{-\alpha_i t})e^{-j\beta_i t} \tag{13-50}$$

Observe that if Re $\lambda_i \triangleq \alpha_i > 0$, $\hat{x}_i(t)$ *decays* to zero in about five time constants, i.e., in $5\tau_i \triangleq 5/\alpha_i$. If Re $\lambda_i \triangleq \alpha_i < 0$, $\hat{x}_i(t)$ *grows* exponentially as time increases. In either case, if Im $\lambda_i \triangleq \beta_i \neq 0$, it is well known that λ_i forms a complex-conjugate pair with another eigenvalue λ_i^*, and the corresponding solution of $y_i(t)$ will contain the exponentially weighted *sinusoidal component*

$$(c_i e^{-\alpha_i t}) \sin \beta_i t \tag{13-51}$$

where c_i is a constant. Let us now examine the different cases assumed by $\lambda_i \triangleq \alpha_i + j\beta_i$.

Case 1: Re $\lambda_i \triangleq \alpha_i \geq 0$. Since $h > 0$, case 1 corresponds to the region Re $\sigma \triangleq (h\lambda_i) \geq 0$, i.e., the closed right half-plane. Suppose that we examine the subregion defined by

$$\text{Re}\,\sigma \triangleq \text{Re}\,(h\lambda) \geq \delta > 0 \tag{13-52}$$

where δ is a real number chosen large enough such that the decaying sinusoidal component defined in Eq. (13-51) becomes insignificant after one time step h; thus,

$$(c_i e^{-\alpha_i h}) \sin \beta_i h \approx 0 \tag{13-53}$$

Since this "fast" transient component of $y_i(t)$ will be completely "swamped" by the remaining components, it is not necessary to obtain an *accurate* representation of Eq. (13-53) by taking *small* step sizes. Instead, h could be chosen as large as possible, as long as the algorithm remains stable. Hence, *it is desirable that the region defined by Eq. (13-52) be stable.* This corresponds to region III in Fig. 13-6.

In the region

$$0 \leq \text{Re}\,\sigma \triangleq \text{Re}\,(h\lambda) < \delta \tag{13-54}$$

the "fast" transient component defined by Eq. (13-53) is no longer negligible. This corresponds therefore to the initial phase of the solution waveform, where the transient has not yet decayed sufficiently in one step size and must therefore be *accurately* determined. Observe that in one time step h the sinusoidal component in Eq. (13-53) goes through an oscillation of

$$N = \frac{|\text{Im}\,\sigma|}{2\pi} \triangleq \frac{|\beta_i h|}{2\pi} \tag{13-55}$$

complete cycles. Since this damped sinusoidal component would not have decayed to zero in the region defined by Eq. (13-54), we must be sure that $N < \frac{1}{8}$ if we are to recover at least eight points per cycle of this decaying sinusoid. Hence, to ensure *accuracy in the initial "fast" transient phase, it is desirable that*

$$\frac{|\beta_i h|}{2\pi} \triangleq \frac{\theta}{2\pi} < \frac{1}{8}, \quad \text{or} \quad \theta < \frac{\pi}{4} \tag{13-56}$$

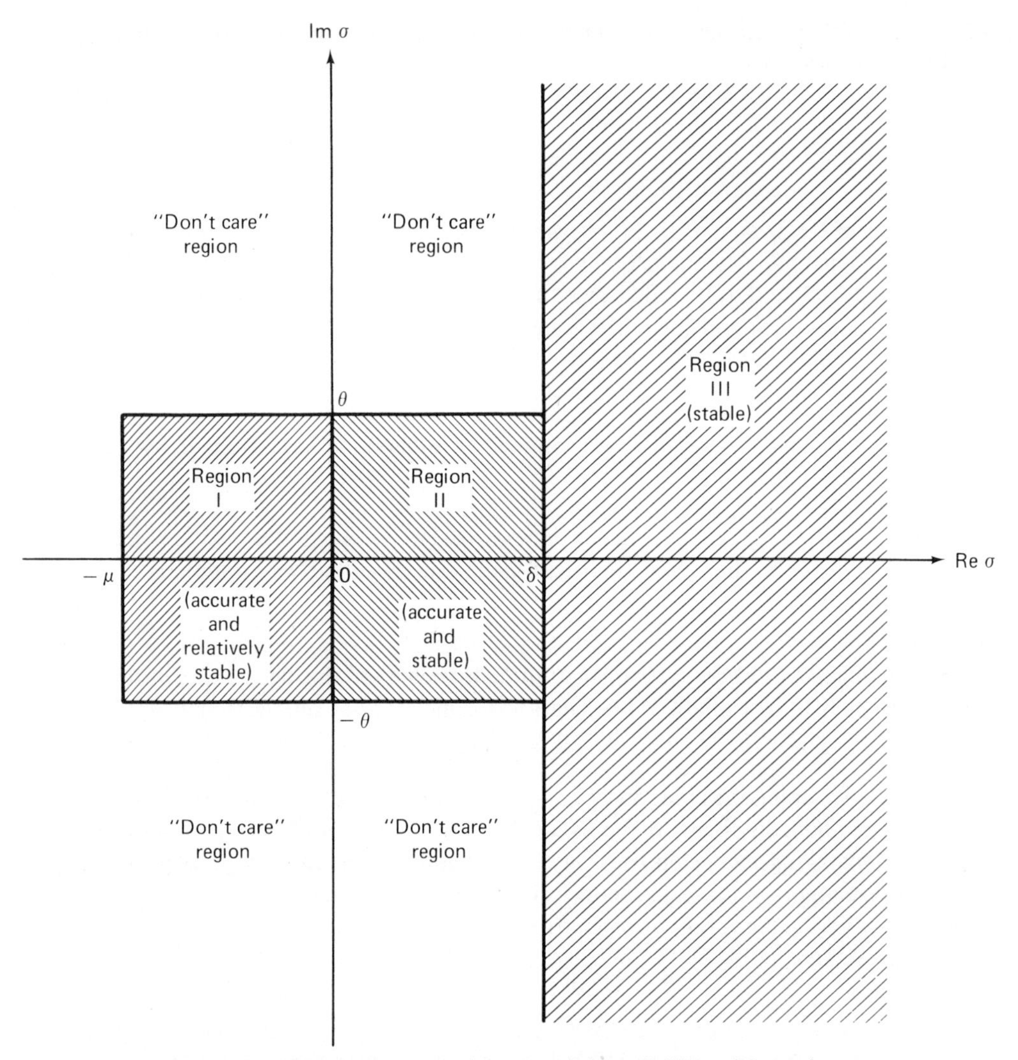

Fig. 13-6. Multistep algorithms for solving stiff differential equations must exhibit the properties indicated in regions I, II, and III.

The region defined by Eqs. (13-54) and (13-56) corresponds to region II in Fig. 13-6. Within this region, the numerical solution must be both *accurate* and *stable.*

Case 2: $\text{Re}\,\lambda_i \triangleq \alpha_i < 0$. Since $h > 0$, case 2 corresponds to the region $\text{Re}\,\sigma \triangleq \text{Re}\,(h\lambda_i) < 0$, i.e., the open left half-plane. In this region, the solution consists of growing exponentials if $\beta_i = 0$, or growing sinusoidal waveforms if $\beta_i \neq 0$. It follows from the general solution[4]

$$y_i = c_1 z_{i_1}^n + c_2 z_{i_2}^n + \ldots + c_{p+1} z_{i_{p+1}}^n \tag{13-57}$$

[4]Equation (13-57) corresponds to Eq. (13-41) for the scalar case.

[$z_{i_1}, z_{i_2}, \ldots, z_{i_{p+1}}$ are the *roots* of the polynomial equation $\mathcal{P}(z) = 0$] that the *principal root* z_{i_1} must satisfy

$$|z_{i_1}| > 1 \tag{13-58}$$

Hence, any multistep algorithm capable of obtaining a growing exponential or sinusoidal solution is necessarily *unstable* in the region defined by $\text{Re}\,\lambda_i < 0$. In the general case where $\beta_i \neq 0$, there will again be N complete cycles over one time step h, where N is defined by Eq. (13-55). Clearly, we must again satisfy Eq. (13-56) in this region for the details of the growing waveform to be *accurately* computed.

Moreover, we must restrict $\text{Re}\,\sigma$ such that

$$-\mu < \text{Re}\,\sigma \triangleq \alpha_i < 0 \tag{13-59}$$

where μ is chosen "small" enough so that the exponential will not grow too much over one time step. It follows from the preceding analysis that, if $\text{Re}\,\lambda_i < 0$, *then it is desirable to choose* $\sigma \triangleq h\lambda$ *within the rectangular region bounded by Eqs. (13-56) and (13-59)*. This corresponds to region I in Fig. 13-6.

We have already shown that the multistep algorithm must be unstable in region I in order for the principal root z_{i_1} to give rise to a growing waveform. The remaining extraneous parasitic roots $z_{i_2}, z_{i_3}, \ldots, z_{i_{p+1}}$ no longer have to lie within the unit circle. However, in order that these parasitic roots will not dominate the principal root, *we must require that the magnitude of all parasitic roots be less than the magnitude of the principal root*. A multistep algorithm that satisfies this property in region I is said to be *relatively stable*.

The remaining regions in Fig. 13-6 are the "*don't care*" *regions*, since we never operate there. Accordingly, the multistep algorithm need not be stable or even relatively stable in these "don't care" regions. We shall now summarize our preceding analysis as follows:

For efficient and accurate solution of stiff differential equations, it is desirable to choose a multistep algorithm having the following three properties:

Stiff Property 1. There exists a $\delta \geq 0$ such that region III is a region of absolute stability.

Stiff Property 2. There exists a $0 < \theta < \frac{1}{4}$ such that region II is accurate and stable.

Stiff Property 3. There exists a $\mu < 0$ such that region I is accurate and relatively stable.

A multistep algorithm satisfying these three properties is said to be *stiffly stable*.

Examination of the regions of absolute stability for the Adams–Bashforth algorithms shown in Fig. 13-2 shows that they all fail to satisfy stiff property 1. It follows that the family of *Adams–Bashforth algorithms are not suitable for solving stiff equations*.

Examination of the regions of absolute stability for the Adams–Moulton algorithms shown in Fig. 13-3 shows that, with the exception of the *backward Euler* and the *trapezoidal* algorithms, they too fail to satisfy stiff property 1.

Are there other multistep algorithms that satisfy stiff property 1? A partial answer to this question is given by Dahlquist [1].

***Theorem 13-1* (*Dahlquist Theorem*).** *A multistep algorithm that is absolutely stable in the region* Re $\sigma \triangleq$ Re $(h\lambda) \geq 0$ *cannot exceed order 2. Moreover, the best (most accurate) method of order 2 that satisfies stiff property 1 is the trapezoidal algorithm.*

Indeed, examination of Fig. 13-3(b) shows that the trapezoidal algorithm satisfies stiff property 1 with $\delta = 0$. The Dahlquist theorem rules out the existence of higher-order "stiffly stable" multistep algorithms *with* $\delta = 0$. Consequently, in searching for higher-order stiffly stable algorithms, it is necessary to choose $\delta > 0$, which we do next.

13-4 DERIVATION OF GEAR'S STIFFLY STABLE ALGORITHMS

To satisfy stiff property 1 with $\delta > 0$, a multistep algorithm must be stable as $\sigma \triangleq h\lambda \longrightarrow \infty$. In this limiting case, the polynomial equation $\mathcal{P}(z) = 0$ given in Eq. (13-7) becomes dominated by the terms multiplying σ, as long as $b_{-1} \neq 0$. Hence, we can write

$$\mathcal{P}(z) \approx (b_{-1}z^{p+1} + b_0 z^p + b_1 z^{p-1} + \ldots + b_p)\sigma = 0 \qquad \text{as } \sigma \longrightarrow \infty \tag{13-60}$$

One simple way to ensure that all roots of $\mathcal{P}(z) = 0$ will lie within the unit circle as $|\sigma| \longrightarrow \infty$ is to choose $b_0 = b_1 = \ldots = b_p = 0$. Under this assumption, $z = 0$ is the only (multiple) root of $\mathcal{P}(z) = 0$ as $|\sigma| \longrightarrow \infty$. Since this root lies within the unit circle, the algorithm is guaranteed to be stable as $|\sigma| \triangleq h|\lambda| \longrightarrow \infty$. Since the roots of the polynomial equation $\mathcal{P}(z) = 0$ are a continuous function of σ, one could hope that, through an appropriate choice of the remaining coefficients, stiffly stable algorithms could be discovered. Such an argument has been advocated by Gear [1], and we shall refer to the resulting multistep formulas as Gear's algorithms.

The kth-order Gear's algorithm is an *implicit* algorithm obtained by setting

$$p = k - 1, \qquad b_0 = b_1 = b_2 = \ldots = b_{k-1} = 0 \tag{13-61}$$

in Eq. (12-1):

$$\boxed{\begin{aligned} x_{n+1} = {} & a_0(k)x_n + a_1(k)x_{n-1} + a_2(k)x_{n-2} \\ & + \ldots + a_{k-1}(k)x_{n-k+1} + h[b_{-1}(k)f(x_{n+1}, t_{n+1})] \end{aligned}} \tag{13-62}$$

where the coefficient a_i is now denoted by $a_i(k)$ to emphasize its dependence on the order k of the method. The $k + 1$ coefficients $a_0(k), a_1(k), a_2(k), \ldots, a_{k-1}(k)$, and $b_{-1}(k)$ are to be determined such that Eq. (13-62) is *exact* for all polynomial solutions of degree k. Since a kth-degree polynomial is uniquely determined by $k + 1$ parameters, we have exactly the correct number of unknown coefficients to be determined using Eq. (12-17); thus,

$$\sum_{i=0}^{k-1} (-i)^j a_i(k) + jb_{-1}(k) = 1, \qquad j = 0, 1, 2, \ldots, k-1 \tag{13-63}$$

Equation (13-63), when expanded, consists of a system of $k+1$ linear equations in the $k+1$ unknowns $a_0(k), a_1(k), \ldots, a_{k-1}(k)$, and $b_{-1}(k)$;

$$\begin{bmatrix} 1 & 1 & 1 & 1 & \cdots & 1 & 0 \\ 0 & -1 & -2 & -3 & \cdots & (k-1) & 1 \\ 0 & 1 & 4 & 9 & \cdots & (k-1)^2 & 2 \\ 0 & -1 & -8 & -27 & \cdots & [-(k-1)]^3 & 3 \\ \cdot & \cdot & \cdot & \cdot & & \cdot & \cdot \\ \cdot & \cdot & \cdot & \cdot & & \cdot & \cdot \\ \cdot & \cdot & \cdot & \cdot & & \cdot & \cdot \\ 0 & (-1)^k & (-2)^k & (-3)^k & \cdots & [-(k-1)]^k & k \end{bmatrix} \begin{bmatrix} a_0(k) \\ a_1(k) \\ a_2(k) \\ a_3(k) \\ \cdot \\ \cdot \\ \cdot \\ b_{-1}(k) \end{bmatrix} = \begin{bmatrix} 1 \\ 1 \\ 1 \\ 1 \\ \cdot \\ \cdot \\ \cdot \\ 1 \end{bmatrix} \tag{13-64}$$

The solution of Eq. (13-64) therefore uniquely specifies the remaining coefficients of the kth-order Gear's algorithm. We shall illustrate this by considering three cases.

CASE 1: $k = 1$. In this case, Eq. (13-64) reduces trivially to $a_0(1) = 1$ and $b_{-1}(1) = 1$. Substituting these coefficients into Eq. (13-62), we obtain the following *first-order Gear's algorithm:*

$$x_{n+1} = x_n + h[f(x_{n+1}, t_{n+1})] \tag{13-65}$$

Observe that Eq. (13-65) is simply the backward Euler algorithm derived earlier in Section 12-3.

CASE 2: $k = 2$. In this case, Eq. (13-64) becomes

$$\begin{bmatrix} 1 & 1 & 0 \\ 0 & -1 & 1 \\ 0 & 1 & 2 \end{bmatrix} \begin{bmatrix} a_0(2) \\ a_1(2) \\ b_{-1}(2) \end{bmatrix} = \begin{bmatrix} 1 \\ 1 \\ 1 \end{bmatrix} \tag{13-66}$$

The solution is given by $a_0(2) = \frac{4}{3}$, $a_1(2) = -\frac{1}{3}$, and $b_{-1}(2) = \frac{2}{3}$. Substituting these coefficients into Eq. (13-62), we obtain the following *second-order Gear's algorithm:*[5]

$$x_{n+1} = \tfrac{4}{3}x_n - \tfrac{1}{3}x_{n-1} + h[\tfrac{2}{3}f(x_{n+1}, t_{n+1})] \tag{13-67}$$

CASE 3: $k = 3$. In this case, Eq. (13-64) becomes

$$\begin{bmatrix} 1 & 1 & 1 & 0 \\ 0 & -1 & -2 & 1 \\ 0 & 1 & 4 & 2 \\ 0 & -1 & -8 & 3 \end{bmatrix} \begin{bmatrix} a_0(3) \\ a_1(3) \\ a_2(3) \\ b_{-1}(3) \end{bmatrix} = \begin{bmatrix} 1 \\ 1 \\ 1 \\ 1 \end{bmatrix} \tag{13-68}$$

The solution is given by $a_0(3) = \frac{18}{11}$, $a_1(3) = -\frac{9}{11}$, $a_2(3) = \frac{2}{11}$, and $b_{-1}(3) = \frac{6}{11}$. Substituting these coefficients into Eq. (13-62), we obtain the following *third-order Gear's algorithm:*

$$x_{n+1} = \tfrac{18}{11}x_n - \tfrac{9}{11}x_{n-1} + \tfrac{2}{11}x_{n-2} + h[\tfrac{6}{11}f(x_{n+1}, t_{n+1})] \tag{13-69}$$

[5]A variable step-size version of Gear's second-order algorithm has been implemented in the simulation program CIRPAC described in reference 2.

For future reference, the formulas for the first- to sixth-order Gear's algorithms are collected in Table 13-1.

TABLE 13-1. Gear's Algorithm

ORDER	
First	$x_{n+1} = x_n + h[f(x_{n+1}, t_{n+1})]$
Second	$x_{n+1} = \frac{4}{3}x_n - \frac{1}{3}x_{n-1} + h[\frac{2}{3}f(x_{n+1}, t_{n+1})]$
Third	$x_{n+1} = \frac{18}{11}x_n - \frac{9}{11}x_{n-1} + \frac{2}{11}x_{n-2}$ $+ h[\frac{6}{11}f(x_{n+1}, t_{n+1})]$
Fourth	$x_{n+1} = \frac{48}{25}x_n - \frac{36}{25}x_{n-1} + \frac{16}{25}x_{n-2} - \frac{3}{25}x_{n-3}$ $+ h[\frac{12}{25}f(x_{n+1}, t_{n+1})]$
Fifth	$x_{n+1} = \frac{300}{137}x_n - \frac{300}{137}x_{n-1} + \frac{200}{137}x_{n-2} - \frac{75}{137}x_{n-3}$ $+ \frac{12}{137}x_{n-4} + h[\frac{60}{137}f(x_{n+1}, t_{n+1})]$
Sixth	$x_{n+1} = \frac{360}{147}x_n - \frac{450}{147}x_{n-1} + \frac{400}{147}x_{n-2} - \frac{225}{147}x_{n-3}$ $+ \frac{72}{147}x_{n-4} - \frac{10}{147}x_{n-5} + h[\frac{60}{147}f(x_{n+1}, t_{n+1})]$

An examination of Table 13-1 shows that the kth-order Gear's algorithm requires k *starting* values; namely, $x_n, x_{n-1}, x_{n-2}, \ldots, x_{n-k+1}$. Hence, a kth-order Gear's algorithm is a k-step algorithm. If we apply Theorem 11-5 from Sec. 11-5-1 to Gear's algorithm, we obtain the following corresponding result:

Theorem 13-2. *The local truncation error ϵ_T for the kth order Gear's algorithm is given by*

$$\epsilon_T = [C_k\hat{x}^{(k+1)}(\hat{\tau})]h^{k+1} = O(h^{k+1}), \qquad t_n < \hat{\tau} < t_{n+1}$$
$$C_1 = -\tfrac{1}{2}, C_2 = -\tfrac{2}{9}, C_3 = -\tfrac{3}{22}, C_4 = -\tfrac{12}{125},$$
$$C_5 = -\tfrac{10}{137}, C_6 = -\tfrac{60}{1029}, \ldots,$$
$$C_k = \frac{1}{(k+1)!}\left[k^{k+1} - \sum_{i=0}^{k-2} a_i(k-1-i)^{k+1} - b_{-1}(k+1)k^k\right]$$

It remains for us to determine the regions of absolute stability for Gear's algorithms.

First-Order Gear's Algorithm

$$k = 1: \quad x_{n+1} = x_n + h[f(x_{n+1}, t_{n+1})]$$

Since this is just the backward Euler algorithm, its region of absolute stability is the exterior of the unit circle centered at $\sigma = -1 + j0$ as derived earlier in Section 13-1-3. See Fig. 13-7(a).

Second-Order Gear's Algorithm

$$k = 2: \quad x_{n+1} = \tfrac{4}{3}x_n - \tfrac{1}{3}x_{n-1} + h[\tfrac{2}{3}f(x_{n+1}, t_{n+1})]$$

The nonzero coefficients are $a_0 = \frac{4}{3}$, $a_1 = -\frac{1}{3}$, and $b_{-1} = \frac{2}{3}$. Substituting these

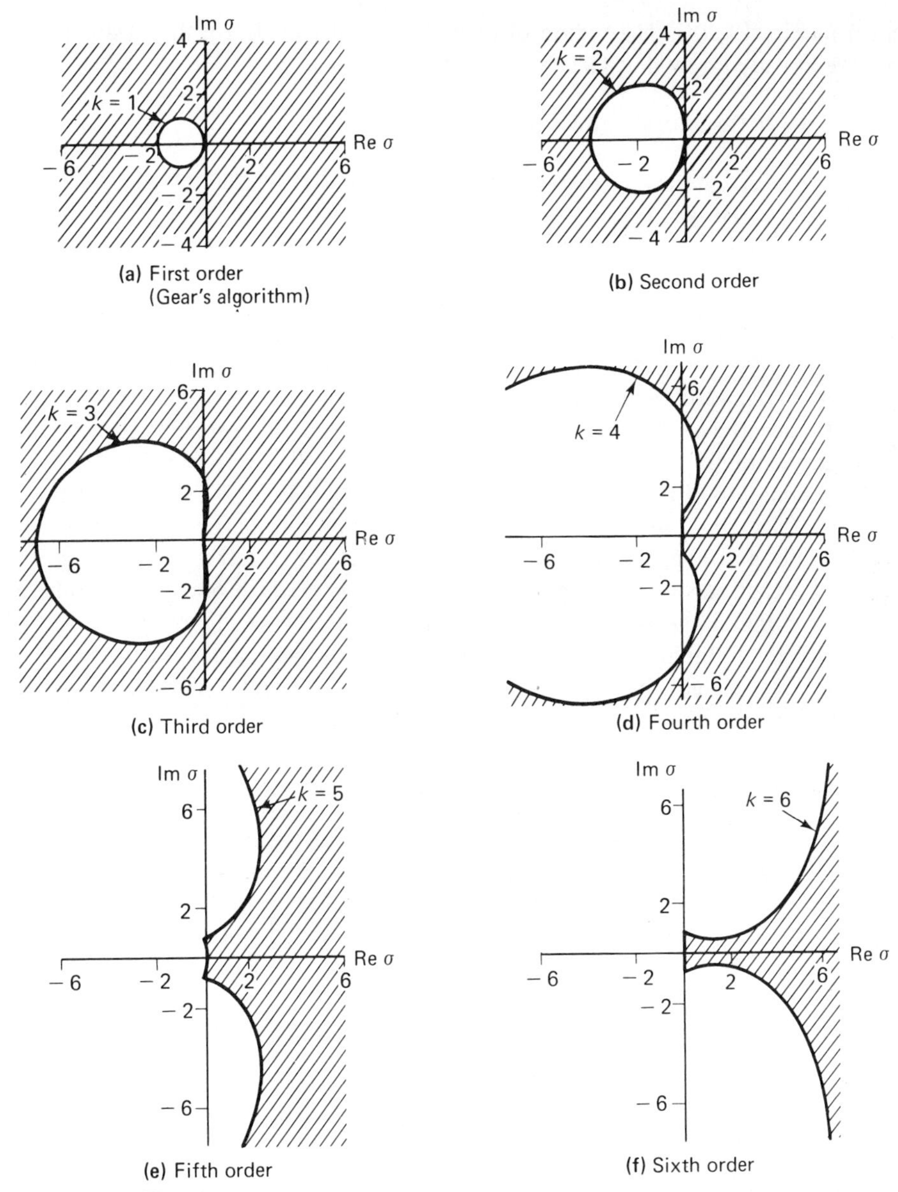

Fig. 13-7. Regions of absolute stability for the family of kth-order Gear's algorithm, $k = 1, 2, \ldots, 6$.

coefficients into Eq. (13-13) and setting $p = k - 1 = 1$, we obtain

$$\sigma(\theta) = \frac{-e^{j2\theta} + \frac{4}{3}e^{j\theta} - \frac{1}{3}}{\frac{2}{3}e^{j2\theta}} = -\frac{3}{2} + 2e^{-j\theta} - \frac{1}{2}e^{-j2\theta}, \qquad 0 \leq \theta \leq 2\pi \tag{13-70}$$

The locus of $\sigma(\theta)$ described by Eq. (13-70) can be found by computing $\sigma(\theta)$ from $\theta = 0$ to $\theta = \pi$, and then applying horizontal symmetry to obtain the unity-root locus Γ_σ

shown in Fig. 13-7(b). The region of absolute stability is clearly the *exterior* of Γ_σ, since we already know that Gear's algorithm is stable as $\sigma \longrightarrow \infty$.

*k*TH-ORDER GEAR'S ALGORITHM

$$x_{n+1} = a_0 x_n + a_1 x_{n-1} + a_2 x_{n-2} + \ldots + a_{k-1} x_{n-k+1} + h[b_{-1} f(x_{n+1}, t_{n+1})]$$

Equations (13-13) and (13-7) reduce to ($p = k - 1$)

$$\sigma(\theta) = \frac{-e^{jk\theta} + a_0 e^{j(k-1)\theta} + a_1 e^{j(k-2)\theta} + \ldots + a_{k-2} e^{j\theta} + a_{k-1}}{b_{-1} e^{jk\theta}}, \qquad 0 \leq \theta \leq 2\pi \tag{13-71}$$

$$\mathcal{P}(z) = (z^k - a_0 z^{k-1} - a_1 z^{k-2} - \ldots - a_{k-1}) + (b_{-1} z^k)\sigma = 0 \tag{13-72}$$

The regions of absolute stability for Gear's algorithms of orders 3, 4, 5, and 6 are shown in the shaded regions in Fig. 13-7(c), (d), (e), and (f), respectively. Due to limits of space, the unity-root locus Γ_σ in Fig. 13-7(e) and (f) are shown open. Needless to say, they actually form a closed curve if traced from $\theta = 0$ to $\theta = 2\pi$.

Examination of the regions of absolute stability in Fig. 13-7 shows that they all satisfy *stiff property 1* with various minimum values of δ. In particular, we find $\delta = 0$ for the first- and second-order algorithms, $\delta = 0.1$ for the third-order algorithm, $\delta = 0.7$ for the fourth-order algorithm, $\delta = 2.4$ for the fifth-order algorithm, and $\delta = 6.1$ for the sixth-order algorithm. Moreover, since the algorithm needs to be only relatively stable in region I, by choosing appropriate values for θ and μ, it is clear that Gear's algorithms also satisfy stiff properties 2 and 3. Hence, we conclude that *Gear's algorithms of orders 1 to 6 are stiffly stable.*

*13-5 CORRECTOR ITERATION FOR GEAR'S ALGORITHM

Since the *k*th-order Gear's algorithm is an *implicit* multistep algorithm ($b_{-1} \neq 0$), it is necessary to solve an implicit equation in each time step; i.e.,

$$x_{n+1} = \sum_{i=0}^{k-1} a_i x_{n-i} + h b_{-1} f(x_{n+1}, t_{n+1}) \tag{13-73}$$

We could apply the fixed-point iteration formula

$$x_{n+1}^{(j+1)} = \sum_{i=0}^{k-1} a_i x_{n-i} + h b_{-1} f(x_{n+1}^{(j)}, t_{n+1}) \tag{13-74}$$

as was done in Section 11-5-2 to obtain

$$x_{n+1} = \lim_{j \to \infty} x_{n+1}^{(j)} \tag{13-75}$$

From Theorem 11-6 we know Eq. (13-75) *converges* if the step size h satisfies the inequality

$$h < \frac{1}{|b_{-1}| L_{n+1}} \tag{13-76}$$

where L_{n+1} is a Lipschitz constant for the function $f(x, t)$ at $t = t_{n+1}$. Since the very purpose for resorting to *stiffly stable* algorithms is to have the freedom to choose *large* step sizes after the "fast" transient component has settled down, the restriction imposed by Eq. (13-76) will clearly defeat our purpose. Consequently, the simple fixed-point iteration technique is not an appropriate *corrector* for implementing

Gear's algorithm. In this section, we shall choose the more general Newton–Raphson technique as the corrector. For the sake of generality, consider a system of state equations

$$\dot{\boldsymbol{x}} = \boldsymbol{f}(\boldsymbol{x}, t) \tag{13-77}$$

The kth-order Gear's algorithm for solving Eq. (13-77) can be recast into the form

$$\boxed{\boldsymbol{x}_{n+1} - hb_{-1}\boldsymbol{f}(\boldsymbol{x}_{n+1}, t_{n+1}) - \sum_{i=0}^{k-1}[a_i\boldsymbol{x}_{n-i}] = \boldsymbol{0}} \tag{13-78}$$

Applying the Newton–Raphson algorithm to Eq. (13-78), we obtain

$$\boxed{\begin{aligned}\boldsymbol{x}_{n+1}^{(j+1)} = \boldsymbol{x}_{n+1}^{(j)} &- [\boldsymbol{1} - hb_{-1}\boldsymbol{J}_f(\boldsymbol{x}_{n+1}^{(j)}, t_{n+1})]^{-1} \\ &\cdot\left\{\boldsymbol{x}_{n+1}^{(j)} - hb_{-1}\boldsymbol{f}(\boldsymbol{x}_{n+1}^{(j)}, t_{n+1}) - \sum_{i=0}^{k-1}[a_i\boldsymbol{x}_{n-i}]\right\}\end{aligned}} \tag{13-79}$$

where $\boldsymbol{1}$ is an $n \times n$ unit matrix and where

$$\boldsymbol{J}_f(\boldsymbol{x}_{n+1}^{(j)}, t_{n+1}) \triangleq \left.\frac{\partial \boldsymbol{f}(\boldsymbol{x}, t)}{\partial \boldsymbol{x}}\right|_{\boldsymbol{x}=\boldsymbol{x}_{n+1}^{(j)}, t=t_{n+1}} \tag{13-80}$$

is the Jacobian matrix of $\boldsymbol{f}(\boldsymbol{x}, t)$ evaluated at $\boldsymbol{x} = \boldsymbol{x}_{n+1}^{(j)}$ and $t = t_{n+1}$. Equation (13-79) is the Newton–Raphson *corrector* for implementing Gear's algorithms. Although one could apply Eq. (13-79) directly to solve for $\boldsymbol{x}_{n+1}^{(j+1)}$, it is possible to simplify the arithmetic considerably by first rewriting Eq. (13-79) into the equivalent form

$$\begin{aligned}[\boldsymbol{1} - hb_{-1}\boldsymbol{J}_f(\boldsymbol{x}_{n+1}^{(j)}, t_{n+1})][\boldsymbol{x}_{n+1}^{(j+1)} - \boldsymbol{x}_{n+1}^{(j)}]& \\ = -\boldsymbol{x}_{n+1}^{(j)} + hb_{-1}\boldsymbol{f}(\boldsymbol{x}_{n+1}^{(j)}, t_{n+1}) &+ \sum_{i=0}^{k-1}[a_i\boldsymbol{x}_{n-i}]\end{aligned} \tag{13-81}$$

Replacing the superscript $(j + 1)$ in Eq. (13-81) by (j), we obtain

$$\begin{aligned}[\boldsymbol{1} - hb_{-1}\boldsymbol{J}_f(\boldsymbol{x}_{n+1}^{(j-1)}, t_{n+1})][\boldsymbol{x}_{n+1}^{(j)} - \boldsymbol{x}_{n+1}^{(j-1)}]& \\ = -\boldsymbol{x}_{n+1}^{(j-1)} + hb_{-1}\boldsymbol{f}(\boldsymbol{x}_{n+1}^{(j-1)}, t_{n+1}) &+ \sum_{i=0}^{k-1}[a_i\boldsymbol{x}_{n-i}]\end{aligned} \tag{13-82}$$

Subtracting Eq. (13-82) from Eq. (13-81), we obtain, upon simplification, the expression

$$\begin{aligned}[\boldsymbol{1} - hb_{-1}\boldsymbol{J}_f(\boldsymbol{x}_{n+1}^{(j)}, t_{n+1})][\boldsymbol{x}_{n+1}^{(j+1)} - \boldsymbol{x}_{n+1}^{(j)}]& \\ = hb_{-1}\{\boldsymbol{f}(\boldsymbol{x}_{n+1}^{(j)}, t_{n+1}) - [\boldsymbol{f}(\boldsymbol{x}_{n+1}^{(j-1)}, t_{n+1})& \\ + \boldsymbol{J}_f(\boldsymbol{x}_{n+1}^{(j-1)}, t_{n+1})(\boldsymbol{x}_{n+1}^{(j)} - \boldsymbol{x}_{n+1}^{(j-1)})]\}&\end{aligned} \tag{13-83}$$

If we define

$$\boldsymbol{d}_{n+1}^{(j)} \triangleq h\boldsymbol{f}(\boldsymbol{x}_{n+1}^{(j-1)}, t_{n+1}) + h\boldsymbol{J}_f(\boldsymbol{x}_{n+1}^{(j-1)}, t_{n+1})[\boldsymbol{x}_{n+1}^{(j)} - \boldsymbol{x}_{n+1}^{(j-1)}] \tag{13-84}$$

and recall that

$$\dot{\boldsymbol{x}}_{n+1}^{(j)} = \boldsymbol{f}(\boldsymbol{x}_{n+1}^{(j)}, t_{n+1}) \tag{13-85}$$

then Eq. (13-83) assumes the following simplified form:

$$\boxed{\boldsymbol{x}_{n+1}^{(j+1)} = \boldsymbol{x}_{n+1}^{(j)} + b_{-1}[\boldsymbol{1} - hb_{-1}\boldsymbol{J}_f(\boldsymbol{x}_{n+1}^{(j)}, t_{n+1})]^{-1}[h\dot{\boldsymbol{x}}_{n+1}^{(j)} - \boldsymbol{d}_{n+1}^{(j)}]} \tag{13-86}$$

If we replace the superscript (j) in Eq. (13-84) by $(j+1)$ and substitute Eq. (13-86) into Eq. (13-84), we obtain

$$\begin{aligned} \boldsymbol{d}_{n+1}^{(j+1)} &= h\dot{\boldsymbol{x}}_{n+1}^{(j)} + h\boldsymbol{J}_f(\boldsymbol{x}_{n+1}^{(j)}, t_{n+1})[\boldsymbol{x}_{n+1}^{(j+1)} - \boldsymbol{x}_{n+1}^{(j)}] \\ &= h\dot{\boldsymbol{x}}_{n+1}^{(j)} + h\boldsymbol{J}_f(\boldsymbol{x}_{n+1}^{(j)}, t_{n+1}) \\ &\quad \cdot\{b_{-1}[\mathbf{1} - hb_{-1}\boldsymbol{J}_f(\boldsymbol{x}_{n+1}^{(j)}, t_{n+1})]^{-1}[h\dot{\boldsymbol{x}}_{n+1}^{(j)} - \boldsymbol{d}_{n+1}^{(j)}]\} \\ &= h\dot{\boldsymbol{x}}_{n+1}^{(j)} + [\mathbf{1} - hb_{-1}\boldsymbol{J}_f(\boldsymbol{x}_{n+1}^{(j)}, t_{n+1})]^{-1}[h\dot{\boldsymbol{x}}_{n+1}^{(j)} - \boldsymbol{d}_{n+1}^{(j)}] \\ &\quad - [\mathbf{1} - hb_{-1}\boldsymbol{J}_f(\boldsymbol{x}_{n+1}^{(j)}, t_{n+1})] \\ &\quad \cdot[\mathbf{1} - hb_{-1}\boldsymbol{J}_f(\boldsymbol{x}_{n+1}^{(j)}, t_{n+1})]^{-1}[h\dot{\boldsymbol{x}}_{n+1}^{(j)} - \boldsymbol{d}_{n+1}^{(j)}] \end{aligned} \tag{13-87}$$

Simplifying, we obtain the following recursive relation for $\boldsymbol{d}_{n+1}^{(j+1)}$:

$$\boxed{\boldsymbol{d}_{n+1}^{(j+1)} = \boldsymbol{d}_{n+1}^{(j)} + [\mathbf{1} - hb_{-1}\boldsymbol{J}_f(\boldsymbol{x}_{n+1}^{(j)}, t_{n+1})]^{-1}[h\dot{\boldsymbol{x}}_{n+1}^{(j)} - \boldsymbol{d}_{n+1}^{(j)}]} \tag{13-88}$$

If we can find an explicit expression for $\boldsymbol{d}_{n+1}^{(0)}$, we can calculate $\boldsymbol{d}_{n+1}^{(j+1)}$ recursively from Eq. (13-88). To do this, we let $j = 0$ in Eq. (13-81) to obtain

$$\begin{aligned} &[\mathbf{1} - hb_{-1}\boldsymbol{J}_f(\boldsymbol{x}_{n+1}^{(0)}, t_{n+1})][\boldsymbol{x}_{n+1}^{(1)} - \boldsymbol{x}_{n+1}^{(0)}] \\ &\quad = -\boldsymbol{x}_{n+1}^{(0)} + hb_{-1}\boldsymbol{f}(\boldsymbol{x}_{n+1}^{(0)}, t_{n+1}) + \sum_{i=0}^{k-1}[a_i\boldsymbol{x}_{n-i}] \\ &\quad = hb_{-1}\dot{\boldsymbol{x}}_{n+1}^{(0)} - b_{-1}\left\{\frac{\boldsymbol{x}_{n+1}^{(0)}}{b_{-1}} - \frac{1}{b_{-1}}\sum_{i=0}^{k-1}[a_i\boldsymbol{x}_{n-i}]\right\} \end{aligned} \tag{13-89}$$

Solving for $\boldsymbol{x}_{n+1}^{(1)}$ from Eq. (13-89), we obtain

$$\begin{aligned} \boldsymbol{x}_{n+1}^{(1)} &= \boldsymbol{x}_{n+1}^{(0)} + b_{-1}[\mathbf{1} - hb_{-1}\boldsymbol{J}_f(\boldsymbol{x}_{n+1}^{(0)}, t_{n+1})]^{-1} \\ &\quad \cdot\left[h\boldsymbol{x}_{n+1}^{(0)} - \left(\frac{\boldsymbol{x}_{n+1}^{(0)}}{b_{-1}} - \frac{1}{b_{-1}}\sum_{i=0}^{k-1}[a_i\boldsymbol{x}_{n-i}]\right)\right] \end{aligned} \tag{13-90}$$

Comparing Eq. (13-90) with Eq. (13-86) [with the superscript (j) replaced by (0)], we identify

$$\boxed{\boldsymbol{d}_{n+1}^{(0)} = \frac{1}{b_{-1}}\boldsymbol{x}_{n+1}^{(0)} - \frac{1}{b_{-1}}\sum_{i=0}^{k-1}[a_i\boldsymbol{x}_{n-i}]} \tag{13-91}$$

Observe that since $\boldsymbol{x}_{n+1}^{(0)}$, the predicted (initial guess) value of $\boldsymbol{x}_{n+1}$, is generally a *linear sum* of $\boldsymbol{x}_n, \boldsymbol{x}_{n-1}, \ldots$, etc., so is $\boldsymbol{d}_{n+1}^{(0)}$. Observe that if the initial guess $\boldsymbol{x}_{n+1}^{(0)}$ turns out to be a correct solution to the corrector equation (13-86), i.e., $\boldsymbol{x}_{n+1}^{(0)} = \boldsymbol{x}_{n+1}$, then Eq. (13-91) can be written as

$$\boldsymbol{d}_{n+1}^{(0)} = \frac{1}{b_{-1}}\left\{\boldsymbol{x}_{n+1} - \sum_{i=0}^{k-1}[a_i\boldsymbol{x}_{n-i}]\right\} \tag{13-92}$$

It follows from Eq. (13-73) and $\boldsymbol{f}(\boldsymbol{x}_{n+1}^{(0)}, t_{n+1}) = \boldsymbol{f}(\boldsymbol{x}_{n+1}, t_{n+1}) = \dot{\boldsymbol{x}}_{n+1}$ that

$$\boldsymbol{d}_{n+1}^{(0)} = h\dot{\boldsymbol{x}}_{n+1} \tag{13-93}$$

Hence, even if $\boldsymbol{x}_{n+1}^{(0)} \neq \boldsymbol{x}_{n+1}$, we can interpret $\boldsymbol{d}_{n+1}^{(0)}$ as the *predicted value* for $h\dot{\boldsymbol{x}}_{n+1}$ such that when Eq. (13-88) converges, we have

$$\lim_{j\to\infty} \boldsymbol{d}_{n+1}^{(j)} \longrightarrow h\dot{\boldsymbol{x}}_{n+1} \tag{13-94}$$

Equations (13-86), (13-88), and (13-91) are the formulas for calculating $\boldsymbol{x}_{n+1}$ using Gear's algorithm. One reason for using three formulas rather than just Eq. (13-79) is that these equations can be programmed more efficiently by the automatic order and step size changing algorithm presented in Section 12-8. This algorithm requires that we work with the *Nordsieck vector* $\boldsymbol{z}_n$ defined in Eq. (12-119).[6] In terms of this vector, the *canonical matrix representation* for the usual predictor–corrector algorithm has been derived in Eqs. (11-161)–(11-165): i.e.,

Predictor Canonical Representation

$$\boldsymbol{z}_{n+1}^{(0)} = \boldsymbol{Z}\boldsymbol{z}_n \tag{13-95}$$

Corrector Canonical Representation

$$\boldsymbol{z}_{n+1}^{(j+1)} = \boldsymbol{z}_{n+1}^{(j)} + F_z(\boldsymbol{z}_{n+1}^{(j)})\boldsymbol{c}_z, \qquad j = 0, 1, 2, \ldots, M-1 \tag{13-96}$$

To implement the kth-order Gear's algorithm, we could devise a kth-order "explicit" algorithm which uses only stored from $\mathbf{y}_n$ and then transform $\mathbf{y}_n$ into $\mathbf{z}_n$. But since all entries in the Nordsieck vector $\boldsymbol{z}_n$ pertain to $t = t_{n+1}$, we could simply apply Taylor expansion as in Section 11-7-2 and obtain the following Pascal triangle matrix $\mathbf{Z}$ as the corrector.

$$\underbrace{\begin{bmatrix} x_{n+1}^{(0)} \\ hx_{n+1}'^{(0)} \\ \dfrac{h^2x_{n+1}''^{(0)}}{2} \\ \cdot \\ \cdot \\ \cdot \\ \dfrac{h^kx_{n+1}^{(k)(0)}}{k!} \end{bmatrix}}_{\boldsymbol{z}_{n+1}^{(0)}} = \underbrace{\begin{bmatrix} 1 & 1 & 1 & 1 & 1 & \cdots & 1 \\ 0 & 1 & 2 & 3 & 4 & \cdots & k \\ 0 & 0 & 1 & 3 & 6 & \cdots & \dfrac{k(k-1)}{2} \\ 0 & 0 & 0 & 1 & 4 & \cdots & \dfrac{k(k-1)(k-2)}{3!} \\ \cdot & \cdot & \cdot & \cdot & 1 & \cdots & \dfrac{k(k-1)(k-2)(k-3)}{4!} \\ \cdot & \cdot & \cdot & \cdot & & \cdots & \\ \cdot & \cdot & \cdot & \cdot & & \cdots & \\ 0 & 0 & 0 & 0 & 0 & 000 & 1 \end{bmatrix}}_{\boldsymbol{Z}} \underbrace{\begin{bmatrix} x_n \\ hx_n' \\ \dfrac{h^2x_n''}{2} \\ \cdot \\ \cdot \\ \cdot \\ \dfrac{h^kx_n^{(k)}}{k!} \end{bmatrix}}_{\boldsymbol{z}_n} \tag{13-97}$$

[6]Recall from Eq. (11-158) that $\boldsymbol{z}_n \triangleq \boldsymbol{T}\boldsymbol{y}_n$, where $\boldsymbol{y}_n$ is a $(2p+2) \times 1$ vector associated with the *predictor-corrector canonical representation* given in Eqs. (11-145)–(11-146) for a *fixed-point iteration corrector*. These equations remain valid for the *Newton-Raphson corrector*, however, if the following changes are made:

1. Replace α_i and β_i in Eqs. (11-140)–(11-141) by a_i and 0, respectively. Note also that $\beta_i^* = 0$ for $i > 0$.
2. Replace the $(p+2)$th component $\sum_{i=0}^{p} [\gamma_i x_{n-i} + \delta_i hx_{n-i}']$ of $\boldsymbol{y}_{n+1}^{(0)}$ in Eq. (11-143) by $d_{n+1}^{(0)}$.
3. Replace the $(p+2)$th component $hf(x_{n+1}^{(j-1)}, t_{n+1})$ of $\boldsymbol{y}_{n+1}^{(j)}$ in Eq. (11-144) by $d_{n+1}^{(j)}$.
4. Set the last "p" components of the vectors defined in Eqs. (11-143)–(11-144) to zero, where $p = k - 1$ for Gear's algorithm.
5. Replace Eq. (11-149) by
$$F_y(\boldsymbol{y}_{n+1}^{(j)}) \triangleq [1 - hb_{-1}J_f(x_{n+1}^{(j)}, t_{n+1})]^{-1}[h\dot{x}_{n+1}^{(j)} - d_{n+1}^{(j)}]$$
for $j = 0, 1, 2, \ldots$, where $J_f \triangleq \partial f(x, t)/\partial x$ and $\dot{x}_{n+1}^{(j)} \triangleq f(x_{n+1}^{(j)}, t_{n+1})$.

Notice that Eqs. (13-95)–(13-97) apply to *only one component* of the state vector $\boldsymbol{x}$. Hence, in programming Gear's algorithm for a system of "N" state equations, each component equation is treated as an independent "scalar" equation and a total of "N" Nordsieck vectors $(z_n)^1, (z_n)^2, \ldots, (z_n)^k, \ldots, (z_n)^N$ must be defined, *mutatis mutandis*, via Eqs. (13-96)–(13-97). To specify the *scalar corrector function* $F_z(z_{n+1}^{(j)})$ for each Nordsieck vector $(z_{n+1})^k$, we observe from footnote 6 that it is simply equal to the kth component of the *vector*

$$\boxed{\boldsymbol{F}_z(z_{n+1}^{(j)}) \triangleq [\mathbf{1} - hb_{-1}\boldsymbol{J}_f(\boldsymbol{x}_{n+1}^{(j)}, t_{n+1})]^{-1}[h\dot{\boldsymbol{x}}_{n+1}^{(j)} - \boldsymbol{d}_{n+1}^{(j)}]} \tag{13-98}$$

where $\dot{\boldsymbol{x}}_{n+1}^{(j)}$ is defined by Eq. (13-85). The vector $\boldsymbol{c}_z$ in Eq. (13-96) can be derived by the same procedure given in Section 11-7-2 for each order k (see Problems 13-24, 13-25, and 13-26), namely:

$$k = 2: \quad \boldsymbol{c}_z = \begin{bmatrix} \frac{2}{3} \\ \frac{3}{3} \\ \frac{1}{3} \end{bmatrix} \qquad k = 3: \quad \boldsymbol{c}_z = \begin{bmatrix} \frac{6}{11} \\ \frac{11}{11} \\ \frac{6}{11} \\ \frac{1}{11} \end{bmatrix} \qquad k = 4: \quad \boldsymbol{c}_z = \begin{bmatrix} \frac{24}{50} \\ \frac{50}{50} \\ \frac{35}{50} \\ \frac{10}{50} \\ \frac{1}{50} \end{bmatrix}$$

$$k = 5: \quad \boldsymbol{c}_z = \begin{bmatrix} \frac{120}{274} \\ \frac{274}{274} \\ \frac{225}{274} \\ \frac{85}{274} \\ \frac{15}{274} \\ \frac{1}{274} \end{bmatrix} \qquad k = 6: \quad \boldsymbol{c}_z = \begin{bmatrix} \frac{720}{1764} \\ \frac{1764}{1764} \\ \frac{1624}{1764} \\ \frac{735}{1764} \\ \frac{175}{1764} \\ \frac{21}{1764} \\ \frac{1}{1764} \end{bmatrix}$$

Observe that Eq. (13-98) is not yet in a form suitable for the Nordsieck vector implementation as defined by Eq. (13-96) because the expression for $\boldsymbol{d}_{n+1}^{(0)}$ in Eq. (13-91) requires the past values x_{n-i} for $0 \leq i \leq k-1$. Our final task in this section is to derive the following *equivalent* expression for $\boldsymbol{d}_{n+1}^{(j)}$ whose value can be obtained directly from the *second component* of the Nordsieck vectors $(z_{n+1}^{(j)})^1, (z_{n+1}^{(j)})^2, \ldots, (z_{n+1}^{(j)}),^N$ as defined earlier in Eq. (11-211):

$$\boxed{\boldsymbol{d}_{n+1}^{(j)} = h\boldsymbol{x}_{n+1}^{\prime(j)}, \quad j = 0, 1, 2, \ldots} \tag{13-99}$$

Let us first derive Eq. (13-99) when $j = 0$. Observe that Eq. (13-91) can be recast as follows:

$$\boldsymbol{d}_{n+1}^{(0)} = \frac{1}{\beta_{-1}}\{\boldsymbol{x}_{n+1}^{(0)} - \sum_{i=0}^{p}[\alpha_i \boldsymbol{x}_{n-i} + \beta_i h\boldsymbol{x}_{n-i}']\} \tag{13-100}$$

where we have defined $\alpha_i \triangleq a_i$, $\beta_{-1} \triangleq b_{-1}$, $p \triangleq k-1$, and $\beta_i \triangleq 0$ for $i = 0, 1, 2, \ldots, p$. Now recall the following general form for a *predictor* given earlier in Eq. (11-136):

$$\boldsymbol{x}_{n+1}^{(0)} = \sum_{i=0}^{p} [\alpha_i^* \boldsymbol{x}_{n-i} + \beta_i^* h\boldsymbol{x}'_{n-i}] \tag{13-101}$$

Substituting Eq. (13-101) into Eq. (13-100) and simplifying, we obtain

$$\begin{aligned}\boldsymbol{d}_{n+1}^{(0)} &= \sum_{i=0}^{p} \left\{\left(\frac{\alpha_i^* - \alpha_i}{\beta_{-1}}\right)\boldsymbol{x}_{n-i} + \left(\frac{\beta_i^* - \beta_i}{\beta_{-1}}\right)h\boldsymbol{x}'_{n-i}\right\} \\ &= \sum_{i=0}^{p} \{\gamma_i \boldsymbol{x}_{n-i} + \delta_i h\boldsymbol{x}'_{n-i}\}\end{aligned} \tag{13-102}$$

where γ_i and δ_i are as defined earlier in Eqs. (11-140)–(11-141). Observe that the right-hand sides of Eqs. (13-102) and (11-214) are identical, and hence we can write

$$\boldsymbol{d}_{n+1}^{(0)} = h\boldsymbol{x}'^{(0)}_{n+1} \tag{13-103}$$

Observe also that the preceding equations represent *predicted values* and are therefore independent of the choice of the corrector algorithm. To derive Eq. (13-99) for $j \geq 1$, let us substitute Eq. (13-98) into Eq. (13-88) and obtain

$$\boldsymbol{d}_{n+1}^{(j)} = \boldsymbol{d}_{n+1}^{(j-1)} + \boldsymbol{F}_z(\boldsymbol{z}_{n+1}^{(j-1)}) \tag{13-104}$$

Applying the same procedure used earlier to derive Eq. (11-216), we found that the second component of $\boldsymbol{c}_z = \boldsymbol{T}\boldsymbol{c}_y$ must be equal to the $(p+2)$th component of $\boldsymbol{c}_y$, namely, 1. It follows from this observation and Eq. (13-96) that

$$h\boldsymbol{x}'^{(j)}_{n+1} = h\boldsymbol{x}'^{(j-1)}_{n+1} + \boldsymbol{F}_z(\boldsymbol{z}_{n+1}^{(j-1)})(1) \tag{13-105}$$

Substituting Eq. (13-104) into Eq. (13-105) and letting $j = 1$, we obtain

$$h\boldsymbol{x}'^{(1)}_{n+1} = h\boldsymbol{x}'^{(0)}_{n+1} + \boldsymbol{d}_{n+1}^{(1)} - \boldsymbol{d}_{n+1}^{(0)} \tag{13-106}$$

It follows from Eqs. (13-103) and (13-106) that

$$\boldsymbol{d}_{n+1}^{(1)} = h\boldsymbol{x}'^{(1)}_{n+1} \tag{13-107}$$

Hence, Eq. (13-99) is valid for $j = 1$. Now suppose it is valid for $j = k-1$ [the *induction hypothesis*]:

$$\boldsymbol{d}_{n+1}^{(k-1)} = h\boldsymbol{x}'^{(k-1)}_{n+1} \tag{13-108}$$

Substituting Eq. (13-104) into Eq. (13-105) and letting $j = k$, we obtain

$$h\boldsymbol{x}'^{(k)}_{n+1} = h\boldsymbol{x}'^{(k-1)}_{n+1} + \boldsymbol{d}_{n+1}^{(k)} - \boldsymbol{d}_{n+1}^{(k-1)} \tag{13-109}$$

Substituting Eq. (13-108) into Eq. (13-109), we obtain

$$\boldsymbol{d}_{n+1}^{(k)} = h\boldsymbol{x}'^{(k)}_{n+1} \tag{13-110}$$

It follows from the *principle of mathematical induction* that Eq. (13-99) is valid for all j. This completes the proof. Q.E.D.

We can now summarize the implementation of Gear's algorithm via the Nordsieck vector as follows:[7]

[7]A subroutine for implementing this algorithm is given in reference 1. A slightly modified version adopting the notation of this book is given in reference 6.

1. Substitute the kth stored Nordsieck vector $(z_n)^k$ associated with the kth component of the state vector $\boldsymbol{x}_n$ into Eq. (13-97) to obtain the kth "predicted" Nordsieck vector $(z_{n+1}^{(0)})^k$, $k = 1, 2, \ldots, N$. The multiplication between the Pascal triangle matrix and the Nordsieck vector $(z_n)^k$ can be programmed very efficiently as indicated in Problem 13-27 [3, 4, 5].
2. For each Nordsieck vector $(z_{n+1}^{(j)})^k$, $k = 1, 2, \ldots, N$, extract the component $(d_{n+1}^{(j)})^k$ to form the vector $\boldsymbol{d}_{n+1}^{(j)}$. Compute the "corrector vector function" $\boldsymbol{F}_z(z_{n+1}^{(j)})$ from Eq. (13-98) by Gaussian elimination, or by any efficient linear equation solution algorithm.
3. Corresponding to each component k of the state vector $\boldsymbol{x}_{n+1}$, compute $(z_{n+1}^{(j+1)})^k$ via Eq. (13-96), where the *scalar* corrector function $F_z(z_{n+1}^{(j)})$ is simply extracted from the kth component of the *corrector vector function* from step 2.
4. Repeat steps (2) and (3) until the iteration converges to within the prescribed error bound.[8] Then go to step 1 for the next time step.
5. Apply the algorithm described in Section 12-8-1 for changing the *order* and *step size* automatically.

REFERENCES

1. **Gear, C. W.** *Numerical Initial Value Problems in Ordinary Differential Equations.* Englewood Cliffs, N.J.: Prentice-Hall, Inc., 1971.
2. **Shichman, H.**, "Integration System of a Nonlinear Network Analysis Program." *IEEE Trans. on Circuit Theory*, pp. 378–386, Aug. 1970.
3. **Gear, C. W.** *The Numerical Integration of Stiff Differential Equations.* Urbana, Ill.: University of Illinois, Department of Computer Science, Rept. No. 221, Jan. 1967.
4. **Gear, C. W.** "The Automatic Integration of Stiff Ordinary Differential Equations." *Information Processing 68.* Amsterdam: North-Holland Publishing Company, 1969, pp. 187–193.
5. **Gear, C. W.** "The Automatic Integration of Ordinary Differential Equations." *Numerical Math.*, Vol. 14, No. 3, pp. 176–179, Mar. 1971.
6. **Stockman, J. F.** *A Critical Examination of Gear's Implicit Integration Algorithm.* Berkeley, Cal.: University of California, Department of Electrical Engineering and Computer Sciences, MS Project Report, December, 1974.
7. **Genin, Y.** "A New Approach to the Synthesis of Stiffly Stable Linear Multistep Formulas." *IEEE Trans. on Circuit Theory*, pp. 352–360, July, 1973.

[8] In iterating Eq. (13-96), only the first two components of $z_{n+1}^{(j+1)}$ need be evaluated since $\boldsymbol{F}_z(z_{n+1}^{(j+1)})$ can be calculated using only these two components. The complete corrected Nordsieck vector can be obtained after the mth iteration, where $m = 1, 2, \ldots$ by accumulating the sum $\sum_{j=0}^{m-1} \boldsymbol{F}_z(z_{n+1}^{(j)})$ as indicated in Problem 13-28. Moreover, it can be shown that

$$\sum_{j=0}^{m-1} \boldsymbol{F}_z(z_{n+1}^{(j)}) \simeq \frac{h^{k+1}\boldsymbol{x}_{n+1}^{(k+1)}}{k!\,(c_z)_{k+1}} \quad \text{for } m \text{ sufficiently large}$$

Therefore, the quantity $\sum_{j=0}^{m-1} \boldsymbol{F}_z(z_{n+1}^{(j)})$ can be used for evaluating the *local truncation error* for testing for the best *step size* and *order*, and for obtaining one higher-order scaled derivative if the order is increased. Even more computing time can be saved by observing that the Jacobian matrix in Eq. (13-98) usually does not change much and therefore need not be reevaluated at each iteration.

PROBLEMS

13-1. Determine with the help of a computer the *region of absolute stability* for the kth-order Adams–Bashforth algorithm, where $k = 2, 3, 4, 5$, and 6.

13-2. Determine with the help of a computer the *region of absolute stability* for the kth-order Adams–Moulton algorithm, where $k = 3, 4, 5$, and 6.

13-3. Determine with the help of a computer the *region of absolute stability* for the kth-order Gear's algorithm, where $k = 2, 3, 4, 5$, and 6.

13-4. Derive the kth-order Gear's algorithm, where $k = 4, 5$, and 6.

13-5. Specify the parameters δ, μ, and θ in Fig. 13-6 for the kth-order Gear's algorithm, where $k = 1, 2, 3, 4, 5$, and 6.

13-6. Derive the seventh-order Gear's algorithm and show that it is *not stiffly stable.*

13-7. Formulate the kth-order Gear's algorithm for a system of n ordinary differential equations $\dot{\mathbf{x}} = \mathbf{f}(\mathbf{x}, t)$, where $k = 1, 2, \ldots, 6$.

13-8. Consider the following system of "stiff" equations:

$$\dot{x}_1 = -2x_1 + x_2 + 100$$
$$\dot{x}_2 = 10^4 x_1 - 10^4 x_2 + 50$$
$$x_1(0) = x_2(0) = 0$$

(a) Determine the *maximum step size* $h_{\max}$ for the forward Euler algorithm to remain numerically stable.
(b) Use the *forward Euler algorithm* with a step size $h = \frac{1}{2}h_{\max}$ to solve for $x_1(t)$ and $x_2(t)$ for $t > 0$.
(c) Repeat part (b) using a step size $h = 2h_{\max}$.

13-9. Use the *backward Euler algorithm* to solve the stiff equations given in Problem 13-8. Choose the following step sizes in terms of the maximum step size $h_{\max}$ determined from Problem 13-8(a).
(a) $h = 10h_{\max}$.
(b) $h = 100h_{\max}$.
(c) $h = 1000h_{\max}$.
(d) $h = 10{,}000h_{\max}$.

13-10. Repeat Problem 13-9 using the *trapezoidal algorithm.*

13-11. Repeat Problem 13-9 using *Gear's second-order algorithm.*

13-12. Repeat Problem 13-9 using *Gear's third-order algorithm.*

13-13. Repeat Problem 13-9 using *Gear's fourth-order algorithm.*

13-14. Repeat Problem 13-9 using *Gear's fifth-order algorithm.*

13-15. Repeat Problem 13-9 using *Gear's sixth-order algorithm.*

13-16. The most general two-step algorithm is given by

$$x_{n+1} = a_0 x_n + a_1 x_{n-1} + h[b_{-1} f(x_{n+1}, t_{n+1}) + b_0 f(x_n, t_n) + b_1 f(x_{n-1}, t_{n-1})]$$

(a) Use the exactness constraints to show that the five coefficients a_0, a_1, b_{-1}, b_0, and b_1 which result in an algorithm of *maximum order* are given by $a_0 = 0$, $a_1 = 1$,

$b_{-1} = \frac{1}{3}$, $b_0 = \frac{4}{3}$, and $b_1 = \frac{1}{3}$. The associated algorithm

$$x_{n+1} = x_{n-1} + \frac{h}{3}[f(x_{n+1}, t_{n+1}) + 4f(x_n, t_n) + f(x_{n-1}, t_{n-1})]$$

is usually called *Milne's algorithm.*

(b) Show that Milne's algorithm is unstable unless Re $\lambda = 0$.

13-17. The most general two-step algorithm is given by

$$x_{n+1} = a_0 x_n + a_1 x_{n-1} + h[b_{-1} f(x_{n+1}, t_{n+1}) + b_0 f(x_n, t_n) + b_1 f(x_{n-1}, t_{n-1})]$$

For a third-order algorithm, we can assign *arbitrary* value to any one of the five coefficients a_0, a_1, b_{-1}, b_0, and b_1, and then determine the remaining coefficients using the exactness constraints.

(a) Determine the coefficients a_0, a_1, b_0, and b_1 as a function of b_{-1}.

(b) Determine the range of values of b_{-1} in which this family of third-order two-step algorithms is stable. Identify the points on this range that coincide with the Adams–Moulton and Gear's third-order algorithm.

(c) Show that when $b_{-1} = \frac{1}{3}$, this third-order two-step algorithm reduces to the fourth-order two-step Milne's algorithm defined in Problem 13-16.

13-18. The most general three-step algorithm is given by

$$x_{n+1} = a_0 x_n + a_1 x_{n-1} + a_2 x_{n-2} + h[b_{-1} f(x_{n+1}, t_{n+1}) + b_0 f(x_n, t_n) + b_1 f(x_{n-1}, t_{n-1}) + b_2 f(x_{n-2}, t_{n-2})]$$

For a fourth-order algorithm, we can assign *arbitrary* values to any two of the seven coefficients a_0, a_1, a_2, b_{-1}, b_0, b_1, and b_2, and then determine the remaining coefficients using the exactness constraints.

(a) Determine the coefficients a_0, a_1, a_2, b_0, and b_1 in terms of b_{-1} and b_2.

(b) Show that when $b_{-1} = \frac{3}{8}$ and $b_2 = \frac{1}{24}$ the resulting algorithm reduces to the fourth-order Adams–Moulton algorithm.

(c) Find the region in the b_2 versus b_{-1} plane in which the above three-step algorithm is stable.

(d) Use the result from part (c) to prove that a stable three-step algorithm cannot exceed order 4, and that there does not exist a stable *explicit* three-step fourth-order algorithm.

13-19. Determine the three-step algorithm of *maximum order* and specify its region of absolute stability.

13-20. Show that there is a $(k - 1)$-dimensional region within which every k-step algorithm of order $k + 1$ is stable. Moreover, if k is odd, show that no points within this region correspond to algorithms of order higher than $k + 1$. If k is even, show that there is a $\frac{1}{2}(k - 2)$-dimensional subregion within which the corresponding algorithm has order $k + 2$ (but none higher).

13-21. Derive the second-order Gear's algorithm with *nonuniform* step sizes $h \triangleq t_{n+1} - t_n$ and $h_n \triangleq t_n - t_{n-1}$.

13-22. Derive the third-order Gear's algorithm with *nonuniform* step sizes $h \triangleq t_{n+1} - t_n$, $h_n \triangleq t_n - t_{n-1}$, and $h_{n-1} \triangleq t_{n-1} - t_{n-2}$.

13-23. Derive the kth-order Gear's algorithm with *nonuniform* step sizes $h \triangleq t_{n+1} - t_n$, $h_k \triangleq t_k - t_{k-1}$, where $k = n, n - 1, \ldots, n - 4$.

13-24. (a) Show that the nonsingular linear transformation matrix $\boldsymbol{T}$ defined in Eq. (11-165) which transforms the vector $\boldsymbol{y}_n$ into the Nordsieck vector $\boldsymbol{z}_n$ for Gear's second-order algorithm is given by

$$\boldsymbol{T} = \begin{bmatrix} 1 & 0 & 0 \\ 0 & 0 & 1 \\ -1 & 1 & 1 \end{bmatrix}$$

Hint: Express $\boldsymbol{y}_n = [x_n x_{n-1} h x'_n]^t$ in terms of

$$\boldsymbol{z}_n = \left[\frac{x_n h x'_n h^2 x''_n}{2}\right]^t$$

and obtain $\boldsymbol{y}_n = \boldsymbol{T}^{-1}\boldsymbol{z}_n$.

(b) Derive the vector $\boldsymbol{c}_z$ associated with Gear's second-order corrector canonical representation.

13-25. Repeat Problem 13-24 for Gear's third-order algorithm with the matrix $\boldsymbol{T}$ given by

$$\boldsymbol{T} = \begin{bmatrix} 1 & 0 & 0 & 0 \\ 0 & 0 & 0 & 1 \\ -\frac{7}{4} & 2 & -\frac{1}{4} & \frac{3}{2} \\ -\frac{3}{4} & 1 & -\frac{1}{4} & \frac{1}{2} \end{bmatrix}$$

13-26. Repeat Problem 13-24 for Gear's fourth-order algorithm with the matrix $\boldsymbol{T}$ given by

$$\boldsymbol{T} = \begin{bmatrix} 1 & 0 & 0 & 0 & 0 \\ 0 & 0 & 0 & 0 & 1 \\ -\frac{85}{36} & 3 & -\frac{3}{4} & \frac{1}{9} & \frac{11}{6} \\ -\frac{5}{3} & \frac{5}{2} & -1 & \frac{1}{6} & 1 \\ -\frac{11}{36} & \frac{1}{2} & -\frac{1}{4} & \frac{1}{18} & \frac{1}{6} \end{bmatrix}$$

13-27. The predictor for Gear's algorithm using the Nordsieck vector representation is given by the Pascal triangle matrix in Eq. (13-97). This equation predicts $z_{n+1}^{(0)}$ for a *scalar* state equation. In the N-dimensional case, each of the "N" Nordsieck vectors $(z_{n+1}^{(0)})^1, (z_{n+1}^{(0)})^2, \ldots, (z_{n+1}^{(0)})^N$ will have to be predicted separately using Eq. (13-97). Show that the following FORTRAN algorithm will effectively predict $z_{n+1}^{(0)}$ for *order K* for each component of the *N*-dimensional state equations:

```
      DO 1    J = 2, K+1
      DO 1    J1 = J, K+1
      J2  = K - J1 + J
      DO 1  I = 1, N
1     Y( J2,I) = Y(J2,I) + Y(J2+1,I)
```

where Y(J,I) for $1 \leq$ J $\leq$ K+1 is the Nordsieck vector $\boldsymbol{z}_n$ for a scalar state equation.

13-28. (a) To compute $\boldsymbol{z}_{n+1}^{(j+1)}$ from Eq. (13-96) efficiently, show that only the first two components of this vector need to be evaluated for each iteration since the remaining components can be corrected via the relation

$$z_{n+1}^{(m)} = z_{n+1}^{(0)} + \sum_{j=0}^{m-1} F_z(z_{n+1}^{(j)})c_z, \quad m = 1, 2, \ldots$$

To derive this relation, show that

$$\frac{h^i x_{n+1}^{(i)^{(m)}}}{i!} = \frac{h^i x_{n+1}^{(i)^{(0)}}}{i!} + \sum_{j=0}^{m-1} F_z(z_{n+1}^{(j)})(c_z)_{i+1}$$

Hint: Observe that the $(i+1)th$ component of each Nordsieck vector $(z_{n+1}^{(j+1)})^k$ satisfies the relation

$$\frac{h^i x_{n+1}^{(i)^{(j+1)}}}{i!} = \frac{h^i x_{n+1}^{(i)^{(j)}}}{i!} + F_z(z_{n+1}^{(j)})(c_z)_{i+1}, \quad j = 0, 1, 2, \ldots$$

(b) Using the approximation

$$\frac{h^k}{k!}[x_{n+1}^{(k)} - x_n^{(k)}] \cong \frac{h^{k+1} x_{n+1}^{(k+1)}}{k!}$$

invoked earlier in Eq. (12-121), and assuming the corrector converges for m sufficiently large, i.e.,

$$\frac{h^k x_{n+1}^{(k)^{(m)}}}{k!} \cong \frac{h^k x_{n+1}^{(k)}}{k!}$$

show that

$$\sum_{j=0}^{m-1} F_z(z_{n+1}^{(j)}) \cong \frac{h^{k+1} x_{n+1}^{(k+1)}}{k!(c_z)_{k+1}} \quad \text{for large } m$$

Hint: Note that the structure of the Pascal triangle matrix implies that the last component of each Nordsieck vector $(z_{n+1}^{(0)})^k$ satisfies the relation

$$\frac{h^k x_{n+1}^{(k)^{(0)}}}{k!} = \frac{h^k x_n^{(k)}}{k!}$$

CHAPTER 14

Algorithms for Generating Symbolic Network Functions

14-1 INTRODUCTION

Up to now we have discussed methods of computer analysis of lumped networks whose branch characteristics are specified by some real or complex *numbers*. In the case of linear networks, it is sometimes very desirable to have branches characterized by *variables* instead of numbers and to obtain symbolic network functions. By a *symbolic network function* we mean an expression for V_{out}/V_{in}, V_{out}/I_{in}, I_{out}/V_{in}, or I_{out}/I_{in} that contains variables. Depending on whether all, some, or none of the network elements are represented by symbols, we have three types of symbolic network functions as illustrated by the following examples:

Type 1. Fully symbolic network functions

$$\frac{V_{out}}{V_{in}} = \frac{s^2LRC}{s^2 2LRC + s(L + R^2C) + R}$$

$$\frac{V_{out}}{V_{in}} = \frac{ZYR}{2ZYR + Z + R^2Y + R}$$

Type 2. Partially symbolic network functions

$$\frac{V_{out}}{V_{in}} = \frac{s^2R}{s^2 2R + s(0.5 \times 10^6 + 150R^2) + 0.75 \times 10^8 R}$$

Type 3. Rational functions of s with numerical coefficients

$$\frac{V_{out}}{V_{in}} = \frac{s^2}{2s^2 + 2 \times 10^4 s + 0.75 \times 10^8}$$

There are many reasons why one may be interested in symbolic network functions [1]. A few of the more important ones are as follows:

INSIGHT. To illustrate the insight that symbolic network functions can provide better than numerical solutions, consider the simple example of a common-collector transistor stage. The voltage gain of such a stage (under the assumption $r_c \longrightarrow \infty$) can be shown to be given by

$$A_V = \frac{V_o}{V_i} = \frac{R_L}{(1-\alpha)r_b + r_e + R_L} \tag{14-1}$$

By inspection of this symbolic network function, it is immediately clear that A_V is positive, less than 1, and very close to 1 (provided that $R_L \gg [(1-\alpha)r_b + r_e]$). Without a symbolic network function, these conclusions could only be reached after the analysis of many numerical cases; and even then some degree of uncertainty still exists.

ERROR CONTROL. In the analysis of networks using digital computers, there are several important sources of numerical errors. Among these are the round-off error and the loss-of-significance error. The former is due to the finite word length of the machine, and the latter occurs during floating-point addition of two numbers of opposite sign but comparable magnitudes. By the proper use of symbolic parameters, the accuracy of the final result of the calculations can be greatly improved. Some simple examples will illustrate the point.

Consider first a current divider consisting of three branches in parallel: $G_1 = 1\ \mho$, $G_2 = 10^{-6}\ \mho$, and $I_s = 1$ A. It is desired to find the current I_1 (through G_1). If a nodal analysis is done and the computer has a word length that gives six significant digits, the answer $I_1 = 1.00000$ will be obtained, because in calculating I_1 from

$$I_1 = \frac{G_1}{G_1 + G_2} I_s$$

the denominator $G_1 + G_2$, whose exact value is 1.000001, will be rounded off to 1.00000. In such a case, better accuracy can be obtained by keeping G_2 as a symbolic parameter, obtaining first

$$I_1 = \frac{1}{1 + G_2} I_s$$

and then computing I_1 by any other more accurate method. For example, $I_1 \cong 1 - G_2 + G_2^2 - \cdots \cong 0.999999$, which is accurate to six significant digits.

As a second example, consider the evaluation of a determinant whose expression is given by

$$\Delta = p + q - p - q$$

Suppose that $p = 5.33333$, $q = 0.0133333$, and all computations retain six significant digits. If Δ is evaluated from the preceding expression by adding numbers in the order indicated, the result will be

$$\begin{aligned}\Delta &= [(5.33333 - 0.0133333) - 5.33333] - 0.0133333 \\ &= -0.0000033\end{aligned}$$

which is obviously not exact. On the other hand, if p (or q) is kept as a symbol, the exact result for Δ will be obtained as follows:

$$\Delta = p + 0.0133333 - p - 0.0133333 = 0$$

SENSITIVITY ANALYSIS. The sensitivity of the system performance due to changes in component characteristics is a very important consideration in the design of systems. In Chapter 15, we shall present a very powerful method, the adjoint network, for solving the sensitivity problem associated with lumped networks.

The symbolic network functions provide another effective method of investigating the sensitivity problems of linear networks (see Section 15-4). The symbolic method has one advantage over the adjoint network method in that, besides giving the exact solution, it very often shows the factors contributing to the sensitivity functions. For example, the voltage gain of a common-collector transistor stage has been shown to be given by Eq. (14-1). Then, one of the sensitivity functions $\partial A_V/\partial \alpha$ may be obtained by differentiation:

$$\frac{\partial A_V}{\partial \alpha} = \frac{R_L r_b}{[(1-\alpha)r_b + r_c + R_L]^2} \tag{14-2}$$

From Eq. (14-2), the roles played by various parameters in $\partial A_V/\partial \alpha$ become obvious. Higher-order sensitivities, such as $\partial^2 A_V/\partial \alpha\, \partial R_L$, may be obtained by repeated differentiation.

PARAMETER ITERATIONS. The sensitivity functions obtained either by the adjoint network method (Chap. 15) or by the symbolic method are applicable only when the changes in system parameters are very small—infinitesimally small, theoretically speaking. When relatively large changes in system parameters are to be considered, other methods have to be used. For example, one may wish to calculate the voltage gain of a certain amplifier when the load resistance R_L takes on successively the values 1000, 2000, 3000, . . . , 10,000 Ω. One brute-force method would be to analyze the amplifier for each value of R_L. Ten analyses of the complete network would then be necessary. But if the gain function is derived with R_L kept as a symbol, it is only necessary to evaluate this gain function for 10 values of R_L, a much simpler task. The same situation exists in the calculation of the frequency response curve. If the gain function $A(s)$ has been obtained with s kept as a variable, it is much simpler to evaluate $|A(j\omega)|$ for different values of ω than to analyze the complete network repeatedly for different values of ω.

Many computer programs with the capability of obtaining type 3 symbolic network functions are now available (see [1, 2] and their references). However, for the applications discussed above, it is *essential* to include variables representing element values, besides the complex frequency variable s. For this reason we shall in this chapter concentrate on computer methods for generating types 1 and 2 symbolic network functions. The problem is of course within the realm of nonnumerical algebra for which a considerable amount of literature exists [3, 4]. Unfortunately, most of the methods, although general and powerful, are unduly complicated for our purpose because they fail to exploit the special properties of linear networks.

In the next three sections we shall present three methods for generating fully (type 1) and partially (type 2) symbolic network functions by a digital computer:

1. Signal-flow-graph method.
2. Tree-enumeration method.
3. Parameter-extraction method.

Other methods which have not been actually implemented and documented are not included in our discussion. The signal-flow-graph method is the most widely used and has resulted in several user-oriented computer programs [5–7]. Therefore, this method will be described in considerable detail. For the remaining two methods, only outlines of the procedures are given. Interested readers may pursue the topic with the aid of the references given.

14-2 SIGNAL-FLOW-GRAPH (SFG) METHOD

The analysis of any linear system has at its foundation the solution of a set of simultaneous linear equations $\boldsymbol{AX} = \boldsymbol{K}$. When the elements of the coefficient matrix $\boldsymbol{A}$ are numbers (real or complex), one may use, among many other methods, the Gaussian elimination method to find $\boldsymbol{X}$, as discussed in Chapter 4. When the elements of $\boldsymbol{A}$ contain symbols, the problem becomes much more difficult. Topological methods become attractive in such cases. By topological methods we mean those techniques which derive the system function from the structure of some graph associated with the system. In this sense, both the signal-flow-graph method of this section and the tree-enumeration method of the next are topological methods.

For the sake of brevity, the term signal-flow graph will sometimes be abbreviated SFG.

14-2-1 Signal-Flow Graph and Mason's Rule

A signal-flow graph is a weighted directed graph representing a system of simultaneous linear equations according to the following three rules:

1. Node weights (node variables) represent variables (known or unknown).
2. Branch weights (branch transmittances) represent coefficients in the relationships among node variables.
3. For every node with some *incoming* branches, there corresponds the equation

   node variable $= \sum$ (incoming branch transmittance $\times$ node variable from which the incoming branch originates)

   where the summation is over all incoming branches (of the node under consideration).

As an example, consider the signal-flow graph of Fig. 14-1. The node variables are (x_0, x_1, x_2, x_3), and the branch transmittances are (a, b, c, d, e, f, g). According to

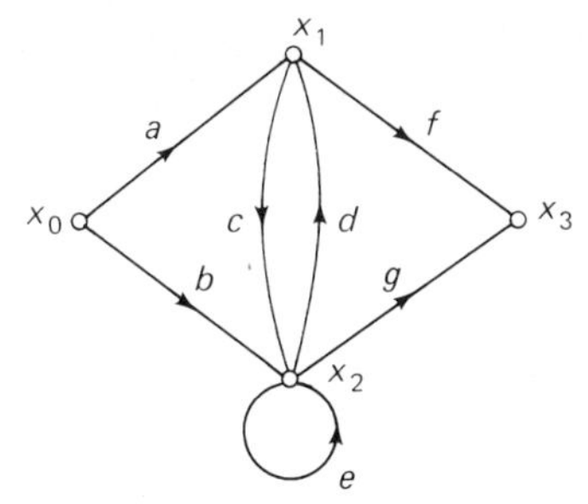

Fig. 14-1. A typical signal-flow graph.

rule 3, the SFG represents the following set of equations:

$$\begin{aligned} x_1 &= ax_0 + \qquad\quad dx_2 \\ x_2 &= bx_0 + cx_1 + ex_2 \\ x_3 &= \qquad\quad fx_1 + gx_2 \end{aligned} \tag{14-3}$$

Thus, from a given SFG, one can write the equations in a *unique* manner according to rule 3.

In a signal-flow graph, a node with only outgoing branches is called a *source node*. A node with some incoming branches is called a *dependent node*. In particular, a dependent node with only incoming branches is called a *sink node*. For example, in Fig. 14-1, x_0 is a source node, (x_1, x_2, x_3) are dependent nodes, and x_3 is a sink node. Dependent-node variables are *treated* as unknown quantities and source-node variables as known quantities in the simultaneous equations. For example, in Eq. (14-3), (x_1, x_2, x_3) are considered as unknowns to be solved for in terms of x_0, which is treated as if it is a known quantity.

Now consider the reverse process of constructing a signal-flow graph to represent a set of simultaneous equations. If the equations are expressed in the form

$$\underset{n\times 1}{\boldsymbol{X}} = \underset{n\times n}{\boldsymbol{A}} \; \underset{n\times 1}{\boldsymbol{X}} + \underset{n\times m}{\boldsymbol{B}} \; \underset{m\times 1}{\boldsymbol{X}_s} \tag{14-4}$$

then, applying rule 3, there is an obvious and *unique* way of constructing the SFG, in which the elements of $\boldsymbol{X}$ appear as dependent nodes, the elements of $\boldsymbol{X}_s$ appear as source nodes, and the elements of $\boldsymbol{A}$ and $\boldsymbol{B}$ appear as branch transmittances as follows: A_{ij} is the transmittance of the branch directed from node X_j to node X_i, and b_{ij} is the transmittance of the branch directed from node X_{sj} to node X_i.

Very often, a set of linear equations is expressed in the form $\boldsymbol{FX} = \boldsymbol{K}$, instead of Eq. (14-4). In such cases, any of several methods may be used to transform $\boldsymbol{FX} = \boldsymbol{K}$ into an equivalent system of the form of Eq. (14-4), from which the corresponding signal-flow graphs may be constructed [8]. Depending on the method used to obtain Eq. (14-4), the resultant signal-flow graphs will be different.

As will be shown in Section 14-3, for active linear networks it is always possible to formulate the equations of motion *directly* in the form of Eq. (14-4). Furthermore, the procedure guarantees $a_{ii} = 0$ for all i. Therefore, in our applications the signal-flow graphs will not contain any "self-loops," such as branch e in Fig. 14-1, even though they are permitted in a general SFG.

Simultaneous linear equations may be solved by the use of Cramer's rule, which requires the evaluation of the determinant and some cofactors of the coefficient matrix. Interestingly enough, the determinant and cofactors can all be obtained from the structure of the signal-flow graph representing the equations. Thus, the SFG provides a graphical or topological way of solving simultaneous linear equations. It is not our intention in this book to derive or prove such a topological rule for signal-flow graphs. Instead, we shall merely state the rule and show its applications.

A few definitions associated with the SFG method are necessary before presenting the topological rule.

Definitions:

Path (forward path). *Imagine each directed branch in the signal-flow graph as a one-way street. A path from node X_i to node X_j is any route leaving node X_i and terminating at node X_j along which no node is encountered more than once.*

Loop (feedback loop). *A path whose initial node and terminal node coincide.*

nth-order loop. *A set of n* ***nontouching*** *loops.*

Path weight. *The product of all branch transmittances in a path.*

Loop weight. *The product of all branch transmittances in a loop.*

For the sake of brevity, we use *path* and *loop* instead of the more precise terms *forward path* and *feedback loop*. The terms "loop" and "loop weight" (also "path" and "path weight") will be used interchangeably. The context should make clear which is intended.

For example, in the SFG of Fig. 14-2, there are two paths from X_1 to X_4, with path weights

$$P_1 = ab, \qquad P_2 = cdfb$$

and three loops with weights

$$L_1 = h, \qquad L_2 = de, \qquad L_3 = fbg$$

There is only one second-order loop whose weight is $L_1 L_2 = hde$. Note that neither $L_1 L_3$ nor $L_2 L_3$ constitutes a second-order loop, because the two loops under consideration touch each other.

From Eq. (14-4), we can solve for $\mathbf{X}$ in terms of $\mathbf{X}_s$ to obtain

$$\mathbf{X} = [\mathbf{1} - A]^{-1}\mathbf{B}\mathbf{X}_s \tag{14-5}$$

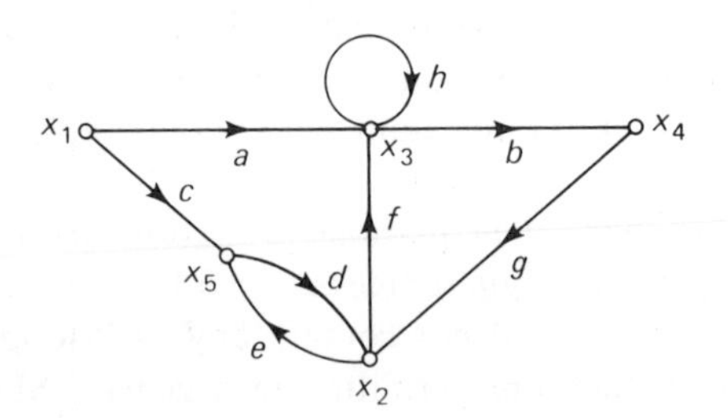

Fig. 14-2. Paths and loops in a signal-flow graph.

provided that the inverse of $[\mathbf{1} - A]$ exists. Thus, any dependent-node variable X_j may be expressed in terms of the source-node variables in the form

$$X_j = T_{j1}X_{s1} + T_{j2}X_{s2} + \cdots + T_{jm}X_{sm} \tag{14-6}$$

Each T_{ji} in Eq. (14-6) is called the *transmission* from the source node X_{si} to the dependent node X_j. The following is a topological rule for evaluating T_{ji} [8]:

$$\begin{aligned} T_{ji} &\triangleq \left(\frac{X_j}{X_{si}}\right)\Bigg|_{X_{si}\neq 0;\ X_{s1}=0,\ X_{s2}=0,\ldots,X_{sm}=0} \\ &= \frac{1}{\Delta}\sum_k P_k\Delta_k \end{aligned} \tag{14-7}$$

where

$$\begin{aligned} \Delta = 1 &- (\text{sum of all loop weights}) \\ &+ (\text{sum of all second-order loop weights}) \\ &- (\text{sum of all third-order loop weights}) + \ldots \end{aligned} \tag{14-8}$$

P_k = weight of the kth path from the source node X_{si} to dependent node X_j

Δ_k = sum of those terms in Δ without any constituent loops touching P_k

and the summation is taken over all paths from X_{si} to X_j.

It can be shown that Δ, given by Eq. (14-8), is actually equal to $|\mathbf{1} - A|$ [referring to Eq. (14-4)]. Therefore, Δ is sometimes called the *graph determinant*.

The preceding rule is commonly called Mason's rule. To illustrate its application, consider some simple examples.

EXAMPLE 14-1. For the signal-flow graph of Fig. 14-1, find the ratio X_3/X_0.

Solution: We enumerate all paths from X_0 to X_3 and all loops.

$$P_1 = af, \qquad P_2 = acg, \qquad P_3 = bg, \qquad P_4 = bdf$$

$$L_1 = e, \qquad L_2 = cd$$

Substituting these values into Eq. (14-7) and paying attention to whether touching occurs among loops or paths, we obtain $\Delta_1 = 1 - e$, $\Delta_2 = \Delta_3 = \Delta_4 = 1$; and Eq. (14-7) becomes

$$\frac{X_3}{X_0} = \frac{af(1-e) + acg + bg + bdf}{1 - e - cd}$$

EXAMPLE 14-2. For the signal-flow graph of Fig. 14-2, find X_4/X_1.

Solution: $P_1 = ab$, $P_2 = cdfb$, $L_1 = h$, $L_2 = ed$, and $L_3 = fbg$. Substituting these values into Eq. (14-7), we have

$$\frac{X_4}{X_0} = \frac{ab(1 - ed) + cdfb}{1 - (h + de + fbg) + (hde)}$$

Mason's formula, as given by Eq. (14-7), requires two sets of rules, one each for the denominator and numerator. From a computer programming point of view, it would be very convenient to evaluate both the denominator and numerator by a single process. This is made possible through the use of the *closed signal-flow graph*, to be described next.

Suppose that we wish to find T_{ji} (X_j due to source-node variable X_{si}). Let us add one more branch to the original signal-flow graph. This additional branch has a symbolic weight $-F$ (distinct from any other branch weights in the SFG), and is directed from node X_j to node X_{si}. Such a derived graph is called a *closed signal-flow graph*. We can evaluate the graph determinant Δ_C for this closed SFG according to the rule given by Eq. (14-8). The terms of Δ_C are next sorted into two groups according to whether each term contains the symbol F or not. The result may be expressed as

$$\Delta_C = D + CF$$

It can be shown fairly easily [9] that $D = \Delta$ and $C = \sum_k P_k \Delta_k$ in Eq. (14-7). In other words,

$$\Delta_C = \Delta + F(\sum_k P_k \Delta_k) \tag{14-9}$$

The result shows that if we evaluate the determinant of the closed signal-flow graph according to Eq. (14-8), and then sort the terms according to Eq. (14-9), we obtain both the denominator and numerator of the desired transmission T_{ji}. Example 14-3 will illustrate the use of a closed signal-flow graph.

EXAMPLE 14-3. Find X_3/X_0 in the signal-flow graph of Fig. 14-1 by the use of Eq. (14-9).

Solution: The closed signal-flow graph for the present problem is shown in Fig. 14-3. We have

$$L_1 = e, \qquad L_2 = cd$$
$$L_3 = -Faf, \qquad L_4 = -Fbg$$
$$L_5 = -Facg, \qquad L_6 = -Fbdf$$

and, from Eq. (14-8),

$$\begin{aligned}\Delta_C &= 1 - (L_1 + L_2 + L_3 + L_4 + L_5 + L_6) + L_1 L_3 \\ &= (1 - e - cd) + F(af + bg + acg + bdf - eaf)\end{aligned}$$

Then

$$T_{30} = \frac{af - eaf + bg + acg + bdf}{1 - e - cd}$$

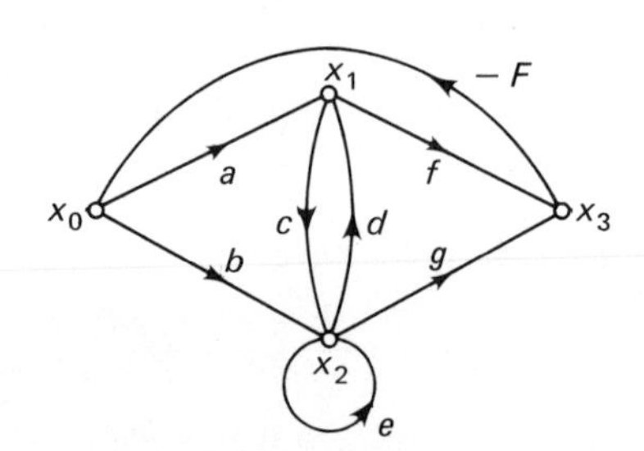

Fig. 14-3. A closed signal-flow graph.

14-2-2 Formulation of the Signal-Flow Graph

In Section 14-2-1 we showed how to find the dependent-node variables in a SFG topologically by the use of Mason's rule. In this section, we shall describe how to systematically formulate a SFG for any linear network consisting of *RLC* elements and independent and controlled sources.

Consider first the simpler case when the network consists of *immittance* (impedance or admittance) elements and independent sources only. We assume that the voltage sources contain no loops, and the current sources contain no cutsets.[1] Then it is possible to select a tree T such that all voltage sources V_E are tree branches, and all current sources I_J are cotree branches. Let the immittance tree branches be characterized by the impedance matrix $\boldsymbol{Z}_T$ and the immittance cotree branches by the admittance matrix $\boldsymbol{Y}_L$. Refer to Fig. 14-4, which depicts the general layout of the SFG (note that ⊙ represents a collection of nodes, and ⇉ represents a collection of branches in the SFG). The SFG is constructed step by step as follows:

1. Apply KVL to express each element of V_L in terms of elements of V_E and V_T.
2. Apply KCL to express each element of $\boldsymbol{I}_T$ in terms of elements of $\boldsymbol{I}_L$ and $\boldsymbol{I}_J$.
3. For immittance tree branches, each voltage is expressed in terms of the current through the branch; i.e., $V_T = \boldsymbol{Z}_T\boldsymbol{I}_T$.
4. For immittance cotree branches, each current is expressed in terms of the voltage across the branch; i.e., $\boldsymbol{I}_L = \boldsymbol{Y}_L V_L$.

A signal-flow graph constructed according to these rules displays the KCL and KVL equations and the branch characteristics in their most primitive form. Consequently, it is called a *primitive signal-flow graph* [8]. For example, the network of Fig. 14-5(a) has a primitive signal-flow graph corresponding to tree $T(ade)$, as shown in Fig. 14-5(b). Note that in a primitive SFG each immittance branch has both its voltage and current represented in the SFG.

The number of nodes in the SFG can very easily be greatly reduced by the use of a *compact signal-flow graph*, which, by definition, is obtained from the primitive

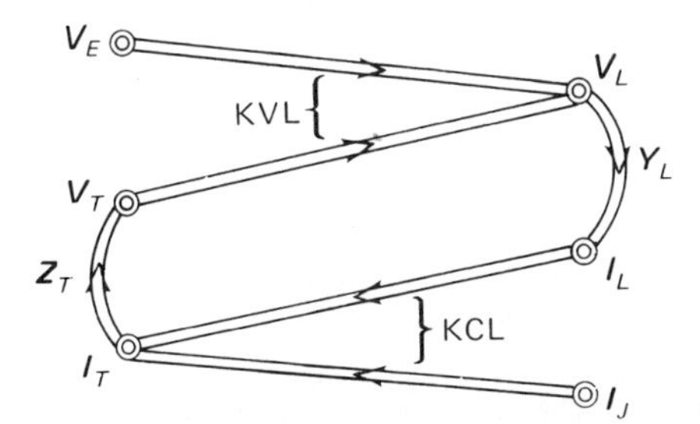

Fig. 14-4. Pattern of a primitive signal-flow graph.

[1]See Section 6-3 for a justification of this assumption.

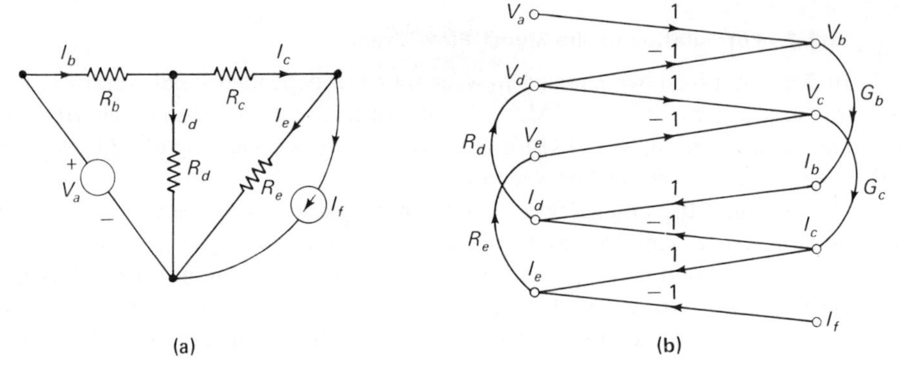

Fig. 14-5. Construction of a primitive signal-flow graph for a network without controlled sources.

signal-flow graph by eliminating the node variables I_T and V_L. Given any network without controlled sources, the compact SFG is formulated after selecting a tree T, as discussed at the beginning of this section, according to the following rules:

1. For each immittance cotree branch Y_k, express its current in terms of tree branch voltages by the use of $I_k = Y_k V_k$ and the KVL equation.
2. For each immittance tree branch Z_j, express its voltage in terms of cotree branch currents by the use of $V_j = Z_j I_j$ and the KCL equation.

The general layout of a compact SFG is depicted in Fig. 14-6(a). For example, for the network of Fig. 14-5(a), by applying the preceding rules we have

$$
\begin{aligned}
I_b &= G_b V_b, & V_b &= V_a - V_d & \therefore\ I_b &= G_b V_a - G_b V_d \\
I_c &= G_c V_c, & V_c &= V_d - V_e & \therefore\ I_c &= G_c V_d - G_c V_e \\
V_d &= R_d I_d, & I_d &= I_b - I_c & \therefore\ V_d &= R_d I_b - R_d I_c \\
V_e &= R_e I_e, & I_e &= I_c - I_f & \therefore\ V_e &= R_e I_c - R_e I_f
\end{aligned}
\tag{14-10}
$$

The last column of equations in Eq. (14-10) is used to construct the compact SFG shown in Fig. 14-6(b), which is considerably simpler in appearance than Fig. 14-5(b).

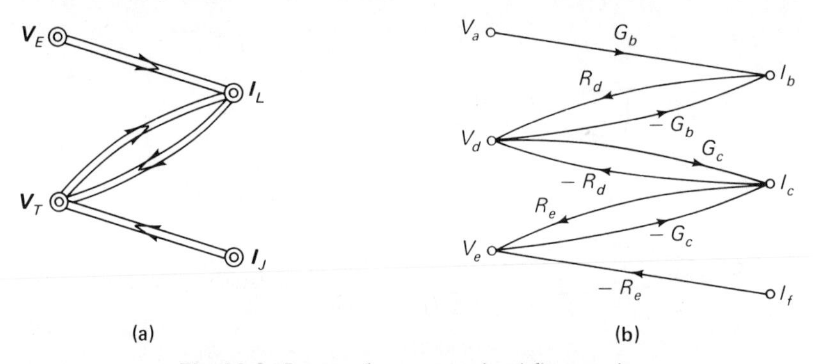

Fig. 14-6. Pattern of a compact signal-flow graph.

Because of its simplicity (as compared to the primitive SFG), the compact SFG will be used from now on.

Next, consider the general case where all four types of controlled sources are permitted in the network. We shall make the following assumptions:[2]

1. Independent and controlled voltage sources form no loops.
2. Independent and controlled current sources form no cutsets.

The following is a step-by-step procedure for formulating the compact SFG for linear networks containing controlled sources:

Step 1. *Temporarily* replace all controlled sources by independent sources of the same type. The resultant network has no controlled sources.

Step 2. Formulate the compact SFG for the network (without controlled sources) obtained in step 1 by the method described previously.

Step 3. Desired outputs and all controlling variables, if not tree branch voltages and cotree currents, should be expressed in terms of the latter quantities and represented in the SFG.

Step 4. Reinstate the constraints of all controlled sources.

An example will now be worked out completely to illustrate the procedure. Consider the feedback amplifier circuit shown in Fig. 14-7(a). It is desired to formulate a compact SFG suitable for the evaluation of the voltage gain function V_o/V_i.

We first replace the controlled current source $g_m V_g$ by an independent current source I_x. A tree is chosen to consist of (V_i, R_1, R_2, R_4). The result is shown in Fig. 14-7(b).

For the cotree branch R_3, the current I_{R3} is expressed in terms of tree branch voltages, resulting in Fig. 14-7(c).

For tree branches R_1, R_2, R_4, the voltages are expressed in terms of cotree branch currents. The result is shown in Fig. 14-7(d).

The desired output voltage V_o and the controlling voltage V_g are expressed in terms of tree branch voltages. The result is shown in Fig. 14-7(e).

Finally, in Fig. 14-7(f), the constraint $I_x = g_m V_g$ is reinstated. A branch with weight $-F$ is added from V_o to V_i to form a closed SFG, as discussed in Section 14-2-1. The method of Section 14-2-1 may now be applied to this compact SFG to find the gain function V_o/V_i.

Even though there are other methods for formulating the SFG that do not rely on the concept of a tree and which may have fewer nodes and branches in the SFG (at the expense of more complicated branch transmittances), they do not lend themselves easily to digital computer programming. On the other hand, the primitive and compact SFG described here have been used successfully in several computer programs.

[2]A network violating these two conditions may still be solvable (see Example 6-3). We shall not attempt to include such extremely rare cases.

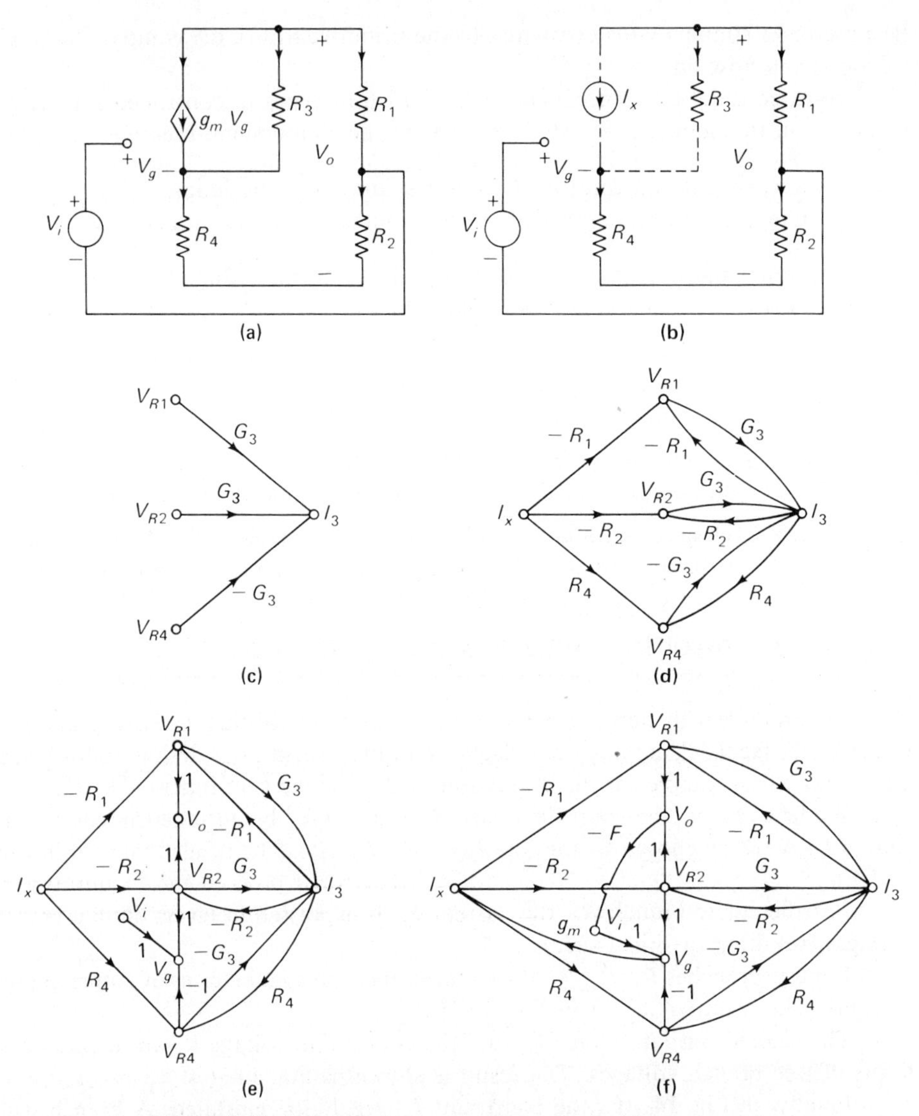

Fig. 14-7. Construction of a compact signal-flow graph for a network with controlled sources.

Note that the compact (or primitive) SFG constructed according to our procedure never contains self-loops or parallel branches (branches with the same initial and terminal nodes).

14-2-3 Enumeration of Paths and Loops

Evaluation of a network function by Eq. (14-7) requires the listing of all paths (from a source node to a dependent node) and all orders of loops. It is important to have an efficient, programmable algorithm for achieving this. We shall first present the basic idea in a very intuitive manner and then describe a path-finding algorithm.

The enumeration of higher-order loops, which is the most crucial part of the SFG method, will be discussed last.

Imagine the directed branches of a SFG to be one-way roads and the nodes to be junctions. At each junction, one may proceed along one of the several roads to reach the next junction. Such choices can be completely described by a *routing table*, such as Table 14-1, for the SFG of Fig. 14-8. The nodes in the second column have been listed in descending order, which is done for convenience rather than necessity.

TABLE 14-1. Routing Table for Fig. 14-8

FROM NODE	TO NODES		
1	5	3	2
2	5		
3	6	2	
4	6	3	
5	6	4	

Suppose that it is desired to find all paths from some initial node I to some destination node L. This may be done systematically as follows:

Starting from the initial node of the path and referring to the routing table (or SFG), proceed to the highest-numbered adjacent node. Everytime a node is reached, the next action to be taken is determined by the following rules.

Rule 1. If the node is the destination node, a path has been found. Record the path, and apply rule 4.

Rule 2. If the node is not the destination node and has not been encountered twice on the present route from the initial node, then proceed on an untraversed branch to the highest-numbered adjacent node. If no such untraversed branch exists, apply rule 4.

Rule 3. If the node is not the destination node but has been encountered for the second time on the present route from the initial node, apply rule 4.

Rule 4. Back up one node on the present route and then proceed on an untraversed branch to the highest-numbered adjacent node. If no such untraversed

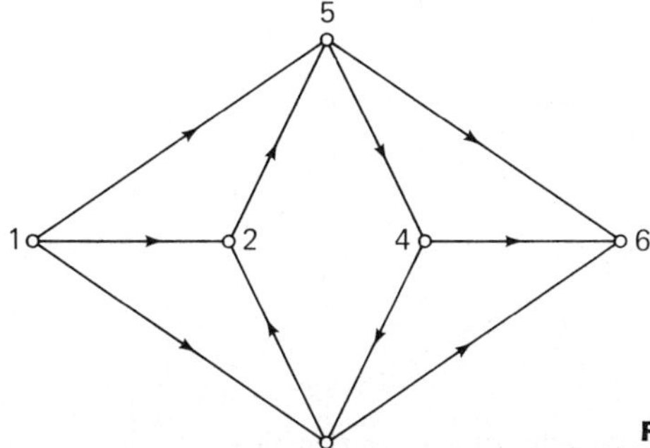

Fig. 14-8. Procedure for finding all paths between two nodes.

branch exists, repeat the "back-up and take next choice" process. If, on backing-up, the initial node I is reached, and all outgoing branches from I have been traversed, then all paths have been found.

We shall now apply the procedure to find all paths from node 1 to node 6 in the SFG of Fig. 14-8. The complete process may be described with the aid of Fig. 14-9.

From node 1, we first proceed to node 5 and then proceed to node 6, and obtain the first path, P_1. Applying rule 4, we back up to node 5 and proceed to node 4. At node 4, rule 2 applies, and we proceed to node 6 to obtain the second path, P_2. Applying rule 4, we back up to node 4 and proceed to node 3. At node 3, rule 2 applies, and we proceed to node 6 to obtain the third path, P_3. Applying rule 4, we back up to node 3 and proceed to node 2. At node 2, rule 2 applies, and we proceed to node 5. At node 5, the condition of rule 3 prevails; therefore, we go to rule 4. We back up from node 5 to node 2, and further to node 3, node 4, node 5, and node 1, and then proceed to node 3.

Continuing the process, a total of nine paths are generated in the following order (see Fig. 14-9):

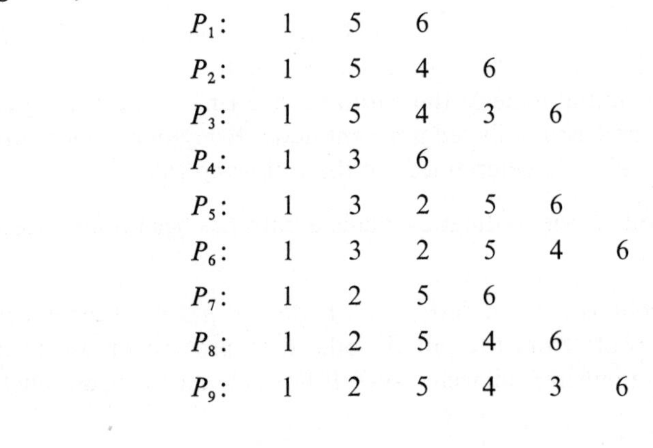

P_1:	1	5	6			
P_2:	1	5	4	6		
P_3:	1	5	4	3	6	
P_4:	1	3	6			
P_5:	1	3	2	5	6	
P_6:	1	3	2	5	4	6
P_7:	1	2	5	6		
P_8:	1	2	5	4	6	
P_9:	1	2	5	4	3	6

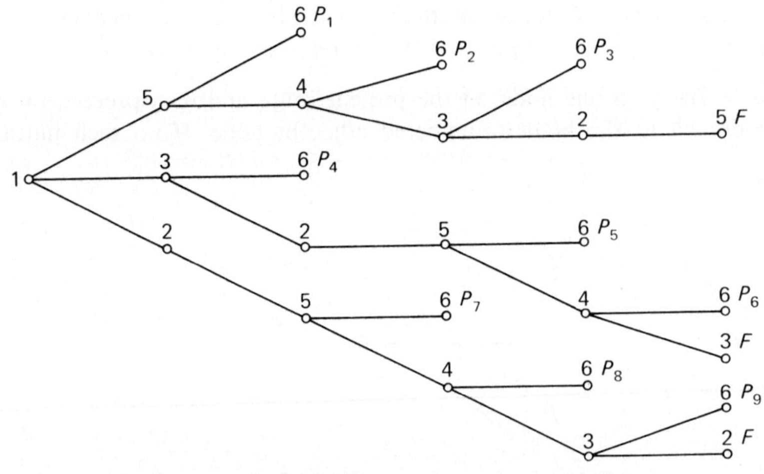

Fig. 14-9. Tree graph depicting the path-finding algorithm.

This example is a brief account of a systematic procedure for listing all paths in a SFG. A precise formulation of the corresponding algorithm is given in Appendix 14A. The algorithm may be readily translated into a digital computer program [6].

14-2-4 Enumeration of First-Order and *n*th-Order Loops

As discussed in Section 14-2-1, the use of a closed signal-flow graph reduces the evaluation of Mason's formula (14-7) to that of enumerating all orders of loops in Eq. (14-9), the total number of which usually places a limit on the size of the network that can be handled by a digital computer.[3] In this section, methods for enumerating loops will be described.

Let the nodes of the SFG be labeled $1, 2, \ldots, N$. All first-order loops that contain node J ($J = 1$ initially) can be found by conceptually splitting node J into two nodes, one containing all incoming branches and the other containing all outgoing branches, and then enumerating all paths between these two nodes, say by the algorithm of Section 14-2-3. All branches connected to node J (including both incoming to and outgoing from node J) are then removed, and the process is repeated for node $J + 1$. Clearly, this procedure will generate all first-order loops without duplication. As an example, let us apply the procedure to the SFG of Fig. 14-10. Then all first-order loops will be generated as given in Table 14-2.

Once the total of L first-order loops has been found, a possible way to find the nth-order loops is to form ${}_LC_n$ combinations of first-order loops and accept only those which are nontouching. Such a brute-force method, however, is not satisfactory in practice because of the very large number of combinations that have to be examined. More efficient methods will now be described.

Let the first-order loops be numbered $1, 2, \ldots, L$, as in Table 14-2. Consider loop K (initially $K = 1$). All second-order loops (or nontouching pairs of loops) containing loop K and another *higher*-numbered loop may be found by comparing the node sequences given in Table 14-2.[4] Next, K is increased by 1, and the process is repeated until $K = L - 1$. All second-order loops generated this way can again be described by a table.

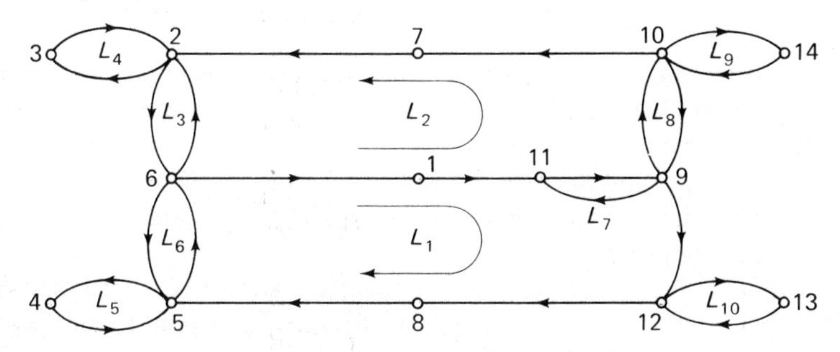

Fig. 14-10. Enumeration of all first-order loops.

[3]By evaluating the "*permanent*" of a certain matrix associated with the SFG, it is possible to determine in advance the total number of all orders of loops (see reference 10).

[4]A special coding technique will make such comparisons very easy. See reference 6.

TABLE 14-2. First-Order Loops in Fig. 14-10

LOOP	NODE SEQUENCE							
1	1	11	9	12	8	5	6	1
2	1	11	9	10	7	2	6	1
3	2	6	2					
4	2	3	2					
5	4	5	4					
6	5	6	5					
7	9	11	9					
8	9	10	9					
9	10	14	10					
10	12	13	12					

As an example, let us apply the procedure to the SFG of Fig. 14-10, whose first-order loops are given by Table 14-2. The second-order loops found are given in Table 14-3. Every loop in the first column together with any loop in the same row constitutes a second-order loop.

TABLE 14-3. Second-Order Loops in Fig. 14-10

LOOP K	HIGHER-NUMBERED LOOPS NOT TOUCHING LOOP K					
1	4	9				
2	5	10				
3	5	7	8	9	10	
4	5	6	7	8	9	10
5	7	8	9	10		
6	7	8	9	10		
7	9	10				
8	10					
9	10					
10	None					

The enumeration of higher-order loops ($n \geq 3$) is a crucial part of any analysis program using the SFG approach. Several algorithms are now in use [5, 6]. We shall describe only one of them [6].

The basic idea is as follows. We choose loop K (initially $K = 1$) and generate all orders of loops consisting of loops *numbered K or higher*. When this is done, we increase K by 1 and repeat the process.

The method is best illustrated with a specific example. Consider the SFG of Fig. 14-10 for which the first- and second-order loops have been found (see Tables 14-2 and 14-3). We desire to find all orders of loops consisting of loops numbered 3 ($K = 3$) or higher. The solution may be described with the aid of the tree graph of Fig. 14-11, which is constructed as follows (we use the terms *root* and *tips* of a tree graph in their intuitive senses).

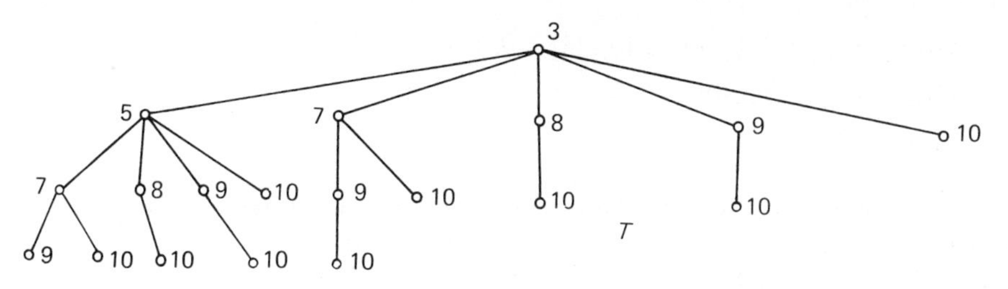

Fig. 14-11. Enumeration of higher-order loops.

The root is numbered K, which is 3 in the present case. Every node (except the root and the tip nodes) in the tree graph has exactly one *immediate ascendent* node and some *immediate descendent* nodes. The immediate descendent nodes of the root K are numbered with those loops not touching loop K. The immediate descendent nodes of any other node corresponding to loop N are those which have the same immediate ascendent node as N *and* which do not touch loop N (as determined by Table 14-3). For example, the immediate descendents of 7 in the third row of Fig. 14-11 are 9 and 10. (7, 8, 9, and 10 have 5 as their immediate ascendent. But loop 8 touches loop 7, from Table 14-3. Therefore, only 9 and 10 are immediate descendents of 7.) A node having no descendents becomes a tip node in the tree graph. The tree graph for the SFG of Fig. 14-10 constructed according to the preceding is shown in Fig. 14-11. Once this tree is constructed, the desired loops are obtained simply by tracing the paths from the root to *all* nodes in the tree. Thus, from Fig. 14-11, we have 18 loops ($n \geq 2$), as follows:

Second-order loops: (3, 5), (3, 7), (3, 8), (3, 9), (3, 10)

Third-order loops: (3, 5, 7), (3, 5, 8), (3, 5, 9), (3, 5, 10)
(3, 7, 9), (3, 7, 10), (3, 8, 10), (3, 9, 10)

Fourth-order loops: (3, 5, 7, 9), (3, 5, 7, 10), (3, 5, 8, 10), (3, 5, 9, 10),
(3, 7, 9, 10)

By performing the process successively for $K = 1, 2, \ldots, (L - 1)$, we have obtained all orders of loops in the SFG.

One immediately raises the question of the computer storage needed for the information about all such trees. Fortunately if we do not insist on listing loops from the tree starting with second-order loops, then third-order loops, etc., only one subtree needs to be stored at any time, and this subtree is updated repeatedly to give the same result as the full tree. For example, let the 18 loops of the previous example be listed sequentially as follows:

Group 1
(3, 5), (3, 7), (3, 8), (3, 9), (3, 10)
(3, 5, 7), (3, 5, 8), (3, 5, 9), (3, 5, 10)
(3, 5, 7, 9), (3, 5, 7, 10)

Group 2 (3, 5, 8, 10)

Group 3 (3, 5, 9, 10)

Group 4 (3, 7, 9), (3, 7, 10), (3, 7, 9, 10)

Group 5 (3, 8, 10)

Group 6 (3, 9, 10)

To generate loop sets of group 1, we need only the subtree T_1 of Fig. 14-12(a). Similarly, to generate loop sets of groups 2–6, we need only the subtrees T_2–T_6, shown respectively in Fig. 14-12(b)–(f). The striking facts about these subtrees are that (1) in any row, only the descendents of one node need be recorded; and (2) T_{k+1} is obtainable from T_k with the aid of Table 14-2. It is seen that the storage of such a subtree requires much less space than for the full tree of Fig. 14-11. The details of the programming technique for updating the subtrees may be found in [6].

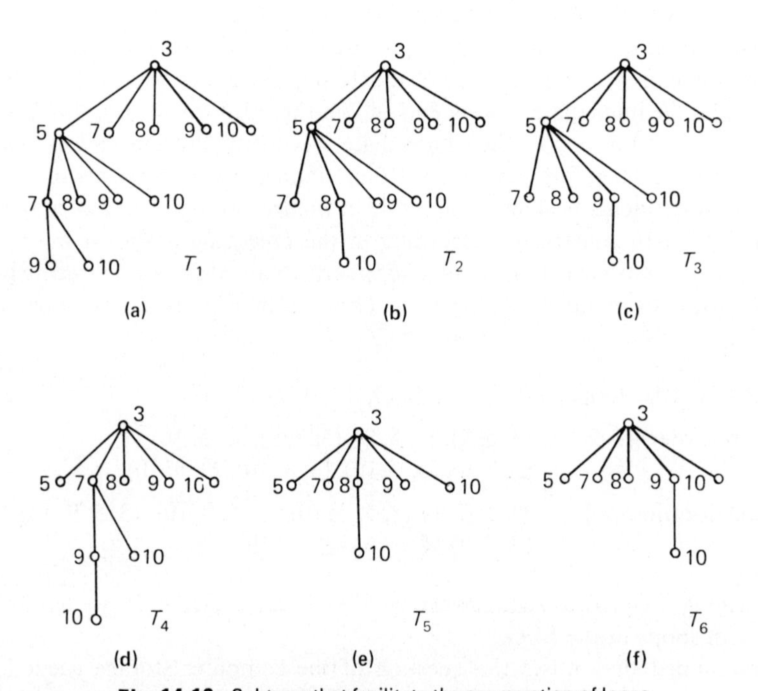

Fig. 14-12. Subtrees that facilitate the enumeration of loops.

14-2-5 Symbol Manipulations in the Signal-Flow-Graph Method

Since the primary objective of this chapter is the generation of *symbolic* network functions, our methods will not be complete without an efficient means of handling the symbolic parameters. The integer code method appears to be the simplest and most widely used, and will therefore be briefly described.

Each branch weight in the SFG is of the form

$$\text{Constant} \cdot \text{symbol} \cdot s^n = b \cdot X \cdot s^n$$

Thus, three parameters (b, X, n) are ascribed to each branch to define its weight—one each for the constant, symbol, and power of s. Each *distinct* symbol in the SFG may be represented by an integer in the following manner:[5]

$$\text{Symbol } X_1 \longleftrightarrow B^0$$
$$\text{Symbol } X_2 \longleftrightarrow B^1$$
$$\text{Symbol } X_3 \longleftrightarrow B^2$$
$$\text{etc.}$$

where the base B is chosen from the set $\{2, 4, \ldots, 2^m\}$ and to satisfy the condition that at most $B - 1$ *network branches* have the same symbolic weight. (For example, if a network has three capacitances with symbolic weight C and all other elements are either numeric or represented by distinct symbols, then we may choose $B = 4$.)

Using this technique, a loop of any order consisting of branches $(p, q, \ldots)$ can be completely described by a constant that is the *product* of the individual constants, a symbol code that is the *sum* of the individual symbol codes, and an exponent of s that is the *sum* of the individual exponents.

To see how the symbol code works, consider the active network of Fig. 14-13(a) and the corresponding closed SFG of Fig. 14-13(b). Since in the network R and C each appear as the symbolic parameter of three elements, we choose $B = 4$. There are four distinct symbols in Fig. 14-13(b), which are coded as follows:

$$R \longleftrightarrow 4^0 = 1$$
$$C \longleftrightarrow 4^1 = 4$$
$$\mu \longleftrightarrow 4^2 = 16$$
$$F \longleftrightarrow 4^3 = 64$$

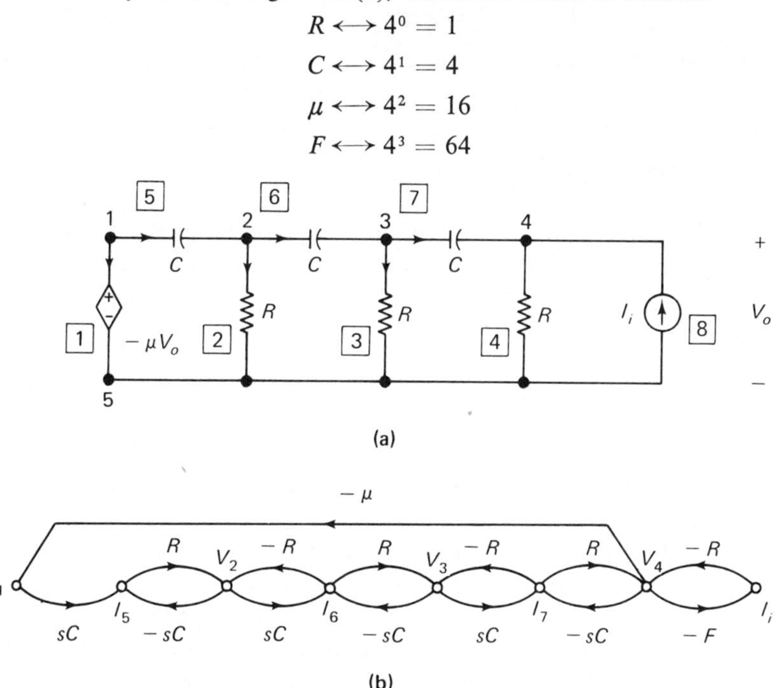

Fig. 14-13. Method of representing symbols.

[5]In reference 11 a method of symbolic network analysis is described in which the variables are represented by prime numbers.

The description of a second-order loop with weight $(sCR)^2 = (1.0)(CCRR)s^2$ will be as follows:

$$\text{Constant} = 1.0$$

$$\text{Symbol code} = 4^1 + 4^1 + 4^0 + 4^0 = 8 + 2 = 10$$

$$\text{Exponent (power of } s) = 2$$

Similarly, a first-order loop with weight $-\mu(SCR)^3$ will have a constant -1.0, an exponent 3, and a symbol code

$$(4^2) + 3(4^1) + 3(4^0) = 16 + 12 + 3 = 31$$

During the intermediate stages of calculations, all symbols are represented by integers, and multiplications of symbols correspond to additions of their symbol codes. It is only in the final output stage that these symbol codes need to be translated or decoded to give the desired symbol combinations. The decomposition of a composite symbol code into its constituent symbols is unique, provided that the stated condition on the choice of B is met. Consider again the example of Fig. 14-13. Suppose that a composite symbol code 74 appears in the final stage, and we wish to know what symbols it represents. We proceed as follows:

$$74 = 1(4^3) + 2(4^1) + 2(4^0)$$

Therefore, the symbol combination is

$$F \cdot C^2 \cdot R^2$$

These coding and decoding operations can be performed very easily with a digital computer.

14-3 TREE-ENUMERATION METHOD

Besides the signal-flow graph method, another topological method widely known is the tree-enumeration method. The basic concepts of this latter method for resistance networks dated back to Kirchhoff (1847). In the 1950s, there was revived interest in the method, and new results were obtained for active linear networks [12, 13]. A few computer programs for evaluating network functions as rational functions of s have been written using the tree-enumeration method [14–17]. We shall give a brief account of the method. Many underlying concepts of the method are also useful in a more powerful method to be described in Section 14-4.

The essence of the method is to first represent a given network by a directed, weighted graph G_d constructed according to certain rules and then evaluate any desired network function by listing the trees (directed trees) of the graph. The details of the method will be described in the next few pages. For the proofs of various asserted properties in this section, the reader may consult the references at the end of this chapter. Throughout this section, we assume that the network consists of RLC elements and voltage-controlled current sources (which we designate as the g_m type) only. All other types of controlled sources must be converted into the g_m type (see Problem 4-19), before the tree-enumeration method can be applied.

14-3-1 Network Functions in Terms of the Determinant and Cofactors of $\boldsymbol{Y}_n$

Let N be an $RLC\text{-}g_m$ network whose n nodes are labeled as shown in Fig. 14-14. If node n is chosen as the reference node, the nodal equation may be written as

$$\boldsymbol{Y}_n \begin{bmatrix} V_1 \\ V_2 \\ V_3 \\ \cdot \\ \cdot \\ \cdot \\ V_{n-1} \end{bmatrix} = \begin{bmatrix} I_1 \\ 0 \\ 0 \\ \cdot \\ \cdot \\ \cdot \\ 0 \end{bmatrix} \tag{14-11}$$

where $\boldsymbol{Y}_n$, the node-admittance matrix, has dimension $(n-1) \times (n-1)$.

Solving Eq. (14-11), we obtain

$$V_1 = \frac{\Delta_{11}}{\Delta} I_1, \qquad V_2 = \frac{\Delta_{12}}{\Delta} I_1, \qquad V_3 = \frac{\Delta_{13}}{\Delta} I_1 \tag{14-12}$$

where Δ is the determinant; and Δ_{ij} is the ijth cofactor of $\boldsymbol{Y}_n$.

From Eq. (14-12), we immediately obtain the following network functions[6] in terms of Δ and Δ_{ij}:

$$Z_{\text{in}} = \frac{V_{\text{in}}}{I_{\text{in}}} = \frac{V_1}{I_1} = \frac{\Delta_{11}}{\Delta} \tag{14-13a}$$

$$\frac{V_o}{I_{\text{in}}} = \frac{V_2}{I_1} = \frac{\Delta_{12}}{\Delta} \tag{14-13b}$$

$$\frac{V_o}{V_{\text{in}}} = \frac{V_2}{V_1} = \frac{\Delta_{12}}{\Delta_{11}} \tag{14-13c}$$

$$\frac{V_o'}{I_{\text{in}}} = \frac{V_2 - V_3}{I_1} = \frac{\Delta_{12} - \Delta_{13}}{\Delta} \tag{14-13d}$$

$$\frac{V_o'}{V_{\text{in}}} = \frac{V_2 - V_3}{V_1} = \frac{\Delta_{12} - \Delta_{13}}{\Delta_{11}} \tag{14-13e}$$

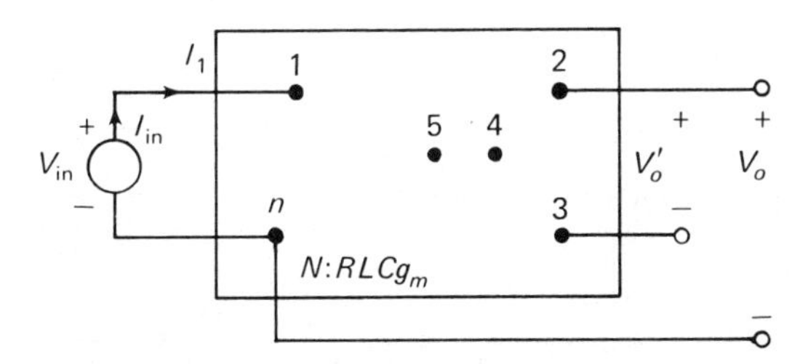

Fig. 14-14. Notations for a two-port network.

[6] See Problem 14-19 for the handling of output currents.

Thus, the evaluation of network functions is basically the evaluation of the determinant and some cofactors. Since we are dealing with symbolic network functions, we can take advantage of the presence of symbols to eliminate the need for evaluating cofactors separately. The method will be described next.

14-3-2 Sorting Scheme

Suppose that in Fig. 14-14 the network functions Z_{in} and V'_o/V_{in} (or V_o/V_{in}) are desired. Let us form an augmented network $\hat{N}$ by adding an admittance $\hat{y}_s$ and a controlled source $\hat{g}_m$ across the input nodes, the controlling voltage being V'_o (or V_o), as shown in Fig. 14-15. It is important that $\hat{y}_s$ and $\hat{g}_m$ be *distinct* from any other variables in N. $\hat{N}$ has the same nodes (and numbering) as N. If, as before, node n is chosen as the reference node, it can readily be shown that $\hat{\boldsymbol{Y}}_n$, the node-admittance matrix of $\hat{N}$, is related to $\boldsymbol{Y}_n$ as follows (for the case of adding $\hat{y}_s$ and $\hat{g}_m V'_o$):

$$\hat{\boldsymbol{Y}}_n = \boldsymbol{Y}_n + \begin{bmatrix} \hat{y}_s & \hat{g}_m & -\hat{g}_m & 0 & \cdots & 0 \\ & & \mathbf{0} & & & \end{bmatrix} \tag{14-14}$$

A straightforward evaluation of the determinant of $\hat{\boldsymbol{Y}}_n$ gives

$$\hat{\boldsymbol{\Delta}}_n = \boldsymbol{\Delta} + \hat{y}_s \boldsymbol{\Delta}_{11} + \hat{g}_m(\boldsymbol{\Delta}_{12} - \boldsymbol{\Delta}_{13}) \tag{14-15a}$$

where $\boldsymbol{\Delta}$ and $\boldsymbol{\Delta}_{ij}$ are associated with $\boldsymbol{Y}_n$. Similarly, if N is augmented with $\hat{y}_s$ and $\hat{g}_m V_o$, then

$$\hat{\boldsymbol{\Delta}}_n = \boldsymbol{\Delta} + \hat{y}_s \boldsymbol{\Delta}_{11} + \hat{g}_m \boldsymbol{\Delta}_{12} \tag{14-15b}$$

From Eq. (14-15), it is seen that, after the evaluation of just one determinant $\hat{\boldsymbol{\Delta}}_n$ (the node determinant of the augmented network), we can obtain $\boldsymbol{\Delta}$, $\boldsymbol{\Delta}_{11}$, $\boldsymbol{\Delta}_{12}$, and $(\boldsymbol{\Delta}_{12} - \boldsymbol{\Delta}_{13})$, which are required in Eq. (14-13), by sorting the terms of $\hat{\boldsymbol{\Delta}}_n$ with respect to the symbols $\hat{y}_s$ and $\hat{g}_m$. If Z_{in} only is required, then N is augmented with $\hat{y}_s$ only. Similarly, if only V_o/I_{in} or V'_o/I_{in} is required, then N is augmented with $\hat{g}_m$ only. The advantage of such a sorting scheme should be obvious, for we need concentrate only on the problem of evaluating a determinant. The cofactors $\boldsymbol{\Delta}_{ii}$ and $\boldsymbol{\Delta}_{ij}$ are obtained through sorting.

We note that the closed signal-flow graph introduced in Section 14-2-1 is actually a sorting scheme for the SFG approach. Convenient as it is, the sorting scheme has

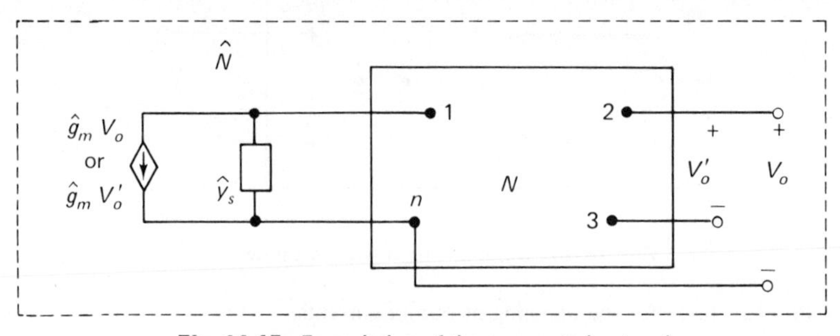

Fig. 14-15. Formulation of the augmented network.

its place only in computer programs that permit symbolic network parameters. In a purely numerical program, the use of sorting is not feasible.

14-3-3 Indefinite Admittance Matrix and Its Graph

In Section 14-3-4 we shall describe how the evaluation of the node determinant of any RLC-g_m network is transformed into a tree-listing problem. To do this, a directed graph associated with the network must be defined first. It is more convenient to define this graph by the concept of an indefinite admittance matrix.

The *indefinite admittance matrix* (IAM) of any n-terminal network N (without independent sources inside N), denoted by $\mathbf{Y}_{\text{ind}}$, is an $n \times n$ matrix relating the terminal currents $\mathbf{I}$ and terminal voltages $\mathbf{V}$ as follows:

$$\mathbf{I} = \mathbf{Y}_{\text{ind}} \mathbf{V} \tag{14-16}$$

where the voltages $\mathbf{V}$ are measured with respect to some point r *external* to N, and the reference directions and node labelings are as shown in Fig. 14-16.

The IAM defined by Eq. (14-16) has many interesting properties, the proofs of which can be found in many textbooks [24]. We shall list a few that are immediately useful for our purposes.

1. *The sum of the elements in each row or each column is zero. Hence the determinant of* $\mathbf{Y}_{\text{ind}}$ *is zero.*
2. *The node admittance matrix* $\mathbf{Y}_n$ *may be obtained from* $\mathbf{Y}_{\text{ind}}$ *by deleting row k and column k, if node k of N is chosen to be the reference node.*
3. *All cofactors of* $\mathbf{Y}_{\text{ind}}$ *are equal.*

Because of properties 1 and 2, we can obtain $\mathbf{Y}_{\text{ind}}$ from a given $\mathbf{Y}_n$ by adding a kth row and a kth column whose entries are determined by the zero-sum property. Also, from properties 2 and 3, we see that the node determinant of any RLC-g_m network is unique, regardless of the reference node chosen to write $\mathbf{Y}_n$.

For RLC-g_m networks, $\mathbf{Y}_{\text{ind}}$ always exists. Moreover, the structure of $\mathbf{Y}_{\text{ind}}$ bears a very simple relationship with the structure of the network. The following two rules,

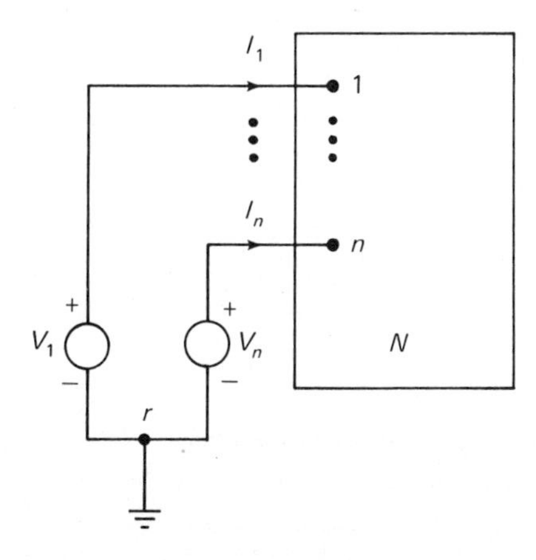

Fig. 14-16. Defining the indefinite admittance matrix of a network.

which can be proved easily, enable one to write $\boldsymbol{Y}_{\text{ind}}$ directly by inspection of the network. (We assume that the nodes of N are numbered $1, 2, \ldots, n$. See also Section 4-6.)

Rule 1. *For each admittance y connected between nodes i and j of N, we have an entry y in the positions (i, i) and (j, j), and an entry $-y$ in the positions (i, j) and (j, i).*

Rule 2. *For each g_m-type controlled source with controlled current source directed from node i to node j and the controlling voltage drop from node k to node m, we have an entry g_m in positions (i, k) and (j, m), and an entry $-g_m$ in positions (i, m) and (j, k).*

As an example, the IAM for the network of Fig. 14-17 may be written directly by the use of these above rules. The result is

$$\boldsymbol{Y}_{\text{ind}} = \begin{bmatrix} y_a + y_b + \hat{y}_s & -y_a + \hat{g}_m & -y_b - \hat{g}_m & -\hat{y}_s \\ -y_a + g_m & y_a + y_c + y_e & -y_e - g_m & -y_c \\ -y_b - g_m & -y_e & g_m + y_b + y_d + y_e & -y_d \\ -\hat{y}_s & -y_c - \hat{g}_m & -y_d + \hat{g}_m & \hat{y}_s + y_c + y_d \end{bmatrix} \tag{14-17}$$

The directed graph G_d associated with $\boldsymbol{Y}_{\text{ind}}$, or more generally any square matrix with the zero-sum property, is by definition to be constructed as follows: (1) For every entry y_{ij} $(i \neq j)$ in $\boldsymbol{Y}_{\text{ind}}$, there is a branch in G_d directed from node i to node j, with a branch weight $-y_{ij}$. (2) Parallel branches (i.e., branches having the same initial node and the same terminal node) may be replaced by a single branch whose weight is the sum of the individual branch weights.

Although the diagonal elements of $\boldsymbol{Y}_{\text{ind}}$ have not been explicitly used in the construction of G_d, no information contained in $\boldsymbol{Y}_{\text{ind}}$ is lost in G_d, because of the zero-sum property.

As an example, G_d for the IAM given by Eq. (14-17) is shown in Fig. 14-18. A closer examination of the rules for constructing G_d from $\boldsymbol{Y}_{\text{ind}}$ and the rules for writing

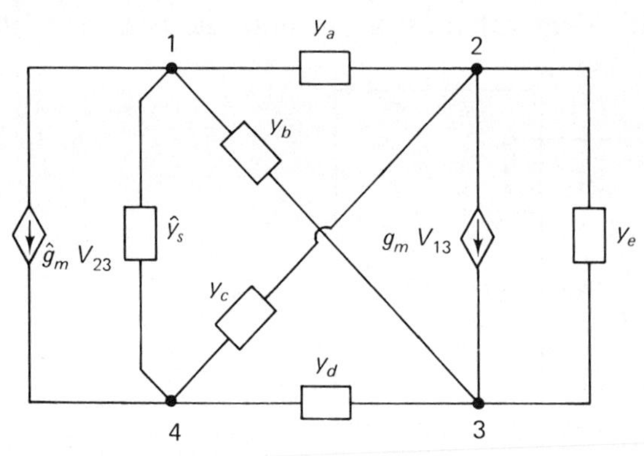

Fig. 14-17. Network with indefinite admittance matrix given by Eq. (14-17).

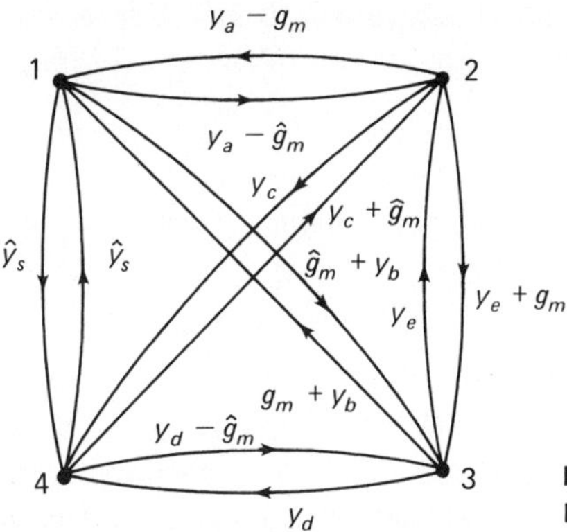

Fig. 14-18. Directed graph associated with Eq. (14-17).

$\boldsymbol{Y}_{\text{ind}}$ from a given *RLC*-g_m network suggests that we may construct G_d directly by inspection of the network, and hence eliminate the intermediate step of writing $\boldsymbol{Y}_{\text{ind}}$. We need only replace each y or g_m element in N by a proper set of directed branches, as shown in Fig. 14-19, and then combine parallel branches.

14-3-4 Node Determinant from Directed Trees of G_d

The concept of a tree was introduced in Chapter 3 for undirected graphs. For *RLC*-g_m networks, the evaluation of network functions by the tree-enumeration method requires the use of *directed trees*, a concept to be defined next.

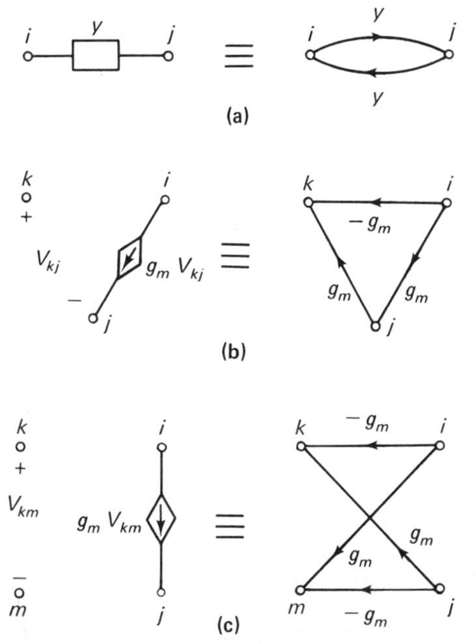

Fig. 14-19. Rules for constructing G_d directly from the given network.

Definition. *Let G_d be a directed graph. A subgraph T_d is said to be a directed tree of G_d,* ***rooted at node*** *r, if and only if*

1. *In T_d, node r has no incoming branches, and each of the remaining nodes has exactly one incoming branch.*[7]
2. *With all branch orientations removed, the resultant undirected graph T is a tree of the resultant undirected graph G.*

As an example, consider the directed graph G_d of Fig. 14-20(a). There are two trees rooted at node 1, shown in Fig. 14-20(b), two trees rooted at node 2, shown in (c), and two trees rooted at node 3, shown in (d); there is no tree rooted at node 4.

In Section 14-3-3 we defined the directed graph G_d associated with a given $\boldsymbol{Y}_{\text{ind}}$ or a given RLC-g_m network. The weight of each branch has the dimension of an admittance (e.g., y_s, $(y_s + g_m)$, or $-g_m$). For any directed tree of G_d, the product of its branch weights is called a *directed-tree admittance product.* The most significant result of this section is now stated in Theorem 14-1.

Theorem 14-1. *For any RLC-g_m network N, the determinant of the node-admittance matrix (with any node as the reference node) is equal to the sum of all directed-tree admittance products of G_d (with any node as the root).*

A proof of this theorem is certainly beyond the scope of this book. The reader is referred to [10]. We shall merely illustrate the use of this theorem, as well as the

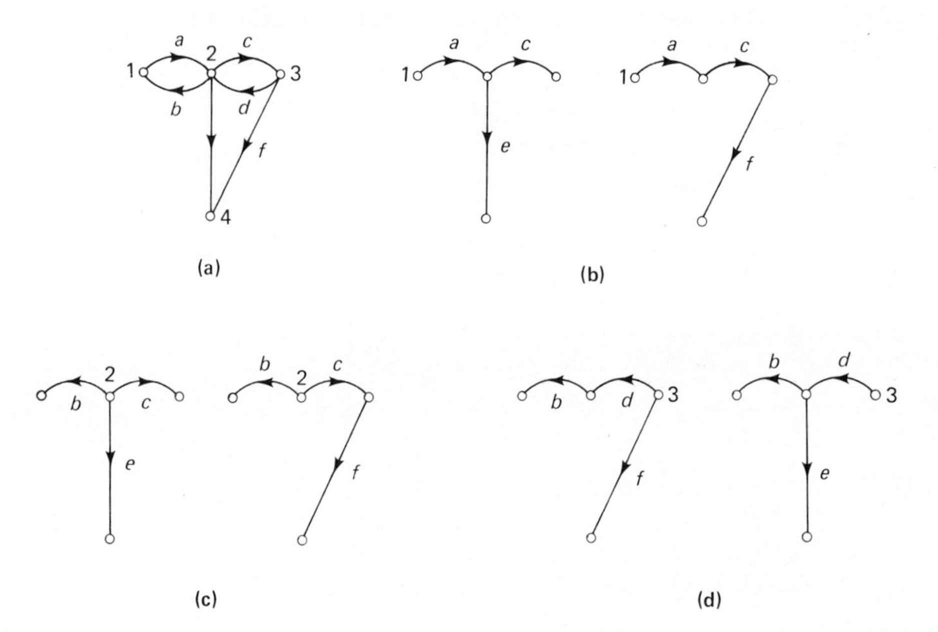

Fig. 14-20. Directed trees of a directed graph.

[7]By changing the word *incoming* to *outgoing* we have another definition of a directed tree as given in reference 10. Theorem 14-1 is valid regardless of which definition is used.

sorting scheme of Section 14-3-2. Consider the simple active network of Fig. 14-21(a). We desire to find the transfer function V_o/I_{in}. The augmented network $\hat{N}$ is shown in Fig. 14-21(b) and the corresponding directed graph in Fig. 14-21(c). Suppose that node 2 is chosen as the root. Then there are three directed trees, which result in

$$\begin{aligned}\hat{\Delta}_n &= \sum \text{tree admittance product} \\ &= [(g_m + Y_L)g_m] + [(g_m + Y_L)(-g_m + Y_f)] + [\hat{g}_m(-g_m + Y_f)] \\ &= g_m^2 + Y_L g_m - g_m g_m - Y_L g_m + (g_m + Y_L)Y_f + \hat{g}_m(-g_m + Y_f) \\ &= (g_m + Y_L)Y_f + \hat{g}_m(-g_m + Y_f)\end{aligned}$$

Comparing with Eq. (14-15b) and using Eq. (14-13b), we have

$$\frac{V_o}{I_{in}} = \frac{-g_m + Y_f}{(g_m + Y_L)Y_f}$$

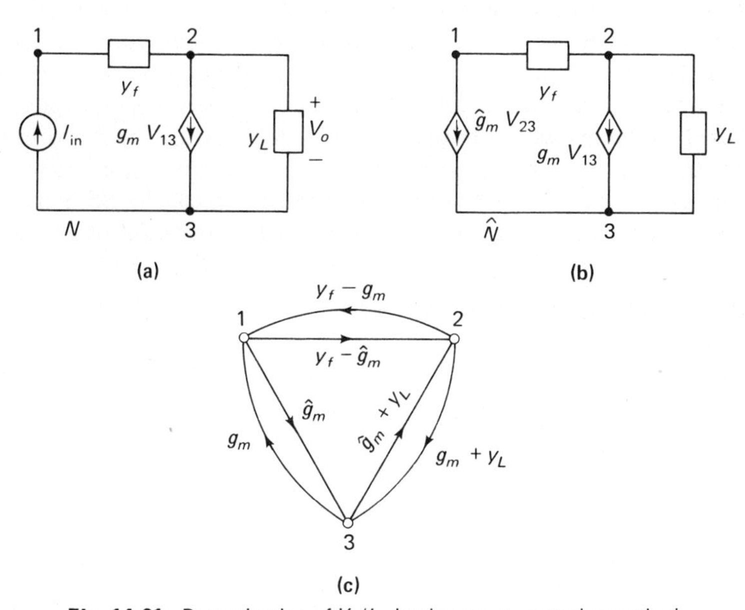

Fig. 14-21. Determination of V_o/I_{in} by the tree-enumeration method.

The reader can easily verify that exactly the same expression will be obtained if node 1 or node 3 is chosen as the root of the directed trees. The third line of $\hat{\Delta}_n$ is shown to emphasize the *cancellation of terms*, an undesirable feature that we would like to avoid as much as possible.

The determinant and cofactors of a node-admittance matrix, when expressed in terms of directed-tree admittance products,[8] are usually referred to as *topological formulas*. For an *RLC* network, the associated directed graph G_d consists only of

[8]Topological formulas for Δ_{ij} require the concept of "directed *k*-trees" (see reference 13). However, with the sorting scheme of Section 14-3-2 ,we do not need this additional concept for evaluating network functions.

branches of the type shown in Fig. 14-19(a). As a simplification in the graphical representation of G_d, an undirected branch with weight y will be substituted for each pair of branches of Fig. 14-19(a). Then we obtain an undirected graph G_n for the *RLC* network *N*. Theorem 14-1 then reduces to the following:

> ***Theorem 14-2.*** *For any RLC network N, the determinant of the node-admittance matrix is equal to the sum of all tree admittance products.*

One advantage of evaluating the node determinant of an *RLC* network by a topological formula (Theorem 14-2) is that the cancellation of terms, which plagues many other conventional methods, can be *completely avoided.*[9] However, for *RLC*-g_m networks, as we have seen in our simple example, the evaluation of the node determinant by a topological formula may still have many cancellations of terms.

Undoubtedly, for the topogical formulas to be of practical value, we must have efficient algorithms for enumerating directed trees; hopefully, the algorithm will not generate many directed-tree admittance products that are only to be cancelled in the final answer. This topic has attracted a great deal of research interest in recent years. However, only a few computer programs for generating network functions have been written based on topological formulas [14-17]. At present, their performance, in regard to both computer running time and storage requirement, appears to be similar to those programs based on the signal-flow-graph approach. Both methods are limited to small networks having, say, fewer than 10 nodes and 20 branches.

14-4 PARAMETER-EXTRACTION METHOD

One serious drawback of the signal-flow-graph and the tree-enumeration methods discussed in the preceding sections is the extremely large number of higher-order loops or number of trees for a network of moderate size. As an indication of the orders of magnitude, let us consider the following results:

1. For a complete undirected graph[10] of n nodes, the number of trees is $n^{(n-2)}$.
2. For a signal-flow graph with L first-order loops that are node-disjoint, the total number of all orders of loops is $2^L - 1$.

Thus, the number of terms that have to be processed increases very fast with n or L, since it appears in the exponent of the expression.

On the other hand, numerical programs based on matrix methods do not experience such a serious difficulty. One reason is that the number of multiplications required for the inversion of an $n \times n$ matrix is only n^3 (see Problem 4-7). This increase is much slower than in the two cases just given.

[9]For an *RLC* network, the number of terms in the node determinant is equal to the number of trees of the associated undirected graph. This number can be determined easily (see Problem 3-8). Various methods for enumerating trees may be found in reference 10.

[10]An undirected graph is said to be "complete" if there is exactly one branch joining every pair of distinct nodes.

In symbolic network analysis, very often we encounter a situation in which only a few elements of a large network are represented by symbolic parameters, whereas the remaining elements all have numerical values assigned. In such a case, it will be most advantageous to solve the problem by a combination of two methods: a parameter-extraction process for removing symbolic parameters from a determinant, and standard numerical methods for evaluating a determinant $|\boldsymbol{G}|$ or $|s\boldsymbol{C} + \boldsymbol{G}|$, the entries of $\boldsymbol{G}$ and $\boldsymbol{C}$ being real constants.

We shall first briefly describe this new method as applied to linear resistive networks. Extension to linear dynamic networks is discussed in Section 14-4-3.

14-4-1 Theorem for Parameter Extraction

From the discussions in Sections 14-3-1 and 14-3-2, we see that in deriving symbolic network functions the crux of the problem is the evaluation of the determinant of a node-admittance matrix $\boldsymbol{Y}_n$ whose entries contain symbolic parameters (variables). For the new method to be discussed, it is more convenient to use the indefinite admittance matrix. Then the problem is to evaluate efficiently the cofactor of an indefinite admittance matrix whose entries contain symbols.

As shown in Section 14-3-3, any admittance y or transconductance g_m (assumed to be a *distinct* variable) must appear exactly in four positions in $\boldsymbol{Y}_{\text{ind}}$. Let α represent $\pm y$ or $\pm g_m$. Then we have, in general,

1. α in positions (i, k) and (j, m).
2. $-\alpha$ in positions (i, m) and (j, k).

See Eq. (14-17) for an example and Eq. (4-79) for the general pattern.

Theorem 14-3 reduces the evaluation of any cofactor of $\boldsymbol{Y}_{\text{ind}}$ (containing α) to two simpler problems, each of which is to evaluate a cofactor of some indefinite admittance matrix not containing α. In other words, the parameter α has been "extracted."

Theorem 14-3. *Let α be a term appearing in an $n \times n$ indefinite admittance matrix $\boldsymbol{Y}_{\text{ind}}$ in four positions: $+\alpha$ in positions (i, k) and (j, m) and $-\alpha$ in positions (i, m) and (j, k). Then*

$$\textit{cofactor of } \boldsymbol{Y}_{\text{ind}} = \textit{cofactor of } (\boldsymbol{Y}_{\text{ind}}|_{\alpha=0}) + (-1)^{j+m}\cdot\alpha\cdot(\textit{cofactor of } \boldsymbol{Y}_\alpha) \tag{14-18}$$

where $\boldsymbol{Y}_\alpha$ is an $(n-1) \times (n-1)$ indefinite admittance matrix not containing α, obtained by modifying $\boldsymbol{Y}_{\text{ind}}$ as follows:[11]

1. *Add row j to row i.*
2. *Add column m to column k.*
3. *Delete row j and column m.*

For a proof of the theorem, see [18].

[11]The cofactor of any 1×1 matrix is defined to be 1.

If $\boldsymbol{Y}_{\text{ind}}|_{\alpha=0}$ and $\boldsymbol{Y}_\alpha$ in Eq. (14-18) contain other symbols, say $\beta, \gamma, \ldots$, the theorem may be applied repeatedly to extract all symbols. Finally, we obtain an expression of a cofactor of $\boldsymbol{Y}_{\text{ind}}$ in the following form:

$$\text{cofactor of } \boldsymbol{Y}_{\text{ind}} = \sum P_j \cdot (\text{cofactor of } \boldsymbol{Y}_j) \tag{14-19}$$

where each P_j is some product of symbolic parameters including the sign, and for resistive networks each $\boldsymbol{Y}_j$ is some indefinite admittance matrix containing only numerical entries. The cofactor of each $\boldsymbol{Y}_j$ may be evaluated by any of the well-known numerical methods, e.g., the Gaussian elimination method.

14-4-2 A Complete Example

We wish to find Z_{in} for the network of Fig. 14-22 by the parameter-extraction process, keeping g_m as a symbolic parameter. According to the sorting scheme of Section 14-3-2, we add an admittance $\hat{y}_s$ across the input terminals. The indefinite admittance of the augmented network $\hat{N}$ is

$$\hat{\boldsymbol{Y}}_{\text{ind}} = \begin{bmatrix} 6+\hat{y}_s & -5 & -1 & -\hat{y}_s \\ g_m - 5 & 15.1 & -g_m - 10 & -0.1 \\ -g_m - 1 & -10 & g_m + 13 & -2 \\ -\hat{y}_s & -0.1 & -2 & \hat{y}_s + 2.1 \end{bmatrix} \tag{14-20}$$

Applying Theorem 14-3 to extract $\hat{y}_s$ first, we obtain

$$\text{cofactor of } \hat{\boldsymbol{Y}}_{\text{ind}} = \text{cofactor of} \begin{bmatrix} 6 & -5 & -1 & 0 \\ g_m - 5 & 15.1 & -g_m - 10 & -0.1 \\ -g_m - 1 & -10 & g_m + 13 & -2 \\ 0 & -0.1 & -2 & 2.1 \end{bmatrix}$$

$$+ \hat{y}_s \cdot \text{cofactor of} \begin{bmatrix} 8.1 & -5.1 & -3 \\ g_m - 5.1 & 15.1 & -g_m - 10 \\ -g_m - 3 & -10 & g_m + 13 \end{bmatrix} \tag{14-21}$$

Next, Theorem 14-3 is applied again to each term of Eq. (14-21) to extract the symbol

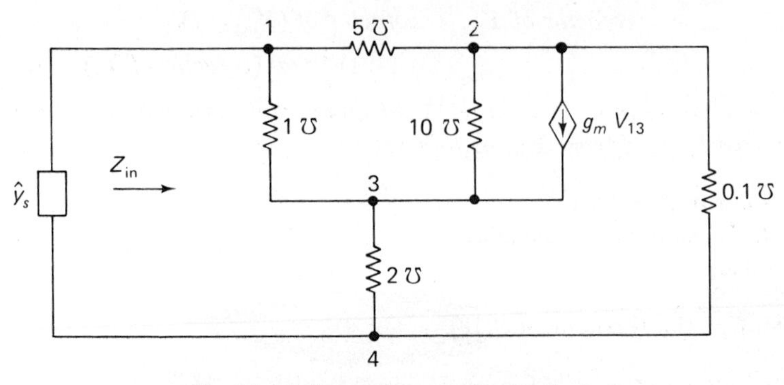

Fig. 14-22. Parameter-extraction method.

g_m. The final result becomes

$$\text{cofactor of } \hat{Y}_{\text{ind}} = \text{cofactor of} \begin{bmatrix} 6 & -5 & -1 & 0 \\ -5 & 15.1 & -10 & -0.1 \\ -1 & -10 & 13 & -2 \\ 0 & -0.1 & -2 & 2.1 \end{bmatrix}$$

$$+ g_m \cdot \text{cofactor of} \begin{bmatrix} 5 & -5 & 0 \\ -3 & 5.1 & -2.1 \\ -2 & -0.1 & 2.1 \end{bmatrix} \tag{14-22}$$

$$+ \hat{y}_s \cdot \text{ cofactor of} \begin{bmatrix} 8.1 & -5.1 & -3 \\ -5.1 & 15.1 & -10 \\ -3 & -10 & 13 \end{bmatrix}$$

$$+ \hat{y}_s g_m \cdot \text{ cofactor of} \begin{bmatrix} 5.1 & -5.1 \\ -5.1 & 5.1 \end{bmatrix}$$

The four cofactors may now be evaluated, say, by the Gaussian elimination method, to yield

$$\text{cofactor of } \hat{Y}_{\text{ind}} = 137.7 + g_m \times 10.5 + \hat{y}_s \times 96.3 + \hat{y}_s g_m \times 5.1 \tag{14-23}$$

Comparing Eq. (14-23) with Eq. (14-15b), we have

$$\Delta = 137.7 + 10.5 g_m$$

$$\Delta_{11} = 96.3 + 5.1 g_m$$

Therefore, from Eq. (14-13a), we have

$$Z_{\text{in}} = \frac{96.3 + 5.1 g_m}{137.7 + 10.5 g_m} \tag{14-24}$$

14-4-3 Extension and Further Remarks

Consider now linear dynamic networks. If a capacitance C_1 is kept as a variable, its admittance sC_1 will be extracted. On the other hand, if C_1 is specified numerically, say $C_1 = 5$, then the admittance $5s$ will be left in $\boldsymbol{Y}_j$ in Eq. (14-19). If the network is modeled with R, C, and g_m elements only,[12] each entry of the indefinite admittance matrix $\boldsymbol{Y}_{\text{ind}}$ is in general a first-degree polynomial in s whose coefficients may contain symbolic parameters. For example, possible entries might be $(3s + 5)$, $(sC_1 + 7)$, and $(sC_2 + g_m)$. After extracting all symbols as discussed above, the cofactor of each $\boldsymbol{Y}_j$ in Eq. (14-19) will be of the form

$$\text{cofactor of } \boldsymbol{Y}_j = |s\boldsymbol{P} + \boldsymbol{Q}| \tag{14-25}$$

where $\boldsymbol{P}$ and $\boldsymbol{Q}$ are square matrices containing only numerical entries, not necessarily symmetric or nonsingular. The determinant, given by Eq. (14-25), is a polynomial in s.

[12]An inductor may be modeled as a gyrator terminated in a capacitor. A gyrator in turn may be modeled with two g_m elements (voltage-controlled current sources). See Problem 4-20.

The evaluation of Eq. (14-25) may be considered as a *generalized* eigenvalue problem for which several methods of solution are available [19–22].

Serious readers will no doubt have these questions in mind: (1) For a network having n symbolic parameters, is it necessary to examine all $2^n - 1$ possible symbol combinations? (2) Is it necessary to store all $\boldsymbol{Y}_j$ in Eq. (14-19) at the same time? If the answers are in the affirmative, the method of parameter extraction will not be practical at all. For as n becomes moderately large, say $n = 20$, both the computer running time and the storage requirement will rule out the use of the method. Fortunately, through some intricate programming techniques, all *valid* symbol products P_j in Eq. (14-19) can be generated very efficiently, and only two IAM's need be stored at any time. The reader is referred to [22] for details. In fact, a computer program using the parameter-extraction approach has demonstrated that with a comparable amount of computer storage, it is now possible to perform symbolic network analysis (provided that only a few, say 10, network elements are represented by symbols) on networks of an order that can be handled by a numerical program such as SPICE.

APPENDIX 14A AN ALGORITHM FOR FINDING ALL PATHS

Finding all paths between any two nodes of a directed graph has many applications. The success of the signal-flow graph as a tool for computer generation of symbolic network functions depends very much on an efficient path-finding algorithm [23]. The following is a step-by-step description of an algorithm, the basic idea of which has been explained in Section 14-2-3.

ALGORITHM PF (PATH-FINDING). Algorithm PF finds all paths between two nodes of a directed graph (without parallel branches) whose nodes are labeled $1, 2, \ldots, N$. The only modification necessary to adapt the algorithm to find all loops through node L is to set $I = L$, where L and I are as defined.

Notations:

- I: Initial path node.
- L: Last path node.
- N: Number of nodes in graph.
- E_J: Number of branches leaving node J.
- $R(J, M)$: Routing table.
- C_J: Column counter for the Jth row of the table R.
- $P(V, W)$: The Vth node in the node sequence of path W.
- $F(K)$: A function used to test whether node K is repeated and whether the last node is reached.

PF1. (Preliminary)

Set $R(J, 1), R(J, 2), \ldots, R(J, E_J)$ to the group of E_J nodes of distance 1 from node J. When using the algorithm to find loops, make the entries of each row decrease as M increases.

Set $R(J, M) \longleftarrow \begin{cases} -1 \text{ for } M = E_J + 1 \text{ and } J = I. \\ 0 \text{ for } M = E_J + 1 \text{ and } J \neq I. \end{cases}$

Set $F(K) \longleftarrow \begin{cases} 1 \text{ for } K = I. \\ 0 \text{ for } K = J \text{ and } J \neq I, L. \\ -1 \text{ for } K = L. \end{cases}$

Set $C_J \longleftarrow 1$ for $J = 1, 2, \ldots, N$.

Set $W \longleftarrow 1$, $V \longleftarrow 2$, $J \longleftarrow I$, $P(1, 1) \longleftarrow I$.

PF2. (*Find the next node*)

Set $P(V, W) \longleftarrow R(J, C_J)$.

PF3. (*Test R*)

IF $R(J, C_J) \begin{cases} < 0 \text{ stop; all paths have been found.} \\ = 0 \text{ set } F(J) \longleftarrow 0\text{; go to step PF6.} \\ > 0 \text{ go to step PF4.} \end{cases}$

PF4. (*Test F*)

IF $F\{R(J, C_J)\} \begin{cases} < 0 \text{ path completed; go to step PF7.} \\ = 0 \text{ go to step PF5.} \\ > 0 \text{ set } C_J \longleftarrow C_J + 1\text{; go to step PF2.} \end{cases}$

PF5. (*Prepare for next node*)

Set $J \longleftarrow P(V, W)$, $F(J) \longleftarrow 1$, $V \longleftarrow V + 1$, go to step PF2.

PF6. (*Back step*)

Set $C_J \longleftarrow 1$, $J \longleftarrow P(V - 2, W)$, $C_J \longleftarrow C_J + 1$, $V \longleftarrow V - 1$; go to step PF2.

PF7. (*Finish path*)

Set $C_J \longleftarrow C_J + 1$; $P(K, W + 1) \longleftarrow P(K, W)$, for $K = 1, 2, \ldots, V - 1$, $W \longleftarrow W + 1$; go to step PF2.

REFERENCES

1. Lin, P. M. "A Survey of Applications of Symbolic Network Functions." *IEEE Trans. Circuit Theory*, Vol. CT-20, pp. 732–737, Nov. 1973.
2. Fidler, J. K., and J. I. Sewell. "Symbolic Analysis for Computer-Aided Circuit Design —the Interpolative Approach." *IEEE Trans. Circuit Theory*, Vol. CT-20, pp. 738–741, Nov. 1973.
3. Brown, W. S. "The ALPAK System for Nonnumerical Algebra on a Digital Computer —I: Polynomials in Several Variables and Truncated Power Series with Polynomial Coefficients." *Bell System Tech. Jour.*, Vol. XLII, pp. 2081–2119, Sept. 1963.
4. *Proceedings of Second Symposium on Symbolic and Algebraic Manipulation*, Los Angeles, Calif., 1971.
5. McNamee, L. P., and H. Potash. *A User's and Programmer's Manual for NASAP.* Los Angeles, Calif.: University of California, Report 68–38, 1968.
6. Lin, P. M., and G. E. Alderson. *SNAP—A Computer Program for Generating Symbolic Network Functions.* Lafayette, Ind.: Purdue University, School of Electrical Engineering, Tech. Report TR-EE-70-16, Aug. 1970.
7. Demari, A. "On-Line Computer Active Network Analysis and Design in Symbolic Form." *Proc. 2nd Cornell Elec. Eng. Conf.*, pp. 94–106, Aug. 1969.
8. Mason, S. J., and H. J. Zimmermann. *Electronic Circuits, Signals and Systems.* New York: John Wiley & Sons, Inc., 1960.

9. ZOBRIST, G. W. "Signal-Flow Graphs as an Aid in the Design of Linear Circuits by Computer," Chap. 5 in G. J. Herskowitz, ed., *Computer-Aided Integrated Circuit Design*. New York: John Wiley & Sons, Inc., 1966.
10. CHEN, W. K. *Applied Graph Theory*. Amsterdam: North-Holland Publishing Company, 1971.
11. ROSKA, T. "Generating Network Functions by Digital Computers Using Prime Numbers." *Summer School on Circuit Theory, Short Contr.*, Vol. 1, pp. 210–217, Prague, 1968.
12. MAYEDA, W., and S. SESHU. *Topological Formulas for Network Functions*. Urbana, Ill.: University of Illinois. Engineering Experiment Station, Bull. No. 446, 1957.
13. CHEN, W. K. "Topological Analysis for Active Networks." *IEEE Trans. Circuit Theory*, Vol. CT-12, pp. 85–91, Mar. 1965.
14. CALAHAN, D. A. *Linear Network Analysis and Realization Digital Computer Programs: An Instruction Manual*. Urbana, Ill.: University of Illinois, Bulletin 472, 1965.
15. GATTS, T. F., Jr., and N. R. MALIK. "Topological Analysis Program for Linear Active Networks (TAPLAN)." *Proc. 13th Midwest Symp. Circuit Theory*, Minneapolis, Minn., May 1970.
16. MANAKTALA, V. K. "A Versatile Small Circuit-Analysis Program." *IEEE Trans. Circuit Theory*, Vol. CT-20, pp. 583–586, Sept. 1973.
17. MCCLENAHAN, J. O., and S. P. CHAN. "Computer Analysis of General Linear Networks Using Digraphs." *Intern. J. Electron.*, No. 22, pp. 153–191, 1972.
18. ALDERSON, G. E., and P. M. LIN. "Integrating Topological and Numerical Methods for Semi-Symbolic Network Analysis." *Proc. 8th Allerton Conf. Circuits and Systems*, pp. 646–654, Urbana, Ill., 1970.
19. PETERS, G., and J. H. WILKINSON. "$Ax = \lambda Bx$ and the Generalized Eigenproblem." *SIAM J. Num. Anal.*, Vol. 7, pp. 470–493, Dec. 1970.
20. MOLER, C. B., and G. STEWART. "An Algorithm for Generalized Matrix Eigenvalue Problems." *SIAM J. Num. Anal.*, Vol. 10, pp. 241–256, Apr. 1973.
21. KAUFMAN, I. "On Poles and Zeros of Linear Systems." *IEEE Trans. Circuit Theory*, Vol. CT-20, pp. 93–101, Mar. 1973.
22. ALDERSON, G. E., and P. M. LIN. "Computer Generation of Symbolic Network Functions—A New Theory and Implementation." *IEEE Trans. Circuit Theory*, Vol. CT-20, pp. 48–56, Jan. 1973.
23. LIN, P. M., and G. E. ALDERSON. "Symbolic Network Functions by a Single Pathfinding Algorithm." *Proc. 7th Allerton Conf. Circuits and Systems*, pp. 196–205, Urbana, Ill., 1969.
24. LEON, B. J., and P. A. WINTZ. *Basic Linear Networks for Electrical and Electronics Engineers*. New York: Holt, Rinehart and Winston, Inc., 1970.

PROBLEMS

14-1. Refer to Fig. P14-1.

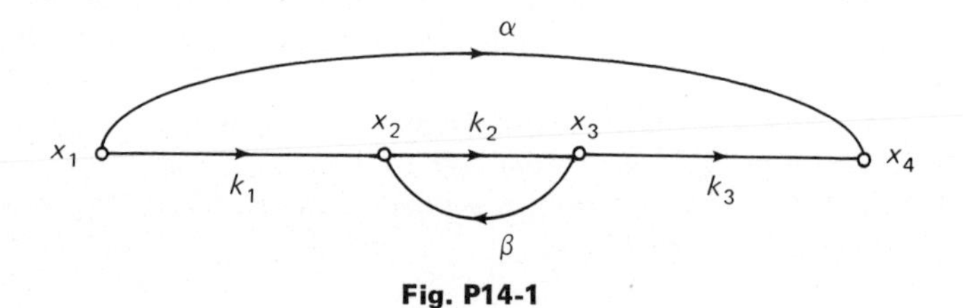

Fig. P14-1

(a) Write the simultaneous equations represented by the signal-flow graph.
(b) Find x_4/x_1 using Cramer's rule.
(c) Find x_4/x_1 by the topological rule of Eq. (14-7).

14-2. Refer to Fig. P14-2.
(a) Write the simultaneous equations represented by the signal-flow graph.
(b) Find x_4/x_1 by the use of Eq. (14-7).

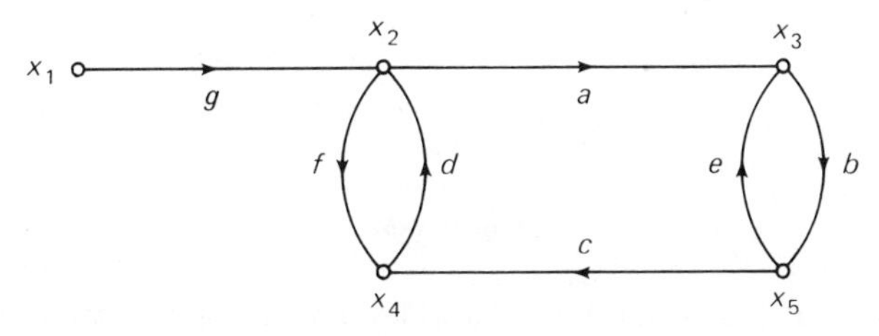

Fig. P14-2

14-3. For the amplifier network shown in Fig. P14-3,
(a) Choose a tree and formulate the corresponding compact signal-flow graph.
(b) Find the transfer function V_o/V_i by the topological rule given by Eq. (14-7).
(c) Add one more branch to the SFG of part (a) to obtain a closed signal-flow graph, and then find V_o/V_i by the sorting rule as given by Eq. (14-9).

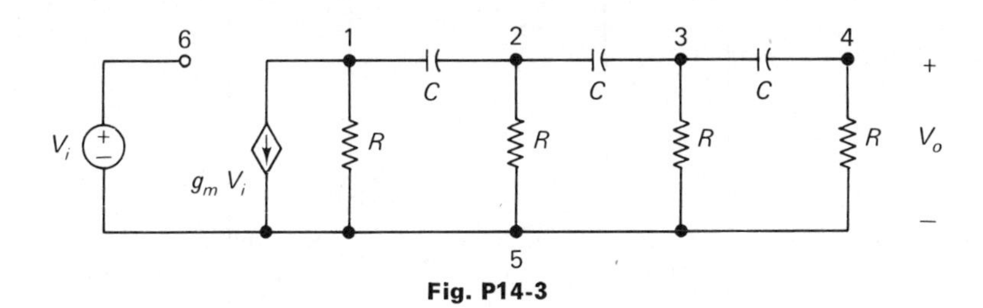

Fig. P14-3

14-4. Repeat Problem 14-3 for the bridged-T network shown in Fig. P14-4.

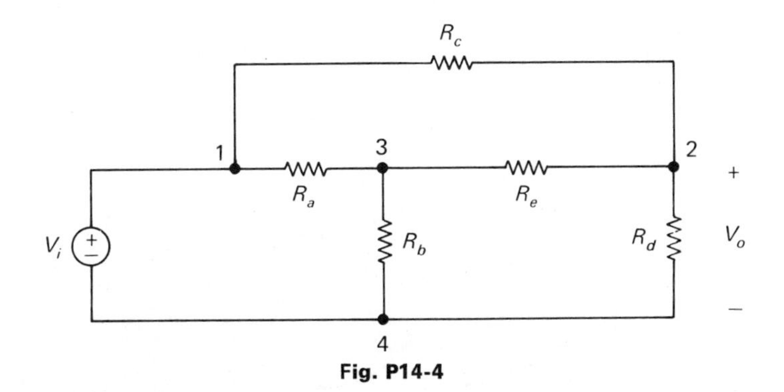

Fig. P14-4

14-5. Using the method described in Section 14-2-3, find all paths from node 4 to node 6 in the graph of Fig. P14-5 (see next page).

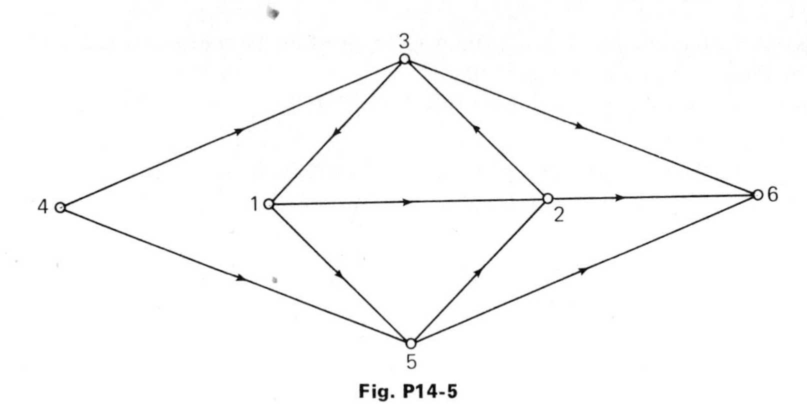

Fig. P14-5

14-6. For the directed graph shown in Fig. P14-6, find all paths from node S to node T, and all loops.

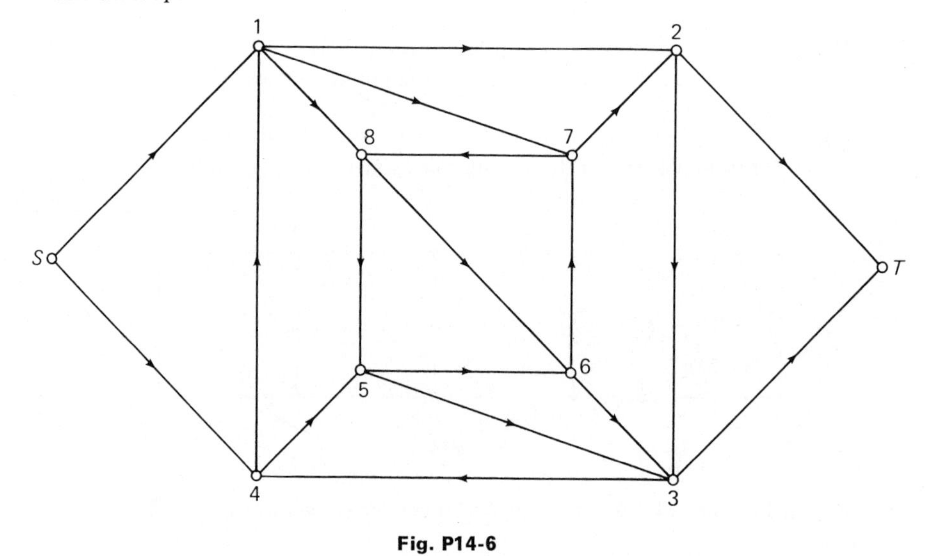

Fig. P14-6

14-7. For the signal-flow graph shown in Fig. P14-7, find the ratio Y/X by the use of Eq. (14-7).

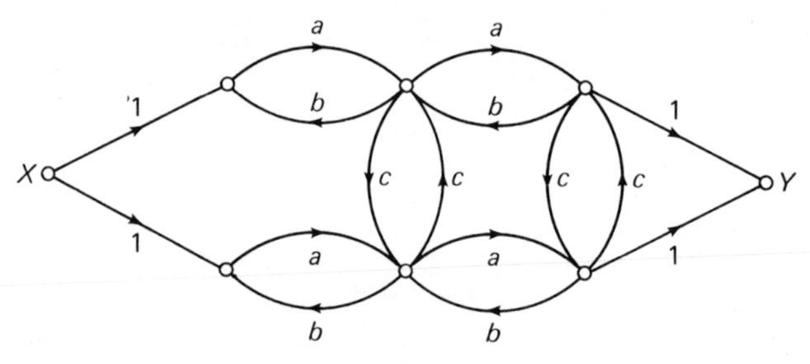

Fig. P14-7

14-8. Repeat Problem 14-7 for the signal-flow graph of Fig. P14-8.

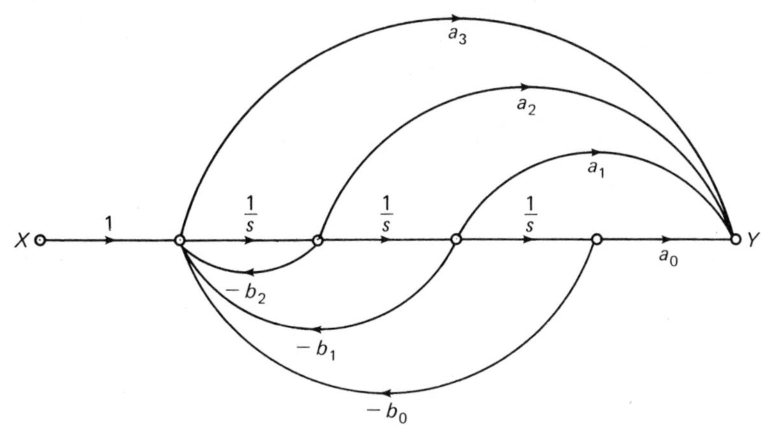

Fig. P14-8

14-9. Repeat Problem 14-7 for the signal-flow graph of Fig. P14-9.

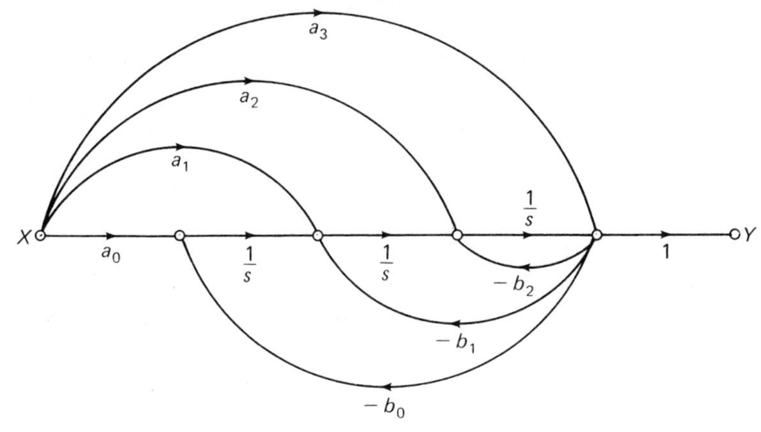

Fig. P14-9

14-10. For the circuit of Fig. P14-10, formulate a suitable signal-flow graph and find R_i.

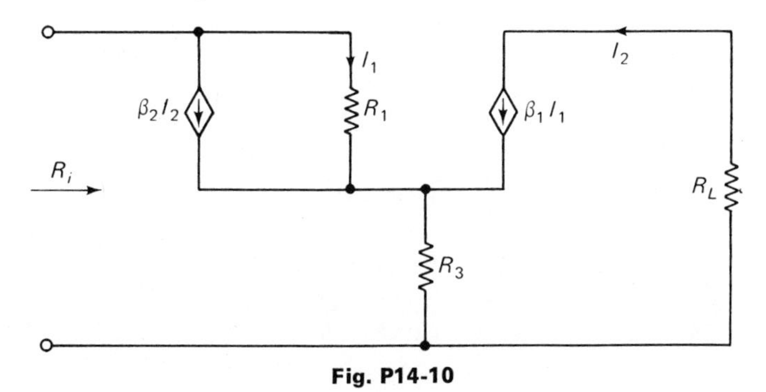

Fig. P14-10

14-11. Repeat Problem 14-10 for Fig. P14-11. Also find I_o/I_i. *Hint:* The branch carrying I_2 must be included in the tree.

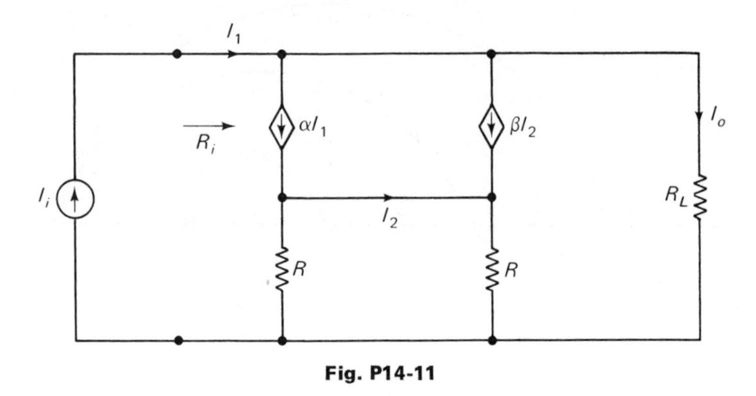

Fig. P14-11

14-12. Repeat Problem 14-10 for Fig. P14-12, but find V_o/I_i instead.

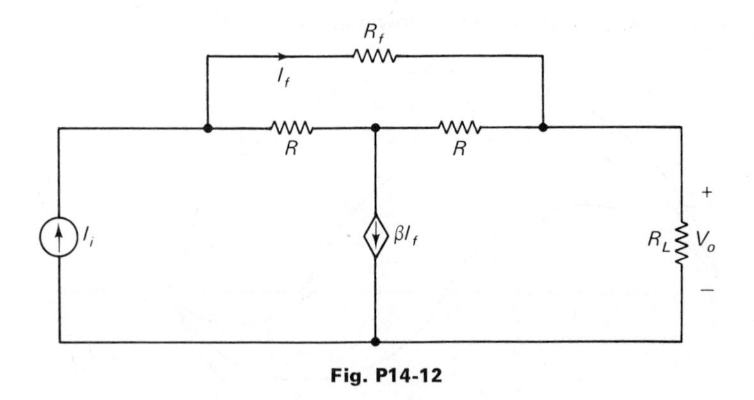

Fig. P14-12

14-13. Repeat Problem 14-10 for Fig. P14-13, but find V_o/V_s instead.

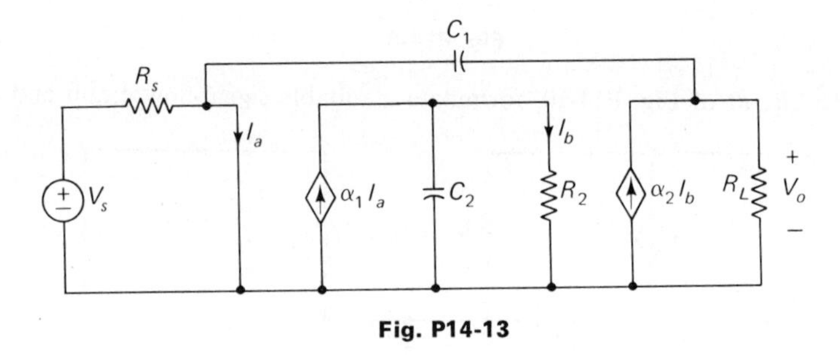

Fig. P14-13

14-14. For the network shown in Fig. P14-14, find Z_{in} by the tree-enumeration method. *Hint:* Augment the network with $\hat{y}_s$ only. Apply Theorem 14-2 and Eqs. (14-15b) and (14-13a) (see references [1–4] of Chapter 3 for methods of enumerating trees).

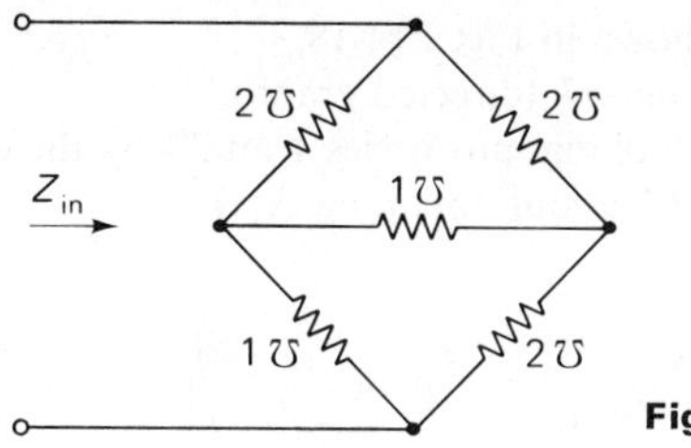

Fig. P14-14

14-15. Repeat Problem 14-10 for the network shown in Fig. P14-15.

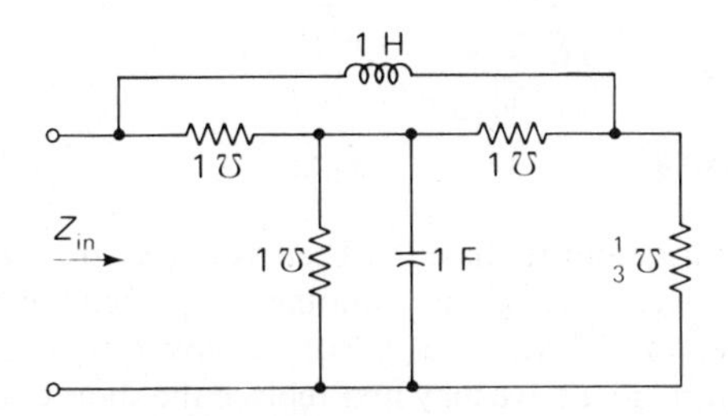

Fig. P14-15

14-16. Repeat Problem 14-10 for the network shown in Fig. P14-16. Note that Theorem 14-2 is also valid for networks containing negative *RLC* elements.

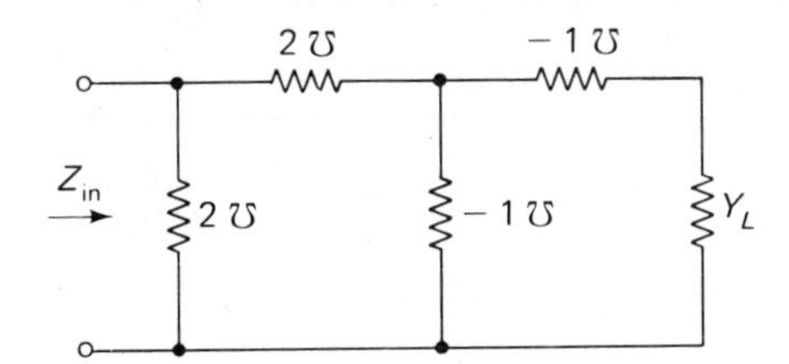

Fig. P14-16

14-17. For the directed graph shown in Fig. P14-17,
(a) Find all directed trees rooted at node 4.
(b) Find all directed trees rooted at node 2.
(c) With nodes 1 and 4 joined, find all directed trees rooted at node 4.

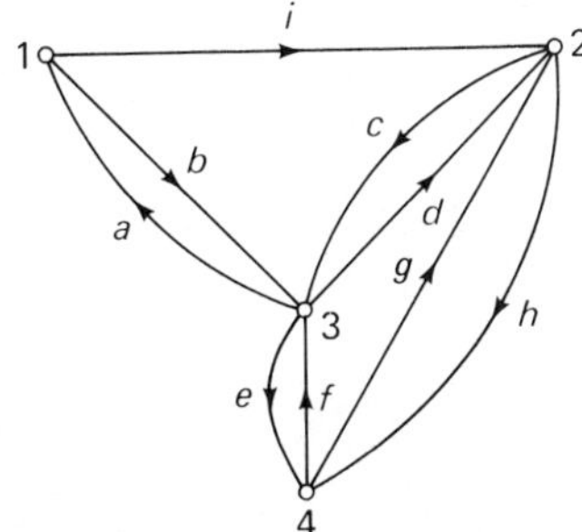

Fig. P14-17

14-18. For the active network shown in Fig. P14-18,
(a) Draw the associated *digraph* (directed graph).
(b) Find the admittance looking into nodes 1 and 4 by the enumeration of directed trees (see Section 14-3-2 about obtaining Δ_{11}).

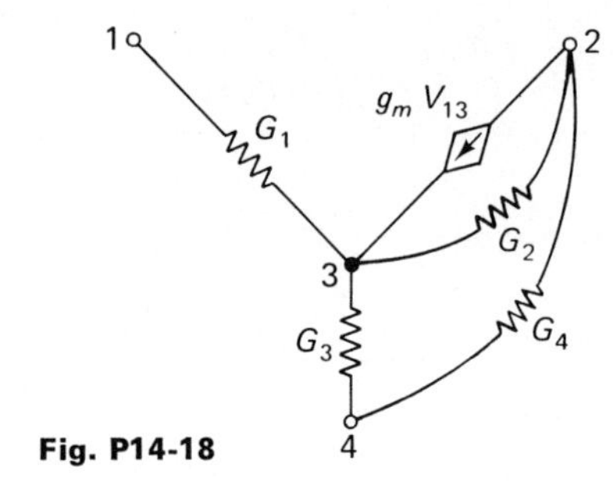

Fig. P14-18

14-19. In Eq. (14-13), the output quantities have all been voltages. If an output current I_o is desired, and I_o is the current through an admittance Y_L, then the problem is essentially the same as determining V_o (across Y_L). For we have $I_o/V_i = Y_L(V_o/V_i)$. If I_o is the current through a short circuit, we may first replace the short circuit by an admittance Y_L, then find V_o (across Y_L), and finally let Y_L approach infinity. Applying this method, find I_o/I_i for the network shown in Fig. P14-19.

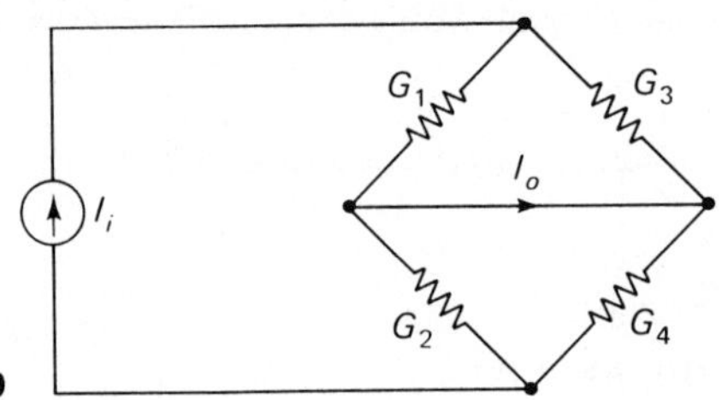

Fig. P14-19

CHAPTER 15

Frequency-Domain and Time-Domain Sensitivity Calculations

15-1 INTRODUCTION

In the design of any system, it is important to know the effect on the system performance due to the variations of some system parameters. In the case of lumped, linear, time-invariant networks, a precise measure of this effect can be expressed in terms of the sensitivity function to be defined next.

Let $T(s)$ be any network function of interest (driving-point immittances, transfer immittances, voltage ratios, or current ratios). At any particular frequency, T is in general a complex number. The value of T of course varies with the frequency ($s = j\omega$) in general, except for resistive networks. Let x be any parameter associated with some network element; x may be the element value (such as the impedance or transadmittance) or some physical parameter (such as temperature or pressure) that affects the element value.

Definition. *The* ***relative sensitivity****, or simply the* ***sensitivity of a network function*** *T, with respect to a parameter x, denoted by S_x^T, is defined as*

$$S_x^T \triangleq \frac{\partial T}{\partial x} \cdot \frac{x}{T} = \frac{\dfrac{\partial T}{T}}{\dfrac{\partial x}{x}} = \frac{\partial \ln T}{\partial \ln x} \tag{15-1}$$

According to Eq. (15-1), we may interpret the sensitivity S_x^T as the ratio of the fractional change in the network function T to the fractional change in the parameter x, provided that all changes are sufficiently small (approaching zero theoretically). The sensitivity, as defined by Eq. (15-1), is just the reciprocal of the sensitivity defined by Bode in his classic work [1]. We use Eq. (15-1) because it is widely accepted in literature nowadays. Sometimes we refer to S_x^T as the *normalized sensitivity*, in contrast with the *unnormalized sensitivity*, which is simply the partial derivative $\partial T/\partial x$.

Sensitivity information is valuable in several applications. An obvious one concerns the tolerance specification of components. Consider the simple example of a resistance voltage divider with the voltage ratio given by

$$T = \frac{R_2}{R_1 + R_2} = \frac{1}{3}$$

From Eq. (15-1), we can easily obtain

$$S_{R_1}^T = \frac{-R_1}{R_1 + R_2} = -\frac{2}{3}$$

$$S_{R_2}^T = \frac{R_1}{R_1 + R_2} = \frac{2}{3}$$

If T is to be accurate within ± 2 per cent, then the required tolerance of R_1 and R_2 (assumed equal) can be determined approximately from

$$\frac{\Delta T}{T} \cong S_{R_1}^T \frac{\Delta R_1}{R_1} + S_{R_2}^T \frac{\Delta R_2}{R_2} = \frac{2}{3}\left(-\frac{\Delta R_1}{R_1} + \frac{\Delta R_2}{R_2}\right)$$

The requirement $|\Delta T/T| \leq 0.02$ will be satisfied if $|\Delta R_1/R_1| \leq 0.015$ and $|\Delta R_2/R_2| \leq 0.015$. Thus, 1 per cent precision resistors will have to be used for the voltage dividers. The reader is referred to [2, 3, 4] for a detailed account of tolerance analysis.

Another important application of sensitivity analysis is in the automated design of linear networks to meet frequency-domain specifications. A brief description of such an application is given in Section 15-6.

A brute-force method to determine S_x^T is to vary x slightly ($\Delta x \to 0$) and calculate the change in T. This amounts to performing numerical differentiation and has some serious drawbacks. First, the accuracy will be poor because of taking the difference between two nearly equal numbers. Second, one analysis of the network is needed for the sensitivity at each frequency and with respect to each parameter x_i. Some general formulas for sensitivity that avoid numerical differentiation have been derived by Bode [1]. His formulas are in terms of the determinant and cofactors of the loop-impedance matrix or node-admittance matrix. For large networks, it is very difficult to obtain these cofactors as *expressions* containing the parameters ($x_1, \ldots, x_n$). Therefore, at each frequency, and for each $S_{x_i}^T$, these cofactors have to be evaluated numerically. In this respect, there is little advantage over the brute-force method. However, numerical differentiation has been avoided, and hence the accuracy is greatly improved.

During the 1960s, tremendous progress was made in computer-aided circuit analysis. We now have many general-purpose computer simulation programs that perform dc, ac, and transient analysis for networks in the 100-branch range. With these powerful tools at hand, several computer-oriented methods have been developed for the calculation of sensitivities. In the next few sections, we shall give an elementary account of the incremental-network method, the adjoint-network method, and the symbolic-function method. Only the calculation of first-order sensitivity in the frequency domain and the time domain will be considered here. Other topics, such as higher-order sensitivities, summed sensitivity invariants, pole–zero sensitivity, and transient sensitivity, will not be treated. Interested readers may refer to [4, 5] for these more advanced topics.

We note that the calculations of $\partial T/\partial x$ and $\partial V/\partial x$ (or $\partial I/\partial x$) are essentially the same problem. For example, let $T = Z_{\text{in}}$ and let I_{in} be an excitation that is independent of the parameter x. Then

$$\frac{\partial T}{\partial x} = \frac{\partial Z_{\text{in}}}{\partial x} = \frac{\partial}{\partial x}\left(\frac{V_{\text{in}}}{I_{\text{in}}}\right) = \frac{1}{I_{\text{in}}} \cdot \frac{\partial V_{\text{in}}}{\partial x}$$

Thus, $\partial Z_{\text{in}}/\partial x$ and $\partial V_{\text{in}}/\partial x$ differ only by a factor $1/I_{\text{in}}$, which is a constant with respect to x.

15-2 INCREMENTAL-NETWORK APPROACH

Figure 15-1(a) shows a portion of a typical linear network N. For fixed topology and excitations, the branch voltages and currents are functions of branch immittances. We wish to determine the changes in branch voltages and currents due to the changes in branch immittances. To this end, we shall vary the impedance of each branch by a *slight amount* and obtain a *perturbed* network N_p, as shown in Fig. 15-1(b). For N, we can write (see Chap. 3)

$$\text{KCL:} \quad \boldsymbol{AI} = \boldsymbol{0} \tag{15-2}$$

$$\text{KVL:} \quad \boldsymbol{BV} = \boldsymbol{0} \tag{15-3}$$

where $\boldsymbol{A}$ and $\boldsymbol{B}$ are the reduced incidence matrix and fundamental loop matrix, respectively, and $\boldsymbol{I}$ and $\boldsymbol{V}$ are the branch current vector and branch voltage vector, respectively. Since the perturbed network N_p has the same topology as N, for N_p we have

$$\text{KCL:} \quad \boldsymbol{A}(\boldsymbol{I} + \Delta\boldsymbol{I}) = \boldsymbol{0} \tag{15-4}$$

$$\text{KVL:} \quad \boldsymbol{B}(\boldsymbol{V} + \Delta\boldsymbol{V}) = \boldsymbol{0} \tag{15-5}$$

From Eqs. (15-2)–(15-5), we immediately have

$$\boldsymbol{A}\Delta\boldsymbol{I} = \boldsymbol{0} \tag{15-6}$$

$$\boldsymbol{B}\Delta\boldsymbol{V} = \boldsymbol{0} \tag{15-7}$$

which indicate that the incremental currents $\Delta\boldsymbol{I}$ and incremental voltages $\Delta\boldsymbol{V}$ have the same constraints as $\boldsymbol{I}$ and $\boldsymbol{V}$. Therefore, $\Delta\boldsymbol{I}$ and $\Delta\boldsymbol{V}$ could possibly be the branch currents and voltages of some network N_i having the same topology as N, provided that the branch characteristics of N_i are properly defined. We shall next investigate how branch characteristics of N_i are to be defined for the purpose of producing $\Delta\boldsymbol{I}$ and $\Delta\boldsymbol{V}$.

Consider an element in N with impedance Z. Then

$$V = ZI \tag{15-8}$$

For the perturbed network, the same element is described by

$$\begin{aligned} V + \Delta V &= (Z + \Delta Z)(I + \Delta I) \\ &= ZI + Z\Delta I + I\Delta Z + \Delta Z \cdot \Delta I \end{aligned} \tag{15-9}$$

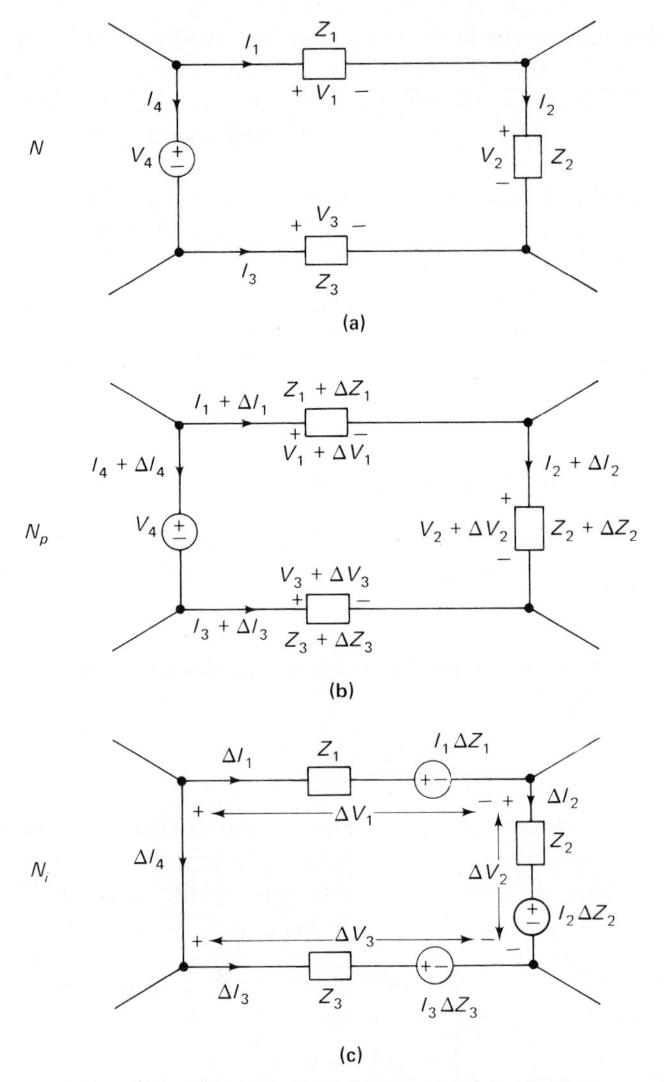

Fig. 15-1. Derivation of the incremental network.

From Eqs. (15-8) and (15-9), we have

$$\Delta V = Z\Delta I + I\Delta Z + \Delta Z \cdot \Delta I \tag{15-10}$$

Provided that ΔZ (and hence ΔI and ΔV) is *infinitesimally small*, we can neglect higher-order terms and rewrite Eq. (15-10) as

$$\Delta V = Z\Delta I + I\Delta Z \tag{15-11}$$

which indicates that in N_i the branch having $(\Delta V, \Delta I)$ consists of an impedance Z (the original impedance in N) in series with a voltage source $I\Delta Z$. This is illustrated in Fig. 15-1(c) and also in the first row of Table 15-1.

TABLE 15-1. Construction of the Incremental Network

	N, ORIGINAL NETWORK	N_i, INCREMENTAL NETWORK
1	I_j Z_j + V_j −	ΔI_j Z_j $I_j\,\Delta Z_j$ + ΔV_j −
2	Y_j I_j + V_j −	ΔI_j Y_j + $V_j\,\Delta Y_j$ − ΔV_j
3	+ V_j − $g_m V_j$ I_k + V_k −	+ ΔV_j − $V_j\,\Delta g_m$ ΔI_k + ΔV_k $g_m\,\Delta V_j$ −
4	I_j $r_m I_j$ I_k + V_k −	ΔI_j $r_m\,\Delta I_j$ $I_j \Delta r_m$ ΔI_k + ΔV_k −
5	+ V_j − μV_j I_k + V_k −	+ ΔV_j − $\mu\,\Delta V_j$ $V_j\;\Delta\mu$ ΔI_k + ΔV_k −
6	I_j βI_j I_k + V_k −	ΔI_j $I_j\;\Delta\beta$ $\beta\,\Delta I_j$ ΔI_k + ΔV_k −
7	I_s + V_s −	$\Delta I_s = 0$ + ΔV_s −
8	I_s + V_s −	ΔI_s + $\Delta V_s = 0$ −

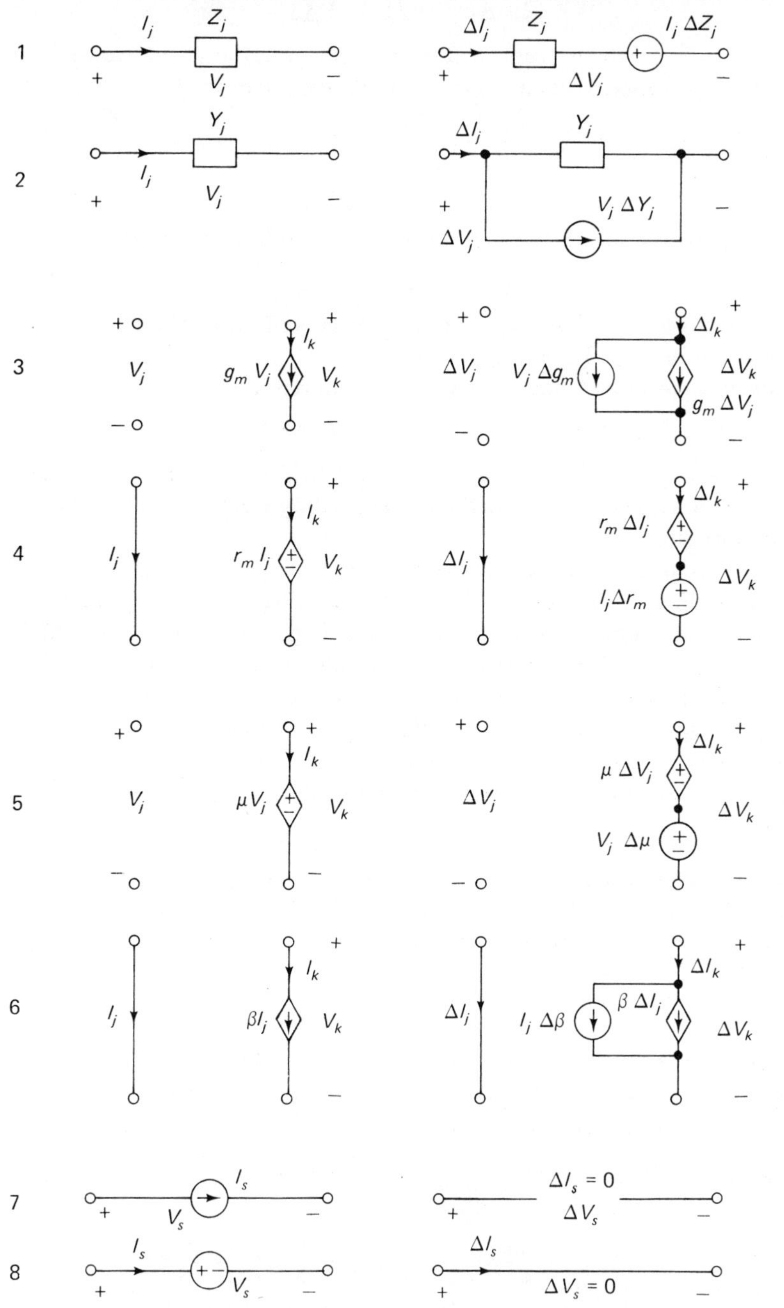

By a similar reasoning, we can use the admittance description of a branch and obtain the following relationship:

$$\Delta I = Y\Delta V + V\Delta Y \tag{15-12}$$

The corresponding network representation is shown in the second row of Table 15-1.

For a voltage-controlled current source, we have, in N,

$$I_k = g_m V_j \tag{15-13}$$

and in N_p,

$$I_k + \Delta I_k = (g_m + \Delta g_m)(V_j + \Delta V_j) \tag{15-14}$$

From Eqs. (15-13) and (15-14), and neglecting second-order effects, we have

$$\Delta I_k = g_m \Delta V_j + \Delta g_m \cdot V_j \tag{15-15}$$

Other types of controlled sources may be treated similarly. The network representations are given in rows 3–6 of Table 15-1.

For an *unperturbed* independent current source, we have

$$I_s = \text{constant in } N \text{ and } N_p$$

Therefore, $\Delta I_s = 0$, indicating an open circuit in N_i. Similarly, for an *unperturbed* independent voltage source in N, the corresponding element in N_i is $\Delta V_s = 0$, indicating a short circuit.

The preceding results are summarized in Table 15-1. The network N_i derived from N according to the rules of Table 15-1, as illustrated in Fig. 15-1(c), is an *incremental network*, since branch currents and voltages in N_i are ΔI's and ΔV's. (This is not to be confused with the linear incremental model discussed in Chapter 2.) It should be emphasized that the rules given in Table 15-1 for the construction of the *incremental network are derived under the assumptions that all parameter changes are very small* (approaching zero, theoretically) and that second-order effects can be neglected. The use of N_i for sensitivity calculations will now be illustrated by two examples.

EXAMPLE 15-1. The active network N shown in Fig. 15-2(a) has nominal element values as follows: $y_1 = 2$ ℧, $y_2 = 4$ ℧, $y_3 = 1$ ℧, and $g_m = 2$ ℧. Find the partial derivatives of V_1 and V_2 with respect to y_1, y_2, y_3, and g_m. Also find $\partial Z_{in}/\partial g_m$.

Solution:

Step 1. Perform an analysis of N. The result is shown in Fig. 15-2(b).

Step 2. Construct the incremental network N_i using Table 15-1 and the result of step 1. The result is shown in Fig. 15-2(c).

Step 3. Perform an analysis of the incremental network. For this particular example, a nodal analysis of N_i yields

$$\Delta V_1 = -\frac{25}{64}\Delta y_1 + \frac{1}{64}\Delta y_2 - \frac{3}{8}\Delta y_3 - \frac{5}{64}\Delta g_m \tag{15-16}$$

$$\Delta V_2 = \frac{5}{64}\Delta y_1 + \frac{3}{64}\Delta y_2 + \frac{3}{8}\Delta y_3 - \frac{15}{64}\Delta g_m \tag{15-17}$$

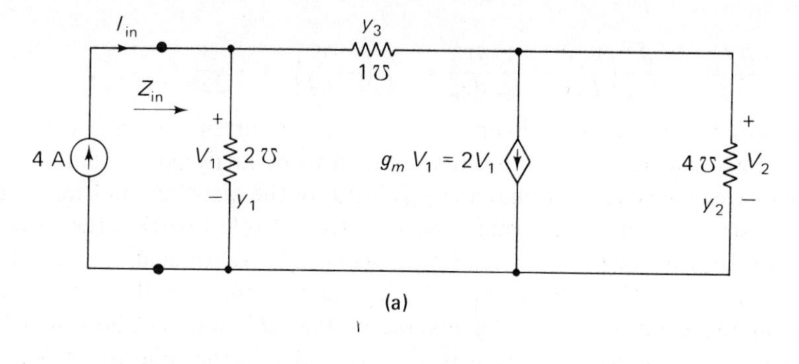

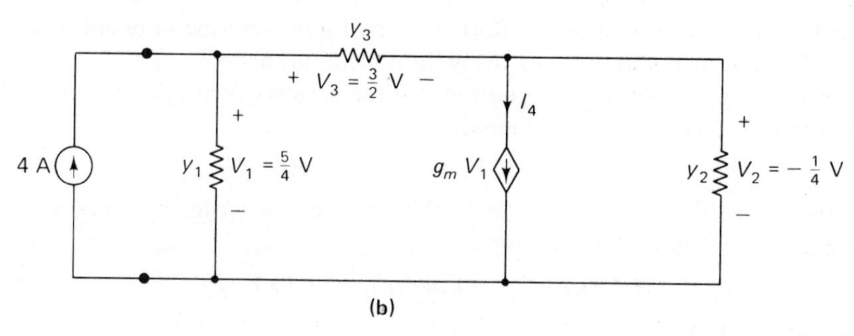

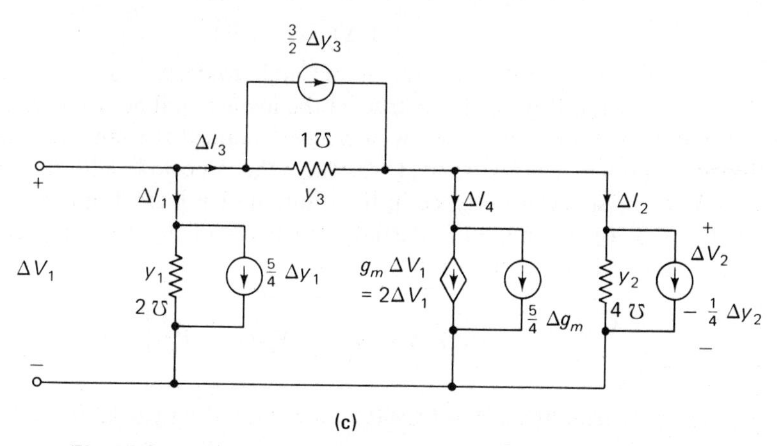

Fig. 15-2. Sensitivity calculation by the incremental-network method.

Step 4. From Eqs. (15-16) and (15-17), we obtain the following partial derivatives (recalling that in deriving N_i, we have assumed $\Delta y_1 \longrightarrow 0$, etc.):

$$\frac{\partial V_1}{\partial y_1} = -\frac{25}{64}, \quad \frac{\partial V_1}{\partial y_2} = \frac{1}{64}, \quad \frac{\partial V_1}{\partial y_3} = -\frac{3}{8}, \quad \frac{\partial V_1}{\partial g_m} = -\frac{5}{64}$$

$$\frac{\partial V_2}{\partial y_1} = \frac{5}{64}, \quad \frac{\partial V_2}{\partial y_2} = \frac{3}{64}, \quad \frac{\partial V_2}{\partial y_3} = \frac{3}{8}, \quad \frac{\partial V_2}{\partial g_m} = -\frac{15}{64}$$

and

$$\frac{\partial Z_{\text{in}}}{\partial g_m} = \frac{\partial}{\partial g_m}\left(\frac{V_1}{I_{\text{in}}}\right) = \frac{1}{I_{\text{in}}}\frac{\partial V_1}{\partial g_m} = \frac{1}{4}\left(-\frac{5}{64}\right) = -\frac{5}{256}$$

In our analysis of N_i, we have kept the independent current sources in symbolic form (e.g., $\frac{5}{4}\,\Delta y_1$) and the network is small enough to be analyzed manually. When large networks are considered, manual analysis is out of the question, and we have to resort to computer simulation programs. We note from Table 15-1 that the networks N and N_i differ only in the placements and the values of independent sources. Thus, if the equation for N is of the form $\boldsymbol{Px} = \boldsymbol{k}$, the equation for N_i will be of the form $\boldsymbol{Px} = \hat{\boldsymbol{k}}$. If, in the solution of N, the inverse or the $\boldsymbol{LU}$ factors (Chap. 4) of the coefficient matrix is obtained, the result is directly usable in the solution of N_i. This is a great saving in computational effort. We shall now describe in detail how this is done in the case of nodal analysis using composite branches.

From Chapter 4, the nodal equations for the network N are given by Eq. (4-73), which is repeated below for convenience:

$$(\boldsymbol{A}\boldsymbol{Y}_b\boldsymbol{A}^t)\boldsymbol{V}_n = \boldsymbol{A}(\boldsymbol{J} - \boldsymbol{Y}_b\boldsymbol{E}) \tag{15-18}$$

When the branch admittance $\boldsymbol{Y}_b$ varies slightly, we can write, to a first-order approximation, the following equation:

$$\boldsymbol{A}\,\Delta\boldsymbol{Y}_b\boldsymbol{A}^t\boldsymbol{V}_n + \boldsymbol{A}\boldsymbol{Y}_b\boldsymbol{A}^t\Delta\boldsymbol{V}_n = -\boldsymbol{A}\Delta\boldsymbol{Y}_b\cdot\boldsymbol{E} \tag{15-19}$$

from which, we have

$$\begin{aligned}(\boldsymbol{A}\boldsymbol{Y}_b\boldsymbol{A}^t)\,\Delta\boldsymbol{V}_n &= -\boldsymbol{A}\,\Delta\boldsymbol{Y}_b\cdot(\boldsymbol{E} + \boldsymbol{A}^t\boldsymbol{V}_n)\\ &= -\boldsymbol{A}\,\Delta\boldsymbol{Y}_b\cdot(\boldsymbol{E} + \hat{\boldsymbol{V}})\end{aligned} \tag{15-20}$$

where $\hat{\boldsymbol{V}}$ and $\boldsymbol{V}_n$ are, respectively, the branch and node-to-datum voltage vectors of N with composite branches. If we had constructed the incremental network N_i and then written the nodal equations for N_i, we would have obtained the same results as Eq. (15-20). Note that both Eqs. (15-19) and (15-20) have the same node-admittance matrix $\boldsymbol{A}\boldsymbol{Y}_b\boldsymbol{A}^t$. If $(\boldsymbol{A}\boldsymbol{Y}_b\boldsymbol{A}^t)^{-1}$ has been obtained in the solution of Eq. (15-18), it may be used directly in Eq. (15-20) to find $\Delta\boldsymbol{V}_n$. Partial derivatives of $\boldsymbol{V}_n$ with respect to any parameter may then be calculated from

$$\boxed{\frac{\partial}{\partial x}\boldsymbol{V}_n = -(\boldsymbol{A}\boldsymbol{Y}_b\boldsymbol{A}^t)^{-1}\boldsymbol{A}\cdot\frac{\partial}{\partial x}\boldsymbol{Y}_b\cdot(\boldsymbol{E} + \hat{\boldsymbol{V}})} \tag{15-21}$$

Such an approach is actually used in ECAP [6] for calculating partial derivatives.

EXAMPLE 15-2. Solve the problem of Example 15-1 by the use of Eq. (15-21).

Solution:

$$\boldsymbol{Y}_b = \begin{bmatrix} y_1 & 0 & 0 & 0\\ 0 & y_2 & 0 & 0\\ 0 & 0 & y_3 & 0\\ g_m & 0 & 0 & 0\end{bmatrix} = \begin{bmatrix} 2 & 0 & 0 & 0\\ 0 & 4 & 0 & 0\\ 0 & 0 & 1 & 0\\ 2 & 0 & 0 & 0\end{bmatrix}$$

$$A = \begin{bmatrix} 1 & 0 & 1 & 0 \\ 0 & 1 & -1 & 1 \end{bmatrix}$$

$$AY_bA^t = \begin{bmatrix} 3 & -1 \\ 1 & 5 \end{bmatrix}$$

(see Section 4-6 for a direct construction of this matrix), and

$$(AY_bA^t)^{-1} = \frac{1}{16}\begin{bmatrix} 5 & 1 \\ -1 & 3 \end{bmatrix}$$

$$\boldsymbol{E} = \boldsymbol{0}, \qquad \boldsymbol{J} = [4, 0, 0, 0]^t$$

The solution of Eq. (15-18) yields

$$V_n = [1.25, -0.25]^t$$

and

$$\hat{V} = A^t V_n = [1.25, -0.25, 1.5, -0.25]^t$$

Substituting these results into Eq. (15-21), we have

$$\frac{\partial}{\partial x} V_n = -\frac{1}{16}\begin{bmatrix} 5 & 1 & 4 & 1 \\ -1 & 3 & -4 & 3 \end{bmatrix} \frac{\partial}{\partial x} Y_b \begin{bmatrix} 1.25 \\ -0.25 \\ 1.5 \\ -0.25 \end{bmatrix} \tag{15-22}$$

If $x = y_1$, then

$$\frac{\partial}{\partial x} Y_b = \frac{\partial}{\partial y_1} Y_b = \begin{bmatrix} 1 & 0 & 0 & 0 \\ 0 & 0 & 0 & 0 \\ 0 & 0 & 0 & 0 \\ 0 & 0 & 0 & 0 \end{bmatrix}$$

and

$$\frac{\partial}{\partial x}\begin{bmatrix} V_1 \\ V_2 \end{bmatrix} = \frac{\partial}{\partial y_1}\begin{bmatrix} V_1 \\ V_2 \end{bmatrix} = \begin{bmatrix} -\frac{5}{16} \times 1.25 \\ \frac{1}{16} \times 1.25 \end{bmatrix} = \begin{bmatrix} -\frac{25}{64} \\ \frac{5}{64} \end{bmatrix}$$

By a similar process, from Eq. (15-22) we obtain

$$\frac{\partial}{\partial x_2}\begin{bmatrix} V_1 \\ V_2 \end{bmatrix} = \begin{bmatrix} \frac{1}{64} \\ \frac{3}{64} \end{bmatrix}, \quad \frac{\partial}{\partial y_3}\begin{bmatrix} V_1 \\ V_2 \end{bmatrix} = \begin{bmatrix} -\frac{3}{8} \\ \frac{3}{8} \end{bmatrix}, \quad \frac{\partial}{\partial g_m}\begin{bmatrix} V_1 \\ V_2 \end{bmatrix} = \begin{bmatrix} -\frac{5}{64} \\ -\frac{15}{64} \end{bmatrix}$$

As shown by Eq. (15-21), the calculation of partial derivatives of all node voltages with respect to all branch admittances requires the following:

1. Solution of the given network N to obtain branch voltages $\hat{V}$.
2. Inversion of the nodal-admittance matrix AY_bA^t.
3. Multiplications as indicated by Eq. (15-21), or operations to the same effect.

As far as computational effort is concerned, the incremental-network approach is less efficient than the adjoint-network approach, to be discussed in Section 15-3. However, the incremental-network approach has the following advantages:

1. It provides better insight into the effect of parameter variations. For example, the effect of ΔZ change may be visualized as that due to a voltage source of value $I \cdot \Delta Z$ (I is the current through Z before the change) in series with Z. The physical picture is easy to grasp.
2. In the analysis of the incremental network, the incremental currents and voltages *of all branches* are available from the calculations. Such incremental quantities are themselves of interest in some applications.

See [7] for additional examples of the applications of incremental networks.

15-3 ADJOINT-NETWORK APPROACH

Referring to Fig. 15-1(a) and (b), we see that both the original network N and the perturbed network N_p have the same topology. More precisely, they have the same incidence matrix (assuming proper branch and node numbering). As will be shown shortly, when two networks possess the same topology, there is an interesting relationship among their branch voltages and currents, commonly referred to as the Tellegen's theorem [8]. Tellegen's theorem has found an abundance of applications in network theory [9]. However, in this book we shall be concerned only with its application in the calculation of sensitivity functions. To do this, we have to digress for a while to introduce the theorem and the concept of adjoint networks. Finally, we shall show by the analysis of two networks, the original network and its adjoint network, how the partial derivatives of any network function T with respect to *all* branch parameters may be determined (at some given frequency).

15-3-1 Tellegen's Theorem

In Chapter 3 we presented some basic concepts and results in network topology. A brief review of Sections 3-1 to 3-5 is expected of the reader at this point so that we may use the previous notations and results. In particular, note that lowercase letters i and v indicate functions of time, and capital letters I and V indicate phasors (for ac analysis).

Kirchhoff's voltage law and current law are algebraic constraints on branch voltages and currents arising from the interconnection of branches and are independent of the branch characteristics. They may be expressed in several equivalent forms. For example, KVL equations may be written as $\boldsymbol{Bv} = \mathbf{0}$ or, equivalently, as $\boldsymbol{v} = \boldsymbol{A}^t\boldsymbol{v}_n$ (see Section 3-5); KCL equations may be written as $\boldsymbol{Ai} = \mathbf{0}$. From these relationships, we immediately have

$$\boldsymbol{v}^t\boldsymbol{i} = (\boldsymbol{A}^t\boldsymbol{v}_n)^t\boldsymbol{i} = \boldsymbol{v}_n^t\boldsymbol{A}\boldsymbol{i} = 0$$

Thus, for any lumped network N we see that

$$\boldsymbol{v}^t\boldsymbol{i} = \boldsymbol{i}^t\boldsymbol{v} = 0 \tag{15-23}$$

The physical meaning of Eq. (15-23) is that the sum of *instantaneous powers* delivered to all branches is equal to zero. This is sometimes described as the conservation of power for electrical networks and can be considered as a consequence of Kirchhoff's voltage and current laws.

Consider now another network $\hat{N}$ having the same topology as N. Then $\boldsymbol{A} = \hat{\boldsymbol{A}}$ after numbering the branches and nodes properly (which we always assume has been done). The networks N and $\hat{N}$ may or may not have the same branch characteristics. Then, from $\hat{\boldsymbol{A}}\hat{\boldsymbol{i}} = \boldsymbol{0}$, $\boldsymbol{v} = \boldsymbol{A}^t\boldsymbol{v}_n$, and $\boldsymbol{A} = \hat{\boldsymbol{A}}$, we immediately have

$$\boldsymbol{v}^t\hat{\boldsymbol{i}} = (\boldsymbol{A}^t\boldsymbol{v}_n)^t\hat{\boldsymbol{i}} = \boldsymbol{v}_n^t\boldsymbol{A}\hat{\boldsymbol{i}} = \boldsymbol{v}_n^t(\hat{\boldsymbol{A}}\hat{\boldsymbol{i}}) = 0$$

By taking the transpose of this equation and exchanging the role of N and $\hat{N}$, we obtain several variations of the equation, as follows:

$$\boxed{\boldsymbol{v}^t\hat{\boldsymbol{i}} = \boldsymbol{i}^t\hat{\boldsymbol{v}} = \hat{\boldsymbol{v}}^t\boldsymbol{i} = \hat{\boldsymbol{i}}^t\boldsymbol{v} = 0} \tag{15-24}$$

Equation (15-24) is called *Tellegen's theorem*. We note that $\boldsymbol{v}^t\hat{\boldsymbol{i}}$ does not have the physical meaning of "sum of branch powers," since $\boldsymbol{v}$ and $\hat{\boldsymbol{i}}$ in general belong to two different networks.

An example will illustrate the use of Eq. (15-24). Consider the network N of Fig. 15-3(a), whose branch currents and voltages are easily shown to be

$$\boldsymbol{i} = [3, 2, 1, -3]^t$$
$$\boldsymbol{v} = [9, 6, 6, 15]^t$$

Figure 15-3(b) shows a network $\hat{N}$ having the same topology N (note the way that the branches in $\hat{N}$ are numbered to agree with N). We find that

$$\hat{\boldsymbol{i}} = [-3, -5, 2, 3]^t$$
$$\hat{\boldsymbol{v}} = [-6, 6, 6, 0]^t$$

It is easy to verify Eq. (15-24), as we have

$$\boldsymbol{v}^t\hat{\boldsymbol{i}} = \hat{\boldsymbol{i}}^t\boldsymbol{v} = 9 \times (-3) + 6 \times (-5) + 6 \times 2 + 15 \times 3 = 0$$

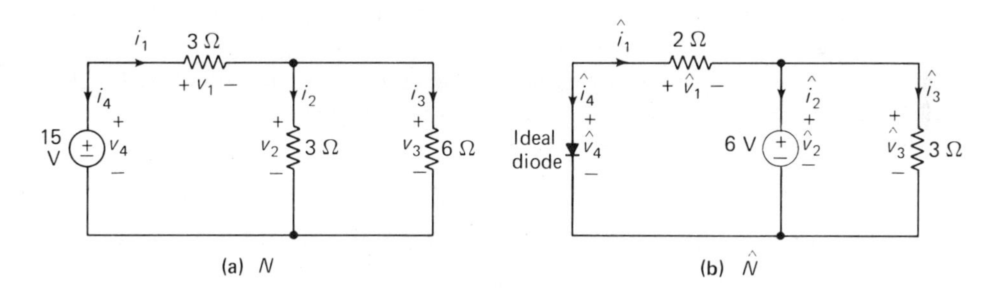

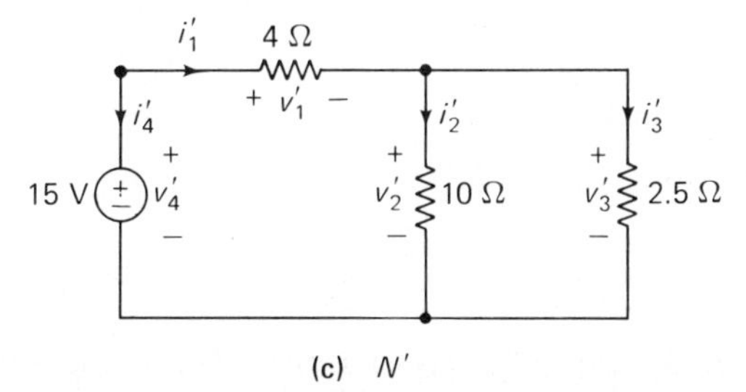

Fig. 15-3. Network for illustrating Tellegen's theorem.

and

$$\boldsymbol{i}^t\hat{\boldsymbol{v}} = \hat{\boldsymbol{v}}^t\boldsymbol{i} = 3 \times (-6) + 2 \times 6 + 1 \times 6 + (-3) \times 0 = 0$$

The only condition for Eq. (15-24) to be valid is that the two networks under consideration have the *same* topology. Now let us consider another network N' that has the same topology as N and $\hat{N}$. (A possible candidate for N' is the perturbed network N_p.) Denote the branch voltage and current vectors of N' by $\boldsymbol{v}'$ and $\boldsymbol{i}'$, respectively, and let

$$\boldsymbol{v}' = \boldsymbol{v} + \Delta\boldsymbol{v}$$

$$\boldsymbol{i}' = \boldsymbol{i} + \Delta\boldsymbol{i}$$

Since the branch characteristics of N' are completely arbitrary, the magnitudes of the elements of $\Delta\boldsymbol{v}$ and $\Delta\boldsymbol{i}$ are not necessarily small. However, N' can also be the perturbed network N_p, which is obtained from N by varying the element values only slightly (see Section 15-2). If a perturbed network N_p is considered, it is implied that the magnitudes of the elements of $\Delta\boldsymbol{v}$ and $\Delta\boldsymbol{i}$ are all very small. Regardless of the sizes of $\Delta\boldsymbol{v}$ and $\Delta\boldsymbol{i}$, by applying Eq. (15-24) to N and N' (or N_p), we have

$$\hat{\boldsymbol{i}}^t\boldsymbol{v}' = \hat{\boldsymbol{i}}^t(\boldsymbol{v} + \Delta\boldsymbol{v}) = \hat{\boldsymbol{i}}^t\boldsymbol{v} + \hat{\boldsymbol{i}}^t\,\Delta\boldsymbol{v} = 0$$

and

$$\hat{\boldsymbol{v}}^t\boldsymbol{i}' = \hat{\boldsymbol{v}}^t(\boldsymbol{i} + \Delta\boldsymbol{i}) = \hat{\boldsymbol{v}}^t\boldsymbol{i} + \hat{\boldsymbol{v}}^t\,\Delta\boldsymbol{i} = 0$$

But $\hat{\boldsymbol{i}}^t\boldsymbol{v} = 0$ and $\hat{\boldsymbol{v}}^t\boldsymbol{i} = 0$ from Eq. (15-24). Therefore, the preceding equations lead to

$$\hat{\boldsymbol{i}}^t\,\Delta\boldsymbol{v} = 0 \tag{15-25a}$$

$$\hat{\boldsymbol{v}}^t\,\Delta\boldsymbol{i} = 0 \tag{15-25b}$$

and

$$\hat{\boldsymbol{i}}^t\,\Delta\boldsymbol{v} - \hat{\boldsymbol{v}}^t\,\Delta\boldsymbol{i} = 0 \tag{15-25c}$$

To illustrate the use of this equation, consider the network N' shown in Fig. 15-3(c). The branch currents and voltages of N' are easily shown to be

$$\boldsymbol{i}' = \boldsymbol{i} + \Delta\boldsymbol{i} = [2.5, 0.5, 2.0, -3]^t$$

$$\boldsymbol{v}' = \boldsymbol{v} + \Delta\boldsymbol{v} = [10, 5, 5, 15]^t$$

Subtracting from $\boldsymbol{i}'$ and $\boldsymbol{v}'$, respectively, the $\boldsymbol{i}$ and $\boldsymbol{v}$ obtained previously for the network N, we have

$$\Delta\boldsymbol{i} = [-0.5, \ -1.5, \quad 1, \quad 0.5]^t$$

$$\Delta\boldsymbol{v} = [\ \ 1, \quad -1, \quad -1, \quad 0]^t$$

To verify Eq. (15-25), we have

$$\hat{\boldsymbol{i}}^t\Delta\boldsymbol{v} = (-3) \times 1 + (-5)(-1) + 2 \times (-1) + 3 \times 0 = 0$$

and

$$\hat{\boldsymbol{v}}^t\Delta\boldsymbol{i} = (-6)(-0.5) + 6 \times (-1.5) + 6 \times 1 + 0 \times (0.5) = 0$$

It is emphasized again that Eq. (15-25) is valid *regardless of the magnitudes of changes in currents and voltages from N to N'*. Although Eq. (15-25c) is just a trivial consequence of Eqs. (15-25a) and (15-25b), it plays a fundamental role in the adjoint-network method of sensitivity analysis to be discussed presently.

In all the preceding derivations, we have assumed currents and voltages to be the instantaneous values at some time instant t. In frequency-domain analysis of linear time-invariant networks, we work with the transforms of currents and voltages (Laplace transforms, Fourier transforms, or phasor transforms). By repeating the derivations, it is easily seen that all previous relationships are still valid when $\boldsymbol{i}$ and $\boldsymbol{v}$ are replaced by their transforms $\boldsymbol{I}$ and $\boldsymbol{V}$, respectively. Two equations of particular importance in later discussions are repeated here.

$$\boldsymbol{V}^t\hat{\boldsymbol{I}} = \boldsymbol{I}^t\hat{\boldsymbol{V}} = \hat{\boldsymbol{V}}^t\boldsymbol{I} = \hat{\boldsymbol{I}}^t\boldsymbol{V} = 0 \tag{15-26}$$

$$\hat{\boldsymbol{I}}^t\Delta\boldsymbol{V} - \hat{\boldsymbol{V}}^t\Delta\boldsymbol{I} = 0 \tag{15-27}$$

15-3-2 Adjoint Network

Two linear time-invariant networks N and $\hat{N}$ are said to be adjoint networks of each other if the following three conditions are satisfied:

Condition 1. Both networks have the same topology; i.e., $\boldsymbol{A} = \hat{\boldsymbol{A}}$. For controlled sources, we consider a controlling voltage as that across an open-circuit branch, and a controlling current as that through a short-circuit branch.

Condition 2. If the nonindependent-source branches of N and $\hat{N}$ possess branch impedance matrices $\boldsymbol{Z}_b$ and $\hat{\boldsymbol{Z}}_b$, respectively, then

$$\boldsymbol{Z}_b^t = \hat{\boldsymbol{Z}}_b \tag{15-28}$$

On the other hand, if branch-admittance matrices $\boldsymbol{Y}_b$ and $\hat{\boldsymbol{Y}}_b$ exist, then

$$\boldsymbol{Y}_b^t = \hat{\boldsymbol{Y}}_b \tag{15-29}$$

In general, the nonsource branches of N and $\hat{N}$ can always be characterized by hybrid matrices $\boldsymbol{H}_b$ and $\hat{\boldsymbol{H}}_b$ in the following manner (see Chap. 6):

$$\begin{bmatrix} \boldsymbol{I}_{b1} \\ \boldsymbol{V}_{b2} \end{bmatrix} = \begin{bmatrix} \boldsymbol{H}_{11b} & \boldsymbol{H}_{12b} \\ \boldsymbol{H}_{21b} & \boldsymbol{H}_{22b} \end{bmatrix} \begin{bmatrix} \boldsymbol{V}_{b1} \\ \boldsymbol{I}_{b2} \end{bmatrix} = \boldsymbol{H}_b \begin{bmatrix} \boldsymbol{V}_{b1} \\ \boldsymbol{I}_{b2} \end{bmatrix} \tag{15-30}$$

and

$$\begin{bmatrix} \hat{\boldsymbol{I}}_{b1} \\ \hat{\boldsymbol{V}}_{b2} \end{bmatrix} = \begin{bmatrix} \hat{\boldsymbol{H}}_{11b} & \hat{\boldsymbol{H}}_{12b} \\ \hat{\boldsymbol{H}}_{21b} & \hat{\boldsymbol{H}}_{22b} \end{bmatrix} \begin{bmatrix} \hat{\boldsymbol{V}}_{b1} \\ \hat{\boldsymbol{I}}_{b2} \end{bmatrix} = \hat{\boldsymbol{H}}_b \begin{bmatrix} \hat{\boldsymbol{V}}_{b1} \\ \hat{\boldsymbol{I}}_{b2} \end{bmatrix} \tag{15-31}$$

For N and $\hat{N}$ to be adjoint networks of each other, we require that[1]

$$\begin{bmatrix} \hat{\boldsymbol{H}}_{11b} & \hat{\boldsymbol{H}}_{12b} \\ \hat{\boldsymbol{H}}_{21b} & \hat{\boldsymbol{H}}_{22b} \end{bmatrix} = \begin{bmatrix} \boldsymbol{H}_{11b}^t & -\boldsymbol{H}_{21b}^t \\ -\boldsymbol{H}_{12b}^t & \boldsymbol{H}_{22b}^t \end{bmatrix} \tag{15-32}$$

[1]Note that Eqs. (15-28) and (15-29) are actually special cases of Eq. (15-32). Furthermore, if the nonsource branches of N can be described by all three matrices $\boldsymbol{Z}_b$, $\boldsymbol{Y}_b$, and $\boldsymbol{H}_b$, it can be shown that any of the three conditions (15-28), (15-29), and (15-32) implies the remaining two. The reasons for imposing these conditions on the nonsource branches of $\hat{N}$ will become clear in Section 15-3-3 when we derive the sensitivity formulas.

Condition 3. Corresponding independent sources in both network are the same in nature (current or voltage sources), but need not have the same values.

The above definition of adjoint networks suggests very clearly how to construct the adjoint network $\hat{N}$ for any given network N: let $\hat{\boldsymbol{A}} = \boldsymbol{A}$ (same topology) and $\hat{\boldsymbol{Z}}_b = \boldsymbol{Z}_b^t$, or make use of Eq. (15-32), from which $\hat{N}$ may be constructed. A few examples will illustrate the procedure.

EXAMPLE 15-3. Construct the adjoint network for network N shown in Fig. 15-4(a).

Solution: Note that branch 5 is an open circuit whose voltage v_5 controls i_6 ($i_6 = g_m v_5$). We have

$$\boldsymbol{Y}_b = \begin{bmatrix} G_1 & 0 & 0 & 0 & 0 & 0 \\ 0 & G_2 & 0 & 0 & 0 & 0 \\ 0 & 0 & G_3 & 0 & 0 & 0 \\ 0 & 0 & 0 & G_4 & 0 & 0 \\ 0 & 0 & 0 & 0 & 0 & 0 \\ 0 & 0 & 0 & 0 & g_m & 0 \end{bmatrix}$$

Then

$$\hat{\boldsymbol{Y}}_b = \boldsymbol{Y}_b^t = \begin{bmatrix} G_1 & 0 & 0 & 0 & 0 & 0 \\ 0 & G_2 & 0 & 0 & 0 & 0 \\ 0 & 0 & G_3 & 0 & 0 & 0 \\ 0 & 0 & 0 & G_4 & 0 & 0 \\ 0 & 0 & 0 & 0 & 0 & g_m \\ 0 & 0 & 0 & 0 & 0 & 0 \end{bmatrix}$$

The corresponding $\hat{N}$ is shown in Fig. 15-4(b).

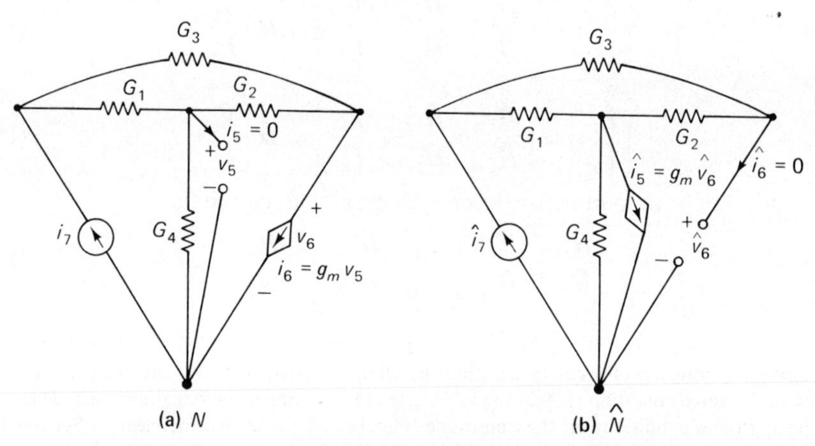

Fig. 15-4. Construction of the adjoint network when $\mathbf{Y}_b$ exists.

EXAMPLE 15-4. Construct the adjoint network for Fig. 15-5(a).

Solution: The nonsource branches of N have no impedance- nor admittance-matrix characterization. Therefore, we resort to Eq. (15-30), the hybrid characterization:

$$\begin{bmatrix} i_1 \\ i_4 \\ \hline v_2 \\ v_3 \\ v_5 \end{bmatrix} = \left[\begin{array}{cc|ccc} 0 & 0 & -2 & 0 & 0 \\ 0 & 0 & 0 & 0 & 0 \\ \hline 2 & 0 & 0 & 0 & 0 \\ 0 & 0 & 0 & 8 & 0 \\ 0 & 3 & 0 & 0 & 0 \end{array}\right] \begin{bmatrix} v_1 \\ v_4 \\ \hline i_2 \\ i_3 \\ i_4 \end{bmatrix}$$

The hybrid branch characterization for the adjoint network $\hat{N}$, according to Eq. (15-32), is

$$\begin{bmatrix} \hat{i}_1 \\ \hat{i}_4 \\ \hline \hat{v}_2 \\ \hat{v}_3 \\ \hat{v}_5 \end{bmatrix} = \left[\begin{array}{cc|ccc} 0 & 0 & -2 & 0 & 0 \\ 0 & 0 & 0 & 0 & -3 \\ \hline 2 & 0 & 0 & 0 & 0 \\ 0 & 0 & 0 & 8 & 0 \\ 0 & 0 & 0 & 0 & 0 \end{array}\right] \begin{bmatrix} \hat{v}_1 \\ \hat{v}_4 \\ \hline \hat{i}_2 \\ \hat{i}_3 \\ \hat{i}_5 \end{bmatrix} \tag{15-33}$$

The adjoint network $\hat{N}$, constructed according to Eq. (15-33) and $\hat{A} = A$, is shown in Fig. 15-5(b).

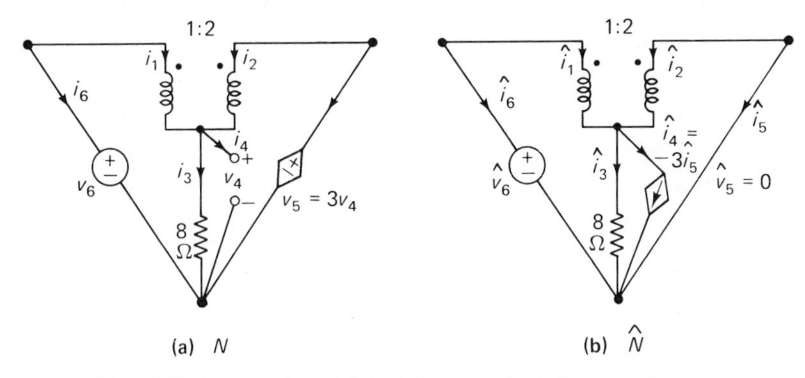

Fig. 15-5. Construction of the adjoint network for the general case.

In the solutions of these two examples, we actually construct the branch-admittance matrix or hybrid matrix for the adjoint network in order to illustrate the principle. Once the principle is understood, it is much more convenient to construct the adjoint network $\hat{N}$ from any given network N by the use of Table 15-2. Note that, for all reciprocal components, their counterparts in the adjoint network are just the components themselves. Table 15-2 is established by the use of Eqs. (15-28), (15-29), or (15-32).

In applications of adjoint networks, it is sometimes convenient to extract all independent sources to form a multiport, as shown in Fig. 15-6. Denote the port currents and voltages by $\boldsymbol{I}_p$ and $\boldsymbol{V}_p$, respectively. Denote the nonsource branch currents

TABLE 15-2. Construction of the Adjoint Network

N, ORIGINAL NETWORK	$\hat{N}$, ADJOINT NETWORK
R	R
L	L
C	C
T, $1:n$	T, $1:n$
1, 2: $+\ V_{12}\ -$; 3, 4: $g_m V_{12}$	1, 2: $g_m V_{34}$; 3, 4: $+\ V_{34}\ -$
1, 2: I_a; 3, 4: $r_m I_a$	1, 2: $r_m I_b$; 3, 4: I_b
1, 2: I_a; 3, 4: kI_a	1, 2: $-kV_{34}$; 3, 4: $+\ V_{34}\ -$
1, 2: $+\ V_{12}\ -$; 3, 4: kV_{12}	1, 2: $-kI_b$; 3, 4: I_b

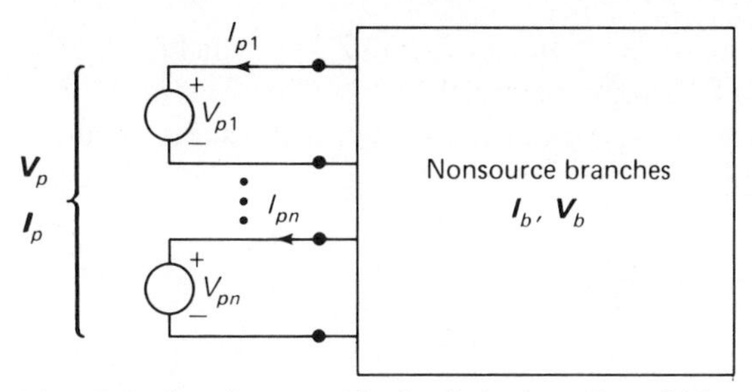

Fig. 15-6. Notations used in the derivation of sensitivity formulas by the adjoint-network method.

and voltages by I_b and V_b, respectively. Then, when the branch impedance matrices for N and its adjoint network $\hat{N}$ exist, we have, by definition,

$$V_b = Z_b I_b \tag{15-34}$$

$$\hat{V}_b = \hat{Z}_b \hat{I}_b \tag{15-35}$$

where $Z_b^t = \hat{Z}_b$. When the open-circuit impedance matrices for the two multiports exist, we have, by definition,

$$V_p = -Z_{oc} I_p \tag{15-36}$$

$$\hat{V}_p = -\hat{Z}_{oc} \hat{I}_p \tag{15-37}$$

where the negative signs are due to the reference directions used in Fig. 15-6. We shall now show that Z_{oc} for N and $\hat{Z}_{oc}$ for its adjoint network $\hat{N}$ are related by

$$\boxed{\hat{Z}_{oc} = Z_{oc}^t} \tag{15-38}$$

From Eq. (15-26) it follows immediately that

$$\hat{I}^t V - \hat{V}^t I = 0 \tag{15-39}$$

In terms of the current and voltage vectors for the nonsource branches and the independent source (port) branches defined earlier, Eq. (15-39) becomes

$$(\hat{I}_p^t V_p + \hat{I}_b^t V_b) - (\hat{V}_p^t I_p + \hat{V}_b^t I_b) = 0$$

or

$$(-\hat{I}_p^t V_p + \hat{V}_p^t I_p) = (\hat{I}_b^t V_b - \hat{V}_b^t I_b) \tag{15-40}$$

Making use of the relationships (15-28), (15-34), and (15-35), we find the right side of Eq. (15-40) to be

$$\boxed{(\hat{I}_b^t V_b - \hat{V}_b^t I_b) = \hat{I}_b^t Z_b I_b - \hat{I}_b^t \hat{Z}_b^t I_b = 0} \tag{15-41a}$$

Similarly, making use of the relationships (15-36) and (15-37), we find the left side of Eq. (15-40) to be

$$(-\hat{I}_p^t V_p + \hat{V}_p^t I_p) = \hat{I}_p^t Z_{oc} I_p - \hat{I}_p^t \hat{Z}_{oc}^t I_p = \hat{I}_p^t (Z_{oc} - \hat{Z}_{oc}^t) I_p \tag{15-41b}$$

Substituting Eqs. (15-41a) and (15-41b) into Eq. (15-40), we have

$$\hat{I}_p^t (Z_{oc} - \hat{Z}_{oc}^t) I_p = 0 \tag{15-42}$$

Since Eq. (15-42) holds for arbitrary $\hat{\boldsymbol{I}}_p$ and $\boldsymbol{I}_p$, we conclude that $\boldsymbol{Z}_{oc} - \hat{\boldsymbol{Z}}^t_{oc} \equiv \boldsymbol{0}$, from which Eq. (15-38) follows immediately.

In a similar manner, we can prove that if the short-circuit admittance matrices exist for the multiports created from N and its adjoint $\hat{N}$, then

$$\boxed{\hat{\boldsymbol{Y}}_{sc} = \boldsymbol{Y}^t_{sc}} \tag{15-43}$$

In general, the multiports created from N and $\hat{N}$ may be characterized by hybrid matrices $\boldsymbol{H}$ and $\hat{\boldsymbol{H}}$, respectively. Then a relationship similar to that given by Eq. (15-32) can be established (see Eq. (15-65) and Problem 15-11).

One important application of Eq. (15-38) is in noise analysis of electronic circuits. Because of the extremely small amplitudes of the voltages and currents involved, noise analysis always uses *linear* network models (with all *signal* sources set to zero). For most common devices such as resistors, diodes, and transistors the noise sources can be modeled as independent current sources each of which is in parallel with a *noiseless* resistor. Suppose that the network model for noise analysis contains $n - 1$ noise current sources denoted by $I_1, \ldots, I_{n-1}$, with the corresponding noiseless resistors denoted by $R_1, \ldots, R_{n-1}$. Let the output noise voltage be the voltage across an open circuit which in turn is considered as a zero-valued current source I_n ($I_n \equiv 0$). These n current sources are extracted to form an n-port. The noise-output voltage due to each I_j alone is then given by $z_{nj}I_j$. Because of the random nature of the noise sources, we cannot add up the individual contributions to obtain the overall output-noise voltage. Instead, the mean-square value of the output-noise voltage is given by (assuming *uncorrelated* noise sources)

$$\bar{V}^2_n = |z_{n1}|^2 \cdot |I_1|^2 + \cdots + |z_{n,n-1}|^2 \cdot |I_{n-1}|^2$$

Each $|I_j|^2$ is known from the device noise model (e.g., $|I_j|^2 = 4kT\Delta f/R_j$ for a resistor of R_j Ω). The question is then how to efficiently calculate $z_{n1}, \ldots, z_{n,n-1}$. If we calculate these transimpedances strictly according to their definitions, then $n - 1$ analyses will be required. However, from Eq. (15-38) we see that if the adjoint network $\hat{N}$ is constructed, then the desired set of transimpedances is identical to

$$\{\hat{z}_{1n}, \ldots, \hat{z}_{n-1,n}\}$$

This latter set of transimpedances can be obtained with just *one* analysis of the adjoint network $\hat{N}$ as follows: Apply a unit current at port n and calculate the voltages at ports $1, 2, \ldots, n - 1$, which are also the voltages across the resistors $R_1, \ldots, R_{n-1}$. Each voltage $\hat{V}_j$ is equal to $\hat{z}_{jn}$. The saving in computational effort is obvious. Further details of this method and its computer implementation may be found in [17].

15-3-3 Calculation of Sensitivities by the Use of Adjoint Networks

For linear n-ports, it is natural to consider first the partial derivatives of the elements of $\boldsymbol{Z}_{oc}$, $\boldsymbol{Y}_{sc}$, or $\boldsymbol{H}$ with respect to branch parameters. As shown in Section 15-1, from the information of $\partial \boldsymbol{Z}_{oc}/\partial x$, we can easily obtain the partial derivatives $\partial \boldsymbol{V}_p/\partial x$.

Similarly, from $\partial \boldsymbol{Y}_{sc}/\partial x$, we can easily obtain $\partial \boldsymbol{I}_p/\partial x$. In the case that a voltage or current whose partial derivative is to be found is not a port quantity, we can always create an additional port such that the quantity of interest becomes a port quantity. This is done by considering the voltage of interest as the voltage across a current port whose current is identically zero. Similarly, the current of interest is considered as the current through a voltage port whose voltage is identically zero. Thus, our main concern will be the calculation of $\partial \boldsymbol{Z}_{oc}/\partial x$, $\partial \boldsymbol{Y}_{sc}/\partial x$, or $\partial \boldsymbol{H}/\partial x$.

As before, let $\boldsymbol{I}$, $(\boldsymbol{I} + \Delta \boldsymbol{I})$, and $\hat{\boldsymbol{I}}$ be the currents associated with the original network N, perturbed network N_p, and adjoint network $\hat{N}$, respectively, and similarly for the voltage vectors. Subscript p indicates port variables, and subscript b indicates variables for nonsource branches. With such partitioning of voltage and current vectors, Eq. (15-27) becomes

$$\hat{\boldsymbol{I}}^t \Delta \boldsymbol{V} - \hat{\boldsymbol{V}}^t \Delta \boldsymbol{I} = (\hat{\boldsymbol{I}}_p^t \Delta \boldsymbol{V}_p + \hat{\boldsymbol{I}}_b^t \Delta \boldsymbol{V}_b) - (\hat{\boldsymbol{V}}_p^t \Delta \boldsymbol{I}_p + \hat{\boldsymbol{V}}_b^t \Delta \boldsymbol{I}_b) = 0$$

It follows that

$$-(\hat{\boldsymbol{I}}_p^t \Delta \boldsymbol{V}_p - \hat{\boldsymbol{V}}_p^t \Delta \boldsymbol{I}_p) = (\hat{\boldsymbol{I}}_b^t \Delta \boldsymbol{V}_b - \hat{\boldsymbol{V}}_b^t \Delta \boldsymbol{I}_b) \tag{15-44}$$

We point out that in Eq. (15-44) $\Delta \boldsymbol{V}_p$ and $\Delta \boldsymbol{I}_p$ need not be small changes.

Now consider the effect due to very small (approaching zero, theoretically) changes in the elements of $\boldsymbol{Z}_b$. Then, *to a first-order approximation*, Eqs. (15-34) and (15-36) yield the following relationships:

$$\Delta \boldsymbol{V}_b = \Delta(\boldsymbol{Z}_b \boldsymbol{I}_b) = \Delta \boldsymbol{Z}_b \cdot \boldsymbol{I}_b + \boldsymbol{Z}_b \Delta \boldsymbol{I}_b \tag{15-45}$$

$$\Delta \boldsymbol{V}_p = -\Delta(\boldsymbol{Z}_{oc} \boldsymbol{I}_p) = -\Delta \boldsymbol{Z}_{oc} \cdot \boldsymbol{I}_p - \boldsymbol{Z}_{oc} \Delta \boldsymbol{I}_p \tag{15-46}$$

Making use of Eq. (15-45) and recalling that $\hat{\boldsymbol{Z}}_b^t = \boldsymbol{Z}_b$, we find the right side of Eq. (15-44) to be

$$\hat{\boldsymbol{I}}_b^t \Delta \boldsymbol{V}_b - \hat{\boldsymbol{V}}_b^t \Delta \boldsymbol{I}_b = (\hat{\boldsymbol{I}}_b^t \Delta \boldsymbol{Z}_b \cdot \boldsymbol{I}_b + \underbrace{\hat{\boldsymbol{I}}_b^t \boldsymbol{Z}_b \Delta \boldsymbol{I}_b) - \hat{\boldsymbol{I}}_b^t \hat{\boldsymbol{Z}}_b^t \Delta \boldsymbol{I}_b}_{\text{cancelled}} = \hat{\boldsymbol{I}}_b^t \Delta \boldsymbol{Z}_b \cdot \boldsymbol{I}_b \tag{15-47}$$

Note the indicated cancellation of terms because of the condition $\hat{\boldsymbol{Z}}_b^t = \boldsymbol{Z}_b$. The final result contains the changes in $\boldsymbol{Z}_b$ only. In fact, this is the basic goal we want to achieve when we define the characteristics of the nonsource branches of the adjoint network [see Eqs. (15-28), (15-29), and (15-32)]; i.e., *the sum of the* $(\hat{I}_k \Delta V_k - \hat{V}_k \Delta I_k)$ *terms for all nonsource branches should result in an expression containing only the changes in network parameters and not the changes in voltages or currents.* The same goal will be adhered to later on when we extend the definition of adjoint network to nonlinear networks.

In a similar manner, by the use of Eq. (15-46) and the fact that $\hat{\boldsymbol{Z}}_{oc}^t = \boldsymbol{Z}_{oc}$, we find the left side of Eq. (15-44) to be

$$-\hat{\boldsymbol{I}}_p^t \Delta \boldsymbol{V}_p + \hat{\boldsymbol{V}}_p^t \Delta \boldsymbol{I}_p = (\hat{\boldsymbol{I}}_p^t \Delta \boldsymbol{Z}_{oc} \cdot \boldsymbol{I}_p + \hat{\boldsymbol{I}}_p^t \boldsymbol{Z}_{oc} \Delta \boldsymbol{I}_p) - \hat{\boldsymbol{I}}_p^t \hat{\boldsymbol{Z}}_{oc}^t \Delta \boldsymbol{I}_p = \hat{\boldsymbol{I}}_p^t \Delta \boldsymbol{Z}_{oc} \cdot \boldsymbol{I}_p \tag{15-48}$$

Putting Eqs. (15-47) and (15-48) into Eq. (15-44), we have

$$\boxed{\hat{\boldsymbol{I}}_p^t \Delta \boldsymbol{Z}_{oc} \cdot \boldsymbol{I}_p = \hat{\boldsymbol{I}}_b^t \Delta \boldsymbol{Z}_b \cdot \boldsymbol{I}_b} \tag{15-49}$$

This is an extremely useful equation for calculating sensitivities, as will be shown shortly. For the special case of one-ports, ($\boldsymbol{Z}_b$ may be asymmetric), $\boldsymbol{Z}_{oc}$ becomes Z_{in}, and we have

$$\boxed{\hat{I}_p \,\Delta Z_{\text{in}} \cdot I_p = \hat{\boldsymbol{I}}_b^t \,\Delta \boldsymbol{Z}_b \cdot \boldsymbol{I}_b} \tag{15-50}$$

By a dual development using the admittance approach, we have for the multiport case

$$\boxed{\hat{\boldsymbol{V}}_p^t \,\Delta \boldsymbol{Y}_{sc} \cdot \boldsymbol{V}_p = \hat{\boldsymbol{V}}_b^t \,\Delta \boldsymbol{Y}_b \cdot \boldsymbol{V}_b} \tag{15-51}$$

and for the one-port case

$$\boxed{\hat{V}_p \,\Delta Y_{\text{in}} \cdot V_p = \hat{\boldsymbol{V}}_b^t \,\Delta \boldsymbol{Y}_b \cdot \boldsymbol{V}_b} \tag{15-52}$$

By a suitable choice of excitations for N and $\hat{N}$, the left side of Eq. (15-49) may produce only one term of interest, Δz_{jk}. For example, let the excitations (for the four-port case) be

$$\hat{\boldsymbol{I}}_p = [0, 0, 1, 0]^t \tag{15-53}$$

$$\boldsymbol{I}_p = [0, 1, 0, 0]^t \tag{15-54}$$

Then the left side of Eq. (15-49) yields

$$\hat{\boldsymbol{I}}_p^t \,\Delta \boldsymbol{Z}_{oc} \cdot \boldsymbol{I}_p = \Delta z_{32}$$

If an analysis of N is performed to obtain nonsource branch currents $\boldsymbol{I}_b$ under condition (15-54), and another analysis is performed on $\hat{N}$ to obtain $\hat{\boldsymbol{I}}_b$ under condition (15-53), then Eq. (15-49) may be used to calculate the partial derivatives of z_{32} with respect to all elements of $\boldsymbol{Z}_b$.

In a similar manner, by a suitable choice of excitations for N and $\hat{N}$, the left side of Eq. (15-51) may produce only one term of interest, Δy_{jk}. Two analyses, one performed on N and another on $\hat{N}$, will be sufficient to determine the partial derivatives of y_{jk} with respect to all elements of $\boldsymbol{Y}_b$.

We shall now illustrate the adjoint-network method of calculating sensitivities with two examples.

EXAMPLE 15-5. Consider the network of Fig. 15-7(a). Find $\partial Y_{\text{in}}/\partial y_1$, $\partial Y_{\text{in}}/\partial y_2$, $\partial Y_{\text{in}}/\partial y_3$, and $\partial Y_{\text{in}}/\partial g_m$ by the adjoint-network method.

Solution: To use the adjoint-network method, we first redraw the circuit diagram, as shown in Fig. 15-7(b). Note the appearance of the voltage source V_p and the open-circuit nonsource branch $y_5 = 0$ in the network N. The solution of branch voltages for N under the condition $V_p = 1$ is next obtained.

$$\boldsymbol{V}_b = [1, -\tfrac{1}{5}, \tfrac{6}{5}, -\tfrac{1}{5}, 1]^t \tag{15-55}$$

Next construct the adjoint network $\hat{N}$. The result is shown in Fig. 15-7(c). The solution of $\hat{\boldsymbol{V}}_b$ corresponding to the condition $\hat{V}_p = 1$ is found to be

$$\hat{\boldsymbol{V}}_b = [1, \tfrac{1}{5}, \tfrac{4}{5}, \tfrac{1}{5}, 1]^t \tag{15-56}$$

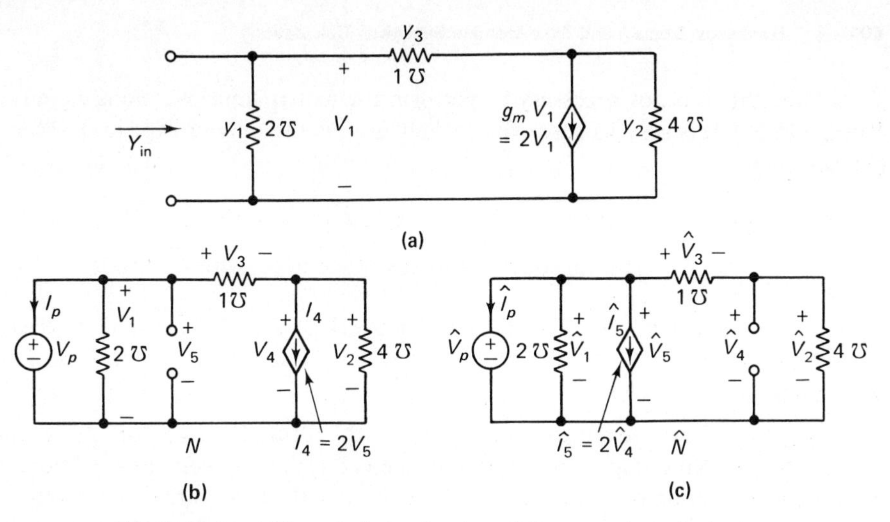

Fig. 15-7. Sensitivity calculation by the adjoint-network method—the one-port case.

The branch-admittance matrix $\boldsymbol{Y}_b$ for N is

$$\boldsymbol{Y}_b = \begin{bmatrix} 2 & 0 & 0 & 0 & 0 \\ 0 & 4 & 0 & 0 & 0 \\ 0 & 0 & 1 & 0 & 0 \\ 2 & 0 & 0 & 0 & 0 \\ 0 & 0 & 0 & 0 & 0 \end{bmatrix} = \begin{bmatrix} y_1 & 0 & 0 & 0 & 0 \\ 0 & y_2 & 0 & 0 & 0 \\ 0 & 0 & y_3 & 0 & 0 \\ g_m & 0 & 0 & 0 & 0 \\ 0 & 0 & 0 & 0 & 0 \end{bmatrix} \tag{15-57}$$

Therefore,

$$\Delta \boldsymbol{Y}_b = \begin{bmatrix} \Delta y_1 & 0 & 0 & 0 & 0 \\ 0 & \Delta y_2 & 0 & 0 & 0 \\ 0 & 0 & \Delta y_3 & 0 & 0 \\ \Delta g_m & 0 & 0 & 0 & 0 \\ 0 & 0 & 0 & 0 & 0 \end{bmatrix} \tag{15-58}$$

Substituting Eqs. (15-55), (15-56), and (15-58) into Eq. (15-52), we have

$$1 \cdot \Delta Y_{\text{in}} \cdot 1 = [1, \tfrac{1}{5}, \tfrac{4}{5}, \tfrac{1}{5}, 1] \begin{bmatrix} \Delta y_1 & 0 & 0 & 0 & 0 \\ 0 & \Delta y_2 & 0 & 0 & 0 \\ 0 & 0 & \Delta y_3 & 0 & 0 \\ \Delta g_m & 0 & 0 & 0 & 0 \\ 0 & 0 & 0 & 0 & 0 \end{bmatrix} \begin{bmatrix} 1 \\ -\frac{1}{5} \\ \frac{6}{5} \\ -\frac{1}{5} \\ 1 \end{bmatrix} \tag{15-59}$$

Recalling that Eqs. (15-49)–(15-52) are obtained under the assumption of infinitesimal changes in the elements of $\boldsymbol{Y}_b$, we can immediately find the desired partial derivatives from Eq. (15-59), as follows:

$$\frac{\partial Y_{\text{in}}}{\partial y_1} = 1, \qquad \frac{\partial Y_{\text{in}}}{\partial y_2} = -\frac{1}{25}$$

$$\frac{\partial Y_{\text{in}}}{\partial y_3} = \frac{24}{25}, \qquad \frac{\partial Y_{\text{in}}}{\partial g_m} = \frac{1}{5}$$

Note that it is not necessary to perform the matrix multiplications as indicated by Eq. (15-59). It is easy to show that, when Δy_{jk} is the only change in $\boldsymbol{Y}_b$, then Eq. (15-52) leads to

$$\frac{\partial Y_{\text{in}}}{\partial y_{jk}} = (j\text{th nonsource branch voltage in } \hat{N}, \text{ with } \hat{V}_p = 1) \cdot (k\text{th nonsource branch voltage in } N, \text{ with } V_p = 1)$$

EXAMPLE 15-6. Consider the network of Fig. 15-8(a). The desired network function is $T = V_o/I_{\text{in}}$. Find $\partial T/\partial R_1$, $\partial T/\partial R_2$, and $\partial T/\partial R_m$ by the adjoint-network method.

Solution: First, we redraw the network in the form of a two-port N, as shown in Fig. 15-8(b). Note that V_o now becomes the voltage of an open-circuited port. Also note the introduction of a short-circuit branch $v_4 = 0$ whose current controls v_3.

The adjoint network $\hat{N}$ may now be constructed with the aid of Table 15-2. The result is Fig. 15-8(c). Since $\partial T/\partial x = \partial z_{21}/\partial x$ is of interest, Eq. (15-49) indicates that we should let

$$\hat{\boldsymbol{I}}_p = [0, 1]^t$$

and

$$\boldsymbol{I}_p = [1, 0]^t$$

resulting in $\hat{\boldsymbol{I}}_p^t \, \Delta \boldsymbol{Z}_{oc} \cdot \boldsymbol{I}_p = \Delta z_{21}$. Analysis of N under the condition $I_{p1} = 1$ and $I_{p2} = 0$ yields the branch currents

$$\boldsymbol{I}_b = [\tfrac{1}{3}, \tfrac{2}{3}, -\tfrac{2}{3}, \tfrac{1}{3}]^t \tag{15-60}$$

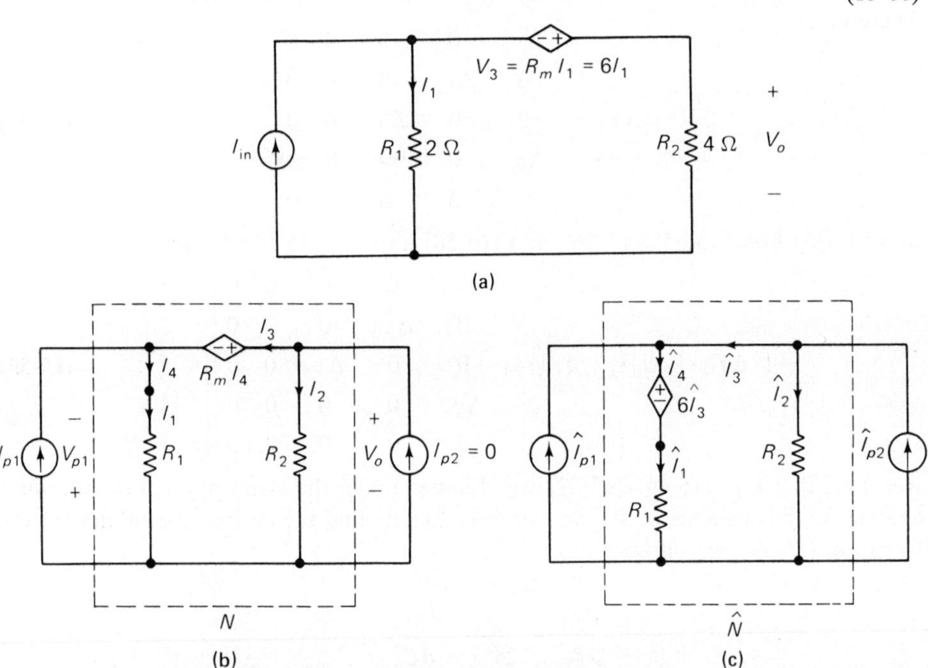

Fig. 15-8. Sensitivity calculation by the adjoint-network method —the two-port case.

Analysis of $\hat{N}$ under the condition $\hat{I}_{p1} = 0$ and $\hat{I}_{p2} = 1$ yields

$$\hat{I}_b = [\tfrac{1}{3}, \tfrac{2}{3}, \tfrac{1}{3}, \tfrac{1}{3}]^t \tag{15-61}$$

The impedance matrix for the nonsource branches of N is

$$Z_b = \begin{bmatrix} R_1 & 0 & 0 & 0 \\ 0 & R_2 & 0 & 0 \\ 0 & 0 & 0 & R_m \\ 0 & 0 & 0 & 0 \end{bmatrix} \tag{15-62}$$

Substituting Eqs. (15-60)–(15-62) into Eq. (15-49), we have

$$\hat{I}_p^t \, \Delta Z_{oc} \cdot I_p = 1 \cdot \Delta z_{21} \cdot 1 = \tfrac{1}{3} \cdot \Delta R_1 \cdot \tfrac{1}{3} + \tfrac{2}{3} \cdot \Delta R_2 \cdot \tfrac{2}{3} + \tfrac{1}{3} \Delta R_m \cdot \tfrac{1}{3}$$

Recalling that Eq. (15-49) is derived under the assumption that ΔZ_b contains infinitesimal changes, we immediately have

$$\frac{\partial z_{21}}{\partial R_1} = \frac{1}{9}, \qquad \frac{\partial z_{21}}{\partial R_2} = \frac{4}{9}, \qquad \frac{\partial z_{21}}{\partial R_m} = \frac{1}{9}$$

So far, all the examples have been resistive linear networks for the sake of easier hand solution. The procedure is applicable to ac analysis (steady state) too. At any given frequency, all the currents, voltages, and immittances will then have complex values. The analyses of N and $\hat{N}$ may be carried out with the aid of a suitable computer simulation program. It should be clear from our examples that the *result of two analyses* (one for the original network and one for its adjoint network) *will yield partial derivatives of z_{jk} or y_{jk} with respect to all branch parameters at one frequency.*

Equations (15-49)–(15-52) contain ΔZ_b and ΔY_b. Care should be exercised to relate ΔZ_b and ΔY_b to changes in network parameters. For example, a network with four nonsource branches, G_1, C_2, Γ_3, and g_m, may have a branch-admittance matrix as follows:

$$Y_b = \begin{bmatrix} G_1 & 0 & 0 & 0 \\ 0 & j\omega C_2 & 0 & 0 \\ 0 & 0 & \dfrac{\Gamma_3}{j\omega} & 0 \\ g_m & 0 & 0 & 0 \end{bmatrix}$$

Then, in terms of ΔG_1, ΔC_2, $\Delta \Gamma_3$, and Δg_m, we have

$$\Delta Y_b = \begin{bmatrix} \Delta G_1 & 0 & 0 & 0 \\ 0 & j\omega \, \Delta C_2 & 0 & 0 \\ 0 & 0 & \dfrac{\Delta \Gamma_3}{j\omega} & 0 \\ \Delta g_m & 0 & 0 & 0 \end{bmatrix}$$

In deriving Eq. (15-49), we have assumed that the nonsource branches of N possess an impedance-matrix description, and that the multiport created from N has an open-circuit impedance matrix. Similar assumptions about the existence of admittance matrices are made in the derivation of Eq. (15-51). These restrictions will now be removed. In general, the nonsource branches of N and $\hat{N}$ can be characterized by hybrid matrices $\boldsymbol{H}_b$ and $\hat{\boldsymbol{H}}_b$, respectively, as shown in Eqs. (15-30) and (15-31). The relationship between $\boldsymbol{H}_b$ and $\hat{\boldsymbol{H}}_b$ is given by Eq. (15-32). The multiports created from N and $\hat{N}$ by extracting all independent sources can be characterized by hybrid matrices $\boldsymbol{H}$ and $\hat{\boldsymbol{H}}$, respectively, as follows:

$$\begin{bmatrix} \boldsymbol{I}_E \\ \boldsymbol{V}_J \end{bmatrix} = \begin{bmatrix} \boldsymbol{H}_{EE} & \boldsymbol{H}_{EJ} \\ \boldsymbol{H}_{JE} & \boldsymbol{H}_{JJ} \end{bmatrix} \begin{bmatrix} \boldsymbol{V}_E \\ \boldsymbol{I}_J \end{bmatrix} = \boldsymbol{H} \begin{bmatrix} \boldsymbol{V}_E \\ \boldsymbol{I}_J \end{bmatrix} \tag{15-63}$$

$$\begin{bmatrix} \hat{\boldsymbol{I}}_E \\ \hat{\boldsymbol{V}}_J \end{bmatrix} = \begin{bmatrix} \hat{\boldsymbol{H}}_{EE} & \hat{\boldsymbol{H}}_{EJ} \\ \hat{\boldsymbol{H}}_{JE} & \hat{\boldsymbol{H}}_{JJ} \end{bmatrix} \begin{bmatrix} \hat{\boldsymbol{V}}_E \\ \hat{\boldsymbol{I}}_J \end{bmatrix} = \hat{\boldsymbol{H}} \begin{bmatrix} \hat{\boldsymbol{V}}_E \\ \hat{\boldsymbol{I}}_J \end{bmatrix} \tag{15-64}$$

where the subscript E indicates independent voltage sources, and J indicates independent current sources. It can be shown (Problem 15-11) that the matrices $\boldsymbol{H}$ and $\hat{\boldsymbol{H}}$ are related in the following manner:

$$\boxed{\begin{bmatrix} \hat{\boldsymbol{H}}_{EE} & \hat{\boldsymbol{H}}_{EJ} \\ \hat{\boldsymbol{H}}_{JE} & \hat{\boldsymbol{H}}_{JJ} \end{bmatrix} = \begin{bmatrix} \boldsymbol{H}^t_{EE} & -\boldsymbol{H}^t_{JE} \\ -\boldsymbol{H}^t_{EJ} & \boldsymbol{H}^t_{JJ} \end{bmatrix}} \tag{15-65}$$

Note that Eqs. (15-38) and (15-43) are special cases of Eq. (15-65).

An equation relating the changes in $\boldsymbol{H}$ to the changes in $\boldsymbol{H}_b$ can be derived. The procedure is basically the same as that used to derive Eq. (15-49), except that the notations are more complicated in the present case because of the further partitioning of the voltage and current vectors. Only the result will be given below; the details of the derivation are left as an exercise (see Problem 15-12). The equations corresponding to Eqs. (15-47) and (15-48) become, respectively,

$$\hat{\boldsymbol{I}}^t_b \,\Delta \boldsymbol{V}_b - \hat{\boldsymbol{V}}^t_b \,\Delta \boldsymbol{I}_b = [-\hat{\boldsymbol{V}}^t_{b1} \quad \hat{\boldsymbol{I}}^t_{b2}] \begin{bmatrix} \Delta \boldsymbol{H}_{11b} & \Delta \boldsymbol{H}_{12b} \\ \Delta \boldsymbol{H}_{21b} & \Delta \boldsymbol{H}_{22b} \end{bmatrix} \begin{bmatrix} \boldsymbol{V}_{b1} \\ \boldsymbol{I}_{b2} \end{bmatrix} \tag{15-66}$$

$$\begin{aligned} -\hat{\boldsymbol{I}}^t_p \,\Delta \boldsymbol{V}_p + \hat{\boldsymbol{V}}^t_p \,\Delta \boldsymbol{I}_p &= -(\hat{\boldsymbol{I}}^t_E \,\Delta \boldsymbol{V}_E + \hat{\boldsymbol{I}}^t_J \,\Delta \boldsymbol{V}_J) + (\hat{\boldsymbol{V}}^t_E \,\Delta \boldsymbol{I}_E + \hat{\boldsymbol{V}}^t_J \,\Delta \boldsymbol{I}_J) \\ &= -\hat{\boldsymbol{I}}^t_J \,\Delta \boldsymbol{V}_J + \hat{\boldsymbol{V}}^t_E \,\Delta \boldsymbol{I}_E \\ &= [\hat{\boldsymbol{V}}^t_E \quad -\hat{\boldsymbol{I}}^t_J] \begin{bmatrix} \Delta \boldsymbol{H}_{EE} & \Delta \boldsymbol{H}_{EJ} \\ \Delta \boldsymbol{H}_{JE} & \Delta \boldsymbol{H}_{JJ} \end{bmatrix} \begin{bmatrix} \boldsymbol{V}_E \\ \boldsymbol{I}_J \end{bmatrix} \end{aligned} \tag{15-67}$$

Substituting Eqs. (15-66) and (15-67) into Eq. (15-44), we obtain the following general relationship, which includes Eqs. (15-49) and (15-51) as special cases:[2]

$$\boxed{\begin{aligned} -\hat{\boldsymbol{I}}^t_J \,\Delta \boldsymbol{V}_J + \hat{\boldsymbol{V}}^t_E \,\Delta \boldsymbol{I}_E &= [\hat{\boldsymbol{V}}^t_E \quad -\hat{\boldsymbol{I}}^t_J] \begin{bmatrix} \Delta \boldsymbol{H}_{EE} & \Delta \boldsymbol{H}_{EJ} \\ \Delta \boldsymbol{H}_{JE} & \Delta \boldsymbol{H}_{JJ} \end{bmatrix} \begin{bmatrix} \boldsymbol{V}_E \\ \boldsymbol{I}_J \end{bmatrix} \\ &= [-\hat{\boldsymbol{V}}^t_{b1} \quad \hat{\boldsymbol{I}}^t_{b2}] \begin{bmatrix} \Delta \boldsymbol{H}_{11b} & \Delta \boldsymbol{H}_{12b} \\ \Delta \boldsymbol{H}_{21b} & \Delta \boldsymbol{H}_{22b} \end{bmatrix} \begin{bmatrix} \boldsymbol{V}_{b1} \\ \boldsymbol{I}_{b2} \end{bmatrix} \end{aligned}} \tag{15-68}$$

[2]Many special cases of Eq. (15-68) can be derived without the use of Tellegen's theorem. See Section 1-10 and reference 10.

TABLE 15-3. Sensitivity Components in Frequency Domain

ELEMENT TYPE	DESCRIPTION IN N	DESCRIPTION IN $\hat{N}$	$\Sigma(\hat{I}\Delta V - \hat{V}\Delta I)$
R	$V = RI$	$\hat{V} = R\hat{I}$	$\hat{I}I\,\Delta R$
G	$I = GV$	$\hat{I} = G\hat{V}$	$-\hat{V}V\,\Delta G$
Z	$V = ZI$	$\hat{V} = Z\hat{I}$	$\hat{I}I\,\Delta Z$
Y	$I = YV$	$\hat{I} = Y\hat{V}$	$-\hat{V}V\,\Delta Y$
C	$I = j\omega CV$	$\hat{I} = j\omega C\hat{V}$	$-j\omega\hat{V}V\,\Delta C$
L	$V = j\omega LI$	$\hat{V} = j\omega L\hat{I}$	$j\omega\hat{I}I\,\Delta L$
μ	$\begin{cases} V_2 = \mu V_1 \\ I_1 = 0 \end{cases}$	$\hat{I}_1 = -\mu\hat{I}_2$ $\hat{V}_2 = 0$	$\hat{I}_2V_1\,\Delta\mu$
g_m	$\begin{cases} I_2 = g_mV_1 \\ I_1 = 0 \end{cases}$	$\hat{I}_1 = g_m\hat{V}_2$ $\hat{I}_2 = 0$	$-\hat{V}_2V_1\,\Delta g_m$
β	$\begin{cases} I_2 = \beta I_1 \\ V_1 = 0 \end{cases}$	$\hat{V}_1 = -\beta\hat{V}_2$ $\hat{I}_2 = 0$	$-\hat{V}_2I_1\,\Delta\beta$
r_m	$\begin{cases} V_2 = r_mI_1 \\ V_1 = 0 \end{cases}$	$\hat{V}_1 = r_m\hat{I}_2$ $\hat{V}_2 = 0$	$\hat{I}_2I_1\,\Delta r_m$
T	$\begin{cases} V_2 = nV_1 \\ I_1 = -nI_2 \end{cases}$	$\hat{V}_2 = n\hat{V}_1$ $\hat{I}_1 = -n\hat{I}_2$	$(\hat{I}_2V_1 + \hat{V}_1I_2)\,\Delta n$

For different types of elements in N and $\hat{N}$, the right side of Eq. (15-68) may be evaluated separately, and the results are given in Table 15-3 for reference purposes.

The procedure for calculating $\partial V_o/\partial x_j$, $\partial I_o/\partial x_j$, or $\partial h_{jk}/\partial x_j$ may be summarized as follows:

> *Step 1.* Perform an analysis of the network N to obtain $\mathbf{V}_{b1}$ and $\mathbf{I}_{b2}$.
>
> *Step 2.* Select the excitations for $\hat{N}$ such that one side of Eq. (15-68) yields only one term, which is ΔV_o, ΔI_o, or Δh_{jk}, of interest. Perform an analysis of the adjoint network $\hat{N}$ to obtain $\hat{\mathbf{V}}_{b1}$ and $\hat{\mathbf{I}}_{b2}$.
>
> *Step 3.* Evaluate the right side of Eq. (15-68), either directly from matrix multiplications or with the aid of Table 15-3. From the resultant expression, obtain the desired partial derivatives.

We shall now illustrate the procedure with a numerical example.

EXAMPLE 15-7. Consider the two-port network shown in Fig. 15-9(a). The nominal element values are $R_1 = 2\,\Omega$, $G_2 = 2\,℧$ and n (transformer turns ratio) $= 2$. Find the partial derivatives of H_{JE} with respect to R_1, G_2, and n.

Solution: An examination of Eq. (15-68) shows that to have ΔH_{JE} only, we may choose $I_J = 0$, $V_E = 1$, $\hat{V}_E = 0$, and $\hat{I}_J = -1$. An analysis of the original network N yields

$$[V_2 \quad V_4 \quad I_1 \quad I_3] = [0.25 \quad 0.25 \quad 0.25 \quad 0.25]$$

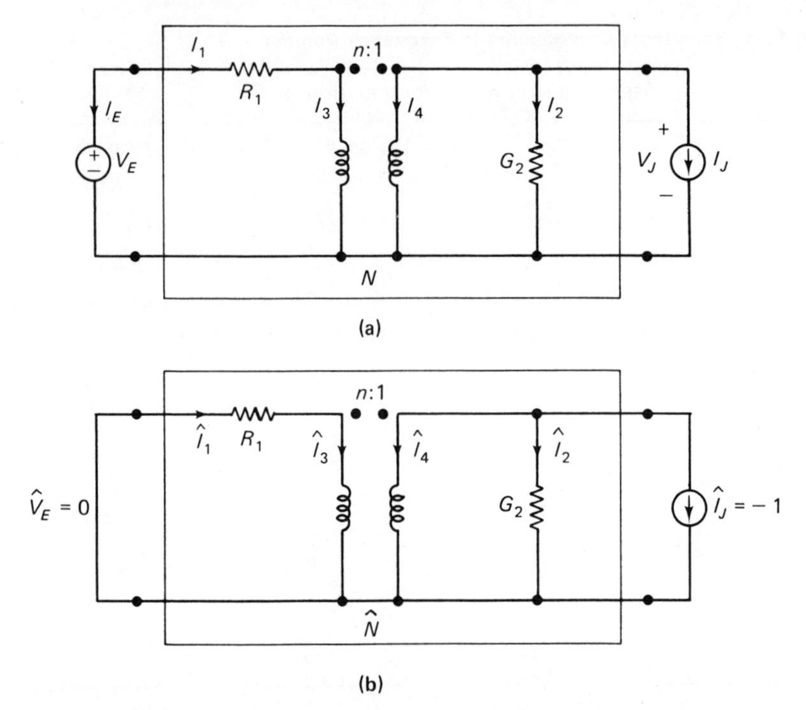

Fig. 15-9. Sensitivity calculation—the hybrid characterization case.

The adjoint network $\hat{N}$, constructed according to Table 15-2, is shown in Fig. 15-9(b). An analysis of $\hat{N}$ yields

$$[\hat{V}_2 \quad \hat{V}_4 \quad \hat{I}_1 \quad \hat{I}_3] = [0.25 \quad 0.25 \quad -0.25 \quad -0.25]$$

Using these values and evaluating the right side of Eq. (15-68) according to Table 15-3, we obtain

$$\Delta H_{JE} = -0.0625\Delta R_1 - 0.0625\Delta G_2 + 0\Delta n \tag{15-69}$$

It follows from Eq. (15-69) that

$$\frac{\partial H_{JE}}{\partial R_1} = -0.0625, \qquad \frac{\partial H_{JE}}{\partial G_2} = -0.0625, \qquad \frac{\partial H_{JE}}{\partial n} = 0 \tag{15-70}$$

The correctness of these results may be verified for this simple network by evaluating the following expressions at the nominal element values:

$$H_{JE} = \frac{n}{R_1 G_2 + n^2} \tag{15-71}$$

$$\frac{\partial H_{JE}}{\partial R_1} = -\frac{nG_2}{(R_1 G_2 + n^2)^2} \tag{15-72a}$$

$$\frac{\partial H_{JE}}{\partial G_2} = -\frac{nR_1}{(R_1 G_2 + n^2)^2} \tag{15-72b}$$

$$\frac{\partial H_{JE}}{\partial n} = \frac{R_1 G_2 - n^2}{(R_1 G_2 + n^2)^2} \tag{15-72c}$$

15-4 SYMBOLIC-NETWORK-FUNCTION APPROACH

In Chapter 14 we described several algorithms for digital computer generation of symbolic network functions. The symbolic network function offers an alternative method of calculating sensitivities that can be more efficient than the adjoint-network method when the number of terms is not excessive and the number of frequency sample points is very large. With the symbolic method, we first obtain the network function T, where *some or all* of the network element values are represented by variables $(x_1, \ldots, x_n)$. The expression for T is then differentiated with respect to each x_i (by computer). Once the expressions for all $\partial T/\partial x_i$ have been obtained, we have only to evaluate these expressions for each set of $(x_1, \ldots, x_n)$ and ω values, as compared to the adjoint-network method, for which we have to perform at each frequency an analysis of the network N and another analysis of its adjoint network $\hat{N}$.

The partial derivative $\partial T/\partial x_i$ can be obtained easily from the expression for T in view of the following basic property of linear networks.

Let a lumped, linear, time-invariant network N consist of impedances, admittances, and all four types of controlled sources (μ, β, g_m, r_m). Let some or all of these network elements be characterized by ***distinct*** *variables $(x_1, \ldots, x_n)$; the remaining elements are assigned numerical values. Then any network function T, which is V_o/V_{in}, V_o/I_{in}, I_o/V_{in}, or I_o/I_{in}, may be expressed as the ratio of two polynomials of* degree *1 in each variable x_i.*

For example, if only two elements are represented by variables, then T can always be expressed as

$$T = \frac{A_o + A_1x_1 + A_2x_2 + A_{12}x_1x_2}{B_o + B_1x_1 + B_2x_2 + B_{12}x_1x_2}$$

One tends to think that the property is obvious until being reminded that the property does not hold if x_i represents the gyration resistance of a gyrator or the turns ratio of a transformer, or if T is not one of the four types of network functions mentioned. A proof of the stated property may be found in [11].

According to this property, we can always write

$$T = \frac{N(x_1, \ldots, x_n)}{D(x_1, \ldots, x_n)} \tag{15-73}$$

where N and D are polynomials of degree 1 in each x_i. Sensitivity function (15-1) can of course be obtained by direct differentiation of Eq. (15-73), once the expression for T has been obtained by hand or by a suitable symbolic analysis program (see Chap. 14). However, for digital computer solution of the problem, it is more convenient to use the method described next.

Equation (15-73) can be rewritten as

$$\begin{aligned} H &\triangleq D - \frac{1}{T}N \\ &= D(x_1, \ldots, x_n) - P \cdot N(x_1, \ldots, x_n) = 0 \end{aligned} \tag{15-74}$$

where $P \triangleq 1/T$, and H is a polynomial of degree 1 in P and in each x_i. Equation (15-74) defines P as an implicit function of the variables $(x_1, \ldots, x_n)$. Let us first find $S^P_{x_i}$.

To simplify the notations, we shall let $Q = x_i$. Because H is of degree 1 in P and Q, we can always write

$$H = A + BQ + P(C + FQ) = 0 \tag{15-75}$$

where A, B, C, and F are polynomials that do not contain the variables P and Q, but may contain other variables. It follows from Eq. (15-75) that

$$P = -\frac{A + BQ}{C + FQ} \tag{15-76}$$

$$Q = -\frac{A + CP}{B + FP} \tag{15-77}$$

$$\frac{\partial P}{\partial Q} = -\frac{\partial H/\partial Q}{\partial H/\partial P} = -\frac{B + FP}{C + FQ} \tag{15-78}$$

and

$$\begin{aligned} S_Q^P = \frac{\partial P}{\partial Q}\frac{Q}{P} &= -\frac{B + FP}{C + FQ} \cdot \frac{A + CP}{B + FP} \cdot \frac{C + FQ}{A + BQ} \\ &= -\frac{A + CP}{A + BQ} \end{aligned} \tag{15-79}$$

Referring to the decomposition of H as shown in Eq. (15-75), we can interpret Eq. (15-79) as follows:

$$\boxed{S_Q^P = -\frac{\text{sum of terms of } H \text{ not containing } Q}{\text{sum of terms of } H \text{ not containing } P}} \tag{15-80}$$

Equation (15-80) contains the variable P, which can be eliminated by the use of Eq. (15-76). The result is

$$S_Q^P = \frac{C}{C + FQ} - \frac{A}{A + BQ} \tag{15-81}$$

Since $P = 1/T$, we have the following simple relationship:

$$S_Q^T = \frac{\partial \ln T}{\partial \ln Q} = -\frac{\partial \ln P}{\partial \ln Q} = -S_Q^P \tag{15-82}$$

The procedure for calculating $S_{x_i}^T$ by the symbolic-function method will now be illustrated with an example.

EXAMPLE 15-8. The transfer function $T = V_o/I_{in}$ for the network of Fig. 15-10 has been found to be

$$T = \frac{R_2 + j\omega C_1 R_2 R_m}{1 + j\omega C_1(R_2 + R_m)} = \frac{N}{D} \tag{15-83}$$

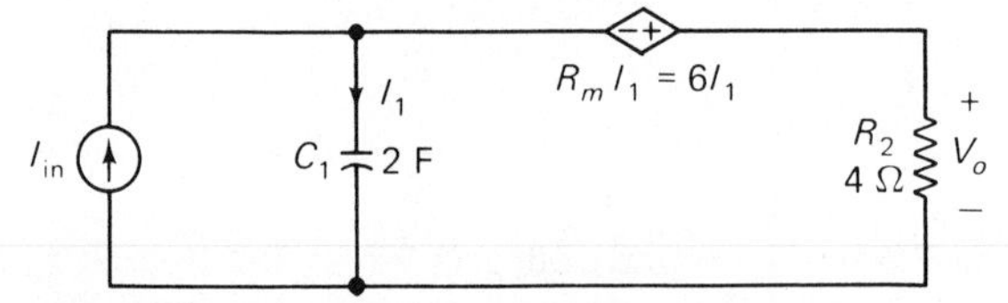

Fig. 15-10. Sensitivity analysis by the symbolic-function method.

Find $S_{R_2}^T$, $S_{R_m}^T$, and $S_{C_1}^T$ under the conditions $C_1 = 2$ F, $R_2 = 4\ \Omega$, $R_m = 6\ \Omega$, and $\omega = \frac{1}{4}, \frac{1}{2}$, and 1 rad/s, successively.

Solution:

$$H = D - \frac{1}{T}N = D - PN$$

$$= 1 + j\omega C_1(R_2 + R_m) - P(R_2 + j\omega C_1 R_2 R_m)$$

From Eqs. (15-80)–(15-82), we immediately have

$$S_{R_2}^T = \frac{1 + j\omega C_1 R_m}{1 + j\omega C_1(R_2 + R_m)} = \frac{1 + j12\omega}{1 + j20\omega}$$

$$S_{R_m}^T = \frac{1 + j\omega C_1 R_2 - PR_2}{1 + j\omega C_1(R_2 + R_m)} = \frac{-96\omega^2}{(1 - 240\omega^2) + j32\omega}$$

$$S_{C_1}^T = \frac{1 - PR_2}{1 + j\omega C_1(R_2 + R_m)} = \frac{-j8\omega}{(1 - 240\omega^2) + j32\omega}$$

where

$$P = \frac{1}{T} = \frac{1 + j20\omega}{4 + j48\omega}$$

At each frequency, we need only evaluate these expressions to obtain the desired sensitivity functions. The results are tabulated as follows:

ω	$P = \frac{1}{T}$	$S_{C_1}^T$	$S_{R_2}^T$	$S_{R_m}^T$
0.25	0.4000 $+j0.0500$	-0.1230 $+j0.2153$	0.6154 $+j0.0769$	0.3230 $+j0.1846$
0.5	0.4122 $+j0.0270$	-0.01712 $+j0.06315$	0.6040 $-j0.0396$	0.3789 $+j0.1207$
1	0.4155 $+j0.0138$	-0.004402 $+j0.03288$	0.6010 $-j0.0200$	0.3946 $+j0.05283$

It appears that to construct H of Eq. (15-74) we would have to obtain the expression for the network function T first. In practice, the contrary is true. Most existing symbolic-analysis programs find H first and then obtain T (or P) from H by sorting terms according to Eq. (15-74). For example, in the signal-flow-graph method, we add a branch with weight P from the output node to the input node to form a closed SFG. The determinant Δ_c for this closed SFG can be shown to be exactly the desired H (see Section 14-2). We apply Mason's rule to find Δ_c, and then obtain the numerator and denominator of T by separating the terms of Δ_c according to Eq. (15-74).

At any given frequency, T and S_x^T are in general complex numbers, as shown in Example 15-8. Let $T = |T|e^{j\theta}$. For some applications, we are interested in $S_x^{|T|}$ and S_x^θ. The following relationships will be useful for such purposes:

$$S_x^T = \frac{\partial \ln T}{\partial \ln x} = \frac{\partial \ln |T|}{\partial \ln x} + j\frac{\partial \theta}{\partial x}x = S_x^{|T|} + jS_x^\theta \theta \tag{15-84}$$

If x is real, then

$$S_x^{|T|} = \text{Re}\,\{S_x^T\} \tag{15-85}$$

and

$$\theta \cdot S_x^\theta = \text{Im}\,\{S_x^T\} \tag{15-86}$$

*15-5 TIME-DOMAIN SENSITIVITY CALCULATIONS

Up to now we have only considered sensitivity calculations in the *frequency domain.* The linear network under consideration is subject to excitations at a single frequency and is in steady state. The phasors (for currents and voltages) and immittances that we deal with are *complex numbers* in general.

The adjoint-network method will now be used for sensitivity calculations in the time domain for linear dynamic networks. In this case, it is usual to talk about the changes in some voltages or currents due to the change in some element parameters. Precisely, the problem to be investigated in this section is the determination of $\partial v_o(t_f)/\partial x_j$ or $\partial i_o(t_f)/\partial x_j$, where x_j is some element parameter (R, L, C, g_m, temperature, etc.), and $v_o(t_f)$ or $i_o(t_f)$ is the output voltage or current at the specified time $t = t_f$. We wish to determine these partial derivatives without "numerical differentiation," the drawbacks of which have been explained in Section 15-1.

The adjoint network defined in Section 15-3-2 will again be used here. It is necessary to generalize several important equations derived previously for the present purpose. By repeating the derivation of Eq. (15-24), but using $\boldsymbol{v}(t)$ and $\boldsymbol{i}(t)$ for N, and $\hat{\boldsymbol{v}}(\tau)$ and $\hat{\boldsymbol{i}}(\tau)$ for $\hat{N}$, we obtain

$$\boxed{\boldsymbol{v}^t(t)\hat{\boldsymbol{i}}(\tau) = \boldsymbol{i}^t(t)\hat{\boldsymbol{v}}(\tau) = \hat{\boldsymbol{i}}^t(\tau)\boldsymbol{v}(t) = \hat{\boldsymbol{v}}^t(\tau)\boldsymbol{i}(t) = 0} \tag{15-87}$$

The noteworthy fact about Eq. (15-87) is that *t and τ are arbitrary.* A product such as $\hat{v}_1(\tau)\cdot i_1(t)$ has no clear physical meaning, because the quantities involved not only belong to different networks, but also belong to different time instants. Nevertheless, it is mathematically a very useful entity, as we shall see shortly.

In a similar manner, we can show that Eq. (15-25) becomes

$$\hat{\boldsymbol{i}}^t(\tau)\,\Delta\boldsymbol{v}(t) = 0 \tag{15-88a}$$

$$\hat{\boldsymbol{v}}^t(\tau)\,\Delta\boldsymbol{i}(t) = 0 \tag{15-88b}$$

$$\hat{\boldsymbol{i}}^t(\tau)\,\Delta\boldsymbol{v}(t) - \hat{\boldsymbol{v}}^t(\tau)\,\Delta\boldsymbol{i}(t) = 0 \tag{15-88c}$$

and Eq. (15-44) becomes

$$-\hat{\boldsymbol{i}}_p^t(\tau)\,\Delta\boldsymbol{v}_p(t) + \hat{\boldsymbol{v}}^t(\tau)\,\Delta\boldsymbol{i}_p(t) = \hat{\boldsymbol{i}}_b^t(\tau)\,\Delta\boldsymbol{v}_b(t) - \hat{\boldsymbol{v}}_b^t(\tau)\,\Delta\boldsymbol{i}_b(t) \tag{15-89}$$

These equations are valid regardless of the values of t and τ. In particular, Eq. (15-89) is valid if we let t and τ be related by $\tau = t_f - t$, where $t_f > 0$ is the time instant at which v_o or i_o is to be investigated. With t and τ so related, we can further integrate both sides of Eq. (15-89) from $t = 0$ to $t = t_f$. The result is

$$\begin{aligned}&\int_0^{t_f} [-\hat{\boldsymbol{i}}_p^t(\tau)\,\Delta\boldsymbol{v}_p(t) + \hat{\boldsymbol{v}}_p^t(\tau)\,\Delta\boldsymbol{i}_p(t)]_{\tau=t_f-t}\,dt \\ &\qquad = \int_0^{t_f} [\hat{\boldsymbol{i}}_b^t(\tau)\,\Delta\boldsymbol{v}_b(t) - \hat{\boldsymbol{v}}_b^t(\tau)\,\Delta\boldsymbol{i}_b(t)]_{\tau=t_f-t}\,dt\end{aligned} \tag{15-90}$$

The reason for introducing the relationship $\tau = t_f - t$ will become clear shortly.

To evaluate the integrals in Eq. (15-90), we first have to perform transient analysis of N and $\hat{N}$ to obtain the required voltages and currents. In the analysis of N, the time variable is designated as t; in the analysis of $\hat{N}$ the time variable is designated as τ. Responses of N are obtained for the interval $0 \leq t \leq t_f$; the responses of $\hat{N}$ are obtained for $0 \leq \tau \leq t_f$. Since $\tau = t_f - t$, as τ *increases* from 0 to t_f, the variable t actually decreases from t_f to 0. For this reason, the adjoint network is sometimes said to be analyzed in *backward time*, whereas the original network is analyzed in *forward time* [12].

Let there be b nonsource branches in N (and hence in $\hat{N}$). We can rewrite the right side of Eq. (15-90) in scalar notations as equal to

$$\sum_{k=1}^{b} \int_0^{t_f} [\hat{i}_{bk}(\tau)\, \Delta v_{bk}(t) - \hat{v}_{bk}(\tau)\, \Delta i_{bk}(t)]_{\tau=t_f-t}\, dt \tag{15-91}$$

Thus, we can evaluate the integrals in Eq. (15-91) for each individual *RLC* branch and for each group of coupled branches. The results are then added together to give the right side of Eq. (15-90). We shall now show in detail the evaluation of the integrals for three types of branches. For other types, only the results will be given. It should be emphasized that in all subsequent derivations we have *neglected second-order effects.*

RESISTANCE BRANCH. $v(t) = Ri(t)$ and $\hat{v}(\tau) = R\hat{i}(\tau)$

$$\begin{aligned}
&\int_0^{t_f} [\hat{i}(\tau)\, \Delta v(t) - \hat{v}(\tau)\, \Delta i(t)]_{\tau=t_f-t}\, dt \\
&\qquad = \int_0^{t_f} \{\hat{i}(\tau)[R\, \Delta i(t) + i(t)\, \Delta R] - R\hat{i}(\tau)\, \Delta i(t)\}_{\tau=t_f-t}\, dt \\
&\qquad = \left\{\int_0^{t_f} [\hat{i}(\tau) i(t)]_{\tau=t_f-t}\, dt\right\} \Delta R
\end{aligned} \tag{15-92}$$

CAPACITANCE BRANCH. $i(t) = dq/dt$, $q(t) = Cv(t)$; and $\hat{i}(\tau) = d\hat{q}/d\tau$ $\hat{q}(\tau) = C\hat{v}(\tau)$

$$\begin{aligned}
&\int_0^{t_f} [\hat{i}(\tau)\, \Delta v(t) - \hat{v}(\tau)\, \Delta i(t)]_{\tau=t_f-t}\, dt \\
&\qquad = \int_0^{t_f} \left\{\hat{i}(\tau)\, \Delta v(t) - \hat{v}(\tau)\left[\Delta \frac{dq}{dt}\right]\right\}_{\tau=t_f-t} dt \\
&\qquad = \int_0^{t_f} \left\{\hat{i}(\tau)\, \Delta v(t) - \hat{v}(\tau) \frac{d}{dt}[C\, \Delta v(t) + v(t)\, \Delta C]\right\}_{\tau=t_f-t} dt \\
&\qquad \triangleq \vartheta_1 + \vartheta_2 + \vartheta_3
\end{aligned} \tag{15-93}$$

The evaluation of ϑ_3 in Eq. (15-93) is straightforward.

$$\begin{aligned}
\vartheta_3 &= -\int_0^{t_f} [\hat{v}(\tau)\, \dot{v}(t)\, \Delta C]_{\tau=t_f-t}\, dt \\
&= \left\{-\int_0^{t_f} [\hat{v}(\tau)\dot{v}(t)]_{\tau=t_f-t}\, dt\right\} \Delta C
\end{aligned} \tag{15-94}$$

The evaluation of $\vartheta_1 + \vartheta_2$ in (15-93) makes use of the method of integration by parts, which states

$$\int_a^b x(t)\frac{dy}{dt}\,dt = -\int_a^b \frac{dx}{dt}y(t)\,dt + [x(t)y(t)]_a^b \tag{15-95}$$

We have

$$\vartheta_1 + \vartheta_2 = \int_0^{t_f} \left\{\hat{i}(\tau)\,\Delta v(t) - \hat{v}(\tau)\frac{d}{dt}[C\,\Delta v(t)]\right\}_{\tau=t_f-t} dt$$

In evaluating ϑ_2, we make use of Eq. (15-95) with $x(t) = \hat{v}(\tau) = \hat{v}(t_f - t)$, and $y(t) = C\,\Delta v(t)$. Then, recalling that $dt = -d\tau$, we have

$$\begin{aligned}\vartheta_1 + \vartheta_2 &= \int_0^{t_f} [\hat{i}(\tau)\,\Delta v(t) - \frac{d\hat{v}}{d\tau}C\,\Delta v(t)]_{\tau=t_f-t}\,dt \\ &\quad - \{[\hat{v}(\tau)C\Delta v(t)]_{\tau=t_f-t}\}_0^{t_f} \\ &= \int_0^{t_f} [\hat{i}(\tau)\,\Delta v(t) - \hat{i}(\tau)\,\Delta v(t)]_{\tau=t_f-t}\,dt \\ &\quad - C\hat{v}(0)\,\Delta v(t_f) + C\hat{v}(t_f)\,\Delta v(0) \\ &= -C\hat{v}(0)\,\Delta v(t_f) + C\hat{v}(t_f)\,\Delta v(0)\end{aligned} \tag{15-96}$$

Since nothing has been said about the initial condition of a capacitor in the adjoint network $\hat{N}$, we can exploit this freedom to achieve some simplification of the analysis of $\hat{N}$. We shall assume that *any capacitor in $\hat{N}$ always has a zero initial condition*; i.e.,

$$\hat{v}_c(\tau) = 0, \qquad \text{when } \tau = 0$$

Then Eq. (15-96) becomes

$$\vartheta_1 + \vartheta_2 = C\hat{v}(t_f)\,\Delta v(0) \tag{15-97}$$

In many situations the initial voltage of the capacitor in N remains fixed when other element parameters vary. Then $\Delta v(0) = 0$, and Eq. (15-93) finally becomes

$$\int_0^{t_f} [\hat{i}(\tau)\,\Delta v(t) - \hat{v}(\tau)\,\Delta i(t)]_{\tau=t_f-t}\,dt = \left\{-\int_0^{t_f} [\hat{v}(\tau)\dot{v}(t)]_{\tau=t_f-t}\,dt\right\}\Delta C \tag{15-98}$$

Coupled Resistive Branches. This category includes gyrators, ideal transformers, and all four types of controlled sources as special cases. To treat the most general case of n coupled resistive branches, we shall first generalize Eqs. (15-30)–(15-32). Any group of coupled resistive branches in N can be characterized by

$$\begin{bmatrix}\boldsymbol{i}_{b1}(t)\\ \boldsymbol{v}_{b2}(t)\end{bmatrix} = \begin{bmatrix}\boldsymbol{H}_{11b} & \boldsymbol{H}_{12b}\\ \boldsymbol{H}_{21b} & \boldsymbol{H}_{22b}\end{bmatrix}\begin{bmatrix}\boldsymbol{v}_{b1}(t)\\ \boldsymbol{i}_{b2}(t)\end{bmatrix} \tag{15-99}$$

The corresponding branches in $\hat{N}$ will then be characterized by

$$\begin{bmatrix}\hat{\boldsymbol{i}}_{b1}(\tau)\\ \hat{\boldsymbol{v}}_{b2}(\tau)\end{bmatrix} = \begin{bmatrix}\boldsymbol{H}^t_{11b} & -\boldsymbol{H}^t_{21b}\\ -\boldsymbol{H}^t_{12b} & \boldsymbol{H}^t_{22b}\end{bmatrix}\begin{bmatrix}\hat{\boldsymbol{v}}_{b1}(\tau)\\ \hat{\boldsymbol{i}}_{b2}(t)\end{bmatrix} \tag{15-100}$$

It is easy to show that Eq. (15-66) then becomes

$$\boxed{\hat{\boldsymbol{i}}^t_b(\tau)\,\Delta\boldsymbol{v}(t) - \hat{\boldsymbol{v}}^t_b(\tau)\,\Delta\boldsymbol{i}(t) = [-\hat{\boldsymbol{v}}^t_{b1}(\tau), \hat{\boldsymbol{i}}^t_{b2}(\tau)]\begin{bmatrix}\Delta\boldsymbol{H}_{11b} & \Delta\boldsymbol{H}_{12b}\\ \Delta\boldsymbol{H}_{21b} & \Delta\boldsymbol{H}_{22b}\end{bmatrix}\begin{bmatrix}\boldsymbol{v}_{b1}(t)\\ \boldsymbol{i}_{b2}(t)\end{bmatrix}} \tag{15-101}$$

For example, a voltage-controlled voltage source $v_2 = \mu v_1$ is characterized by

$$\begin{bmatrix} i_1(t) \\ v_2(t) \end{bmatrix} \begin{bmatrix} 0 & 0 \\ \mu & 0 \end{bmatrix} \begin{bmatrix} v_1(t) \\ i_2(t) \end{bmatrix}$$

Then, applying Eq. (15-101), we have

$$\sum_{k=1}^{2} \int_0^{t_f} [\hat{i}_k(\tau)\, \Delta v_k(t) - \hat{v}_k(\tau)\, \Delta i_k(t)]_{\tau=t_f-t}\, dt = \left\{ \int_0^{t_f} [\hat{i}_2(\tau) v_1(t)]_{\tau=t_f-t}\, dt \right\} \Delta\mu \qquad (15\text{-}102)$$

All other types of coupled branches can be handled in a similar manner. For an inductance branch, the derivation is a complete dual to the preceding derivation for a capacitance branch. For reference purposes, these results are given in Table 15-4.

TABLE 15-4. Sensitivity Components—Time Domain

ELEMENT TYPE	DESCRIPTION	$\int_0^{t_f} [\hat{i}(\tau)\, \Delta v(t) - \hat{v}(\tau)\, \Delta i(t)]_{\tau=t_f-t}\, dt$
R	$v_R = Ri_R$	$\left\{ \int_0^{t_f} [\hat{i}_R(\tau) i_R(t)]_{\tau=t_f-t}\, dt \right\} \Delta R$
C	$\begin{cases} i_C = \dfrac{dq}{dt} \\ q = Cv_C \end{cases}$	$\left\{ -\int_0^{t_f} [\hat{v}_C(\tau) \dot{v}_C(t)]_{\tau=t_f-t}\, dt \right\} \Delta C$
G	$i_G = Gv_G$	$\left\{ -\int_0^{t_f} [\hat{v}_G(\tau) v_G(t)]_{\tau=t_f-t}\, dt \right\} \Delta G$
L	$\begin{cases} v_L = \dfrac{d\lambda}{dt} \\ \lambda = Li_L \end{cases}$	$\left\{ -\int_0^{t_f} [\hat{i}_L(\tau) \dot{i}_L(t)]_{\tau=t_f-t}\, dt \right\} \Delta L$
μ	$\begin{cases} v_2 = \mu v_1 \\ i_1 = 0 \end{cases}$	$\left\{ \int_0^{t_f} [\hat{i}_2(\tau) v_1(t)]_{\tau=t_f-t}\, dt \right\} \Delta \mu$
β	$\begin{cases} i_2 = \beta i_1 \\ v_1 = 0 \end{cases}$	$\left\{ -\int_0^{t_f} [\hat{v}_2(\tau) i_1(t)]_{\tau=t_f-t}\, dt \right\} \Delta \beta$
g_m	$\begin{cases} i_2 = g_m v_1 \\ v_1 = 0 \end{cases}$	$\left\{ -\int_0^{t_f} [\hat{v}_2(\tau) v_1(t)]_{\tau=t_f-t}\, dt \right\} \Delta g_m$
r_m	$\begin{cases} v_2 = r_m i_1 \\ v_1 = 0 \end{cases}$	$\left\{ \int_0^{t_f} [\hat{i}_2(\tau) i_1(t)]_{\tau=t_f-t}\, dt \right\} \Delta r_m$
T	$\begin{cases} v_2 = nv_1 \\ i_1 = -ni_2 \end{cases}$	$\left\{ \int_0^{t_f} [\hat{i}_2(\tau) v_1(t) + \hat{v}_1(\tau) i_2(t)]_{\tau=t_f-t}\, dt \right\} \Delta n$

Notes: (1) See Table 15-2 for the description of adjoint-network elements. (2) In column 3, the integrals are evaluated under the assumptions that $\hat{v}_C(0) = 0$, $\hat{i}_L(0) = 0$, $\Delta v_L(0) = 0$, and $\Delta i_C(0) = 0$.

Having considered the evaluation of the right side of Eq. (15-90), let us next turn our attention to the left side, which is concerned with the independent sources in N and $\hat{N}$. As before, we consider any output voltage in N as the voltage across an independent current source. Similarly, any output current in N is considered to be the current through an independent voltage source. By a proper choice of the excitations for $\hat{N}$, the left side of Eq. (15-90) may yield only one term, which is precisely the desired change in output (Δv_o or Δi_o) at $t = t_f$. For example, if $v_{p3}(t_f)$ is of interest, we make use of the unit impulse function $\delta(\tau)$ and let

$$\hat{i}_{p3}(\tau) = -\delta(\tau)$$

and all other independent sources in $\hat{N}$ be zero valued. The left side of Eq. (15-90) becomes

$$\int_0^{t_f} [-\hat{\mathbf{i}}_p^t(\tau)\,\Delta \mathbf{v}_p(t) + \hat{\mathbf{v}}_p^t(\tau)\,\Delta \mathbf{i}_p(t)]_{\tau=t_f-t}\,dt = -\int_0^{t_f} [\hat{i}_{p3}(\tau)\,\Delta v_{p3}(t)]_{\tau=t_f-t}\,dt$$

$$= \int_0^{t_f} \delta(t - t_f)\,\Delta v_{p3}(t)\,dt = \Delta v_{p3}(t_f)$$

From the preceding discussion, we can summarize the procedure for calculating $\partial v_o(t_f)/\partial x$ or $\partial i_o(t_f)/\partial x$ as follows:

Step 1. Perform a transient analysis of the original network N for the time interval $t = [0, t_f]$. Obtain $i(t)$ or $v(t)$ for resistive branches, $\dot{v}(t)$ for capacitive branches, and $\dot{i}(t)$ for inductive branches.

Step 2. Construct the adjoint network $\hat{N}$ according to Eq. (15-32) or Table 15-2, and then impose the following two conditions:

1. All capacitor voltages and inductor currents in $\hat{N}$ are zero at $\tau = 0$.
2. All independent sources in $\hat{N}$ are set to zero, except for $\hat{i}_{pk}(\tau) = -\delta(\tau)$ (impulse function), when $\Delta v_{pk}(t_f)$ is of interest, or $\hat{v}_{pk}(\tau) = \delta(\tau)$ when $\Delta i_{pk}(t_f)$ is of interest.

Step 3. Perform a transient analysis of the adjoint network $\hat{N}$ for the time interval $\tau = [0, t_f]$. Obtain $\hat{i}(\tau)$ or $\hat{v}(\tau)$ for resistive branches, $\hat{v}(\tau)$ for capacitive branches, and $\hat{i}(\tau)$ for inductive branches.

Step 4. Evaluate Eq. (15-90) to obtain the sensitivity components. With the excitations for $\hat{N}$ chosen as described in step 3, the left side of Eq. (15-90) should be exactly $\Delta v_o(t_f)$ or $\Delta i_o(t_f)$. The right side may be evaluated with the aid of Table 15-4.

Step 5. Find $\partial v_o(t_f)/\partial x_j$ or $\partial i_o(t_f)/\partial x_j$ from the result of step 4, using the fact that $\partial f/\partial x_j = \Delta f/\Delta x_j$ when x_j is the only parameter that varies, and $\Delta x_j \longrightarrow 0$.

We shall now illustrate the procedure with a simple numerical example. In each step, analytical solutions of voltages and currents will be obtained. Of course, for any nontrivial networks, solutions of voltages and currents will have to be obtained numerically with a computer simulation program.

EXAMPLE 15-9. Consider the simple RC network shown in Fig. 15-11(a). $v_C(0) = 10$ V and $v_{p1} = 30$ V. Find $\partial v_o(t_f)/\partial R$ and $\partial v_o(t_f)/\partial C$. Note that the output voltage $v_o(t)$ has been shown in the diagram as the voltage $v_{p2}(t)$ across the zero-valued independent current source $i_{p2}(t)$.

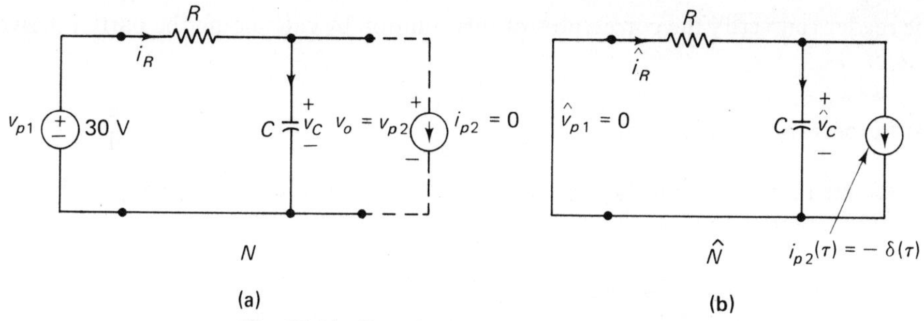

Fig. 15-11. Time-domain sensitivity calculation.

Solution:

Step 1. By ordinary circuit analysis, we obtain, for $t \geq 0$,

$$i_R(t) = \frac{20}{R} e^{-t/RC}$$

$$v_C(t) = 30 - 20e^{-t/RC}$$

$$\dot{v}_C(t) = \frac{20}{RC} e^{-t/RC}$$

Step 2. The adjoint network $\hat{N}$ is shown in Fig. 15-11(b) with proper excitations.

Step 3. By ordinary circuit analysis, we obtain, for $\tau \geq 0$,

$$\hat{i}_R(\tau) = -\frac{1}{RC} e^{-\tau/RC}$$

$$\hat{v}_C(\tau) = \frac{1}{C} e^{-\tau/RC}$$

Step 4. Evaluation of sensitivity components (see Table 15-4).
For R,

$$\int_0^{t_f} \left[\frac{20}{R} e^{-t/RC}\left(-\frac{1}{RC} e^{-\tau/RC}\right)\right]_{\tau = t_f - t} dt \cdot \Delta R = -\frac{20t_f}{R^2C} e^{-t_f/RC} \cdot \Delta R$$

For C,

$$-\int_0^{t_f} \left[\frac{20}{RC} e^{-t/RC} \cdot \frac{1}{C} e^{-\tau/RC}\right]_{\tau = t_f - t} dt \cdot \Delta C = -\frac{20t_f}{RC^2} e^{-t_f/RC} \cdot \Delta C$$

Step 5. Putting the results of step 4 into Eq. (15-90), we have

$$\Delta v_o(t_f) = -\frac{20t_f}{R^2C} e^{-t_f/RC} \cdot \Delta R - \frac{20t_f}{RC^2} e^{-t_f/RC} \cdot \Delta C$$

from which

$$\frac{\partial v_o(t_f)}{\partial R} = -\frac{20t_f}{R^2C} e^{-t_f/RC}$$

and

$$\frac{\partial v_o(t_f)}{\partial C} = -\frac{20t_f}{RC^2} e^{-t_f/RC}$$

The reader can verify the correctness of this solution by calculating the partial derivatives of

$$v_o(t_f) = 30 - 20e^{-t_f/RC}$$

with respect to R and C.

Some remarks about the adjoint-network method of calculating $\partial v_o(t_f)/\partial x_j$ and $\partial i_o(t_f)/\partial x_j$ are now in order.

1. Two transient analyses, one performed on N in forward time $(0 \leq t \leq t_f)$ and another performed on $\hat{N}$ in backward time $(0 \leq \tau \leq t_f)$ are sufficient for the determination of the partial derivatives of $v_o(t_f)$ or $i_o(t_f)$ with respect to all element parameters $x_1, \ldots, x_n$.
2. Although the solution of N at $t = 0, h, 2h, \ldots, t_f = mh$ needs to be stored, the solution of $\hat{N}$ at $\tau = 0, h, 2h, \ldots, t_f = mh$ *need not be stored.* To illustrate this point, suppose that the integral in Eq. (15-92) is approximated by the following summation (with $t_f = 5h$):

$$\begin{aligned} \vartheta &= \int_0^{t_f} [\hat{i}(\tau)i(t)]_{\tau=t_f-t}\, dt \\ &\cong h[\hat{i}(5h)i(0) + \hat{i}(4h)i(h) + \hat{i}(3h)i(2h) \\ &\quad + \hat{i}(2h)i(3h) + \hat{i}(h)i(4h) + \hat{i}(0)i(5h)] \end{aligned} \tag{15-103}$$

Now the values of $i(0)$ to $i(5h)$ are calculated and *stored.* Initially, we set $\vartheta = 0$. As $\hat{i}(0)$ is determined from the analysis of $\hat{N}$, we update ϑ to be $\vartheta = h[\hat{i}(0)i(5h)]$. The value of $\hat{i}(0)$ is no longer needed as far as the determination of ϑ is concerned [although it may be needed in the numerical-integration algorithm to obtain $\hat{i}(h)$]. As $\hat{i}(h)$ is next determined, we update ϑ to be $\vartheta = h[\hat{i}(0)i(5h) + \hat{i}(\tau)i(4h)]$. The process is continued until $\hat{i}(5h)$ is finally obtained, and ϑ is updated to that given by Eq. (15-103).
3. The derivatives $\dot{v}(t)$ and $\dot{i}(t)$ are required for capacitive branches and inductive branches, respectively. If the state-variable method is used for the transient analysis, these derivatives are either given directly by the state equation (8-4), or are related to the state variables $\boldsymbol{x}$ and the forcing functions $\boldsymbol{u}$ by some algebraic equation, possibly involving the derivatives of the forcing function $\boldsymbol{u}$ also.

15-6 CALCULATION OF ERROR GRADIENT BY THE ADJOINT-NETWORK METHOD

We have investigated several methods for the calculation of the partial derivatives of $z_{jk}, y_{jk}, h_{jk}, V_o, I_o, v_o(t_f), i_o(t_f)$, etc., with respect to network element parameters $(x_1, \ldots, x_n)$. When the sensitivity information is used for the purpose of network optimization, the more pertinent information is actually the partial derivatives of a properly selected error function $\mathcal{E}$ with respect to $(x_1, \ldots, x_n)$. To illustrate the basic idea, consider a resistive linear network subject to constant excitations. The desired values of some output voltages and currents are specified. We would like to adjust the parameter values $(x_1, \ldots, x_n)$ in an efficient manner, until either the desired outputs are obtained or some properly defined error function $\mathcal{E}$ is reduced to an acceptable low value. The error function most commonly used in network optimization is the weighted sum of square-of-errors, given next for the case of m outputs:

$$\mathcal{E} \triangleq \frac{1}{2} \sum_{k=1}^{m} W_k(\phi_k - \bar{\phi}_k)^2 \tag{15-104}$$

where ϕ_k, $\bar{\phi}_k$, and W_k are the kth *actual* output, kth *desired* output, and kth weighting factor, respectively. Since we wish to vary the element parameters $(x_1, \ldots, x_n)$ so as to reduce $\mathcal{E}$ as much as possible, it is important to know the effect on $\mathcal{E}$ due to the change in each x_j alone. In other words, we need the partial derivatives $\partial\mathcal{E}/\partial x_j$, $j = 1, 2, \ldots, n$. Now the *gradient* of any scalar function $\mathcal{E}(x_1, \ldots, x_n)$, denoted by $\nabla\mathcal{E}$, is defined to be the *vector*

$$\nabla\mathcal{E} \triangleq \left[\frac{\partial\mathcal{E}}{\partial x_1}, \ldots, \frac{\partial\mathcal{E}}{\partial x_n}\right]^t \tag{15-105}$$

The change in $\mathcal{E}$, *to a first order*, may be expressed as

$$\begin{aligned} \Delta\mathcal{E} &= \frac{\partial\mathcal{E}}{\partial x_1}\Delta x_1 + \ldots + \frac{\partial\mathcal{E}}{\partial x_n}\Delta x_n \\ &= \nabla\mathcal{E}\cdot\Delta\boldsymbol{x} \end{aligned} \tag{15-106}$$

For any given set of parameter values, once the gradient $\nabla\mathcal{E}$ is ascertained, we can alter the parameter values $(x_1, \ldots, x_n)$ according to some gradient technique [13] to efficiently reduce the error. (For example, we may let $\Delta\boldsymbol{x} = -\lambda\cdot\nabla\mathcal{E}$ for a suitable value of $\lambda > 0$.) After arriving at an improved set of parameter values, the gradient $\nabla\mathcal{E}$ is evaluated again, and the parameter values adjusted again in accordance with the new gradient. The process is repeated and terminated when the error function $\mathcal{E}$ is acceptably small, or when no appreciable reduction in $\mathcal{E}$ can be achieved, or when the number of iterations exceeds some predetermined maximum.

A detailed discussion of various optimization techniques is clearly beyond the scope of this book. Those interested should refer to other sources for this topic [13, 14]. Here, we shall only outline the procedure for calculating the gradients by the adjoint-network method. Three cases will be considered: linear resistive networks, linear dynamic networks in the frequency domain, and linear dynamic networks in the time domain. As always with the adjoint-network method, we consider any output voltage to be the voltage across an independent current source, and any output current to be the current through an independent voltage source (see the remarks at the beginning of Section 15-3-3).

15-6-1 Linear Resistive Networks with Constant Excitations

The error function $\mathcal{E}$, defined by Eq. (15-104), may be rewritten with matrix notation as

$$\mathcal{E} = \tfrac{1}{2}(\boldsymbol{V}_J - \bar{\boldsymbol{V}}_J)^t\boldsymbol{W}_J(\boldsymbol{V}_J - \bar{\boldsymbol{V}}_J) + \tfrac{1}{2}(\boldsymbol{I}_E - \bar{\boldsymbol{I}}_E)^t\boldsymbol{W}_E(\boldsymbol{I}_E - \bar{\boldsymbol{I}}_E) \tag{15-107}$$

where the subscripts J and E indicate independent current and voltage sources, respectively. If the voltage across some independent current source is not among the output variables, then the corresponding weighting factor in the diagnoal matrix $\boldsymbol{W}_J$ is set to zero. Similar use of the zero diagonal elements in the diagonal matrix $\boldsymbol{W}_E$ is obvious.

To a first-order approximation, we obtain from Eq. (15-107)

$$\Delta \mathcal{E} = (\boldsymbol{V}_J - \bar{\boldsymbol{V}}_J)^t \boldsymbol{W}_J \Delta \boldsymbol{V}_J + (\boldsymbol{I}_E - \bar{\boldsymbol{I}}_E)^t \boldsymbol{W}_E \, \Delta \boldsymbol{I}_E \tag{15-108}$$

Since all independent sources in N are constant and independent of the parameters $(x_1, \ldots, x_2)$, we have $\Delta \boldsymbol{V}_E = \boldsymbol{0}$ and $\Delta \boldsymbol{I}_J = 0$. Then Eq. (15-44) becomes

$$-\hat{\boldsymbol{I}}_J^t \, \Delta \boldsymbol{V}_J + \hat{\boldsymbol{V}}_E^t \, \Delta \boldsymbol{I}_E = \hat{\boldsymbol{I}}_b^t \, \Delta \boldsymbol{V}_b - \hat{\boldsymbol{V}}_b^t \, \Delta \boldsymbol{I}_b \tag{15-109}$$

A comparison between the right side of Eq. (15-108) and the left side of Eq. (15-109) shows that if we let the excitations in $\hat{N}$ be related to output errors by

$$\hat{\boldsymbol{I}}_J = -\boldsymbol{W}_J(\boldsymbol{V}_J - \bar{\boldsymbol{V}}_J) \tag{15-110a}$$

and

$$\hat{\boldsymbol{V}}_E = \boldsymbol{W}_E(\boldsymbol{I}_E - \bar{\boldsymbol{I}}_E) \tag{15-110b}$$

then Eq. (15-108) becomes

$$\Delta \mathcal{E} = (\hat{\boldsymbol{I}}_b^t \, \Delta \boldsymbol{V}_b - \hat{\boldsymbol{V}}_b^t \, \Delta \boldsymbol{I}_b) \tag{15-111}$$

The right side of Eq. (15-111) may be related to the changes in $(x_1, \ldots, x_n)$ by the use of Eq. (15-67) or Table 15-3. From the resultant expression, the gradient $\nabla \mathcal{E}$ is readily obtained.

In summary, the adjoint-network method of calculating the gradient of the error function $\mathcal{E}$, defined by Eq. (15-107) for a linear resistive network with constant excitations, consists of the following steps:

Step 1. Perform an analysis of N. Obtain those elements of $\boldsymbol{I}_E$ and $\boldsymbol{V}_J$ that are considered outputs. For each nonsource branch in N, also obtain the voltage or current, whichever is the independent variable in the branch description.

Step 2. For any chosen set of weighting factors for output errors, let the excitations in $\hat{N}$ be given by Eq. (15-110). Perform an analysis of $\hat{N}$. For each nonsource branch in $\hat{N}$ obtain the voltage or current, whichever is the independent variable in the branch description.

Step 3. Calculate the right side of Eq. (15-111) with the aid of Eq. (15-67) or Table 15-3. Obtain the gradient $\nabla \mathcal{E}$ by comparing the resultant equation with Eq. (15-106).

EXAMPLE 15-10. Consider the network shown in Fig. 15-12(a). Let the desired output be $\bar{V}_1 = 1$ V and $\bar{V}_2 = 2$ V, and the weighting factors be $W_1 = W_2 = 1$. The adjustable parameters are the resistances R_1 and R_2, whose nominal values are 2 Ω and 4 Ω, respectively. Find the gradient $\nabla \mathcal{E}$ by the adjoint-network method.

Solution: The network is first redrawn as shown in Fig. 15-12(b). Note that a short-circuit element $V_4 = 0$ and a zero-valued current source $I_{J2} = 0$ have been introduced. Also note the reference directions for I_{J1}, V_{J1}, I_{J2}, and V_{J2}.

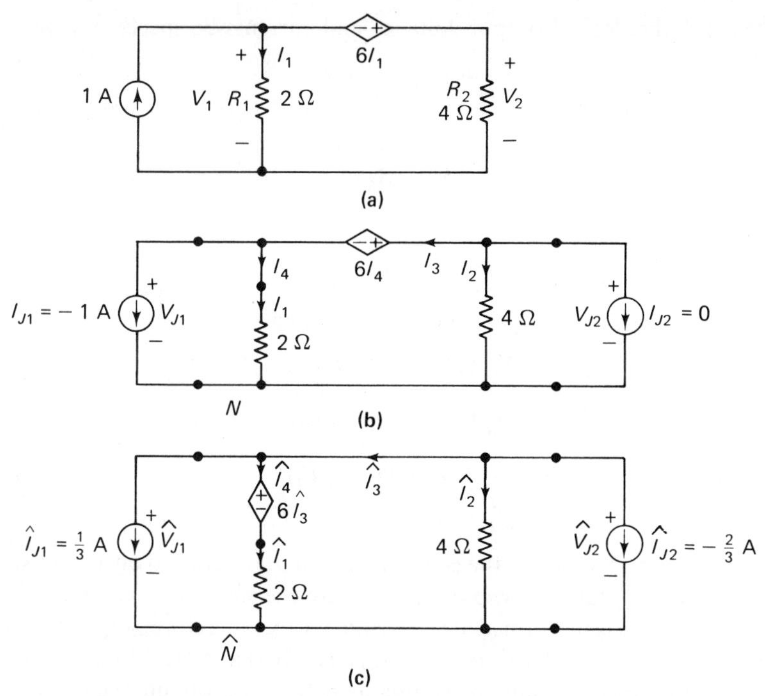

Fig. 15-12. Calculation of the gradient by the adjoint-network method.

Step 1. With $I_{J1} = -1$ and $I_{J2} = 0$, the network N is solved to give

$$\boldsymbol{I}_b = [\tfrac{1}{3}, \tfrac{2}{3}, -\tfrac{2}{3}, \tfrac{1}{3}]^t$$

$$\boldsymbol{V}_J = [\tfrac{2}{3}, \tfrac{8}{3}]^t$$

Step 2. The adjoint network $\hat{N}$ is shown in Fig. 15-12(c). The excitations in $\hat{N}$ are, according to Eq. (15-110),

$$\hat{I}_{J1} = -1 \times (\tfrac{2}{3} - 1) = \tfrac{1}{3}$$

$$\hat{I}_{J2} = -1 \times (\tfrac{8}{3} - 2) = -\tfrac{2}{3}$$

The adjoint network $\hat{N}$ is solved to give

$$\hat{\boldsymbol{I}}_b = [-\tfrac{1}{18}, \tfrac{7}{18}, \tfrac{5}{18}, -\tfrac{1}{18}]^t$$

Step 3. Using Table 15-3 and Eq. (15-111), we obtain

$$\Delta\mathcal{E} = (-\tfrac{1}{18} \times \tfrac{1}{3})\,\Delta R_1 + (\tfrac{7}{18} \times \tfrac{2}{3})\,\Delta R_2$$

Therefore,

$$\nabla\mathcal{E} = [-\tfrac{1}{54}, \tfrac{7}{27}]^t$$

★15-6-2 Calculation of Error Gradients for Linear Dynamic Networks—The Frequency-Domain Case

The procedure is basically the same as that presented in Section 15-6-1. Since currents and voltages are now complex-valued phasors, and since outputs may be specified for a number of frequency points, some minor changes in notations are to be expected. Let the desired output voltages and currents be specified at frequencies

$$\omega_1, \omega_2, \ldots, \omega_m$$

Designate as $\mathcal{E}_k$ the contribution at frequency ω_k to the error function $\mathcal{E}$. That is,

$$\mathcal{E} = \sum_{k=1}^{m} \mathcal{E}_k \tag{15-112a}$$

and

$$\nabla\mathcal{E} = \sum_{k=1}^{m} \nabla\mathcal{E}_k \tag{15-112b}$$

At each frequency ω_k, the error component is defined to be

$$\begin{aligned}\mathcal{E}_k &\triangleq \tfrac{1}{2}\sum_i W_{J_i}(j\omega_k)\,|V_{J_i}(j\omega_k) - \bar{V}_{J_i}(j\omega_k)|^2 \\ &\quad + \tfrac{1}{2}\sum_i W_{E_i}(j\omega_k)\,|I_{E_i}(j\omega_k) - \bar{I}_{E_i}(j\omega_k)|^2 \\ &= \tfrac{1}{2}(\boldsymbol{V}_J - \bar{\boldsymbol{V}}_J)^* \boldsymbol{W}_J (\boldsymbol{V}_J - \bar{\boldsymbol{V}}_J) \\ &\quad + \tfrac{1}{2}(\boldsymbol{I}_E - \bar{\boldsymbol{I}}_E)^* \boldsymbol{W}_E (\boldsymbol{I}_E - \bar{\boldsymbol{I}}_E)\end{aligned} \tag{15-113}$$

where the superscript * indicates the *conjugate transpose* operation on a matrix. $\boldsymbol{W}_J$ and $\boldsymbol{W}_E$ are real diagonal matrices whose elements are the weighting factors. It is understood that the analysis is performed at the frequency ω_k; therefore, $j\omega_k$ will be dropped for simplicity. Note the similarity between Eqs. (15-107) and (15-113).

Following exactly the same reasoning as before, we obtain the equations corresponding to Eqs. (15-108), (15-110), and (15-111):

$$\Delta\mathcal{E}_k = \operatorname{Re}\{(\boldsymbol{V}_J - \bar{\boldsymbol{V}}_J)^* \boldsymbol{W}_J\, \Delta\boldsymbol{V}_J + (\boldsymbol{I}_E - \bar{\boldsymbol{I}}_E)^* \boldsymbol{W}_E\, \Delta\boldsymbol{I}_E\} \tag{15-114}$$

$$\hat{\boldsymbol{I}}_J = -\boldsymbol{W}_J(\boldsymbol{V}_J - \bar{\boldsymbol{V}}_J)_{\text{conj}} \tag{15-115a}$$

$$\hat{\boldsymbol{V}}_E = \boldsymbol{W}_E(\boldsymbol{I}_E - \bar{\boldsymbol{I}}_E)_{\text{conj}} \tag{15-115b}$$

$$\Delta\mathcal{E}_k = \operatorname{Re}(\hat{\boldsymbol{I}}_b^t\, \Delta\boldsymbol{V}_b - \hat{\boldsymbol{V}}_b^t\, \Delta\boldsymbol{I}_b) \tag{15-116}$$

where $\boldsymbol{A}_{\text{conj}}$ means the conjugate of $\boldsymbol{A}$.

The procedure for calculating the gradient component $\nabla\mathcal{E}_k$ may be summarized as follows:

Step 1. At the frequency ω_k, with given excitations, perform an ac analysis of N to obtain $\boldsymbol{V}_J$ and $\boldsymbol{I}_E$. For each nonsource branch, also obtain the voltage or current, whichever is the independent variable in the branch description.

Step 2. For any chosen set of weighting factors for output errors, let the excitation for $\hat{N}$ be given by Eq. (15-115). Perform an ac analysis of $\hat{N}$. For each nonsource branch in $\hat{N}$, obtain the voltage or current.

Step 3. Calculate the right side of Eq. (15-116) with the aid of Eq. (15-67) or Table 15-3. Obtain $\nabla\mathcal{E}_k$ by comparing the resultant equation with Eq. (15-106).

This procedure is repeated at each frequency of interest, and finally the gradient $\nabla\mathcal{E}$ is the sum of the contributions at all frequencies as given by Eq. (15-112b).

*15-6-3 Calculation of Error Gradients for Linear Dynamic Networks—The Time-Domain Case

In the time-domain case, the error function corresponding to Eq. (15-107) becomes

$$\begin{aligned}\mathcal{E} &\triangleq \frac{1}{2}\int_0^{t_f}\{\sum_i w_{J_i}(t)[v_{J_i}(t) - \bar{v}_{J_i}(t)]^2 + \sum_i w_{E_i}(t)[i_{E_i}(t) - \bar{i}_{E_i}(t)]^2\}\,dt \\ &= \frac{1}{2}\int_0^{t_f}\{[\boldsymbol{v}_J(t) - \bar{\boldsymbol{v}}_J(t)]^t \boldsymbol{w}_J(t)[\boldsymbol{v}_J(t) - \bar{\boldsymbol{v}}_J(t)] \\ &\qquad + \tfrac{1}{2}[\boldsymbol{i}_E(t) - \bar{\boldsymbol{i}}_E(t)]^t \boldsymbol{w}_E(t)[\boldsymbol{i}_E(t) - \bar{\boldsymbol{i}}_E(t)]\}\,dt\end{aligned} \tag{15-117}$$

where $0 \leq t \leq t_f$ is the time interval over which the excitations and desired outputs are given as functions of t. Except for using lowercase letters to indicate time functions, all the variables, subscripts, and superscripts in this section have the same meanings as those in Section 15-6-1. In fact, the reasonings used in proving various equations for the determination of $\nabla\mathcal{E}$ are also the same as in Section 15-6-1. (The reader is urged to understand the material of Section 15-6-1 perfectly before proceeding with the present section.) Thus, corresponding to Eqs. (15-108)–(15-111), we have the following equations, respectively:

$$\Delta\mathcal{E} = \int_0^{t_f}\{[\boldsymbol{v}_J(t) - \bar{\boldsymbol{v}}_J(t)]^t \boldsymbol{w}_J(t)\,\Delta\boldsymbol{v}_J(t) + [\boldsymbol{i}_E(t) - \bar{\boldsymbol{i}}_E(t)]^t \boldsymbol{w}_E(t)\,\Delta\boldsymbol{i}_E(t)\}\,dt \tag{15-118}$$

$$\begin{aligned}&\int_0^{t_f}[-\hat{\boldsymbol{i}}_J^t(\tau)\,\Delta\boldsymbol{v}_J(t) + \hat{\boldsymbol{v}}_E^t(\tau)\,\Delta\boldsymbol{i}_E(t)]_{\tau=t_f-t}\,dt \\ &\qquad = \int_0^{t_f}[\hat{\boldsymbol{i}}_b^t(\tau)\,\Delta\boldsymbol{v}_b(t) - \hat{\boldsymbol{v}}_b^t(\tau)\,\Delta\boldsymbol{i}_b(t)]_{\tau=t_f-t}\,dt\end{aligned} \tag{15-119}$$

$$\hat{\boldsymbol{i}}_J(\tau) = \{-\boldsymbol{w}_J(t)[\boldsymbol{v}_J(t) - \bar{\boldsymbol{v}}_J(t)]\}_{t=t_f-\tau} \tag{15-120a}$$

$$\hat{\boldsymbol{v}}_E(\tau) = \{\boldsymbol{w}_E(t)[\boldsymbol{i}_E(t) - \bar{\boldsymbol{i}}_E(t)]\}_{t=t_f-\tau} \tag{15-120b}$$

$$\Delta\mathcal{E} = \int_0^{t_f}[\hat{\boldsymbol{i}}_b^t(\tau)\,\Delta\boldsymbol{v}_b(t) - \hat{\boldsymbol{v}}_b^t(\tau)\,\Delta\boldsymbol{i}_b(t)]_{\tau=t_f-t}\,dt \tag{15-121}$$

The procedure for calculating the gradient $\nabla\mathcal{E}$ in the time domain may be summarized as follows:

Step 1. Perform a transient analysis of N over the time interval $0 \leq t \leq t_f$. Obtain those elements of $\boldsymbol{i}_E(t)$ and $\boldsymbol{v}_J(t)$ that are considered as outputs. Obtain $v(t)$ or $i(t)$ (whichever is the independent variable in the branch description) for each resistive branch. Obtain $\dot{v}(t)$ for each capacitive branch and $\dot{i}(t)$ for each inductive branch.

Step 2. Perform a transient analysis of the adjoint network $\hat{N}$ over the time interval $0 \leq \tau \leq t_f$, with $\hat{v}_C(\tau) = 0$ and $\hat{i}_L(\tau) = 0$ at $\tau = 0$, and under the excitations given by Eq. (15-120). As before, if any elements of $\boldsymbol{i}_E(t)$ and $\boldsymbol{v}_J(t)$ are not output variables, then the corresponding weighting factors are set to zero. Obtain $\hat{v}(\tau)$ or $\hat{i}(\tau)$ for each resistive branch. Obtain $\hat{v}(\tau)$ for each capacitive branch and $\hat{i}(\tau)$ for each inductive branch.

Step 3. Calculate the right side of Eq. (15-121) according to Table 15-4. From the resultant expression, find $\nabla\mathcal{E}$ by comparing with Eq. (15-106).

Example 15-11 will illustrate the procedure.

EXAMPLE 15-11. Consider the *RC* network shown in Fig. 15-13(a). The adjustable parameters are R_1, R_2, C_3, and C_4 with nominal values $R_1 = R_2 = 2\,\Omega$, and $C_3 = C_4 = 2$ F. The capacitors are initially uncharged. An excitation $v_E(t) = 4e^{-2t}$ is applied, and an output voltage $\bar{v}_o(t) = 3e^{-2t}$ is desired over the time interval $0 \leq t \leq 1$. Find the gradient

$$\nabla\mathcal{E} = \left[\frac{\partial\mathcal{E}}{\partial R_1}, \frac{\partial\mathcal{E}}{\partial R_2}, \frac{\partial\mathcal{E}}{\partial C_3}, \frac{\partial\mathcal{E}}{\partial C_4}\right]^t$$

where $\mathcal{E}$ is defined by Eq. (15-117).

Solution: A zero-valued independent current source i_J has been added in Fig. 15-13(a) so that the output becomes $v_J(t)$. The initial time is $t = 0$, and the final time is $t_f = 1$ s.

Step 1. An analysis of N yields

$$i_{R1}(t) = i_{R2}(t) = e^{-2t}$$
$$v_{C3}(t) = v_{C4}(t) = 2e^{-2t}$$
$$\dot{v}_{C3}(t) = \dot{v}_4(t) = -4e^{-2t}$$

Step 2. The adjoint network is shown in Fig. 15-13(b). Since i_E is not an output, we let $w_E = 0$. Arbitrarily choose $w_J = 1$. Then, according to Eq. (15-120), the excitations in $\hat{N}$ are

$$\hat{i}_J(\tau) = -[2e^{-2t} - 3e^{-2t}]_{t=1-\tau} = e^{-2+2\tau}$$

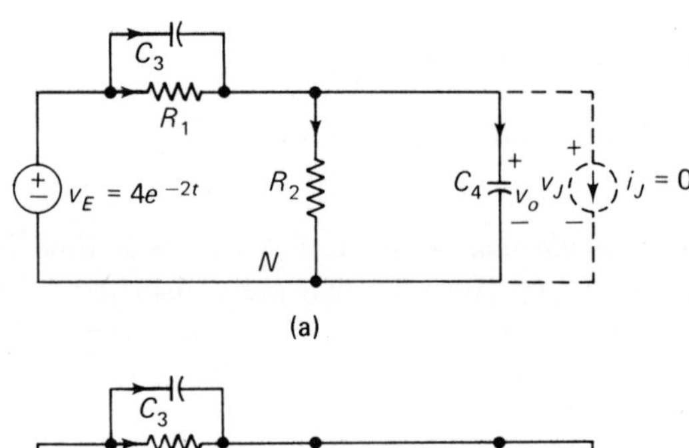

(a)

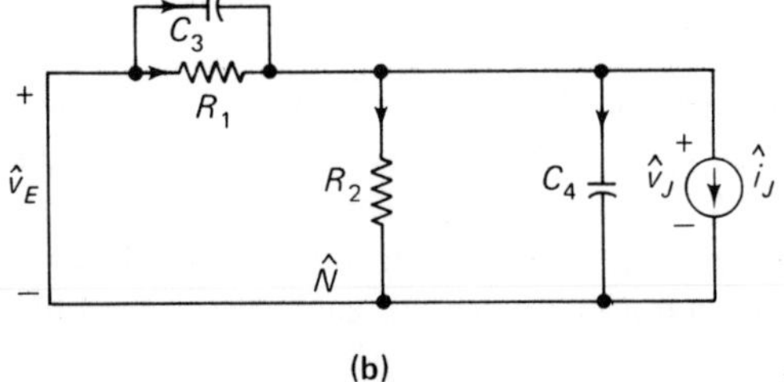

(b)

Fig. 15-13. Time-domain gradient calculation.

and

$$\hat{v}_E(\tau) = 0$$

A transient analysis of $\hat{N}$ [under the conditions $\hat{v}_{C3}(0) = \hat{v}_{C4}(0) = 0$] yields

$$\hat{i}_{R1}(\tau) = -\hat{i}_{R2}(\tau) = \frac{e^{-2}}{18}[-e^{-(1/4)\tau} + e^{2\tau}]$$

$$\hat{v}_{C3}(\tau) = -\hat{v}_{C4}(\tau) = \frac{e^{-2}}{9}[-e^{-(1/4)\tau} + e^{2\tau}]$$

Step 3. From the third column of Table 15-4, the following sensitivity components are obtained:
For R_1,

$$\left\{\int_0^1 \left[\frac{e^{-2}}{18}(-e^{-(1/4)\tau} + e^{2\tau})e^{-2t}\right]_{\tau=1-t} dt\right\} \Delta R_1 = 0.01087\,\Delta R_1$$

For C_3,

$$\left\{-\int_0^1 \left[\frac{e^{-2}}{9}(-e^{-(1/4)\tau} + e^{2\tau})(-4)e^{-2t}\right]_{\tau=1-t} dt\right\} \Delta C_3 = 0.08696\,\Delta C_3$$

In a similar manner, we find the sensitivity components for R_2 and C_4 to be $-0.01087\,\Delta R_2$ and $-0.08696\,\Delta C_4$, respectively. Therefore, we have

$$\begin{aligned}\nabla\mathcal{E} &= \left[\frac{\partial\mathcal{E}}{\partial R_1}, \frac{\partial\mathcal{E}}{\partial R_2}, \frac{\partial\mathcal{E}}{\partial C_3}, \frac{\partial\mathcal{E}}{\partial C_4}\right]^t \\ &= [0.01087, -0.01087, 0.08696, -0.08696]^t\end{aligned}$$

Some remarks about the procedure for calculating the gradient in the time domain are in order.

1. From Eq. (15-120), the excitations in $\hat{N}$ are just the weighted output errors *in reverse time.*
2. When the gradient calculation is done with a digital computer simulation program, all integrals in Eq. (15-121) will be approximated by summations. For the original network N, the transient responses are obtained at

   $$t = 0, h, 2h, \ldots, t_f = mh$$

   and the results *must be stored.* For the adjoint network $\hat{N}$, the transient responses are obtained at

   $$\tau = 0, h, 2h, \ldots, t_f = mh$$

   but the results *need not be stored.* Each time the transient responses at $\tau = kh$ $(k = 0, 1, 2, \ldots)$ are obtained, they are used immediately to update the integrals in Eq. (15-121); and their values are no longer needed (see Section 15-5 for a more detailed explanation).
3. As pointed out earlier, the derivatives $\dot{v}_C(t)$ and $\dot{i}_L(t)$ are readily available from the transient analysis program if the state-variable approach is used.

15-7 SENSITIVITY CALCULATION FOR NONLINEAR RESISTIVE NETWORKS

When a nonlinear resistive network is subject to dc excitations, there may be a unique solution, several solutions, or no solution at all. In Chapters 5 and 7 we studied several methods for finding the solutions, or the operating points as they are sometimes called. Obviously, when the parameters of some elements vary, so will the operating points. In this section we shall describe how the adjoint-network method can be used to determine the sensitivity of an operating point due to the changes in parameters of *linear and nonlinear* resistive elements. To this end, we must first generalize the definition of adjoint networks to include nonlinear resistive elements.

For the network under consideration, we shall assume that any nonlinear two-terminal resistor is either *current-controlled*, such that

$$V = f(I, \mathbf{x}) \tag{15-122a}$$

or *voltage-controlled*, such that

$$I = g(V, \mathbf{x}) \tag{15-122b}$$

where the x_i's are some physical parameters affecting the I–V characteristics. For example, a junction diode may be characterized by (see Section 2-4-1)

$$I_d = I_s(e^{(q/kMT)V_d} - 1)$$

Then we have three parameters, $x_1 = I_s$, $x_2 = M$, and $x_3 = T$ (q and k are constants).

How shall we define the characteristic of the adjoint element of a nonlinear resistor? In Section 15-3-3, it is pointed out that in defining adjoint elements, our goal is to make ($\hat{I}\,\Delta V - \hat{V}\,\Delta I$) *independent of* ΔV *and* ΔI, but *dependent* on Δx *only*. Consider first Eq. (15-122a). To a first-order approximation, we can write

$$\Delta V = \frac{\partial f}{\partial I}\Delta I + \frac{\partial f}{\partial x_1}\Delta x_1 + \ldots + \frac{\partial f}{\partial x_m}\Delta x_m$$

Then we have

$$\begin{aligned}\hat{I}\,\Delta V - \hat{V}\,\Delta I &= \hat{I}\Big(\frac{\partial f}{\partial I}\Delta I + \frac{\partial f}{\partial x_1}\Delta x_1 + \ldots + \frac{\partial f}{\partial x_m}\Delta x_m\Big) - \hat{V}\,\Delta I \\ &= \Big(\hat{I}\frac{\partial f}{\partial I} - \hat{V}\Big)\Delta I + \hat{I}\Big(\frac{\partial f}{\partial x_1}\Delta x_1 + \ldots + \frac{\partial f}{\partial x_m}\Delta x_m\Big)\end{aligned} \tag{15-123}$$

To make the coefficients of ΔI identically zero, we must have

$$\boxed{\hat{V} = \frac{\partial f}{\partial I}\hat{I}} \tag{15-124}$$

which shows that the adjoint element of a current-controlled nonlinear resistor R is a *linear* resistor whose resistance is equal to the *incremental resistance* of R at the operating point and that

$$\boxed{\hat{I}\,\Delta V - \hat{V}\,\Delta I = \hat{I}\Big(\frac{\partial f}{\partial x_1}\Delta x_1 + \ldots + \frac{\partial f}{\partial x_m}\Delta x_m\Big)} \tag{15-125}$$

For the special case of a linear resistor of value R, Eq. (15-124) defines the adjoint element as a linear resistor also of value R. This of course agrees with the previous definition given in Table 15-2.

In a similar manner, it can be shown that the adjoint element of a voltage-controlled nonlinear resistor G, characterized by Eq. (15-122b), is a *linear resistor* characterized by

$$\boxed{\hat{I} = \frac{\partial g}{\partial V}\hat{V}} \tag{15-126}$$

and that

$$\boxed{\hat{I}\,\Delta V - \hat{V}\,\Delta I = -\hat{V}\left(\frac{\partial g}{\partial x_1}\Delta x_1 + \ldots + \frac{\partial g}{\partial x_m}\Delta x_m\right)} \tag{15-127}$$

Now consider a nonlinear resistive network whose only nonlinearities appear in two-terminal resistors characterized by Eq. (15-122). The independent voltage and current sources have constant values and are designated by $\boldsymbol{V}_E$ and $\boldsymbol{I}_J$, respectively. As before, we consider any output voltage V_o to be an element of $\boldsymbol{V}_J$ and any output current I_0 to be an element of $\boldsymbol{I}_E$. The procedure for calculating $\partial V_o/\partial x_j$ or $\partial I_o/\partial x_j$ for all x's may be summarized as follows:

Step 1. Perform a dc analysis of the nonlinear resistive network N to ascertain the operating point. For each nonsource branch, obtain the voltage or current, whichever is the independent variable in the branch description. For each nonlinear resistor, determine the *incremental* resistance or conductance at the operating point.

Step 2. Construct the adjoint network $\hat{N}$, making use of Table 15-2 for linear elements, and Eq. (15-124) or (15-126) for nonlinear resistors.

Step 3. Perform a dc analysis of the *linear* network $\hat{N}$, with all excitations except one set to zero. The single excitation is

$$\hat{I}_{Jk} = -1, \qquad \text{when } V_{jk} \text{ is the output}$$

or

$$\hat{V}_{Ek} = 1, \qquad \text{when } I_{Jk} \text{ is the output}$$

For each nonsource branch, obtain the voltage or current, whichever is the independent variable in the branch description.

Step 4. Evaluate Eq. (15-44). With the excitations chosen in step 2, the left side of Eq. (15-44) should yield only one term, which is precisely the desired ΔV_o or ΔI_o. The right side of Eq. (15-44) is evaluated with the aid of Table 15-2 (for linear elements) and Eqs. (15-125) and (15-127) for nonlinear resistors. Finally, from the resultant expressions, we obtain $\partial V_o/\partial x_j$ or $\partial I_o/\partial x_j$.

EXAMPLE 15-12. Consider the nonlinear resistive network shown in Fig. 15-14(a). The two nonlinear resistors are characterized by $I_1 = k_1 V_1^3$ and $I_2 = k_2 V_2^3$, with nominal values $k_1 = k_2 = 1$. Let $V_o = V_2$. Find V_o, $\partial V_o/\partial k_1$, and $\partial V_o/\partial k_2$.

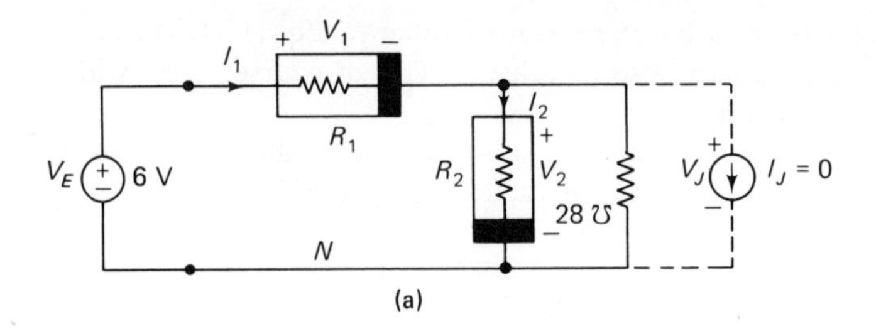

(a)

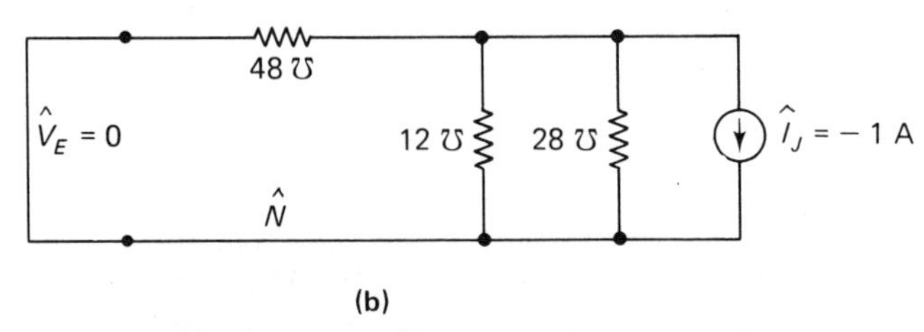

(b)

Fig. 15-14. Operating sensitivity analysis.

Solution: A zero-valued current source $I_J = 0$ has been added to Fig. 15-14(a) so that the output is $V_o = V_J$.

Step 1. A dc analysis of the network N yields

$$V_1 = 4, \qquad V_2 = 2$$

$$\frac{\partial I_1}{\partial V_1} = 3V_1^2 = 48, \qquad \frac{\partial I_2}{\partial V_2} = 3V_2^2 = 12$$

Step 2. The adjoint network $\hat{N}$ is shown in Fig. 15-14(b).

Step 3. The excitations in $\hat{N}$ are

$$\hat{V}_E = 0, \qquad \hat{I}_J = -1$$

a dc analysis of $\hat{N}$ yields

$$\hat{V}_1 = -\frac{1}{88}, \qquad \hat{V}_2 = \frac{1}{88}$$

Step 4. From Eqs. (15-44) and (15-127), we obtain

$$\Delta V_J = -\hat{V}_1 V_1^3 \, \Delta k_1 - \hat{V}_2 V_2^3 \, \Delta k_2$$
$$= \tfrac{8}{11} \Delta k_1 - \tfrac{1}{11} \Delta k_2$$

Therefore, the answers are $V_o = 2$ V, $\partial V_o / \partial k_1 = \frac{8}{11}$, and $\partial V_o / \partial k_2 = -\frac{1}{11}$.

We have extended the definition of the adjoint networks to include only nonlinear two-terminal resistors of the voltage-controlled and current-controlled types. Further extension of the adjoint networks to include nonlinear, time-varying *RLC* and coupled elements is also possible. Parametric characterizations of these elements will be necessary. A discussion of this most general case of all adjoint networks and applications is beyond the scope of this book. Interested readers should consult appropriate references [12, 15, 16].

REFERENCES

1. BODE, H. W. *Network Analysis and Feedback Amplifier Design.* New York: Van Nostrand Reinhold Company, 1945.
2. CALAHAN, D. A. *Computer-Aided Network Design.* New York: McGraw-Hill Book Company, 1972.
3. KARAFIN, B. J. "Optimum Assignment of Component Tolerances for Electrical Networks." *Bell System Tech. J.*, Vol. 50, No. 4, pp. 1225–1242, Apr. 1971.
4. GEHER, K. *Theory of Network Tolerances.* Budapest: Akademiai, Kiado, 1971.
5. BUTLER, W. J., and S. S. HAYKIN. "Multiparameter Sensitivity Problems in Network Theory." *Proc. IEE (London)*, Vol. 117, pp. 2228–2236, Dec. 1970.
6. JENSEN, R. W., and M. D. LIEBERMAN. *IBM Electronic Circuit Analysis Program.* Englewood Cliffs, N.J.: Prentice-Hall, Inc., 1968.
7. VALLESE, L. M. "Incremental Versus Adjoint Models for Network Sensitivity Analysis." *IEEE Trans.*, CAS-21, pp. 46–49, Jan. 1974.
8. TELLEGEN, B. D. H. *A General Network Theorem, with Applications.* Phillips Res. Rept., No. 7, pp. 259–269, 1952.
9. PENFIELD, P., Jr., R. SPENCE, and S. DUINKER. *Tellegen's Theorem and Electrical Networks.* Cambridge, Mass.: The MIT Press, 1970.
10. DESOER, C. A. "Teaching Adjoint Networks to Juniors." *IEEE Trans. Education*, Vol. E-16, pp. 10–14, Feb. 1973.
11. LIN, P. M. "A Survey of Applications of Symbolic Network Functions." *IEEE Trans. Circuit Theory*, Vol. CT-20, pp. 732–737, Nov. 1973.
12. ROHRER, R. A. "Fully Automated Network Design by Digital Computer: Preliminary Consideration." *Proc. IEEE*, Vol. 55, pp. 1929–1939, Nov. 1967.
13. TEMES, G. C. "Optimization Methods in Circuit Design." Chap. 5 in F. F. Kuo and W. G. Magnuson, Jr., eds., *Computer Oriented Circuit Design.* Englewood Cliffs, N.J.: Prentice-Hall, Inc., 1969.
14. DIRECTOR, S. W. "Survey of Circuit Oriented Optimization Techniques." *IEEE Trans. Circuit Theory*, Vol. CT-18, pp. 3–10, Jan. 1971.
15. DIRECTOR, S. W., and R. A. ROHRER. "The Generalized Adjoint Network and Network Sensitivity." *IEEE Trans. Circuit Theory*, Vol. CT-16, pp. 318–323, Aug. 1969.
16. DIRECTOR, S. W., and R. A. ROHRER. "Automated Network Design—The Frequency Domain Case." *IEEE Trans. Circuit Theory*, Vol. CT-16, pp. 330–337, Aug. 1969.
17. ROHRER, R. A., L. NAGEL, R. MEYER, and L. WEBER. "Computationally Efficient Electronic—Circuit Noise Calculations." *IEEE J. Solid State Circuits*, Vol. 5, pp. 204–213, Aug. 1971.

PROBLEMS

15-1. Consider the network of Fig. 15-3(a).

(a) Find all branch currents.

(b) Find $\partial i_1/\partial R_3$, $\partial i_2/\partial R_3$, and $\partial i_3/\partial R_3$ (numerical answers) by the use of the incremental-network method.

15-2. For the circuit of Fig. P15-2, find the partial derivative of V_o with respect to β by the incremental-network method.

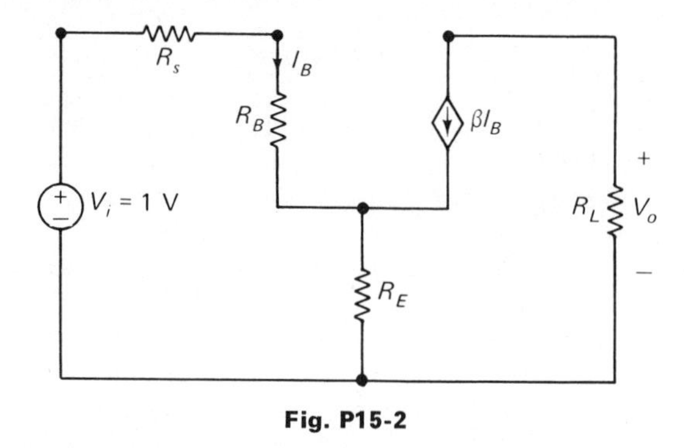

Fig. P15-2

15-3. For the circuit of Fig. P15-3, find $\partial V_{o1}/\partial \mu$ and $\partial V_{o2}/\partial \mu$ by the incremental-network method.

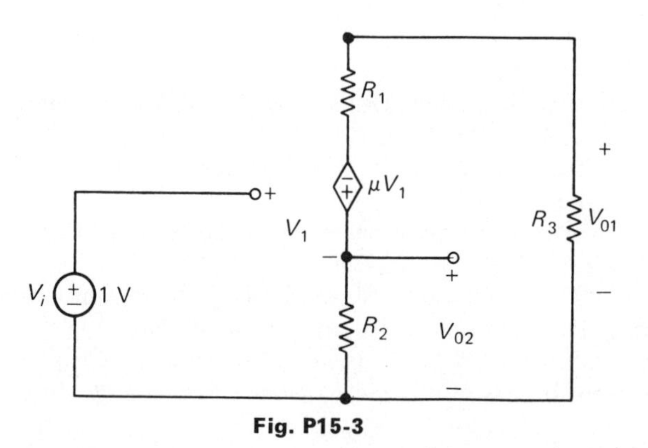

Fig. P15-3

15-4. For the two networks N and N' shown in Fig. P15-4, find all branch currents and voltages and verify Eq. (15-27).

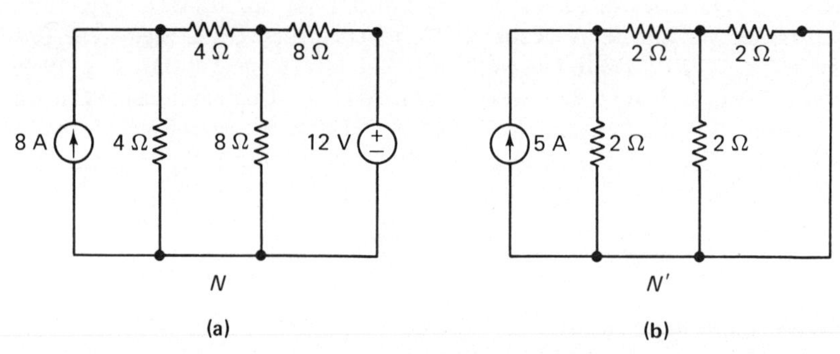

Fig. P15-4

15-5. Construct the adjoint network for the network shown in Fig. P15-5.

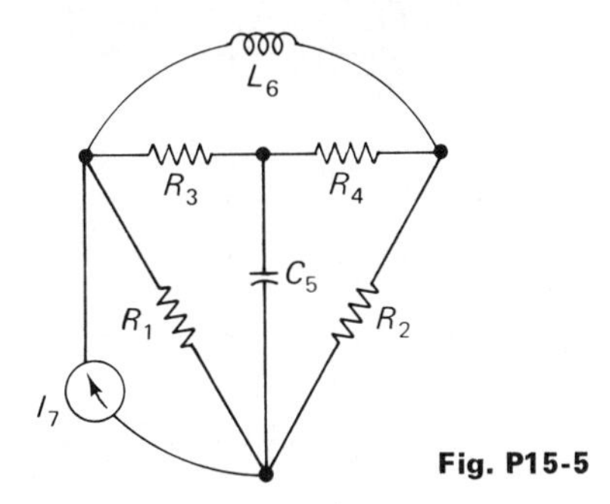

Fig. P15-5

15-6. Repeat Problem 15-5 for Fig. P15-6.

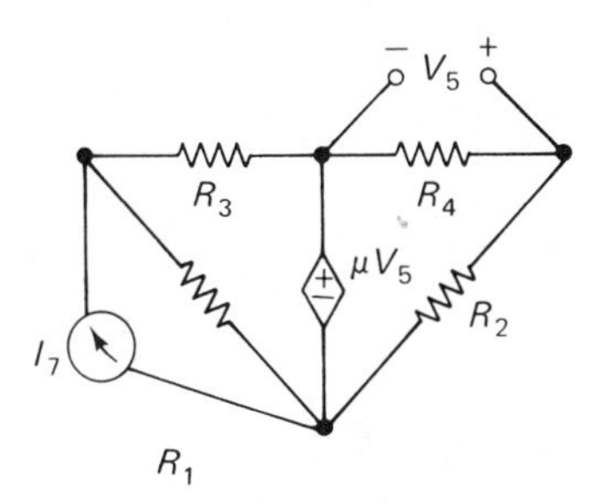

Fig. P15-6

15-7. Repeat Problem 15-5 for Fig. P15-7.

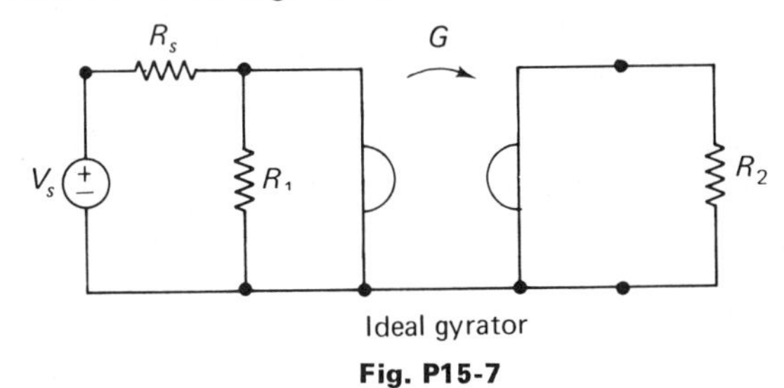

Ideal gyrator

Fig. P15-7

15-8. Find $\partial Z_{in}/\partial R_a$, $\partial Z_{in}/\partial R_b$, and $\partial Z_{in}/\partial R_c$ for the network of Fig. P15-8 by the adjoint-network method.

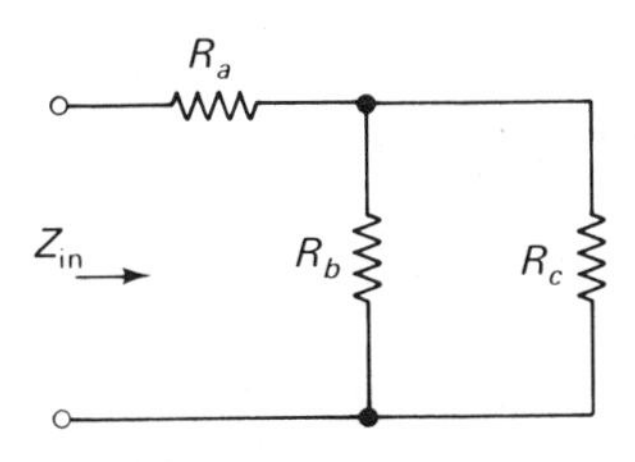

Fig. P15-8

15-9. Find $\partial Y_{in}/\partial Y_f$, $\partial Y_{in}/\partial Y_L$, and $\partial Y_{in}/\partial g_m$ for the network of Fig. P15-9 by the adjoint-network method (see art on next page).

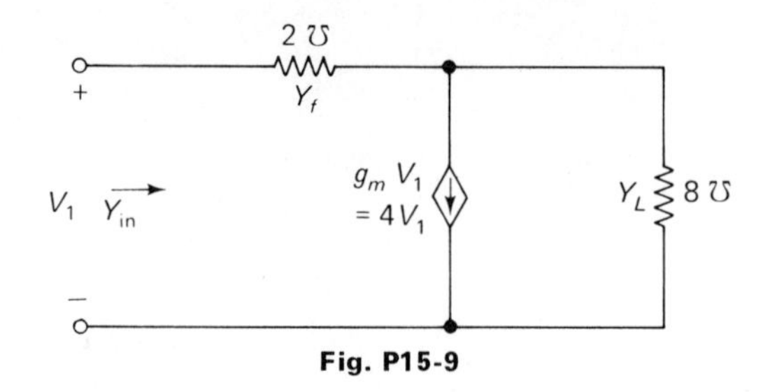

Fig. P15-9

15-10. In Fig. P15-10, let $T = V_o/I_{in}$. Find $\partial T/\partial R_i$, $i = 1, 2, 3, 4$, by the adjoint-network method.

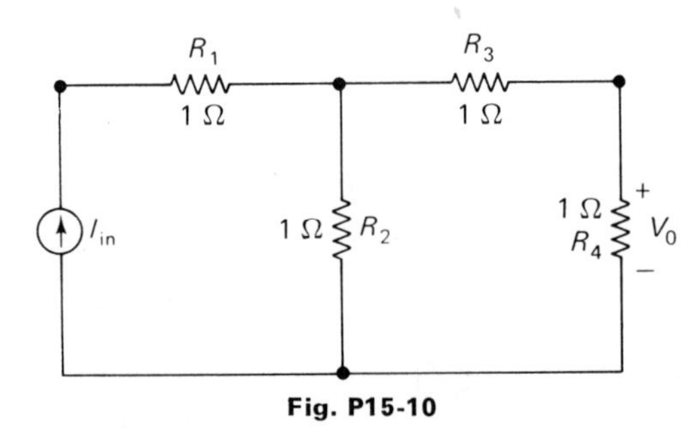

Fig. P15-10

15-11. Prove the relationship given by Eq. (15-65). *Hint:* Follow the method used to prove Eq. (15-38).

15-12. Prove the relationship given by Eq. (15-68). *Hint:* Follow the method used to prove Eq. (15-49).

15-13. For the circuit of Fig. P15-13, find the partial derivatives of I_o with respect to R_1, R_2, and r_m by the adjoint-network method.

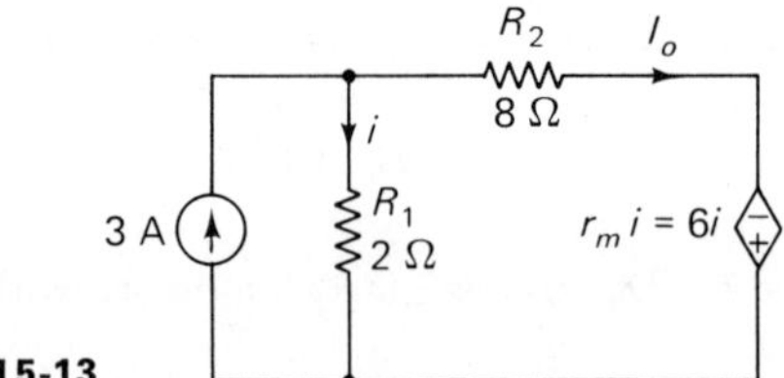

Fig. P15-13

15-14. For the circuit of Fig. P15-14, find the partial derivatives of V_o with respect to R_1, R_2, and β by the adjoint-network method.

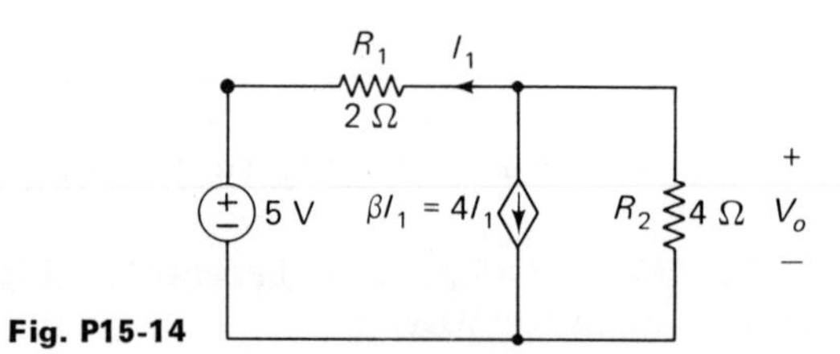

Fig. P15-14

15-15. For the circuit of Fig. P15-15, find the partial derivatives of the current ratio I_o/I_i with respect to R_1, R_2, and n by the adjoint-network method.

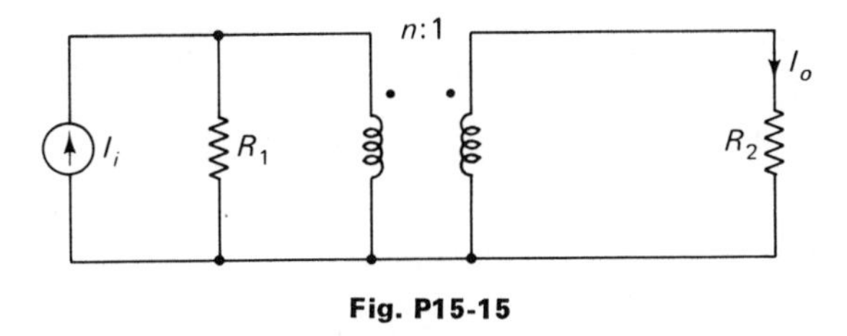

Fig. P15-15

15-16. Solve Problem 15-2 again by the adjoint-network method.

15-17. Solve Problem 15-3 again by the adjoint-network method.

15-18. Find the expression for Z_{in} for Fig. P15-8. Then find the expressions for $\partial Z_{in}/\partial R_a$, $\partial Z_{in}/\partial R_b$, and $\partial Z_{in}/\partial R_c$ by the symbolic method discussed in Section 15-4. Formulate the closed signal-flow graph (see Chap. 14), and verify that the expression for H is simply the determinant of the closed SFG.

15-19. Solve Problem 15-9 again by the symbolic method.

15-20. Consider the RL network shown in Fig. P15-20. $i_o = 0$ at $t = 0$. Find the partial derivatives of $i_o(t_f)$ with respect to R and L by the adjoint-network method (E and $t_f > 0$ are constants).

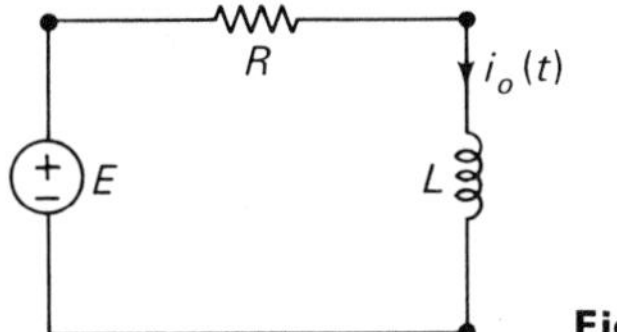

Fig. P15-20

15-21. For the circuit of Fig. P15-21, $v_C(0)$ is known to be constant. Find $\partial v_o(t_f)/\partial R$ and $\partial v_o(t_f)/\partial C$ by the adjoint-network method.

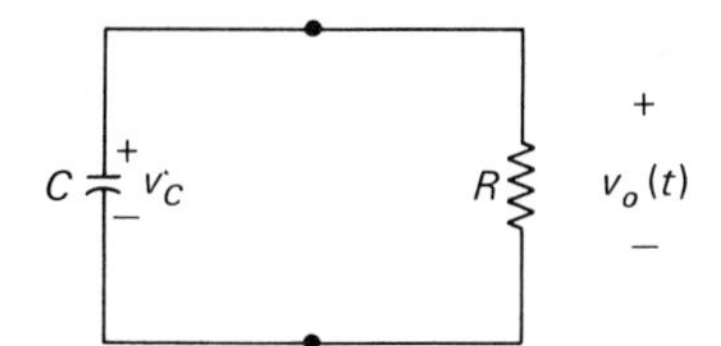

Fig. P15-21

15-22. The adjustable parameters in the circuit of Fig. P15-22 are R_1, R_2, R_3, and R_4. The desired outputs are $V_1 = 4$ V and $V_2 = 2$ V. Find the gradient $\nabla\mathcal{E}$ as defined by Eq. (15-105). Let all nonzero weighting factors be 1.

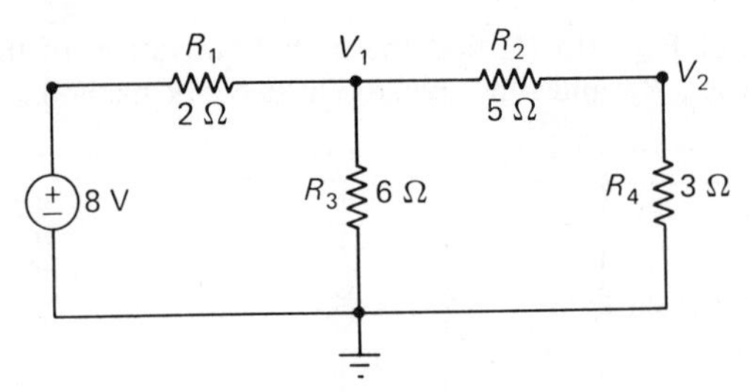

Fig. P15-22

15-23. Repeat Problem 15-22 for the circuit of Fig. P15-23. The parameters are now G_1, G_2, G_3, and g_m. The desired outputs are $V_1 = 5$ V and $V_2 = 3$ V.

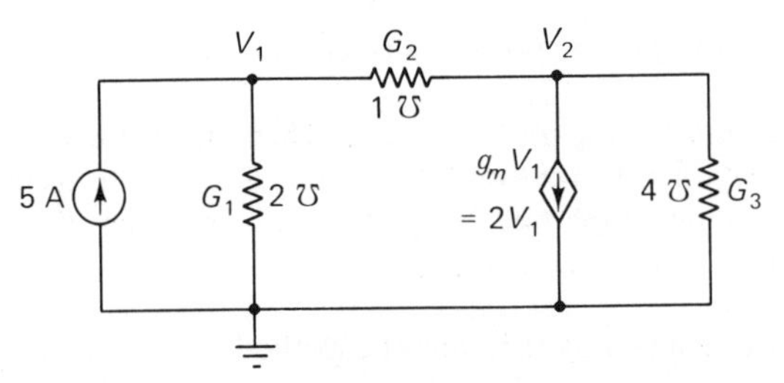

Fig. P15-23

15-24. For the circuit of Fig. P15-24, verify that $V_o = 1$ V is a solution. Then find $\partial V_o/\partial k_1$, $\partial V_o/\partial k_2$, and $\partial V_o/\partial R_3$ by the adjoint-network method.

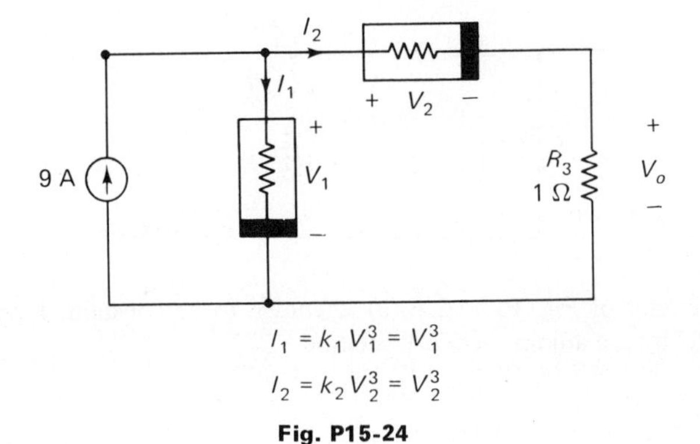

$$I_1 = k_1 V_1^3 = V_1^3$$
$$I_2 = k_2 V_2^3 = V_2^3$$

Fig. P15-24

CHAPTER 16

Introduction to Sparse-Matrix Techniques for Circuit Analysis

16-1 INTRODUCTION

In the preceding chapters we studied a number of methods for digital computer analysis of electrical networks. At the core of any general circuit-analysis method is the solution of n simultaneous linear equations

$$Ax = \mu \tag{16-1}$$

Two general methods for solving Eq. (16-1), the Gaussian elimination method and the ***LU*** factorization method, have been described in Chapter 4. Both methods require approximately $n^3/3$ operations[1] for large n (see Problem 4-5 for the exact count). In power networks and integrated electronic circuits, one can easily encounter networks with hundreds of nodes. If nodal equations are written for a 200-node network, then $n = 200$; and the number of operations will be approximately $(200)^3/3 \cong 2.67 \times 10^6$. Assume that each multiplication requires 6 μs (CDC 6500). Then the time required for performing multiplications alone will be 16 s. The total CPU (central processor unit) time will be considerably longer, because we have yet to add the time required for all other operations such as additions, subtractions, and memory addressing. The computing time goes up very rapidly as n increases. The situation becomes more serious when Eq. (16-1) has to be solved repeatedly (with different numerical coefficients), as in the Newton–Raphson method of solving nonlinear functional equations (see Chap. 5). Besides the difficulty of excessive computing time, the storage of the

[1] In this chapter, the count of operations refers to multiplications and divisions only.

coefficient matrix A as a two-dimensional array of size $n \times n$ may present difficulty for a moderate-sized computer when n is large. Even with the present-day digital computers, it is clear that direct application of the *full-matrix methods*[2] of Chapter 4 to solve a large number of simultaneous equations is very uneconomical, if not impossible.

It happens that the large systems encountered in network analysis (power networks or electronic circuits) are usually characterized by simultaneous equations having a sparse coefficient matrix. A square matrix A is considered to be *sparse* if only a small percentage (say, below 30 per cent) of its entries are nonzero. In electronic circuits, the nodal-admittance matrix Y_n for a typical 10-node circuit contains fewer than 50 per cent nonzero entries. For a typical 100-node circuit, Y_n has about 5 per cent nonzero entries [1]. In such cases, the number of operations required for solving Eq. (16-1) could be reduced to very much below $n^3/3$ if a suitable scheme is used. Example 16-1 will illustrate the idea.

EXAMPLE 16-1. Consider first the following equation:

$$\begin{bmatrix} 2 & 3 & -1 & -1 \\ 1 & 3 & 2 & 1 \\ 2 & -2 & 4 & 1 \\ 3 & -1 & 1 & 2 \end{bmatrix} \begin{bmatrix} x_1 \\ x_2 \\ x_3 \\ x_4 \end{bmatrix} = \begin{bmatrix} 1 \\ -2 \\ 3 \\ 8 \end{bmatrix} \tag{16-2}$$

One can apply the Gaussian elimination method or the LU factorization method to obtain the solution

$$x = [2 \quad -1 \quad -1 \quad 1]^t$$

In either method the exact number of operations (multiplications and divisions) is given by (see Problems 4-4 and 16-2)

$$\frac{n^3}{3} + n^2 - \frac{n}{3} = \frac{4^3}{3} + 4^2 - \frac{4}{3} = 36$$

Next consider the following equation whose coefficient matrix has only 50 per cent nonzero elements.

$$\begin{bmatrix} 2 & 0 & 0 & 6 \\ 0 & 4 & 0 & 8 \\ 0 & 0 & 2 & 4 \\ 4 & 0 & 0 & 6 \end{bmatrix} \begin{bmatrix} x_1 \\ x_2 \\ x_3 \\ x_4 \end{bmatrix} = \begin{bmatrix} 8 \\ 4 \\ 8 \\ 10 \end{bmatrix} \tag{16-3}$$

Suppose that the same digital computer program used to solve Eq. (16-2) is also used to solve Eq. (16-3). Assume that the program has no provision for taking advantage of zero elements. Then the number of operations required to solve Eq. (16-3) is also 36. The computer time required to solve Eq. (16-3) will be the same as that for Eq. (16-2), despite the fact that Eq. (16-3) is sparse, because it takes the same amount of time

[2]The two methods of solving $Ax = \mu$, discussed in Chapter 4, are applicable whether some elements of A are zero or all elements of A are nonzero. Such methods are said to be *full-matrix methods*.

for a digital computer to calculate 0.0×0.0 as it takes to calculate 2.5×3.2. Thus, sparsity of the coefficient matrix is of little help if we use full-matrix techniques that do not take advantage of the zero elements.

Now let us solve Eq. (16-3) manually with perhaps the aid of a pocket calculator. Whenever we encounter operations involving zero elements that produce no changes, e.g., $0 \times a = 0$, or $a + 0 = a$, we simply get the results without any computational effort. Then the actual operations required are fewer than for the full-matrix case. For Eq. (16-3), nine operations are required in the forward-elimination step to obtain

$$\begin{bmatrix} 1 & 0 & 0 & 3 \\ 0 & 1 & 0 & 2 \\ 0 & 0 & 1 & 2 \\ 0 & 0 & 0 & 1 \end{bmatrix} \begin{bmatrix} x_1 \\ x_2 \\ x_3 \\ x_4 \end{bmatrix} = \begin{bmatrix} 4 \\ 1 \\ 4 \\ 1 \end{bmatrix} \tag{16-4}$$

Three operations are required in the back-substitution step to obtain

$$\begin{bmatrix} 1 & 0 & 0 & 0 \\ 0 & 1 & 0 & 0 \\ 0 & 0 & 1 & 0 \\ 0 & 0 & 0 & 1 \end{bmatrix} \begin{bmatrix} x_1 \\ x_2 \\ x_3 \\ x_4 \end{bmatrix} = \begin{bmatrix} 1 \\ -1 \\ 2 \\ 1 \end{bmatrix} \tag{16-5}$$

Therefore, a total of 12 operations are required to obtain the solution of Eq. (16-3),

$$\boldsymbol{x} = [1 \quad -1 \quad 2 \quad 1]^t$$

The number of operations is only one third that for the full-matrix case. The saving will be much more impressive and significant when n is large and the matrix is sparser.

In Example 16-1, we have illustrated one aspect of sparse matrix techniques: *to reduce computing time by omitting trivial operations involving zero elements.* The important point here is the ability to *skip* the trivial operations completely. In manual solutions, this presents no difficulty. For example, in calculating $a_{21} \times a_{34}$, we look up the values of a_{21} and a_{34}. If either a_{21} or a_{34} is seen (literally, by the eyes) to be zero, we simply give the product a zero value *without any computational effort.* With digital computer solutions, "looking up" a_{21} means addressing a cell in the computer memory and transferring its content to a proper register. To check whether a_{21} is zero also requires central processor time (about the same as that required for an addition). Obviously, we very much desire to avoid such addressing and checking of an element that has been given to be zero. Thus, a sparse-matrix technique intended for digital computer implementation should have the following two features:

1. *Only nonzero elements with necessary indexing information are stored.*
2. *Operations involving zeros are not performed.*

Although our main concern has been the reduction of computational effort, it should be pointed out that the availability of limited core memory of a digital computer may be the more compelling reason for requiring the first feature.

A detailed study of various sparse-matrix techniques and their actual implementations is clearly beyond our scope, as it involves considerable knowledge of digital computer software [2]. Nevertheless, we can illustrate the basic principles and advantages of sparse-matrix techniques with problems that can be solved manually without the knowledge of computer programming, as we have done in Example 16-1. This is what we shall do in Sections 16-2 to 16-4. In studying these sections, the reader may imagine himself confronting the task of solving a large sparse system of equations, equipped only with paper, pencil, and a pocket calculator. Then he will find the concepts and techniques discussed immensely useful. Finally, in Sections 16-5 and 16-6, we briefly discuss some aspects of digital computer implementation of sparse-matrix techniques. Our discussion in this chapter is introductory. For an in-depth study of sparse-matrix techniques and applications, the reader is referred to [7, 11].

16-2 EFFECT OF ORDERING OF EQUATIONS

In Section 16-1 we showed how the sparsity of the **A** matrix may reduce the computational effort in solving $Ax = \boldsymbol{\mu}$. We shall now demonstrate that in the case of a sparse matrix A the order in which the unknowns are eliminated in the forward-elimination step also has an important effect on the solution effort.

Consider the equation

$$\begin{bmatrix} 2 & 4 & -2 & -6 \\ 3 & 9 & 0 & 0 \\ -2 & 0 & 8 & 0 \\ 2 & 0 & 0 & -12 \end{bmatrix} \begin{bmatrix} x_1 \\ x_2 \\ x_3 \\ x_4 \end{bmatrix} = \begin{bmatrix} 6 \\ -6 \\ 14 \\ 26 \end{bmatrix} \tag{16-6}$$

Let us apply the Gaussian elimination method described in Chapter 4. The forward-elimination step requires 30 multiplications and divisions to yield

$$\begin{bmatrix} 1 & 2 & -1 & -3 \\ 0 & 1 & 1 & 3 \\ 0 & 0 & 1 & -9 \\ 0 & 0 & 0 & 1 \end{bmatrix} \begin{bmatrix} x_1 \\ x_2 \\ x_3 \\ x_4 \end{bmatrix} = \begin{bmatrix} 3 \\ -5 \\ 20 \\ -2 \end{bmatrix} \tag{16-7}$$

The back-substitution step requires 6 operations [assuming that the first two off-diagonal elements in column 3 of Eq. (16-7) are treated as general nonzero elements instead of ± 1] to yield

$$\begin{bmatrix} 1 & 0 & 0 & 0 \\ 0 & 1 & 0 & 0 \\ 0 & 0 & 1 & 0 \\ 0 & 0 & 0 & 1 \end{bmatrix} \begin{bmatrix} x_1 \\ x_2 \\ x_3 \\ x_4 \end{bmatrix} = \begin{bmatrix} 1 \\ -1 \\ 2 \\ -2 \end{bmatrix}$$

Thus, a total of 36 operations is required for the solution, the same as that required for a full A matrix. The sparsity of A with zero elements distributed as shown in Eq. (16-6) produces no benefit at all as far as the number of multiplications and divisions is concerned.

However, Eq. (16-6) can also be rewritten in a different order, as follows:

$$\begin{bmatrix} -12 & 0 & 0 & 2 \\ 0 & 9 & 0 & 3 \\ 0 & 0 & 8 & -2 \\ -6 & 4 & -2 & 2 \end{bmatrix} \begin{bmatrix} x_4 \\ x_2 \\ x_3 \\ x_1 \end{bmatrix} = \begin{bmatrix} 26 \\ -6 \\ 14 \\ 6 \end{bmatrix} \tag{16-8}$$

Now if the unknowns are eliminated in the order x_4, x_2, and x_3, the forward-elimination step requires only 13 operations to yield

$$\begin{bmatrix} 1 & 0 & 0 & -\frac{1}{6} \\ 0 & 1 & 0 & \frac{1}{3} \\ 0 & 0 & 1 & -\frac{1}{4} \\ 0 & 0 & 0 & 1 \end{bmatrix} \begin{bmatrix} x_4 \\ x_2 \\ x_3 \\ x_1 \end{bmatrix} = \begin{bmatrix} -\frac{13}{6} \\ -\frac{2}{3} \\ \frac{7}{4} \\ 1 \end{bmatrix} \tag{16-9}$$

The back-substitution step requires only 3 operations to yield the solution

$$\begin{bmatrix} 1 & 0 & 0 & 0 \\ 0 & 1 & 0 & 0 \\ 0 & 0 & 1 & 0 \\ 0 & 0 & 0 & 1 \end{bmatrix} \begin{bmatrix} x_4 \\ x_2 \\ x_3 \\ x_1 \end{bmatrix} = \begin{bmatrix} -2 \\ -1 \\ 2 \\ 1 \end{bmatrix}$$

Thus, by merely rewriting Eq. (16-6) in a different order, we have reduced the number of operations from 36 to 16. The saving will be much more impressive for larger, sparser matrices. Obviously, then, for a given sparse system of equations, we are anxious to know the best order to eliminate the unknowns so that the computational effort is minimized. The problem has been referred to as the *optimal-ordering* problem [3]. Of course, it is not necessary to actually rewrite the equations as we have done in Eq. (16-8). The same reduction of computational effort may be achieved by working with the original equation (16-6) but choosing the pivot elements (see Chap. 4 for notations) in the following order: a_{44}, $a_{22}^{(2)}$, $a_{33}^{(3)}$, $a_{11}^{(4)}$. Thus, the optimal-ordering problem can be viewed as a problem of choosing the pivot elements (not necessarily restricted to diagonal elements).

Although the Gaussian elimination method has been used in all the examples given so far, we shall in the next few sections discuss the optimal ordering problem for the case of solving Eq. (16-1) by the $\boldsymbol{LU}$ factorization method (see Chap. 4). It is pertinent here to compare both methods and see why the $\boldsymbol{LU}$ decomposition method is preferable for the sparse-matrix case.

If the equation $\boldsymbol{Ax} = \boldsymbol{\mu}$ is to be solved just once, both the Gaussian elimination method and the $\boldsymbol{LU}$ decomposition method require $(n^3 + 3n^2 - n)/3$ multiplications–divisions (see Problems 4-5 and 16-2). To obtain $\boldsymbol{A}^{-1}$ by Gaussian elimination requires n^3 operations; $\boldsymbol{LU}$ factorization requires $(n^3 - n)/2$ operations. To obtain a new solution $\boldsymbol{x}$ for a new $\boldsymbol{\mu}$ requires n^2 further operations, whether $\boldsymbol{x} = \boldsymbol{A}^{-1}\boldsymbol{\mu}$ is used or $\boldsymbol{x}$ is calculated from $\boldsymbol{Ly} = \boldsymbol{\mu}$ and $\boldsymbol{Ux} = \boldsymbol{y}$. Thus, for the full-matrix case, there is really no decisive advantage of the $\boldsymbol{LU}$ factorization method over the Gaussian elimination method, aside from the fact that more operations are required for finding $\boldsymbol{A}^{-1}$ than with $\boldsymbol{LU}$ factorization.

This situation is quite different, however, for the sparse-matrix case. Often, the inverse of a sparse matrix may be full, whereas the $\boldsymbol{LU}$ factors may retain the sparsity [4]. Consider, for example, the coefficient matrix $\boldsymbol{A}$ in Eq. (16-8). Straightforward calculations show that (see Section 4-4)

$$\boldsymbol{L} = \begin{bmatrix} -12 & 0 & 0 & 0 \\ 0 & 9 & 0 & 0 \\ 0 & 0 & 8 & 0 \\ -6 & 4 & -2 & -\frac{5}{6} \end{bmatrix}$$

$$\boldsymbol{U} = \begin{bmatrix} 1 & 0 & 0 & -\frac{1}{6} \\ 0 & 1 & 0 & \frac{1}{3} \\ 0 & 0 & 1 & -\frac{1}{4} \\ 0 & 0 & 0 & 1 \end{bmatrix}$$

and

$$\boldsymbol{A}^{-1} = \begin{bmatrix} \frac{1}{60} & \frac{4}{45} & -\frac{1}{20} & -\frac{1}{5} \\ -\frac{1}{5} & -\frac{1}{15} & \frac{1}{10} & \frac{2}{5} \\ \frac{3}{20} & \frac{2}{15} & \frac{1}{20} & -\frac{3}{10} \\ \frac{3}{5} & \frac{8}{15} & -\frac{3}{10} & -\frac{6}{5} \end{bmatrix}$$

Note that $\boldsymbol{L}$ and $\boldsymbol{U}$ are sparse, whereas $\boldsymbol{A}^{-1}$ is full. If Eq. (16-8) is to be solved many times for different $\boldsymbol{\mu}$, then each new solution requires 16 operations by the $\boldsymbol{x} = \boldsymbol{A}^{-1}\boldsymbol{\mu}$ method. In contrast, the $\boldsymbol{LU}$ factorization requires only 6 operations for each new solution. The advantage is clearly seen.

As a second example, consider the tridiagonal matrix ($\times$ stands for a nonzero element)

$$\boldsymbol{A} = \begin{bmatrix} \times & \times & & & \\ \times & \times & \times & & \\ & \times & \times & \times & \\ & & \times & \times & \times \\ & & & \times & \times \end{bmatrix}$$

It can easily be shown that

$$\boldsymbol{A}^{-1} = \begin{bmatrix} \times & \times & \times & \times & \times \\ \times & \times & \times & \times & \times \\ \times & \times & \times & \times & \times \\ \times & \times & \times & \times & \times \\ \times & \times & \times & \times & \times \end{bmatrix}, \quad \boldsymbol{L} = \begin{bmatrix} \times & & & & \\ \times & \times & & & \\ & \times & \times & & \\ & & \times & \times & \\ & & & \times & \times \end{bmatrix},$$

$$\boldsymbol{U} = \begin{bmatrix} 1 & \times & & & \\ & 1 & \times & & \\ & & 1 & \times & \\ & & & 1 & \times \\ & & & & 1 \end{bmatrix}$$

Again, A^{-1} is full, and L and U are sparse. The tridiagonal structure is important because the nodal-admittance matrix of an *RLC* ladder network is of such form.

A problem of great practical importance is then the following: Given a sparse matrix A, find a suitable reordering of the rows and columns such that the LU factors of the permuted matrix retain the sparsity as much as possible.

If A represents the nodal-admittance matrix of a general lumped linear network, the problem may be restated as follows: Given a linear network, find the best numbering of nodes such that the node-admittance matrix $Y_n = LU$ has LU factors as sparse as possible. This problem will be explored in the next two sections.

16-3 DETERMINATION OF FILLS IN *LU* FACTORIZATION

As shown in Eq. (4-61), the LU factors of an $n \times n$ matrix A may be recorded as a single matrix Q, where

$$Q = L + (U - 1) \quad \text{and} \quad A = LU$$

Whenever $a_{ij} = 0$ but $q_{ij} \neq 0$, we say that a "fill" has been created. We wish to reduce the number of fills in the LU factorization by a suitable permutation of rows and columns of A. But first, we must have a convenient way of determining the number of fills when rows and columns of A are not interchanged or reordered. Two methods, one using matrices and the other using graphs, will be described in this section. The optimal-ordering problem will be discussed in the next section.

We shall make some assumptions on the matrix A to simplify our investigation. Although we are here considering only a restricted class of matrices, it covers the cases most often encountered in computer-aided circuit analysis.

Assumption 1. A is *structurally* symmetric (but not necessarily symmetric).

By *structural symmetry* we mean the property that a_{ij} and a_{ji} are both zero or both nonzero, as illustrated by the two matrices

$$\begin{bmatrix} 3 & 0 & 2 & 0 \\ 0 & 4 & 0 & 0 \\ 4 & 0 & 8 & 5 \\ 0 & 0 & 5 & 6 \end{bmatrix}, \quad \begin{bmatrix} \times & 0 & \times & 0 \\ 0 & \times & 0 & 0 \\ \times & 0 & \times & \times \\ 0 & 0 & \times & \times \end{bmatrix}$$

Assumption 1 is reasonable because the nodal-admittance matrix Y_n of an *RLC* network is always symmetric. When g_m-type controlled sources are included, it is possible that $a_{ij} = 0$ and $a_{ji} \neq 0$. However, a more accurate modeling of the active devices usually leads to $a_{ij} \neq 0$, too, although $|a_{ij}|$ may be very small. Besides, even if $a_{ij} = 0$ and $a_{ji} \neq 0$, the present state of the art does not permit an *easy* exploitation of this asymmetrical case.[3] Therefore, admitting the fact that the method is not the best we shall *process both a_{ij} and a_{ji} as nonzero elements whenever one of them is nonzero.*

[3]One method that takes full advantage of zero elements is described in Section 16-6, the implementation of which requires more advanced programming techniques than that assuming structural symmetry.

Assumption 2. $a_{ii} \neq 0$ for all i.

Assumption 3. As the ***LU*** factorization algorithm is carried out, $l_{ii} \neq 0$ for all i.

Both assumptions 2 and 3 are certainly valid for $\boldsymbol{Y}_n$ of an *RLC* network, and also for almost all *practical* active circuits. Of course, it is easy to create an active circuit on paper that violates these assumptions.

Assumption 4. Complete numerical cancellation leading to zero does not occur in the process of ***LU*** factorization.

One may object to assumption 4 at first, because we can conceivably encounter a situation as follows: $a_{43} = 0$, $l_{41} = 2$, $l_{42} = 3$, $u_{13} = -6$, $u_{23} = 4$, and by Eq. (4-63)

$$\begin{aligned} l_{43} = q_{43} &= a_{43} - l_{41}u_{13} - l_{42}u_{23} \\ &= 0 - 2 \times (-6) - 3 \times 4 = 0 \end{aligned}$$

Thus, no fill has been created in this step. Yet, according to assumption 4, we have to say that one fill has been created, because the cancellation as shown is not something that we can count on to happen when the nonzero elements are given different values. Since we are investigating the number of fills for the general case when only the structure of $\boldsymbol{A}$ is known, assumption 4 is a reasonable one.

It can easily be shown that if $\boldsymbol{A}$ is structurally symmetric, so is $\boldsymbol{Q}$. Thus, the total number of fills is twice the fills in $\boldsymbol{U}$ or $\boldsymbol{L}$. Furthermore, on account of assumption 4, $q_{ij} \neq 0$ whenever $a_{ij} \neq 0$.

In Section 4-4, we described Crout's algorithm for ***LU*** factorization. A careful examination of the expressions given by Eq. (4-63) for the $n = 4$ case reveals that the computations can be carried out in a slightly different order as shown next. This variation, usually credited to Doolittle [5], is more convenient from the programming point of view (see Section 16-5).

LU FACTORIZATION ALGORITHM.

Start with the given matrix $\boldsymbol{A}$.

Step 1. *Copy:* Copy column 1 (no computation).

Step 2. *Divide:* In row 1, divide all nondiagonal elements by the diagonal element.

Step 3. *Subtract:* For each element (i, j), $i > 1$ and $j > 1$, subtract from it the product of $(i, 1)$ and $(1, j)$ elements (see Fig. 16-1).

Step 4. *Check:* If the order of the submatrix subject to subtractions in step 3 is 2 or higher, *refer to this submatrix* and go back to step 1. Otherwise, the algorithm is completed, and the updated matrix is the desired $\boldsymbol{Q}$ matrix.

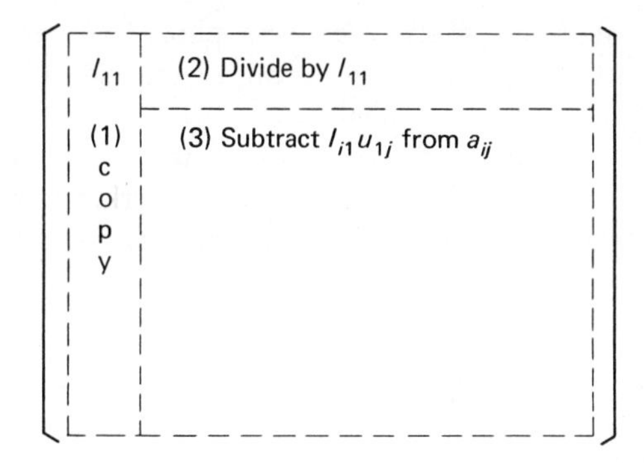

Fig. 16-1. Steps in ***LU*** factorization.

EXAMPLE 16-2. Find the ***LU*** factors for the matrix A of Example 4-1.

Solution:

$$A = \begin{bmatrix} 2 & 4 & 4 & 2 \\ 3 & 3 & 12 & 6 \\ 2 & 4 & -1 & 2 \\ 4 & 2 & 1 & 1 \end{bmatrix} \xrightarrow{\text{steps 1 and 2}} \begin{bmatrix} 2 & 2 & 2 & 1 \\ 3 & 3 & 12 & 6 \\ 2 & 4 & -1 & 2 \\ 4 & 2 & 1 & 1 \end{bmatrix}$$

$$\xrightarrow{\text{step 3}} \begin{bmatrix} 2 & 2 & 2 & 1 \\ 3 & -3 & 6 & 3 \\ 2 & 0 & -5 & 0 \\ 4 & -6 & -7 & -3 \end{bmatrix} \xrightarrow{\text{steps 4, 1, and 2}} \begin{bmatrix} 2 & 2 & 2 & 1 \\ 3 & -3 & -2 & -1 \\ 2 & 0 & -5 & 0 \\ 4 & -6 & -7 & -3 \end{bmatrix}$$

$$\xrightarrow{\text{step 3}} \begin{bmatrix} 2 & 2 & 2 & 1 \\ 3 & -3 & -2 & -1 \\ 2 & 0 & -5 & 0 \\ 4 & -6 & -19 & -9 \end{bmatrix} \xrightarrow{\text{steps 4, 1, and 2}} \begin{bmatrix} 2 & 2 & 2 & 1 \\ 3 & -3 & -2 & -1 \\ 2 & 0 & -5 & 0 \\ 4 & -6 & -19 & -9 \end{bmatrix}$$

$$\xrightarrow{\text{step 3}} \begin{bmatrix} 2 & 2 & 2 & 1 \\ 3 & -3 & -2 & -1 \\ 2 & 0 & -5 & 0 \\ 4 & -6 & -19 & -9 \end{bmatrix} = \boldsymbol{Q}, \quad \boldsymbol{L} = \begin{bmatrix} 2 & 0 & 0 & 0 \\ 3 & -3 & 0 & 0 \\ 2 & 0 & -5 & 0 \\ 4 & -6 & -19 & -9 \end{bmatrix}$$

$$\boldsymbol{U} = \begin{bmatrix} 1 & 2 & 2 & 1 \\ 0 & 1 & -2 & -1 \\ 0 & 0 & 1 & 0 \\ 0 & 0 & 0 & 1 \end{bmatrix}$$

The result of course agrees with that obtained in Example 4-1. Note that in the algorithm we can change the word "column," to "row," and vice versa, to obtain another form of the algorithm, which is actually used in [1].

In the present study we are only interested in knowing whether each element in $\boldsymbol{Q}$ is zero or nonzero. The actual value of the element is unimportant. Thus, the preceding algorithm can be carried out very rapidly, even with paper and pencil, since no actual arithmetic operations are involved. Fills in $\boldsymbol{Q}$ can be created only in step 3, and only under the conditions that the (i, j) element is zero and both $(i, 1)$ and $(1, j)$ elements are nonzero. Example 16-3 illustrates the use of the procedure to determine the number of fills. Blank or 0 indicates a zero element. A $\times$ indicates a nonzero element, and $\otimes$ indicates a nonzero element created in the factorization process, and hence a fill.

EXAMPLE 16-3.

$$\boldsymbol{A}_1 = \begin{bmatrix} \times & \times & 0 & \times & 0 \\ \times & \times & 0 & 0 & \times \\ 0 & 0 & \times & \times & 0 \\ \times & 0 & \times & \times & \times \\ 0 & \times & 0 & \times & \times \end{bmatrix}, \qquad \boldsymbol{Q}_1 = \begin{bmatrix} \times & \times & 0 & \times & 0 \\ \times & \times & 0 & \otimes & \times \\ 0 & 0 & \times & \times & 0 \\ \times & \otimes & \times & \times & \times \\ \times & \times & 0 & \times & \times \end{bmatrix}$$

$$\boldsymbol{A}_2 = \begin{bmatrix} \times & & & & \times \\ & \times & & & \times \\ & & \times & & \times \\ & & & \times & \times \\ \times & \times & \times & \times & \times \end{bmatrix}, \qquad \boldsymbol{Q}_2 = \begin{bmatrix} \times & & & & \times \\ & \times & & & \times \\ & & \times & & \times \\ & & & \times & \times \\ \times & \times & \times & \times & \times \end{bmatrix}$$

$$\boldsymbol{A}_3 = \begin{bmatrix} \times & \times & \times & \times & \times \\ \times & \times & & & \\ \times & & \times & & \\ \times & & & \times & \\ \times & & & & \times \end{bmatrix}, \qquad \boldsymbol{Q}_3 = \begin{bmatrix} \times & \times & \times & \times & \times \\ \times & \times & \otimes & \otimes & \otimes \\ \times & \otimes & \times & \otimes & \otimes \\ \times & \otimes & \otimes & \times & \otimes \\ \times & \otimes & \otimes & \otimes & \times \end{bmatrix}$$

There are 2 fills in $\boldsymbol{Q}_1$, no fills in $\boldsymbol{Q}_2$, and 12 fills in $\boldsymbol{Q}_3$. Note that $\boldsymbol{A}_2$ and $\boldsymbol{A}_3$ are related structurally by row and column permutations or, in other words, by reordering of rows and columns. The tremendous effect of ordering on fills is seen in $\boldsymbol{Q}_2$ and $\boldsymbol{Q}_3$.

The number of fills can also be determined easily with the aid of a graph. This alternative method provides, perhaps, better insight into the problem. Define an undirected graph G associated with a matrix $\boldsymbol{A}$ (satisfying assumptions 1 to 4) according to the rule that, for each pair of nonzero, nondiagonal elements (a_{ij}, a_{ji}) in $\mathbf{A}$, there corresponds an undirected branch joining nodes i and j in G. For example, the graphs associated with $\boldsymbol{A}_1$, $\boldsymbol{A}_2$, and $\boldsymbol{A}_3$ are shown in Fig. 16-2 as G_1, G_2, and G_3, respectively. Note that the diagonal elements of $\boldsymbol{A}$ are assumed to be nonzero and are not used in the construction of G.

Since $\boldsymbol{Q}$ is structurally symmetric, it suffices to determine the number of fills in the $\boldsymbol{U}$ matrix only. As shown in Chapter 4, the $\boldsymbol{U}$ matrix in the $\boldsymbol{LU}$ factorization is exactly the upper triangular matrix obtained in the forward-elimination step of the Gaussian elimination method. Thus the problem is the same as determining the fills created by the elimination of unknowns in $\boldsymbol{Ax} = \boldsymbol{\mu}$.

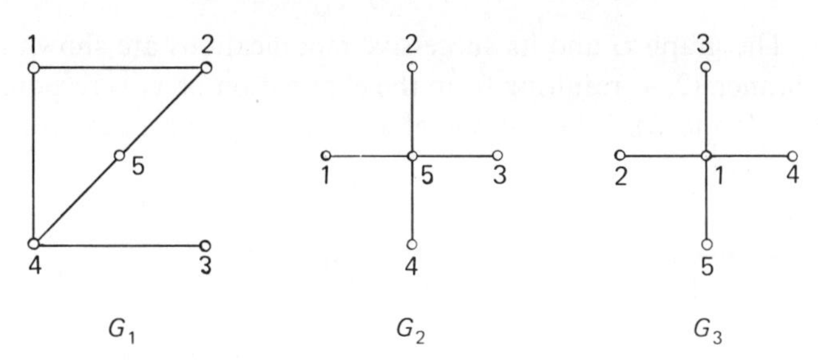

Fig. 16-2. Graphs associated with the matrices of Example 16-3.

Starting with the equation $\boldsymbol{Ax} = \boldsymbol{\mu}$, let us arbitrarily eliminate one unknown x_i by solving from the ith equation to obtain

$$x_i = -\frac{1}{a_{ii}}\left[\sum_{\substack{k=1\\k\neq i}}^{n} a_{ik}x_k - \mu_i\right] \tag{16-10}$$

and then substituting Eq. (16-10) into the remaining equations. The result will be a system of equations having one fewer unknowns than previously. It is not difficult to show (see [6]) that the graph $\hat{G}$ associated with the reduced system may be derived very simply from the graph G associated with the original system, as follows:

1. *For* every *pair of branches* (i, j) *and* (i, k) *in* G *incident at node* i, *create a branch joining nodes* j *and* k *(see Fig. 16-3). If a newly created branch is connected in parallel with an existing branch, represent the two as a single branch.*
2. *Delete node* i *and all branches incident at node* i. *The resultant graph, having one fewer nodes, is the desired graph* $\hat{G}$.

In step 1, if any newly created branch is *not* in parallel with an existing branch, a fill in the $\boldsymbol{U}$ factor has been created. As nodes $x_i, x_j, \ldots$ are eliminated successively, the number of fills can be tallied and the positions of fills can be recorded if so desired. Some examples will illustrate the graphical procedure.

EXAMPLE 16-4. Determine the number of fills in the $\boldsymbol{LU}$ factorization of $\boldsymbol{A}_1$ given in Example 16-3, if the elimination is done in the natural order.

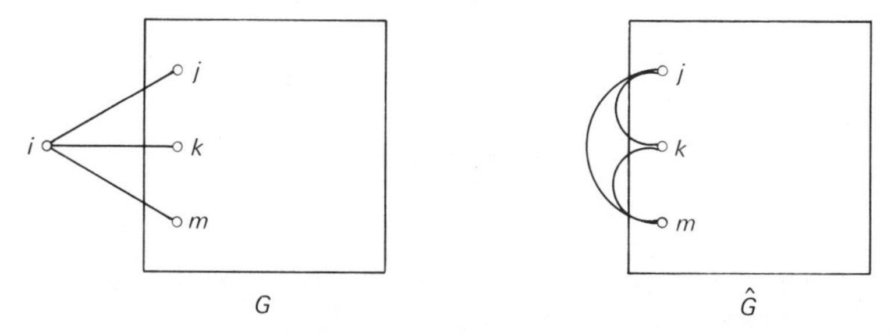

Fig. 16-3. Rule for eliminating a node.

Solution: The graph G and its successive modifications are shown in Fig. 16-4. Note that the branch (2, 4) resulting from the elimination of x_1 is responsible for two fills (one each in $\boldsymbol{U}$ and $\boldsymbol{L}$). Eliminations of x_2, x_3, and x_4 produce no fills. The total number of fills is two.

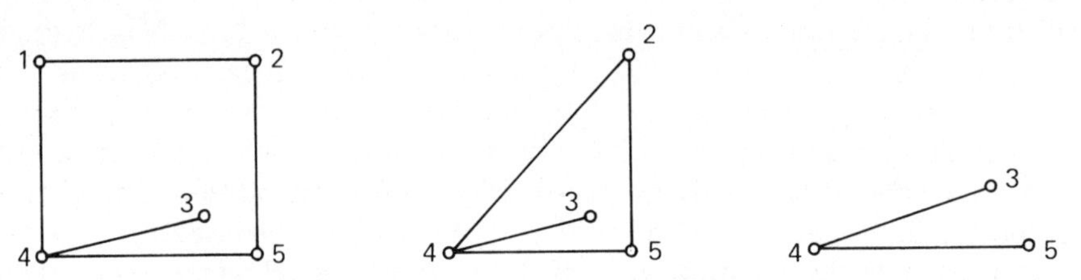

Fig. 16-4. Use of the graphical procedure to determine the number of fills.

EXAMPLE 16-5. Determine the number of fills in reducing $A_2 x = \boldsymbol{\mu}$ to the triangular form for each of the following orders of eliminating unknowns. The matrix A_2 is given in Example 16-3.

Case 1. x_1, x_2, x_3, x_4.

Case 2. x_5, x_1, x_2, x_3.

Case 3. x_1, x_5, x_2, x_3.

Solution: The graphs and their successive modifications are shown in Fig. 16-5. We see the number of fills in $\boldsymbol{U}$ is zero for case 1, six for case 2 and three for case 3. The $\boldsymbol{L}$ matrix will of course have the same number of fills.

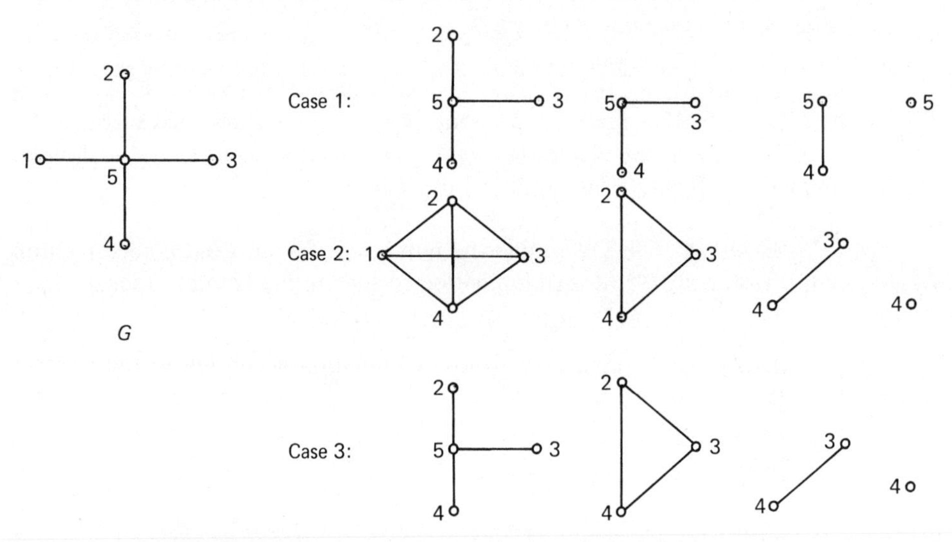

Fig. 16-5. Number of fills depends on order in which the nodes are eliminated.

16-4 A NEAR-OPTIMUM ORDERING ALGORITHM

The problem of optimum ordering of equations to minimize the number of fills in the ***LU*** factorization has not yet been solved. What we shall describe in this section may be called a near-optimum algorithm in the sense that it cannot be guaranteed to produce the minimum number of fills, although in almost all practical cases it will. The algorithm is essentially that given in [1]. Unlike [1], in which the algorithm is described in terms of array manipulations, the description given here is based completely on graph manipulations. This latter approach is believed to provide a clearer picture of the algorithm and is more comprehensible to those whose knowledge of computer programming is limited. However, when it comes to actual implementation on a digital computer, these graph manipulations will again have to be converted into array manipulations. Several other near-optimum ordering algorithms may be found in [8].

Consider the nodal equation

$$\mathbf{Y}_n \mathbf{V}_n = \mathbf{J}_n$$

corresponding to a given network N, whose nodes have been numbered $1, 2, \ldots, n$ (the reference node is always numbered 0). Let us assume that the matrix $\mathbf{Y}_n$ satisfies the four assumptions of Section 16-3. We wish to find a way of renumbering the nodes (except the reference node) so as to greatly reduce the number of fills in the ***LU*** factorization. We shall indicate the new node numbering by circled integers ①, ②, . . . , ⓝ.

To describe the algorithm, we first construct the undirected graph G associated with $\mathbf{Y}_n$ according to the rules described in Section 16-3. All operations will now be referred to the graph G and its successive subgraphs. The algorithm consists of four steps. The first three steps search for the proper node and renumber it with the next higher integer. Steps 1 and 2 produce no fills. All fills are results of step 3. Step 4 modifies the graph and makes it ready for the next search.

ALGORITHM RO (REORDERING). Initially, the graph G_s under consideration is the original graph G, and let $J = 1$.

Step 1. In G_s, search for a node with only one branch connected to it. If no such branch exists, go to step 2. Otherwise, renumber this node J. Go to step 4 and return here. Then go back to the beginning of step 1.

Step 2. In G_s, search for a node whose elimination according to the rules of Section 16-3 creates only branches in parallel with some existing branches (and hence no fills). If no such node exists, go to step 3. Otherwise, renumber this node J. Go to step 4 and return here. Then go to step 1.

Step 3. For each node in G_s determine the number of fills if that *particular node only* is to be eliminated. Choose a node whose elimination will produce the smallest number of fills and renumber this node J. (If several nodes have the same number of fills, choose the one having the largest number of branches connected to it.) Go to step 4 and return here. Then go to step 2.

Step 4. If all nodes in G_s except one have been renumbered, renumber this last node n and stop. Otherwise, remove the newly numbered node J and all branches connected to it. Combine branches in parallel and represent as a single branch. Call the resultant graph G_s. Increment J by 1 and return to step 1 or step 2, as the case may be.

We shall now illustrate the use of the algorithm with two examples.

EXAMPLE 16-6. The graph G associated with $\mathbf{Y}_n$ is shown in Fig. 16-6(a). Renumber the nodes by the preceding algorithm and determine the number of fills.

Solution: Step 1 produces no renumbering. From step 2, node 4 is renumbered ①. Step 4 results in a new G_s, as shown in Fig. 16-6(b). Continuing with the algorithm, we obtain the final result:

$$4 \longrightarrow ①, \quad 1 \longrightarrow ②, \quad 5 \longrightarrow ③$$
$$2 \longrightarrow ④, \quad 3 \longrightarrow ⑤, \quad 6 \longrightarrow ⑥$$

The number of fills is zero, since step 3 is never entered in this example.

EXAMPLE 16-7. Given the graph G associated with a system of 14 equations as shown in Fig. 16-7(a), renumber the nodes by the preceding algorithm and determine the number of fills.

Solution: One possible sequence of events is as follows:

Step 1. results in the renumbering 1 → ①, 4 → ②.

Step 2. 14 → ③, 3 → ④.

Step 1. 2 → ⑤.

Step 2. 7 → ⑥, 6 → ⑦, 12 → ⑧.

At this time, the graph G_s is as shown in Fig. 16-7(b). No renumbering is possible with step 1 or 2. Therefore, we proceed to step 3.

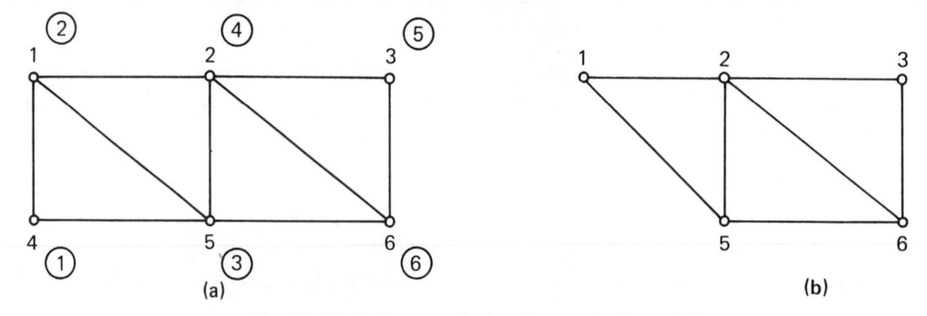

Fig. 16-6. Node renumbering that results in no fill.

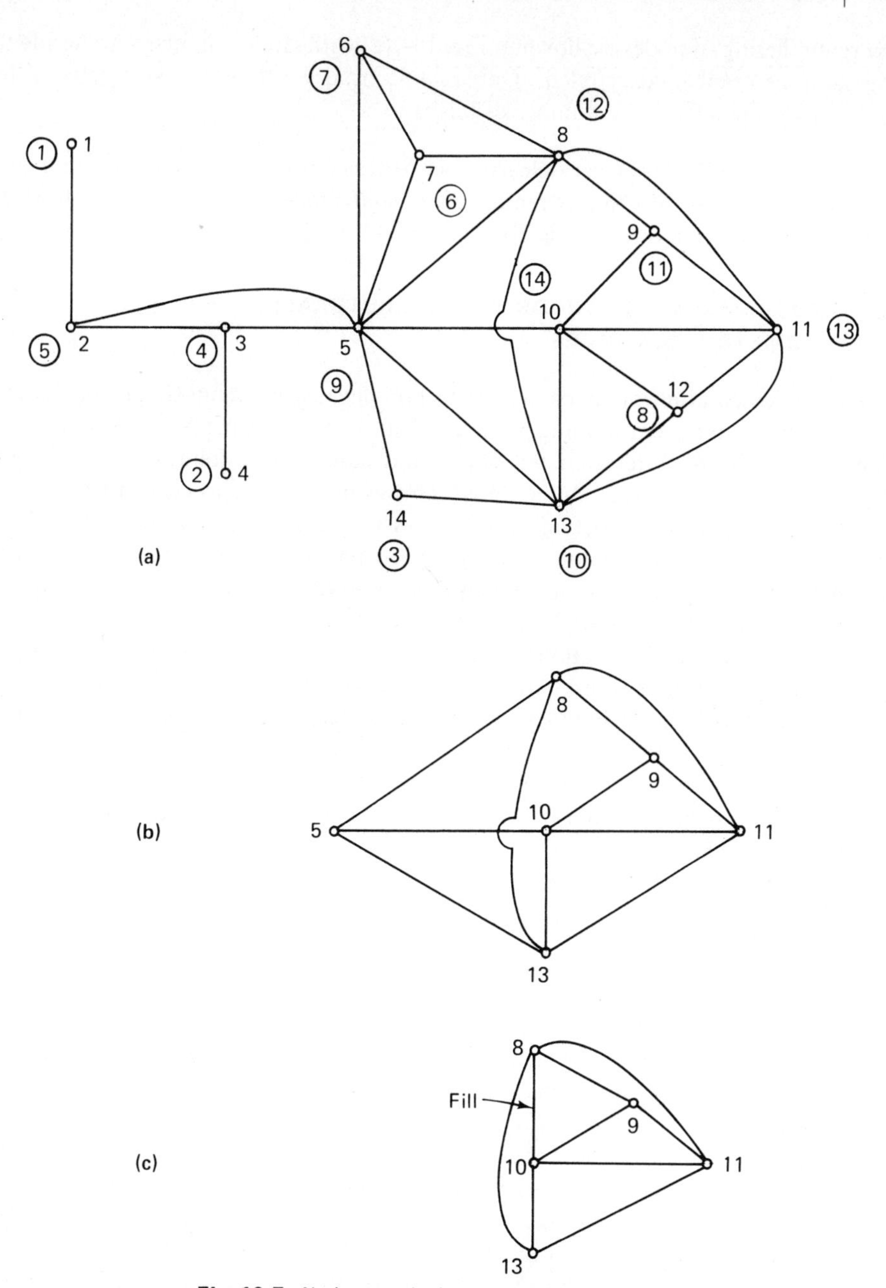

Fig. 16-7. Node renumbering that results in two fills.

Step 3. 5 ⟶ ⑨. One fill is created by the branch connecting nodes 8 and 10. After step 4, the graph G_s is as shown in Fig. 16-7(c).

Step 2. 13 ⟶ ⑩, 9 ⟶ ⑪, 8 ⟶ ⑫, 11 ⟶ ⑬.

Step 4. 10 ⟶ ⑭. All nodes have been renumbered. Stop.

The renumbering of nodes is shown in Fig. 16-7(a) with circled numbers alongside the original node numbers (uncircled). Only one fill occurs in $\boldsymbol{U}$. From symmetry, $\boldsymbol{L}$ has one fill also. Therefore, the number of fills is two.

Example 16-7 shows very clearly the necessity for going back to step 1 after step 2 if the latter results in any renumbering, and similarly for going back to step 2 after step 3 if any renumbering is done in step 3.

*16-5 PROGRAMMING METHODS FOR STRUCTURALLY SYMMETRIC MATRICES[4]

Actual implementation of sparse-matrix techniques involves considerable knowledge of digital computer programming. A detailed study of this topic is clearly beyond the scope of this book. Interested readers should consult the literature on this topic [2, and 7 with its extensive bibliography]. However, for the special and yet practical case of a sparse matrix satisfying the four assumptions stated in Section 16-3, the implementation is simple enough that only elementary knowledge of FORTRAN programming is required to understand the process. Therefore, we shall in this section describe the techniques applicable to this special case.

As pointed out in Section 16-1, a digital computer implementation of any sparse-matrix technique must store only nonzero elements and operate only on nonzero elements. The techniques for achieving these goals will be described in the next two subsections; the actual FORTRAN programs implementing these techniques are given in the Appendix 16A. We draw very heavily from [1] as far as the basic implementation techniques are concerned. Minor changes from [1] are mainly for the sake of simpler exposition.

16-5-1 Storage of Nonzero Elements

When a square matrix $\boldsymbol{A}$ of order n is sparse, it is inefficient to store $\boldsymbol{A}$ as a two-dimensional array that requires n^2 memory cells. One obvious method to reduce the storage is to store, for each nonzero element a_{ij}, the triplet (i, j, a_{ij}). Although many other more efficient storage schemes are available for sparse matrices [7], we shall describe one scheme [1] that is particularly suitable for the $\boldsymbol{LU}$ factorization process described in Section 16-3. We assume that $\boldsymbol{A}$ is structurally symmetrical.

Since in the factorization of $\boldsymbol{A}$ into $\boldsymbol{A} = \boldsymbol{LU}$, new nonzero elements called fills may be created, it is necessary to store these fills also. In other words, what we should plan to cope with is the storage of the nonzero elements of the $\boldsymbol{Q}$ matrix [$\boldsymbol{Q} = \boldsymbol{L} + \boldsymbol{U} - \mathbf{1}$, see Eq. (4-61)], not just the $\boldsymbol{A}$ matrix. Given a sparse matrix $\boldsymbol{A}$, it is easy to determine the number of fills and their locations in the $\boldsymbol{Q}$ matrix, as described in Section 16-3. If a near-optimum reordering of the nodes is desired, the algorithm of Section 16-4 may be used, the implementation of which, however, involves consider-

[4]Readers not familiar with FORTRAN programming may omit this section. Computer programs based on these techniques are also applicable to structurally asymmetric matrices, but with lowered computational efficiency (see Example 16-8).

able programming effort, (see [1] for details). We shall assume in this section that *a certain ordering of the nodes has been decided upon and the structure of* $\boldsymbol{Q}$ *(locations of nonzero elements) has been determined.* Our aim is to efficiently store the nonzero elements in $\boldsymbol{Q}$.

We recall from Section 16-3 that $\boldsymbol{Q}$ is obtained from $\boldsymbol{A}$ by a series of operations. The matrices $\boldsymbol{Q}$ and $\boldsymbol{A}$ are not stored as two separate matrices. Rather, $\boldsymbol{Q}$ is rewritten over $\boldsymbol{A}$. At the beginning of the $\boldsymbol{LU}$ factorization, the matrix is $\boldsymbol{A}$. At the end of the algorithm, the matrix becomes $\boldsymbol{Q}$. The nonzero elements of $\boldsymbol{Q}$ are to be stored in three one-dimensional arrays, as follows:

1. Array DIAG stores the nonzero diagonal elements of $\boldsymbol{Q}$ in the order of $q_{11}, q_{22}, \ldots, q_{nn}$.
2. Array UPACK stores in packed form the nondiagonal, nonzero elements of $\boldsymbol{U}$ in the order of row 1, . . . , row n.
3. Array LPACK stores in packed form the nondiagonal, nonzero elements of $\boldsymbol{L}$ in the order of column 1, . . . , column n.

A specific example will illustrate the storage scheme best. Consider the matrix

$$\boldsymbol{A} = \begin{bmatrix} 4 & 3 & 0 & 7 & 0 \\ 2 & 8 & 0 & 0 & -2 \\ 0 & 0 & 6 & 3 & 0 \\ -1 & 0 & 3 & 7 & -3 \\ 0 & -5 & 0 & 4 & 5 \end{bmatrix} \tag{16-11}$$

Using the method of Section 16-3, we can easily determine the structure of $\boldsymbol{Q}$.

$$\boldsymbol{Q} = \begin{bmatrix} \times & \times & 0 & \times & 0 \\ \times & \times & 0 & \otimes & \times \\ 0 & 0 & \times & \times & 0 \\ \times & \otimes & \times & \times & \times \\ 0 & \times & 0 & \times & \times \end{bmatrix} \tag{16-12}$$

Note that there are two fills indicated by $\otimes$ in Eq. (16-12). The *initial* contents of the three arrays are as follows:

$$\begin{array}{lll} \text{DIAG}(1) = 4, & \text{UPACK}(1) = 3, & \text{LPACK}(1) = 2 \\ \text{DIAG}(2) = 8, & \text{UPACK}(2) = 7, & \text{LPACK}(2) = -1 \\ \text{DIAG}(3) = 6, & \text{UPACK}(3) = 0, & \text{LPACK}(3) = 0 \\ \text{DIAG}(4) = 7, & \text{UPACK}(4) = -2, & \text{LPACK}(4) = -5 \\ \text{DIAG}(5) = 5, & \text{UPACK}(5) = 3, & \text{LPACK}(5) = 3 \\ & \text{UPACK}(6) = -3, & \text{LPACK}(6) = 4 \end{array} \tag{16-13}$$

Note that UPACK(3) and LPACK(3) are fills. They are initially given the value zero. Other elements of the arrays are *initially* taken from the $\boldsymbol{A}$ matrix.

Having stored the nonzero elements, we must have a simple way of addressing these elements. To put it differently, if we want to use a_{jj}, where can we locate it in Eq. (16-13)? The diagonal elements present no difficulty, since DIAG(J) is a_{jj} at the beginning and q_{jj} at the end of the factorization algorithm. To locate a nonzero nondiagonal element a_{ij} (or q_{ij}), we use two more one-dimensional arrays, UCOL and UROWST.

The UCOL array provides the information for the column positions of elements in UPACK. In general,

$$\text{UCOL(K)} = m \tag{16-14}$$

means that the kth element in UPACK is in the mth column of $\boldsymbol{A}$ (or $\boldsymbol{Q}$). Because of the structural symmetry of $\boldsymbol{A}$ (or $\boldsymbol{Q}$), Eq. (16-14) also means that the kth element in LPACK is in the mth row of $\boldsymbol{A}$ (or $\boldsymbol{Q}$). For example UCOL(4) = 5 means that UPACK(4) = a_{i5} and LPACK(4) = a_{5i}.

The array UROWST tells from which element of the UPACK array each row of the $\boldsymbol{U}$ matrix starts. In general

$$\text{UROWST(I)} = m$$

and (16-15)

$$\text{UROWST(I+1)} = m_1$$

means that there are $(m_1 - m)$ nondiagonal, nonzero elements of the ith row of $\boldsymbol{U}$ which are stored in UPACK, starting with the mth element of UPACK. From symmetry, Eq. (16-15) also means that the ith column of $\boldsymbol{L}$ starts with the mth element in LPACK. For example, two consecutive elements of UROWST,

$$\text{UROWST(3)} = 18$$

$$\text{UROWST(4)} = 21$$

will imply the following: The eighteenth, nineteenth, and twentieth elements of UPACK are in the third row of $\boldsymbol{A}$ (or $\boldsymbol{Q}$); and the corresponding elements in LPACK are in the third column of $\boldsymbol{A}$ (or $\boldsymbol{Q}$). In the case when the ith row of $\boldsymbol{U}$ has no nonzero, nondiagonal elements, UROWST(I) is assigned the same value as UROWST(I+1).

For the $\boldsymbol{A}$ matrix of Eq. (16-11), the indexing arrays UCOL and UROWST would be as follows:

$$\begin{aligned} &\text{UCOL(1)} = 2, && \text{UROWST(1)} = 1 \\ &\text{UCOL(2)} = 4, && \text{UROWST(2)} = 3 \\ &\text{UCOL(3)} = 4, && \text{UROWST(3)} = 5 \\ &\text{UCOL(4)} = 5, && \text{UROWST(4)} = 6 \\ &\text{UCOL(5)} = 4, && \text{UROWST(5)} = 7 \\ &\text{UCOL(6)} = 5, && \end{aligned} \tag{16-16}$$

A total of five arrays have been defined: DIAG, UPACK, LPACK, UCOL, and UROWST. Some estimates of the proper lengths of these arrays are in order. Consider a nodal-analysis program for handling 50-node, 200-branch networks. Then obviously DIAG and UROWST may be dimensioned 50. On the average, 200 branches (RLC and g_m) will produce about 400 nonzero nondiagonal elements in $\boldsymbol{Y}_n$ (see Section 4-6 for the reasons). Thus, UPACK, LPACK, and UCOL must each be dimensioned 200. A total of 700 cells are needed for storage and indexing purposes, compared with the 600 cells required for storing the triplets (i, j, a_{ij}) and the 2500 cells required for storing $\boldsymbol{Y}_n$ as a two-dimensional array. Although the present scheme requires slightly more storage than the triplet method, the disadvantage is completely outweighed by the ease of $\boldsymbol{LU}$ factorization using the present storage scheme.

Let us now consider the following addressing problem: given an element in $\boldsymbol{A}$ (or $\boldsymbol{Q}$), determine its location in the DIAG, UPACK, and LPACK arrays, and vice versa.

We shall illustrate the solution with a specific example. Refer to the arrays given in Eqs. (16-13) and (16-16). The correspondence for diagonal elements of $\boldsymbol{A}$ is obvious, since DIAG(J) $= a_{jj}$. Let us consider the locating of a nondiagonal element such as a_{24}. Since a_{24} is above the diagonal, it is located in UPACK array. From

$$\text{UROWST}(2) = 3, \qquad \text{UROWST}(3) = 5$$

we know that only the third and fourth elements of UPACK are in the second row of $\boldsymbol{A}$. From

$$\text{UCOL}(3) = 4, \qquad \text{UCOL}(4) = 5$$

we see that $a_{24} =$ UPACK(3) $= 0$. Because of structural symmetry, we have $a_{42} =$ LPACK(3) $= 0$.

Next, consider the reverse problem. We desire to find the element of $\boldsymbol{A}$ corresponding to UPACK(2). From UCOL(2) $= 4$, we know UPACK(2) is in column 4 of the $\boldsymbol{A}$ matrix. From

$$\text{UROWST}(1) = 1 < 2 < 3 = \text{UROWST}(2)$$

we see that UPACK(2) is in row 1 of $\boldsymbol{A}$. Therefore, UPACK(2) $= 7 = a_{14}$. By structural symmetry, LPACK(2) $= -1 = a_{41}$.

16-5-2 *LU* Factorization and the Solution of $LUx = \mu$

Once the structure of the $\boldsymbol{Q}$ matrix is ascertained (Section 16-3) and the arrays DIAG, UPACK, LPACK, UCOL, and UROWST have been set up as described in Section 16-5-1, we can proceed with the $\boldsymbol{LU}$ factorization, *operating only on nonzero elements*. The procedure is exactly as described in Section 16-3, except that in the present case some extra effort is needed to access any element a_{ij}. The storage scheme of Section 16-5-1 has been devised mainly with the ease of $\boldsymbol{LU}$ factorization in mind.

We shall illustrate the factorization process by referring to the specific matrix of Eq. (16-11), whose associated arrays are given in Eqs. (16-13) and (16-16) at the beginning of the factorization algorithm. Various steps of the first iterations are explained below in detail.

Step 1. Copy column 1. This step requires no action, since the desired contents are already stored in UPACK.

Step 2. Divide. Since UROWST(1) $= 1$ and UROWST(2) $= 3$, the elements UPACK(1) $= 3$ and UPACK(2) $= 7$ are in row 1 and are divided by DIAG(1) $= 4$.

Step 3. Subtract. From the structural symmetry of A and Q matrices, we know LPACK(1) = 2 and LPACK(2) = −1 are in column 1 of the A matrix. Then the following products must be subtracted from appropriate elements of A:

$$\text{LPACK}(1) \times \text{UPACK}(1) = 2 \times 3$$
$$\text{LPACK}(1) \times \text{UPACK}(2) = 2 \times 7$$
$$\text{LPACK}(2) \times \text{UPACK}(1) = -1 \times 3$$
$$\text{LPACK}(2) \times \text{UPACK}(2) = -1 \times 7$$

Consider the first product. From UCOL(1) = 2, we see that the product 2 × 3 must be subtracted from a_{22}, which is stored in DIAG(2). Similarly, the fourth product −1 × 7 must be subtracted from DIAG(4), because UCOL(2) = 4.

Now consider the second product, LPACK(1) × UPACK(2). From UCOL(1) = 2 and UCOL(2) = 4, we see that the product 2 × 7 must be subtracted from $a_{\text{UCOL}(1),\text{UCOL}(2)} = a_{24}$. Following the addressing procedure discussed earlier, we find that a_{24} is stored in UPACK(3). Now a_{24} is a fill, and therefore UPACK(3) = 0 initially. After this subtraction, we have

$$\text{UPACK}(3) = 0 - 2 \times 7 = -14$$

Note that the fill element is now nonzero. From symmetry considerations we see that the third product LPACK(2) × UPACK(1) = −1 × 3 must be subtracted from $a_{\text{UCOL}(2),\text{UCOL}(1)} = a_{42}$, which is stored in LPACK(3).

Step 4. The submatrix to be considered next is completely described by the latter parts of various arrays as follows:

DIAG and UROWST: Starting from the second element.

UPACK, LPACK, and UCOL: Starting from the third element, since UROWST(2) = 3.

This completes the first iteration. The second iteration again starts with step 1, but the matrix is now a submatrix defined in step 4 of the first iteration. The process is repeated until the $(n - 1)$th iteration is completed. The last updated matrix (described by the five arrays) is the desired $\boldsymbol{Q}$ matrix, which contains $\boldsymbol{L}$ and $\boldsymbol{U}$.

A complete FORTRAN subroutine called LUFACT for this factorization process using sparse-matrix techniques is given in Appendix 16A.

Having obtained the $\boldsymbol{LU}$ factors of A, we rewrite the equation $\boldsymbol{Ax} = \boldsymbol{LUx} = \boldsymbol{\mu}$ as two equations:

$$\boldsymbol{Ly} = \boldsymbol{\mu} \tag{16-17}$$

and

$$\boldsymbol{Ux} = \boldsymbol{y} \tag{16-18}$$

First, we solve for $\boldsymbol{y}$ from Eq. (16-17) and then $\boldsymbol{x}$ from Eq. (16-18). Since both $\boldsymbol{L}$ and $\boldsymbol{U}$ are triangular matrices, the solutions can be obtained by forward substitution for Eq. (16-17) and back substitution for Eq. (16-18). It can easily be shown that for the *full-matrix* case the solution of $\boldsymbol{Ly} = \boldsymbol{\mu}$ requires $n(n + 1)/2$ operations, and the solution of $\boldsymbol{Ux} = \boldsymbol{y}$ requires $n(n - 1)/2$ operations (see Problem 16-5). Thus, a total of n^2 operations are required for each solution when a new $\boldsymbol{\mu}$ is given.

To avoid operations on zero elements by the use of the five arrays defined previously, we have to organize the forward and back substitutions in such a way that coefficients from the $\boldsymbol{U}$ matrix are used row after row, since they are stored in the UPACK array in such manner. Similarly, the coefficients from the $\boldsymbol{L}$ matrix are used column after column. We shall illustrate the procedures with $\boldsymbol{L}$ and $\boldsymbol{U}$ matrices of order 4.

Forward Substitution:

$$\begin{bmatrix} l_{11} & 0 & 0 & 0 \\ l_{21} & l_{22} & 0 & 0 \\ l_{31} & l_{32} & l_{33} & 0 \\ l_{41} & l_{42} & l_{43} & l_{44} \end{bmatrix} \begin{bmatrix} y_1 \\ y_2 \\ y_3 \\ y_4 \end{bmatrix} = \begin{bmatrix} \mu_1 \\ \mu_2 \\ \mu_3 \\ \mu_4 \end{bmatrix}$$

The first column reduction step yields

$$\begin{bmatrix} 1 & 0 & 0 & 0 \\ 0 & l_{22} & 0 & 0 \\ 0 & l_{32} & l_{33} & 0 \\ 0 & l_{42} & l_{43} & l_{44} \end{bmatrix} \begin{bmatrix} y_1 \\ y_2 \\ y_3 \\ y_4 \end{bmatrix} = \begin{bmatrix} \mu_1^{(1)} \\ \mu_2^{(1)} \\ \mu_3^{(1)} \\ \mu_4^{(1)} \end{bmatrix} \tag{16-19}$$

where

$$\begin{aligned} \mu_1^{(1)} &= \frac{\mu_1}{l_{11}} \\ \mu_2^{(1)} &= \mu_2 - l_{21}\mu_1^{(1)} \\ \mu_3^{(1)} &= \mu_3 - l_{31}\mu_1^{(1)} \\ \mu_4^{(1)} &= \mu_4 - l_{41}\mu_1^{(1)} \end{aligned} \tag{16-20}$$

The second column reduction step yields

$$\begin{bmatrix} 1 & 0 & 0 & 0 \\ 0 & 1 & 0 & 0 \\ 0 & 0 & l_{33} & 0 \\ 0 & 0 & l_{43} & l_{44} \end{bmatrix} \begin{bmatrix} y_1 \\ y_2 \\ y_3 \\ y_4 \end{bmatrix} = \begin{bmatrix} \mu_1^{(1)} \\ \mu_2^{(2)} \\ \mu_3^{(2)} \\ \mu_4^{(2)} \end{bmatrix} \tag{16-21}$$

where

$$\begin{aligned} \mu_2^{(2)} &= \frac{\mu_2^{(1)}}{l_{22}} \\ \mu_3^{(2)} &= \mu_3^{(1)} - l_{32}\mu_2^{(2)} \\ \mu_4^{(2)} &= \mu_4^{(1)} - l_{42}\mu_2^{(2)} \end{aligned} \tag{16-22}$$

The third column reduction step yields

$$\begin{bmatrix} 1 & 0 & 0 & 0 \\ 0 & 1 & 0 & 0 \\ 0 & 0 & 1 & 0 \\ 0 & 0 & 0 & l_{44} \end{bmatrix} \begin{bmatrix} y_1 \\ y_2 \\ y_3 \\ y_4 \end{bmatrix} = \begin{bmatrix} \mu_1^{(1)} \\ \mu_2^{(2)} \\ \mu_3^{(3)} \\ \mu_4^{(3)} \end{bmatrix} \tag{16-23}$$

where

$$\mu_3^{(3)} = \frac{\mu_3^{(2)}}{l_{33}}$$
$$\mu_4^{(3)} = \mu_4^{(2)} - l_{43}\mu_3^{(3)} \tag{16-24}$$

The fourth column reduction step is simply to divide the last equation by l_{44} to obtain

$$\begin{bmatrix} 1 & 0 & 0 & 0 \\ 0 & 1 & 0 & 0 \\ 0 & 0 & 1 & 0 \\ 0 & 0 & 0 & 1 \end{bmatrix} \begin{bmatrix} y_1 \\ y_2 \\ y_3 \\ y_4 \end{bmatrix} = \begin{bmatrix} y_1 \\ y_2 \\ y_3 \\ y_4 \end{bmatrix} = \begin{bmatrix} \mu_1^{(1)} \\ \mu_2^{(2)} \\ \mu_3^{(3)} \\ \mu_4^{(4)} \end{bmatrix} \tag{16-25}$$

where

$$\mu_4^{(4)} = \frac{\mu_4^{(3)}}{l_{44}}$$

The solution of $\boldsymbol{y}$ is given by Eq. (16-25).

For the general case of the $\boldsymbol{L}$ matrix of order n, the solution algorithm is as follows (with $\boldsymbol{\mu}^{(0)} = \boldsymbol{\mu}$):

For $i = 1, 2, \ldots, n$, successively,

$$\mu_i^{(i)} = \frac{\mu_i^{(i-1)}}{l_{ii}} \tag{16-26}$$

and

$$\mu_k^{(i)} = \mu_k^{(i-1)} - l_{ki}\mu_i^{(i)}, \quad k = i+1, i+2, \ldots, n \tag{16-27}$$

This algorithm is for the full-matrix case. If $\boldsymbol{L}$ is sparse, in applying the algorithm we shall obtain the elements of $\boldsymbol{L}$ from the array LPACK. This is very easy now since, as indicated by Eq. (16-27), only elements in the *same column* of $\boldsymbol{L}$ are used in each column reduction step.

Back Substitution:

$$\begin{bmatrix} 1 & u_{12} & u_{13} & u_{14} \\ 0 & 1 & u_{23} & u_{24} \\ 0 & 0 & 1 & u_{34} \\ 0 & 0 & 0 & 1 \end{bmatrix} \begin{bmatrix} x_1 \\ x_2 \\ x_3 \\ x_4 \end{bmatrix} = \begin{bmatrix} y_1 \\ y_2 \\ y_3 \\ y_4 \end{bmatrix}$$

The unknowns are solved successively in the following order:

$$x_4 = y_4$$
$$x_3 = y_3 - u_{34}x_4$$
$$x_2 = y_2 - u_{23}x_3 - u_{24}x_4$$
$$x_1 = y_1 - u_{12}x_2 - u_{13}x_3 - u_{14}x_4$$

For the general case of the U matrix of order n, the solution algorithm is as follows:

Calculate the unknowns in the order $x_n, x_{n-1}, \ldots, x_1$ from

$$\begin{aligned} x_j &= y_j - u_{j,j+1}x_{j+1} - u_{j,j+2}x_{j+2} - \ldots - u_{j,n}x_n \\ &= y_j - \sum_{k=j+1}^{n} u_{jk}x_k \end{aligned} \tag{16-28}$$

The algorithm is for the full-matrix case. If U is sparse, in applying the algorithm we obtain the elements of U from the UPACK array and therefore operate only on nonzero elements. Equation (16-28) shows that only elements in the *same row* of the U matrix are used in the calculation of x_j. That is precisely the way the UPACK array is structured—elements in the same row of U are stored in consecutive locations of the UPACK array. This feature makes the addressing of elements much simpler.

The subroutine SOLV in Appendix 16A shows the actual FORTRAN codes for solving $Ly = \mu$ and $Ux = y$ using a sparse-matrix technique. A main program called SPARSE, complete with READ and WRITE statements, is also given.[5] The complete program, consisting of SPARSE, LUFACT, and SOLV, embodies all the techniques discussed in this section. The program is dimensioned to solve a system of 50 equations, provided that the number of nonzero elements (including the fills) does not exceed 400. The six arrays used in the program are dimensioned as follows:

DIAG(50), UPACK(200), LPACK(200)

UCOL(200), UROWST(50) X(50)

The input to SPARSE consists of the following data:

1. One card containing the title of the problem to be printed verbatim.
2. One card containing 3 integers:
   NNODE, the order of A.
   LENGTH, the number of elements (including fills) in UPACK.
   NTIMES, the number of μ_k vectors.
3. Cards containing the element values of the arrays DIAG, UPACK, LPACK, UCOL, UROWST, and the constant vectors $\mu_1, \mu_2, \ldots, \mu_m$.

See Example 16-8 for further explanations of the input data. We point out that it is necessary first to determine the number of fills and their locations before using the program SPARSE. If a near-optimal ordering algorithm is implemented, the information about the fills is part of the output. If no ordering algorithm is implemented, the information on the fills is either provided by the user (as in Example 16-8) or by another subroutine written for that purpose.

The main program SPARSE then calls the subroutine LUFACT to perform the LU factorization. At entry to LUFACT, the five arrays, DIAG, UPACK, LPACK, UCOL, and UROWST contain the information about A. At exit from LUFACT, the same five arrays contain the information about Q, and hence about L and U. Note

[5]The first FORTRAN statement in SPARSE is for CDC 6000 series computers.

how the algorithm of Section 16-3 is actually carried out. Step 1 requires no action. Step 2 is carried out by the statements between comments C1 and C2. One DO statement is used. Step 3 is carried out by the statements between C2 and C4. Three DO statements are used. Step 4 is carried out by the DO statement preceding comment C1. The addressing scheme described in Section 16-5-1 can be seen in the statements between C3 and C4. With the generous use of DO loops, the source program is short, but the execution time is somewhat slowed owing to the addressing operations required (e.g., see comment C3 in the program).

Having obtained the $\boldsymbol{LU}$ factors, the program SPARSE then transfers the information, together with one constant vector $\boldsymbol{\mu}$, to subroutine SOLV for the solutions of $\boldsymbol{Ly} = \boldsymbol{\mu}$ and $\boldsymbol{Ux} = \boldsymbol{y}$. At entry to SOLV, the array X contains the constant vector $\boldsymbol{\mu}$. At exit from SOLV, the array X contains the solution vector $\boldsymbol{x}$. SOLV uses Eq. (16-26) for $\boldsymbol{Ly} = \boldsymbol{\mu}$ and Eq. (16-28) for $\boldsymbol{Ux} = \boldsymbol{y}$, and carries out the iterations with DO statements.

Example 16-8 will illustrate the use of SPARSE.

EXAMPLE 16-8. Given $\boldsymbol{Ax} = \boldsymbol{\mu}$ as

$$\begin{bmatrix} 4 & 0 & 0 & -3 \\ 2 & 6 & 0 & 0 \\ 0 & 0 & -1 & 3 \\ 5 & 0 & -2 & 1 \end{bmatrix} \begin{bmatrix} x_1 \\ x_2 \\ x_3 \\ x_4 \end{bmatrix} = \begin{bmatrix} -5 \\ -4 \\ 7 \\ 4 \end{bmatrix} \tag{16-29}$$

find $\boldsymbol{x}$ by the use of SPARSE. The complete listing of SPARSE is given in Appendix 16A.

Solution: Although $a_{12} = 0$, we have to treat it as if it were nonzero because $a_{21} \neq 0$ (see assumption 1 of Section 16-3).

Working out a *dummy* $\boldsymbol{LU}$ factorization as described in Section 16-3, we find that there are two fills, in locations (2, 4) and (4, 2), as follows:

$$\boldsymbol{Q} = \begin{bmatrix} \times & \times & 0 & \times \\ \times & \times & 0 & \otimes \\ 0 & 0 & \times & \times \\ \times & \otimes & \times & \times \end{bmatrix} \tag{16-30}$$

Even though a_{12}, a_{24}, and a_{42} in Eq. (16-29) are zeros, they have to be included in the input to SPARSE along with other nonzero elements. Therefore, referring to Eq. (16-29), we have the following input data for SPARSE:

N = 4, LENGTH = 4, NTIMES = 1

DIAG:	4,	6,	−1,	1
UPACK:	0,	−3,	0,	3
LPACK:	2,	5,	0,	−2
UCOL:	2,	4,	4,	4
UROWST:	1,	3,	4	
X:	−5,	−4,	7,	4

A listing of the data cards and the output of SPARSE is given in Figs. 16-8 and 16-9, respectively. The answers obtained from SPARSE are $x_1 = 1$, $x_2 = -1$, $x_3 = 2$, and $x_4 = 3$.

```
SOLUTION OF EQ.(16-29) BY SPARSE MATRIX TECHNIQUE.
     4     4     1
4.0          6.0          -1.0          1.0
0.0          -3.0         0.0           3.0
2.0          5.0          0.0           -2.0
      2     4     4     4
      1     3     4     -0
-5.          -4.          7.0           4.0
```

Fig. 16-8. Input to SPARSE for Example 16-8.

```
SOLUTION OF EQ.(16-29) BY SPARSE MATRIX TECHNIQUE.

INPUT DATA
           4               4               1

           4.0000          6.0000          -1.0000          1.0000
           0.0000          -3.0000         0.0000           3.0000
           2.0000          5.0000          0.0000           -2.0000
      2    4       4       4
      1    3       4

RESULT OF LU FACTORIZATION
ELEMENTS OF DIAG,UPACK,LPACK,UCOL, AND UROWST

           4.0000          6.0000          -1.0000          -1.2500
           0.0000          -.7500          .2500            -3.0000
           2.0000          5.0000          0.0000           -2.0000
      2    4       4       4
      1    3       4       5

CONSTANT VECTOR

           -5.0000         -4.0000         7.0000           4.0000

SOLUTION VECTOR

           1.0000          -1.0000         2.0000           3.0000
```

Fig. 16-9. Output from SPARSE for Example 16-8.

★16-6 OPTIMAL CROUT ALGORITHM[6]

The word *optimal* has different meanings in different situations. Here, any algorithm for solving a sparse system of linear equations will be said to be *optimal* if only nonzero elements are stored and operated on. The Crout algorithm described in Section 16-5 is not optimal because a fraction of the operations may still involve zero-valued elements. Example 16-8 illustrates this point. The element a_{12} in Eq. (16-29) is zero. Yet because $a_{21} \neq 0$, the algorithm requires the storage of $a_{12} = 0$ and operates on a_{12} twice during the first iteration, in steps 2 and 3. However, it should be pointed out that, for the special case of a *structurally symmetric* matrix A, the implementation of the Crout algorithm of Section 16-5 is optimal in the sense just defined.

[6]Readers not familiar with FORTRAN programming may omit this section.

We shall now consider an *optimal Crout algorithm* for solving $\boldsymbol{Ax} = \boldsymbol{\mu}_k$, $k = 1, 2, \ldots, m$. The $n \times n$ matrix $\boldsymbol{A}$ need not be structurally symmetric. Thus, we remove assumption 1 of Section 16-3. But we shall retain the remaining three assumptions about $\boldsymbol{A}$ for the same reasons given in Section 16-3. If a reordering of the equations is desired to reduce the number of fills, *we assume that it has been done*. The ideas presented in this section are basically those of [2]. Both the storage scheme for nonzero elements and the $\boldsymbol{LU}$ factorization process are different from the method of Section 16-5. The reader is asked to review Section 4-4 before proceeding.

The *locations* of the nonzero elements of the $n \times n$ matrix $\boldsymbol{A}$ can be specified by a binary number $\hat{A}$ of n^2 digits as follows:

$$\hat{A} = \hat{a}_1\hat{a}_2 \ldots \hat{a}_{n^2} \tag{16-31}$$

where (1) each digit $\hat{a}_i$ corresponds to a specific location in $\boldsymbol{A}$, as follows,

$$\begin{bmatrix} \hat{a}_1 & \hat{a}_{n+1} & \cdots & \hat{a}_{(2n-1)} \\ \hat{a}_2 & \hat{a}_{2n} & \cdots & \\ \cdot & \cdot & & \\ \cdot & \cdot & & \\ \cdot & \cdot & & \\ \hat{a}_n & \hat{a}_{3n-2} & & \hat{a}_{n^2} \end{bmatrix} \tag{16-32}$$

and (2) $\hat{a}_i = 1$ if the corresponding element in $\boldsymbol{A}$ is nonzero, and $\hat{a}_i = 0$ otherwise.

The order of elements shown in Eq. (16-32) has been used earlier in Chapter 4 in performing $\boldsymbol{LU}$ factorization. Equation (4-62) is a special case of Eq. (16-32) for $n = 4$. As an example of using Eq. (16-31), the number $\hat{A}$ associated with the $\boldsymbol{A}$ matrix of Eq. (16-29) is easily seen to be

$$\hat{A} = 1101001100001111$$

Suppose that the computer to be used has a word length of p bits. Then the storage of $\hat{A}$ requires only $[n^2/p] + 1$ computer words.[7]

Another binary number $\hat{Q}$ will be used to specify the *locations* of the nonzero elements of $\boldsymbol{Q}$ ($\boldsymbol{Q} = \boldsymbol{L} + \boldsymbol{U} - \boldsymbol{1}$) in exactly the same manner as $\hat{A}$ is used for $\boldsymbol{A}$. For example, if $n = 5$ and

$$\hat{Q} = 1101110101010011110101111$$

then the structure of $\boldsymbol{Q}$ is as shown in Eq. (16-12). The storage of $\hat{Q}$ requires another $[n^2/p] + 1$ computer words.

We shall perform $\boldsymbol{A} = \boldsymbol{LU}$ factorization using the Crout algorithm described in Section 4-4. The general formulas for q_{ij} are given in steps 4 and 6 of the algorithm. For the special case of $n = 4$, the 16 formulas for q_{ij} are given by Eq. (4-63). Such a set of n^2 formulas constitutes what is called a *full Crout algorithm.*

Let us examine the formulas of the full Crout algorithm *one by one*. Considerable simplifications of these formulas can be achieved if $\boldsymbol{A}$ is sparse. For the purpose of simplifying the formulas, we need only $\hat{A}$, which contains the information about the *locations* of nonzero elements of $\boldsymbol{A}$. The actual values of these nonzero elements are immaterial. As we examine each formula, in the order specified by Eq. (16-32), if any term on the right side is found to be zero, that term will be omitted from the formula.

[7]The notation $[h]$ here means the largest integer that is smaller than h.

If all terms on the right side are zero, then the left side is zero, and the complete formula will be omitted from the set of n^2 formulas. After such a "clean-up" operation, the remaining formulas constitute what is called a *reduced Crout algorithm* (for $\boldsymbol{LU}$ factorization). These formulas of course will be valid for all matrices having the same locations of nonzero elements as specified by the number $\hat{A}$.

The *values* of the nonzero elements of $\boldsymbol{A}$ are stored in a one-dimensional array A(I) whose length is equal to the number of nonzero elements in $\boldsymbol{A}$. [Should the notations $\boldsymbol{A}$ for a matrix, and A(I) for a linear array cause confusion, the reader may change A(I) to D(I) from this point on.] The values of the *nonzero* elements are taken from $\boldsymbol{A}$ in the order indicated by Eq. (16-32), and stored successively in A(1), A(2), ... For example, applying the rule to the $\boldsymbol{A}$ matrix in Eq. (16-29), we have

$$\boldsymbol{A} = \begin{bmatrix} 4 & 0 & 0 & -3 \\ 2 & 6 & 0 & 0 \\ 0 & 0 & -1 & 3 \\ 5 & 0 & -2 & 1 \end{bmatrix} = \begin{bmatrix} \text{A}(1) & 0 & 0 & \text{A}(4) \\ \text{A}(2) & \text{A}(5) & 0 & 0 \\ 0 & 0 & \text{A}(6) & \text{A}(8) \\ \text{A}(3) & 0 & \text{A}(7) & \text{A}(9) \end{bmatrix} \tag{16-33}$$

Therefore, the array A(I) has nine elements:

$$\text{A}(1) = 4, \quad \text{A}(2) = 2, \quad \text{A}(3) = 5, \quad \text{A}(4) = -3, \quad \text{A}(5) = 6,$$
$$\text{A}(6) = -1, \quad \text{A}(7) = -2, \quad \text{A}(8) = 3, \quad \text{A}(9) = 1$$

In exactly the same manner, the *values* of the nonzero elements in $\boldsymbol{Q}$ will be stored in a one-dimensional array Q(J). The length of Q(J) is equal to the length of A(I) plus the number of *fills* (see Section 16-3), and becomes known only after the number of formulas in the reduced Crout algorithm is counted. These formulas also contain the information about the *locations* of the nonzero elements of $\boldsymbol{Q}$, from which the binary number $\hat{Q}$ may be generated.

After the $\boldsymbol{LU}$ factors are obtained by the use of the reduced Crout algorithm, we proceed to solve the triangular systems of equations $\boldsymbol{Ly} = \boldsymbol{\mu}$ and $\boldsymbol{Ux} = \boldsymbol{y}$. For the solution of $\boldsymbol{Ly} = \boldsymbol{\mu}$, we use

$$y_i = \frac{\mu_i - \sum_{k=1}^{i-1} l_{ik} y_k}{l_{ii}} \tag{16-34}$$

The derivation of this formula is straightforward and is left as an exercise. For the solution of $\boldsymbol{Ux} = \boldsymbol{y}$, we use Eq. (16-28). Again, these formulas may be *reduced*, meaning that any term found to be zero may be omitted from the formula. When this is done, we have a set of formulas that represents an optimal Crout algorithm for solving $\boldsymbol{Ax} = \boldsymbol{\mu}_k$, $k = 1, 2, \ldots, m$.

EXAMPLE 16-9. The structure of a system of four linear equations $\boldsymbol{Ax} = \boldsymbol{\mu}$ is as follows [note that Eq. (16-29) is of such form]:

$$\begin{bmatrix} a_{11} & 0 & 0 & a_{14} \\ a_{21} & a_{22} & 0 & 0 \\ 0 & 0 & a_{33} & a_{34} \\ a_{41} & 0 & a_{43} & a_{44} \end{bmatrix} \begin{bmatrix} x_1 \\ x_2 \\ x_3 \\ x_4 \end{bmatrix} = \begin{bmatrix} \mu_1 \\ \mu_2 \\ \mu_3 \\ \mu_4 \end{bmatrix} \tag{16-35}$$

(a) Find $\hat{A}$.
(b) Show the correspondence between the elements of A(I) and $\boldsymbol{A}$.
(c) Find the complete set of formulas of the optimal Crout algorithm.
(d) Show the correspondence between the elements of Q(J) and $\boldsymbol{Q}$.
(e) Find $\hat{Q}$.

Solution:

(a) By comparing Eqs. (16-35) and (16-32) or Eq. (4-62), we have $\hat{A} =$ 1101001100001111.
(b) See the right side of Eq. (16-33): $A(1) = a_{11}$, $A(2) = a_{21}$, $A(3) = a_{41}$, $A(4) = a_{14}$, $A(5) = a_{22}$, $A(6) = a_{33}$, $A(7) = a_{43}$, $A(8) = a_{34}$, and $A(9) = a_{44}$.
(c) The result is given in Table 16-1.

TABLE 16-1. Formulas for Solving Eq. (16-35)

FULL CROUT ALGORITHM	OPTIMAL CROUT ALGORITHM
$q_{11} = l_{11} = a_{11}$	$Q(1) = A(1)$
$q_{21} = l_{21} = a_{21}$	$Q(2) = A(2)$
$q_{31} = l_{31} = a_{31}$	Omitted
$q_{41} = l_{41} = a_{41}$	$Q(3) = A(3)$
$q_{12} = u_{12} = \dfrac{a_{12}}{l_{11}}$	Omitted
$q_{13} = u_{13} = \dfrac{a_{13}}{l_{11}}$	Omitted
$q_{14} = u_{14} = \dfrac{a_{14}}{l_{11}}$	$Q(4) = \dfrac{A(4)}{A(1)}$
$q_{22} = l_{22} = a_{22} - l_{21}u_{12}$	$Q(5) = A(5)$
$q_{32} = l_{32} = a_{32} - l_{31}u_{12}$	Omitted
$q_{42} = l_{42} = a_{42} - l_{41}u_{12}$	Omitted
$q_{23} = u_{23} = \dfrac{a_{23} - l_{21}u_{13}}{l_{22}}$	Omitted
$q_{24} = u_{24} = \dfrac{a_{24} - l_{21}u_{14}}{l_{22}}$	$Q(6) = -\dfrac{Q(2)Q(4)}{Q(5)}$
$q_{33} = l_{33} = a_{33} - l_{31}u_{13} - l_{32}u_{23}$	$Q(7) = A(6)$
$q_{43} = l_{43} = a_{43} - l_{41}u_{13} - l_{42}u_{23}$	$Q(8) = A(7)$
$q_{34} = u_{34} = \dfrac{a_{34} - l_{31}u_{14} - l_{32}u_{24}}{l_{33}}$	$Q(9) = \dfrac{A(8)}{Q(7)}$
$q_{44} = l_{44} = a_{44} - l_{41}u_{14} - l_{42}u_{24} - l_{43}u_{34}$	$Q(10) = A(9) - Q(3)Q(4) - Q(8)Q(9)$
$y_1 = \dfrac{\mu_1}{l_{11}}$	$Y(1) = \dfrac{\mu(1)}{Q(1)}$
$y_2 = \dfrac{\mu_2 - l_{21}y_1}{l_{22}}$	$Y(2) = \dfrac{\mu(2) - Q(2)Y(1)}{Q(5)}$
$y_3 = \dfrac{\mu_3 - l_{31}y_1 - l_{32}y_2}{l_{33}}$	$Y(3) = \dfrac{\mu(3)}{Q(7)}$
$y_4 = \dfrac{\mu_4 - l_{41}y_1 - l_{42}y_2 - l_{43}y_3}{l_{44}}$	$Y(4) = \dfrac{\mu(4) - Q(3)Y(1) - Q(8)Y(3)}{Q(10)}$
$x_4 = y_4$	$X(4) = Y(4)$
$x_3 = y_3 - u_{34}y_4$	$X(3) = Y(3) - Q(9)X(4)$
$x_2 = y_2 - u_{23}x_3 - u_{24}x_4$	$X(2) = Y(2) - Q(6)X(4)$
$x_1 = y_1 - u_{12}x_2 - u_{13}x_3 - u_{14}x_4$	$X(1) = Y(1) - Q(4)X(4)$

(d) From Table 16-1, we have

$$
\boldsymbol{Q} = \begin{bmatrix} Q(1) & 0 & 0 & Q(4) \\ Q(2) & Q(5) & 0 & Q(6) \\ 0 & 0 & Q(7) & Q(9) \\ Q(3) & 0 & Q(8) & Q(10) \end{bmatrix}
$$

(e) From the above $\boldsymbol{Q}$ matrix, we have $\hat{Q} = 1101001100011111$.

Using the result of part (c), we can write a complete FORTRAN program for solving Eq. (16-35) that completely avoids operations on zero-valued elements. The listing of this program is shown in Fig. 16-10.

So far, we have only shown how to obtain the optimal Crout formulas by hand for solving a given system of equations $\boldsymbol{Ax} = \boldsymbol{\mu}$. *The whole process can be automated.* The FORTRAN program given in Fig. 16-10 can be generated by another FORTRAN program. The FORTRAN program GNSO described in [2] accomplishes this task through the use of some intricate programming techniques.

The input to GNSO consists of (1) N, the order of the matrix A; (2) NT, the number of nonzero off-diagonal elements of A (recall that all diagonal elements of A are assumed to be nonzero); (3) binary digits of the number $\hat{A}$, which specifies the *locations* of the nonzero elements of A.

The output of GNSO contains (1) a FORTRAN program called SOLVE (such as the one shown in Fig. 16-10), which represents the optimal Crout algorithm and is

```
      DIMENSION A(9),B(4),Q(10),Y(4),X(4)
    1 READ(5,2)(A(I),I=1,9)
    2 FORMAT(8F10.4)
      WRITE(6,3)(A(I),I=1,9)
    3 FORMAT(1X,10E12.4)
    4 READ(5,2)(B(I),I=1,4)
      WRITE(6,3)(B(I),I=1,4)
      Q(1)=A(1)
      Q(2)=A(2)
      Q(3)=A(3)
      Q(4)=A(4)/Q(1)
      Q(5)=A(5)
      Q(6)=-Q(2)*Q(4)/Q(5)
      Q(7)=A(6)
      Q(8)=A(7)
      Q(9)=A(8)/Q(7)
      Q(10)=A(9)-Q(3)*Q(4)-Q(8)*Q(9)
      Y(1)=B(1)/Q(1)
      Y(2)=(B(2)-Q(2)*Y(1))/Q(5)
      Y(3)=B(3)/Q(7)
      Y(4)=(B(4)-Q(3)*Y(1)-Q(8)*Y(3))/Q(10)
      X(4)=Y(4)
      X(3)=Y(3)-Q(9)*X(4)
      X(2)=Y(2)-Q(6)*X(4)
      X(1)=Y(1)-Q(4)*X(4)
      WRITE(6,3)(Q(I),I=1,10)
      WRITE(6,3)(X(I),I=1,4)
      GO TO 4
      END
```

Fig. 16-10. FORTRAN Program for Solving Eq. (16-35).

punched on cards or stored on tape;[8] (2) the information about $\hat{A}$ and $\hat{Q}$, printed on paper; (3) control cards for running the program.

The FORTRAN program GNSO uses 14 sequences of integers of lengths of at most n in generating the optimal Crout algorithm. The details of the indexing techniques used to generate the formulas in the optimal Crout algorithm are beyond the scope of this book. The reader is referred to [2].

The program SOLVE is then compiled and run with data cards that consist of (1) numerical values of the nonzero elements of $\boldsymbol{A}$, in the order shown by Eq. (16-32); (2) numerical values of the elements of $\boldsymbol{\mu}$, whether zero or nonzero, in the natural order. The output of SOLVE contains the solution vector $\boldsymbol{x}$, and other information, such as Q(I).

Since the program representing the optimal Crout algorithm *contains no DO loops*, the source program may be very long. Two factors must be considered in using the present approach: (1) the time for compiling SOLVE may be a significant part of the total computer time that is to be charged. (2) The object program may be too long to reside in the computer memory as a whole. This will necessitate an exchange between the computer main memory and slow-speed peripheral storage devices. As a consequence, program execution will be slowed. On the other hand, once the program has been compiled so that the object program can fit in the computer memory, the execution of the algorithm is extremely fast. This is mainly due to the special structure of the program, which makes the indexing of elements very efficient. See [2] for actual performance data.

In contrast, the FORTRAN program based on the technique of Section 16-5 is very short, owing to the use of DO loops. But the scheme for addressing elements, as described in Section 16-5, is much more complicated. Thus, one expects the execution time to be longer. See [1] for actual performance data.

The choice of a sparse-matrix technique involves many factors, such as speed, storage, and program transferability. It appears that, between the two methods discussed in the present chapter, neither is definitely better than the other. The program SPICE2 [9] uses the optimal Crout algorithm.[9] The program ASTAP [10] offers the user a choice of three methods. The slowest method contains DO loops and also uses the least amount of storage. The fastest method (in regard to execution time) uses the optimal Crout formulas discussed in the present section.

APPENDIX 16A LISTING OF SPARSE

```
      PROGRAM SPARSE(INPUT,OUTPUT,TAPE5=INPUT,TAPE6=OUTPUT)
C     THIS PROGRAM SOLVES THE  SIMULTANEOUS EQUATIONS AX=B REPEATEDLY
C     FOR DIFFERENT B VECTORS.  FOR AN EXPLANATION OF THE SPARSE MATRIX
C     TECHNIQUE USED, AND THE LIMITATIONS OF THE PROGRAM, SEE SEC. 16-5.
      COMMON DIAG(50),UPACK(200),LPACK(200),UCOL(200),UROWST(50),X(50)
      INTEGER HIGH,HI,UCOL,UROWST
```

[8]The program SOLVE is not to be confused with SOLV in Appendix 16A.

[9]The program SPICE2 (see reference 9) uses a subroutine named CODGEN, similar in purpose to GNSO, but which directly generates machine codes to perform the optimal Crout algorithm.

```
      REAL LPACK
 7    READ (5,8)(UPACK(J),J=1,80)
      WRITE(6,9)(UPACK(J),J=1,80)
      READ(5,1) NNODE,LENGTH,NTIMES
      READ(5,2)(DIAG(J),J=1,NNODE)
      READ(5,2)(UPACK(J),J=1,LENGTH)
      READ(5,2)(LPACK(J),J=1,LENGTH)
      READ(5,1)( UCOL(J),J=1,LENGTH)
      READ(5,1)(UROWST(J),J=1,NNODE)
      WRITE(6,3)NNODE,LENGTH,NTIMES
      WRITE(6,4)(DIAG(J),J=1,NNODE)
      WRITE(6,4)(UPACK(J),J=1,LENGTH)
      WRITE(6,4)(LPACK(J),J=1,LENGTH)
      WRITE(6,6)(UCOL(J),J=1,LENGTH)
      NM1=NNODE-1
      WRITE(6,6)(UROWST(J),J=1,NM1)
      CALL LUFACT(NNODE,LENGTH)
      WRITE(6,5)
      WRITE(6,4)(DIAG(J),J=1,NNODE)
      WRITE(6,4)(UPACK(J),J=1,LENGTH)
      WRITE(6,4)(LPACK(J),J=1,LENGTH)
      WRITE(6,6)(UCOL(J),J=1,LENGTH)
      WRITE(6,6)(UROWST(J),J=1,NM1)
      IF(NTIMES.EQ.0) GO TO 7
      DO 10 J=1,NTIMES
      READ(5,2)(X(J),J=1,NNODE)
      WRITE(6,11)
      WRITE(6,4)(X(J),J=1,NNODE)
      CALL SOLV (NNODE,LENGTH)
      WRITE(6,12)
      WRITE(6,4)(X(J),J=1,NNODE)
10    CONTINUE
      GO TO 7
 1    FORMAT(16I5)
 2    FORMAT(8F10.0)
 3    FORMAT(   /11H INPUT DATA/  3I10/)
 4    FORMAT(1H ,8F12.4)
 5    FORMAT(  //27H RESULT OF LU FACTORIZATION/
     +  46H ELEMENTS OF DIAG,UPACK,LPACK,UCOL, AND UROWST/)
 6    FORMAT( 16I5)
 8    FORMAT(80A1)
 9    FORMAT(1H1,80A1)
11    FORMAT(   /16H CONSTANT VECTOR/ )
12    FORMAT(  /16H SOLUTION VECTOR/ )
      END



      SUBROUTINE LUFACT(NNODE,LENGTH)
      COMMON    DIAG(50),UPACK(200),LPACK(200),UCOL(200),UROWST(50)
      INTEGER HIGH,HI,UCOL,UROWST
      REAL LPACK
      NM1=NNODE-1
      UROWST(NNODE)=LENGTH+1
      DO 6 N=1,NM1
C1    OBTAIN ONE ROW OF THE U MATRIX
      LOW=UROWST(N)
      HIGH=UROWST(N+1) -1
      IF(HIGH.LT.LOW) GO TO 6
      DO 1 MU=LOW,HIGH
 1    UPACK(MU)=UPACK(MU)/DIAG(N)
C2    SUBTRACTION OF PRODUCTS
      DO 6 ML=LOW,HIGH
      I=UCOL(ML)
      DO 6 MU=LOW,HIGH
      J=UCOL(MU)
      IF(I-J)3,2,6
 2    DIAG(J)=DIAG(J)-UPACK(MU)*LPACK(ML)
```

```
      GO TO 6
C3    SEARCH FOR LOCATION IN UPACK CORRESPONDING TO (I,J) ELEMENT IN Q.
 3    LO  =UROWST(I)
      HI   =UROWST(I+1)-1
      IF(HI.LT.LO) GO TO 6
      DO 4 LOC=LO,HI
      IF(UCOL(LOC).EQ.J) GO TO 5
 4    CONTINUE
 5    UPACK(LOC)=UPACK(LOC)-UPACK(MU)*LPACK(ML)
      LPACK(LOC)=LPACK(LOC) -LPACK(MU)*UPACK(ML)
C4
 6    CONTINUE
      RETURN
      END
```

```
      SUBROUTINE SOLV (NNODE,LENGTH)
      COMMON     DIAG(50),UPACK(200),LPACK(200),UCOL(200),UROWST(50)
     1,X(50)
      INTEGER HIGH,UCOL,UROWST
      REAL LPACK
C SOLVE THE LOWER TRIANGULAR SYSTEM
      NM1=NNODE-1
      UROWST(NNODE)=LENGTH+1
      DO 1 J=1,NM1
      X(J)=X(J)/DIAG(J)
      LOW=UROWST(J)
      HIGH=UROWST(J+1)-1
      IF(HIGH.LT.LOW) GO TO 1
      DO 3 K=LOW,HIGH
      MCOL=UCOL(K)
      X(MCOL)=X(MCOL)-LPACK(K)*X(J)
 3    CONTINUE
 1    CONTINUE
      X(NNODE)=X(NNODE)/DIAG(NNODE)
C SOLVE THE UPPER TRIANGULAR SYSTEM
      DO 2 J=2,NNODE
      K=NNODE-J+1
      LOW=UROWST(K)
      HIGH=UROWST(K+1)-1
      IF(HIGH.LT.LOW) GO TO 2
      DO 4 M=LOW,HIGH
      MCOL=UCOL(M)
      X(K)=X(K)-UPACK(M)*X(MCOL)
 4    CONTINUE
 2    CONTINUE
      RETURN
      END
```

REFERENCES

1. BERRY, R. D. "An Optimum Ordering of Electronic Circuit Equations for a Sparse Matrix Solution." *IEEE Trans. Circuit Theory*, Vol. CT-18, pp. 40–50, Jan. 1971.
2. GUSTAVSON, F. G., W. LINIGER, and R. A. WILLOUGHBY. "Symbolic Generation of an Optimal Crout Algorithm for Sparse Systems of Equations." *J. ACM*, Vol. 17, pp. 87–109, 1970.
3. TINNEY, W. F., and J. W. WALKER. "Direct Solutions of Sparse Network Equations by Optimally Ordered Triangular Factorization." *Proc. IEEE*, Vol. 55, pp. 1801–1809, Nov. 1967.

4. CALAHAN, D. A. *Computer-Aided Network Design.* New York: McGraw-Hill Book Company, 1972.
5. FOX, L. *An Introduction to Numerical Linear Algebra.* New York: Oxford University Press, Inc., 1965.
6. PARTER, S. "The Use of Linear Graphs in Gauss Elimination." *SIAM Rev.*, Vol. 3, pp. 119–131, Apr. 1961.
7. TEWARSON, R. P. *Sparse Matrices.* New York: Academic Press, Inc., 1973.
8. MARKOWITZ, H. M. "The Elimination Form of the Inverse and Its Application to Linear Programming." *Management Sci.*, Vol. 3, pp. 255–269, 1957.
9. NAGEL, L. W. *SPICE2: A Computer Program to Simulate Semiconductor Circuits.* Berkeley Calif.: University of California, Ph.D. dissertation, May 1975.
10. WEEKS, W. T., A. J. JIMENEZ, G. W. MAHONEY, D. MEHTA, H. H. QASSEMZADEH, and T. R. SCOTT. "Algorithms for ASTAP—A Network-Analysis Program." *IEEE Trans. Circuit Theory*, Vol. CT-20, pp. 628–634, Nov. 1973.
11. ROSS, D. J., and R. A. WILLOUGHBY, eds. *Sparse Matrices and Their Applications.* New York: Plenum Press, 1972.

PROBLEMS

16-1. Show that the number of multiplications and divisions required in the LU factorization of a square matrix of order n is $n(n^2 - 1)/3$.

16-2. Show that the number of operations (multiplications and divisions) required for solving Eq. (16-1) by the LU factorization method is $(n^3 + 3n^2 - n)/3$.

16-3. Verify that the number of operations required for solving Eq. (16-3) is 12 if the Gaussian elimination method is used and trivial operations of the type $a \times 0 = 0$ are avoided.

16-4. Using the same method as Problem 16-3, show that the number of operations required for solving Eq. (16-8) is 16.

16-5. Assume that the LU factors and the inverse of A have been obtained. Show that for every new $\boldsymbol{\mu}$ the solution of $Ax = \boldsymbol{\mu}$ requires n^2 operations, whether x is obtained by $x = A^{-1}\boldsymbol{\mu}$ or by the LU factorization method.

16-6. Apply the algorithm described in Section 16-3 to obtain the LU factors of the A matrix in Eq. (16-8).

16-7. Repeat Problem 16-6 for the matrix in Eq. (16-2).

16-8. Verify the number and locations of fills shown in Q_1, Q_2, and Q_3 of Example 16-3 by applying the LU factorization algorithm described in Section 16-3.

16-9. Construct the undirected graph associated with a matrix A of the following structure:

$$\begin{bmatrix} \times & \times & 0 & \times & 0 \\ \times & \times & \times & 0 & \times \\ 0 & \times & \times & 0 & 0 \\ \times & 0 & 0 & \times & 0 \\ 0 & \times & 0 & 0 & \times \end{bmatrix}$$

16-10. Using the graph of Problem 16-9, determine the number of fills in the LU factorization if the pivot elements are chosen in the natural order along the diagonal.

16-11. Using the graph of Problem 16-9, find an ordering of the nodes that will minimize the number of fills in the $\boldsymbol{LU}$ factorization.

16-12. Refer to Example 16-7. Find an alternative reordering of nodes that yields no more fills than the solution given.

16-13. It is desired to solve Eq. (16-8) by the use of the program SPARSE, which is based on the method of Section 16-5. Determine the *initial* contents of the arrays DIAG, UPACK, LPACK, UCOL, and UROWST.

16-14. Repeat Problem 16-13 for Eq. (16-6).

16-15. Prove the formula given by Eq. (16-28) for solving $\boldsymbol{Ux} = \boldsymbol{y}$.

16-16. Prove the formula given by Eq. (16-34) for solving $\boldsymbol{Ly} = \boldsymbol{\mu}$.

16-17. Repeat parts (a) to (e) of Example 16-9 for a system of four equations having zero elements distributed as shown in Eq. (16-8).

16-18. Repeat Problem 16-17 for Eq. (16-3).

CHAPTER 17

Advanced Algorithms and Computational Techniques for Computer Simulation Programs

17-1 GENERALIZED ASSOCIATED DISCRETE CIRCUIT MODEL APPROACH

We have already shown in Section 1-9 that the transient analysis of a *dynamic network* can be transformed into a sequence of dc analyses of resistive *networks* upon replacing each capacitor and each inductor by a resistive *discrete circuit model* associated with a given integration algorithm. In particular, the discrete circuit models associated with the *backward Euler algorithm, trapezoidal algorithm, and Gear's second-order algorithm* were derived in Section 1-9. To make use of more sophisticated higher-order *implicit algorithms*, such as those presented in Chapter 13, as well as those yet to be discovered, we shall now derive a *generalized discrete circuit model* associated with any *implicit* multistep algorithm

$$x_{n+1} = \sum_{i=0}^{p} a_i x_{n-i} + h \sum_{i=-1}^{p} b_i f(x_{n-i}, t_{n-i}), \qquad b_{-1} \neq 0 \tag{17-1}$$

as presented earlier in Eq. (13-2). To emphasize the implicit nature of this equation, let us rewrite it as follows:

$$x_{n+1} = \hat{x}_n + hb_{-1} f(x_{n+1}, t_{n+1}), \qquad b_{-1} \neq 0 \tag{17-2}$$

where

$$\hat{x}_n \triangleq \sum_{i=0}^{p} [a_i x_{n-i} + hb_i f(x_{n-i}, t_{n-i})] \tag{17-3}$$

17-1-1 Generalized Associated Discrete Circuit Model for Capacitors

Applying Eq. (17-2) to solve the first-order differential equation $\dot{v} = f(v, t)$ with a *step size h*, we obtain

$$v_{n+1} = \hat{v}_n + hb_{-1}\dot{v}_{n+1}, \qquad b_{-1} \neq 0 \tag{17-4}$$

where

$$\hat{v}_n \triangleq \sum_{j=0}^{p} [a_j v_{n-j} + hb_j \dot{v}_{n-j}] \quad \text{and} \quad \dot{v}_k \triangleq f(v_k, t_k) \tag{17-5}$$

Since

$$i(t) = C(v(t))\frac{dv(t)}{dt} \tag{17-6}$$

where

$$C(v) \triangleq \frac{d\hat{q}(v)}{dv} \tag{17-7}$$

for a nonlinear voltage-controlled capacitor characterized by $q = \hat{q}(v)$, it follows that

$$\dot{v}(t_{n+1}) = \frac{1}{C(v(t_{n+1}))} i(t_{n+1}) \tag{17-8}$$

Observe that, whereas Eq. (17-8) is an exact relationship, Eq. (17-4) represents only an "approximate solution" owing to the inherent local truncation error associated with the multistep numerical-integration algorithm. If we approximate the exact solution $v(t_{n+1})$ and $i(t_{n+1})$ by v_{n+1} and i_{n+1}, respectively, then Eq. (17-8) becomes

$$\dot{v}_{n+1} = \frac{1}{C(v_{n+1})} i_{n+1} \tag{17-9}$$

Substituting Eq. (17-9) into Eq. (17-4) and solving for i_{n+1}, we obtain

$$i_{n+1} = \left[\frac{C(v_{n+1})}{hb_{-1}}\right][v_{n+1} - \hat{v}_n] \triangleq g_C(v_{n+1}) \tag{17-10}$$

where

$$\hat{v}_n \triangleq \sum_{j=0}^{p} [a_j v_{n-j} + hb_j \dot{v}_{n-j}] \tag{17-11}$$

Observe that, since $\hat{v}_n$ depends only on the *previously stored* values $v_n, v_{n-1}, \ldots, v_{n-p}$ and $\dot{v}_n, \dot{v}_{n-1}, \ldots, \dot{v}_{n-p}$, it is a *constant*. Hence, Eq. (17-10) specifies i_{n+1} as a function only of the variable v_{n+1}; i.e., $i_{n+1} = g_C(v_{n+1})$. But this function can be interpreted as the v-i curve of a nonlinear resistor with terminal voltage v_{n+1} and terminal current i_{n+1}, as shown in Fig. 17-1(a). If a dynamic network contains only resistors and capacitors, and if each capacitor is replaced by this equivalent nonlinear resistor model, the solution voltage across each of these nonlinear resistors would be equal to the corresponding capacitor voltage at $t = t_{n+1}$. Hence, *the generalized discrete circuit model for a nonlinear capacitor* [*characterized by* $C(v) \triangleq d\hat{q}(v)/dv$] *associated with the implicit multistep integration algorithm of Eq. (17-1), at time* $t = t_{n+1}$, *is simply the nonlinear resistor shown in Fig. 17-1(a)*. In the special case when the capacitor is *linear*

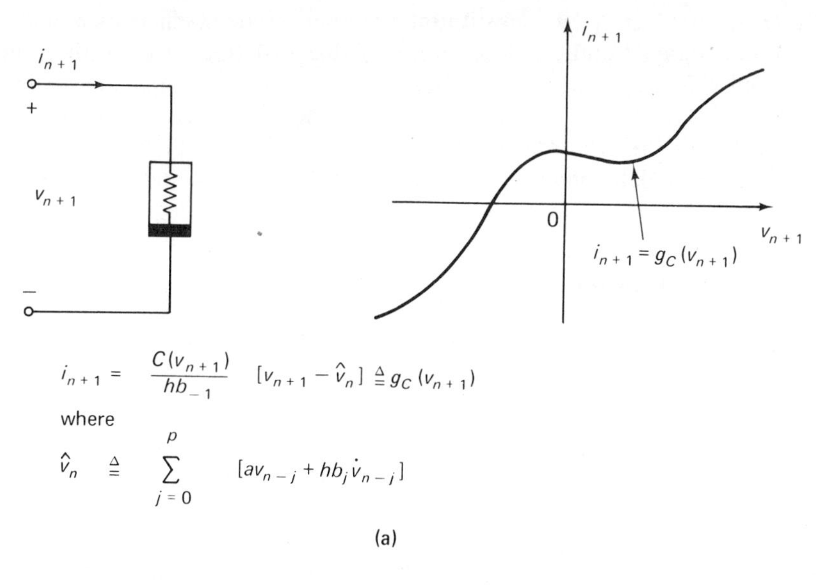

(a)

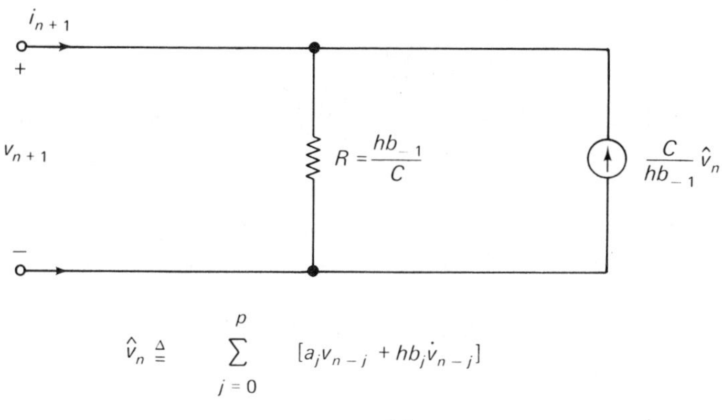

(b)

Fig. 17-1. (a) Generalized associated discrete circuit model for a nonlinear capacitor with incremental capacitance $C(v) \triangleq dq(v)/dv$; (b) generalized associated discrete circuit model for a linear capacitor with constant capacitance C.

with a constant capacitance C, Eq. (17-10) reduces to

$$i_{n+1} = \left[\frac{C}{hb_{-1}}\right] v_{n+1} - \left[\frac{C}{hb_{-1}} \hat{v}_n\right] \tag{17-12}$$

where

$$\hat{v}_n \triangleq \sum_{j=0}^{p} [a_j v_{n-j} + hb_j \dot{v}_{n-j}] \tag{17-13}$$

Observe that Eq. (17-12) is the terminal equation describing the *linear resistive circuit* shown in Fig. 17-1(b). Hence, *this equivalent circuit is the generalized discrete circuit model for a linear capacitor associated with the implicit multistep integration algorithm of Eq. (17-1)* at $t = t_{n+1}$. By substituting the appropriate coefficients a_i and b_i from Table 12-2 (Adams–Moulton algorithm) or Table 13-1 (Gear's algorithm) into Eqs. (17-10) and (17-11), or into Eqs. (17-12) and (17-13) if the capacitor is linear, the result would be the associated discrete circuit model for the kth-order Adams–Moulton algorithm or the kth-order Gear's algorithm. In particular, the three associated discrete circuit models shown in Figs. 1-27, 1-28, and 1-29 would follow trivially.

17-1-2 Generalized Associated Discrete Circuit Model for Inductors

Applying Eq. (17-2) to solve the first-order differential equation $\dot{i} = f(i, t)$ with a *step size h*, we obtain

$$i_{n+1} = \hat{i}_n + hb_{-1}\dot{i}_{n+1}, \qquad b_{-1} \neq 0 \tag{17-14}$$

where

$$\hat{i}_n \triangleq \sum_{j=0}^{p} [a_j i_{n-j} + hb_j \dot{i}_{n-j}] \quad \text{and} \quad \dot{i}_k \triangleq f(i_k, t_k) \tag{17-15}$$

Since

$$v(t) = L(i(t))\frac{di(t)}{dt} \tag{17-16}$$

where

$$L(i) \triangleq \frac{d\hat{\phi}(i)}{di} \tag{17-17}$$

for a *nonlinear current-controlled inductor* characterized by $\phi = \hat{\phi}(i)$, it follows that

$$\dot{i}_{n+1} = \frac{1}{L(i_{n+1})} v_{n+1} \tag{17-18}$$

Substituting Eq. (17-18) into Eq. (17-14) and solving for v_{n+1}, we obtain

$$v_{n+1} = \left[\frac{L(i_{n+1})}{hb_{-1}}\right][i_{n+1} - \hat{i}_n] \triangleq g_L(i_{n+1}) \tag{17-19}$$

where

$$\hat{i}_n \triangleq \sum_{j=0}^{p} [a_j i_{n-j} + hb_j \dot{i}_{n-j}] \tag{17-20}$$

Following the same reasoning as in Section 17-1-1, we now conclude that the *generalized discrete circuit model for a nonlinear inductor [characterized by L(i) $\triangleq d\hat{\phi}(i)/di$] associated with the implicit multistep integration algorithm of Eq. (17-1) is simply the nonlinear resistor shown in Fig. 17-2(a)*. In the special case when the inductor is *linear* with a constant inductance L, Eq. (17-19) reduces to

$$i_{n+1} = \left[\frac{hb_{-1}}{L}\right] v_{n+1} + \hat{i}_n \tag{17-21}$$

where

$$\hat{i}_n \triangleq \sum_{j=0}^{p} [a_j i_{n-j} + hb_j \dot{i}_{n-j}] \tag{17-22}$$

Observe that Eq. (17-21) is the terminal equation describing the *linear resistive circuit* shown in Fig. 17-2(b). Hence, *this equivalent circuit is the generalized discrete circuit*

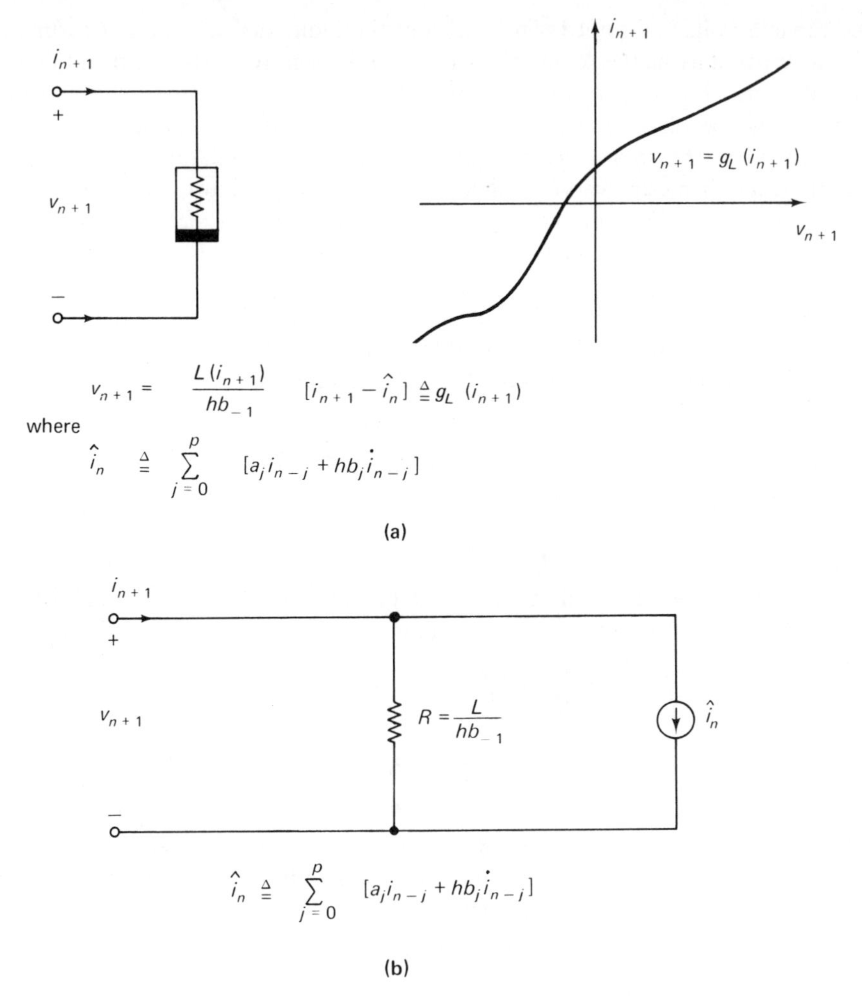

Fig. 17-2. (a) Generalized associated discrete circuit model for a nonlinear inductor with incremental inductance $L(i) \triangleq d\phi(i)/di$; (b) generalized associated discrete circuit model for a linear inductor with constant inductance L.

model for a linear inductor associated with the implicit multistep integration algorithm of Eq. (17-1) at $t = t_{n+1}$. Observe that the three associated discrete circuit models shown in Figs. 1-30, 1-31, and 1-32 now follow as special cases of this generalized model. Also, both models shown in Figs. 17-1(b) and 17-2(b) are particularly suited for *nodal analysis* [see Problem 17-25].

17-1-3 Transforming a Dynamic Network into a Generalized Associated Discrete Resistive Network

If we replace each capacitor and inductor in a dynamic network N by the associated discrete circuit models given in Figs. 17-1 and 17-2, the result is a nonlinear resistive circuit, henceforth called the *generalized associated discrete network* N_R. If the capacitors in N do not form loops and the inductors in N do not form cutsets, and if N contains neither controlled sources nor negative resistors, then the number of state variables is equal to the sum of the number of capacitors and inductors, and solving the resistive network N_R at $t = t_{n+1}$ would be identical to solving the network N by the *state equation approach*. Hence, by updating the parameters in N_R after each time step, we would indeed obtain the same solution voltage across each capacitor model and solution current entering each inductor model as those obtained by integrating the state equations of N, as long as the same numerical integration algorithm is used.

If the dynamic network N contains some loops of capacitors and some cutsets of inductors, the number of state variables will be less than the total number of capacitors and inductors. Let us denote the state equation by

$$\dot{\boldsymbol{v}}_C = \boldsymbol{f}_C(\boldsymbol{v}_C, \boldsymbol{i}_L, t) \tag{17-23}$$

$$\dot{\boldsymbol{i}}_L = \boldsymbol{f}_L(\boldsymbol{v}_C, \boldsymbol{i}_L, t) \tag{17-24}$$

where

$$\boldsymbol{v}_C \triangleq [v_{C_1} \quad v_{C_2} \quad \cdots \quad v_{C\alpha}]^t \tag{17-25}$$

$$\boldsymbol{i}_L \triangleq [i_{L_1} \quad i_{L_2} \quad \cdots \quad i_{L\beta}]^t \tag{17-26}$$

are the capacitor voltage and inductor current *state variables*. If we denote the *remaining* capacitor voltages and inductor currents by

$$\hat{\boldsymbol{v}}_C \triangleq [v_{C\alpha+1} \quad v_{C\alpha+2} \quad \cdots \quad v_{C\gamma}]^t \tag{17-27}$$

$$\hat{\boldsymbol{i}}_L \triangleq [i_{L\beta+1} \quad i_{L\beta+2} \quad \cdots \quad i_{L\delta}]^t \tag{17-28}$$

then

$$\hat{\boldsymbol{v}}_C = \boldsymbol{K}_C \boldsymbol{v}_C \tag{17-29}$$

$$\hat{\boldsymbol{i}}_L = \boldsymbol{K}_L \boldsymbol{i}_L \tag{17-30}$$

where the elements of the matrices $\boldsymbol{K}_C$ and $\boldsymbol{K}_L$ are 1, 0, or -1. If we differentiate both sides of Eqs. (17-29) and (17-30) and then substitute the expression for $\dot{\boldsymbol{v}}_C$ and $\dot{\boldsymbol{i}}_L$ from Eqs. (17-23) and (17-24), we obtain

$$\dot{\hat{\boldsymbol{v}}}_C = \boldsymbol{K}_C \boldsymbol{f}_C(\boldsymbol{v}_C, \boldsymbol{i}_L, t) \tag{17-31}$$

$$\dot{\hat{\boldsymbol{i}}}_L = \boldsymbol{K}_L \boldsymbol{f}_L(\boldsymbol{v}_C, \boldsymbol{i}_L, t) \tag{17-32}$$

Equations (17-23), (17-24), (17-31), and (17-32) can be combined into the following compact form:

$$\dot{\boldsymbol{x}} = \boldsymbol{f}(\boldsymbol{x}, t) \tag{17-33}$$

where

$$\boldsymbol{x} \triangleq \begin{bmatrix} \boldsymbol{v}_C \\ \hat{\boldsymbol{v}}_C \\ \boldsymbol{i}_L \\ \hat{\boldsymbol{i}}_L \end{bmatrix}, \qquad \boldsymbol{f}(\boldsymbol{x}, t) \triangleq \begin{bmatrix} \boldsymbol{f}_C(\boldsymbol{v}_C, \boldsymbol{i}_L, t) \\ \boldsymbol{K}_C \boldsymbol{f}_C(\boldsymbol{v}_C, \boldsymbol{i}_L, t) \\ \boldsymbol{f}_L(\boldsymbol{v}_C, \boldsymbol{i}_L, t) \\ \boldsymbol{K}_L \boldsymbol{f}_L(\boldsymbol{v}_C, \boldsymbol{i}_L, t) \end{bmatrix} \tag{17-34}$$

Observe that each capacitor voltage and each inductor current is now included in the augmented vector $\boldsymbol{x}$. Hence, if we replace each capacitor and inductor in N by a generalized associated discrete circuit model chosen from Figs. 17-1 and 17-2, the resulting solution voltage across each capacitor model and solution current entering each inductor model are identical to that obtained by numerically solving Eq. (17-33), as long as the same integration algorithm is used. Since the vectors $\hat{\boldsymbol{v}}_C$ and $\hat{\boldsymbol{i}}_L$ are constrained by the *state vectors* $\boldsymbol{v}_C$ and $\boldsymbol{i}_L$ by Eqs. (17-29) and (17-30), their *initial* values $\hat{\boldsymbol{v}}_C(0)$ and $\hat{\boldsymbol{i}}_L(0)$ must be chosen to be consistent with these constraints when solving Eq. (17-33) numerically. However, this precautionary step is automatically taken care of by the generalized associated discrete circuit N_R, because the vectors $\hat{\boldsymbol{v}}_C$ and $\hat{\boldsymbol{i}}_L$ must satisfy KVL and KCL around each loop and cutset in N_R corresponding to each capacitor loop and inductor cutset in N. Hence, our approach is indeed a general one. Since formulating state equations for a nonlinear dynamic network is a formidable task, as we have demonstrated in Chapter 10, it is clearly advantageous to develop computer simulation programs using the approach presented in this section. However, the approach presented in Chapter 10 would still be preferred if the nonlinear dynamic network N is very large but contains only a few nonlinear resistors. The state-variable approach is also useful if it is necessary to obtain the state equations in explicit form, as in the case when a stability analysis is desired. Perhaps the most serious objection to the associated discrete circuit approach is that it is extremely difficult to implement an automatic control of order and step size, a feature easily achieved with the state-variable approach.

17-2 TABLEAU APPROACH

We showed in Section 17-1 that the *transient analysis* of a nonlinear dynamic network can be reduced to the *dc analysis* of a sequence of *nonlinear resistive networks* by replacing each capacitor and inductor by a discrete circuit model associated with a preselected *implicit* numerical-integration algorithm. Since the network to be analyzed is resistive, the Newton–Raphson algorithm presented in Chapter 5 can be used to find the solutions. In particular, we can invoke the "discrete circuit equivalent of the Newton–Raphson nodal analysis theorem" from Chapter 5 to transform the nonlinear resistive network into an equivalent discrete *linear resistive* circuit. In view of this observation, the efficient analysis of linear resistive networks becomes of crucial importance. Since such networks are usually fairly large, and since a complete transient analysis would often require solving the same networks (albeit with different parameter values) many times, the sparse-matrix techniques presented in Chapter 16 should be brought to bear here. Since the sparser the associated matrix, the better the computational efficiency, it is important that the optimal equation-formulation algorithm be selected. Our objective in this section is to present such an algorithm.

Choosing the same *composite branch* shown in Fig. 4-1(b) and using the same symbols and notations as presented in Section 4-2, we obtain three sets of equations:

$$\text{KCL } (n-1 \text{ equations}): \quad \boldsymbol{Ai} = \boldsymbol{AJ} \tag{17-35}$$

$$\text{KVL } (b \text{ equations}):^{1} \quad \boldsymbol{v} = \boldsymbol{A}^t\boldsymbol{v}_n + \boldsymbol{E} \tag{17-36}$$

$$\text{Element constitutive relations } (b \text{ equations}): \quad \boldsymbol{K}_i\boldsymbol{i} + \boldsymbol{K}_v\boldsymbol{v} = \boldsymbol{s} \tag{17-37}$$

[1]To show that Eq. (17-36) represents KVL equations, let us premultiply both sides of Eq. (17-36) by the *fundamental loop matrix* $\boldsymbol{B}$ to obtain

$$\boldsymbol{Bv} = (\boldsymbol{BA}^t)\boldsymbol{v}_n + \boldsymbol{BE}$$

Substituting $\boldsymbol{BA}^t = \boldsymbol{0}$ (Corollary 3-3) into this equation, we obtain $\boldsymbol{Bv} = \boldsymbol{BE}$, the familiar KVL equations for composite branches.

where

$$\boldsymbol{i} \triangleq \begin{bmatrix} i_1 \\ i_2 \\ \cdot \\ \cdot \\ \cdot \\ i_b \end{bmatrix}, \quad \boldsymbol{v} \triangleq \begin{bmatrix} v_1 \\ v_2 \\ \cdot \\ \cdot \\ \cdot \\ v_b \end{bmatrix}, \quad \boldsymbol{J} \triangleq \begin{bmatrix} J_1 \\ J_2 \\ \cdot \\ \cdot \\ \cdot \\ J_b \end{bmatrix}, \quad \boldsymbol{E} \triangleq \begin{bmatrix} E_1 \\ E_2 \\ \cdot \\ \cdot \\ \cdot \\ E_b \end{bmatrix} \tag{17-38}$$

denote the *branch current vector*, the *branch voltage vector*, the *independent current source vector*, and the *independent voltage source vector*, respectively. The vector $\boldsymbol{v}_n$ denotes the $(n-1) \times 1$ *node-to-datum* voltage vector, where n is the total number of nodes. The matrix $\boldsymbol{A}$ is the *reduced incidence matrix*, and the matrices $\boldsymbol{K}_i$ and $\boldsymbol{K}_v$ are *branch parameter matrices* for describing the constitutive relation of the linear elements. The vector **s** is a constant $b \times 1$ vector for allowing a constant to be added to the constitutive relation of a resistor. For example, if all elements are two-terminal linear resistors with resistances $R_1, R_2, \ldots, R_b$, then $\boldsymbol{K}_i = -\boldsymbol{Z}_b$, $\boldsymbol{K}_v = \boldsymbol{1}$, and $\boldsymbol{s} = \boldsymbol{0}$, where $\boldsymbol{Z}_b$ is the $b \times b$ diagonal *branch resistance matrix*. If all elements are two-terminal linear resistors with conductances $G_1, G_2, \ldots, G_b$, then $\boldsymbol{K}_i = \boldsymbol{1}$, $\boldsymbol{K}_v = -\boldsymbol{Y}_b$, and $\boldsymbol{s} = \boldsymbol{0}$, where $\boldsymbol{Y}_b$ is the $b \times b$ diagonal *branch conductance matrix*. In the most general case, $\boldsymbol{K}_i$ and $\boldsymbol{K}_v$ are not necessarily $b \times b$ diagonal matrices, and $\boldsymbol{s}$ is not necessarily a zero vector. For example, let

$$\boldsymbol{K}_i = \begin{bmatrix} 0 & -10 \\ 1 & 0 \\ 0 & 0 \\ -1 & \alpha \end{bmatrix}, \quad \boldsymbol{K}_v = \begin{bmatrix} 0 & 1 \\ 0 & 0 \\ 0 & 1 \\ g_m & 0 \end{bmatrix}, \quad \boldsymbol{s} = \begin{bmatrix} 0 \\ 6 \\ 4 \\ 10 \end{bmatrix} \tag{17-39}$$

Substituting Eq. (17-39) into Eq. (17-37), we obtain four constitutive relations:

$$v_2 = 10 i_2 \tag{17-40}$$

$$i_1 = 6 \tag{17-41}$$

$$v_2 = 4 \tag{17-42}$$

$$i_1 = \alpha i_2 + g_m v_1 + 10 \tag{17-43}$$

Observe that Eq. (17-40) represents a 10-Ω resistor, Eq. (17-41) represents a 6-A current source, Eq. (17-42) represents a 4-V voltage source, and Eq. (17-43) represents a controlled current source that depends on i_2, v_1, and a constant equal to 10.

Equations (17-35)–(17-37) constitute a system of $2b + n - 1$ linear equations in $2b + n - 1$ unknowns, i.e., b branch currents, b branch voltages, and $n - 1$ node-to-datum voltages. Even though we now have many more equations than other formulation systems, such as nodal analysis, these equations are *extremely sparse*, and it is often more efficient computationally to solve a much sparser, albeit larger, system of linear equations. It is convenient to recast Eqs. (17-35)–(17-37) into the following *matrix tableau* form:

$$\begin{array}{l} n-1 \text{ KCL equations} \\ b \text{ KVL equations} \\ b \text{ element constitutive relations} \end{array} \left\{ \begin{bmatrix} \boldsymbol{A} & \boldsymbol{0} & \boldsymbol{0} \\ \boldsymbol{0} & \boldsymbol{1} & -\boldsymbol{A}^t \\ \boldsymbol{K}_i & \boldsymbol{K}_v & \boldsymbol{0} \end{bmatrix} \begin{bmatrix} \boldsymbol{i} \\ \boldsymbol{v} \\ \boldsymbol{v}_n \end{bmatrix} = \begin{bmatrix} \boldsymbol{A}\boldsymbol{J} \\ \boldsymbol{E} \\ \boldsymbol{s} \end{bmatrix} \right. \tag{17-44}$$

Observe that, unlike other methods of analysis, the solution to Eq. (17-44) automatically yields all branch current and voltage variables, as well as all node-to-datum voltages. The computational efficiency of course depends on the sparsity of the *tableau matrix*

$$T \triangleq \begin{bmatrix} A & 0 & 0 \\ 0 & 1 & -A^t \\ K_i & K_v & 0 \end{bmatrix} \tag{17-45}$$

Since the tableau matrix equation contains KCL, KVL, and the element constitutive relations, it should be possible to derive from it any other equation-formulation method. In particular, we shall now show that the well-known nodal analysis can be interpreted as solving the matrix tableau equation

$$\begin{bmatrix} A & 0 & 0 \\ 0 & 1 & -A^t \\ 1 & -Y_b & 0 \end{bmatrix} \begin{bmatrix} i \\ v \\ v_n \end{bmatrix} = \begin{bmatrix} AJ \\ E \\ 0 \end{bmatrix} \tag{17-46}$$

via Gaussian elimination with a *preselected pivoting order*. For example, if we pivot along the diagonal elements of the lower-left-corner unit matrix, we obtain

$$\begin{bmatrix} 1 & -Y_b & 0 \\ 0 & 1 & -A^t \\ A & 0 & 0 \end{bmatrix} \begin{bmatrix} i \\ v \\ v_n \end{bmatrix} = \begin{bmatrix} 0 \\ E \\ AJ \end{bmatrix} \tag{17-47}$$

These "row pivoting" operations are of course equivalent to simply reordering the last $n - 1$ equations first and the first b equations last. Now let us carry out a forward Gaussian elimination cycle on Eq. (17-47) to reduce the lower-left-corner submatrix A to a zero matrix. In terms of matrix operations, this step is equivalent to premultiplying the first row of Eq. (17-47) by $-A$ and then adding the result to the third row to obtain

$$\begin{bmatrix} 1 & -Y_b & 0 \\ 0 & 1 & -A^t \\ 0 & AY_b & 0 \end{bmatrix} \begin{bmatrix} i \\ v \\ v_n \end{bmatrix} = \begin{bmatrix} 0 \\ E \\ AJ \end{bmatrix} \tag{17-48}$$

The next forward Gaussian elimination cycle would reduce the submatrix AY_b in Eq. (17-48) to a zero matrix. This step is equivalent to premultiplying the second row of Eq. (17-48) by $-AY_b$ and then adding the result to the third row to obtain

$$\begin{bmatrix} 1 & -Y_b & 0 \\ 0 & 1 & -A^t \\ 0 & 0 & AY_bA^t \end{bmatrix} \begin{bmatrix} i \\ v \\ v_n \end{bmatrix} = \begin{bmatrix} 0 \\ E \\ A(J - Y_bE) \end{bmatrix} \tag{17-49}$$

We can now identify the last row in Eq. (17-49) to be the familiar nodal equations [see Eq. (4-73)]

$$(AY_bA^t)v_n = A(J - Y_bE) \tag{17-50}$$

Hence, if we continue the forward Gaussian elimination cycle to reduce the matrix in Eq. (17-49) to an upper triangular matrix, and then obtain the solution v_n by back substitution, we are precisely solving the nodal equations by Gaussian elimination!

The preceding development makes use of row pivoting only. This corresponds simply to reordering the equations. If we wish also to reorder variables, then we can also perform *column pivoting*. For example, by column pivoting Eq. (17-44) from the right to the left, we obtain the following equivalent matrix tableau equation:

$$\begin{bmatrix} \mathbf{0} & \mathbf{0} & A \\ -A^t & \mathbf{1} & \mathbf{0} \\ \mathbf{0} & K_v & K_i \end{bmatrix} \begin{bmatrix} v_n \\ v \\ i \end{bmatrix} = \begin{bmatrix} AJ \\ E \\ s \end{bmatrix} \tag{17-51}$$

We shall leave it as an exercise for the reader to verify that *any general topological method of circuit analysis*—loop analysis, hybrid analysis, or even state-variable analysis—*can be interpreted as simply solving the matrix tableau equations by Gaussian elimination with a particular combination of row and column pivoting order.* Recall from Chapter 16 that for each sparse matrix an optimal pivoting order should be chosen to minimize computation. Since, except in contrived examples, this optimal pivoting order will be different from the *particular combination* of pivoting order required by a particular method of circuit analysis, it is clear that, from the computational efficiency point of view, the tableau approach is superior to most other methods of network analysis. Moreover, unlike other topological equation-formulation algorithms, it is a trivial matter to program the tableau equations. The only task left is to develop a highly efficient sparse-matrix Gaussian elimination algorithm.

From the theoretical point of view, the tableau approach has three shortcomings. First, if the network does not have a solution (such as networks containing a loop of independent voltage sources that do not add to zero), the numerical calculation will obviously terminate prematurely without any indication of where the source of difficulty is. Second, if the network has indeterminate solutions (such as networks containing norators that result in infinitely many solutions), the forward Gaussian elimination cycle will also terminate prematurely owing to the presence of a row of "zero" pivots. Thus, the original tableau matrix is singular; otherwise, a unique solution would exist. Again, it may not be easy to pinpoint the source of the difficulty numerically. Finally, the tableau matrix has a much larger dimension than any other topological matrix and is therefore cumbersome to handle. This objection, however, is not too serious from a programming standpoint, since the tableau matrix is extremely sparse and only the nonzero elements need to be stored. Of course, the tableau approach requires the development of a highly efficient sparse-matrix Gaussian-elimination-solution software package. Without such software, the tableau approach would lose its computational advantage.

17-3 VARIABLE STEP-SIZE VARIABLE-ORDER ALGORITHM FOR SOLVING IMPLICIT DIFFERENTIAL-ALGEBRAIC SYSTEMS

There are many occasions in computer-aided circuit design when one is confronted with the task of solving a system of implicit equations of the form

$$f_1(x_1, x_2, \ldots, x_n, \dot{x}_1, \dot{x}_2, \ldots, \dot{x}_m, t) = 0 \tag{17-52a}$$

$$f_2(x_1, x_2, \ldots, x_n, \dot{x}_1, \dot{x}_2, \ldots, \dot{x}_m, t) = 0 \tag{17-52b}$$

$$\vdots$$

$$f_n(x_1, x_2, \ldots, x_n, \dot{x}_1, \dot{x}_2, \ldots, \dot{x}_m, t) = 0 \tag{17-52n}$$

where $m \leq n$. If we define the vectors

$$\boldsymbol{x} \triangleq [x_1 \quad x_2 \quad \cdots \quad x_n]^t \tag{17-53}$$

and

$$\dot{\boldsymbol{x}} \triangleq [\dot{x}_1 \quad \dot{x}_2 \quad \cdots \quad \dot{x}_n]^t \tag{17-54}$$

then Eq. (17-52) can be expressed compactly as

$$\boxed{\boldsymbol{f}(\boldsymbol{x}, \dot{\boldsymbol{x}}, t) = \boldsymbol{0}} \tag{17-55}$$

Observe that, if $m < n$, the functions $f_i(\boldsymbol{x}, \dot{\boldsymbol{x}}, t)$ will not contain the variables $\dot{x}_{m+1}$, $\dot{x}_{m+2}, \ldots, \dot{x}_n$. Also, since $m \leq n$, Eq. (17-55) always contains as many equations as there are unknowns. We shall henceforth refer to Eq. (17-55) as an *implicit differential-algebraic system*. This system clearly includes the *explicit* state equations

$$\dot{\boldsymbol{x}} = \boldsymbol{h}(\boldsymbol{x}, t) \tag{17-56}$$

of Chapter 10 as a special case; i.e.,

$$\boldsymbol{f}(\boldsymbol{x}, \dot{\boldsymbol{x}}, t) \triangleq \boldsymbol{h}(\boldsymbol{x}, t) - \dot{\boldsymbol{x}} = \boldsymbol{0} \tag{17-57}$$

If an efficient algorithm can be developed for solving Eq. (17-55) directly, it would be unnecessary to go through the ordeal of transforming it first into the normal form of Eq. (17-56). Our objective in this section is to derive such an algorithm.

Suppose that the solution $\boldsymbol{x}(t)$ of Eq. (17-55) had been found at $t = t_n$, $t = t_{n-1}$, $\ldots$, and t_{n+1-k}, where the *step size* $h_j \triangleq t_{j+1} - t_j$ *need not be uniform*. If we denote $\boldsymbol{x}(t_j)$ by $\boldsymbol{x}_j$, the solution $\boldsymbol{x}_{n+1}$ of Eq. (17-55) at $t = t_{n+1}$ must satisfy

$$\boldsymbol{f}(\boldsymbol{x}_{n+1}, \dot{\boldsymbol{x}}(t_{n+1}), t_{n+1}) = \boldsymbol{0} \tag{17-58}$$

We shall show in Section 17-3-1 that it is possible to derive a general formula, henceforth called the *backward-differentiation formula* (BDF), which *approximates* to within any prescribed accuracy the present value $\dot{\boldsymbol{x}}(t_{n+1})$ of the time derivative of $\boldsymbol{x}(t)$ at $t = t_{n+1}$ in terms of $\boldsymbol{x}_{n+1}$ and k *past values* $\boldsymbol{x}_n, \boldsymbol{x}_{n-1}, \ldots, \boldsymbol{x}_{n-k+1}$. If we let $\dot{\boldsymbol{x}}_{n+1}$ denote the approximate value of $\dot{\boldsymbol{x}}(t_{n+1})$, then the BDF of order k will be shown to be given by

$$\boxed{\dot{\boldsymbol{x}}_{n+1} = -\frac{1}{h}\sum_{i=0}^{k} \alpha_i \boldsymbol{x}_{n+1-i} \triangleq \boldsymbol{g}(\boldsymbol{x}_{n+1})} \tag{17-59}$$

where $\alpha_0, \alpha_1, \ldots, \alpha_k$ are *constants* and $h \triangleq t_{n+1} - t_n$ is the *present* step size.[2] Observe that, although the right side of Eq. (17-59) depends on both the present value $\boldsymbol{x}_{n+1}$ and k past values $\boldsymbol{x}_n, \boldsymbol{x}_{n-1}, \ldots, \boldsymbol{x}_{n-k+1}$, we have denoted it as a function $\boldsymbol{g}(\boldsymbol{x}_{n+1})$ of only $\boldsymbol{x}_{n+1}$ to emphasize that $\boldsymbol{x}_{n+1}$ is still unknown. If we substitute Eq. (17-59) into Eq.

[2]Since the step size need not be uniform, we should really write $h_{n+1} = t_{n+1} - t_n$. However, for convenience we shall omit the subscript $n + 1$, and let h denote the "present" step size.

(17-58), we obtain

$$\boxed{\boldsymbol{f}(\boldsymbol{x}_{n+1}, \boldsymbol{g}(\boldsymbol{x}_{n+1}), t_{n+1}) = \boldsymbol{0}} \tag{17-60}$$

Observe that Eq. (17-60) is now a *system of nonlinear algebraic equations* in terms of the unknown variable $\boldsymbol{x}_{n+1}$. Hence, the solution $\boldsymbol{x}_{n+1}$ of the *implicit differential-algebraic system* in Eq. (17-55) can be found by solving Eq. (17-60) by any efficient algorithm for solving nonlinear algebraic equations. In particular, if we use the *Newton–Raphson algorithm* of Chapter 5, then $\boldsymbol{x}_{n+1}$ is found by using the following recursive relation:

$$\boldsymbol{x}_{n+1}^{(j+1)} = \boldsymbol{x}_{n+1}^{(j)} - [\boldsymbol{J}(\boldsymbol{x}_{n+1}^{(j)})]^{-1}\boldsymbol{f}(\boldsymbol{x}_{n+1}^{(j)}, \boldsymbol{g}(\boldsymbol{x}_{n+1}^{(j)}), t_{n+1}) \tag{17-61}$$

where

$$\boldsymbol{J}(\boldsymbol{x}_{n+1}^{(j)}) \triangleq \left.\frac{\partial \boldsymbol{f}(\boldsymbol{x}, \dot{\boldsymbol{x}}, t)}{\partial \boldsymbol{x}}\right|_{\substack{\boldsymbol{x}=\boldsymbol{x}_{n+1}^{(j)} \\ \dot{\boldsymbol{x}}=\boldsymbol{g}(\boldsymbol{x}_{n+1}^{(j)})}} - \frac{\alpha_0}{h}\left.\frac{\partial \boldsymbol{f}(\boldsymbol{x}, \dot{\boldsymbol{x}}, t)}{\partial \dot{\boldsymbol{x}}}\right|_{\substack{\boldsymbol{x}=\boldsymbol{x}_{n+1}^{(j)} \\ \dot{\boldsymbol{x}}=\boldsymbol{g}(\boldsymbol{x}_{n+1}^{(j)})}} \tag{17-62}$$

is the *Jacobian matrix* of $\boldsymbol{f}(\boldsymbol{x}_{n+1}, \boldsymbol{g}(\boldsymbol{x}_{n+1}), t_{n+1})$ evaluated at $\boldsymbol{x}_{n+1} = \boldsymbol{x}_{n+1}^{(j)}$. After $\boldsymbol{x}_{n+1}$ is found, we simply repeat the process and compute $\boldsymbol{x}_{n+2}, \boldsymbol{x}_{n+3}, \ldots$, etc.

Since only $\boldsymbol{x}_0 \triangleq \boldsymbol{x}(t_0)$ is given at the initial time $t = t_0$, it is necessary to choose the order $k = 1$ in Eq. (17-59) for computing $\boldsymbol{x}_1$. Knowing $\boldsymbol{x}_0$ and $\boldsymbol{x}_1$, we can compute for $\boldsymbol{x}_2$ using $k \leq 2$. Knowing $\boldsymbol{x}_0$, $\boldsymbol{x}_1$, and $\boldsymbol{x}_2$, we can compute for $\boldsymbol{x}_3$ using $k \leq 3$. For most applications, the order k seldom exceeds 6. As an example, Table 17-1 shows the pertinent formulas when the maximum order for k is chosen in finding the first six values $\boldsymbol{x}_1, \boldsymbol{x}_2, \ldots, \boldsymbol{x}_6$. After the first six values are obtained, any subsequent $\boldsymbol{x}_{n+1}$ is usually found by using the BDF of order $k \leq 6$.

TABLE 17-1. Pertinent Equations for Solving Implicit Differential-Algebraic Systems Using BDF of Maximum Order *k*

$k = 1$	$\boldsymbol{f}(\boldsymbol{x}_1, \boldsymbol{g}(\boldsymbol{x}_1), t_1) = \boldsymbol{0}$ $\boldsymbol{g}(\boldsymbol{x}_1) \triangleq -\frac{1}{h}[\alpha_0\boldsymbol{x}_1 + \alpha_1\boldsymbol{x}_0] = \dot{\boldsymbol{x}}_1$
$k = 2$	$\boldsymbol{f}(\boldsymbol{x}_2, \boldsymbol{g}(\boldsymbol{x}_2), t_2) = \boldsymbol{0}$ $\boldsymbol{g}(\boldsymbol{x}_2) \triangleq -\frac{1}{h}[\alpha_0\boldsymbol{x}_2 + \alpha_1\boldsymbol{x}_1 + \alpha_2\boldsymbol{x}_0] = \dot{\boldsymbol{x}}_2$
$k = 3$	$\boldsymbol{f}(\boldsymbol{x}_3, \boldsymbol{g}(\boldsymbol{x}_3), t_3) = \boldsymbol{0}$ $\boldsymbol{g}(\boldsymbol{x}_3) \triangleq -\frac{1}{h}[\alpha_0\boldsymbol{x}_3 + \alpha_1\boldsymbol{x}_2 + \alpha_2\boldsymbol{x}_1 + \alpha_3\boldsymbol{x}_0] = \dot{\boldsymbol{x}}_3$
$k = 4$	$\boldsymbol{f}(\boldsymbol{x}_4, \boldsymbol{g}(\boldsymbol{x}_4), t_4) = \boldsymbol{0}$ $\boldsymbol{g}(\boldsymbol{x}_4) \triangleq -\frac{1}{h}[\alpha_0\boldsymbol{x}_4 + \alpha_1\boldsymbol{x}_3 + \alpha_2\boldsymbol{x}_2 + \alpha_3\boldsymbol{x}_1 + \alpha_4\boldsymbol{x}_0] = \dot{\boldsymbol{x}}_4$
$k = 5$	$\boldsymbol{f}(\boldsymbol{x}_5, \boldsymbol{g}(\boldsymbol{x}_5), t_5) = \boldsymbol{0}$ $\boldsymbol{g}(\boldsymbol{x}_5) \triangleq -\frac{1}{h}[\alpha_0\boldsymbol{x}_5 + \alpha_1\boldsymbol{x}_4 + \alpha_2\boldsymbol{x}_3 + \alpha_3\boldsymbol{x}_2 + \alpha_4\boldsymbol{x}_1 + \alpha_5\boldsymbol{x}_0] = \dot{\boldsymbol{x}}_5$
$k = 6$	$\boldsymbol{f}(\boldsymbol{x}_6, \boldsymbol{g}(\boldsymbol{x}_6), t_6) = \boldsymbol{0}$ $\boldsymbol{g}(\boldsymbol{x}_6) \triangleq -\frac{1}{h}[\alpha_0\boldsymbol{x}_6 + \alpha_1\boldsymbol{x}_5 + \alpha_2\boldsymbol{x}_4 + \alpha_3\boldsymbol{x}_3 + \alpha_4\boldsymbol{x}_2 + \alpha_5\boldsymbol{x}_1 + \alpha_6\boldsymbol{x}_0] = \dot{\boldsymbol{x}}_6$

17-3-1 Deriving the Backward-Differentiation Formula (BDF)

Since the same α_i and k are used for each component of the *vector equation* in Eq. (17-59), it suffices to derive the BDF for the scalar case: i.e.,

$$\boxed{\dot{x}_{n+1} = -\frac{1}{h}\sum_{i=0}^{k} \alpha_i x_{n+1-i}} \tag{17-63}$$

Our strategy for deriving the BDF is similar to that used in Chapter 12 for deriving the multistep integration algorithms: we shall specify the coefficients α_i such that Eq. (17-63) gives the *exact value* of $\dot{x}(t)$ at $t = t_{n+1}$ whenever $x(t)$ is a polynomial of degree k. To do this, let $x(t)$ be the *unique* kth-degree polynomial passing through the $k + 1$ points $x_{n+1}, x_n, \ldots, x_{n+1-k}$ at $t = t_{n+1}, t_n, \ldots, t_{n+1-k}$:

$$x(t) = a_0 + a_1 t + a_2 t^2 + a_3 t^3 + \ldots + a_k t^k \tag{17-64}$$

where the coefficients $a_0, a_1, \ldots, a_k$ are the solutions of the matrix equation

$$\begin{bmatrix} 1 & t_{n+1} & t_{n+1}^2 & \cdots & t_{n+1}^k \\ 1 & t_n & t_n^2 & \cdots & t_n^k \\ 1 & t_{n-1} & t_{n-1}^2 & \cdots & t_{n-1}^k \\ \cdot & \cdot & \cdot & & \cdot \\ \cdot & \cdot & \cdot & & \cdot \\ \cdot & \cdot & \cdot & & \cdot \\ 1 & t_{n+1-k} & t_{n+1-k}^2 & \cdots & t_{n+1-k}^k \end{bmatrix} \begin{bmatrix} a_0 \\ a_1 \\ a_2 \\ \cdot \\ \cdot \\ \cdot \\ a_k \end{bmatrix} = \begin{bmatrix} x_{n+1} \\ x_n \\ x_{n-1} \\ \cdot \\ \cdot \\ \cdot \\ x_{n+1-k} \end{bmatrix} \tag{17-65}$$

To display explicitly that Eq. (17-64) passes through the $k + 1$ points $x_{n+1}, x_n, \ldots, x_{n+1-k}$, we can recast this equation into the equivalent form

$$x(t) = x_{n+1}p_{n+1}(t) + x_n p_n(t) + \ldots + x_{n+1-k}p_{n+1-k}(t) \tag{17-66}$$

where $p_{n+1}(t), p_n(t), \ldots, p_{n+1-k}(t)$ are kth-*degree polynomials* passing through the following $k + 1$ points:[3]

$$\begin{array}{llll} p_{n+1}(t_{n+1}) = 1, & p_{n+1}(t_n) = 0, & p_{n+1}(t_{n-1}) = 0, & \ldots, p_{n+1}(t_{n+1-k}) = 0 \\ p_n(t_{n+1}) = 0, & p_n(t_n) = 1, & p_n(t_{n-1}) = 0, & \ldots, p_n(t_{n+1-k}) = 0 \\ \cdot & \cdot & \cdot & \cdot \\ \cdot & \cdot & \cdot & \cdot \\ \cdot & \cdot & \cdot & \cdot \\ p_{n+1-k}(t_{n+1}) = 0, & p_{n+1-k}(t_n) = 0, & p_{n+1-k}(t_{n-1}) = 0, & \ldots, p_{n+1-k}(t_{n+1-k}) = 1 \end{array}$$

It follows from Eq. (17-66) that

$$\dot{x}(t) = x_{n+1}\dot{p}_{n+1}(t) + x_n\dot{p}_n(t) + \ldots + x_{n+1-k}\dot{p}_{n+1-k}(t) \tag{17-67}$$

Substituting $t = t_{n+1}$ in Eq. (17-67), we obtain

$$\begin{aligned} \dot{x}(t_{n+1}) &= \dot{p}_{n+1}(t_{n+1})x_{n+1} + \dot{p}_n(t_{n+1})x_n + \ldots + \dot{p}_{n+1-k}(t_{n+1})x_{n+1-k} \\ &= -\frac{1}{h}\left\{\sum_{i=0}^{k} [-h\dot{p}_{n+1-i}(t_{n+1})x_{n+1-i}]\right\} \end{aligned} \tag{17-68}$$

[3]The coefficients for specifying each kth-degree polynomial $p_j(t)$ can be found by solving a matrix equation similar to Eq. (17-65).

where $h \triangleq t_{n+1} - t_n$ is the present step size. If we define the *constant*

$$\alpha_i \triangleq -h\dot{p}_{n+1-i}(t_{n+1}) \tag{17-69}$$

then Eq. (17-68) simplifies to

$$\dot{x}(t_{n+1}) = -\frac{1}{h}\sum_{i=0}^{k}\alpha_i x_{n+1-i} \tag{17-70}$$

Equation (17-70) gives the *exact* value of $\dot{x}(t)$ at $t = t_{n+1}$ in terms of the present value x_{n+1} and k past values $x_n, x_{n-1}, \ldots, x_{n+1-k}$, *provided* $x(t)$ is a polynomial of degree less than or equal to k.[4] Now, if $x(t)$ is an arbitrary function, we can apply the Weierstrass theorem and approximate it arbitrarily closely by a polynomial of sufficiently high degree k. Hence, if we let $\dot{x}_{n+1}$ denote the *approximate value* of $\dot{x}(t_{n+1})$, we obtain the kth-*order backward-differentiation formula* given in Eq. (17-63).

The coefficient α_i given by Eq. (17-69) depends on the time $t = t_{n+1}$ and must be reevaluated for each new time $t_{n+2}, t_{n+3}, \ldots$, etc. Moreover, each polynomial $p_{n+1-i}(t)$ requires the solution of a $(k+1) \times (k+1)$ matrix equation similar to Eq. (17-65). Hence, computing α_i from Eq. (17-69) would represent a rather time-consuming process. Let us now derive a much more efficient method for computing α_i for the BDF

$$\dot{x}_{n+1} = -\frac{1}{h}[\alpha_0 x_{n+1} + \alpha_1 x_n + \alpha_2 x_{n-1} + \ldots + \alpha_k x_{n+1-k}] \tag{17-71}$$

We shall choose $\alpha_0, \alpha_1, \ldots, \alpha_k$ such that Eq. (17-71) gives the *exact* value $\dot{x}_{n+1} = \dot{x}(t_{n+1})$ whenever $x(t)$ is *any* polynomial of degree $m = 0, 1, 2, \ldots, k$. In particular, we pick the following $k+1$ polynomials:

$$m = 0: \quad x(t) = \frac{1}{h^0}[t_{n+1} - t]^0, \qquad \dot{x}(t) = 0 \tag{17-72a}$$

$$m = 1: \quad x(t) = \frac{1}{h^1}[t_{n+1} - t]^1, \qquad \dot{x}(t) = -\frac{1}{h} \tag{17-72b}$$

$$m = 2: \quad x(t) = \frac{1}{h^2}[t_{n+1} - t]^2, \qquad \dot{x}(t) = -\frac{2}{h^2}[t_{n+1} - t] \tag{17-72c}$$

$$\vdots$$

$$m = k: \quad x(t) = \frac{1}{h^k}[t_{n+1} - t]^k, \qquad \dot{x}(t) = -\frac{k}{h^k}[t_{n+1} - t]^{k-1} \tag{17-72k+1}$$

If we let $x(t_j) = x_j$ and $\dot{x}(t_j) = \dot{x}_j$ in Eq. (17-72), where $j = n+1, n, \ldots, n+1-k$, then Eq. (17-71) assumes the following form:

$$m = 0: \qquad 0 = -\frac{1}{h}[\alpha_0 + \alpha_1 + \alpha_2 + \ldots + \alpha_k] \tag{17-73a}$$

$$m = 1: \quad -\frac{1}{h} = -\frac{1}{h}\left\{\alpha_0(0) + \alpha_1\left[\frac{t_{n+1} - t_n}{h}\right] + \alpha_2\left[\frac{t_{n+1} - t_{n-1}}{h}\right] + \ldots + \alpha_k\left[\frac{t_{n+1} - t_{n+1-k}}{h}\right]\right\} \tag{17-73b}$$

[4]Observe that any polynomial of degree $m < k$ can be considered as a kth-degree polynomial with zero coefficients multiplying the missing higher-degree terms.

$$m = 2: \qquad 0 = -\frac{1}{h}\left\{\alpha_0(0) + \alpha_1\left[\frac{t_{n+1} - t_n}{h}\right]^2 + \alpha_2\left[\frac{t_{n+1} - t_{n-1}}{h}\right]^2 + \cdots + \alpha_k\left[\frac{t_{n+1} - t_{n+1-k}}{h}\right]^2\right\} \tag{17-73c}$$

$$\vdots$$

$$m = k: \qquad 0 = -\frac{1}{h}\left\{\alpha_0(0) + \alpha_1\left[\frac{t_{n+1} - t_n}{h}\right]^k + \alpha_2\left[\frac{t_{n+1} - t_{n-1}}{h}\right]^k + \cdots + \alpha_k\left[\frac{t_{n+1} - t_{n+1-k}}{h}\right]^k\right\} \tag{17-73k+1}$$

Equations (17-73) constitute a system of $k + 1$ equations in the $k + 1$ unknown coefficients $\alpha_0, \alpha_1, \ldots, \alpha_k$. Hence, these coefficients can be determined by solving the following $(k + 1) \times (k + 1)$ matrix equation:

$$\begin{bmatrix} 1 & 1 & 1 & \cdots & 1 \\ 0 & \left[\frac{t_{n+1} - t_n}{h}\right] & \left[\frac{t_{n+1} - t_{n-1}}{h}\right] & \cdots & \left[\frac{t_{n+1} - t_{n+1-k}}{h}\right] \\ 0 & \left[\frac{t_{n+1} - t_n}{h}\right]^2 & \left[\frac{t_{n+1} - t_{n-1}}{h}\right]^2 & \cdots & \left[\frac{t_{n+1} - t_{n+1-k}}{h}\right]^2 \\ \vdots & \vdots & \vdots & \cdots & \vdots \\ 0 & \left[\frac{t_{n+1} - t_n}{h}\right]^k & \left[\frac{t_{n+1} - t_{n-1}}{h}\right]^k & \cdots & \left[\frac{t_{n+1} - t_{n+1-k}}{h}\right]^k \end{bmatrix} \begin{bmatrix} \alpha_0 \\ \alpha_1 \\ \alpha_2 \\ \vdots \\ \alpha_k \end{bmatrix} = \begin{bmatrix} 0 \\ 1 \\ 0 \\ \vdots \\ 0 \end{bmatrix} \tag{17-74a}$$

Observe again from Eq. (17-74a) that the coefficients $\alpha_0, \alpha_1, \ldots, \alpha_k$ for the BDF depend on the *nonuniform* discrete times $t_{n+1}, t_n, \ldots, t_{n+1-k}$ and must be *reevaluated* at each new time step $t_{n+2}, t_{n+3}, \ldots$, etc. However, if the discrete times are all uniformly spaced, i.e., $t_{n+1} - t_n = t_n - t_{n-1} = \ldots = t_{n-k+2} - t_{n-k+1} = h$, then Eq. (17-74a) reduces to

$$\begin{bmatrix} 1 & 1 & 1 & 1 & \cdots & 1 \\ 0 & 1 & 2 & 3 & \cdots & k \\ 0 & 1 & 2^2 & 3^2 & \cdots & k^2 \\ \vdots & \vdots & \vdots & \vdots & \cdots & \vdots \\ 0 & 1 & 2^k & 3^k & \cdots & k^k \end{bmatrix} \begin{bmatrix} \alpha_0 \\ \alpha_1 \\ \alpha_2 \\ \vdots \\ \alpha_k \end{bmatrix} = \begin{bmatrix} 0 \\ 1 \\ 0 \\ \vdots \\ 0 \end{bmatrix} \tag{17-74b}$$

In this case, it is clear that the coefficients $\alpha_0, \alpha_1, \ldots, \alpha_k$ do not depend on the discrete times. Nor do they depend on the step size h. Hence, the coefficients $\alpha_0, \alpha_1, \ldots, \alpha_k$ in this case do not have to be reevaluated at each new time step $t_{n+2}, t_{n+3}, \ldots$, etc.

17-3-2 Predicting the Initial Guess for Newton–Raphson Iteration

Let us return to the system of nonlinear algebraic equations given by Eq. (17-61). Recall that this iteration converges whenever the initial guess $\boldsymbol{x}_{n+1}^{(0)}$ is sufficiently close to the solution $\boldsymbol{x}_{n+1}$ of Eq. (17-60). Although the initial guess is usually arbitrarily chosen when no advanced knowledge of a neighborhood of the solution is known, in the present situation we shall take advantage of the past values of $\boldsymbol{x}_{n+1}$ to *predict* a good initial guess. Again, to simplify notation, we shall consider the scalar case of Eq. (17-60); i.e.,

$$f(x_{n+1}, g(x_{n+1}), t_{n+1}) = 0 \tag{17-75a}$$

where

$$g(x_{n+1}) \triangleq -\frac{1}{h}[\alpha_0 x_{n+1} + \alpha_1 x_n + \alpha_2 x_{n-1} + \ldots + \alpha_k x_{n+1-k}] \tag{17-75b}$$

represents a kth-order approximation to $\dot{x}(t_{n+1})$. Let $x(t)$ denote the *unique* kth-degree polynomial passing through $k+1$ points $x_{n+1}, x_n, \ldots, x_{n+1-k}$, where x_{n+1} is the exact solution of Eq. (17-75a). Let $x^P(t)$ denote the *unique* kth-degree polynomial passing through $k+1$ points $x_n, x_{n-1}, \ldots, x_{n-k}$. A typical curve representing the polynomial $x(t)$ (solid curve) and a typical curve representing the polynomial $x^P(t)$ (dashed curve) are shown in Fig. 17-3. Since each polynomial is of degree k, we can only force it to pass through $k+1$ *prescribed* points. Hence, we expect $x_{n+1}^P \triangleq x^P(t_{n+1}) \neq x_{n+1}$. However, since both polynomials pass through k *common* points, $x_{n+1-k}, x_{n+2-k}, \ldots, x_{n-1}, x_n$, we can minimize the error $x_{n+1} - x_{n+1}^P$ by decreasing the step size $h \triangleq t_{n+1} - t_n$. In other words, rather than choosing an arbitrary initial guess $x_{n+1}^{(0)}$, we might as well choose

$$x_{n+1}^{(0)} = x_{n+1}^P \triangleq x^P(t_{n+1}) \tag{17-76}$$

where x_{n+1}^P is called the *predicted* value of x_{n+1}. With this initial guess, the Newton–Raphson algorithm will usually converge in a few iterations. Now the unique kth-degree polynomial $x^P(t)$ that passes through the $k+1$ points $x_n, x_{n-1}, \ldots, x_{n-k}$ can

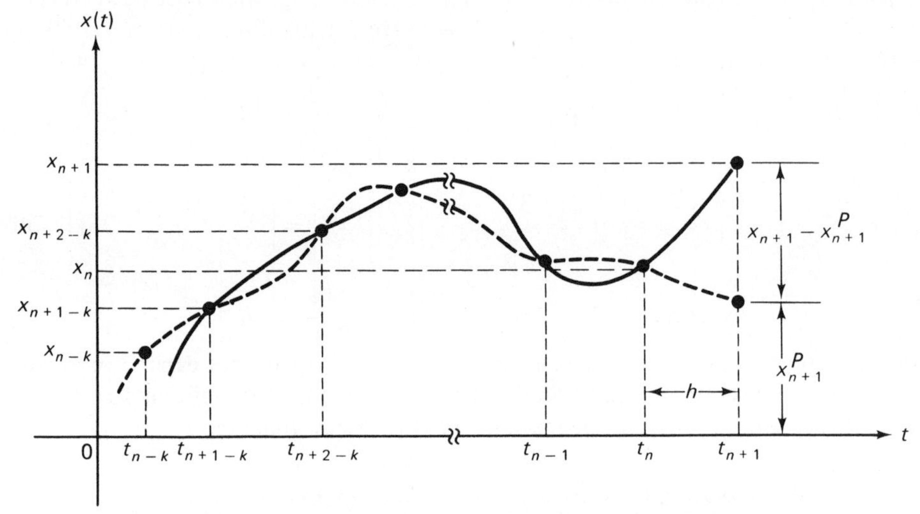

Fig. 17-3. Two kth-degree polynomials passing through $k+1$ prescribed points. The polynomial $x(t)$ (solid curve) passes through $x_{n+1}, x_n, \ldots, x_{n+1-k}$; the polynomial $x^P(t)$ (dashed curve) passes through $x_n, x_{n-1}, \ldots, x_{n-k}$.

be written in a form similar to Eq. (17-66):

$$x^P(t) = x_n p_n(t) + x_{n-1} p_{n-1}(t) + \ldots + x_{n-k} p_{n-k}(t) \quad (17\text{-}77)$$

where

$$\begin{array}{lllll} p_n(t_n) = 1, & p_n(t_{n-1}) = 0, & p_n(t_{n-2}) = 0, & \ldots, & p_n(t_{n-k}) = 0 \\ p_{n-1}(t_n) = 0, & p_{n-1}(t_{n-1}) = 1, & p_{n-1}(t_{n-2}) = 0, & \ldots, & p_{n-1}(t_{n-k}) = 0 \\ \cdot & \cdot & \cdot & \cdots & \cdot \\ \cdot & \cdot & \cdot & \cdots & \cdot \\ \cdot & \cdot & \cdot & \cdots & \cdot \\ p_{n-k}(t_n) = 0, & p_{n-k}(t_{n-1}) = 0, & p_{n-k}(t_{n-2}) = 0, & \ldots, & p_{n-k}(t_{n-k}) = 1 \end{array}$$

If we let

$$\gamma_i \triangleq p_{n+1-i}(t_{n+1})$$

then at $t = t_{n+1}$, Eq. (17-77) assumes the form

$$\boxed{x_{n+1}^P = \sum_{i=1}^{k+1} \gamma_i x_{n+1-i}} \quad (17\text{-}78)$$

The coefficients $\gamma_1, \gamma_2, \ldots, \gamma_{k+1}$ can be determined efficiently by demanding that Eq. (17-78) give the exact value $x_{n+1}^P = x^P(t_{n+1})$ whenever $x^P(t)$ is a polynomial of degree $m = 0, 1, 2, \ldots, k$. In particular, suppose that we pick the following $k + 1$ polynomials:

$$m = 0: \quad x^P(t) = \frac{1}{h^0}[t_{n+1} - t]^0 \quad (17\text{-}79a)$$

$$m = 1: \quad x^P(t) = \frac{1}{h^1}[t_{n+1} - t]^1 \quad (17\text{-}79b)$$

$$m = 2: \quad x^P(t) = \frac{1}{h^2}[t_{n+1} - t]^2 \quad (17\text{-}79c)$$

$$\vdots \qquad \vdots$$

$$m = k: \quad x^P(t) = \frac{1}{h^k}[t_{n+1} - t]^k \quad (17\text{-}79k{+}1)$$

where $h \triangleq t_{n+1} - t_n$. If we let $x^P(t_j) = x_j^P$ in Eq. (17-79), where $j = n + 1, n, \ldots, n - k$, then Eq. (17-78) assumes the form

$$m=0: \quad 1 = \gamma_1 + \gamma_2 + \ldots + \gamma_{k+1} \quad (17\text{-}80a)$$

$$m=1: \quad 0 = \gamma_1\left[\frac{t_{n+1} - t_n}{h}\right] + \gamma_2\left[\frac{t_{n+1} - t_{n-1}}{h}\right] + \ldots + \gamma_{k+1}\left[\frac{t_{n+1} - t_{n-k}}{h}\right] \quad (17\text{-}80b)$$

$$m=2: \quad 0 = \gamma_1\left[\frac{t_{n+1} - t_n}{h}\right]^2 + \gamma_2\left[\frac{t_{n+1} - t_{n-1}}{h}\right]^2 + \ldots + \gamma_{k+1}\left[\frac{t_{n+1} - t_{n-k}}{h}\right]^2 \quad (17\text{-}80c)$$

$$\vdots \qquad \vdots$$

$$m=k: \quad 0 = \gamma_1\left[\frac{t_{n+1} - t_n}{h}\right]^k + \gamma_2\left[\frac{t_{n+1} - t_{n-1}}{h}\right]^k + \ldots + \gamma_{k+1}\left[\frac{t_{n+1} - t_{n-k}}{h}\right]^k \quad (17\text{-}80k{+}1)$$

Equation (17-80) constitutes a system of $k+1$ equations in the $k+1$ coefficients $\gamma_1, \gamma_2, \ldots, \gamma_{k+1}$. Hence, these coefficients can be determined by solving the following $(k+1) \times (k+1)$ matrix equation:

$$\begin{bmatrix} 1 & 1 & \cdots & 1 \\ \left[\frac{t_{n+1}-t_n}{h}\right] & \left[\frac{t_{n+1}-t_{n-1}}{h}\right] & \cdots & \left[\frac{t_{n+1}-t_{n-k}}{h}\right] \\ \left[\frac{t_{n+1}-t_n}{h}\right]^2 & \left[\frac{t_{n+1}-t_{n-1}}{h}\right]^2 & \cdots & \left[\frac{t_{n+1}-t_{n-k}}{h}\right]^2 \\ \vdots & \vdots & & \vdots \\ \left[\frac{t_{n+1}-t_n}{h}\right]^k & \left[\frac{t_{n+1}-t_{n-1}}{h}\right]^k & \cdots & \left[\frac{t_{n+1}-t_{n-k}}{h}\right]^k \end{bmatrix} \begin{bmatrix} \gamma_1 \\ \gamma_2 \\ \gamma_3 \\ \vdots \\ \gamma_{k+1} \end{bmatrix} = \begin{bmatrix} 1 \\ 0 \\ 0 \\ \vdots \\ 0 \end{bmatrix} \tag{17-81}$$

An examination of both Eqs. (17-74a) and (17-81) shows that columns $2, 3, \ldots, k+1$ of the matrix in Eq. (17-74a) are identical to columns $1, 2, 3, \ldots, k$ of the matrix in Eq. (17-81). Both belong to a general class of matrices called *Vandermonde* matrices. By exploiting the special properties of Vandermonde matrices, it is possible to relate the two sets of coefficients $\alpha_0, \alpha_1, \ldots, \alpha_k$ and $\gamma_1, \gamma_2, \ldots, \gamma_{k+1}$. Hence, even though these coefficients must be reevaluated after each time step, a highly efficient algorithm can be developed for determining the new coefficients at t_{n+1} from the coefficients at t_n in $10(k+1)$ operations. The interested reader is referred to [1] for the details of this algorithm. See also [13] for a related algorithm.

17-3-3 Local Truncation Error of Backward-Differentiation Formula

The kth-order BDF defined by Eq. (17-59) is exact only if the function $\dot{x}(t)$ is a *polynomial* of degree $m \leq k$. For any other function, we have $\dot{x}(t_{n+1}) \neq \dot{x}_{n+1}$, and it would be important to know the size of the error.

If we define the *local truncation error* ϵ_T at $t = t_{n+1}$ by

$$\epsilon_T \triangleq h[\dot{x}(t_{n+1}) - \dot{x}_{n+1}] \tag{17-82}$$

a remarkable theorem results.

Theorem 17-1. *The local truncation error of the kth-order backward-differentiation formula*

$$\dot{x}_{n+1} = -\frac{1}{h}\sum_{i=0}^{k} \alpha_i x_{n+1-i}$$

is given by

$$\epsilon_T = E_k = O(h^{k+2}) \tag{17-83}$$

where

$$\boxed{E_k \triangleq \left[\frac{h}{t_{n+1} - t_{n-k}}\right](x_{n+1} - x^P_{n+1})} \tag{17-84}$$

and x^P_{n+1} is the predicted *value of x_{n+1} as defined in Eq. (17-78). The term $O(h^{k+2})$ denotes higher-order terms in the* step size $h \triangleq t_{n+1} - t_n$ *of degrees greater than or equal to $k+2$.*

For a proof of this theorem, see reference 1. This theorem is remarkable because, unlike the local truncation error for other numerical integration algorithms, Eq. (17-84) is extremely easy to calculate. Moreover, observe that *the same* x^P_{n+1} is used for the Newton–Raphson iteration and for computing the error!

17-3-4 Backward-Differentiation Formula in Terms of Backward Differences

To implement the BDF given by Eq. (17-63) and the prediction formula given by Eq. (17-78) in a computer, it would be necessary to store the past values $\mathbf{x}_n, \mathbf{x}_{n-1}, \ldots, \mathbf{x}_{n-k}$. It has been found in [1] that to reduce *round-off errors* it is advantageous to store the *backward differences*.[5]

$$\boxed{\Delta x_{n-i} \triangleq x_{n+1-i} - x_{n-i}, \qquad i = 1, 2, \ldots, k} \tag{17-85}$$

In terms of Δx_{n-i}, Eqs. (17-63) and (17-78) assume the following equivalent forms:

1. *BDF in terms of backward differences*

$$\dot{x}_{n+1} = -\frac{1}{h}\sum_{i=0}^{k-1} \hat{\alpha}_i \,\Delta x_{n-i} \tag{17-86}$$

where

$$\hat{\alpha}_i \triangleq \sum_{j=0}^{i} \alpha_j \tag{17-87}$$

2. *Initial guess prediction in terms of backward differences*

$$x^P_{n+1} = x_n + \sum_{i=1}^{k} \hat{\gamma}_i \,\Delta x_{n-i} \tag{17-88}$$

where

$$\hat{\gamma}_i \triangleq -1 + \sum_{j=1}^{i} \gamma_j \tag{17-89}$$

To show that Eqs. (17-86) and (17-87) are equivalent to Eq. (17-63), let us substitute Eqs. (17-85) and (17-87) into Eq. (17-86) to obtain

$$\begin{aligned}
\dot{x}_{n+1} &= -\frac{1}{h}\{\hat{\alpha}_0 \,\Delta x_n + \hat{\alpha}_1 \,\Delta x_{n-1} + \hat{\alpha}_2 \,\Delta x_{n-2} + \ldots + \hat{\alpha}_{k-1} \,\Delta x_{n+1-k}\} \\
&= -\frac{1}{h}\{\alpha_0(x_{n+1} - x_n) + (\alpha_0 + \alpha_1)(x_n - x_{n-1}) \\
&\quad + (\alpha_0 + \alpha_1 + \alpha_2)(x_{n-1} - x_{n-2}) + \ldots \\
&\quad + (\alpha_0 + \alpha_1 + \ldots + \alpha_{k-1})(x_{n+2-k} - x_{n+1-k})\} \\
&= -\frac{1}{h}\{(\alpha_0 x_{n+1} + \alpha_1 x_n + \alpha_2 x_{n-1} + \ldots + \alpha_{k-1} x_{n+2-k}) \\
&\quad - (\alpha_0 + \alpha_1 + \ldots + \alpha_{k-1}) x_{n+1-k}\}
\end{aligned} \tag{17-90}$$

[5]Since the material in this section is based entirely on reference 1, we have adopted the backward-difference notation used in this paper, as opposed to that defined earlier in Section 11-7-1, to facilitate further study of the paper.

Now let us invoke the exactness constraint given by Eq. (17-73a) to obtain

$$\alpha_0 + \alpha_1 + \ldots + \alpha_{k-1} = -\alpha_k \tag{17-91}$$

Substituting Eq. (17-91) into Eq. (17-90), we obtain

$$\dot{x}_{n+1} = -\frac{1}{h}\{\alpha_0 x_{n+1} + \alpha_1 x_n + \alpha_2 x_{n-1} + \ldots + \alpha_{k-1} x_{n+2-k} + \alpha_k x_{n+1-k}\} \tag{17-92}$$

which is seen to be identical to Eq. (17-63).

To show that Eqs. (17-88) and (17-89) are equivalent to Eq. (17-78), let us substitute Eqs. (17-85) and (17-89) into Eq. (17-88) to obtain

$$\begin{aligned} x_{n+1}^P &= x_n + \{\hat{\gamma}_1 \, \Delta x_{n-1} + \hat{\gamma}_2 \, \Delta x_{n-2} + \ldots + \hat{\gamma}_k \, \Delta x_{n-k}\} \\ &= x_n + \{(-1 + \gamma_1)(x_n - x_{n-1}) + (-1 + \gamma_1 + \gamma_2)(x_{n-1} - x_{n-2}) \\ &\quad + \ldots + (-1 + \gamma_1 + \gamma_2 + \ldots + \gamma_k)(x_{n+1-k} - x_{n-k})\} \\ &= (\gamma_1 x_n + \gamma_2 x_{n-1} + \ldots + \gamma_{k-1} x_{n+2-k} + \gamma_k x_{n+1-k}) \\ &\quad - (-1 + \gamma_1 + \gamma_2 + \ldots + \gamma_k) x_{n-k} \end{aligned} \tag{17-93}$$

Now let us invoke the exactness constraint given by Eq. (17-80a) to obtain

$$-1 + \gamma_1 + \gamma_2 + \ldots + \gamma_k = -\gamma_{k+1} \tag{17-94}$$

Substituting Eq. (17-94) into Eq. (17-93), we obtain

$$x_{n+1}^P = \gamma_1 x_n + \gamma_2 x_{n-1} + \ldots + \gamma_k x_{n+1-k} + \gamma_{k+1} x_{n-k} \tag{17-95}$$

which is seen to be identical to Eq. (17-78).

17-3-5 Algorithm for Variable Step-Size Variable-Order Backward-Differentiation Formula

We can now combine the preceding results to arrive at the following highly efficient algorithm for solving any system of implicit differential-algebraic equations:

BDF 1. Set $t_{n+1} = t_n + h$.

BDF 2. Determine the coefficients $\gamma_1, \gamma_2, \ldots, \gamma_{k+1}$ at $t = t_{n+1}$ and compute the *predicted value*

$$\mathbf{x}_{n+1}^P = \sum_{i=1}^{k+1} \gamma_i \mathbf{x}_{n+1-i}$$

BDF 3. Determine the coefficients $\alpha_0, \alpha_1, \ldots, \alpha_k$ at $t = t_{n+1}$ and substitute the BDF

$$\dot{\mathbf{x}}_{n+1} = -\frac{1}{h}\sum_{i=0}^{k} \alpha_i \mathbf{x}_{n+1-i} \triangleq \mathbf{g}(\mathbf{x}_{n+1})$$

in place of $\dot{\mathbf{x}}(t_{n+1})$ in the equation

$$\mathbf{f}(\mathbf{x}_{n+1}, \dot{\mathbf{x}}(t_{n+1}), t_{n+1}) = \mathbf{0}$$

BDF 4. With $\mathbf{x}_{n+1}^P$ as the *initial guess*, solve

$$\mathbf{f}(\mathbf{x}_{n+1}, \mathbf{g}(\mathbf{x}_{n+1}), t_{n+1}) = \mathbf{0}$$

for $\mathbf{x}_{n+1}$ using the Newton–Raphson algorithm.

BDF 5. Compute the error term

$$E_k \triangleq \left[\frac{h}{t_{n+1} - t_{n-k}}\right] \| \boldsymbol{x}_{n+1} - \boldsymbol{x}^P_{n+1} \|$$

BDF 6. Compare E_k with the user-prescribed error-control criterion.[6] Select a *new* time *step* h and new *order* k such that h is the maximum value for which E_k is commensurate with the user-specified maximum error. If E_k is too large, reject the old step.

BDF 7. Set $n = n + 1$ and go to BDF 1.

17-4 GENERALIZED TABLEAU APPROACH WITH VARIABLE ORDER AND VARIABLE STEP SIZE

The tableau approach presented in Section 17-2 shows that it is advantageous, from a computational point of view, to solve the original set of "unreduced" circuit equations made up of KCL, KVL, and element constitutive relations. This is because the matrices associated with any "reduced" system—such as nodal, loop, or hybrid equations—are much less sparse than the tableau matrix. In Section 17-2, each capacitor and inductor is first replaced by a *resistive* associated discrete circuit model. This somewhat *ad hoc procedure* automatically constrains the variables in the equations to be the *terminal voltage* and *terminal current* of the circuit elements. However, it is sometimes desirable to include *capacitor charges* and *inductor flux linkages* as additional variables in the equations.[7] Hence, in the most general case, the *unreduced* system of equations for a circuit containing n nodes and b composite branches would consist of the following:[8]

KCL ($n - 1$ equations)		$\boldsymbol{A i} - \boldsymbol{A J} = \boldsymbol{0}$	(17-96a)
KVL (b equations)		$\boldsymbol{v} - \boldsymbol{A}^t \boldsymbol{v}_n - \boldsymbol{E} = \boldsymbol{0}$	(17-96b)
Constitutive relations:			
b_R resistors	(b_R equations)	$\boldsymbol{f}_R(\boldsymbol{v}_R, \boldsymbol{i}_R) = \boldsymbol{0}$	(17-96c)
b_C capacitors	(b_C equations)	$\boldsymbol{f}_C(\boldsymbol{v}_C, \boldsymbol{q}_C) = \boldsymbol{0}$	(17-96d)
	(b_C equations)	$\boldsymbol{i}_C - \dot{\boldsymbol{q}}_C = \boldsymbol{0}$	(17-96e)
b_L inductors	(b_L equations)	$\boldsymbol{f}_L(\boldsymbol{i}_L, \boldsymbol{\phi}_L) = \boldsymbol{0}$	(17-96f)
	(b_L equations)	$\boldsymbol{v}_L - \dot{\boldsymbol{\phi}}_L = \boldsymbol{0}$	(17-96g)

[6]There is no optimum error-control criterion that applies for all problems. However, a satisfactory, albeit somewhat ad hoc, criterion is given in reference 1.

[7]Recall from Section 10-5 that, to minimize error propagation in the numerical solution of state equations, it is desirable to choose capacitor charges and inductor flux linkages as the state variables.

[8]Since we are dealing with composite branches here, all sources are included in the generalized branch, and n is the number of nodes in the "composite" graph. This is reflected by the terms $\boldsymbol{A J}$ and $\boldsymbol{E}$ in Eqs. (17-96a) and (17-96b). Observe also that $\boldsymbol{i} = [\boldsymbol{i}_R \;\; \boldsymbol{i}_C \;\; \boldsymbol{i}_L]^t$ and $\boldsymbol{v} = [\boldsymbol{v}_R \;\; \boldsymbol{v}_C \;\; \boldsymbol{v}_L]^t$.

Observe that Eq. (17-96) contains a total of

$$N \triangleq (n-1) + b + (b_R + b_C + b_L) + (b_C + b_L) = (n-1) + 2b + b_C + b_L$$

equations involving the following variables:

NAME	VARIABLE	DIMENSION
Node-to-datum voltage vector	$\boldsymbol{v}_n$	$(n-1) \times 1$
Resistor voltage vector	$\boldsymbol{v}_R$	$b_R \times 1$
Resistor current vector	$\boldsymbol{i}_R$	$b_R \times 1$
Capacitor voltage vector	$\boldsymbol{v}_C$	$b_C \times 1$
Capacitor current vector	$\boldsymbol{i}_C$	$b_C \times 1$
Capacitor charge vector	$\boldsymbol{q}_C$	$b_C \times 1$
Inductor voltage vector	$\boldsymbol{v}_L$	$b_L \times 1$
Inductor current vector	$\boldsymbol{i}_L$	$b_L \times 1$
Inductor flux-linkage vector	$\boldsymbol{\phi}_L$	$b_L \times 1$

Observe that there are a total of

$$(n-1) + 2(b_R + b_C + b_L) + b_C + b_L = (n-1) + 2b + b_C + b_L = N$$

variables in Eq. (17-96). Since we have the same number of equations as there are variables, and since all equations in Eq. (17-96) are independent, we have a system of N well-defined *implicit algebraic-differential equations* of the form

$$\boldsymbol{f}(\boldsymbol{x}, \dot{\boldsymbol{x}}, t) = \boldsymbol{0} \tag{17-97}$$

where

$$\boldsymbol{x} \triangleq [\boldsymbol{v}_n \quad \boldsymbol{v}_R \quad \boldsymbol{i}_R \quad \boldsymbol{v}_C \quad \boldsymbol{i}_C \quad \boldsymbol{q}_C \quad \boldsymbol{v}_L \quad \boldsymbol{i}_L \quad \boldsymbol{\phi}_L]^t \tag{17-98}$$

Observe that Eqs. (17-97) and Eq. (17-55) are identical; hence, the *variable step-size, variable-order algorithm* presented in Section 17-3 is directly applicable here. In particular, the solution $\boldsymbol{x}(t)$ at any time $t = t_{n+1}$ is obtained by the Newton–Raphson iteration given in Eq. (17-61). In practice, we never invert the Jacobian matrix $\boldsymbol{J}$. Instead, we recast Eq. (17-61) into the following equivalent system of linear equations,

$$\boldsymbol{J}(\boldsymbol{x}_{n+1}^{(j)})\boldsymbol{y}_{n+1}^{(j+1)} = \boldsymbol{u}_{n+1}^{(j)} \tag{17-99}$$

where

$$\boldsymbol{y}_{n+1}^{(j+1)} \triangleq \boldsymbol{x}_{n+1}^{(j+1)} - \boldsymbol{x}_{n+1}^{(j)} \tag{17-100}$$

$$\boldsymbol{u}_{n+1}^{(j)} \triangleq -\boldsymbol{f}(\boldsymbol{x}_{n+1}^{(j)}, \boldsymbol{g}(\boldsymbol{x}_{n+1}^{(j)}), t_{n+1}) \tag{17-101}$$

and solve for $\boldsymbol{y}_{n+1}^{(j+1)}$. The Jacobian matrix $\boldsymbol{J}(\boldsymbol{x}_{n+1}^{(j)})$ can be interpreted as a *generalized tableau matrix* and is *extremely sparse*. Hence, once again we have a tableau approach, except that the solution now includes capacitor charges and inductor flux linkages in addition to the voltage and current variables obtained by the *ad hoc tableau approach* presented in Section 17-2. In fact, the generalized tableau approach could include any number of additional variables

$$\boldsymbol{z} \triangleq [z_1 \quad z_2 \quad \cdots \quad z_p]^t$$

where

$$\dot{\mathbf{z}} = \mathbf{w}(\mathbf{x}, \mathbf{z}) \tag{17-102}$$

because we can combine Eqs. (17-97) and (17-102) into the following equivalent *augmented* system of *implicit algebraic equations:*

$$\mathbf{f}(\mathbf{x}, \dot{\mathbf{x}}, t) = \mathbf{0} \tag{17-103}$$

$$\dot{\mathbf{z}} - \mathbf{w}(\mathbf{x}, \mathbf{z}) = \mathbf{0} \tag{17-104}$$

The flexibility of the generalized tableau approach allows one to perform computer optimization in a very efficient manner. The interested reader should consult reference 2.

17-5 ALGORITHM FOR DETERMINING STEADY-STATE PERIODIC SOLUTIONS OF NONLINEAR CIRCUITS WITH PERIODIC INPUTS

In the analysis of many communication circuits driven by periodic inputs, one is interested only in determining the steady-state periodic response. If conventional numerical techniques are employed, one has no alternative but to integrate the differential equations over a sufficiently long interval of time for the transient waveform to die out. This procedure is satisfactory if the transient decays rapidly. However, for many lightly damped or high Q circuits, the transient will decay very slowly, and it becomes prohibitively expensive for the computer to integrate over such a long transient regime. Before we attempt to develop an algorithm for obviating this difficulty, it is instructive to examine a simple example, which provides the clue for overcoming the problem. Consider the *RLC* circuit shown in Fig. 17-4. The state equations are given by

$$\dot{v}_C = -\frac{i_L}{C} \tag{17-105a}$$

$$\dot{i}_L = \frac{1}{L}(v_C - Ri_L - V_s \cos \omega t) \tag{17-105b}$$

The complete response due to initial states $v_C(0)$ and $i_L(0)$ is given by

$$v_C(t) = k_1 e^{-p_1 t} + k_2 e^{-p_2 t} + V_C \cos(\omega t + \theta) \tag{17-106a}$$

$$i_L(t) = (p_1 C)k_1 e^{-p_1 t} + (p_2 C)k_2 e^{-p_2 t} - (\omega C)V_C \sin(\omega t + \theta) \tag{17-106b}$$

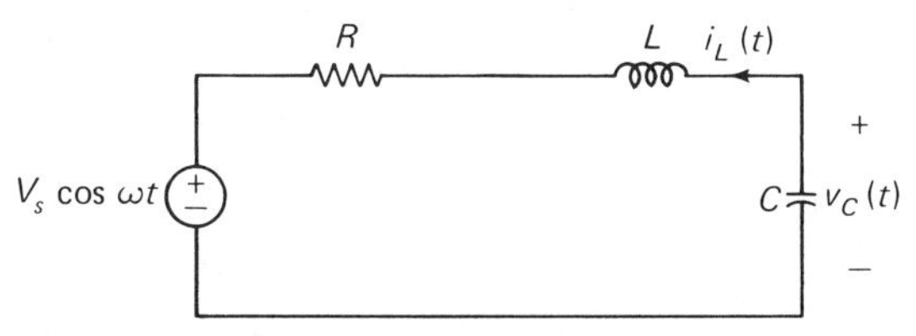

Fig. 17-4. Simple circuit for illustrating how the transient component may be suppressed by appropriate choice of initial conditions.

where

$$V_C \triangleq \frac{V_s}{[(1-\omega^2 LC)^2 + (\omega RC)^2]^{1/2}}, \qquad \theta \triangleq -\tan^{-1}\left[\frac{\omega RC}{1-\omega^2 LC}\right] \tag{17-106c}$$

and where $s_1 = -p_1$ and $s_2 = -p_2$ are the natural frequencies. The two arbitrary constants k_1 and k_2 are determined by the initial states $v_C(0)$ and $i_L(0)$:

$$k_1 + k_2 = v_C(0) - V_C \cos\theta \tag{17-107a}$$

$$(p_1 C)k_1 + (p_2 C)k_2 = i_L(0) + (\omega C)V_C \sin\theta \tag{17-107b}$$

It follows from Eq. (17-107) that $k_1 = k_2 = 0$ if, and only if,

$$v_C(0) = V_C \cos\theta \tag{17-108a}$$

$$i_L(0) = -(\omega C)V_C \sin\theta \tag{17-108b}$$

With this choice of initial states, the *complete solution* is given by $v_C(t) = V_C \cos(\omega t + \theta)$ and $i_L(t) = -(\omega C)V_C \sin(\omega t + \theta)$, both of which are periodic waveforms of period $T = 2\pi/\omega$. Hence, we have shown that it is possible to suppress the transient component if the initial states are chosen as specified by Eq. (17-108). Our objective in this section will be to generalize this concept and devise an algorithm for determining the appropriate *initial-state* vector $\boldsymbol{x}(0) \triangleq \boldsymbol{x}_0$ such that the complete solution to the state equations

$$\dot{\boldsymbol{x}} = \boldsymbol{f}(\boldsymbol{x}, t) \tag{17-109}$$

has no transient component. Throughout this section, we assume that all inputs are periodic of *period T*; hence,

$$\boldsymbol{f}(\boldsymbol{x}, t) = \boldsymbol{f}(\boldsymbol{x}, t + T) \tag{17-110}$$

Moreover, we assume that the circuit has a unique and continuously differentiable periodic solution $\boldsymbol{x}(t)$ of the same period T.[9]

17-5-1 Formulating the Fixed-Point Problem

Let $\boldsymbol{x}(t)$ be the solution to Eq. (17-109) with initial state $\boldsymbol{x}(0) = \boldsymbol{x}_0$. Then integrating both sides of Eq. (17-109) from time 0 to time t, we obtain

$$\boldsymbol{x}(t) = \int_0^t \boldsymbol{f}(\boldsymbol{x}(t), t)\, dt + \boldsymbol{x}_0 \triangleq \boldsymbol{x}(t, \boldsymbol{x}_0) \tag{17-111}$$

Observe that we have denoted the right side of this equation by $\boldsymbol{x}(t, \boldsymbol{x}_0)$ to emphasize that the solution $\boldsymbol{x}(t)$ at any time t depends on the initial state $\boldsymbol{x}_0$. Our objective is to find $\boldsymbol{x}_0$ such that at $t = T$, $\boldsymbol{x}(T, \boldsymbol{x}_0) = \boldsymbol{x}_0$. Since this relationship is seldom satisfied for an arbitrary choice of $\boldsymbol{x}_0$, let us define the function

$$\boldsymbol{F}(\boldsymbol{x}_0) \triangleq \boldsymbol{x}(T, \boldsymbol{x}_0) = \int_0^T \boldsymbol{f}(\boldsymbol{x}(t), t)\, dt + \boldsymbol{x}_0 \tag{17-112}$$

Since we are fixing the time $t = T$, $\boldsymbol{F}(\boldsymbol{x}_0)$ is a function *only* of $\boldsymbol{x}_0$ and is independent of t. In fact, in the scalar case, the function $F(x_0)$ can be represented by a curve in the

[9]The algorithm to be developed in this section is not applicable if the solution has a different period, say, a *subharmonic solution* of period T'/m. However, if the period T' of the solution is known, then the algorithm still holds if T is replaced by T' in Eq. (17-112) and in all subsequent equations.

F versus x_0 plane! It follows from Eq. (17-112) that the initial state $\boldsymbol{x}_0 = \boldsymbol{p}_0$ which gives rise to a periodic solution $\boldsymbol{p}(t)$ (with no transient component) must satisfy the equation

$$\boxed{\boldsymbol{x}_0 = \boldsymbol{F}(\boldsymbol{x}_0)} \tag{17-113}$$

Observe that Eq. (17-113) is identical to the standard form given by Eq. (5-20) for a *fixed-point iteration.* Hence, if the function $\boldsymbol{F}(\boldsymbol{x}_0)$ is a *contraction,* the solution $\boldsymbol{x}_0 = \boldsymbol{p}_0$ can be found by applying the fixed-point iteration algorithm of Section 5-3; i.e.,

$$\boldsymbol{x}_0^{(n+1)} = \boldsymbol{F}(\boldsymbol{x}_0^{(n)}) \tag{17-114}$$

The first two iterates are

$$\boldsymbol{x}_0^{(1)} = \boldsymbol{F}(\boldsymbol{x}_0^{(0)}) = \int_0^T \boldsymbol{f}(\boldsymbol{x}, t)\, dt + \boldsymbol{x}_0^{(0)} \tag{17-115}$$

$$\begin{aligned} \boldsymbol{x}_0^{(2)} = \boldsymbol{F}(\boldsymbol{x}_0^{(1)}) &= \int_0^T \boldsymbol{f}(\boldsymbol{x}, t)\, dt + \boldsymbol{x}_0^{(1)} \\ &= \int_0^T \boldsymbol{f}(\boldsymbol{x}, t)\, dt + \int_0^T \boldsymbol{f}(\boldsymbol{x}, t)\, dt + \boldsymbol{x}_0^{(0)} \\ &= \int_T^{2T} \boldsymbol{f}(\boldsymbol{x}, t)\, dt + \int_0^T \boldsymbol{f}(\boldsymbol{x}, t)\, dt + \boldsymbol{x}_0^{(0)} \\ &= \int_0^{2T} \boldsymbol{f}(\boldsymbol{x}, t)\, dt + \boldsymbol{x}_0^{(0)} \end{aligned} \tag{17-116}$$

In this manipulation we have invoked the assumption $\boldsymbol{f}(\boldsymbol{x}, t) = \boldsymbol{f}(\boldsymbol{x}, t + T)$ in changing the limits of integration from $[0, T]$ to $[T, 2T]$. By a similar procedure, it is easy to see that the $(j + 1)$th iterate is given by

$$\boldsymbol{x}_0^{(j+1)} = \boldsymbol{F}(\boldsymbol{x}_0^{(j)}) = \int_0^{(j+1)T} \boldsymbol{f}(\boldsymbol{x}, t)\, dt + \boldsymbol{x}_0^{(0)} \tag{17-117}$$

This iteration must be repeated until $\boldsymbol{x}_0^{(j+1)}$ differs from the preceding iterated value $\boldsymbol{x}_0^{(j)}$ by a small prespecified quantity. Observe that Eq. (17-117) can be interpreted as integrating the state equation with an initial state $\boldsymbol{x}_0^{(0)}$ over a sufficiently long time interval $[0, (j + 1)T]$ such that all transient components have decayed to zero; i.e.,

$$\int_{jT}^{(j+1)T} \boldsymbol{f}(\boldsymbol{x}, t)\, dt \longrightarrow \boldsymbol{0} \tag{17-118}$$

In other words, the *fixed-point iteration algorithm* is really the *conventional method* in a disguised form! To devise a much more efficient algorithm, let us rewrite Eq. (17-113) as

$$\hat{\boldsymbol{F}}(\boldsymbol{x}_0) \triangleq \boldsymbol{x}_0 - \boldsymbol{F}(\boldsymbol{x}_0) \tag{17-119}$$

and solve for $\boldsymbol{x}_0$ by the Newton–Raphson algorithm:

$$\boldsymbol{x}_0^{(j+1)} = \boldsymbol{x}_0^{(j)} - [\boldsymbol{1} - \boldsymbol{F}'(\boldsymbol{x}_0^{(j)})]^{-1}[\boldsymbol{x}_0^{(j)} - \boldsymbol{F}(\boldsymbol{x}_0^{(j)})] \tag{17-120}$$

where

$$\boldsymbol{F}'(\boldsymbol{x}_0^{(j)}) \triangleq \left.\frac{\partial \boldsymbol{F}(\boldsymbol{x}_0)}{\partial \boldsymbol{x}_0}\right|_{\boldsymbol{x}_0 = \boldsymbol{x}_0^{(j)}} \tag{17-121}$$

17-5-2 Evaluating the Jacobian Matrix $F'(x_0^{(j)})$ by Numerical Differentiation

Since it is necessary to determine the Jacobian matrix $\boldsymbol{F}'(\boldsymbol{x}_0^{(j)})$ before we can evaluate Eq. (17-120), let us study the interpretation of this matrix and devise some efficient numerical techniques for computing $\boldsymbol{F}'(\boldsymbol{x}_0^{(j)})$. Recall from Eq. (17-112) that the vector $\boldsymbol{F}(\boldsymbol{x}_0)$ is given by

$$\boldsymbol{F}(\boldsymbol{x}_0) \triangleq \begin{bmatrix} x_1(T, \boldsymbol{x}_0) \\ x_2(T, \boldsymbol{x}_0) \\ \cdot \\ \cdot \\ \cdot \\ x_n(T, \boldsymbol{x}_0) \end{bmatrix} = \begin{bmatrix} \int_0^T f_1(\boldsymbol{x}(t), t)\, dt + x_{0_1} \\ \int_0^T f_2(\boldsymbol{x}(t), t)\, dt + x_{0_2} \\ \cdot \\ \cdot \\ \cdot \\ \int_0^T f_n(\boldsymbol{x}(t), t)\, dt + x_{0_n} \end{bmatrix} \tag{17-122}$$

where $\boldsymbol{x}_0 \triangleq [x_{0_1} \quad x_{0_2} \quad \cdots \quad x_{0_n}]^t$ is the initial state of $\boldsymbol{x}$ at $t = 0$. Since T remains constant, each component

$$F_i(\boldsymbol{x}_0) = x_i(T, x_{0_1}, x_{0_2}, \ldots, x_{0_n}) \triangleq x_i(T, \boldsymbol{x}_0) \tag{17-123}$$

depends only on $x_{0_1}, x_{0_2}, \ldots, x_{0_n}$. The Jacobian matrix of $\boldsymbol{F}(\boldsymbol{x}_0)$ is therefore given by

$$\boldsymbol{F}'(\boldsymbol{x}_0) = \begin{bmatrix} \frac{\partial x_1(T, \boldsymbol{x}_0)}{\partial x_{0_1}} & \frac{\partial x_1(T, \boldsymbol{x}_0)}{\partial x_{0_2}} & \cdots & \frac{\partial x_1(T, \boldsymbol{x}_0)}{\partial x_{0_n}} \\ \frac{\partial x_2(T, \boldsymbol{x}_0)}{\partial x_{0_1}} & \frac{\partial x_2(T, \boldsymbol{x}_0)}{\partial x_{0_2}} & \cdots & \frac{\partial x_2(T, \boldsymbol{x}_0)}{\partial x_{0_n}} \\ \cdot & \cdot & & \cdot \\ \cdot & \cdot & & \cdot \\ \cdot & \cdot & & \cdot \\ \frac{\partial x_n(T, \boldsymbol{x}_0)}{\partial x_{0_1}} & \frac{\partial x_n(T, \boldsymbol{x}_0)}{\partial x_{0_2}} & \cdots & \frac{\partial x_n(T, \boldsymbol{x}_0)}{\partial x_{0_n}} \end{bmatrix} \tag{17-124}$$

To obtain the elements of this matrix with $\boldsymbol{x}_0 = \boldsymbol{x}_0^{(j)}$ numerically, we must first solve the state equations (17-109) from $t = 0$ to $t = T$ with initial states $\boldsymbol{x}_0^{(j)}$ and $\boldsymbol{x}_0^{(j)} + \Delta\boldsymbol{x}_0$, where $\Delta\boldsymbol{x}_0 \triangleq [\Delta x_{0_1} \quad \Delta x_{0_2} \quad \cdots \quad \Delta x_{0_n}]^t$ is a *small* perturbation vector. Let $\boldsymbol{x}(T, \boldsymbol{x}_0^{(j)})$ and $\boldsymbol{x}(T, \boldsymbol{x}_0^{(j)} + \Delta\boldsymbol{x}_0)$ be the corresponding solution evaluated at the end of one period, $t = T$. Then the ikth element of the preceding matrix can be computed as follows:

$$\left.\frac{\partial x_i(T, \boldsymbol{x}_0)}{\partial x_{0_k}}\right|_{\boldsymbol{x}_0 = \boldsymbol{x}_0^{(j)}} = \frac{x_i(T, \boldsymbol{x}_0^{(j)} + \Delta\boldsymbol{x}_0) - x_i(T, \boldsymbol{x}_0^{(j)})}{\Delta x_{0_k}} \tag{17-125}$$

Observe that since there are n^2 elements in $\boldsymbol{F}'(\boldsymbol{x}_0)$, we need to carry out the numerical differentiation a total of n^2 times. Moreover, we have to integrate the state equation *twice* from $t = 0$ to $t = T$, first with $\boldsymbol{x}_0^{(j)}$ as the initial state and then with $\boldsymbol{x}_0^{(j)} + \Delta\boldsymbol{x}_0$ as the "perturbed" initial state. Only after these computations do we have the complete Jacobian matrix $\boldsymbol{F}'(\boldsymbol{x}_0^{(j)})$. Substituting this matrix and the relationship $\boldsymbol{F}(\boldsymbol{x}_0^{(j)}) =$

$\boldsymbol{x}(T, \boldsymbol{x}_0^{(j)})$ into Eq. (17-120), we finally obtain the next Newton–Raphson iterate:

$$\boxed{\boldsymbol{x}_0^{(j+1)} = \boldsymbol{x}_0^{(j)} - [\mathbf{1} - \boldsymbol{F}'(\boldsymbol{x}_0^{(j)})]^{-1}[\boldsymbol{x}_0^{(j)} - \boldsymbol{x}(T, \boldsymbol{x}_0^{(j)})]} \tag{17-126}$$

17-5-3 Evaluating the Jacobian Matrix $\boldsymbol{F}'(\boldsymbol{x}_0^{(j)})$ by Transient Analysis of Sensitivity Networks

An examination of Eq. (17-125) reveals that the ikth element $\partial x_i(T, \boldsymbol{x}_0)/\partial x_{0_k}$ of the Jacobian matrix $\boldsymbol{F}'(\boldsymbol{x}_0)$ can be interpreted as the *sensitivity* of the state variable x_i at time T due to a change in the *initial state* x_{0_k} of the state variable x_k at $t = 0$. For this reason, we shall henceforth refer to $\boldsymbol{F}'(\boldsymbol{x}_0)$ as the *sensitivity matrix*. We shall now show that this matrix can be determined by computing the *transient response* of *n time-varying linear incremental circuits* from $t = 0$ to $t = T$, after the transient response of the original circuit has been determined over the same time interval with initial state $\boldsymbol{x}_0$. In other words, a total of $n + 1$ separate transient analyses over the interval $[0, T]$ will suffice to determine the sensitivity matrix.

For complete generality, we shall allow the network $\mathfrak{N}$ to contain nonlinear resistors, inductors, and capacitors, as well as any system of *linearly coupled* resistive circuit elements, such as linear controlled sources (all four types), ideal transformers, and gyrators.

Our first task will be to determine the dependence of each branch current $i_j(t)$ and branch voltage $v_j(t)$ on the *initial state* $\boldsymbol{x}_0$. To emphasize that both $i_j(t)$ and $v_j(t)$ depend on $\boldsymbol{x}_0$, we let $\boldsymbol{i}(t, \boldsymbol{x}_0)$ and $\boldsymbol{v}(t, \boldsymbol{x}_0)$ denote the branch current solution vector and branch voltage solution vector at time t of all circuit elements in $\mathfrak{N}$, except *independent* current sources and *independent* voltage sources, which are denoted by the composite branch symbols $\boldsymbol{J}(t)$ and $\boldsymbol{E}(t)$, respectively. In terms of these explicit notations, Eqs. (17-35) and (17-36) now assume the form

$$\text{KCL:} \quad \boldsymbol{A}\boldsymbol{i}(t, \boldsymbol{x}_0) = \boldsymbol{A}\boldsymbol{J}(t) \tag{17-127a}$$

$$\text{KVL:} \quad \boldsymbol{v}(t, \boldsymbol{x}_0) = \boldsymbol{A}^t\boldsymbol{v}_n(t, \boldsymbol{x}_0) + \boldsymbol{E}(t) \tag{17-127b}$$

The constitutive relations of the circuit elements are given by the following:

Linear Coupled Elements:[10]

$$\boldsymbol{K}_i\boldsymbol{i}_\alpha(t, \boldsymbol{x}_0) + \boldsymbol{K}_v\boldsymbol{v}_\alpha(t, \boldsymbol{x}_0) = \mathbf{0} \tag{17-127c}$$

Nonlinear Resistors:

$$\boldsymbol{v}_R(t, \boldsymbol{x}_0) = \hat{\boldsymbol{v}}_R(\boldsymbol{i}_R(t, \boldsymbol{x}_0)), \quad \text{for current-controlled resistors} \tag{17-127d}$$

$$\boldsymbol{i}_G(t, \boldsymbol{x}_0) = \hat{\boldsymbol{i}}_G(\boldsymbol{v}_G(t, \boldsymbol{x}_0)), \quad \text{for voltage-controlled resistors} \tag{17-127e}$$

Nonlinear Capacitors:

$$\boldsymbol{q}_C(t, \boldsymbol{x}_0) = \hat{\boldsymbol{q}}_C(\boldsymbol{v}_C(t, \boldsymbol{x}_0)) \tag{17-127f}$$

$$\boldsymbol{v}_C(0) = \boldsymbol{v}_{C_0}, \quad \text{initial capacitor voltage} \tag{17-127g}$$

[10]Equation (17-127c) is simply Eq. (17-37) with $\boldsymbol{s} = \mathbf{0}$ and includes all *linear* resistors, linear controlled sources (all four types), and linear resistive n-ports, such as ideal transformers and gyrators.

Nonlinear Inductors:

$$\boldsymbol{\phi}_L(t, \boldsymbol{x}_0) = \hat{\boldsymbol{\phi}}_L(\boldsymbol{i}_L(t, \boldsymbol{x}_0)) \tag{17-127h}$$

$$\boldsymbol{i}_L(0) = \boldsymbol{i}_{L_0}, \qquad \text{initial inductor current} \tag{17-127i}$$

It follows from these notations that

$$\boldsymbol{i}(t, \boldsymbol{x}_0) = [\boldsymbol{i}_\alpha(t, \boldsymbol{x}_0) \quad \boldsymbol{i}_R(t, \boldsymbol{x}_0) \quad \boldsymbol{i}_G(t, \boldsymbol{x}_0) \quad \boldsymbol{i}_C(t, \boldsymbol{x}_0) \quad \boldsymbol{i}_L(t, \boldsymbol{x}_0)]^t \tag{17-128a}$$

$$\boldsymbol{v}(t, \boldsymbol{x}_0) = [\boldsymbol{v}_\alpha(t, \boldsymbol{x}_0) \quad \boldsymbol{v}_R(t, \boldsymbol{x}_0) \quad \boldsymbol{v}_G(t, \boldsymbol{x}_0) \quad \boldsymbol{v}_C(t, \boldsymbol{x}_0) \quad \boldsymbol{v}_L(t, \boldsymbol{x}_0)]^t \tag{17-128b}$$

Now suppose that we differentiate both sides of Eq. (17-127) with respect to the initial states $x_{0_1}, x_{0_2}, \ldots, x_{0_n}$:

$$\text{KCL:} \quad A\frac{\partial \boldsymbol{i}(t, \boldsymbol{x}_0)}{\partial \boldsymbol{x}_0} = \boldsymbol{0} \tag{17-129a}$$

$$\text{KVL:} \quad \frac{\partial \boldsymbol{v}(t, \boldsymbol{x}_0)}{\partial \boldsymbol{x}_0} = A^t\frac{\partial \boldsymbol{v}_n(t, \boldsymbol{x}_0)}{\partial \boldsymbol{x}_0} \tag{17-129b}$$

The constitutive relations are then given by the following:

Linear Coupled Elements:

$$\boldsymbol{K}_i\frac{\partial \boldsymbol{i}_\alpha(t, \boldsymbol{x}_0)}{\partial \boldsymbol{x}_0} + \boldsymbol{K}_v\frac{\partial \boldsymbol{v}_\alpha(t, \boldsymbol{x}_0)}{\partial \boldsymbol{x}_0} = \boldsymbol{0} \tag{17-129c}$$

Nonlinear Resistors:

$$\frac{\partial \boldsymbol{v}_R(t, \boldsymbol{x}_0)}{\partial \boldsymbol{x}_0} = \left.\frac{\partial \hat{\boldsymbol{v}}_R(\boldsymbol{i}_R)}{\partial \boldsymbol{i}_R}\right|_{\boldsymbol{v}_R=\boldsymbol{v}_R(t,\boldsymbol{x}_0)} \cdot \frac{\partial \boldsymbol{i}_R(t, \boldsymbol{x}_0)}{\partial \boldsymbol{x}_0}, \quad \text{for current-controlled resistors} \tag{17-129d}$$

$$\frac{\partial \boldsymbol{i}_G(t, \boldsymbol{x}_0)}{\partial \boldsymbol{x}_0} = \left.\frac{\partial \hat{\boldsymbol{i}}_G(\boldsymbol{v}_G)}{\partial \boldsymbol{v}_G}\right|_{\boldsymbol{v}_G=\boldsymbol{v}_G(t,\boldsymbol{x}_0)} \cdot \frac{\partial \boldsymbol{v}_C(t, \boldsymbol{x}_0)}{\partial \boldsymbol{x}_0}, \quad \text{for voltage-controlled resistors} \tag{17-129e}$$

Nonlinear Capacitors:

$$\frac{\partial \boldsymbol{q}_C(t, \boldsymbol{x}_0)}{\partial \boldsymbol{x}_0} = \left.\frac{\partial \hat{\boldsymbol{q}}_C(\boldsymbol{v}_C)}{\partial \boldsymbol{v}_C}\right|_{\boldsymbol{v}_C=\boldsymbol{v}_C(t,\boldsymbol{x}_0)} \cdot \frac{\partial \boldsymbol{v}_C(t, \boldsymbol{x}_0)}{\partial \boldsymbol{x}_0} \tag{17-129f}$$

$$\frac{\partial v_{C_j}(0)}{\partial x_{0_k}} = \begin{cases} 1, & \text{if } x_{0_k} = v_{C_j}(0) \\ 0, & \text{if } x_{0_k} \neq v_{C_j}(0) \end{cases} \tag{17-129g}$$

Nonlinear Inductors:

$$\frac{\partial \boldsymbol{\phi}_L(t, \boldsymbol{x}_0)}{\partial \boldsymbol{x}_0} = \left.\frac{\partial \hat{\boldsymbol{\phi}}_L(\boldsymbol{i}_L)}{\partial \boldsymbol{i}_L}\right|_{\boldsymbol{i}_L=\boldsymbol{i}_L(t,\boldsymbol{x}_0)} \cdot \frac{\partial \boldsymbol{i}_L(t, \boldsymbol{x}_0)}{\partial \boldsymbol{x}_0} \tag{17-129h}$$

$$\frac{\partial i_{L_j}(0)}{\partial x_{0_k}} = \begin{cases} 1, & \text{if } x_{0_k} = i_{L_j}(0) \\ 0, & \text{if } x_{0_k} \neq i_{L_j}(0) \end{cases} \tag{17-129i}$$

If we introduce the abbreviated notation

$$\bar{y}(t, \boldsymbol{x}_0) \triangleq \frac{\partial y(t, \boldsymbol{x}_0)}{\partial \boldsymbol{x}_0} \tag{17-130}$$

then Eq. (17-129) assumes the following simplified form:

$$\text{KCL:} \quad A\bar{\boldsymbol{i}}(t, \boldsymbol{x}_0) = \boldsymbol{0} \tag{17-131a}$$

$$\text{KVL:} \quad \bar{\boldsymbol{v}}(t, \boldsymbol{x}_0) = A^t\bar{\boldsymbol{v}}_n(t, \boldsymbol{x}_0) \tag{17-131b}$$

The constitutive relations are then given by the following:

Linear Coupled Elements:

$$\mathbf{K}_i\bar{\mathbf{i}}_\alpha(t, \mathbf{x}_0) + \mathbf{K}_v\bar{\mathbf{v}}_\alpha(t, \mathbf{x}_0) = \mathbf{0} \tag{17-131c}$$

Nonlinear Resistors:

$$\bar{\mathbf{v}}_R(t, \mathbf{x}_0) = \left.\frac{\partial \hat{\mathbf{v}}_R(\mathbf{i}_R)}{\partial \mathbf{i}_R}\right|_{\mathbf{i}_R=\mathbf{i}_R(t,\mathbf{x}_0)} \cdot \bar{\mathbf{i}}_R(t, \mathbf{x}_0), \quad \text{for current-controlled resistors} \tag{17-131d}$$

$$\bar{\mathbf{i}}_G(t, \mathbf{x}_0) = \left.\frac{\partial \hat{\mathbf{i}}_G(\mathbf{v}_G)}{\partial \mathbf{v}_G}\right|_{\mathbf{v}_G=\mathbf{v}_G(t,\mathbf{x}_0)} \cdot \bar{\mathbf{v}}_G(t, \mathbf{x}_0), \quad \text{for voltage-controlled resistors} \tag{17-131e}$$

Nonlinear Capacitors:

$$\bar{\mathbf{q}}_C(t, \mathbf{x}_0) = \left.\frac{\partial \hat{\mathbf{q}}_C(\mathbf{v}_C)}{\partial \mathbf{v}_C}\right|_{\mathbf{v}_C=\mathbf{v}_C(t,\mathbf{x}_0)} \cdot \bar{\mathbf{v}}_C(t, \mathbf{x}_0) \tag{17-131f}$$

$$\bar{v}_{C_j}(0) = \begin{cases} 1, & \text{if } x_{0_k} = v_{C_j}(0) \\ 0, & \text{if } x_{0_k} \neq v_{C_j}(0) \end{cases} \tag{17-131g}$$

Nonlinear Inductors:

$$\bar{\boldsymbol{\phi}}_L(t, \mathbf{x}_0) = \left.\frac{\partial \hat{\boldsymbol{\phi}}_L(\mathbf{i}_L)}{\partial \mathbf{i}_L}\right|_{\mathbf{i}_L=\mathbf{i}_L(t,\mathbf{x}_0)} \bar{\mathbf{i}}_L(t, \mathbf{x}_0) \tag{17-131h}$$

$$\bar{i}_{L_j}(0) = \begin{cases} 1, & \text{if } x_{0_k} = i_{L_j}(0) \\ 0, & \text{if } x_{0_k} \neq i_{L_j}(0) \end{cases} \tag{17-131i}$$

Examination of these equations reveals the following important observations:

1. Equations (17-131a) and (17-131b) can be interpreted as the KCL and KVL equations of an associated network $\bar{\mathfrak{N}}$ having *identical* topology as the original network $\mathfrak{N}$, with all *independent current sources* in $\mathfrak{N}$ replaced by *open circuits* $[\mathbf{J}(t) = \mathbf{0}]$ and with all *independent voltage sources* in $\mathfrak{N}$ replaced by *short circuits* $[\mathbf{E}(t) = \mathbf{0}]$. Observe that an *overbar* is attached to each branch current and voltage variable in the new network $\bar{\mathfrak{N}}$.
2. Equation (17-131c) shows that the location, type, and circuit parameter value of all linear coupled and uncoupled elements remain unchanged. In particular, a linear resistor of value R in $\mathfrak{N}$ remains a linear resistor of value R in $\bar{\mathfrak{N}}$, a linear controlled source in $\mathfrak{N}$ remains a linear controlled source in $\bar{\mathfrak{N}}$, an ideal transformer in $\mathfrak{N}$ remains an ideal transformer in $\bar{\mathfrak{N}}$, etc. Moreover, all locations and parameter values remain invariant.
3. If we examine a typical equation representing resistor m from Eqs. (17-131d) and (17-131e), we obtain

$$\bar{v}_{R_m}(t, \mathbf{x}_0) = R_m(t, \mathbf{x}_0)\bar{i}_{R_m}(t, \mathbf{x}_0) \tag{17-132a}$$

where

$$R_m(t, \mathbf{x}_0) \triangleq \left.\frac{\partial \hat{v}_{R_m}(i_{R_m})}{\partial i_{R_m}}\right|_{i_{R_m}=i_{R_m}(t,\mathbf{x}_0)} \tag{17-132b}$$

and

$$\bar{i}_{G_m}(t, \mathbf{x}_0) = G_m(t, \mathbf{x}_0)\bar{v}_{G_m}(t, \mathbf{x}_0) \tag{17-133a}$$

where

$$G_m(t, \boldsymbol{x}_0) \triangleq \left.\frac{\partial \hat{i}_{G_m}(v_{R_m})}{\partial v_{R_m}}\right|_{v_{R_m}=v_{R_m}(t,\boldsymbol{x}_0)} \tag{17-133b}$$

It follows from Eqs. (17-132) and (17-133) that each *nonlinear resistor in* $\mathfrak{N}$ *should be replaced by a time-varying linear resistor* defined by Eq. (17-132b) or (17-133b), depending on whether the nonlinear resistor is current controlled or voltage controlled (see Table 17-2). Observe that, in the special case of a *linear* resistor, $R_m(t, \boldsymbol{x}_0) = R_m$ and $G_m(t, \boldsymbol{x}_0) = G_m$, as they should.

4. If we examine a typical equation representing capacitor m from Eq. (17-131f), we obtain

$$\bar{q}_{C_m}(t, \boldsymbol{x}_0) = C_m(t, \boldsymbol{x}_0)\bar{v}_{C_m}(t, \boldsymbol{x}_0) \tag{17-134a}$$

where

$$C_m(t, \boldsymbol{x}_0) \triangleq \left.\frac{\partial \hat{q}_{C_m}(v_{C_m})}{\partial v_{C_m}}\right|_{v_{C_m}=v_{C_m}(t,\boldsymbol{x}_0)} \tag{17-134b}$$

$$\bar{v}_{C_m}(0) = \begin{cases} 1, & \text{if } x_{0_k} = v_{C_m}(0) \\ 0, & \text{if } x_{0_k} \neq v_{C_m}(0) \end{cases} \tag{17-134c}$$

It follows from Eq. (17-134) that each *nonlinear capacitor m in* $\mathfrak{N}$ *with initial voltage* $v_{C_m}(0)$ *should be replaced by an uncharged but time-varying linear capacitor* [defined by Eq. (17-134b)] *in series* with a 1-V (resp., 0-V) voltage source if the initial state x_{0_k} being perturbed is (resp., is not) $v_{C_m}(0)$. If the capacitor is linear, then $C_m(t, \boldsymbol{x}_0) = C_m$, a time-invariant linear capacitor (see Table 17-2).

5. If we examine a typical equation representing inductor m from Eq. (17-131h), we obtain

$$\bar{\phi}_{L_m}(t, \boldsymbol{x}_0) = L_m(t, \boldsymbol{x}_0)\bar{i}_{L_m}(t, \boldsymbol{x}_0) \tag{17-135a}$$

where

$$L_m(t, \boldsymbol{x}_0) \triangleq \left.\frac{\partial \hat{\phi}_{L_m}(i_{L_m})}{\partial i_{L_m}}\right|_{i_{L_m}=i_{L_m}(t,\boldsymbol{x}_0)} \tag{17-135b}$$

$$\bar{i}_{L_m}(0) = \begin{cases} 1, & \text{if } x_{0_k} = i_{L_m}(0) \\ 0, & \text{if } x_{0_k} \neq i_{L_m}(0) \end{cases} \tag{17-135c}$$

It follows from Eq. (17-135) that each *nonlinear inductor m in* $\mathfrak{N}$ *with initial current* $i_{L_m}(0)$ *should be replaced by a time-varying linear inductor* [defined by Eq. (17-135b)] *with zero initial current in parallel* with a 1-A (resp., 0-A) current source if the initial state x_{0_k} being perturbed is (resp., is not) $i_{L_m}(0)$. If the inductor is linear, then $L_m(t, \boldsymbol{x}_0) = L_m$, a time-invariant linear inductor (see Table 17-2).

We shall henceforth refer to the circuit $\overline{\mathfrak{N}}$ constructed in accordance with the preceding recipe as a *sensitivity network*. Observe that, since the *initial-state vector* is $\boldsymbol{x}_0 = [x_{0_1} \quad x_{0_2} \quad \cdots \quad x_{0_n}]^t$, we must perturb the n initial states $x_{0_1}, x_{0_2}, \ldots, x_{0_n}$ one at a time, and that to each perturbed initial state x_{0_k} there corresponds an associated linear time-varying sensitivity network $\overline{\mathfrak{N}}(x_{0_k})$. Hence, there are a total of n sensitivity networks whose transient response must be computed from $t = 0$ to $t = T$. In par-

TABLE 17-2. Models and Constitutive Relations for Sensitivity Network

CIRCUIT ELEMENT	CONSTITUTIVE RELATION FOR CIRCUIT ELEMENTS IN THE ORIGINAL NETWORK	CONSTITUTIVE RELATION FOR CIRCUIT ELEMENT MODEL FOR THE SENSITIVITY NETWORK
Linear-coupled elements	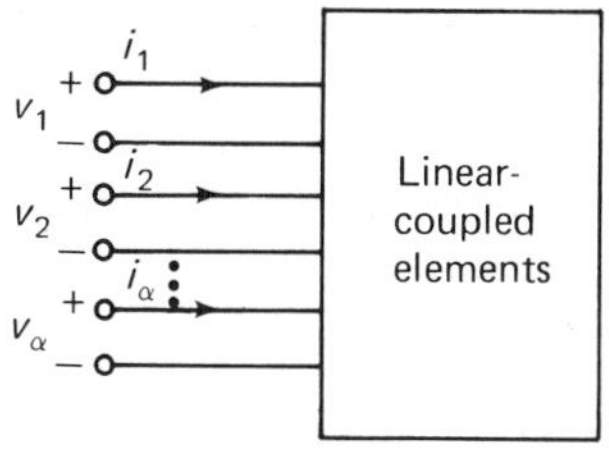 Linear-coupled elements; $K_i i_\alpha + K_v v_\alpha = 0$	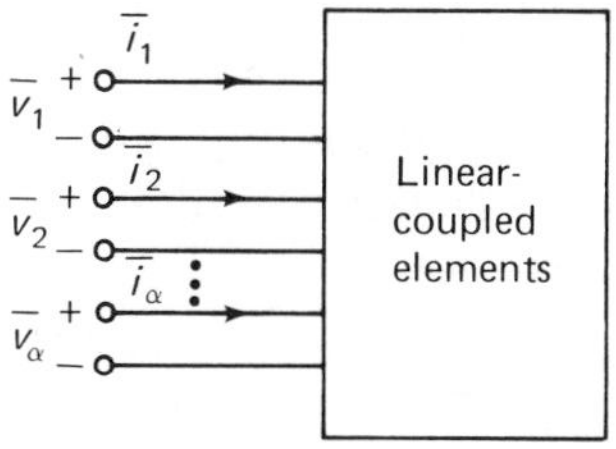 Linear-coupled elements; $k_i \bar{i}_\alpha + K_v \bar{v}_\alpha = 0$
Nonlinear resistor (current-controlled)	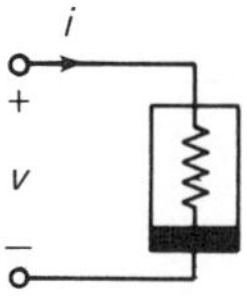 $v = \hat{v}(i)$	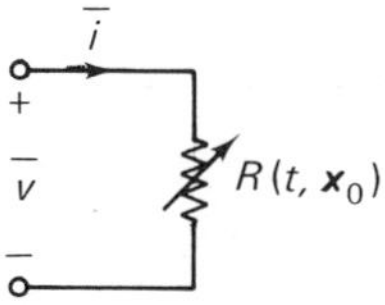 $\bar{v} = R(t, \mathbf{x}_0)\,\bar{i}$; $R(t, \mathbf{x}_0) \triangleq \left.\frac{d\hat{v}(i)}{di}\right\|_{i = i(t, \mathbf{x}_0)}$
Nonlinear resistor (voltage-controlled)	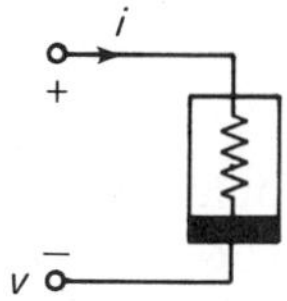 $i = \hat{i}(v)$	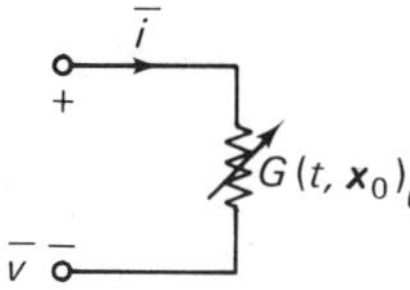 $\bar{i} = G(t, \mathbf{x}_0)\,\bar{v}$; $G(t, \mathbf{x}_0) \triangleq \left.\frac{d\hat{i}(v)}{dv}\right\|_{v = v(t, \mathbf{x}_0)}$
Nonlinear capacitor	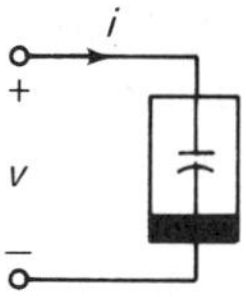 $q = \hat{q}(v)$; $v(0) = v_0$	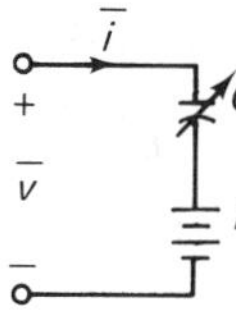 $C(t, \mathbf{x}_0) \triangleq \left.\frac{\partial\hat{q}(v)}{\partial v}\right\|_{v = v(t, \mathbf{x}_0)}$; $E = 1$, if $x_{0_k} = v_0$; 0, if $x_{0_k} \neq v_0$
Nonlinear inductor	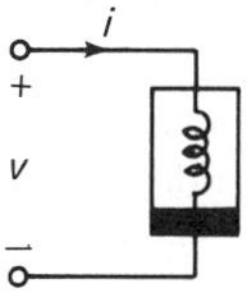 $\varphi = \hat{\varphi}(i)$; $i(0) = i_0$	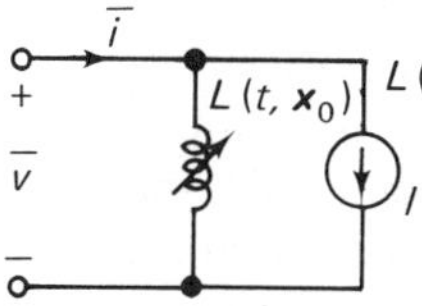 $L(t, \mathbf{x}_0) = \left.\frac{\partial\hat{\varphi}(i)}{\partial i}\right\|_{i = i(t, \mathbf{x}_0)}$; $I = 1$, if $x_{0_k} = i_0$; 0, if $x_{0_k} \neq i_0$
Linear resistor	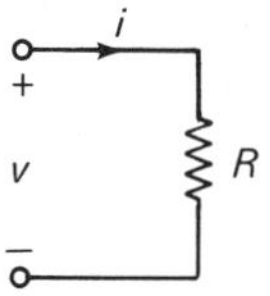 $v = Ri$	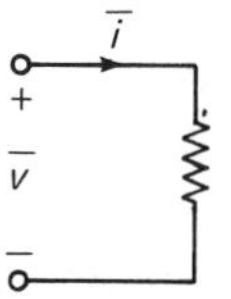 $\bar{v} = R\bar{i}$

TABLE 17-2. (cont.)

CIRCUIT ELEMENT	CONSTITUTIVE RELATION FOR CIRCUIT ELEMENTS IN THE ORIGINAL NETWORK	CONSTITUTIVE RELATION FOR CIRCUIT ELEMENT MODEL FOR THE SENSITIVITY NETWORK
Linear capacitor	$i = C\dfrac{dv}{dt}$	$\bar{i} = C\dfrac{d\bar{v}}{dt}$
Linear inductor	$v = L\dfrac{di}{dt}$	$\bar{v} = L\dfrac{d\bar{i}}{dt}$

ticular, the solution vector

$$\bar{\boldsymbol{x}}(T, \boldsymbol{x}_0) \triangleq \left.\frac{\partial \boldsymbol{x}(t, \boldsymbol{x}_0)}{\partial x_{0_k}}\right|_{t=T} \tag{17-136}$$

of the kth sensitivity network $\overline{\mathfrak{N}}(x_{0_k})$ evaluated at $t = T$ would give us precisely the elements of column k of the *sensitivity matrix* given in Eq. (17-124)! Hence, by computing n transient analyses of the n sensitivity networks $\overline{\mathfrak{N}}(x_{0_k})$, $k = 1, 2, \ldots, n$, from $t = 0$ to $t = T$, we would obtain the desired sensitivity matrix $\boldsymbol{F}'(\boldsymbol{x}_0^{(j)})$. Since the transient response $\boldsymbol{x}(t, \boldsymbol{x}_0^{(j)})$ of the original network $\mathfrak{N}$ must be computed from $t = 0$ to $t = T$ in order to construct the sensitivity network, a total of $n + 1$ transient analyses over the time interval $[0, T]$ must be computed before the Newton–Raphson iteration given by Eq. (17-126) can be carried out. Observe that in the special case where $\mathfrak{N}$ is a linear time-invariant network, the n associated sensitivity networks are also linear and time invariant.

The sensitivity networks $\overline{\mathfrak{N}}(x_{0_k})$ associated with the perturbed states x_{0_k}, $k = 1, 2, \ldots, n$, have identical topology (except that all independent current sources in $\mathfrak{N}$ are replaced by open circuits and all independent voltage sources in $\mathfrak{N}$ are replaced by short circuits). Corresponding to each nonlinear element in $\mathfrak{N}$, an incremental linear time-varying element must be determined for $\overline{\mathfrak{N}}$ in accordance with Eqs. (17-132b), (17-133b), (17-134b), and (17-135b). For many existing computer simulation programs employing the *generalized associated discrete circuit model approach*, as described in Section 17-1, the partial derivatives required by these equations are already available and can be simply retrieved. Hence, for such programs the sensitivity networks could be constructed and analyzed rather efficiently. For details on how this can be done, the reader is referred to [3–5].

17-5-4 Convergence of the Iteration Algorithm

After the sensitivity matrix $\boldsymbol{F}'(\boldsymbol{x}_0^{(j)})$ is computed with $\boldsymbol{x}_0 = \boldsymbol{x}_0^{(j)}$, we can proceed to evaluate $\boldsymbol{x}_0^{(j+1)}$ using the Newton–Raphson iterate given in Eq. (17-126). This procedure is then iterated until $\boldsymbol{x}_0^{(j+1)}$ converges to the *desired* initial state $\boldsymbol{x}_0 = \boldsymbol{p}_0$. On the strength of the *Newton–Raphson–Kantorovich theorem* stated in Chapter 5, or the convergence proof given in Section 5A.3, we can be assured that as long as the initial guess $\boldsymbol{x}_0^{(0)}$ is close to $\boldsymbol{p}_0$ the preceding algorithm will converge quadratically to $\boldsymbol{p}_0$. Of course, if our initial guess $\boldsymbol{x}_0^{(0)}$ is too far off from $\boldsymbol{p}_0$, the algorithm may diverge and another initial guess would have to be chosen. If the network has more than one periodic solution of the same period T, the algorithm may converge to any one of these multiple solutions, depending on the choice of the initial guess. Moreover, it is entirely possible for the algorithm to converge to an *unstable* periodic solution, i.e., a solution which exists in theory but can never be observed in practice owing to the unavoidable presence of noise.

We shall now summarize the *algorithm for determining the steady-state response of nonlinear circuits with periodic inputs*, as follows:

Step 1. Compute the transient response $\boldsymbol{x}(t, \boldsymbol{x}_0^{(j)})$ of the network from $t = 0$ to $t = T$ with initial state $\boldsymbol{x} = \boldsymbol{x}_0^{(j)}$.

Step 2. Compute the *sensitivity matrix*

$$\boldsymbol{F}'(\boldsymbol{x}_0^{(j)}) \triangleq \left.\frac{\partial \boldsymbol{F}(\boldsymbol{x}_0)}{\partial \boldsymbol{x}_0}\right|_{\boldsymbol{x}=\boldsymbol{x}_0^{(j)}}$$

by using either the numerical differentiation formula given in Eq. (17-125), or by analyzing n associated *sensitivity networks* $\overline{\mathfrak{N}}(x_{0_m})$, $m = 1, 2, \ldots, n$.

Step 3. Compute $\boldsymbol{x}_0^{(j+1)}$ from Eq. (17-126).

Step 4. Return to step 1 unless $\|\boldsymbol{x}(T, \boldsymbol{x}_0^{(j)}) - \boldsymbol{x}_0^{(j)}\| < \varepsilon$ and $\|\boldsymbol{x}_0^{(j+1)} - \boldsymbol{x}_0^{(j)}\| < \delta$, where ε and δ are (user-specified) arbitrarily small positive numbers.

Step 5. Stop.

17-6 ALGORITHM FOR DETERMINING STEADY-STATE PERIODIC SOLUTIONS OF NONLINEAR OSCILLATORS

The state equations $\dot{\boldsymbol{x}} = \boldsymbol{f}(\boldsymbol{x}, t)$ considered in Section 17-5 are said to be *nonautonomous* because the time variable t is present explicitly in the function $\boldsymbol{f}(\boldsymbol{x}, t)$ owing to the presence of periodic input time functions. In sharp contrast to this observation, the state equations of a nonlinear oscillator assume the form

$$\dot{\boldsymbol{x}} = \boldsymbol{f}(\boldsymbol{x}) \tag{17-137}$$

This system of equations is said to be *autonomous*, because it can oscillate by itself without any external input. For most practical applications, one is interested only in determining the *steady-state periodic waveform* of the nonlinear oscillator, and once again it is computationally inefficient to determine this solution by conventional numerical-integration methods. The algorithm presented in Section 17-5 also is *not* applicable here for two fatal reasons:

1. The period of oscillation T is not known and hence Eq. (17-112) cannot be evaluated.
2. Even if the oscillation period T is given a priori, the algorithm is still not applicable in view of a standard result from the theory of ordinary differential equations [6] which asserts that the sensitivity matrix $\boldsymbol{F}'(\boldsymbol{x}_0) \triangleq \boldsymbol{A}$, as defined in Eq. (17-124), always has an eigenvalue $\lambda = 1$ whenever $\boldsymbol{x}_0$ lies on a periodic orbit. This implies that $\boldsymbol{A}\boldsymbol{x} = \lambda\boldsymbol{x} = \boldsymbol{x}$ and $(\boldsymbol{1} - \boldsymbol{A})\boldsymbol{x} = \boldsymbol{0}$ for all $\boldsymbol{x}$. Hence, the matrix $\boldsymbol{H}' \triangleq (\boldsymbol{1} - \boldsymbol{A}) = \boldsymbol{1} - \boldsymbol{F}'(\boldsymbol{x}_0)$ is *singular*. Therefore, Eq. (17-126) cannot be evaluated and the algorithm is not directly applicable.

To salvage this algorithm, however, let us rewrite Eqs. (17-111)–(17-113) as follows:

$$\boldsymbol{x}(t) = \int_0^t \boldsymbol{f}(\boldsymbol{x}(t))\,dt + \boldsymbol{x}_0 \triangleq \boldsymbol{x}(t, \boldsymbol{x}_0) \tag{17-111'}$$

$$\boldsymbol{F}(T, \boldsymbol{x}_0) \triangleq \boldsymbol{x}(T, \boldsymbol{x}_0) = \int_0^T \boldsymbol{f}(\boldsymbol{x}(t))\,dt + \boldsymbol{x}_0 \tag{17-112'}$$

$$\boldsymbol{x}_0 = \boldsymbol{F}(T, \boldsymbol{x}_0) \tag{17-113'}$$

Observe that the period T is now added to the arguments of the function $\boldsymbol{F}(T, \boldsymbol{x}_0)$ because it is an unknown variable that must be determined along with the unknown initial state $\boldsymbol{x}_0 = \boldsymbol{p}_0$. In other words, we now have $n + 1$ variables $\{x_{0_1}, x_{0_2}, \ldots, x_{0_n}, T\}$ to be determined from Eq. (17-113′), which is a system of only n equations. To have a unique solution, it is necessary that we either *assume* an appropriate value for one of these $n + 1$ variables, or we must attempt to find another independent equation relating these variables. To exploit the first alternative, we observe that, if $\boldsymbol{x} = \boldsymbol{p}(t)$ is a periodic solution of the *autonomous* system $\dot{\boldsymbol{x}} = \boldsymbol{f}(\boldsymbol{x})$, then so is $\boldsymbol{x} = \boldsymbol{p}(t + t_0)$, where t_0 is any constant, because $\boldsymbol{f}(\boldsymbol{x})$ does not contain t as an argument; hence, $\dot{\boldsymbol{p}}(t) = \boldsymbol{f}(\boldsymbol{p}(t))$ implies $\dot{\boldsymbol{p}}(t + t_0) = \boldsymbol{f}(\boldsymbol{p}(t + t_0))$. In other words, Eq. (17-137) admits an infinite number of periodic solutions, each one differing from the others by a translation in time. To illustrate this observation, consider the typical family of *periodic* solutions $x_1 = p_1(t)$, $x_2 = p_2(t), \ldots, x_k = p_k(t), \ldots, x_n = p_n(t)$ of period T, as shown in Fig. 17-5. Observe that since the system $\dot{\boldsymbol{x}} = \boldsymbol{f}(\boldsymbol{x})$ is autonomous (no time-varying inputs), if we pick another set of points $x'_{0_1}, x'_{0_2}, \ldots, x'_{0_k}, \ldots, x'_{0_n}$ on these solution waveforms *at any time* $t = t_0$ as a *new* initial state, the corresponding solutions will be given by the same family of solutions, but with the time origin translated to $t = t_0$. Hence, we are free to assume a value for one of the n initial states, say $x_{0_k} = p_{k_0}$, and determine the remaining initial states $x_{0_1}, x_{0_2}, \ldots, x_{0_{k-1}}, x_{0_{k+1}}, \ldots, x_n$ which fall on the same time instant $t = t_0$ as that of $x_{0_k} = p_{k_0}$. Hence, our procedure is equivalent to that of choosing a new time origin for the periodic solutions. The only

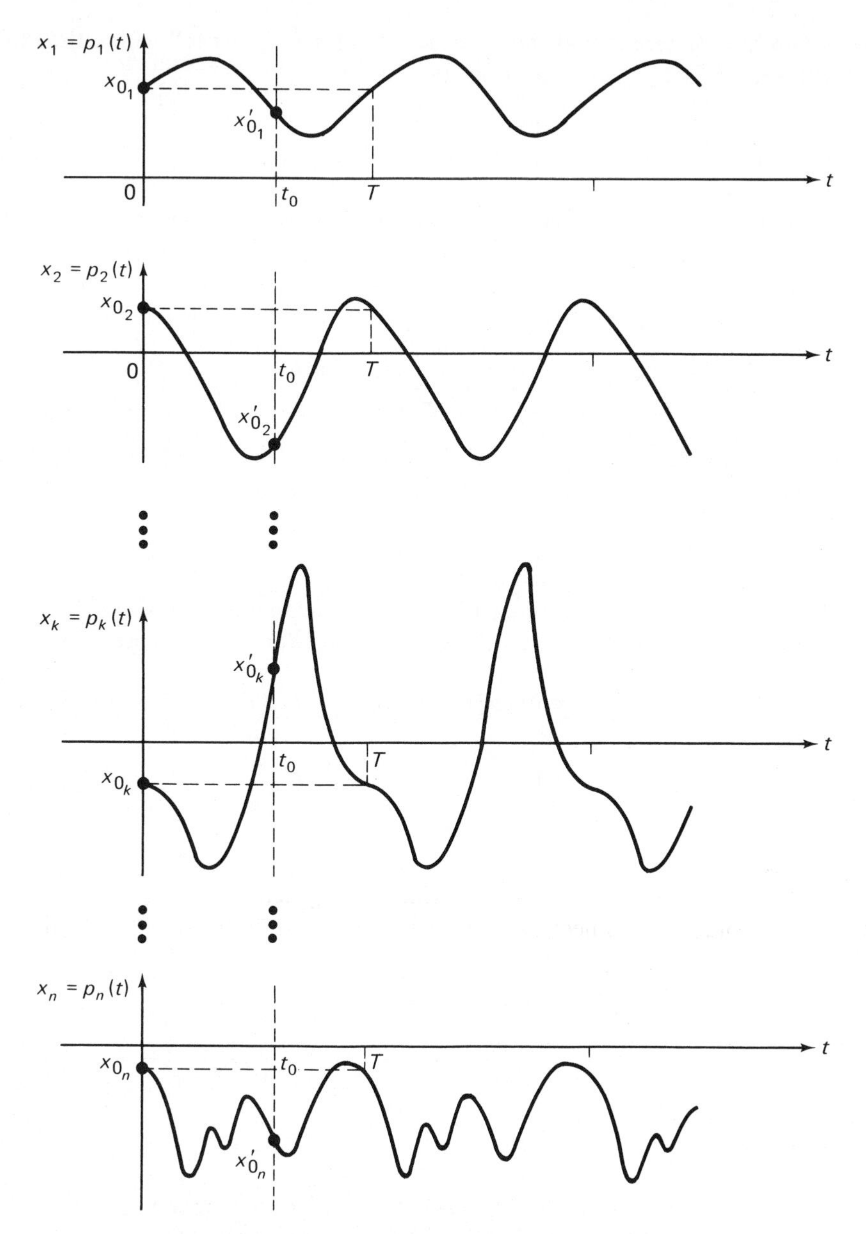

Fig. 17-5. Typical family of periodic solutions $x_1 = p_1(t)$, $x_2 = p_2(t)$, $\ldots$, $x_n = p_n(t)$ for an *autonomous system* $\dot{\mathbf{x}} = \mathbf{f}(\mathbf{x})$ with initial states $x_{0_1}, x_{0_2}, \ldots, x_{0_n}$. A different set of initial states, $x'_{0_1}, x'_{0_2}, \ldots, x'_{0_n}$ corresponding to points located at any time t_0 will give rise to the same family of periodic solution waveforms, apart from a translation by t_0 seconds.

important precaution here is that the *assumed* value for p_{k_0} must fall within the range of values taken on by $x_k = p_k(t)$ over one period; i.e.,

$$p_{k_{\min}} < p_{k_0} < p_{k_{\max}} \tag{17-138}$$

where $p_{k_{\min}}$ and $p_{k_{\max}}$ denote, respectively, the minimum and maximum values of $p_k(t)$, over one period.

Once the value of the initial state x_{0_k} is fixed, by assumption, we have exactly n unknowns left, i.e., $x_{0_1}, x_{0_2}, \ldots, x_{0_{k-1}}, x_{0_{k+1}}, \ldots, x_{0_n}$, and T. If we define the new $n \times 1$ unknown vector

$$\boldsymbol{y} \triangleq [x_{0_1} \quad x_{0_2} \quad \cdots \quad x_{0_{k-1}} \quad T \quad x_{0_{k+1}} \quad \cdots \quad x_n]^t \tag{17-139}$$

then Eq. (17-113′) can be recast into the equivalent form

$$\boldsymbol{H}(\boldsymbol{y}) \triangleq \boldsymbol{x}_0 - \boldsymbol{F}(T, \boldsymbol{x}_0) = \boldsymbol{0} \tag{17-140}$$

Equation (17-116) represents a system of n nonlinear algebraic equations in n unknowns and can be solved by the Newton–Raphson iteration

$$\boxed{\boldsymbol{y}^{(j+1)} = \boldsymbol{y}^{(j)} - [\boldsymbol{1}(-k) - \boldsymbol{F}'(T^{(j)}, \boldsymbol{x}_0^{(j)})]^{-1}[\boldsymbol{x}_0^{(j)} - \boldsymbol{x}(T^{(j)}, \boldsymbol{x}_0^{(j)})]} \tag{17-141}$$

where $\boldsymbol{1}(-k)$ denotes a diagonal matrix with *unit* diagonal elements except for the kth element, which is *zero*.[11] The matrix $\boldsymbol{F}'(T, \boldsymbol{x}_0) \triangleq \partial \boldsymbol{F}(T, \boldsymbol{x}_0)/\partial \boldsymbol{y}$ is given by

$$\boldsymbol{F}'(T, \boldsymbol{x}_0) = \begin{bmatrix} \dfrac{\partial x_1(T, \boldsymbol{x}_0)}{\partial x_{0_1}} & \cdots & \dfrac{\partial x_1(T, \boldsymbol{x}_0)}{\partial x_{0_{k-1}}} & \dfrac{\partial x_1(T, \boldsymbol{x}_0)}{\partial T} & \dfrac{\partial x_1(T, \boldsymbol{x}_0)}{\partial x_{0_{k+1}}} & \cdots & \dfrac{\partial x_1(T, \boldsymbol{x}_0)}{\partial x_{0_n}} \\ \dfrac{\partial x_2(T, \boldsymbol{x}_0)}{\partial x_{0_1}} & \cdots & \dfrac{\partial x_2(T, \boldsymbol{x}_0)}{\partial x_{0_{k-1}}} & \dfrac{\partial x_2(T, \boldsymbol{x}_0)}{\partial T} & \dfrac{\partial x_2(T, \boldsymbol{x}_0)}{\partial x_{0_{k+1}}} & \cdots & \dfrac{\partial x_2(T, \boldsymbol{x}_0)}{\partial x_{0_n}} \\ \vdots & \cdots & \vdots & \vdots & \vdots & & \\ \dfrac{\partial x_k(T, \boldsymbol{x}_0)}{\partial x_{0_1}} & \cdots & \dfrac{\partial x_k(T, \boldsymbol{x}_0)}{\partial x_{0_{k-1}}} & \dfrac{\partial x_k(T, \boldsymbol{x}_0)}{\partial T} & \dfrac{\partial x_k(T, \boldsymbol{x}_0)}{\partial x_{0_{k+1}}} & \cdots & \dfrac{\partial x_k(T, \boldsymbol{x}_0)}{\partial x_{0_n}} \\ \vdots & \cdots & \vdots & \vdots & \vdots & & \\ \dfrac{\partial x_n(T, \boldsymbol{x}_0)}{\partial x_{0_1}} & \cdots & \dfrac{\partial x_n(T, \boldsymbol{x}_0)}{\partial x_{0_{k-1}}} & \dfrac{\partial x_n(T, \boldsymbol{x}_0)}{\partial T} & \dfrac{\partial x_n(T, \boldsymbol{x}_0)}{\partial x_{0_{k+1}}} & \cdots & \dfrac{\partial x_n(T, \boldsymbol{x}_0)}{\partial x_{0_n}} \end{bmatrix} \tag{17-142}$$

Comparing Eq. (17-142) with Eq. (17-124), we observe that, except for column k, $\boldsymbol{F}'(T, \boldsymbol{x}_0)$ is identical to $\boldsymbol{F}'(\boldsymbol{x}_0)$ and hence can be determined by the methods already presented in Section 17-5. To evaluate the kth column, we observe from Eq. (17-112) that

$$\frac{\partial x_j(T, \boldsymbol{x}_0)}{\partial T} = f_j(\boldsymbol{x}(T, \boldsymbol{x}_0)), \qquad j = 1, 2, \ldots, n \tag{17-143}$$

[11] Observe that the Jacobian matrix $\boldsymbol{H}'(\boldsymbol{y}) \triangleq \partial \boldsymbol{H}(\boldsymbol{y})/\partial \boldsymbol{y}$ in question here requires taking partial derivatives with respect to the variables in $\boldsymbol{y}$, and the kth element of $\boldsymbol{y}$ is T, not x_{0_k}!

Now if x_j represents the voltage v_{C_j} of a *nonlinear capacitor* with incremental capacitance $C_j(v_{C_j})$, then

$$f_j(\boldsymbol{x}(T, \boldsymbol{x}_0)) = \dot{v}_{C_j}(T, \boldsymbol{x}_0) = \frac{i_{C_j}(T, \boldsymbol{x}_0)}{C_j(v_{C_j}(T, \boldsymbol{x}_0))} \tag{17-144}$$

Similarly, if x_j represents the current i_{L_j} of a *nonlinear inductor* with incremental inductance $L_j(i_{L_j})$, then

$$f_j(\boldsymbol{x}(T, \boldsymbol{x}_0)) = \dot{i}_{L_j}(T, \boldsymbol{x}_0) = \frac{v_{L_j}(T, \boldsymbol{x}_0)}{L_j(i_{L_j}(T, \boldsymbol{x}_0))} \tag{17-145}$$

To initiate the Newton–Raphson iteration given by Eq. (17-141), we must as usual choose an initial guess $\boldsymbol{y}^{(0)} = [x_{0_1}^{(0)} \quad x_{0_2}^{(0)} \quad \cdots \quad x_{0_{k-1}}^{(0)} \quad T^{(0)} \quad x_{0_{k+1}}^{(0)} \quad \cdots \quad x_n^{(0)}]^t$ in order to compute $\boldsymbol{F}'(T^{(0)}, \boldsymbol{x}_0^{(0)})$ from Eq. (17-142) and then evaluate $\boldsymbol{y}^{(1)}$ from Eq. (17-141). Substituting $T^{(1)}$ and $\boldsymbol{x}_0^{(1)}$ from $\boldsymbol{y}^{(1)}$ into Eq. (17-142), we can compute $\boldsymbol{F}'(T^{(1)}, \boldsymbol{x}_0^{(1)})$, and hence $\boldsymbol{y}^{(2)}$, etc. Observe that, although the vector $\boldsymbol{x}_0^{(j)}$ denotes the jth iterated vector for $\boldsymbol{x}_0$, the kth component remains unchanged throughout all iterations; i.e.,[12]

$$x_{0_k}^{(j)} = p_{k_0}, \qquad j = 0, 1, 2, \ldots \tag{17-146}$$

So far we have implicitly assumed that the *inverse* of the matrix in Eq. (17-141) exists. It is shown in [4] that this assumption is valid as long as

$$\boxed{f_k(\boldsymbol{x}(T^{(j)}, \boldsymbol{x}_0^{(j)})) \neq 0, \qquad j = 0, 1, 2, \ldots} \tag{17-147}$$

This condition furnishes us with a strategy for choosing our particular kth component x_{0_k}, which so far has been an arbitrary choice; let us choose that component k such that

$$|f_k(\boldsymbol{x}(T^{(j)}, \boldsymbol{x}_0^{(j)}))| \geq |f_m(\boldsymbol{x}(T^{(j)}, \boldsymbol{x}_0^{(j)}))|, \quad m = 1, 2, \ldots, k-1, k+1, \ldots, n \tag{17-148}$$

Moreover, if we choose

$$x_{0_k}^{(j)} = x_{0_k}^{(j)}(T^{(j)}, \boldsymbol{x}_0^{(j)}) \triangleq p_{k_0}^{(j)} \tag{17-149}$$

then not only are we guaranteed that Eq. (17-147) is satisfied, but also $p_{k_0}^{(j)}$ will almost be certain to satisfy Eq. (17-138). This is because $p_{k_{\min}}$ and $p_{k_{\max}}$ occur at $\dot{x}_k(t) \triangleq f_k(\boldsymbol{x}(t)) = 0$, and we know that $f_k(\boldsymbol{x}(T^{(j)}, \boldsymbol{x}_0^{(j)})) \neq 0$. Hence, rather than choosing the component k and the initial state x_{0_k} arbitrarily, the preceding strategy allows us to determine them automatically. Using this strategy, the component k is no longer fixed (as was originally assumed), and the variable T in the vector $\boldsymbol{y}$ defined in Eq. (17-139) will shift in accordance with the location k. Similarly, Eq. (17-146) no longer stays invariant and must be replaced by Eq. (17-149). The Newton–Raphson iteration given by Eq. (17-141) still applies and will converge whenever the initial guess is sufficiently close to the correct solution. Observe that the final solution consists of not

[12] An alternative interpretation is to augment the system given by Eq. (17-113′) with the trivial but independent equation $x_{0_k} = p_{n_0}$, thereby obtaining a system of $n+1$ independent equations in $n+1$ unknowns.

only a desired *initial state* $\boldsymbol{p}_0 = [p_{1_0} \quad p_{2_0} \quad \cdots \quad p_{n_0}]$ on the periodic orbit, but also the exact value of the period T of oscillation.

We shall now summarize the *algorithm for determining the periodic solution of a nonlinear oscillator*, as follows:

Step 1. Compute the transient response $\boldsymbol{x}(t, \boldsymbol{x}_0^{(j)})$ of the network $\mathfrak{N}$ from $t = 0$ to $t = T^{(j)}$ with initial state $\boldsymbol{x}_0 = \boldsymbol{x}_0^{(j)}$.

Step 2. Choose k such that Eq. (17-148) is satisfied, and choose $x_{0_k}^{(j)}$ as given by Eq. (17-149).

Step 3. Compute the *sensitivity matrix*

$$\boldsymbol{F}'(T^{(j)}, \boldsymbol{x}_0^{(j)}) \triangleq \left.\frac{\partial \boldsymbol{F}(T, \boldsymbol{x}_0)}{\partial \boldsymbol{x}_0}\right|_{\boldsymbol{x}=\boldsymbol{x}_0^{(j)}}$$

by using either the numerical differentiation formula given by Eq. (17-125), or by analyzing $n - 1$ associated sensitivity networks $\overline{\mathfrak{N}}(x_{0_m})$, $m = 1, 2, \ldots, k-1, k+1, \ldots, n$. The kth column of this matrix is computed using Eqs. (17-144) and (17-145).

Step 4. Compute $\boldsymbol{y}^{(j+1)}$, and hence $\boldsymbol{x}_0^{(j+1)}$ and $T^{(j+1)}$, from Eq. (17-141).

Step 5. Return to step 1 unless $\|\boldsymbol{x}(T^{(j)}, \boldsymbol{x}_0^{(j)}) - \boldsymbol{x}_0^{(j)}\| < \varepsilon$ and $\|\boldsymbol{x}_0^{(j+1)} - \boldsymbol{x}_0^{(j)}\| < \delta$, where ε and δ are (user-specified) arbitrarily small positive numbers.

Step 6. Stop.

17-7 SPECTRUM AND DISTORTION ANALYSIS OF NONLINEAR COMMUNICATION CIRCUITS

Communication circuits such as mixers and modulators often have two or more inputs. In analyzing these circuits, the inputs are often assumed to be periodic sine waves of two or more distinct frequencies. For simplicity, let us suppose that there are only two frequencies ω_1 and ω_2. If these two frequencies are *commensurable*, i.e., if there exist two integers n_1 and n_2 having no common factors such that $\omega_2/\omega_1 = n_2/n_1$, then the inputs have a common period, $T = n_1 T_1 = n_2 T_2$, where $T_1 = 2\pi/\omega_1$ and $T_2 = 2\pi/\omega_2$. If the circuit under consideration is *linear*, it follows from the superposition principle that the voltage and current associated with each element in the circuit are also sinusoid with period T. In general, the response of a *linear circuit to two or more sinusoidal inputs of commensurable frequencies is always a sinusoid, and can therefore be represented by a phasor at frequency* $\omega = 2\pi/T$, where T is the least common period of the sinusoidal inputs. This phasor can be computed by at least two methods. One method applies the *periodic steady-state algorithm* presented in Section 17-4 with T as the period. Since the circuit is linear, the algorithm will converge in one iteration. The second method would be to carry out an ac frequency-domain analysis at each frequency ω_i of the sinusoidal inputs and then use superposition to find the phasor at the output frequency ω.

Communication circuits, however, are *nonlinear* by nature. In general, *harmonic frequencies* $m\omega_i$ and *combination or beat frequencies* $\omega = m_1\omega_1 \pm m_2\omega_2 \pm \cdots \pm m_k\omega_k$ of all orders will be generated. If all detailed aspects of the solution waveform are desired, or if the input frequencies are *not* commensurable, one has no choice but to analyze the circuit as an initial-value problem, and numerically integrate the associated system of *state equations* using the algorithm presented in Chapter 13, or to solve an associated system of *implicit algebraic-differential equations* using the techniques presented in Sections 17-2 and 17-3. On many occasions, however, one is interested only in determining the *spectrum*, i.e., the sinusoidal frequency components, of the periodic output waveform in steady state. In this case, the *time-domain* periodic response can be determined first using the algorithm presented in Section 17-4, and then the associated spectrum can be determined by *fast Fourier transform* (FFT) techniques [7].[13]

17-7-1 Distortion Analysis of Quasi-Linear Communication Circuits

There is an important class of communication circuits that ideally should be linear, but is never completely linear because of the inherent nonlinearity of the devices used in the circuits. For example, we know from Chapter 2 that the Ebers–Moll model for bipolar transistors contains among other things, *nonlinear resistors, nonlinear controlled sources, and nonlinear capacitors.* In the design of small-signal amplifiers, we often replace this model by a small-signal *linear incremental hybrid-pi model.* This approach is valid, because for sufficiently small signals, the higher harmonic components (for the single-frequency input case) as well as the *beat-frequency* (for the multifrequency input case) components are much smaller than the fundamental component and are therefore negligible. For long-distance communication circuits, such as frequency-division wideband coaxial analog systems or cable TV systems, however, a *repeater amplifier* must be inserted every few miles to maintain signal strength, and it has been observed that the *undesirable* frequency components have a tendency to become additive along the line and hence can no longer be neglected. These undesirable components are generally called *distortions* and must be kept to an extremely low level, e.g., -80 to -100 dB relative to the fundamental signal component. For this class of problems, only the first few harmonics and beat-frequency components are appreciable. For example, if the input consists of three sinusoids at frequencies ω_1, ω_2, and ω_3, then the distortion is usually dominated by the harmonic components $2\omega_i$, $3\omega_i$, $i = 1, 2, 3$, and the beat-frequency components $\omega_1 \pm \omega_2$, $\omega_1 \pm \omega_3$, $\omega_2 \pm \omega_3$, and $\omega_1 \pm \omega_2 \pm \omega_3$. The *harmonic components are usually called distortion products;* the *beat-frequency components are usually called intermodulation distortion products.* Since each distortion product pertains to a *sinusoidal* waveform at a harmonic

[13]The *fast Fourier transform* is a highly efficient numerical technique for signal analysis, and programs for implementing this technique are available as subroutines at most computing centers.

or beat frequency ω_d, it suffices to determine the *phasor* associated with this sinusoid. Since the magnitude of each distortion-product phasor is much smaller compared to the magnitude of the fundamental-signal phasors, the circuit is said to be *quasi-linear*. For this class of circuits, there are at least three special techniques for computing distortion-product phasors: the *perturbation approach* [8], the *Volterra series approach* [9, 10], and the *Picard iteration approach* [11]. We shall present only the perturbation approach since it can be easily implemented with the help of an *ac linear circuit analysis* program. In contrast to this, the Volterra series approach is much more *algebraic* and somewhat circuit dependent in nature, and is therefore difficult to implement into a general simulation program.[14] The Picard iteration approach requires the use of an efficient fast Fourier transform package, and the iteration may not converge in some cases. However, unlike the perturbation approach in which one distortion product is obtained at a time, the Picard iteration approach yields the entire spectrum of interest in one analysis.

17-7-2 Low-Distortion Analysis by the Perturbation Approach

To present the main ideas of this approach, let us consider a network containing, in addition to *linear elements* and *independent sources*, a *nonlinear resistor* characterized by $I_R = \hat{I}_R(V_R)$, a *nonlinear inductor* characterized by $I_L = \hat{I}_L(\Phi_L)$, and a *nonlinear capacitor* characterized by $Q_C = \hat{Q}_C(V_C)$.[15] Let the network be driven by two sinusoidal sources of frequencies ω_1 and ω_2. Let us extract all *independent* sources and all nonlinear elements as shown in Fig. 17-6(a). The *n*-port $\mathfrak{N}$ contains all *linear* resistors, inductors, capacitors, and *linear coupled* elements, such as linear controlled sources and ideal transformers. Observe that we also added an output port across which is the output voltage $V_o(t)$ of main interest. The dc voltage and current sources are denoted by a single battery and a single current source. The magnitudes V_{s1} and V_{s2} of the two sinusoidal sources are assumed to be *very small* throughout this section in order that our circuit can be considered to be *quasi-linear*. Let us determine first the dc operating point of the circuit by carrying out a *dc analysis* with $V_{s1} = V_{s2} = 0$. Let us next expand the nonlinear resistor characteristic $I_R = \hat{I}_R(V_R)$ about its dc operating point $(I_{R_{dc}}, V_{R_{dc}})$ as follows:

$$\begin{aligned} I_R = I_{R_{dc}} &+ \left.\frac{\partial \hat{I}_R(V_R)}{\partial V_R}\right|_{V_R = V_{R_{dc}}} (V_R - V_{R_{dc}}) \\ &+ \frac{1}{2} \left.\frac{\partial^2 \hat{I}_R(V_R)}{\partial V_R^2}\right|_{V_R = V_{R_{dc}}} (V_R - V_{R_{dc}})^2 + \text{higher-order terms} \end{aligned} \tag{17-150a}$$

[14]Because of its algebraic nature, the Volterra series approach does have the advantage that, for a specific simple circuit, it is often possible to derive *closed-form* expressions for the lower-order distortion products.

[15]Once the main concept is well understood, it becomes a trivial matter to generalize the perturbation approach to allow more than one *nonlinear* resistor, inductor, and capacitor, as well as more than two sinusoidal input sources.

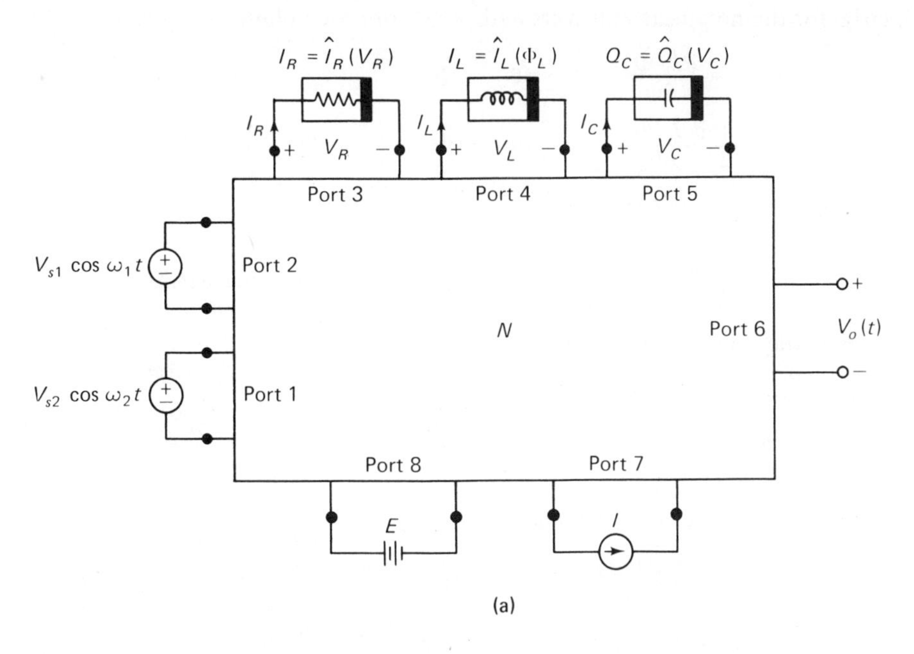

(a)

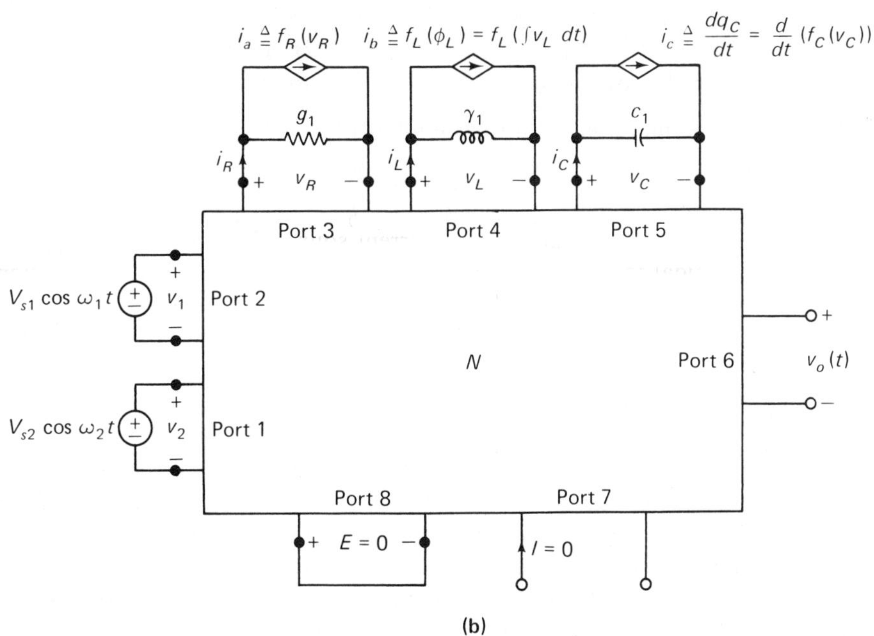

(b)

Fig. 17-6. Quasi-linear circuit shown in (a) can be represented by the *incremental circuit* shown in (b).

Similarly, for the nonlinear inductors and capacitors, we obtain

$$\begin{aligned} I_L = I_{L_{dc}} &+ \left.\frac{\partial \hat{I}_L(\Phi_L)}{\partial \Phi_L}\right|_{\Phi_L=\Phi_{L_{dc}}} (\Phi_L - \Phi_{L_{dc}}) \\ &+ \frac{1}{2} \left.\frac{\partial^2 \hat{I}_L(\Phi_L)}{\partial \Phi_L^2}\right|_{\Phi_L=\Phi_{L_{dc}}} (\Phi_L - \Phi_{L_{dc}})^2 + \text{higher-order terms} \end{aligned} \tag{17-150b}$$

$$\begin{aligned} Q_C = Q_{C_{dc}} &+ \left.\frac{\partial \hat{Q}_C(V_C)}{\partial V_C}\right|_{V_C=V_{C_{dc}}} (V_C - V_{C_{dc}}) \\ &+ \frac{1}{2} \left.\frac{\partial^2 \hat{Q}_C(V_C)}{\partial V_C^2}\right|_{V_C=V_{C_{dc}}} (V_C - V_{C_{dc}})^2 + \text{higher-order terms} \end{aligned} \tag{17-150c}$$

If we define the following *incremental ac variables*,

$$\begin{aligned} &i_R \triangleq I_R - I_{R_{dc}}, \quad &&v_R \triangleq V_R - V_{R_{dc}} \\ &i_L \triangleq I_L - I_{L_{dc}}, \quad &&v_L \triangleq V_L - V_{L_{dc}}, \quad &&\phi_L \triangleq \Phi_L - \Phi_{L_{dc}} \\ &i_C \triangleq I_C - I_{C_{dc}}, \quad &&v_C \triangleq V_C - V_{C_{dc}}, \quad &&q_C \triangleq Q_C - Q_{C_{dc}} \end{aligned}$$

and the following *incremental parameters*,

$$\begin{aligned} g_1 &\triangleq \left.\frac{\partial \hat{I}_R(V_R)}{\partial V_R}\right|_{V_R=V_{R_{dc}}}, \quad && g_2 \triangleq \frac{1}{2} \left.\frac{\partial^2 \hat{I}_R(V_R)}{\partial V_R^2}\right|_{V_R=V_{R_{dc}}} \\ \gamma_1 &\triangleq \left.\frac{\partial \hat{I}_L(\Phi_L)}{\partial \Phi_L}\right|_{\Phi_L=\Phi_{L_{dc}}}, \quad && \gamma_2 \triangleq \frac{1}{2} \left.\frac{\partial^2 \hat{I}_L(\Phi_L)}{\partial \Phi_L^2}\right|_{\Phi_L=\Phi_{L_{dc}}} \\ c_1 &\triangleq \left.\frac{\partial \hat{Q}_C(V_C)}{\partial V_C}\right|_{V_C=V_{C_{dc}}}, \quad && c_2 \triangleq \frac{1}{2} \left.\frac{\partial^2 \hat{Q}_C(V_C)}{\partial V_C^2}\right|_{V_C=V_{C_{dc}}} \end{aligned}$$

then Eq. (17-150) assumes the following simplified form:

$$i_R = g_1 v_R + f_R(v_R) \tag{17-151a}$$

$$i_L = \gamma_1 \phi_L + f_L(\phi_L) \tag{17-151b}$$

$$q_C = c_1 v_C + f_C(v_C) \tag{17-151c}$$

where

$$f_R(v_R) \triangleq g_2 v_R^2 + \text{higher-order terms in } v_R \tag{17-152a}$$

$$f_L(\phi_L) \triangleq \gamma_2 \phi_L^2 + \text{higher-order terms in } \phi_L \tag{17-152b}$$

$$f_C(v_C) \triangleq c_2 v_C^2 + \text{higher-order terms in } v_C \tag{17-152c}$$

It follows from Eq. (17-151) that about the dc operating point the circuit shown in Fig. 17-6(a) can be represented by the *incremental circuit* shown in Fig. 17-6(b). Observe that the *nonlinear* components in Eq. (17-151) are lumped together and are represented by a *nonlinear voltage-controlled current source* in Fig. 17-6(b).

To develop the perturbation method, let us introduce a dimensionless variable ε in Eq. (17-151) and in the circuit of Fig. 17-6(b) to obtain

$$i_R = g_1 v_R + \varepsilon f_R(v_R) \tag{17-153a}$$

$$i_L = \gamma_1 \phi_L + \varepsilon f_L(\phi_L) \tag{17-153b}$$

$$q_C = c_1 v_C + \varepsilon f_C(v_C) \tag{17-153c}$$

The solution $v_R(t)$, $v_L(t)$, $v_C(t)$, and $v_o(t)$ for the circuit shown in Fig. 17-6(b) (with the parameter ε multiplying the controlled current sources) will of course depend on ε, and should be denoted by $v_R(t, \varepsilon)$, $v_L(t, \varepsilon)$, $v_C(t, \varepsilon)$, and $v_o(t, \varepsilon)$. The correct solution is obtained by simply setting $\varepsilon = 1$ in these expressions. Now, at each time instant t, we can expand solution waveforms as a *power series* in ε. To see this, let us expand $v_R(t, \varepsilon)$ by a Taylor series about $\varepsilon = 1$ to obtain

$$v_R(t, \varepsilon) = v_R(t, 1) + \frac{\partial v_R(t, \varepsilon)}{\partial \varepsilon}\bigg|_{\varepsilon=1} (\varepsilon - 1) + \frac{1}{2}\frac{\partial^2 v_R(t, \varepsilon)}{\partial \varepsilon^2}\bigg|_{\varepsilon=1} (\varepsilon - 1)^2 \qquad (17\text{-}154)$$
$$+ \text{ higher-order terms in } (\varepsilon - 1)$$

This series converges as long as ε is close to 1. If we expand each term in Eq. (17-154) and rearrange them such that all terms multiplying ε^j are collected together, $j = 0, 1, 2, \ldots$, then Eq. (17-154) can be written as follows:

$$v_R(t, \varepsilon) = v_{R1}(t) + \varepsilon v_{R2}(t) + \varepsilon^2 v_{R3}(t) \ldots \qquad (17\text{-}155a)$$

where $v_{R_1}(k)$, $v_{R_2}(t)$, and $v_{R_3}(t)$ denote the collection of all terms in the expansion which are multiplied with ε^0, ε^1, and ε^2, respectively. Similarly, we can write

$$v_L(t, \varepsilon) = v_{L1}(t) + \varepsilon v_{L2}(t) + \varepsilon^2 v_{L3}(t) + \ldots \qquad (17\text{-}155b)$$

$$v_C(t, \varepsilon) = v_{C1}(t) + \varepsilon v_{C2}(t) + \varepsilon^2 v_{C3}(t) + \ldots \qquad (17\text{-}155c)$$

$$v_o(t, \varepsilon) = v_{o1}(t) + \varepsilon v_{o2}(t) + \varepsilon^2 v_{o3}(t) + \ldots \qquad (17\text{-}155d)$$

Now if we consider the controlled sources in Fig. 17-6(b) as *independent sources*, then the network can be considered *linear* and the output voltage $v_o(t, \varepsilon)$ can be obtained by superposition:

$$\begin{aligned} v_o(t, \varepsilon) = &\int_{-\infty}^{t} h_{61}(t - \tau)v_1(\tau)\,d\tau + \int_{-\infty}^{t} h_{62}(t - \tau)v_2(\tau)\,d\tau \\ &- \int_{-\infty}^{t} h_{63}(t - \tau)i_a(\tau)\,d\tau - \int_{-\infty}^{t} h_{64}(t - \tau)i_b(\tau)\,d\tau \\ &- \int_{-\infty}^{t} h_{65}(t - \tau)i_c(\tau)\,d\tau \end{aligned} \qquad (17\text{-}156)$$

where $v_1(t) \triangleq V_{s1} \cos \omega_1 t$, $v_2(t) \triangleq V_{s2} \cos \omega_2 t$, and where $h_{6j}(t)$ denotes the *impulse response* from port j to port 6.[16] Observe that the last three terms in Eq. (17-156) cannot yet be evaluated, because $i_a(t) \triangleq \varepsilon f_R(v_R(t))$, $i_b(t) \triangleq \varepsilon f_L(\phi_L(t))$, and $i_c(t) \triangleq \varepsilon(d/dt)(f_c(v_c(t))$ depend on the *unknown* time functions $v_R(t)$, $\phi_L(t) \triangleq \int v_L(t)\,dt$, and $v_c(t)$. However, when $\varepsilon = 0$, Eq. (17-155) reduces to $v_R(t) = v_{R1}(t)$, $v_L(t) = v_{L1}(t)$, $v_C(t) = v_{C1}(t)$, and $v_o(t) = v_{o1}(t)$. These time functions can be obtained by solving the *linear circuit* in Fig. 17-6(b) with the controlled current sources left out (since $i_a = i_b = i_c = 0$ when $\varepsilon = 0$). By superposition, we know these solutions must assume the form

$$v_{R1}(t) = V_{R,\omega_1} \cos(\omega_1 t + \theta_{R,\omega_1}) + V_{R,\omega_2} \cos(\omega_2 t + \theta_{R,\omega_2}) \qquad (17\text{-}157a)$$

$$v_{L1}(t) = V_{L,\omega_1} \cos(\omega_1 t + \theta_{L,\omega_1}) + V_{L,\omega_2} \cos(\omega_2 t + \theta_{L,\omega_2}) \qquad (17\text{-}157b)$$

[16]The last three terms in Eq. (17-156) are preceded by a *negative sign* because the reference current direction for the controlled sources is defined *leaving* the ports, whereas the impulse response is defined in terms of current entering the ports.

$$v_{C1}(t) = V_{C,\omega_1} \cos(\omega_1 t + \theta_{C,\omega_1}) + V_{C,\omega_2} \cos(\omega_2 t + \theta_{C,\omega_2}) \tag{17-157c}$$

$$v_{o1}(t) = V_{o,\omega_1} \cos(\omega_1 t + \theta_{o,\omega_1}) + V_{o,\omega_2} \cos(\omega_2 t + \theta_{o,\omega_2}) \tag{17-157d}$$

We shall now show that it is possible to compute $v_{R2}(t)$, $v_{L2}(t)$, $v_{C2}(t)$ and $v_{o2}(t)$ from $v_{R1}(t)$, $v_{L1}(t)$, $v_{C1}(t)$, and $v_{o1}(t)$. To do this, we *neglect* the higher-order terms in Eq. (17-152) and substitute Eq. (17-155) into Eq. (17-152) to obtain

$$\begin{aligned} i_a \triangleq \varepsilon f_R(v_R) &= \varepsilon g_2\{v_{R1}(t) + \varepsilon v_{R2}(t) + \varepsilon^2 v_{R3}(t) + \ldots\}^2 \\ &= \varepsilon g_2\{v_{R1}^2 + \varepsilon[2v_{R1}v_{R2}] + \varepsilon^2[v_{R2}^2 + 2v_{R1}v_{R3}] + \varepsilon^3[\ldots]\} \end{aligned} \tag{17-158a}$$

$$\begin{aligned} i_b \triangleq \varepsilon f_L(\phi_L) &= \varepsilon \gamma_2 \left\{ \int v_{L1}(t)\,dt + \varepsilon \int v_{L2}(t)\,dt + \varepsilon^2 \int v_{L3}(t)\,dt + \ldots \right\}^2 \\ &= \varepsilon \gamma_2 \left\{ \left[\int v_{L1}\,dt \right]^2 + \varepsilon \left[2\left(\int v_{L1}\,dt \right)\left(\int v_{L2}\,dt \right) \right] \right. \\ &\quad \left. + \varepsilon^2 \left[\left(\int v_{L2}\,dt \right)^2 + 2\left(\int v_{L1}\,dt \right)\left(\int v_{L3}\,dt \right) \right] + \varepsilon^3[\ldots] \right\} \end{aligned} \tag{17-158b}$$

$$\begin{aligned} q_C \triangleq \varepsilon f_C(v_C) &= \varepsilon c_2\{v_{C1}(t) + \varepsilon v_{C2}(t) + \varepsilon^2 v_{C3}(t) + \ldots\}^2 \\ &= \varepsilon c_2\{v_{C1}^2 + \varepsilon[2v_{C1}v_{C2}] + \varepsilon^2[v_{C2}^2 + 2v_{C1}v_{C3}] + \varepsilon^3[\ldots]\} \end{aligned} \tag{17-158c}$$

If we substitute Eq. (17-158) for i_a, i_b, and $i_C \triangleq \dot{q}_C(t)$ in Eq. (17-156) and collect the terms, multiplying ε, ε^2, ε^3, etc., together, we obtain

$$\begin{aligned} v_o(t, \varepsilon) &= \int_{-\infty}^{t} h_{61}(t-\tau)v_1(\tau)\,d\tau + \int_{-\infty}^{t} h_{62}(t-\tau)v_2(\tau)\,d\tau \\ &\quad - \varepsilon \left\{ \int_{-\infty}^{t} h_{63}(t-\tau)[g_2 v_{R1}^2(\tau)]\,d\tau \right. \\ &\quad + \int_{-\infty}^{t} h_{64}(t-\tau)\gamma_2 \left[\int_0^{\tau} v_{L1}(t)\,dt \right]^2 d\tau \\ &\quad \left. + \int_{-\infty}^{t} h_{65}(t-\tau)\frac{d}{d\tau}[c_2 v_{C1}^2(\tau)]\,d\tau \right\} \\ &\quad - \varepsilon^2 \left\{ \int_{-\infty}^{t} h_{63}(t-\tau)[2g_2 v_{R1}(\tau)v_{R2}(\tau)]\,d\tau \right. \\ &\quad + \int_{-\infty}^{t} h_{64}(t-\tau)\left[2\gamma_2 \left(\int_0^{\tau} v_{L1}(t)\,dt \right)\left(\int_0^{\tau} v_{L2}(t)\,dt \right) \right] d\tau \\ &\quad \left. + \int_{-\infty}^{t} h_{65}(t-\tau)\frac{d}{d\tau}[2c_2 v_{C1}(\tau)v_{C2}(\tau)]\,d\tau \right\} \\ &\quad - \varepsilon^3\{\ldots\} - \ldots \end{aligned} \tag{17-159}$$

Now recall that $v_o(t, \varepsilon)$ is given earlier by Eq. (17-155d) as a power series in ε. Hence, Eqs. (17-155d) and (17-159) are identical *for all values of* ε. In particular, when $\varepsilon = 0$, the corresponding terms not multiplied by ε in these equations must certainly be equal. If we increase ε by an infinitesimal amount from zero, the terms multiplied by ε will begin to become important while those multiplied by ε^2, ε^3, etc., will remain negligible. Hence, the coefficients of ε in Eqs. (17-155d) and (17-159) must also be equal. By the same reasoning, it follows that in fact the coefficients of ε^2, ε^3, etc., in

Eqs. (17-155d) and (17-159) must be equal. Hence, equating the respective coefficients of ε^0, ε^1, ε^2, etc., we obtain

$$\varepsilon^0: \quad v_{o1}(t) = \int_{-\infty}^{t} h_{61}(t-\tau)v_1(\tau)\,d\tau + \int_{-\infty}^{t} h_{62}(t-\tau)v_2(\tau)\,d\tau \tag{17-160a}$$

where

$$v_1(t) \triangleq V_{s1}\cos\omega_1 t \quad \text{and} \quad v_2(t) \triangleq V_{s2}\cos\omega_2 t.$$

$$\begin{aligned}\varepsilon^1: \quad v_{o2}(t) = &-\int_{-\infty}^{t} h_{63}(t-\tau)[g_2 v_{R1}^2(\tau)]\,d\tau \\ &-\int_{-\infty}^{t} h_{64}(t-\tau)\gamma_2\left[\int_0^{\tau} v_{L1}(t)\,dt\right]^2 d\tau \\ &-\int_{-\infty}^{t} h_{65}(t-\tau)\frac{d}{d\tau}[c_2 v_{C1}^2(\tau)]\,d\tau\end{aligned} \tag{17-160b}$$

$$\begin{aligned}\varepsilon^2: \quad v_{o3}(t) = &-\int_{-\infty}^{t} h_{63}(t-\tau)[2g_2 v_{R1}(\tau)v_{R2}(\tau)]\,d\tau \\ &-\int_{-\infty}^{t} h_{64}(t-\tau)\left[2\gamma_2\left(\int_0^{\tau} v_{L1}(t)\,dt\right)\left(\int_0^{\tau} v_{L2}(t)\,dt\right)\right]d\tau \\ &-\int_{-\infty}^{t} h_{65}(t-\tau)\frac{d}{d\tau}[2c_2 v_{C1}(\tau)v_{C2}(\tau)]\,d\tau\end{aligned} \tag{17-160c}$$

Let us now focus our attention on Eq. (17-160b). Observe that it depends only on $v_{R1}^2(t)$, $[\int v_{L1}(t)\,dt]^2$, and $(d/dt)[v_{C1}^2(t)]$. These expressions can be determined from Eq. (17-157) with the help of the trigonometric identities $\cos^2 x = \frac{1}{2} + \frac{1}{2}\cos 2x$, $(\cos x)(\cos y) = \frac{1}{2}\cos(x+y) + \frac{1}{2}\cos(x-y)$, $\sin^2 x = \frac{1}{2} - \frac{1}{2}\cos 2x$, and $(\sin x)(\sin y) = \frac{1}{2}\cos(x-y) - \frac{1}{2}\cos(x+y)$; thus,

$$\begin{aligned}v_{R1}^2(t) &= [V_{R,\omega_1}\cos(\omega_1 t + \theta_{R,\omega_1})]^2 + [V_{R,\omega_2}\cos(\omega_2 t + \theta_{R,\omega_2})]^2 \\ &\quad + 2[V_{R,\omega_1}\cos(\omega_1 t + \theta_{R,\omega_1})][V_{R,\omega_2}\cos(\omega_2 t + \theta_{R,\omega_2})] \\ &= \tfrac{1}{2}(V_{R,\omega_1}^2 + V_{R,\omega_2}^2) + \tfrac{1}{2}[V_{R,\omega_1}^2\cos 2(\omega_1 t + \theta_{R,\omega_1})] \\ &\quad + \tfrac{1}{2}[V_{R,\omega_2}^2\cos 2(\omega_2 t + \theta_{R,\omega_2})] \\ &\quad + \{V_{R,\omega_1}V_{R,\omega_2}\cos[(\omega_1+\omega_2)t + (\theta_{R,\omega_1} + \theta_{R,\omega_2})]\} \\ &\quad + \{V_{R,\omega_1}V_{R,\omega_2}\cos[(\omega_1-\omega_2)t + (\theta_{R,\omega_1} - \theta_{R,\omega_2})]\}\end{aligned} \tag{17-161a}$$

$$\begin{aligned}\left[\int V_{L1}(t)\,dt\right]^2 &= \frac{1}{2}\left\{\left[\frac{V_{L,\omega_1}}{\omega_1}\right]^2 + \left[\frac{V_{L,\omega_2}}{\omega_2}\right]^2\right\} - \frac{1}{2}\left\{\left[\frac{V_{L,\omega_1}}{\omega_1}\right]^2\cos 2(\omega_1 t + \theta_{L,\omega_1})\right\} \\ &\quad - \frac{1}{2}\left\{\left[\frac{V_{L,\omega_2}}{\omega_2}\right]^2\cos 2(\omega_2 t + \theta_{L,\omega_2})\right\} \\ &\quad + \left\{\left[\frac{V_{L,\omega_1}}{\omega_1}\right]\left[\frac{V_{L,\omega_2}}{\omega_2}\right]\cos[(\omega_1-\omega_2)t + (\theta_{L,\omega_1} - \theta_{L,\omega_2})]\right\} \\ &\quad - \left\{\left[\frac{V_{L,\omega_1}}{\omega_1}\right]\left[\frac{V_{L,\omega_2}}{\omega_2}\right]\cos[(\omega_1+\omega_2)t + (\theta_{L,\omega_1} + \theta_{L,\omega_2})]\right\}\end{aligned} \tag{17-161b}$$

$$\frac{d}{dt}[v_{C1}^2(t)] = \frac{d}{dt}\Big(\frac{1}{2}(V_{C,\omega_1}^2 + V_{C,\omega_2}^2) + \frac{1}{2}[V_{C,\omega_1}^2 \cos 2(\omega_1 t + \theta_{C,\omega_1})]$$
$$+ \frac{1}{2}[V_{C,\omega_2}^2 \cos 2(\omega_2 t + \theta_{C,\omega_2})] \tag{17-161c}$$
$$+ \{V_{C,\omega_1} V_{C,\omega_2} \cos[(\omega_1 + \omega_2)t + (\theta_{C,\omega_1} + \theta_{C,\omega_2})]\}$$
$$+ \{V_{C,\omega_1} V_{C,\omega_2} \cos[(\omega_1 - \omega_2)t + (\theta_{C,\omega_1} - \theta_{C,\omega_2})]\}\Big)$$

Examination of Eq. (17-161) shows that each of the expressions $v_{R1}^2(t)$, $[\int v_{L1}(t)\,dt]^2$, and $(d/dt)[v_{C1}^2(t)]$ contains a constant term and four sinusoids of frequencies $2\omega_1$, $2\omega_2$, $(\omega_1 + \omega_2)$, and $(\omega_1 - \omega_2)$. Since the time integral or the time derivative of a sinusoid is again a sinusoid *of the same frequency*, it follows that if we substitute Eq. (17-161) into Eq. (17-160) and collect all terms having identical frequencies, we obtain

$$v_{o2}(t) = -\Big[k_{R1}\int_{-\infty}^{t} h_{63}(t-\tau)\,d\tau + k_{L1}\int_{-\infty}^{t} h_{64}(t-\tau)\,d\tau\Big]$$
$$-\Big\{\int_{-\infty}^{t} h_{63}(t-\tau) I_{R_1}\cos(2\omega_1\tau + \theta_{R_1})\,d\tau$$
$$-\int_{-\infty}^{t} h_{64}(t-\tau) I_{L_1}\cos(2\omega_1\tau + \theta_{L_1})\,d\tau$$
$$+\int_{-\infty}^{t} h_{65}(t-\tau) I_{C_1}\cos(2\omega_1\tau + \theta_{C_1})\,d\tau\Big\}$$
$$-\Big\{\int_{-\infty}^{t} h_{63}(t-\tau) I_{R_2}\cos(2\omega_2\tau + \theta_{R_2})\,d\tau$$
$$-\int_{-\infty}^{t} h_{64}(t-\tau) I_{L_2}\cos(2\omega_2\tau + \theta_{L_2})\,d\tau$$
$$+\int_{-\infty}^{t} h_{65}(t-\tau) I_{C_2}\cos(2\omega_2\tau + \theta_{C_2})\,d\tau\Big\} \tag{17-162}$$
$$-\Big\{\int_{-\infty}^{t} h_{63}(t-\tau) I_{R_3}\cos[(\omega_1+\omega_2)\tau + \theta_{R_3}]\,d\tau$$
$$-\int_{-\infty}^{t} h_{64}(t-\tau) I_{L_3}\cos[(\omega_1+\omega_2)\tau + \theta_{L_3}]\,d\tau$$
$$+\int_{-\infty}^{t} h_{65}(t-\tau) I_{C_3}\cos[(\omega_1+\omega_2)\tau + \theta_{C_3}]\,d\tau\Big\}$$
$$-\Big\{\int_{-\infty}^{t} h_{63}(t-\tau) I_{R_4}\cos[(\omega_1-\omega_2)\tau + \theta_{R_4}]\,d\tau$$
$$+\int_{-\infty}^{t} h_{64}(t-\tau) I_{L_4}\cos[(\omega_1-\omega_2)\tau + \theta_{L_4}]\,d\tau$$
$$+\int_{-\infty}^{t} h_{65}(t-\tau) I_{C_4}\cos[(\omega_1-\omega_2)\tau + \theta_{C_4}]\,d\tau\Big\}$$

where k_{R_1} and k_{L_1} are constants resulting from the dc terms in Eq. (17-161) and where the *amplitude* and *phase* for each sinusoid of frequency ω_a in Eq. (17-162) represents the *resulting* amplitude and phase of the phasor obtained by adding all phasors of the same frequency ω_d. Since we are interested in the periodic steady-state response, we can ignore the first two transient terms enclosed by brackets in Eq. (17-162) and focus our attention on the remaining terms which tend to a periodic waveform in steady state. Observe that *each* remaining term can be interpreted as the response due to a sinusoid of a particular frequency, as shown in Fig. 17-7. For obvious reasons, we shall call these *independent* current sources *distortion current sources.* The periodic steady-state output voltage $v_{o2}(t)$ can then be obtained by calculating its periodic

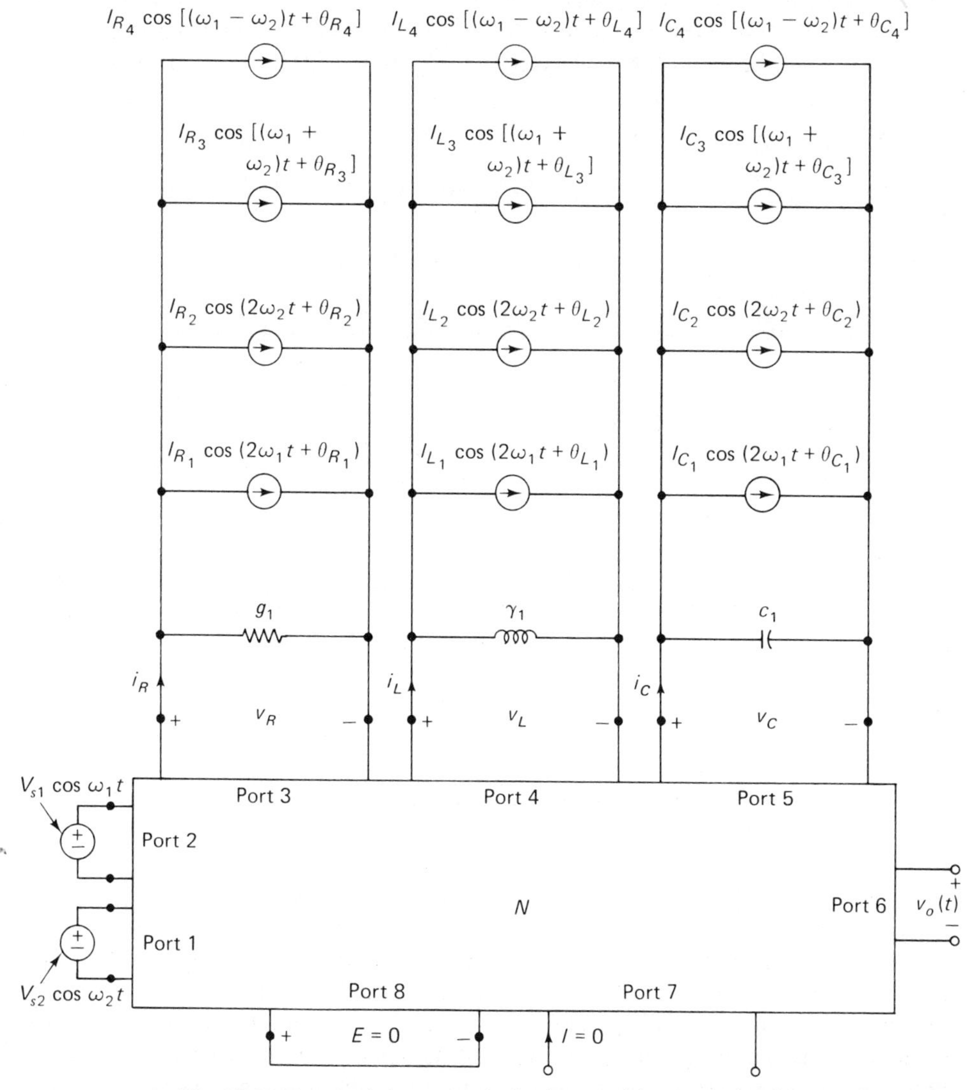

Fig. 17-7. Linearized incremental circuit containing second-order distortion current sources.

response due to each distortion current source separately and then adding them. Conversely, the harmonic or intermodulation distortion at a particular frequency ω_d can be determined by setting all independent voltage and current sources of frequency $\omega \neq \omega_d$ to zero. For example, to compute the intermodulation distortion at frequency $\omega_1 + \omega_2$, we "*short*" the two input voltage sources and "*open*" all distortion current sources except those with frequency $\omega_d = \omega_1 + \omega_2$, as shown in Fig. 17-8. Each distortion component of $v_{o2}(t)$ as computed from Fig. 17-7 is called a *second-order harmonic distortion* if $\omega_d = 2\omega_1$ or $2\omega_2$, and a *second-order intermodulation distortion* if $\omega_d = \omega_1 \pm \omega_2$. Observe that if we compute for the periodic steady voltage $v_R(t)$, $v_L(t)$, and $v_C(t)$ associated with the circuit shown in Fig. 17-7, but with the two input voltage sources shorted, we would obtain precisely $v_{R2}(t)$, $v_{L2}(t)$, and $v_{C2}(t)$.

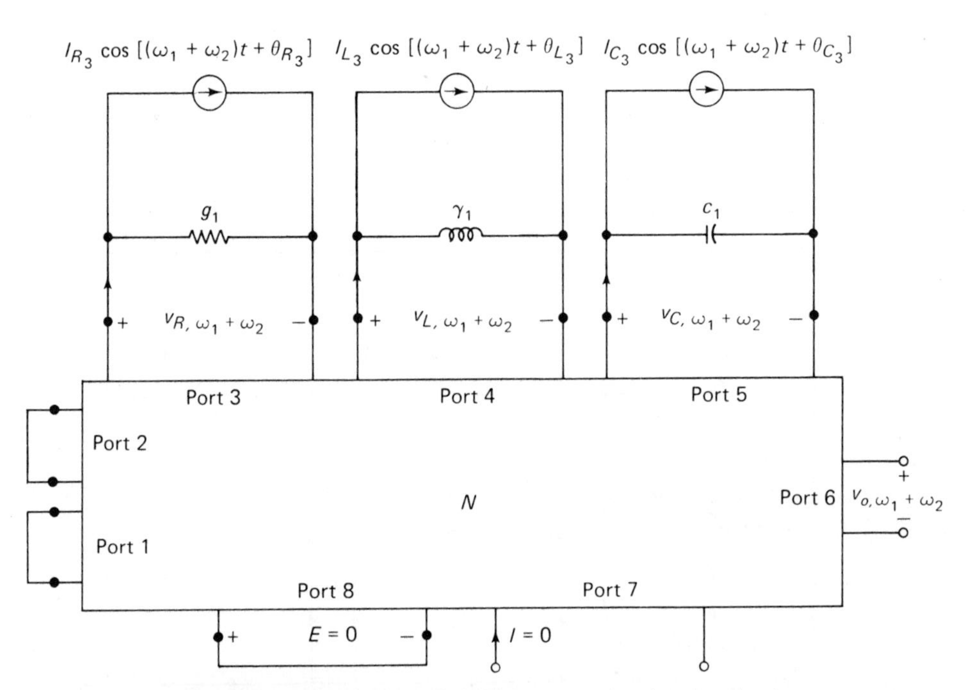

Fig. 17-8. Linear incremental circuit for computing the second-order intermodulation distortion component of frequency $\omega_d = \omega_1 + \omega_2$.

Now that we have computed $v_{R1}(t)$, $v_{L1}(t)$, $v_{C1}(t)$, $v_{R2}(t)$, $v_{L2}(t)$, and $v_{C2}(t)$, we can return to Eq. (17-159c) and repeat the procedure to determine $v_{o3}(t)$, as well as $v_{R3}(t)$, $v_{L3}(t)$, and $v_{C3}(t)$. Again, $v_{o3}(t)$ can be interpreted as the response due to *distortion current* sources at *third-order* harmonic and intermodulation frequencies. In view of our quasi-linear assumption, usually only one or two of these third-order distortion components is significant enough to warrant its determination. Otherwise, the computational task will be so enormous that this approach is no longer appealing. We shall now summarize the *low-distortion analysis algorithm* as follows:

Step 1. Find the dc operating point and expand each nonlinear function by a Taylor series about the operating point.
Step 2. Compute the periodic steady-state, first-order voltages $v_{o1}(t)$, $v_{R1}(t)$, $v_{L1}(t)$, and $v_{C1}(t)$ by neglecting all nonlinearities of degree $n \geq 2$.
Step 3. Determine the phasors associated with each second-order distortion current source.[17]
Step 4. Compute the second-order harmonic or intermodulation distortion component of $v_{o2}(t)$, $v_{R2}(t)$, $v_{L2}(t)$, and $v_{C2}(t)$ by analyzing the *linear circuit* with the appropriate distortion current sources at the prescribed frequency.
Step 5. To determine third-order harmonic or intermodulation distortion component of $v_{o3}(t)$, $v_{R3}(t)$, $v_{L3}(t)$, and $v_{C3}(t)$, calculate the phasors associated with each third-order distortion current source and go to step 4, with 2 replaced by 3.
Step 6. Stop.

This algorithm has been successfully applied to the distortion analysis of bipolar transistor circuits. The interested reader is referred to references 8 and 12 for the details.

REFERENCES

1. BRAYTON, R. K., F. G. GUSTAVSON, and G. D. HACHTEL. "A New Efficient Algorithm for Solving Differential-Algebraic Systems Using Implicit Backward-Differentiation Formulas." *Proc. IEEE,* Vol. 60, No. 1, pp. 98–108, Jan. 1972.
2. HACHTEL, G. D., R. K. BRAYTON, and F. G. GUSTAVSON. "The Sparse Tableau Approach to Network Analysis and Design." *IEEE Trans. Circuit Theory,* Vol. CT-18, No. 1, pp. 101–113, Jan. 1971.
3. APRILLE, T. J., Jr., and T. N. Trick. "Steady State Analysis of Nonlinear Circuits with Periodic Inputs." *Proc. IEEE,* Vol. 60, pp. 108–114, Jan. 1972.
4. APRILLE, T. J., Jr., and T. N. TRICK. "A Computer Algorithm to Determine the Steady-State Response of Nonlinear Oscillators." *IEEE Trans. Circuit Theory,* Vol. CT-19, pp. 354–360, July 1972.
5. COLLÒN, F. R., and T. N. Trick. "Fast Periodic Steady-State Analysis for Large-Signal Electronic Circuits." *IEEE J. Solid State Circuits,* Vol. SC-8, No. 4, pp. 260–269, Aug. 1973.
6. PONTRYAGIN, L. S. *Ordinary Differential Equations.* Reading, Mass.: Addison-Wesley Publishing Company, Inc., 1962.
7. BRIGHAM, E. O. *The Fast Fourier Transform.* Englewood Cliffs, N.J.: Prentice-Hall, Inc., 1974.
8. KUO, Y. L. "Distortion Analysis of Bipolar Transistor Circuits." *IEEE Trans. Circuit Theory,* Vol. CT-20, pp. 709–716, Nov. 1973.

[17]The phasors associated with the distortion current sources of a given order can usually be derived in explicit algebraic form; hence, step 3 in general requires very little computing time.

9. NARAYANAN, S. "Transistor Distortion Analysis Using Volterra Series Representations." *Bell System Tech. J.*, Vol. 46, No. 5, pp. 991–1024, May–June 1967.
10. NARAYANAN, S. "Application of Volterra Series to Intermodulation Distortion Analysis of Transistor Feedback Amplifier." *IEEE Trans. Circuit Theory*, Vol. CT-17, pp. 518–527, Nov. 1970.
11. NEILL, T. B. M. "Improved Method of Analyzing Nonlinear Electrical Networks." *Electron. Letters*, Vol. 5, No. 1, pp. 13–15, Jan. 9, 1969.
12. CHISHOLM, S. H., and L. W. NAGEL. "Efficient Computer Simulation of Distortion in Electronic Circuits." *IEEE Trans. Circuit Theory*, Vol. CT-20, pp. 742–745, Nov. 1973.
13. VAN BOKHOVEN, W. M. G. "Linear Implicit Differentiation Formulas of Variable Step and Order." *IEEE Trans. on Circuits and Systems*, Vol. CAS-22, pp. 109–115, Feb. 1975.

PROBLEMS

17-1. Construct the discrete circuit models associated with the *Adams–Moulton algorithm* of orders 3–6 for a nonlinear capacitor and a nonlinear inductor.

17-2. Construct the discrete circuit models associated with *Gear's algorithm* of orders 3–6 for a nonlinear capacitor and a nonlinear inductor.

17-3. Modify Gear's second-order algorithm so that the formula will allow *nonuniform* step sizes to be chosen. Construct an associated discrete circuit model for a nonlinear capacitor and a nonlinear inductor.

17-4. Construct a simple *linear* circuit containing a *loop* of three capacitors and derive its discrete circuit model associated with the backward Euler algorithm. Find the solution using two methods.
(a) Derive the state equation and integrate using the backward Euler formula.
(b) Solve the associated discrete circuit using the same step size and initial conditions $v_{C1}(0) = v_{C2}(0) = 1$.

17-5. Construct a simple *linear* circuit containing a *cutset* of three inductors and derive its discrete circuit model associated with the trapezoidal algorithm. Find the solution using two methods.
(a) Derive the state equation and integrate using the trapezoidal algorithm.
(b) Solve the associated discrete circuit using the same step size and initial conditions $i_{L1}(0) = i_{L2}(0) = 1$.

17-6. Construct a circuit containing a linear resistor, a linear inductor, a linear capacitor, an ideal two-port transformer, a current-controlled voltage source, an independent voltage source, and an independent current source. Construct the *tableau matrix* associated with this circuit.

17-7. Show that by choosing a particular row and column pivoting order the tableau approach reduces to the familiar *loop analysis.*

17-8. Show that by choosing a particular row and column pivoting order the tableau approach reduces to the familiar *cutset analysis.*

17-9. Show that by choosing a particular row and column pivoting order the tableau approach reduces to the familiar *state-variable analysis.*

17-10. Derive the coefficients $\alpha_0, \alpha_1, \ldots, \alpha_k$ for the kth-order *backward-differentiation formula* with a *uniform* step size h. Let $k = 1, 2, \ldots, 6$. Compare these coefficients with those for the *Adams–Moulton algorithm* of the same order.

17-11. The matrix defined by

$$V(x_1, x_2, \ldots, x_n) \triangleq \begin{bmatrix} 1 & 1 & 1 & \cdots & 1 \\ x_1 & x_2 & x_3 & \cdots & x_n \\ x_1^2 & x_2^2 & x_3^2 & \cdots & x_n^2 \\ \cdot & & & & \\ \cdot & & & & \\ \cdot & & & & \\ x_1^{n-1} & x_2^{n-1} & x_3^{n-1} & \cdots & x_n^{n-1} \end{bmatrix}$$

is called a *Vandermonde matrix* and is uniquely determined by specifying $x_1, x_2, \ldots, x_n$.

(a) Show that all Vandermonde matrices obey the following *shifting property*:

$$V(x_1 - a, x_2 - a, \ldots, x_{k+1} - a)^{-1} V(x_1, x_2, \ldots, x_{k+1}) = \boldsymbol{D}^{-1}\boldsymbol{P}_k\boldsymbol{D}$$

where $\boldsymbol{D}$ is a diagonal matrix with diagonal elements $1, a, a^2, \ldots, a^k$, and $\boldsymbol{P}_k$ is a $(k + 1) \times (k + 1)$ square matrix.

(b) Show that the matrix defining the BDF given in Eq. (17-73) is a Vandermonde matrix, and show that two such matrices corresponding to $t = t_{n+1}$ and $t = t_n$ are simply related by a shifting factor $a = 1$.

17-12. The polynomial defined by

$$p_k^n(t) \triangleq \sum_{i=0}^{k} \frac{\prod_{j=0}^{k} (t - t_{n-j}) \times (t_{n-i})}{(t - t_{n-i}) \prod_{\substack{j=0 \\ i \neq j}}^{k} (t_{n-i} - t_{n-j})}$$

is called the *Lagrange interpolation formula*. Show that this polynomial has the property that

$$p_k^n(t_j) = x(t_j), \qquad j = n, n - 1, \ldots, n - k$$

17-13. Show that a kth-degree polynomial $p_k^{n+1}(t)$ in the variable t which passes through $k + 1$ prescribed points $x(t_{n+1}), x(t_n), \ldots, x(t_{n+1-k})$ is given by

$$\begin{aligned} p_k^{n+1}(t) = a_0 &+ a_1(t - t_{n+1}) + a_2(t - t_{n+1})(t - t_n) \\ &+ a_3(t - t_{n+1})(t - t_n)(t - t_{n-1}) + \cdots \\ &+ a_k(t - t_{n+1})(t - t_n) \ldots (t - t_{n+2-k}) \end{aligned}$$

where

$$a_j \triangleq \sum_{i=0}^{j} \frac{x(t_{n+1-i})}{f'_j(t_{n+1-i})}$$

and

$$f_j(t) \triangleq \prod_{i=0}^{j} (t - t_{n+1-i})$$

In particular, show that

$$p_{k+1}^{n+1}(t) = p_k^{n+1}(t) + a_{k+1} \prod_{i=0}^{k} (t - t_{n+1-i})$$

where $p_{k+1}^{n+1}(t)$ is a polynomial which passes through the $k + 2$ points $x(t_{n+1}), x(t_n), \ldots, x(t_{n-k})$.

17-14. If we define the local truncation error

$$E_k \triangleq h[\dot{x}_{n+1}^{(k+1)} - \dot{x}_{n+1}^{(k)}]$$

show with the help of the relations from Problems 17-12 and 17-13 that we have

$$\begin{aligned} E_k &= h a_{k+1} \prod_{i=1}^{k} (t_{n+1} - t_{n+1-i}) \\ &= \left[\frac{h}{t_{n+1} - t_{n-k}}\right][x(t_{n+1}) - p_k^n(t_{n+1})] \\ &= \left[\frac{h}{t_{n+1} - t_{n-k}}\right][x_{n+1} - x_{n+1}^P] \end{aligned}$$

17-15. Construct a circuit containing a nonlinear resistor, a linear inductor, and a linear capacitor, and apply the *variable step-size, variable-order BDF algorithm* given in Section 17-3-5 to find the solution.

17-16. Construct a circuit containing a nonlinear resistor, a linear inductor, and a linear capacitor, and apply the *generalized tableau approach* given in Section 17-4 to find the solution.

17-17. Show that for a *linear time-varying* system described by the *state equation*

$$\dot{x}(t) = A(t)x + u(t), \qquad A(t + T) = A(t)$$

the *periodic steady-state algorithm* presented in Section 17-5 converges in one step; i.e.,

$$\mathbf{P}_0 = \mathbf{x}_0 - \left[\mathbf{1} - \frac{\partial \mathbf{x}(t)}{\partial \mathbf{x}_0}\bigg|_{t=T}\right]^{-1} [\mathbf{x}_0 - \mathbf{x}(T)]$$

where $\mathbf{x}_0$ is any *initial guess.*

17-18. Show that the Jacobian matrix $\partial \mathbf{x}(t)/\partial \mathbf{x}_0$ given in Problem 17-17 is equal to the *state transition matrix* associated with the matrix $A(t)$.

17-19. Construct the sensitivity circuit model associated with the following circuit elements: linear controlled sources (all four types), ideal two-port transformers, gyrators, nonlinear voltage-controlled current sources, and nonlinear current-controlled voltage sources.

17-20. Derive the *phasors* associated with the *second-order* distortion current sources shown in Fig. 17-7.

17-21. Derive the circuit shown in Fig. 17-6(a) with three sinusoids of frequency ω_1, ω_2, and ω_3, and determine the *phasors* associated with the *third-order intermodulation* distortion current sources at frequency $\omega_d = \omega_1 \pm \omega_2 \pm \omega_3$.

17-22. Derive the phasors associated with the *second-order* harmonic and intermodulation distortion current sources for a circuit containing a nonlinear two-port resistor characterized by

$$i_a = \hat{i}_a(v_a, v_b)$$
$$i_b = \hat{i}_b(v_a, v_b)$$

17-23. Derive the phasors associated with the *second-order* harmonic and intermodulation distortion *voltage* and *current* sources for a circuit containing a nonmonotonic, voltage-controlled nonlinear resistor $i_a = \hat{i}_a(v_a)$ and a nonmonotonic, current-controlled nonlinear resistor $v_b = \hat{v}_b(i_b)$.

17-24. Derive the *phasors* associated with the *second-order* harmonic and intermodulation distortion current sources associated with a two-port characterized by

$$i_a = \hat{i}_a(v_a, v_b) + C_a(v_a)\frac{dv_a}{dt}$$

$$i_b = \hat{i}_b(v_a, v_b) + C_b(v_b)\frac{dv_b}{dt}$$

17-25. To *initiate* a conventional *nodal analysis* on an associated discrete circuit obtained by replacing each capacitor (resp., inductor) in a *dynamic* network by the discrete capacitor (resp., inductor) circuit model shown in Fig. 17-1 (resp., Fig. 17-2), it is necessary to be given not only the initial capacitor voltages (resp., inductor currents) but also the initial capacitor currents (resp., inductor voltages) in order that $\dot{v}_{C_k}(0)$ (resp., $\dot{i}_{L_k}(0)$) may be computed.

(a) Explain why the initial capacitor currents (resp., inductor voltages) cannot be arbitrarily prescribed.

(b) Given the initial capacitor voltages (resp., inductor currents), devise an algorithm for computing the associated initial capacitor currents (resp., inductor voltages) by nodal analysis.

Index